The Indian Ocean and its Role in the Global Climate System

The Indian Ocean and its Role in the Global Climate System

Edited by

Caroline C. Ummenhofer
Department of Physical Oceanography, Woods Hole Oceanographic Institution, Woods Hole, MA, United States

Raleigh R. Hood
Horn Point Laboratory, University of Maryland Center for Environmental Science, Cambridge, MD, United States

Elsevier
Radarweg 29, PO Box 211, 1000 AE Amsterdam, Netherlands
125 London Wall, London EC2Y 5AS, United Kingdom
50 Hampshire Street, 5th Floor, Cambridge, MA 02139, United States

Copyright © 2024 Elsevier Inc. All rights are reserved, including those for text and data mining, AI training, and similar technologies.

Publisher's note: Elsevier takes a neutral position with respect to territorial disputes or jurisdictional claims in its published content, including in maps and institutional affiliations.

No part of this publication may be reproduced or transmitted in any form or by any means, electronic or mechanical, including photocopying, recording, or any information storage and retrieval system, without permission in writing from the publisher. Details on how to seek permission, further information about the Publisher's permissions policies and our arrangements with organizations such as the Copyright Clearance Center and the Copyright Licensing Agency, can be found at our website: www.elsevier.com/permissions.

This book and the individual contributions contained in it are protected under copyright by the Publisher (other than as may be noted herein).

Notices
Knowledge and best practice in this field are constantly changing. As new research and experience broaden our understanding, changes in research methods, professional practices, or medical treatment may become necessary.

Practitioners and researchers must always rely on their own experience and knowledge in evaluating and using any information, methods, compounds, or experiments described herein. In using such information or methods they should be mindful of their own safety and the safety of others, including parties for whom they have a professional responsibility.

To the fullest extent of the law, neither the Publisher nor the authors, contributors, or editors, assume any liability for any injury and/or damage to persons or property as a matter of products liability, negligence or otherwise, or from any use or operation of any methods, products, instructions, or ideas contained in the material herein.

ISBN: 978-0-12-822698-8

For information on all Elsevier publications
visit our website at https://www.elsevier.com/books-and-journals

Publisher: Candice Janco
Acquisitions Editor: Maria Elekidou
Editorial Project Manager: Rupinder Heron
Production Project Manager: Sruthi Satheesh
Cover Designer: Christian Bilbow

Typeset by STRAIVE, India

Contents

Contributors	xiii
Preface	xvii
Glossary	xix
Acronyms	xxv

1. Introduction to the Indian Ocean

Raleigh R. Hood, Caroline C. Ummenhofer, Helen E. Phillips, and Janet Sprintall

1. Introduction	1
2. Research history	2
3. Geology	3
4. Oceanography	5
4.1 Ocean circulation	5
4.2 Upper-ocean structure	7
4.3 Interocean basin connections and heat transport	9
5. Atmosphere	10
6. Hydrology and hydrography	11
7. Biogeochemistry, productivity, and fisheries	13
7.1 Nutrients, phytoplankton, and zooplankton	13
7.2 Oxygen, carbon, and pH	16
7.3 Productivity and fisheries	17
8. Summary and conclusions	20
9. Educational resources	20
Acknowledgments	21
Author contributions	21
References	21

2. A brief historical overview of the maritime Indian Ocean World (ancient times to 1950)

Timothy D. Walker

1. Monsoon winds, fishing, and coastwise trade: The earliest maritime peoples of the Indian Ocean rim	33
2. Arab traders and the spread of Islamic culture by sea (c. 800–1400 CE)	34
3. The Chola kingdom: Maritime power of South Asia (9th–12th centuries CE)	36
4. Indian Ocean world seaborn trade from China (13th–15th centuries CE): Merchants of the Song and Ming dynasties	36
5. Sea change: Arrival of the Portuguese, armed trade, and colonization (c. 1498–1600)	37
6. Advent of the Dutch: Dutch East India Company ambitions and activities in the Indian Ocean world (c. 1600–1800)	39
7. The English East India company in the Indian Ocean world (c. 1600–1858)	40
8. Arrival of the French (18th–20th centuries)	42
9. Slave trading in the Indian Ocean	43
10. Whaling voyages in the Indian Ocean (c. 1785–1920)	44
11. The Indian Ocean world during the late colonial era (1860–1950)	44
12. Conclusion	45
13. Educational resources	46
Acknowledgments	46
References	46

3. Past, present, and future of the South Asian monsoon

Caroline C. Ummenhofer, Ruth Geen, Rhawn F. Denniston, and Mukund Palat Rao

1. Introduction	49
2. Monsoon dynamics	50
2.1 Climatological dynamics	50
2.2 Subseasonal dynamics	53
3. Drivers of South Asian monsoon variability	54
3.1 Pacific Ocean drivers	55
3.2 Indian Ocean drivers	56
3.3 Atlantic Ocean drivers	57
3.4 High-latitude influences	58

4. Past variability in the South Asian monsoon 58
   4.1 Late Holocene monsoon variability from stalagmite and lacustrine proxies 58
   4.2 Last millennium variability from tree rings 61
5. Changes in the South Asian monsoon in a warming world 63
6. Seasonal forecasting 66
7. Conclusions 67
8. Educational resources 68
Acknowledgments 68
Author contributions 68
References 68

4. Intraseasonal variability in the Indian Ocean region

Charlotte A. DeMott, James H. Ruppert, Jr., and Adam Rydbeck

1. Introduction to intraseasonal ocean-atmosphere coupling 79
2. The intraseasonal oscillation 81
   2.1 Overview 81
   2.2 Atmospheric processes and their dynamic interpretation 81
   2.3 ISO forcing of the ocean 83
   2.4 Differences between boreal winter and boreal summer 84
   2.5 Model limitations to ISO simulation 84
3. Intraseasonal oceanic variability 84
   3.1 1D perspective 84
   3.2 2D perspective 85
   3.3 Ocean layers 86
4. Ocean feedbacks to the atmosphere 87
   4.1 Intraseasonal SST feedbacks to maintenance and propagation of atmospheric ISOs 87
   4.2 Other modes of SST variability that affect atmospheric ISV 87
5. ISV and the maritime continent prediction barrier 90
6. Conclusions 92
7. Educational resources 92
Acknowledgments 93
Author contributions 93
References 93

5. Climate phenomena of the Indian Ocean

Toshio Yamagata, Swadhin Behera, Takeshi Doi, Jing-Jia Luo, Yushi Morioka, and Tomoki Tozuka

1. Introduction 103
2. IOD and its flavors 103
3. Indian Ocean Basin mode and interbasin connections 106
4. Indian Ocean subtropical dipole 106
5. Ningaloo Niño/Niña 108
6. Predictability and prediction of IOD, Indian Ocean subtropical dipole, and Ningaloo Niño 110
7. Conclusions 113
8. Educational resources 114
Author contributions 114
References 114

6. Extreme events in the Indian Ocean: Marine heatwaves, cyclones, and tsunamis

Ming Feng, Matthieu Lengaigne, Sunanda Manneela, Alex Sen Gupta, and Jérôme Vialard

1. Introduction 121
2. Marine heatwaves 123
   2.1 Introduction 123
   2.2 Mechanisms and relationships with climate variability and change 123
   2.3 Impacts of marine heatwaves 126
3. Tropical cyclones 127
   3.1 TC mechanisms and control by the ocean-atmosphere background state 127
   3.2 Present climate 128
   3.3 Future TC projections 130
4. Tsunami 131
   4.1 Introduction 131
   4.2 Mechanisms of tsunami generation 131
   4.3 Impact of tsunamis 133
   4.4 Tsunami monitoring system 133
5. Conclusions 134
6. Educational resources 135
Acknowledgments 135
Author contributions 136
References 136

7. Impacts of the Indian Ocean on regional and global climate

Caroline C. Ummenhofer,
Andréa S. Taschetto, Takeshi Izumo,
and Jing-Jia Luo

1. Introduction 145
2. Impacts on regional (hydro)climate 146
   - 2.1 Impacts in Indian Ocean rim countries 146
   - 2.2 Remote teleconnections 150
3. Indian Ocean-ENSO interactions 152
   - 3.1 Impacts of ENSO on the Indian Ocean 152
   - 3.2 Impacts of the Indian Ocean on ENSO 153
   - 3.3 Impacts of Indian Ocean-ENSO interactions on climate predictability 155
4. The effect of long-term warming of the Indian Ocean on regional and global climate 156
   - 4.1 Indian Ocean long-term warming 156
   - 4.2 Effects on Indian Ocean regional climate 157
   - 4.3 Remote effects on global climate 158
5. Conclusions 159
6. Educational resources 160
   Acknowledgments 160
   Author contributions 160
   References 160

8. Indian Ocean circulation

Helen E. Phillips, Viviane V. Menezes,
Motoki Nagura, Michael J. McPhaden,
P.N. Vinayachandran, and Lisa M. Beal

1. Introduction 169
2. Monsoon circulation 171
   - 2.1 Introduction 171
   - 2.2 Cross-equatorial gyre circulations 172
   - 2.3 Somali current system and western Arabian Sea 174
   - 2.4 Eastern Arabian Sea and Bay of Bengal 174
   - 2.5 Southwest/Northeast Monsoon Currents 175
   - 2.6 Marginal Seas 178
3. Equatorial regime 179
   - 3.1 Introduction 179
   - 3.2 Mean circulation 179
   - 3.3 Wyrtki jets 179
   - 3.4 Equatorial undercurrents 180
   - 3.5 Equatorial waves 181
   - 3.6 Equatorial deep jets 181
4. Southern hemisphere circulation 181
   - 4.1 Introduction 181
   - 4.2 Subtropical gyre circulation 182
   - 4.3 Western boundary 183
   - 4.4 Eastern boundary 184
5. Overturning circulations 185
   - 5.1 Introduction 185
   - 5.2 Shallow overturning cells 186
   - 5.3 Deep circulation 186
   - 5.4 Abyssal circulation 187
6. Conclusions 190
7. Educational resources 191
   Acknowledgments 191
   Author contributions 191
   References 192

9. Oceanic basin connections

Janet Sprintall, Arne Biastoch,
Laura K. Gruenburg, and Helen E. Phillips

1. Introduction 205
2. The Indonesian Throughflow 206
   - 2.1 Introduction 206
   - 2.2 Pathways through the Indonesian seas 207
   - 2.3 Biogeochemistry within the Indonesian seas 208
   - 2.4 The ITF influence on the properties and currents within the Indian Ocean 208
3. Southeastern boundary exchanges 209
   - 3.1 Introduction 209
   - 3.2 Exports from the Indian Ocean 210
   - 3.3 Imports to the Indian Ocean 212
4. The Agulhas leakage 213
   - 4.1 Introduction to the greater Agulhas current system 213
   - 4.2 Temporal variability of the Agulhas current and Agulhas leakage 215
   - 4.3 Impact of the Agulhas leakage on the Atlantic Ocean 216
5. Southern Ocean water mass exchanges 216
6. Predicted changes to Interbasin boundary current connections 218
7. Conclusions 219
8. Educational resources 220
   Acknowledgments 220
   Author contributions 220
   References 220

10. Decadal variability of the Indian Ocean and its predictability

Tomoki Tozuka, Lu Dong, Weiqing Han, Matthieu Lengaigne, and Lei Zhang

1. Introduction 229
2. Observational datasets 230
3. Internal decadal climate variability 232
   3.1 Remote forcing from other regions 232
   3.2 Intrinsic Indian Ocean decadal variability 233
4. Externally forced signals 234
   4.1 Detection and attribution of Indian Ocean warming 234
   4.2 Nonuniform warming patterns 236
5. Predictability 240
6. Conclusions 240
7. Educational resources 241
Acknowledgments 241
Author contributions 241
References 241

11. Indian Ocean primary productivity and fisheries variability

Francis Marsac, Bernadine Everett, Umair Shahid, and Peter G. Strutton

1. Introduction 245
2. Indian Ocean productivity: Variability and trends 246
3. Trends in coastal fisheries 249
   3.1 The western Indian Ocean 249
   3.2 North Indian Ocean 251
   3.3 The Eastern Indian Ocean 254
4. Trends in tuna fisheries 255
   4.1 Outlook for tuna fisheries in the Indian Ocean 255
   4.2 Status and management of tuna and billfish stocks 256
   4.3 Vulnerability to climate change 256
5. Conclusion 259
6. Educational resources 260
Acknowledgments 260
Authors contribution 260
References 260

12. Oxygen, carbon, and pH variability in the Indian Ocean

Raleigh R. Hood, Timothy Rixen, Marina Levy, Dennis A. Hansell, Victoria J. Coles, and Zouhair Lachkar

1. Introduction 265
   1.1 The northern Indian Ocean oxygen minimum zones (OMZs) 265
   1.2 Role of the Indian Ocean in the global carbon cycle 267
2. Oxygen concentrations and the biogeochemical impacts of the OMZs 271
   2.1 Oxygen distributions, sources, and sinks 271
   2.2 Biogeochemical impacts of the northern Indian Ocean OMZs 272
   2.3 Recent (decadal) changes in oxygen concentrations 273
   2.4 Future changes in oxygen concentrations 274
3. Carbon concentrations and fluxes 274
   3.1 Inorganic carbon distributions and fluxes 274
   3.2 Spatial and temporal variability in pH 275
   3.3 Large-scale DOC and POC distribution and fluxes 276
   3.4 Regional DOC and POC distributions and fluxes 277
4. Summary and conclusions 284
5. Educational resources 285
Acknowledgments 285
Author contributions 285
References 285

13. Nutrient, phytoplankton, and zooplankton variability in the Indian Ocean

Raleigh R. Hood, Victoria J. Coles, Jenny A. Huggett, Michael R. Landry, Marina Levy, James W. Moffett, and Timothy Rixen

1. Introduction 293
2. Nutrient, phytoplankton and zooplankton variability 294
   2.1 Arabian Sea and the Northwestern Indian Ocean 294
   2.2 Bay of Bengal 304
   2.3 Equatorial Indian Ocean, including the Java upwelling and the SCTR 307
   2.4 Eastern Indian Ocean and the Leeuwin Current 310
   2.5 Southwestern Indian Ocean and the Agulhas Current 312
   2.6 The southern subtropical gyre 314
3. Summary and conclusions 315
4. Educational resources 316
Acknowledgments 316
Author contributions 316
References 317

14. Air-sea exchange and its impacts on biogeochemistry in the Indian Ocean

Hermann W. Bange, Damian L. Arévalo-Martínez, Srinivas Bikkina, Christa A. Marandino, Manmohan Sarin, Susann Tegtmeier, and Vinu Valsala

1. Introduction	329
2. Greenhouse gases	330
2.1 Carbon dioxide	330
2.2 Methane	331
2.3 Nitrous oxide	332
3. Reactive gases	334
3.1 Biogenic gases: Dimethylsulfide, isoprene, and very short-lived halocarbons	334
3.2 Pollutants: Carbon monoxide (CO), sulfur dioxide (SO_2), nitrogen oxides (NO_x), ozone (O_3), and mercury (Hg0)	335
4. Aerosols	337
4.1 Sources	337
4.2 Transport	338
4.3 Air-sea deposition	339
4.4 Biogeochemical significance	340
5. Conclusions and outlook	341
6. Educational resources	342
Acknowledgments	343
Author contributions	343
References	343

15. Microbial ecology of the Indian Ocean

Carolin Regina Löscher and Christian Furbo Reeder

1. Microbial diversity in the Indian Ocean	351
1.1 Introduction	351
1.2 Microbial research in the Indian Ocean: Early expeditions and methods	351
2. Microbial diversity in the Indian Ocean	352
2.1 Prokaryotic microbes in the Arabian Sea and the bay of Bengal—Who is out there?	352
2.2 Primary producers—Who is fixing the carbon?	353
3. Functional diversity of microbes in the Indian Ocean	354
3.1 Microbes of the nitrogen (N) cycle	354
3.2 The microbial sulfur cycle	358
4. Microbial feedback loops impacting OMZ intensity and expansion	358
5. Conclusions and outlook	359
6. Educational resources	360
Acknowledgments	360
Author contributions	360
References	360

16. Physical and biogeochemical characteristics of the Indian Ocean marginal seas

Faiza Y. Al-Yamani, John A. Burt, Joaquim I. Goes, Burton Jones, Ramaiah Nagappa, V.S.N. Murty, Igor Polikarpov, Maria Saburova, Mohammed Alsaafani, Alkiviadis Kalampokis, Helga do R. Gomes, Sergio de Rada, Dale Kiefer, Turki Al-Said, Manal Al-Kandari, Khalid Al-Hashmi, and Takahiro Yamamoto

1. Introduction	365
2. The Andaman Sea	365
2.1 Physical setting	365
2.2 Oceanographic setting	366
2.3 Biogeochemical features and biological characteristics	367
3. The Arabian/Persian Gulf	369
3.1 Physical setting	369
3.2 Oceanographic setting	371
3.3 Biogeochemical features and biological characteristics	371
4. The Gulf of Aden	372
4.1 Physical setting	372
4.2 Oceanographic setting	373
4.3 Biogeochemical features and biological characteristics	374
5. The Red Sea	375
5.1 Physical setting	375
5.2 Oceanographic setting	375
5.3 Biogeochemical features and biological characteristics	377
6. The Sea of Oman	377
6.1 Physical setting	377
6.2 Oceanographic setting	378
6.3 Biogeochemical features and biological characteristics	379
7. Conclusions	381
8. Educational resources	382
Acknowledgments	382
Authors contributions	383
References	383

17. The Indian Ocean Observing System (IndOOS)

Michael J. McPhaden, Lisa M. Beal, T.V.S. Udaya Bhaskar, Tong Lee, Motoki Nagura, Peter G. Strutton, and Lisan Yu

1. Introduction	393
2. Historical background	394
3. IndOOS design	395
3.1 In situ components	395
3.2 Satellites	397
4. Scientific highlights	398
4.1 Ocean circulation	398
4.2 Equatorial waves	401
4.3 Air-sea fluxes and mixed layer processes	402
4.4 Oceanic processes involved in IOD development	404
4.5 Sea level variability and rise	405
4.6 Ocean biogeochemistry	407
5. Challenges and future directions	408
6. Educational resources	409
Acknowledgments	409
Author contributions	409
References	409

18. Modeling the Indian Ocean

Toshiaki Shinoda, Tommy G. Jensen, Zouhair Lachkar, Yukio Masumoto, and Hyodae Seo

1. Introduction	421
2. Ocean circulation and upper-ocean structure	422
2.1 Boundary currents	422
2.2 Equatorial currents	424
2.3 Indonesian Throughflow	424
2.4 Temperature and salinity upper-ocean structure	424
2.5 Future perspectives	425
3. Climate variability	425
3.1 Modeling of climate variations	425
3.2 Ocean dynamics	426
3.3 Heat budget analyses and salinity impacts for the IOD	426
3.4 Dynamical prediction of the IOD variations	427
3.5 Future perspectives	427
4. Regional coupled climate modeling	427
4.1 Processes and regions suitable for regional coupled climate model applications	428
4.2 Synthesis, issues, and outlook	431
5. Biogeochemistry	431
5.1 Modeling phytoplankton bloom dynamics in the Indian Ocean	431
5.2 Modeling Indian Ocean oxygen minimum zones (OMZs)	433
5.3 Modeling Indian Ocean ecosystems under a warmer climate	433
6. Conclusions and discussion	434
7. Educational resources	435
Acknowledgments	436
Author contributions	436
References	436

19. Paleoclimate evidence of Indian Ocean variability across a range of timescales

Mahyar Mohtadi, Nerilie J. Abram, Steven C. Clemens, Miriam Pfeiffer, James M. Russell, Stephan Steinke, and Jens Zinke

1. Introduction	445
2. Interannual to decadal variability	445
2.1 Sea surface temperature (SST) and ocean circulation	445
2.2 Rainfall, monsoon, and the intertropical convergence zone (ITCZ)	449
3. Centennial to millennial variability	450
3.1 Monsoon and the ITCZ	450
3.2 IOD and ENSO	453
3.3 Ocean circulation and conditions	454
4. Orbital and glacial-interglacial variability	454
4.1 Monsoon and the ITCZ	454
4.2 Sea-level and ocean conditions	459
5. Conclusions	460
6. Educational resources	460
Acknowledgments	460
Author contributions	460
References	460

20. Future projections for the tropical Indian Ocean

M.K. Roxy, J.S. Saranya, Aditi Modi, A. Anusree, Wenju Cai, Laure Resplandy, Jérôme Vialard, and Thomas L. Frölicher

1. Introduction	469
2. Projected changes in sea surface temperature, Indian Ocean Dipole, and heat content	471

3. Projected changes in marine heatwaves 474
4. Projected changes in the biogeochemistry of the Indian Ocean 474
5. Summary and discussion 477
6. Educational resources 478
Author contributions 478
References 478

Index 483

Contributors

Numbers in parentheses indicate the pages on which the authors' contributions begin.

Nerilie J. Abram (445), Research School of Earth Sciences; ARC Centre of Excellence for Climate Extremes, The Australian National University, Canberra, ACT, Australia

Khalid Al-Hashmi (365), Sultan Qaboos University, Al-Khod, Muscat, Oman

Manal Al-Kandari (365), Kuwait Institute for Scientific Research, Kuwait City, Kuwait

Mohammed Alsaafani (365), King Abdulaziz University, Jeddah, Saudi Arabia

Turki Al-Said (365), Kuwait Institute for Scientific Research, Kuwait City, Kuwait

Faiza Y. Al-Yamani (365), Kuwait Institute for Scientific Research, Kuwait City, Kuwait

A. Anusree (469), Centre for Climate Change Research, Indian Institute of Tropical Meteorology, Ministry of Earth Sciences, Pune, India

Damian L. Arévalo-Martínez (329), GEOMAR Helmholtz Centre for Ocean Research Kiel, Kiel, Germany

Hermann W. Bange (329), GEOMAR Helmholtz Centre for Ocean Research Kiel, Kiel, Germany

Lisa M. Beal (169,393), Rosenstiel School of Marine, Atmospheric, and Earth Science, University of Miami, Coral Gables, Miami, FL, United States

Swadhin Behera (103), Application Lab, Japan Agency for Marine-Earth Science and Technology, Yokohama, Japan

Arne Biastoch (205), GEOMAR Helmholtz Centre for Ocean Research and Kiel University, Kiel, Germany

Srinivas Bikkina (329), Chubu University, Kasugai, Japan; National Institute of Oceanography, Dona Paula, India

John A. Burt (365), Mubadala ACCESS Center, New York University Abu Dhabi, Abu Dhabi, United Arab Emirates

Wenju Cai (469), Frontier Science Centre for Deep Ocean Multispheres and Earth System and Physical Oceanography Laboratory, Ocean University of China, Qingdao, China; Center for Southern Hemisphere Oceans Research (CSHOR), CSIRO Oceans and Atmosphere, Hobart, TAS, Australia

Steven C. Clemens (445), Department of Earth, Environmental, and Planetary Sciences, Brown University, Providence, RI, United States

Victoria J. Coles (265,293), Horn Point Laboratory, University of Maryland Center for Environmental Science, Cambridge, MD, United States

Sergio de Rada (365), US Naval Research Laboratory, Stennis Space Centre, MS, United States

Charlotte A. DeMott (79), Department of Atmospheric Science, Colorado State University, Fort Collins, CO, United States

Rhawn F. Denniston (49), Department of Geology, Cornell College, Mount Vernon, IA, United States

Takeshi Doi (103), Application Lab, Japan Agency for Marine-Earth Science and Technology, Yokohama, Japan

Lu Dong (229), Frontier Science Centre for Deep Ocean Multispheres and Earth System and Physical Oceanography Laboratory, Ocean University of China, Qingdao, People's Republic of China

Bernadine Everett (245), Oceanographic Research Institute, South African Association for Marine Biological Research, Durban, South Africa

Ming Feng (121), CSIRO Environment, Indian Ocean Marine Research Centre, Crawley, WA; Center for Southern Hemisphere Oceans Research, CSIRO Oceans and Atmosphere, Hobart, TAS, Australia

Thomas L. Frölicher (469), Climate and Environmental Physics, Physics Institute; Oeschger Centre for Climate Change Research, University of Bern, Bern, Switzerland

Ruth Geen (49), School of Geography, Earth and Environmental Sciences, University of Birmingham, Birmingham, United Kingdom

Joaquim I. Goes (365), Lamont-Doherty Earth Observatory at Columbia University, Palisades, NY, United States

Helga do R. Gomes (365), Lamont-Doherty Earth Observatory at Columbia University, Palisades, NY, United States

Laura K. Gruenburg (205), School of Marine and Atmospheric Sciences, Stony Brook University, Stony Brook, NY, United States

Alex Sen Gupta (121), University of New South Wales, Sydney, NSW; Australian Research Council Centre of Excellence for Climate Extremes, Sydney, Australia

Weiqing Han (229), Department of Atmospheric and Oceanic Sciences, University of Colorado, Boulder, CO, United States

Dennis A. Hansell (265), Rosenstiel School of Marine and Atmospheric Science, University of Miami, Miami, FL, United States

Raleigh R. Hood (1,265,293), Horn Point Laboratory, University of Maryland Center for Environmental Science, Cambridge, MD, United States

Jenny A. Huggett (293), Oceans and Coasts, Department of Forestry, Fisheries and the Environment, Cape Town, South Africa

Takeshi Izumo (145), Institut de Recherche pour le Développement (IRD), EIO Laboratory, University of French Polynesia, Puna'auia, French Polynesia

Tommy G. Jensen (421), Ocean Sciences Division, US Naval Research Laboratory, Stennis Space Center, MS, United States

Burton Jones (365), King Abdullah University of Science and Technology, Thuwal, Saudi Arabia

Alkiviadis Kalampokis (365), King Abdullah University of Science and Technology, Thuwal, Saudi Arabia

Dale Kiefer (365), University of Southern California, Los Angeles, CA, United States

Zouhair Lachkar (265,421), Arabian Center for Climate and Environmental Sciences; Center for Prototype Climate Modeling, New York University Abu Dhabi, Abu Dhabi, United Arab Emirates

Michael R. Landry (293), Scripps Institution of Oceanography, University of California San Diego, La Jolla, CA, United States

Tong Lee (393), Jet Propulsion Laboratory, California Institute of Technology, Pasadena, CA, United States

Matthieu Lengaigne (121,229), MARBEC, University of Montpellier, CNRS, IFREMER, IRD, Sète, France

Marina Levy (265,293), LOCEAN-IPSL, Sorbonne Universités (UPMC, Univ Paris 06)-CNRS-IRD-MNHN, Paris, France

Carolin Regina Löscher (351), University of Southern Denmark, Odense, Denmark

Jing-Jia Luo (103,145), Institute of Climate and Application Research, Nanjing University of Information Science and Technology, Nanjing, China

Sunanda Manneela (121), Indian National Centre for Ocean Information Services (INCOIS), Hyderabad, India

Christa A. Marandino (329), GEOMAR Helmholtz Centre for Ocean Research Kiel, Kiel, Germany

Francis Marsac (245), MARBEC, Univ Montpellier, CNRS, Ifremer, IRD, Sète, France; IRD, Seychelles Fishing Authority, Victoria, Seychelles

Yukio Masumoto (421), Department of Earth and Planetary Science, University of Tokyo, Tokyo, Japan

Michael J. McPhaden (169,393), NOAA/Pacific Marine Environmental Laboratory, Seattle, WA, United States

Viviane V. Menezes (169), Woods Hole Oceanographic Institution, Woods Hole, MA, United States

Aditi Modi (469), Centre for Climate Change Research, Indian Institute of Tropical Meteorology, Ministry of Earth Sciences, Pune; Interdisciplinary Programme in Climate Studies, Indian Institute of Technology Bombay, Mumbai, India

James W. Moffett (293), Department of Biological Sciences, Earth Sciences and Civil and Environmental Engineering, University of Southern California, Los Angeles, CA, United States

Mahyar Mohtadi (445), MARUM-Center for Marine Environmental Sciences; Faculty of Geosciences, University of Bremen, Bremen, Germany

Yushi Morioka (103), Application Lab, Japan Agency for Marine-Earth Science and Technology, Yokohama, Japan

V.S.N. Murty (365), CSIR – National Institute of Oceanography, Dona Paula, Goa, India

Ramaiah Nagappa (365), CSIR – National Institute of Oceanography, Dona Paula, Goa, India

Motoki Nagura (169,393), Japan Agency for Marine-Earth Science and Technology, Yokosuka, Kanagawa, Japan

Miriam Pfeiffer (445), Institut für Geowissenschaften, Christian-Albrechts-Universität zu Kiel, Kiel, Germany

Helen E. Phillips (1,169,205), Institute for Marine and Antarctic Studies and Australian Antarctic Program Partnership, University of Tasmania, Hobart, TAS, Australia

Igor Polikarpov (365), Kuwait Institute for Scientific Research, Kuwait City, Kuwait

Mukund Palat Rao (49), Centre de Recerca Ecològica i Aplicacions Forestals, Barcelona, Catalonia, Spain; Department of Plant Science, University of California Davis, Davis, CA; Tree Ring Laboratory, Lamont-Doherty Earth Observatory of Columbia University, Palisades, NY; Cooperative Programs for the Advancement of Earth System Science, University Corporation for Atmospheric Research, Boulder, CO, United States

Christian Furbo Reeder (351), University of Southern Denmark, Odense, Denmark; Marseille Institute for Ocean Science, Marseille, France

Laure Resplandy (469), Department of Geosciences and High Meadows Environmental Institute, Princeton University, Princeton, NJ, United States

Timothy Rixen (265,293), Leibniz Centre for Tropical Marine Research, University of Bremen, Bremen, Germany

M.K. Roxy (469), Centre for Climate Change Research, Indian Institute of Tropical Meteorology, Ministry of Earth Sciences, Pune, India

James H. Ruppert, Jr. (79), School of Meteorology, The University of Oklahoma, Norman, OK, United States

James M. Russell (445), Department of Earth, Environmental, and Planetary Sciences, Brown University, Providence, RI, United States

Adam Rydbeck (79), Ocean Sciences Division, US Naval Research Laboratory, Stennis Space Center, Hancock County, MS, United States

Maria Saburova (365), Kuwait Institute for Scientific Research, Kuwait City, Kuwait

J.S. Saranya (469), Centre for Climate Change Research, Indian Institute of Tropical Meteorology, Ministry of Earth Sciences, Pune, India; School of Earth and Environmental Sciences, College of Natural Sciences, Seoul National University, Seoul, Republic of Korea

Manmohan Sarin (329), Physical Research Laboratory, Ahmedabad, India

Hyodae Seo (421), Department of Physical Oceanography, Woods Hole Oceanographic Institution, Woods Hole, MA, United States

Umair Shahid (245), WWF, Karachi, Pakistan

Toshiaki Shinoda (421), Department of Physical and Environmental Sciences, Texas A&M University-Corpus Christi, Corpus Christi, TX, United States

Janet Sprintall (1,205), Climate, Atmospheric Sciences and Physical Oceanography, Scripps Institution of Oceanography, University of California San Diego, La Jolla, CA, United States

Stephan Steinke (445), Department of Geological Oceanography & State Key Laboratory of Marine Environmental Science, College of Ocean and Earth Sciences, Xiamen University, Xiamen, People's Republic of China

Peter G. Strutton (245,393), Institute for Marine and Antarctic Studies; Australian Research Council Centre of Excellence for Climate Extremes, University of Tasmania, Hobart, TAS, Australia

Andréa S. Taschetto (145), Climate Change Research Centre and ARC Centre of Excellence for Climate Extremes, University of New South Wales, Sydney, NSW, Australia

Susann Tegtmeier (329), University of Saskatchewan, Saskatoon, SK, Canada

Tomoki Tozuka (103,229), Application Lab, Japan Agency for Marine-Earth Science and Technology, Yokohama; Department of Earth and Planetary Science, Graduate School of Science, The University of Tokyo, Tokyo, Japan

T.V.S. Udaya Bhaskar (393), ESSO-Indian National Centre for Ocean Information Services, Hyderabad, India

Caroline C. Ummenhofer (1,49,145), Department of Physical Oceanography, Woods Hole Oceanographic Institution, Woods Hole, MA, United States

Vinu Valsala (329), Indian Institute of Tropical Meteorology, Pune, India

Jérôme Vialard (121,469), LOCEAN-IPSL, Sorbonne Universités (UPMC, Univ Paris 06)-CNRS-IRD-MNHN, Paris, France

P.N. Vinayachandran (169), Centre for Atmospheric and Oceanic Sciences, Indian Institute of Science, Bangalore, India

Timothy D. Walker (33), Department of History, University of Massachusetts Dartmouth, North Dartmouth; Woods Hole Oceanographic Institution, Woods Hole, MA, United States

Toshio Yamagata (103), Application Lab, Japan Agency for Marine-Earth Science and Technology, Yokohama, Japan; Institute of Climate and Application Research, Nanjing University of Information Science and Technology, Nanjing, China

Takahiro Yamamoto (365), Kuwait Institute for Scientific Research, Kuwait City, Kuwait

Lisan Yu (393), Woods Hole Oceanographic Institution, Woods Hole, MA, United States

Lei Zhang (229), State Key Laboratory of Tropical Oceanography, South China Sea Institute of Oceanology, Chinese Academy of Sciences; Global Ocean and Climate Research Center, South China Sea Institute of Oceanology, Guangzhou, Guangdong, People's Republic of China

Jens Zinke (445), School of Geography, Geology and the Environment, University of Leicester, Leicester, United Kingdom

Preface

The Indian Ocean is unique and different and represents one of the great frontiers in climate science and oceanography. The Indian Ocean and its surrounding countries stand out globally as the region with the highest risk of natural hazards, with coastal communities vulnerable to weather and climate extremes. The vagaries of the Asian monsoon directly affect more than a billion people and a third of the global population lives in the vicinity of the Indian Ocean. The Indian Ocean is also particularly susceptible to human-induced climate change, with robust warming trends and pronounced changes in heat and freshwater observed in recent decades. Paired with a sparse and relatively short instrumental record, the Indian Ocean thus represents challenges to predict and forecast environmental conditions and their effects on climate and weather in the surrounding countries.

The Indian Ocean is unusual among tropical ocean basins due to its lack of steady easterlies and relatively deep thermocline, seasonal reversal of monsoon winds and concurrent ocean currents, and lack of northward heat export due to the Asian continent to the north. These characteristics shape the very dynamic intraseasonal and seasonal variability, air-sea interactions, and biological responses of the Indian Ocean. However, the advent of new technologies, an expanded observation system, and rapid advances in environmental predictions and forecasting capabilities in recent decades open new and exciting opportunities for improved environmental and climate risk management in a region particularly vulnerable to changing conditions.

Given the unique characteristics of the Indian Ocean and that it has traditionally been understudied, understanding of the Indian Ocean as a wholistic system is still limited. This book provides a rare interdisciplinary synthesis of recent advances in knowledge and understanding of the physical climate system in the Indian Ocean, interlinked with interactions with its biogeochemistry and ecology, and impacts on human and natural systems in surrounding countries. Recent trends and future projections of the Indian Ocean, including warming and extreme events—both in the physical and biogeochemical realm—such as marine heatwaves, climate and weather extremes, ocean acidification, and deoxygenation are detailed. The textbook identifies recent new understanding and technologies to provide stakeholders with relevant knowledge for more informed decision-making and highlights knowledge gaps to encourage students, practitioners, and researchers to help overcome these.

With a total of 20 peer-reviewed chapters, more than 175 figures, 3000 references, 100 educational resources and links directing readers to more in-depth information, and 100 glossary entries of key concepts, the book delivers a comprehensive overview of our current understanding of the Indian Ocean from an interdisciplinary perspective. Contributions by more than 90 authors from around the world with expertise across a wide range of fields (e.g., atmospheric and climate science, biogeochemistry, ecology, environmental science, fisheries, geology, history, meteorology, microbial ecology, numerical modeling, all fields of oceanography, paleoclimate, remote sensing, statistics, and weather forecasting) underpin the content of this textbook. Furthermore, all chapters were peer-reviewed by at least two experts in the field: constructive and valuable feedback from ~40 reviewers ensured that the material in the individual chapters and the book as a whole provided a comprehensive, interconnected, and up-to-date review of Indian Ocean science. The time and efforts of all members of the Indian Ocean research community that helped in completing this book are greatly appreciated.

Caroline C. Ummenhofer
Raleigh R. Hood

Glossary

Aerosols Fine solid particles or liquid droplets in the atmosphere. Aerosol particles have diameters typically less than 1 μm (10^{-6} m).

Agulhas Current Strongest western boundary current in the Southern Hemisphere that is fed by the waters flowing east and west of Madagascar to flow poleward along the eastern bank of the African continent.

Agulhas leakage Relatively small component of the Agulhas Current that leaks westward into the Atlantic Ocean through eddies that are generated southwest of the Cape of Good Hope, South Africa. The warm and salty eddies derived from the Indian Ocean play a significant role in the upper branch of the global meridional overturning circulation.

Agulhas Return Current Part of the poleward flowing Agulhas Current that retroflects south of the Agulhas Bank to flow eastward within the northern boundary of the Antarctic Circumpolar Current.

Air-sea gas exchange Transfer (flux) of gases across the ocean/atmosphere interface. The flux is bidirectional and can lead to a release of gas from the surface ocean to the overlying atmosphere (=emission) or to an uptake of gas from the atmosphere into the surface ocean. The flux is mainly driven by the gas partial pressure difference between the surface ocean and the overlying atmosphere and the wind speed.

Amount effect Phenomenon observed in many tropical settings in which the ratio of $^{18}O/^{16}O$ in rainwater is inversely proportional to the rainfall amount.

Atlantic Meridional Overturning Circulation (AMOC) Zonally integrated meridional transport of surface and deep currents in the Atlantic. The AMOC transports up to 35 Sv ($1\,\text{Sv} = 10^6\,\text{m}^3\,\text{s}^{-1}$) and consists primarily of a northward-flowing, warm, and saline upper limb, and a southward-flowing, cold, and fresh lower limb.

Anoxia Conditions in which the environment is completely devoid of oxygen (i.e., anaerobic or anoxic conditions).

Antarctic Oscillation Also known as the Southern Annular Mode. Leading mode of large-scale atmospheric variability in the Southern Hemisphere, characterized by an anomalous pressure center over Antarctica and a zonally symmetric pressure anomaly of opposite sign at mid-latitudes. The positive (negative) phases of the mode are associated with poleward (equatorward) displacement of the midlatitude westerly winds.

Anthropogenic carbon dioxide CO_2 generated by human activities, such as fossil fuel combustion, that enters the natural carbon cycle (usually through the atmosphere).

Biogeochemistry Studies the cycling of crucial elements in natural systems. Thus biogeochemical refers to the processes that involve biological, geological, and chemical transformations of key elements such as carbon and nitrogen.

Biological pump Transfer of carbon from the ocean surface to the deep ocean by sinking particles.

Bjerknes feedback Positive feedback loop named after Jacob Bjerknes describing ocean-atmosphere interactions over the tropical Pacific that govern the development of El Niño-Southern Oscillation (ENSO) events. It involves reinforcing variations between surface winds, sea surface temperatures, and thermocline depth.

Boreal summer intraseasonal oscillation Intraseasonal oscillation characterized by both eastward and northward propagating components most frequently observed in the Asian boreal summer monsoon systems.

Carbon biomass Mass of organic carbon molecules/compounds in living biological organisms in each area or ecosystem at a given time.

Carbonate pump Physicochemical processes that transport carbon as dissolved inorganic carbon (DIC) to the ocean's interior from the surface. The carbonate pump is the "hard tissue" component of the biological pump. It is propelled by calcium carbonate shell-forming surface-inhabiting marine organisms, like coccolithophores. Upon death or by their cast-off shells, the DIC formed is an important part of the oceanic carbon cycle.

Carbon isotopes Isotopes are atoms of the same element but with different masses and may be stable or radioactive; two carbon isotopes (^{12}C and ^{13}C) are stable and one (^{14}C) is radioactive. Carbon isotopes may be selectively enriched or depleted relative to the others owing to differences in mass by processes related to climate and/or metabolism. Carbon isotopes are also used to measure primary production (^{13}C or ^{14}C uptake rate). $\delta^{13}C$ ratio is a common paleo proxy.

Central Pacific El Niño-Southern Oscillation (ENSO) Type of ENSO event characterized by maximum surface anomalies in the central equatorial Pacific, distinct from canonical ENSO in terms of its spatial and temporal characteristics, as well as its teleconnection patterns. Also referred to as "ENSO Modoki", where "Modoki" is a Japanese word for "similar but different."

Chlorophyll-a Green pigment absorbing most energy from wavelengths of violet-blue and orange-red light and used in oxygenic photosynthesis to generate chemical energy by photosynthetic plants.

Cloud radiative feedbacks Effect of cloud fields on warming or cooling the atmosphere by altering vertical profiles of solar and/or infrared heating, and the changes in weather induced by these heating anomalies

Cold pool Short-lived (up to one day) locally cooler zone of air (O(10–100 km)) in the atmospheric boundary layer formed by the sinking of negatively buoyant air (as a downdraft) that has been cooled by evaporation of cloud and rainwater (O(1 K)).

Convective quasi-equilibrium Theoretical framework for the tropical atmosphere that assumes the atmospheric lapse rate is maintained close to a moist adiabat due to the occurrence of frequent, intense moist convection.

Coral bleaching Whitening of coral resulting from the loss of a coral's symbiotic microalgae or the degradation of the algae's photosynthetic pigment during exposure to elevated temperatures.

Cyclogenesis Broad term encompassing different processes leading to the formation of some sort of cyclone of any size.

Dark Ages Cold Period Period spanning ~450–750 CE that was characterized by anomalously cool conditions across many parts of the Northern Hemisphere and may also have been associated with reduced rainfall across parts of Asia.

Denitrification Microbial-driven stepwise reduction of nitrate (NO_3^-) to dinitrogen (N_2) gas. During this metabolic pathway (which occurs under low oxygen (suboxic) conditions), N_2O is produced as an obligate intermediate. However, when O_2 concentrations drop further and are on the verge of anoxia or sulfidic conditions (i.e., anoxia with the presence of hydrogen sulfide), N_2O is consumed by reduction to N_2.

Deposition (wet/dry) Transfer of particles from the atmosphere to the surface ocean. Wet deposition is synonymous with the wash-out of atmospheric particles by rain.

Diabatic heating Warming of air parcels through the addition of radiative heating or latent heat release associated with water phase change.

Diapycnal Occurring across a surface of constant density in the ocean. When the density surface is horizontal, diapycnal exchange can be interpreted as vertical exchange.

Diazotrophs Microbes (bacteria and archaea) that fix atmospheric nitrogen (N_2) gas into a more usable form, such as ammonia, and are able to grow without external sources of fixed nitrogen.

Diurnal warm layer Thin (O(1–10 m)) stably stratified layer of the upper ocean that warms as much as 1–3 K per day through absorption of solar radiation and cools by a similar amount through infrared surface cooling and downward mixing at night.

Dust storm A meteorological phenomenon common in arid and semi-arid regions, arising when a strong wind blows loose sand and dirt from a dry surface and moves fine sediment particles from one place and deposits them in another.

Eccentricity Measure of the amount by which the Earth's orbit about the sun deviates from a circle, with periods of 405, 124, and 95 thousand years. The 124 and 95 thousand-year eccentricity bands are referred to as the ~100,000-year band.

El Niño The warm phase of the El Niño-Southern Oscillation (ENSO), characterized by anomalous surface warming and weaker trade winds in the eastern equatorial Pacific Ocean.

El Niño-Southern Oscillation (ENSO) Strong year-to-year climate variability originating in the equatorial Pacific Ocean through coupled ocean-atmosphere interactions. ENSO manifests itself in anomalous surface warming (El Niño) or cooling (La Niña) in the central to eastern equatorial Pacific that typically peaks in boreal winter.

Endemism Ecological state of a biological species being native and restricted to a particular geographic region as a result of isolation or in response to environmental conditions.

Endosymbiotic dinoflagellates of corals Photoautotrophic unicellular algae of the family Symbiodiniaceae (colloquially called zooxanthellae) residing within host coral cells, where the products of their photosynthetic processing provide most of the coral host's metabolic energy requirements.

Eutrophic environments Surface ocean with high nutrient concentrations leading to high biological productivity (high photosynthesis). Eutrophic conditions are usually found in coastal areas due to nutrient input by rivers or groundwater.

Euphotic zone Surface layer of the ocean that receives sunlight, allowing phytoplankton to perform photosynthesis (biological production), typically defined as extending down to the light level where there is only 1% of the flux compared to the surface.

Free troposphere Portion of the atmosphere above the marine atmospheric boundary layer and below the tropopause.

Hadley circulation Thermally driven meridional circulation in the atmosphere consisting of poleward flow in the upper troposphere, subsiding air over high-pressure regions of the subtropics, a return surface flow as part of the trade winds, and rising air in the Intertropical Convergence Zone.

Halocline A sharp change in oceanic salinity at a particular depth.

Holocene Climate Optimum Interval generally dated to 9500–5500 years ago and representing the first half of the Holocene (the climate period since the termination of the last ice age, 11,600 years ago); characterized by anomalous warmth in the high northern latitudes, likely owing to orbitally driven changes in summer sunlight.

Hypersaline Water characterized by high salt concentration.

Hypoxia Phenomenon that occurs in aquatic environments when dissolved oxygen in seawater is reduced in concentration to a point where it becomes detrimental to aquatic organisms living in the system. An aquatic system completely devoid of dissolved oxygen is termed anoxic but a system with low concentration (in the range between 1% and 30% saturation) is called hypoxic or dysoxic. Most fish cannot live below 30% saturation and hypoxia and the shoaling of hypoxic waters often leads to the mortality of smaller pelagic fish.

Indian Ocean Dipole (IOD) Coupled ocean-atmosphere phenomenon in the tropical Indian Ocean peaking in boreal fall. Its positive phase is characterized by anomalous tropical southeastern Indian Ocean surface cooling and anomalous warming in the western tropical Indian Ocean.

Indonesian Throughflow (ITF) Exchange of water from the Pacific Ocean into the Indian Ocean driven by the large-scale pressure gradient between the two ocean basins that weaves its way through the multitude of islands within the Indonesian archipelago.

Indo-Pacific Warm Pool Broad region of sea surface temperatures warmer than about 28°C that spans the tropical Indian and western Pacific Oceans.

Interdecadal Pacific Oscillation (IPO) Decadal mode of climate variability occurring in the Pacific with a period of 15–30 years. Its positive phase features a meridionally broad El Niño-like surface temperature anomaly pattern with cooling at higher latitudes in the North and South Pacific; similar to the Pacific Decadal Oscillation, which exhibits a more prominent Northern Hemisphere signal than the IPO.

Intertropical Convergence Zone (ITCZ) Basin-scale near-equatorial belt of low pressure where the northeast trade winds meet the southeast trade winds. As these winds converge, uplifted moist air forms a band of heavy precipitation that migrates latitudinally with the seasons.

Intraseasonal variability (ISV) Temporal oscillations with a dominant timescale falling within 20–100 days and peaking near 30–60 days.

Intraseasonal oscillation (ISO) All-season slowly eastward moving ($\sim 5\,\mathrm{m\,s^{-1}}$), large-scale ($O(10^4\,\mathrm{km})$) tropical cloud complex with a period ranging from 20 to 100 days.

Kelvin wave Large-scale gravity wave in the atmosphere and ocean in balance with the Coriolis force, which exists either along a topographic boundary or an equatorial waveguide; Kelvin waves propagate eastward along the equator, but can also propagate along coastlines in the ocean and along mountain ranges in the atmosphere.

Lacustrine Relating to lakes. Lacustrine sediments are composed of mineral and biological materials often linked to climate.

La Niña The cold phase of the El Niño-Southern Oscillation (ENSO), characterized by anomalous surface cooling and stronger trade winds in the eastern equatorial Pacific Ocean.

Leeuwin Current Poleward flowing eastern boundary current off the coast of Western Australia that transports relatively warm and fresh waters southward. This current is fed by the Indonesian Throughflow (ITF) and zonal flows from the subtropical southern Indian Ocean.

Little Ice Age Period of global cooling spanning $\sim$1450–1850 CE that was likely triggered by reduced solar irradiance and several large volcanic eruptions.

Madden-Julian Oscillation (MJO) Boreal winter manifestation of the intraseasonal oscillation (ISO), characterized by predominantly eastward propagation across the Indo-Pacific Warm Pool. A traveling large-scale pattern of tropical deep convection that is flanked to the east and west by suppressed tropical rainfall. It is usually first observed near the east coast of Africa and travels eastward at 4 to $8\,\mathrm{m\,s^{-1}}$.

Maritime continent Region of large (Sumatra, Java, Papua New Guinea) and small islands and Indonesian Seas that act as a leaky boundary between the tropical Indian Ocean and the western Pacific Ocean and trigger frequent diurnal thunderstorm activity.

Mascarene High Semi-permanent subtropical anticyclone located over the southern Indian Ocean.

Medieval Climate Anomaly Period of anomalous climate spanning $\sim$850–1250 CE characterized by enhanced warmth in the North Atlantic region.

Mesoscale In oceanography, mesoscale refers to processes that occur at scales of eddies, typically from 10 to 100 km or larger. Timescales are generally on the order of about one month.

Mesoscale eddies Energetic, swirling, time-dependent water circulations occurring almost everywhere in the ocean with space scales of 50–500 km and timescales of 10–100 days.

Methanogenesis Final step of microbial organic matter decomposition (respiration), which takes place under strictly anaerobic (anoxic) conditions and results in the release of CH_4. This process represents the major natural production pathway of CH_4.

Microbial loop Trophic pathway in the marine microbial food web where dissolved organic carbon is returned to higher trophic levels via its incorporation into bacterial biomass, and then coupled with the classic food chain formed by phytoplankton-zooplankton-nekton.

Mode of variability Natural, recurrent climate phenomenon with an underlying space-time structure that displays a preferred spatial pattern and temporal variation in components of the climate system (e.g., ocean, atmosphere, cryosphere).

Moist static energy Energy of an air parcel due to its internal and gravitational potential energy and latent heat content. This can be calculated as $h = c_p T + gz + Lq$, where h is moist static energy, c_p is the specific heat capacity of dry air at constant pressure, T is temperature, g is gravitational acceleration, z is height, L is the latent heat of vaporization of water, and q is specific humidity.

Moisture mode Type of tropical weather disturbance whose existence, scale, and propagation characteristics depend critically and predominantly on the evolution of water vapor within and above the marine atmospheric boundary layer.

Monsoon Seasonally reversing winds coupled with the seasonal cycle of precipitation. Over South Asia, these winds are typically from the southwest during the months of May–September and from the northeast from October through early May.

Monsoon depression Low pressure, westward propagating system observed in a monsoon trough. Over India, these have traditionally been categorized by the India Meteorological Department as monsoon lows, monsoon depressions, deep depressions, and cyclonic storms based on their wind speed and sea level pressure anomaly strengths.

Monsoon Intraseasonal Oscillation Dominant mode of tropical intraseasonal variability observed over Asia in boreal summer. See also boreal summer intraseasonal oscillation.

Nitrification Microbially-driven stepwise oxidation of ammonia (NH_4^+) to nitrate (NO_3^-). This is a chemoautotrophic (carbon fixation using energy derived from chemical reactions) metabolic pathway that occurs under oxic to suboxic conditions that produce N_2O as a by-product. This reaction is mediated by nitrifying bacteria and archaea. Due to its widespread occurrence, it is an important process for N_2O production in the global ocean.

Noctiluca scintillans Mixotrophic dinoflagellate that meets its energy requirements through ingestion of external prey and via photosynthesis by green endosymbionts *Protoeuglena noctilucae* within its central cytoplasm.

Obliquity The tilt of the spin axis with respect to the plane of the ecliptic that varies between a minimum of 22.05 degree and a maximum of 24.45 degree. Changes in the obliquity with a period of about 41,000 years affect the seasonality of incoming solar radiation equally in both hemispheres with a greater signal at high latitudes.

Oligotrophic environments Surface ocean with depleted nutrient concentrations leading to low biological productivity (low photosynthesis). Oligotrophic conditions are usually found in the central gyres of the major ocean basins.

Orography Portion of topography dealing with mountains. In coastal areas, mountains can steer or channel the winds.

Oxygen isotope ($\delta^{18}O$) Isotopes are atoms of the same element but with different masses that may be stable or radioactive; all oxygen isotopes (^{16}O, ^{17}O, and ^{18}O) are stable. Oxygen isotopes are selectively enriched or depleted by processes related to climate and/or metabolism.

Oxygen Minimum Zone (OMZ) Zone in which oxygen saturation in seawater in the ocean is at its lowest. This zone typically exists at depths of about 200 to 1500 m, depending on local circumstances. OMZs are found worldwide, especially in areas where the interplay of physical and biological processes concurrently lower oxygen concentrations and restricts the water from mixing with surrounding waters.

Pacific Decadal Oscillation (PDO) Sea surface temperature pattern that varies on decadal timescales, characterized by pronounced sea surface temperature anomalies in the North Pacific and sea surface temperature anomalies of opposite signs in the tropical Pacific, and is closely related to the strength of the wintertime Aleutian low-pressure system; similar to the Interdecadal Pacific oscillation (IPO).

Phenological Biological processes that undergo periodic cycles in nature generally related to seasonal and longer term variability.

Precession Axial and apsidal precession are the wobble of Earth's axis and rotation of Earth's elliptical orbit over time with periods of about 19 and 23 kyr, respectively, which affect the seasonal cycle of incoming solar radiation and its hemispheric distribution.

Primary production Synthesis of organic molecules by fixing atmospheric or aqueous carbon dioxide using light as the energy source (photosynthesis) or by oxidizing or reducing inorganic chemical compounds as the energy source (chemosynthesis).

Rectifier effect Imprint of variability of one timescale onto another, usually longer, timescale.

Rossby wave Planetary waves in the atmosphere and ocean that arise from the meridional variation of the effective Coriolis force due to the Earth's rotation, typically propagating westward with phase speeds that decrease with increasing latitude.

Scleractinian corals Reef-building stony or hard corals in the phylum Cnidaria that build calcium carbonate skeletons.

Sea-breeze Mesoscale circulation observed in coastal regions featuring an onshore near-surface wind and return flow aloft. This is driven by the daytime warming of land relative to the ocean and the resultant pressure gradients. Monsoons have historically been described as large-scale versions of the sea breeze, but recent advances in dynamics suggest this is not the case.

Shared Socioeconomic Pathways (SSP) Global emissions pathways under different socioeconomic scenarios for use in Coupled Model Intercomparison Project Phase 6 (CMIP6), and widely used in the Intergovernmental Panel on Climate Change Sixth Assessment Report for future projections.

Solubility pump Uptake and transfer of CO_2 from the atmosphere to the ocean driven by the CO_2 solubility and sinking of water masses to the deep ocean.

South Australia Current system Coupled system of surface and deeper currents in the region south of Australia that permit exchange with the southeast Indian Ocean. The complex circulation includes shelf break currents flowing predominantly eastward and the counterflowing Flinders Current beneath that can rise up to the sea surface further offshore.

Southern Annular Mode Leading mode of large-scale atmospheric variability in the Southern Hemisphere, characterized by an anomalous pressure center over Antarctica and a zonally symmetric pressure anomaly of opposite sign at midlatitudes. The positive (negative) phases of the mode are associated with poleward (equatorward) displacement of the midlatitude westerly winds. Also known as the Antarctic oscillation.

Speleothem Deposits formed in caves through the precipitation of minerals (typically calcium carbonate) from drip water. Speleothems are a commonly used proxy for past climates, particularly in (sub)tropical regions.

Spring predictability barrier Refers to the fact that El Niño-Southern Oscillation (ENSO) forecast skill is significantly reduced during boreal spring, and any time a forecast is made for the other side of spring, because of the tendency for the ENSO phase to shift as El Niño and La Niña episodes decay after their usual winter peak.

Submesoscale Submesoscale processes are those small-scale processes that occur at space scales of less than 10 km and have timescales from hours to days.

Subsidence Large-scale ($O(10^2$–10^4 km)) descent of atmospheric column mass coupled to rising motion elsewhere and sustained by column infrared cooling ($O(1$ hPa/day or 10^{-1} m s^{-1})).

Subtropical Geographic and climatic zone located between 30 to 40 latitudes in the Northern and Southern Hemispheres.

Tasman leakage Large-scale current system that links the East Australian Current western boundary current of the South Pacific to a broad westward flow south of Tasmania into the southern Indian Ocean subtropical gyre.

Teleconnection Changes in atmospheric or oceanic circulation over widely separated, geographically fixed spatial locations; often a consequence of large-scale wave motions, whereby energy is transferred from source regions along preferred atmospheric/oceanic paths.

Thermocline Zone of maximum vertical temperature gradient, separating warm and cold layers of water. The 20°C isotherm is often used as an indicator of thermocline depth in tropical oceans.

Trace gases Gaseous constituents of the Earth's atmosphere with atmospheric mixing ratios (mole fractions) less than 1%.

Troposphere Lowest layer of the Earth's atmosphere, with a depth of $\sim$10 km near the poles to nearly 20 km in the deep tropics, where most clouds and weather occur.

Walker circulation The time-mean, thermally driven zonally oriented overturning circulation in the tropics with rising branches situated over the maritime continent and Indo-Pacific Warm Pool, and sinking branches found over eastern tropical Africa and the eastern Pacific Ocean.

Wyrtki jets Wind-driven eastward jets in the surface layer of the Indian Ocean that occur twice per year (April–May and October–November) along the equator during the transition between the Northeast and Southwest Monsoons.

Acronyms

AMOC	Atlantic Meridional Overturning Circulation
BCE	Before Common Era
BP	Before Present
CE	Common Era
CLIVAR	Climate and Ocean: Variability, Predictability, and Change
CMIP	Coupled Model Intercomparison Project
DIC	dissolved inorganic carbon
DJF	December-January-February
DMI	Dipole Mode index
DMS	dimethyl sulfide
DOC	dissolved organic carbon
ECMWF	European Centre for Medium Range Forecasting
EEZ	exclusive economic zone
EKE	eddy kinetic energy
ENSO	El Niño-Southern Oscillation
EOF	empirical orthogonal function
FAO	Food and Agriculture Organization (of the United Nations)
GCM	general circulation model
GDP	gross domestic product
ICOADS	International Comprehensive Ocean-Atmosphere Data Set
IIOE	International Indian Ocean Expedition
IMOS	Integrated Marine Observing System
IndOOS	Indian Ocean Observing System
IOD	Indian Ocean Dipole
IPCC	Intergovernmental Panel on Climate Change
IPO	Interdecadal Pacific Oscillation
ISO	intraseasonal oscillation
ISV	intraseasonal variability
ITCZ	Intertropical Convergence Zone
ITF	Indonesian Throughflow
JGOFS	Joint Global Ocean Flux Study
JJA	June-July-August
JMA	Japan Meteorological Agency
MAM	March-April-May
MHW	marine heatwave
MJO	Madden-Julian Oscillation
NAO	North Atlantic Oscillation
NPP	net primary productivity
OHC	ocean heat content
OMZ	oxygen minimum zone
PDO	Pacific Decadal Oscillation
POC	particulate organic carbon
RAMA	Research Moored Array for African-Asian-Australian Monsoon Analysis and Prediction

SCOR	Scientific Committee on Ocean Research
SCTR	Seychelles-Chagos Thermocline Ridge
SLP	sea level pressure
SON	September-October-November
SSH	sea surface height
SSP	Shared Socioeconomic Pathway
SSS	sea surface salinity
SST	sea surface temperature
TC	tropical cyclone
TOGA	Tropical Ocean Global Atmosphere
TRMM	Tropical Rainfall Measuring Mission
WCRP	World Climate Research Program
WOCE	World Ocean Circulation Experiment
XBT	eXpendable BathyThermograph
XRF	X-ray fluorescence

Chapter 1

Introduction to the Indian Ocean

Raleigh R. Hood[a], Caroline C. Ummenhofer[b], Helen E. Phillips[c], and Janet Sprintall[d]

[a]Horn Point Laboratory, University of Maryland Center for Environmental Science, Cambridge, MD, United States, [b]Department of Physical Oceanography, Woods Hole Oceanographic Institution, Woods Hole, MA, United States, [c]Institute for Marine and Antarctic Studies and Australian Antarctic Program Partnership, University of Tasmania, Hobart, TAS, Australia, [d]Climate, Atmospheric Sciences and Physical Oceanography, Scripps Institution of Oceanography, University of California San Diego, La Jolla, CA, United States

1 Introduction

The Indian Ocean is a remarkable place. Unlike the Pacific and Atlantic, the Indian Ocean has a low-latitude land boundary to the north and the Indian subcontinent partitions the northern basin (Fig. 1). The partitioning of the northern basin effectively creates two subbasins consisting of the Arabian Sea and the Bay of Bengal, where the differences in evaporation, precipitation, and river runoff give rise to pronounced differences in salinity and stratification between subbasins. As a result of the proximity of the high mountainous terrain of the Eurasian land mass and the heating and cooling of air masses over it, the tropical Indian Ocean is subject to strong monsoonal wind forcing that reverses seasonally (for reviews, see Schott & McCreary, 2001; Ummenhofer et al., 2024a). In the northern hemisphere, these are referred to as the Southwest Monsoon winds (blowing from the southwest toward the northeast during the boreal summer) and the Northeast Monsoon winds (blowing from the northeast toward the southwest in the boreal winter). The Coriolis effect causes the winds to change their direction at the equator, and in the Southern Hemisphere, they become the Southeast Monsoon winds (blowing from the southeast toward the northwest in the austral winter) and the Northwest Monsoon winds (blowing from the northwest toward the southeast in the austral summer), respectively (Fig. 1). These winds, combined with the large differences in salinity and stratification, profoundly impact marine physical, biogeochemical, and ecosystem dynamics in the Arabian Sea and the Bay of Bengal that extend into the southern tropical Indian Ocean (Brewin et al., 2012; Hood et al., 2009, 2017, 2024a, 2024b; Marsac et al., 2024; Phillips et al., 2021; Schott & McCreary, 2001; Sprintall et al., 2024).

The winds along the equator in the Indian Ocean are stronger during the westerlies of the Southwest Monsoon and, therefore, downwelling favorable on average (Schott et al., 2009; Wang & McPhaden, 2017). As a result, the near-surface waters along the equator tend to be oligotrophic, i.e., typically undetectable nitrate, phosphate, and silicate concentrations down to ∼100 m (see Hood et al., 2024a). Upwelling centers in the Indian Ocean are generally found in off-equatorial regions. Coastal upwelling occurs in the Arabian Sea and the Bay of Bengal in response to the Southwest Monsoon winds. Local Southwest Monsoon-induced wind stress curl drives upwelling in the southern tropical Indian Ocean between ∼5–15°S and ∼50–80°E at the Seychelles-Chagos Thermocline Ridge (SCTR) (Hermes & Reason, 2008; McPhaden & Nagura, 2014; Nyadjro et al., 2017; Xie et al., 2002; Yokoi et al., 2008) and the east of Sri Lanka between ∼6–10°N and ∼82–87°E in Sri Lanka Dome (Burns et al., 2017; Schott & McCreary, 2001; Shankar et al., 2002; Vinayachandran & Yamagata, 1998). In addition, upwelling occurs along the coasts of Sumatra, Java, and Bali in Indonesia in response to the Southeast Monsoon winds (Hood et al., 2017, 2024a; Sprintall et al., 1999; Susanto et al., 2001).

In the southeastern tropical Indian Ocean, the Indonesian Throughflow (ITF) connects the Pacific and Indian Ocean basins. The ITF influences water mass properties of the Indian Ocean through exchanges of heat and freshwater (Schott & McCreary, 2001) and biogeochemical properties through exchanges of nutrients (Ayers et al., 2014; Hood et al., 2024a; Sprintall et al., 2024; Talley & Sprintall, 2005) and presumably also planktonic organisms (Hood et al., 2017).

The southern Indian Ocean subtropical gyre is an oligotrophic ocean habitat. To the north, the South Equatorial Current transports warm, nutrient-enriched freshwater from the ITF across the basin (Schott & McCreary, 2001; Sprintall et al., 2024). To the west, the Agulhas Current is an unusually large, poleward-flowing, upwelling-favorable western boundary

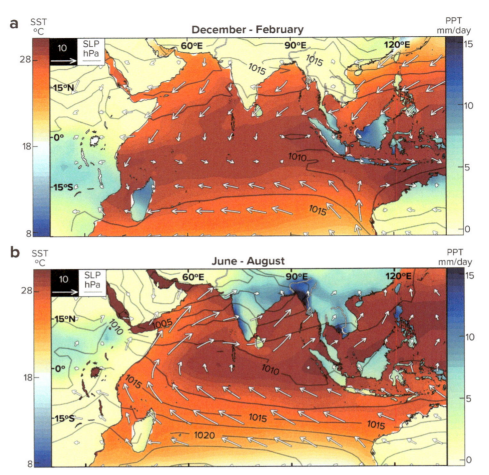

FIG. 1 Average sea surface temperature (SST in °C) with surface wind vectors (m/s), sea level pressure (SLP contours in hPa), and land precipitation (*blue shading* over land in mm/day) for (a) December through February and (b) June through August. Climatological conditions of SLP and winds based on NCEP/NCAR reanalysis (Kalnay, 1996), precipitation based on CMAP (Xie & Arkin, 1997), and SST on NOAA OISST version 2 (Reynolds et al., 2002) for the period 1982–2019.

current (Beal et al., 2011, 2015; Bryden et al., 2005; Hood et al., 2017, 2024a, 2024b; Sprintall et al., 2024). To the east, the Leeuwin Current is a small current that is unusual in being a southward-flowing eastern boundary current and in having the largest eddy kinetic energy among all midlatitude eastern boundary current systems (Feng et al., 2005, 2007, 2010; Hood et al., 2017, 2024a, 2024b; Phillips et al., 2021, 2024; Sprintall et al., 2024).

Finally, a significant fraction of the world's population lives in the coastal and interior regions of Indian Ocean rim countries, and they are directly impacted by the variability of the monsoons and associated rains. Moreover, many of these populations reside in low-lying coastal zones and island nations and are therefore threatened by tropical cyclones and sea level rise, both of which accelerate coastal erosion and the degradation of coastal ecosystems. Other Indian Ocean processes—such as seasonal variations in oceanic circulation and their associated biogeochemical and ecological responses—also directly and indirectly impact these populations through their influence on fisheries and food supplies (Feng et al., 2024; Hood et al., 2015; Marsac et al., 2024; Walker, 2024).

This chapter provides a brief overview of the history, geology, and physical and biogeochemical variability in the Indian Ocean. The reader is referred to the individual chapters in this book for greater detail.

2 Research history

The Indian Ocean was largely overlooked in the early days of oceanography, e.g., the Challenger expedition (1872–76) made only a single leg from Cape Town to Melbourne (see the review by Benson & Rehbock, 2002). The John Murray Expedition was the first major expedition to the Indian Ocean. This effort was focused on the Arabian Sea in 1933–34 (Sewell, 1934) and led to the discovery of the Arabian Sea Oxygen Minimum Zone (OMZ). During the International Geophysical Year (1957–58), oceanographic exploration of the southern Indian Ocean was carried out by Australian, French,

Japanese, New Zealand, and Soviet researchers. The International Indian Ocean Expedition (IIOE) in the early 1960s was the second major expedition to the Indian Ocean. The IIOE was a basin-wide interdisciplinary study that involved 46 research vessels from 14 different countries (Behrman, 1981; Hood et al., 2015). Among its many legacies, the IIOE led to the publication of the first oceanographic atlas of the basin (Wyrtki et al., 1971) and a detailed map of the Indian Ocean bathymetry (Heezen & Tharp, 1966). The latter provided crucial information that contributed to the development of the theory of plate tectonics (Hood et al., 2015).

Subsequent measurements were made in the Indian Ocean under the Indian Ocean Experiment (INDEX 1979) and the Geochemical Ocean Section Study (Moore, 1984) in the 1970s, the Tropical Ocean Global Atmosphere (TOGA) program (1985–94; McPhaden et al., 1998; Webster et al., 1998), the Joint Global Ocean Flux Study (JGOFS; Fasham, 2003), and the World Ocean Circulation Experiment (WOCE) in the 1990s (Woods, 1985). The Geochemical Ocean Section Study was a global survey of the three-dimensional distributions of chemical, isotopic, and radiochemical tracers in the ocean. A key objective was to investigate the deep thermohaline circulation. The TOGA program did not do as much in the Indian Ocean as it did in the Pacific. Yet, it still led to some research progress related to physical variability, ocean-atmosphere interactions, and the influence of the El Niño-Southern Oscillation (ENSO) in the Indian Ocean. The JGOFS Arabian Sea Process Study was an interdisciplinary effort that contributed greatly to the current understanding of monsoon-forced biogeochemical variability in the Arabian Sea. WOCE provided the first systematic large-scale surface to bottom sampling of the Indian Ocean along multiple zonal and meridional transects.

The CLIVAR Repeat Hydrography project (Gould et al., 2013) in the first decade of the 21st century maintained selected transects from the WOCE sampling, and this global effort has continued through to the present under the ongoing GO-SHIP program (Talley et al., 2017). These large-scale survey programs have been consolidated with the implementation of the Indian Ocean Observing System (IndOOS) starting in 2006 (Beal et al., 2020; McPhaden et al., 2024). IndOOS is a multinational network of sustained oceanic measurements that is composed of five in situ observing networks: profiling floats (Argo), a moored tropical array (Research Moored Array for African-Asian-Australian Monsoon Analysis and Prediction; RAMA), repeat lines of temperature profiles (expendable bathythermograph, XBT, network), surface drifters, and tide gauges. The Second International Indian Ocean Expedition (IIOE-2; Hood et al., 2015) was launched in 2015. Like the IIOE, IIOE-2 is a basin-wide interdisciplinary study involving many countries. It is slated to continue through 2030. In addition to these major international efforts, there have been numerous national individual expeditions and programs that have contributed substantially to the inventory of measurements in the Indian Ocean.

Apart from the Geochemical Ocean Section Study, JGOFS, IIOE, and IIOE-2, most of the international programs have been primarily focused on atmospheric and physical oceanographic processes. As a result, understanding of the complex physical dynamics of the Indian Ocean has advanced much more rapidly compared to other disciplines. The WOCE, CLIVAR, and GO-SHIP transects have had an increasing emphasis on collecting biogeochemical measurements providing crucial information about carbon and nutrient concentrations and distributions in the Indian Ocean. Nonetheless, the complex physical dynamics that give rise to compound biogeochemical responses are not well sampled or understood in many subregions of the Indian Ocean. This is particularly true for the equatorial and Southern Hemisphere regions where much of the current understanding of biogeochemical variability is based on satellite remote sensing and numerical modeling (Hood et al., 2015, 2024a, 2024b).

3 Geology

The Indian Ocean was formed as a result of the breakup of the southern supercontinent Gondwana about 150 million years ago (Norton & Sclater, 1979). This breakup was due to the movement to the northeast of the Indian subcontinent about 125 million years ago. The Indian subcontinent collided with Eurasia about 50 million years ago, which coincided with the western movement of Africa and the separation of Australia from Antarctica. The approximate present configuration of the Indian Ocean was established by 36 million years ago. Most of the Indian Ocean basin is less than 80 million years old (Hood et al., 2015).

The oceanic ridges in the Indian Ocean are part of the worldwide oceanic ridge system where seafloor spreading occurs. The ridges form a remarkable triple junction in the shape of an inverted "Y" on the ocean floor of the Indian Ocean (Hood et al., 2015; Kennett, 1982; Talley et al., 2011; Turcotte & Schubert, 2002; Fig. 2). Starting in the upper northwest with the Carlsberg Ridge in the Arabian Sea, the ridge turns due south past the Chagos-Laccadive Plateau and becomes the Mid-Indian (or Central Indian) Ridge. Southeast of Madagascar, the ridge branches into the Southwest Indian Ridge, which continues to the southwest until it merges into the Atlantic-Indian Ridge south of Africa, and the Southeast Indian Ridge which extends to the east until it joins the Pacific-Antarctic Ridge south of Tasmania. These tectonically complex and active spreading centers have significant impacts on deep ocean chemical distributions (Hood et al., 2015; Nishioka et al., 2013; Vu & Sohrin, 2013), and they support diverse hydrothermal vent deep-sea communities (see http://www.interridge.org/WG/VentEcology).

4 The Indian Ocean and its role in the global climate system

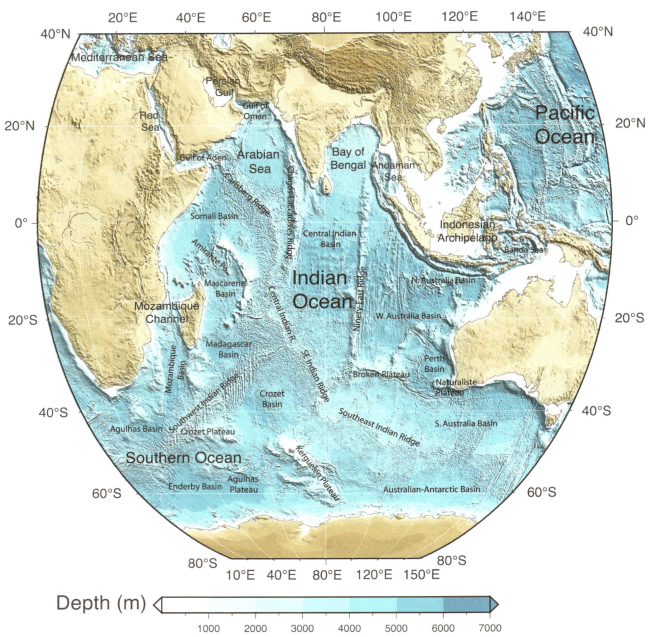

FIG. 2 Map of Indian Ocean topography, derived from ETOPO2 bathymetry data from NOAA NGDC. *(From Talley et al. (2011).)*

In addition, there are several prominent aseismic ridges in the Indian Ocean (Fig. 2). Perhaps the most striking is the Ninety East Ridge. It is the straightest and longest ridge in the world ocean (Hood et al., 2015; Kennett, 1982). It runs northward along the 90° E meridian for 4500 km from the zonal Broken Ridge at latitudes 31°S to 9°N. Other important aseismic ridges in the Indian Ocean include the Madagascar, Chagos-Laccadive, and Mascarene plateaus (Fig. 2). These ridges, which extend to the surface to form island chains in many places, can have a profound impact on both surface and deep circulations in the Indian Ocean, and they can dramatically enhance biological productivity in the surface waters in regions where surface currents interact with topographic features and island chains, which can lead to nutrient and/or trace metal fertilization and pronounced island-wake effects (e.g., Strutton et al., 2015).

The deep Indian Ocean basins are characterized by relatively flat plains of thick sediment that extend from the flanks of the oceanic ridges (Divins, 2003; Hood et al., 2015). The Indian Ocean's ridge topography defines several separate basins that range in width from 320 to 9000 km across. They include the Arabian, Somali, Mascarene, Madagascar, Mozambique, Agulhas, and Crozet basins in the west and the Central Indian Ocean, and the Wharton and South Australia basins in the east (Fig. 2).

The continental shelves of the Indian Ocean are, for the most part, relatively narrow, extending to an average width of only 120 km (Talley et al., 2011; Fig. 2). The widest shelves are found off India near Mumbai (Bombay) and off northwestern Australia. The shelf break is typically found at a depth of about 150 m. The Ganges, Indus, and Zambezi rivers have carved particularly large canyons into the shelf breaks and slopes in the Bay of Bengal, Arabian Sea, and off Mozambique, respectively.

Finally, it should be noted that the tectonic activity associated with subduction zones of the Java Trench and the Sunda Arc trench system has generated numerous tsunamis and volcanic eruptions over geologic time, which have had widespread impacts in the Indian Ocean (Feng et al., 2024; Shen-Tu, 2016). Most recently, the Indian Ocean Tsunami of December 26, 2004 claimed more than 283,000 human lives in 14 countries, inundating coastal communities with waves up to 30 m high (Feng et al., 2024; Lay et al., 2005; Shen-Tu, 2016). This tsunami was one of the deadliest natural disasters in recorded history. Indonesia was the hardest-hit country, followed by Sri Lanka, India, and Thailand. Moreover, the geography of Indonesia is dominated by volcanoes that are generated by these subduction zones. As of 2012, Indonesia has 127 active volcanoes, and about 5 million people live and work within the volcanic danger zones (Hood et al., 2015).

4 Oceanography

4.1 Ocean circulation

4.1.1 Near-surface currents

The atmospheric and oceanic circulation of the Indian Ocean are different from those in the Pacific and Atlantic, largely due to geography. The Asian landmass limits the northern extent of the Indian Ocean to only ~25°N, so there is no dense water formation at high northern latitudes as observed in the Atlantic and Pacific. As discussed earlier, the intense seasonal heating and cooling of air masses over Asia drives the seasonal monsoons (Fig. 1). Nonetheless, the timing of the onset and relaxation of the monsoons, their strength, and the associated wet and dry periods in the northern Indian Ocean rim countries are somewhat variable due to the influence of multiple larger-scale climate modes and smaller-scale ocean-atmosphere interactions (Phillips et al., 2024). The seasonally reversing winds drive corresponding surface coastal ocean currents in the northern Indian Ocean (Fig. 3), i.e., anticyclonic upwelling circulations during the Southwest Monsoon and cyclonic downwelling circulations during the Northeast Monsoon (Hood et al., 2017; Phillips et al., 2024). During the spring and fall inter-monsoon transitions, the winds relax and drive the equatorial Wyrtki Jets that flow rapidly from the west to the east along the equator (McPhaden et al., 2015; Strutton et al., 2015; Wyrtki, 1973).

The connection with the Pacific Ocean through the ITF in the southeastern tropical Indian Ocean also strongly influences Indian Ocean circulation patterns. The warm and fresh ITF into the Indian Ocean sets up a north-south pressure gradient that drives westward transport in the South Equatorial Current (Phillips et al., 2024) and southward transport in the Leeuwin Current (Hood et al., 2017; Fig. 3). The Leeuwin Current is a relatively small (<5 Sv) eastern boundary current (Godfrey & Ridgway, 1985; Feng et al., 2003; Hood et al., 2017; Sprintall et al., 2024) with elevated kinetic energy that sheds relatively high chlorophyll-*a* (Chl*a*), warm-core, downwelling eddies westward into open ocean waters (Feng et al., 2005, 2007, 2010; Hood et al., 2017; Waite et al., 2007). The South Equatorial Current crosses the southern tropical Indian Ocean between ~10°S and 20°S and feeds into the northward-flowing East African Coastal Current and the southward-flowing Agulhas Current (Fig. 3). The Agulhas is the largest Southern Hemisphere western boundary current (60–85 Sv) with sources derived from the Mozambique Channel, the Southeast Madagascar Current, and the southwest Indian Ocean subgyre (Hood et al., 2017; Phillips et al., 2024; Stramma & Lutjeharms, 1997). It flows southwestward along the coast of southern Africa from ~25°S to 40°S where most of it retroflects sharply eastward back into the Indian Ocean (becoming the eastward-flowing Agulhas Return Current), while shedding eddies ("Agulhas Rings") that propagate into the Atlantic (Hood et al., 2017; Lutjeharms, 2006; Sprintall et al., 2024; Fig. 2). Meanders and eddies in the Agulhas Current propagate alongshore and interact with winds and topographic features which gives rise to seasonally variable localized upwelling and downwelling (Hood et al., 2017; Lutjeharms, 2006; Phillips et al., 2024).

In the central-eastern South Indian Ocean, the surface flow is generally eastward between 20°S and 30°S (Fig. 3; Godfrey & Ridgway, 1985; Phillips et al., 2024; Schott et al., 2009; Sharma, 1976; Sharma et al., 1978). This flow is driven by the large-scale, poleward drop in sea surface height (Godfrey & Ridgway, 1985; Phillips et al., 2024; Schott et al., 2009) that is associated with the transition from the warm and fresh South Equatorial Current waters to the cooler, saltier, and

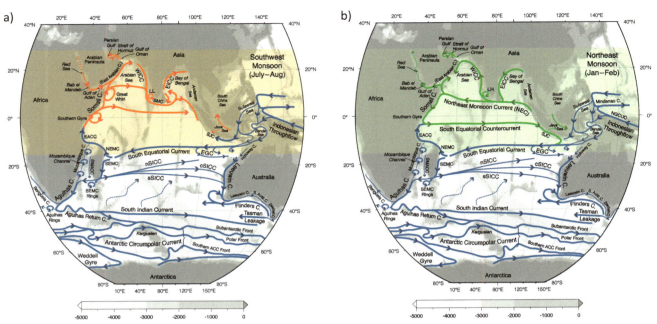

FIG. 3 Schematic near-surface circulation during the Southwest Monsoon (July–August, left panel). Schematic near-surface circulation during the Northeast Monsoon (January–February, right panel). *Blue*: year-round mean flows with no seasonal reversals. *Orange*: monsoonally reversing circulation (after Schott & McCreary, 2001). The ACC fronts are taken directly from Orsi et al. (1995). Acronyms: *EACC*, East African Coastal Current; *NEMC*, Northeast Madagascar Current; *SEMC*, Southeast Madagascar Current; *SMACC*, Southwest MAdagascar Coastal Current; *WICC*, West Indian Coastal Current; *EICC*, East Indian Coastal Current; *LH and LL*, Lakshadweep high and low; *SJC*, South Java Current; *EGC*, Eastern Gyral Current; *SICC*, South Indian Countercurrent (south, central, and southern branches); *NEC*, Northeast Monsoon Current. Updated from Talley et al. (2011), originally based on Schott and McCreary (2001). The *light gray* shading shows seafloor bathymetry. *(Modified from Phillips et al. (2021).)*

denser waters to the south. Narrower eastward jets are embedded in this general eastward flow (Fig. 3; Divakaran & Brassington, 2011; Maximenko et al., 2009; Menezes et al., 2014; Phillips et al., 2024) that starts as a single flow near the southern tip of Madagascar around 25°S and then divide into separate jets at the Central Indian Ridge (65°E–68°E) (Fig. 3; Menezes et al., 2014; Phillips et al., 2024).

Variability in the atmospheric and oceanic circulation of the Indian Ocean is the result of complex interactions that are both internal and external to the Indian Ocean. The major drivers of this variability are still not fully understood (Beal et al., 2019, 2020; Phillips et al., 2024).

4.1.2 The overturning circulation

The Indian Ocean overturning cells are important for the redistribution and transport of heat and other properties southward from the tropical Indian Ocean (Figs. 4 and 5; Beal et al., 2020; Phillips et al., 2024; Schott et al., 2002). The shallow overturning (<500m) in the Indian Ocean (Figs. 4 and 5) consists of the cross-equatorial cell and the southern cell (Beal et al., 2020; Miyama et al., 2003; Phillips et al., 2024; Schott et al., 2004). The ascending branches of these cells, which connect to different upwelling zones in the southern and northern Indian Ocean (Figs. 4 and 5), play an important role in regulating the heat balance in the tropical Indian Ocean (Beal et al., 2020; Lee, 2004; Lee & McPhaden, 2008; Sprintall et al., 2024). In a deeper cell (>500m), highly oxygenated mode and intermediate waters of various density classes enter the Indian Ocean from the Southern Ocean at lower thermocline to intermediate depths and move northward. The upper part of this cell mixes with lighter water above and upwells to the sea surface, and then returns southward via wind-driven Ekman transport of near-surface waters (Fig. 4; Beal et al., 2020; Schott et al., 2009; Sprintall et al., 2024). Deep and abyssal water masses flow equatorward in deep western boundary currents along the Madagascar and African coasts in the western Indian Ocean, and along the Southeast Indian and Ninety East Ridges in the eastern Indian Ocean. These deep waters are upwelled within the Indian Ocean and then mix together with the lower part of the mode and intermediate waters and return south to exit the basin at shallower depths via the Agulhas Current as part of the return flow within the meridional overturning circulation (Beal et al., 2020; Sprintall et al., 2024).

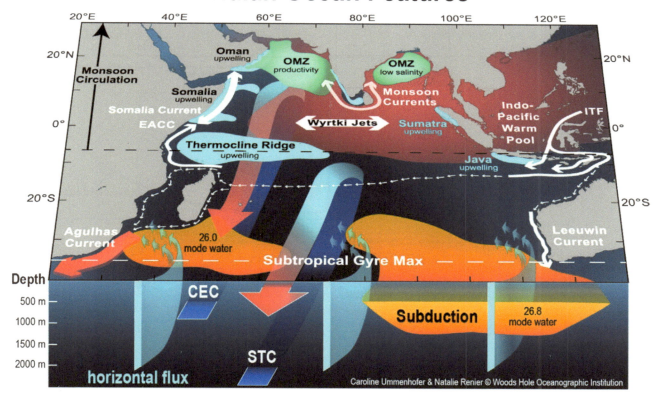

FIG. 4 Indian Ocean main oceanographic features and phenomena. The surface circulation seasonally reverses north of 10°S under the influence of monsoons. The summer monsoon also promotes the intense Somali current as well as upwelling and high productivity in the western Arabian Sea. High surface layer productivity, sinking of biomass, and its remineralization at depth also lead to the formation of subsurface oxygen minimum zones (OMZs) in the Arabian Sea and Bay of Bengal. The Indo-Pacific warm pool is a region of intense air-sea interactions, where the Madden-Julian oscillation, monsoon intraseasonal oscillation, and Indian Ocean dipole develop. The Indian Ocean is a gateway of the global oceanic circulation, with inputs of heat and freshwater through the Indonesian Throughflow, which exit the basin though boundary currents, mainly the Agulhas Current along Africa, but also the Leeuwin Current along Australia. There are two vertical overturning cells connecting subducted waters south of 30°S to the tropical Indian Ocean: the shallow subtropical overturning cell where water upwells in the "thermocline ridge" open-ocean upwelling region, and the cross-equatorial cell where water upwells farther north in the Arabian Sea of the coast of Somalia and Oman. These cells are the main source of subsurface ventilation due to the presence of continents to the north. *(From Beal et al. (2020).)*

4.2 Upper-ocean structure

4.2.1 Sea surface temperature (SST)

The tropical Indian Ocean includes the largest fraction of SST warmer than 28°C (relative to the size of the basin) in any ocean basin and has warmed faster than either the tropical Pacific or Atlantic (Fox-Kemper et al., 2021; Han et al., 2014; Phillips et al., 2024), with implications for primary productivity (Roxy et al., 2014, 2016, 2024). Over the past decade, the Indian Ocean has accounted for 50%–70% of the total global upper-ocean (700m) heat uptake, due to global warming (Beal et al., 2020; Lee et al., 2015; Phillips et al., 2024; Ummenhofer et al., 2021). The patterns of SST variability in the Indian Ocean are largely driven by the spatial and seasonal variability in net heat fluxes at the sea surface (Beal et al., 2020; Phillips et al., 2024; Yu et al., 2007). In the southern Indian Ocean, this results in a seasonally driven SST pattern with the lowest SST during the austral winter and the highest SST during the austral summer (Fig. 1). In contrast, in the northern Indian Ocean, there is cooling and lower SST during the Northeast Monsoon due to the influence of the monsoon winds. Warming and higher SST occur there during the rest of the year and especially during spring and fall inter-monsoon periods when the winds are weak (Beal et al., 2020; Phillips et al., 2024; Yu et al., 2007). One important exception to this general pattern is the SCTR, where the seasonal cycle of SST is also strongly influenced by substantial intraseasonal variations in upwelling

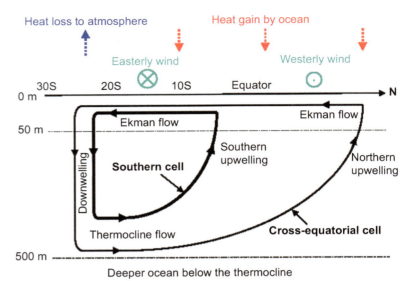

FIG. 5 Conceptual illustration of the time-mean meridional overturning circulation of the upper Indian Ocean that consists of a southern and a cross-equatorial cell. The time-mean zonal wind and surface heat flux are also shown schematically. This flow is believed to partially supply the cross-equatorial thermocline flow. *(From Lee (2004).)*

(Beal et al., 2020; Foltz et al., 2010; Hermes & Reason, 2008; Phillips et al., 2024; Ummenhofer et al., 2021; Vialard et al., 2008). The seasonally reversing monsoon winds also drive upwelling and downwelling patterns that lead to more complex SST variability in the western Arabian Sea and in the coastal zones of Java and Bali (Beal et al., 2020; Chowdary et al., 2015; Hood et al., 2017; Phillips et al., 2024; Sprintall et al., 2024; Yu et al., 2007).

4.2.2 Upper-ocean stratification

Mixed layer depth in the northern Indian Ocean exhibits a strong seasonal cycle that is associated with the seasonally reversing monsoon winds (Beal et al., 2020; de Boyer Montegut et al., 2007; McCreary & Kundu, 1989; Phillips et al., 2024; Prasad, 2004; Rao et al., 1989; Rao & Sivakumar, 2003; Schott et al., 2002; Sreenivas et al., 2008). In the Arabian Sea, the winds induce Ekman transport that generates convergences and divergences that vary with both the monsoon phase and from the coastal to offshore regions. The Bay of Bengal exhibits weaker seasonal mixed layer depth variations due to the strong salinity stratification induced by river runoff that prevents entrainment mixing, particularly in the northern part of the Bay (Babu et al., 2004; Beal et al., 2020; Gopalakrishna et al., 2002; Narvekar & Kumar, 2006; Phillips et al., 2024; Prasad, 2004; Rao et al., 1989; Shenoi et al., 2002). In addition, temperature inversions are generated by river runoff and wintertime surface cooling near the surface in the northern Bay of Bengal. In this region, the surface water becomes cooler than the subsurface in winter owing to surface cooling, but still, the stratification is stable, due to freshwater supply by the rivers (Girishkumar et al., 2013; Nagura et al., 2015; Shetye et al., 1996; Thadathil et al., 2002). At the equator, the predominantly eastward flow causes a deepening of mixed layer depth from the west to the east (Ali & Sharma, 1994; Beal et al., 2020; O'Brien & Hurlburt, 1974; Phillips et al., 2024), although heavy precipitation results in shoaling of the mixed layer depth west of Sumatra during the Northwest Monsoon (Beal et al., 2020; Du et al., 2005; Masson et al., 2002; Phillips et al., 2024; Qu & Meyers, 2005). Thin mixed layers are associated with the SCTR due to upwelling (Beal et al., 2020; McCreary et al., 1993; Phillips et al., 2024; Resplandy et al., 2009; Vialard et al., 2009). In general, seasonal mixed layer depth shoaling and deepening in the southern Indian Ocean is related to the annual cycle of net surface heat flux and wind, with mixed layers deepening during the austral winter due to increased surface heat loss and increased winds and vice versa during the austral summer (Beal et al., 2020; Foltz et al., 2010; McCreary et al., 1993; Phillips et al., 2024).

Fresher surface layers can result in shallow mixed layers overlying a strongly stable salt-stratified pycnocline within a deeper isothermal layer: the region between the mixed layer depth and the isothermal layer is referred to as the barrier layer (Lukas & Lindstrom, 1991; Sprintall & Tomczak, 1992). The thickest barrier layers in the Indian Ocean are in the northern Bay of Bengal as a result of river runoff and in the equatorial waters off Sumatra as a result of precipitation and perhaps advection of fresh water from the Bay of Bengal (Fig. 6; Felton et al., 2014; Qu & Meyers, 2005; Sprintall & Tomczak, 1992). Significant seasonal variability of the barrier layer thickness in the Indian Ocean arises due to the interplay of surface water convergence, wave propagation, and precipitation associated with monsoon forcing (Felton et al., 2014; Qu & Meyers, 2005). Barrier layer thickness can impact the summer monsoon, development of the Indian Ocean Dipole (IOD), and tropical cyclone-induced upwelling in the Indian Ocean (Balaguru et al., 2012; DeMott et al., 2024; Feng

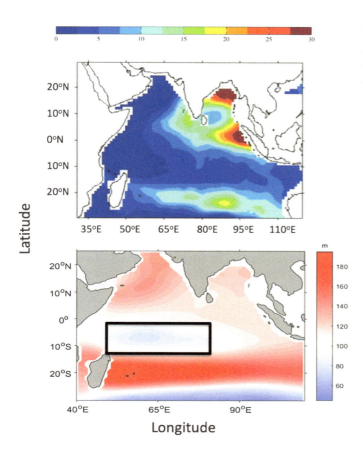

FIG. 6 Top panel: Annual mean barrier layer thickness in the Indian Ocean estimated from Argo data. Bottom panel: Climatological distribution of the thermocline depth in the Indian Ocean. The box denotes the approximate location of the SCTR. *(Top panel: Modified from Felton et al. (2014). Bottom panel: Modified from Yuan (2020).)*

et al., 2024; Guo et al., 2013; Kumari et al., 2018; Masson et al., 2005; Qiu et al., 2012; Sengupta et al., 2008; Tozuka et al., 2024; Yamagata et al., 2024).

As with mixed layer depth, the thermocline depth is strongly influenced by the monsoon winds and Ekman pumping in the northern Indian Ocean, with coastal upwelling during the Southwest Monsoon causing shoaling of the thermocline in coastal waters and deepening of the thermocline due to convergence and downwelling further offshore (Beal et al., 2020; Phillips et al., 2024; Fig. 6). However, remotely wind-forced planetary waves also impact thermocline depth in the northern Indian Ocean. For example, coastal Kelvin waves that originate in the eastern equatorial Indian Ocean can alter coastal currents, upwelling and downwelling signatures, and thermocline depth in both the Bay of Bengal and Arabian Sea (Beal et al., 2020; Hood et al., 2017; Phillips et al., 2024). The shallowest thermocline in the tropical Indian Ocean is associated with upwelling in the SCTR (Beal et al., 2020; McCreary et al., 1993; Phillips et al., 2024; Resplandy et al., 2009; Vialard et al., 2009; Fig. 6). The thermocline depth in the SCTR varies seasonally, shoaling in May and December and deepening in February and September, in response to the interaction between local wind-driven Ekman pumping and remotely forced downwelling Rossby waves that also influence interannual thermocline variability (Beal et al., 2020; Nyadjro et al., 2017; Phillips et al., 2024; Tozuka et al., 2010; Trenary & Han, 2012; Yu et al., 2005). At higher latitudes, the wind-induced Ekman pumping drives convergence in the subtropics deepening the thermocline and divergence further south leads to a shoaling thermocline (Fig. 6; Phillips et al., 2024; Sprintall & Tomczak, 1992).

4.3 Interocean basin connections and heat transport

The signature of the ITF water masses is readily identifiable as a low-salinity surface layer (Gordon et al., 1997; Hood et al., 2024a; Sprintall et al., 2024) separated from a low-salinity, high-silicate intermediate depth core that stretches across nearly the entire Indian Ocean within the South Equatorial Current (Sprintall et al., 2024; Talley & Sprintall, 2005). The ITF water masses then feed into the East African Coastal Current and the Agulhas Current, the latter via the Mozambique Channel and the Southeast Madagascar Current (Sprintall et al., 2024; Fig. 3). A smaller ITF contribution exports fresh water and heat into the poleward-flowing Leeuwin Current. The ITF nutrient inputs support a substantial amount of new production in the

Indian Ocean and significantly impact basin-wide biogeochemical cycles (Ayers et al., 2014; Hood et al., 2024a). The ITF itself is influenced by regional climate forcing from intraseasonal to interannual to decadal timescales. Yet, there remains a relatively poor understanding of the water mass mixing within the Indonesian seas and nutrient fluxes via the ITF due to the lack of in situ physical observations, and particularly the lack of biogeochemical measurements (Sprintall et al., 2024).

A portion of the western boundary East Australian Current Extension that is not reflected back into the Pacific Ocean turns south around Tasmania, Australia, and finds its way into the Indian Ocean as the "Tasman leakage" (Fig. 3; Ridgway & Dunn, 2007; Sprintall et al., 2024) and the Flinders Current that follows the southern Australian shelf-break (Fig. 3; Duran et al., 2020; Sprintall et al., 2024). Together with the ITF from the Pacific to the north of Australia, the Pacific water leakage south of Australia constitute important interocean contributions to the surface return flow of the global thermohaline circulation. The water masses derived from the ITF and Tasman leakage undergo significant changes along their pathway through the Indian Ocean, ultimately feeding into the Atlantic via the Agulhas Current ("Agulhas Rings") (Fig. 3; Speich et al., 2002; Sprintall et al., 2024).

The Indian Ocean has the world's largest southward meridional heat transport (Lumpkin & Speer, 2007), with an estimated 1.5 PW exiting the basin across 32°S. This export of heat balances the net heat gain from the atmosphere and the heat transport into the Indian Ocean from the Pacific via the ITF (Fig. 3) that is then advected out of the basin via horizontal currents and the aforementioned shallow overturning cells (Sprintall et al., 2024). The South Equatorial Current transports heat gained from the atmosphere and the ITF westward and the Agulhas Current then moves this heat poleward and into the Atlantic at surface and intermediate depths (Bryden & Beal, 2001; Sprintall et al., 2024). The Leeuwin Current makes a much smaller contribution to the poleward flow of heat from the ITF (Feng et al., 2003; Furue et al., 2017; Smith et al., 1991; Sprintall et al., 2024). However, the mesoscale eddies generated by the Leeuwin Current carry heat into the interior of the Indian Ocean, which ultimately contribute to heat export across 32°S (Dilmahamod et al., 2018; Domingues et al., 2006, Feng et al., 2007; Phillips et al., 2024; Sprintall et al., 2024).

5 Atmosphere

The South Asian monsoon represents the largest monsoon system on Earth. It is also a lifeline for nearly 2 billion people in the region, who receive 80% of their annual precipitation during the summer monsoon (Fig. 1), setting the rhythm of agricultural production (Singh, 2016; Ummenhofer et al., 2024a). Even small perturbations in the timing of onset, seasonal distribution, and monsoon intensity can have disproportionate effects on agriculture, most of which is rain-fed (Kumar et al., 2005; Parthasarathy et al., 1988). Historically, droughts have occasioned widespread famine and even today trigger large-scale impoverishment and rural-to-urban migration. For example, a 20% reduction in monsoon rainfall in 2002 resulted in billions of dollars in damage to the Indian economy (Gadgil et al., 2004); weakening of the monsoons and ensuing drought have had major societal effects and even upheavals throughout Asia (e.g., Cook et al., 2010; Mohtadi et al., 2024; Ummenhofer et al., 2024a).

Traditionally characterized to first order as continent-scale "sea breezes," where in summer the land heats faster than the ocean, causing warm air to rise over the continent and moist air to be drawn in from the ocean, the South Asian monsoon exhibits seasonally reversing wind and precipitation patterns (Geen et al., 2020; Wang, 2006; Wang et al., 2005; Webster et al., 1998). Yet, more recently, theoretical advances have led to the emergence of the monsoon as a regional manifestation of the seasonal variation of the global tropical atmospheric overturning and migration of the associated convergence zones (Geen et al., 2020). Monsoon wind strength and precipitation vary on intraseasonal timescales, with so-called active and break spells, on interannual timescales in response to modes of climate variability, such as ENSO, as well as on multidecadal and longer timescales.

The Pacific Ocean's influence on the Asian monsoon is well established, going back to the late 19th century, when a catastrophic drought and famine in India in 1877 coincided with unusually high pressure over Australia (Taschetto et al., 2020). The atmospheric pressure seesaw between the Pacific and Indian Oceans, first reported by Hildebrandsson (1897) and coined the Southern Oscillation by Walker (1925), became one of the key predictors of the strength of the South Asian monsoon. The pressure changes are indicative of zonal displacements of the ascending and descending branches of the tropical Walker Circulation; El Niño events are associated with anomalous descent of dry air over South Asia that tends to reduce monsoonal precipitation (e.g., Kumar et al., 1999), while an enhanced South Asian summer monsoon is typically observed during La Niña events when an intensified ascending branch is located over Indonesia (Ummenhofer et al., 2024a).

This traditional model of ENSO-South Asian monsoon interaction has worked well in the past, yet forecasting centers failed to predict the wet monsoon season in 1997 during one of the strongest El Niño events on record, and the failure of the monsoon in 2002 during a weak El Niño (Taschetto et al., 2020). Recent studies suggest that the well-established relationships between ENSO and the Asian monsoon seem to be changing (e.g., Kumar et al., 1999), or that ENSO diversity plays a significant role. Regarding the latter, the location of maximum warming in the Pacific during an El Niño event, and the associated shift of the Walker circulation, appear to be important factors in determining whether or not an Indian summer monsoon failure materializes. Additionally, the role of other climate modes, such as the IOD and Indian Ocean Basin Mode, also modulate Indian summer monsoon rainfall (Anil et al., 2016; Ashok et al., 2001; Behera et al., 1999; Chang & Li, 2000; Li et al., 2001; Li & Zhang, 2002; Yamagata et al., 2024) and can modulate ENSO's influence on multi-decadal timescales (Ummenhofer et al., 2011, 2024a). Recent advances in regional modeling not only produce improved skill in simulating Indian summer monsoon rainfall but also better capture the transition of the active/break cycles and remote impacts by ENSO (Shinoda et al., 2024).

The Indian Ocean is characterized by very active intraseasonal variability (ISV, see DeMott et al., 2024), with far-reaching implications for regional climate. ISV of the atmosphere and ocean refers to disturbances with characteristic timescales spanning approximately 30–90 days. In the tropics, atmospheric ISV is chiefly driven by the intraseasonal oscillation (ISO)—a large-scale atmospheric disturbance often featuring organized cloud clusters so large that they span nearly the entire Indian Ocean basin. The ISO remotely affects global weather through numerous teleconnections across timescales, from mesoscale to interannual (Zhang, 2005, 2013). Perturbations in the atmosphere introduced by ISV modify the ocean via atmosphere-ocean coupled feedbacks, and vice versa (DeMott et al., 2015, 2024). These feedbacks act across the ocean-atmosphere interface through the fluxes of heat, freshwater, and momentum, which are themselves regulated by processes that control the wind speed, temperature, and moisture content of the atmospheric boundary layer and the heat content of the upper-ocean mixed layer. Theories and global modeling studies have demonstrated that local air-sea coupling is critical for a realistic simulation of the northward propagating ISV (DeMott et al., 2015, 2024; Fu et al., 2007, 2008): in the tropics, local air-sea coupling is effectively communicated to the deep troposphere via their impacts on convection, leading to well-defined regional air-sea coupled effects on monsoon precipitation and its ISV (DeMott et al., 2015, 2024; Shinoda et al., 2024).

Changes in the broader Indian Ocean region as seen in recent decades (Roxy et al., 2024), including a warming Indian Ocean, contribute to increasing monsoon droughts and floods, and premonsoon heatwaves over South Asia (Rohini et al., 2016; Roxy et al., 2015, 2017, 2024; Ummenhofer et al., 2024a; Wang et al., 2021). Yet, there is still considerable uncertainty about the future of the South Asian monsoon under anthropogenic forcing. Observations and climate models indicate a decreased Indian summer monsoon in the second half of the 20th century, primarily due to anthropogenic aerosol forcing, while in the long term, the South Asian monsoon is projected to increase over the 21st century and exhibit enhanced interannual variability (Doblas-Reyes et al., 2021; Douville et al., 2021; and references therein). The strong warming of the Indian Ocean in recent decades also has far-reaching implications for regional and global climate and weather patterns (Ummenhofer et al., 2024a, 2024b) and for Indian Ocean productivity (Roxy, 2014; Roxy et al., 2016, 2024).

Paleoclimate proxies provide a long-term context for Asian monsoon variability (Mohtadi et al., 2024; Ummenhofer et al., 2024a). A number of proxies have been used to reconstruct summer-monsoon wind strength in the Bay of Bengal and Arabian Sea where winds are 90% steady from the southwest at $\sim 15\,\mathrm{m\,s^{-1}}$ during the summer-monsoon months. Although the close association between the onset of summer monsoon rains over India and the abrupt strengthening of the southwesterly Somali Jet over the Arabian Sea is well established in the modern climatology (Boos & Emanuel, 2009), none of these wind-related proxies directly record rainfall. More broadly, millennial-scale variability in temperature and productivity of the Indian Ocean and rainfall over the surrounding continents appear to be controlled by changes in the strength of the oceanic global thermohaline circulation. Cooling of the North Atlantic (e.g., during Heinrich Stadials and the Younger Dryas) provokes rapid changes in the atmospheric circulation by displacing the Intertropical Convergence Zone (ITCZ) to the warmer Southern Hemisphere, inducing a weaker Northern Hemisphere monsoon and a drier and warmer northern Indian Ocean. As such, periods of a less vigorous Atlantic Meridional Overturning Circulation (AMOC) during the last glacial period are generally associated with anomalously dry conditions for the South Asian monsoon (Berkelhammer et al., 2012; Dutt et al., 2015). In contrast, periods of stronger AMOC and a relatively warmer North Atlantic result in generally wetter conditions at or north of the equator and drier conditions over the Southern Hemisphere monsoon regions (Mohtadi et al., 2014, 2024; Wurtzel et al., 2018).

6 Hydrology and hydrography

The Indo-Pacific Warm Pool (Fig. 7b), defined by the 28°C isotherm and associated with the ascending branch of the Walker Circulation, represents a key regional and global feature (Ummenhofer et al., 2024b). The Indo-Pacific Warm Pool is characterized by high climatological mean rainfall rates, with the precipitation maximum tapering off toward the western part of the basin (Fig. 7a). Lower precipitation is found in the subtropics, with minima near $\sim 20°$–$30°$S off the coast of

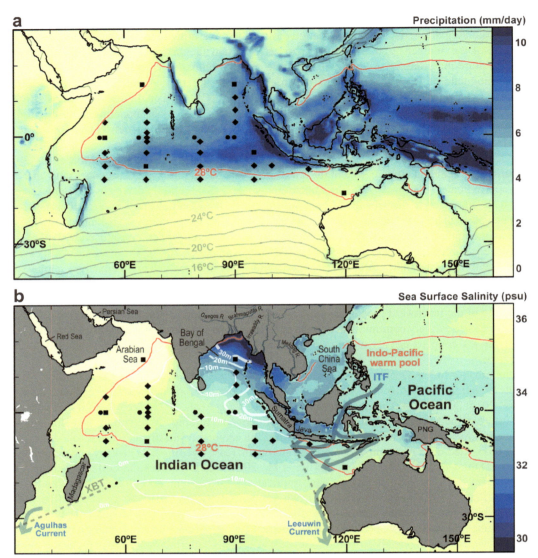

FIG. 7 (a) Average precipitation (shaded; in mm/day) and surface ocean temperature contours superimposed (in °C); (b) sea surface salinity (SSS, shaded; in psu) and barrier layer thickness (in m). The 28°C isotherm defines the Indo-Pacific Warm Pool. The *black symbols* denote positions of the RAMA array moorings in 2018, *gray dashed lines* key XBT lines; and major geographical features, rivers, and currents are indicated as well. *(Adapted from Ummenhofer et al. (2021) and barrier layer thickness from Felton et al. (2014).)*

Western Australia and in the northwestern Arabian Sea (Adler et al., 2017; Ummenhofer et al., 2024b). Both areas are also associated with high evaporation (not shown). There are strong spatial variations in precipitation, particularly in the tropics, near orographic features such as the west coasts of India, Myanmar, northern Thailand, Sumatra, and Borneo (Fig. 7a). Similarly, large variability in precipitation is caused by the seasonal march of the moisture transport of the monsoon system and the interannual movement of the Walker circulation in both zonal extent and magnitude associated with ENSO and IOD variability (Ummenhofer et al., 2024b; Yang et al., 2010).

To first order, the observed sea surface salinity (SSS; Fig. 7b) is expected to reflect the evaporation-minus-precipitation patterns (Ummenhofer et al., 2021). Using the same color scheme, Fig. 7 illustrates where the time-mean SSS pattern corresponds to the time-mean precipitation pattern; locations with a mismatch between the two fields are indicative of regions where effects from river discharge and ocean dynamics dominate. The maritime continent region is relatively fresh due to the direct input from extensive convection and precipitation as part of the rising branch of the Walker circulation, as well as Southeast Asian riverine input, for example, from the Mekong River system entering the South China Sea (Fig. 7b).

Freshwater from the Indonesian Seas lowers the salinity from the surface to ~600 m depth from where the ITF enters the southeast Indian Ocean at ~10°–15°S all the way across the Indian Ocean to the coast of Madagascar (Fig. 7b; Gordon, 1986; Hu et al., 2019; Sprintall et al., 2024; Talley & Sprintall, 2005). To the south of the ITF influence, the low surface salinity along 15°S is located about 6°–8° south of the rainfall maximum, and so is likely largely contributed through southward Ekman transport in response to the dominant easterly winds found south of 10°S in the Indian Ocean (Sengupta et al., 2006). On the other hand, the part of the fresh ITF that enters the poleward flowing Leeuwin Current acts to erode the effect of the local evaporation-precipitation maximum in this region such that relatively low SSS is found against the coast of Western Australia (Fig. 7b).

The northern Indian Ocean is characterized by stark salinity contrasts between the Arabian Sea and Bay of Bengal. River runoff from the Ganges, Brahmaputra, and Irrawaddy River systems (Fig. 7b) accounts for 60% of total riverine input north of 30°S (Sengupta et al., 2006). This produces the freshest surface waters of all the global tropical oceans (Fig. 7b) and is associated with strong salinity stratification in the upper 50–80 m in the Bay of Bengal (Mahadevan et al., 2016a, 2016b). However, the uncertainty in the freshwater distribution and mixing pathways for riverine input is high, and shallow, salinity-controlled mixed layers and the presence of barrier layers significantly affect the upper-ocean temperature (Sengupta et al., 2006; Thadathil et al., 2002; Wijesekera et al., 2016). In contrast, the saltiest waters of the Indian Ocean basin are found in the northern Arabian Sea influenced by the high-salinity inflow from the marginal Red and Persian Seas (Al-Yamani et al., 2024) that is largely evaporatively driven (Zhai et al., 2015). Subsurface salinity patterns in the upper 200 m within the Indian Ocean basin primarily reflect the SSS patterns (Hu et al., 2019).

7 Biogeochemistry, productivity, and fisheries

7.1 Nutrients, phytoplankton, and zooplankton

In the Arabian Sea, the strongest upwelling occurs during the Southwest Monsoon, driving nutrient enrichment and elevated phytoplankton productivity in coastal waters off Somalia, Yemen, and Oman and along western India (Fig. 8; see also Hood et al., 2017, 2024a, 2024b, and references cited therein). Multiple lines of evidence indicate that production in these upwelled waters is limited by silicate (Si) and iron (Fe) (Hood et al., 2024a; Moffett & Landry, 2020; Moffett et al., 2015; Naqvi et al., 2010; Wiggert et al., 2006; Fig. 9). In contrast, the Northeast Monsoon drives downwelling except in the northern central Arabian Sea where primary production increases due to wind-driven nutrient entrainment (Hood et al., 2017, 2024a; Wiggert et al., 2000, 2005). Yet, the seasonal and spatial variability in mesozooplankton biomass in the Arabian Sea is surprisingly weak (Baars, 1999; Hood et al., 2024a, 2024b; Madhupratap et al., 1992). The relative grazing contributions of meso- and microzooplankton vary spatially and seasonally in a way that is consistent with spatially separated co-regulation by grazing and iron limitation (Hood et al., 2024a; Moffett & Landry, 2020).

The biogeochemistry and productivity in the Bay of Bengal are influenced by the same factors that impact the Arabian Sea, but seasonal wind effects are less pronounced due to weaker winds and strong freshwater stratification (Gomes et al., 2000; Hood et al., 2017, 2024a, 2024b; Kumar et al., 2002, 2007; Thushara et al., 2019; Vinayachandran et al., 2002, 2005; Vinayachandran, 2009; Wijesekera et al., 2016; Fig. 8). Although Bay of Bengal surface waters are generally oligotrophic, there are regions of high Chl*a* and production that are associated with river plumes, wind-induced nutrient entrainment, and upwelling eddies (Hood et al., 2024a; Vinayachandran, 2009; Fig. 8). This variability in productivity drives significant fluctuations in mesozooplankton biomass (Fernandes & Ramaiah, 2019; Hood et al., 2024a; Muraleedharan et al., 2007; Ramaiah et al., 2010). As a result of the low oxygen concentrations in waters beneath the pycnocline, mesozooplankton in the Bay of Bengal are likely forced to inhabit a thin mixed layer where trophic interactions are more concentrated compared to other regions in the Indian Ocean (Hood et al., 2024a).

There are strong nutrient, Chl*a*, production and zooplankton responses to monsoon wind forcing in the SCTR. Vertical sections show that the nutricline and deep Chl*a* maximum shoal sharply in the SCTR region (Hood et al., 2024a). In the SCTR, the highest Chl*a* and primary production is observed in the austral winter (June–August) due to the strong Southeast Monsoon winds that increase wind stirring and induce upwelling (Dilmahamod, 2014; Hood et al., 2024a; Resplandy et al., 2009; Fig. 8c). Zooplankton biomass is relatively low for most of the year in the SCTR region with a pronounced peak during the Southeast Monsoon upwelling in August (austral winter) (Hood et al., 2024a). Southeast Monsoon winds along the southern coasts of the Indonesian island chains, combined with the upwelling-favorable South Java Current, drive nutrient inputs that give rise to substantial increases in Chl*a* concentration and primary production (Hood et al., 2017, 2024a; Fig. 8c). This elevated productivity during the Southeast Monsoon supports order of magnitude increases in zooplankton biomass compared to open-ocean waters further south (Hood et al., 2024a; Tranter & Kerr, 1969, 1977).

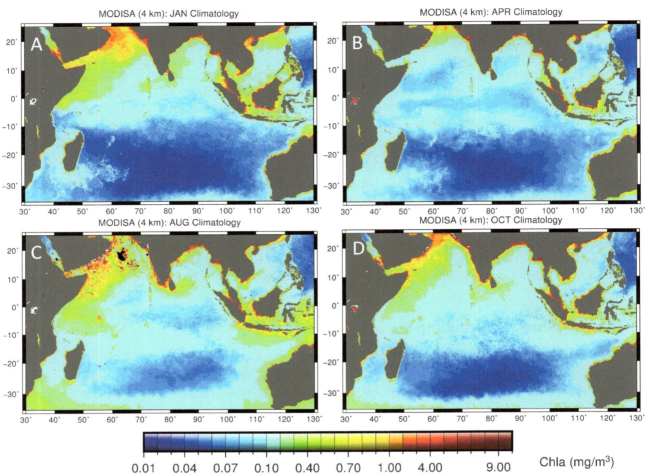

FIG. 8 Monthly climatology of MODIS-Aqua (4-km resolution) chlorophyll: (a) January, (b) April, (c) August, and (d) October. The climatology fields were obtained from the Goddard DAAC (http://daac.gsfc.nasa.gov). *(From Hood et al. (2017).)*

The ITF has a strong influence on biogeochemical properties and productivity in the eastern Indian Ocean and in the Leeuwin Current (Hood et al., 2017; Sprintall et al., 2024). The tropical origins of the Leeuwin Current combine with its southward-flowing downwelling favorable tendency to create a warm oligotrophic current (Hood et al., 2017, 2024a). However, local wind forcing during the austral summer can override this general tendency and drive localized upwelling (Gersbach et al., 1999; Hanson et al., 2005a, 2005b; Hood et al., 2017, 2024a; Pearce & Pattiaratchi, 1999; Rossi et al., 2013a, 2013b). Seasonal nutrient climatologies from the southwestern Australian shelf, the Leeuwin Current, and offshore suggest nitrogen (N) limitation of primary production (Lourey et al., 2006). Primary production is lowest during the austral summer (Fig. 8a) when the water column is most stratified (Hanson et al., 2007). However, primary production in nearshore upwelling regions can attain very high levels (Furnas, 2007; Hood et al., 2017). In the poleward-flowing Leeuwin Current, tropical zooplankton are transported southward (Buchanan & Beckley, 2016; Sutton & Beckley, 2016) and warm-core anticyclonic eddies carry moderately high Chl*a* coastal water offshore (Hood et al., 2017, 2024a; Paterson et al., 2008; Waite et al., 2007, 2015). In open ocean waters of the southeastern Indian Ocean, the highest mesozooplankton biomass is observed in tropical waters south of Java and to the north of northwest Australia driven by upwelling during the Southeast Monsoon (Hood et al., 2024a; Tranter & Kerr, 1969, 1977; Fig. 8).

In the southwest Indian Ocean meanders and topographic interactions of alongshore-propagating eddies in the Mozambique Channel and Agulhas Current drive upwelling and downwelling and lateral transport which have a strong influence on nutrient concentrations and primary production (Hood et al., 2017, 2024a; Jose et al., 2014, 2016; Lamont et al., 2014; Lutjeharms, 2006; Meyer et al., 2002; Roberts et al., 2014). The Agulhas is an oligotrophic current that is derived from oligotrophic surface waters from the southwestern tropical Indian Ocean (Lutjeharms, 2006). Nutrient concentrations, Chl*a* and production in surface waters of the Agulhas are particularly low during the austral summer and higher in the

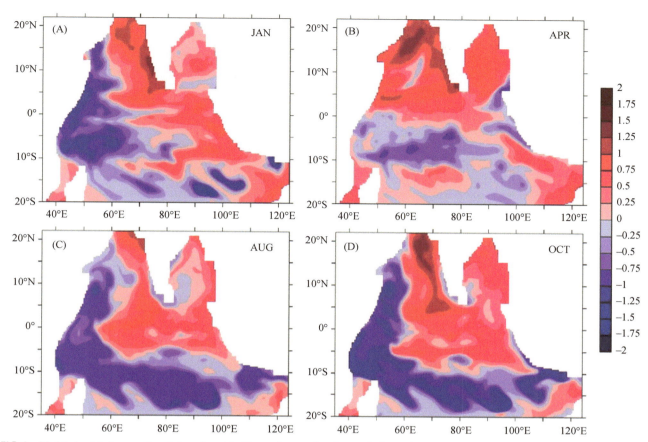

FIG. 9 Model-simulated seasonal evolution of most limiting surface nutrient for net plankton with *blue* (*red*) indicating Fe (N) limited growth (i.e., *red* is iron replete). The four seasons consist of (a) January (NEM); (b) April (Spring Intermonsoon); (c) August (SWM); and (d) October (Fall Intermonsoon). *(Modified from Wiggert et al. (2006).)*

austral winter (Hood et al., 2017, 2024a; Machu & Garçon, 2001; Fig. 8a, c, respectively). However, the Agulhas Current can elevate primary production in the coastal zone through meandering and topographic interactions that drive upwelling and conversely can suppress primary production when the current impinges onto the shelf (Hood et al., 2017, 2024a; Schumann et al., 2005). In general, zooplankton biomass is also low in these waters, except in coastal/shelf regions (Huggett & Kyewalyanga, 2017). The Agulhas Current transports Indo-Pacific zooplankton southward (De Decker, 1973; Hood et al., 2024a). Over the Agulhas Bank, primary production is somewhat elevated, and the zooplankton community is dominated by the large calanoid copepod *Calanus agulhensis* in the austral spring (De Decker & Marska, 1991; Hood et al., 2024a; Huggett & Richardson, 2000).

In general, the southern subtropical gyre of the Indian Ocean is extremely oligotrophic (Harms et al., 2019; Hood et al., 2024a; Fig. 8). However, during the austral summer and fall, the South-East Madagascar bloom often occurs off the southeastern tip of Madagascar extending eastward into the subtropical gyre (Longhurst, 2001; Fig. 8a). It has been suggested that the bloom is fueled by N_2 fixation, with eddies and/or the South Indian Counter Current advecting and dispersing it eastward (Dilmahamod et al., 2019; Hood et al., 2024a; Huhn et al., 2012; Poulton et al., 2009; Raj et al., 2010; Srokosz & Quartly, 2013; Srokosz et al., 2004, 2015; Uz, 2007). Phytoplankton production (Harms et al., 2019) and mesozooplankton abundance (Rochford, 1969, 1977) are particularly low in the western subtropical gyre, and variability is associated with specific water masses that have different Chl*a* concentrations (Cedras et al., 2020; Hood et al., 2024a). Zooplankton biomass is highest in the eastern subtropical gyre, likely due to the influence of increased winds and eddy activity associated with Leeuwin Current and proximity to the Australian land mass (Hood et al., 2024a). Grazing rates estimated from the southern subtropical gyre of the Indian Ocean indicate that the dominant pico-sized phytoplankton largely escape direct feeding by mesozooplankton, and microzooplankton are the major grazers as is typical for highly oligotrophic waters (Hood et al., 2024a; Jaspers et al., 2009).

7.2 Oxygen, carbon, and pH

The Arabian Sea has the thickest OMZ in the world where oxygen concentrations in intermediate water (~200–800 m) decline to nearly zero (e.g., Hood et al., 2024b, and references cited therein; Morrison et al., 1999; Naqvi et al., 2005). These low oxygen concentrations have significant impacts on nitrogen cycling, which include denitrification-driven depletion of subsurface nitrate (NO_3^-) concentrations, generation of nitrite (NO_2^-) maxima, production of greenhouse gases (nitrous oxide, N_2O), and globally significant fluxes of N_2 gas into the atmosphere from the ocean (Codispoti et al., 2001; Naqvi et al., 2005; Ramaswamy et al., 2001). In contrast, in the Bay of Bengal, these biogeochemical impacts are much less pronounced where dissolved oxygen concentrations in intermediate waters remain just above the threshold for denitrification (Hood et al., 2024b; Naqvi et al., 2005). In both the Arabian Sea and the Bay of Bengal, the low oxygen conditions in the water column are associated with high biological oxygen demand and slow ventilation. It appears that the more intense OMZ in the Arabian Sea is due to a weaker mineral (e.g., plankton shell) ballast effect and a higher biological production that support a higher biological oxygen demand compared to the Bay of Bengal, which, in turn, balances the higher physical oxygen supply (Hood et al., 2024b; Rixen et al., 2019). Elsewhere in the Indian Ocean, oxygen concentrations generally remain above the hypoxic threshold of 2 mg/L with concentrations increasing southward due to the influence of northward transport of well-oxygenated intermediate waters from the Southern Hemisphere. Highly oxygenated, intermediate water is formed along the northern edge of the Antarctic Circumpolar Current, and subsequently spreads throughout the Indian Ocean (McCreary et al., 2013).

It is anticipated that global warming will lead to the intensification of monsoon winds (Goes et al., 2005). Regional modeling studies have suggested that stronger Southwest Monsoon winds intensify the upwelling, increasing biological productivity and respiration in excess of the increase in wind-driven ventilation and are also responsible for deepening of the OMZ (Hood et al., 2024b; Lachkar et al., 2018). Increases in Arabian Sea denitrification are expected to cause increases in the N_2 and N_2O production. However, studies of long-term OMZ variability in the global ocean reveal only a small decline in oxygen concentrations in the Arabian Sea and the Bay of Bengal OMZs compared to in OMZs in the Pacific and Atlantic Oceans (Hood et al., 2024b; Ito et al., 2017; Naqvi, 2019; Schmidtko et al., 2017; Stramma et al., 2008). Moreover, outbreaks of hydrogen sulfide (H_2S) have so far not been reported in the northern Indian Ocean during the past 50 years (Rixen et al., 2020), other than in bottom waters on the Indian shelf (Naqvi et al., 2000). The absence of H_2S suggests that the balance of physical oxygen supply and biological oxygen demand has been fairly constant in the Arabian Sea and the Bay of Bengal OMZs and has maintained the current hypoxic/anoxic conditions (Hood et al., 2024b; Rixen et al., 2020). However, some of the most recent studies suggest that oxygen levels are now declining in the Arabian Sea and that these declines are giving rise to major biogeochemical and ecological impacts (Goes & Gomes, 2016; Goes et al., 2020; Gomes et al., 2014; Lachkar et al., 2019, 2020). Earth System Model-projected future changes in the northern Indian Ocean OMZs are highly uncertain (Bopp et al., 2013; Cabré et al., 2015; Kwiatkowski et al., 2020; Rixen et al., 2020). This uncertainty largely arises from differences in the magnitude and timing of the ventilation and biological oxygen demand in the models and how strongly they offset each other (Hood et al., 2024b; Resplandy, 2018; Rixen et al., 2020).

Inorganic and organic carbon pools and air-sea carbon fluxes in the Indian Ocean are sparsely sampled compared to other ocean regions (Hood et al., 2024b; Fig. 10). The Indian Ocean is responsible for ~20% of the oceanic uptake of atmospheric CO_2 (Takahashi et al., 2002). The Arabian Sea is a significant source of CO_2 to the atmosphere due to Southwest Monsoon-driven upwelling of water with elevated pCO_2 (Fig. 10; see also De Verneil et al., 2021; Takahashi et al., 2009, 2014). In contrast, whether the Bay of Bengal is a source or sink of CO_2 to the atmosphere is still uncertain due to the lack of measurements (Fig. 10; Bates et al., 2006; Hood et al., 2024b). The Indian Ocean appears to be a strong net sink of atmospheric CO_2 south of 14°S. This sink is due to the combined effects of both the biological pump (biologically mediated carbon export) and the solubility pump (CO_2 dissolution and physical transport and mixing) (Fig. 10; Hood et al., 2024b; Valsala et al., 2012). Some of the lowest pH values in the world have been measured in surface waters of the Indian Ocean (Feely et al., 2009). Further acidification is anticipated in the Indian Ocean over the coming decades given projected increases in oceanic CO_2 concentrations. This acidification will likely have severe negative impacts on coral reefs and other calcifying organisms (Bednarsek et al., 2012; de Moel et al., 2009; Hoegh-Guldberg et al., 2007; Hood et al., 2015, 2024b; IPCC, 2019).

As observed globally, dissolved organic carbon (DOC) concentrations in the Indian Ocean are high in stratified near-surface tropical and subtropical waters because these are regions where DOC is produced and accumulates. In contrast, the lowest DOC concentrations are observed in the deep ocean where heterotrophic consumption of DOC is much greater than autotrophic production (Hansell, 2009; Hansell et al., 2009). Furthermore, as observed globally, near-surface particulate organic carbon (POC) concentrations in the Indian Ocean are elevated in coastal regions (Gardner et al., 2006), with the

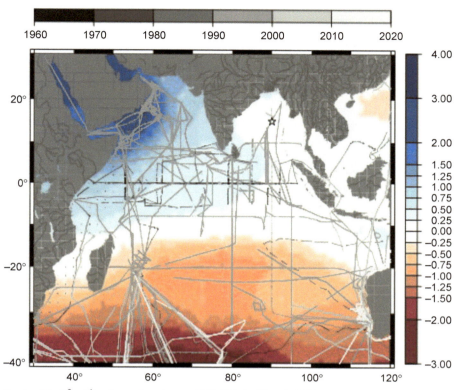

FIG. 10 Annual CO_2 flux ($mol\,C\,m^{-2}\,yr^{-1}$) referenced to the year 2000 (Takahashi et al., 2009) over the Indian Ocean. Data points colored by year of collection are overlaid as points. Major rivers are also delineated over the continents. *(From Hood et al. (2024a).)*

highest POC concentrations observed in the northwestern part of the Arabian Sea and off the coast of Somalia (Hood et al., 2024b; Fig. 11). These high POC values are coincident with high Chl*a* concentrations (Gardner et al., 2006; Fig. 11). A tongue of elevated POC and Chl*a* concentration is also associated with the Agulhas retroflection (Gardner et al., 2006; Fig. 11). The lowest POC (and Chl*a*) concentrations in the Indian Ocean are found in the oligotrophic southern subtropical gyre (Gardner et al., 2006; Figs. 8 and 11).

The spatial and temporal variability in POC export flux in the Indian Ocean is similar to the spatial and temporal variability in POC and Chl*a* concentration, which is consistent with the idea that organic carbon fluxes are largely controlled by primary production (Hood et al., 2024b; Rixen et al., 2019). However, the spatial variability of POC flux is also strongly influenced by lithogenic matter (ballast) in regions like the Bay of Bengal that have large river inputs (Fig. 7; Hood et al., 2024b; Rixen et al., 2019). In the Arabian Sea, the highest POC fluxes are observed during the Southwest Monsoon. In the Bay of Bengal, POC export flux is generally lower and the seasonal variability is weaker compared to the Arabian Sea (Hood et al., 2024b; Rixen et al., 2019). In eastern-central equatorial waters of the Indian Ocean, ^{234}Th-estimates of POC export flux have revealed elevated values (Subha Anand et al., 2017), which is surprising given the downwelling-favorable equatorial circulation (Hood et al., 2024b; Strutton et al., 2015) and the observed low burial rates of organic carbon (Jahnke, 1996). There are very few POC export flux measurements off western Australia (Hood et al., 2024b). The highest organic carbon burial rates in the southwestern Indian Ocean are observed near the coast off southeastern Africa and in the open ocean in the Agulhas retroflection (Jahnke, 1996). Export fluxes are extremely low in the southern subtropical gyre of the Indian Ocean (Harms et al., 2021; Hood et al., 2024b).

7.3 Productivity and fisheries

The spatial distribution and temporal variability of primary productivity, from regional to basin-scale and seasonal to interannual, have been quantified in the Indian Ocean using satellite ocean color data (Hood et al., 2017, 2024a;

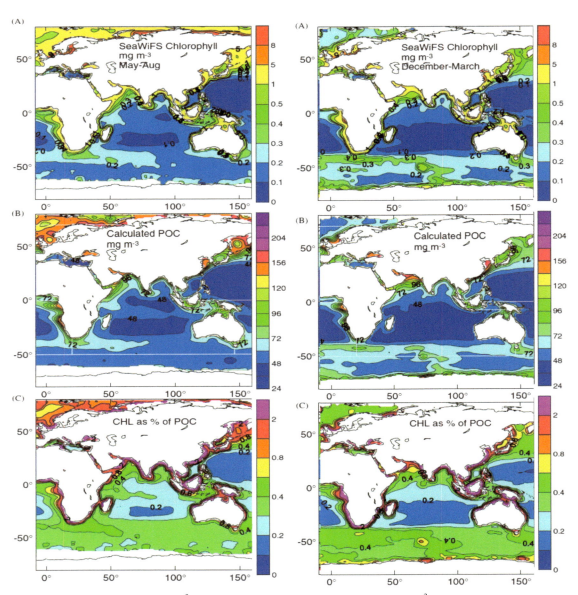

FIG. 11 Distribution of: (a) SeaWiFS Chl (mg m^{-3}, level 3, reprocessing 4 data); (b) average POC (mg m^{-3}) over one attenuation depth calculated from K_{490}:c_p:POC; (c) Chl as a percentage of POC for: (left panels) summer season (1997–2002, May–August, 20 months); (right panels) winter season (1997–2002, December–March, 20 months). *White lines* in (b) mark boundaries separating ocean basins. *(Modified from Gardner et al. (2006).)*

Marsac et al., 2024). The spatial patterns in Chl*a* and POC concentration and primary production are described in the previous sections. Goes et al. (2005) used a seven-year (1997–2004) satellite ocean color time series to document an increase in Chl*a* in the Arabian Sea that they attributed to global warming-induced strengthening of upwelling-favorable Southwest Monsoon winds in the western Arabian Sea. Others have argued that the increase in Chl*a* observed by Goes et al. (2005) reflects decadal variability (Prakash et al., 2012; Prakash & Ramesh, 2007; Rixen et al., 2014). In contrast, Roxy et al. (2016) examined a broader area of the northwestern Indian Ocean (including the Arabian Sea) and a longer time series (1997–2013) and came to the opposite conclusion that Chl*a* and primary production has been decreasing due to greater stratification in a warming ocean. Based on a 23-year time series, Marsac et al. (2024) concluded that there is no evidence of any long-term trends in the mean Chl*a* concentration or the areal extent of extremely oligotrophic waters. Clearly, there are conflicting assessments of long-term changes in Chl*a* concentration and primary production in the Indian Ocean that need to be resolved (Hood et al., 2024a).

Several studies have investigated the impact of interannual ENSO and IOD variability on the primary productivity of the basin (Currie et al., 2013; Keerthi et al., 2017; McCreary et al., 2009; Wiggert et al., 2009). For example, the IOD generates significant increases in Chl*a* concentration and primary production in September along the eastern boundary of the Indian Ocean in tropical waters that are normally highly oligotrophic driven by IOD-induced increases in upwelling off Sumatra (Wiggert et al., 2009). Positive Chl*a* and productivity anomalies are also apparent in the southeastern Bay of Bengal due to increased upwelling and/or mixing, while negative anomalies are observed over much of the Arabian Sea and SCTR where upwelling is suppressed (Marsac et al., 2024; Wiggert et al., 2009).

In 2019, tuna catches in the Indian Ocean represented 23% of the world tuna and billfish catch (FAO, 2020), and 90% of the neritic tunas are caught by the coastal/artisanal fisheries. The spatial distribution of actively feeding tuna schools is very patchy (Fonteneau, 1986; Marsac et al., 2024; Ravier & Fromentin, 2001, 2004), which reflects the patchiness in the distribution of prey. Patchiness in the distribution of tuna prey can often be related to mesoscale eddies (Tew-Kai & Marsac, 2009), indicating that mesoscale features often promote aggregations of tuna (Digby et al., 1999). Several studies have shown linkages between large-scale climate signals such as the ENSO and IOD and the Pacific Decadal Oscillation and changes in mesoscale activity (e.g., Palastanga et al., 2006; Tew-Kai & Marsac, 2009; Zheng et al., 2018). However, predictions on the future state of mesoscale activity in the ocean are highly uncertain because coupled climate models do not resolve the mesoscale. Based on the present understanding, it seems likely that tropical tunas will face challenging conditions in the open ocean if climate change results in a reduction of mesoscale activity (Marsac et al., 2024).

The distribution of tuna fleets in the Indian Ocean have been observed to shift at interannual timescales, moving from the west to the east in the equatorial Indian Ocean during the 1997–98 IOD (Marsac, 2017; Marsac & Le Blanc, 1998, 1999; Marsac et al., 2024; Fig. 12). As discussed earlier, a positive IOD disrupts the typical seasonal evolution of phytoplankton bloom dynamics over the whole Indian Ocean basin. The purse seiner fleet could not operate efficiently in the western Indian Ocean during the 1997–1998 positive IOD due to the absence of tuna schools and therefore had to move to areas with higher biological productivity in the eastern equatorial Indian Ocean (Fig. 12; Marsac, 2017; Marsac et al., 2024). It appears that the IOD-modulated prey enhancement is a major driver of tuna and tuna fleet relocation in the Indian Ocean.

Multiple studies suggest that tuna fisheries will be significantly impacted on multidecadal timescales by climate change (e.g., Lehodey et al., 2006; Ravier & Fromentin, 2001, 2004). Roxy et al. (2016) predict dramatic declines in net primary production in the northwestern Indian Ocean in the future due to climate change. These changes are also projected to expand over the 21st century (Dueri et al., 2014; Roxy et al., 2024). Thanks to their ability to adapt to changing conditions, tunas will likely survive these environmental changes, unless poor management results in severe depletion of tuna stocks. Regardless, tuna fishers in the Indian Ocean will very likely have to adapt to a significant reshaping of the tuna habitat, both horizontally and in depth (Moustahfid et al., 2018). These changes will almost certainly impact socioeconomic conditions in Indian Ocean rim countries that are highly dependent on their domestic tuna fisheries (Marsac et al., 2024).

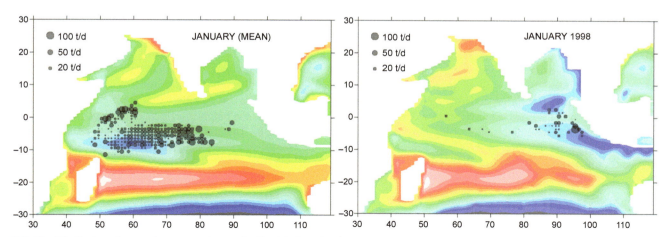

FIG. 12 Distribution of purse seine tuna catch per unit effort (circles, t day^{-1}) on free swimming schools. Left panel: January mean, catch per unit effort 1991–2002 and 20°C isothermal depth (m, *color shaded*), 1980–2005. Right panel: Catch per unit effort and 20°C depth in January 1998 (positive dipole). *(From Marsac et al. (2024).)*

8 Summary and conclusions

The seasonally reversing winds drive corresponding ocean surface currents in the northern Indian Ocean, with anticyclonic upwelling circulations during the Southwest Monsoon and cyclonic downwelling circulations during the Northeast Monsoon. And, unlike the Atlantic and Pacific, these winds drive equatorial currents that are, on average, downwelling favorable. The ITF forces and feeds the Leeuwin Current, the only poleward-flowing eastern boundary current in the world. Near-surface waters derived from the ITF and the Tasman leakage south of Australia cross the Indian Ocean basin before feeding into the poleward-flowing Agulhas Current, which sheds eddies into the Atlantic to complete the global thermohaline circulation. Variability in the circulation of the Indian Ocean is the result of complex interactions that are both internal and external to the Indian Ocean on many different timescales, and these interactions are not fully understood.

The monsoon winds also influence SST. As is observed globally, there is lower SST at higher latitudes in the Southern Hemisphere of the Indian Ocean where there is strong surface cooling, whereas in the northern Indian Ocean, there is cooling and lower SST during the Northeast Monsoon due to the influence of the monsoon winds, and warming and higher SST during the rest of the year, especially during the spring and fall inter-monsoon seasons when winds are weak. In contrast, salinity in the Indian Ocean is controlled not only by net air-sea fluxes (evaporation minus precipitation), but also freshwater inflow from large rivers, inflow of relatively fresh waters from the Pacific Ocean via the ITF, and the inflow of saltier waters from the Red Sea and the Persian and Arabian Gulfs. These different drivers combine to give the Indian Ocean a unique salinity distribution, i.e., the Arabian Sea is relatively salty and the Bay of Bengal is relatively fresh, whereas the tropics are relatively fresh, and the southern subtropics are relatively salty. Fresher surface layers also give rise to the formation of barrier layers—especially in the northern Bay of Bengal—that can influence the summer monsoon, development of the IOD, and tropical cyclone-induced upwelling. However, the uncertainty in the freshwater distribution and mixing pathways for riverine input is high, and shallow, salinity-controlled mixed layers and the presence of barrier layers can significantly affect upper-ocean processes in ways that are not fully understood.

The monsoonal wind forcing is also a major driver of biogeochemical and ecological variability with both stimulatory (Southwest Monsoon) and suppressing (Northeast Monsoon) effects throughout many Indian Ocean regions (Arabian Sea, Bay of Bengal, SCTR, Java, central/eastern subtropical gyre). In addition, there are regionally specific processes that significantly modulate the biogeochemical and ecological responses. For example, Fe/Si limitation in the Arabian Sea; freshwater and stratification in the Bay of Bengal; the influence of ITF nutrient inputs, poleward transport, downwelling, and seaward-propagating eddies in the Leeuwin Current; and alongshore-propagating eddies, meanders, upwelling, and poleward transport in the Agulhas Current. The southern subtropical gyre of the Indian Ocean is extremely oligotrophic, yet the Southeast Madagascar bloom occurs during the most stratified season (austral summer/fall). All of these regionally specific processes have significant biogeochemical consequences, yet many have received relatively little research attention.

And don't forget that the South Asian monsoon represents the largest monsoon system on Earth, the Indian Ocean has the world's largest southward meridional heat transport, and the Indian Ocean is characterized by very active intraseasonal variability with far-reaching implications for regional climate. Moreover, the Arabian Sea has the thickest OMZ in the world with globally significant biogeochemical impacts, and some of the lowest pH values in the world have been measured in surface waters of the Indian Ocean. And in 2019, tuna catches in the Indian Ocean represented 23% of the world tuna and billfish catch, with 90% of the neritic tunas caught by the coastal/artisanal fisheries.

Adding to all this complexity (or perhaps a consequence of it) are the conflicting reports as to whether or not primary production is changing in the Indian Ocean in response to global warming, and highly uncertain future projections of climate change impacts on biogeochemistry. Moreover, both interannual studies in the Indian Ocean and multidecadal studies in other ocean basins suggest that fisheries in the Indian Ocean will be significantly impacted by climate change and that these changes will impact socioeconomic conditions in Indian Ocean rim countries that are highly dependent on their domestic tuna fisheries. Clearly, more work needs to be done to reassess the potential effects of climate variability and change on the biogeochemistry and ecology of the entire Indian Ocean basin. This need is made even more pressing by the fact that a significant fraction of the world's population lives in the coastal and interior regions of Indian Ocean rim countries and they are directly impacted by this atmospheric and oceanic variability.

9 Educational resources

- Ocean Data View, free software for plotting oceanographic data. Available at: https://odv.awi.de.
- World Ocean Atlas, a collection of objectively analyzed, quality controlled temperature, salinity, oxygen, phosphate, silicate, and nitrate means based on profile data from the World Ocean Database. Available at: https://www.ncei.noaa.gov/products/world-ocean-atlas.

- Surface Ocean CO$_2$ Atlas (SOCAT) is a synthesis of quality-controlled, surface ocean fCO$_2$ (fugacity of carbon dioxide) observations by the international marine carbon research community. Available at: https://www.socat.info.
- Satellite ocean color data. Available at: https://oceancolor.gsfc.nasa.gov.
- GFDL Sea Surface Temperature Simulation for the Indian Ocean. Available at: https://www.youtube.com/watch?v=ZVssbK0K4wc.
- Educational video of the Asian monsoon. Available at: https://www.youtube.com/watch?v=RQkOhKEtC20.
- Wikipedia Page on the Joint Global Ocean Flux Study: https://en.wikipedia.org/wiki/Joint_Global_Ocean_Flux_Study.
- Information on and resources related to the GO-SHIP Program: https://www.go-ship.org/About.html.

Acknowledgments

The development of this article was supported by the Scientific Committee for Oceanic Research via direct funding to the Second International Indian Ocean Expedition and indirect funding through the Integrated Marine Biosphere Research regional program SIBER (Sustained Indian Ocean Biogeochemistry and Ecosystem Research). Additional support was provided by NASA Grant Number 80NSSC17K0258 49A37A, NOAA Grant Number NA15NMF4570252 NCRS-17, and NSF Grant Number 2009248 to R. Hood. H. Phillips acknowledges support from the Earth Systems and Climate Change and Climate Systems Hubs of the Australian government's National Environmental Science Programme. J. Sprintall acknowledges funding to support her effort from the National Science Foundation under Grant Number OCE-1851316. C.C. Ummenhofer acknowledges support from NSF under AGS-2002083 and AGS-2102844 and the James E. and Barbara V. Moltz Fellowship for Climate-Related Research at WHOI. This article also benefitted from comments provided by Pete Strutton and one anonymous reviewer. This is UMCES contribution 6345.

Author contributions

RRH conceived and led the chapter overall, with contributions in particular to Sections 1, 2, 3, and 7. HP wrote Section 4.1 and drafted Fig. 3. JS wrote Section 4.3. CCU wrote Sections 5 and 6 and drafted Fig. 7. All authors contributed to the discussion of content and overall chapter structure and provided feedback on the entire chapter.

References

Adler, R. F., Gu, G., Sapiano, M., Wang, J.-J., & Huffman, G. J. (2017). Global precipitation: Means, variations and trends during the satellite era (1979–2014). *Surveys in Geophysics*, *38*(4), 679–699.

Ali, M., & Sharma, R. (1994). Estimation of mixed layer depth in the equatorial Indian Ocean using Geosat altimeter data. *Marine Geodesy*, *17*(1), 63–72.

Al-Yamani, F. Y., Burt, J. A., Goes, J. I., Jones, B., Nagappa, R., Murty, V. S. N., … Yamamoto, T. (2024). Chapter 16: Physical and biogeochemical characteristics of the Indian Ocean marginal seas. In C. C. Ummenhofer, & R. R. Hood (Eds.), *The Indian Ocean and its role in the global climate system* (pp. 365–391). Amsterdam: Elsevier. https://doi.org/10.1016/B978-0-12-822698-8.00008-1.

Anil, N., Ramesh Kumar, M., Sajeev, R., & Saji, P. (2016). Role of distinct flavours of IOD events on Indian summer monsoon. *Natural Hazards*, *82*(2), 1317–1326.

Ashok, K., Guan, Z., & Yamagata, T. (2001). Impact of the Indian Ocean dipole on the relationship between the Indian monsoon rainfall and ENSO. *Geophysical Research Letters*, *28*(23), 4499–4502.

Ayers, J. M., Strutton, P. G., Coles, V. J., Hood, R. R., & Matear, R. J. (2014). Indonesian Throughflow nutrient fluxes and their potential impact on Indian Ocean productivity. *Geophysical Research Letters*, *41*(14), 5060–5067.

Baars, M. A. (1999). On the paradox of high mesozooplankton biomass, throughout the year in the western Arabian Sea: Re-analysis of IIOE data and comparison with newer data. *International Journal of Marine Science*, *28*, 125–127.

Babu, K., Sharma, R., Agarwal, N., Agarwal, V. K., & Weller, R. (2004). Study of the mixed layer depth variations within the north Indian Ocean using a 1-D model. *Journal of Geophysical Research: Oceans*, *109*(C8).

Balaguru, K., Chang, P., Saravanan, R., Leung, L. R., Xu, Z., Li, M., & Hsieh, J.-S. (2012). Ocean barrier layers' effect on tropical cyclone intensification. *Proceedings of the National Academy of Sciences*, *109*(36), 14343–14347.

Bates, N. R., Pequignet, A. C., & Sabine, C. L. (2006). Ocean carbon cycling in the Indian Ocean: 1. Spatiotemporal variability of inorganic carbon and air-sea CO2gas exchange. *Global Biogeochemical Cycles*, *20*(3).

Beal, L. M., De Ruiter, W. P. M., Biastoch, A., Zahn, R., & SCOR/WCRP/IAPSO_Working_Group_136. (2011). On the role of the Agulhas system in ocean circulation and climate. *Nature*, *472*, 429–436.

Beal, L. M., Elipot, S., Houk, A., & Leber, G. M. (2015). Capturing the transport variability of a western boundary jet: Results from the Agulhas Current time-series experiment (ACT). *Journal of Physical Oceanography*, *45*, 1302–1324.

Beal, L., Vialard, J., Roxy, M., Li, J., Andres, M., Annamalai, H., Feng, M., Han, W., Hood, R., & Lee, T. (2020). A road map to IndOOS-2: Better observations of the rapidly warming Indian Ocean. *Bulletin of the American Meteorological Society*, *101*(11), E1891–E1913.

Beal, L. M., Vialard, J., Roxy, M. K., et al. (2019). *A roadmap to sustained observations of the Indian Ocean for 2020–2030*. CLIVAR-4/2019, GOOS-237, 204 pp.

Bednarsek, N., Tarling, G. A., Bakker, D. C. E., Fielding, S., Jones, E. M., Venables, H. J., Ward, P., Kuzirian, A., Leze, B., Feely, R. A., & Murphy, E. J. (2012). Extensive dissolution of live pteropods in the Southern Ocean. *Nature Geoscience, 5*, 881–885.

Behera, S., Krishnan, R., & Yamagata, T. (1999). Unusual ocean-atmosphere conditions in the tropical Indian Ocean during 1994. *Geophysical Research Letters, 26*(19), 3001–3004.

Behrman, D. (1981). *Assault on the largest unknown: The International Indian Ocean Expedition, 1959–1965* (p. 96). Paris: UNESCO Press.

Benson, K. R., & Rehbock, P. F. (2002). *Oceanographic history the pacific and beyond*. Seattle and London: University of Washington Press. 556 p.

Berkelhammer, M., Sinha, A., Stott, L., Cheng, H., Pausata, F. S., & Yoshimura, K. (2012). An abrupt shift in the Indian monsoon 4000 years ago. *Geophysical Monograph Series, 198*(7).

Boos, W. R., & Emanuel, K. A. (2009). Annual intensification of the Somali jet in a quasi-equilibrium framework: Observational composites. *Quarterly Journal of the Royal Meteorological Society: A Journal of the Atmospheric Sciences, Applied Meteorology and Physical Oceanography, 135*(639), 319–335.

Bopp, L., Resplandy, L., Orr, J. C., Doney, S. C., Dunne, J. P., Gehlen, M., Halloran, P., Heinze, C., Ilyina, T., & Seferian, R. (2013). Multiple stressors of ocean ecosystems in the 21st century: projections with CMIP5 models. *Biogeosciences, 10*(10), 6225–6245.

Brewin, R. J., Hirata, T., Hardman-Mountford, N. J., Lavender, S. J., Sathyendranath, S., & Barlow, R. (2012). The influence of the Indian Ocean Dipole on interannual variations in phytoplankton size structure as revealed by Earth Observation. *Deep-Sea Rearch, Part II, 77*, 117–127.

Bryden, H. L., & Beal, L. M. (2001). Role of the Agulhas Current in Indian Ocean circulation and associated heat and freshwater fluxes. *Deep Sea Research Part I, 48*(8), 1821–1845.

Bryden, H. L., Beal, L. M., & Duncan, L. M. (2005). Structure and transport of the Agulhas Current and its temporal variability. *Journal of Oceanography, 61*, 479–492.

Buchanan, P., & Beckley, L. (2016). Chaetognaths of the Leeuwin Current system: Oceanographic conditions drive epi-pelagic zoogeography in the southeast Indian Ocean. *Hydrobiologia, 763*(1), 81–96.

Burns, J. M., Subrahmanyam, B., & Murty, V. (2017). On the dynamics of the Sri Lanka Dome in the Bay of Bengal. *Journal of Geophysical Research: Oceans, 122*(9), 7737–7750.

Cabré, A., Marinov, I., Bernardello, R., & Bianchi, D. (2015). Oxygen minimum zones in the tropical Pacific across CMIP5 models: Mean state differences and climate change trends. *Biogeosciences, 12*(18), 5429–5454.

Cedras, R., Halo, I., & Gibbons, M. (2020). Biogeography of pelagic calanoid copepods in the Western Indian Ocean. *Deep Sea Research Part II: Topical Studies in Oceanography, 179*, 104740.

Chang, C., & Li, T. (2000). A theory for the tropical tropospheric biennial oscillation. *Journal of the Atmospheric Sciences, 57*(14), 2209–2224.

Chowdary, J. S., Bandgar, A. B., Gnanaseelan, C., & Luo, J. J. (2015). Role of tropical Indian Ocean air–sea interactions in modulating Indian summer monsoon in a coupled model. *Atmospheric Science Letters, 16*(2), 170–176.

Codispoti, L., Brandes, J. A., Christensen, J., Devol, A., Naqvi, S., Paerl, H. W., & Yoshinari, T. (2001). The oceanic fixed nitrogen and nitrous oxide budgets: Moving targets as we enter the anthropocene? *Scientia Marina (Barcelona), 65*(Suppl. 2), 85–105.

Cook, E. R., Anchukaitis, K. J., Buckley, B. M., D'Arrigo, R. D., Jacoby, G. C., & Wright, W. E. (2010). Asian monsoon failure and megadrought during the last millennium. *Science, 328*, 486–489.

Currie, J. C., Lengaigne, M., Vialard, J., Kaplan, D. M., Aumont, O., Naqvi, S., & Maury, O. (2013). Indian Ocean dipole and El Nino/southern oscillation impacts on regional chlorophyll anomalies in the Indian Ocean. *Biogeosciences, 10*(10), 6677–6698.

de Boyer Montegut, C., Vialard, J., Shenoi, S. S. C., Shankar, D., Durand, F., Ethe, C., & Madec, G. (2007). Simulated seasonal and interannual variability of the mixed layer heat budget in the northern Indian Ocean. *Journal of Climate, 20*(13), 3249–3268.

De Decker, A. H. B. (1973). Agulhas Bank plankton. In B. Zeitzschel (Ed.), *The biology of the Indian Ocean* (pp. 189–219). Berlin: Springer-Verlag.

De Decker, A., & Marska, G. (1991). *A new species of Calanus (Copepoda, Calanoida) from South African waters*. South African Museum.

de Moel, H., Ganssen, G. M., Peeters, F. J. C., Jung, S. J. A., Kroon, D., Brummer, G. J. A., & Zeebe, R. E. (2009). Planktonic foraminiferal shell thinniing in the Arabian Sea due to anthropogenic ocean acidification? *Biogeosciences, 6*, 1917–1925.

De Verneil, A., Lachkar, Z., Smith, S., & Lévy, M. (2021). Evaluating the Arabian Sea as a regional source of atmospheric CO_2: Seasonal variability and drivers. *Biogeosciences Discussions*, 1–38.

DeMott, C. A., Klingaman, N. P., & Woolnough, S. J. (2015). Atmosphere-ocean coupled processes in the Madden-Julian oscillation. *Reviews of Geophysics, 53*(4), 1099–1154.

DeMott, C. A., Ruppert, J. H., Jr., & Rydbeck, A. (2024). Chapter 4: Intraseasonal variability in the Indian Ocean region. In C. C. Ummenhofer, & R. R. Hood (Eds.), *The Indian Ocean and its role in the global climate system* (pp. 79–101). Amsterdam: Elsevier. https://doi.org/10.1016/B978-0-12-822698-8.00006-8.

Digby, S., Antczak, T., Leben, R., Born, G., Barth, S., Cheney, R., Foley, D., Goni, G. J., Jacobs, G., & Shay, N. (1999). Altimeter data for operational use in the marine environment. In *Vol. 2. Proceedings Oceans' 99. MTS/IEEE. Riding the Crest into the 21st Century. Conference and Exhibition. Conference Proceedings (IEEE Cat. No. 99CH37008)* (pp. 605–613). IEEE.

Dilmahamod, A. F. (2014). *Links between the Seychelles-Chagos thermocline ridge and large scale climate modes and primary productivity; and the annual cycle of chlorophyll-a* (Ph.D.). University of Cape Town.

Dilmahamod, A. F., Aguiar-González, B., Penven, P., Reason, C., De Ruijter, W., Malan, N., & Hermes, J. (2018). SIDDIES corridor: a major east-west pathway of long-lived surface and subsurface eddies crossing the subtropical South Indian Ocean. *Journal of Geophysical Research: Oceans, 123*(8), 5406–5425.

Dilmahamod, A. F., Penven, P., Aguiar-González, B., Reason, C., & Hermes, J. (2019). A new definition of the South-East Madagascar Bloom and analysis of its variability. *Journal of Geophysical Research: Oceans, 124*(3), 1717–1735.

Divakaran, P., & Brassington, G. B. (2011). Arterial ocean circulation of the southeast Indian Ocean. *Geophysical Research Letters, 38*(1).

Divins, D. (2003). *Total sediment thickness of the world's oceans & marginal seas*. Boulder, CO: NOAA National Geophysical Data Center.

Doblas-Reyes, F. J., Sörensson, A. A., Almazroui, M., Dosio, A., Gutowski, W. J., Haarsma, R., Hamdi, R., Hewitson, B., Kwon, W.-T., Lamptey, B. L., Maraun, D., Stephenson, T. S., Takayabu, I., Terray, L., Turner, A., & Zuo, Z. (2021). Linking global to regional climate change. In V. Masson-Delmotte, P. Zhai, A. Pirani, S. L. Connors, C. Péan, S. Berger, ... B. Zhou (Eds.), *Climate change 2021: The physical science basis. Contribution of working group I to the sixth assessment report of the intergovernmental panel on climate change* (pp. 1363–1512). Cambridge, UK: Cambridge University Press.

Domingues, C. M., Wijffels, S. E., Maltrud, M. E., Church, J. A., & Tomczak, M. (2006). Role of eddies in cooling the Leeuwin Current. *Geophysical Research Letters*, *33*(5).

Douville, H., Raghavan, K., Renwick, J., Allan, R. P., Arias, P. A., Barlow, M., Cerezo-Mota, R., Cherchi, A., Gan, T. Y., Gergis, J., Jiang, D., Khan, A., Pokam Mba, W., Rosenfeld, D., Tierney, J., & Zolina, O. (2021). Water cycle changes. In V. Masson-Delmotte, P. Zhai, A. Pirani, S. L. Connors, C. Péan, S. Berger, ... B. Zhou (Eds.), *Climate change 2021: The physical science basis. Contribution of working group I to the sixth assessment report of the intergovernmental panel on climate change* (pp. 1055–1210). Cambridge, UK: Cambridge University Press.

Du, Y., Qu, T., Meyers, G., Masumoto, Y., & Sasaki, H. (2005). Seasonal heat budget in the mixed layer of the southeastern tropical Indian Ocean in a high-resolution ocean general circulation model. *Journal of Geophysical Research: Oceans*, *110*(C4).

Dueri, S., Bopp, L., & Maury, O. (2014). Projecting the impacts of climate change on skipjack tuna abundance and spatial distribution. *Global Change Biology*, *20*(3), 742–753.

Duran, E. R., Phillips, H. E., Furue, R., Spence, P., & Bindoff, N. L. (2020). Southern Australia Current System based on a gridded hydrography and a high-resolution model. *Progress in Oceanography*, *181*, 102254.

Dutt, S., Gupta, A. K., Clemens, S. C., Cheng, H., Singh, R. K., Kathayat, G., & Edwards, R. L. (2015). Abrupt changes in Indian summer monsoon strength during 33,800 to 5500 years BP. *Geophysical Research Letters*, *42*(13), 5526–5532.

FAO. (2020). *FishStatJ-Software for fishery and aquaculture statistical time series* (online). Rome: FAO Fisheries Division. Updated, v. 22.

Fasham, M. J. (2003). JGOFS: A retrospective view. In *Ocean Biogeochemistry* (pp. 269–277). Springer.

Feely, R. A., Doney, S. C., & Cooley, S. R. (2009). Ocean acidification: Present conditions and future changes in a high-CO_2 world. *Oceanography*, *22*(4), 36–47.

Felton, C. S., Subrahmanyam, B., Murty, V., & Shriver, J. F. (2014). Estimation of the barrier layer thickness in the Indian Ocean using Aquarius Salinity. *Journal of Geophysical Research: Oceans*, *119*(7), 4200–4213.

Feng, M., Majewski, L., Fandry, C., & Waite, A. (2007). Characteristics of two counter-rotating eddies in the Leeuwin Current system off the Western Australian coast. *Deep-Sea Research, Part II*, *54*, 961–980.

Feng, M., Meyers, G., Pearce, A., & Wijffels, S. (2003). Annual and interannual variations of the Leeuwin Current. *Journal of Geophysical Research, Oceans*, *108*(C11), 3355–3376.

Feng, M., Lengaigne, M., Manneela, S., Gupta, A. S., & Vialard, J. (2024). Chapter 6: Extreme events in the Indian Ocean: Marine heatwaves, cyclones, and tsunamis. In C. C. Ummenhofer, & R. R. Hood (Eds.), *The Indian Ocean and its role in the global climate system* (pp. 121–144). Amsterdam: Elsevier. https://doi.org/10.1016/B978-0-12-822698-8.00011-1.

Feng, M., Slawinski, D., Beckley, L. E., & Keesing, J. K. (2010). Retention and dispersal of shelf waters influenced by interactins of ocean boundary current and coastal geography. *Marine and Freshwater Research*, *61*, 1259–1267.

Feng, M., Wijffels, S., Godfrey, S., & Meyers, G. (2005). Do eddies play a role in the momentum balance of the Leeuwin Current? *Journal of Physical Oceanography*, *35*, 964–975.

Fernandes, V., & Ramaiah, N. (2019). Spatial structuring of zooplankton communities through partitioning of habitat and resources in the Bay of Bengal during spring intermonsoon. *Turkish Journal of Zoology*, *43*(1), 68–93.

Foltz, G., Vialard, J., Kumar, P., & McPhaden, M. J. (2010). Seasonal mixed layer heat balance of the southwestern tropical Indian Ocean. *Journal of Climate*, *23*, 947–965.

Fonteneau, A. (1986). Analysis of exploitation of some yellowfin tuna concentrations by purse seiners during the period 1980–1983 in the East Atlantic. *ICCAT Collective Volumes of Scientific Papers*, *25*, 81–98.

Fox-Kemper, B., et al. (2021). Ocean, cryosphere and sea level change. In V. Masson-Delmotte, et al. (Eds.), *Climate change 2021: The physical science basis. Contribution of working group I to the sixth assessment report of the intergovernmental panel on climate change*. Cambridge: Cambridge University Press.

Fu, X., Wang, B., Waliser, D. E., & Tao, L. (2007). Impact of atmosphere–ocean coupling on the predictability of monsoon intraseasonal oscillations. *Journal of the Atmospheric Sciences*, *64*(1), 157–174.

Fu, X., Yang, B., Bao, Q., & Wang, B. (2008). Sea surface temperature feedback extends the predictability of tropical intraseasonal oscillation. *Monthly Weather Review*, *136*(2), 577–597.

Furnas, M. (2007). Intra-seasonal and inter-annual variations in phytoplankton biomass, primary production and bacterial production at North West Cape, Western Australia: Links to the 1997–1998 El Niño event. *Continental Shelf Research*, *27*(7), 958–980.

Furue, R., Guerreiro, K., Phillips, H. E., McCreary, J. P., Jr., & Bindoff, N. L. (2017). On the Leeuwin Current System and its linkage to zonal flows in the south Indian Ocean as inferred from a gridded hydrography. *Journal of Physical Oceanography*, *47*(3), 583–602.

Gadgil, S., Vinayachandran, P., Francis, P., & Gadgil, S. (2004). Extremes of the Indian summer monsoon rainfall, ENSO and equatorial Indian Ocean oscillation. *Geophysical Research Letters*, *31*(12).

Gardner, W., Mishonov, A., & Richardson, M. (2006). Global POC concentrations from in-situ and satellite data. *Deep Sea Research Part II: Topical Studies in Oceanography*, *53*(5–7), 718–740.

Geen, R., Bordoni, S., Battisti, D. S., & Hui, K. (2020). Monsoons, ITCZs, and the concept of the global monsoon. *Reviews of Geophysics*, *58*, e2020RG000700. https://doi.org/10.1029/2020RG000700.

Gersbach, G. H., Pattiaratchi, C., Ivey, G. N., & Cresswell, G. R. (1999). Upwelling on the south-west coast of Australia—Source of the Capes Current? *Continental Shelf Research*, *19*, 363–400.

Girishkumar, M., Ravichandran, M., & McPhaden, M. (2013). Temperature inversions and their influence on the mixed layer heat budget during the winters of 2006–2007 and 2007–2008 in the Bay of Bengal. *Journal of Geophysical Research: Oceans*, *118*(5), 2426–2437.

Godfrey, J. S., & Ridgway, K. R. (1985). The large-scale environment of the poleward-flowing Leeuwin Current, Western Australia: Longshore steric height gradients, wind stresses, and geostrophic flow. *Journal of Physical Oceanography*, *15*, 481–495.

Goes, J. I., & Gomes, H.d. R. (2016). An ecosystem in transition: The emergence of mixotrophy in the Arabian Sea. In *Aquatic microbial ecology and biogeochemistry: A dual perspective* (pp. 155–170). Springer.

Goes, J. I., Thoppil, P. G., Gomes, H. D., & Fasullo, J. T. (2005). Warming of the Eurasian landmass is making the Arabian Sea more productive. *Science*, *308*, 545–547.

Goes, J. I., Tian, H., do Rosario Gomes, H., Anderson, O. R., Al-Hashmi, K., de Rada, S., Luo, H., Al-Kharusi, L., Al-Azri, A., & Martinson, D. G. (2020). Ecosystem state change in the Arabian Sea fuelled by the recent loss of snow over the Himalayan-tibetan plateau region. *Scientific Reports*, *10*(1), 1–8.

Gomes, H. D., Goes, J. I., Motondkar, S. G. P., Buskey, E. J., Basu, S., Parab, S., & Thoppil, P. G. (2014). Massive outbreaks of Notiluca scintillans blooms in teh Arabian Sea due to spread of hypoxia. *Nature Communications*, *5*(4863). https://doi.org/10.1038/ncomms5862.

Gomes, H. R., Goes, J. I., & Saino, T. (2000). Influence of physical processes and freshwater discharge on the seasonality of phytplankton regime in the Bay of Bengal. *Continental Shelf Research*, *20*, 313–330.

Gopalakrishna, V., Murty, V., Sengupta, D., Shenoy, S., & Araligidad, N. (2002). Upper ocean stratification and circulation in the northern Bay of Bengal during southwest monsoon of 1991. *Continental Shelf Research*, *22*(5), 791–802.

Gordon, A. L. (1986). Interocean exchange of thermocline water. *Journal of Geophysical Research: Oceans*, *91*(C4), 5037–5046.

Gordon, A. L., Ma, S., Olson, D. B., Hacker, P., Ffield, A., Talley, L. D., Wilson, D., & Baringer, M. (1997). Advection and diffusion of Indonesian Throughflow Water within the Indian Ocean South Equatorial Current. *Geophysical Research Letters*, *24*(21), 2573–2576.

Gould, J., Sloyan, B., & Visbeck, M. (2013). In situ ocean observations: A brief history, present status, and future directions. In *Vol. 103. International Geophysics* (pp. 59–81). Elsevier.

Guo, F., Liu, Q., Zheng, X.-T., & Sun, S. (2013). The role of barrier layer in southeastern Arabian Sea during the development of positive Indian Ocean Dipole events. *Journal of Ocean University of China*, *12*(2), 245–252.

Han, W., Vialard, J., McPhaden, M. J., Lee, T., Masumoto, Y., Feng, M., & de Ruiter, W. P. (2014). Indian Ocean decadal variability: A review. *Bulletin of the American Meteorological Society*, *95*(11), 1679–1703.

Hansell, D. A. (2009). Dissolved organic carbon in the carbon cycle of the Indian Ocean. In *Geophysical Monograph Series: Vol. 185. Indian Ocean Biogeochemical Processes and Ecological Variability* (pp. 217–230).

Hansell, D. A., Carlson, C. A., Repeta, D. J., & Schlitzer, R. (2009). Dissolved organic matter in the ocean: A controversy stimulates new insights. *Oceanography*, *22*(4), 202–211.

Hanson, C. E., Pattiaratchi, C. B., & Waite, A. M. (2005a). Sporadic upwelling on a downwelling coast: Phytoplankton responses to spatially variable nutrient dynamics off the Gascoyne region of Western Australia. *Continental Shelf Research*, *25*, 1561–1582.

Hanson, C. E., Pattriaratchi, C. B., & Waite, A. M. (2005b). Seasonal production regimes off south-western Australia: Influence of the Capes and Leeuwin currents on phytoplankton dynamics. *Marine and Freshwater Research*, *56*, 1011–1026.

Hanson, C. E., Pesant, S., Waite, A. M., & Pattiaratchi, C. B. (2007). Assessing the magnitude and significance of deep chlorophyll maxima of the coastal eastern Indian Ocean. *Deep-Sea Research, Part II*, *54*, 884–901.

Harms, N. C., Lahajnar, N., Gaye, B., Rixen, T., Dähnke, K., Ankele, M., Schwarz-Schampera, U., & Emeis, K.-C. (2019). Nutrient distribution and nitrogen and oxygen isotopic composition of nitrate in water masses of the subtropical southern Indian Ocean. *Biogeosciences*, *16*(13), 2715–2732.

Harms, N. C., Lahajnar, N., Gaye, B., Rixen, T., Schwarz-Schampera, U., & Emeis, K.-C. (2021). Sediment trap-derived particulate matter fluxes in the oligotrophic subtropical gyre of the South Indian Ocean. *Deep Sea Research Part II: Topical Studies in Oceanography*, 104924.

Heezen, B. C., & Tharp, M. (1966). Physiography of the Indian Ocean. *Philosophical Transactions of the Royal Society of London Series A—Mathematical, Physical and Engineering Sciences*, *259*, 137–149.

Hermes, J. C., & Reason, C. J. C. (2008). Annual cycle of the South Indian Ocean (Seychelles-Chagos) thermocline ridge in a regional ocean model. *Journal of Geophysical Research*, *113*, C04035. https://doi.org/10.1029/2007JC004363.

Hildebrandsson, H. H. (1897). *Quelques recherches sur les centres d'action de l'atmosphère*. Norstedt & Söner.

Hoegh-Guldberg, O., Mumby, P. J., Hooten, A. J., Steneck, R. S., Greenfield, P., Gomez, E., Harvell, C. D., Sale, P. F., Edwards, A. J., Caldeira, K., Knowlton, N., Eakin, C. M., Iglesias-Prieto, R., Muthiga, N., Bradbury, R. H., Dubi, A., & Hatziolos, M. E. (2007). Coral reefs under rapid climate change and oceand acidification. *Science*, *318*(5857), 1737–1742.

Hood, R. R., Bange, H. W., Beal, L., Beckley, L. E., Burkill, P., Cowie, G. L., D'Adamo, N., Ganssen, G., Hendon, H., Hermes, J., Honda, M., McPhaden, M., Roberts, M., Singh, S., Urban, E., & Yu, W. (2015). *Science plan of the second International Indian Ocean Expedition (IIOE-2): A basin-wide research program*. Scientific Committee on Oceanic Research.

Hood, R. R., Beckley, L. E. B., & Wiggert, J. D. (2017). Biogeochemical and ecological impacts of boundary currents in the Indian Ocean. *Progress in Oceanography*, *156*, 290–325.

Hood, R. R., Coles, V. J., Huggett, J. A., Landry, M. R., Levy, M., Moffett, J. W., & Rixen, T. (2024a). Chapter 13: Nutrient, phytoplankton, and zooplankton variability in the Indian Ocean. In C. C. Ummenhofer, & R. R. Hood (Eds.), *The Indian Ocean and its role in the global climate system* (pp. 293–327). Amsterdam: Elsevier. https://doi.org/10.1016/B978-0-12-822698-8.00020-2.

Hood, R. R., Rixen, T., Levy, M., Hansell, D. A., Coles, V. J., & Lachkar, Z. (2024b). Chapter 12: Oxygen, carbon, and pH variability in the Indian Ocean. In C. C. Ummenhofer, & R. R. Hood (Eds.), *The Indian Ocean and its role in the global climate system* (pp. 265–291). Amsterdam: Elsevier. https://doi.org/10.1016/B978-0-12-822698-8.00017-2.

Hood, R. R., Wiggert, J. D., & Naqvi, S. W. A. (2009). Indian Ocean research: Opportunities and challenges. *Geophysical Monograph Series*, 409–429. https://doi.org/10.1029/2008GM000714.

Hu, S., Zhang, Y., Feng, M., Du, Y., Sprintall, J., Wang, F., Hu, D., Xie, Q., & Chai, F. (2019). Interannual to decadal variability of upper-ocean salinity in the southern Indian Ocean and the role of the Indonesian Throughflow. *Journal of Climate*, *32*(19), 6403–6421.

Huggett, J., & Kyewalyanga, M. (2017). Ocean productivity: The RV Dr Fridtjof Nansen in the Western Indian Ocean. In J. C. Groeneveld, & K. A. Koranteng (Eds.), *The RV Dr Fridtjof Nansen in the Western Indian Ocean: Voyages of marine research and capacity development* (pp. 55–80). Rome: FAO. Appendix (pp. 189–216). ISBN 978-92-5-109872-1, p. 55 http://www.fao.org/3/a-i7652e.pdf.

Huggett, J., & Richardson, A. (2000). A review of the biology and ecology of Calanus agulhensis off South Africa. *ICES Journal of Marine Science*, *57*(6), 1834–1849.

Huhn, F., Von Kameke, A., Pérez-Muñuzuri, V., Olascoaga, M. J., & Beron-Vera, F. J. (2012). The impact of advective transport by the South Indian Ocean Countercurrent on the Madagascar plankton bloom. *Geophysical Research Letters*, *39*(6).

IPCC. (2019). *IPCC special report on the ocean and cryosphere in a changing climate*.

Ito, T., Minobe, S., Long, M. C., & Deutsch, C. (2017). Upper ocean O2 trends: 1958–2015. *Geophysical Research Letters*, *44*(9), 4214–4223.

Jahnke, R. A. (1996). The global ocean flux of particulate organic carbon: Areal distribution and magnitude. *Global Biogeochemical Cycles*, *10*(1), 71–88.

Jaspers, C., Nielsen, T. G., Carstensen, J., Hopcroft, R. R., & Møller, E. F. (2009). Metazooplankton distribution across the Southern Indian Ocean with emphasis on the role of Larvaceans. *Journal of Plankton Research*, *31*(5), 525–540.

Jose, Y. S., Aumont, O., Machu, E., Penven, P., Moloney, C. L., & Maury, O. (2014). Influence of mesoscale eddies on biological production in the Mozambique Channel: Severl contrasted examples from a coupled ocean-biogeochemistry model. *Deep-Sea Rearch, Part II*, *100*, 79–93.

Jose, Y. S., Penven, P., Aumont, O., Machu, E., Moloney, C., Shillington, F., & Maury, O. (2016). Suppressing and enhancing effects of mesoscale dynamics on biological production in the Mozambique Channel. *Journal of Marine Systems*, *158*, 129–139.

Kalnay, E., et al. (1996). The NCEP/NCAR 40-year reanalysis project. *Bulletin of the American Meteorological Society*, *77*, 437–470.

Keerthi, M. G., Lengaigne, M., Levy, M., Vialard, J., Parvathi, V., Boyer Montégut, C. D., Ethé, C., Aumont, O., Suresh, I., & Akhil, V. P. (2017). Physical control of interannual variations of the winter chlorophyll bloom in the northern Arabian Sea. *Biogeosciences*, *14*(15), 3615–3632.

Kennett, J. P. (1982). *Marine geology*. New Jersey, Prentice-Hall: Englewood Cliffs. 813 p.

Kumar, S., Muraleedharan, P., Prasad, T., Gauns, M., Ramaiah, N., De Souza, S., Sardesai, S., & Madhupratap, M. (2002). Why is the Bay of Bengal less productive during summer monsoon compared to the Arabian Sea? *Geophysical Research Letters*, *29*(24), 88-1–88-4.

Kumar, S. P., Nuncio, M., Ramaiah, N., Sardesai, S., Narvekar, J., Fernandes, V., & Paul, J. T. (2007). Eddy-mediated biological productivity in the Bay of Bengal during fall and spring intermonsoons. *Deep Sea Research Part I: Oceanographic Research Papers*, *54*(9), 1619–1640.

Kumar, K. K., Rajagopalan, B., & Cane, M. A. (1999). On the weakening relationship between the Indian monsoon and ENSO. *Science*, *284*(5423), 2156–2159.

Kumar, R., Singh, R., & Sharma, K. (2005). Water resources of India. *Current Science*, 794–811.

Kumari, A., Kumar, S. P., & Chakraborty, A. (2018). Seasonal and interannual variability in the barrier layer of the Bay of Bengal. *Journal of Geophysical Research: Oceans*, *123*(2), 1001–1015.

Kwiatkowski, L., Torres, O., Bopp, L., Aumont, O., Chamberlain, M., Christian, J. R., Dunne, J. P., Gehlen, M., Ilyina, T., & John, J. G. (2020). Twenty-first century ocean warming, acidification, deoxygenation, and upper-ocean nutrient and primary production decline from CMIP6 model projections. *Biogeosciences*, *17*(13), 3439–3470.

Lachkar, Z., Lévy, M., & Smith, S. (2018). Intensification and deepening of the Arabian Sea oxygen minimum zone in response to increase in Indian monsoon wind intensity. *Biogeosciences*, *15*(1), 159–186.

Lachkar, Z., Lévy, M., & Smith, K. S. (2019). Strong intensification of the Arabian Sea oxygen minimum zone in response to Arabian Gulf warming. *Geophysical Research Letters*, *46*(10), 5420–5429.

Lachkar, Z., Mehari, M., Al Azhar, M., Lévy, M., & Smith, S. (2020). Fast local warming of sea-surface is the main factor of recent deoxygenation in the Arabian Sea. *Biogeosciences Discussions*, 1–27.

Lamont, T., Barlow, R. G., Morris, T., & van den Berg, M. A. (2014). Characterization of mesoscale features and phytoplankton variability in the Mozambique Channel. *Deep-Sea Rearch, Part II*, *100*, 94–105.

Lay, T., Kanamori, H., Ammon, C. J., Nettles, M., Ward, S. N., Aster, R. C., Beck, S. L., Bilek, S. L., Brudzinski, M. R., Butler, R., DeShon, H. R., Ekstrom, G., Satake, K., & Sipkin, S. (2005). The great Sumatra-Andaman earthquake of 26 December 2004. *Science*, *308*(5725), 1127–1133.

Lee, T. (2004). Decadal weakening of the shallow overturning circulation in the South Indian Ocean. *Geophysical Research Letters*, *31*(18).

Lee, T., & McPhaden, M. J. (2008). Decadal phase change in large-scale sea level and winds in the Indo-Pacific region at the end of the 20th century. *Geophysical Research Letters*, *35*, 1.

Lee, S.-K., Park, W., Baringer, M. O., Gordon, A. L., Huber, B., & Liu, Y. (2015). Pacific origin of the abrupt increase in Indian Ocean heat content during the warming hiatus. *Nature Geoscince*, *8*(6), 445–449.

Lehodey, P., Alheit, J., Barange, M., Baumgartner, T., Beaugrand, G., Drinkwater, K., Fromentin, J.-M., Hare, S., Ottersen, G., & Perry, R. (2006). Climate variability, fish, and fisheries. *Journal of Climate*, *19*(20), 5009–5030.

Li, T., & Zhang, Y. (2002). Processes that determine the quasi-biennial and lower-frequency variability of the South Asian monsoon. *Journal of the Meteorological Society of Japan. Series II*, *80*(5), 1149–1163.

Li, T., Zhang, Y., Chang, C. P., & Wang, B. (2001). On the relationship between Indian Ocean sea surface temperature and Asian summer monsoon. *Geophysical Research Letters*, *28*(14), 2843–2846.

Longhurst, A. (2001). A major seasonal phytoplankton bloom in Madagascar. *Deep Sea Research Part I*, *48*(11), 2413–2422.

Lourey, M. J., Dunn, J. R., & Waring, J. (2006). A mixed layer nutrient climatology of Leeuwin Current and Western Australian shelf waters: seasonal nutrient dynamics and biomass. *Journal of Marine Systems*, *59*, 25–51.

Lukas, R., & Lindstrom, E. (1991). The mixed layer of the western equatorial Pacific Ocean. *Journal of Geophysical Research: Oceans*, *96*(S01), 3343–3357.

Lumpkin, R., & Speer, K. (2007). Global ocean meridional overturning. *Journal of Physical Oceanography*, *37*, 2550–2562.

Lutjeharms, J. R. E. (2006). *The Agulhas Current* (p. 329). Berlin, Heidelbert, New York: Springer.

Machu, E., & Garçon, V. (2001). Phytoplankton seasonal distribution from SeaWiFS data in the Agulhas Current system. *Journal of Marine Research*, *59*(5), 795–812.

Madhupratap, M., Haridas, P., Ramaiah, N., & Achuthankutty, C. (1992). Zooplankton of the southwest coast of India: Abundance, composition, temporal and spatial variability in 1987. In B. N. Desai (Ed.), *Oceanography of the Indian Ocean* (pp. 99–112). New Delhi: Oxford and IBH Publishing Co.

Mahadevan, A., Jaeger, G. S., Freilich, M., Omand, M. M., Shroyer, E. L., & Sengupta, D. (2016a). Freshwater in the Bay of Bengal: Its fate and role in air-sea heat exchange. *Oceanography*, *29*(2), 72–81.

Mahadevan, A., Paluszkiewicz, T., Ravichandran, M., Sengupta, D., & Tandon, A. (2016b). Introduction to the special issue on the Bay of Bengal: From monsoons to mixing. *Oceanography*, *29*(2), 14–17.

Marsac, F. (2017). The Seychelles Tuna fishery and climate change. In B. Phillips, & M. Perez-Ramirez (Eds.), *Vol. 2. Climate change impacts on fisheries and aquaculture* (pp. 523–568). Wiley Blackwell.

Marsac, F., Everett, B., Shahid, U., & Strutton, P. G. (2024). Chapter 11: Indian Ocean primary productivity and fisheries variability. In C. C. Ummenhofer, & R. R. Hood (Eds.), *The Indian Ocean and its role in the global climate system* (pp. 245–264). Amsterdam: Elsevier. https://doi.org/10.1016/B978-0-12-822698-8.00019-6.

Marsac, F., & Le Blanc, J. L. (1998). Interannual and ENSO-associated variability of the coupled ocean–atmosphere system with possible impacts on the yellowfin tuna fisheries of the Indian and Atlantic oceans. In J. S. Beckett (Ed.), *Vol. L(1). ICCAT Tuna symposium. ICCAT collective volumes of scientific papers* (pp. 345–377).

Marsac, F., & Le Blanc, J. L. (1999). Oceanographic changes during the 1997–1998 El Niño in the Indian ocean and their impact on the purse seine fishery. In *Vol. 2. Proceedings first session of the IOTC working party on tropical tunas, Mahe, Seychelles* (pp. 147–157). IOTC Proceedings.

Masson, S., Delecluse, P., Boulanger, J. P., & Menkes, C. (2002). A model study of the seasonal variability and formation mechanisms of the barrier layer in the eastern equatorial Indian Ocean. *Journal of Geophysical Research: Oceans*, *107*(C12). SRF 18-1–SRF 18-20.

Masson, S., Luo, J. J., Madec, G., Vialard, J., Durand, F., Gualdi, S., Guilyardi, E., Behera, S., Delécluse, P., & Navarra, A. (2005). Impact of barrier layer on winter-spring variability of the southeastern Arabian Sea. *Geophysical Research Letters*, *32*(7).

Maximenko, N., Niiler, P., Centurioni, L., Rio, M.-H., Melnichenko, O., Chambers, D., Zlotnicki, V., & Galperin, B. (2009). Mean dynamic topography of the ocean derived from satellite and drifting buoy data using three different techniques. *Journal of Atmospheric and Oceanic Technology*, *26*(9), 1910–1919.

McCreary, J. P., & Kundu, P. K. (1989). A numerical investigation of sea surface temperature variability in the Arabian Sea. *Journal of Geophysical Research: Oceans*, *94*(C11), 16097–16114.

McCreary, J. P., Kundu, P. K., & Molinari, R. L. (1993). A numerical investigation of dynamics, thermodynamics and mixed-layer processes in the Indian Ocean. *Progress in Oceanography*, *31*, 181–244.

McCreary, J. P., Murtugudde, R., Vialard, J., Vinayachandran, P. N., Wiggert, J. D., Hood, R. R., Shankar, D., & Shetye, S. R. (2009). Biophysical processes in the Indian Ocean. In J. Wiggert, R. R. Hood, S. W. A. Naqvi, K. H. Brink, & S. L. Smith (Eds.), *Vol. 185. Indian ocean biogeochemical processes and ecological variability* (pp. 9–32). Washington D.C.: American Geophysical Union.

McCreary, J. P., Yu, Z., Hood, R. R., Vinayachandran, P. N., Furue, R., Ishida, A., & Richards, K. J. (2013). Dynamcis of the Indian-Ocean oxygen minumum zones. *Progress in Oceanography*, *112–113*, 15–37.

McPhaden, M. J., Beal, L. M., Bhaskar, T. V. S. U., Lee, T., Nagura, M., Strutton, P. G., & Yu, L. (2024). Chapter 17: The Indian Ocean Observing System (IndOOS). In C. C. Ummenhofer, & R. R. Hood (Eds.), *The Indian Ocean and its role in the global climate system* (pp. 393–419). Amsterdam: Elsevier. https://doi.org/10.1016/B978-0-12-822698-8.00002-0.

McPhaden, M. J., Busalacchi, A. J., Cheney, R., Donguy, J. R., Gage, K. S., Halpern, D., Ji, M., Julian, P., Meyers, G., & Mitchum, G. T. (1998). The Tropical Ocean-Global Atmosphere observing system: A decade of progress. *Journal of Geophysical Research: Oceans*, *103*(C7), 14169–14240.

McPhaden, M. J., & Nagura, M. (2014). Indian Ocean dipole interpreted in terms of recharge oscillator theory. *Climate Dynamics*, *42*, 1569–1586.

McPhaden, M. J., Wang, Y., & Ravichandran, M. (2015). Volume transports of the Wyrtki jets and their relationship to the Indian Ocean Dipole. *Journal of Geophysical Research: Oceans*, *120*(8), 5302–5317.

Menezes, V. V., Phillips, H. E., Schiller, A., Bindoff, N. L., Domingues, C. M., & Vianna, M. L. (2014). South Indian Countercurrent and associated fronts. *Journal of Geophysical Research: Oceans*, *119*(10), 6763–6791.

Meyer, A. A., Lutjeharms, J. R. E., & de Villiers, S. (2002). The nutrient characteristics of the Natal Bight, South Africa. *Journal of Marine Systems*, *35*, 11–37.

Miyama, T., McCreary, J. P., Jr., Jensen, T. G., Loschnigg, J., Godfrey, S., & Ishida, A. (2003). Structure and dynamics of the Indian-Ocean cross-equatorial cell. *Deep Sea Research Part II: Topical Studies in Oceanography*, *50*(12–13), 2023–2047.

Moffett, J. W., & Landry, M. R. (2020). Grazing control and iron limitation of primary production in the Arabian Sea: Implications for anticipated shifts in Southwest Monsoon intensity. *Deep Sea Research Part II: Topical Studies in Oceanography*, *179*, 104687.

Moffett, J. W., Vedamati, J., Goepfert, T. J., Pratihary, A., Gauns, M., & Naqvi, S. W. A. (2015). Biogeochemistry of iron in the Arabian Sea. *Limnology and Oceanography*, *60*(5), 1671–1688.

Mohtadi, M., Abram, N. J., Clemens, S. C., Pfeiffer, M., Russell, J. M., Steinke, S., & Zinke, J. (2024). Chapter 19: Paleoclimate evidence of Indian Ocean variability across a range of timescales. In C. C. Ummenhofer, & R. R. Hood (Eds.), *The Indian Ocean and its role in the global climate system* (pp. 445–467). Amsterdam: Elsevier. https://doi.org/10.1016/B978-0-12-822698-8.00007-X.

Mohtadi, M., Prange, M., Oppo, D. W., De Pol-Holz, R., Merkel, U., Zhang, X., Steinke, S., & Luckge, A. (2014). North Atlantic forcing of tropical Indian Ocean climate. *Nature, 509*(7498), 76–80.

Moore, W. S. (1984). Review of the geosecs project. *Nuclear Instruments and Methods in Physics Research, 223*(2–3), 459–465.

Morrison, J. M., Codispoti, L. A., Smith, S. L., Wishner, K., Flagg, C., Gardner, W. D., Gaurin, S., Naqvi, S. W. A., Manghnani, V., Prosperie, L., & Gundersen, J. S. (1999). The oxygen minimum zone in the Arabian Sea during 1995. *Deep-Sea Research (Part II, Topical Studies in Oceanography), 46*(8–9), 1903–1931.

Moustahfid, H., Marsac, F., & Gangopadhyay, A. (2018). Chapter 12: Climate change impacts, vulnerabilities and adaptations: Western Indian Ocean marine fisheries. In M. Barange, T. Bahri, M. C. M. Beveridge, K. L. Cochrane, & S. Funge-Smith (Eds.), *Impacts of climate change on fisheries and aquaculture: synthesis of current knowledge, adaptation and mitigation options* (pp. 251–279). Rome: FAO. FAO Fish Aquac Tech Paper 627.

Muraleedharan, K., Jasmine, P., Achuthankutty, C., Revichandran, C., Kumar, P. D., Anand, P., & Rejomon, G. (2007). Influence of basin-scale and mesoscale physical processes on biological productivity in the Bay of Bengal during the summer monsoon. *Progress in Oceanography, 72*(4), 364–383.

Nagura, M., Terao, T., & Hashizume, M. (2015). The role of temperature inversions in the generation of seasonal and interannual SST variability in the far northern Bay of Bengal. *Journal of Climate, 28*(9), 3671–3693.

Naqvi, S. W. A. (2019). Evidence for ocean deoxygenation and its patterns: Indian Ocean. In D. Laffoley, & J. M. Baxter (Eds.), *Ocean deoxygenation: Everyone's problem.* Gland, Switzerland: IUCN.

Naqvi, S. W. A., Bange, H. W., Gibb, S. W., Goyet, C., Hatton, A. D., & Upstill-Goddard, R. C. (2005). Biogeochemical ocean-atmosphere transfers in the Arabian Sea. *Progress in Oceanography, 65*, 116–144.

Naqvi, S. W. A., Jayakumar, D. A., Narvekar, P. V., Naik, H., Sarma, V., D'Souza, W., Joseph, S., & George, M. D. (2000). Increased marine production of N_2O due to intensifying anoxia on the Indian continental shelf. *Nature, 408*(6810), 346–349.

Naqvi, S. W. A., Moffett, J. W., Gauns, M. U., Narvekar, P. V., Pratihary, A. K., Naik, H., Shenoy, D. M., Jayakumar, D. A., Goepfert, T. J., Patra, P. K., Al-Azri, A., & Ahmed, S. I. (2010). The Arabian Sea as a high-nutrient, low-chlorophyll region during the late Southwest Monsoon. *Biogeosciences, 7*, 2091–2100.

Narvekar, J., & Kumar, S. P. (2006). Seasonal variability of the mixed layer in the central Bay of Bengal and associated changes in nutrients and chlorophyll. *Deep Sea Research Part I: Oceanographic Research Papers, 53*(5), 820–835.

Nishioka, J., Obata, H., & Tsumune, D. (2013). Evidence of an extensive spread of hydrothermal dissolved iron in the Indian Ocean. *Earth and Planetary Science Letters, 361*, 26–33.

Norton, I. O., & Sclater, J. G. (1979). A model for the evolution of the Indian Ocean and the breakup of Gondwanaland. *Journal of Geophysical Research, 84*(B12), 6803–6830.

Nyadjro, E. S., Jensen, T. G., Richman, J. G., & Shriver, J. F. (2017). On the relationship between wind, SST, and the thermocline in the Seychelles–Chagos thermocline ridge. *IEEE Geoscience and Remote Sensing Letters, 14*(12), 2315–2319.

O'Brien, J. J., & Hurlburt, H. (1974). Equatorial jet in the Indian Ocean: Theory. *Science, 184*(4141), 1075–1077.

Orsi, A. H., Whitworth, T., III, & Nowlin, W. D., Jr. (1995). On the meridional extent and fronts of the Antarctic Circumpolar Current. *Deep Sea Research Part I: Oceanographic Research Papers, 42*(5), 641–673. https://doi.org/10.1016/0967-0637(95)00021-W.

Palastanga, V., Van Leeuwen, P., & De Ruijter, W. (2006). A link between low-frequency mesoscale eddy variability around Madagascar and the large-scale Indian Ocean variability. *Journal of Geophysical Research: Oceans, 111*(C9).

Parthasarathy, B., Munot, A., & Kothawale, D. (1988). Regression model for estimation of Indian foodgrain production from summer monsoon rainfall. *Agricultural and Forest Meteorology, 42*(2–3), 167–182.

Paterson, H. L., Feng, M., Waite, A. M., Gomis, D., Beckley, L. E., Holliday, D., & Thompson, P. A. (2008). Physical and chemical signatures of a developing anticyclonic eddy in the Leeuwin Current, eastern Indian Ocean. *Journal of Geophysical Research: Oceans, 113*(C7).

Pearce, A., & Pattiaratchi, C. (1999). The Capes Current: A summer countercurrent flowing past Cape Leeuwin and Cape Naturaliste, Western Australia. *Continental Shelf Research, 19*(3), 401–420.

Phillips, H. E., et al. (2021). Progress in understanding of Indian Ocean circulation, variability, air–sea exchange, and impacts on biogeochemistry. *Ocean Science, 17*, 1677–1751.

Phillips, H. E., Menezes, V. V., Nagura, M., McPhaden, M. J., Vinayachandran, P. N., & Beal, L. M. (2024). Chapter 8: Indian Ocean circulation. In C. C. Ummenhofer, & R. R. Hood (Eds.), *The Indian Ocean and its role in the global climate system* (pp. 169–203). Amsterdam: Elsevier. https://doi.org/10.1016/B978-0-12-822698-8.00012-3.

Poulton, A. J., Stinchcombe, M. C., & Quartly, G. D. (2009). High numbers of *Trichodesmium* and diazotrophic diatoms in the southwest Indian Ocean. *Geophysical Research Letters, 36*(15). https://doi.org/10.1029/2009GL039719.

Prakash, P., Prakash, S., Rahaman, H., Ravichandran, M., & Nayak, S. (2012). Is the trend in chlorophyll-a in the Arabian Sea decreasing? *Geophysical Research Letters, 39*(23).

Prakash, S., & Ramesh, R. (2007). Is the Arabian Sea getting more productive? *Currrent Science, 92*(5), 667–671.

Prasad, T. (2004). A comparison of mixed-layer dynamics between the Arabian Sea and Bay of Bengal: One-dimensional model results. *Journal of Geophysical Research: Oceans, 109*(C3).

Qiu, Y., Cai, W., Li, L., & Guo, X. (2012). Argo profiles variability of barrier layer in the tropical Indian Ocean and its relationship with the Indian Ocean Dipole. *Geophysical Research Letters, 39*(8).

Qu, T., & Meyers, G. (2005). Seasonal variation of barrier layer in the southeastern tropical Indian Ocean. *Journal of Geophysical Research: Oceans*, *110*(C11).

Raj, R. P., Peter, B. N., & Pushpadas, D. (2010). Oceanic and atmospheric influences on the variability of phytoplankton bloom in the Southwestern Indian Ocean. *Journal of Marine Systems*, *82*(4), 217–229.

Ramaiah, N., Fernandes, V., Paul, J., Jyothibabu, R., Gauns, M., & Jayraj, E. (2010). Seasonal variability in biological carbon biomass standing stocks and production in the surface layers of the Bay of Bengal. *Indian Journal of Marine Sciences*, *39*(3), 369–379.

Ramaswamy, V., Boucher, O., Haigh, J., Hauglustaine, D., Haywood, J., Myhre, G., Nakajima, T., Shi, G. Y., & Solomon, S. (2001). Radiative forcing of climate change. In J. T. Houghton, Y. Ding, D. J. Griggs, M. Noguer, P. J. Van der Linden, X. Dai, K. Maskell, & C. A. Johnson (Eds.), *Climate change 2001: The scientific basis: Contribution of working group I to the third assessment report of the intergovernmental panel on climate change* (pp. 349–416). Cambridge, UK: Cambridge University Press.

Rao, R. R., Molinari, R. L., & Festa, J. F. (1989). Evolution of the climatological near-surface thermal structure of the tropical Indian Ocean: 1. Description of mean monthly mixed layer depth, and sea surface temperature, surface current, and surface meteorological fields. *Journal of Geophysical Research: Oceans*, *94*(C8), 10801–10815.

Rao, R., & Sivakumar, R. (2003). Seasonal variability of sea surface salinity and salt budget of the mixed layer of the north Indian Ocean. *Journal of Geophysical Research: Oceans*, *108*(C1), 9-1–9-14.

Ravier, C., & Fromentin, J.-M. (2001). Long-term fluctuations in the eastern Atlantic and Mediterranean bluefin tuna population. *ICES Journal of Marine Science*, *58*(6), 1299–1317.

Ravier, C., & Fromentin, J. M. (2004). Are the long-term fluctuations in Atlantic bluefin tuna (*Thunnus thynnus*) population related to environmental changes? *Fisheries Oceanography*, *13*(3), 145–160.

Resplandy, L. (2018). Climate change and oxygen in the ocean. *Nature: International Weekly Journal of Science*, 7705.

Resplandy, L., Vialard, J., Lévy, M., Aumont, O., & Dandoneau, Y. (2009). Seasonal and intraseasonal biogeochemical variability in the thermocline ridge of the Indian Ocean. *Journal of Geophysical Research*, *114*, C07024. https://doi.org/10.1029/2008JC005246.

Reynolds, R. W., Rayner, N. A., Smith, T. M., Stokes, D. C., & Wang, W. (2002). An improved in situ and satellite SST analysis for climate. *Journal of Climate*, *15*, 1609–1625.

Ridgway, K., & Dunn, J. (2007). Observational evidence for a Southern Hemisphere oceanic supergyre. *Geophysical Research Letters*, *34*(13).

Rixen, T., Baum, A., Gaye, B., & Nagel, B. (2014). Seasonal and interannual variations in the nitrogen cycle in the Arabian Sea. *Biogeosciences*, *11*(20), 5733–5747.

Rixen, T., Cowie, G., Gaye, B., Goes, J., do Rosário Gomes, H., Hood, R. R., Lachkar, Z., Schmidt, H., Segschneider, J., & Singh, A. (2020). Reviews and syntheses: Present, past, and future of the oxygen minimum zone in the northern Indian Ocean. *Biogeosciences*, *17*(23), 6051–6080.

Rixen, T., Gaye, B., Emeis, K.-C., & Ramaswamy, V. (2019). The ballast effect of lithogenic matter and its influences on the carbon fluxes in the Indian Ocean. *Biogeosciences*, *16*(2), 485–503.

Roberts, M. J., Ternon, J.-F., & Morris, T. (2014). Interaction of dipole eddies with the western continental slope of the Mozambique Channel. *Deep-Sea Rearch, Part II*, *100*, 54–67.

Rochford, D. (1969). Seasonal variations in the Indian Ocean along 110 EI Hydrological structure of the upper 500 m. *Marine and Freshwater Research*, *20*(1), 1–50.

Rochford, D. (1977). Further studies of plankton ecosystems in the eastern Indian Ocean. II. Seasonal variations in water mass distribution (upper 150 m) along 110° E. (August 1962–August 1963). *Marine and Freshwater Research*, *28*(5), 541–555.

Rohini, P., Rajeevan, M., & Srivastava, A. (2016). On the variability and increasing trends of heat waves over India. *Scientific Reports*, *6*(1), 1–9.

Rossi, V., Feng, M., Pattiaratchi, C., Roughan, M., & Waite, A. M. (2013a). On the factors influencing the development of sporadic upwelling in the Leeuwin Current system. *Journal of Geophysical Research*, *118*(7), 3608–3621.

Rossi, V., Feng, M., Pattiaratchi, C., Roughan, M., & Waite, A. M. (2013b). Linking synoptic forcing and local mesoscale processes with biological dynamics off Ningaloo Reef. *Journal of Geophysical Research*, *118*(3), 1211–1225.

Roxy, M. (2014). Sensitivity of precipitation to sea surface temperature over the tropical summer monsoon region—and its quantification. *Climate Dynamics*, *43*(5–6), 1159–1169.

Roxy, M. K., Ghosh, S., Pathak, A., Athulya, R., Mujumdar, M., Murtugudde, R., Terray, P., & Rajeevan, M. (2017). A threefold rise in widespread extreme rain events over central India. *Nature Communications*, *8*(1), 1–11.

Roxy, M. K., Modi, A., Murtugudde, R., Valsala, V., Panickal, S., Kumar, S. P., Ravichandran, M., Vichi, M., & Levy, M. (2016). A reduction in marine primary productivity driven by rapid warming over the tropical Indian Ocean. *Geophysical Research Letters*, *43*, 826–833.

Roxy, M. K., Ritika, K., Terray, P., & Masson, S. (2014). The curious case of Indian Ocean warming. *Journal of Climate*, *27*(22), 8501–8509.

Roxy, M. K., Ritika, K., Terray, P., Murtugudde, R., Ashok, K., & Goswami, B. (2015). Drying of Indian subcontinent by rapid Indian Ocean warming and a weakening land-sea thermal gradient. *Nature Communications*, *6*(1), 1–10.

Roxy, M. K., Saranya, J. S., Modi, A., Anusree, A., Cai, W., Resplandy, L., … Frölicher, T. L. (2024). Chapter 20: Future projections for the tropical Indian Ocean. In C. C. Ummenhofer, & R. R. Hood (Eds.), *The Indian Ocean and its role in the global climate system* (pp. 469–482). Amsterdam: Elsevier. https://doi.org/10.1016/B978-0-12-822698-8.00004-4.

Schmidtko, S., Stramma, L., & Visbeck, M. (2017). Decline in global oceanic oxygen content during the past five decades. *Nature*, *542*(7641), 335–339.

Schott, F. A., Dengler, M., & Schoenefeldt, R. (2002). The shallow overturning circulation of the Indian Ocean. *Progress in Oceanography*, *53*, 57–103.

Schott, F. A., & McCreary, J. P. (2001). The monsoon circulation in the Indian Ocean. *Progress in Oceanography*, *51*(1), 1–123.

Schott, F. A., McCreary, J. P., Jr., & Johnson, G. C. (2004). Shallow overturning circulations of the tropical-subtropical oceans, in Earth's Climate: The Ocean-Atmosphere Interaction. In C. Wang, S.-P. Xie, & J. A. Carton (Eds.), *Vol. 147. Geophysical monograph* (pp. 261–304). Washington, D.C.: American Geophysical Union.

Schott, F. A., Xie, S. P., & McCreary, J. P. (2009). Indian Ocean circulation and climate variability. *Reviews of Geophysics, 47*, RG1002.

Schumann, E. H., Churchill, J. R. S., & Zaayman, H. J. (2005). Oceanic variability in the western sector of Algoa Bay, South Africa. *African Journal of Marine Science, 27*(1), 65–80.

Sengupta, D., Bharath Raj, G., & Shenoi, S. (2006). Surface freshwater from Bay of Bengal runoff and Indonesian Throughflow in the tropical Indian Ocean. *Geophysical Research Letters, 33*(22).

Sengupta, D., Goddalehundi, B. R., & Anitha, D. (2008). Cyclone-induced mixing does not cool SST in the post-monsoon north Bay of Bengal. *Atmospheric Science Letters, 9*(1), 1–6.

Sewell, R. B. S. (1934). The John Murray Expedition to the Arabian Sea. *Nature, 133*, 669–672.

Shankar, D., Vinayachandran, P., & Unnikrishnan, A. (2002). The monsoon currents in the north Indian Ocean. *Progress in Oceanography, 52*(1), 63–120.

Sharma, G. S. (1976). Water characteristics and current structure at 65°E during the southwest monsoon. *Journal of the Oceanographical Society of Japan, 32*, 284–296.

Sharma, G., Gouveia, A., & Satyendranath, S. (1978). Incursion of the Pacific Ocean water into the Indian Ocean. *Proceedings of the Indian Academy of Sciences-Section A, Earth and Planetary Sciences, 87*(3), 29–45.

Shenoi, S., Shankar, D., & Shetye, S. (2002). Differences in heat budgets of the near-surface Arabian Sea and Bay of Bengal: Implications for the summer monsoon. *Journal of Geophysical Research: Oceans, 107*(C6), 5-1–5-14.

Shen-Tu, B. (2016). *Java subduction zone earthquake: The worst is yet to come*. Verisk. https://www.air-worldwide.com/publications/air-currents/2016/Java-Subduction-Zone-Earthquake- -The-Worst-Is-Yet-to-Come-/.

Shetye, S. R., Gouveia, A. D., Shankar, D., Shenoi, S. S. C., Vinayachandran, P. N., Sundar, D., Michael, G. S., & Nampoothiri, G. (1996). Hydrography and circulation in the western Bay of Bengal during the northeast monsoon. *Journal of Geophysical Research, Oceans, 101*(C6), 14011–14025.

Shinoda, T., Jenson, T. G., Lachkar, Z., Masumoto, Y., & Seo, H. (2024). Chapter 18: Modeling the Indian Ocean. In C. C. Ummenhofer, & R. R. Hood (Eds.), *The Indian Ocean and its role in the global climate system* (pp. 421–443). Amsterdam: Elsevier. https://doi.org/10.1016/B978-0-12-822698-8.00001-9.

Singh, D. (2016). Tug of war on rainfall changes. *Nature Climate Change, 6*(1), 20–22.

Smith, R. L., Huyer, A., Godfrey, J. S., & Church, J. (1991). The Leeuwin Current off Western Australia, 1986–1987. *Journal of Physical Oceanography, 21*, 323–345.

Speich, S., Blanke, B., de Vries, P., Drijfhout, S., Döös, K., Ganachaud, A., & Marsh, R. (2002). Tasman leakage: A new route in the global ocean conveyor belt. *Geophysical Research Letters, 29*(10), 55-1–55-4.

Sprintall, J., Biastoch, A., Gruenburg, L. K., & Phillips, H. E. (2024). Chapter 9: Oceanic basin connections. In C. C. Ummenhofer, & R. R. Hood (Eds.), *The Indian Ocean and its role in the global climate system* (pp. 205–227). Amsterdam: Elsevier. https://doi.org/10.1016/B978-0-12-822698-8.00003-2.

Sprintall, J., Chong, J., Syamsudin, F., Morawitz, W., Hautala, S., Bray, N., & Wijffels, S. (1999). Dynamics of the South Java Current in the Indo-Australian Basin. *Geophysical Research Letters, 26*(16), 2493–2496.

Sprintall, J., & Tomczak, M. (1992). Evidence of the barrier layer in the surface layer of the tropics. *Journal of Geophysical Research: Oceans, 97*(C5), 7305–7316.

Sreenivas, P., Patnaik, K., & Prasad, K. (2008). Monthly variability of mixed layer over Arabian Sea using ARGO data. *Marine Geodesy, 31*(1), 17–38.

Srokosz, M. A., & Quartly, G. D. (2013). The Madagascar bloom: A serendipitous study. *Journal of Geophysical Research, 118*(1), 14–25.

Srokosz, M. A., Quartly, G. D., & Buck, J. J. (2004). A possible plankton wave in the Indian Ocean. *Geophysical Research Letters, 31*(13).

Srokosz, M., Robinson, J., McGrain, H., Popova, E., & Yool, A. (2015). Could the Madagascar bloom be fertilized by Madagascan iron? *Journal of Geophysical Research: Oceans, 120*(8), 5790–5803.

Stramma, L., Johnson, G. C., Sprintall, J., & Mohrholz, V. (2008). Expanding oxygen-minimum zones in the tropical oceans. *Science, 320*, 655–658.

Stramma, L., & Lutjeharms, J. R. E. (1997). The flow field of the subtropical gyre of the South Indian Ocean. *Journal of Geophysical Research, Oceans, 102*(C3), 5513–5530.

Strutton, P. G., Coles, V. J., Hood, R. R., Matear, R. J., McPhaden, M. J., & Phillips, H. E. (2015). Biogeochemical variability in the central equatorial Indian Ocean during the monsoon transition. *Biogeosciences, 12*(8), 2367–2382.

Subha Anand, S., Rengarajan, R., Sarma, V., Sudheer, A., Bhushan, R., & Singh, S. (2017). Spatial variability of upper ocean POC export in the Bay of Bengal and the Indian Ocean determined using particle-reactive 234 Th. *Journal of Geophysical Research: Oceans, 122*(5), 3753–3770.

Susanto, R. D., Gordon, A. L., & Zheng, Q. (2001). Upwelling along the coasts of Java and Sumatra and its relation to ENSO. *Geophysical Research Letters, 28*(8), 1599–1602.

Sutton, A. L., & Beckley, L. E. (2016). Influence of the Leeuwin Current on the epipelagic euphausiid assemblages of the south-east Indian Ocean. *Hydrobiologia, 779*(1), 193–207.

Takahashi, T., Sutherland, S. C., Chipman, D. W., Goddard, J. G., Ho, C., Newberger, T., Sweeney, C., & Munro, D. R. (2014). Climatological distributions of pH, pCO_2, total CO_2, alkalinity, and $CaCO_3$ saturation in the global surface ocean, and temporal changes at selected locations. *Marine Chemistry, 164*, 95–125.

Takahashi, T., Sutherland, S. C., Sweeney, C., Poisson, A., Metzl, N., Tilbrook, B., Bates, N., Wanninkhof, R., Feely, R. A., Sabine, C., Olafsson, J., & Nojiri, Y. (2002). Global sea-air CO_2 flux based on climatological surface ocean pCO_2, and seasonal biological and temperature effects. *Deep-Sea Research Part II, 49*(9–10), 1601–1623.

Takahashi, T., et al. (2009). Climatological mean and decadal change in surface ocean pCO_2, and net sea-air CO_2 flux over the global ocean. *Deep-Sea Rearch, Part II, 56*, 554–577.

Talley, L. D., Johnson, G. C., Purkey, S., Feely, R. A., & Wanninkhof, R. (2017). *Global Ocean Ship-based Hydrographic Investigations Program (GO-SHIP) provides key climate-relevant deep ocean observations* (p. 15). US CLIVAR Variations.

Talley, L. D., Pickard, G. L., Emery, W. J., & Swift, J. H. (2011). *Descriptive physical oceanography: An introduction*. New York: Academic Press, Elsevier.

Talley, L. D., & Sprintall, J. (2005). Deep expression fo the Indonesian Throughflow: Indonesian intermediate water in the south equatorial current. *Journal of Geophysical Research, Oceans, 110*, C10009. https://doi.org/10.1029/2004JC002826.

Taschetto, A. S., Ummenhofer, C. C., Stuecker, M. F., Dommenget, D., Ashok, K., Rodrigues, R. R., & Yeh, S. W. (2020). ENSO atmospheric teleconnections. In M. McPhaden, A. Santoso, & W. Cai (Eds.), *El Niño Southern Oscillation in a changing climate* (pp. 309–335). American Geophysical Union.

Tew-Kai, E., & Marsac, F. (2009). Patterns of variability of sea surface chlorophyll in the Mozambique Channel: A quantitative approach. *Journal of Marine Systems, 77*(1–2), 77–88.

Thadathil, P., Gopalakrishna, V., Muraleedharan, P., Reddy, G., Araligidad, N., & Shenoy, S. (2002). Surface layer temperature inversion in the Bay of Bengal. *Deep Sea Research Part I: Oceanographic Research Papers, 49*(10), 1801–1818.

Thushara, V., Vinayachandran, P. N. M., Matthews, A. J., Webber, B. G., & Queste, B. Y. (2019). Vertical distribution of chlorophyll in dynamically distinct regions of the southern Bay of Bengal. *Biogeosciences, 16*(7), 1447–1468.

Tozuka, T., Dong, L., Han, W., Lengainge, M., & Zhang, L. (2024). Chapter 10: Decadal variability of the Indian Ocean and its predictability. In C. C. Ummenhofer, & R. R. Hood (Eds.), *The Indian Ocean and its role in the global climate system* (pp. 229–244). Amsterdam: Elsevier. https://doi.org/10.1016/B978-0-12-822698-8.00014-7.

Tozuka, T., Yokoi, T., & Yamagata, T. (2010). A modeling study of interannual variations of the Seychelles Dome. *Journal of Geophysical Research: Oceans, 115*, C4.

Tranter, D., & Kerr, J. (1969). Seasonal variations in the Indian Ocean along 110 EV Zooplankton biomass. *Marine and Freshwater Research, 20*(1), 77–84.

Tranter, D. J., & Kerr, J. D. (1977). Further studies of plankton ecosystems in the eastern Indian Ocean. III. Numerical abundance and biomass. *Australian Journal of Marine and Freshwater Research, 28*, 557–583.

Trenary, L. L., & Han, W. (2012). Intraseasonal-to-interannual variability of South Indian Ocean sea level and thermocline: Remote versus local forcing. *Journal of Physical Oceanography, 42*(4), 602–627.

Turcotte, D. L., & Schubert, G. (2002). *Geodynamics*. Cambridge University Press.

Ummenhofer, C. C., Geen, R., Denniston, R. F., & Rao, M. P. (2024a). Chapter 3: Past, present, and future of the South Asian monsoon. In C. C. Ummenhofer, & R. R. Hood (Eds.), *The Indian Ocean and its role in the global climate system* (pp. 49–78). Amsterdam: Elsevier. https://doi.org/10.1016/B978-0-12-822698-8.00013-5.

Ummenhofer, C. C., Gupta, A. S., Li, Y., Taschetto, A. S., & England, M. H. (2011). Multi-decadal modulation of the El Niño–Indian monsoon relationship by Indian Ocean variability. *Environmental Research Letters, 6*(3), 034006.

Ummenhofer, C. C., Murty, S. A., Sprintall, J., Lee, T., & Abram, N. J. (2021). Heat and freshwater changes in the Indian Ocean region. *Nature Reviews Earth & Environment, 2*, 1–17.

Ummenhofer, C. C., Taschetto, A. S., Izumo, T., & Luo, J.-J. (2024b). Chapter 7: Impacts of the Indian Ocean on regional and global climate. In C. C. Ummenhofer, & R. R. Hood (Eds.), *The Indian Ocean and its role in the global climate system* (pp. 145–168). Amsterdam: Elsevier. https://doi.org/10.1016/B978-0-12-822698-8.00018-4.

Uz, B. M. (2007). What causes the sporadic phytoplankton bloom southeast of Madagascar? *Journal of Geophysical Research, 112*(C9). https://doi.org/10.1029/2006JC003685.

Valsala, V., Maksyutov, S., & Murtugudde, R. (2012). A window for carbon uptake in the southern subtropical Indian Ocean. *Geophysical Research Letters, 39*, L17605.

Vialard, J., Duvel, J. P., McPhaden, M., Bouruet-Aubertot, P., Ward, B., Key, E., Bourras, D., Weller, R., Minnett, P., Weill, A., Cassou, C., Eymard, L., Fristedt, T., Basdevant, C., Dandoneau, Y., Duteil, O., Izumo, T., de Boyer Montégut, C., Masson, S., ... Kennan, S. (2009). Cirene: Air-sea interactions in the Seychelles-Chagos thermocline ridge region. *Bulletin of the American Meteorological Society, 90*(1), 45–61.

Vialard, J., Foltz, G., McPhaden, M. J., Duvel, J. P., & de Boyer Montegut, C. (2008). Strong Indian Ocean sea surface temperature signals associated with the Madden-Julian oscillation in late 2007 and early 2008. *Geophysical Research Letters, 35*, L19608. https://doi.org/10.1029/2008GL035238.

Vinayachandran, P. N. (2009). Impact of physical processes on chlorophyll distribution in the Bay of Bengal. In J. D. Wiggert, R. R. Hood, S. W. A. Naqvi, K. H. Brink, & S. L. Smith (Eds.), *Vol. 185. Indian Ocean biogeochemical processes and ecological variability*. Washington D.C.: American Geophysical Union.

Vinayachandran, P. N., McCreary, J. P., Hood, R. R., & Kohler, K. E. (2005). A numerical investigation of phytoplankton blooms in the Bay of Bengal during the Northeast Monsoon. *Journal of Geophysical Research, 110*(C12001). https://doi.org/10.1029/2005JC002966.

Vinayachandran, P., Murty, V., & Ramesh Babu, V. (2002). Observations of barrier layer formation in the Bay of Bengal during summer monsoon. *Journal of Geophysical Research: Oceans, 107*(C12). SRF 19-1–SRF 19-9.

Vinayachandran, P., & Yamagata, T. (1998). Monsoon response of the sea around Sri Lanka: Generation of thermal domesand anticyclonic vortices. *Journal of Physical Oceanography, 28*(10), 1946–1960.

Vu, H. T. D., & Sohrin, Y. (2013). Diverse stoichiometry of dissolved trace metals in the Indian Ocean. *Scientific Reports, 3*(1), 1–5.

Waite, A. M., Beckley, L. E., Guidi, L., Landrum, J., Holliday, D., Montoya, J., Paterson, H., Feng, M., Thompson, P. A., & Raes, E. J. (2015). Cross-shelf transport, oxygen depletion and nitrate release within a forming mesoscale eddy in the eastern Indian Ocean. *Limnology and Oceanography, 61*(1), 103–121.

Waite, A. M., Thompson, P. A., Pesant, S., Feng, M., Beckley, L. E., Domingues, C. M., Gaughan, D., Hanson, C. E., Holl, C. M., Koslow, T., Meuleners, M., Montoya, J. P., Moore, T., Muhling, B. A., Paterson, H., Rennie, S., Strzelecki, J., & Twomey, L. (2007). The Leeuwin Current and its eddies: An introductory overview. *Deep-Sea Research, Part II, 54*, 789–796.

Walker, G. T. (1925). Correlation in seasonal variations of weather—A further study of world weather. *Monthly Weather Review*, *53*(6), 252–254.

Walker, T. D. (2024). Chapter 2: A brief historical overview of the maritime Indian Ocean World (ancient times to 1950). In C. C. Ummenhofer, & R. R. Hood (Eds.), *The Indian Ocean and its role in the global climate system* (pp. 33–48). Amsterdam: Elsevier. https://doi.org/10.1016/B978-0-12-822698-8.00005-6.

Wang, B. (2006). *The Asian Monsoon*. Chichester, UK: Praxis Publishing Limited. 788 pp.

Wang, B., Biasutti, M., Byrne, M. P., Castro, C., Chang, C.-P., Cook, K., Fu, R., Grimm, A. M., Ha, K.-J., & Hendon, H. (2021). Monsoons climate change assessment. *Bulletin of the American Meteorological Society*, *102*(1), E1–E19.

Wang, P., Clemens, S., Beaufort, L., Braconnot, P., Ganssen, G., Jian, Z., Kershaw, P., & Sarnthein, M. (2005). Evolution and variability of the Asian monsoon system: State of the art and outstanding issues. *Quaternary Science Reviews*, *24*, 595–629.

Wang, Y., & McPhaden, M. J. (2017). Seasonal cycle of cross-equatorial flow in the central Indian Ocean. *Journal of Geophysical Research: Oceans*, *122*(5), 3817–3827.

Webster, P. J., Magana, V. O., Palmer, T., Shukla, J., Tomas, R., Yanai, M., & Yasunari, T. (1998). Monsoons: Processes, predictability, and the prospects for prediction. *Journal of Geophysical Research: Oceans*, *103*(C7), 14451–14510.

Wiggert, J. D., Hood, R. R., Banse, K., & Kindle, J. C. (2005). Monsoon-driven biogeochemical processes in the Arabian Sea. *Progress in Oceanography*, *65*(2–4), 176–213.

Wiggert, J. D., Jones, B. H., Dickey, T. D., Weller, R. A., Brink, K. H., Marra, J., & Codispoti, L. A. (2000). The northeast monsoon's impact on mixing, phytoplankton biomass and nutrient cycling in the Arabian Sea. *Deep-Sea Research Part II*, *47*, 1353–1385.

Wiggert, J. D., Murtugudde, R. G., & Christian, J. R. (2006). Annual ecosystem variability in the tropical Indian Ocean: Results of a coupled bio-physical ocean general circulation model. *Deep-Sea Research, Part II*, *53*, 644–676.

Wiggert, J. D., Vialard, J., & Behrenfeld, M. (2009). Basinwide modification of dynamical and biogeochemical processes by the positive phase of the Indian Ocean Dipole during the SeaWiFS era. In J. D. Wiggert, R. R. Hood, S. W. A. Naqvi, S. L. Smith, & K. H. Brink (Eds.), *Indian Ocean biogeochemical processes and ecological variability* (pp. 385–407). Washington D.C.: American Geophysical Union.

Wijesekera, H. W., Shroyer, E., Tandon, A., Ravichandran, M., Sengupta, D., Jinadasa, S., Fernando, H. J., Agrawal, N., Arulananthan, K., & Bhat, G. (2016). ASIRI: An ocean–atmosphere initiative for Bay of Bengal. *Bulletin of the American Meteorological Society*, *97*(10), 1859–1884.

Woods, J. D. (1985). The world ocean circulation experiment. *Nature*, *314*(6011), 501–511.

Wurtzel, J. B., Abram, N. J., Lewis, S. C., Bajo, P., Hellstrom, J. C., Troitzsch, U., & Heslop, D. (2018). Tropical Indo-Pacific hydroclimate response to North Atlantic forcing during the last deglaciation as recorded by a speleothem from Sumatra, Indonesia. *Earth and Planetary Science Letters*, *492*, 264–278.

Wyrtki, K. (1973). An equatorial jet in the Indian Ocean. *Science*, *181*, 262–264.

Wyrtki, K., Bennett, E. B., & Rochford, D. J. (1971). *Oceanographic atlas of the International Indian Ocean expeditions*. National Science Foundation. OCE/NSF 86-00-001.

Xie, S. P., Annamalai, H., Schott, F. A., & McCreary, J. P. (2002). Structure and mechanisms of South Indian Ocean climate variability. *Journal of Climate*, *15*, 874–878.

Xie, P., & Arkin, P. A. (1997). Global precipitation: A 17-year monthly analysis based on gauge observations, satellite estimates, and numerical model outputs. *Bulletin of the American Meteorological Society*, *78*, 2539–2558.

Yamagata, T., Behera, S., Doi, T., Luo, J.-J., Morioka, Y., & Tozuka, T. (2024). Chapter 5: Climate phenomena of the Indian Ocean. In C. C. Ummenhofer, & R. R. Hood (Eds.), *The Indian Ocean and its role in the global climate system* (pp. 103–119). Amsterdam: Elsevier. https://doi.org/10.1016/B978-0-12-822698-8.00009-3.

Yang, J., Liu, Q., & Liu, Z. (2010). Linking observations of the Asian monsoon to the Indian Ocean SST: Possible roles of Indian Ocean basin mode and dipole mode. *Journal of Climate*, *23*(21), 5889–5902.

Yokoi, T., Tozuka, T., & Yamagata, T. (2008). Seasonal variation of the Seychelles Dome. *Journal of Climate*, *21*(15), 3740–3754.

Yu, L., Jin, X., & Weller, R. A. (2007). Annual, seasonal, and interannual variability of air–sea heat fluxes in the Indian Ocean. *Journal of Climate*, *20*(13), 3190–3209.

Yu, W., Xiang, B., Liu, L., & Liu, N. (2005). Understanding the origins of interannual thermocline variations in the tropical Indian Ocean. *Geophysical Research Letters*, *32*(24).

Yuan, X. (2020). *Sea surface silinity and the ocean structure in the tropical Indian Ocean* (Ph.D.). University Twente. 109 p.

Zhai, P., Bower, A. S., Smethie, W. M., Jr., & Pratt, L. J. (2015). Formation and spreading of Red Sea Outflow Water in the Red Sea. *Journal of Geophysical Research: Oceans*, *120*(9), 6542–6563.

Zhang, C. (2005). Madden-Julian oscillation. *Reviews in Geophysics*, *43*. RG 2003.

Zhang, C. (2013). Madden–Julian oscillation: Bridging weather and climate. *Bulletin of the American Meteorological Society*, *94*, 1849–1870.

Zheng, S., Feng, M., Du, Y., Meng, X., & Yu, W. (2018). Interannual variability of Eddy Kinetic energy in the subtropical Southeast Indian Ocean associated with the El Niño-Southern oscillation. *Journal of Geophysical Research: Oceans*, *123*(2), 1048–1061.

Chapter 2

A brief historical overview of the maritime Indian Ocean World (ancient times to 1950)[★]

Timothy D. Walker[a,b]

[a]Department of History, University of Massachusetts Dartmouth, North Dartmouth, MA, United States, [b]Woods Hole Oceanographic Institution, Woods Hole, MA, United States

1 Monsoon winds, fishing, and coastwise trade: The earliest maritime peoples of the Indian Ocean rim

From the earliest days of settled human habitation in the lands washed by the Indian Ocean, people have ventured into this body of water for sustenance, transportation, and trade. It has been navigated intensely and systematically for over 5000 years—far longer than any other of the world's seas—and is the first of the Earth's oceans to have been regularly traversed by humans on a deep-water offshore passage. The proximity of prototypical city states in the Indus River valley, Mesopotamia, and Egypt, all with fluvial or overland links to its waters, meant that the Indian Ocean figured as a key conduit of exploration, migration, and commerce, allowing for interaction and exchange between the earliest human civilizations (MacPherson, 2004). And yet, of all the world's ocean basins, the Indian Ocean, along with its neighbor, the Southern Ocean bordering Antarctica, is the least studied and understood by modern scientists and scholars.

The Indian Ocean ranks as the third largest of Earth's oceans, accounting for 27% of the world's maritime space or 14% of the total globe (Hattendorf et al., 2007). From its western limit at Cape Agulhas (roughly 20°E), it extends to 146°E, encompassing the southern tip of Tasmania. 60°S marks the Indian Ocean's southern boundary; adjacent lies the Southern Ocean. Its northern contours follow the coasts of Africa and the Middle East (including key maritime links to the Red Sea and Persian Gulf), India, the Bay of Bengal and the Malay Peninsula, and the islands of Indonesia. In all these regions, since long before recorded time, coastwise seafaring has been a common economic pursuit, with indigenous peoples developing varied efficient sailing or fishing technologies to exploit the sea.

Unique climatic conditions made the Indian Ocean a nursery for long-distance, wind-driven navigation. By the second millennia BCE (Before Common Era), sailors of the Indian Ocean basin well understood the patterns of monsoon winds of the Arabian Sea, which would reliably reverse its direction twice each year in the tropical zone (~20°N–20°S). Predictable seasonal changes in the direction of the prevailing monsoon winds provided an obvious benefit for traveling by sail between ports in coastal eastern Africa, South Asia, the Persian Gulf, and the Red Sea. The winter monsoon blows steadily from the northeast to the southwest between November and March. Summer monsoon winds blow from the southwest to the northeast between May and September (Pearson, 2003). This dependable regional weather pattern has set the rhythm for seaborne trade and travel in the Indian Ocean basin throughout most of human history, until the advent of steam-powered navigation. By about 900 BCE, mariners in India used the monsoons to sail seasonally between Gujarat and Bengal; a few centuries later, they made regular transoceanic passages across the southern Bay of Bengal between the Tamil Nadu coast or Sri Lanka and ports in Malaysia, Sumatra, and Java (Hattendorf et al., 2007).

Also important for Indian Ocean navigation under sail is a belt of reliable westerly trade winds that blow consistently for most of the year in an area to the south of the monsoon wind region, between 15° and 30°S. Still further south is a region known to early modern mariners as the "Roaring Forties" (between 40° and 50°S)—an area of strong, steady westerly winds that build up massive seas. Beginning in the 16th and 17th centuries, eastbound sailing masters found conditions in these

---

[★]This book has a companion website hosting complementary materials. Visit this URL to access it: https://www.elsevier.com/books-and-journals/book-companion/9780128226988.

latitudes useful for making rapid transits from the Atlantic to the easternmost Indian Ocean or to the Pacific (Pearson, 2003).

The earliest trade routes in the Indian Ocean world were established by merchant mariners of Egypt and Mesopotamia who, beginning in the fourth millennium BCE, carried goods in vessels that hugged the coast of the Red Sea, continued along the Arabian Peninsula to Hormuz and into the Persian Gulf. Somewhat later, c. 2900–2500 BCE, coastwise trade evolved between Mesopotamia and the Indus River valley, thus linking three archetypal civilization zones by maritime commerce (MacPherson, 2004). The ancient dock and harbor works constructed at Lothal, India, c. 2200 BCE, is the world's oldest known seaport; it served as an entrepôt for cities of the Harappan civilization further inland upriver (Rao & Sarkar, 1985). New influences that penetrated the Indian Ocean world from eastern waters began between approximately 500 BCE and 500 CE (Common Era), when Austronesian mariners, sailing westward from the Melaka and Sunda Straits via the Maldives and Chagos Archipelagos south of the Indian subcontinent, created a maritime trade network linking the islands of Indonesia with coastal South Asia, ports of the Persian Gulf and Mesopotamian interior via the Tigris and Euphrates rivers, the Red Sea, East Africa, and Madagascar (Manguin, 2016).

During the classical period in the West, following the Roman conquest of Egypt in 30 BCE, Roman sailors operating from Red Sea ports conducted regular seaborne expeditions into the Indian Ocean, plying trade routes south along Africa's east coast and eastward to the Persian Gulf and the Gulf of Khambhat (Cambay) in India (Young, 2001). Key goods sought were silk fabrics, spices, and incense. Roman imperial trade in the Indian Ocean expanded during the first to third centuries CE to include ports along the entire west Indian coast, around the southern tip of the subcontinent to trade with the southeast Tamil coast, and north all the way to the mouth of the Ganges River. Roman ships on the east coast of Africa ventured as far south as Rhapta, a trade emporium in modern Tanzania or Mozambique. These maritime routes are described in detail in a contemporary Greco-Roman navigational guide, the *Periplus Maris Erythraei* (Casson, 2012). Constantinople's rise between the fourth and seventh centuries CE helped sustain the trade between the Mediterranean Sea and the Indian Ocean world via the Red Sea and the Persian Gulf, linking maritime trade communities in India, East Africa, and the eastern margins of the Bay of Bengal with Europe (Paine, 2013).

2 Arab traders and the spread of Islamic culture by sea (c. 800–1400 CE)

Religious and cultural developments along the eastern margins of the Red Sea had an enormous impact on the Indian Ocean world. With the rapid spread of Islam eastward and south out of the Arabian Peninsula beginning in the seventh century CE, the nature and of objectives of Indian Ocean navigation started to change as, over the course of a few hundred years, much of the coastline came to be influenced politically by Islamic powers. Because regional rulers often converted to Islam of their own volition, the religion spread readily and peacefully around the Indian Ocean rim, rather than being propagated by military conquest (Reid, 1988). Consequently, much of the Indian Ocean world trade activity came to be monopolized by practicing Muslims but was generally carried on without armed competition. The advent of this new religion affected maritime traffic in a number of ways, with the introduction and dissemination of innovative seafaring technologies, a shifting and recentering of trade routes, and an expansion of voyages to the *Hejaz* region of the western Arabian Peninsula as religious pilgrims made annual *hajj* journeys in fulfillment of their Islamic faith (Chaudhuri, 1985). The cosmopolitan nature of open, unfortified trade ports, through which most maritime commerce was conducted across the region, is a notable feature of the Indian Ocean world during this era (Reid, 1993) (Fig. 1).

Lateen-rigged *dhow* vessels, common to the waters of the northern and western Arabian Sea by 600 CE, came to dominate shipping throughout the Indian Ocean during this period of Muslim ascendancy, prior to European arrival c. 1500 CE. The *dhow*'s triangular lateen sail, loose-footed and laced to a long spar, or yard, typically on a single mast, provided great versatility and maneuverability for shifting Indian Ocean wind conditions. Though *dhows* used for commercial navigation could range in size from 30 up to 500 tons' burden, they had no keel and so tended to be shallow-draft vessels well adopted for short near-coastal hops between ports, as well as long deep-water downwind passages. Because of their durability and rot resistance, teak from Malabar in southwest India and hardwoods from tropical Africa were the preferred material for hull frames and planking (Tripati et al., 2016). Coconut fiber, called *coir*, provided the tough cordage used for sail-handling and anchor lines, but also to lace together the planks of the carvel-built hulls, which contained no metal fastenings (Agius, 2010).

Long-distance trade networks in this Muslim-dominated maritime world naturally tended to direct cargoes and profits through major Islamic trade entrepôts, including key nodal ports on the edges of the Indian Ocean world, like Calicut (Kozhikode), Hormuz, Aden near the mouth of the Red Sea, Mombasa in East Africa, and Melaka, which in time grew to be the essential maritime waypoint between the Indian Ocean and eastern Asia. By the early 15th century, trade from the eastern Indian Ocean, carried by Islamic merchants and ships of other regional seafaring peoples, extended through the Straits of Melaka to the South China Sea, the Java Sea, and further to the Spice Islands (Prange, 2018).

FIG. 1 The Ottoman Turkish Fleet protecting shipping in the Gulf of Aden; Indian Ocean (16th century). *(Anonymous, Turkish School (Public Domain).)*

Muslim merchant communities became established in most major trade ports of the Indian Ocean. Spices constituted the most valuable of the bulk commodities traded, with horses from the Arabian Peninsula, pepper from Malabar and Sumatra, cinnamon from Ceylon (Sri Lanka), and cloves, nutmeg, and mace from the Spice Islands of eastern Indonesia (the Moluccas, modern Maluku) filling the holds of countless vessels (Chaudhuri, 1985). Islamic mercantile distribution networks carried these goods eastward to China and northward to Persian or Mughal regions, or northwest to Ottoman lands, there to be transshipped onward to the Mediterranean and Europe (Casale, 2006). Other important Indian Ocean world trade goods included African ivory and gold, Asian cotton and silk textiles or decorative carpets, medicines, semiprecious stones, and Chinese lacquer ware and porcelains (Prange, 2018).

3 The Chola kingdom: Maritime power of South Asia (9th–12th centuries CE)

During the 9th to 12th centuries CE, roughly contemporary with the rise of Islamic power in the Indian Ocean world region, the Hindu Chola dynasty, centered in the Kaveri River valley of southeastern India with their kingdom's capital near modern Tiruchchirappalli, became the first major maritime power indigenous to South Asia. In 1025 CE, the Chola, contesting access to direct seaborne trade links with southern China, conducted a successful naval campaign against the Srivijaya empire, based on the northeast coast of Sumatra (Shaffer, 2015). By the mid-11th century, the Chola kings had expanded their land holdings to most of southern India, including the cultural center and inland river port Kanchipuram, northern Sri Lanka, part of the Deccan plain, and were conducting seaborne raids along the Coromandel coast, north to the Ganges River delta and across the Bay of Bengal to Burma and the Malay peninsula (Paine, 2013).

Expanded maritime trade followed, with Chola ships venturing through the Melaka Strait to spread goods and South Asian cultural influence as far as coastal Annam (Vietnam) and ports in southern Song-era China. For nearly a century and a half, from 1010 to 1153 CE, Chola territories included the Maldives archipelago, and they maintained seaborn trade links as far as the Abbasid caliphate in Baghdad. The central geographic position of Chola lands was essential to their control over trade routes to Southeast Asia, the northern Arabian Sea, and East Africa. Sri Lanka and the ports of southern India were transit areas, ports of call for African, North Indian, Arab, and East Asian traders, whose ships voyaged to and from lands to the east of the Strait of Melaka, especially China, and throughout the Indian Ocean world (Subbarayalu, 2009).

The Chola fleet of this period represented the height of shipbuilding and navigational skill in ancient India. Open ocean navigation and piloting techniques included basic celestial pathfinding, knowledge and use of a rudimentary compass, and the imprecise processes of deduced reckoning. Like other seafaring peoples of coastal India, the Chola used coconut-fiber *coir* lines and sewn-plank vessel construction (Deloche & Walker, 1994). However, recent finds by the Archeological Survey of India suggest that the Chola may have created more strongly built craft, clinker-planked over light internal frames, that were better suited to longer voyages.

4 Indian Ocean world seaborn trade from China (13th–15th centuries CE): Merchants of the Song and Ming dynasties

While there had been very early Chinese voyages into the Indian Ocean—one group of envoys reportedly reached Kanchipuram in eastern India during the second century BCE—these had been irregular, limited to a few occasional vessels. During the Tang and Song dynasties, beginning in the 7th century and continuing until the 12th century, merchant junks sailed regularly, though in small numbers, from ports in southern China—Guangzhou principal among them—to conduct trade in the established entrepôts of Southeast Asia and the eastern Indian Ocean (Hall, 2019). Commercial voyages returned to China with a significant proportion of the pepper, ginger, ivory, and gold trade of the contemporary Indian Ocean world, but from India, they also imported Buddhist texts and practices that would influence eastern Asian cultures profoundly.

Direct Chinese links with the Indian Ocean world were disrupted temporarily in the mid-14th century due to the rise of a powerful maritime state in Java and Bali, the Majapahit kingdom, which monopolized seaborne trade in the key entrepôt ports between south and east Asia. The Majapahit ascendency was short-lived, and Ming dynasty expeditions began to resume trade ties with India in 1368 (Hall, 2019). Only later, for a brief window of time during the early 15th century CE, did powerful Chinese fleets make forays into the western Indian Ocean. Between 1405 and 1433, experienced soldier and Ming courtier Zheng He commanded a series of seven major expeditions from China. Their mission was to conduct commerce, explore, and, most important, carry out diplomacy through intimidation, creating a network of tribute-paying trade partners—nominally client states that recognized Chinese superiority. Selected in part because of his conversion to Islam, admiral Zheng He was a political ally and favorite of the Yongle Emperor, whose foreign policy was built around supporting these unprecedented expeditions (Dreyer, 2006).

The largest of Zheng He's vessels would have dwarfed contemporary ships from any other seafaring people. Called "treasure ships," these enormous craft were approximately 90–130 m long (their precise dimensions are a subject of much historical debate), had up to four decks, and could carry a thousand people with ease. Nine masts ranked three abreast diagonally across the vessel's broad beam, up to 40 m wide, carried junk-rigged sails, which were highly adaptable to many wind and weather conditions (Dreyer, 2006). A handful of treasure ships formed the core of Zheng He's fleets, but the total number of vessels, including supply and troop ships, may have numbered in the hundreds. Altogether they carried tens of thousands of sailors, soldiers, diplomats, and merchants. When they hove into view off any shore, such intimidating armadas made a powerful impression.

Over a quarter century, Zheng He's repeated expeditions called at ports in Borneo, Java, Sumatra, the Malay Peninsula, Thailand, India, the Horn of Africa, the Arabian Peninsula, and the Persian Gulf, successfully establishing a Chinese presence throughout the Indian Ocean world. Following the death of the Yongle Emperor in 1424, however, the Chinese made an abrupt strategic decision to withdraw their imperial interest in political or territorial influence within the Indian Ocean region, and instead opted to be content with the voluminous trade carried to Chinese ports by a broad range of domestic and foreign merchants (Li, 2010). Thus, when the first Portuguese ships arrived two generations later, the only Asian maritime power strong enough to oppose the Europeans militarily at sea was effectively absent from the Indian Ocean world.

Noteworthy, too, is the rise of the Arakan maritime kingdom on the eastern coast of the Bay of Bengal during the first third of the 15th century, contemporary with the withdrawal of Chinese seaborne power from the Indian Ocean world. For three and a half centuries, until 1785, the Arakan empire, centered at the capital, Mrauk-U on the Burma littoral, exercised strong commercial and cultural influence across the eastern Indian Ocean region through its active trading networks. Cosmopolitan Mrauk-U was home to many ethnic and religious communities—Buddhist, Hindu, Islamic, and Christian—and was frequented by Arab, Portuguese, Dutch, and Danish traders. Weakened by decline during the 18th century, Arakan was conquered by the Burmese Konbaung dynasty; the region later fell under British control in 1826 (Charney, 1998).

5 Sea change: Arrival of the Portuguese, armed trade, and colonization (c. 1498–1600)

Portuguese attempts to reach the Indian Ocean—and Asian commercial ports—by sea began soon after their conquest of Ceuta, a Muslim-held trade port opposite Gibraltar in North Africa, in 1415. A regular program of reconnaissance and commercial voyages southward along the West African coast followed. This effort gained a new urgency after 1453, when the Ottoman Turks conquered Constantinople, thereby making almost all European trade with Asia dependent on cooperation with, and according to terms set by, Muslim middlemen (Costa et al., 2016). By 1488, a small Portuguese squadron under Bartolomeu Dias rounded Africa's Cape of Good Hope, sailing south of the near-coastal Agulhas Current, and became the first Europeans to enter the Indian Ocean from the Atlantic. Nearly a decade later, Vasco da Gama embarked on the voyage that reached Calicut, Southwest India, and finally linked Europe to Asia for direct seaborne commerce (Disney, 2009) (Fig. 2).

Upon arrival in the Indian Ocean world, European powers used their maritime military advantages—heavily-built ships carrying powerful cannon—to ward off the region's endemic piracy, insinuate themselves into preexisting Asian trade networks, and then manipulate those perennial commercial systems for their own advantage. Prior to the late 15th-century Portuguese incursion, there had been no organized attempt by any political power to assert comprehensive control over Asia's sea lanes and long-distance trade. Instead, inter-port trade had been conducted largely without recourse to violence, managed through free competition for commerce between ports, the levying of modest customs duties, and welcoming merchants of all nations. The ports that thrived did not restrict trade by enacting discriminating regulations (Prakash, 1999).

European newcomers, by contrast, early on leveraged their asymmetrical sea power, as well as access to gold and silver bullion extracted from the Americas, to force admittance and conduct trade in Asian ports that resisted them. Paying for goods with precious metals was vital to early European trade in the Indian Ocean world, because demand for products from the West was generally very weak in Asia. Western colonial powers, beginning with the Portuguese, imposed monopolies or discriminatory regulations in the ports and seaways they controlled to restrict trade only to favored groups. Armed trade necessitated that Europeans fortify their conquered enclaves against attack from rival powers, or from uprisings by aggrieved indigenous populations. Due to insufficient military power on land, centuries would pass before European rule would extend deep into the hinterland, beyond the range of their cannon mounted aboard ships or in seaport fortresses (Prakash, 1999).

Portugal entered the Indian Ocean world at a fortuitous time. The Ottoman Turks, the Ming Chinese, Safavid Iran, and the Mughal sultanate in Delhi were all too preoccupied with internal political upheaval and managing their terrestrial empires to counter seaborne European interlopers. These established land-based powers in Asia were little concerned with who managed trade along their coasts, so long as tax revenues and trade goods continued to flow inland. Moreover, the limited scale of Portuguese activities initially threatened only local rulers of small coastal kingdoms (Goldstone, 2008). Exploiting rivalries among port city-states on the Indian Ocean world rim enabled the Portuguese to develop a foothold.

FIG. 2 Fleet of Vasco da Gama that departed in 1497 (First Portuguese Fleet to India), in the manuscript Memoria Das Armadas Que De Portugal Pasaram Ha India E Esta Primeira E Ha Com Que Vasco Da Gama Partio Ao Descobrimento Dela Por Mamdado De El Rei Dom Manuel No Segundo Anno De Seu Reinado E No Do Nascimento De Cristo De 1497. *(Unknown author; circa 1567 (Academia Das Ciências De Lisboa) (Public Domain).)*

First Vasco da Gama, followed by Afonso de Albuquerque, who commanded the fleet sent in 1503, established trading factories and a network of alliances with potentates on Mozambique Island and Malindi in East Africa, and in the Indian ports of Calicut (Kozhikode), Cochin (Kochi), and Cannanore (Kannur) (Subrahmanyam, 2012).

With an uncanny grasp of strategic maritime geography, between 1505 and 1535, the Portuguese quickly established fortified seaport bases in key locations to consolidate their territorial gains. Their fortresses at Malacca (1509), Goa (1510), Hormuz (1515), and Diu (1535) protected critical choke points in Asia's seaborne trade networks and were extremely effective in maintaining Portuguese imperial influence (Newitt, 2005). Portuguese outposts soon extended from East Africa to Southeast Asia, Indonesia, the Spice Islands, China, and Japan. The conquest of Goa, midway between Kochi and Bombay (Mumbai) on India's west coast, established this strategic port as the administrative and cultural hub of the *Estado*

FIG. 3 Early representation of Portuguese-held Goa, India, with exaggerated fortifications to deter potential European rivals. *(From Braun and Hogenberg, 1572. Civitates Orbis Terrarum I, 54. After an unidentified Portuguese manuscript. (Public Domain).)*

da Índia (Portugal's eastern empire), and as a dissemination point for missionaries intent on spreading Roman Catholicism (Borges, 1996). In East Africa, the Portuguese held fortified coastal harbors at Mozambique Island beginning in 1498, Zanzibar (1503–1698), Malindi (1499–1593), and Mombasa, the main port for the Kenyan hinterland (1593–1698) (Pearson, 2010). For over a century, Portuguese warships became the dominant naval power operating in the region, able to unilaterally assert the principle of *mare clausum*—theoretical legal authority over the Indian Ocean, with the right to close the sea to potential rivals—and enforce their imperial *cartaz* system: licensing and collecting fees for all waterborne trade (Disney, 2009). Although the Portuguese empire in Asia suffered numerous setbacks in the 17th century, they maintained key territories in mainland India until driven out by independent Indian national forces in 1961 (Fig. 3).

6 Advent of the Dutch: Dutch East India Company ambitions and activities in the Indian Ocean world (c. 1600–1800)

The earliest forays of Dutch *fluyt* ships into the Indian Ocean were made in the context of the Protestant Reformation, ambitions to create a network of lucrative colonial trading enclaves in Asia modeled on, and in competition with, the Portuguese *Estado da Índia*, and the Calvinist Netherlands' 80-year struggle for independence from rule by the staunchly Catholic Spanish Hapsburgs. After the Dutch Revolt began in 1568, they started to extend their maritime commerce beyond Europe, striking at Spanish colonial interests in the Atlantic. When young king Sebastian of Portugal was lost without heir in 1578 at the battle of Ksar-El-Kebir in Morocco, Felipe II of Spain asserted his claim to the Portuguese crown. The two kingdoms and their maritime colonies were thus joined in an Iberian Union lasting 60 years, from 1580 until 1640. These circumstances gave the Dutch a justifiable *casus belli* to attack Portuguese overseas possessions in Asia and the Atlantic, which under Spanish rule were left relatively weakly defended (Boxer, 1973; 1990). Trade and religion motivated the global rivalry between the Protestant Dutch and the Catholic Iberian realms (Fig. 4).

Two seminal books, published in Amsterdam in 1595 and 1596 by Jan Huygen van Linschoten, a Dutch merchant adventurer who had spent 7 years traveling and working in the *Estado da Índia*, opened the door for the Netherlands to exploit Portuguese vulnerability in the East. These works, entitled *Travel Accounts of Portuguese Navigation in the Orient* and *Itinerario*, a diary of Linschoten's voyage to the Portuguese East Indies, contained strategic commercial and geographic information, including detailed instructions for sailing between Lisbon and the Indian Ocean colonies, as well as for navigation routes between Goa, China, and Japan (van Linschoten, 1595, 1596). Quickly translated into English (1598) and other languages, Linschoten's writing catalyzed efforts across Europe to establish direct trade links with Asia, which led to the chartering of joint stock East India companies in England (1600) and the Netherlands (1602), followed by a steadily increasing wave of European navigation and colonization in the Indian Ocean (Saldanha, 2011).

During the first two thirds of the 17th century, the Dutch VOC (*Vereenigde Nederlandsche Oost Indische Compagnie*, or United Dutch East India Company) established and consolidated their maritime empire in the Indian Ocean and further east, largely at the expense of the Portuguese. The Dutch attacked Goa in 1603 and again in 1610, hoping to displace the Portuguese from their profitable main base in Asia but were twice repelled. Ambon Island in the Moluccas, the source for lucrative cloves, nutmeg, and mace, was the next objective; the Dutch managed to oust the Portuguese in 1605. Contesting control of the Malacca Strait, the key strategic maritime bottleneck between India and China began the following year, though it would not fall to the Dutch until 1641. In need of a strategic headquarters for their Asian colonies, VOC forces

FIG. 4 Dutch *fluyt* ships; East India Company (VOC) Insignia (17th Century); Warehouse in Hoorm, Netherlands. Wikimedia Commons. *(Public Domain.)*

under Jan Coen conquered a port near the western tip of Java in 1619 and founded Batavia, which remained the center of Dutch commercial activity and administration in Asia until 1946 (Parthesius, 2010).

In 1609, Dutch legal scholar and counsel to the VOC Hugo Grotius published his treatise entitled *Mare Liberum* ("The Free Sea"), which refuted Portuguese claims to unilateral control of regional navigation and laid the theoretical foundations of a new principle in maritime law: that all the world's oceans were international territory, and therefore ships of all nations had the immutable right to freely ply the seas for trade, without tax, toll, or molestation (Vervliet, 2009).

In 1638, the Dutch established a stepping-stone base on Mauritius, a promising island in the central Indian Ocean named for Prince Mauritz of Nassau. They founded Cape Town in 1652 at a natural harbor near Africa's Cape of Good Hope, as a mid-voyage waypoint to resupply VOC vessels sailing between the Netherlands and Batavia (often taking advantage of the Indian Ocean's strong westerly winds in the southern "roaring forties" latitudes). Settlers soon expanded into the broader Cape Colony. During most of the mid-17th century, the Dutch and Portuguese waged a protracted war for Ceylon and its cinnamon production; VOC forces finally took Colombo in 1656, and consolidated control over Ceylon in 1658. A Dutch attack in 1662 wrested Nagapattinam, a base for missionary activity and trade on India's eastern Coromandel Coast, from Portuguese control; Cranganore (modern Kodungallur) and Cochin, important commercial harbors and sources of highly valuable pepper on the Malabar Coast of southwest India, were taken the following year, in 1663 (Prakash, 1999). Though Dutch fortunes waned significantly in Asia, as in Europe, during the 18th and 19th centuries, they asserted colonial sovereignty over the Indonesia archipelago until after World War II.

7 The English East India company in the Indian Ocean world (c. 1600–1858)

Toward the end of Queen Elizabeth I's long reign, powerful mercantile interests in London persuaded her that breaking into the East Indies trade would be a good way to counter hostile Catholic Iberian power in Europe while profiting handsomely in the process. In 1600, therefore, the final Tudor monarch chartered the English East India Company, granting permission to trade in a vast expanse of the Indian and Pacific oceans, though the precise geography was little understood in the British Isles at the time (Lawson, 2014). That would change over the next two and a half centuries, as the British became the dominant naval and political power in the region, with colonies and bases ringing the Indian Ocean perimeter. The first East India Company voyage in 1601 succeeded in establishing trade bases at Bantam, Java, to trade for pepper, and in the Spice Islands to obtain nutmeg, mace, and cloves (Lawson, 2014).

In India, the English established early footholds with a factory at Masulipatnam on the eastern coast in 1611, and the following year at Surat, a portal for precious stones and textiles on the Gulf of Khambhat (Cambay) north of Bombay. A joint Anglo-Persian attack took Hormuz Island from the Portuguese in 1622, giving the East India Company and Safavid Shah Abbas the Great control of the Persian Gulf trade. By 1640, the East India Company had become well established in India, with a convenient trade port at Madras (Chennai) midway along India's East coast. Two decades later, they gained Bombay, a key West coast port, from the Portuguese as part of the dowery when princess Catarina of Bragança married Charles II in 1662. Calcutta (Kolkata) near the mouth of the Ganges River in Northeast India completed this trinity in 1690, at the end of the first Anglo-Mughal War (Black, 2004) (Fig. 5).

FIG. 5 Stylized map of Hormuz Island and New Hormuz City in Johann Caspar Arkstee and Henricus Merkus' *Allgemeine Historie der Reisen zu Wasser und Lande, oder Sammlung aller Reisebeschreibungen*, Leipzig, 1747. *(Public Domain.)*

In the Indian subcontinent, the British extended their control from the coastal ports inland during the second half of the 18th century, exploiting their superior weaponry and rivalries among fractious regional rulers to defeat and subordinate them to East India Company administration. By 1820, military successes against Napoleon's French forces and indigenous Maratha warriors in India put more key ports and bases across the Indian Ocean world into British hands, from Cape Town (1806) to Singapore (1819), with Colombo in Sri Lanka (1796), Mauritius Island (1810), and Kochi in Kerala (1814) in between. British forces took Aden, a strategic Ottoman-dominated port controlling access to the Red Sea, in 1839. On the East African coast, British interests would come to control Cape Colony (South Africa), annexing the harbor at Natal Bay (modern Durban) in 1844, Mombasa, Kenya, in 1887, and Zanzibar Island in 1890 (Black, 2004). By the end of the 19th century, the United Kingdom's maritime, political, and commercial dominance in the Indian Ocean world was complete (Fig. 6).

FIG. 6 Zanzibar Harbor and Dhows. Photographer: Sir John Kirk [1r] (1/1), British Library: Visual Arts, Photo 355/1(120), in Qatar Digital Library [Accessed 11 August 2021]. *(Public Domain.)*

8 Arrival of the French (18th–20th centuries)

French attempts to create a maritime empire to rival those of other European powers began in the mid-16th century but were hampered by domestic turmoil, so colonization in the Indian Ocean did not begin until the early 17th century, following the Wars of Religion and the Edict of Nantes. With encouragement from newly crowned king Henri IV, a French trading company chartered in December 1600 sent two ships into the Indian Ocean the following spring; their objective was to obtain spices in the Molucca Islands and to trade with Japan (Quinn, 2000). Neither ship returned—the Dutch captured one and the other was wrecked—but two survivors of the venture, François Martin de Vitré and François Pyrard de Laval, published accounts of their experiences that helped to catalyze French colonialism in Asia (Martin de Vitré, 1604; Pyrard de Laval, 1615). Only in 1664 did Louis XIV's minister Jean-Baptiste Colbert establish a French East India Company with sufficient capital and resources to compete with rival European interests.

The *Compagnie des Indes Orientales* initially targeted Madagascar unsuccessfully for colonization, but instead established its first port base on a small Indian Ocean island to the east, Bourbon (Réunion), in 1664; in 1715, the French would occupy nearby Mauritius, abandoned as unprofitable by the Dutch in 1710, and rename it Île-de-France. French settlers developed both islands into brutal but lucrative plantation economies, where thousands of enslaved laborers from Madagascar, East Africa, and other points around the Indian Ocean world produced sugar and some tobacco (Prakash, 1998). In 1673, the French contested the port of Trincomalee in Ceylon, hoping to break into the cinnamon trade, but the Dutch successfully drove them out within a year.

Meanwhile, French commercial efforts on the Indian mainland focused on establishing trade enclaves at Pondicherry (Puducherry), an open roadstead harbor on India's Southeast Coromandel Coast, in 1674, and at Chandernagore, inland on the Hooghly River near Calcutta in Northeast India, in 1675. French holdings in mainland India expanded during the second quarter of the 18th century, with bases established in 1725 at the Godavari River port of Yanaon on the East coast for the export of textiles, and at Mahé on the southwestern Malabar Coast, near Cannanore, for exporting pepper and other goods. In 1739, the French established themselves at Karikal, a coastal enclave south of Pondicherry (Ray, 1999). France maintained legal jurisdiction over these Indian mainland territories until the middle 1950s.

In 1756, at the outset of the Seven Years' War, French colonists began to occupy the Seychelles Islands, about 1750 km east of Mombasa. However, French setbacks during major global imperial conflicts between 1756 and 1815 resulted in the permanent loss of Île-de-France, the Seychelles, and other ports to Britain (Baugh, 2014). French occupation and dominance of the Comoros Islands, about midway between northern Mozambique and Madagascar, began in the 1840s and lasted until 1975, though one island in the archipelago, Mayotte, remains a French overseas department. In Madagascar, two French military interventions between 1883 and 1896 resulted in Madagascar becoming a protectorate and colony of France until 1958. Under international treaty, notwithstanding contested discovery by rival European explorers dating to the 16th century and exploitation by whalers, fishermen, and seal hunters, France administers the Kerguelen Islands, Crozet Islands, Saint Paul's Island, and Amsterdam Island, all in the southernmost latitudes of the Indian Ocean, as well as several small islands scattered across the Mozambique Channel (Aldrich & Connell, 2006) (Fig. 7).

FIG. 7 British ships the Resolution and the Discovery at Christmas Harbor in the Kerguelen Islands, under command of Captains James Cook and James King, 1776, in *A Voyage to the Pacific Ocean […]*, first edition (London, 1784) *(Public Domain)*.

9 Slave trading in the Indian Ocean

Enslavement and slave trading, common in the Indian Ocean world by 2500 BCE, was practiced for 4500 years—far longer than in the trans-Atlantic context, though historically it was undertaken on a less intensive scale. The volume of human bondage in the Indian Ocean world, especially persons captured in East Africa, exploded during the final century of legal slave holding, which ended only in the early 20th century (Drescher & Engerman, 1998). In ancient times, merchants crisscrossed the Indian Ocean carrying relatively small numbers of enslaved people to and from every bordering shore and beyond—young women held as concubines; children as household servants; young men as footmen or guards. Such conspicuous human property displayed the power of, and brought prestige to, the owners and their households; slaves were valued principally for their rarity and exoticism rather than for their productive manual labor (Freamon, 2019).

Islamic expansion, with its consequent *jihadi* wars fought against infidel nonbelievers, occasioned a vast increase in slave trafficking across the Indian Ocean, Red Sea, and Gulf of Aden between approximately 800 and 1400 CE. Indian Ocean world slave trading conducted by Muslims expanded in the 19th century, just as the practice was being outlawed in Europe and the Western Hemisphere. Traders based in Zanzibar, Pemba, Sofala, Quelimane, Kilwa, Malindi, and Mogadishu shipped hundreds of thousands of enslaved people from the Swahili Coast—a region stretching between northern Mozambique and Madagascar to southern Somalia—to destinations in Yemen, the Hejaz on the Arabian Peninsula, Basra in the Persian Gulf, and Oman (Walvin, 2005). The rise of a chain of independent polities along the Swahili Coast influenced the commercial culture of the western Indian Ocean and created opportunities for wealth and power to accrue in that region, outside the ambit of European colonialism (Pearson, 2010) (Fig. 8).

Among the European powers, the Portuguese, Dutch, British, Danish, and French interests expanded both the enslavement of indigenous colonial populations and long-distance trading across Asian waters between 1500 and 1800, for purposes of prestige slaveholding, as well as for agricultural and construction labor. The Dutch in particular exploited bonded human labor, trafficking tens of thousands of enslaved workers to their colonial enclaves, especially in Java, Ceylon, and the South African Cape Colony (Vink, 2003).

Training and arming slaves as elite warriors was a common custom in many regions around the Indian Ocean hinterland for over a millennium, from before the Islamic period until the 20th century. During the 18th and 19th centuries, "most slaves in the Persian Gulf were employed as soldiers, household servants, sailors and dock hands, and pearl divers" (Machado, 2004). In East Africa toward the end of the 19th century, the powerful merchant Tippu Tip's principal warriors in Zanzibar and the adjacent mainland were predominantly slaves (Miers, 1975). Well into the 1930s, the Sultans of the Qu'ayti state in what is now central Yemen relied on specialized units of enslaved African soldiers, universally recruited from East Africa and procured from Zanzibar (Alpers, 2004).

FIG. 8 East African slaving dhow diagram demonstrating how enslaved cargo was transported. *(Slaving Dhow East Africa. London Illustrated News March 1, 1873 (Public Domain).)*

SECTION OF VESSEL, SHOWING THE MANNER OF STOWING SLAVES ON BOARD.

Definitive and determined legal curtailment of slave trafficking across the Indian Ocean began only in the mid-19th century. In 1842, under treaties negotiated in Europe between the British, Portuguese, and other regional Indian Ocean world governments, the movement of enslaved individuals between colonies of the Indian Ocean rim was prohibited—especially the transport of forced laborers from Africa to India (Walker, 2006). However, slave holding in Portuguese colonies in India remained legal, and contraband slave smuggling continued virtually unchecked and with little discouragement from local colonial officials (Pinto, 1994). For their part, French colonial elites were in the habit of masking their commerce in African slaves with "legitimate" labor schemes, providing "freely contracted" African workers to plantations on Réunion and Mayotte islands following the French abolition of slavery in 1848 (Allen, 2008). During the latter decades of the 1800s, the British Navy began to patrol the traditional slave-trading ports along the East African coast systematically to enforce international laws proscribing maritime transport of human chattel (Doulton, 2013).

10 Whaling voyages in the Indian Ocean (c. 1785–1920)

Native peoples in Madagascar, as elsewhere in the Indian Ocean world, traditionally captured whales close to shore; the Flemish artist Theodor de Bry published illustrated accounts of Madagascan whaling in 1598 (Dow, 2013). Beginning in the late 18th century, however, Europeans and Americans in the whale fishery regularly made the Indian Ocean a scene of marine megafauna exploitation on an industrial scale. As cetacean populations in the South Atlantic were depleted by overhunting, whalemen based in the United Kingdom and France, followed by larger numbers of Yankee whalers from New England, began to venture into the Indian Ocean world, in search of humpback and sperm whales, valuable for the oil their blubber provided when rendered, as well as for their bone, ivory, and baleen (Stackpole, 1972). Approximately 15% of all U.S.-based whaling voyages between 1785 and 1910 targeted the Indian Ocean as their principal destination. Whalers intending to hunt in the Pacific Ocean often rounded the Cape of Good Hope and used prevailing westerly winds to transit the Indian Ocean in southern latitudes, as this was considerably easier on a vessel's rigging and crew than sailing west against the wind around Cape Horn (Sherman, 1986).

Primary Indian Ocean whaling areas for sperm whales included the Zanzibar Grounds and the Coast of Arabia Grounds in the western margins of the Indian Ocean; the Mahe Banks and the Mauritius Ground in mid-ocean; and the Western Australia Ground in the Southeast. Humpback whales were found in latitudes further south, in the Crozet Islands Ground, the Desolation Island Ground, and the New Holland (Australia) Ground. Whale hunters learned that both types of whales frequented the Mozambique Channel near Delagoa Bay, and the northern Madagascar Strait (Townsend, 1935).

Nineteenth-century U.S. whaling voyages traveled routinely to some of the most remote waters on the Southern and Indian Oceans, and carefully charted unfamiliar seas that previously were poorly conceptualized by mariners of North America and Europe. Over more than a century of maritime activity, New England whalemen played an exceptionally valuable role in refining human geographic and spatial knowledge of the Indian Ocean world through navigation records made while underway. Their daily logbook observations and latitude/longitude sightings gradually improved the accuracy of ocean charts, by recording with increasing precision the positions of islands, safe havens, currents, winds, and maritime hazards. In the process, they also gathered precious information about regional weather conditions that is required for secure sailing, as well as ethnographic information that helped mariners, scholars, governments, and commercial interests to better understand the human geography of regions where the whalers traveled, which in turn facilitated provisioning and trade. Whaling voyages were a critical training ground in cartography and navigation for generations of mariners (Dyer, 2017).

11 The Indian Ocean world during the late colonial era (1860–1950)

Industrialization in the West had powerful consequences in the Indian Ocean world, as imperial navies and commercial vessels became larger, faster, and far more powerful relative to traditional indigenous maritime capacities. First iron and then steel-hulled ships, armed with rifled, breech-loading cannon, propelled by steam and then diesel engines, were capable of navigation independent of the monsoon weather patterns that had forever dictated wind-driven commerce or seaborne military campaigns (MacPherson, 2004). Britain emerged during this period as the dominant power afloat in the Indian Ocean, though French, Portuguese, and Dutch vessels continued to assert their colonial presence. Increasingly, industrialists improved Indian Ocean world port facilities and shipped raw materials from Asia to Europe for manufacturing. After 1860, submarine telegraph cables, laid mainly by British companies, improved communications between Europe and the Indian Ocean world (Hattendorf et al., 2007). The opening of the Suez Canal in 1869 provided a much more direct water route between the Indian Ocean world and Europe, cutting navigation time and costs dramatically, which in turn spurred increased trade and travel, including colonial immigration from and to Europe. Consequently, maritime traffic around the Cape of Good Hope declined markedly in the late 19th century (Paine, 2013).

During World War I, naval operations contesting Indian Ocean shipping lanes were minimal. At the war's outset, a German cruiser, the *Emden*, raided several poorly defended colonial ports, disrupting commerce, diverting vital Allied naval assets, and capturing or sinking 30 merchant and military vessels before being destroyed by an Australian warship near the Cocos (Keeling) Islands. Indian Ocean world colonies were of tremendous strategic importance to the belligerent European nations, especially the British and French; maritime transport delivered not only vital raw materials and foodstuffs, but also hundreds of thousands of men brought from Indochina, Madagascar, India, and east Africa to Europe for military service (Koller, 2008).

The interwar years saw a consolidation of British power in the Indian Ocean world following the withdrawal of Germany's colonial presence in east Africa. At the same time, incipient anticolonial and independence movements took root in various regions of the Indian Ocean world, notably in India, Indonesia, Burma, and eastern Africa; these initiatives would help to hasten decolonization in the coming decades. Oil exports from the Persian Gulf through the Straits of Hormuz, increasingly central to industrialized Western economies, gained strategic importance beginning in the 1930s. Diego Garcia island, part of the Chagos Archipelago south of the Maldives, recognized for its strategic mid-ocean location, became a fortified British base at the outset of the Second World War (Lesser, 1991).

The Indian Ocean world witnessed significant maritime military operations during World War II. Because of the critical material and personnel resources that could flow from Asia and East Africa to Britain, France, and the Netherlands, Axis naval forces aggressively sought to disrupt Allied shipping in the Indian Ocean. German, Italian, and Japanese submarines carried out unrestricted attacks on Allied vessels, while surface warships raided colonial enclaves. Allied fleets conducted multiple amphibious landings to secure strategic islands and the Burmese mainland, while military aircraft operating from rival Allied and Axis carrier groups or island bases played a crucial role in the contest for dominance of this key maritime region (Jackson, 2018).

Rapid decolonization following World War II, especially as the United Kingdom and the Netherlands, under pressure from powerful pro-independence social movements, withdrew from their roles in South and Southeast Asia, meant that the navies of the newly independent nations of India, Pakistan, and Indonesia emerged as important seaborne powers in the region. By the mid-20th century, seaways transiting the Indian Ocean from Melaka in the East to Hormuz and Suez in the West remained among the most important strategic and commercial seaways of the world, carefully monitored by warships of the emerging Cold War superpowers and their allies (Lee, 2013).

12 Conclusion

Europe's colonial presence in the Indian Ocean world contracted markedly during the third quarter of the 20th century. Madagascar island emerged as an independent nation in 1960, while Indian military force drove the Portuguese from their South Asian enclaves in late 1961—shortly after the French left voluntarily in 1956. Between 1963 and 1977, former colonies newly recognized as independent countries accounted for virtually the entire east African coastline. On the Indian Ocean's eastern margin, Malaysia and Singapore gained independence from Britain in 1963 and 1965, respectively. In the early 21st century, only a handful of remote but strategic island territories remain under French and British administration in the Indian Ocean.

As colonial claims of sovereignty in the Indian Ocean world diminished, maritime trade expanded, along with the regional population. Nowadays, an enormous volume of global maritime trade either originates in or transits the Indian Ocean—especially petroleum products carried through the Hormuz Strait—while coastwise trade in East Africa, South Asia, the Bay of Bengal, and the Andaman Sea maintains its importance serving regional markets. About 40% of contemporary global seaborne commerce passes through the Strait of Melaka, entering or exiting the Indian Ocean (Zulkifli et al., 2020). Tracking recent shifts in weather, wind patterns, and extreme storm events in the Indian Ocean basin is essential to understanding how such developments will impact human populations of the region, as well as ocean shipping activities on which national economies depend.

While the Indian remains the ocean least studied by scholars and scientists, there has been an enormous growth of regional interest in the Indian Ocean world over the past two decades, across many academic disciplines—the humanities, social sciences, and earth sciences. As a result, an annual flurry of scholarly publications has begun to roll back the fog of unfamiliarity that has existed for far too long about all aspects of this key global region.

13 Educational resources

This selected resource list provides key links for the history of trade and cultural connections, as well as maps and images of sites throughout the Indian Ocean world:

- https://www.asianstudies.org/publications/eaa/archives/the-indian-ocean-in-world-history/
- https://www.indianoceanhistory.org/
- https://theconversation.com/exploring-the-indian-ocean-as-a-rich-archive-of-history-above-and-below-the-water-line-133817
- https://worldhistoryconnected.press.uillinois.edu/11.1/anderson.html
- https://www.thoughtco.com/indian-ocean-trade-routes-195514
- https://www.colonialvoyage.com/
- Indian Ocean Maritime History Atlas: https://www.jstor.org/stable/41563180

Acknowledgments

The composition of this chapter was accomplished with the essential aid of a Publication Completion Support Grant from the Office of the Dean of the College of Arts and Sciences, University of Massachusetts Dartmouth, for which I am deeply grateful. Support for this research was provided by NSF under BCS-1852647. For their critical reading of the text and indispensable suggestions, I would like to extend heartfelt thanks to my colleagues, historians Len Travers and Lincoln P. Paine.

References

Agius, D. A. (2010). *In the wake of the dhow: The Arabian gulf and Oman*. Reading, UK: Garnet & Ithaca Press. pp. 31–36; 49–58.

Aldrich, R., & Connell, J. (2006). *France's overseas frontier*. Cambridge, UK: Cambridge University Press. pp. 17–33; 262–269.

Allen, R. B. (2008). The constant demand of the French: The Mascarene slave trade and the worlds of the Indian Ocean and Atlantic during the eighteenth and nineteenth centuries. *The Journal of African History, 49*(1) 43–72.

Alpers, E. (2004). Flight to freedom: Escape from slavery among bonded Africans in the Indian Ocean world, c. 1750–1962. In G. Campbell (Ed.), *The structure of slavery in the Indian Ocean, Africa and Asia* (p. 57). London: Frank Cass.

Baugh, D. (2014). *The global seven years war 1754–1763: Britain and France in a great power contest*. London and New York: Routledge Publishing. pp. 282–304; 461–481.

Black, J. (2004). *The British seaborne empire*. New Haven, Connecticut: Yale University Press. pp. 50–51; 104–106; 114–115; 118–119; 130–133; 138–144; 161–163; 180–181; 193–195.

Borges, C. (1996). Intercultural movements in the Indian Ocean region: Churchmen, chroniclers and travellers in voyage and in action. In K. S. Mathew (Ed.), *Indian Ocean and cultural interaction, a.D. 1400–1800* (pp. 21–34). Pondicherry, India: Pondicherry University Press.

Boxer, C. R. (1973). *The Dutch seaborne empire, 1600–1800*. New York: Penguin Books. (reprinted 1990). pp. 1–26; 95–100.

Braun, G., & Hogenberg, F. (1572). Goa fortissima India urbs in Christianorum potestatem anno salutis 1509 deuenit. Civitates Orbis Terrarum I (54). After an unidentified Portuguese manuscript.

Casale, G. (2006). The ottoman administration of the spice trade in the sixteenth-century Red Sea and Persian gulf. *Journal of the Economic and Social History of the Orient, 49*(2) 170–198.

Casson, L. (2012). *The periplus Maris Erythraei: Text with introduction, translation, and commentary* (pp. 51–93). Princeton, NJ: Princeton University Press.

Charney, M. (1998). Crisis and reformation in a maritime kingdom of Southeast Asia: Forces of instability and political disintegration in Western Burma (Arakan), 1603–1701. *Journal of the Economic and Social History of the Orient, 41*(2) 185–219.

Chaudhuri, K. N. (1985). *Trade and civilisation in the Indian Ocean: An economic history from the rise of Islam to 1750*. Cambridge, UK: Cambridge University Press. pp. 34–57; 15–20; 25–30.

Costa, L. F., Lains, P., & Miranda, S. M. (2016). *An economic history of Portugal, 1143–2010* (pp. 45–50). Cambridge, UK: Cambridge University Press.

Deloche, J., & Walker, J. (1994). *Transport and communications in India prior to steam locomotion, vol. II, water transport*. Delhi, India: Oxford University Press. pp. 179–186; 200–205.

Disney, A. R. (2009). *A history of Portugal and the Portuguese empire. Vol. 2*. Cambridge, UK: Cambridge University Press. pp. 37–39; 119–125; 156–157.

Doulton, L. (2013). 'The flag that sets us free:' Antislavery, Africans, and the Royal Navy in the Western Indian Ocean. In R. W. Harms, B. K. Freamon, & D. W. Blight (Eds.), *Indian Ocean slavery in the age of abolition* (pp. 101–118). New Haven, CT: Yale University Press.

Dow, G. F. (2013). *Whale ships and whaling: A pictorial history* (p. 240). Mineola, NY: Dover Publications.

Drescher, S., & Engerman, S. L. (Eds.). (1998). *A historical guide to world slavery* (pp. 41–46). New York and Oxford, UK: Oxford University Press.

Dreyer, E. L. (2006). *Zheng he: China and the oceans in the early Ming dynasty, 1405–1433*. Harlow, UK: Pearson Longman Publishing. pp. 1–10; 22–34; 102–115.

Dyer, M. P. (2017). *O'er the wide and tractless sea: Original art of the Yankee whale hunt*. New Bedford, MA: Old Dartmouth Historical Society/New Bedford Whaling Museum. pp. 13–27; 91–98.

Freamon, B. K. (2019). *Possessed by the right hand: The problem of slavery in Islamic law and Muslim cultures* (pp. 78–83). Leiden and Boston: Brill Academic Publishing.

Goldstone, J. (2008). *Why Europe? The rise of the west in world history 1500–1850* (p. 56). New York: McGraw-Hill.

Hall, K. R. (2019). *Maritime trade and state development in early Southeast Asia*. Honolulu: University of Hawaii Press. pp. 28–42; 252–270.

Hattendorf, J. B., et al. (2007). *The Oxford encyclopedia of maritime history. Vol. 2*. Oxford, UK: Oxford University Press. pp. 190–191; 201; 207–208.

Jackson, A. (2018). *Of islands, ports, and sea lanes: Africa and the Indian Ocean in the second world war* (pp. 18–53). Warwick, UK: Helion and Company.

Koller, C. (2008). The recruitment of colonial troops in Africa and Asia and their deployment in Europe during the first world war. *Immigrants & Minorities, 26*(1–2) 111–128.

Lawson, P. (2014). *The East India company: A history* (pp. 13–36). London and New York: Routledge Publishing.

Lee, C. J. (2013). The Indian Ocean during the cold war: Thinking through a critical geography. *History Compass, 11*(7) 524–530.

Lesser, I. O. (1991). *Oil, the Persian Gulf, and grand strategy—Contemporary issues in historical perspective* (pp. 5–26). Santa Monica, CA: RAND Corporation.

Li, K. (2010). *The Ming maritime trade policy in transition, 1368 to 1567. Vol. 8* (pp. 7–9). Weißbaden, Germany: Otto Harrassowitz Verlag.

London. (1784). Capt. James Cook & Capt. James King, A Voyage to the Pacific Ocean. Undertaken, by the Command of his Majesty, for making Discoveries in the Northern Hemisphere. To determine The Position and Extent of the West Side of North America; its Distance from Asia; and the Practicability of a Northern Passage to Europe. Performed under the direction of Captains Cook, Clerke, and Gore. *His Majestys Ships the Resolution and Discovery. In the Years 1776, 1777, 1778, 1779, and 1780*. First edition.

Machado, P. (2004). A forgotten corner of the Indian Ocean: Gujarati merchants, Portuguese India and the Mozambique slave trade, c. 1730–1830. In G. Campbell (Ed.), *The structure of slavery in the Indian Ocean, Africa and Asia* (p. 19). London: Frank Cass.

MacPherson, K. (2004). *The Indian Ocean: A history of people and the sea*. New Delhi, India: Oxford University Press. pp. 3–8; 24–28.

Manguin, P.-Y. (2016). Austronesian shipping in the Indian Ocean: From outrigger boats to trading ships. In G. Campbell (Ed.), *Early exchange between Africa and the wider Indian Ocean world* (pp. 51–76). London: Palgrave Macmillan.

Martin de Vitré, F. (1604). *La Description du premier voyage fait aux Indes orientales par les Français en l'an 1603, contenant les mœurs, les lois, façon de vivre, religions et habits des Indiens*. Paris: Chez Laurens Sonnius.

Miers, S. (1975). *Britain and the ending of the slave trade*. New York: Africana Publishing Company. pp. 58; 76.

Newitt, M. (2005). *A history of Portuguese overseas expansion, 1400–1668*. London and New York: Routledge Publishing. pp. 68; 90.

Paine, L. (2013). *The sea and civilization: A maritime history of the world*. New York: Alfred A. Knopf. pp. 203–207; 215–219; 276–278; 522–528.

Parthesius, R. (2010). *Dutch ships in tropical waters: The development of the Dutch East India company (VOC) shipping network in Asia, 1595–1660* (pp. 31–47). Amsterdam, The Netherlands: Amsterdam University Press.

Pearson, M. N. (2003). *The Indian Ocean*. London and New York: Routledge Publishing. pp. 19–23; 23–24.

Pearson, M. N. (2010). *Port cities and intruders: The Swahili coast, India, and Portugal in the early modern era* (pp. 129–135). Baltimore, Maryland: Johns Hopkins University Press.

Pinto, C. (1994). *Trade and finance in Portuguese India; a study of the Portuguese country trade, 1770–1840* (pp. 170–171). New Delhi, India: Concept Publishing Company.

Prakash, O. (1998). *European commercial enterprise in pre-colonial India*. Cambridge, UK: Cambridge University Press. pp. 1–14; 23–30; 60–65; 79–80; 109–110.

Prakash, O. (1999). The Dutch East India company in the trade of the Indian Ocean. In *Ashin das Gupta and Michael Naylor Pearson, India and the Indian Ocean, 1500–1800* (pp. 185–198). New Delhi, India: Oxford University Press.

Prange, S. R. (2018). *Monsoon Islam: Trade and faith on the medieval Malabar Coast* (pp. 26–42). Cambridge, UK: Cambridge University Press.

Pyrard de Laval, F. (1615). *Voyage de François Pyrard de Laval [...] navigation aux Indes Orientales, aux Moluques et au Brésil [...] Avec la description des pays, moeurs, loix, façons de vivre, police et gouvernement: du trafic et commerce qui s'y fait, et plusieurs autres singularitez. Vol. I*. Paris: Samuel Thiboust.

Quinn, F. (2000). *The French overseas empire* (pp. 39–48). Westport, CT: Praeger Publishing.

Rao, S. R., & Sarkar, S. S. (1985). *Lothal, a Harappan port town* (pp. 17–29). New Delhi, India: Archaeological Survey of India.

Ray, I. (1999). The French company and the merchants of Bengal (1680–1730). In I. Ray, & L. Subramanian (Eds.), *The French East India company and the trade of the Indian Ocean: A collection of essays* (pp. 77–87). New Delhi, India: Munshiram Manoharlal Publishers.

Reid, A. (1988). *Southeast Asia in the age of commerce, 1450–1680: Volume one: The lands below the winds*. New Haven, CT, and London, UK: Yale University Press.

Reid, A. (1993). *Southeast Asia in the age of commerce 1450–1680: Volume 2: Expansion and crisis*. New Haven, CT, and London, UK: Yale University Press.

Saldanha, A. (2011). The itineraries of geography: Jan Huygen van Linschoten's Itinerario and Dutch expeditions to the Indian Ocean, 1594–1602. *Annales of the Association of American Geographers, 101*, 149–177.

Shaffer, L. N. (2015). *Maritime Southeast Asia to 1500* (pp. 37–49). London and New York: Routledge Publishing.

Sherman, S. C. (1986). In J. M. Downey, & V. M. Adams (Eds.), *Whaling logbooks and journals, 1613–1927: An inventory of manuscript records in public collections* (pp. 437–444). New York & London: Garland Publishing.

Stackpole, E. A. (1972). *Whales and destiny: The rivalry between America, France, and Britain for control of the southern whale fishery, 1785–1825* (pp. 113–146). Amherst: University of Massachusetts Press.

Subbarayalu, Y. (2009). A note on the navy of the Chola state. In H. Kulke, K. Kesavapany, & V. Sakhuja (Eds.), *Nagapattinam to Suvarnadwipa: Reflections on the Chola naval expeditions to Southeast Asia* (pp. 91–95). Singapore: Institute of Southeast Asian Studies.

Subrahmanyam, S. (2012). *The Portuguese empire in Asia, 1500–1700: A political and economic history* (pp. 60–82). Hoboken, NJ: John Wiley & Sons.

Townsend, C. H. (1935). "The Townsend whaling charts," four charts published with "the distribution of certain whales as shown by logbook records of American whale ships". In *Zoologica*. New York: The New York Zoological Society (un-numbered).

Tripati, S., Shukla, S. R., Shashikala, S., & Sardar, A. (2016). Role of teak and other hardwoods in shipbuilding as evidenced from literature and shipwrecks. *Current Science*, *111*(7) 1262–1268.

van Linschoten, J. H. (1595). *Reys-gheschrift vande navigatien der Portugaloysers in Orienten*. Amsterdam: Cornelius Claesz.

van Linschoten, J. H. (1596). *Itinerario: Voyage ofte schipvaert van Jan Huygen van Linschoten naer Oost ofte Portugaels Indien, 1579–1592*. Amsterdam: Cornelius Claesz.

Vervliet, J. (2009). Introduction. In *Feenstra, Robert, Hugo Grotius Mare liberum 1609–2009: Original latin text and English translation* (pp. ix–xi). Leiden, The Netherlands: Brill Academic Publishers.

Vink, M. (2003). The World's oldest trade: Dutch slavery and slave trade in the Indian Ocean in the seventeenth century. *Journal of World History*, *14*(2) 134–144.

Walker, T. (2006). Slaves or soldiers? African conscripts in Portuguese India, 1857–1860. In R. Eaton, & I. Chatterjee (Eds.), *Slavery in the Indian Ocean region* (pp. 243–247). Indiana University Press.

Walvin, J. (2005). *Atlas of slavery*. London and New York: Routledge Publishing. pp. 23–26; 127–130.

Young, G. K. (2001). *Rome's eastern trade: International commerce and imperial policy, 31 BC–AD 305* (pp. 19–29). London and New York: Routledge Publishing.

Zulkifli, N., Ibrahim, R. I. R., Rahman, A. A. A., & Yasid, A. F. M. (2020). Maritime cooperation in the straits of Malacca (2016–2020): Challenges and recommend for a new framework. *Asian Journal of Research in Education and Social Sciences*, *2*(2) 10–32.

Chapter 3

Past, present, and future of the South Asian monsoon

Caroline C. Ummenhofer[a], Ruth Geen[b], Rhawn F. Denniston[c], and Mukund Palat Rao[d,e,f,g]

[a]Department of Physical Oceanography, Woods Hole Oceanographic Institution, Woods Hole, MA, United States, [b]School of Geography, Earth and Environmental Sciences, University of Birmingham, Birmingham, United Kingdom, [c]Department of Geology, Cornell College, Mount Vernon, IA, United States, [d]Centre de Recerca Ecològica i Aplicacions Forestals, Barcelona, Catalonia, Spain, [e]Department of Plant Science, University of California Davis, Davis, CA, United States, [f]Tree Ring Laboratory, Lamont-Doherty Earth Observatory of Columbia University, Palisades, NY, United States, [g]Cooperative Programs for the Advancement of Earth System Science, University Corporation for Atmospheric Research, Boulder, CO, United States

1 Introduction

The South Asian monsoon is a key component of the global climate system and manifests as seasonal changes in the atmospheric and oceanic circulation over the Indian Ocean basin, with widespread implications for regional water resources in surrounding countries. As the largest of the Earth's monsoon systems, the South Asian monsoon is a lifeline for the nearly 2 billion people in the region, who receive more than 75% of their annual precipitation during the summer monsoon (June–September; Doblas-Reyes et al., 2021; IPCC, 2021). Its domain is defined here as a broad region encompassing Bangladesh, Bhutan, Cambodia, India, Laos, Maldives, Myanmar, Nepal, Pakistan, Sri Lanka, and Vietnam (Fig. 1), in accordance with the regional South Asian monsoon domain defined by the IPCC (2021, and references therein). Even small perturbations in the timing of onset, seasonal distribution, and intensity of monsoon rainfall can have disproportionate effects on agriculture, most of which is rain-fed (Kumar et al., 2005a; Parthasarathy et al., 1988). Historically, a weakened monsoon and ensuing drought often had major societal effects throughout Asia, occasioning widespread famine and even today triggering large-scale impoverishment and rural-to-urban migration. For example, a 20% shortage in monsoon rainfall in 2002 resulted in billions of dollars in damage to the Indian economy (Gadgil et al., 2004).

Traditionally characterized to first order as continent-scale "sea breezes," where in summer the land heats faster than the ocean, causing warm air to rise over the continent and moist air to be drawn in from the ocean, the South Asian monsoon exhibits seasonally reversing wind and precipitation patterns that impact the most populated regions on Earth (Clift & Plumb, 2008; Geen et al., 2020; Wang, 2006; Wang et al., 2005; Webster et al., 1998). Yet, more recently, theoretical advances have led to the emergence of the monsoon as a regional manifestation of the seasonal variation of the global tropical atmospheric overturning and migration of the associated convergence zones (Geen et al., 2020). Monsoon wind strength and precipitation vary on intraseasonal timescales, with so-called active and break spells, on interannual timescales in response to modes of climate variability, such as the El Niño-Southern Oscillation (ENSO), as well as on multidecadal and longer timescales (e.g., Lau et al., 2000; Webster et al., 1998). Yet, high-quality instrumental records of hydroclimate variability in many regions across South Asia, with the exception of India, are relatively short and often do not extend prior to 1950, posing challenges to understanding South Asian summer monsoon variability at multidecadal to centennial timescales. Hydroclimatically sensitive paleo proxies (e.g., stalagmite, lacustrine, and tree-ring records) can provide long-term context for monsoon variability and provide links to societal impacts in response to sustained and widespread drought periods over the last millennium (e.g., Cook et al., 2010). Proxy evidence across the region suggests a weakening of the South Asian monsoon over the past 200 years, with observations and climate models also indicating a decreased Indian summer monsoon in the second half of the 20th century primarily due to anthropogenic aerosol forcing, while in the long-term, the South Asian monsoon is projected to increase over the 21st century and exhibit enhanced interannual variability (Doblas-Reyes et al., 2021; Douville et al., 2021; and references therein). Despite enormous progress in

---

[★]This book has a companion website hosting complementary materials. Visit this URL to access it: https://www.elsevier.com/books-and-journals/book-companion/9780128226988.

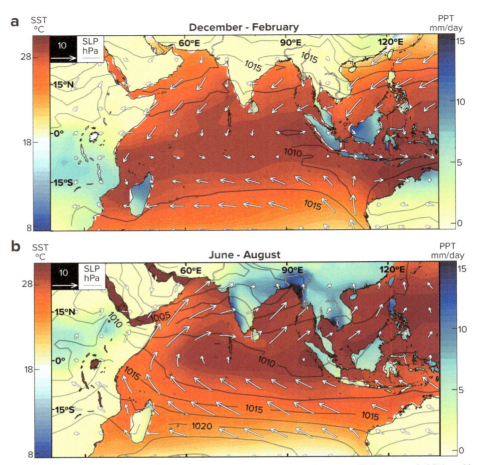

FIG. 1 Average sea surface temperature (SST in °C) with surface wind vectors (m/s), sea level pressure (SLP contours in hPa), and land precipitation (*blue* shading over land in mm/day) for (a) December-February and (b) June-August. Climatological conditions of SLP and winds based on NCEP/NCAR reanalysis (Kalnay et al., 1996), precipitation based on CMAP (Xie & Arkin, 1997), and SST on NOAA OISST v2 (Reynolds et al., 2002) for the period 1982–2019.

monsoon predictions since the early 1900s, operational monsoon forecasts in recent decades have had limited skill. As such, successful predictions of the South Asian monsoon, its onset, duration, intensity, and active and break cycles and regional variations still represent considerable challenges for weather and seasonal forecasts (e.g., Webster et al., 1998).

2 Monsoon dynamics

The summer monsoon season over South Asia is marked by intensification of precipitation and a reversal of the prevailing wind from easterly to westerly over the Arabian Sea, India, Bay of Bengal, and into Southeast Asia (Fig. 1). Monsoon rain begins over the southern Bay of Bengal between late April and mid-May, reaches Kerala between mid-May and mid-June, and then progresses northwestward over India. Over India, this season has historically been referred to as the Southwest monsoon, in reference to the prevailing wind associated with this summer rain (Rao, 1976). Withdrawal occurs through September to November, with the rainband retreating southward. Tamil Nadu and Sri Lanka receive most annual rain during this withdrawal phase, which has historically been referred to as the Northeast monsoon (Dhar & Rakhecha, 1983).

2.1 Climatological dynamics

The South Asian monsoon forms part of a global monsoon system (Wang & Ding, 2008). The tropical rainband migrates north and south throughout the year on a global scale, with particularly pronounced excursions over continents (Gadgil, 2018). Monsoons in individual regions have their own specific characteristics arising from local geography. The dynamics governing the South Asian monsoon climatology can broadly be understood by considering three factors.

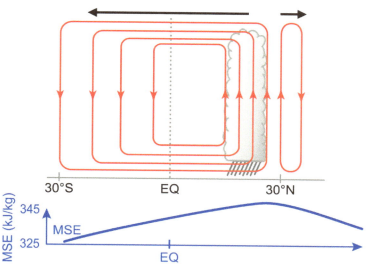

FIG. 2 Schematic illustrating the summertime meridional overturning (Hadley) circulation. The *gray* cloud denotes deep convection, clouds, and precipitation. *Red* contours and arrows indicate the mass streamfunction. The divide between the cross-equatorial winter Hadley cell and summer Hadley cell coincides with the maximum in subcloud moist static energy, shown in *blue*. The *black* arrows indicate the direction of vertically integrated moist static energy transport. *(Adapted from Geen et al. (2020); illustration by Beth Tully.)*

2.1.1 Warming of the summer hemisphere relative to the winter hemisphere

A useful concept for understanding monsoon circulations is moist static energy. This describes the energy of an air parcel due to its enthalpy, gravitational potential energy, and latent heat content. Moist static energy can be calculated as

$$h = c_p T + gz + Lq,$$

where h is the moist static energy, c_p is the specific heat capacity of dry air at constant pressure, T is the temperature, g is the gravitational acceleration, z is the height, L is the latent heat of vaporization of water, and q is the specific humidity. Warm and humid, high-moist-static-energy air provides fuel for deep convection in the monsoons, and the transport and spatial distribution of the moist static energy provide helpful constraints and insights into the location of the tropical rainband, including monsoon rain.

The atmosphere and ocean can be considered as a heat engine, transporting moist static energy from regions warmed by insolation to cooler parts of the globe. Monsoon circulations form an important part of this heat engine, and changes in drivers of atmospheric energy transport have been linked to altered monsoon intensity and extent (e.g., Biasutti et al., 2018; Kang, 2020, and references therein). In the tropical atmosphere, energy is predominantly transported in the upper branches of the Hadley circulation. This energy transport is illustrated by black arrows in Fig. 2. As the Northern Hemisphere warms through summer, its moist static energy increases relative to the Southern Hemisphere. The tropical convergence zone shifts northward so that the upper branch of the Southern Hemisphere Hadley cell transports energy across the Equator, from the higher to lower moist static energy hemisphere (e.g., Adam et al., 2016; Kang et al., 2008). The shift of the convergence zone is particularly pronounced over land, where monsoon regions form with strong, cross-equatorial Hadley circulations (Bordoni & Schneider, 2008; Gadgil, 2018).

In addition to these links between the monsoon and energy transport, arguments based on the upper-level (e.g., at 200 hPa) angular momentum budget and the assumption of convective quasi-equilibrium indicate that the division between Hadley cells should coincide with the maximum in subcloud moist static energy (Privé & Plumb, 2007). Convective quasi-equilibrium posits that where moist convection is frequent and intense, the atmospheric lapse rate is close to moist adiabatic. The rainband then sits just equatorward of the moist static energy maximum, where convergence is strongest, illustrated by the blue line and cloud in Fig. 2. These arguments were built from consideration of the zonally averaged circulation, but monsoon rain has been shown to follow this rule of thumb on a regional scale (Nie et al., 2010).

2.1.2 Warming of land relative to ocean

During summer, land warms faster than the ocean due to its lower heat capacity and the ocean's vertical redistribution of heat. The South Asian monsoon has often been described as a large-scale sea-breeze driven by this summer land-sea contrast (Halley, 1686). However, the observed behavior of the system disagrees with the sea-breeze perspective (Gadgil, 2003, 2018). For example, climatological land temperatures over India are highest before the onset of the monsoon (Simpson,

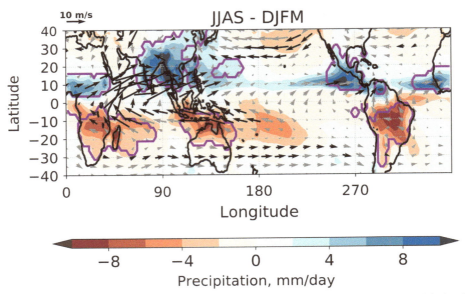

FIG. 3 Figure illustrating the localization of monsoons over land. Shading and arrows show the difference between precipitation (GPCP; Huffman et al., 2001) and 850-hPa wind velocity (JRA-55; Kobayashi et al., 2015) in June–September compared with December–March. *Magenta* contours indicate where the difference in precipitation exceeds 2 mm/day and where over 55% of the annual total rainfall falls in the hemisphere's summer season (cf. Wang & Ding, 2008). *Black* arrows show where the wind changes its direction seasonally by more than 90 degree; *gray* arrows show the windspeed difference where this criterion is not met. *(Adapted from Geen et al. (2020).)*

1921), while interannual variability indicates anomalously high land temperatures are associated with drought years (Kothawale & Kumar, 2002). Reanalysis and observational data indicate that the rainband is tightly linked to moist static energy on a regional scale, rather than temperature (Hurley & Boos, 2013). In addition, idealized modeling studies in which there is a seasonal cycle but no land generate monsoon-like circulations, emphasizing that land-sea contrast is a secondary ingredient (Bordoni & Schneider, 2008).

While the land-sea temperature contrast does not appear to be the primary physical driver of monsoons, it is certainly fundamental to their localization and regional characteristics. This is illustrated in Fig. 3; regions with a monsoon, i.e., a significant seasonal difference between summer and winter precipitation and circulation, are collocated with land masses. Land warms quickly relative to ocean, shifting the low-level moist static energy maximum to higher latitudes over land and supporting the development of a localized cross-equatorial Hadley circulation (e.g., Nie et al., 2010). Warming of land relative to ocean additionally sets up stationary-wave patterns, with pressure lows over the continents and highs over the ocean (Rodwell & Hoskins, 2001). Circulations develop around these pressure systems, with summertime southwesterly low-level winds over South Asia and southeasterly winds over the Western Pacific (Figs. 1 and 3). These winds advect moisture from the Indian Ocean over land, fueling the monsoon. The summertime stationary-wave circulations and their interactions with the land-surface and orography strongly influence the moist static energy distribution and the monsoon's northward extent over land (Chou & Neelin, 2003).

2.1.3 Asian orography

The Tibetan Plateau sits to the north of India and the Bay of Bengal. Simulations in which components of the orography are removed have highlighted its influence on the extent and progression of the monsoon over South Asia. For example, in simulations in which the Tibetan Plateau is removed, the monsoon rainband is weakened and confined to southern India and Southeast Asia (Boos & Kuang, 2010). Removing the orography allows ventilation by cooler, drier air from the north (cf. Chou & Neelin, 2003), which shifts the maximum in low-level moist static energy southward. Model simulations show that blocking this northerly flow, even with only a narrow orographic barrier, allows the monsoon rain to regain its observed northward extent (Boos & Kuang, 2010).

Effects of ventilation are observed during the onset of the South Asian monsoon (Fig. 4). Although the prevailing summertime low-level wind over India is westerly, the onset front progresses from the south to the northwest of the continent. At mid-levels (~600–750 hPa), the subtropical westerly jet is diverted southward along the westward side of the plateau (e.g., Sato & Kimura, 2007). This generates a northwesterly mid-level flow that brings dry, continental air over India and suppresses convection (red arrows, Fig. 4a). During the monsoon onset, this air mass is undercut by a moist low-level flow

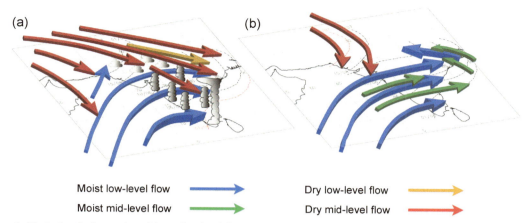

FIG. 4 Schematic illustrating the interaction of dry and moist airflows during South Asian summer monsoon. Panel (a) shows the situation around the time of onset at Kerala (around 1 June). *Red* and *green* arrows indicate dry and moist mid-level air, respectively, and *orange* and *blue* arrows indicate dry and moist low-level air, respectively. Mid-level dry air is advected from the northwest and is undercut by the low-level flow. The dry air suppresses convection, until the humid low-level air provides sufficient moistening of the dry layer to allow onset of moist convection and monsoon rain. Panel (b) shows the situation of around 15 July when the onset has progressed to the northwest. Typical clouds are represented in (a), but the more extensive cloud cover in (b) is omitted for clarity. *(Adapted from Parker et al. (2016).)*

(blue arrows, Fig. 4a). Monsoon onset can only occur once this flow has sufficiently moistened the mid-level dry air from below so that deep, moist convection can occur, as illustrated by the deep cloud bank over Kerala in this figure. Thus, while the prevailing summertime low-level wind over India is southwesterly, the monsoon front progresses northwestward through the season, delayed by the need to moisten the mid-levels from below (Parker et al., 2016). Later in the season, as the monsoon onset has progressed northwestward, the dry layer retreats and the mid-level flow changes the direction to moist the monsoon westerlies (Fig. 4b).

2.2 Subseasonal dynamics

Although the climatological picture might suggest a relatively smooth progression of monsoon rain across South Asia, rain does not fall steadily over the continent through the season. Each year the monsoon features several active and break phases in which rain intensifies or weakens, with the sum of these influencing the overall annual rainfall (Fig. 5; Gadgil, 2003). These active and break phases have been observed to occur in cycles with a period of 30–60 days (e.g., Hartmann & Michelsen, 1989). This summertime variability over Asia is influenced by propagating patterns of tropical deep convection, known as intraseasonal oscillations (Goswami & Mohan, 2001). Research has revealed that rainfall is modulated by the eastward propagating Madden Julian Oscillation (MJO; Madden & Julian, 1971), as well as a northward and eastward propagating mode known as the boreal summer intraseasonal oscillation or monsoon intraseasonal oscillation (e.g., Kikuchi, 2021). The dynamical mechanisms of both the MJO and boreal summer intraseasonal oscillation are the subject of ongoing research; recent advances were reviewed by Jiang et al. (2020), Zhang et al. (2020), and Kikuchi (2021); see also DeMott et al. (2024).

Stepping downward again to consider even smaller spatial scales and shorter timescales, within the monsoon rainband precipitation falls within short-lived weather systems. In particular, the summer monsoon over South Asia features monsoon lows and depressions: low-pressure systems of diameter 1500–2000 km that form over the Bay of Bengal and Arabian Sea, lasting 3 to 5 days (e.g., Godbole, 1977). Monsoon lows are categorized by the Indian Meteorological Department as having surface wind speeds less than 8.5 m/s, while monsoon depressions are categorized by wind speeds between 8.5 and 13.5 m/s. Such systems have been estimated to be responsible for approximately 40% of monsoon precipitation over northeast and central India (e.g., Hurley & Boos, 2015). Recent work indicates that these systems are generated by moist barotropic instability due to the meridional shear of the monsoon trough (Diaz & Boos, 2019a, 2019b). A study of extreme precipitation events over Pakistan and northern India from 1951 to 2007 identified that approximately ¾ of events occurred with a monsoon low in the vicinity, either directly driving the rain, or enhancing the monsoon circulation (Hunt et al., 2018).

Interactions with the extratropics additionally contribute to South Asian rainfall extremes, particularly over Pakistan and northern India. In 2010, a blocking high drove heatwaves over Europe and Russia. Coupling of active periods of monsoon rain and extratropical wave activity downstream of this block drove intense precipitation over Pakistan, resulting in record-breaking flooding (Hong et al., 2011). Similar midlatitude-tropical interactions have since been identified as

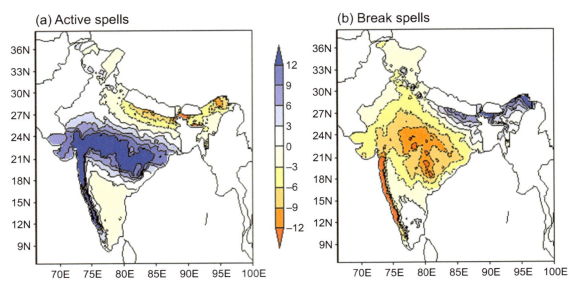

FIG. 5 Composite daily rainfall anomaly map of (a) active spells and (b) break spells identified in the India Meteorological Department high resolution (0.25 × 0.25 degree) gridded daily rainfall dataset for the period 1901–2014. *(Reprinted by permission from Springer Nature, Pai et al. (2016).)*

modulating rainfall over Pakistan (Ullah et al., 2021). Precipitation over northern India and Pakistan is also influenced by western disturbances, which are cyclonic storms embedded in the subtropical westerly jet, arising from interactions between extratropical cyclones propagating along the jet with stationary low-level circulations (Dimri et al., 2015). These are related to extreme precipitation, particularly in winter, although they are also observed during some summer extreme precipitation events (Hunt et al., 2018).

The role of atmospheric rivers over Asia has recently been investigated (Liang & Yong, 2021). Atmospheric rivers are synoptic systems featuring "river-like" filaments of enhanced lower-troposphere water vapor transport. These have been suggested to contribute up to 36% of total summer precipitation and 52%–56% of extreme summer precipitation in South Asia (Liang & Yong, 2021), and appear strongly linked to flood events (Mahto et al., 2023). Further work may elucidate their broader driving dynamics over Asia and connections to the monsoon and associated synoptic systems.

Recent years have seen a number of extreme flooding events across South Asia, triggered by extreme rainfall in systems as described earlier. For example, an analysis of the catastrophic floods in Pakistan in August 2022 identified that multiday extreme precipitation arose from two atmospheric rivers passing over southern Pakistan (Nanditha et al., 2023). Rain fell onto wet antecedent soil moisture conditions, caused by prolonged rainfall in July. Similarly, flooding in Kerala in 2018 occurred when heavy rain in a low-pressure system in early August saturated soils and filled reservoirs, followed by further intense rain in a monsoon depression later that month (Hunt & Menon, 2020). Consistent with these examples, an analysis of high river flow events identified multiday precipitation, resulting in high rainfall on wet antecedent soil conditions, as the prominent driver of flooding in Indian river basins (Nanditha & Mishra, 2022).

3 Drivers of South Asian monsoon variability

Variations of the South Asian monsoon can be related to changes in the forcing mechanisms of the mean monsoon circulation (Kucharski & Abid, 2017) detailed in Section 2. Variations in the monsoon are partly linked with internal atmospheric variability on synoptic to (sub)seasonal timescales, such as rainfall variations in active and break phases of the monsoon, and are hence not predictable at seasonal timescales. On the other hand, the predictable component of monsoon variability is linked to more slowly varying boundary conditions in the climate system (Charney & Shukla, 1981; Kucharski & Abid, 2017), such as ocean temperatures, soil moisture, and Eurasian snow cover. The focus of this section is on the drivers of the predictable component of South Asian monsoon variability, and their interactions, across a range of timescales. The role of external drivers, such as greenhouse gases, aerosols, and land-use change, will be addressed in Section 5, while for a discussion of orbital and solar forcing on South Asian monsoon variability on millennial and longer timescales, the reader is referred to Mohtadi et al. (2024).

3.1 Pacific Ocean drivers

The influence of the Pacific Ocean on the Asian monsoon is well established, going back to the late 19th century when catastrophic drought and famine in India coincided with anomalous high pressure over Australia (Taschetto et al., 2020; and references therein). This out-of-phase atmospheric pressure relationship between the Pacific and Indian Oceans, termed the "Southern Oscillation" by Walker (1924) and first reported by Hildebrandsson (1897), became a key predictor for the strength of the South Asian monsoon. The pressure seesaw reflects shifts in the location of the ascending and descending branches of the Walker circulation; i.e., El Niño events are typically associated with anomalous descent over South Asia that tends to reduce monsoonal precipitation (Fig. 6; e.g., Kumar et al., 1999; Rasmusson & Carpenter, 1983).

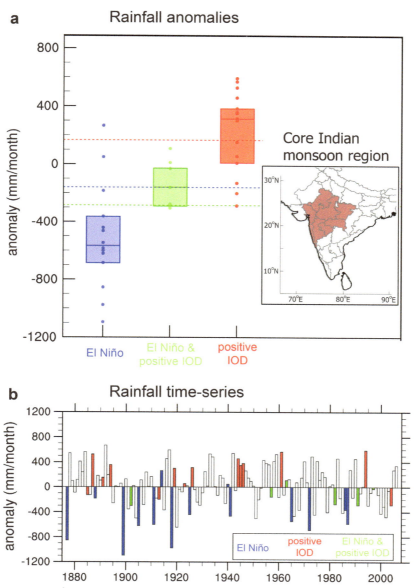

FIG. 6 Rainfall anomalies (mm month^{-1}) for the different ENSO/IOD categories for the core Indian monsoon region (see the map inset) during the June–September months for the period 1877–2006: (a) rainfall anomalies shown as dots for El Niño (*blue*), co-occurring El Niño and positive IOD (*green*), and positive IOD (*red*) events. The colored boxes are delimited by the upper and lower quartiles, with the middle bar denoting the median rainfall in the respective category. Dashed lines indicate the 90% confidence level (as estimated by Monte Carlo testing) for the medians for the different categories (indicated in color). (b) Time series of rainfall anomalies with the associated ENSO/IOD categories indicated in color. *(Adapted from Ummenhofer et al. (2011).)*

Several additional mechanisms appear to be at play during El Niño events due to adjustments of the Walker circulation and the resultant descending motion over the Indo-Pacific warm pool (Kucharski & Abid, 2017). The response to the anomalous subsidence over the maritime continent includes, for example, modulation of the Indian Ocean regional Hadley cell resulting in anomalous ascent over the central equatorial Indian Ocean and descent over the Indian subcontinent (Krishnamurthy & Goswami, 2000). Furthermore, a Matsuno-Gill-type (Gill, 1980; Matsuno, 1966) equatorial Rossby wave anticyclone response is excited to the west across the Indian Ocean that leads to a tilted ridge with suppressed rainfall extending toward India (Wang et al., 2003). In addition, warming of the tropical troposphere reduces the land-sea temperature gradient and induces an opposing tropospheric temperature gradient that counteracts the monsoon circulation; in turn, this weakens the South Asian monsoon trough and shortens the length of the rainy season (Goswami & Xavier, 2005) while also increasing the frequency of longer break and shorter active spells (Dwivedi et al., 2015). In contrast, to first order, an enhanced South Asian summer monsoon with broadly opposite climate anomaly patterns and active/break periods occurs during La Niña events (Taschetto et al., 2020, and references therein).

This traditional view of the ENSO-South Asian monsoon relationship worked well for much of the 20th century. However, during the strong 1997 El Niño, the suppression of convection over the maritime continent due to the shift in the Walker circulation was proposed to be strong enough to substantially change the local Hadley circulation over the Indian Ocean leading to large-scale convergence over the monsoon trough leading to deep monsoon depressions over northern India. Hence, the intertropical convergence zone (ITCZ) was preferentially located to the north over the Asian monsoon domain, resulting in above-average rainfall and subseasonal synoptic activity over the region (Slingo & Annamalai, 2000). It has also been suggested that the well-established relationship between ENSO and the South Asian monsoon is changing (e.g., Kumar et al., 1999), or that the location of maximum warming/cooling in the equatorial Pacific during ENSO events are critical factors in determining the extent to which the Indian summer monsoon is affected: central Pacific warming appears more likely to result in monsoon drought than the canonical eastern Pacific El Niño events (Kumar et al., 1999, 2006; Weng et al., 2007). Furthermore, the strength of the relationship between Pacific conditions and the South Asian monsoon is nonstationary over time (e.g., Webster & Yang, 1992) and varies on decadal to multidecadal timescales (Kripalani & Kulkarni, 1997) due to various modulating influences (Kucharski & Abid, 2017), including the Indian Ocean (Ashok et al., 2001; Krishnaswamy et al., 2015; Kucharski et al., 2006; Ummenhofer et al., 2011, 2013), Pacific decadal variability (Krishnamurthy & Goswami, 2000; Krishnamurthy & Krishnamurthy, 2014, 2016), Atlantic variability (Chang et al., 2001; Kucharski et al., 2007, 2008), or due to stochastic processes (Gershunov et al., 2001).

On decadal timescales, the Interdecadal Pacific Oscillation (IPO) exerts considerable influence on the atmospheric circulation and hydroclimate across the Indo-Pacific and can modulate ENSO's impact on interannual timescales (e.g., Kucharski et al., 2021; Power et al., 2021; Salinger et al., 2014). The IPO is a decadal mode of Pacific variability (similar to the Pacific Decadal Oscillation), but with a meridionally broader tropical El Niño-like warm temperature anomaly pattern and cool extratropical Pacific during its positive phase. Much like El Niño, positive IPO phases are characterized by weakened trade winds and warmer SST in the eastern Pacific, cooler SST in the Indo-Pacific warm pool region, as well as a weakening and eastward shift of the Walker circulation from its climatological position, resulting in anomalous subsidence over the Indo-Pacific warm pool and extending into South Asia (e.g., Meehl & Hu, 2006).

The positive IPO phase exhibits anomalously strong subsidence north of the equator over the Indian Ocean sector associated with an intensified (meridional) Hadley circulation over South Asia leading to reduced moisture flow over western and central India and an anomalously dry summer monsoon (Joshi & Kucharski, 2017; Joshi & Rai, 2015; Krishnamurthy & Goswami, 2000; Krishnamurthy & Krishnamurthy, 2014, 2016; Krishnan & Sugi, 2003; Salzmann & Cherian, 2015; Sen Roy, 2006). Hydroclimatic effects of El Niño events coinciding with the positive IPO phase are amplified over the region. Broadly opposite circulation and hydroclimatic anomalies occur during the negative (cool) IPO phase, which resembles La Niña-like conditions. Monsoon variability on decadal timescales linked to the IPO have also been reported in other regions of Southeast Asia, such as for Myanmar (D'Arrigo & Ummenhofer, 2015; Sen Roy & Sen Roy, 2011), Thailand (Xu et al., 2019), and Vietnam (Buckley et al., 2010, 2014, 2019; Duc et al., 2018).

3.2 Indian Ocean drivers

At (sub)seasonal timescales, variations related to the boreal summer intraseasonal oscillation modulate active and break periods of the South Asian summer monsoon (see Section 2.2 for more details, as well as DeMott et al., 2024). At interannual timescales, even before the Indian Ocean Basin Mode (Chambers et al., 1999) and the IOD (Saji et al., 1999; Webster et al., 1999) were more widely adopted in the community, tropical Indian Ocean variability was recognized as a driver of regional hydroclimate, including for the South Asian monsoon (e.g., Chang & Li, 2000; Clark et al., 2000; Li et al., 2001).

The Indian Ocean Basin Mode manifests as uniform warming (cooling) across the tropical Indian Ocean during El Niño (La Niña) events, developing mostly in boreal winter following the peak of ENSO, though anomalies can persist until the following summer and influence the South Asian monsoon (Kucharski & Abid, 2017). As such, the Indian Ocean Basin Mode has been interpreted as prolonging ENSO's impact in the broader Indian Ocean region (e.g., Kucharski & Abid, 2017; Taschetto et al., 2011; Xie et al., 2009; Yang et al., 2007). Specifically, Indian Ocean Basin Mode warming leads to an enhanced South Asian monsoon over the Arabian Sea, likely resulting from the creation of a secondary heating source forcing a Matsuno-Gill response in the upper troposphere (Yang et al., 2007) that generates significant circumglobal teleconnections over the Northern Hemisphere during boreal summer (Yang et al., 2009). On the other hand, ENSO-induced changes in SST in the Southwest Indian Ocean in boreal spring, in conjunction with anomalous subsidence over the maritime continent, prevent the northwestward migration of the ITCZ and the associated deep moist layer, causing the Indian summer monsoon onset in June to be delayed by a week (Annamalai et al., 2005). Furthermore, local SST in the Arabian Sea and Bay of Bengal modulates Indian monsoon rainfall, likely through influences on the moisture flow to South Asia (e.g., Ashok et al., 2001; Vecchi et al., 2004; Izumo et al., 2008; Ummenhofer et al., 2024).

The IOD also exerts considerable influence on the South Asian monsoon (e.g., Ashok et al., 2001, 2004; Ashok & Saji, 2007; Behera et al., 1999; Behera & Ratnam, 2018; Cherchi et al., 2007, 2021; Cherchi & Navarra, 2013; Gadgil et al., 2004; Kucharski & Abid, 2017; Saji & Yamagata, 2003; Ummenhofer et al., 2011, 2013). While there is considerable covariability between ENSO and IOD through the Walker circulation facilitating tropical basin interactions (e.g., Cai et al., 2019), and debate continues about the degree of (in)dependence of the IOD (e.g., Stuecker et al., 2017), there exists increasing evidence that the Indian Ocean exhibits substantial interannual variability distinct from Pacific influences. For details, see Yamagata et al. (2024) and Ummenhofer et al. (2024) and references therein. Characterized in its positive phase by anomalous cold SST in the Southeast Indian Ocean and warm anomalies in the equatorial western Indian Ocean, which develop in boreal summer and peak in the fall, positive IOD events are associated with enhanced Indian summer monsoon rainfall (Fig. 6), while a weakened monsoon occurs during negative IOD events (Kucharski & Abid, 2017). As summarized in a recent review by Cherchi et al. (2021), the direct effect of the IOD on the Indian monsoon occurs through changes in moisture transport over the western Indian Ocean or modulations of the local Hadley cell, with stronger ascending motion and a northward shift of its ascending branch during positive IOD events, both leading to a stronger summer monsoon (Ashok et al., 2001, 2004; Ashok & Saji, 2007; Behera & Ratnam, 2018; Gadgil et al., 2004). However, the response across the Indian subcontinent is not spatially uniform: for a positive IOD, the anomalous moisture transport toward the Indian subcontinent helps strengthen the monsoon trough through an intensified monsoon-Hadley circulation, with anomalously enhanced rainfall bordered to the north and south of the trough by reduced precipitation (Anil et al., 2016; Behera & Ratnam, 2018; Cherchi et al., 2021).

In addition, the IOD can also modulate the relationship between the South Asian monsoon and ENSO (Ashok et al., 2001, 2004; Behera & Yamagata, 2003; Cherchi & Navarra, 2013; Gadgil et al., 2004; Ummenhofer et al., 2011, 2013). Variations in equatorial Indian Ocean SST on decadal timescales have also been linked to low-frequency modulations of the strength of the Indian summer monsoon (Kucharski et al., 2006) and can modulate ENSO's influence on multidecadal timescales (Fig. 6b; Ummenhofer et al., 2011, 2013). Several studies furthermore identify a strengthening influence of Indian Ocean conditions associated with the IOD over ENSO for the Indian summer monsoon in recent decades (Feba et al., 2019; Krishnaswamy et al., 2015). A relative strengthening of cross-equatorial flow from the Southern Hemisphere into the equatorial Indian Ocean has been reported during the summer monsoon season in recent decades due to an intensification of the subtropical high pressure system in the southern Indian Ocean, which in turn weakens the teleconnection from the Pacific to the Indian monsoon during El Niño events (Feba et al., 2019).

3.3 Atlantic Ocean drivers

Atlantic influences on the South Asian monsoon exist across a range of timescales and are transmitted via several different pathways: Rossby wave trains emanating from the midlatitude North Atlantic are associated with abrupt declines in late season monsoon rainfall through the interaction of upper-level winds and episodic North Atlantic vorticity anomalies (Borah et al., 2020). This link to the North Atlantic explains half of the 20th century Indian summer monsoon droughts, largely coinciding with those monsoon failures that occurred when equatorial Pacific Ocean temperatures were near-neutral (Borah et al., 2020). On interannual timescales, an inverse relationship between the winter/spring North Atlantic Oscillation (NAO) and South Asian monsoon has been described (e.g., Dugam et al., 1997), likely transmitted via associated shifts in storm-track patterns over the North Atlantic affecting northern Eurasian surface air temperatures (Chang et al., 2001). An anomalously warm (cold) North Atlantic results in weakening (strengthening) of the meridional gradient of tropospheric temperature due to a cool (warm) tropospheric temperature anomaly over Eurasia during boreal late summer/fall leading to an early (late) withdrawal of the Indian summer monsoon and decrease (increase) of monsoon rainfall (e.g., Goswami et al., 2006a). On longer timescales, a strengthening and poleward shift of the jet stream over the North Atlantic in recent decades appeared to affect the meridional temperature contrast over Eurasia into boreal spring and disrupt ENSO's impact on the monsoon (Chang et al., 2001). However, there remain unresolved uncertainties in the causality of the NAO-South Asian monsoon relationship on interannual to interdecadal timescales (Cui et al., 2014).

Tropical Atlantic SST variability can also influence the strength of the Indian summer monsoon through the excitation of Rossby waves, modulating the atmospheric circulation and convective activity over the Sahel, and this in turn affecting western Indian Ocean winds and SST (e.g., Kucharski et al., 2007, 2008; Vittal et al., 2020; Zhang & Delworth, 2006). For example, cooler than normal SST in the southeastern Atlantic lead to upper-level convergence over the tropical Atlantic and Africa, compensated by divergent motion over the Indo-Pacific warm pool and Pacific, while a near-surface cyclone forms over the Indian subcontinent that results in enhanced southwesterly moisture flow and an anomalously wet summer monsoon (Kucharski et al., 2008). Broadly opposite climate anomalies are seen for warm spring and summer South Atlantic SST and an ensuing anomalously dry Indian monsoon (Kucharski et al., 2008). This mechanism was also suggested to contribute to the apparent weakening of the ENSO-Indian summer monsoon relationship in recent decades, with the anomalous cool SST in the tropical Atlantic of the 1980s–1990s exciting a Rossby wave train response and a remote teleconnection with increased convergence over India opposing drought-inducing subsidence during El Niño events of the late 20th century (Kucharski et al., 2007).

On multidecadal timescales, Indian summer monsoon variability has been seen to covary with low-frequency modulation of North Atlantic SST associated with Atlantic multidecadal variability over the course of the 20th century in observations and climate model simulations (Goswami et al., 2006a; Joshi & Rai, 2015; Luo et al., 2018b; Zhang & Delworth, 2006). The mechanism invoked resembles what is observed following strong NAO influences on seasonal to interannual timescales and is likely induced by a clustering of NAO events during decades with anomalous Atlantic multidecadal variability conditions (Goswami et al., 2006a). However, the transition to the North Atlantic warm phase in 1995 did not produce the expected secular increase in Indian summer monsoon rainfall (Luo et al., 2018a). Whether or not this decoupling is due to natural variability or instead reflects anthropogenic influences (e.g., aerosols) is the subject of debate (e.g., Luo et al., 2018a; Roxy et al., 2015), but evidence suggests that the link between Atlantic multidecadal variability and Indian summer monsoon rainfall may have varied through time (Goswami et al., 2006a), linked in part also to constructive/destructive influences of Atlantic and Pacific decadal variability (e.g., Joshi & Rai, 2015).

3.4 High-latitude influences

Eurasian snow cover and soil moisture are additional factors that can modulate the South Asian monsoon, with enhanced (reduced) snow cover preceding a weak (strong) summer monsoon (e.g., Barnett et al., 1989; Vernekar et al., 1995). As such, wintertime Northern Hemisphere circulation anomalies affect the extent of Eurasian snow cover and via influences on springtime hemispheric temperatures and soil moisture bridge the gap between boreal winter Northern Hemisphere conditions to early summer South Asian monsoon variations: for example, quasi-stationary tropospheric circulation anomalies during a negative Arctic Oscillation displace the mid-Eastern jet poleward, leading to anomalous surface heating and drying (Buermann et al., 2005). Such surface conditions lead to southwesterly anomalies to the east of the lower-troposphere monsoon jet, which enhances the airmass convergence and moisture transport into the monsoon region at the start of the season. In contrast, the opposite occurs in response to anomalously high Eurasian snow cover and cooler surface temperatures over the Tibetan Plateau, reducing the mid-tropospheric meridional temperature gradient over the Indian subcontinent and resulting in a weaker monsoon circulation (Vernekar et al., 1995). However, this inverse relationship between Eurasian snow cover and the South Asian summer monsoon appears weaker post-1990 due to reduced Eurasian snow cover no longer being able to regulate the summer mid-tropospheric temperature gradient (Zhang et al., 2019). On the other hand, the association between springtime Siberian snow extent and Indian summer monsoon rainfall appears stronger in recent decades, with increased (reduced) accumulated snow during spring providing wetter (drier) soil during the following summer, which results in a weakened (strengthened) meridional tropospheric temperature gradient driving corresponding weakening (strengthening) of the subtropical westerly jet, and summer monsoon Hadley cell (Singh et al., 2021).

4 Past variability in the South Asian monsoon

4.1 Late Holocene monsoon variability from stalagmite and lacustrine proxies

Oxygen isotope ratios in stalagmites are widely used as paleomonsoon proxies (Figs. 7 and 8) because stalagmite carbonate preserves precipitation $\delta^{18}O$ variability, which is often tied in (sub)tropical settings to the amount effect (i.e., an inverse

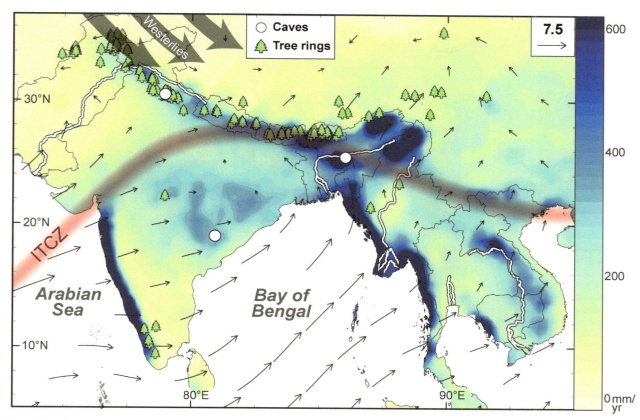

FIG. 7 Key climatic features of the South Asian monsoon region, including summertime (June–August) rainfall (colored in mm/year) and winds (vectors in m/s), northernmost location of the intertropical convergence zone (ITCZ), and position of the Westerlies; locations of paleoclimate proxy records (speleothems and tree rings) mentioned in this chapter are indicated as well.

relationship between rainwater $^{18}O/^{16}O$ ratios and rainfall amount) in monsoonal rainfall. However, across South Asia, the amount effect varies markedly in strength. In general, the amount effect is most pronounced where seasonal aridity is a main regional feature: the drier and longer the dry season, the more likely that a classical amount effect will be present (Dansgaard, 1964). Within the core monsoon zone of India (Fig. 7; Sinha et al., 2011) and just to the north of it (Kathayat et al., 2017), the amount effect can explain most of the variance in precipitation $\delta^{18}O$. However, outside the core monsoon zone, this effect may be weak to absent, as is the case in Kathmandu, Nepal, and Cherrapunji, northeastern India (Adhikari et al., 2020; Breitenbach et al., 2010), and thus oxygen isotope time series must be viewed in a more holistic paleoenvironmental sense.

Across many portions of South Asia, precipitation isotope ratios appear to be dominated by source effects. The controls of atmospheric circulation dynamics on oxygen isotope ratios in Indian precipitation were explored using isotope-enabled climate model simulations (Kathayat et al., 2021). These analyses revealed that moisture is increasingly sourced from the isotopically heavier Arabian Sea during weak Indian summer monsoon years and from the isotopically lighter southern Indian Ocean when summer monsoon strength is enhanced (Kathayat et al., 2021). Further complicating the interpretation of stalagmite $\delta^{18}O$ as a paleo-Indian summer monsoon proxy are precipitation/evaporation feedbacks, such as those hypothesized in northeastern India, where secular decreases in rainwater $\delta^{18}O$ over the course of the monsoon season were ascribed, in part, to a freshening (and thus decreasing of $\delta^{18}O$) across Bay of Bengal surface waters from early monsoon runoff (Breitenbach et al., 2010).

In addition, while the majority of South Asian precipitation falls during the summer monsoon, up to 20% of precipitation across the Himalayan front is derived from the midlatitude Westerlies, which are characterized by generally higher $\delta^{18}O$ values than Indian summer monsoon rains (Fig. 7; Adhikari et al., 2020). Central Himalayan glacial ice records reflect this complexity. For example, although the Dasuopu and East Rongbuk glaciers are proximally located, with Dasuopu only 125 km further west, their oxygen isotope time series are ascribed to different processes: at Dasuopu, $\delta^{18}O$ variability is generally regarded as a paleotemperature signal (Thompson et al., 2000), while at East Rongbuk $\delta^{18}O$ is tied to precipitation amount (Zhang et al., 2005). These seemingly contradictory interpretations were reconciled by Pang et al. (2014)

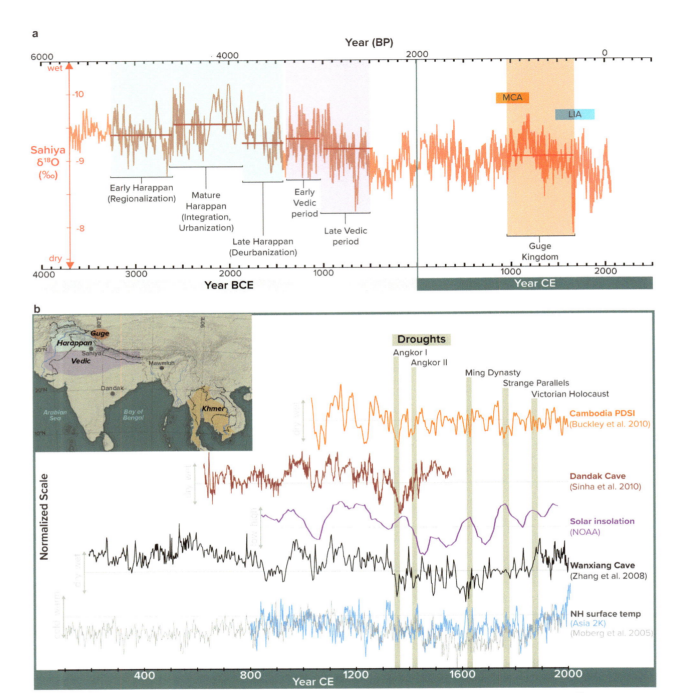

FIG. 8 South Asian monsoon variability over past millennia and links to past civilizations and iconic drought periods. (a) Sahiya δ¹⁸O speleothem record and cultural history for past civilizations (shown in the inset in b); key climatic periods are indicated for Medieval Climate Anomaly (MCA) and Little Ice Age (LIA); horizontal red bars indicate the mean δ¹⁸O for vertical bars. (b) Speleothem δ¹⁸O records from China and India, tree-ring-derived Palmer Drought Severity Index (PDSI), Northern Hemispheric surface air temperature reconstructions, and solar insolation data as indicated. Beige bars indicate key drought periods discussed. Inset: schematic of regional extent of past civilizations. *(Panel (a) adapted from Kathayat et al. (2017) and (b) adapted from Buckley et al. (2014).)*

who argued that Dasuopu receives a markedly larger contribution of precipitation from the Westerlies, whereas East Rongbuk is blocked from this moisture source by topographic highs, including Mt. Qomolangma (Mt. Everest), and thus supplied more substantially by the Indian summer monsoon.

Where an amount effect is identifiable, changes in stalagmite $\delta^{18}O$ values have been linked to South Asian summer monsoon rainfall variability that, in turn, has been tied to societal shifts (Fig. 8). These include a sustained period of enhanced monsoon rainfall during the late mid-Holocene Climate Optimum (~4.5–4.0ka) that is linked with expansion of agriculture that supported urbanization of the Harappan civilization (Fig. 8). A severe drought from ~1593 to 1623 CE is also associated with collapse of the Guge Kingdom in Tibet (Fig. 8; Kathayat et al., 2017). Some late Holocene drought intervals identified in central (Sinha et al., 2011) and northern India coincide with events documented in paleo-South Asian summer monsoon reconstructions from Oman (Fleitmann et al., 2007) and beyond, suggesting they were geographically expansive. Drought associated with the "4.2 ka event", which has been linked to the collapse of the Akkadian Empire in Mesopotamia (Weiss et al., 1993) and the Old Kingdom in Egypt (Stanley et al., 2003), is evident across many parts of India. Lake Rara, western Nepal (Nakamura et al., 2016), and Kotla Dahar Lake, north-central India, record pronounced drought activity associated at this time (Dixit et al., 2014), and stalagmites from Mawmluh cave in northeastern India are used as the type locality for the Meghalayan stage (Berkelhammer et al., 2012). However, this signal does not appear in the Sahiya record from northern India (Kathayat et al., 2016), raising questions about its relationship to the Indian summer monsoon.

Longer-term shifts in the strength of the Indian summer monsoon track Northern Hemisphere temperature, possibly through modulation of the ITCZ and positioning of the westerly jet (e.g., Kang, 2020). Likely as a result, well-recognized, multicentennial climate intervals such as the Little Ice Age (~1450–1850 CE), Medieval Climate Anomaly (~850–1250 CE), and Dark Ages Cold Period (~400–750 CE) coincide with secular shifts in Indian summer monsoon rainfall in northern Indian stalagmite records (Kathayat et al., 2017). More complete spatial coverage of Indian summer monsoon activity could be developed from lacustrine sediments, but most published lake records preserve evidence of late Holocene Indian summer monsoon variability at multidecadal to centennial scales, and thus lack the temporal resolution required for analysis of annual to decadal monsoon variations.

4.2 Last millennium variability from tree rings

Instrumental records of hydroclimate and streamflow variability in South Asia are relatively short and generally do not extend prior to 1950 (Cook et al., 2010). This limitation has made it challenging to understand variability in the South Asian summer monsoon at decadal to centennial timescales. Dendroclimatology, the study of tree rings as natural paleoclimate proxies, has contributed significantly to furthering our understanding of historical monsoon variability prior to the short observational record. The growth of trees is influenced by local environmental conditions (Fritts, 1976). Trees generally grow more and put on wider rings in favorable years and conversely grow less and have narrower rings during years with unfavorable climate conditions (Fritts, 1976). Moreover, many tree species can live for centuries (Liu et al., 2019; Piovesan & Biondi, 2021). Globally, tree growth is commonly controlled by growing season soil moisture variability with trees exhibiting higher growth in wet years relative to drought years (Brienen et al., 2016; St. George & Ault, 2014). Additionally, various wood properties of tree rings such as their density, anatomical characteristics, and their isotopic composition (primarily $\delta^{13}C$ and $\delta^{18}O$ stable isotopes) are also influenced by environmental conditions (Brienen et al., 2016; Fritts, 1976; Loader et al., 2007). Although many tropical tree species do not produce annual rings, the monsoonal climate of South Asia with pronounced wet and dry season causes some tree species to produce annual growth rings facilitating their use in dendroclimatology (e.g., Bhattacharyya et al., 1992; Bhattacharyya & Shah, 2009; Brienen et al., 2016; Grant, 1992).

The relationship between monsoonal moisture and tree growth has facilitated the use of tree rings for the reconstruction of South Asian summer monsoon variability and helped better characterize the severity of recent and past extremes events such as flood and droughts (e.g., Cook et al., 2010; Gaire et al., 2019; Nguyen et al., 2020; Sano et al., 2012; Shah et al., 2007; Shi et al., 2018). In the South Asian context, tree-ring studies have documented and validated drought events such as the Strange Parallels Drought (1756–1768 CE), East India Drought (1790–1796 CE), and the Great Drought (1876–1877 CE) that were linked to monsoon failure (Fig. 9) and were previously known from documentary and oral records (Cook et al., 2010; Davis, 2001; Shah et al., 2014; Shi et al., 2014; Zaw et al., 2020). Further, these studies have also documented the extreme severity of these events relative to the instrumental and paleoclimate record (Cook et al., 2022). Tree rings have also helped elucidate the large subcontinental scale spatial extent of these extreme events. This has been aided by the development of spatially resolved gridded hydroclimate reconstructions such as the Monsoon Asia Drought Atlas (Cook et al., 2010), and more recently, the Paleo Hydrodynamics Data Assimilation (Steiger et al., 2018) and Last Millennium Reanalysis (Tardif et al., 2019) products. The latter two datasets combine

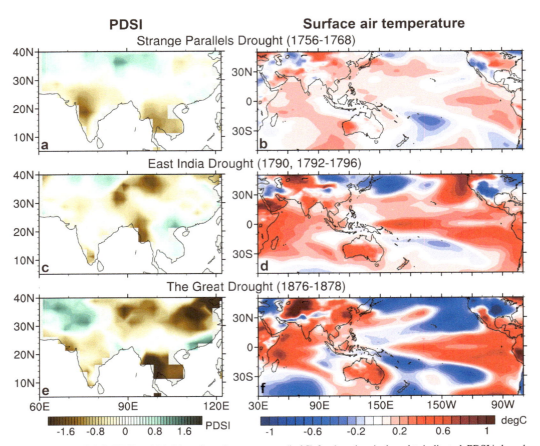

FIG. 9 Composite anomalies of (*left*) PDSI and (*right*) surface air temperature (in °C) for three iconic droughts indicated. PDSI is based on hydroclimate reconstructions from tree-ring width records from the Monsoon Asia Drought Atlas (Cook et al., 2010), and surface air temperatures from the Paleo Hydrodynamics Data Assimilation product (Steiger et al., 2018); all composites based on the June–August season and relative to 1700–1900 baseline conditions.

paleo-proxy data with climate models in a data assimilation approach to reconstruct multiple climate fields (Steiger et al., 2018; Tardif et al., 2019). These gridded spatial reconstruction products (e.g., Cook et al., 2010; Shi et al., 2018; Steiger et al., 2018; Fig. 9) harness the large network of publicly available tree-ring data archived in the International Tree Ring Databank shared by tree-ring scientists around the world and the spatial autocorrelation inherent in hydroclimate to reconstruct spatial explicitly climate anomaly fields (Zhao et al., 2019).

South Asia's hydroclimate exhibits strong linkages (or teleconnections) with large-scale ocean-atmospheric patterns such as ENSO, the IOD, and IPO (Fig. 6; e.g., Lenssen et al., 2020; Saji & Yamagata, 2003; Ummenhofer et al., 2013), and to abrupt natural forcing events such as volcanic eruptions (Shi et al., 2014). Tree rings have been utilized to examine the ocean-atmospheric drivers of extreme events like the Strange Parallels Drought (Fig. 9a and b) and the Great Drought (Fig. 9e and f) and to contextualize the calamitous societal consequences of these events (Fig. 8; D'Arrigo et al., 2011; D'Arrigo & Ummenhofer, 2015; Hernandez et al., 2015; Singh et al., 2018b; Mishra & Aadhar, 2021). Buckley et al. (2014) suggested that the late 14th century ITCZ movement over the tropical western Pacific (e.g., Denniston et al., 2016; Sachs et al., 2009; Yan et al., 2015) may have contributed to the climate instability that led to the demise of Cambodia's expansive Khmer Kingdom by the early 15th century (Fig. 8). At the same time, a southward shift in the trajectory of fall/winter tropical storms brought heavy rains to Southeast Asia. Similar issues of high-volume water damage to dams and levees in northern Vietnam's Red River Valley caused significant disruption to agricultural systems (Buckley et al., 2014; Penny et al., 2019). The Strange Parallels Drought gripped much of monsoon South Asia in the latter half of the 18th century, contributed to the collapse of regional polities, and was connected to the tropical Pacific Ocean and regions of ENSO variability (Fig. 9a and b; Sano et al., 2008; Buckley et al., 2010, 2014; Cook et al., 2010; Lieberman & Buckley, 2012). Persistent central Pacific ENSO conditions in the tropical Pacific may have accounted for anomalous hydroclimate during this period (Hernandez et al., 2015). In the aftermath of the Great Drought (1876–1877), 12.3 to 29.3 million people died in colonial India alone (Davis, 2001).

The extreme severity, duration, and extent of the Great Drought is linked to a rare combination of a strong positive IOD (1877), a record-breaking El Niño (1877–1878), as well as unusually warm North Atlantic ocean (1878) conditions (Fig. 9e and f; Singh et al., 2018b). In addition, the severe societal consequences for these deficits were likely exacerbated by the mismanagement of colonial British Administration with policies such as the continued export of grains from South Asia to England and the promotion of cash crops over food crops (Davis, 2001).

While the South Asian summer monsoon is an important driver of the climate of South Asia, its influence on regional climate and tree growth varies spatially. For example, the hydroclimate of the western Himalayas and the Karakorum region of northwestern South Asia is primarily influenced by wintertime precipitation in the form of western disturbances (Rao, 2020; Thapa et al., 2020, 2022). Therefore, tree-ring studies in northwestern South Asia have predominantly focused on wintertime hydroclimate variability (Ahmed et al., 2013; Rao et al., 2018; Thapa et al., 2020). In many parts of western and central Nepal, tree growth is more sensitive to premonsoon season precipitation than monsoon season precipitation (Dawadi et al., 2013; Panthi et al., 2017). Tree-ring work in western Nepal has highlighted the extreme severity of the ongoing premonsoon and monsoon season drought in the region over the past three to four centuries (Bhandari et al., 2019; Panthi et al., 2017; Sano et al., 2012). Further, some regions of peninsular South India and Sri Lanka receive a large fraction of their annual precipitation from the South Asian winter monsoon, and tree isotopic content and wood anatomy has been shown to capture these influences (Bhattacharyya et al., 2007; Managave et al., 2011).

The expected changes in the South Asian summer monsoon (e.g., Turner & Annamalai, 2012) enhance flood risk in large river basins such as the Ganges, Brahmaputra, and Indus (Ali et al., 2019; Uhe et al., 2019). The annual growth of trees is often controlled by the same environmental drivers as streamflow in rivers (Meko et al., 1995; Shekhar et al., 2022). These drivers commonly include monsoon precipitation (D'Arrigo et al., 2011), summer drought (Cook & Jacoby, 1983), snowpack, and summer temperatures (Rao et al., 2018; Woodhouse & Lukas, 2006). Multiple tree-ring reconstructions of past streamflow have been developed in South Asia to better understand long-term decadal to centennial streamflow variability (Chen et al., 2019; Nguyen et al., 2020; Shah et al., 2014, 2019; Singh et al., 2022; Speer et al., 2019). These reconstructions allow for comparisons between recent observed hydroclimate trends in relation to preinstrumental period streamflow and those projected under continued climate change due to an intensified monsoon (Rao et al., 2020). For example, tree-ring reconstructions of streamflow suggest that flood risk estimates developed using reconstructed streamflow records are higher than those developed using the observational record in the lower Brahmaputra River in Bangladesh and northeastern India, likely due to decreased precipitation during the observational period from aerosol forcing (Rao et al., 2020), pointing to an underestimated likelihood of flood hazard in coming decades.

5 Changes in the South Asian monsoon in a warming world

While some hydroclimatically sensitive tree-ring records and speleothems across the region suggest a long-term weakening of the South Asian monsoon over the past 200 years (e.g., Sinha et al., 2015; Xu et al., 2013, 2018), quantifying centennial monsoon precipitation changes in observations at a regional scale is often hampered by limitations in the observational network and multidecadal variability (Doblas-Reyes et al., 2021; Douville et al., 2021; Knutson & Zeng, 2018; Singh et al., 2019b; Sontakke et al., 2008; Turner & Annamalai, 2012; Ummenhofer et al., 2021). The length of the Indian rain-gauge networks is an exception that enables various observational products (cf. Khouider et al. (2020) for an intercomparison) and that reveal pronounced decadal variability back to the 19th century (Sontakke et al., 2008). While these products are useful for mesoscale analysis, inhomogeneities of gauge distribution and reporting in India still present challenges in assessing long-term precipitation trends (Rajeevan & Bhate, 2009; Singh et al., 2019b). On the other hand, to quantify long-term rainfall changes, only gauges that have been carefully quality-controlled can be used, which reduces the time period covered and number of available stations to only ~2000 stations post-1950 (Doblas-Reyes et al., 2021; Rajeevan et al., 2006), though a newer gridded observational product extends back to 1901 (Pai et al., 2015).

As summarized in the IPCC Sixth Assessment Report (Douville et al., 2021), there is high confidence that the South Asian summer monsoon has weakened in the second half of the 20th century (Figs. 10 and 11). However, over India and Southeast Asia, the observed rainfall trends are not consistent across datasets and seasons, nor spatially coherent (Doblas-Reyes et al., 2021; Douville et al., 2021). The observed decrease in the South Asian monsoon precipitation counters the expected strengthening of the monsoon due to thermodynamic constraints on the hydrological cycle in a warming world (Goswami et al., 2006b). This has primarily been linked to anthropogenic aerosol forcing, with the predominant Northern Hemisphere emissions affecting the inter-hemispheric energy transports, with additional contributions from volcanic aerosols from El Chichón and Mount Pinatubo eruptions (e.g., Bollasina et al., 2011; Krishnan et al., 2016; Patil et al., 2019; Polson et al., 2014; Seth et al., 2019; Undorf et al., 2018). However, the large spread in the

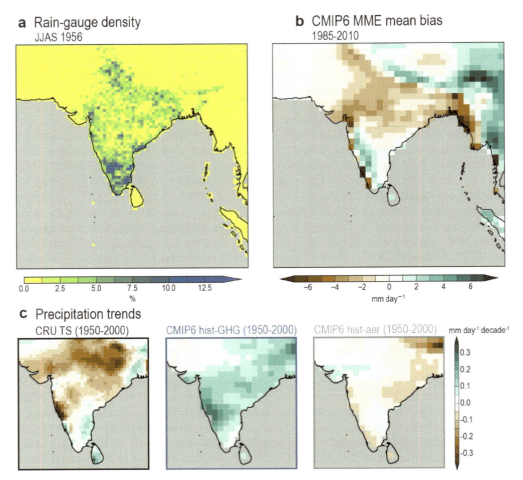

FIG. 10 Changes in the Indian summer monsoon (June–September) in the historical period. (a) Observational uncertainty demonstrated by a snapshot of rain-gauge density (% of 0.05 degree subgrid boxes containing at least one gauge) in the APHRO-MA 0.5 degree daily precipitation dataset for 1956. (b) Multi-model ensemble (MME) mean bias of CMIP6 models for precipitation (mm/day) compared to CRU TS observations for the 1985–2010 period. (c) Maps of rainfall trends (mm/day per decade) for the 1950–2000 period for CRU TS observations (1950–2000), and the CMIP6 model runs using greenhouse gas forcing and aerosol forcing for the historic period. *(Adapted from Doblas-Reyes et al. (2021), Fig. 10.19.)*

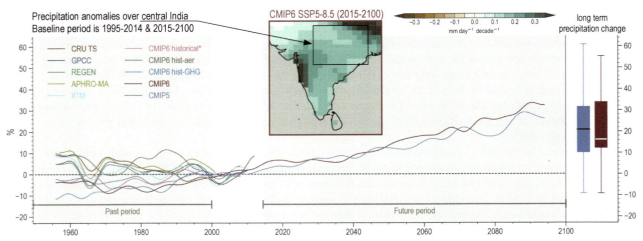

FIG. 11 Changes in the Indian summer monsoon (June–September) in the historical and future periods. Low-pass filtered time series of precipitation anomalies (%, relative to 1995–2014 baseline) averaged over central India (cf. inset). Time series are shown for different observational estimates in comparison with the CMIP6 mean for the all-forcings historical (*pink*) the aerosol-only (hist-aer, *gray*) and greenhouse gas-only (hist-GHG, *blue*). *Dark red* and *blue* lines show low-pass filtered multi-model ensemble-mean change in the CMIP6 historical/SSP5-8.5 and CMIP5 historical/RCP8.5 experiments for future projections to 2100. To the right, box-and-whisker plots show the 2081–2100 change averaged over the CMIP5 (*blue*) and CMIP6 (*dark red*) ensembles. *(Adapted from Doblas-Reyes et al. (2021), Fig. 10.19.)*

magnitude of the precipitation response to anthropogenic forcing in CMIP5 models suggests considerable internal variability of the regional monsoon (Douville et al., 2021; Saha et al., 2014; Salzmann et al., 2014; Sinha et al., 2015; Wang et al., 2021).

Over the second half of the 20th century, observations and climate models show a significant decrease in Indian summer monsoon precipitation and soil moisture, while at the same time, drought severity and the frequency and duration of monsoon breaks (or "dry spells") have increased (Fig. 10; e.g., Bollasina et al., 2011; Doblas-Reyes et al., 2021; Guhathakurta et al., 2017; Krishnan et al., 2016; Mishra et al., 2012; Niranjan Kumar et al., 2013; Ramarao et al., 2019; Singh et al., 2014, 2019b; Wang et al., 2021). These hydroclimatic changes have coincided with a concurrent weakening of the large-scale monsoon circulation, and reduced frequency of monsoon depressions forming over the Bay of Bengal (Douville et al., 2021; Prajeesh et al., 2013; Roxy et al., 2015; Vishnu et al., 2016). Observed trends in the Indian summer monsoon precipitation have also been linked to atmospheric circulation changes associated with low-level divergence and the Somali jet (Turner & Hannachi, 2010), and aerosols inducing changes in radiative forcing (Lin et al., 2018; Seth et al., 2019; Singh et al., 2019a) and in the local Hadley circulation (Ji et al., 2011). The rapid warming of the equatorial Indian Ocean with effects on the meridional temperature gradient, as well as decadal Pacific variability associated with IPO phasing are seen as likely contributing factors as well (Huang et al., 2020; Krishnan et al., 2016; Roxy et al., 2015).

In addition, a growing body of evidence also highlights the role of agricultural intensification, irrigation, regional land use and management changes, as well as urbanization in driving changes in mean and extreme South Asian monsoon precipitation (e.g., Devanand et al., 2019; McDermid et al., 2023; Niyogi et al., 2010; Shastri et al., 2015; Singh et al., 2016, 2019b, 2018a). South Asia is among the most heavily irrigated regions of the world, with the Indo-Gangetic basin among the most heavily cultivated hotspots, with extensive irrigation capable of modifying the surface energy balance and conversely regional precipitation responses (e.g., Devanand et al., 2019; McDermid et al., 2023). Over South Asia, expanding irrigation appears to have contributed to reduced monsoon precipitation due to a decrease in total column moisture over India, a weakening of the regional circulation and hence reduced low-level moisture convergence, which—together with lower moist static energy—has a drying effect on precipitation, despite the increase in moisture availability (Singh et al., 2018a). However, spatiotemporal variations exist, with excess irrigation over northern India in regional process-based model experiments leading to a shift in rainfall to northwestern India during the late monsoon season, while it was shown to enhance extreme rainfall over central India, consistent with observed precipitation trends (Devanand et al., 2019).

There is some indication that the intensity of daily rainfall and extremes have increased over the Indian subcontinent in some regions, especially in urbanized settings (Shastri et al., 2015; Singh et al., 2016) in the late 20th century (Deshpande et al., 2016; Pai et al., 2015; Roxy et al., 2017; Seneviratne et al., 2021; Singh et al., 2019a), but inhomogeneity in rain gauge stations may have introduced artificial jumps in extreme rainfall (Lin & Huybers, 2019), making such assessments of extreme rainfall changes in observations challenging (Doblas-Reyes et al., 2021). In contrast, according to climate model simulations, anthropogenic aerosols dominate late 20th century changes in wet and dry event frequencies over India compared to other climate forcings (Singh et al., 2019a): in particular, local aerosols suppress (enhance) the frequency of wet (dry) events over eastern-central India, where the largest increases in aerosol loading have occurred; in contrast, remote aerosols result in fewer rainless days over western India (Singh et al., 2019a).

While broad-scale features of the South Asian monsoon are simulated in general circulation models, several large biases have persisted for generations of climate models (IPCC, 2021; Singh et al., 2019b; Wang et al., 2021), including a large dry bias over India (Fig. 10b) that has been associated with too weak lower-tropospheric circulation (Sperber et al., 2013), as well as with cold SST biases in the Arabian Sea limiting the available moisture (Levine et al., 2013; Sandeep & Ajayamohan, 2015). It was also suggested that biases in moist static energy and upper-tropospheric temperature associated with the South Asian monsoon occurred due to overly smoothed model topography (Boos & Hurley, 2013; Eyring et al., 2021). In CMIP6, improvements have been made regarding robustness across models and reduced uncertainties in spatial patterns, especially near orography, though features of the boreal summer intraseasonal oscillation that controls most of subseasonal South Asian monsoon rainfall variations are still poorly simulated (Gusain et al., 2020; IPCC, 2021; Katzenberger et al., 2021; Sabeerali et al., 2013). Furthermore, assessments of whether aerosols enhance or suppress extreme heavy rainfall at subseasonal timescales are challenging and results are often spatially heterogeneous, as they depend on complex physical processes of aerosol-cloud interactions, i.e., indirect aerosol effects, to be included in climate models, not just direct aerosol or cloud-albedo effects (Lin et al., 2018).

Some recent studies have suggested an apparent recovery in the Indian summer monsoon since the 2000s (Hari et al., 2020; Jin & Wang, 2017), but the analysis period is short especially when considering multidecadal variations in the monsoon system (e.g., Doblas-Reyes et al., 2021; Krishnamurthy & Krishnamurthy, 2014; Sontakke et al., 2008). As such, according to the modeling evidence summarized in the IPCC Sixth Assessment Report (Doblas-Reyes et al., 2021), precipitation variability associated with the South Asian summer monsoon in the near-term is likely to be dominated by the effects of internal variability. In contrast, long-term projections see a weakened upper-tropospheric meridional temperature gradient induced by upper-level heating over the tropical Indo-Pacific (Sooraj et al., 2015) and increases in both the mean and extreme South Asian summer monsoon precipitation (Fig. 11), with a stronger increase in monsoon rainfall in CMIP6 over CMIP5 models, dominated by thermodynamic mechanisms leading to enhancement of the available moisture (Doblas-Reyes et al., 2021; Katzenberger et al., 2021; Seneviratne et al., 2021; and references therein). In addition, a projected northward movement of the lower-tropospheric monsoon winds and related genesis of monsoon depressions has been suggested (Sandeep et al., 2018).

6 Seasonal forecasting

The beginning of the South Asian monsoon precipitation season over land is traditionally defined by the "monsoon onset": a dramatic, easily observed phenomenon characterized by a sharp increase in rainfall that is readily distinguishable from premonsoonal rainfall (Ananthakrishnan & Soman, 1988; Bombardi et al., 2020). The climatological onset of the South Asian monsoon typically occurs around May 20 over the Nicobar and Andaman Islands to then extend across the Bay of Bengal, while simultaneously advancing into the Indian subcontinent from the south (IPCC, 2021; Pai et al., 2020). Commencing typically around June 1 in Kerala, it progresses south to north and east to west to cover the entire subcontinent by July 15; conversely, monsoon retreat commences in northwest India around September 1 and withdraws from the subcontinent by mid-October (IPCC, 2021; Pai et al., 2020). However, large interannual and decadal variations exist around both monsoon onset and retreat dates (Ghanekar et al., 2019), highlighting the need for skillfully predicting the monsoon. Yet, successful predictions of the South Asian monsoon, its onset, duration, intensity, as well as active and break cycles and regional variations still represent considerable challenges for seasonal forecasts (e.g., Bombardi et al., 2020; Webster et al., 1998).

The skill of seasonal Indian summer monsoon rainfall predictions remains low and does not approach the potential predictability limit (e.g., Kang & Shukla, 2006; Kumar et al., 2005b) for seasonal monsoon forecasts at a 1-month lead time (Rao et al., 2019). Hence, despite enormous progress in monsoon predictions since the early 1900s, operational monsoon forecasts in recent decades have had limited skill (Chevuturi et al., 2021; Rao et al., 2019; Zhang et al., 2022).

While the traditional view of the ENSO-South Asian monsoon relationship worked well for much of the 20th century, forecasts widely failed to predict the wet monsoon season in 1997 during one of the strongest El Niño events on record, and they failed to predict the failure of the monsoon in 2002 during a weak El Niño (Taschetto et al., 2020). While epoch-specific skillful monsoon precursors have been recognized (e.g., Kumar et al., 1999), it has been suggested that failures in monsoon forecasts in the 1990s and 2000s are due to the models' inability to capture new predictability sources emerging in a warming climate (Wang et al., 2015). In particular, central Pacific ENSO events, strengthening subtropical high-pressure systems in the Pacific during boreal spring, and a deepening Asian Low are not well captured in the models and could represent sources for low overall skill (Wang et al., 2015). The ENSO teleconnection represents one of the main prospects for seasonal prediction; limited skill of climate and seasonal forecasting models in representing and forecasting South Asian monsoon precipitation is often attributed to errors in the ENSO-monsoon teleconnection (IPCC, 2021; Jain et al., 2019; Sperber et al., 2013; Takaya et al., 2023; Webster et al., 1998), though biases in Indian Ocean SST and teleconnections in response to the IOD are also limiting factors (Johnson et al., 2017).

Several recent reviews (e.g., Bombardi et al., 2020; Madolli et al., 2022; Mohanty et al., 2019; Takaya et al., 2021, 2023) summarize the current status and recent developments of South Asian monsoon predictions. Limitations in statistical models led the Indian Meteorological Department to focus increasingly on dynamical forecasting for their operational seasonal forecasting as part of the Monsoon Mission since 2012, resulting in improved prediction skill for the Indian summer monsoon (Madolli et al., 2022). Evaluating the seasonal prediction skill of the Asian summer monsoon for two generations of models in hindcast data, it was found that the more recent system had superior prediction skill likely due to a better representation of the seasonal distribution of Asian monsoon precipitation and of the teleconnections to key Pacific and Indian Ocean climate drivers in the models (Takaya et al., 2023). Recent advances in long-range predictions with skillful predictions up to a year ahead (e.g., Takaya et al., 2021) also provide promising avenues for South Asian monsoon forecasting. For further details on seasonal forecasting of Indian Ocean conditions, including intraseasonal variability and Indian Ocean climate modes, such as the IOD, the reader is referred to other chapters by DeMott et al. (2024), Shinoda et al. (2024), Yamagata et al. (2024). Overall, the prospects for skillful forecasts of South Asian large-scale circulation features are better than for actual monsoon precipitation (Chevuturi et al., 2021; Johnson et al., 2017; Zhang et al., 2022).

7 Conclusions

The South Asian monsoon is a key component of the global climate system and manifests as seasonal changes in the atmospheric and oceanic circulation over the Indian Ocean basin, with widespread implications for regional water resources in surrounding countries. Traditionally characterized to first order as continent-scale "sea breezes," where in summer the land heats faster than the ocean, causing warm air to rise over the continent and moist air to be drawn in from the ocean, the South Asian monsoon exhibits seasonally reversing wind and precipitation patterns. Yet, more recent theoretical advances have led to the emergence of the monsoon as a regional manifestation of the seasonal variation of the global tropical atmospheric overturning and migration of the associated convergence zones (Geen et al., 2020). The dynamics governing the South Asian monsoon climatology can broadly be understood by considering three factors: (1) warming of the summer hemisphere relative to the winter hemisphere, (2) warming of land relative to ocean, and (3) Asian orography. Variations of the South Asian monsoon across a range of timescales can hence be related to changes in the forcing mechanisms of the mean monsoon circulation. Monsoon variations linked with internal atmospheric variability on synoptic to (sub)seasonal timescales, such as rainfall variations in active and break phases of the monsoon, are not predictable at seasonal timescales. In contrast, the predictable component of monsoon variability is linked to more slowly varying boundary conditions in the climate system (Charney & Shukla, 1981; Kucharski & Abid, 2017), such as ocean temperatures, soil moisture, and Eurasian snow cover. As such, interannual to multidecadal variability in the Atlantic (e.g., associated with the NAO, tropical SST, and Atlantic multidecadal variability), in the Pacific (e.g., associated with ENSO and Pacific decadal variability), and Indian Ocean (e.g., related to MJO and IOD), as well as high-latitude influences affecting Eurasian snow cover, all play a substantial role in modulating South Asian monsoon variability across a range of timescales. However, instrumental records of hydroclimate variability in South Asia are relatively short, posing challenges to understanding South Asian summer monsoon variability at decadal to centennial timescales.

As the largest of the Earth's monsoon systems, the South Asian monsoon is a lifeline for the nearly 2 billion people in the region, who receive 80%–85% of their annual precipitation during the summer monsoon (June–September). Even small perturbations in the timing of onset, seasonal distribution, and intensity of monsoon rainfall can have disproportionate effects on agriculture, most of which is rain-fed. Historically, a weakened monsoon and ensuing widespread drought, as recorded by hydroclimatically sensitive paleo proxies (e.g., stalagmite, lacustrine, and tree-ring records) that provide long-term context for such monsoon variability, often resulted in severe societal impacts throughout the late Holocene. Oxygen isotope ratios in stalagmites are widely used as paleomonsoon proxies that have in turn been tied to societal shifts: for example, a sustained period of enhanced monsoon rainfall during the late mid-Holocene Climate Optimum (~4.5–4.0 ka) linked with expansion of agriculture that supported urbanization of the Harappan civilization; a severe drought from ~1593–1623 CE associated with the collapse of the Guge Kingdom in Tibet (Kathayat et al., 2017); tree-ring based drought reconstructions of climate instability that led to the demise of Cambodia's expansive Khmer Kingdom by the early 15th century (Buckley et al., 2014), as well as the Strange Parallels Drought (1756–1768 CE), East India Drought (1790–1796 CE), and the Great Drought (1876–1877 CE) reflected in a regional drought atlas (Cook et al., 2010). Despite the valuable contributions of dendrochronology for a better understanding of South Asian climate and summer monsoon variability, some challenges remain. Key among these challenges is the bias in the current tree-ring network toward conifer sites in the mountainous Himalaya, Karakorum, Tibetan Plateau regions (Borgaonkar et al., 2010; Dawadi et al., 2013; Shah et al., 2007; Thapa et al., 2017; Zhao et al., 2019), and the rapid rates of deforestation of old-growth forests across much of South Asia (Sudhakar Reddy et al., 2018). A wider and more geographically distributed network of tree-ring studies across the subcontinent that expands beyond the traditionally used conifer species and ring-width measurements is needed to better resolve spatial variability and nuances of the South Asian summer monsoon variability across the subcontinent.

Proxy evidence across the region suggests a weakening of the South Asian monsoon over the past 200 years, with observations and climate models also indicating a decreased Indian summer monsoon in the second half of the 20th century primarily due to anthropogenic aerosol forcing, while in the long term, the South Asian monsoon is projected to increase over the 21st century and exhibit enhanced interannual variability (Doblas-Reyes et al., 2021; Douville et al., 2021). While broad-scale features of the South Asian monsoon are simulated in general circulation models, several large biases have persisted for generations of models, affecting not just climate model projections for the 21st century, but also seasonal forecasting skill. Despite enormous progress in monsoon predictions since the early 1900s, operational monsoon forecasts in recent decades have had limited skill. As such, successful predictions of the South Asian monsoon, its onset, duration, intensity, as well as active and break cycles and regional variations still represent considerable challenges for weather and seasonal forecasts.

8 Educational resources

- Educational video of the Asian monsoon. Available at: https://www.youtube.com/watch?v=RqkOhKEtC20.
- Books, such as Wang (2006), Clift and Plumb (2008).
- Review papers, such as Webster et al. (1998), Kucharski and Abid (2017), Biasutti et al. (2018), Hill (2019), Geen et al. (2020), Cherchi et al. (2021) and web resources, such as Gottschalck (2014).
- Drought atlas portal: http://drought.memphis.edu/.
- Paleo Hydrodynamics Data Assimilation (PHYDA) product: https://zenodo.org/record/1198817#.YpUJZGDMIUN.
- Last Millennium Reanalysis (LMR) project global climate reconstructions, version 2: https://www.ncei.noaa.gov/access/paleo-search/study/27850.

Acknowledgments

Use of the following datasets is gratefully acknowledged: Global Precipitation Climatology Center dataset by the German Weather Service (DWD), as well as NCEP/NCAR reanalysis, CMAP, and NOAA Optimum Interpolation SST version 2 provided by NOAA/OAR/ESRL PSD, Boulder, Colorado, through their website http://www.cdc.noaa.gov/psd, and supported by the U.S. Department of Energy, Office of Science Innovative and Novel Computational Impact on Theory and Experiment program, and Office of Biological and Environmental Research, and by the NOAA Climate Program Office. Constructive comments by two anonymous reviewers helped improve an earlier version of the chapter and are gratefully acknowledged. We acknowledge support from the U.S. National Science Foundation under AGS-2002083 (to CCU), AGS-2102844 (to CCU), and AGS-2102864 (to RFD). CCU also acknowledges support from the *James E. and Barbara V. Moltz Fellowship for Climate-Related Research*, and MPR from European Union Horizon 2020 Marie Skłodowska-Curie Grant #101031748 and NOAA Climate and Global Change Postdoctoral Fellowship (UCAR-CPAESS) award #NA18NWS4620043B. Thanks to Natalie Renier (WHOI Graphics) for help with the design of Figs. 1, 7, 8, 10, and 11.

Author contributions

CCU conceived and led the chapter overall, with contributions, in particular, in Sections 1, 3, 5, and 6, and accompanying figures (Figs. 1, 6–11). RG wrote Section 2 and provided accompanying Figs. 2–5, RFD wrote Section 4.1, and MPR wrote Section 4.2. All authors contributed to Sections 7 and 8, discussion of content, and overall chapter structure and provided feedback on the entire chapter.

References

Adam, O., Bischoff, T., & Schneider, T. (2016). Seasonal and interannual variations of the energy flux equator and ITCZ. Part I: Zonally averaged ITCZ position. *Journal of Climate, 29*, 3219–3230.

Adhikari, N., Gao, J., Yao, T., Yang, Y., & Dai, D. (2020). The main controls of the precipitation stable isotopes at Kathmandu, Nepal. *Tellus B: Chemical and Physical Meteorology, 72*, 1–17.

Ahmed, M., Zafar, M. U., Hussain, A., Akbar, M., Wahab, M., & Khan, N. (2013). Dendroclimatic and dendrohydrological response of two tree species from Gilgit valleys. *Pakistan Journal of Botany, 45*, 987–992.

Ali, H., Modi, P., & Mishra, V. (2019). Increased flood risk in Indian sub-continent under the warming climate. *Weather and Climate Extremes, 25*, 100212.

Ananthakrishnan, R., & Soman, M. K. (1988). The onset of the southwest monsoon over Kerala: 1901-1980. *Journal of Climatology, 8*, 283–296.

Anil, N., Ramesh Kumar, M. R., Sajeev, R., & Saji, P. K. (2016). Role of distinct flavors of IOD events on Indian summer monsoon. *Natural Hazards, 82*, 1317–1326.

Annamalai, H., Liu, P., & Xie, S.-P. (2005). Southwest Indian Ocean SST variability: Its local effect and remote influence on Asian monsoons. *Journal of Climate, 18*, 4150–4167.

Ashok, K., Guan, Z., Saji, N. H., & Yamagata, T. (2004). Individual and combined influences of ENSO and the Indian Ocean dipole on the Indian summer monsoon. *Journal of Climate, 17*, 3141–3155.

Ashok, K., Guan, Z., & Yamagata, T. (2001). Impact of the Indian Ocean dipole on the relationship between the Indian monsoon rainfall and ENSO. *Geophysical Research Letters, 26*, 4499–4502.

Ashok, K., & Saji, N. H. (2007). On the impacts of ENSO and Indian Ocean dipole events on sub-regional Indian summer monsoon rainfall. *Natural Hazards, 42*, 273–285.

Barnett, T. P., Dümenil, L., Schlese, U., Roeckner, E., & Latif, M. (1989). The effect of Eurasian snow cover on regional and global climate variations. *Journal of Atmospheric Sciences, 46*, 661–686.

Behera, S. K., Krishnan, R., & Yamagata, T. (1999). Unusual Ocean–atmosphere conditions in the tropical Indian Ocean during 1994. *Geophysical Research Letters, 26*, 3001–3004.

Behera, S. K., & Ratnam, J. V. (2018). Quasi-asymmetric response of the Indian summer monsoon rainfall to opposite phases of the IOD. *Scientific Reports*. https://doi.org/10.1038/s41598-017-18396-6.

Behera, S. K., & Yamagata, T. (2003). Influence of the Indian Ocean dipole on the southern oscillation. *Journal of the Meteorological Society of Japan, Series II, 81*, 169–177.

Berkelhammer, M., Sinha, A., Stott, L., Cheng, H., Pausata, F. S., & Yoshimura, K. (2012). An abrupt shift in the Indian monsoon 4000 years ago. *Climates, Landscapes, and Civilizations, 198*, 75–88.

Bhandari, S., Gaire, N. P., Shah, S. K., Speer, J. H., Bhuju, D. R., & Thapa, U. K. (2019). A 307-year tree-ring SPEI reconstruction indicates modern drought in Western Nepal Himalayas. *Tree-Ring Research, 75*, 73–85.

Bhattacharyya, A., Eckstein, D., Shah, S. K., & Chaudhary, V. (2007). Analyses of climatic changes around Perambikulum, South India, based on early wood mean vessel area of teak. *Current Science, 93*, 1159–1164.

Bhattacharyya, A., & Shah, S. K. (2009). Tree-ring studies in India: Past appraisal, present status, and future prospects. *IAWA Journal, 30*, 361–370.

Bhattacharyya, A., Yadav, R. R., Borgaonkar, H. P., & Pant, G. B. (1992). Growth-ring analysis of Indian tropical trees: Dendroclimatic potential. *Current Science, 62*, 736–741.

Biasutti, M., Voigt, A., Boos, W. R., Braconnot, P., Hargreaves, J. C., Harrison, S. P., Kang, S. M., Mapes, B. E., Scheff, J., Schumacher, C., Sobel, A. H., & Xie, S. P. (2018). Global energetics and local physics as drivers of past, present and future monsoons. *Nature Geoscience, 11*, 392–400.

Bollasina, M. A., Ming, Y., & Ramaswamy, V. (2011). Anthropogenic aerosols and the weakening of the South Asian summer monsoon. *Science, 334*, 502–505.

Bombardi, R. J., Moron, V., & Goodnight, J. S. (2020). Detection, variability, and predictability of monsoon onset and withdrawal dates: A review. *International Journal of Climatology, 40*, 641–667.

Boos, W. R., & Hurley, J. V. (2013). Thermodynamic bias in the multimodel mean boreal summer monsoon. *Journal of Climate, 26*, 2279–2287.

Boos, W. R., & Kuang, Z. (2010). Dominant control of the South Asian monsoon by orographic insulation versus plateau heating. *Nature, 463*, 218–222.

Borah, P. J., Venugopal, V., Sukhatme, J., Muddebihal, P., & Goswami, B. N. (2020). Indian monsoon derailed by a North Atlantic wavetrain. *Science, 370*, 1335–1338.

Bordoni, S., & Schneider, T. (2008). Monsoons as eddy-mediated regime transitions of the tropical overturning circulation. *Nature Geoscience, 1*, 515–519.

Borgaonkar, H. P., Sikder, A. B., Ram, S., & Pant, G. B. (2010). El Niño and related monsoon drought signals in 523-year-long ring width records of teak (Tectona grandis L.F.) trees from South India. *Palaeogeography, Palaeoclimatology, Palaeoecology, 285*, 74–84.

Breitenbach, S. F. M., Adkins, J. F., Meyer, H., Marwan, N., Kumar, K. K., & Haug, G. H. (2010). Strong influence of water vapor source dynamics on stable isotopes in precipitation observed in southern Meghalaya, NE India. *Earth and Planetary Science Letters, 292*, 212–220.

Brienen, R. J. W., Schöngart, J., & Zuidema, P. A. (2016). Tree rings in the tropics: Insights into the ecology and climate sensitivity of tropical trees. In G. Goldstein, & L. S. Santiago (Eds.), *Tropical tree physiology: Adaptations and responses in a changing environment* (pp. 439–461). Cham: Springer International Publishing.

Buckley, B. M., Fletcher, R., Wang, S.-Y. S., Zottoli, B., & Pottier, C. (2014). Monsoon extremes and society over the past millennium on mainland Southeast Asia. *Quaternary Science Reviews, 95*, 1–19.

Buckley, B. M., Ummenhofer, C. C., D'Arrigo, R. D., Hansen, K. G., Truong, L. H., Le, C. N., & Stahle, D. K. (2019). Interdecadal Pacific oscillation reconstructed from trans-Pacific tree rings: 1350-2004 CE. *Climate Dynamics, 53*, 3181–3196.

Buckley, B. M., et al. (2010). Climate as a contributing factor in the demise of Angkor, Cambodia. *Proceedings of the National Academy of Sciences of the United States of America, 107*, 6748–6752.

Buermann, W., Lintner, B., & Bonfils, C. (2005). A wintertime Arctic oscillation signature on early-season Indian Ocean monsoon intensity. *Journal of Climate, 18*, 2247–2269.

Cai, W., et al. (2019). Pantropical climate interactions. *Science, 363*, eaav4236.

Chambers, D. P., Tapley, B. D., & Stewart, R. H. (1999). Anomalous warming in the Indian Ocean coincident with El Niño. *Journal of Geophysical Research, 104*, 3035–3047.

Chang, C.-P., Harr, P., & Ju, J. (2001). Possible roles of Atlantic circulations on the weakening Indian monsoon rainfall–ENSO relationship. *Journal of Climate, 14*, 2376–2380.

Chang, C.-P., & Li, T. (2000). A theory for the tropical tropospheric biennial oscillation. *Journal of Atmospheric Sciences, 57*, 2209–2224.

Charney, J. G., & Shukla, J. (1981). Predictability of monsoons. In *Monsoon dynamics* (pp. 99–109). Cambridge, UK: Cambridge University Press.

Chen, F., Shang, H., Panyushkina, I., Meko, D., Li, J., Yuan, Y., Yu, S., Chen, F., He, D., & Luo, X. (2019). 500-year tree-ring reconstruction of Salween River streamflow related to the history of water supply in Southeast Asia. *Climate Dynamics, 53*, 6595–6607.

Cherchi, A., Gualdi, S., Behera, S., Luo, J. J., Masson, S., Yamagata, T., et al. (2007). The influence of tropical Indian Ocean SST on the Indian summer monsoon. *Journal of Climate, 20*, 3083–3105.

Cherchi, A., & Navarra, A. (2013). Influence of ENSO and of the Indian Ocean dipole on the Indian summer monsoon variability. *Climate Dynamics, 41*, 81–103.

Cherchi, A., Terray, P., Ratna, S. B., Sankar, S., Sooraj, K. P., & Behera, S. K. (2021). Chapter 8: Indian Ocean Dipole influence on Indian summer monsoon and ENSO: A review. In J. Chowdary, A. Parekh, & C. Gnanaseelan (Eds.), *Indian summer monsoon variability: El Nino teleconnections and beyond* (pp. 157–182). Elsevier.

Chevuturi, A., et al. (2021). Forecast skill of the Indian monsoon and its onset in the ECMWF seasonal forecasting system 5 (SEAS5). *Climate Dynamics, 56*, 2941–2957.

Chou, C., & Neelin, J. D. (2003). Mechanisms limiting the northward extent of the northern summer monsoons over North America, Asia, and Africa. *Journal of Climate, 16*, 406–425.

Clark, C. O., Cole, J. E., & Webster, P. J. (2000). Indian Ocean SST and Indian summer rainfall: Predictive relationships and their decadal variability. *Journal of Climate, 13*, 2503–2519.

Clift, P. D., & Plumb, R. A. (2008). *The Asian monsoon: Causes, history and effects. Vol. 270*. Cambridge: Cambridge University Press.

Cook, E. R., Anchukaitis, K. J., Buckley, B. M., D'Arrigo, R. D., Jacoby, G. C., & Wright, W. E. (2010). Asian monsoon failure and megadrought during the last millennium. *Science, 328*, 486–489.

Cook, E. R., & Jacoby, G. C. (1983). Potomac river streamflow since 1730 as reconstructed by tree rings. *Journal of Applied Meteorology and Climatology, 22*, 1659–1672.

Cook, B. I., et al. (2022). Megadroughts in the common era and the anthropocene. *Nature Reviews Earth and Environment, 3*, 741–757.

Cui, X., Gao, Y., Sun, J., Guo, D., Li, S., & Johannessen, O. M. (2014). Role of natural external forcing factors in modulating the Indian summer monsoon rainfall, the winter North Atlantic oscillation and their relationship on inter-decadal timescale. *Climate Dynamics, 43*, 2283–2295.

D'Arrigo, R., Palmer, J., Ummenhofer, C. C., Kyaw, N. N., & Krusic, P. (2011). Three centuries of Myanmar monsoon climate variability inferred from teak tree rings. *Geophysical Research Letters, 38*. https://doi.org/10.1029/2011GL049927.

D'Arrigo, R. D., & Ummenhofer, C. C. (2015). The climate of Myanmar: Evidence for effects of the Pacific decadal oscillation. *International Journal of Climatology, 35*, 634–640.

Dansgaard, W. (1964). Stable isotopes in precipitation. *Tellus, 16*, 436–468.

Davis, M. (2001). *Late Victorian holocausts: El Niño famines and the making of the third world, London* (p. 2001). New York: Verso.

Dawadi, B., Liang, E., Tian, L., Devkota, L. P., & Yao, T. (2013). Pre-monsoon precipitation signal in tree rings of timberline Betula utilis in the Central Himalayas. *Quaternary International, 283*, 72–77.

DeMott, C. A., Ruppert, J. H., Jr., & Rydbeck, A. (2024). Chapter 4: Intraseasonal variability in the Indian Ocean region. In C. C. Ummenhofer, & R. R. Hood (Eds.), *The Indian Ocean and its role in the global climate system* (pp. 79–101). Amsterdam: Elsevier. https://doi.org/10.1016/B978-0-12-822698-8.00006-8.

Denniston, R. F., et al. (2016). Expansion and contraction of the Indo-Pacific tropical rain belt over the last three millennia. *Scientific Reports, 6*. https://doi.org/10.1038/srep34485.

Deshpande, N. R., Kothawale, D. R., & Kulkarni, A. (2016). Changes in climate extremes over major river basins of India. *International Journal of Climatology, 36*, 4548–4559.

Devanand, A., Huang, M., Ashfaq, M., Barik, B., & Ghosh, S. (2019). Choice of irrigation water management practice affects Indian summer monsoon rainfall and its extremes. *Geophysical Research Letters, 46*, 9126–9135.

Dhar, O. N., & Rakhecha, P. R. (1983). Foreshadowing Northeast summer rainfall over Tamil Nadu, India. *Monthly Weather Review, 111*, 109–112.

Diaz, M., & Boos, W. R. (2019a). Barotropic growth of monsoon depressions. *Quarterly Journal of the Royal Meteorological Society, 145*, 824–844. https://doi.org/10.1002/qj.3467.

Diaz, M., & Boos, W. R. (2019b). Monsoon depression amplification by moist barotropic instability in a vertically sheared environment. *Quarterly Journal of the Royal Meteorological Society, 145*, 2666–2684. https://doi.org/10.1002/qj.3585.

Dimri, A. P., Niyogi, D., Barros, A. P., Ridley, J., Mohanty, U. C., Yasunari, T., & Sikka, D. R. (2015). Western disturbances: A review. *Reviews of Geophysics, 53*, 225–246. https://doi.org/10.1002/2014RG000460.

Dixit, Y., Hodell, D. A., & Petrie, C. A. (2014). Abrupt weakening of the summer monsoon in Northwest India ~4100 yr ago. *Geology, 42*, 339–342.

Doblas-Reyes, F. J., Sörensson, A. A., Almazroui, M., Dosio, A., Gutowski, W. J., Haarsma, R., Hamdi, R., Hewitson, B., Kwon, W.-T., Lamptey, B. L., Maraun, D., Stephenson, T. S., Takayabu, I., Terray, L., Turner, A., & Zuo, Z. (2021). Linking global to regional climate change. In V. Masson-Delmotte, P. Zhai, A. Pirani, S. L. Connors, C. Péan, S. Berger, … B. Zhou (Eds.), *Climate change 2021: The physical science basis. Contribution of working group I to the sixth assessment report of the Intergovernmental Panel on Climate Change* (pp. 1363–1512). Cambridge, UK: Cambridge University Press.

Douville, H., Raghavan, K., Renwick, J., Allan, R. P., Arias, P. A., Barlow, M., Cerezo-Mota, R., Cherchi, A., Gan, T. Y., Gergis, J., Jiang, D., Khan, A., Mba, W. P., Rosenfeld, D., Tierney, J., & Zolina, O. (2021). Water cycle changes. In V. Masson-Delmotte, P. Zhai, A. Pirani, S. L. Connors, C. Péan, S. Berger, … B. Zhou (Eds.), *Climate change 2021: The physical science basis. contribution of working group I to the sixth assessment report of the Intergovernmental Panel on Climate Change* (pp. 1055–1210). Cambridge, UK: Cambridge University Press.

Duc, H. N., Bang, H. Q., & Quang, N. X. (2018). Influence of the Pacific and Indian Ocean climate drivers on the rainfall in Vietnam. *International Journal of Climatology, 38*, 5717–5732.

Dugam, S. S., Kakade, S. B., & Verma, R. K. (1997). Interannual and long-term variability in the North Atlantic oscillation and Indian summer monsoon rainfall. *Theoretical and Applied Climatology, 58*, 21–29.

Dwivedi, S., Goswami, B. N., & Kucharski, F. (2015). Unraveling the missing link of ENSO control over the Indian monsoon rainfall. *Geophysical Research Letters, 42*, 8201–8207.

Eyring, V., Gillett, N. P., Rao, K. M. A., Barimalala, R., Parrillo, M. B., Bellouin, N., Cassou, C., Durack, P. J., Kosaka, Y., McGregor, S., Min, S., Morgenstern, O., & Sun, Y. (2021). Human influence on the climate system. In V. Masson-Delmotte, P. Zhai, A. Pirani, S. L. Connors, C. Péan, S. Berger, … B. Zhou (Eds.), *Climate change 2021: The physical science basis. Contribution of working group I to the sixth assessment report of the Intergovernmental Panel on Climate Change* (pp. 423–552). Cambridge, UK: Cambridge University Press.

Feba, F., Ashok, K., & Ravichandran, M. (2019). Role of changed Indo-Pacific atmospheric circulation in the recent disconnect between the Indian summer monsoon and ENSO. *Climate Dynamics, 52*, 1461–1470.

Fleitmann, D., Burns, S. J., Mangini, A., Mudelsee, M., Kramers, J., Villa, I., Neff, U., Al-Subbary, A. A., Buettner, A., Hippler, D., & Matter, A. (2007). Holocene ITCZ and Indian monsoon dynamics recorded in stalagmites from Oman and Yemen (Socotra). *Quaternary Science Reviews, 26*, 170–188.

Fritts, H. (1976). *Tree rings and climate*. 567 pp San Diego, CA: Academic.

Gadgil, S. (2003). The Indian monsoon and its variability. *Annual Reviews Earth Planetary Science, 31*, 429–467. https://doi.org/10.1146/annurev.earth.31.100901.141251.

Gadgil, S. (2018). The monsoon system: Land-sea breeze or the ITCZ? *Journal of Earth System Science, 127*. https://doi.org/10.1007/s12040-017-0916-x.

Gadgil, S., Vinaychandran, P. N., Francis, P. A., & Gadgil, S. (2004). Extremes of the Indian summer monsoon rainfall, ENSO and equatorial Indian Ocean oscillation. *Geophysical Research Letters, 31*, L12213.

Gaire, N. P., Dhakal, Y. R., Shah, S. K., Fan, Z.-X., Bräuning, A., Thapa, U. K., Bhandari, S., Aryal, S., & Bhuju, D. R. (2019). Drought (scPDSI) reconstruction of trans-Himalayan region of central Himalaya using *Pinus wallichiana* tree-rings. *Palaeogeography, Palaeoclimatology, Palaeoecology, 514*, 251–264.

Geen, R., Bordoni, S., Battisti, D. S., & Hui, K. (2020). Monsoons, ITCZs, and the concept of the global monsoon. *Reviews of Geophysics, 58*. https://doi.org/10.1029/2020RG000700. e2020RG000700.

Gershunov, A., Schneider, N., & Barnett, T. (2001). Low-frequency modulation of the ENSO–Indian monsoon rainfall relationship: Signal or noise? *Journal of Climate, 14*, 2486–2492.

Ghanekar, S. P., Bansod, S. D., Narkhedkar, S. G., & Kulkarni, A. (2019). Variability of Indian summer monsoon onset over Kerala during 1971-2018. *Theoretical and Applied Climatology, 138*, 729–742.

Gill, A. (1980). Some simple solutions for heat-induced tropical circulations. *Quarterly Journal of the Royal Meteorological Society, 106*, 447–462.

Godbole, R. V. (1977). The composite structure of the monsoon depression. *Tellus, 29*, 25–40. https://doi.org/10.1111/j.2153-3490.1977.tb00706.x.

Goswami, B. N., Madhusoodanan, M. S., Neema, C. P., & Sengupta, D. (2006a). A physical mechanism for North Atlantic SST influence on the Indian summer monsoon. *Geophysical Research Letters, 33*, L02706. https://doi.org/10.1029/2005GL024803.

Goswami, B. N., et al. (2006b). Increasing trend of extreme rain events over India in a warming environment. *Science, 314*, 1442–1445.

Goswami, B. N., & Mohan, R. S. A. (2001). Intraseasonal oscillations and interannual variability of the Indian summer monsoon. *Journal of Climate, 14*, 1180–1198.

Goswami, B. N., & Xavier, P. K. (2005). ENSO control on the south Asian monsoon through the length of the rainy season. *Geophysical Research Letters, 32*, L18717. https://doi.org/10.1029/2005GL023216.

Gottschalck, J. (2014). *What is the MJO, and why do we care*. https://www.climate.gov/news-features/blogs/enso/what-mjo-and-why-do-we-care.

Grant, M. E. (1992). Dendrochronology in South Asia, an update. *Southeast Asian Studies, 8*(1), 115–123.

Guhathakurta, P., Menon, P., Inkane, P. M., Krishnan, U., & Sable, S. T. (2017). Trends and variability of meteorological drought over the districts of India using standardized precipitation index. *Journal of Earth System Science, 126*, 120. https://doi.org/10.1007/s12040-017-0896-x.

Gusain, A., Ghosh, S., & Karmakar, S. (2020). Added value of CMIP6 over CMIP5 models in simulating Indian summer monsoon rainfall. *Atmospheric Research, 232*, 104680. https://doi.org/10.1016/j.atmosres.2019.104680.

Halley, E. (1686). An historical account of the trade winds, and monsoons, observable in the seas between and near the Tropicks, with an attempt to assign the physical cause of the said winds. *Philosophical Transactions of the Royal Society, 16*, 153–168. https://doi.org/10.1098/rstl.1686.0026.

Hari, V., Villarini, G., Karmakar, S., Wilcox, L. J., & Collins, M. (2020). Northward propagation of the intertropical convergence zone and strengthening of Indian summer monsoon rainfall. *Geophysical Research Letters, 47*(23). https://doi.org/10.1029/2020gl089823. e2020GL089823.

Hartmann, D. L., & Michelsen, M. L. (1989). Intraseasonal periodicities in Indian rainfall. *Journal of the Atmospheric Sciences, 46*(18), 2838–2862.

Hernandez, M., Ummenhofer, C. C., & Anchukaitis, K. J. (2015). Multi-scale drought and ocean-atmosphere variability in monsoon Asia. *Environmental Research Letters*. https://doi.org/10.1088/1748-9326/10/7/074010.

Hildebrandsson, H. H. (1897). *Quelques recherches sur les centres d'action de l' atmosphere*. Junglica Svenska Vetenskaps-Akademiens Handlingar. XXIX.

Hill, S. A. (2019). Theories for past and future monsoon rainfall changes. *Current Climate Change Reports, 5*, 160–171. https://doi.org/10.1007/s40641-019-00137-8.

Hong, C.-C., Hsu, H.-H., Lin, N.-H., & Chiu, H. (2011). Roles of European blocking and tropical-extratropical interaction in the 2010 Pakistan flooding. *Geophysical Research Letters, 38*, L13806. https://doi.org/10.1029/2011GL047583.

Huang, X., et al. (2020). The recent decline and recovery of Indian summer monsoon rainfall: Relative roles of external forcing and internal variability. *Journal of Climate, 33*, 5035–5060.

Huffman, G. J., Adler, R. F., Morrissey, M. M., Bolvin, D. T., Curtis, S., Joyce, R., McGavock, B., & Susskind, J. (2001). Global precipitation at one-degree daily resolution from multisatellite observations. *Journal of Hydrometeorology, 2*, 36–50.

Hunt, K. M. R., & Menon, A. (2020). The 2018 Kerala floods: A climate change perspective. *Climate Dynamics, 54*, 2433–2446. https://doi.org/10.1007/s00382-020-05123-7.

Hunt, K. M. R., Turner, A. G., & Shaffrey, L. C. (2018). Extreme daily rainfall in Pakistan and North India: Scale interactions, mechanisms, and precursors. *Monthly Weather Review, 146*, 1005–1022. https://doi.org/10.1175/MWR-D-17-0258.1.

Hurley, J. V., & Boos, W. R. (2013). Interannual variability of monsoon precipitation and local subcloud equivalent potential temperature. *Journal of Climate, 26*, 9507–9527.

Hurley, J. V., & Boos, W. R. (2015). A global climatology of monsoon low-pressure systems. *Quarterly Journal of the Royal Meteorological Society, 141*, 1049–1064.

IPCC. (2021). Annex V: Monsoons (Cherchi, A., A. Turner (eds.)). In V. Masson-Delmotte, P. Zhai, A. Pirani, S. L. Connors, C. Péan, S. Berger, … B. Zhou (Eds.), *Climate change 2021: The physical science basis. Contribution of working group I to the sixth assessment report of the Intergovernmental Panel on Climate Change* (pp. 2193–2204). Cambridge, UK: Cambridge University Press. https://doi.org/10.1017/9781009157896.019.

Izumo, T., de Boyer Montegut, C., Luo, J.-J., Behera, S. K., Masson, S., & Yamagata, T. (2008). The role of the western Arabian Sea upwelling in Indian monsoon rainfall variability. *Journal of Climate, 21*, 5603–5623.

Jain, S., Scaife, A. A., & Mitra, A. K. (2019). Skill of Indian summer monsoon rainfall prediction in multiple seasonal prediction systems. *Climate Dynamics*, *52*, 5291–5301.

Ji, Z., Kang, S., Zhang, D., Zhu, C., Wu, J., & Xu, Y. (2011). Simulation of the anthropogenic aerosols over South Asia and their effects on Indian summer monsoon. *Climate Dynamics*, *36*, 1633–1647.

Jiang, X., Adames, Á. F., Kim, D., Maloney, E. D., Lin, H., Kim, H., et al. (2020). Fifty years of research on the Madden-Julian oscillation: Recent progress, challenges, and perspectives. *Journal of Geophysical Research: Atmospheres*, *125*. https://doi.org/10.1029/2019JD030911. e2019JD030911.

Jin, Q., & Wang, C. (2017). A revival of Indian summer monsoon rainfall since 2002. *Nature Climate Change*, *7*, 587–594.

Johnson, S. J., et al. (2017). An assessment of Indian monsoon seasonal forecasts and mechanisms underlying monsoon interannual variability in the Met Office GloSea5-GC2 system. *Climate Dynamics*, *48*, 1447–1465.

Joshi, M. K., & Kucharski, F. (2017). Impact of interdecadal Pacific oscillation on Indian summer monsoon rainfall: An assessment from CMIP5 climate models. *Climate Dynamics*, *48*, 2375–2391.

Joshi, M. K., & Rai, A. (2015). Combined interplay of the Atlantic multidecadal oscillation and the interdecadal Pacific oscillation on rainfall and its extremes over Indian subcontinent. *Climate Dynamics*, *44*, 3339–3359.

Kalnay, E., et al. (1996). The NCEP/NCAR 40-year reanalysis project. *Bulletin of the American Meteorological Society*, *77*, 437–470.

Kang, S. M. (2020). Extratropical influence on the tropical rainfall distribution. *Current Climate Change Reports*, *6*, 24–36.

Kang, S. M., Held, I. M., Frierson, D. M., & Zhao, M. (2008). The response of the ITCZ to extratropical thermal forcing: Idealized slab-ocean experiments with a GCM. *Journal of Climate*, *21*, 3521–3532. https://doi.org/10.1175/2007JCLI2146.1.

Kang, I.-S., & Shukla, J. (2006). Dynamic seasonal prediction and predictability of the monsoon. In B. Wang (Ed.), *The Asian monsoon* (pp. 585–612). Springer.

Kathayat, G., Sinha, A., Tanoue, M., Yoshimura, K., Li, H., Zhang, H., & Cheng, H. (2021). Interannual oxygen isotope variability in Indian summer monsoon precipitation reflects changes in moisture sources. *Communications Earth & Environment*, *2*, 96.

Kathayat, G., et al. (2016). Indian monsoon variability on millennial-orbital timescales. *Scientific Reports*, *6*. https://doi.org/10.1038/srep24374.

Kathayat, G., et al. (2017). The Indian monsoon variability and civilization changes in the Indian subcontinent. *Science Advances*, *3*, e1701296.

Katzenberger, A., Schewe, J., Pongratz, J., & Levermann, A. (2021). Robust increase of Indian monsoon rainfall and its variability under future warming in CMIP6 models. *Earth System Dynamics*, *12*, 367–386.

Khouider, B., et al. (2020). A novel method for interpolating daily station rainfall data using a stochastic lattice model. *Journal of Hydrometeorology*, *21*, 909–933.

Kikuchi, K. (2021). The boreal summer Intraseasonal oscillation (BSISO): A review. *Journal of the Meteorological Society of Japan Series II*, *99*, 933–972. https://doi.org/10.2151/jmsj.2021-045.

Knutson, T. R., & Zeng, F. (2018). Model assessment of observed precipitation trends over land regions: Detectable human influences and possible low bias in model trends. *Journal of Climate*, *31*, 4617–4637.

Kobayashi, S., Ota, Y., Harada, Y., Ebita, A., Moriya, M., Onoda, H., Onogi, K., Kamahori, H., Kobayashi, C., Endo, H., Miyaoka, K., & Takahashi, K. (2015). The JRA-55 reanalysis: General specifications and basic characteristics. *Journal of the Meteorological Society of Japan*, *93*, 5–48.

Kothawale, D., & Kumar, K. R. (2002). Tropospheric temperature variation over India and links with the Indian summer monsoon: 1971-2000. *Mausam*, *53*, 289–308.

Kripalani, R. H., & Kulkarni, A. (1997). Rainfall variability over Southeast Asia—Connections with Indian monsoon and ENSO extremes: New perspectives. *International Journal of Climatology*, *17*, 1155–1168.

Krishnamurthy, V., & Goswami, B. N. (2000). Indian monsoon–ENSO relationship on interdecadal timescale. *Journal of Climate*, *13*, 579–595.

Krishnamurthy, L., & Krishnamurthy, V. (2014). Influence of PDO on South Asian summer monsoon and monsoon-ENSO relation. *Climate Dynamics*, *42*, 2397–2410.

Krishnamurthy, L., & Krishnamurthy, V. (2016). Indian monsoon's relation with the decadal part of PDO in observations and NCAR CCSM4. *International Journal of Climatology*, *37*, 1824–1833.

Krishnan, R., & Sugi, M. (2003). Pacific decadal oscillation and variability of the Indian summer monsoon rainfall. *Climate Dynamics*, *21*, 233–242.

Krishnan, R., et al. (2016). Deciphering the desiccation trend of the South Asian monsoon hydroclimate in a warming world. *Climate Dynamics*, *47*, 1007–1027.

Krishnaswamy, J., et al. (2015). Non-stationary and non-linear influence of ENSO and Indian Ocean dipole on the variability of Indian monsoon rainfall and extreme rain events. *Climate Dynamics*, *45*, 175–184.

Kucharski, F., & Abid, M. A. (2017). Interannual variability of the Indian monsoon and its link to ENSO. In *Oxford Research Encyclopedias, Climate Science*. https://doi.org/10.1093/acrefore/9780190228620.013.615.

Kucharski, F., Biastoch, A., Ashok, K., & Yuan, D. (2021). Indian Ocean variability and interactions. In C. R. Mechoso (Ed.), *Interacting climates of ocean basins: Observations, mechanisms, predictability, and impacts* (pp. 153–185). Cambridge University Press.

Kucharski, F., Bracco, A., Yoo, J. H., & Molteni, F. (2007). Low-frequency variability of the Indian monsoon–ENSO relationship and the tropical Atlantic: The "weakening" of the 1980s and 1990s. *Journal of Climate*, *20*, 4255–4266.

Kucharski, F., Bracco, A., Yoo, J. H., & Molteni, F. (2008). Atlantic forced component of the Indian monsoon interannual variability. *Geophysical Research Letters*, *35*, L04706. https://doi.org/10.1029/2007GL033037.

Kucharski, F., Molteni, F., & Yoo, J. H. (2006). SST forcing of decadal Indian monsoon rainfall variability. *Geophysical Research Letters*, *33*, L03709. https://doi.org/10.1029/2005GL025371.

Kumar, K. K., Hoerling, M., & Rajagopalan, B. (2005a). Advancing dynamical prediction of Indian monsoon rainfall. *Geophysical Research Letters, 32*, L08704. https://doi.org/10.1029/2004GL021979.

Kumar, K. K., Rajagopalan, B., & Cane, M. A. (1999). On the weakening relationship between the Indian monsoon and ENSO. *Science, 284*, 2156–2159.

Kumar, K. K., Rajagopalan, B., Hoerling, M., Bates, G., & Cane, M. (2006). Unraveling the mystery of Indian monsoon failure during El Niño. *Science, 314*, 115–119.

Kumar, R., Singh, R., & Sharma, K. (2005b). Water resources of India. *Current Science, 89*, 794–811.

Lau, K.-M., Kim, K.-M., & Yang, S. (2000). Dynamical and boundary forcing characteristics of regional components of the Asian summer monsoon. *Journal of Climate, 13*, 2461–2482.

Lenssen, N. J. L., Goddard, L., & Mason, S. (2020). Seasonal forecast skill of ENSO teleconnection maps. *Weather and Forecasting, 35*, 2387–2406.

Levine, R. C., Turner, A. G., Marathayil, D., & Martin, G. M. (2013). The role of northern Arabian Sea surface temperature biases in CMIP5 model simulations and future projections of Indian summer monsoon rainfall. *Climate Dynamics, 41*, 155–172.

Li, T., Zhang, Y., Chang, C.-P., & Wang, B. (2001). On the relationship between Indian Ocean Sea surface temperature and Asian summer monsoon. *Geophysical Research Letters, 28*, 2843–2846.

Liang, J., & Yong, Y. (2021). Climatology of atmospheric rivers in the Asian monsoon region. *International Journal of Climatology, 41*(Suppl. 1), E801–E818. https://doi.org/10.1002/joc.6729.

Lieberman, V., & Buckley, B. (2012). The impact of climate on Southeast Asia, circa 950–1820: New findings. *Modern Asian Studies, 46*, 1049–1096.

Lin, M., & Huybers, P. (2019). If rain falls in India and no one reports it, are historical trends in monsoon extremes biased? *Geophysical Research Letters, 46*, 1681–1689.

Lin, L., Xu, Y., Wang, Z., Diao, C., Dong, W., & Xie, S.-P. (2018). Changes in extreme rainfall over India and China attributed to regional aerosol-cloud interaction during the late 20th century rapid industrialization. *Geophysical Research Letters, 45*, 7857–7865.

Liu, J., Yang, B., & Lindenmayer, D. B. (2019). The oldest trees in China and where to find them. *Frontiers in Ecology and the Environment, 17*, 319–322.

Loader, N. J., McCarroll, D., Gagen, M., Robertson, I., & Jalkanen, R. (2007). Extracting climatic information from stable isotopes in tree rings. In *Terrestrial ecology* (pp. 25–48). Elsevier.

Luo, F.-F., Li, S., & Furevik, T. (2018a). Weaker connection between the Atlantic multidecadal oscillation and Indian summer rainfall since the mid-1990s. *Atmospheric and Oceanic Science Letters, 11*, 37–43.

Luo, F.-F., Li, S., Gao, Y., Keenlyside, N., Svendsen, L., & Furevik, T. (2018b). The connection between the Atlantic multidecadal oscillation and the Indian summer monsoon in CMIP5 models. *Climate Dynamics, 51*, 3023–3039.

Madden, R. A., & Julian, P. R. (1971). Detection of a 40–50 day oscillation in the zonal wind in the tropical Pacific. *Journal of Atmospheric Sciences, 28*, 702–708.

Madolli, M. J., Himanshu, S. K., Patro, E. R., & DeMichele, C. (2022). Past, present and future perspectives of seasonal prediction of Indian summer monsoon rainfall: A review. *Asia-Pacific Journal of Atmospheric Sciences, 58*, 591–615.

Mahto, S. S., et al. (2023). Atmospheric rivers that make landfall in India are associated with flooding. *Communications Earth & Environment, 4*, 120. https://doi.org/10.1038/s43247-023-00775-9.

Managave, S. R., Sheshshayee, M. S., Bhattacharyya, A., & Ramesh, R. (2011). Intra-annual variations of teak cellulose $\delta^{18}O$ in Kerala, India: Implications to the reconstruction of past summer and winter monsoon rains. *Climate Dynamics, 37*, 555–567.

Matsuno, T. (1966). Quasi-geostrophic motions in the equatorial area. *Journal of the Meteorological Society of Japan, 44*, 25–43.

McDermid, S., et al. (2023). Irrigation in the earth system. *Nature Reviews Earth and Environment, 4*, 435–453.

Meehl, G. A., & Hu, A. (2006). Megadroughts in the Indian monsoon region and Southwest North America and a mechanism for associated multidecadal Pacific Sea surface temperature anomalies. *Journal of Climate, 19*, 1605–1623.

Meko, D. M., Stockton, C. W., & Boggess, W. R. (1995). The tree-ring record of severe sustained drought. *JAWRA Journal of the American Water Resources Association, 31*(5), 789–801.

Mishra, V., & Aadhar, S. (2021). Famines and likelihood of consecutive megadroughts in India. *npj Climate and Atmospheric Science, 4*. https://doi.org/10.1038/s41612-021-00219-1.

Mishra, V., Smoliak, B., Lettenmaier, D. P., & Wallace, J. M. (2012). A prominent pattern of year-to-year variability in Indian summer monsoon rainfall. *Proceedings of the National Academy of Sciences, 109*, 7213–7217.

Mohanty, U. C., et al. (2019). A review on the monthly and seasonal forecast of the Indian summer monsoon. *Mausam, 70*, 425–442.

Mohtadi, M., Abram, N. J., Clemens, S. C., Pfeiffer, M., Russell, J. M., Steinke, S., & Zinke, J. (2024). Chapter 19: Paleoclimate evidence of Indian Ocean variability across a range of timescales. In C. C. Ummenhofer, & R. R. Hood (Eds.), *The Indian Ocean and its role in the global climate system* (pp. 445–467). Amsterdam: Elsevier. https://doi.org/10.1016/B978-0-12-822698-8.00007-X.

Nakamura, A., et al. (2016). Weak monsoon event at 4.2 ka recorded in sediment from Lake Rara, Himalayas. *Quaternary International, 397*, 349–359.

Nanditha, J. S., & Mishra, V. (2022). Multiday precipitation is a prominent driver of floods in Indian river basins. *Water Resources Research, 58*. https://doi.org/10.1029/2022WR032723. e2022WR032723.

Nanditha, J. S., et al. (2023). The Pakistan flood of August 2022: Causes and implications. *Earth's Future, 11*. https://doi.org/10.1029/2022EF003230. e2022EF003230.

Nguyen, H. T. T., Turner, S. W. D., Buckley, B. M., & Galelli, S. (2020). Coherent streamflow variability in monsoon Asia over the past eight centuries—Links to oceanic drivers. *Water Resources Research, 56*(12). e2020WR027883.

Nie, J., Boos, W. R., & Kuang, Z. (2010). Observational evaluation of a convective quasi-equilibrium view of monsoons. *Journal of Climate, 23*, 4416–4428. https://doi.org/10.1175/2010JCLI3505.1.

Niranjan Kumar, K., et al. (2013). On the observed variability of monsoon droughts over India. *Weather and Climate Extremes, 1*, 42–50.

Niyogi, D., Kishtawal, C., Tripathi, S., & Govindaraju, R. S. (2010). Observational evidence that agricultural intensification and land use change may be reducing the Indian summer monsoon rainfall. *Water Resources Research, 46*, W03533.

Pai, D. S., Sridhar, L., Badwaik, M. R., & Rajeevan, M. (2015). Analysis of the daily rainfall events over India using a new long period (1901–2010) high resolution (0.25° × 0.25°) gridded rainfall data set. *Climate Dynamics, 45*, 755–776.

Pai, D. S., Sridhar, L., & Ramesh Kumar, M. R. (2016). Active and break events of Indian summer monsoon during 1901–2014. *Climate Dynamics, 46*, 3921–3939.

Pai, D. S., et al. (2020). Normal dates of onset/progress and withdrawal of southwest monsoon over India. *Mausam, 71*, 553–570.

Pang, H., Hou, S., Kaspari, S., & Mayewski, P. A. (2014). Influence of regional precipitation patterns on stable isotopes in ice cores from the Central Himalayas. *The Cryosphere, 8*, 289–301.

Panthi, S., Bräuning, A., Zhou, Z.-K., & Fan, Z.-X. (2017). Tree rings reveal recent intensified spring drought in the central Himalaya, Nepal. *Global and Planetary Change, 157*, 26–34.

Parker, D. J., et al. (2016). The interaction of moist convection and mid-level dry air in the advance of the onset of the Indian monsoon. *Quarterly Journal of the Royal Meteorological Society, 142*, 2256–2272. https://doi.org/10.1002/qj.2815.

Parthasarathy, B., Munot, A., & Kothawale, D. (1988). Regression model for estimation of Indian foodgrain production from summer monsoon rainfall. *Agricultural and Forest Meteorology, 42*, 167–182.

Patil, N., Venkataraman, C., Muduchuru, K., Ghosh, S., & Mondal, A. (2019). Disentangling sea-surface temperature and anthropogenic aerosol influences on recent trends in South Asian monsoon rainfall. *Climate Dynamics, 52*, 2287–2302.

Penny, D., Hall, T., Evans, D., & Polkinghorne, M. (2019). Geoarchaeological evidence from Angkor, Cambodia, reveals a gradual decline rather than a catastrophic 15th-century collapse. *Proceedings of the National Academy of Sciences, 116*, 4871–4876.

Piovesan, G., & Biondi, F. (2021). On tree longevity. *New Phytology, 231*, 1318–1337.

Polson, D., Bollasina, M., Hegerl, G. C., & Wilcox, L. J. (2014). Decreased monsoon precipitation in the Northern Hemisphere due to anthropogenic aerosols. *Geophysical Research Letters, 41*, 6023–6029.

Power, S., et al. (2021). Decadal climate variability in the tropical Pacific: Characteristics, causes, predictability, and prospects. *Science, 374*. https://doi.org/10.1126/science.aay9165.

Prajeesh, A. G., Ashok, K., & Rao, D. V. B. (2013). Falling monsoon depression frequency: A Gray-Sikka conditions perspective. *Scientific Reports, 3*, 2989. https://doi.org/10.1038/srep02989.

Privé, N. C., & Plumb, R. A. (2007). Monsoon dynamics with interactive forcing. Part I: Axisymmetric studies. *Journal of the Atmospheric Sciences, 64*, 1417–1430.

Rajeevan, M., & Bhate, J. (2009). A high resolution daily gridded rainfall dataset (1971-2005) for mesoscale meteorological studies. *Current Science, 96*, 558–562.

Rajeevan, M., Bhate, J., Kale, J. D., & Lal, B. (2006). High resolution daily gridded rainfall data for the Indian region: Analysis of break and active monsoon spells. *Current Science, 91*(3), 296–306.

Ramarao, M. V. S., et al. (2019). On observed aridity changes over the semiarid regions of India in a warming climate. *Theoretical and Applied Climatology, 136*, 693–702.

Rao, Y. P. (1976). *Southwest monsoon. Meteorolical monograph, no.1*. New Delhi: India Meteorol. Dept. 366 pp.

Rao, M. P. (2020). *Hydroclimate variability and environmental change in Eurasia over the past millennium and its impacts*. Columbia University.

Rao, M. P., et al. (2018). Six centuries of upper Indus basin streamflow variability and its climatic drivers. *Water Resources Research, 54*, 5687–5701.

Rao, S. A., et al. (2019). Monsoon Mission: A targeted activity to improve monsoon prediction across scales. *Bulletin of the American Meteorological Society, 100*, 2509–2532.

Rao, M. P., et al. (2020). Seven centuries of reconstructed Brahmaputra River discharge demonstrate underestimated high discharge and flood hazard frequency. *Nature Communications, 11*, 6017.

Rasmusson, E. M., & Carpenter, T. H. (1983). The relationship between the eastern Pacific Sea surface temperature and rainfall over India and Sri Lanka. *Monthly Weather Review, 111*, 517–528.

Reynolds, R. W., Rayner, N. A., Smith, T. M., Stokes, D. C., & Wang, W. (2002). An improved in situ and satellite SST analysis for climate. *Journal of Climate, 15*, 1609–1625.

Rodwell, M. J., & Hoskins, B. J. (2001). Subtropical anticyclones and summer monsoons. *Journal of Climate, 14*, 3192–3211.

Roxy, M. K., Ritika, K., Terray, P., Murtugudde, R., Ashok, K., & Goswami, B. N. (2015). Drying of Indian subcontinent by rapid Indian Ocean warming and a weakening land-sea thermal gradient. *Nature Communications, 6*. https://doi.org/10.1038/ncomms8423.

Roxy, M. K., et al. (2017). A threefold rise in widespread extreme rain events over Central India. *Nature Communications, 8*, 708. https://doi.org/10.1038/s41467-017-00744-9.

Sabeerali, C. T., et al. (2013). Simulation of boreal summer intraseasonal oscillations in the latest CMIP5 coupled GCMs. *Journal of Geophysical Research: Atmospheres, 118*, 4401–4420.

Sachs, J., et al. (2009). Southward movement of the Pacific intertropical convergence zone AD 1400-1850. *Nature Geoscience, 2*, 519–525.

Saha, A., Ghosh, S., Sahana, A. S., & Rao, E. P. (2014). Failure of CMIP5 climate models in simulating post-1950 decreasing trend of Indian monsoon. *Geophysical Research Letters, 41*, 7323–7330.

Saji, N. H., Goswami, B. N., Vinaychandran, P. N., & Yamagata, T. (1999). A dipole mode in the tropical Indian Ocean. *Nature, 401*, 360–363.

Saji, N. H., & Yamagata, T. (2003). Possible impacts of Indian Ocean dipole mode events on global climate. *Climate Research, 25*, 151–169.

Salinger, M. J., Shrestha, M. L., Ailikun, D. W., McGregor, J. L., & Wang, S. (2014). Climate in Asia and the Pacific: Climate variability and change. In M. Manton, & L. Stevenson (Eds.), *Vol. 56. Climate in Asia and the Pacific. Advances in global change research* (pp. 17–57). Dordrecht: Springer.

Salzmann, M., & Cherian, R. (2015). On the enhancement of the Indian summer monsoon drying by Pacific multidecadal variability during the latter half of the twentieth century. *Journal of Geophysical Research-Atmospheres, 120*, 9103–9118.

Salzmann, M., Weser, H., & Cherian, R. (2014). Robust response of Asian summer monsoon to anthropogenic aerosols in CMIP5 models. *Journal of Geophysical Research: Atmospheres, 119*, 11321–11337.

Sandeep, S., & Ajayamohan, R. S. (2015). Origin of cold bias over the Arabian Sea in climate models. *Scientific Reports, 4*, 6403. https://doi.org/10.1038/srep06403.

Sandeep, S., Ajayamohan, R. S., Boos, W. R., Sabin, T. P., & Praveen, V. (2018). Decline and poleward shift in Indian summer monsoon synoptic activity in a warming climate. *Proceedings of the National Academy of Sciences, 115*, 2681–2686.

Sano, M., Buckley, B. M., & Sweda, T. (2008). Tree-ring based hydroclimate reconstruction over northern Vietnam from *Fokienia hodginsii*: Eighteenth century mega-drought and tropical Pacific influence. *Climate Dynamics, 33*, 331–340.

Sano, M., Ramesh, R., Sheshshayee, M., & Sukumar, R. (2012). Increasing aridity over the past 223 years in the Nepal Himalaya inferred from a tree-ring $\delta 18O$ chronology. *The Holocene, 22*, 809–817.

Sato, T., & Kimura, F. (2007). How does the Tibetan plateau affect the transition of Indian monsoon rainfall? *Monthly Weather Review, 135*, 2006–2015. https://doi.org/10.1175/MWR3386.1.

Sen Roy, S. (2006). The impacts of ENSO, PDO, and local SSTs on winter precipitation in India. *Physical Geography, 5*, 464–474.

Sen Roy, S., & Sen Roy, N. (2011). Influence of Pacific Decadal Oscillation and El Niño-Southern Oscillation on the summer monsoon precipitation in Myanmar. *International Journal of Climatology, 31*, 14–21.

Seneviratne, S. I., Zhang, X., Adnan, M., Badi, W., Dereczynski, C., Di Luca, A., Ghosh, S., Iskandar, I., Kossin, J., Lewis, S., Otto, F., Pinto, I., Satoh, M., Vicente-Serrano, S. M., Wehner, M., & Zhou, B. (2021). Weather and climate extreme events in a changing climate. In V. Masson-Delmotte, P. Zhai, A. Pirani, S. L. Connors, C. Péan, S. Berger, … B. Zhou (Eds.), *Climate change 2021: The physical science basis. Contribution of working group I to the sixth assessment report of the Intergovernmental Panel on Climate Change* (pp. 1513–1766). Cambridge, United Kingdom and New York, NY, USA: Cambridge University Press.

Seth, A., Giannini, A., Rojas, M., Rauscher, S. A., Bordoni, S., Singh, D., & Camargo, S. J. (2019). Monsoon responses to climate changes—Connecting past, present and future. *Current Climate Change Reports, 5*, 63–79.

Shah, S. K., Bhattacharyya, A., & Chaudhary, V. (2007). Reconstruction of June–September precipitation based on tree-ring data of teak (Tectona grandis L.) from Hoshangabad, Madhya Pradesh, India. *Dendrochronologia, 25*(1), 57–64.

Shah, S. K., Bhattacharyya, A., & Chaudhary, V. (2014). Streamflow reconstruction of Eastern Himalaya River, Lachen 'Chhu', North Sikkim, based on tree-ring data of Larix griffithiana from Zemu Glacier basin. *Dendrochronologia, 32*(2), 97–106.

Shah, S. K., Singh, R., Mehrotra, N., & Thomte, L. (2019). River flow reconstruction of the Lohit River Basin, North–east India based on tree–rings of *Pinus merkusii* (Merkus pine). *Journal of Palaeosciences, 68*.

Shastri, H., Paul, S., Ghosh, S., & Karmakar, S. (2015). Impacts of urbanization on Indian summer monsoon rainfall extremes. *Journal of Geophysical Research-Atmospheres, 120*, 495–516.

Shekhar, M., Ranhotra, P. S., Bhattacharyya, A., Singh, A., Dhyani, R., & Singh, S. (2022). Tree-ring-based hydrological records reconstructions of the Himalayan Rivers: Challenges and opportunities. In S. Rani, & R. Kumar (Eds.), *Climate change: Impacts, responses and sustainability in the Indian Himalaya* (pp. 47–72). Cham: Springer International Publishing.

Shi, F., Li, J., & Wilson, R. J. S. (2014). A tree-ring reconstruction of the South Asian summer monsoon index over the past millennium. *Scientific Reports, 4*(1), 6739.

Shi, H., Wang, B., Cook, E. R., Liu, J., & Liu, F. (2018). Asian summer precipitation over the past 544 years reconstructed by merging tree rings and historical documentary records. *Journal of Climate, 31*(19), 7845–7861.

Shinoda, T., Jenson, T. G., Lachkar, Z., Masumoto, Y., & Seo, H. (2024). Chapter 18: Modeling the Indian Ocean. In C. C. Ummenhofer, & R. R. Hood (Eds.), *The Indian Ocean and its role in the global climate system* (pp. 421–443). Amsterdam: Elsevier. https://doi.org/10.1016/B978-0-12-822698-8.00001-9.

Simpson, G. C. (1921). The South-West monsoon. *Quarterly Journal of the Royal Meteorological Society, 47*, 151–171. https://doi.org/10.1002/qj.49704719901.

Singh, D., Bollasina, M., Ting, M., & Diffenbaugh, N. S. (2019a). Disentangling the influence of local and remote anthropogenic aerosols on south Asian monsoon daily rainfall characteristics. *Climate Dynamics*. https://doi.org/10.1007/s00382-018-4512-9.

Singh, R., Kishtawal, C. M., & Singh, C. (2021). The strengthening association between Siberian snow and Indian summer monsoon rainfall. *Journal of Geophysical Research-Atmospheres, 126*. https://doi.org/10.1029/2020JD033779.

Singh, D., Gosh, S., Roxy, M. K., & McDermid, S. (2019b). Indian summer monsoon: Extreme events, historical changes, and role of anthropogenic forcings. *WIREs Climate Change, 10*. https://doi.org/10.1002/wcc.571.

Singh, V., Misra, K. G., Yadava, A. K., & Yadav, R. R. (2022). Application of tree rings in understanding long-term variability in river discharge of high Himalayas, India. In N. Kumaran, & P. Damodara (Eds.), *Holocene climate change and environment* (pp. 247–264). Elsevier.

Singh, D., McDermid, S. P., Cook, B. I., Puma, M. J., Nazarenko, L., & Kelley, M. (2018a). Distinct influences of land cover and land management on seasonal climate. *Journal of Geophysical Research – Atmospheres, 123*, 12017–12039.

Singh, D., Seager, R., Cook, B. I., Cane, M., Ting, M., Cook, E., & Davis, M. (2018b). Climate and the global famine of 1876–78. *Journal of Climate, 31*, 9445–9467.

Singh, D., Tsiang, M., Rajaratnam, B., & Diffenbaugh, N. S. (2014). Observed changes in extreme wet and dry spells during the south Asian summer monsoon season. *Nature Climate Change, 4*, 456–461.

Singh, J., Vittal, H., Karmakar, S., Ghosh, S., & Niyogi, D. (2016). Urbanization causes nonstationarity in Indian summer monsoon rainfall extremes. *Geophysical Research Letters, 43*, 11269–11277.

Sinha, A., Berkelhammer, M., Stott, L., Mudelsee, M., Cheng, H., & Biswas, J. (2011). The leading mode of Indian summer monsoon precipitation variability during the last millennium. *Geophysical Research Letters, 38*, L15703.

Sinha, A., et al. (2015). Trends and oscillations in the Indian summer monsoon rainfall over the last two millennia. *Nature Communications, 6*(1), 6309. https://doi.org/10.1038/ncomms7309.

Slingo, J. M., & Annamalai, H. (2000). The El Niño of the century and the response of the Indian summer monsoon. *Monthly Weather Review, 128*, 1778–1796.

Sontakke, N. A., Singh, N., & Singh, H. N. (2008). Instrumental period rainfall series of the Indian region (AD 1813–2005): Revised reconstruction, update and analysis. *The Holocene, 18*, 1055–1066.

Sooraj, K. P., Terray, P., & Mujumdar, M. (2015). Global warming and the weakening of the Asian summer monsoon circulation: Assessments from the CMIP5 models. *Climate Dynamics, 45*, 233–252.

Speer, J. H., Shah, S. K., Truettner, C., Pacheco, A., Bekker, M. F., Dukpa, D., Cook, E. R., & Tenzin, K. (2019). Flood history and river flow variability recorded in tree rings on the Dhur River, Bhutan. *Dendrochronologia, 56*, 125605.

Sperber, K. R., et al. (2013). The Asian summer monsoon: An intercomparison of CMIP5 vs. CMIP3 simulations of the late 20th century. *Climate Dynamics, 41*, 2711–2744.

St. George, S., & Ault, T. R. (2014). The imprint of climate within northern hemisphere trees. *Quaternary Science Reviews, 89*, 1–4.

Stanley, J.-D., Krom, M. D., Cliff, R. A., & Woodward, J. C. (2003). Short contribution: Nile flow failure at the end of the Old Kingdom, Egypt: Strontium isotopic and petrologic evidence. *Geoarchaeology, 18*, 395–402.

Steiger, N. J., Smerdon, J. E., Cook, E. R., & Cook, B. I. (2018). A reconstruction of global hydroclimate and dynamical variables over the common era. *Scientific Data, 5*, 180086.

Stuecker, M. F., et al. (2017). Revisiting ENSO/Indian Ocean dipole phase relationships. *Geophysical Research Letters, 44*, 2481–2492.

Sudhakar Reddy, C., Saranya, K. R. L., Vazeed Pasha, S., Satish, K. V., Jha, C. S., Diwakar, P. G., Dadhwal, V. K., Rao, P. V. N., & Krishna Murthy, Y. V. N. (2018). Assessment and monitoring of deforestation and forest fragmentation in South Asia since the 1930s. *Global and Planetary Change, 161*, 132–148.

Takaya, Y., Kosaka, Y., Watanabe, M., & Maeda, S. (2021). Skilful predictions of the Asian summer monsoon one year ahead. *Nature Communications, 12*. https://doi.org/10.1038/s41467-021-22299-6.

Takaya, Y., Ren, H.-L., Vitart, F., & Robertson, A. W. (2023). Current status and progress in the seasonal prediction of the Asian summer monsoon. *Mausam, 74*, 455–466.

Tardif, R., Hakim, G. J., Perkins, W. A., Horlick, K. A., Erb, M. P., Emile-Geay, J., & Noone, D. (2019). Last millennium reanalysis with an expanded proxy database and seasonal proxy modeling. *Climate of the Past, 15*, 1251–1273.

Taschetto, A. S., Sen Gupta, A., Hendon, H. H., Ummenhofer, C. C., & England, M. H. (2011). The contribution of Indian Ocean Sea surface temperature anomalies on Australian summer rainfall during El Niño events. *Journal of Climate, 24*, 3734–3747.

Taschetto, A. S., Ummenhofer, C. C., Stuecker, M., Dommenget, D., Ashok, K., Rodrigues, R., & Yeh, S.-W. (2020). El Niño-Southern Oscillation atmospheric teleconnections. In M. J. McPhaden, A. Santoso, & W. Cai (Eds.), *El Niño Southern Oscillation in a changing climate*. https://doi.org/10.1002/9781119548164.ch14.

Thapa, U., St, S., George, D. K., & Gaire, N. (2017). Tree growth across the Nepal Himalaya during the last four centuries. *Progress in Physical Geography: Earth and Environment, 41*(4), 478–495.

Thapa, U. K., St. George, S., & Trouet, V. (2020). Poleward excursions by the Himalayan subtropical jet over the past four centuries. *Geophysical Research Letters, 47*(22). e2020GL089631.

Thapa, U. K., Stevenson, S., & Midhun, M. (2022). Orbital forcing strongly influences the poleward shift of the spring Himalayan jet during the past millennium. *Geophysical Research Letters, 49*(3). e2021GL095955.

Thompson, L. G., Yao, T., Mosley-Thompson, E., Davis, M. E., Henderson, K. A., & Lin, P.-N. (2000). A high-resolution millennial record of the South Asian monsoon from Himalayan ice cores. *Science, 289*, 1916–1919.

Turner, A. G., & Annamalai, H. (2012). Climate change and the South Asian summer monsoon. *Nature Climate Change, 2*, 587.

Turner, A. G., & Hannachi, A. (2010). Is there regime behaviour in monsoon convection in the late 20th century? *Geophysical Research Letters, 37*, L16706.

Uhe, P. F., Mitchell, D. M., Bates, P. D., Sampson, C. C., Smith, A. M., & Islam, A. S. (2019). Enhanced flood risk with 1.5°C global warming in the Ganges–Brahmaputra–Meghna basin. *Environmental Research Letters, 14*(7), 074031.

Ullah, W., Wang, G., Lou, D., Ullah, S., Bhatti, A. S., Ullah, S., Karim, A., Hagan, D. F. T., & Ali, G. (2021). Large-scale atmospheric circulation patterns associated with extreme monsoon precipitation in Pakistan during 1981-2018. *Atmospheric Research, 253*, 105489.

Ummenhofer, C. C., D'Arrigo, R. D., Anchukaitis, K. J., Buckley, B. M., & Cook, E. R. (2013). Links between Indo-Pacific climate variability and drought in the monsoon Asia drought atlas. *Climate Dynamics, 40*(5), 1319–1334.

Ummenhofer, C. C., Murty, S. A., Sprintall, J., Lee, T., & Abram, N. J. (2021). Heat and freshwater changes in the Indian Ocean region. *Nature Reviews Earth and Environment, 2*. https://doi.org/10.1038/s43017-021-00192-6.

Ummenhofer, C. C., Sen Gupta, A., Li, Y., Taschetto, A. S., & England, M. H. (2011). Multi-decadal modulation of the El Nino–Indian monsoon relationship by Indian Ocean variability. *Environmental Research Letters, 6*(3), 034006.

Ummenhofer, C. C., Taschetto, A. S., Izumo, T., & Luo, J.-J. (2024). Chapter 7: Impacts of the Indian Ocean on regional and global climate. In C. C. Ummenhofer, & R. R. Hood (Eds.), *The Indian Ocean and its role in the global climate system* (pp. 145–168). Amsterdam: Elsevier. https://doi.org/10.1016/B978-0-12-822698-8.00018-4.

Undorf, S., Bollasina, M. A., & Hegerl, G. C. (2018). Impacts of the 1900-74 increase in anthropogenic aerosol emissions from North America and Europe on Eurasian summer climate. *Journal of Climate, 31*, 8381–8399.

Vecchi, G., Harrison, D., & Bond, D. (2004). Interannual Indian rainfall variability and Indian Ocean sea surface temperature anomalies. In C. Wang, S.-P. Xie, & J. A. Carton (Eds.), *Earth climate: The ocean-atmosphere interaction* (pp. 247–260). Washington, DC: American Geophysical Union, Geophysical Monograph 147.

Vernekar, A. D., Zhou, J., & Shukla, J. (1995). The effect of Eurasian snow cover on the Indian monsoon. *Journal of Climate, 8*, 248–266.

Vishnu, S., Francis, P. A., Shenoi, S. S. C., & Ramakrishna, S. S. V. S. (2016). On the decreasing trend of the number of monsoon depressions in the Bay of Bengal. *Environmental Research Letters, 11*, 014011. https://doi.org/10.1088/1748-9326/11/1/014011.

Vittal, H., Villarini, G., & Zhang, W. (2020). Early prediction of the Indian summer monsoon rainfall by the Atlantic Meridional Mode. *Climate Dynamics, 54*, 2337–2346.

Walker, G. T. (1924). Correlation in seasonal variations of weather—A further study of world weather. *Memoirs of the Indian Meteorological Department, 24*, 275–332.

Wang, B. (2006). *The Asian monsoon*. Chichester, UK: Praxis Publishing Limited. 788 pp.

Wang, B., Biasutti, M., Byrne, M. P., Castro, C., Chang, C.-P., Cook, K., Fu, R., Grimm, A. M., Ha, K.-J., & Hendon, H. (2021). Monsoons climate change assessment. *Bulletin of the American Meteorological Society, 102*(1), E1–E19.

Wang, P., Clemens, S., Beaufort, L., Braconnot, P., Ganssen, G., Jian, Z., Kershaw, P., & Sarnthein, M. (2005). Evolution and variability of the Asian monsoon system: State of the art and outstanding issues. *Quaternary Science Reviews, 24*, 595–629.

Wang, B., & Ding, Q. (2008). Global monsoon: Dominant mode of annual variation in the tropics. *Dynamics of Atmosphere and Oceans, 44*, 165–183. https://doi.org/10.1016/j.dynatmoce.2007.05.002.

Wang, B., Wu, R., & Li, T. (2003). Atmosphere-warm ocean interaction and its impacts on Asian-Australian monsoon variation. *Journal of Climate, 16*, 1195–1211.

Wang, B., et al. (2015). Rethinking Indian monsoon rainfall prediction in the context of recent global warming. *Nature Communications, 6*. https://doi.org/10.1038/ncomms8154.

Webster, P. J., Magana, V. O., Palmer, T. N., Shukla, J., Tomas, R. A., Yanai, M., & Yasunari, T. (1998). Monsoons: Processes, predictability, and the prospects for prediction. *Journal of Geophysical Research, 103*, 14451–14510.

Webster, P. J., Moore, A. M., Loschnigg, J. P., & Leben, R. R. (1999). Coupled ocean-atmosphere dynamics in the Indian Ocean during 1997-1998. *Nature, 401*, 356–360.

Webster, P. J., & Yang, S. (1992). Monsoon and ENSO: Selectively interactive systems. *Quarterly Journal of the Royal Meteorological Society, 118*, 877–926.

Weiss, H., Courty, M.-A., Wetterstrom, W., Guichard, F., Senior, L., Meadown, R., & Curnow, A. (1993). The genesis and collapse of third millennium north Mesopotamian civilization. *Science, 261*, 995–1004.

Weng, H., Ashok, K., Behera, S. K., Rao, S. A., & Yamagata, T. (2007). Impacts of recent El Niño Modoki on dry/wet conditions in the Pacific rim during boreal summer. *Climate Dynamics, 29*, 113–129.

Woodhouse, C. A., & Lukas, J. J. (2006). Multi-century tree-ring reconstructions of Colorado streamflow for water resource planning. *Climate Change, 78*, 293–315.

Xie, P., & Arkin, P. A. (1997). Global precipitation: A 17-year monthly analysis based on gauge observations, satellite estimates, and numerical model outputs. *Bulletin of the American Meteorological Society, 78*, 2539–2558.

Xie, S.-P., Hu, K., Hafner, J., Tokinaga, H., Du, Y., Huang, G., & Sampe, T. (2009). Indian Ocean capacitor effect on Indo-Western Pacific climate during the summer following El Niño. *Journal of Climate, 22*, 730–747.

Xu, C., Buckley, B. M., Promchote, P., Wang, S.-Y. S., An, W., Pumijumnong, N., Sano, M., Nakatsuka, T., & Guo, Z. (2019). Increased variability of Chao Phraya River peak-season flow in Thailand is associated with strong ENSO activity: Evidence from tree ring $\delta^{18}O$. *Geophysical Research Letters, 46*. https://doi.org/10.1029/2018GL081458.

Xu, C., Sano, M., & Nakatsuka, T. (2013). A 400-year record of hydroclimate variability and local ENSO history in northern Southeast Asia inferred from tree-ring $\delta^{18}O$. *Palaeogeography, Palaeoclimatology, Palaeoecology, 386*, 588–598.

Xu, C., et al. (2018). Decreasing Indian summer monsoon on the northern Indian sub-continent during the last 180 years: Evidence from five tree-ring cellulose oxygen isotope chronologies. *Climate of the Past, 14*, 653–664.

Yamagata, T., Behera, S., Doi, T., Luo, J.-J., Morioka, Y., & Tozuka, T. (2024). Chapter 5: Climate phenomena of the Indian Ocean. In C. C. Ummenhofer, & R. R. Hood (Eds.), *The Indian Ocean and its role in the global climate system* (pp. 103–119). Amsterdam: Elsevier. https://doi.org/10.1016/B978-0-12-822698-8.00009-3.

Yan, H., et al. (2015). Dynamics of the intertropical convergence zone over the western Pacific during the little ice age. *Nature Geoscience, 8*, 315–320.

Yang, J., Liu, Q., Liu, Z., Wu, L., & Huang, F. (2009). Basin mode of Indian Ocean Sea surface temperature and northern hemisphere circumglobal teleconnection. *Geophysical Research Letters, 36*, L19705. https://doi.org/10.1029/2009GL039559.

Yang, J., Liu, Q., Xie, S.-P., Liu, Z., & Wu, L. (2007). Impact of the Indian Ocean SST basin mode on the Asian summer monsoon. *Geophysical Research Letters, 34*, L02708. https://doi.org/10.1029/2006GL028571.

Zaw, Z., Fan, Z.-X., Bräuning, A., Xu, C.-X., Liu, W.-J., Gaire, N. P., Panthi, S., & Than, K. Z. (2020). Drought reconstruction over the past two centuries in southern Myanmar using teak tree-rings: Linkages to the Pacific and Indian oceans. *Geophysical Research Letters, 47*(10). e2020GL087627.

Zhang, C., Adames, Á. F., Khouider, B., Wang, B., & Yang, D. (2020). Four theories of the Madden-Julian oscillation. *Reviews of Geophysics, 58*. https://doi.org/10.1029/2019RG000685. e2019RG000685.

Zhang, R., & Delworth, T. L. (2006). Impact of Atlantic multidecadal oscillations on India/Sahel rainfall and Atlantic hurricanes. *Geophysical Research Letters, 33*. https://doi.org/10.1029/2006GL026267.

Zhang, T., Jiang, X., Yang, S., Chen, J., & Li, Z. (2022). A predictable prospect of the South Asian summer monsoon. *Nature Communications, 13*. https://doi.org/10.1038/s41467-022-34881-7.

Zhang, D., Qin, D., Hou, S., Kang, S., Ren, J., & Mayewski, P. A. (2005). Climatic significance of $\delta^{18}O$ records from an 80.36 m ice core in the East Rongbuk Glacier, Mount Qomolangma (Everest). *Science China Series D, 48*, 266–272.

Zhang, T., et al. (2019). The weakening relationship between Eurasian spring snow cover and Indian summer monsoon rainfall. *Science Advances, 5*. https://doi.org/10.1126/sciadv.aau893.

Zhao, S., Pederson, N., D'Orangeville, L., HilleRisLambers, J., Boose, E., Penone, C., Bauer, B., Jiang, Y., & Manzanedo, R. D. (2019). The International Tree-Ring Data Bank (ITRDB) revisited: Data availability and global ecological representativity. *Journal of Biogeography, 46*(2), 355–368.

Chapter 4

Intraseasonal variability in the Indian Ocean region

Charlotte A. DeMott[a], James H. Ruppert, Jr.[b], and Adam Rydbeck[c]

[a]Department of Atmospheric Science, Colorado State University, Fort Collins, CO, United States, [b]School of Meteorology, The University of Oklahoma, Norman, OK, United States, [c]Ocean Sciences Division, US Naval Research Laboratory, Stennis Space Center, Hancock County, MS, United States

1 Introduction to intraseasonal ocean-atmosphere coupling

Intraseasonal variability (ISV) of the atmosphere and ocean refers to disturbances with characteristic timescales spanning approximately 30–90 days. In the tropics, atmospheric ISV includes convectively coupled equatorial waves (e.g., Kelvin and equatorial Rossby waves; Kiladis et al., 2009), but it is chiefly driven by the intraseasonal oscillation (ISO)—a large-scale atmospheric disturbance often featuring organized cloud clusters so large that they span nearly the entire Indian Ocean basin (Fig. 1). The ISO remotely affects global weather through numerous teleconnections across timescales, from mesoscale to interannual (Zhang, 2005, 2013). Perturbations in the atmosphere introduced by ISV modify the ocean via atmosphere-ocean coupled feedbacks, and vice versa (cf. DeMott et al., 2015). These feedbacks act across the ocean-atmosphere interface through the fluxes of heat, freshwater, and momentum, which are themselves regulated by processes that control the wind speed, temperature, and moisture content of the atmospheric boundary layer and the heat content of the upper ocean mixed layer. We begin with a discussion of processes that regulate these layers.

Processes within the tropical atmosphere that regulate the exchange of heat, mass, and momentum across the air-sea interface act primarily through large-scale wind and convective cloud activity, as illustrated in Fig. 2. Convective activity is strongly coupled to free tropospheric humidity, which fundamentally relies on the detrainment of moisture across the boundary layer top, vertical moisture transport out of the boundary layer by clouds, and the re-evaporation of detrained cloud droplets to the free troposphere. Moist convection also responds to variations in humidity caused by large-scale advection. In suppressed conditions, when large-scale subsidence inhibits deep convection, drying of the boundary layer by detrainment and cloud vertical transport is offset by surface evaporation. In active conditions, when deep convection is widespread, rainfall is enhanced and frequent convectively generated cold pools cool the boundary layer (de Szoeke et al., 2017). Additionally, convectively active regimes in the Indian Ocean often coincide with enhanced westerly winds due to the associated large-scale circulation patterns (Section 2) (Moum et al., 2014). Consequently, transitions from suppressed to active regimes are characterized by an associated change from very quiescent conditions to a highly disturbed atmospheric boundary layer.

A similar set of ocean processes affect the temperature of the ocean mixed layer, as well as the stratification of the upper ocean, and help regulate exchanges across the air-sea interface. For example, changes in ocean currents near the mixed layer base promote shear-driven mixing of surface and deep ocean waters, while sharp discontinuities in salinity and temperature, including those linked to biological activity, drive diffusive mixing. Ocean currents may also advect water masses with different salinities into a region, sometimes favoring the formation of ocean barrier layers (Cronin & McPhaden, 2002) that resist the vertical mixing of cool thermocline water into the mixed layer (Section 3.3). Shifts in mixed layer properties are also sensitive to variations of upper ocean heat content partly controlled by upwelling and downwelling oceanic Kelvin and Rossby waves that modulate the depth of the thermocline.

The processes that regulate the energetics of the atmospheric boundary layer and the upper ocean mixed layer mediate the coupled feedbacks between the ocean and atmosphere, which are communicated by the fluxes between these two layers, as shown in Fig. 3. Atmospheric cloudiness regulates the amount of solar radiation reaching the ocean surface, while wind and rainfall provide fluxes of momentum and freshwater. The ocean mixing and stratification response to these inputs

---

[⊛]This book has a companion website hosting complementary materials. Visit this URL to access it: https://www.elsevier.com/books-and-journals/book-companion/9780128226988.

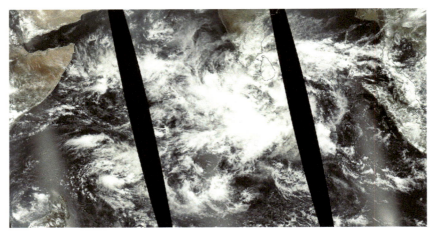

FIG. 1 Visible satellite image showing the ISO (also referred to as the Madden-Julian Oscillation) convectively active phase over the Indian Ocean on November 24, 2011 during the DYNAMO field campaign. *(Image courtesy of NASA Worldview.)*

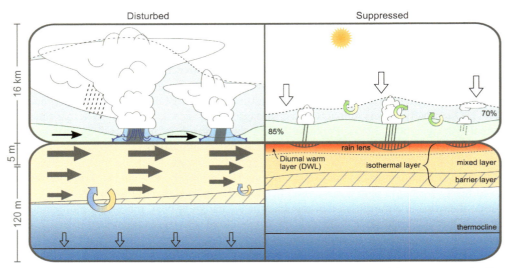

FIG. 2 Processes affecting kinematic and thermodynamic states of the atmospheric boundary layer (*top*) and the upper ocean mixed layer (*bottom*) during convectively disturbed (*left*) and suppressed conditions (*right*). Large-scale sinking motion, horizontal winds or currents, and entrainment mixing processes are denoted by open vertical block arrows, solid horizontal arrows, and shaded rotor arrows, respectively. Atmospheric humidity profiles are illustrated as layers with 85% and 70% relative humidity.

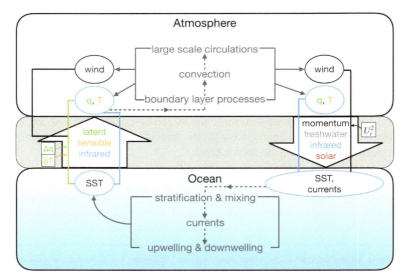

FIG. 3 Conceptual illustration of how atmospheric and oceanic processes shown in Fig. 2 determine fluxes of momentum, mass, and heat across the air-sea interface. Near-surface vertical gradients of specific humidity and temperature (the *green-orange* dashed line) are denoted by Δq and ΔT, respectively. Variables influencing infrared fluxes are connected by *blue* lines. Momentum fluxes are regulated by the difference in wind speed and surface currents, U_r.

affects the vertical transport of momentum that impacts ocean currents and the mixed layer depth, all of which regulate the sea surface temperature (SST). The SST response is communicated to the atmosphere through surface turbulent and longwave radiation fluxes. The SST feedback to the atmosphere is strongly modulated by the temperature and humidity of the atmospheric boundary layer and low-level wind speed. Energy fluxed to the atmospheric boundary layer subsequently influences convection and large-scale circulations, closing the feedback loop through their influences on momentum, freshwater, and solar radiation fluxes to the ocean.

Diagnosis of the leading processes modulating ocean-atmosphere coupled feedbacks on intraseasonal timescales is complicated by the inherently multiscale nature of these feedbacks, where forcing on one timescale can influence the response on another through the rectifier effect (e.g., Kessler & Kleeman, 2000). Examples of multiscale coupled feedbacks include large-amplitude diurnal SST variations, which both rectify intraseasonal SST changes (Webster et al., 1996) and invigorate low-level atmospheric moistening (Ruppert & Johnson, 2015; Sui et al., 1997; Zhao & Nasuno, 2020); synoptic scale low-level winds forcing subseasonal-to-seasonal oceanic Kelvin and equatorial Rossby waves (Kessler et al., 1995; Schott et al., 2009); and low-frequency Indian Ocean SST patterns associated with the Indian Ocean Dipole (IOD; Saji et al., 1999) affecting the zonal migration of intraseasonal cloudiness (Izumo et al., 2010; Shinoda & Han, 2005; Wilson et al., 2013).

The goals of this chapter are to provide an overview of atmospheric and oceanic intraseasonal processes and their coupled feedbacks rooted both at intraseasonal and other timescales, and to review recent findings regarding their manifestations in the Indian Ocean.

2 The intraseasonal oscillation

2.1 Overview

Atmospheric ISV is chiefly regulated by the ISO, a large-scale tropical convective system that varies by season. The boreal winter (November–April) ISO is usually referred to as the Madden-Julian oscillation (MJO) after Madden and Julian (1972). The MJO resembles an eastward-propagating Walker circulation, with a large-scale ($O(10,000 km)$) envelope of active deep moist convection in its ascending phase and suppressed convection in its descending phase (Fig. 4a). For a given MJO event, deep convection usually initiates at the western edge of the Warm Pool in the western Indian Ocean and then propagates across the Indian and western Pacific oceans with a phase speed of about $5 m s^{-1}$ before diminishing near the eastern edge of the Warm Pool near the dateline.

Throughout the boreal winter season, MJO propagation is predominantly eastward, with a southward "detour" of the convective signal around the maritime continent (Kim et al., 2017), where it interacts with the Australian summer monsoon (Wheeler et al., 2009). During the boreal summer (May–October), deep easterly wind shear (Jiang et al., 2004) and strong ocean surface warming (e.g., Fu et al., 2003; Yang et al., 2020) add a pronounced northward component to ISO propagation. This boreal summer ISO interacts with the active and break periods of the Indian and Southeast Asian summer monsoons. Descriptions of the ISO in this chapter predominantly refer to the boreal winter ISO (i.e., the MJO), except where noted otherwise.

The ISO modulates the upper Indian Ocean through exchanges of internal energy, mass, and momentum, and drives ocean ISV. Ocean ISV includes warming, freshening, accelerating, and stabilizing of the upper ocean, which are regulated by ISO cloud, precipitation, and wind anomalies, as well as the generation of oceanic equatorial shallow water waves (i.e., Kelvin and equatorial Rossby waves) that subsequently evolve quasi-independently of ISO forcing and have a preferred basin resonance on intraseasonal timescales (Han et al., 1999; Jensen, 1993). Other modes of ocean ISV include variations in the surface current and upwelling systems near Somalia in the western Indian Ocean (Shinoda et al., 2017) and Sri Lanka in the central Indian Ocean, equatorial jets (Wyrtki, 1973; Yoshida, 1959), coastal Kelvin waves (Arief & Murray, 1996), ocean barrier layers (Sprintall & Tomczak, 1992), and Seychelles-Chagos thermocline ridge (SCTR) perturbations (Vialard et al., 2008). The ocean responses to ISO forcing in heat, stratification, and circulation can affect both ongoing and subsequent ISO events.

2.2 Atmospheric processes and their dynamic interpretation

In an Eulerian framework, the ISO lifecycle over the Warm Pool transitions between large-scale suppressed conditions, which feature light winds and shallow trade cumulus clouds, and large-scale convectively active conditions, which feature strong winds and deep, widespread cloudiness (Fig. 2). These regimes are modulated by slow eastward propagation of the ISO dipole heating structure, with enhanced diabatic heating (dominated by condensation heating) in the active phase and reduced diabatic heating in the suppressed phase (Fig. 4a). The ISO active phase features low-level easterly and upper-level

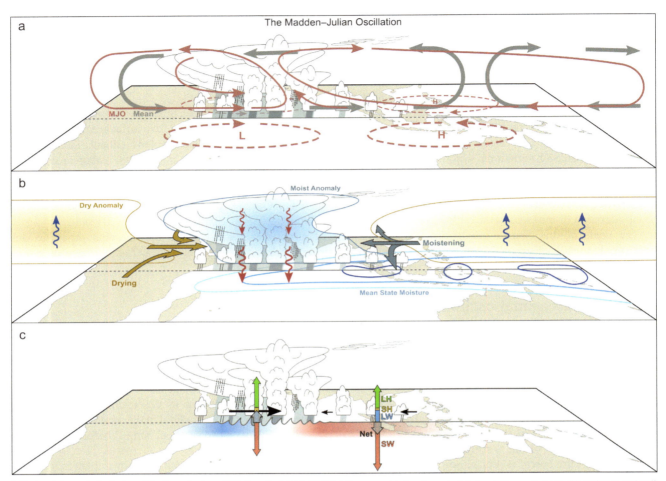

FIG. 4 Schematic depiction of the boreal winter ISO (i.e., the MJO), highlighting associated patterns in (a) circulation and cloud, (b) moisture and radiative forcing, and (c) surface fluxes and ocean mixed layer temperature. In the top-panel, both the time-mean Walker circulation (*gray*) and intraseasonal circulation (*red*) are depicted. SW denotes the net shortwave, LW net longwave, LH latent, and SH sensible heat fluxes, while Net denotes their sum.

westerly[a] wind anomalies to its east, and the opposite pattern to its west. Conditions following the passage of the ISO active phase are marked by extensive stratiform anvil clouds and strong low-level equatorward and westerly winds. Continued eastward ISO propagation leads to a gradual return to suppressed conditions. While the ISO convective signal diminishes near the dateline, its wind signal often continues eastward and may circumnavigate the globe. Upon reaching the western Indian Ocean, the wind anomalies often contribute to the initiation of the next MJO event (Chen & Zhang, 2019; Matthews, 2008; Powell & Houze, 2015; Zhang et al., 2017).

The processes that initiate, destabilize, maintain, and propagate the ISO have challenged the tropical meteorology community for decades (Jiang et al., 2020; Zhang et al., 2013), motivating competing theories to explain its existence and behavior (Zhang, 2005; Zhang et al., 2020). The need for progress on this issue culminated in an international effort to observe ISO convective initiation in the central Indian Ocean during the boreal winter of 2011–2012: the Cooperative Indian Ocean Experiment on Intraseasonal Variability in the Year 2011 (CINDY2011)–Dynamics of the MJO (DYNAMO) (Yoneyama et al., 2013). This campaign (denoted DYNAMO hereafter) yielded unprecedented observations of the ISO and shed new light on its underlying nature.

In efforts to understand the ISO, observations and modeling studies increasingly highlight the importance of dynamic coupling between moisture, the cloud population, and circulation on intraseasonal timescales and large spatial scales (Fig. 4a). This coupling is consistent with "moisture mode theory" (Adames & Kim, 2016; Raymond, 2001; Sobel & Maloney, 2012; Zhang et al., 2020). During DYNAMO, the transition from suppressed to active ISO conditions was

---

a. Here, we use the atmospheric convention to describe wind directions. An easterly wind blows from the east; a westerly wind blows from the west.

met with a roughly coinciding reduction of large-scale subsidence, deepening of the cloud population, and moistening of the free troposphere (Johnson et al., 2015; Powell & Houze, 2015; Ruppert & Johnson, 2015). The increase in cloud coverage during this transition greatly reduced net radiative cooling in the atmosphere. Growing evidence indicates that the anomalous longwave heating of the atmosphere due to cloud trapping (Fig. 4b) is crucial to maintaining the ISO against dissipation (Andersen & Kuang, 2012; Ciesielski et al., 2017; Sobel et al., 2014).

Eastward propagation of the ISO active phase is promoted by anomalous moistening of the environment to its east and drying to its west (Kiranmayi & Maloney, 2011; Maloney, 2009; Wang et al., 2017). A primary mechanism for generating anomalous moistening in the free troposphere to the east of the ISO active phase is the advection of mean state moisture (contours in Fig. 4b) by ISO horizontal wind anomalies. Zonal and meridional gradients of mean state moisture, therefore, are crucial for the ISO's eastward propagation (Gonzalez & Jiang, 2017). The mean state moisture pattern, in turn, is tightly coupled with the mean Walker circulation in this region (Fig. 4a).

Recent evidence suggests that seasonal and interannual variations in ISO propagation characteristics are modulated by low-frequency SST patterns that regulate convective activity and atmospheric humidity patterns and their spatial gradients (Jiang, 2017; Klingaman & Demott, 2020; Roxy et al., 2019; Wang et al., 2019). Other processes that contribute to east-of-convection moistening include vertical moisture advection forced by boundary layer frictional convergence (Wang & Li, 1994; Wang & Rui, 1990) or SST gradient-driven convergence (de Szoeke & Maloney, 2020; Hsu & Li, 2012), and evaporation of condensate detrained from trade cumuli (DeMott et al., 2019; Itterly et al., 2021; Ruppert & Johnson, 2016; Sui et al., 1997).

Moisture mode theory is arguably the most successful paradigm at explaining both the existence and behavior of the ISO, which is based on two key aspects of large-scale tropical dynamics: (1) the intimate coupling between moisture and convection and (2) the rapid action of gravity waves, which homogenize temperature and promote a strong coupling between vertical motion and diabatic heating (Adames & Maloney, 2021). "Recharge-discharge theory" (Bladé & Hartmann, 1993) interprets the Eulerian variation caused by the ISO as a build-up (recharging) of moisture or convective instability over several weeks and its discharging during the active phase. While this evolution is indeed observed in an Eulerian sense (cf. also Kemball-Cook & Weare, 2001; Benedict & Randall, 2007), this theory fails to account for the above-noted processes that act on large spatial scales and promote horizontal moisture advection, which are now recognized as crucial to ISO propagation and maintenance.

2.3 ISO forcing of the ocean

The large-scale changes to winds, cloudiness, and rainfall throughout the ISO lifecycle strongly regulate exchanges of heat, water, and momentum across the air-sea interface, yielding SST perturbations of ±0.2°C on average, but as large as ±1°C for individual events (Fig. 4c). Because ocean dynamics modulate the heating and cooling of the upper ocean, SST anomaly patterns are not as spatially coherent as their atmospheric drivers (Section 3). Typical changes to the ocean net surface energy flux during active and suppressed phases of the ISO are shown in Fig. 5 and highlight the dominant roles of the net surface shortwave and surface latent heat fluxes in setting the net surface heating or cooling of the ocean throughout the ISO.

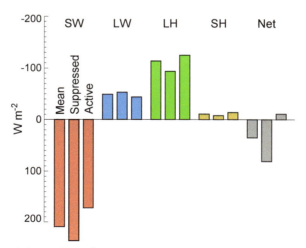

FIG. 5 Typical surface radiative and turbulent enthalpy fluxes over the tropical Indian Ocean (approximately 10°S–10°N and 30°E–90°E) for November–April mean conditions, and for ISO suppressed and active conditions. Magnitudes represent a blended average from in situ measurements and objectively analyzed gridded surface flux products.

The local transition from the ISO convectively suppressed to the active phase is marked by a gradual increase in cloud population, cloud height, and wind speed. Net surface heating of the ocean during this transition period is reduced, but typically remains positive, with daily mean SSTs increasing right up until the onset of deep convection and strong winds (Anderson et al., 1996; de Szoeke et al., 2015; Hendon & Glick, 1997; Zhang & McPhaden, 2000).

The arrival of the ISO active phase is marked by an abrupt transition from ocean warming to ocean cooling (de Szoeke et al., 2015; Zhang & McPhaden, 2000) initiated by the strong wind mixing that accompanies ISO convection, and sustained by the development of broad stratiform cloud decks that inhibit surface heating by solar radiation (Hendon & Glick, 1997; Lau & Wu, 2010; Moum et al., 2014). Much of the energy accumulated in the upper ocean during ISO suppressed and transition phases is rapidly transferred back to the atmosphere during the period of increased westerly winds and ocean-to-atmosphere surface turbulent fluxes associated with the ISO active phase. A portion of the energy is also mixed to the deeper ocean as wind mixing and ocean current-induced shear mixing erode the ocean mixed layer (Duvel et al., 2004; Duvel & Vialard, 2007; Halkides et al., 2015; Jayakumar et al., 2011; Lloyd & Vecchi, 2010; McPhaden & Foltz, 2013; Vialard et al., 2008; Vinayachandran & Saji, 2008). The ocean response to ISO active phase wind forcing may be tempered by the stratifying effects of copious and widespread precipitation falling onto the ocean surface, thus favoring development of strong surface currents versus deep ocean mixing during portions of the wind event (e.g., Moum et al., 2014). These processes are discussed in more detail in Section 3.

2.4 Differences between boreal winter and boreal summer

Coupled ocean-atmosphere feedbacks during the boreal summer season over the Indian Ocean are broadly similar to those observed during the boreal winter in that they are primarily driven by intraseasonal variations in surface solar heating and surface turbulent heat fluxes (Weller et al., 1998; Zhang et al., 2018, 2019). There is a general consensus, however, that the ocean response and, potentially, the ocean feedbacks to the atmosphere (see Section 4.1) during the boreal summer are stronger than during the boreal winter. These differences are particularly noteworthy in the northernmost Indian Ocean in the Arabian Sea and the Bay of Bengal, where intraseasonal SST perturbations are as large as 1 K (Fu et al., 2003; Gao et al., 2018). The greater surface warming during the boreal summer compared to boreal winter likely results from the smaller sun declination and therefore larger surface insolation per negative cloud anomaly, and from differences in the background ocean stability. The Arabian Sea and northern Bay of Bengal are both significantly freshened by springtime snow melt carried by the Indus and Ganges river systems, respectively, and sometimes freshened by heavy rainfall from tropical cyclones. Both of these processes can contribute to the formation of an ocean barrier layer (e.g., Sprintall & Tomczak, 1992), whereby vertical offsets in the upper ocean halocline and thermocline insulate the intervening layer from mixing (Section 3.3).

2.5 Model limitations to ISO simulation

Realistic simulation of observed intraseasonal ocean-atmosphere coupled processes in the Indian Ocean hinges upon the fidelity of many parameterized subgrid-scale processes. In particular, the sensitivity of parameterized atmospheric convection to environmental moisture (e.g., Randall, 2013; Thayer-Calder & Randall, 2009) and boundary layer enthalpy perturbations (DeMott et al., 2019) are strongly linked to ISO simulation skill, while ocean stratification and temperature response is sensitive to turbulent mixing parameterization in the ocean model (Shinoda et al., 2024).

Apart from parameterized processes, ISO simulation may also be hindered by coarse grid resolution and coupling frequency that impede a model's ability to simulate ISO-supportive processes, such as oceanic Kelvin and equatorial Rossby waves, equatorial Yoshida-Wyrtki jets (Jensen et al., 2015), and the formation of ocean diurnal warm layers and barrier layers (Bernie et al., 2008; Klingaman & Woolnough, 2014; Seo et al., 2014; Zhao & Nasuno, 2020). Model resolution also affects the representation of topography, and thus coupled ocean-land-atmosphere interactions, especially within the maritime continent. These processes are discussed further in the following sections, while more details concerning their simulation can be found in Shinoda et al. (2024).

3 Intraseasonal oceanic variability

3.1 1D perspective

3.1.1 Effects of surface fluxes on the upper ocean

The mixed layer is a region of relatively homogenous temperature and salinity in the upper ocean that is generally between 30 and 80 m deep in the Indian Ocean (de Boyer Montégut et al., 2004). Surface fluxes are significantly correlated with the intraseasonal mixed layer temperature tendency and are the dominant process forcing intraseasonal mixed layer temperature variability in much of the Indian Ocean (Drushka et al., 2012; Duvel & Vialard, 2007; Jayakumar et al., 2011; Lloyd & Vecchi, 2010; Vialard et al., 2012), particularly in regions poleward of 5 degree latitude (Halkides et al., 2015). Along the equator, however, the intraseasonal mixed layer temperature tendency is primarily forced by ocean dynamics such as advection or entrainment, with notable secondary contributions by surface fluxes.

In the Bay of Bengal, mixed layer temperature anomalies associated with the northward-propagating boreal summer ISO are primarily forced by changes in the shortwave and surface latent heat fluxes whose amplitude generally increases to the north (Duncan & Han, 2009; Girishkumar et al., 2011; Parampil et al., 2010; Sanchez-Franks et al., 2018; Sengupta & Ravichandran, 2001; Waliser et al., 2004). The relative significance of surface fluxes versus other mixed layer temperature forcings increases to the north because barrier layers formed by precipitation and river runoff in the northern Bay of Bengal (Rao & Sivakumar, 2003; Zeng et al., 2009) suppress vertical advection and entrainment. Net surface flux heating is also dominant over ocean dynamics in the central and eastern Arabian Sea (Halkides et al., 2015; Li et al., 2016). However, mixed layer temperature variability in those regions is relatively modest compared to that in the western Arabian Sea, which is predominantly forced by horizontal advection. The SCTR is also characterized by strong intraseasonal mixed layer temperature variations, with ∼70% of the mixed layer temperature forcing generated by surface fluxes (Duvel et al., 2004; Halkides et al., 2015; Vialard et al., 2008; Vinayachandran & Saji, 2008).

3.1.2 Vertical Ocean dynamics

Turbulent mixing, vertical advection, and vertical diffusion force exchanges along the mixed layer base between the warm upper-ocean and cooler thermocline (see Fig. 2). Shear-induced turbulent flux cooling, commonly referred to as entrainment, often becomes important in the central Indian Ocean during periods of enhanced ISO convection (Drushka et al., 2012). During these conditions, episodes of very strong mixing can last for several hours at a time over many days (Pujiana et al., 2018). Argo profiling data suggest that entrainment cooling accounts for ∼50% of the mixed layer temperature tendency during enhanced ISO convection (Drushka et al., 2012). During DYNAMO, mixing from below accounted for ∼33% of the 1°C mixed layer cooling in the days following a strong westerly wind burst. The winds forced strong eastward surface currents, which penetrated down to 100m and generated shear-induced entrainment that persisted for several weeks after the wind relaxed (Moum et al., 2014).

3.2 2D perspective

In certain cases, horizontal advection may be the strongest forcing of the intraseasonal mixed layer temperature tendency. Harrison and Vecchi (2001) used satellite data to determine that surface heat fluxes were insufficient to explain a large-magnitude ISO cooling event in the central Indian Ocean during the winter of 1999, leading them to hypothesize that horizontal advection was responsible for up to 80% of the mixed layer temperature anomaly. Indeed, using the ECCO-JPL ocean modeling system, it was recently demonstrated that horizontal advection often dominates intraseasonal mixed layer temperature variations along the equator (Halkides et al., 2015). Using reanalysis data in the central equatorial Indian Ocean, horizontal advection was found to be responsible for ∼67% of the intraseasonal mixed layer temperature tendency while net surface heat fluxes account for ∼33% (Halkides et al., 2015). However, in situ observations indicate a reduced role for horizontal advection and greater contribution by surface heat fluxes (Drushka et al., 2012; Duvel et al., 2004; Parampil et al., 2010; Sengupta & Ravichandran, 2001; Vialard et al., 2008). Further study is therefore necessary on this issue.

Equatorial jets, such as the Yoshida-Wyrtki jet, are characterized by strong eastward currents in the upper 100m of the ocean, and are featured most prominently between 60 and 90°E (Wyrtki, 1973; Yoshida, 1959). McPhaden and Foltz (2013) used Research Moored Array for African-Asian-Australian Monsoon Analysis and Prediction (RAMA) data in the central Indian Ocean to demonstrate the importance of strong zonal advective cooling forced by an enhanced Yoshida-Wyrtki jet during late October to mid-November of 2005. During this event, zonal advection cooled the mixed layer by 50–100 W m^{-2}, almost double that of vertical ocean dynamics. The Yoshida-Wyrtki jet during the same time of year in DYNAMO had current speeds of 1.5 m s^{-1} and was associated with an eastward transport of 24 Sv (1 Sv = 10^6 m^3 s^{-1}) in the upper 100m (Moum et al., 2014).

Horizontal advection associated with a Rossby wave contributed to the abrupt cooling in the south-central Indian Ocean during November of 2011 (Seiki et al., 2013). This cooling preceded the convective onset of the second DYNAMO ISO event. The rapid cooling was hypothesized to have delayed the convective onset of the ISO and diminished its intensity in the region of strongest cooling. Intraseasonal equatorial Rossby waves generated by ISO events have also been observed to shift the western edge of the equatorial Indian Ocean warm pool westward, sometimes generating warm intraseasonal SST anomalies of 0.15°C in the western Indian Ocean (Rydbeck et al., 2017).

The cycle of intraseasonal sea surface salinity anomalies associated with the ISO as first documented in Matthews et al. (2010) using Argo data is strongly modulated by horizontal advection associated with ISO wind stress (e.g., Cronin & McPhaden, 2002), as observed using Aquarius satellite data (Shinoda et al., 2013), Hybrid Coordinate Ocean Model (HYCOM) reanalysis (Nyadjro et al., 2020), and RAMA moorings (Horii et al., 2016). These intraseasonal anomalies of 0.05 psu are comparable to the annual cycle of salinity (Matthews et al., 2010). Horizontal advection dominates the intraseasonal salinity tendency in the eastern Indian Ocean, while evaporation minus precipitation is dominant in the western Indian Ocean and Bay of Bengal (Guan et al., 2014; Horii et al., 2016; Li et al., 2015). River runoff is a leading contributor in the Bay of Bengal (Trott et al., 2019).

3.3 Ocean layers

The intraseasonal shoaling/deepening of the mixed layer occurs at a rate of $\sim 1\,\mathrm{m\,day^{-1}}$ (Drushka et al., 2012) with amplitudes of $\pm 10\,\mathrm{m}$ (Halkides et al., 2015; Keerthi et al., 2016). In rare instances, rapid deepening can occur, such as when the mixed layer deepened from 10 to 50 m in approximately 24 h following the arrival of a westerly wind burst (Moum et al., 2014). In a recent coupled modeling study, the mixed layer depth of the Indian Ocean was varied through the use of different Langmuir turbulence and submesoscale mixed layer eddy restratification parameterizations (Orenstein et al., 2021). The patterns and variability of rainfall in the model notably shifted due to the different mixed layer depth dynamics in the model.

The mixed layer depth is important because it regulates the mixed layer temperature tendency response to the net surface heat fluxes through two effects: (1) the scaling effect and (2) the penetrative effect (Shinoda & Hendon, 1998). In the mixed layer temperature tendency calculation, the surface heat fluxes are scaled by the depth of the mixed layer. For a given surface flux, the mixed layer temperature tendency magnitude is greater for shallower versus deep mixed layers. The penetrative effect describes the increase of shortwave radiation that penetrates the base of the mixed layer as it shoals. Not accounting for the penetrative effect when calculating the mixed layer temperature results in a 20%–70% overestimate of SST across the Indian Ocean (Keerthi et al., 2016). Drushka et al. (2012) showed that the scaling and penetrative effects largely cancel one another during ISO suppressed phases but can result in a net 40% cooling during ISO enhanced phases. However, Keerthi et al. (2016) found that compensation between penetrative and scaling effects largely cancel one another regardless of the ISO phase.

During the ISO suppressed phase, the superposition of low-level easterly anomalies onto mean state low-level westerlies during the boreal winter leads to a dramatic reduction in near-surface winds and evaporative cooling (Fig. 4a), which, together with enhanced solar surface heating, promote reduced mechanical mixing and enhanced warming of the upper ocean. The intense daytime warming of the upper ocean shoals (thins) the mixed layer, thus reducing its heat capacity and enabling a pronounced diurnal cycle of SST (Fig. 2). The near-surface stable diurnal warm layer that forms in response to daytime solar heating can be $<1\,\mathrm{m}$ thick under weak winds, with temperature differences reaching $\sim 5^\circ\mathrm{C}$ between the surface and underlying mixed layer (Kawai & Wada, 2007; Soloviev & Lukas, 1997). The cycle between daytime diurnal warm layer formation and nocturnal cooling and mixing causes a diurnal cycle of SST—namely, of skin SST (Fairall et al., 1996; Kawai & Wada, 2007)—that often reaches amplitudes of $2^\circ\mathrm{C}$, with much greater extrema localized in space and/or time (Bellenger et al., 2010; Kawai & Wada, 2007; Moum et al., 2014; Ruppert & Johnson, 2015). The diurnal increase in SST increases surface latent and sensible heat fluxes by up to $\sim 60\,\mathrm{W\,m^{-2}}$ around midday or $10\,\mathrm{W\,m^{-2}}$ in the daily average (Kawai & Wada, 2007; Ruppert & Johnson, 2015). Given the high sensitivity to vertical mixing, diurnal warm layer formation generally only occurs under wind speeds weaker than $\sim 6\text{–}7\,\mathrm{m\,s^{-1}}$ (Thompson et al., 2019). At stronger wind speeds, wind-driven turbulence and enhanced evaporative cooling prevent diurnal warm layer formation (Moum et al., 2014). Chlorophyll concentrations (Matthews et al., 2014) and rainfall (Thompson et al., 2019) can additionally act to increase the near-surface stratification and suppress vertical mixing, thereby acting to concentrate and enhance near-surface heating and cooling (Shackelford et al., 2022; Soloviev & Lukas, 1997; Thompson et al., 2019; Webster et al., 1996).

Turbulent mixing during ISO events is also dependent on the presence of barrier layers, which are quantified in terms of thickness and/or potential energy (e.g., Chi et al., 2014; Lukas & Lindstrom, 1991; McPhaden & Foltz, 2013; Sprintall & Tomczak, 1992) in the Indian Ocean. The barrier layer restricts the impact of surface fluxes on the thermocline and the impact of turbulent processes at the base of the mixed layer on the upper ocean. Barrier layers can be up to 48 m thick (Alappattu et al., 2017), as observed under a band of convective precipitation using an Airborne eXpendable Conductivity Temperature Depth during DYNAMO, but are more commonly 0–10 m thick. During DYNAMO, the barrier layer was thickest during the suppressed phase and was generally reduced during the active phases (Pujiana et al., 2018). However, observations using Airborne eXpendable Conductivity Temperature Depths show that $\sim 2\%$ of barrier layers are thicker than 30 m during the suppressed phase, while the percentage of thick barrier layers jumps to 15% during the enhanced phase (Alappattu et al., 2017). These variations are important because the intraseasonal mixed layer temperature anomaly is double the magnitude for thin barrier layers versus thick barrier layers ($\pm 0.2^\circ\mathrm{C}$ versus $\pm 0.1^\circ\mathrm{C}$) (Drushka et al., 2014;

Girishkumar et al., 2011). Large barrier layer thickness variations (>20m) are generally coincident with thick (>30m) mean barrier layers primarily found in the northern Bay of Bengal and equatorial eastern Indian Ocean (Drushka et al., 2012).

The leading mechanisms responsible for intraseasonal barrier layer formation and dissipation are an area of active research. Early studies suggested that interleaving of ocean layers is responsible for barrier layer formation. In these frameworks, differential horizontal advection can either transport a salty warm layer below the warm, fresh mixed layer (Lukas & Lindstrom, 1991) or a fresh surface layer over the top of a relatively salty layer (Cronin & McPhaden, 2002). The modeling study of Schiller and Godfrey (2003) demonstrated that a barrier layer can develop during the suppressed phase of the ISO, when weak winds permit the mixed layer to shoal. The barrier layer then thins when the mixed layer deepens during the active phase.

4 Ocean feedbacks to the atmosphere

4.1 Intraseasonal SST feedbacks to maintenance and propagation of atmospheric ISOs

Understanding how intraseasonal SST variability affects the ISO is challenging for several reasons. First, although the observed ISO always involves ocean-atmosphere coupled feedbacks, some ISO events appear to be more sensitive to ocean coupling than others (Fu et al., 2015; Gottschalck et al., 2013). Model experiments designed to test the sensitivity of the ISO to ocean coupling by comparing ISO simulation in coupled general circulation models and atmospheric-only general circulation models with prescribed SSTs may suffer from differences in mean state conditions (Klingaman & Woolnough, 2014; Zhang et al., 2006) or erroneous phasing of SST and rainfall when high-frequency (i.e., daily) SSTs are prescribed to the atmospheric general circulation model (Fu et al., 2003; Pegion & Kirtman, 2008). These issues may be less evident with regional models, where convection is strongly constrained by boundary forcing, and especially if regional simulations are performed at convection-permitting resolutions (i.e., ≲2km in the horizontal).

DeMott et al. (2016) estimated that intraseasonal SST fluctuations directly contribute up to about 2% of column moistening associated with ISO maintenance, and about 10% of moistening associated with ISO propagation. The former, although small, offsets about 10% of the column moisture lost through precipitation, and thus helps sustain the ISO against dissipation (Bui et al., 2020; Riley Dellaripa & Maloney, 2015). The latter is consistent with the idea that warm SST anomalies east of ISO convection promote ISO propagation by reducing static stability of the lower atmosphere, reducing surface pressure, and enhancing boundary layer convergence (Wang & Rui, 1990). ISO propagation may also be aided by intraseasonal SST feedbacks to local circulations that encourage ISO convection to "detour" south of the maritime continent during December through February. Zhang and Ling (2017) showed that ISO events that propagate beyond the maritime continent (i.e., Sumatra, Borneo, Indonesia, and surrounding islands) are associated with warmer intraseasonal SST anomalies, specifically in the Java Sea, Banda Sea, and Timor Sea, than nonpropagating events. Zhou and Murtugudde (2020) found that these large warm SST anomalies (nearly +1 K) that develop during the ISO suppressed phase (e.g., Vialard et al., 2013) induce low-level cyclonic (clockwise) flow near 120°E and 12°S that moistens the atmosphere south of the maritime continent by advecting high column water vapor near the equator southward.

Other work suggests that cross-scale feedbacks that link the intraseasonal scale with shorter or longer timescales are important to the ISO. For example, several studies have shown the importance of diurnal SST warming for simulating the MJO, either through its rectification onto the intraseasonal SST (Bernie et al., 2005, 2008; Klingaman et al., 2011; Zhao & Nasuno, 2020) or by diurnal moistening of the lower atmosphere (Ruppert, 2016; Seo et al., 2014). Another cross-scale feedback was identified by DeMott et al. (2019) who prescribed monthly SSTs from four coupled general circulation models to their respective atmospheric simulations. Despite identical SST mean state and low-frequency SST variability in the coupled and atmospheric-only general circulation model pairs, mean state column water vapor was more peaked on the equator in the coupled general circulation models, which promoted ISO propagation in the coupled general circulation models by enhancing east-of-convection moistening by meridional advection of mean state moisture by anomalous poleward flow (Fig. 4b). The more peaked mean state moisture in coupled general circulation models was attributed to SST-driven changes to the magnitude and altitude of convective moistening. Understanding the role of SST-modulated surface fluxes for mean state moisture and MJO propagation remains an area of active research.

4.2 Other modes of SST variability that affect atmospheric ISV

4.2.1 The SST diurnal cycle

In addition to the intraseasonal cycle of SST, the SST diurnal cycle associated with diurnal warm layers is also often prominently superimposed onto the ISV mode. Given the intraseasonal variability of both wind speeds and cloudiness, diurnal SST variability is most pronounced during the suppressed phase of the ISO and leading up to the active phase (Figs. 2 and 6)

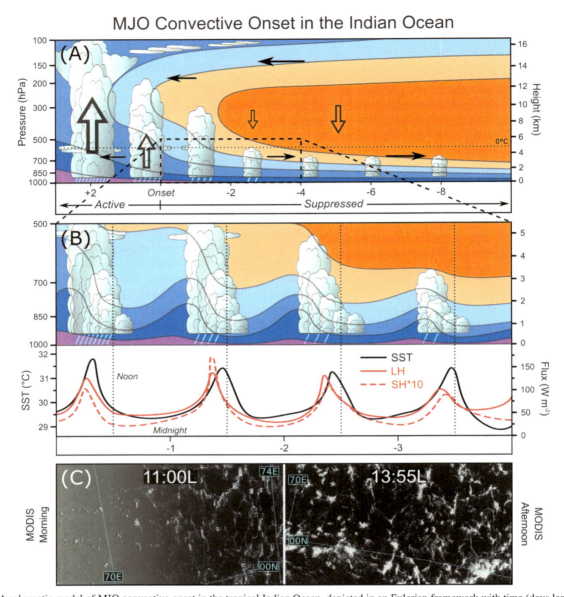

FIG. 6 A schematic model of MJO convective onset in the tropical Indian Ocean, depicted in an Eulerian framework with time (days leading up to MJO onset) from the right to the left. (a) Convective clouds, relative humidity (shading; *brown* is drier), and circulation (arrows) from the suppressed to the early-active phase. (b) (*top*) Subset of (a) emphasizing diurnal variability during the late suppressed phase; (*bottom*) SST (*black*), LH (*red*, solid), and SH (multiplied by 10; *red*, dashed). (c) Characteristic cloud field during the suppressed phase as depicted by MODIS true-color imagery in the (*left*) morning (0600 UTC, 1100L) and (*right*) afternoon (0855 UTC, 1355L) (from October 12, 2011), highlighting the prominence of mesoscale cloud organization (i.e., open cells and horizontal convective rolls). *(From Ruppert & Johnson, 2015. © American Meteorological Society. Used with permission.)*

(Bellenger et al., 2010; Itterly et al., 2021; Matthews et al., 2014; Ruppert & Johnson, 2015; Thompson et al., 2019). The stratification that results from strong diurnal heating in this regime generally stabilizes the upper ocean. When integrated over many days, this heating has the effect of "rectifying" a stronger cycle of SST on intraseasonal timescales (Bernie et al., 2005; Matthews et al., 2014; Shinoda, 2005; Webster et al., 1996).

The overlying atmosphere strongly responds to this rectified SST forcing (Fig. 6). The diurnally increased latent and sensible heat fluxes into the troposphere due to solar heating during the suppressed phase of the ISO cause both diurnal and day-to-day variations of atmospheric mixed layer depth during the ISO suppressed phase (Johnson & Ciesielski, 2017). In response, the shallow moist convection that prevails during the suppressed phase of the ISO exhibits a prominent afternoon deepening (Chen & Houze, 1997; Ruppert & Johnson, 2015; Sakaeda et al., 2017, 2018; Sui et al., 1997). On longer timescales, this diurnal invigoration superimposes onto the ISO, causing a diurnally-varying, stepwise moistening of the troposphere, over a progressively deeper layer in advance of the ISO active phase (Johnson et al., 1999; Kikuchi & Takayabu, 2004; Ruppert, 2016; Ruppert & Johnson, 2015). While the direct cloud-radiation interaction is also a key driver of the diurnal cycle of tropical convection (Gray & Jacobson, 1977; Randall et al., 1991; Ruppert et al., 2018), the formation of oceanic diurnal warm layers is likely important to the afternoon signature of moistening and cloud deepening.

4.2.2 Interannual SST variability

The IOD (Saji et al., 1999) is the primary mode of interannual variability in the Indian Ocean. The positive phase of the IOD is characterized by warm SST anomalies in the western tropical Indian Ocean, and cold anomalies in the east; the pattern is reversed for negative IOD. The effect of these SST patterns on the ISO can be interpreted with the moisture mode theory through SST regulation of mean state moisture and background low-level convergence (Section 2.2). During positive IOD, the typical eastward zonal moisture gradient is weakened or reversed, thereby reducing moistening by zonal moisture advection (e.g., Adames & Wallace, 2015), and inhibiting ISO propagation (Benedict et al., 2015; Wilson et al., 2013). Warm SST anomalies and increased low-level convergence in the eastern Indian Ocean also contribute to enhanced ISO propagation during negative IOD (Shinoda & Han, 2005).

4.2.3 SST gradients

The horizontal SST distribution in the tropical Indian Ocean is hypothesized to modulate both the initiation and propagation of the ISO. Localized warm SST anomalies hydrostatically induce relatively low pressure in the overlying atmospheric boundary layer while cold SST anomalies induce relatively high pressure. The low-level wind speed is proportional to the magnitude of the horizontal SST gradient, while the divergence is proportional to the Laplacian of the SST (∇^2SST), with divergence (convergence) favored over cold (warm) anomalies. In the tropics, SST "hotspots" were hypothesized to drive solenoidal circulations, like a sea-breeze circulation, that could initiate atmospheric convection (Malkus, 1957). Lindzen and Nigam (1987) and Back and Bretherton (2005) demonstrated that the distribution and amplitude of SST gradients in the Pacific Intertropical Convergence Zone (ITCZ) play an essential role in determining the patterns of low-level atmospheric convergence in the ITCZ, and hence deep convection there. In fact, Li and Carbone (2012) observed that ~75% of rainfall events in the western Pacific Ocean initiated near warm patches where the SST gradient was strong, demonstrating that these events can be attributed to the SST gradients through their influence on boundary layer convergence.

For intraseasonal phenomena such as the ISO, Hsu and Li (2012) estimated that SST-induced pressure gradients were responsible for 10%–25% of the low-level convergence that leads ISO convection, assisting its eastward propagation and westward vertical tilt with height. This horizontal convergence serves as both an impetus for upward motion and a mechanism of moisture aggregation in the atmospheric boundary layer. Using a simple model to estimate the amount of moist static energy convergence forced by warm SST anomalies ahead of ISO convection in the Indian Ocean, de Szoeke and Maloney (2020) observed that SST gradient-driven boundary layer convergence contributed approximately two to three times more moist static energy to the lower atmosphere than enhanced surface latent heat fluxes generated by anomalously warm SSTs which also lead ISO convection (DeMott et al., 2016).

The impact of the SST spatial distribution on the convective onset of ISO events in the Indian Ocean has also been observed (Webber et al., 2010, 2012a, 2012b). Rydbeck and Jensen (2017) found that oceanic downwelling equatorial Rossby waves augmented the intraseasonal humidification of the atmospheric boundary layer preceding the ISO onset by generating persistent warm intraseasonal SST anomalies, which in turn enhanced the horizontal SST gradient in the western Indian Ocean (see also Rydbeck et al., 2017). The SST gradient-induced atmospheric boundary layer convergence from this effect accounted for up to 45% of the intraseasonal signal of convergence. Using an idealized convection-resolving model, Rydbeck et al. (2019a) further demonstrated that the enhanced SST gradients in the western Indian Ocean also tend to accelerate the low-level winds, which increase the surface latent heat fluxes and assist the humidification of the atmospheric boundary layer prior to and during the ISO convective onset. Carbone and Li (2015) observed that SST-induced boundary layer convergence tends to lead ISO convection by ~10 days, especially in the western Indian Ocean, further emphasizing the likely importance of SST-induced boundary layer convergence during ISO convective initiation.

4.2.4 Oceanic equatorial waves

Intraseasonal low-level wind variations in the tropical Indian Ocean drive large-scale shifts in oceanic mass that generate large-scale, intraseasonal oceanic responses (e.g., Moum et al., 2016). Westerly surface wind anomalies associated with the ISO drive Ekman convergence along the equator, which depresses the thermocline, increases the sea surface height, and generates eastward propagating downwelling equatorial Kelvin waves with strong eastward currents (i.e., Yoshida jets). In the neighboring off-equatorial regions, upwelling equatorial Rossby waves are also generated by the same wind stress, similar to a Gill (1980) response. Because the eastward propagation of ISO winds tends to dampen short-wavelength and high-frequency ocean waves, wind work associated with the ISO preferentially rectifies onto the low-frequency, long wavelength spectrum of the ocean (Kessler et al., 1995). Oceanic Rossby waves can also be forced by Kelvin wave energy that reflects off an eastern boundary, such as Sumatra in the case of the Indian Ocean. First (second) baroclinic mode equatorial Kelvin waves propagate to the east with phase speeds near $2.7\,\mathrm{m\,s^{-1}}$ ($1.5\,\mathrm{m\,s^{-1}}$), requiring 25–30 days (50 days) to cross the equatorial basin width of the Indian Ocean. Pujiana and McPhaden (2020) observed that second baroclinic mode Kelvin waves dominate eastward propagating intraseasonal variability along the equatorial Indian Ocean.

Prominent variations of horizontal currents and sea surface heights at periods of 180 days (Fu, 2007; Han et al., 1999; Jensen, 1993; Luyten & Roemmich, 1982; McPhaden, 1982; Reppin et al., 1999; Reverdin & Luyten, 1986) and 90 days (Brandt et al., 2003; Fu, 2007; Han, 2005; Han et al., 2001; Iskandar & McPhaden, 2011; McPhaden, 1982; Nagura & McPhaden, 2012; Rydbeck et al., 2017, 2021; Rydbeck & Jensen, 2017; Suresh et al., 2013; West et al., 2018) manifest both in models and observations of the Indian Ocean. Resonant basin modes have previously been hypothesized to explain much of this variance. Such modes occur when counterpropagating oceanic equatorial waves are excited by intraseasonal wind forcing and subsequently reflect off the basin boundaries, such that wind-generated and reflected waves harmonically and constructively interfere with one another. While early theories of the 90-day resonant basin mode suggested the importance of constructive interference between Kelvin waves reflected from the western boundary and Rossby waves reflected from the eastern boundary (Fu, 2007; Han, 2005; Han et al., 2001, 2011), more recent work indicates that only Rossby reflections at the eastern boundary, which constructively interfere with wind-generated waves, are necessary to explain the observed 90-day variability (Nagura & McPhaden, 2012).

Rossby and Kelvin waves generated by the ISO are an important source of intraseasonal SST and ocean heat content variability and are hypothesized to also assist the initiation and maintenance of the ISO in the Indian Ocean. Using satellite observations, Webber et al. (2010, 2012a, 2012b) first demonstrated that the initiation of primary ISO events (i.e., those with no predecessor intraseasonal signal in the atmosphere) in the western Indian Ocean was, in fact, preceded by intraseasonal downwelling oceanic Rossby waves that warmed the SST by ∼0.15°C prior to ISO convective onset. They hypothesized a positive feedback loop in which the ISO generates a downwelling Kelvin wave that propagates eastward and reflects at the eastern boundary as downwelling Rossby waves (Fig. 7). These Rossby waves propagate westward, depress the thermocline, and generate warm SST anomalies in the western Indian Ocean that subsequently initiate convection in the next ISO event. Rydbeck et al. (2017) showed that these waves primarily warmed the western Indian Ocean by advecting the equatorial warm pool to the west. However, data suggest that this feedback loop of ISO initiation is relatively rare, likely occurring less than once per year (Rydbeck & Jensen, 2017; Webber et al., 2012b). However, a more common interaction is observed to occur whereby oceanic Rossby waves depress the thermocline and increase ocean heat content by 10%–15% in the central Indian Ocean during the enhanced convective phase of already existing ISO events (Rydbeck et al., 2021, 2019b; Shinoda et al., 2013).

Intraseasonal equatorial waves impact ocean dynamics across the entire tropical Indian Ocean. In the SCTR, Rossby waves have been observed to modulate the upper ocean temperature and salinity profiles (Shinoda et al., 2013, 2017). This is important because the climatologically shallow mixed layer of the region exhibits a greater sensitivity to surface flux forcing (Duvel et al., 2004). Rossby waves propagating into the region are associated with a $20\,\mathrm{cm\,s^{-1}}$ acceleration of the northward flowing Somali current (Shinoda et al., 2017). These northward currents are a precursor to the spin up of the Great Whirl (Beal & Donohue, 2013; Melzer et al., 2019), implying important downstream impacts by Rossby waves. The Yoshida jets associated with these Kelvin waves also transport high-salinity water originating in the Arabian Sea into the eastern equatorial Indian Ocean, affecting the ocean stratification in the region (Jensen et al., 2015; Shinoda et al., 2013).

5 ISV and the maritime continent prediction barrier

The maritime continent is one of the rainiest places on earth, and the complex meteorology there is greatly affected by linkages across both spatial and temporal scales (Houze et al., 2015; Krishnamurti et al., 1973; Liu et al., 2007; Neale & Slingo, 2003; Ramage, 1968; Yamanaka et al., 2018). The diurnal cycle is the leading mode of rainfall production in the maritime continent. The overall character of the diurnal cycle is the consequence of land-sea breezes and mountain-valley circulations, which serve as the primary rainfall triggering mechanisms (Houze et al., 1981; Johnson & Priegnitz, 1981; Kikuchi & Wang, 2008). In addition, the offshore propagation of rainfall is greatly modulated and promoted by gravity waves (Coppin & Bellon, 2019; Fujita et al., 2011; Ichikawa & Yasunari, 2006, 2007; Mapes et al., 2003;

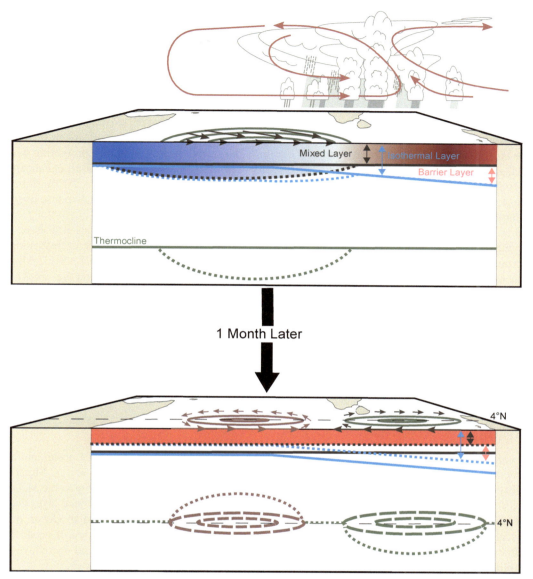

FIG. 7 Schematic of large-scale ocean Kelvin and Rossby wave response to ISO wind forcing. The top panel shows the quasi-immediate response, and the bottom panel shows the delayed response after wave energy has reflected off the eastern boundary. Sea surface height is shown in solid *green/red* contours, and mean (perturbations to) ocean layers are shown using solid (dashed) contours.

Ruppert & Zhang, 2019). On intraseasonal timescales, propagation of the ISO across the maritime continent drives pronounced changes to this diurnal cycle through its influence on large-scale background horizontal wind, vertical motion, moisture, and the preferred mode of the convective cloud population (Peatman et al., 2014; Ruppert et al., 2020; Sakaeda et al., 2017, 2020; Vincent & Lane, 2016).

These prominently diurnal precipitation mechanisms are extremely challenging to accurately represent in the numerical models employed for both operational weather and climate prediction, in turn causing errors at intraseasonal timescales (Baranowski et al., 2019; Dai & Trenberth, 2004; Kim et al., 2014b; Neale & Slingo, 2003; Slingo et al., 2003). The long-standing prediction challenges related to this issue have come to be referred to as the maritime continent "prediction barrier" (Fu et al., 2011; Inness & Slingo, 2006; Kim et al., 2009, 2014a; Lin et al., 2006; Seo et al., 2009; Vitart et al., 2007; Weaver et al., 2011). The prediction barrier takes the form of large errors in representing both if and how the eastward-propagating ISO transits the maritime continent (Ahn et al., 2020; Jiang, 2017). While the root cause for the barrier effect remains an open subject, mounting evidence suggests a key role of horizontal moisture advection, in a manner consistent with the prevailing moisture mode theory for ISO propagation (see Section 2.2). Recent research implicates background moisture as a key signature of a model's ability to accurately represent MJO propagation in relation to the maritime continent. Namely, a given model is found to accurately predict eastward ISO propagation across—or its blocking by—the maritime continent if it accurately predicts the background humidity state (Gonzalez & Jiang, 2017; Jiang, 2017; Ling et al., 2017).

The foregoing discussion implies that understanding the barrier effect of the maritime continent equates to understanding the fundamental controls over the background humidity state in the region. These findings further imply that the accurate model treatment of any local or regional process that affects the mean humidity state of the maritime continent is crucial to accurate prediction of ISV. Numerical model experiments highlight that the existence of the islands themselves plays a critical role in controlling this humidity state. Namely, when the islands are present, much of the solar forcing received at the surface is communicated to the atmosphere through sensible heat flux, in turn promoting deep convection and rainfall (Hagos et al., 2016; Ruppert & Chen, 2020; Tseng et al., 2017). This "island rainfall enhancement" (Cronin et al., 2015; Sobel et al., 2011) dramatically increases moisture over the region (Ruppert & Chen, 2020). Experiments using a global ocean-atmosphere coupled model corroborate that the increased time-mean humidity response over the maritime continent promotes a stronger ISO (Tseng et al., 2017).

The maritime continent is additionally subject to intraseasonal ocean-atmosphere feedback processes unique to the complex bathymetry of this region. Copious rainfall freshens the region's inner seas through direct freshening of the ocean surface and through freshwater input from the islands' many river systems (e.g., Matthews et al., 2013). Upper ocean stabilization by these surface freshening processes is strongly countered by the vigorous tidal mixing that takes place along the northeast-to-southwest water trajectories that form the branches of the Indonesian Throughflow (e.g., Ffield & Gordon, 1996; Gordon et al., 2010; Susanto et al., 2000; see Sprintall et al., 2024). Sensitivity tests with global ocean-atmosphere coupled models indicate a prominent role for tidal mixing in reducing SSTs and climatological rainfall over the Indonesian seas compared to adjacent open ocean regions (Jochum & Potemra, 2008; Koch-Larrouy et al., 2010; Napitu et al., 2015; Sprintall et al., 2014). This regional ocean response to rainfall has implications for the relative freshness of water entering the Indian Ocean from the Indonesian Throughflow exit passages, as well as for the regional distribution of atmospheric water vapor surrounding the maritime continent and its role in regulating ISO propagation.

6 Conclusions

Intraseasonal variability in the Indian Ocean is the result of the continuous adjustments of the ocean and atmosphere to fluxes of momentum, freshwater, and heat whose variability spans sub-daily to interannual scales. Much of the Indian Ocean variability on intraseasonal timescales is forced by the atmosphere's ISO (often referred to as the MJO), and is characterized by ISV in upper ocean currents, sea surface height, salinity, SST, thermocline, and mixed layer depth. ISO wind anomalies force oceanic Kelvin and equatorial Rossby waves that propagate across the Indian Ocean on monthly-to-seasonal timescales; abundant rainfall during the ISO active phase contributes to the formation of ocean barrier layers; reduced cloudiness during the ISO suppressed phase promotes the formation of extensive diurnal warm layers in the upper ocean that amplify anomalous SST perturbations on 30- to 60-day timescales.

The multiscale ocean response to ISO forcing in turn regulates the SST and the ocean-to-atmosphere surface turbulent and infrared radiative fluxes. Warming and moistening of the atmospheric boundary layer by these fluxes contribute to the initiation and maintenance of atmospheric convection. The complexity of the ocean response to ISO forcing supports a multitude of local and remote ocean feedback processes to the atmosphere that span temporal scales, and thus contributes to the diversity of ISO events observed in nature.

7 Educational resources

Additional information on the MJO and ocean-atmosphere coupling may be found at the following links:

- What is the MJO and why do we care? https://www.climate.gov/news-features/blogs/enso/what-mjo-and-why-do-we-care.
- MJO realtime forecasts: https://www.cpc.ncep.noaa.gov/products/precip/CWlink/MJO/CLIVAR/clivar_wh.shtml.
- Realtime sea surface height anomaly data: https://sos.noaa.gov/datasets/sea-surface-height-anomaly-real-time/.
- https://www.meted.ucar.edu/training_module.php?id=13#.X-0Ug8ZKjOQ.
- https://www.meted.ucar.edu/training_module.php?id=996&tab=03#.X-0UgulKhTY.
- Ocean Literacy Portal: https://oceanliteracy.unesco.org/?post-types=all&sort=popular.

Acknowledgments

C. DeMott was supported by DOE RGMA award DE-SC0020092, NSF Physical Oceanography award AGS-1924659, and NOAA CVP award NA18OAR4310405. J. Ruppert was supported by NSF Physical & Dynamic Meteorology award AGS-1712290. A. Rydbeck was supported by NRL base funding and ONR DRIs MISO-BoB and PISTON.

Author contributions

All authors contributed to all sections of this chapter. CAD led Section 2, AR led Section 3, and JR led Section 5. JR created Figs. 2, 4, and 5, AR created Fig. 7, and CAD created Fig. 3.

References

Adames, Á. F., & Kim, D. (2016). The MJO as a dispersive, convectively coupled moisture wave: Theory and observations. *Journal of the Atmospheric Sciences, 73*, 913–941. https://doi.org/10.1175/JAS-D-15-0170.1.

Adames, Á. F., & Maloney, D. E. D. (2021). Moisture mode theory's contribution to advances in our understanding of the Madden-Julian oscillation and other tropical disturbances. *Current Climate Change Reports, 7*, 72–85. https://doi.org/10.1007/s40641-021-00172-4.

Adames, Á. F., & Wallace, J. M. (2015). Three-dimensional structure and evolution of the moisture field in the MJO. *Journal of the Atmospheric Sciences, 72*, 3733–3754. https://doi.org/10.1175/JAS-D-15-0003.1.

Ahn, M., et al. (2020). MJO propagation across the maritime continent: Are CMIP6 models better than CMIP5 models? *Geophysical Research Letters, 47*, 1–9. https://doi.org/10.1029/2020GL087250.

Alappattu, D. P., Wang, Q., Kalogiros, J., Guy, N., & Jorgensen, D. P. (2017). Variability of upper ocean thermohaline structure during a MJO event from DYNAMO aircraft observations. *Journal of Geophysical Research, Oceans, 122*, 1122–1140. https://doi.org/10.1002/2016JC012137.

Andersen, J. A., & Kuang, Z. (2012). Moist static energy budget of MJO-like disturbances in the atmosphere of a zonally symmetric Aquaplanet. *Journal of Climate, 25*, 2782–2804. https://doi.org/10.1175/JCLI-D-11-00168.1.

Anderson, S. P., Weller, R. A., & Lukas, R. B. (1996). Surface buoyancy forcing and the mixed layer of the Western Pacific warm pool: Observations and 1D model results. *Journal of Climate, 9*, 3056–3085. https://doi.org/10.1175/1520-0442(1996)009<3056:SBFATM>2.0.CO;2.

Arief, D., & Murray, S. P. (1996). Low-frequency fluctuations in the Indonesian throughflow through Lombok Strait. *Journal of Geophysical Research, Oceans, 101*, 12455–12464. https://doi.org/10.1029/96JC00051.

Back, L. E., & Bretherton, C. S. (2005). The relationship between wind speed and precipitation in the Pacific ITCZ. *Journal of Climate, 18*, 4317–4328. https://doi.org/10.1175/JCLI3519.1.

Baranowski, D. B., Waliser, D. E., Jiang, X., Ridout, J. A., & Flatau, M. K. (2019). Contemporary GCM fidelity in representing the diurnal cycle of precipitation over the maritime continent. *Journal of Geophysical Research – Atmospheres, 124*, 747–769. https://doi.org/10.1029/2018JD029474.

Beal, L. M., & Donohue, K. A. (2013). The Great Whirl: Observations of its seasonal development and interannual variability. *Journal of Geophysical Research, Oceans, 118*, 1–13. https://doi.org/10.1029/2012JC008198.

Bellenger, H., Takayabu, Y. N., Ushiyama, T., & Yoneyama, K. (2010). Role of diurnal warm layers in the diurnal cycle of convection over the tropical Indian Ocean during MISMO. *Monthly Weather Review, 138*, 2426–2433. https://doi.org/10.1175/2010MWR3249.1.

Benedict, J. J., Pritchard, M. S., & Collins, W. D. (2015). Sensitivity of MJO propagation to a robust positive Indian Ocean dipole event in the superparameterized CAM. *Journal of Advances in Modeling Earth Systems, 7*, 1901–1917. https://doi.org/10.1002/2015MS000530.

Benedict, J. J., & Randall, D. A. (2007). Observed characteristics of the MJO relative to maximum rainfall. *Journal of the Atmospheric Sciences, 64*, 2332–2354. https://doi.org/10.1175/JAS3968.1.

Bernie, D. J., Guilyardi, E., Madec, G., Slingo, J. M., Woolnough, S. J., & Cole, J. (2008). Impact of resolving the diurnal cycle in an ocean–atmosphere GCM. Part 2: A diurnally coupled CGCM. *Climate Dynamics, 31*, 909–925. https://doi.org/10.1007/s00382-008-0429-z.

Bernie, D. J., Woolnough, S. J., Slingo, J. M., & Guilyardi, E. (2005). Modeling diurnal and intraseasonal variability of the ocean mixed layer. *Journal of Climate, 18*, 1190–1202. https://doi.org/10.1175/JCLI3319.1.

Bladé, I., & Hartmann, D. L. (1993). Tropical intraseasonal oscillations in a simple nonlinear model. *Journal of the Atmospheric Sciences, 50*, 2922–2939. https://doi.org/10.1175/1520-0469(1993)050<2922:TIOIAS>2.0.CO;2.

Brandt, P., Dengler, M., Rubino, A., Quadfasel, D., & Schott, F. (2003). Intraseasonal variability in the southwestern Arabian Sea and its relation to the seasonal circulation. *Deep-Sea Research Part II: Topical Studies in Oceanography, 50*, 2129–2141. https://doi.org/10.1016/S0967-0645(03)00049-3.

Bui, H. X., Maloney, E. D., Riley Dellaripa, E. M., & Singh, B. (2020). Wind speed, surface flux, and intraseasonal convection coupling from CYGNSS data. *Geophysical Research Letters, 47*. https://doi.org/10.1029/2020GL090376.

Carbone, R. E., & Li, Y. (2015). Tropical oceanic rainfall and sea surface temperature structure: Parsing causation from correlation in the MJO. *Journal of the Atmospheric Sciences, 72*, 2703–2718. https://doi.org/10.1175/JAS-D-14-0226.1.

Chen, S. S., & Houze, R. A. (1997). Diurnal variation and life-cycle of deep convective systems over the tropical pacific warm pool. *Quarterly Journal of the Royal Meteorological Society, 123*, 357–388. https://doi.org/10.1002/qj.49712353806.

Chen, X., & Zhang, F. (2019). Relative roles of preconditioning moistening and global circumnavigating mode on the MJO convective initiation during DYNAMO. *Geophysical Research Letters, 46*, 1079–1087. https://doi.org/10.1029/2018GL080987.

Chi, N.-H., Lien, R.-C., D'Asaro, E. A., & Ma, B. B. (2014). The surface mixed layer heat budget from mooring observations in the Central Indian Ocean during Madden–Julian oscillation events. *Journal of Geophysical Research, Oceans, 119*, 4638–4652. https://doi.org/10.1002/2014JC010192.

Ciesielski, P. E., Johnson, R. H., Jiang, X., Zhang, Y., & Xie, S. (2017). Relationships between radiation, clouds, and convection during DYNAMO. *Journal of Geophysical Research – Atmospheres*, 1–20. https://doi.org/10.1002/2016JD025965.

Coppin, D., & Bellon, G. (2019). Physical mechanisms controlling the offshore propagation of convection in the tropics: 1. Flat Island. *Journal of Advances in Modeling Earth Systems, 11*, 3042–3056. https://doi.org/10.1029/2019MS001793.

Cronin, T. W., Emanuel, K. A., & Molnar, P. (2015). Island precipitation enhancement and the diurnal cycle in radiative-convective equilibrium. *Quarterly Journal of the Royal Meteorological Society, 141*, 1017–1034. https://doi.org/10.1002/qj.2443.

Cronin, M. F., & McPhaden, M. J. (2002). Barrier layer formation during westerly wind bursts. *Journal of Geophysical Research, Oceans, 107*, SRF 21-1. https://doi.org/10.1029/2001jc001171.

Dai, A., & Trenberth, K. E. (2004). The diurnal cycle and its depiction in the community climate system model. *Journal of Climate, 17*, 930–951. https://doi.org/10.1175/1520-0442(2004)017<0930:TDCAID>2.0.CO;2.

de Boyer Montégut, C., Madec, G., Fischer, A. S., Lazar, A., & Iudicone, D. (2004). Mixed layer depth over the global ocean: An examination of profile data and a profile-based climatology. *Journal of Geophysical Research, C: Oceans, 109*, 1–20. https://doi.org/10.1029/2004JC002378.

de Szoeke, S. P., Edson, J. B., Marion, J. R., Fairall, C. W., & Bariteau, L. (2015). The MJO and air–sea interaction in TOGA COARE and DYNAMO. *Journal of Climate, 28*, 597–622. https://doi.org/10.1175/JCLI-D-14-00477.1.

de Szoeke, S. P., & Maloney, E. D. (2020). Atmospheric mixed layer convergence from observed MJO sea surface temperature anomalies. *Journal of Climate, 33*, 547–558. https://doi.org/10.1175/JCLI-D-19-0351.1.

de Szoeke, S. P., Skyllingstad, E. D., Zuidema, P., & Chandra, A. S. (2017). Cold pools and their influence on the tropical marine boundary layer. *Journal of the Atmospheric Sciences, 74*, 1149–1168. https://doi.org/10.1175/JAS-D-16-0264.1.

DeMott, C. A., Benedict, J. J., Klingaman, N. P., Woolnough, S. J., & Randall, D. A. (2016). Diagnosing Ocean feedbacks to the MJO: SST-modulated surface fluxes and the moist static energy budget. *Journal of Geophysical Research, 121*, 8350–8373. https://doi.org/10.1002/2016JD025098.

DeMott, C. A., Klingaman, N. P., Tseng, W., Burt, M. A., Gao, Y., & Randall, D. A. (2019). The convection connection: How ocean feedbacks affect tropical mean moisture and MJO propagation. *Journal of Geophysical Research – Atmospheres, 124*, 11910–11931. https://doi.org/10.1029/2019JD031015.

DeMott, C. A., Klingaman, N. P., & Woolnough, S. J. (2015). Atmosphere-ocean coupled processes in the Madden-Julian oscillation. *Reviews of Geophysics, 53*, 1099–1154. https://doi.org/10.1002/2014RG000478.

Drushka, K., Sprintall, J., & Gille, S. T. (2014). Subseasonal variations in salinity and barrier-layer thickness in the eastern equatorial Indian Ocean. *Journal of Geophysical Research, Oceans, 119*, 805–823. https://doi.org/10.1002/2013JC009422.

Drushka, K., Sprintall, J., Gille, S. T., & Wijffels, S. (2012). In situ observations of Madden-Julian oscillation mixed layer dynamics in the Indian and western Pacific oceans. *Journal of Climate, 25*, 2306–2328. https://doi.org/10.1175/JCLI-D-11-00203.1.

Duncan, B., & Han, W. (2009). Indian Ocean intraseasonal sea surface temperature variability during boreal summer: Madden-Julian oscillation versus submonthly forcing and processes. *Journal of Geophysical Research, Oceans, 114*. https://doi.org/10.1029/2008JC004958.

Duvel, J. P., Roca, R., & Vialard, J. (2004). Ocean mixed layer temperature variations induced by intraseasonal convective perturbations over the Indian Ocean. *Journal of the Atmospheric Sciences, 61*, 1004–1023. https://doi.org/10.1175/1520-0469(2004)061<1004:OMLTVI>2.0.CO;2.

Duvel, J. P., & Vialard, J. (2007). Indo-Pacific Sea surface temperature perturbations associated with intraseasonal oscillations of tropical convection. *Journal of Climate, 20*, 3056–3082.

Fairall, C. W., Bradley, E. F., Godfrey, J. S., Wick, G. A., Edson, J. B., & Young, G. S. (1996). Cool-skin and warm-layer effects on sea surface temperature. *Journal of Geophysical Research, 101*, 1295–1308.

Ffield, A., & Gordon, A. L. (1996). Tidal mixing signatures in the Indonesian seas. *Journal of Physical Oceanography, 26*, 1924–1937. https://doi.org/10.1175/1520-0485(1996)026<1924:TMSITI>2.0.CO;2.

Fu, L. L. (2007). Intraseasonal variability of the equatorial Indian ocean observed from sea surface height, wind, and temperature data. *Journal of Physical Oceanography, 37*, 188–202. https://doi.org/10.1175/JPO3006.1.

Fu, X., Wang, B., Lee, J.-Y., Wang, W., & Gao, L. (2011). Sensitivity of dynamical intraseasonal prediction skills to different initial conditions. *Monthly Weather Review, 139*, 2572–2592. https://doi.org/10.1175/2011MWR3584.1.

Fu, X., Wang, W., Lee, J.-Y., Wang, B., Kikuchi, K., Xu, J., Li, J., & Weaver, S. (2015). Distinctive roles of air–sea coupling on different MJO events: A new perspective revealed from the DYNAMO/CINDY field campaign. *Monthly Weather Review, 143*, 794–812. https://doi.org/10.1175/MWR-D-14-00221.1.

Fu, X., Wang, B., Li, T., & McCreary, J. P. (2003). Coupling between northward-propagating, intraseasonal oscillations and sea surface temperature in the Indian Ocean. *Journal of the Atmospheric Sciences, 60*, 1733–1753. https://doi.org/10.1175/1520-0469(2003)060<1733:CBNIOA>2.0.CO;2.

Fujita, M., Yuneyama, K., Mori, S., Nasuno, T., & Satoh, M. (2011). Diurnal convection peaks over the Eastern Indian Ocean off Sumatra during different MJO phases. *Journal of the Meteorological Society of Japan, 89A*, 317–330. https://doi.org/10.2151/jmsj.2011-A22.

Gao, Y., Klingaman, N. P., DeMott, C. A., & Hsu, P. (2018). Diagnosing ocean feedbacks to the BSISO: SST-modulated surface fluxes and the moist static energy budget. *Journal of Geophysical Research – Atmospheres*. https://doi.org/10.1029/2018JD029303.

Gill, A. E. (1980). Some simple solutions for heat-induced tropical circulation. *Quarterly Journal of the Royal Meteorological Society, 106*, 447–462. https://doi.org/10.1002/qj.49710644905.

Girishkumar, M. S., Ravichandran, M., McPhaden, M. J., & Rao, R. R. (2011). Intraseasonal variability in barrier layer thickness in the south central Bay of Bengal. *Journal of Geophysical Research, Oceans, 116*, C03009. https://doi.org/10.1029/2010JC006657.

Gonzalez, A. O., & Jiang, X. (2017). Winter mean lower tropospheric moisture over the maritime continent as a climate model diagnostic metric for the propagation of the Madden-Julian oscillation. *Geophysical Research Letters, 44*, 2588–2596. https://doi.org/10.1002/2016GL072430.

Gordon, A. L., et al. (2010). The Indonesian throughflow during 2004–2006 as observed by the INSTANT program. *Dynamics of Atmospheres and Oceans, 50*, 115–128. https://doi.org/10.1016/j.dynatmoce.2009.12.002.

Gottschalck, J., Roundy, P. E., Schreck, C. J., III, Vintzileos, A., & Zhang, C. (2013). Large-scale atmospheric and oceanic conditions during the 2011–12 DYNAMO field campaign. *Monthly Weather Review, 141*, 4173–4196. https://doi.org/10.1175/MWR-D-13-00022.1.

Gray, W. M., & Jacobson, R. W. (1977). Diurnal variation of deep cumulus convection. *Monthly Weather Review, 105*, 1171–1188.

Guan, B., Lee, T., Halkides, D. J., & Waliser, D. E. (2014). Aquarius surface salinity and the Madden-Julian oscillation: The role of salinity in surface layer density and potential energy. *Geophysical Research Letters, 41*, 2858–2869. https://doi.org/10.1002/2014GL059704.

Hagos, S. M., Zhang, C., Feng, Z., Burleyson, C. D., De Mott, C., Kerns, B., Benedict, J. J., & Martini, M. N. (2016). The impact of the diurnal cycle on the propagation of Madden-Julian oscillation convection across the maritime continent. *Journal of Advances in Modeling Earth Systems, 8*, 1552–1564. https://doi.org/10.1002/2016MS000725.

Halkides, D. J., Waliser, D. E., Lee, T., Menemenlis, D., & Guan, B. (2015). Quantifying the processes controlling intraseasonal mixed-layer temperature variability in the tropical Indian Ocean. *Journal of Geophysical Research, Oceans, 120*, 692–715. https://doi.org/10.1002/2014JC010139.

Han, W. (2005). Origins and dynamics of the 90-day and 30-60-day variations in the equatorial Indian Ocean. *Journal of Physical Oceanography, 35*, 708–728. https://doi.org/10.1175/JPO2725.1.

Han, W., Lawrence, D. M., & Webster, P. J. (2001). Dynamical response of equatorial Indian Ocean to intraseasonal winds: Zonal flow. *Geophysical Research Letters, 28*, 4215–4218. https://doi.org/10.1029/2001GL013701.

Han, W., McCreary, J. P., Anderson, D. L. T., & Mariano, A. J. (1999). Dynamics of the eastern surface jets in the equatorial Indian Ocean. *Journal of Physical Oceanography, 29*, 2191–2209. https://doi.org/10.1175/1520-0485(1999)029<2191:DOTESJ>2.0.CO;2.

Han, W., McCreary, J. P., Masumoto, Y., Vialard, J., & Duncan, B. (2011). Basin resonances in the equatorial Indian Ocean. *Journal of Physical Oceanography, 41*, 1252–1270. https://doi.org/10.1175/2011JPO4591.1.

Harrison, D. E., & Vecchi, G. A. (2001). January 1999 Indian ocean cooling event. *Geophysical Research Letters, 28*, 3717–3720. https://doi.org/10.1029/2001GL013506.

Hendon, H. H., & Glick, J. (1997). Intraseasonal air–sea interaction in the tropical Indian and Pacific Oceans. *Journal of Climate, 10*, 647–661. https://doi.org/10.1175/1520-0442(1997)010<0647:IASIIT>2.0.CO;2.

Horii, T., Ueki, I., Ando, K., Hasegawa, T., Mizuno, K., & Seiki, A. (2016). Impact of intraseasonal salinity variations on sea surface temperature in the eastern equatorial Indian Ocean. *Journal of Oceanography, 72*, 313–326. https://doi.org/10.1007/s10872-015-0337-x.

Houze, R. A., Geotis, S. G., Marks, F. D., & West, A. K. (1981). Winter monsoon convection in the vicinity of North Borneo. Part I: Structure and time variation of the clouds and precipitation. *Monthly Weather Review, 109*, 1595–1614. https://doi.org/10.1175/1520-0493(1981)109<1595:WMCITV>2.0.CO;2.

Houze, R. A., Rasmussen, K. L., Zuluaga, M. D., & Brodzik, S. R. (2015). The variable nature of convection in the tropics and subtropics: A legacy of 16 years of the tropical rainfall measuring Mission satellite. *Reviews of Geophysics, 53*, 994–1021. https://doi.org/10.1002/2015RG000488.

Hsu, P. C., & Li, T. (2012). Role of the boundary layer moisture asymmetry in causing the eastward propagation of the Madden-Julian oscillation. *Journal of Climate, 25*, 4914–4931. https://doi.org/10.1175/JCLI-D-11-00310.1.

Ichikawa, H., & Yasunari, T. (2006). Time–space characteristics of diurnal rainfall over Borneo and surrounding oceans as observed by TRMM-PR. *Journal of Climate, 19*, 1238–1260. https://doi.org/10.1175/JCLI3714.1.

Ichikawa, H., & Yasunari, T. (2007). Propagating diurnal disturbances embedded in the Madden-Julian oscillation. *Geophysical Research Letters, 34*, L18811. https://doi.org/10.1029/2007GL030480.

Inness, P. M., & Slingo, J. M. (2006). The interaction of the Madden–Julian oscillation with the maritime continent in a GCM. *Quarterly Journal of the Royal Meteorological Society, 132*, 1645–1667. https://doi.org/10.1256/qj.05.102.

Iskandar, I., & McPhaden, M. J. (2011). Dynamics of wind-forced intraseasonal zonal current variations in the equatorial Indian Ocean. *Journal of Geophysical Research, Oceans, 116*, C06019. https://doi.org/10.1029/2010JC006864.

Itterly, K., Taylor, P., & Roberts, J. B. (2021). Satellite perspectives of sea surface temperature diurnal warming on atmospheric moistening and radiative heating during MJO. *Journal of Climate, 34*, 1203–1226. https://doi.org/10.1175/JCLI-D-20-0350.1.

Izumo, T., Masson, S., Vialard, J., de Boyer Montegut, C., Behera, S. K., Madec, G., Takahashi, K., & Yamagata, T. (2010). Low and high frequency Madden–Julian oscillations in austral summer: Interannual variations. *Climate Dynamics, 35*, 669–683. https://doi.org/10.1007/s00382-009-0655-z.

Jayakumar, A., Vialard, J., Lengaigne, M., Gnanaseelan, C., McCreary, J. P., & Kumar, B. P. (2011). Processes controlling the surface temperature signature of the Madden-Julian oscillation in the thermocline ridge of the Indian Ocean. *Climate Dynamics, 37*, 2217–2234. https://doi.org/10.1007/s00382-010-0953-5.

Jensen, T. G. (1993). Equatorial variability and resonance in a wind-driven Indian Ocean model. *Journal of Geophysical Research, 98*, 22533–22552. https://doi.org/10.1029/93JC02565.

Jensen, T. G., Shinoda, T., Chen, S., & Flatau, M. (2015). Ocean response to CINDY/DYNAMO MJOs in air-sea-coupled COAMPS. *Journal of the Meteorological Society of Japan. Series II, 93A*, 157–178. https://doi.org/10.2151/jmsj.2015-049.

Jiang, X. (2017). Key processes for the eastward propagation of the Madden-Julian oscillation based on multimodel simulations. *Journal of Geophysical Research – Atmospheres, 122*, 755–770. https://doi.org/10.1002/2016JD025955.

Jiang, X., Li, T., & Wang, B. (2004). Structures and mechanisms of the northward propagating boreal summer intraseasonal oscillation. *Journal of Climate*, 1022–1039.

Jiang, X., et al. (2020). Fifty years of research on the Madden-Julian oscillation: Recent progress, challenges, and perspectives. *Journal of Geophysical Research – Atmospheres*, *125*, 1–64. https://doi.org/10.1029/2019JD030911.

Jochum, M., & Potemra, J. (2008). Sensitivity of tropical rainfall to Banda Sea diffusivity in the community climate system model. *Journal of Climate*, *21*, 6445–6454. https://doi.org/10.1175/2008JCLI2230.1.

Johnson, R. H., & Ciesielski, P. E. (2017). Multiscale variability of the atmospheric boundary layer during DYNAMO. *Journal of the Atmospheric Sciences*, *74*, 4003–4021. https://doi.org/10.1175/JAS-D-17-0182.1.

Johnson, R. H., Ciesielski, P. E., Ruppert, J. H., & Katsumata, M. (2015). Sounding-based thermodynamic budgets for DYNAMO. *Journal of the Atmospheric Sciences*, *72*, 598–622. https://doi.org/10.1175/JAS-D-14-0202.1.

Johnson, R. H., & Priegnitz, D. L. (1981). Winter monsoon convection in the vicinity of North Borneo. Part II: Effects on large-scale fields. *Monthly Weather Review*, *109*, 1615–1628. https://doi.org/10.1175/1520-0493(1981)109<1615:WMCITV>2.0.CO;2.

Johnson, R. H., Rickenbach, T. M., Rutledge, S. A., Ciesielski, P. E., & Schubert, W. H. (1999). Trimodal characteristics of tropical convection. *Journal of Climate*, *12*, 2397–2418. https://doi.org/10.1175/1520-0442(1999)012<2397:TCOTC>2.0.CO;2.

Kawai, Y., & Wada, A. (2007). Diurnal sea surface temperature variation and its impact on the atmosphere and ocean: A review. *Journal of Oceanography*, *63*, 721–744.

Keerthi, M. G., Lengaigne, M., Drushka, K., Vialard, J., Montegut, C. D. B., Pous, S., Levy, M., & Muraleedharan, P. M. (2016). Intraseasonal variability of mixed layer depth in the tropical Indian Ocean. *Climate Dynamics*, *46*, 2633–2655. https://doi.org/10.1007/s00382-015-2721-z.

Kemball-Cook, S. R., & Weare, B. C. (2001). The onset of convection in the Madden–Julian oscillation. *Journal of Climate*, *14*, 780–793. https://doi.org/10.1175/1520-0442(2001)014<0780:TOOCIT>2.0.CO;2.

Kessler, W. S., & Kleeman, R. (2000). Rectification of the Madden–Julian oscillation into the ENSO cycle. *Journal of Climate*, *13*, 3560–3575. https://doi.org/10.1175/1520-0442(2000)013<3560:ROTMJO>2.0.CO;2.

Kessler, W. S., McPhaden, M. J., & Weickmann, K. M. (1995). Forcing of intraseasonal Kelvin waves in the equatorial Pacific. *Journal of Geophysical Research*, *100*, 10613–10631. https://doi.org/10.1029/95jc00382.

Kikuchi, K., & Takayabu, Y. N. (2004). The development of organized convection associated with the MJO during TOGA COARE IOP: Trimodal characteristics. *Geophysical Research Letters*, *31*, L10101. https://doi.org/10.1029/2004GL019601.

Kikuchi, K., & Wang, B. (2008). Diurnal precipitation regimes in the global tropics. *Journal of Climate*, *21*, 2680–2696. https://doi.org/10.1175/2007JCLI2051.1.

Kiladis, G. N., Wheeler, M. C., Haertel, P. T., Straub, K. H., & Roundy, P. E. (2009). Convectively coupled equatorial waves. *Reviews of Geophysics*, *47*, RG2003. https://doi.org/10.1029/2008RG000266.

Kim, D., et al. (2009). Application of MJO simulation diagnostics to climate models. *Journal of Climate*, *22*, 6413–6436. https://doi.org/10.1175/2009JCLI3063.1.

Kim, D., Kim, H., & Lee, M.-I. (2017). Why does the MJO detour the maritime continent during austral summer? *Geophysical Research Letters*, *44*, 2579–2587. https://doi.org/10.1002/2017GL072643.

Kim, D., Kug, J.-S., & Sobel, A. H. (2014a). Propagating versus nonpropagating Madden–Julian oscillation events. *Journal of Climate*, *27*, 111–125. https://doi.org/10.1175/JCLI-D-13-00084.1.

Kim, H.-M., Webster, P. J., Toma, V. E., & Kim, D. (2014b). Predictability and prediction skill of the MJO in two operational forecasting systems. *Journal of Climate*, *27*, 5364–5378. https://doi.org/10.1175/JCLI-D-13-00480.1.

Kiranmayi, L., & Maloney, E. D. (2011). Intraseasonal moist static energy budget in reanalysis data. *Journal of Geophysical Research – Atmospheres*, *116*, 1–12. https://doi.org/10.1029/2011JD016031.

Klingaman, N. P., & Demott, C. A. (2020). Mean state biases and interannual variability affect perceived sensitivities of the Madden-Julian oscillation to air-sea coupling. *Journal of Advances in Modeling Earth Systems*, *12*. https://doi.org/10.1029/2019MS001799.

Klingaman, N. P., & Woolnough, S. J. (2014). The role of air-sea coupling in the simulation of the Madden-Julian oscillation in the Hadley Centre model. *Quarterly Journal of the Royal Meteorological Society*, *140*, 2272–2286. https://doi.org/10.1002/qj.2295.

Klingaman, N. P., Woolnough, S. J., Weller, H., & Slingo, J. M. (2011). The impact of finer-resolution air-sea coupling on the intraseasonal oscillation of the Indian summer monsoon. *Journal of Climate*, *24*, 2451–2468.

Koch-Larrouy, A., Lengaigne, M., Terray, P., Madec, G., & Masson, S. (2010). Tidal mixing in the Indonesian seas and its effect on the tropical climate system. *Climate Dynamics*, *34*, 891–904. https://doi.org/10.1007/s00382-009-0642-4.

Krishnamurti, T. N., Kanamitsu, M., Koss, W. J., & Lee, J. D. (1973). Tropical east–west circulations during the Northern winter. *Journal of the Atmospheric Sciences*, *30*, 780–787. https://doi.org/10.1175/1520-0469(1973)030<0780:TECDTN>2.0.CO;2.

Lau, K.-M., & Wu, H.-T. (2010). Characteristics of precipitation, cloud, and latent heating associated with the Madden–Julian oscillation. *Journal of Climate*, *23*, 504–518. https://doi.org/10.1175/2009JCLI2920.1.

Li, Y., & Carbone, R. E. (2012). Excitation of rainfall over the tropical Western Pacific. *Journal of the Atmospheric Sciences*, *69*, 2983–2994. https://doi.org/10.1175/JAS-D-11-0245.1.

Li, Y., Han, W., & Lee, T. (2015). Intraseasonal sea surface salinity variability in the equatorial Indo-Pacific Ocean induced by Madden-Julian oscillations. *Journal of Geophysical Research, C: Oceans*, *120*, 2233–2258. https://doi.org/10.1002/2014JC010647.

Li, Y., Han, W., Wang, W., & Ravichandran, M. (2016). Intraseasonal variability of SST and precipitation in the Arabian Sea during the Indian summer monsoon: Impact of ocean mixed layer depth. *Journal of Climate*, *29*, 7889–7910. https://doi.org/10.1175/JCLI-D-16-0238.1.

Lin, J.-L., et al. (2006). Tropical Intraseasonal variability in 14 IPCC AR4 climate models. Part I: Convective signals. *Journal of Climate*, *19*, 2665–2690. https://doi.org/10.1175/JCLI3735.1.

Lindzen, R. R. S., & Nigam, S. (1987). On the role of sea surface temperature gradients in forcing low-level winds and convergence in the tropics. *Journal of the Atmospheric Sciences, 44*, 2418–2436. https://doi.org/10.1175/1520-0469(1987)044<2418:OTROSS>2.0.CO;2.

Ling, J., Zhang, C., Wang, S., & Li, C. (2017). A new interpretation of the ability of global models to simulate the MJO. *Geophysical Research Letters, 44*, 5798–5806. https://doi.org/10.1002/2017GL073891.

Liu, C., Zipser, E. J., & Nesbitt, S. W. (2007). Global distribution of tropical deep convection: Different perspectives from TRMM infrared and radar data. *Journal of Climate, 20*, 489–503. https://doi.org/10.1175/JCLI4023.1.

Lloyd, I. D., & Vecchi, G. A. (2010). Submonthly Indian ocean cooling events and their interaction with large-scale conditions. *Journal of Climate, 23*, 700–716. https://doi.org/10.1175/2009JCLI3067.1.

Lukas, R., & Lindstrom, E. (1991). The mixed layer of the western equatorial Pacific Ocean. *Journal of Geophysical Research, 96*, 3343. https://doi.org/10.1029/90jc01951.

Luyten, J. R., & Roemmich, D. H. (1982). Equatorial currents at semi-annual period in the Indian Ocean. *Journal of Physical Oceanography, 12*, 406–413. https://doi.org/10.1175/1520-0485(1982)012<0406:ecasap>2.0.co;2.

Madden, R. A., & Julian, P. R. (1972). Description of global-scale circulation cells in the tropics with a 40–50 day period. *Journal of the Atmospheric Sciences, 29*, 1109–1123. https://doi.org/10.1175/1520-0469(1972)029<1109:DOGSCC>2.0.CO;2.

Malkus, J. S. (1957). Trade cumulus cloud groups: Some observations suggesting a mechanism of their origin. *Tellus, 9*, 33–44. https://doi.org/10.1111/j.2153-3490.1957.tb01851.x.

Maloney, E. D. (2009). The moist static energy budget of a composite tropical intraseasonal oscillation in a climate model. *Journal of Climate, 22*, 711–729. https://doi.org/10.1175/2008JCLI2542.1.

Mapes, B. E., Warner, T. T., & Xu, M. (2003). Diurnal patterns of rainfall in northwestern South America. Part III: Diurnal gravity waves and nocturnal convection offshore. *Monthly Weather Review, 131*, 830–844. https://doi.org/10.1175/1520-0493(2003)131<0830:DPORIN>2.0.CO;2.

Matthews, A. J. (2008). Primary and successive events in the Madden-Julian oscillation. *Quarterly Journal of the Royal Meteorological Society, 134*, 439–453. https://doi.org/10.1002/qj.

Matthews, A. J., Baranowski, D. B., Heywood, K. J., Flatau, P. J., & Schmidtko, S. (2014). The surface diurnal warm layer in the Indian Ocean during CINDY/DYNAMO. *Journal of Climate, 27*, 9101–9122. https://doi.org/10.1175/JCLI-D-14-00222.1.

Matthews, A. J., Pickup, G., Peatman, S. C., Clews, P., & Martin, J. (2013). The effect of the Madden-Julian oscillation on station rainfall and river level in the Fly River system, Papua New Guinea. *Journal of Geophysical Research – Atmospheres, 118*, 10926–10935. https://doi.org/10.1002/jgrd.50865.

Matthews, A. J., Singhruck, P., & Heywood, K. J. (2010). Ocean temperature and salinity components of the Madden-Julian oscillation observed by Argo floats. *Climate Dynamics, 35*, 1149–1168. https://doi.org/10.1007/s00382-009-0631-7.

McPhaden, M. J. (1982). Variability in the central equatorial Indian Ocean. Part I: Ocean dynamics. *Journal of Marine Research, 40*, 157–176.

McPhaden, M. J., & Foltz, G. R. (2013). Intraseasonal variations in the surface layer heat balance of the central equatorial Indian Ocean: The importance of zonal advection and vertical mixing. *Geophysical Research Letters, 40*, 2737–2741. https://doi.org/10.1002/grl.50536.

Melzer, B. A., Jensen, T. G., & Rydbeck, A. V. (2019). Evolution of the Great Whirl using an altimetry-based Eddy tracking algorithm. *Geophysical Research Letters, 46*, 4378–4385. https://doi.org/10.1029/2018GL081781.

Moum, J. N., Pujiana, K., Lien, R.-C., & Smyth, W. D. (2016). Ocean feedback to pulses of the Madden–Julian oscillation in the equatorial Indian ocean. *Nature Communications, 7*, 13203. https://doi.org/10.1038/ncomms13203.

Moum, J. N., et al. (2014). Air-sea interactions from westerly wind bursts during the November 2011 MJO in the Indian Ocean. *Bulletin of the American Meteorological Society, 95*, 1185–1199. https://doi.org/10.1175/BAMS-D-12-00225.1.

Nagura, M., & McPhaden, M. J. (2012). The dynamics of wind-driven intraseasonal variability in the equatorial Indian Ocean. *Journal of Geophysical Research, Oceans, 117*. https://doi.org/10.1029/2011JC007405.

Napitu, A. M., Gordon, A. L., & Pujiana, K. (2015). Intraseasonal sea surface temperature variability across the Indonesian seas. *Journal of Climate, 28*, 8710–8727. https://doi.org/10.1175/JCLI-D-14-00758.1.

Neale, R., & Slingo, J. (2003). The maritime continent and its role in the global climate: A GCM study. *Journal of Climate, 16*, 834–848. https://doi.org/10.1175/1520-0442(2003)016<0834:TMCAIR>2.0.CO;2.

Nyadjro, E. S., Rydbeck, A. V., Jensen, T. G., Richman, J. G., & Shriver, J. F. (2020). On the generation and salinity impacts of intraseasonal Westward Jets in the equatorial Indian Ocean. *Journal of Geophysical Research, Oceans*. https://doi.org/10.1029/2020JC016066.

Orenstein, P., Fox-Kemper, B., Johnson, L., Li, Q., & Sane, A. (2021). Evaluating coupled climate model parameterizations via skill at reproducing the monsoon intraseasonal oscillation. *Journal of Climate, 35*. published online ahead of print 2021).

Parampil, S. R., Gera, A., Ravichandran, M., & Sengupta, D. (2010). Intraseasonal response of mixed layer temperature and salinity in the Bay of Bengal to heat and freshwater flux. *Journal of Geophysical Research, Oceans, 115*, C05002. https://doi.org/10.1029/2009JC005790.

Peatman, S. C., Matthews, A. J., & Stevens, D. P. (2014). Propagation of the Madden-Julian oscillation through the maritime continent and scale interaction with the diurnal cycle of precipitation. *Quarterly Journal of the Royal Meteorological Society, 140*, 814–825. https://doi.org/10.1002/qj.2161.

Pegion, K., & Kirtman, B. P. (2008). The impact of air–sea interactions on the simulation of tropical intraseasonal variability. *Journal of Climate, 21*, 6616–6635. https://doi.org/10.1175/2008JCLI2180.1.

Powell, S. W., & Houze, R. A. (2015). Effect of dry large-scale vertical motions on initial MJO convective onset. *Journal of Geophysical Research – Atmospheres, 120*, 4783–4805. https://doi.org/10.1002/2014JD022961.

Pujiana, K., & McPhaden, M. J. (2020). Intraseasonal Kelvin waves in the equatorial Indian Ocean and their propagation into the Indonesian seas. *Journal of Geophysical Research, Oceans, 125*, 1–18. https://doi.org/10.1029/2019JC015839.

Pujiana, K., Moum, J. N., & Smyth, W. D. (2018). The role of turbulence in redistributing upper-ocean heat, freshwater, and momentum in response to the MJO in the equatorial Indian Ocean. *Journal of Physical Oceanography*, 48, 197–220. https://doi.org/10.1175/JPO-D-17-0146.1.

Ramage, C. S. (1968). Role of a tropical "maritime continent" in the atmospheric circulation. *Monthly Weather Review*, 96, 365–370. https://doi.org/10.1175/1520-0493(1968)096<0365:ROATMC>2.0.CO;2.

Randall, D. A. (2013). Beyond deadlock. *Geophysical Research Letters*, 40, 5970–5976. https://doi.org/10.1002/2013GL057998.

Randall, D. A., Harshvardhan, & Dazlich, D. A. (1991). Diurnal variability of the hydrologic cycle in a general circulation model. *Journal of the Atmospheric Sciences*, 48, 40–62. https://doi.org/10.1175/1520-0469(1991)048<0040:DVOTHC>2.0.CO;2.

Rao, R. R., & Sivakumar, R. (2003). Seasonal variability of sea surface salinity and salt budget of the mixed layer of the North Indian Ocean. *Journal of Geophysical Research, Oceans*, 108, 3009. https://doi.org/10.1029/2001jc000907.

Raymond, D. J. (2001). A new model of the Madden–Julian oscillation. *Journal of the Atmospheric Sciences*, 58, 2807–2819. https://doi.org/10.1175/1520-0469(2001)058<2807:ANMOTM>2.0.CO;2.

Reppin, J., Schott, F. A., Fischer, J., & Quadfasel, D. (1999). Equatorial currents and transports in the upper Central Indian Ocean: Annual cycle and interannual variability. *Journal of Geophysical Research, Oceans*, 104, 15495–15514. https://doi.org/10.1029/1999jc900093.

Reverdin, G., & Luyten, J. (1986). Near-surface meanders in the equatorial Indian Ocean. *Journal of Physical Oceanography*, 16, 1088–1100. https://doi.org/10.1175/1520-0485(1986)016<1088:nsmite>2.0.co;2.

Riley Dellaripa, E. M., & Maloney, E. D. (2015). Analysis of MJO wind-flux feedbacks in the Indian Ocean using RAMA Buoy observations. *Journal of the Meteorological Society of Japan. Series II*, 93A, 1–20. https://doi.org/10.2151/jmsj.2015-021.

Roxy, M. K., Dasgupta, P., McPhaden, M. J., et al. (2019). Twofold expansion of the Indo-Pacific warm pool warps the MJO life cycle. *Nature*, 575, 647–651. https://doi.org/10.1038/s41586-019-1764-4.

Ruppert, J. H. (2016). Diurnal timescale feedbacks in the tropical cumulus regime. *Journal of Advances in Modeling Earth Systems*, 8, 1483–1500. https://doi.org/10.1002/2016MS000713.

Ruppert, J. H., & Chen, X. (2020). Island rainfall enhancement in the maritime continent. *Geophysical Research Letters*, 47. https://doi.org/10.1029/2019GL086545.

Ruppert, J. H., Chen, X., & Zhang, F. (2020). Convectively forced diurnal gravity waves in the maritime continent. *Journal of the Atmospheric Sciences*, 77, 1119–1136. https://doi.org/10.1175/JAS-D-19-0236.1.

Ruppert, J. H., & Johnson, R. H. (2015). Diurnally modulated cumulus moistening in the preonset stage of the Madden-Julian oscillation during DYNAMO. *Journal of the Atmospheric Sciences*, 72, 1622–1647. https://doi.org/10.1175/JAS-D-14-0218.1.

Ruppert, J. H., & Johnson, R. H. (2016). On the cumulus diurnal cycle over the tropical warm pool. *Journal of Advances in Modeling Earth Systems*, 8, 669–690. https://doi.org/10.1002/2015MS000610.

Ruppert, J. H., Johnson, R. H., & Hohenegger, C. (2018). Diurnal circulation adjustment and organized deep convection. *Journal of Climate*, 31, 4899–4916. https://doi.org/10.1175/JCLI-D-17-0693.1.

Ruppert, J. H., & Zhang, F. (2019). Diurnal forcing and phase locking of gravity waves in the maritime continent. *Journal of the Atmospheric Sciences*, 76, 2815–2835. https://doi.org/10.1175/JAS-D-19-0061.1.

Rydbeck, A. V., & Jensen, T. G. (2017). Oceanic impetus for convective onset of the Madden-Julian oscillation in the western Indian ocean. *Journal of Climate*, 30, 4299–4316. https://doi.org/10.1175/JCLI-D-16-0595.1.

Rydbeck, A. V., Jensen, T. G., & Flatau, M. K. (2021). Reciprocity in the Indian Ocean: Intraseasonal oscillation and ocean planetary waves. *Journal of Geophysical Research: Oceans*, 126. https://doi.org/10.1029/2021JC017546.

Rydbeck, A. V., Jensen, T. G., & Igel, M. R. (2019a). Idealized modeling of the atmospheric boundary layer response to SST forcing in the Western Indian Ocean. *Journal of the Atmospheric Sciences*, 76, 2023–2042. https://doi.org/10.1175/JAS-D-18-0303.1.

Rydbeck, A. V., Jensen, T. G., & Nyadjro, E. S. (2017). Intraseasonal sea surface warming in the western Indian Ocean by oceanic equatorial Rossby waves. *Geophysical Research Letters*, 44, 4224–4232. https://doi.org/10.1002/2017GL073331.

Rydbeck, A. V., Jensen, T. G., Smith, T. A., Flatau, M. K., Janiga, M. A., Reynolds, C. A., & Ridout, J. A. (2019b). Ocean heat content and the intraseasonal oscillation. *Geophysical Research Letters*. https://doi.org/10.1029/2019GL084974.

Saji, N. H., Goswami, B. N., Vinayachandran, P. N., & Yamagata, T. (1999). A dipole mode in the tropical Indian Ocean. *Nature*, 401, 360–363. https://doi.org/10.1038/43854.

Sakaeda, N., Kiladis, G., & Dias, J. (2017). The diurnal cycle of tropical cloudiness and rainfall associated with the Madden–Julian oscillation. *Journal of Climate*, 30, 3999–4020. https://doi.org/10.1175/JCLI-D-16-0788.1.

Sakaeda, N., Kiladis, G., & Dias, J. (2020). The diurnal cycle of rainfall and the convectively coupled equatorial waves over the maritime continent. *Journal of Climate*, 33, 3307–3331. https://doi.org/10.1175/JCLI-D-19-0043.1.

Sakaeda, N., Powell, S. W., Dias, J., & Kiladis, G. N. (2018). The diurnal variability of precipitating cloud populations during DYNAMO. *Journal of the Atmospheric Sciences*, 75, 1307–1326. https://doi.org/10.1175/JAS-D-17-0312.1.

Sanchez-Franks, A., Kent, E. C., Matthews, A. J., Webber, B. G. M., Peatman, S. C., & Vinayachandran, P. N. (2018). Intraseasonal variability of air-sea fluxes over the Bay of Bengal during the Southwest Monsoon. *Journal of Climate*, 31, 7087–7109. https://doi.org/10.1175/JCLI-D-17-0652.1.

Schiller, A., & Godfrey, J. S. (2003). Indian Ocean intraseasonal variability in an ocean general circulation model. *Journal of Climate*, 16, 21–39. https://doi.org/10.1175/1520-0442(2003)016<0021:IOIVIA>2.0.CO;2.

Schott, F. A., Xie, S.-P., & McCreary, J. P. (2009). Indian Ocean circulation and climate variability. *Reviews of Geophysics*, 47, RG1002. https://doi.org/10.1029/2007RG000245.

Seiki, A., Katsumata, M., Horii, T., Hasegawa, T., Richards, K. J., Yoneyama, K., & Shirooka, R. (2013). Abrupt cooling associated with the oceanic Rossby wave and lateral advection during CINDY2011. *Journal of Geophysical Research, Oceans, 118*, 5523–5535. https://doi.org/10.1002/jgrc.20381.

Sengupta, D., & Ravichandran, M. (2001). Oscillations of bay of Bengal Sea surface temperature during the 1998 summer monsoon. *Geophysical Research Letters, 28*, 2033–2036. https://doi.org/10.1029/2000GL012548.

Seo, H., Subramanian, A. C., Miller, A. J., & Cavanaugh, N. R. (2014). Coupled impacts of the diurnal cycle of sea surface temperature on the Madden-Julian oscillation. *Journal of Climate, 27*, 8422–8443.

Seo, K.-H., Wang, W., Gottschalck, J., Zhang, Q., Schemm, J.-K. E., Higgins, W. R., & Kumar, A. (2009). Evaluation of MJO forecast skill from several statistical and dynamical forecast models. *Journal of Climate, 22*, 2372–2388. https://doi.org/10.1175/2008JCLI2421.1.

Shackelford, K., DeMott, C. A., Leeuwen, P. J., Thompson, E., & Hagos, S. (2022). Rain-induced stratification of the equatorial Indian ocean and its potential feedback to the atmosphere. *Journal of Geophysical Research, Oceans, 127*. https://doi.org/10.1029/2021JC018025.

Shinoda, T. (2005). Impact of the diurnal cycle of solar radiation on intraseasonal SST variability in the Western Equatorial Pacific. *Journal of Climate, 18*, 2628–2636. https://doi.org/10.1175/JCLI3432.1.

Shinoda, T., & Han, W. (2005). Influence of the Indian Ocean dipole on atmospheric subseasonal variability. *Journal of Climate, 18*, 3891–3909. https://doi.org/10.1175/JCLI3510.1.

Shinoda, T., Han, W., Zamudio, L., Lien, R. C., & Katsumata, M. (2017). Remote Ocean response to the Madden-Julian oscillation during the DYNAMO field campaign: Impact on Somali current system and the Seychelles-Chagos thermocline ridge. *Atmosphere (Basel), 8*, 171. https://doi.org/10.3390/atmos8090171.

Shinoda, T., & Hendon, H. H. (1998). Mixed layer modeling of intraseasonal variability in the tropical Western Pacific and Indian oceans. *Journal of Climate, 11*, 2668–2685. https://doi.org/10.1175/1520-0442(1998)011<2668:MLMOIV>2.0.CO;2.

Shinoda, T., Jensen, T. G., Flatau, M., Chen, S., Han, W., & Wang, C. (2013). Large-scale oceanic variability associated with the Madden-Julian oscillation during the CINDY/DYNAMO field campaign from satellite observations. *Remote Sensing, 5*, 2072–2092. https://doi.org/10.3390/rs5052072.

Shinoda, T., Jenson, T. G., Lachkar, Z., Masumoto, Y., & Seo, H. (2024). Chapter 18: Modeling the Indian Ocean. In C. C. Ummenhofer, & R. R. Hood (Eds.), *The Indian Ocean and its role in the global climate system* (pp. 421–443). Amsterdam: Elsevier. https://doi.org/10.1016/B978-0-12-822698-8.00001-9.

Slingo, J., Inness, P., Neale, R., Woolnough, S., & Yang, G. Y. (2003). Scale interactions on diurnal to seasonal timescales and their relevance to model systematic errors. *Annales de Geophysique, 46*, 139–156. https://doi.org/10.4401/ag-3383.

Sobel, A. H., Burleyson, C. D., & Yuter, S. E. (2011). Rain on small tropical islands. *Journal of Geophysical Research, 116*, D08102. https://doi.org/10.1029/2010JD014695.

Sobel, A., & Maloney, E. (2012). An idealized semi-empirical framework for modeling the Madden–Julian oscillation. *Journal of the Atmospheric Sciences, 69*, 1691–1705. https://doi.org/10.1175/JAS-D-11-0118.1.

Sobel, A., Wang, S., & Kim, D. (2014). Moist static energy budget of the MJO during DYNAMO. *Journal of the Atmospheric Sciences, 71*, 4276–4291. https://doi.org/10.1175/JAS-D-14-0052.1.

Soloviev, A., & Lukas, R. (1997). Observation of large diurnal warming events in the near-surface layer of the western equatorial Pacific warm pool. *Deep-Sea Research Part I: Oceanographic Research Papers, 44*, 1055–1076. https://doi.org/10.1016/S0967-0637(96)00124-0.

Sprintall, J., Biastoch, A., Gruenburg, L. K., & Phillips, H. E. (2024). Chapter 9: Oceanic basin connections. In C. C. Ummenhofer, & R. R. Hood (Eds.), *The Indian Ocean and its role in the global climate system* (pp. 205–227). Amsterdam: Elsevier. https://doi.org/10.1016/B978-0-12-822698-8.00003-2.

Sprintall, J., Gordon, A. L., Koch-Larrouy, A., Lee, T., Potemra, J. T., Pujiana, K., & Wijffels, S. E. (2014). The Indonesian seas and their role in the coupled ocean–climate system. *Nature Geoscience, 7*, 487–492. https://doi.org/10.1038/ngeo2188.

Sprintall, J., & Tomczak, M. (1992). Evidence of the barrier layer in the surface layer of the tropics. *Journal of Geophysical Research, 97*, 7305. https://doi.org/10.1029/92JC00407.

Sui, C.-H., Lau, K.-M., Takayabu, Y. N., & Short, D. A. (1997). Diurnal variations in tropical oceanic cumulus convection during TOGA COARE. *Journal of the Atmospheric Sciences, 54*, 639–655.

Suresh, I., Vialard, J., Lengaigne, M., Han, W., McCreary, J., Durand, F., & Muraleedharan, P. M. (2013). Origins of wind-driven intraseasonal sea level variations in the North Indian Ocean coastal waveguide. *Geophysical Research Letters, 40*, 5740–5744. https://doi.org/10.1002/2013GL058312.

Susanto, R. D., Gordon, A. L., Sprintall, J., & Herunadi, B. (2000). Intraseasonal variability and tides in Makassar Strait. *Geophysical Research Letters, 27*, 1499–1502. https://doi.org/10.1029/2000GL011414.

Thayer-Calder, K., & Randall, D. A. (2009). The role of convective moistening in the Madden–Julian oscillation. *Journal of the Atmospheric Sciences, 66*, 3297–3312. https://doi.org/10.1175/2009JAS3081.1.

Thompson, E. J., Moum, J. N., Fairall, C. W., & Rutledge, S. A. (2019). Wind limits on rain layers and diurnal warm layers. *Journal of Geophysical Research, Oceans, 124*, 897–924. https://doi.org/10.1029/2018JC014130.

Trott, C. B., Subrahmanyam, B., Murty, V. S. N., & Shriver, J. F. (2019). Large-scale fresh and salt water exchanges in the Indian Ocean. *Journal of Geophysical Research, Oceans, 124*, 6252–6269. https://doi.org/10.1029/2019JC015361.

Tseng, W.-L., Hsu, H.-H., Keenlyside, N., June Chang, C.-W., Tsuang, B.-J., Tu, C.-Y., & Jiang, L.-C. (2017). Effects of surface orography and land–sea contrast on the Madden–Julian oscillation in the maritime continent: A numerical study using ECHAM5-SIT. *Journal of Climate, 30*, 9725–9741. https://doi.org/10.1175/JCLI-D-17-0051.1.

Vialard, J., Drushka, K., Bellenger, H., Lengaigne, M., Pous, S., & Duvel, J. P. (2013). Understanding Madden-Julian-induced sea surface temperature variations in the North Western Australian Basin. *Climate Dynamics, 41*, 3203–3218. https://doi.org/10.1007/s00382-012-1541-7.

Vialard, J., Foltz, G. R., McPhaden, M. J., Duvel, J. P., & de Boyer Montégut, C. (2008). Strong Indian Ocean sea surface temperature signals associated with the Madden-Julian oscillation in late 2007 and early 2008. *Geophysical Research Letters, 35*, L19608. https://doi.org/10.1029/2008GL035238.

Vialard, J., Jayakumar, A., Gnanaseelan, C., Lengaigne, M., Sengupta, D., & Goswami, B. N. (2012). Processes of 30-90 days sea surface temperature variability in the northern Indian Ocean during boreal summer. *Climate Dynamics, 38*, 1901–1916. https://doi.org/10.1007/s00382-011-1015-3.

Vinayachandran, P. N., & Saji, N. H. (2008). Mechanisms of South Indian Ocean intraseasonal cooling. *Geophysical Research Letters, 35*, L23607. https://doi.org/10.1029/2008GL035733.

Vincent, C. L., & Lane, T. P. (2016). Evolution of the diurnal precipitation cycle with the passage of a Madden–Julian oscillation event through the maritime continent. *Monthly Weather Review, 144*, 1983–2005. https://doi.org/10.1175/MWR-D-15-0326.1.

Vitart, F., Woolnough, S., Balmaseda, M. A., & Tompkins, A. M. (2007). Monthly forecast of the Madden-Julian oscillation using a coupled GCM. *Monthly Weather Review, 135*, 2700–2715. https://doi.org/10.1175/MWR3415.1.

Waliser, D. E., Murtugudde, R., & Lucas, L. E. (2004). Indo-Pacific Ocean response to atmospheric intraseasonal variability: 2. Boreal summer and the intraseasonal oscillation. *Journal of Geophysical Research, Oceans, 109*. https://doi.org/10.1029/2003jc002002.

Wang, B., Chen, G., & Liu, F. (2019). Diversity of the Madden-Julian oscillation. *Science Advances, 5*, eaax0220. https://doi.org/10.1126/sciadv.aax0220.

Wang, B., & Li, T. (1994). Convective interaction with boundary-layer dynamics in the development of a tropical intraseasonal system. *Journal of the Atmospheric Sciences, 51*, 1386–1400. https://doi.org/10.1175/1520-0469(1994)051<1386:CIWBLD>2.0.CO;2.

Wang, L., Li, T., Maloney, E., & Wang, B. (2017). Fundamental causes of propagating and nonpropagating MJOs in MJOTF/GASS models. *Journal of Climate, 30*, 3743–3769. https://doi.org/10.1175/JCLI-D-16-0765.1.

Wang, B., & Rui, H. (1990). Dynamics of the coupled Moist Kelvin–Rossby wave on an equatorial β-plane. *Journal of the Atmospheric Sciences, 47*, 397–413. https://doi.org/10.1175/1520-0469(1990)047<0397:DOTCMK>2.0.CO;2.

Weaver, S. J., Wang, W., Chen, M., & Kumar, A. (2011). Representation of MJO variability in the NCEP climate forecast system. *Journal of Climate, 24*, 4676–4694. https://doi.org/10.1175/2011JCLI4188.1.

Webber, B. G. M., Matthews, A. J., & Heywood, K. J. (2010). A dynamical ocean feedback mechanism for the Madden-Julian oscillation. *Quarterly Journal of the Royal Meteorological Society, 136*, 740–754. https://doi.org/10.1002/qj.604.

Webber, B. G. M., Stevens, D. P., Matthews, A. J., & Heywood, K. J. (2012a). Dynamical Ocean forcing of the Madden-Julian oscillation at lead times of up to five months. *Journal of Climate, 25*, 2824–2842. https://doi.org/10.1175/JCLI-D-11-00268.1.

Webber, B. G. M. M., Matthews, A. J., Heywood, K. J., & Stevens, D. P. (2012b). Ocean Rossby waves as a triggering mechanism for primary Madden-Julian events. *Quarterly Journal of the Royal Meteorological Society, 138*, 514–527. https://doi.org/10.1002/qj.936.

Webster, P. J., Clayson, C. A., & Curry, J. A. (1996). Clouds, radiation, and the diurnal cycle of sea surface temperature in the tropical Western Pacific. *Journal of Climate, 9*, 1712–1730. https://doi.org/10.1175/1520-0442(1996)009<1712:CRATDC>2.0.CO;2.

Weller, R. A., Baumgartner, M. F., Josey, S. A., Fischer, A. S., & Kindle, J. C. (1998). Atmospheric forcing in the Arabian Sea during 1994–1995: Observations and comparisons with climatology and models. *Deep-Sea Research Part II: Topical Studies in Oceanography, 45*, 1961–1999. https://doi.org/10.1016/S0967-0645(98)00060-5.

West, B. J., Han, W., & Li, Y. (2018). The role of oceanic processes in the initiation of Indian summer monsoon intraseasonal oscillations over the Indian Ocean. *Journal of Geophysical Research, Oceans, 123*, 3685–3704. https://doi.org/10.1029/2017JC013564.

Wheeler, M. C., Hendon, H. H., Cleland, S., Meinke, H., & Donald, A. (2009). Impacts of the Madden–Julian oscillation on Australian rainfall and circulation. *Journal of Climate, 22*, 1482–1498. https://doi.org/10.1175/2008JCLI2595.1.

Wilson, E. A., Gordon, A. L., & Kim, D. (2013). Observations of the Madden Julian oscillation during Indian Ocean dipole events. *Journal of Geophysical Research – Atmospheres, 118*, 2588–2599. https://doi.org/10.1002/jgrd.50241.

Wyrtki, K. (1973). An equatorial jet in the Indian Ocean. *Science, 181*, 262–264. https://doi.org/10.1126/science.181.4096.262.

Yamanaka, M. D., Ogino, S.-Y., Wu, P.-M., Jun-Ichi, H., Mori, S., Matsumoto, J., & Syamsudin, F. (2018). Maritime continent coastlines controlling Earth's climate. *Progress in Earth and Planetary Science, 5*, 21. https://doi.org/10.1186/s40645-018-0174-9.

Yang, Y., Lee, J., & Wang, B. (2020). Dominant process for northward propagation of boreal summer intraseasonal oscillation over the Western North Pacific. *Geophysical Research Letters, 47*. https://doi.org/10.1029/2020GL089808.

Yoneyama, K., Zhang, C., & Long, C. N. (2013). Tracking pulses of the Madden- -Julian oscillation. *Bulletin of the American Meteorological Society, 94*, 1871–1891. https://doi.org/10.1175/BAMS-D-12-00157.1.

Yoshida, K. (1959). A theory of the Cromwell current (the equatorial undercurrent) and of the equatorial upwelling. *Journal of the Oceanographic Society of Japan, 15*, 159–170. https://doi.org/10.5928/kaiyou1942.15.159.

Zeng, L., Yan, D., Xie, S.-P., & Wang, D. (2009). Barrier layer in the South China Sea during summer 2000. *Dynamics of Atmospheres and Oceans, 47*(1–3), 38–54. ISSN 0377-0265 https://doi.org/10.1016/j.dynatmoce.2008.08.001.

Zhang, C. (2005). Madden-Julian oscillation. *Reviews of Geophysics, 43*, RG2003. https://doi.org/10.1029/2004RG000158.

Zhang, C. (2013). Madden–Julian oscillation: Bridging weather and climate. *Bulletin of the American Meteorological Society, 94*, 1849–1870. https://doi.org/10.1175/BAMS-D-12-00026.1.

Zhang, C., Adames, Á. F., Khouider, B., Wang, B., & Yang, D. (2020). Four theories of the Madden-Julian oscillation. *Reviews of Geophysics, 58*. https://doi.org/10.1029/2019RG000685.

Zhang, C., Dong, M., Gualdi, S., Hendon, H. H., Maloney, E. D., Marshall, A., Sperber, K. R., & Wang, W. (2006). Simulations of the Madden–Julian oscillation in four pairs of coupled and uncoupled global models. *Climate Dynamics, 27*, 573–592. https://doi.org/10.1007/s00382-006-0148-2.

Zhang, C., Gottschalck, J., Maloney, E. D., Moncrieff, M. W., Vitart, F., Waliser, D. E., Wang, B., & Wheeler, M. C. (2013). Cracking the MJO nut. *Geophysical Research Letters, 40*, 1223–1230. https://doi.org/10.1002/grl.50244.

Zhang, L., Han, W., Li, Y., & Maloney, E. D. (2018). Role of North Indian Ocean air–sea interaction in summer monsoon intraseasonal oscillation. *Journal of Climate, 31*, 7885–7908. https://doi.org/10.1175/JCLI-D-17-0691.1.

Zhang, Q., Li, T., & Liu, J. (2019). Contrast of evolution characteristics of boreal summer and winter intraseasonal oscillations over tropical Indian Ocean. *Journal of Meteorological Research, 33*, 678–694. https://doi.org/10.1007/s13351-019-9015-z.

Zhang, C., & Ling, J. (2017). Barrier effect of the Indo-Pacific maritime continent on the MJO: Perspectives from tracking MJO precipitation. *Journal of Climate, 30*, 3439–3459. https://doi.org/10.1175/JCLI-D-16-0614.1.

Zhang, C., & McPhaden, M. J. (2000). Intraseasonal surface cooling in the equatorial Western Pacific. *Journal of Climate, 13*, 2261–2276. https://doi.org/10.1175/1520-0442(2000)013<2261:ISCITE>2.0.CO;2.

Zhang, F., Taraphdar, S., & Wang, S. (2017). The role of global circumnavigating mode in the MJO initiation and propagation. *Journal of Geophysical Research – Atmospheres, 122*, 5837–5856. https://doi.org/10.1002/2016JD025665.

Zhao, N., & Nasuno, T. (2020). How does the air-sea coupling frequency affect convection during the MJO passage? *Journal of Advances in Modeling Earth Systems, 12*, 1–20. https://doi.org/10.1029/2020MS002058.

Zhou, L., & Murtugudde, R. (2020). Oceanic impacts on MJOs detouring near the maritime continent. *Journal of Climate, 33*, 2371–2388. https://doi.org/10.1175/JCLI-D-19-0505.1.

Chapter 5

Climate phenomena of the Indian Ocean

Toshio Yamagata[a,b], Swadhin Behera[a], Takeshi Doi[a], Jing-Jia Luo[b], Yushi Morioka[a], and Tomoki Tozuka[a,c]

[a]Application Lab, Japan Agency for Marine-Earth Science and Technology, Yokohama, Japan, [b]Institute of Climate and Application Research, Nanjing University of Information Science and Technology, Nanjing, China, [c]Department of Earth and Planetary Science, Graduate School of Science, The University of Tokyo, Tokyo, Japan

1 Introduction

Climate variations rooted in the tropical oceans play an important role in the global climate system. All three tropical basins (Pacific, Indian, and Atlantic Oceans) give rise to intrinsic modes of climate variations, and extreme events associated with them affect human lives and the environment. Rooted in the Indian Ocean, the Indian Ocean Dipole (IOD) is recognized as one of the most influential climate phenomena besides the El Niño-Southern Oscillation (ENSO) of the tropical Pacific Ocean. Many research studies over the past couple of decades have shown IOD's large impacts in many parts of the world. Hence, a better prediction of IOD is essential for improving climate predictability. While considerable progress has been made in the understanding and prediction of ENSO variability over the years, we have just started to understand IOD variability and the intrinsic processes associated with its evolution. In this chapter, we have reviewed some of those studies and tried to provide the present state of IOD research and future perspectives.

The subtropical oceans also give rise to modes of climate variations. However, the ocean dynamics and the coupling processes in the subtropics are different from those of the tropics. It is now known that the Indian Ocean subtropical dipole is a prominent subtropical climate phenomenon. Here, we have reviewed several articles that have studied this phenomenon in depth and tried to understand its predictability based on coupled general circulation model results.

There is also another climate phenomenon discovered off the western coast of Australia. Being confined to the coastal waters of the Ningaloo Reef, the phenomenon is called the Ningaloo Niño/Niña. The phenomenon is phase-locked to austral summer as sea surface temperature (SST) anomalies start to develop off the northwest coast in September, evolve to peak off the west coast during its mature phase in December-February, and then start to decline in March. Significant advances have been made in our understanding of the Ningaloo Niño/Niña since the 2011 event was first investigated. We have reviewed these studies and provided possible ways to deepen our understanding of some open questions.

2 IOD and its flavors

The IOD has emerged as a dominant ocean-atmosphere coupled mode in the tropical climate system. Until its discovery as an intrinsic mode in 1999 (Saji et al., 1999; Yamagata et al., 2004), the Indian Ocean has long been considered as a passive element in the tropical climate system essentially controlled by ENSO through an atmospheric bridge (Alexander et al., 2002; Cadet, 1985; Klein et al., 1999; Lau & Nath, 2003) and by the Asian summer monsoon via air-sea fluxes associated with the monsoon flow (Webster et al., 1998). Indeed, ENSO influences the Indian Ocean basin particularly during its mature phase, as discussed in Section 3. However, during boreal fall, east-west dipole mode events can develop in the Indian Ocean with or without the influence of ENSO (e.g., the 1997 and the 1994 events, respectively), which demonstrates the intrinsic nature of ocean–atmosphere coupled variability in the basin (Saji et al., 1999). One such typical positive IOD event with devastating climatic impacts appeared recently in 2019 in the absence of a canonical El Niño event, notwithstanding some lingering claims of ENSO as a trigger. The SST anomalies associated with the 2019 positive IOD showed characteristic cold anomalies clearly off the Sumatra coast and warm anomalies off the Somali coast (Fig. 1, upper panel). The associated easterly wind anomalies in between those SST anomalies demonstrate the presence of an intrinsic ocean-atmosphere coupling process in the basin, through a Bjerknes feedback mechanism, as discussed first for the 1994 IOD event (Behera et al., 1999; Saji et al., 1999; Vinayachandran et al., 1999). The nature of the coupling was further appreciated when opposite-signed anomalies

---

This book has a companion website hosting complementary materials. Visit this URL to access it: https://www.elsevier.com/books-and-journals/book-companion/9780128226988.

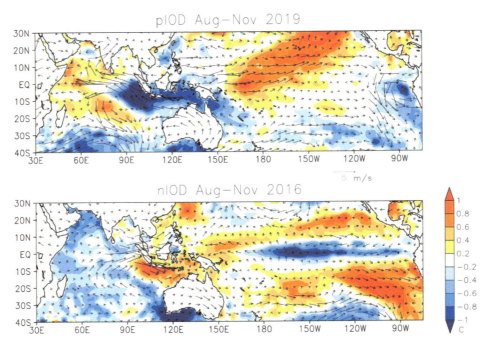

FIG. 1 Detrended anomalies of SST (°C) and 850 hPa wind (m/s) for August-November derived from NOAA NCEP Global SST analysis (Reynolds et al., 2002) and NCEP/NCAR reanalysis dataset (Kalnay et al., 1996), respectively. The upper panel is for the 2019 positive IOD, and the bottom panel is for the 2016 negative IOD. *(Adapted from Behera et al., 2020.)*

persisted over the tropical Indian Ocean during the negative IOD event in 2016 (Fig. 1, lower panel). Opposite polarity SST anomalies prevailed on both sides of the basin with strong westerly wind anomalies in between them through the Bjerknes feedback mechanism. These surface conditions are associated with remarkable variations in the subsurface ocean, as well.

The subsurface ocean variability manifests itself as a dipole structure in sea-level anomalies (Fig. 2). The dipole pattern is formed after the adjustment of the thermocline through the propagation of oceanic Rossby and Kelvin waves (Chen et al., 2019; Feng & Meyers, 2003; Rao et al., 2002; Rao & Behera, 2005; Yamagata et al., 2004, 2015). The wind stress curl anomalies associated with the positive IOD force westward propagating downwelling long Rossby waves north of 10°S and eastward propagating upwelling Kelvin waves at the equator. The downwelling Rossby waves suppress the thermocline near the Somali coast, while the equatorial Kelvin waves raise the thermocline near the Sumatra coast at the peak of the event. These Rossby and Kelvin waves further play a dominant role in shaping the subsurface ocean variability. As

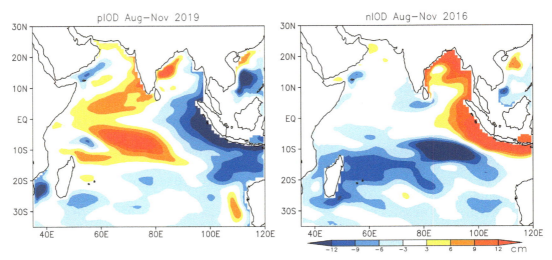

FIG. 2 Sea level anomalies (cm) derived from the NCEP global ocean data assimilation system. The left panel is for the 2019 positive IOD, and the right panel is for the 2016 negative IOD. *(Adapted from Behera et al., 2020.)*

a part of the subsurface processes, the downwelling Rossby waves that propagate to the Somali coast during positive IOD events pass to the equator as coastal Kelvin waves. The coastal Kelvin waves then transform into equatorial Kelvin waves and propagate eastward along the equator. After reaching the Sumatra coast, these downwelling Kelvin waves weaken the upwelling there and help the positive IOD event turn to a negative IOD event (Rao et al., 2002). This turnabout process, often called a delayed oscillator (Suarez & Schopf, 1988), is clearly due to the subsurface ocean dynamics, and gives rise to the quasi-biennial nature of the IOD variability (Feng & Meyers, 2003). In recent years, however, the change in the Indian Ocean in response to global warming seems to favor evolution of more consecutive positive IODs and hence weakens the biennial nature of the IOD (cf., Cai et al., 2013).

Just like the ENSO events, each of the IOD events varies from event to event. What stands out in the spatial structure of some IOD events is that the western pole anchors in the central tropical Indian Ocean rather than in the western Indian Ocean. Endo and Tozuka (2016) named these events IOD Modoki events, following the ENSO Modoki events in the tropical Pacific (Ashok et al., 2007; Ashok & Yamagata, 2009; Behera & Yamagata, 2018; Weng et al., 2007). Although the characteristics of the IOD Modoki bear a close resemblance to the canonical IODs discussed earlier, the longitudinal difference in the location of the warm (cold) SST anomalies has different climatic impacts on the surrounding regions, especially over East Africa. This is because of different atmospheric circulation anomalies and teleconnection patterns compared to the canonical IOD (Endo & Tozuka, 2016).

How the IOD event is triggered has been of great interest among researchers. One suggested mechanism is related to cold SST anomalies associated with enhanced upwelling in the eastern Indian Ocean. Anomalous southeasterlies and suppressed convection in this region work together with those SST anomalies in a feedback loop to trigger a positive IOD event (e.g., Behera et al., 1999; Saji et al., 1999). Other suggested mechanisms are related to the atmospheric pressure variability in the eastern Indian Ocean (e.g., Gualdi et al., 2003; Li et al., 2003), favorable changes in winds in relation to the Pacific ENSO and the Indian monsoon (e.g., Annamalai et al., 2003), oceanic conditions of the Arabian Sea related to the Indian monsoon (Prasad & McClean, 2004), and teleconnection from the southern extratropical region (e.g., Lau & Nath, 2004; Zhang et al., 2020). Feng et al. (2014) suggested that the Indian Ocean subtropical dipole discovered by Behera and Yamagata (2001) could help the development of the equatorial easterly anomalies through the modulation of the Mascarene High and Indian monsoon trough leading to the development of some IOD events. Nevertheless, these mechanisms may not work sometimes; there are cases where the IOD did not form despite favorable preconditions (e.g., 1979), but formed in spite of unfavorable preconditions (e.g., 2007) (Behera et al., 2006, 2008; Gualdi et al., 2003; Rao & Yamagata, 2004).

The IOD's termination is dominantly linked to the seasonal change in the monsoon winds (Saji et al., 1999). However, we note that it may be linked to intraseasonal oscillations (ISOs)/Madden Julian Oscillations (MJOs) originating in the tropical Indian Ocean (Rao & Yamagata, 2004). Rao and Yamagata (2004) and Rao et al. (2007) observed strong 30- to 60-day oscillations of equatorial zonal winds prior to the termination of all IOD events. The strong westerlies associated with the ISO excite anomalous downwelling Kelvin waves that deepen the thermocline in the eastern Indian Ocean and terminate the coupled feedback processes, as discussed by Fischer et al. (2005) for the 1994 positive IOD event. Similarly, anomalously high ISO activities in the boreal summer might explain the aborted positive IOD event of 1974 (Gualdi et al., 2003).

IOD impacts are observed in various parts of the globe through either direct or indirect teleconnections (Saji & Yamagata, 2003), as shown in the schematic diagram (Fig. 3). We observe its direct impact on the rainfall in East Africa (Behera et al., 2005; Black et al., 2003; Clark et al., 2003; Gebregiorgis et al., 2019; Manatsa et al., 2008; Manatsa & Behera, 2013; Marchant et al., 2006) because of its proximity to the western pole. The IOD also influences the climatic conditions directly near the eastern pole; the Pacific ENSO itself is known to be affected (Behera & Yamagata, 2003; Izumo et al., 2010; Saji et al., 2018). Many other cases of the IOD impacts are reported now such as the Sri Lankan Maha rainfall (Zubair et al., 2003), the Brazilian rainfall (Chan et al., 2008), Indian summer monsoon rainfall (Anil et al., 2016; Ashok et al., 2001; Behera et al., 1999), the summer conditions over East Asia (Guan et al., 2003; Guan & Yamagata, 2003), European summer temperature (Behera et al., 2012), and the Australian rainfall (Ashok et al., 2003; Ummenhofer et al., 2009, 2013). In particular, the summer conditions of southeastern Australia are much affected by some of the strong IOD events; the positive IOD-induced dryness causes severe forest fires in the region (Cai et al., 2009; Harris & Lucas, 2019). In a recent study, Abram et al. (2021) showed that IOD preconditions the bushfire season in southeast Australia, and they found that the recent 2019 positive IOD contributed to the severe 2019 Black Summer bushfires. In Japan, the number of heatstroke-related deaths in the Kanto region (Akihiko et al., 2014) rises when the daily maximum temperature exceeds 35°C in some of the positive IOD years. Especially, positive IOD and/or La Niña years like that of 1994, 2007, and 2010 caused stronger heatwaves, which resulted in a large number of heatstroke (hyperthermia)-related deaths among people over 65 years old. In an expected future warmer climate, even a small temperature perturbation related to the tropical climate variability may lead to many more hot days and associated fatalities. According to coral data, positive IOD frequency increased in the 20th century (Abram et al., 2008) and the Indian Ocean's long-term warming trend resembles the

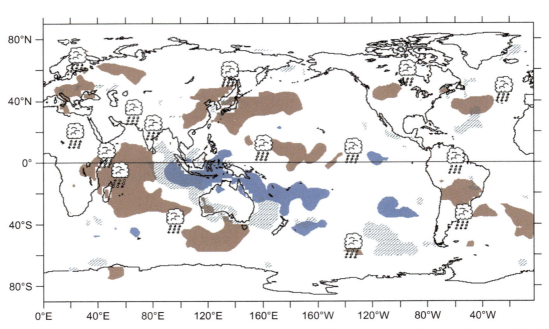

FIG. 3 Schematic of the positive IOD-related global surface temperature and rainfall anomalies during September-November. Blue (*brown*) color means colder (warmer) than normal temperature, and clouds (hashed lines) indicate wetter (drier) than normal condition. *(Adapted from Yamagata et al., 2015.)*

positive IOD (Dhame et al., 2020). Moreover, both IOD's frequency and intensity are expected to change in the future climate as shown by both paleoclimate data (Abram et al., 2008; Nakamura et al., 2009) and climate model projections (Cai et al., 2013; Zheng, 2019).

3 Indian Ocean Basin mode and interbasin connections

The Indian Ocean basin mode appears statistically as the dominant mode of SST anomalies in the empirical orthogonal function (EOF) analysis. As shown in Fig. 4, it is a basin-wide mode associated with El Niño/La Niña events in the tropical Pacific (Behera et al., 2003; Cadet, 1985; Klein et al., 1999; Venzke et al., 2000; Wallace et al., 1998) and is discussed in further details by Ummenhofer et al. (2024). It has a strong impact on the South Asian summer monsoon (Chang & Li, 2000; Li et al., 2001; Li & Zhang, 2002), and it creates a low-level anticyclonic anomaly over the Philippines, the latter of which appears frequently during the El Niño decaying summer (e.g., Du et al., 2009; Wu et al., 2009; Xie et al., 2009; Yang et al., 2007). South of the equator, the Indian Ocean basin mode warming affects Australia by prolonging the El Niño-related dry conditions over northern Australia (Taschetto et al., 2011). In the northwest Pacific, the Indian Ocean basin mode induces a delayed influence on the regional climate (e.g., Chowdary et al., 2010) through an enhancement of the western North Pacific subtropical high and the South Asian high. In particular, Wu et al. (2009) suggested that a low-level anomalous cyclone induced by the warm SST anomaly of a warm Indian Ocean basin mode can enhance southerlies and promote deep convective precipitation to the east of the Indian Ocean basin mode, as confirmed later by model simulation experiments (Chowdary et al., 2011; Hu & Duan, 2015; Huang et al., 2010; Li et al., 2008; Wu et al., 2010). This, in turn, causes an anomalous anticyclonic circulation over the tropical Northwest Pacific and South Asia, prolonging the influences of ENSO like charging/discharging (Yang et al., 2007). Thus, we may infer that the climate variations associated with Indian Ocean basin mode are due to the Indian Ocean capacitor effect (Xie et al., 2009).

4 Indian Ocean subtropical dipole

Interannual SST variations in the southern Indian Ocean have a close link with rainfall variations over southern Africa through changes in location and intensity of the Mascarene High (Mason, 1995; Richard et al., 2001). The SST anomalies tend to have a meridional dipole structure of positive and negative SST anomalies tilting from northwest to southeast in the subtropical Indian Ocean. The overall ocean and atmosphere structure was called the Indian Ocean subtropical dipole by

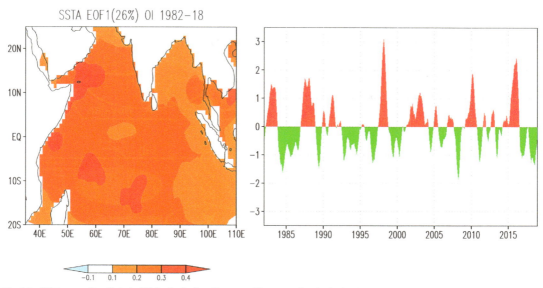

FIG. 4 EOF1 of the SST anomalies (left, in °C) in the Indian Ocean and its normalized principal component (right). The SST anomalies are derived from the NOAA NCEP Global SST analysis.

Behera and Yamagata (2001). The positive phase of the Indian Ocean subtropical dipole is characterized by positive and negative SST anomalies in the southwestern and northeastern parts of the southern Indian Ocean, respectively. It is also accompanied by a southward shift and strengthening of the Mascarene High. The Indian Ocean subtropical dipole shows a strong seasonal evolution; it starts to develop in austral spring (September-November), reaches its peak in austral summer (December-February), and decays in austral autumn (March-May). One of the strongest positive Indian Ocean subtropical dipoles occurred together with La Niña in 2010 when southern Africa received an excessive amount of summer rainfall, and the consecutive floods caused loss of human lives and serious damages to infrastructure, and socioeconomic activities such as agriculture (UN Office for the Coordination of Humanitarian Affairs (OCHA), 2011). Indian Ocean subtropical dipole-related anomalies over the southeastern Indian Ocean are also thought to influence the Indian summer monsoon through the Mascarene High. Following a positive (negative) Indian Ocean subtropical dipole, the Mascarene High shifts southeastward (northwestward) from austral to boreal summer causing a weakening (strengthening) of the monsoon circulation system, by modulating the local Hadley cell, and a weak (strong) Indian Summer Monsoon (Terray et al., 2003).

Since its introduction, several studies have discussed physical mechanisms underlying the Indian Ocean subtropical dipole (Behera & Yamagata, 2001; Fauchereau et al., 2003; Hermes & Reason, 2005; Suzuki et al., 2004). During a positive Indian Ocean subtropical dipole, the Mascarene High tends to shift southward and strengthen anomalously. In the southwestern part of the southern Indian Ocean, anomalous northerly winds, as a result of weakening of the mean circulation, reduce evaporation from the ocean surface. Anomalous decrease in the evaporation was once considered to induce the positive SST anomalies directly. However, subsequent research (Morioka et al., 2010) identified that the suppressed evaporation rather leads to a thinner-than-normal surface ocean mixed layer and that the warming of the mixed-layer is enhanced by the climatological shortwave radiation. Thus, the interannual variation of the mixed-layer thickness plays a key role in development of the Indian Ocean subtropical dipole during austral summer when the mixed layer is shallowest (Fig. 5). This point warns, in general, about an easy application of the wind-evaporation-SST feedback mechanism to the formation of climate modes (cf., Xie & Philander, 1994). After reaching its peak, the Indian Ocean subtropical dipole starts to decay owing to both entrainment and evaporation processes (Morioka et al., 2012). In the vicinity of the positive SST anomaly pole, the temperature differences between the mixed layer and subsurface ocean become larger and enhance cooling of the mixed layer by entrainment. Also, the positive SST anomalies tend to increase evaporation from the ocean surface, which contributes to decay of the positive SST anomalies. A similar but opposite mechanism operates in the vicinity of the negative SST anomaly pole of the Indian Ocean subtropical dipole.

The interannual variations of the Mascarene High play a crucial role in triggering the Indian Ocean subtropical dipole through mixed-layer variations. Therefore, it is important to understand potential sources of the Mascarene High variations for realistic simulation and skillful prediction of the Indian Ocean subtropical dipole. One of the key factors controlling the Mascarene High variations is certainly ENSO; the positive (negative) Indian Ocean subtropical dipole tends to co-occur with La Niña (El Niño) (Behera & Yamagata, 2001; Zinke et al., 2004). It is known that ENSO modifies both the Walker

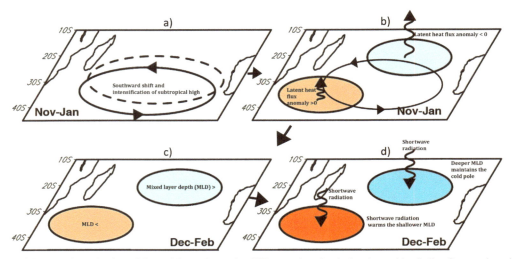

FIG. 5 Schematic of the growth mechanism of the positive and negative SST anomaly poles during the positive Indian Ocean subtropical dipole event. The *color red* (*blue*) indicates the positive (negative) mixed layer temperature anomaly.

and Hadley circulations over the Indian Ocean, affecting the Mascarene High. Also, ENSO induces an atmospheric teleconnection pattern extending from the South Pacific to the South Atlantic, called the Pacific-South American mode (Mo & Paegle, 2001). The Pacific-South America pattern has downstream influences on the Mascarene High through eastward propagation of atmospheric Rossby waves (Morioka et al., 2013). On the other hand, atmospheric internal variability in the middle and high latitudes of the Southern Hemisphere, i.e., the Antarctic Oscillation (Gong & Wang, 1999) or the Southern Annular Mode (Thompson & Wallace, 2000), also have impacts on the Mascarene High (Fauchereau et al., 2003; Morioka et al., 2014). A positive phase of the Southern Annular Mode is accompanied by negative sea level pressure (SLP) anomalies over Antarctica and positive SLP anomalies in mid-latitudes. The positive Southern Annular Mode causes a southward shift and strengthening of the Mascarene High, which is responsible for the positive Indian Ocean subtropical dipole. Besides those external influences, in situ local air-sea interaction contributes to the Mascarene High variability as well (Morioka et al., 2015). During the positive Indian Ocean subtropical dipole, the meridional gradient of surface air temperature, i.e., baroclinicity, becomes larger to the south of the positive SST anomalies in the southwestern Indian Ocean. This provides a favorable condition for shifting storm tracks poleward and maintaining the southward shift and strengthening of the Mascarene High. Therefore, both the remote forcing and local air-sea interaction are important for the Mascarene High variability during the Indian Ocean subtropical dipole. Accurate representation of the Mascarene High variability using coupled general circulation models is necessary for realistic simulation and skillful prediction of the Indian Ocean subtropical dipole (Doi et al., 2016; Kataoka et al., 2012) and hence southern African rainfall (Dieppois et al., 2019).

5 Ningaloo Niño/Niña

Recently, several warm events observed near the western coast of Australia have been referred to as Ningaloo Niño (Feng et al., 2013; Kataoka et al., 2014). As a typical marine heatwave, these events profoundly affect coral reefs and the marine ecosystem of the Ningaloo reef, which is a biodiversity hotspot (Tittensor et al., 2010; Zinke et al., 2014). Several recent studies discussed the marine heatwave mechanisms associated with the Ningaloo Niño. For example, Holbrook et al. (2019) and Sen Gupta et al. (2020) discussed the marine heatwave mechanisms including the Ningaloo Niño event. Hayashida et al. (2020) highlighted the importance of high-resolution climate models to simulate local coastal processes to represent this marine heatwave correctly. Feng et al. (2019) showed that these types of marine heatwaves have increased since 1990, and marine heatwaves are projected to increase in a warmer future (Frölicher et al., 2018).

The Ningaloo Niño is one of the five coastal Niños, together with Benguela Niño, California Niño, Dakar Niño, and Chile Niño located along the eastern boundaries of the global oceans (cf. Oettli et al., 2016, 2020; Xue et al., 2020; Yuan & Yamagata, 2014). They are called Niños as they appear like El Niños confined to respective coastal regions and just like ENSO, all those coastal phenomena have positive (Niño) and negative (Niña) phases. During the positive (negative) phase, the northwestern part of Australia receives more (less) rainfall (Kataoka et al., 2014; Tozuka et al., 2014). Thus, coastal climate phenomena are important even for regional climate. Here, we focus on Ningaloo Niño/Niña events evolving through either local and/or nonlocal processes (Kataoka et al., 2014; Kusunoki et al., 2020, 2021; Tozuka & Oettli, 2018).

Some of the Ningaloo Niño events develop through the local air-sea interactions and are called locally-amplified Ningaloo Niños. Kataoka et al. (2018) demonstrated that the Ningaloo Niño events could develop through such an intrinsic air-sea interaction process in which warm SST anomalies and northerly wind anomalies are generated off the west coast of Australia in a positive feedback loop (the upper left panel of Fig. 6). The air-sea feedback is similar to the conventional Bjerknes feedback, but it occurs in the coastal region instead of the equator, and may well be referred to as "coastal Bjerknes feedback." In this process, the positive SST anomalies generate cyclonic anomalies in the overlying atmosphere slightly displaced to the west (Tozuka et al., 2014), and alongshore northerly wind anomalies are induced. Since these wind anomalies reduce the surface wind speed, the transport of cool and dry air from the higher latitudes is also reduced. In due course, latent and sensible heat loss to the atmosphere is reduced (Guo et al., 2020; Marshall et al., 2015; Zhang et al., 2018). These heat flux anomalies lead to anomalously shallow mixed layers, which, in turn, results in more effective warming of the surface mixed layer even by the climatological shortwave radiation (Kataoka et al., 2017). The northerly wind anomalies, at the same time, accelerate the warm coastal Leeuwin Current and contribute to anomalous warming in the coastal area (Benthuysen et al., 2014; Kataoka et al., 2017; Marshall et al., 2015).

Remote atmospheric and oceanic forcings from the Pacific may also contribute to development of some of the Ningaloo Niños (Feng et al., 2013; Kido et al., 2016) through a nonlocally amplifying process (the upper right panel of Fig. 6). The Matsuno-Gill type response (Gill, 1980; Matsuno, 1966) to positive diabatic heating anomalies over the western equatorial Pacific during La Niña (Feng et al., 2013; Tozuka et al., 2014) enhances negative SLP anomalies off the west coast of Australia with poleward wind anomalies. In addition, downwelling coastal waves originating from the western tropical Pacific penetrate through the maritime continent and propagate poleward along the Australian coast, which is referred to as the Clarke-Meyers effect (Clarke, 1991; Clarke & Liu, 1994; Meyers, 1996; Yamagata et al., 2004). They contribute to strengthening the warm Leeuwin Current (Feng et al., 2003, 2008; Kusunoki et al., 2020, 2021). It is also known that the advection of anomalously fresher water may further enhance the positive heat advection by strengthening the cross-shelf density gradient of the Leeuwin Current in some events (Feng, 2015b). Almost like a mirror image, Ningaloo Niña events evolve through just the opposite feedback/response processes (lower panels of Fig. 6). However, only a few studies have focused on the Ningaloo Niña (Feng et al., 2021).

The La Niña (El Niño) could be a major trigger of the Ningaloo Niño (Niña) as mentioned earlier, but a recent study (Zhang & Han, 2018) suggested that SLP anomalies associated with the Ningaloo Niño/Niña may generate SLP gradient in

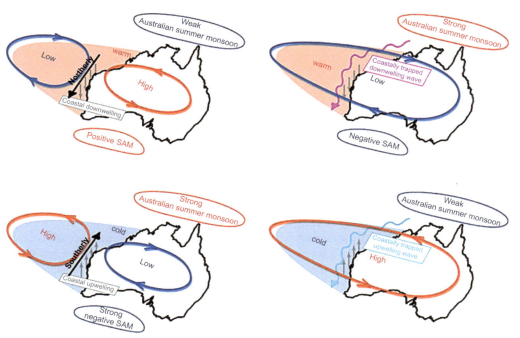

FIG. 6 Schematic diagrams of the locally amplified (left panels) and the nonlocally amplified (right) Ningaloo Niño (upper panels) and Ningaloo Niña (lower panels). *(Adapted from Kataoka et al. (2014).)*

the western equatorial Pacific. The resulting equatorial zonal wind stress anomalies trigger equatorial Kelvin waves and contribute to the development of ENSO events. In addition to the ENSO in the Pacific, the IOD in the Indian Ocean sometimes may act as a trigger of the Ningaloo Niño, because the anomalously cool SST in the eastern pole of the IOD may induce northerly wind stress anomalies along the west coast of Australia (Zhang et al., 2018). However, this relation does not seem to be stable; the correlation coefficient between the Ningaloo Niño index and dipole mode index is almost negligible when calculated over 1960–2011 (Marshall et al., 2015). Instead, the Mascarene High variability seems to play an important role in the evolution of both Ningaloo Niño/Niña and Indian Ocean subtropical dipole. More research is certainly needed on the role of the subtropical high over the Indian Ocean.

The Ningaloo Niño seems to occur more frequently during the negative phase of the Interdecadal Pacific Oscillation (Feng et al., 2015a; Tanuma & Tozuka, 2020) owing to the anomalous warming of the southeastern Indian Ocean on interdecadal timescales (Doi et al., 2015; Li et al., 2017, 2019) related to the Pacific influences. The interaction with the basinwide climate modes at such a longer timescale and the global warming trend may provide a rich area of investigation.

6 Predictability and prediction of IOD, Indian Ocean subtropical dipole, and Ningaloo Niño

Studies of the climate phenomena through routine frameworks of experimental predictions will not only benefit society with reliable climate services but also further improve understanding of the dynamics of those phenomena. This is because the verification process is essential to the advancement of prediction science. The IOD is shown to be predictable at least a season ahead (e.g., Doi et al., 2017, 2019; Liu et al., 2017; Luo et al., 2007, 2008; Wajsowicz, 2005; Wang et al., 2009; Zhu et al., 2015) using coupled general circulation models. Nevertheless, the IOD predictability is limited by several factors, in particular, by the so-called "winter predictability barrier" (Feng et al., 2017; Mu et al., 2017). This predictability barrier is rooted in IOD's strong phase locking to the annual reversal of the monsoon winds (Luo et al., 2007). In spite of such a predictability barrier, several IOD events, such as the 2006 and 2019 events, were predicted almost a year ahead as shown in Fig. 7a (Doi et al., 2020; Luo et al., 2007, 2008) using the SINTEX-F coupled general circulation model (Luo et al., 2005). A detailed analysis of why this succeeded leads to new insights.

Event-to-event diversity in predictability of IOD is shown to be rooted in differing development mechanisms of IOD (Tanizaki et al., 2017). Some studies have shown better IOD predictions while co-occurring with canonical ENSO (Song et al., 2008; Yang et al., 2015; Zhao & Hendon, 2009). Recently, Doi et al. (2020) have shown that the super positive IOD in 2019 was predictable a few seasons ahead by overcoming the winter predictability barrier (Fig. 8); success in capturing the precondition of El Niño Modoki in the Pacific was key. Some IOD events appear to be strongly influenced by weather noise and intraseasonal disturbances, leading to low potential predictability (Baba, 2020; Rao et al., 2009; Rao & Yamagata, 2004). Considering the stochasticity, a large ensemble prediction system seems to be useful to predict an extreme event (Doi et al., 2019). The Indian Ocean basin mode is shown to be more predictable (Luo et al., 2016; Wu & Tang, 2019; Zhu et al., 2015) compared to the IOD (Figs. 7a, b and 9a, b). This is mainly because ENSO-induced basin-scale warming/cooling is responsible for the Indian Ocean basin mode (Kug et al., 2006; Schott et al., 2009; Wu & Tang, 2019).

Compared to those tropical climate variations, predicting extratropical climate phenomena is more challenging due to both large atmospheric internal variability and weak air-sea coupling. Despite those limitations, Yuan et al. (2014) showed that some strong Indian Ocean subtropical dipole events are predictable one season ahead by the SINTEX-F family ensemble mean (Doi et al., 2020; Figs. 7c and 9c). This is partly due to their link with IOD and ENSO events. However, the prediction skill of Indian Ocean subtropical dipole events is still low particularly because of the lower predictability of the SST anomaly in its southwestern pole. Further understanding of the relative roles of local wind forcing and remote forcing originated from ENSO and IOD events will be the key to further understanding the potential predictability of SST in that region (Yuan et al., 2014; Zhang et al., 2019).

In contrast to the aforementioned basin-scale climate modes, the Ningaloo Niño (Niña), a coastal climate mode, has a much smaller lateral scale of a few hundred kilometers. Even so, the prediction of such regional climate modes is vital for the local socioeconomy as well as the marine ecosystem (Doi et al., 2015). Doi et al. (2013) found some skill in the prediction of Ningaloo Niño/Niña using the SINTEX-F; the model was skillful in predicting the Ningaloo Niño (anomaly correlation coefficient > 0.5) up to 4 months ahead when initialized on the first day of each month between August and November (Figs. 7d and 9d). For some of the strong events, they found good prediction skill even at longer lead times. In particular, the prediction skill was good for events that were concurrent with ENSO. On the other hand, it remains a challenge to predict the locally amplified Ningaloo Niño/Niña events. This is, needless to say, related to the fact that the present global models have difficulties in resolving coastal fine-scale physical processes due to their coarse resolution.

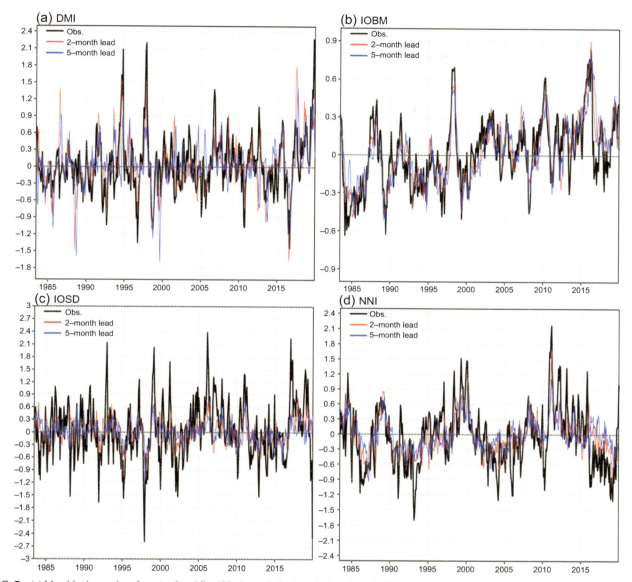

FIG. 7 (a) Monthly time series of two (*red*) and five (*blue*) months lead predictions from the SINTEX-F Family ensemble mean (Doi et al., 2020) with the observation (*black*) for the Dipole Mode Index (DMI) from 1984 to 2019 (°C). DMI is first introduced by Saji et al. (1999) as the SST anomaly difference between the western pole off East Africa (50–70°E, 10°S–10°N) and the eastern pole off Sumatra (90–110°E, 10°S to equator). For instance, when the model was integrated from January 1, March is 2-month lead prediction. (b) Same as (a), but for the Indian Ocean basin mode (IOBM) index defined as the SST anomaly averaged over the tropical Indian Ocean (40°E–100°E, 20°S–20°N) (Wu & Tang, 2019). (c) Same as (a), but for the Indian Ocean subtropical dipole mode index (IOSD) defined as the SST anomaly difference between the domains (55°E–65°E, 37°S–27°S) and (90°E–100°E, 28°S–18°S) (Behera & Yamagata, 2001). (d) Same as (a), but for the Ningaloo Niño index (NNI) is defined by Kataoka et al. (2014) as the SST anomaly averaged over the domain (108°E–116°E, 28°S–22°S). The SST anomalies are derived from the NOAA OISSTv2.

Since many coupled general circulation models have biases in simulating the Indian Ocean subtropical dipole (Kataoka et al., 2012) and the Ningaloo Niño/Niña (Kido et al., 2016), further model improvement may lead to better prediction of the subtropical modes. For example, a high-resolution version of SINTEX-F with a dynamical sea-ice model (called SINTEX-F2, Masson et al., 2012; Sasaki et al., 2013) is more skillful in predicting the Indian Ocean subtropical dipole and the Ningaloo Niño (Doi et al., 2016). In addition to those dynamical research directions in improving the predictability, a statistical approach such as machine learning may provide a useful tool in climate prediction (Ratnam et al., 2020). We expect that a combined dynamical/statistical approach may lead to a dramatic improvement in prediction skill. This, in turn, will contribute to a better understanding of these tropical and extra-tropical phenomena and the interactions among different scale phenomena.

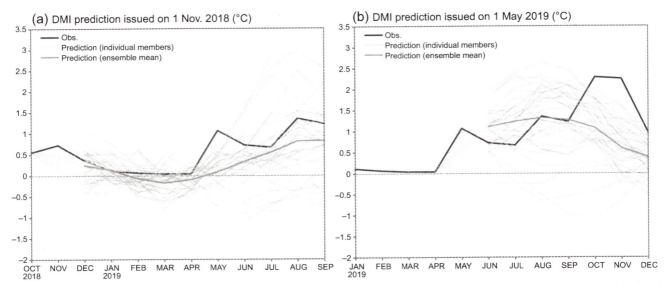

FIG. 8 (a) Monthly Dipole Mode Index (DMI) from October 2018 to September 2019 (°C) from the observational data of NOAA OISSTv2 (*black*) and the prediction issued on November 1, 2018 by SINTEX-F Family system (thin *gray*: each ensemble member; thick *gray*: ensemble mean). (b) Same as (a), but for the May 1, 2019 initialization. *(Adapted from Doi et al. (2020).)*

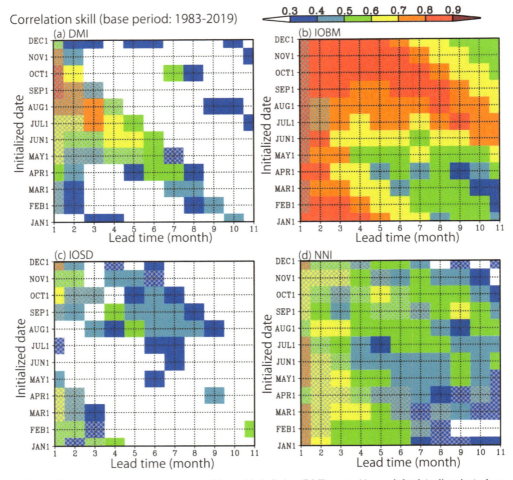

FIG. 9 (a) Seasonally stratified correlation skills for monthly Dipole Mode Index (DMI) up to 11-month lead (*x*-direction) along a fixed start time (*y*-direction) in 1984–2019. For instance, when the model was integrated from January 1, March is 2-month lead prediction. Values lower than the persistence (lag auto-correlation of observation) are shown by *gray* hatching. (b, c, d) Same as (a), but for Indian Ocean basin mode (IOBM), Indian Ocean subtropical dipole (IOSD), and Ningaloo Niño index (NNI), respectively.

7 Conclusions

The first wave of interest in understanding the Indian Ocean surged in the 1960s among oceanographers who wished to raise knowledge of this basin to the same level as the Atlantic and the Pacific. The understanding of the time-dependent ocean circulation forced by the monsoonal winds was much progressed during this period owing to the first international collaboration supported by the Scientific Committee on Ocean Research (SCOR) of the International Council of Scientific Unions at the time. Since then, the Indian Ocean was considered a passive ocean dominated by the Asian monsoon, and later by the atmospheric and oceanic remote forcing by the Pacific ENSO. This view has changed drastically in recent decades after the discovery of the IOD (Saji et al., 1999), and the study of natural variability of the Indian Ocean has received renewed interest. Understanding the ocean-atmosphere coupled variability has become a major research topic and several new phenomena have been catalogued as new climate modes. We have focused on those new climate phenomena in this chapter.

The positive (negative) IOD appears interannually in the tropical Indian Ocean as a zonally oriented SST dipole in association with the anomalous easterlies (westerlies). The IOD events occur sometimes with the Pacific ENSO. In other words, the Pacific ENSO sometimes occurs with the IOD. How those basin-wide climate modes interact with each other is still an active research topic, together with examination of initiation and termination processes involving phenomena of different space and time scales (Luo et al., 2010).

The IOBM associated with the Pacific ENSO appears statistically as the dominant mode of SST variability in the Indian Ocean despite that a clear SST dipole is seen during strong IOD years. The Indian Ocean basin mode is particularly important from a socioeconomic viewpoint because it prolongs the ENSO influences in Asia as a heat capacitor of the Indian Ocean.

At the mid-latitudes, we find a slightly slanted meridional SST dipole in the southern Indian Ocean. This interannual phenomenon was named Indian Ocean subtropical dipole by Behera and Yamagata (2001) during a study to understand the rainfall variability in southern Africa. The Indian Ocean subtropical dipole is closely related to the variability of the Mascarene High and local anomalous air-sea fluxes with mixed-layer variations. Interactions with ENSO and IOD in the tropics as well as the Southern Annular Mode in the high-latitudes through the atmospheric bridge are important in triggering the Indian Ocean subtropical dipole and await more investigation.

It is known that there is a close analogy between equatorial upwelling and coastal upwelling in the Indian Ocean (Yoshida, 1959). The wind-induced equatorial Kelvin waves play a key role in the former, and the wind-induced coastal Kelvin waves play a key role in the latter. It should not, therefore, be unreasonable to expect a coastal version of El Niño. The unprecedented coastal heatwave that occurred off the west coast of Australia in the 2010/2011 austral summer caused serious negative impacts on local marine ecosystems. The phenomenon was named Ningaloo Niño (Feng et al., 2013; Kataoka et al., 2014). Since it involves air-sea coupled processes, it has received much attention as a local interannual climate phenomenon. The remote atmospheric and oceanic forcing from the Pacific certainly contribute to the evolution of some strong events as seen in the austral summer of 2010/2011 during La Niña. Just like the IOD right after its discovery, Ningaloo Niño/Niña's relation with the Pacific ENSO was a very active research topic at the beginning. The current consensus is that some events, called locally-amplified Ningaloo Niños, can develop without ENSO, owing to the coastal Bjerknes feedback process in which warm SST anomalies and northerly wind anomalies develop off the west coast of Australia. Interestingly, Zhang and Han (2018) even suggest that the atmospheric condition associated with Ningaloo Niño/Niña may contribute to the development of ENSO events. Elucidation of the interrelationships among various climate modes always gives us new stimuli in climate research. Possible interactions among Ningaloo Niño, IOD, Indian Ocean subtropical dipole, and others await further study.

Efforts to predict the climate phenomena of different space and time scales are useful not only to society to prepare adaptation and mitigation measures but also to climate research communities to improve their understanding of those phenomena for better prediction. This is because the verification process is essential to the advancement of science and climate science is not an exception. IOD prediction is not easy compared to ENSO prediction, but it is currently predictable at least a season ahead using coupled general circulation models despite the winter prediction barrier. The IOBM is more predictable because it is directly linked with ENSO. The prediction of the Indian Ocean subtropical dipole in the extra-tropical region is challenging except for some cases with links to tropical climate modes like IOD and ENSO. The low predictability is rooted in both large atmospheric internal variability and weak nonlinear air-sea coupling.

In contrast to the aforementioned basin-scale climate modes, prediction of the Ningaloo Niño/Niña is challenging because of its small horizontal scale. We need to develop a high-resolution coupled general circulation model that is able to capture the evolution of a locally amplified mode. However, some phenomena related to ENSO through oceanic and atmospheric bridges are again predictable to some extent using coupled general circulation models. We expect that a new combined dynamical/statistical approach may lead to a dramatic improvement in the prediction skill even for such local climate modes.

As described in this chapter, the climate research of the Indian Ocean has received vigorous interests in recent decades and led to the international IIOE-2 (Second International Indian Ocean Expedition for 2015–2025). In contrast to the IIOE-1, which was initiated at a high international executive level, the IIOE-2 is based on enthusiasm at the researcher level that led to the discovery of various new phenomena and the recognition of interbasin interactions. In addition, the social and ecological importance of Indian Ocean climate variability is increasing under climate change as shown by the devastating super IOD impacts in 2019 (Hermes et al., 2019; Zhou et al., 2021). Therefore, the situation has much changed from the days when "The Indian Ocean Bubble" was edited by Henry Stommel half a century ago (https://darchive.mblwhoilibrary.org/handle/1912/218?show=full). As expected by Gene LaFond (in the fourth issue of The Indian Ocean Bubble published in July 1959), the active participation of researchers from countries bordering the Indian Ocean has introduced strong incentives to promote research and development in this interesting ocean. It is not a bubble any more.

8 Educational resources

Here, we provide a list of resources and materials which may be useful for educational purposes.

IOD forecasts:

Japan Agency for Marine-Earth Science and Technology (JAMSTEC), Japan: http://www.jamstec.go.jp/virtualearth/general/en/

Bureau of Meteorology (BOM), Australia: http://www.bom.gov.au/climate/enso/indices.shtml?bookmark=iod

Japan Meteorological Agency (JMA), Japan: https://ds.data.jma.go.jp/tcc/tcc/products/model/indices/3-mon/indices1/shisu_forecast.php

IOD time series:

PSL, NOAA: https://psl.noaa.gov/gcos_wgsp/Timeseries/DMI/

Observed SST data sets for IOD index:

Available data sets at PSD/ESRL/NOAA: COBE-SST, ERSST, NOAA OIV2: https://psl.noaa.gov/data/gridded/index.html

Model SST data for deriving the IOD index:

North American Multi Model Ensemble (NMME): https://iridl.ldeo.columbia.edu/SOURCES/.Models/.NMME/

Software useful for deriving the IOD index:

Climate Data Operators (CDO): https://code.mpimet.mpg.de/projects/cdo/
NCAR Command Language (NCL): https://www.ncl.ucar.edu/index.shtml
GrADS: http://cola.gmu.edu/grads/
R: https://www.r-project.org/

Author contributions

SB and TY wrote Sections 1 and 3. SB and TT wrote Section 2. YM wrote Section 4. TT wrote Section 5. J-JL and TD wrote Section 6. TY wrote Section 7. All authors contributed to Section 8. TY, together with SB, organized the overall structure of the chapter.

References

Abram, N. J., Gagan, M. K., Cole, J. E., Hantoro, W. S., & Mudelsee, M. (2008). Recent intensification of tropical climate variability in the Indian Ocean. *Nature Geoscience*, *1*, 849–853.

Abram, N. J., et al. (2021). Connections of climate change and variability to large and extreme forest fires in Southeast Australia. *Communications Earth & Environment*, *2*, 1–17. https://doi.org/10.1038/s43247-020-00065-8.

Akihiko, T., Morioka, Y., & Behera, S. K. (2014). Role of climate variability in the heatstroke death rates of Kanto region in Japan. *Scientific Reports*, *4*, 5655.

Alexander, M. A., Bladé, I., Newman, M., Lanzante, J. R., Lau, N.-C., & Scott, J. D. (2002). The atmospheric bridge: The influence of ENSO teleconnections on air-sea interaction over the global oceans. *Journal of Climate*, *15*, 2205–2231.

Anil, N., Ramesh Kumar, M. R., Sajeev, R., & Saji, P. K. (2016). Role of distinct flavours of IOD events on Indian summer monsoon. *Natural Hazards*, *82*, 1317–1326.

Annamalai, H., Murtugudde, R., Potemra, J., Xie, S. P., Liu, P., & Wang, B. (2003). Coupled dynamics over the Indian Ocean: Spring initiation of the zonal mode. *Deep-Sea Research II*, *50*, 2305–2330.

Ashok, K., Behera, S. K., Rao, S. A., Weng, H., & Yamagata, T. (2007). El Niño Modoki and its possible teleconnection. *Journal of Geophysical Research, 112*, C11007.

Ashok, K., Guan, Z., & Yamagata, T. (2001). Impact of the Indian Ocean dipole on the relationship between the Indian monsoon rainfall and ENSO. *Geophysical Research Letters, 26*, 4499–4502.

Ashok, K., Guan, Z., & Yamagata, T. (2003). Influence of the Indian Ocean dipole on the Australian winter rainfall. *Geophysical Research Letters, 30*, 1821.

Ashok, K., & Yamagata, T. (2009). The El Niño with a difference. *Nature, 461*, 481–484.

Baba, Y. (2020). Roles of atmospheric variabilities in the formation of the Indian Ocean dipole. *Ocean Dynamics, 70*, 21–39.

Behera, S. K., Doi, T., & Ratnam, J. V. (2020). Air-sea interactions in tropical Indian Ocean: The Indian Ocean dipole. In S. K. Behera (Ed.), *Tropical and extratropical air-sea interactions* (pp. 115–139). Amsterdam: Elsevier Inc.

Behera, S. K., Krishnan, R., & Yamagata, T. (1999). Unusual ocean-atmosphere conditions in the tropical Indian Ocean during 1994. *Geophysical Research Letters, 26*, 3001–3004.

Behera, S. K., Luo, J.-J., Masson, S., Delecluse, P., Gualdi, S., Navarra, A., & Yamagata, T. (2005). Paramount impact of the Indian Ocean dipole on the East African short rain: A CGCM study. *Journal of Climate, 18*, 4514–4530.

Behera, S. K., Luo, J.-J., Mason, S., Rao, S. A., Sakuma, H., & Yamagata, T. (2006). A CGCM study on the interaction between IOD and ENSO. *Journal of Climate, 19*, 1688–1705.

Behera, S. K., Luo, J.-J., & Yamagata, T. (2008). The unusual IOD event of 2007. *Geophyical Research Letters, 35*, L14S11.

Behera, S. K., Rao, S. A., Saji, H. N., & Yamagata, T. (2003). Comments on "A cautionary note on the interpretation of EOFs". *Journal of Climate, 16*, 1087–1093.

Behera, S. K., Ratnam, J. V., Masumoto, Y., & Yamagata, T. (2012). Origin of extreme summers in Europe – The Indo-Pacific connection. *Climate Dynamics, 41*, 663–676.

Behera, S. K., & Yamagata, T. (2001). Subtropical SST dipole events in the southern Indian Ocean. *Geophysical Research Letters, 28*, 327–330.

Behera, S., & Yamagata, T. (2003). Influence of the Indian Ocean dipole on the southern oscillation. *Journal of the Meteorological Society of Japan, Ser. II, 81*, 169–177.

Behera, S. K., & Yamagata, T. (2018). Climate dynamics of ENSO Modoki phenomenon. In *Oxford Research Encyclopedia of Climate Science*. https://doi.org/10.1093/acrefore/9780190228620.013.612.

Benthuysen, J., Feng, M., & Zhong, L. (2014). Spatial patterns of warming off Western Australia during the 2011 Ningaloo Niño: Quantifying impacts of remote and local forcing. *Continental Shelf Research, 91*, 232–246.

Black, E., Slingo, J., & Sperber, K. R. (2003). An observational study of the relationship between excessively strong short rains in coastal East Africa and Indian Ocean SST. *Monthly Weather Review, 31*, 74–94.

Cadet, D. L. (1985). The southern oscillation over the Indian Ocean. *Journal of Climatology, 5*, 189–212.

Cai, W., Sullivan, A., & Cowan, T. (2009). How rare are the 2006-2008 positive Indian Ocean dipole events? An IPCC AR4 climate model perspective. *Geophysical Research Letters, 36*, L08702.

Cai, W., Zheng, X. T., Weller, E., Collins, M., Cowan, T., Lengaigne, M., Yu, W., & Yamagata, T. (2013). Projected response of the Indian Ocean dipole to greenhouse warming. *Nature Geoscience, 6*, 999–1007.

Chan, S., Behera, S. K., & Yamagata, T. (2008). The Indian Ocean dipole teleconnection to South America. *Geophysical Research Letters, 35*, L14S12.

Chang, C.-P., & Li, T. (2000). A theory of the tropical tropospheric biennial oscillation. *Journal of the Atmospheric Sciences, 57*, 2209–2224.

Chen, Z., Du, Y., Wen, Z., Wu, R., & Xie, S.-P. (2019). Evolution of south tropical Indian ocean warming and the climatic impacts following strong El Niño events. *Journal of Climate, 32*, 7329–7347. https://doi.org/10.1175/jcli-d-18-0704.1.

Chowdary, J. S., Xie, S.-P., Lee, J.-Y., Kosaka, Y., & Wang, B. (2010). Predictability of summer northwest Pacific climate in 11 coupled model hindcasts: Local and remote forcing. *Journal of Geophysical Research: Atmospheres, 115*, D22121. https://doi.org/10.1029/2010JD014595.

Chowdary, J. S., Xie, S.-P., Luo, J.-J., Hafner, J., Behera, S., Masumoto, Y., & Yamagata, T. (2011). Predictability of Northwest Pacific climate during summer and the role of the tropical Indian Ocean. *Climate Dynamics, 36*, 607–621.

Clark, C. O., Webster, P. J., & Cole, J. E. (2003). Interdecadal variability of the relationship between the Indian Ocean zonal mode and East African coastal rainfall anomalies. *Journal of Climate, 16*, 548–554.

Clarke, A. J. (1991). On the reflection and transmission of low-frequency energy at the irregular western Pacific Ocean boundary. *Journal of Geophysical Research, 96*, 3289–3305.

Clarke, A. J., & Liu, X. (1994). Interannual sea level in the northern and eastern Indian Ocean. *Journal of Physical Oceanography, 24*, 1224–1235.

Dhame, S., Taschetto, A. S., Santoso, A., & Meissner, K. J. (2020). Indian Ocean warming modulates global atmospheric circulation trends. *Climate Dynamics, 55*, 2053–2073.

Dieppois, B., Pohl, B., Crétat, J., Eden, J., Sidibe, M., New, M., Rouault, M., & Lawler, D. (2019). Southern African summer-rainfall variability, and its teleconnections, on interannual to interdecadal timescales in CMIP5 models. *Climate Dynamics, 53*, 3505–3527.

Doi, T., Behera, S. K., & Yamagata, T. (2013). Predictability of the Ningaloo Niño/Niña. *Scientific Reports, 3*, 2892.

Doi, T., Behera, S. K., & Yamagata, T. (2015). An interdecadal regime shift in rainfall predictability related to the Ningaloo Niño in the late 1990s. *Journal of Geophysical Research Oceans, 120*, 1388–1396.

Doi, T., Behera, S. K., & Yamagata, T. (2016). Improved seasonal prediction using the SINTEX-F2 coupled model. *Journal of Advances in Modeling Earth Systems, 8*, 1847–1867.

Doi, T., Behera, S. K., & Yamagata, T. (2019). Merits of a 108-member ensemble system in ENSO and IOD predictions. *Journal of Climate, 32*, 957–972.

Doi, T., Behera, S. K., & Yamagata, T. (2020). Predictability of the super IOD event in 2019 and its link with El Niño Modoki. *Geophysical Research Letters, 47*, e2019GL086713.

Doi, T., Storto, A., Behera, S. K., Navarra, A., & Yamagata, T. (2017). Improved prediction of the Indian Ocean dipole mode by use of subsurface ocean observations. *Journal of Climate, 30*, 7953–7970.

Du, Y., Xie, S.-P., Huang, G., & Hu, K. (2009). Role of air–sea interaction in the long persistence of El Niño-induced North Indian Ocean warming. *Journal of Climate*, *22*, 2023–2038.

Endo, S., & Tozuka, T. (2016). Two flavors of the Indian Ocean dipole. *Climate Dynamics*, *46*, 3371–3385.

Fauchereau, N., Trzaska, S., Richard, Y., Roucou, P., & Camberlin, P. (2003). Sea-surface temperature co-variability in the southern Atlantic and Indian oceans and its connections with the atmospheric circulation in the southern hemisphere. *International Journal of Climatology*, *23*, 663–677.

Feng, M., Benthuysen, J., Zhang, N., & Slawinski, D. (2015b). Freshening anomalies in the Indonesian throughflow and impacts on the Leeuwin current during 2010–2011. *Geophysical Research Letters*, *42*, 8555–8562.

Feng, M., Biastoch, A., Böning, C., Caputi, N., & Meyers, G. (2008). Seasonal and interannual variations of upper ocean heat balance off the west coast of Australia. *Journal of Geophysical Research*, *113*, C12025.

Feng, M., Caputi, N., Chandrapavan, A., Chen, M., Hart, A., & Kangas, M. (2021). Multi-year marine cold-spells off the west coast of Australia and effects on fisheries. *Journal of Marine Systems*, *214*, 103473.

Feng, R., Duan, W., & Mu, M. (2017). Estimating observing locations for advancing beyond the winter predictability barrier of Indian Ocean dipole event predictions. *Climate Dynamics*, *48*, 1173–1185.

Feng, M., Hendon, H. H., Xie, S. P., Marshall, A. G., Schiller, A., Kosaka, Y., Caputi, N., & Pearce, A. (2015a). Decadal increase in Ningaloo Niño since the late 1990s. *Geophysical Research Letters*, *42*, 104–112.

Feng, M., Hendon, H., Xie, S.-P., Marshall, A. G., Schiller, A., Kosaka, Y., … Pearce, A. (2019). Decadal increase in Ningaloo Niño since the late 1990s. *Geophysical Research Letters*, *42*. https://doi.org/10.1002/2014GL062509.

Feng, J., Hu, D., & Yu, L. (2014). How does the Indian Ocean subtropical dipole trigger the tropical Indian Ocean dipole via the Mascarene high? *Acta Oceanologica Sinica*, *33*, 64–76.

Feng, M., McPhaden, M. J., Xie, S. P., & Hafner, J. (2013). La Niña forces unprecedented Leeuwin current warming in 2011. *Scientific Reports*, *3*, 1277.

Feng, M., & Meyers, G. (2003). Interannual variability in the tropical Indian Ocean: A two-year time scale of IOD. *Deep-Sea Research II*, *50*, 2263–2284.

Feng, M., Meyers, G., Pearce, A., & Wijffels, S. (2003). Annual and interannual variations of the Leeuwin current at 32°S. *Journal of Geophysical Research*, *108*, 3355.

Fischer, A., Terray, P., Guilyardi, E., Gualdi, S., & Delecluse, P. (2005). Two independent triggers for the Indian Ocean dipole/zonal mode in a coupled GCM. *Journal of Climate*, *18*, 3428–3449.

Frölicher, T. L., Fischer, E. M., & Gruber, N. (2018). Marine heatwaves under global warming. *Nature*, *560*, 360–364.

Gebregiorgis, D., Rayner, D., Linderholm, W., & H. (2019). Does the IOD independently influence seasonal monsoon patterns in northern Ethiopia? *Atmosphere*, *10*, 432.

Gill, A. E. (1980). Some simple solutions for heat-induced tropical circulation. *Quarterly Journal of the Royal Meteorological Society*, *106*, 447–462.

Gong, D., & Wang, S. (1999). Definition of Antarctic oscillation index. *Geophysical Research Letters*, *26*, 459–462.

Gualdi, S., Guilyardi, E., Navarra, A., Masina, S., & Delecluse, P. (2003). The interannual variability in the tropical Indian Ocean as simulated by a CGCM. *Climate Dynamics*, *20*, 567–582.

Guan, Z., Ashok, K., & Yamagata, T. (2003). Summer-time response of the tropical atmosphere to the Indian Ocean dipole sea surface temperature anomalies. *Journal of the Meteorological Society of Japan*, *81*, 531–561.

Guan, Z., & Yamagata, T. (2003). The unusual summer of 1994 in East Asia: IOD teleconnections. *Geophysical Research Letters*, *30*, 1544.

Guo, Y., Li, Y., Wang, F., Wei, Y., & Rong, Z. (2020). Processes controlling sea surface temperature variability of Ningaloo Niño. *Journal of Climate*, *33*, 4369–4389.

Harris, S., & Lucas, C. (2019). Understanding the variability of Australian fire weather between 1973 and 2017. *PLoS ONE*, *14*, e0222328.

Hayashida, H., Matear, R. J., Strutton, P. G., & Zhang, X. (2020). Insights into projected changes in marine heatwaves from a high-resolution ocean circulation model. *Nature Communications*, *11*, 1–9. https://doi.org/10.1038/s41467-020-18241-x.

Hermes, J. C., Masumoto, Y., Beal, L. M., Roxy, M. K., Vialard, J., et al. (2019). A sustained ocean observing system in the Indian Ocean for climate related scientific knowledge and societal needs. *Frontiers in Marine Science*, *6*, 355.

Hermes, J. C., & Reason, C. J. C. (2005). Ocean model diagnosis of interannual coevolving SST variability in the South Indian and South Atlantic Oceans. *Journal of Climate*, *18*, 2864–2882.

Holbrook, N. J., et al. (2019). A global assessment of marine heatwaves and their drivers. *Nature Communications*, *10*, 1–13. https://doi.org/10.1038/s41467-019-10206-z.

Hu, J., & Duan, A. (2015). Relative contributions of the Tibetan Plateau thermal forcing and the Indian Ocean sea surface temperature basin mode to the interannual variability of the East Asian summer monsoon. *Climate Dynamics*, *45*, 2697–2711.

Huang, G., Qu, X., & Hu, K. (2010). The impact of the tropical Indian Ocean on South Asian High in boreal summer. *Advances in Atmocpheric Sciences*, *28*, 421–432.

Izumo, T., Vialard, J., Lengaigne, M., de Boyer Montegut, C., Behera, S. K., Luo, J. J., et al. (2010). Influence of the state of the Indian Ocean dipole on the following years El Niño. *Nature Geoscience*, *3*, 168–172.

Kataoka, T., Masson, S., Izumo, T., Tozuka, T., & Yamagata, T. (2018). Can Ningaloo Niño/Niña develop without El Niño/southern oscillation? *Geophysical Research Letters*, *45*, 7040–7048.

Kataoka, T., Tozuka, T., Behera, S. K., & Yamagata, T. (2014). On the Ningaloo Niño/Niña. *Climate Dynamics*, *43*, 1463–1482.

Kataoka, T., Tozuka, T., Masumoto, Y., & Yamagata, T. (2012). The Indian Ocean subtropical dipole mode simulated in the CMIP3 models. *Climate Dynamics*, *39*, 1385–1399.

Kataoka, T., Tozuka, T., & Yamagata, T. (2017). Generation and decay mechanisms of Ningaloo Niño/Niña. *Journal of Geophysical Research Oceans*, *122*, 8913–8932.

Kalnay, E., Kanamitsu, M., Kister, R., Collins, W., Deaven, D., Gandin, L., ... Joseph, D. (1996). The NCEP/NCAR 40-year reanalysis project. *Bulletin of the American Meteorological Society, 77*, 437–471. https://doi.org/10.1175/1520-0477(1996)077<0437:TNYRP>2.0CO;2.

Kido, S., Kataoka, T., & Tozuka, T. (2016). Ningaloo Niño simulated in the CMIP5 models. *Climate Dynamics, 47*, 1469–1484.

Klein, S. A., Soden, B. J., & Lau, N. C. (1999). Remote sea surface temperature variations during ENSO: Evidence for a tropical atmospheric bridge. *Journal of Climate, 12*, 917–932.

Kug, J.-S., Li, T., An, S.-I., Kang, I.-S., Luo, J.-J., Masson, S., et al. (2006). Role of the ENSO–Indian Ocean coupling on ENSO variability in a coupled GCM. *Geophysical Research Letters, 33*, L09710.

Kusunoki, H., Kido, S., & Tozuka, T. (2020). Contribution of oceanic wave propagation from the tropical Pacific to asymmetry of the Ningaloo Niño/Niña. *Climate Dynamics, 54*, 4865–4875.

Kusunoki, H., Kido, S., & Tozuka, T. (2021). Air-sea interaction in the western tropical Pacific and its impact on asymmetry of the Ningaloo Niño/Niña. *Geophysical Research Letters, 48*, e2021GL093370.

Lau, N.-C., & Nath, M. J. (2003). Atmosphere–ocean variations in the Indo-Pacific sector during ENSO episodes. *Journal of Climate, 16*, 3–20.

Lau, N.-C., & Nath, M. J. (2004). Coupled GCM simulation of atmosphere-ocean variability associated with zonally asymmetric SST changes in the tropical Indian Ocean. *Journal of Climate, 17*, 245–265.

Li, Y., Han, W., & Zhang, L. (2017). Enhanced decadal warming of the Southeast Indian Ocean during the recent global surface warming slowdown. *Geophysical Research Letters, 44*, 9876–9884.

Li, Y., Han, W., Zhang, L., & Wang, F. (2019). Decadal SST variability in the Southeast Indian Ocean and its impact on regional climate. *Journal of Climate, 32*, 6299–6318.

Li, S. L., Lu, J., Huang, G., & Hu, K. M. (2008). Tropical Indian Ocean basin warming and East Asian summer monsoon: A multiple AGCM study. *Journal of Climate, 21*, 6080–6088.

Li, T., Wang, B., Chang, C. P., & Zhang, Y. (2003). A theory for the Indian Ocean dipole–zonal mode. *Journal of the Atmospheric Sciences, 60*, 2119–2135.

Li, T., & Zhang, Y. S. (2002). Processes that determine the quasi-biennial and lower-frequency variability of the South Asian monsoon. *Journal of the Meteorological Society of Japan, 80*, 1149–1163.

Li, T., Zhang, Y. S., Chang, C.-P., & Wang, B. (2001). On the relationship between Indian Ocean SST and Asian summer monsoon. *Geophysical Research Letters, 28*, 2843–2846.

Liu, H., Tang, Y., Chen, D., & Lian, T. (2017). Predictability of the Indian Ocean dipole in the coupled models. *Climate Dynamics, 48*, 2005–2024.

Luo, J. J., Behera, S., Masumoto, Y., Sakuma, H., & Yamagata, T. (2008). Successful prediction of the consecutive IOD in 2006 and 2007. *Geophysical Research Letters, 35*, L14S02.

Luo, J. J., Masson, S., Behera, S., Shingu, S., & Yamagata, T. (2005). Seasonal climate predictability in a coupled OAGCM using a different approach for ensemble forecasts. *Journal of Climate, 18*, 4474–4497.

Luo, J. J., Masson, S., Behera, S., & Yamagata, T. (2007). Experimental forecasts of the Indian Ocean dipole using a coupled OAGCM. *Journal of Climate, 20*, 2178–2190.

Luo, J. J., Yuan, C., Sasaki, W., Behera, S. K., Masumoto, Y., Yamagata, T., et al. (2016). Current status of intraseasonal-seasonal-to-interannual prediction of the Indo-Pacific climate. *World Scientific Series on Asia-Pacific Weather and Climate, 7*, 63–107.

Luo, J. J., Zhang, R., Behera, S. K., Masumoto, Y., Jin, F. F., Lukas, R., et al. (2010). Interaction between El Niño and extreme Indian Ocean dipole. *Journal of Climate, 23*, 726–742.

Manatsa, D., & Behera, S. K. (2013). On the epochal strengthening in the relationship between rainfall of East Africa and IOD. *Journal of Climate, 26*, 5655–5673.

Manatsa, D., Chingombe, W., & Matarira, C. H. (2008). The impact of the positive Indian Ocean dipole on Zimbabwe droughts. *International Journal of Climatology, 28*, 2011–2029.

Marchant, R., Mumbi, C., Behera, S., & Yamagata, T. (2006). The Indian Ocean dipole – The unsung driver of climatic variability in East Africa. *African Journal of Ecology, 45*, 4–16.

Matsuno, T. (1966). Quasi-geostrophic motions in the equatorial area. *Journal of the Meteorological Society of Japan., 44*, 25–43.

Marshall, A. G., Hendon, H. H., Feng, M., & Schiller, A. (2015). Initiation and amplification of the Ningaloo Niño. *Climate Dynamics, 45*, 2367–2385.

Mason, S. J. (1995). Sea-surface temperature—South African rainfall associations, 1910–1989. *International Journal of Climatology, 15*, 119–135.

Masson, S., Terray, P., Madec, G., Luo, J. J., Yamagata, T., & Takahashi, K. (2012). Impact of intra-daily SST variability on ENSO characteristics in a coupled model. *Climate Dynamics, 39*, 681–707. https://doi.org/10.1007/s00382-011-1247-2.

Meyers, G. (1996). Variation of the Indonesian throughflow and the El Niño Southern Oscillation. *Journal of Geophysical Research, 101*, 12255–12263.

Mo, K. C., & Paegle, J. N. (2001). The Pacific–South American modes and their downstream effects. *International Journal of Climatology, 21*, 1211–1229.

Morioka, Y., Masson, S., Terray, P., Prodhomme, C., Behera, S. K., & Masumoto, Y. (2014). Role of tropical SST variability on the formation of subtropical dipoles. *Journal of Climate, 27*, 4486–4507.

Morioka, Y., Takaya, K., Behera, S. K., & Masumoto, Y. (2015). Local SST impacts on the summertime Mascarene high variability. *Journal of Climate, 28*, 678–694.

Morioka, Y., Tozuka, T., Masson, S., Terray, P., Luo, J. J., & Yamagata, T. (2012). Subtropical dipole modes simulated in a coupled general circulation model. *Journal of Climate, 25*, 4029–4047.

Morioka, Y., Tozuka, T., & Yamagata, T. (2010). Climate variability in the southern Indian Ocean as revealed by self-organizing maps. *Climate Dynamics, 35*, 1059–1072.

Morioka, Y., Tozuka, T., & Yamagata, T. (2013). How is the Indian Ocean subtropical dipole excited? *Climate Dynamics, 41*, 1955–1968.

Mu, M., Feng, R., & Duan, W. (2017). Relationship between optimal precursors for Indian Ocean dipole events and optimally growing initial errors in its prediction. *Journal of Geophysical Research Oceans, 122*, 1141–1153.

Nakamura, N., Kayanne, H., Iijima, H., McClanahan, T. R., Behera, S. K., & Yamagata, T. (2009). Mode shift in the Indian Ocean climate under global warming stress. *Geophysical Research Letters, 36*, L23708.

Oettli, P., Morioka, Y., & Yamagata, T. (2016). A regional climate mode discovered in the North Atlantic: Dakar Niño/Niña. *Scientific Reports, 6*, 18782.

Oettli, P., Yuan, C., & Richter, I. (2020). The other coastal Niño/Niña-the Benguela, California and Dakar Niños/Niñas. In S. K. Behera (Ed.), *Tropical and extratropical air-sea interactions* (pp. 237–266). Amsterdam: Elsevier Inc.

Prasad, T. G., & McClean, J. L. (2004). Mechanisms for anomalous warming in the western Indian Ocean during dipole mode events. *Journal of Geophysical Research Oceans, 109*, C02019.

Rao, S. A., & Behera, S. K. (2005). Subsurface influence on SST in the tropical Indian Ocean structure and interannual variabilities. *Dynamics of Atmospheres and Oceans, 39*, 103–135.

Rao, S. A., Behera, S. K., Masumoto, Y., & Yamagata, T. (2002). Interannual subsurface variability in the tropical Indian Ocean with a special emphasis on the Indian Ocean dipole. *Deep-Sea Research II, 49*, 1549–1572.

Rao, S. A., Luo, J. J., Behera, S. K., & Yamagata, T. (2009). Generation and termination of Indian Ocean dipole events in 2003, 2006 and 2007. *Climate Dynamics, 33*, 751–767.

Rao, S. A., Masson, S., Luo, J.-J., Behera, S. K., & Yamagata, T. (2007). Termination of Indian Ocean dipole events in a general circulation model. *Journal of Climate, 20*, 3018–3035.

Rao, S. A., & Yamagata, T. (2004). Abrupt termination of Indian Ocean dipole events in response to intraseasonal disturbances. *Geophysical Research Letters, 31*, L19306.

Ratnam, J. V., Dijkstra, H. A., & Behera, S. K. (2020). A machine learning based prediction system for the Indian Ocean dipole. *Scientific Reports, 10*, 284.

Reynolds, R. W., Rayner, N. A., Smith, T. M., Stokes, D. C., & Wang, W. (2002). An improved in situ and satellite SST analysis for climate. *Journal of Climate, 15*, 1609–1625.

Richard, Y., Fauchereau, N., Poccard, I., Rouault, M., & Trzaska, S. (2001). 20th century droughts in southern Africa: Spatial and temporal variability, teleconnections with oceanic and atmospheric conditions. *International Journal of Climatology, 21*, 873–885.

Saji, N. H., Goswami, B. N., Vinayachandran, P. N., & Yamagata, T. (1999). A dipole mode in the tropical Indian Ocean. *Nature, 401*, 360–363.

Saji, N. H., Jin, D., & Thilakan, V. A. (2018). Model for super El Niños. *Nature Communicactions, 9*, 2528.

Saji, N. H., & Yamagata, T. (2003). Possible impacts of Indian Ocean Dipole Mode events on global climate. *Climate Research, 25*, 151–169.

Sasaki, W., Richards, K. J., & Luo, J. J. (2013). Impact of vertical mixing induced by small vertical scale structures above and within the equatorial thermocline on the tropical Pacific in a CGCM. *Climate Dynamics, 41*, 443–453.

Schott, F. A., Xie, S.-P., & McCreary, J. P. (2009). Indian Ocean circulation and climate variability. *Reviews in Geophysics, 47*, RG1002.

Sen Gupta, A., et al. (2020). Drivers and impacts of the most extreme marine heatwaves events. *Scientific Reports, 10*, 1–15. https://doi.org/10.1038/s41598-020-75445-3.

Song, Q., Vecchi, G. A., & Rosati, A. J. (2008). Predictability of the Indian Ocean sea surface temperature anomalies in the GFDL coupled model. *Geophysical Research Letters, 35*, L02701.

Suarez, M. J., & Schopf, P. S. (1988). A delayed action oscillator for ENSO. *Journal of the Atmospheric Sciences, 45*, 3283–3287.

Suzuki, R., Behera, S. K., Iizuka, S., & Yamagata, T. (2004). Indian Ocean subtropical dipole simulated using a coupled general circulation model. *Journal of Geophysical Research: Oceans, 109*, C09001.

Tanizaki, C., Tozuka, T., Doi, T., & Yamagata, T. (2017). Relative importance of the processes contributing to the development of SST anomalies in the eastern pole of the Indian Ocean Dipole and its implication for predictability. *Climate Dynamics, 49*, 1289–1304.

Tanuma, N., & Tozuka, T. (2020). Influences of the interdecadal Pacific oscillation on the locally amplified Ningaloo Niño. *Geophysical Research Letters, 47*, e2020GL088712.

Taschetto, A. S., Sen Gupta, A., Hendon, H. H., Ummenhofer, C. C., & England, M. H. (2011). The contribution of Indian Ocean sea surface temperature anomalies on Australian summer rainfall during El Niño events. *Journal of Climate, 24*, 3734–3747.

Terray, P., Delecluse, P., Labattu, S., & Terray, L. (2003). Sea surface temperature associations with the late Indian summer monsoon. *Climate Dynamics, 21*, 593–618.

Thompson, D. W., & Wallace, J. M. (2000). Annular modes in the extratropical circulation. Part I: Month-to-month variability. *Journal of Climate, 13*, 1000–1016.

Tittensor, D. P., Mora, C., Jetz, W., Lotze, L. K., Ricard, D., Berghe, E. V., & Worm, B. (2010). Global patterns and predictors of marine biodiversity across taxa. *Nature, 466*, 1098–1101.

Tozuka, T., Kataoka, T., & Yamagata, T. (2014). Locally and remotely forced atmospheric circulation anomalies of Ningaloo Niño/Niña. *Climate Dynamics, 43*, 2197–2205.

Tozuka, T., & Oettli, P. (2018). Asymmetric cloud-shortwave radiation-sea surface temperature feedback of Ningaloo Niño/Niña. *Geophysical Research Letters, 45*, 9870–9879.

Ummenhofer, C. C., England, M. H., McIntosh, P. C., Meyers, G. A., Pook, M. J., Risbey, J. S., Sen Gupta, A., & Taschetto, A. S. (2009). What causes Southeast Australia's worst droughts? *Geophysical Research Letters, 36*, L04706.

Ummenhofer, C. C., Schwarzkopf, F. U., Meyers, G. A., Behrens, E., Biastoch, A., & Böning, C. W. (2013). Pacific Ocean contribution to the asymmetry in eastern Indian Ocean variability. *Journal of Climate, 26*, 1152–1171.

Venzke, S., Latif, M., & Villwock, A. (2000). The coupled GCM ECHO-2. Part II: Indian Ocean response to ENSO. *Journal of Climate, 13*, 1371–1383.

Vinayachandran, P. N., Saji, N. H., & Yamagata, T. (1999). Response of the equatorial Indian Ocean to an anomalous wind event during 1994. *Geophysical Research Letters, 26*, 1613–1616.

Ummenhofer, C. C., Taschetto, A. S., Izumo, T., & Luo, J.-J. (2024). Chapter 7: Impacts of the Indian Ocean on regional and global climate. In C. C. Ummenhofer, & R. R. Hood (Eds.), The Indian Ocean and its role in the global climate system (pp. 145–168). Amsterdam: Elsevier. https://doi.org/10.1016/B978-0-12-822698-8.00018-4.

UN Office for the Coordination of Humanitarian Affairs (OCHA). (2011). *Southern African floods and cyclones: Overview of 2010/2011 rainfall season, December 2010 to May 2011*. 12 pp.

Wajsowicz, R. C. (2005). Potential predictability of tropical Indian Ocean SST anomalies. *Geophysical Research Letters, 32*, L24702.

Wallace, J. M., Rasmusson, E. M., Mitchell, T. P., Kousky, V. E., Sarachik, E. S., & von Storch, H. (1998). On the structure and evolution of ENSO-related climate variability in the tropical Pacific: Lessons from TOGA. *Journal of Geophysical Research, 103*, 14241–14259.

Wang, B., Lee, J.-Y., Kang, I.-S., Shukla, J., Park, C.-K., Kumar, A., … Yamagata, T. (2009). Advance and prospectus of seasonal prediction: assessment of the APCC/CliPAS 14-model ensemble retrospective seasonal prediction (1980-2004). *Climate Dynamics, 33*, 93–117.

Webster, P. J., Magaña, V. O., Palmer, T. N., Shukla, J., Tomas, R. A., Yanai, M., & Yasunari, T. (1998). Monsoons: Processes, predictability, and the prospects for prediction. *Journal of Geophysical Research, 103*, 14451–14510.

Weng, H., Ashok, K., Behera, S. K., Rao, S. A., & Yamagata, T. (2007). Impacts of recent El Niño Modoki on dry/wet conditions in the Pacific Rim during boreal summer. *Climate Dynamics, 29*, 113–129.

Wu, Y., & Tang, Y. (2019). Seasonal predictability of the tropical Indian Ocean SST in the North American multimodel ensemble. *Climate Dynamics, 53*, 3361–3372.

Wu, R., Yang, S., Liu, S., Sun, L., Lian, Y., & Gao, Z. (2010). Changes in the relationship between Northeast China summer temperature and ENSO. *Journal of Geophysical Research, 115*. https://doi.org/10.1029/2010D014422.

Wu, B., Zhou, T. J., & Li, T. (2009). Contrast of rainfall–SST relationships in the western North Pacific between the ENSO developing and decaying summers. *Journal of Climate, 22*, 4398–4405.

Xie, S.-P., Hu, K., Hafner, J., Tokinaga, H., Du, Y., Huang, G., & Sampe, T. (2009). Indian Ocean capacitor effect on Indo-western Pacific climate during the summer following El Niño. *Journal of Climate, 22*, 730–747.

Xie, S.-P., & Philander, G. (1994). A coupled ocean-atmosphere model of relevance to the ITCZ in the eastern Pacific. *Tellus, 46A*, 340–350.

Xue, J., Luo, J. J., Yuan, C., & Yamagata, T. (2020). Discovery of Chile Niño/Niña. *Geophysical Research Letters, 47*, e2019GL086468.

Yamagata, T., Behera, S. K., Luo, J.-J., Masson, S., Jury, M. R., & Rao, S. A. (2004). Coupled ocean-atmosphere variability in the tropical Indian Ocean. In C. Wang, S. P. Xie, & J. A. Carton (Eds.), *Geophysical Monograph Series, 147. Earth's climate: The ocean-atmosphere interaction* (pp. 189–211). Washington: American Geophysical Union.

Yamagata, T., Morioka, Y., & Behera, S. K. (2015). Longstanding and new modes of Indo-Pacific climate variations. In S. K. Behera, & T. Yamagata (Eds.), *Vol. 7. Indo-Pacific climate variability and predictability. World scientific series on Asia-Pacific weather and climate* (pp. 1–23). World Scientific.

Yang, J., Liu, Q., Xie, S.-P., Liu, Z., & Wu, L. (2007). Impact of the Indian Ocean SST basin mode on the Asian summer monsoon. *Geophysical Research Letters, 34*, L02708.

Yang, Y., Xie, S. P., Wu, L., Kosaka, Y., Lau, N. C., & Vecchi, G. A. (2015). Seasonality and predictability of the Indian Ocean dipole mode: ENSO forcing and internal variability. *Journal of Climate, 28*, 8021–8036.

Yoshida, K. (1959). A theory of the Cromwell current and of the equatorial upwelling—An interpretation in a similarity to a coastal circulation. *Journal of the Oceanographic Society of Japan, 15*, 154–170.

Yuan, C., Tozuka, T., Luo, J. J., & Yamagata, T. (2014). Predictability of the subtropical dipole modes in a coupled ocean-atmosphere model. *Climate Dynamics, 42*, 1291–1308.

Yuan, C., & Yamagata, T. (2014). California Niño/Niña. *Scientific Reports, 4*, 4801.

Zhang, L., Du, Y., Cai, W., Chen, Z., Tozuka, T., & Yu, J.-Y. (2020). Triggering the Indian Ocean dipole from the southern hemisphere. *Geophysical Research Letters, 47*, e2020GL088648.

Zhang, L., & Han, W. (2018). Impact of Ningaloo Niño on tropical Pacific and an interbasin coupling mechanism. *Geophysical Research Letters, 45*, 11300–11309. https://doi.org/10.1029/2018GL078579.

Zhang, L., Han, W., Li, Y., & Lovenduski, N. S. (2019). Variability of sea level and upper-ocean heat content in the Indian Ocean: Effects of Subtropical Indian Ocean Dipole and ENSO. *Journal of Climate, 32*, 7227–7245.

Zhang, L., Han, W., Li, Y., & Shinoda, T. (2018). Mechanisms for generation and development of the Ningaloo Niño. *Journal of Climate, 31*, 9239–9259.

Zhao, M., & Hendon, H. H. (2009). Representation and prediction of the Indian Ocean Dipole in the POAMA seasonal forecast model. *Quarterly Journal of the Royal Meteorological Society, 135*, 337–352.

Zheng, X.-T. (2019). Indo-Pacific climate modes in warming climate: Consensus and uncertainty across model projections. *Current Climate Change Reports, 5*, 308–321.

Zhou, Z.-Q., Xie, S.-P., & Zhang, R. (2021). Historic Yangtze flooding of 2020 tied to extreme Indian Ocean conditions. *Proceedings of the National Academy of Sciences of the United States of America, 118*, e202225518.

Zhu, J., Huang, B., Kumar, A., & Kinter, J. L. (2015). Seasonality in prediction skill and predictable pattern of tropical Indian Ocean SST. *Journal of Climate, 28*, 7962–7984.

Zinke, J., Dullo, W. C., Heiss, G. A., & Eisenhauer, A. (2004). ENSO and Indian Ocean subtropical dipole variability is recorded in a coral record off Southwest Madagascar for the period 1659 to 1995. *Earth and Planetary Science Letters, 228*, 177–194.

Zinke, J., Rountrey, A., Feng, M., Xie, S.-P., Dissard, D., Rankenburg, K., Lough, J. M., & McCulloch, M. T. (2014). Corals record long-term Leeuwin current variability including Ningaloo Niño/Niña since 1795. *Nature Communications, 5*, 3607.

Zubair, L., Rao, S. A., & Yamagata, T. (2003). Modulation of Sri Lankan Maha rainfall by the Indian Ocean dipole. *Geophysical Research Letters, 30*, 1063.

Chapter 6

Extreme events in the Indian Ocean: Marine heatwaves, cyclones, and tsunamis[☆]

Ming Feng[a,b], Matthieu Lengaigne[c], Sunanda Manneela[d], Alex Sen Gupta[e,f], and Jérôme Vialard[g]

[a]CSIRO Environment, Indian Ocean Marine Research Centre, Crawley, WA, Australia, [b]Center for Southern Hemisphere Oceans Research, CSIRO Oceans and Atmosphere, Hobart, TAS, Australia, [c]MARBEC, University of Montpellier, CNRS, IFREMER, IRD, Sète, France, [d]Indian National Centre for Ocean Information Services (INCOIS), Hyderabad, India, [e]University of New South Wales, Sydney, NSW, Australia, [f]Australian Research Council Centre of Excellence for Climate Extremes, Sydney, Australia, [g]LOCEAN-IPSL, Sorbonne Universités (UPMC, Univ Paris 06)-CNRS-IRD-MNHN, Paris, France

1 Introduction

Extreme events are, by definition, rare and characterized by physical environmental conditions that strongly deviate from the norm. These events are often associated with negative impacts on ecosystems, cause physical damage to natural or man-made infrastructures, and can lead to heavy human casualties. In the marine environment, extreme events associated with marine heatwaves (MHWs), tropical cyclones (TCs) and accompanying storm surges, and tsunamis have received particular interest due to their large potential socioeconomic impacts. This is particularly the case for countries bordering the Indian Ocean, which are home to one-third of the world's population. Low-lying coastal areas with high population density and poor disaster management capabilities make these countries particularly vulnerable to marine extremes.

MHWs occur when ocean temperatures are extremely warm for an extended period, which can have significant impacts on marine life. For example, an intense MHW off western Australia in early 2011 irrevocably altered the ecosystems along ~2000 km of coastline (Wernberg et al., 2013; Fig. 1a and b). Anomalously warm sea surface temperature (SST) persisted for 4 months, with peak warming of 5°C. This led to large shifts in species distributions as well as high levels of invertebrate, fish, and bird mortalities (Pearce et al., 2011). The extreme temperatures also caused severe destruction of habitat-forming species, with unprecedented levels of coral bleaching, mass die-off of seagrasses, and collapse of kelp forests. The MHW had lasting effects on fisheries production, and the marine ecosystem functions had only partially recovered after 7 years (Caputi et al., 2019).

TCs are characterized by a deep low-pressure system, surrounded by strong cyclonic winds, counterclockwise (clockwise) in the Northern (Southern) Hemisphere, and heavy precipitation bands. A hotspot for TC devastation is the Bay of Bengal, where the shallow continental shelf amplifies storm surges (Antony et al., 2014). The Bay of Bengal only hosts 4% of TCs globally (Singh & Roxy, 2022), but accounts for 80%–90% of fatalities and 14 of the 20 deadliest TCs (Needham et al., 2015). In particular, the Ganges-Brahmaputra Delta can amplify storm surges to up to 14m, leading to some of the highest fatalities ever recorded, with an estimated 700,000 human losses since 1960 (Needham et al., 2015). The densely populated Irrawaddy delta, another very vulnerable Bay of Bengal area, was struck by the category 4 TC Nargis in May 2008, inducing a storm surge of up to 4m that inundated areas as far as 40km inland (Fig. 1c and d). This caused more than 130,000 deaths, 1 million homeless, and over $10 billion in economic losses (Webster, 2008).

Climate variability associated with the El Niño-Southern Oscillation (ENSO) in the Pacific Ocean, the Indian Ocean Dipole (IOD), or the Madden-Julian Oscillation (MJO) in the Indian Ocean can modulate the frequency and intensity of TCs and MHWs in the Indian Ocean. ENSO influences the Indian Ocean through oceanic planetary waves that propagate through the Indonesian Seas (Meyers, 1996; Yamagata et al., 2024) and through atmospheric teleconnections that modulate the basin-wide tropical Indian Ocean SST until the following boreal summer (Sprintall et al., 2020; Taschetto et al., 2020).

[☆]This book has a companion website hosting complementary materials. Visit this URL to access it: https://www.elsevier.com/books-and-journals/book-companion/9780128226988.

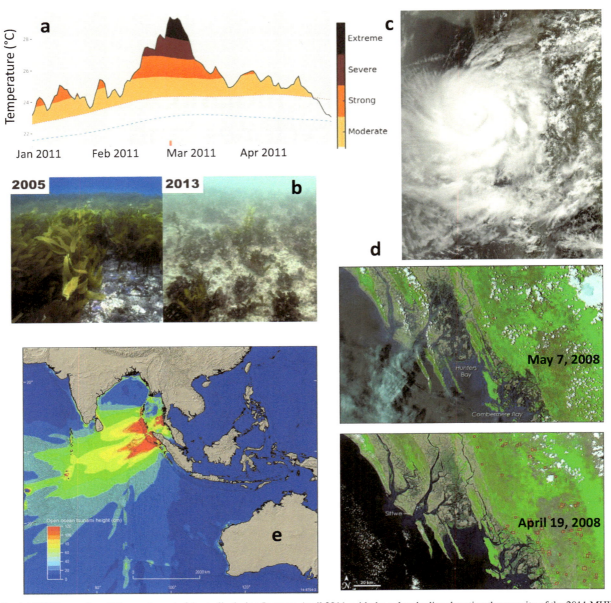

FIG. 1 (a) SST time series off the west coast of Australia during January–April 2011, with the color shading denoting the severity of the 2011 MHW, in terms of moderate, strong, severe, and extreme categories; (b) kelp forest before and after the 2011 MHW off the west coast of Australia (images provided by Thomas Wernberg, University of Western Australia); (c) very Severe Cyclonic Storm Nargis, seen from MODIS on the Terra satellite at 0645Z May 2 near landfall on Myanmar, and (d) satellite images on April 19 and May 7, 2008, before and after the flooding related to Nargis—courtesy the MODIS Rapid Response Team at NASA GSF (see https://earthobservatory.nasa.gov/images/19876/cyclone-nargis-floods-burma-myanmar); (e) Amplitude of tsunami waves generated by the 2004 Sumatra-Andaman earthquake on 26 December 2004, obtained from Geoscience Australia (www.ga.gov.au).

The IOD, which can be triggered by the development of ENSO events but can also occur independently, is associated with warming in the west and cooling in the east-southeast tropical Indian Ocean during its positive phase (Yamagata et al., 2024). The MJO (Madden & Julian, 1971, 1972), characterized by eastward-propagating centers of enhanced and reduced convection separated by more than 10,000 km and with a periodicity of around 30–90 days (DeMott et al., 2024; Zhang, 2005), is the dominant mode of variability in the Indian Ocean at subseasonal timescales. At longer timescales, the Indian Ocean has experienced one of the strongest surface warming trends globally over recent decades (e.g., Cheng et al., 2017; Han et al., 2014), which can also modulate TC and MHW statistics.

Tsunamis are fast-traveling oceanic surface waves that are triggered by a variety of underwater geological processes. The subduction of the Indian and Australian plates below the Eurasian plate creates hotspots for marine earthquakes that generate tsunamis. Tsunami waves cause coastal flooding and inland inundation, which is one of the major threats to the communities living near the coast. The 2004 Sumatra-Andaman earthquake, the largest underwater earthquake for over

40 years globally, triggered the December 26, 2004, tsunami (Fig. 1e). This was the most devastating tsunami recorded in the Indian Ocean, causing unprecedented damage along the coast of 14 countries and around 230,000 fatalities (Fritz & Borrero, 2006; Okal et al., 2006a, 2006b, 2006c). The estimated economic damage from this disaster was more than US$9900 million with Indonesia accounting for almost half of the total (Tsunami Evaluation Coalition, 2006). Major impacts were also felt in Thailand, Sri Lanka, and India. The economy of the Maldives, which is based on tourism and fishing was most heavily impacted, with a loss of US$470 million after the 2004 tsunami—about 62% of its gross domestic product. The fast onset of tsunamis and the unpredictability of geophysical processes that trigger tsunamis make tsunami risk management a unique challenge compared to other hazards.

In the three following sections, we review the research of MHWs, TCs, and tsunamis in the Indian Ocean, followed by a summary and outstanding challenges.

2 Marine heatwaves

2.1 Introduction

The term "marine heatwave" was initially used to describe an extreme surface warming event off the west coast of Australia during 2010–11 austral summer (Fig. 2; Pearce et al., 2011; Pearce & Feng, 2013). MHW is now a generic name for periods of extreme high ocean temperatures persisting for days to months, which can have a spatial extent of hundreds to thousands of kilometers and penetrate into the subsurface ocean (Benthuysen et al., 2018; Hobday et al., 2016). Hobday et al. (2016) developed a framework for defining MHWs based on prolonged periods of anomalously high SST exceeding the seasonally varying 90th percentile of SST. This was extended by Hobday et al. (2018) to categorize moderate (Category I) to extreme (Category IV) events (e.g., Fig. 2). However, other well-established definitions also exist, for example, the degree heating weeks metric which has been widely used to assess bleaching risk for coral reefs (Donner, 2011). Recurring MHWs off the west coast of Australia have been well studied and are referred to as a climate mode Ningaloo Niño, in reference to their deleterious impacts on the Ningaloo coral reef (Feng et al., 2013, 2015; Kataoka et al., 2014). According to Hobday et al. (2018) categorization, the 2011 Ningaloo Niño was classified as a Category IV MHW, and strong to severe MHWs have also been studied in the tropical Indian Ocean off north and northwest Australia, in the Bay of Bengal and the Andaman Sea, and around the Seychelles Islands (Fig. 2).

2.2 Mechanisms and relationships with climate variability and change

Upper ocean temperature variability is governed by fluxes of heat across the air-sea interface and between different parts of the ocean, related to both atmospheric and oceanic processes. Modified air-sea heat fluxes are related to changes in insolation, longwave radiation, and latent and sensible heat losses. These are in turn affected by synoptic conditions including

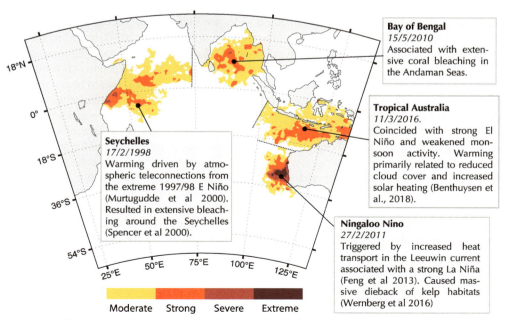

FIG. 2 Daily snapshots of MHW categories, for events that have been described in the literature.

cloud cover, wind speed, and atmospheric temperature and moisture levels (Holbrook et al., 2019). Oceanic fluxes of heat are associated with horizontal advection by deep geostrophic or shallow Ekman currents, vertical advection, associated with divergent surface currents, eddy transport, and ocean mixing processes. For MHWs, the magnitude of warming for a given air-sea heat flux is also modulated by the depth to which the heating extends, often related to the depth of the ocean surface mixed layer (Sen Gupta et al., 2020; Tozuka et al., 2021). These local processes and synoptic conditions that initiate, maintain, and eventually terminate MHW events are often modulated by remote and regional climate phenomena.

2.2.1 Subseasonal to interannual variability

El Niño events have played an important role in triggering a number of MHW events outside of the tropical Pacific through their teleconnection, including the 2014–2016 "Blob" MHWs in the northeast Pacific (Bond et al., 2015; Di Lorenzo & Mantua, 2016) and the extensive central south Pacific MHW in 2009 (Lee et al., 2010). El Niño is known to induce warming of the tropical Indian Ocean, via changes in the Walker Circulation (reducing evaporative heat loss and increasing solar insolation). El Niño also induces anticyclonic wind anomalies in the southern tropical Indian Ocean, which generate westward-propagating downwelling Rossby waves that sustain the warming in the western Indian Ocean (Schott et al., 2009; Xie et al., 2002). The wind-evaporation-SST feedback has been shown to promote warming in the Seychelles Dome region and to extend the warming into the northern Indian Ocean (Du et al., 2009). Thus, both remote and local processes promote Indian Ocean Basin warming in the following year after the peak of El Niño events (Du et al., 2009; Yang et al., 2007), as denoted by the Indian Ocean Basin Mode index (average SST anomalies within 20°S-20°N, 40–100°E; Fig. 3a).

As such, the El Niño-driven teleconnection has facilitated MHW development in the Indian Ocean (Fig. 3b–d), with more frequent and more intense MHWs during the Indian Ocean Basin Mode warming years, especially after the 1997–98 extreme El Niño, strong 2009–10 central Pacific El Niño, and the 2014–16 *Godzilla* El Niño (Kintisch, 2016). For example, MHW events tend to occur off northern Australia in austral summer during strong El Niño, associated with weakening of the Australian monsoon (Zhang et al., 2017), with the most severe MHW occurring during the 2015–16 El Niño (Benthuysen et al., 2018), and in the tropical Indian Ocean in general (Saranya et al., 2022).

Following Holbrook et al. (2019), climate modes that cause an increase in MHW activity are mapped over the Indian Ocean (Fig. 4). Note that the role of ocean warming (or Indian Ocean Basin Mode) on MHWs is not considered here.

Our analysis indicates that canonical El Niño is typically associated with increased MHW occurrences in many regions in the tropical Indian Ocean, the equatorial western and eastern Indian Ocean, southern tropical Indian Ocean, and the Bay of Bengal (Fig. 4). Positive IOD tends to increase the likelihood of MHWs in the rest of the western tropical Indian Ocean and negative IOD increases MHWs off Sumatra-Java in the eastern Indian Ocean. Severe MHWs in the Bay of Bengal in May 1998 and April–May 2010 were associated with strong El Niños (Lix et al., 2016; Saranya et al., 2022). There was substantial warming in the western Indian Ocean during and after the extreme 1997–98 El Niño and synchronous positive IOD event (Spencer et al., 2000; Fig. 2). The 2016 negative IOD sustains MHW events in the tropical southeastern Indian Ocean, primarily in austral spring (Benthuysen et al., 2018).

Consistent with previous studies related to the Ningaloo Niño, we find that MHWs off the west coast of Australia are strongly related to La Niña events that peak in the central Pacific (Fig. 4; Feng et al., 2021), including the 2010–11 Ningaloo Niño (Feng et al., 2013). Strong equatorial easterlies over the Pacific during La Niña deepen the thermocline in the western Pacific, which is transmitted through the Indonesian Seas and along the coast of Western Australia by oceanic Kelvin Waves. This strengthens southward heat transport of the Leeuwin Current, the warm eastern boundary current off the west coast of Australia (Feng et al., 2013). Subdued latent heat loss from the ocean due to weaker wind anomalies also contributes to the MHW development (Benthuysen et al., 2014). Two-way interaction between warm SST in the southeast Indian Ocean and the cooling anomalies in the central tropical Pacific has been proposed as another mechanism to amplify Ningaloo Niño (Zhang & Han, 2018). Many Ningaloo Niño events develop in the absence of La Niña (Kataoka et al., 2014; Zhang & Han, 2018). Regional air-sea coupling, such as coastal Bjerknes feedback (Kataoka et al., 2014), wind-evaporation-SST feedback (Marshall et al., 2015), and shortwave radiation-mixed layer depth feedback (Kataoka et al., 2017), can also induce and influence the Ningaloo Niño evolution (Tozuka et al., 2021; Yamagata et al., 2024). The MJO also appears to enhance tropical MHW development during its suppressed phase before the onset of MJO deep convection (Zhang et al., 2017). For example, weakened MJO activity and suppressed MJO convection during austral summer helped to prolong the peak of the 2015–16 MHW event off north Australia (Benthuysen et al., 2018). The influence of climate modes is more mooted in the extratropics, although there is some increase in MHW activity in the high latitudes associated with the Southern Annular Mode (Fig. 4).

Modulation of MHWs by large-scale climate modes may offer some predictability of the likelihood of MHWs on seasonal timescales (Yamagata et al., 2024). While ENSO forecasting has relatively high skill at leads of one-to-two seasons (Barnston et al., 2019), there is still limited predictability of IOD (Doi et al., 2016), and regional air-sea coupling is yet to be fully understood. The intraseasonal MJO forecasts have skill out to a few weeks (Lim et al.,

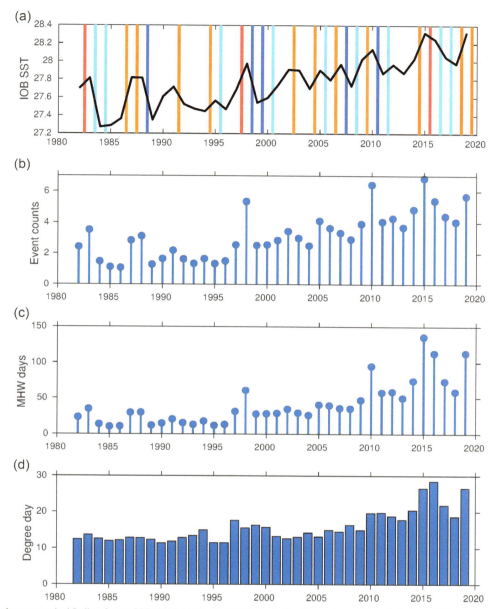

FIG. 3 (a) Annual mean tropical Indian Ocean SST (20°S-20°N, 40–100°E, the Indian Ocean Basin Mode index), and (b) average annual MHW event counts, (c) days, and (d) cumulative intensity over all 0.25 degree grid points north of 40°S in the Indian Ocean. In (a), the *red* and *blue* shadings denote the El Niño and La Niña events, respectively (*dark red* and *blue* colors denote strong events).

2018). As an intrinsically atmospheric climate mode, the Southern Annular Mode, which describes north-south migration and intensity of the subtropical westerlies, does not offer the prospect of any long-term predictability. However, the record-negative Southern Annular Mode associated with a rare stratospheric sudden warming in 2019 was predicted about one season ahead (Lim et al., 2021).

2.2.2 Historical trends and future projection

Globally, the frequency and duration of MHWs have increased by 34% and 17%, respectively, over the last century, largely in response to anthropogenic climate change; combined, this corresponds to an increase in annual MHW days by over 50% globally (Oliver et al., 2018). The Indian Ocean witnessed a faster warming rate since the 1860s (Roxy et al., 2014), resulting in an even larger increase in the occurrence probability of MHWs in this ocean, with a more than twofold increase in the annual number and total duration of MHWs and a substantial increase in their cumulative intensity over the past four decades (Fig. 3). While these extreme events are obviously worsened by the increase in background temperatures

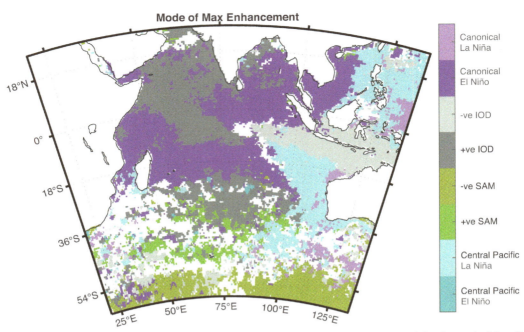

FIG. 4 Climate mode/phase that causes the largest significant enhancement in the likelihood of MHW days (following method described in Holbrook et al., 2019). The number of MHW days during the positive and negative phases of a number of climate indices (|index| > 0.5) are counted: canonical El Niño and La Niña:Niño34 index; Indian Ocean Dipole-Dipole Mode Index; Southern Annular Mode—SAM index; central Pacific ENSO-Niño Modoki Index. Note that the local linear trend was removed from the SST and climate mode time series prior to analysis to isolate relationships associated with interannual variability.

(Laufkötter et al., 2020; Oliver, 2019), MHW statistics are also modulated by low-frequency natural variability. The negative phase of the Interdecadal Pacific Oscillation (Power et al., 1999), for instance, is associated with the dominance of La Niña over El Niño in the Pacific Ocean (Kosaka & Xie, 2013), resulting in strengthened Indonesian Throughflow transport (Lee et al., 2015), increased upper ocean heat content in the southern Indian Ocean and an increased likelihood of Ningaloo Niño during the 1998–2013 period (Feng et al., 2015).

Anthropogenic warming has increased the occurrence probabilities of the largest iconic MHWs around the world by over 20 times (Frölicher et al., 2018; Laufkötter et al., 2020). Under the influence of anthropogenic greenhouse gas forcing, global SST is very likely to continue to rise over the coming decades (Pachauri et al., 2014), and the MHWs are projected to further increase in frequency, duration, spatial extent, and intensity, particularly in the northern Indian Ocean (Collins et al., 2019; Oliver, 2019). Future MHW characteristics may also change in relation to changes in climate variability (Roxy et al., 2024). In particular, climate models suggest a doubling of extreme La Niña under business as usual emission scenarios by the end of the century (Cai et al., 2015), which could have important implications for the occurrence of Ningaloo Niño events. It is noted, however, that the response of ENSO to global warming is still under debate (Wengel et al., 2021).

2.3 Impacts of marine heatwaves

Marine species function within thermal ranges that are aligned with the temperature variability of the environment they inhabit. MHWs that take organisms outside of their thermal tolerances can have major impacts on the performance and mortality of species, particularly on habitat-forming species, e.g., corals, seagrasses, and kelps form the basis of major marine ecosystems and are unable to escape excessive thermal stress (Smale et al., 2019).

Many studies have examined the impact of MHWs on coral reefs. These thermally sensitive species bleach (i.e., expel their symbiotic microbes) if temperatures increase to a threshold above their normal summer peak. If high temperatures persist for an extended period, it leads to coral mortality. Widespread increases in tropical SST after the peak of 1997–98 El Niño led to bleaching across the Seychelles (Spencer et al., 2000) and the Andaman Islands in the Bay of Bengal (Lix et al., 2016). The severe MHWs in the Bay of Bengal in May 2010 (Fig. 2) also caused widespread coral bleaching in the Andaman Seas (Krishnan et al., 2011). Extensive MHW-associated bleaching was also reported in the Andaman Sea and the Gulf of Mannar in 2016 (Edward et al., 2018; Wouthuyzen et al., 2018), on the northwest shelf of Australia in

2016 (Gilmour et al., 2019; Hughes et al., 2018), and the Persian Gulf in 2017 (Burt et al., 2019), associated with the 2015–16 El Niño.

Associated with the 2011 Ningaloo Niño, kelp forests were destroyed over at least 200 km along the west coast of Australia during the extensive warming. The rapid loss of kelp habitat on a background of long-term warming triggered a tropicalization of the region with the invasion of tropical and subtropical seaweeds, corals, and fishes that have suppressed the recovery of kelp habitats (Wernberg et al., 2016). The loss of habitats may have cascading effects on higher trophic level predators (Wild et al., 2019). After the 2011 Ningaloo Niño, the impact on the ecosystem of Western Australia was exacerbated by more MHWs during the following two austral summers, causing significant fish mortalities and triggering management adaptation such as fishing and spawning stock closures (Caputi et al., 2016). Ecosystem models predict that the widespread mortality of habitat-forming taxa will have long-term and, in some cases, irreversible consequences, especially if MHWs become more frequent or severe (Babcock et al., 2019).

3 Tropical cyclones

3.1 TC mechanisms and control by the ocean-atmosphere background state

The most accepted theoretical paradigm for TC intensification is the wind-induced surface heat exchange mechanism (Emanuel, 2004), a positive feedback between convection in the TC eyewall and energy exchanges with the ocean (Fig. 5). Indeed, TCs only develop over the ocean, where very strong surface winds extract heat and moisture, through evaporation. The resulting atmospheric boundary layer warming and moistening destabilize the atmosphere in the vicinity of the TC eye and maintain deep atmospheric convection and strong rain rates. The associated mid-tropospheric latent heating maintains the convergent surface winds, closing the feedback loop and allowing the TC to grow (Fig. 5). TCs thus generally decay when they move over land, where surface evaporation is much smaller.

The efficiency of the abovementioned mechanism depends on the background atmospheric environmental conditions (Fig. 5), which exert a strong control on the TC activity (e.g., Emanuel et al., 2004; Frank, 1987; Gray, 1968, 1975). Warm SST enhances evaporation and destabilizes the atmosphere, favoring TC growth. This is also the case for anomalously high mid-tropospheric relative humidity, which enhances convective buoyancy and large-scale low-level cyclonic vorticity (i.e.,

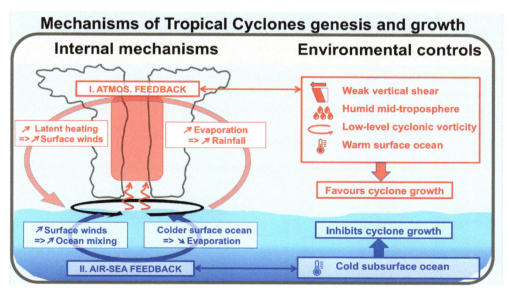

FIG. 5 Sketch of the processes at play in Tropical Cyclone (TC) development. *(Left)* Internal processes. The *red* arrows symbolize the Wind-Induced Surface Heat Exchange positive feedback loop: surface winds favor strong surface evaporation, which sustains the TC rainfall and associated latent heat release. This heat release maintains the strong TC winds, closing the feedback loop. The *blue* arrows symbolize the feedback loop associated with air-sea coupling, that limits the growth of TCs. TC winds induce vertical oceanic mixing that cools the ocean surface and weakens evaporation, reducing the efficiency of the wind-induced surface heat exchange feedback. The amount of cooling is strongly modulated by the subsurface oceanic structure. *(Right)* Control of large-scale environmental parameters. A warm surface ocean, and weak vertical shear, high humidity, and enhanced low-level vorticity in the atmosphere all favor the development of TCs. A cold subsurface ocean favors a strong surface cooling, hence inhibiting TC intensification. Climate variability and change influence those parameters, and hence influence TC genesis.

counterclockwise/clockwise background winds in the Northern/Southern Hemisphere) that favors the low-level cyclone spin up. Strong vertical wind shear in the troposphere on the other hand tends to decouple the elevated heating from the surface winds, making TC growth less efficient. Stochastic processes, such as wind surges or mesoscale interactions also play a strong role in TC genesis, so favorable environmental conditions do not guarantee TC development but increase its probability (Gray, 1998).

Coupling between the TC and the upper ocean reduces the efficiency of the wind-induced surface heat exchange feedback (e.g., Vincent et al., 2014; Fig. 5). The strong wind stress at the air-sea interface under TCs generates intense oceanic vertical mixing (Price, 1981), which can cool the ocean under the TCs by several degrees (e.g., Vincent et al., 2012). TC intensity is sensitive to the cooling under the storm eye (e.g., Schade, 2000), which reduces evaporation and hence the energy supply of cyclones (e.g., Lengaigne et al., 2019; Singh & Roxy, 2022). The amount of cooling depends on the subsurface oceanic stratification, with a much larger cooling in presence of a shallow thermocline. As a result, the growth rate of the cyclone can depend on subsurface oceanic conditions, which are used in combination with the atmospheric environmental controls described above in some TC intensity statistical forecasts. In the following section, we will describe the Indian Ocean TC climatology, and how it is influenced by climate variability and change (see Yamagata et al., 2024; Roxy et al., 2024, for a description of Indian Ocean climate variability and change).

3.2 Present climate

In the southern Indian Ocean, TCs preferentially occur over a zonally elongated band centered around 15°S (Fig. 6a) during austral summer (from November to April), when the Intertropical Convergence Zone is well established and high midtropospheric humidity, low vertical shear, and warm SST combine to favor cyclogenesis (Fig. 6b; e.g., Menkes et al., 2012). More intense TCs occur in the eastern basin (Ramsay et al., 2012).

The unique geographical setting of the northern Indian Ocean results in several unique climatological TC features. Most of the northern Indian Ocean TCs occur over the Bay of Bengal, which hosts four times more TCs than the Arabian Sea (Fig. 6a; Evan & Camargo, 2011). The relatively small Arabian Sea and Bay of Bengal basins limit TC life span, and hence their maximum intensity, relative to other basins. The northern Indian Ocean also exhibits a unique bimodal TC seasonal distribution, with TC-prone seasons before and after the Indian summer monsoon (Fig. 6c). This is largely due to the strong vertical shear that prevents TC formation at the height of the summer monsoon in the Bay of Bengal (e.g., Menkes et al., 2012) and Arabian Sea (e.g., Evan & Camargo, 2011; Gray, 1968). Decreased absolute vorticity and colder SSTs also contribute to suppressing TCs during the summer monsoon (Fig. 6c), counteracting the high ambient relative humidity (Li et al., 2013). While cyclogenesis occurs more frequently after the monsoon in the Bay of Bengal, strong TCs are more frequent before the monsoon due to reduced vertical wind shear and higher SST relative to the postmonsoon (Fig. 6c, Neetu et al., 2019). Recent modeling studies also suggest that the negative feedback associated with air-sea coupling alleviates this background state influence, by reducing the number of intense premonsoon TCs (Neetu et al., 2019) and reduces cyclogenesis by ~20% over the Indian Ocean as well as the proportion of intense TCs (Lengaigne et al., 2019).

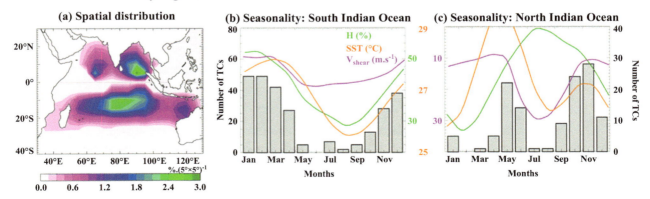

FIG. 6 (a) Climatological distribution of the annual number of TCs in the Indian Ocean from the IBTRacS v3.1 dataset. (b) Southern and (c) Northern Indian Ocean seasonal cycle of the number of TCs (bars) and of the main background state favorable parameters: vertical shear in the atmosphere (m/s, *purple*), SST (°C, *orange*) and mid-tropospheric relative humidity (%, *green*). Vertical axis for vertical shear is inverted to allow all parameters to be favorable when upward.

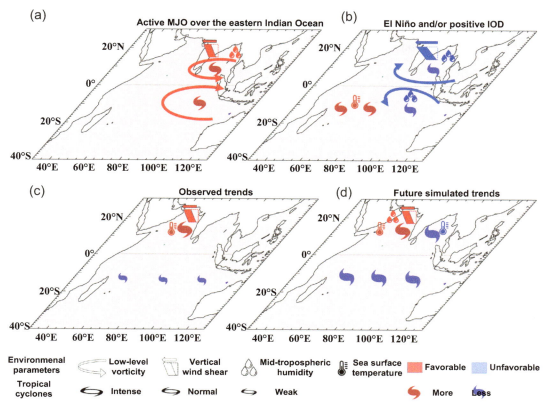

FIG. 7 Sketches of TC changes and controlling large-scale environmental factors related to (a) Madden-Julian Oscillation (MJO), (b) El Niño or/and positive Indian Ocean Dipole (IOD), (c) observed long-term trends, and (d) projected future trends. Various symbols indicate the main environmental factors (low-level vorticity, tropospheric vertical shear, humidity, and SST) that have been brought forward to influence TCs. Those changes are represented in *red* if they are favorable to the TC development (i.e., weak shear, moist atmosphere, strong low-level vorticity) and in *blue* if they impede TC development (i.e., strong shear, dry atmosphere, weak low-level vorticity). A *red* cyclonic symbol stands for more TCs and a *blue* for less. The size of the cyclonic symbol stands for its intensity.

As depicted in Fig. 7a, the MJO (DeMott et al., 2024) and convectively coupled equatorial Rossby waves influence the Indian Ocean TC genesis and development at intraseasonal time scales, through its effect on large-scale convection and the atmospheric background state (e.g., Bessafi & Wheeler, 2006; Frank & Roundy, 2006; Landu et al., 2020). For instance, TCs are two to four times more likely to occur in the southern Indian Ocean during an active MJO phase (e.g., Bessafi & Wheeler, 2006; Hall et al., 2001), primarily through modulation of low-level vorticity (Wang & Moon, 2017). Active MJO phases also increase the probability for a TC to undergo rapid intensification (Klotzbach, 2014). Over the Bay of Bengal, increased cyclonic vorticity induced by equatorial westerlies, reduced vertical wind shear, and increased mid-tropospheric humidity favor TC development during an active MJO phase over the eastern Indian Ocean (Bhardwaj et al., 2019; Kikuchi & Wang, 2010; Krishnamohan et al., 2012), especially when combined with equatorial Rossby wave activity (Landu et al., 2020). This was for instance the case for the deadly TC Nargis, which was associated with a Rossby wave vortex that formed when an MJO reached the maritime continent (Kikuchi et al., 2009).

As first uncovered by Nicholls (1979) for the Australian region, ENSO is a major driver of interannual TC variations in the Indian Ocean (e.g., Ho et al., 2006; Camargo et al., 2007; Kuleshov et al., 2008; Ramsay et al., 2012; see Fig. 7b). In the Southern Hemisphere, changes in low-level vorticity, relative humidity, and SSTs across the basin generally shift TC activity westward during El Niño events and eastward during La Niña (Kuleshov et al., 2009; Ramsay et al., 2012). The strong relationship between Australian TC activity and ENSO has however weakened noticeably since the late 20th century (e.g., Dowdy, 2014), and an increasing influence of Indian Ocean SST related to the IOD (Wijnands et al., 2015) or Indian Ocean subtropical dipole (Ramsay et al., 2017) has been reported. In the northern Indian Ocean, El Niño conditions are also thought to reduce TC activity over the Bay of Bengal after the monsoon (Ng & Chan, 2012; Girishkumar & Ravichandran, 2012; see Fig. 3b), as a result of reduced low-level vorticity, increased vertical wind shear, and reduced relative humidity (Felton et al., 2013). The opposite picture generally occurs during La Niña. Other studies (e.g., Li et al., 2016; Yuan & Cao, 2013) have, however, argued that this modulation of the postmonsoon TC activity in the Bay of Bengal is related to the IOD rather than directly to ENSO. In reality, it is difficult to separate the respective ENSO and IOD influences on Indian Ocean TC activity, because these phenomena are connected (e.g., Annamalai et al., 2003), and because of the strong intrinsic variability in TC activity. Some authors propose that the IOD and ENSO jointly influence Indian Ocean TCs (e.g., Mahala et al., 2015; Yuan et al., 2019) while others point to interannual variations in the monsoon intensity as a possible driver of TC interannual variability (Evan & Camargo, 2011; Sattar & Cheung, 2019). ENSO is often accounted for in seasonal statistical forecasts of the TC activity for both the southern (e.g., Liu & Chan, 2012; Werner & Holbrook, 2011) and northern Indian Ocean (e.g., Wahiduzzaman et al., 2020). Additional predictors in the Indian Ocean may further improve the skill of seasonal TC prediction models (Ramsay et al., 2017; Wijnands et al., 2015), especially over the Australian region where the ENSO-TCs relationship recently weakened.

A growing body of literature has examined the detection and attribution of TC activity changes over the Indian Ocean for the past two decades (see Walsh et al., 2019; Knutson et al., 2019 for recent reviews). Despite inconclusive seminal studies (e.g., Kuleshov et al., 2010), recent attempts applied to longer records and/or accounting for the effects of ENSO variability (Dowdy, 2014) have found a significant decline in the number of TCs and severe TCs over the past four decades (Fig. 7c), both for the entire southern Indian Ocean (Kuleshov et al., 2020) and for the Western Australian region (Chand et al., 2019). Recent studies also reported an increasing number of very intense TCs over the Arabian Sea during the postmonsoon season (Fig. 7c), including the occurrence of five very severe TCs since 1998 (e.g., Evan et al., 2011; Murakami et al., 2017), but there are no significant trends in the adjacent Bay of Bengal (Bhardwaj & Singh, 2020; Deshpande et al., 2021). Rapid intensification of TCs is associated with the increasing TC activities in the Arabian Sea (Deshpande et al., 2021; Singh & Roxy, 2022). While the increasing TCs intensity in the Arabian Sea is likely attributable to climate change in response to enhanced warming and reduced vertical wind shear (Murakami et al., 2017), the attribution of these trends in the rest of the basin to anthropogenic climate change and their related controlling large-scale environmental factors are still uncertain for two reasons (Knutson et al., 2019). First, observed TC historical datasets are often sparse and limited by data quality, particularly in the pre-satellite era (Kossin et al., 2013). Second, natural climate variations drive large low-frequency TC activity variations, making it difficult to disentangle the climate change signal from internal variability "noise" (Knutson et al., 2019). A high-resolution climate model study shows that the declining trend of TC numbers in the southern Indian Ocean is unlikely due to internal climate variability alone (Murakami et al., 2020).

3.3 Future TC projections

Because of observational limitations, most studies investigating the influence of climate change on TC activity have been performed using theoretical and modeling tools. Almost all projections performed with TC-resolving models project a future increase in the maximum intensity of TCs at the global scale (see Walsh et al., 2019 and Knutson et al., 2020 for recent reviews). There is a high confidence in this projection by climate models, as it is consistent with expectations from potential intensity theory (e.g., Sobel et al., 2016). Projections also indicate more intense TCs for individual basins, including in the southern and northern Indian Ocean, where a 5% intensity increase per 2°C warming is predicted (Fig. 7d). These models also project a robust increase in TC rainfall rate of 20% per 2°C warming in all basins (Knutson et al., 2020), including the Indian Ocean, in response to the tropospheric water vapor content increase in a warmer climate. The most confident TC-related projection is that sea level rise will lead to higher storm inundation levels, irrespective of changes in TC characteristics. The plausible increase in TC intensity may further intensify extreme wind waves and aggravate the TC's impacts on coastal structures (Timmermans et al., 2018).

A large majority of TC-resolving simulations also project a future decrease in TC numbers globally (Knutson et al., 2020). Unlike TC intensity projections, the lack of a well-accepted theory for cyclogenesis strongly limits the confidence in this projected change. In addition, projections of TC numbers for individual basins are in general not consistent across models. An exception is the southern Indian Ocean (Fig. 7d), where most climate models simulate a robust decrease in TC frequency of ~10%–20% per 2°C warming (e.g., Cattiaux et al., 2020; Knutson et al., 2020; Muthige et al., 2018; Roberts et al., 2020). TC frequency changes in the northern Indian Ocean have received less attention owing to the small TC sample and the inability of most climate models to capture the complex TC climatology there. As recently confirmed by low-resolution models' results (Bell et al., 2020), ensemble simulations from a high-resolution model with a reasonable TC representation in that region indicate that TC frequency may significantly increase over the Arabian Sea (~+45%) and decrease over the Bay of Bengal (~−30%) by the late 21st century (Murakami et al., 2013). The Arabian Sea TC frequency increase may be related to the larger projected SST warming (Murakami et al., 2012) in response to climate change than in other tropical ocean regions (Gopika et al., 2020). A decrease in vertical shears and an increase in relative humidity may also contribute (Bell et al., 2020).

4 Tsunami

4.1 Introduction

The world has witnessed many devastating tsunamis recently, including Sumatra (2004), South of Java (2006), Solomon Islands (2007), Samoa Islands (2009), Chile (2010), Mentawai, Indonesia (2010), Tohoku, Japan (2011), Chile (2015), New Zealand (2016) and Palu tsunami and Sunda strait tsunami, Indonesia (2018). Although tsunamis are less frequent in the Indian Ocean than in the Pacific Ocean, many of the highest-impact events have occurred in this region. In particular, the devastation caused by the 2004 tsunami shook the world.

4.2 Mechanisms of tsunami generation

Tsunamis are usually triggered by two different natural processes, viz., geological and meteorological. In this chapter, we focus on the more frequent and higher-impact tsunamis, which are generated by the displacement of a substantial volume of water due to undersea geological processes such as earthquakes, submarine landslides, coastal subsidence, and flank collapse triggered by volcanic eruptions. The impact of anthropogenic underwater nuclear explosions and large meteorites in the ocean also has the potential to generate tsunamis (Blades & Siracusa, 2014; Lowe et al., 2003; Wünnemann & Weiss, 2015).

The surface of the earth is broken into large tectonic plates. A subduction zone is formed at boundaries between convergent plates, where the dense oceanic plates are forced under the less dense continental plates. In such regions, the connecting surfaces between those two plates are called megathrust faults, where most of the great tsunamigenic earthquakes occur. In addition to plate convergence regions, seamounts and aseismic ridges that take part in the subduction process may also enhance the intensity of earthquakes (Kelleher & McCann, 1976).

Submarine/subaerial landslides along steep continental slopes are also important tsunami generation mechanisms (Walder et al., 2003), such as the Palu tsunami in Sulawesi, Indonesia on the 28th September 2018 (Frederik et al., 2019; Sunny et al., 2019). The collapse of a volcano flank or magma chamber due to coastal or underwater eruption is another possible mechanism. The 1883 Krakatau volcano explosion and collapse generated a tsunami reaching a height of about 41 m that destroyed parts of neighboring Java and Sumatra (Yokoyama, 1987). On 22nd December 2018, a devastating tsunami with a run-up of ~13 m struck Sunda Strait, Indonesia, caused by a volcanic flank collapse discharging massive amounts of pyroclastic material into the sea (Muhari et al., 2019).

4.2.1 Subduction zones in the Indian Ocean

Seismically active Indian Ocean regions occur where the Indian and Australian plates are moving north and eastward relative to the Eurasian plate forming a convergent boundary to the north and east, and along the divergent Indo-African boundary (Delescluse et al., 2012; Yue et al., 2012). The Indian Ocean has two major tsunamigenic zones (Fig. 8). The

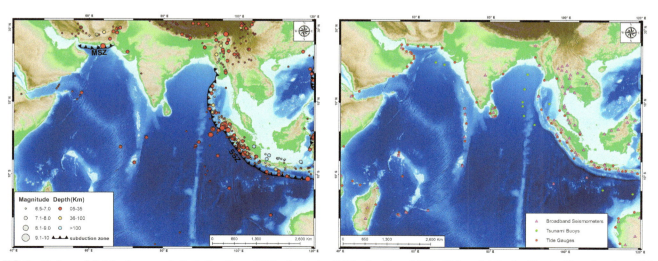

FIG. 8 (*Left panel*) Subduction zones in the Indian Ocean (SSZ—Sunda Arc Subduction Zone, MSZ—Makran Subduction Zone) with earthquakes ≥6.5 magnitude for the period 1900–2020 from USGS catalogue. (*Right panel*) Observation networks (open data) in the Indian Ocean region.

Sunda Arc subduction zone extends along the edge of the northeastern Indian Ocean for 6000 km. It was associated with the 9.1 magnitude Sumatra earthquake of 26th December 2004 and Nias earthquake of 8.6 magnitude on 28th March 2005. In contrast with other subduction zones, the seismicity associated with the second, Makran subduction zone in the northwestern Arabian Sea is low. However, it was associated with the 27th November 1945 earthquake of magnitude 8.1 that caused a devastating tsunami that affected coastal areas of Pakistan, Iran, Oman, and India (Hoffmann et al., 2013; Neetu et al., 2011). The Makran subduction zone has one of the world's largest accretionary wedges, approximately 900 km in length (Mokhtari et al., 2008).

4.2.2 Tsunami wave dynamics

The massive permanent vertical deformation (uplift or subsidence) caused by undersea earthquakes forces sudden vertical displacement of a huge volume of water from the seafloor to the surface. The energy travels outwards from the source region in all directions as surface gravity waves (Abe, 1973). Detecting the source energy transferred to the ocean is the key to the determination of tsunami hazard potential (Morgan, 2011). Particularly, far-field tsunamis are sensitive to the location and energy of the initial ocean-surface displacement (Titov et al., 2017). In the deep ocean, a tsunami's potential is proportional to the total source energy which is related to both potential energy from the vertical seafloor and kinetic energy due to the horizontal displacement of the continental slope (Song, 2007). The wave-amplitude e-folding timescale is twice as long because available potential energy is proportional to the squared wave amplitude (Davies et al., 2020). In general, for a far-field tsunami, deep ocean wave amplitudes are typically a few tens of centimeters or less and for near-source regions wave amplitudes sometimes exceed more than 1 m. Thus, the wave amplitudes depend on parameters such as the magnitude of the earthquake, fault dimensions, initial surface elevation, distance from the source, and local bathymetry near shore, etc. (Ben-Menahem & Rosenman, 1972; Davies & Griffin, 2020; Gica et al., 2007; Sannikova et al., 2021; Yolsal & Taymaz, 2010). The speed of a tsunami depends on the depth of the water, with faster speeds in deeper water. The velocity of the tsunami wave, v, is given by

$$v = \sqrt{gd}$$

where g is the acceleration of gravity (9.8 m/s^2) and d is the depth of water in meters.

For an average ocean depth of 4 km, a tsunami wave travels about 720 km per hour, the speed of a jet plane (Fig. 9). As the wave energy decays exponentially with time, a tsunami that can cross the entire basin within hours experiences little energy loss (Davies et al., 2020; Rabinovich et al., 2013). A stronger directivity of tsunamis also slows down their decay (Sannikova et al., 2021). Due to their wavelengths of hundreds of kilometers, tsunamis are barely noticeable in the open ocean for far-field sources.

The amplification of tsunamis as they travel from the deep ocean onto a continental shelf is approximated using Green's Law (Green, 1838) which describes the progression of the wave height of periodic waves on plane beaches. As the wave approaches the shallow water of the continental shelf and the shore, the wave speed and wavelength decrease so that the conservation of total energy within the wave causes a dramatic increase in wave height. The waves in shallow water are estimated using Green's law that the tsunami amplitude is inversely proportional to the fourth root of water depth (Synolakis, 1991; Wang et al., 2019; Yamanaka et al., 2019). For example, a tsunami in 4000 m of water with a height of 1 m will transform into a wave of 4 to 5 m height when the water depth is 10 m. This is a good approximation when the wave amplitude is small compared to the water depth and the continental slope is sufficiently gentle that the width of the slope region is large compared to the wavelength (George et al., 2020). In the case of steeper slopes, the amplification is less, and the reflection of waves is more. Close to the coast, the wave height varies spatially due to complex topography, with some tsunamis amplified to 10 m above sea level within bays, harbors, or lagoons due to reduced crest width. However, the presence of complex bottom topography and possibly discontinuity near shore, significantly contribute to the attenuation of tsunami energy and waves in some cases (Heidarzadeh et al., 2016; Husrin et al., 2012; Piché et al., 2020). The impact of tsunami waves may be more destructive if they arrive at high tide or along with a storm surge (Lee et al., 2015; Mofjeld et al., 2007; Shelby et al., 2016; Wang & Liu, 2021; Winckler et al., 2017). The strength of the tsunami not only depends on earthquake magnitude but also on earthquake characteristics such as variations in rupture, slip distribution, the thickness of sediments, slab dip angle, and strain rate (Comer, 1980; Davies, 2019; Geist, 1998; Geist & Yoshioka, 1996; Gibbons et al., 2022; Hirshorn et al., 2019; Rasmussen et al., 2015). For example, a 2012 earthquake of magnitude 8.5 Mw off Sumatra, which was only 300 km away from the December 2004 earthquake, produced a much smaller run-up than the 2004 event, as a result of strike-slip faulting rather than thrust faulting (Fig. 9) (Kumar et al., 2012).

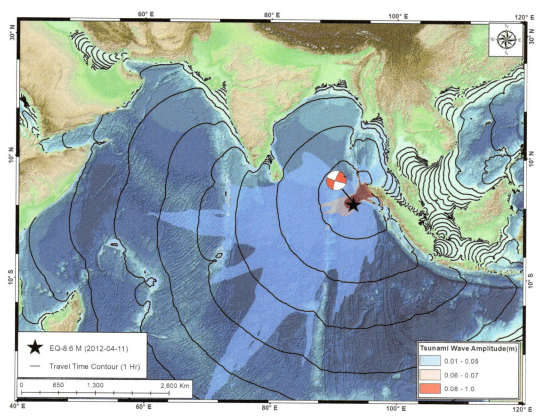

FIG. 9 The location of the April 11, 2012, earthquake of magnitude 8.5 Mw, the tsunami wave travel times (contours, every 60 min), and the maximum wave directivity (shading).

4.3 Impact of tsunamis

Tsunamis typically cause the most severe damage and casualties in low-lying areas such as lagoons, harbors, river mouths, bays, and beaches (ADB, 2005; IUCN, 2005; Sheth et al., 2006; Tadibaght et al., 2022; Tomita et al., 2006; Yamanaka & Shimozono, 2022). Even small tsunamis associated with strong currents can cause severe damage to coastal structures, ports, and harbors (Benazir et al., 2019; Chua et al., 2021; Muhari et al., 2015). Debris in the water can also amplify tsunami danger and the ensuing damage (Matsuba et al., 2020; Nistor et al., 2018; Suppasri et al., 2019). Impacts from a large tsunami often trigger secondary hazards such as fires, water contamination by hazardous chemicals or sewage, and post-tsunami epidemics (Hokugo, 2013; Kume et al., 2009; Suppasri et al., 2021; Villholth & Neupane, 2011). These secondary impacts can be more devastating than direct destruction. Other consequences include enduring changes to coastal areas, as in the case of the Andaman and Nicobar Islands after the 2004 tsunami, where the subsidence of the land inundated with salt water made agricultural land useless (Telford & Cosgrave, 2006). Tsunamis can damage coral reefs, resulting in long-term impacts on marine biodiversity and ecosystem function (Foster et al., 2006; Kumaraguru et al., 2005). It can take years or decades to bring back normality and rebuild the physical and social infrastructures. In addition to the associated economic loss, it leaves an unquantifiable amount of psychological and emotional trauma among the affected people (Jayasuriya et al., 2006; Musa et al., 2014; Neuner et al., 2006). Cascade disasters are still a significant problem that need to be investigated together with other dependencies, vulnerabilities, amplifying factors, and secondary catastrophes.

4.4 Tsunami monitoring system

Unique to tsunamis among other ocean extremes is that we can neither prevent their generation nor make an accurate prediction of when they occur. A tsunami warning system was established in the Pacific Ocean region, following the 1946 Alaska tsunami. However, the Indian Ocean region was completely lacking in tsunami detection and warning systems until

2004. The objective of the tsunami warning system is to provide early warnings on potentially destructive tsunamis to disaster management officials and other stakeholders. The three pillars of the warning system are (i) hazard and risk assessment, (ii) monitoring, detection, and dissemination, and (iii) community awareness and preparedness (IOC-Tsunami Website, 2022). Assessment of tsunami hazards and risks is an important measure to support mitigation and preparedness activities. Generally, the worst-case scenarios were used for assessments, though Probabilistic Tsunami Hazard Assessment or hybrid methods are currently taking over as new approaches for estimating hazard levels (Davies & Griffin, 2020; Grezio et al., 2017; Lorito et al., 2015; Selva et al., 2021). However, tsunami hazard assessments are available to only a few regions due to the unavailability of high-resolution bathymetry and topography data which is needed to model them precisely.

The tsunami warning system uses real-time seismic data from local and global networks to monitor seismicity around the potential tsunamigenic zones. A dense network of seismic stations helps determine the preliminary earthquake parameters such as epicenter, magnitude, and depth of an earthquake event within a few minutes (Hanka et al., 2010; Kanamori & Rivera, 2008). The earthquake focal mechanism comprising the direction of slip, the orientation of the fault, the depth at which the rupture occurred, the length of the rupture, and the magnitude of the displacements are reconstructed for large events rapidly using the available seismic data (Honda, 1962; Khattri, 1973). The preliminary earthquake parameters are surmised into seafloor deformations using simplified approximation models such as Mansinha and Smylie (1971) and Okada (1985, 1992).

After detecting the potential tsunamigenic earthquake using observation networks, numerical models are employed to estimate the impact on the coast using worst-case scenarios of possible tsunami sources. Numerical models such as TUNAMI-N2 (Imamura et al., 2006) and HySea (González-Vida et al., 2021; Macías et al., 2016) have been developed to simulate tsunami propagation and inundation. With the help of these models, and based on the inputs from the observational network, the tsunami can be forecasted for any specific location, to inform local disaster managers of tsunami arrival time, wave heights, and extent of inundation (Tang et al., 2009). The forecasted information and sea-level observations are disseminated to disaster management officials for their subsequent action. However, current practices do not quantify uncertainty in forecasting and define alert levels deterministically due to high data deficiency at the early stage of a tsunami (Selva et al., 2021). Also, the current operational warning systems are unable to provide early warnings for local tsunamis, which can reach within a few minutes the nearest coast (Carvajal et al., 2019; Muhari et al., 2018), and warning systems are still developing procedures for tsunamis triggered by submarine/coastal landslides and volcanic sources.

The sea level observation network of tsunami buoys and tide gauges provides information about the triggering, progression, and severity of a tsunami (Greenslade & Warne, 2012; Omira et al., 2009). The sea-level data is also used for source modeling, which seeks to understand the mechanism of earthquake generation (Hossen et al., 2015; Satake, 1987; Satake et al., 2013). During the last decade, the number of seismic and sea level monitoring stations in the Indian Ocean has grown substantially, however, the openly available data is still limited (Fig. 8b).

Community awareness and preparedness are given equal importance in the tsunami early warning system. In the case of local tsunamis where warning systems are still in progress to provide timely warnings, developing self-evacuation culture plays a crucial role in saving lives. The coastal communities at risk need to be prepared to respond quickly to official warnings as well as natural signs of a possible tsunami.

5 Conclusions

Occurrences of MHWs in the Indian Ocean are modulated by climate variability and change. Anthropogenic warming already induced a massive increase in MHWs frequency and intensity, which will undoubtedly be amplified in the future with projected unabated warming. Variability of MHWs is generally associated with El Niño-induced basin warming, although strong La Niña events are crucial in driving MHWs off the west coast of Australia. MHW research is still in its infancy. There is a need to understand the physical processes leading to MHW development in various hotspots in the Indian Ocean. State-of-the-art models are unable to skillfully predict the strength and duration of MHWs, as predictability depends on both the accurate prediction of the modes of climate variability and on complex regional processes that amplify MHWs. Combined dynamic and statistical forecasting, with the aid of machine learning, may provide additional prediction skills. While most MHW knowledge is based on SST analysis, in situ observing systems and event-based sampling are crucial to understand the depth structure of the MHWs and their compound events with other marine extremes.

TCs are modulated by ENSO, IOD, and MJO variability through changes in the large-scale environmental parameters favoring TC genesis and growth. Observed TCs trends over the past decades, their attribution to anthropogenic climate change, and the identification of their environmental drivers are still uncertain because of uncertainties in TC historical datasets and their more chaotic nature. Future model projections however indicate more intense TCs (high confidence), fewer TCs in the southern Indian Ocean (high agreement between models, moderate confidence), more TCs in the Arabian Sea and fewer in the Bay of Bengal (low confidence).

TC forecasts in the Indian Ocean have improved drastically from infancy a few decades ago to accurate and advanced predictions of TC track and intensity in recent years, which greatly lessens TC-induced damages. However, there are new challenges due to climate change. Cyclones are now undergoing rapid intensification, due to high ocean heat content (e.g., Singh et al., 2021). Cyclone forecasts are unable to predict the rapid intensification due to the gaps in incorporating high-resolution subsurface ocean data.

Climate model projections of TCs are limited by spatial resolution and the sensitivity of TCs to atmospheric convection schemes in climate models. In addition, confidence in future regional projections of TC numbers is hampered by uncertainties in projected SSTs patterns (e.g., Dutheil et al., 2020; Murakami et al., 2012). Regional SST projections vary considerably between climate models and are sensitive to common present-day model biases. In downscaled atmospheric projections forced by specified SSTs, that explicitly resolve high-resolution TC processes, the reliability of TC projections is limited by the lack of air-sea interactions (Trenberth et al., 2018), especially for intense TCs for which air-sea feedbacks play a strong role. TC intensity increase could indeed be overestimated when not accounting for air-sea coupling as the increased thermal stratification in a warming climate could boost the negative air-sea coupling feedback on TC intensity (Huang et al., 2015; Ogata et al., 2016). The uncertainties in future TC projections could hence be lowered by reducing model biases to provide more credible regional SST projections, and by accounting for air-sea coupling in TC-resolving simulations. The valuable and short-term socioeconomic benefits are, however, likely to arise from improving TC intensity and trajectory prediction on weather timescales.

The high vulnerability of the Indian Ocean rim countries to tsunamis has led to the rapid development of both tsunami science (including a better understanding of how earthquake characteristics influence tsunami amplitude) and tsunami warning systems (with both in situ measurements and tsunami real-time forecast systems), particularly since the 2004 event. Despite this, there are still major challenges related to real-time data sharing for early detection, quick characterization of earthquake source mechanism, atypical and local tsunamis that may strike within a few minutes of generation, the lack of high-density coastal sea-level stations, improving real-time propagation and inundation models with advanced technology (Fauzi & Mizutani, 2020; Wang et al., 2019) and the availability of high-resolution bathymetry and topography for accurate inundation forecasts. Emerging techniques such as Global Navigation Satellite Systems (Babeyko et al., 2010; Foster et al., 2012; Hoechner et al., 2013; Sobolev et al., 2007), submarine optical cable instrumentation (Kawaguchi et al., 2015; Matias et al., 2021), high-frequency coastal radars, infrasound sensors, satellite altimetry, and ionospheric monitoring, the multisensor approach will be important for improving the monitoring of tsunamis (Angove et al., 2019; Behrens et al., 2010). However, the success of a warning system depends on how individuals in the hazard zone recognize warning signals, make correct decisions, and act quickly. As a result, public awareness and preparedness also play a crucial role in tsunami risk reduction.

6 Educational resources

Marine heatwaves
MHW Tracker: https://www.phys.ocean.dal.ca/~schlegel/NOAA Reef watch: https://coralreefwatch.noaa.gov/satellite/index.php

Tropical cyclones
https://www.metlink.org/resource/tropical-cyclone-teaching-resources/
https://gpm.nasa.gov/education/keywords/tropical-cycloneshttps://www.nhc.noaa.gov/outreach/

Tsunamis
Seismic Data Source: http://ds.iris.edu/ds/nodes/dmc/data/
Education material: https://www.iris.edu/hq/programs/epo/about
Latest earthquakes: https://www.usgs.gov/natural-hazards/earthquake-hazards/earthquakes

Acknowledgments

MF was supported by the Centre for Southern Hemisphere Oceans Research (CSHOR), which is a joint initiative between the Qingdao National Laboratory for Marine Science and Technology (QNLM), CSIRO, University of New South Wales and the University of Tasmania. ASG was supported by the Australian Research Council Future Fellowship (FT220100475). We thank Helen Phillips, Roxy Matthew Koll, Gareth Davies, and one anonymous reviewer for the review comments.

Author contributions

MF and ASG drafted the marine heatwave section, ML and JV drafted the tropical cyclone section, and SM drafted the tsunami section. MF coordinated the writing, and ASG, ML, JV, and SM commented on the whole chapter.

References

Abe, K. (1973). Tsunami and mechanism of great earthquakes. *Physics of the Earth and Planetary Interiors, 7*, 143–153.

ADB. (2005). *An initial assessment of the impact of the earthquake and tsunami of December 2004*. Manila: Asian Development Bank.

Angove, M., Arcas, D., Bailey, R., Carrasco, P., Coetzee, D., Fry, B., Gledhill, K., Harada, S., von Hillebrandt-Andrade, C., Kong, L., & McCreery, C. (2019). Ocean observations required to minimize uncertainty in global tsunami forecasts, warnings, and emergency response. *Frontiers in Marine Science, 6*, 350.

Annamalai, H., Murtugudde, R., Potemra, J., Xie, S. P., Liu, P., & Wang, B. (2003). Coupled dynamics over the Indian Ocean: Spring initiation of the zonal mode. *Deep Sea Research Part II: Topical Studies in Oceanography, 50*(12–13), 2305–2330.

Antony, C., Testut, L., & Unnikrishnan, A. S. (2014). Observing storm surges in the Bay of Bengal from satellite altimetry. *Estuarine, Coastal and Shelf Science, 151*, 131–140.

Babcock, R. C., Bustamante, R. H., Fulton, E. A., Fulton, D. J., Haywood, M. D., Hobday, A. J., Kenyon, R., Matear, R. J., Plaganyi, E. E., Richardson, A. J., & Vanderklift, M. A. (2019). Severe continental-scale impacts of climate change are happening now: Extreme climate events impact marine habitat forming communities along 45% of Australia's coast. *Frontiers in Marine Science, 6*, 411.

Babeyko, A. Y., Hoechner, A., & Sobolev, S. V. (2010). Source modeling and inversion with near real-time GPS: A GITEWS perspective for Indonesia. *Natural Hazards and Earth System Sciences, 10*.

Barnston, A. G., Tippett, M. K., Ranganathan, M., & L'Heureux, M. L. (2019). Deterministic skill of ENSO predictions from the North American Multi-model Ensemble. *Climate Dynamics, 53*, 7215–7234. https://doi.org/10.1007/s00382-017-3603-3.

Behrens, J., Androsov, A., Babeyko, A. Y., Harig, S., Klaschka, F., & Mentrup, L. (2010). A new multi-sensor approach to simulation assisted tsunami early warning. *Natural Hazards and Earth System Sciences, 10*(6), 1085–1100.

Bell, S. S., Chand, S. S., Tory, K. J., Ye, H., & Turville, C. (2020). North Indian Ocean tropical cyclone activity in CMIP5 experiments: Future projections using a model-independent detection and tracking scheme. *International Journal of Climatology, 40*(15), 6492–6505.

Benazir, Syamsidik, & Luthfi, M. (2019). Assessment on damages of harbor complexes due to impacts of the 2018 Palu-Donggala Tsunami, Indonesia. In T. V. Nguyen, X. Dou, & T. Tran (Eds.), *International conference on Asian and Pacific Coasts*.

Ben-Menahem, A., & Rosenman, M. (1972). Amplitude patterns of tsunami waves from submarine earthquakes. *Journal of Geophysical Research, Space Physics, 77*, 3097–3128.

Benthuysen, J., Feng, M., & Zhong, L. (2014). Spatial patterns of warming off Western Australia during the 2011 Ningaloo Niño: Quantifying impacts of remote and local forcing. *Continental Shelf Research, 91*, 232–246.

Benthuysen, J. A., Oliver, E. C. J., Feng, M., & Marshall, A. G. (2018). Extreme marine warming across tropical Australia during austral summer 2015–2016. *Journal of Geophysical Research: Oceans, 123*, 1301–1326. https://doi.org/10.1002/2017JC013326.

Bessafi, M., & Wheeler, M. C. (2006). Modulation of South Indian Ocean tropical cyclones by the Madden–Julian oscillation and convectively coupled equatorial waves. *Monthly Weather Review, 134*(2), 638–656.

Bhardwaj, P., & Singh, O. (2020). Climatological characteristics of Bay of Bengal tropical cyclones: 1972–2017. *Theoretical and Applied Climatology, 139*(1), 615–629.

Bhardwaj, P., Singh, O., Pattanaik, D. R., & Klotzbach, P. J. (2019). Modulation of Bay of Bengal tropical cyclone activity by the Madden-Julian oscillation. *Atmospheric Research, 229*, 23–38.

Blades, D. M., & Siracusa, J. M. (2014). A history of U.S. nuclear testing and its influence on nuclear thought, 1945–1963. *Journal of American History, 102*(1). https://doi.org/10.1093/jahist/jav293.

Bond, N. A., Cronin, M. F., Freeland, H., & Mantua, N. (2015). Causes and impacts of the 2014 warm anomaly in the NE Pacific. *Geophysical Research Letters, 42*(9), 3414–3420.

Burt, J. A., Paparella, F., Al-Mansoori, N., Al-Mansoori, A., & Al-Jailani, H. (2019). Causes and consequences of the 2017 coral bleaching event in the southern Persian/Arabian Gulf. *Coral Reefs, 38*(4), 567–589.

Cai, W., Wang, G., Santoso, A., McPhaden, M. J., Wu, L., Jin, F. F., Timmermann, A., Collins, M., Vecchi, G., Lengaigne, M., & England, M. H. (2015). Increased frequency of extreme La Niña events under greenhouse warming. *Nature Climate Change, 5*(2), 132–137.

Camargo, S. J., Emanuel, K. A., & Sobel, A. H. (2007). Use of a genesis potential index to diagnose ENSO effects on tropical cyclone genesis. *Journal of Climate, 20*(19), 4819–4834.

Caputi, N., Kangas, M., Chandrapavan, A., Hart, A., Feng, M., Marin, M., & Lestang, S. D. (2019). Factors affecting the recovery of invertebrate stocks from the 2011 Western Australian extreme marine heatwave. *Frontiers in Marine Science, 6*, 484.

Caputi, N., Kangas, M., Denham, A., Feng, M., Pearce, A., Hetzel, Y., & Chandrapavan, A. (2016). Management adaptation of invertebrate fisheries to an extreme marine heat wave event at a global warming hot spot. *Ecology and Evolution, 6*, 3583–3593.

Carvajal, M., Araya-Cornejo, C., Sepúlveda, I., Melnick, D., & Haase, J. S. (2019). Nearly instantaneous tsunamis following the Mw 7.5 2018 Palu earthquake. *Geophysical Research Letters, 46*(10), 5117–5126.

Cattiaux, J., Chauvin, F., Bousquet, O., Malardel, S., & Tsai, C. L. (2020). Projected changes in the Southern Indian Ocean cyclone activity assessed from high-resolution experiments and CMIP5 models. *Journal of Climate, 33*(12), 4975–4991.

Chand, S. S., Dowdy, A. J., Ramsay, H. A., Walsh, K. J., Tory, K. J., Power, S. B., & Kuleshov, Y. (2019). Review of tropical cyclones in the Australian region: Climatology, variability, predictability, and trends. *Wiley Interdisciplinary Reviews: Climate Change, 10*(5), e602.

Cheng, L., Trenberth, K. E., Fasullo, J., Boyer, T., Abraham, J., & Zhu, J. (2017). Improved estimates of ocean heat content from 1960 to 2015. *Science Advances, 3*(3), e1601545.

Chua, C. T., Switzer, A. D., Suppasri, A., Li, L., Pakoksung, K., Lallemant, D., Jenkins, S. F., Charvet, I., Chua, T., Cheong, A., & Winspear, N. (2021). Tsunami damage to ports: Cataloguing damage to create fragility functions from the 2011 Tohoku event. *Natural Hazards and Earth System Sciences, 21*, 1887–1908. https://doi.org/10.5194/nhess-21-1887-2021.

Collins, M., et al. (2019). Extremes, abrupt changes and managing risks. et al. In H.-O. Pörtner (Ed.), *IPCC special report on the ocean and cryosphere in a changing climate*.

Comer, R. P. (1980). Tsunami height and earthquake magnitude: Theoretical basis of an empirical relation. *Geophysical Research Letters, 7*, 445–448. https://doi.org/10.1029/GL007i006p00445.

Davies, G. (2019). Tsunami variability from uncalibrated stochastic earthquake models: Tests against deep ocean observations 2006–2016. *Geophysical Journal International, 218*(3), 1939–1960. https://doi.org/10.1093/gji/ggz260.

Davies, G., & Griffin, J. (2020). Sensitivity of probabilistic tsunami hazard assessment to far-field earthquake slip complexity and rigidity depth-dependence: Case study of Australia. *Pure and Applied Geophysics, 177*(3), 1521–1548.

Davies, G., Romano, F., & Lorito, S. (2020). Global dissipation models for simulating tsunamis at far-field coasts up to 60 hours post-earthquake: Multi-site tests in Australia. *Frontiers in Earth Science, 8*, 497.

Delescluse, M., Chamot-Rooke, N., Cattin, R., et al. (2012). April 2012 intra-oceanic seismicity off Sumatra boosted by the Banda-Aceh megathrust. *Nature, 490*, 240–244. https://doi.org/10.1038/nature11520.

DeMott, C. A., Ruppert, J. H., Jr., & Rydbeck, A. (2024). Chapter 4: Intraseasonal variability in the Indian Ocean region. In C. C. Ummenhofer, & R. R. Hood (Eds.), *The Indian Ocean and its role in the global climate system* (pp. 79–101). Amsterdam: Elsevier. https://doi.org/10.1016/B978-0-12-822698-8.00006-8.

Deshpande, M., Singh, V. K., Kranthi, G. M., Roxy, M. K., Emmanuel, R., & Kumar, U. (2021). Changing status of tropical cyclones over the North Indian Ocean. *Climate Dynamics*, 1–23. https://link.springer.com/article/10.1007/s00382-021-05880-z.

Di Lorenzo, E., & Mantua, N. (2016). Multi-year persistence of the 2014/15 North Pacific marine heatwave. *Nature Climate Change, 6*(11), 1042–1047.

Doi, T., Behera, S. K., & Yamagata, T. (2016). Improved seasonal prediction using the SINTEX-F2 coupled model. *Journal of Advances in Modeling Earth Systems, 8*, 1847–1867. https://doi.org/10.1002/2016MS000744.

Donner, S. D. (2011). An evaluation of the effect of recent temperature variability on the prediction of coral bleaching events. *Ecological Applications, 21*(5), 1718–1730.

Dowdy, A. J. (2014). Long-term changes in Australian tropical cyclone numbers. *Atmospheric Science Letters, 15*(4), 292–298.

Du, Y., Xie, S. P., Huang, G., & Hu, K. (2009). Role of air–sea interaction in the long persistence of El Niño–induced North Indian Ocean warming. *Journal of Climate, 22*(8), 2023–2038.

Dutheil, C., Lengaigne, M., Bador, M., Vialard, J., Lefèvre, J., Jourdain, N. C., & Menkès, C. (2020). Impact of projected sea surface temperature biases on tropical cyclones projections in the South Pacific. *Scientific Reports, 10*(1), 1–12.

Edward, J. K. P., Mathews, G., Raj, K. D., Laju, R. L., Bharath, M. S., Arasamuthu, A., Kumar, P. D., Bilgi, D. S., & Malleshappa, H. (2018). Coral mortality in the Gulf of Mannar, southeastern India, due to bleaching caused by elevated sea temperature. *Current Science, 114*(9), 1967–1972.

Emanuel, K. A. (2004). Tropical cyclone energetics and structure. *Atmospheric Turbulence and Mesoscale Meteorology, 165*, 192.

Emanuel, K., DesAutels, C., Holloway, C., & Korty, R. (2004). Environmental control of tropical cyclone intensity. *Journal of the Atmospheric Sciences, 61*, 843–858.

Evan, A. T., & Camargo, S. J. (2011). A climatology of Arabian Sea cyclonic storms. *Journal of Climate, 24*, 140–158.

Evan, A. T., Kossin, J. P., & Ramanathan, V. (2011). Arabian Sea tropical cyclones intensified by emissions of black carbon and other aerosols. *Nature, 479*(7371), 94–97.

Fauzi, A., & Mizutani, N. (2020). Machine learning algorithms for real-time tsunami inundation forecasting: A case study in Nankai region. *Pure and Applied Geophysics, 177*(3), 1437–1450.

Felton, C. S., Subrahmanyam, B., & Murty, V. S. N. (2013). ENSO-modulated cyclogenesis over the Bay of Bengal. *Journal of Climate, 26*(24), 9806–9818.

Feng, M., Hendon, H. H., Xie, S. P., Marshall, A. G., Schiller, A., Kosaka, Y., Caputi, N., & Pearce, A. (2015). Decadal increase in Ningaloo Niño since the late 1990s. *Geophysical Research Letters, 42*, 104–112.

Feng, M., McPhaden, M. J., Xie, S. P., & Hafner, J. (2013). La Niña forces unprecedented Leeuwin Current warming in 2011. *Scientific Reports, 3*, 1277.

Feng, M., Zhang, Y., Hendon, H. H., McPhaden, M. J., & Marshall, A. G. (2021). Niño 4 West (Niño−4W) sea surface temperature variability. *Journal of Geophysical Research: Oceans, 126*(9).

Foster, J. H., Brooks, B. A., Wang, D., Carter, G. S., & Merrifield, M. A. (2012). Improving tsunami warning using commercial ships. *Geophysical Research Letters, 39*(9).

Foster, R., Hagan, A., Perera, N., Gunawan, C. A., Silaban, I., Yaha, Y., Manuputty, Y., Hazam, I., & Hodgson, G. (2006). *Tsunami and earthquake damage to coral reefs of Aceh, Indonesia*. Pacific Palisades, California, USA: Reef Check Foundation. 33 pp.

Frank, W. M. (1987). Tropical cyclone formation. In *A global view of tropical cyclones* (pp. 53–90).

Frank, W. M., & Roundy, P. E. (2006). The role of tropical waves in tropical cyclogenesis. *Monthly Weather Review, 134*(9), 2397–2417.

Frederik, M. C. G., Udrekh, Adhitama, R., et al. (2019). First results of a bathymetric survey of Palu Bay, Central Sulawesi, Indonesia following the tsunamigenic earthquake of 28 September 2018. *Pure and Applied Geophysics Yokoyam, 176,* 3277–3290. https://doi.org/10.1007/s00024-019-02280-7.

Fritz, H. M., & Borrero, J. C. (2006). Somalia field survey after the December 2004 Indian Ocean tsunami. *Earthquake Spectra, 22*(S3), 219–233. https://doi.org/10.1193/1.2201972.

Frölicher, T. L., Fischer, E. M., & Gruber, N. (2018). Marine heatwaves under global warming. *Nature, 560,* 360–364. https://doi.org/10.1038/s41586-018-0383-9.

Geist, E. L. (1998). Local tsunamis and earthquake source parameters. *Advances in Geophysics, 39,* 117–209.

Geist, E., & Yoshioka, S. (1996). Source parameters controlling the generation and propagation of potential local tsunamis along the Cascadia margin. *Natural Hazards, 13,* 151–177 (15).

George, J., Ketcheson, D. I., & LeVeque, R. J. (2020). Shoaling on steep continental slopes: Relating transmission and reflection coefficients to Green's law. *Pure and Applied Geophysics, 177,* 1659–1674. https://doi.org/10.1007/s00024-019-02316-y.

Gibbons, S. J., Lorito, S., De La Asunción, M., Volpe, M., Selva, J., Macías, J., Sánchez-Linares, C., Brizuela, B., Vöge, M., Tonini, R., & Lanucara, P. (2022). The sensitivity of tsunami impact to earthquake source parameters and manning friction in high-resolution inundation simulations. *Frontiers in Earth Science, 9.*

Gica, E., Teng, M. H., Liu, P. L.-F., Titov, V., & Zhou, H. (2007). Sensitivity analysis of source parameters for earthquake-generated distant tsunamis. *Journal of Waterway Port, Coastal, and Ocean Engineering, 133,* 429–441.

Gilmour, J. P., Cook, K. L., Ryan, N. M., Puotinen, M. L., Green, R. H., Shedrawi, G., Hobbs, J. P. A., Thomson, D. P., Babcock, R. C., Buckee, J., & Foster, T. (2019). The state of Western Australia's coral reefs. *Coral Reefs, 38*(4), 651–667.

Girishkumar, M. S., & Ravichandran, M. (2012). The influences of ENSO on tropical cyclone activity in the Bay of Bengal during October–December. *Journal of Geophysical Research: Oceans, 117*(C2), C02033.

González-Vida, J. M., Castro, M. J., Macías, J., de la Asunción, M., Ortega, S., & Parés, C. (2021). Tsunami-HySEA: A numerical model developed for tsunami early warning systems (TEWS). In *Progress in industrial mathematics: Success stories* (pp. 209–226). Cham: Springer.

Gopika, S., Izumo, T., Vialard, J., Lengaigne, M., Suresh, I., & Kumar, M. R. (2020). Aliasing of the Indian Ocean externally-forced warming spatial pattern by internal climate variability. *Climate Dynamics, 54*(1), 1093–1111.

Gray, W. M. (1968). Global view of the origin of tropical disturbances and storms. *Monthly Weather Review, 96,* 669–700.

Gray, W. M. (1975). *Tropical cyclone genesis (doctoral dissertation).* Colorado State University. Libraries.

Gray, W. M. (1998). The formation of tropical cyclones. *Meteorology and Atmospheric Physics, 67*(1), 37–69.

Green, G. (1838). On the motion of waves in a variable canal of small depth and width. *Transactions of the Cambridge Philosophical Society, 6,* 457.

Greenslade, D. J. M., & Warne, J. O. (2012). Assessment of the effectiveness of a sea-level observing network for tsunami warning. *Journal of Waterway, Port, Coastal, and Ocean Engineering, 138*(3), 246–255.

Grezio, A., Babeyko, A., Baptista, M. A., Behrens, J., Costa, A., Davies, G., Geist, E. L., Glimsdal, S., González, F. I., Griffin, J., & Harbitz, C. B. (2017). Probabilistic tsunami hazard analysis: Multiple sources and global applications. *Reviews of Geophysics, 55*(4), 1158–1198.

Sen Gupta, A., Thomsen, M., Benthuysen, J. A., Hobday, A. J., Oliver, E., Alexander, L. V., Burrows, M. T., Donat, M. G., Feng, M., Holbrook, N. J., & Perkins-Kirkpatrick, S., Moore, P. J., Rodrigues, R., Scannell, H. A., Taschetto, A. S., Ummenhofer, C. C., Wernberg, T., & Smale, D. A. (2020). Drivers and impacts of the most extreme marine heatwaves events. *Scientific Reports, 10*(1), 1–15. https://doi.org/10.1038/s41598-020-75445-3.

Hall, J. D., Matthews, A. J., & Karoly, D. J. (2001). The modulation of tropical cyclone activity in the Australian region by the Madden–Julian oscillation. *Monthly Weather Review, 129*(12), 2970–2982.

Han, W., Vialard, J., McPhaden, M. J., Lee, T., Masumoto, Y., Feng, M., & De Ruijter, W. P. (2014). Indian Ocean decadal variability: A review. *Bulletin of the American Meteorological Society, 95*(11), 1679–1703.

Hanka, W., Saul, J., Weber, B., Becker, J., & Harjadi, P. (2010). Real-time earthquake monitoring for tsunami warning in the Indian Ocean and beyond. *Natural Hazards and Earth System Sciences, 10*(12), 2611–2622.

Heidarzadeh, M., Harada, T., Satake, K., Ishibe, T., & Gusman, A. (2016). Comparative study of two tsunamigenic earthquakes in the Solomon Islands: 2015 Mw 7.0 normal-fault and 2013 Santa Cruz Mw 8.0 megathrust earthquakes. *Geophysical Research Letters, 43*(9), 4340–4349.

Hirshorn, B., Weinstein, S., Wang, D., Koyanagi, K., Becker, N., & McCreery, C. (2019). Earthquake source parameters, rapid estimates for tsunami forecasts and warnings. In R. Meyers (Ed.), *Encyclopedia of complexity and systems science.* Berlin, Heidelberg: Springer. https://doi.org/10.1007/978-3-642-27737-5_160-2.

Ho, C. H., Kim, J. H., Jeong, J. H., Kim, H. S., & Chen, D. (2006). Variation of tropical cyclone activity in the South Indian Ocean: El Niño–Southern Oscillation and Madden-Julian Oscillation effects. *Journal of Geophysical Research: Atmospheres, 111*(D22).

Hobday, A. J., Alexander, L. V., Perkins, S. E., Smale, D. A., Straub, S. C., Oliver, E. C. J., Benthuysen, J. A., et al. (2016). A hierarchical approach to defining marine heatwaves. *Progress in Oceanography, 141,* 227–238.

Hobday, A. J., Oliver, E. C. J., Gupta, A. S., Benthuysen, J. A., Burrow, M. T., Donat, M. G., Hollbrook, N. J., et al. (2018). Categorizing and naming marine heatwaves. *Oceanography, 31*(2), 1–13.

Hoechner, A., Ge, M., Babeyko, A. Y., & Sobolev, S. V. (2013). Instant tsunami early warning based on real-time GPS–Tohoku 2011 case study. *Natural Hazards and Earth System Sciences, 13*(5), 1285–1292.

Hoffmann, G., Rupprechter, M., Al Balushi, N., Grützner, C., & Reicherter, K. (2013). The impact of the 1945 Makran tsunami along the coastlines of the Arabian Sea (northern Indian Ocean)—A review. *Zeitschrift für Geomorphologie, 57*(Suppl 4), 257–277. https://doi.org/10.1127/0372-8854/2013/S-00134.

Hokugo, A. (2013). Mechanism of tsunami fires after the Great East Japan Earthquake 2011 and evacuation from the tsunami fires. *Procedia Engineering, 62,* 140–153.

Holbrook, N. J., Scannell, H. A., Gupta, A. S., Benthuysen, J. A., Feng, M., Oliver, E. C., Alexander, L. V., Burrows, M. T., Donat, M. G., Hobday, A. J., & Moore, P. J. (2019). A global assessment of marine heatwaves and their drivers. *Nature Communications, 10*(1), 1–13.

Honda, H. (1962). Earthquake mechanism and seismic waves. *Journal of Physics of the Earth, 10*(2), 1–97.

Hossen, M. J., Cummins, P. R., Dettmer, J., & Baba, T. (2015). Tsunami waveform inversion for sea surface displacement following the 2011 Tohoku earthquake: Importance of dispersion and source kinematics. *Journal of Geophysical Research: Solid Earth, 120*(9), 6452–6473.

Huang, P., Lin, I. I., Chou, C., & Huang, R. H. (2015). Change in ocean subsurface environment to suppress tropical cyclone intensification under global warming. *Nature Communications, 6*(1), 1–9.

Hughes, T. P., Anderson, K. D., Connolly, S. R., Heron, S. F., Kerry, J. T., Lough, J. M., Baird, A. H., Baum, J. K., Berumen, M. L., Bridge, T. C., & Claar, D. C. (2018). Spatial and temporal patterns of mass bleaching of corals in the Anthropocene. *Science, 359*(6371), 80–83.

Husrin, S., Strusiska, A., & Oumeraci, H. (2012). Experimental study on tsunami attenuation by mangrove forest. *Earth, Planets and Space, 64*(10), 973–989. https://doi.org/10.5047/eps.2011.11.008.

Imamura, F., Yalciner, A., & Ozyurt, G. (2006). *Tsunami modeling manual.* http://www.tsunami.civil.tohoku.ac.jp/hokusai3/E/projects/manual-ver-3.1.pdf.

IOC-Tsunami Website. (2022). http://www.ioc-tsunami.org/index.php?option=com_content&view=article&id=413:indian-ocean-tsunami&catid=20:latest-news&lang=en&Itemid=68. Accessed 28 Oct 2022.

IUCN. (2005). *Rapid environmental and socio-economic assessment of tsunami-damage in terrestrial and marine coastal ecosystems of Ampara and Batticaloa districts of eastern Sri Lanka.* Available online at: http://www.iucn.org/tsunami/resources/iucn-reports.htm.

Jayasuriya, S., Steele, P., & Weerakoon, D. (2006). Post-tsunami recovery: Issues and challenges in Sri Lanka. *ADB Institute Research Paper Series No. 71.*

Kanamori, H., & Rivera, L. (2008). Source inversion of phase: Speeding up seismic tsunami warning. *Geophysical Journal International, 175*(1), 222–238.

Kataoka, T., Tozuka, T., Behera, S., & Yamagata, T. (2014). On the Ningaloo Niño/Niña. *Climate Dynamics, 43*, 1463–1482.

Kataoka, T., Tozuka, T., & Yamagata, T. (2017). Generation and decay mechanisms of Ningaloo Niño/Niña. *Journal of Geophysical Research: Oceans, 122*, 8913–8932.

Kawaguchi, K., Kaneko, S., Nishida, T., & Komine, T. (2015). Construction of the DONET real-time seafloor observatory for earthquakes and tsunami monitoring. In *Seafloor observatories* (pp. 211–228). Berlin, Heidelberg: Springer.

Kelleher, J., & McCann, W. (1976). Buoyant zones, great earthquakes, and unstable boundaries of subduction. *Journal of Geophysical Research, 81*, 4885–4898.

Khattri, K. (1973). Earthquake focal mechanism studies—A review. *Earth-Science Reviews, 9*, 19–63.

Kikuchi, K., & Wang, B. (2010). Formation of tropical cyclones in the northern Indian Ocean associated with two types of tropical intraseasonal oscillation modes. *Journal of the Meteorological Society of Japan Series II, 88*(3), 475–496.

Kikuchi, K., Wang, B., & Fudeyasu, H. (2009). Genesis of tropical cyclone Nargis revealed by multiple satellite observations. *Geophysical Research Letters, 36*(6).

Kintisch, E. (2016). How a "Godzilla" El Niño shook up weather forecasts. *Science, 352*, 1501–1502. https://doi.org/10.1126/science.352.6293.1501.

Klotzbach, P. J. (2014). The Madden–Julian oscillation's impacts on worldwide tropical cyclone activity. *Journal of Climate, 27*(6), 2317–2330.

Knutson, T., Camargo, S. J., Chan, J. C., Emanuel, K., Ho, C. H., Kossin, J., & Wu, L. (2019). Tropical cyclones and climate change assessment: Part I: Detection and attribution. *Bulletin of the American Meteorological Society, 100*(10), 1987–2007.

Knutson, T., Camargo, S. J., Chan, J. C., Emanuel, K., Ho, C. H., Kossin, J., & Wu, L. (2020). Tropical cyclones and climate change assessment: Part II: Projected response to anthropogenic warming. *Bulletin of the American Meteorological Society, 101*(3), E303–E322.

Kosaka, Y., & Xie, S. P. (2013). Recent global-warming hiatus tied to equatorial Pacific surface cooling. *Nature, 501*(7467), 403–407.

Kossin, J. P., Olander, T. L., & Knapp, K. R. (2013). Trend analysis with a new global record of tropical cyclone intensity. *Journal of Climate, 26*(24), 9960–9976.

Krishnamohan, K. S., Mohanakumar, K., & Joseph, P. V. (2012). The influence of Madden–Julian oscillation in the genesis of North Indian Ocean tropical cyclones. *Theoretical and Applied Climatology, 109*(1–2), 271–282.

Krishnan, P., Roy, S. D., George, G., Srivastava, R. C., Anand, A., Murugesan, S., Kaliyamoorthy, M., Vikas, N., & Soundararajan, R. (2011). Elevated sea surface temperature during May 2010 induces mass bleaching of corals in the Andaman. *Current Science*, 111–117.

Kuleshov, Y., Chane Ming, F., Qi, L., Chouaibou, I., Hoareau, C., & Roux, F. (2009). Tropical cyclone genesis in the southern hemisphere and its relationship with the ENSO. *Annales Geophysicae, 27*(6), 2523–2538. Copernicus GmbH.

Kuleshov, Y., Fawcett, R., Qi, L., Trewin, B., Jones, D., McBride, J., & Ramsay, H. (2010). Trends in tropical cyclones in the South Indian Ocean and the South Pacific Ocean. *Journal of Geophysical Research: Atmospheres, 115*(D1).

Kuleshov, Y., Gregory, P., Watkins, A. B., & Fawcett, R. J. (2020). Tropical cyclone early warnings for the regions of the Southern Hemisphere: strengthening resilience to tropical cyclones in small island developing states and least developed countries. *Natural Hazards, 104*, 1295–1313.

Kuleshov, Y., Qi, L., Fawcett, R., & Jones, D. (2008). On tropical cyclone activity in the Southern Hemisphere: Trends and the ENSO connection. *Geophysical Research Letters, 35*(14). https://doi.org/10.1029/2007GL032983.

Kumar, T. S., Nayak, S., Kumar, C. P., Yadav, R. B. S., Kumar, B. A., Sunanda, M. V., Devi, E. U., Kumar, N. K., Kishore, S. A., & Shenoi, S. S. C. (2012). Successful monitoring of the 11 April 2012 tsunami off the coast of Sumatra by Indian Tsunami Early Warning Centre. *Current Science, 102*(11), 1519–1526. JSTOR 24084760.

Kumaraguru, A. K., Jayakumar, K., Wilson, J. J., & Ramakritinan, C. M. (2005). Impact of the tsunami of 26 December 2004 on the coral reef environment of Gulf of Mannar and Palk Bay in the southeast coast of India. *Current Science*, 1729–1741.

Kume, T., Umetsu, C., & Palanisami, K. (2009). Impact of the December 2004 tsunami on soil, groundwater and vegetation in the Nagapattinam district, India. *Journal of Environmental Management, 90*(10), 3147–3154.

Landu, K., Goyal, R., & Keshav, B. S. (2020). Role of multiple equatorial waves on cyclogenesis over Bay of Bengal. *Climate Dynamics, 54*(3), 2287–2296.

Laufkötter, C., Zscheischler, J., & Frölicher, T. L. (2020). High-impact marine heatwaves attributable to human-induced global warming. *Science, 369*(6511), 1621–1625.

Lee, H. S., Shimoyama, T., & Popinet, S. (2015). Impacts of tides on tsunami propagation due to potential Nankai Trough earthquakes in the Seto Inland Sea, Japan. *Journal of Geophysical Research, Oceans, 120*, 6865–6883.

Lee, T., et al. (2010). Record warming in the South Pacific and western Antarctica associated with the strong Central-Pacific El Niño in 2009–10. *Geophysical Research Letters, 37*(19), L19704.

Lengaigne, M., Neetu, S., Samson, G., Vialard, J., Krishnamohan, K. S., Masson, S., & Menkès, C. E. (2019). Influence of air–sea coupling on Indian Ocean tropical cyclones. *Climate Dynamics, 52*(1), 577–598.

Li, Z., Li, T., Yu, W., Li, K., & Liu, Y. (2016). What controls the interannual variation of tropical cyclone genesis frequency over Bay of Bengal in the post-monsoon peak season? *Atmospheric Science Letters, 17*(2), 148–154.

Li, Z., Yu, W., Li, T., Murty, V. S. N., & Tangang, F. (2013). Bimodal character of cyclone climatology in the Bay of Bengal modulated by monsoon seasonal cycle. *Journal of Climate, 26*(3), 1033–1046.

Lim, E. P., Hendon, H. H., Butler, A. H., Thompson, D. W., Lawrence, Z. D., Scaife, A. A., … Comer, R. (2021). The 2019 Southern Hemisphere stratospheric polar vortex weakening and its impacts. *Bulletin of the American Meteorological Society, 102*(6), E1150–E1171.

Lim, Y., Son, S., & Kim, D. (2018). MJO prediction skill of the subseasonal-to-seasonal prediction models. *Journal of Climate, 31*, 4075–4094. https://doi.org/10.1175/JCLI-D-17-0545.1.

Liu, K. S., & Chan, J. C. (2012). Interannual variation of Southern Hemisphere tropical cyclone activity and seasonal forecast of tropical cyclone number in the Australian region. *International Journal of Climatology, 32*(2), 190–202.

Lix, J. K., Venkatesan, R., George Grinson, R. R., Rao, V. K., Jineesh, M. M., Arul, G., Vengatesan, S., Ramasundaram, R. S., & Atmanand, M. A. (2016). Differential bleaching of corals based on El Niño type and intensity in the Andaman Sea, southeast bay of Bengal. *Environmental Monitoring and Assessment, 188*(3), 175.

Lorito, S., Selva, J., Basili, R., Romano, F., Tiberti, M. M., & Piatanesi, A. (2015). Probabilistic hazard for seismically induced tsunamis: Accuracy and feasibility of inundation maps. *Geophysical Journal International, 200*(1), 574–588.

Lowe, D. R., Byerly, G. R., Kyte, F. T., Shukolyukov, A., Asaro, F., & Krull, A. (2003). Spherule beds 3.47-3.24 billion years old in the Barberton Greenstone Belt, South Africa: A record of large meteorite impacts and their influence on early crustal and biological evolution. *Astrobiology, 3*(1), 7–48. https://doi.org/10.1089/153110703321632408.

Macías, J., Mercado, A., González-Vida, J. M., Ortega, S., & Castro, M. J. (2016). Comparison and computational performance of Tsunami-HySEA and MOST models for LANTEX 2013 scenario: Impact assessment on Puerto Rico coasts. In *Vol. I. Global tsunami science: Past and future* (pp. 3973–3997). Cham: Birkhäuser.

Madden, R. A., & Julian, P. R. (1971). Detection of a 40–50 day oscillation in the zonal wind in the tropical Pacific. *Journal of the Atmospheric Sciences, 28*(5). 702–702.

Madden, R. A., & Julian, P. R. (1972). Description of global-scale circulation cells in the tropics with a 40–50 day period. *Journal of the Atmospheric Sciences, 29*(6), 1109–1123.

Mahala, B. K., Nayak, B. K., & Mohanty, P. K. (2015). Impacts of ENSO and IOD on tropical cyclone activity in the bay of Bengal. *Natural Hazards, 75*(2), 1105–1125.

Mansinha, L. A., & Smylie, D. E. (1971). The displacement fields of inclined faults. *Bulletin of the Seismological Society of America, 61*(5), 1433–1440.

Marshall, A. G., Hendon, H. H., Feng, M., & Schiller, A. (2015). Initiation and amplification of the Ningaloo Niño. *Climate Dynamics, 45*, 2367–2385.

Matias, L., Carrilho, F., Sá, V., Omira, R., Niehus, M., Corela, C., Barros, J., & Omar, Y. (2021). The contribution of submarine optical fiber telecom cables to the monitoring of earthquakes and tsunamis in the NE Atlantic. *Frontiers in Earth Science*, 611.

Matsuba, M., Tanaka, Y., Yamakita, T., Ishikawa, Y., & Fujikura, K. (2020). Estimation of tsunami debris on seafloors towards future disaster preparedness: Unveiling spatial varying effects of combined land use and oceanographic factors. *Marine Pollution Bulletin, 157*, 111289.

Menkes, C. E., Lengaigne, M., Marchesiello, P., Jourdain, N. C., Vincent, E. M., Lefèvre, J., & Royer, J. F. (2012). Comparison of tropical cyclogenesis indices on seasonal to interannual timescales. *Climate Dynamics, 38*(1–2), 301–321.

Meyers, G. (1996). Variation of the Indonesian throughflow and the El Niño Southern Oscillation. *Journal of Geophysical Research, 101*, 12255–12263.

Mofjeld, H. O., González, F. I., Titov, V. V., Venturato, A. J., & Newman, J. C. (2007). Effects of tides on maximum tsunami wave heights: Probability distributions. *Journal of Atmospheric and Oceanic Technology, 24*(1), 117–123.

Mokhtari, M., Abdollahie Fard, I., & Hessami, K. (2008). Structural elements of the Makran region, Oman Sea and their potential relevance to tsunamigenisis. *Natural Hazards, 47*, 185–199. https://doi.org/10.1007/s11069-007-9208-0.

Morgan, R. (2011). *Top 100 stories of 2010, #84: Yardstick for killer waves*. Discover. http://discovermagazine.com/2011/jan-feb/84/.

Muhari, A., Charvet, I., Tsuyoshi, F., et al. (2015). Assessment of tsunami hazards in ports and their impact on marine vessels derived from tsunami models and the observed damage data. *Natural Hazards, 78*, 1309–1328. https://doi.org/10.1007/s11069-015-1772-0.

Muhari, A., Heidarzadeh, M., Susmoro, H., et al. (2019). The December 2018 Anak Krakatau volcano tsunami as inferred from post-tsunami field surveys and spectral analysis. *Pure and Applied Geophysics, 176*, 5219–5233. https://doi.org/10.1007/s00024-019-02358-2.

Muhari, A., Imamura, F., Arikawa, T., Hakim, A. R., & Afriyanto, B. (2018). Solving the puzzle of the September 2018 Palu, Indonesia, tsunami mystery: Clues from the tsunami waveform and the initial field survey data. *Journal of Disaster Research, 13*(Scientific Communication), sc20181108.

Murakami, H., Delworth, T. L., Cooke, W. F., Zhao, M., Xiang, B., & Hsu, P. C. (2020). Detected climatic change in global distribution of tropical cyclones. *Proceedings of the National Academy of Sciences, 117*, 10706–10714.

Murakami, H., Mizuta, R., & Shindo, E. (2012). Future changes in tropical cyclone activity projected by multi-physics and multi-SST ensemble experiments using the 60-km-mesh MRI-AGCM. *Climate Dynamics, 39*(9–10), 2569–2584.

Murakami, H., Sugi, M., & Kitoh, A. (2013). Future changes in tropical cyclone activity in the North Indian Ocean projected by high-resolution MRI-AGCMs. *Climate Dynamics, 40*(7–8), 1949–1968.

Murakami, H., Vecchi, G. A., & Underwood, S. (2017). Increasing frequency of extremely severe cyclonic storms over the Arabian Sea. *Nature Climate Change, 7*(12), 885–889.

Musa, R., Draman, S., Jeffrey, S., Jeffrey, I., Abdullah, N., Halim, N. A. M., Wahab, N. A., Mukhtar, N. Z. M., Johari, S. N. A., Rameli, N., & Midin, M. (2014). Post tsunami psychological impact among survivors in Aceh and West Sumatra, Indonesia. *Comprehensive Psychiatry, 55*, S13–S16.

Muthige, M. S., Malherbe, J., Englebrecht, F. A., Grab, S., Beraki, A., Maisha, T. R., & Van der Merwe, J. (2018). Projected changes in tropical cyclones over the South West Indian Ocean under different extents of global warming. *Environmental Research Letters, 13*(6), 065019.

Needham, H. F., Keim, B. D., & Sathiaraj, D. (2015). A review of tropical cyclone-generated storm surges: Global data sources, observations, and impacts. *Reviews of Geophysics, 53*(2), 545–591.

Neetu, S., Lengaigne, M., Vialard, J., Samson, G., Masson, S., Krishnamohan, K. S., & Suresh, I. (2019). Premonsoon/postmonsoon Bay of Bengal tropical cyclones intensity: Role of air-sea coupling and large-scale background state. *Geophysical Research Letters, 46*(4), 2149–2157.

Neetu, S., Suresh, I., Shankar, R., Nagarajan, B., Sharma, R., Shenoi, S. S. C., Unnikrishnan, A. S., & Sundar, D. (2011). Trapped waves of the 27 November 1945 Makran tsunami: Observations and numerical modeling. *Natural Hazards, 59*, 1609–1618. https://doi.org/10.1007/s11069-011-9854-0.

Neuner, F., Schauer, E., Catani, C., Ruf, M., & Elbert, T. (2006). Post-tsunami stress: A study of posttraumatic stress disorder in children living in three severely affected regions in Sri Lanka. *Journal of Traumatic Stress: Official Publication of The International Society for Traumatic Stress Studies, 19*(3), 339–347.

Ng, E. K., & Chan, J. C. (2012). Interannual variations of tropical cyclone activity over the North Indian Ocean. *International Journal of Climatology, 32*(6), 819–830.

Nicholls, N. (1979). A possible method for predicting seasonal tropical cyclone activity in the Australian region. *Monthly Weather Review, 107*(9), 1221–1224.

Nistor, I., Palermo, D., Nouri, Y., Murty, T., & Saatcioglu, M. (2018). Tsunami-induced forces on structures. In *Handbook of coastal and ocean engineering* (pp. 481–506).

Ogata, T., Mizuta, R., Adachi, Y., Murakami, H., & Ose, T. (2016). Atmosphere-ocean coupling effect on intense tropical cyclone distribution and its future change with 60 km-AOGCM. *Scientific Reports, 6*(1), 1–8.

Okada, Y. (1985). Surface deformation due to shear and tensile faults in a half-space. *Bulletin of the Seismological Society of America, 75*, 1135–1154.

Okada, Y. (1992). Internal deformation due to shear and tensile faults in a half-space. *Bulletin of the Seismological Society of America, 82*(2), 1018–1040.

Okal, E. A., Fritz, H. M., Raad, P. E., Synolakis, C. E., Al-Shijbi, Y., & Al-Saifi, M. (2006a). Oman field survey after the December 2004 Indian Ocean tsunami. *Earthquake Spectra, 22*, S203–S218.

Okal, E. A., Fritz, H. M., Raveloson, R., Joelson, G., Pančošková, P., & Rambolamanana, G. (2006b). Madagascar field survey after the December 2004 Indian Ocean tsunami. *Earthquake Spectra, 22*, S263–S283.

Okal, E. A., Sladen, A., & Okal, E. A.-S. (2006c). Rodrigues, Mauritius and Réunion Islands, field survey after the December 2004 Indian Ocean tsunami. *Earthquake Spectra, 22*, S241–S261.

Oliver, E. C. J. (2019). Mean warming not variability drives marine heatwave trends. *Climate Dynamics*. https://doi.org/10.1007/s00382-019-04707-2.

Oliver, E. C. J., et al. (2018). Longer and more frequent marine heatwaves over the past century. *Nature Communications, 9*(1), 1324.

Omira, R., Baptista, M. A., Matias, L., Miranda, J. M., Catita, C., Carrilho, F., & Toto, E. (2009). Design of a sea-level tsunami detection network for the Gulf of Cadiz. *Natural Hazards and Earth System Sciences, 9*(4), 1327–1338.

Pachauri, R. K., Allen, M. R., Barros, V. R., Broome, J., Cramer, W., Christ, R., Church, J. A., Clarke, L., Dahe, Q., Dasgupta, P., & Dubash, N. K. (2014). Climate change 2014: Synthesis report. In *Contribution of working groups I, II and III to the fifth assessment report of the intergovernmental panel on climate change* (p. 151). IPCC.

Pearce, A. F., & Feng, M. (2013). The rise and fall of the "marine heat wave" off Western Australia during the summer of 2010/2011. *Journal of Marine Systems, 111-112*, 139–156.

Pearce, A., Lenanton, R., Jackson, G., et al. (2011). The "marine heat wave" off Western Australia during the summer of 2010/2011. In *Vol. 222. Fisheries research report*. Western Australia: Department of Fisheries. 36 pp.

Piché, S., Nistor, I., & Murty, T. (2020). Modeling tsunami attenuation and impacts on coastal communities. In P. K. Srivastava, S. K. Singh, U. C. Mohanty, & T. Murty (Eds.), *Techniques for disaster risk management and mitigation*. https://doi.org/10.1002/9781119359203.ch19.

Power, S., Casey, T., Folland, C., Colman, A., & Mehta, V. (1999). Inter-decadal modulation of the impact of ENSO on Australia. *Climate Dynamics, 15*(5), 319–324.

Price, J. F. (1981). Upper Ocean response to a hurricane. *Journal of Physical Oceanography, 11*, 153–175. https://doi.org/10.1175/1520. 0485(1981) 011<0153:UORTAH>2.0.CO;2.

Rabinovich, A. B., Candella, R. N., & Thomson, R. E. (2013). The open ocean energy decay of three recent trans-Pacific tsunamis. *Geophysical Research Letters, 40*(12), 3157–3162. https://doi.org/10.1002/grl.50625.

Ramsay, H. A., Camargo, S. J., & Kim, D. (2012). Cluster analysis of tropical cyclone tracks in the Southern Hemisphere. *Climate Dynamics, 39*(3–4), 897–917.

Ramsay, H. A., Richman, M. B., & Leslie, L. M. (2017). The modulating influence of Indian Ocean Sea surface temperatures on Australian region seasonal tropical cyclone counts. *Journal of Climate, 30*, 4843–4856.

Rasmussen, L., Bromirski, P. D., Miller, A. J., Arcas, D., Flick, R. E., & Hendershott, M. C. (2015). Source location impact on relative tsunami strength along the U.S. West Coast. *Journal of Geophysical Research, Oceans, 120*(7), 4945–4961.

Roberts, M. J., Camp, J., Seddon, J., Vidale, P. L., Hodges, K., Vannière, B., & Caron, L. P. (2020). Projected future changes in tropical cyclones using the CMIP6 HighResMIP multimodel ensemble. *Geophysical Research Letters, 47*(14). e2020GL088662.

Roxy, M. K., Ritika, K., Terray, P., & Masson, S. (2014). The curious case of Indian Ocean warming. *Journal of Climate, 27*(22), 8501–8509.

Roxy, M. K., Saranya, J. S., Modi, A., Anusree, A., Cai, W., Resplandy, L., … Frölicher, T. L. (2024). Chapter 20: Future projections for the tropical Indian Ocean. In C. C. Ummenhofer, & R. R. Hood (Eds.), *The Indian Ocean and its role in the global climate system* (pp. 469–482). Amsterdam: Elsevier. https://doi.org/10.1016/B978-0-12-822698-8.00004-4.

Sannikova, N. K., Segur, H., & Arcas, D. (2021). Influence of tsunami aspect ratio on near and far-field tsunami amplitude. *Geosciences, 11*, 178. https://doi.org/10.3390/geosciences11040178.

Saranya, J. S., Roxy, M. K., Dasgupta, P., & Anand, A. (2022). Genesis and trends in marine heatwaves over the tropical Indian Ocean and their interaction with the Indian summer monsoon. *JGR Oceans*. In Press https://www.essoar.org/doi/10.1002/essoar.10506673.3.

Satake, K. (1987). Inversion of tsunami waveforms for the estimation of a fault heterogeneity: Method and numerical experiments. *Journal of Physics of the Earth, 35*(3), 241–254.

Satake, K., Fujii, Y., Harada, T., & Namegaya, Y. (2013). Time and space distribution of coseismic slip of the 2011 Tohoku Earthquake as inferred from Tsunami waveform data. *Bulletin of the Seismological Society of America, 103*(2B), 1473–1492. https://doi.org/10.1785/0120120122.

Sattar, A. M., & Cheung, K. K. (2019). Comparison between the active tropical cyclone seasons over the Arabian Sea and Bay of Bengal. *International Journal of Climatology, 39*(14), 5486–5502.

Schade, L. R. (2000). Tropical cyclone intensity and sea surface temperature. *Journal of the Atmospheric Sciences, 57*, 3122–3130. https://doi.org/10.1175/1520-0469(2000).

Schott, F. A., Xie, S. P., & McCreary, J. P., Jr. (2009). Indian Ocean circulation and climate variability. *Reviews of Geophysics, 47*(1).

Selva, J., Lorito, S., Volpe, M., Romano, F., Tonini, R., Perfetti, P., Bernardi, F., Taroni, M., Scala, A., Babeyko, A., & Løvholt, F. (2021). Probabilistic tsunami forecasting for early warning. *Nature Communications, 12*(1), 1–14.

Shelby, M., Grilli, S. T., & Grilli, A. R. (2016). Tsunami hazard assessment in the Hudson River Estuary based on dynamic tsunami-tide simulations. *Pure and Applied Geophysics, 173*(12), 3999–4037. https://doi.org/10.1007/s00024-016-1315-y.

Sheth, A., Sanyal, S., Jaiswal, A., & Gandhi, P. (2006). Effects of the December 2004 Indian Ocean Tsunami on the Indian Mainland. *Earthquake Spectra, 22*(S3), S435–S473.

Singh, V. K., & Roxy, M. K. (2022). *A review of the ocean-atmosphere interactions during tropical cyclones in the North Indian Ocean*. https://arxiv.org/abs/2012.04384.

Singh, V. K., Roxy, M. K., & Deshpande, M. (2021). Role of warm ocean conditions and the MJO in the genesis and intensification of extremely severe cyclone Fani. *Scientific Reports, 11*, 3607. https://doi.org/10.1038/s41598-021-82680-9.

Smale, D. A., et al. (2019). Marine heatwaves threaten global biodiversity and the provision of ecosystem services. *Nature Climate Change, 9*, 306–312.

Sobel, A. H., Camargo, S. J., Hall, T. M., Lee, C. Y., Tippett, M. K., & Wing, A. A. (2016). Human influence on tropical cyclone intensity. *Science, 353*(6296), 242–246.

Sobolev, S. V., Babeyko, A. Y., Wang, R., Hoechner, A., Galas, R., Rothacher, M., Sein, D. V., Schröter, J., Lauterjung, J., & Subarya, C. (2007). Tsunami early warning using GPS-Shield arrays. *Journal of Geophysical Research: Solid Earth, 112*(B8).

Song, Y. T. (2007). Detecting tsunami genesis and scales directly from coastal GPS stations. *Geophysical Research Letters, 34*, L19602. https://doi.org/10.1029/2007GL031681.

Spencer, T., Teleki, K. A., Bradshaw, C., & Spalding, M. D. (2000). Coral bleaching in the southern Seychelles during the 1997–1998 Indian Ocean warm event. *Marine Pollution Bulletin, 40*(7), 569–586.

Sprintall, J., Cravatte, S., Dewitte, B., Yan, D., & Gupta, A. S. (2020). ENSO oceanic teleconnections. In *El Niño southern oscillation in a changing climate* (pp. 337–359). American Geophysical Union (AGU). https://doi.org/10.1002/9781119548164.ch15.

Sunny, R. C., Cheng, W., & Horrillo, J. (2019). Video content analysis of the 2018 Sulawesi Tsunami, Indonesia: Impact at Palu Bay. *Pure and Applied Geophysics, 176*, 4127–4138. https://doi.org/10.1007/s00024-019-02325-x.

Suppasri, A., Al'Ala, M., Luthfi, M., & Comfort, L. K. (2019). Assessing the tsunami mitigation effectiveness of the planned Banda Aceh Outer Ring Road (BORR), Indonesia. *Natural Hazards and Earth System Sciences*, 299–312.

Suppasri, A., Maly, E., Kitamura, M., Pescaroli, G., Alexander, D., & Imamura, F. (2021). Cascading disasters triggered by tsunami hazards: A perspective for critical infrastructure resilience and disaster risk reduction. *International Journal of Disaster Risk Reduction, 66*, 102597.

Synolakis, C. E. (1991). Green's law and the evolution of solitary waves. *Physics of Fluids A, 3*, 490.

Tadibaght, A., M'rini, A., Siame, L., & Bellier, O. (2022). Tsunami impact assessment for low-lying cities along the Northern Atlantic coast of Morocco using MIRONE software. *Journal of African Earth Sciences, 192*, 104580. ISSN 1464-343X https://doi.org/10.1016/j.jafrearsci.2022.104580.

Tang, L., Titov, V. V., & Chamberlin, C. D. (2009). Development, testing, and applications of site-specific tsunami inundation models for real-time forecasting. *Journal of Geophysical Research: Oceans, 114*(C12).

Taschetto, A. S., Ummenhofer, C. C., Stuecker, M. F., Dommenget, D., Ashok, K., Rodrigues, R. R., & Yeh, S.-W. (2020). ENSO atmospheric teleconnections. In *El Niño Southern Oscillation in a changing climate* (pp. 309–335). American Geophysical Union (AGU). https://doi.org/10.1002/9781119548164.ch14.

Telford, J., & Cosgrave, J. (2006). *Joint evaluation of the international response to the Indian Ocean tsunami: Synthesis report*. Tsunami Evaluation Coalition (TEC).

Timmermans, B., Patricola, C., & Wehner, M. (2018). Simulation and analysis of hurricane-driven extreme wave climate under two ocean warming scenarios. *Oceanography, 31*(2), 88–99.

Titov, V., Song, Y. T., Tang, L., Bernard, E. N., Bar-Sever, Y., & Wei, Y. (2017). Consistent estimates of tsunami energy show promise for improved early warning. *Pure and Applied Geophysics*, 3863–3880. https://doi.org/10.1007/s00024-016-1312-1. 173.

Tomita, T., Imamura, F., Arikawa, T., Yasuda, T., & Kawata, Y. (2006). Damage caused by the 2004 Indian Ocean tsunami on the Southwestern Coast of Sri Lanka. *Coastal Engineering Journal, 48*(2), 99–116. https://doi.org/10.1142/S0578563406001362Frontiers.

Tozuka, T., Feng, M., Han, W., Kido, S., & Zhang, L. (2021). The Ningaloo Niño/Niña: Mechanisms, relation with other climate modes and impacts. In *Tropical and extratropical air-sea interactions* (pp. 207–219).

Trenberth, K. E., Cheng, L., Jacobs, P., Zhang, Y., & Fasullo, J. (2018). Hurricane Harvey links to ocean heat content and climate change adaptation. *Earth's Future, 6*(5), 730–744.

Tsunami Evaluation Coalition. (2006). *Joint evaluation of the international response to the Indian Ocean tsunami—Synthesis report*.

Villholth, K. G., & Neupane, B. (2011). Tsunamis as long-term hazards to coastal groundwater resources and associated water supplies. In *Tsunami—A growing disaster* (pp. 87–104). Shanghai, China: Intech.

Vincent, E. M., Emanuel, K. A., Lengaigne, M., Vialard, J., & Madec, G. (2014). Influence of upper ocean stratification interannual variability on tropical cyclones. *Journal of Advances in Modeling Earth Systems, 6*(3), 680–699.

Vincent, E. M., Lengaigne, M., Madec, G., Vialard, J., Samson, G., Jourdain, N., Menkes, C. E., & Jullien, S. (2012). Processes setting the characteristics of sea surface cooling induced by tropical cyclones. *Journal of Geophysical Research, 117*, C02020. https://doi.org/10.1029/2011JC007396.

Wahiduzzaman, M., Oliver, E. C., Wotherspoon, S. J., & Luo, J. J. (2020). Seasonal forecasting of tropical cyclones in the North Indian Ocean region: The role of El Niño-southern oscillation. *Climate Dynamics, 54*(3), 1571–1589.

Walder, J. S., Watts, P., Sorensen, O. E., & Janssen, K. (2003). Tsunamis generated by subaerial mass flows. *Journal of Geophysical Research, 108*(B5), 2236.

Walsh, K. J., Camargo, S. J., Knutson, T. R., Kossin, J., Lee, T. C., Murakami, H., & Patricola, C. (2019). Tropical cyclones and climate change. *Tropical Cyclone Research and Review, 8*(4), 240–250.

Wang, J., & Liu, P. L.-F. (2021). Numerical study on impacts of a concurrent storm-tide-tsunami event in Macau and Hong Kong. *Coastal Engineering, 170*, 104000.

Wang, B., & Moon, J. Y. (2017). An anomalous genesis potential index for MJO modulation of tropical cyclones. *Journal of Climate, 30*(11), 4021–4035.

Wang, Y., Satake, K., Cienfuegos, R., Quiroz, M., & Navarrete, P. (2019). Far-field tsunami data assimilation for the 2015 Illapel earthquake. *Geophysical Journal International, 219*(1), 514–521. https://doi.org/10.1093/gji/ggz309.

Webster, P. J. (2008). Myanmar's deadly daffodil. *Nature Geoscience, 1*(8), 488–490.

Wengel, C., Lee, S. S., Stuecker, M. F., et al. (2021). Future high-resolution El Niño/Southern Oscillation dynamics. *Nature Climate Change, 11*, 758–765. https://doi.org/10.1038/s41558-021-01132-4.

Wernberg, T., et al. (2013). An extreme climatic event alters marine ecosystem structure in a global biodiversity hotspot. *Nature Climate Change, 3*, 78–82.

Wernberg, T., Bennett, S., Babcock, R. C., et al. (2016). Climate-driven regime shift of a temperate marine ecosystem. *Science, 353*, 169–172.

Werner, A., & Holbrook, N. J. (2011). A Bayesian forecast model of Australian region tropical cyclone formation. *Journal of Climate, 24*(23), 6114–6131.

Wijnands, J. S., Qian, G., Shelton, K. L., Fawcett, R. J. B., Chan, J. C. L., & Kuleshov, Y. (2015). Seasonal forecasting of tropical cyclone activity in the Australian and the South Pacific Ocean regions. *Mathematics of Climate and Weather Forecasting, 1*(1).

Wild, S., Krützen, M., Rankin, R. W., Hoppitt, W. J., Gerber, L., & Allen, S. J. (2019). Long-term decline in survival and reproduction of dolphins following a marine heatwave. *Current Biology, 29*(7), R239–R240.

Winckler, P., Sepúlveda, I., Aron, F., & Contreras-López, M. (2017). How do tides and tsunamis interact in a highly energetic channel? The case of Canal Chacao, Chile. *Journal of Geophysical Research: Oceans, 122*, 9605–9624.

Wouthuyzen, S., Abrar, M., & Lorwens, J. (2018). A comparison between the 2010 and 2016 El-Nino induced coral bleaching in the Indonesian waters. In *Vol. 118. IOP conference series: Earth and environmental science* (p. 012051). IOP Publishing. No. 1.

Wünnemann, K., & Weiss, R. (2015). The meteorite impact-induced tsunami hazard. *Philosophical Transactions of the Royal Society. A*.3732014038120140381.

Xie, S. P., Annamalai, H., Schott, F. A., & McCreary, J. P., Jr. (2002). Structure and mechanisms of South Indian Ocean climate variability. *Journal of Climate, 15*(8), 864–878.

Yamagata, T., Behera, S., Doi, T., Luo, J.-J., Morioka, Y., & Tozuka, T. (2024). Chapter 5: Climate phenomena of the Indian Ocean. In C. C. Ummenhofer, & R. R. Hood (Eds.), *The Indian Ocean and its role in the global climate system* (pp. 103–119). Amsterdam: Elsevier. https://doi.org/10.1016/B978-0-12-822698-8.00009-3.

Yamanaka, Y., Saito, S., Shimozono, T., & Tajima, Y. (2019). A numerical study on nearshore behavior of Japan Sea tsunamis using Green's functions for Gaussian sources based on linear Boussinesq theory. *Coastal Engineering Journal, 61*, 187–198.

Yamanaka, Y., & Shimozono, T. (2022). Tsunami inundation characteristics along the Japan Sea coastline: Effect of dunes, breakwaters, and rivers. *Earth, Planets and Space, 74*, 19. https://doi.org/10.1186/s40623-022-01579-5.

Yang, J., Liu, Q., Xie, S. P., Liu, Z., & Wu, L. (2007). Impact of the Indian Ocean SST basin mode on the Asian summer monsoon. *Geophysical Research Letters, 34*(2).

Yokoyama, I. (1987). A scenario of the 1883 Krakatau tsunamis. *Journal of Volcanology and Geothermal Research, 34*, 123.

Yolsal, S., & Taymaz, T. (2010). Sensitivity analysis on relations between earthquake source rupture parameters and far-field tsunami waves: Case studies in the eastern Mediterranean region. *Turkish Journal of Earth Sciences*, *19*(3). https://doi.org/10.3906/yer-0902-8. Article 2.

Yuan, J., & Cao, J. (2013). North Indian Ocean tropical cyclone activities influenced by the Indian Ocean Dipole mode. *Science China Earth Sciences*, *56*(5), 855–865.

Yuan, J., Gao, Y., Feng, D., & Yang, Y. (2019). The zonal dipole pattern of tropical cyclone genesis in the Indian Ocean influenced by the tropical Indo-Pacific Ocean sea surface temperature anomalies. *Journal of Climate*, *32*(19), 6533–6549.

Yue, H., Lay, T., & Koper, K. D. (2012). En échelon and orthogonal fault ruptures of the 11 April 2012 great intraplate earthquakes. *Nature*. https://doi.org/10.1038/nature11492.

Zhang, C. (2005). Madden-Julian oscillation. *Reviews of Geophysics*, *43*(2).

Zhang, N., Feng, M., Hendon, H. H., Hobday, A. J., & Zinke, J. (2017). Opposite polarities of ENSO drive distinct patterns of coral bleaching potentials in the Southeast Indian Ocean. *Scientific Reports*, *7*, 2443.

Zhang, L., & Han, W. (2018). Impact of Ningaloo Niño on tropical Pacific and an inter-basin coupling mechanism. *Geophysical Research Letters*, *45*, 11300–11309.

Chapter 7

Impacts of the Indian Ocean on regional and global climate

Caroline C. Ummenhofer[a], Andréa S. Taschetto[b], Takeshi Izumo[c], and Jing-Jia Luo[d]

[a]Department of Physical Oceanography, Woods Hole Oceanographic Institution, Woods Hole, MA, United States, [b]Climate Change Research Centre and ARC Centre of Excellence for Climate Extremes, University of New South Wales, Sydney, NSW, Australia, [c]Institut de Recherche pour le Développement (IRD), EIO Laboratory, University of French Polynesia, Puna'auia, French Polynesia, [d]Institute of Climate and Application Research, Nanjing University of Information Science and Technology, Nanjing, China

1 Introduction

Upper-ocean conditions in the Indian Ocean exert considerable influence on regional hydroclimate across a range of timescales, from subseasonal to interannual, decadal, and beyond. The impact of Indian Ocean variability on regional and global weather and climate is increasingly recognized, and large strides have been made in recent decades in its understanding and prediction. Regional impacts can have considerable societal relevance, especially in the densely populated Indian Ocean rim with agrarian societies that are vulnerable to changes in weather and climate. Furthermore, the Indian Ocean exhibits strong warming trends and appears particularly vulnerable to anthropogenic climate change (Beal et al., 2020; see Roxy et al., 2024). With large portions of the basin being part of the Indo-Pacific Warm Pool, the Indian Ocean is particularly sensitive to warming trends affecting precipitation, as the climatological sea surface temperature (SST) are close to the ~28°C threshold temperature required for deep convection to occur (Deser et al., 2010; Zhang et al., 2018). The robust Indian Ocean warming can also further modulate or exacerbate how the Indian Ocean impacts regional and global weather patterns on seasonal to interannual timescales, as well as future projections beyond the ocean basin in the 21st century.

On subseasonal timescales, both northward- and eastward-propagating intraseasonal oscillations in the ocean-atmosphere system are active in the Indian Ocean (Schott et al., 2009). The Madden-Julian Oscillation (MJO) features eastward-propagating anomalies of large-scale atmospheric circulation and deep convection that interact with the warm underlying SST over the Indian Ocean and the maritime continent. The MJO greatly influences weather and climate events (Zhang, 2005, 2013; Zhang et al., 2020). For details on intraseasonal variability over the Indian Ocean, refer to DeMott et al. (2024).

On interannual timescales, variability in the Indian Ocean is characterized by several climate phenomena (Fig. 1), namely the Indian Ocean Dipole (IOD), the Indian Ocean Basin Mode, the Indian Ocean Subtropical Dipole, and the Ningaloo Niño/Niña. The Indian Ocean Basin Mode is not an independent mode of variability but a response to remote forcing from the Pacific related to the El Niño-Southern Oscillation (ENSO; Klein et al., 1999), although some authors argue that this pattern can manifest without the influence of ENSO (Kosaka et al., 2021). For further details, Yamagata et al. (2024) provide an overview of the characteristics, mechanisms, and evolution of these climate phenomena. Variability and change in marine heatwaves, including Ningaloo Niño/Niña events, are summarized in Feng et al. (2024).

Indian Ocean modes were initially considered mainly as an ocean-atmosphere response to the Pacific's ENSO variability. However, numerous studies have recently demonstrated the integrated nature and extensive two-way interactions among the tropical ocean basins (e.g., Cai et al., 2019). These interactions include significant feedback from the Indian Ocean on co-occurring ENSO events, as well as on the following year's ENSO development. A better understanding of such interbasin interactions aids seasonal prediction efforts. Indian Ocean-ENSO interactions will be described here, focusing predominantly on atmospheric teleconnections. For details of oceanic teleconnections via the Indonesian Throughflow (ITF), see Sprintall et al. (2024).

---

[*]This book has a companion website hosting complementary materials. Visit this URL to access it: https://www.elsevier.com/books-and-journals/book-companion/9780128226988.

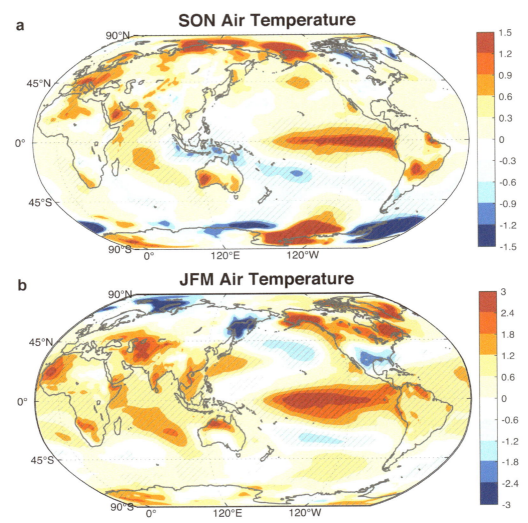

FIG. 1 Regression of (a) Dipole Mode Index and (b) Indian Ocean Basin Mode index on air temperature anomalies (°C/std) for the September–November (SON) and January–March (JFM) seasons, respectively. Indices and temperature fields detrended prior to the analysis. Hatched areas encompass statistically significant regressions at the 95% level based on a Student's *t*-test. Two-meter air temperature from NCEP/NCAR Reanalysis (Kalnay et al., 1996) and Dipole Mode Index and Indian Ocean Basin Mode index based on the December 1948 to November 2019 HadISST data (Rayner et al., 2003).

On longer timescales, the Indian Ocean has experienced considerable surface warming over the past century, with warming rates since the 1950s exceeding those of other tropical ocean basins (Han et al., 2014; Ihara et al., 2008), likely due to a combination of natural and anthropogenic causes (e.g., Alory et al., 2007; Du & Xie, 2008; Lee et al., 2015; Roxy et al., 2014, 2024; Tozuka et al., 2024; Zhang et al., 2018). The strong centennial warming trend in Indian Ocean SST has important implications for interbasin interactions with the Pacific and Atlantic mean state and variability, and the potential to offset the effects of anthropogenic warming in the Pacific Ocean via a thermostat mechanism (Zhang et al., 2019a). It has also been shown to influence regional climate in Indian Ocean rim countries and modulate extratropical teleconnections (Dhame et al., 2020).

The chapter focuses on the regional impacts that modes of climate variability exert in Indian Ocean rim countries and further afield (Section 2). Feedback from the Indian Ocean to the tropical Pacific, especially in relation to ENSO, is detailed in Section 3. Section 4 summarizes our current understanding of the effect of long-term Indian Ocean warming on regional and global climate.

2 Impacts on regional (hydro)climate

2.1 Impacts in Indian Ocean rim countries

The Indian Ocean is characterized by very active intraseasonal variability (see DeMott et al., 2024), with far-reaching implications for regional hydroclimate. For example, flooding events in Indonesia and Malaysia were implicated with unusual conditions in air-sea interactions over the eastern Indian Ocean associated with MJO-related variability (Aldrian, 2008; Tangang et al., 2008). Northward-propagating intraseasonal oscillations over the Indian Ocean also play a critical role in modulating active and break periods in the monsoon and thus impact boreal summer rainfall across southern Asia from India to the Philippines (Annamalai & Sperber, 2005; Schott et al., 2009). In particular, strong intraseasonal SST signals over the shallow thermocline of the Seychelles-Chagos Thermocline Ridge (SCTR; 5°S–10°S) appear to interact with northward-propagating mesoscale instabilities to affect Indian summer monsoon precipitation (Zhou et al., 2017) or with active MJO phases (Vialard et al., 2008), with possible interannual modulation (Izumo et al., 2010b). The SCTR has thus importance for seasonal forecasts of Indian summer monsoon rainfall: Persistence of warm SST anomalies there can delay the north-northwest migration of the Intertropical Convergence Zone (ITCZ) to the northern Indian Ocean and thus postpone the monsoon onset over southern India (Annamalai et al., 2005a), thereby decreasing Somalia-Oman upwelling, raising local SST and evaporation, and resulting in stronger rainfall along the Western Ghats during the monsoon mature phase (Izumo et al., 2008). Interactions between Indian Ocean climate modes also exist across timescales, as seen for the Bay of Bengal, an area of particularly active northward-propagating intraseasonal oscillations. SST warming in the northern Bay of Bengal led to the onset of local intraseasonal rainfall by 5 days, with this rainfall-SST relationship strengthened (through latent heating) in years with anomalous warm SST in the Bay of Bengal, resulting in stronger low-level moisture convergence, as occurs during the negative phase of the IOD (Jongaramrungruang et al., 2017).

Even before the Indian Ocean Basin Mode (Chambers et al., 1999) and the IOD (Saji et al., 1999; Webster et al., 1999) were more widely adopted in the community, tropical Indian Ocean variability was recognized as a driver of regional hydroclimate, such as for East Africa (Goddard & Graham, 1999; Hastenrath et al., 1993; Latif et al., 1999), South Africa (Goddard & Graham, 1999; Landman & Mason, 1999; Mason, 1995; Reason & Lutjeharms, 1998; Reason & Mulenga, 1999), India (Clark et al., 2000), and Australia (Ansell et al., 2000a, 2000b; Smith et al., 2000). SST in the Arabian Sea, central Indian Ocean, and the northwest shelf of Australia in the preceding boreal fall-winter were found to positively correlate with Indian monsoon strength the following summer offering predictive value (Clark et al., 2000). For southern African rainfall, in addition to local SST, tropical western Indian Ocean SST were thought to be important, likely through their role in modulating moisture convergence from the Indian Ocean (Mason, 1995; Reason & Lutjeharms, 1998; Reason & Mulenga, 1999).

The unusual conditions in the ocean-atmosphere system over the broader Indian Ocean region during the boreal fall and into the winter of 1997–98 garnered considerable attention (Saji et al., 1999; Webster et al., 1999), confirming the ocean-atmosphere coupling between zonal wind, atmospheric convection, and SST peaking in fall described by Reverdin et al. (1986). In addition to observed anomalies in SST and the subsurface thermocline across the equatorial Indian Ocean, changes in the seasonal winds led to shifts in rain-bearing convective systems that resulted in enhanced rainfall and widespread flooding across East Africa, while drought and wildfires ensued around Indonesia during 1997–98. In the 20 years since the formal description of this coupled ocean-atmosphere mode in the Indian Ocean, which is now widely known as the IOD (also sometimes called the Indian Ocean dipole zonal mode), the knowledge about its formation, behavior, evolution, trends, and impacts has grown exponentially (Kucharski et al., 2021; Saji, 2018; see Yamagata et al., 2024). Successful attempts have been made to predict the IOD (e.g., Luo et al., 2008), and seasonal IOD predictions have further increased in accuracy and lead time (Doi et al., 2019), including improved skill for multiyear IOD predictions (Feba et al., 2021). Furthermore, for affected countries such as Australia, the IOD is now regarded as an important climate mode for anticipating regional impacts on the seasonal forecasting timescale as is ENSO and operational seasonal forecasting centers around the world feature the state of the IOD alongside ENSO.

The interest is explained by the IOD's extensive impacts in surrounding countries (Kucharski et al., 2021; Saji, 2018; Saji & Yamagata, 2003). While the IOD starts to develop during boreal summer and already exerts influence on regional climate during this season (e.g., Ashok et al., 2003b, 2004, 2007; Kucharski et al., 2021; and references therein), its peak is reached during boreal fall and thus regional rainfall anomalies are shown for that season (Fig. 2). The impacts of the positive phase of IOD with anomalously warm SST in the western Indian Ocean and cool anomalies in the east include heavy rainfall and flooding in East Africa (e.g., Behera et al., 1999, 2005; Birkett et al., 1999; Black et al., 2003; Manatsa & Behera, 2013; Manatsa et al., 2012; Preethi et al., 2015; Ummenhofer et al., 2009b; Wainwright et al., 2021; Webster et al., 1999); droughts, wildfires, and streamflow reductions in Indonesia (e.g., Abram et al., 2003; D'Arrigo et al., 2011; D'Arrigo & Wilson, 2008; Hendon, 2003); droughts and bushfires in Australia (e.g., Abram et al., 2021; Ashok et al., 2003a; Cai et al., 2009a, 2009b, 2011; Marshall et al., 2021; Trewin, 2020; Ummenhofer et al., 2008, 2009a, 2009c, 2011a; Wang & Cai, 2020); a strengthening of the Asian monsoon (e.g., Ashok et al., 2001, 2004; Gadgil et al., 2004; Guan et al., 2003; Ummenhofer et al., 2011b) and enhanced typhoon activity and rainfall in the Northwest Pacific (Chowdary et al., 2011; Kosaka et al., 2021; Pradhan et al., 2011; and references therein). The mechanisms behind these remote teleconnections are further discussed in Section 2.2.

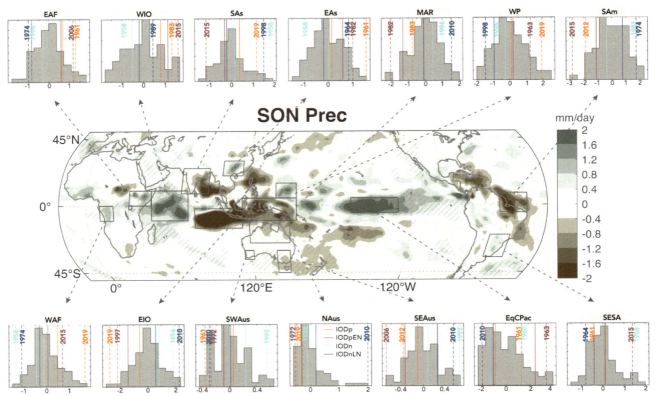

FIG. 2 Regression of Dipole Mode Index onto the September–November (SON) precipitation anomalies (mm day^{-1}/std) and probability density functions of September–November precipitation anomalies (mm day^{-1}) averaged over *gray boxes*. *Boxes* are selected according to larger regression values and existing literature on IOD precipitation effects. Hatched areas encompass statistically significant regressions at the 95% level based on a Student's t-test. *Vertical lines* in histograms indicate the 3-month mean September–November precipitation anomalies when maximum regression occurred during (pure) positive IOD *(orange line)*, positive IOD combined with El Niño *(red line)* (pure) negative IOD *(light blue line)*, and negative IOD combined with La Niña *(dark blue line)*. *Dashed lines* represent the largest 3-month mean September–November precipitation anomaly that occurred in those events. Precipitation data from NCEP/NCAR Reanalysis (Kalnay et al., 1996), and Dipole Mode Index based on the HadISST data (Rayner et al., 2003), all for December 1948 to November 2019. IOD and ENSO years based on the Australian Bureau of Meteorology (http://www.bom.gov.au/climate/iod/) and Ummenhofer et al. (2009a).

While the IOD features prominently during boreal summer and fall, the annual average Indian Ocean SST variability is actually dominated by uniform fluctuations of SST across the basin, commonly known as the Indian Ocean Basin Mode pattern (Fig. 3; Chambers et al., 1999), which peaks during boreal winter-spring. Although the Indian Ocean Basin Mode pattern is a forced response to ENSO (e.g., Ohba & Ueda, 2005, 2009b), nowadays its importance for modulating local and remote climate over land (e.g., Kosaka et al., 2021; and references therein) and for ENSO's life cycle (Cai et al., 2019) has been widely recognized. The influence of ENSO on the Indian Ocean Basin Mode is asymmetric, with stronger warming developing after El Niño events, and cooling of lower magnitude following La Niña (e.g., Hong et al., 2010; for details see Yamagata et al., 2024). Indian Ocean Basin Mode warming impacts regional climate and generates atmospheric teleconnections to the extratropics, while Indian Ocean cooling events are much less likely to significantly affect local and remote climate (e.g., Ohba & Watanabe, 2012; Okumura & Deser, 2010).

One of the challenges in understanding the contribution of Indian Ocean SST to regional climate is their partial co-variation with Pacific SST anomalies. Disentangling the contributions of individual ocean basins can be difficult but is possible in principle with a modeling approach. Using atmospheric general circulation model experiments, previous studies demonstrated that the Indian Ocean Basin Mode warming amplifies the effects of El Niño in Australia (Ashok et al., 2014; Taschetto et al., 2011). More specifically, the anomalous Indian Ocean warming after the El Niño peak induces anomalous ascending motion over the basin and consequent anomalous subsidence over the Australian tropics. This reinforces the Walker circulation changes initiated by El Niño and prolongs the dry conditions over Australia during austral fall. The Indian Ocean Basin Mode warming also affects the Arabian Peninsula, the South Asian monsoon, and the western Pacific climate (Annamalai et al., 2005a, 2005b; Dasari et al., 2021; Kosaka et al., 2021; Watanabe & Jin, 2002; Xie et al., 2009).

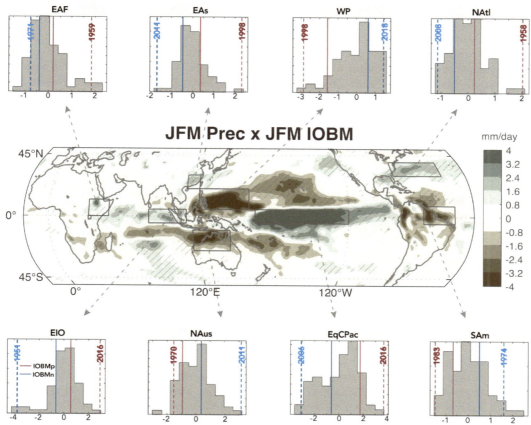

FIG. 3 Regression of Indian Ocean Basin Mode index onto the January–March (JFM) precipitation anomalies (mm day^{-1}/std) and probability density functions of January–March precipitation anomalies (mm day^{-1}) averaged over *gray boxes*. *Boxes* are selected according to larger regression values and existing literature on IOBM precipitation effects. Hatched areas encompass statistically significant regressions at the 95% level based on a Student's *t*-test. *Vertical lines* in histograms indicate the 3-month mean January–March precipitation anomalies when maximum regression occurred during positive *(red line)* and negative Indian Ocean Basin Mode *(blue line)*. *Dashed lines* represent the largest 3-month mean January–March precipitation anomaly that occurred in those events. Precipitation data from NCEP/NCAR Reanalysis (Kalnay et al., 1996), and Indian Ocean Basin Mode index based on the HadISST data (Rayner et al., 2003), all for December 1948 to November 2019 and calculated as the averaged SST anomalies over the tropical Indian Ocean basin (20°N–20°S). Indian Ocean Basin Mode years are selected when the normalized index exceeds one standard deviation.

The enhanced South Asian monsoon due to Indian Ocean Basin Mode warming likely results from the creation of a secondary heating source that generates significant circumglobal teleconnections over the Northern Hemisphere during boreal summer (Yang et al., 2009). While initially triggered by ENSO, the Indian Ocean Basin Mode thus plays an active role in affecting the Northern Hemisphere climate in the western Pacific through the so-called Indo-western Pacific Ocean capacitor effect (Kosaka et al., 2021). This capacitor effect is sustained by wind-evaporation-SST feedback in the western Pacific and Indian Ocean from northern spring to summer (Kosaka et al., 2021). Furthermore, local SST in the Arabian Sea and Bay of Bengal modulate Indian monsoon rainfall, likely through influences on the moisture flow to South Asia (e.g., Annamalai et al., 2005b; Ashok et al., 2001; Izumo et al., 2008; Vecchi et al., 2004). The Indian Ocean SST warming also affects climatic conditions in Africa (e.g., Bader & Latif, 2003; Giannini et al., 2003; Goddard & Graham, 1999; Latif et al., 1999; Rodríguez-Fonseca et al., 2015).

In the subtropical Indian Ocean, a characteristic dipolar SST pattern occurs that is characterized in its positive phase by warm anomalies in the southwestern and cool anomalies in the southeastern Indian Ocean (Behera & Yamagata, 2001; Hermes & Reason, 2005; Suzuki et al., 2004). A positive Indian Ocean subtropical dipole leads to enhanced low-level moisture advection onto land and anomalous wet conditions in southern Africa (Behera & Yamagata, 2001; Hoell et al., 2017; Reason, 2001, 2002). The Indian Ocean subtropical dipole also affects rainfall over southwestern Australia (England et al., 2006; Ummenhofer et al., 2008, 2009c), the Indian summer monsoon (Terray et al., 2003), and tropical cyclone trajectories in the southern Indian Ocean (Ash & Matyas, 2012; Ramsay et al., 2017).

Furthermore, the Indian Ocean was found to be particularly influential for regional hydroclimate on decadal timescales (e.g., Clark et al., 2003), as shown for prolonged droughts and pluvial periods in Australia and East Africa (Ummenhofer et al., 2009a, 2011a, 2018). A prolonged absence of negative IOD events was linked to the iconic multiyear droughts that southeast Australia experienced during the past 120 years (Ummenhofer et al., 2009a). Negative IOD events (as well as La Niñas) thus play an important role in breaking multiyear droughts in Australia (King et al., 2020). Changes in the tropical atmospheric circulation across the Indo-Pacific on multidecadal timescales (L'Heureux et al., 2013; Vecchi & Soden, 2007) affect the relationship between Indian Ocean SST and regional rainfall. When the Pacific Walker cell weakened and the Indian Ocean one strengthened post-1961, the East African short rains became more variable and wetter (Nicholson, 2015). Similarly, an epochal strengthening in the relationship between the IOD and East African short rain was observed post-1961 compared to previous decades, with further strengthening after 1997, explaining spatially coherent events across the region and frequent rainfall extremes (Jiang et al., 2021b; Manatsa & Behera, 2013).

2.2 Remote teleconnections

Variations in the Indian Ocean SST associated with both the IOD and Indian Ocean Basin Mode can perturb the troposphere and generate changes in circulation and precipitation in the remote tropics and extratropics. Warmer SST generate a local increase in moisture, and thus in surface moist static energy. This, in turn, favors deep convection, rainfall, and diabatic heating (Neelin & Held, 1987). Increased diabatic heating triggers equatorial and off-equatorial waves (Gill, 1980; Matsuno, 1966). The characteristics of the baroclinic Kelvin and Rossby waves can be modified by the tropospheric background flow. Background vertical shear plays a key role in converting energy from the heat-induced baroclinic flow anomalies into barotropic wind anomalies near the heating source in the tropics (Kasahara & da Silva Dias, 1986; Lee et al., 2009). This can excite prominent barotropic Rossby waves to the extratropics of both hemispheres.

In the Southern Hemisphere, the propagation of stationary Rossby waves from the Indian Ocean varies seasonally due to the mean background flow. In austral winter when the IOD develops, the climatological subtropical jet creates a zone of zero absolute vorticity across Australian longitudes that inhibits the propagation of the wave train from the tropics into the extratropics (McIntosh & Hendon, 2018). This zone is alleviated in spring when the subtropical jet weakens, allowing the propagation of stationary Rossby waves from the Indian Ocean to the southern extratropics. Thus, during austral spring when the IOD peaks, a stationary Rossby wave train emanates from the Indian Ocean sector across southern Australia and toward South America (Ashok et al., 2007; Cai et al., 2011, 2020; Chan et al., 2008; Saji et al., 2005). This teleconnection pattern affects the mean-state mid-latitude westerlies, thus altering baroclinicity and extratropical storm systems reaching southern regions of Australia (Ashok et al., 2003a; Risbey et al., 2009). This extratropical teleconnection to Australia resembles that from the Indian Ocean Basin Mode warming during austral late summer and fall (Taschetto et al., 2011). It extends across the Pacific mid-latitudes influencing Antarctic sea ice (Nuncio & Yuan, 2015) and reaches the South American subtropics. Over South America, the IOD teleconnection leads to a precipitation dipole (Sena & Magnusdottir, 2021). It should be noted that these relationships are sensitive to whether the IOD event occurs in conjunction with ENSO or not, accounting for the differences seen in Fig. 2.

The stationary Rossby wave train associated with increased diabatic heating during Indian Ocean Basin Mode warming events extends across the Pacific (Fig. 4) and can induce anomalous conditions in the mid-latitudes of both hemispheres akin to the Pacific North/South America patterns. For South America, the Rossby wave train produces an anomalous high-pressure center off the southeast coast of South America. The anomalous high inhibits rainfall over the region where the South Atlantic Convergence Zone forms, while enhancing circulation associated with the South American low-level jet to the south. Consequently, an anomalous dipole rainfall pattern is generated in association with the Rossby wave train pattern, with decreased precipitation over the South American tropics and enhanced rainfall in the La Plata region (Chan et al., 2008; Taschetto & Ambrizzi, 2012).

The impact of the Indian Ocean Basin Mode warming can also extend to Northern Hemisphere land areas (e.g., Yang et al., 2009). Variability in the SCTR can influence the Pacific North American region in boreal winter (Annamalai et al., 2007). The anomalous tropospheric geopotential height caused by the typical Indian Ocean Basin Mode warming during and after strong El Niño events also leads to higher pressure and reduced clouds over North Africa and South Asia, causing these surrounding land masses to cool (Herold & Santoso, 2018). Similarly, the strong 2019 positive IOD event contributed to an anomalously positive North Atlantic Oscillation (NAO) during the boreal winter of 2019–20 (Hardiman et al., 2020). Two teleconnection pathways exist to the North Atlantic, namely a tropospheric pathway via a Rossby wave train emanating from the Indian Ocean via the Pacific to the Atlantic, as well as a stratospheric teleconnection via the Aleutian region and the stratospheric polar vortex (Hardiman et al., 2020); both pathways are similar to the teleconnection patterns described for ENSO impacts on the North Atlantic sector (Taschetto et al., 2020). Furthermore, the IOD has been reported

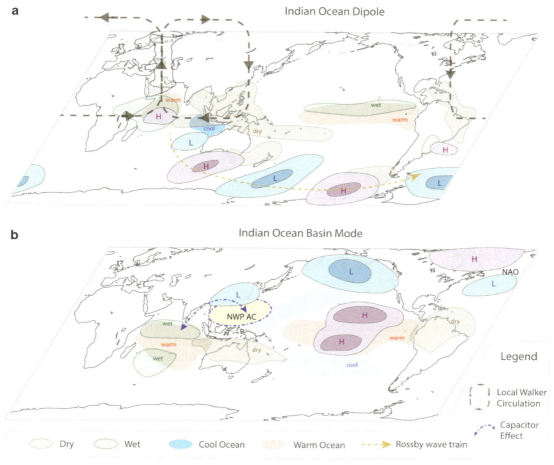

FIG. 4 (a, b) Teleconnection mechanisms to a positive IOD and Indian Ocean Basin Mode. Illustrated in the schematic are changes in local Walker circulation *(dashed brown cells)*, areas of above *(green patch)* and below *(brown patch)* precipitation, warm *(red patch)* and cool *(blue patch)* SST anomalies, Rossby wave train emanating from the Indian Ocean (high and low centers and *dashed yellow arrow*), and the capacitor effect onto the Northwest Pacific anticyclone *(purple dashed arrow and yellow circle)*.

to trigger Atlantic Niño events through an atmospheric teleconnection that leads to a weakening of the easterly trade winds in the tropical Atlantic (Zhang & Han, 2021).

Similarly to the IOD and Indian Ocean Basin Mode, MJO events can trigger Rossby wave trains from the Indian Ocean toward both North and South America, akin to the Pacific North/South America patterns on intraseasonal timescales (Carvalho et al., 2004; De Souza & Ambrizzi, 2006; Seo & Lee, 2017; Tseng et al., 2019), modulating winter temperatures in North America (e.g., Zheng et al., 2018). The MJO-related teleconnections have been associated with precipitation variability in the South Atlantic Convergence Zone and temperature anomalies in southeast South America (Alvarez et al., 2016; Barreiro et al., 2019). The impacts of the wave-train atmospheric circulation can go beyond rainfall effects: in the austral summer of 2013–14, a persistent blocking high was established in the southeast South Atlantic as part of a Rossby wave train induced by MJO convective activity originating in the Indian Ocean that resulted in intense drought conditions in southeast Brazil (Rodrigues et al., 2019); the blocking circulation inhibited convection, cloud, and precipitation over the South Atlantic Convergence Zone, while at the same time increasing shortwave radiation into the ocean and reducing evaporative cooling at the surface due to weaker winds. Intense dry conditions, heatwaves, and drought over land resulted, as well as a severe marine heatwave in the adjacent ocean (Rodrigues et al., 2019). What determines the severity of the impacts and location of the anomalous high center over southeast South America in response to the anomalous Indian Ocean forcing is not fully understood, but model experiments and Rossby wave theory indicate that the location, intensity, and zonal extent of the heating source in the tropics, in combination with the background atmospheric flow, are key players in this wave train pathway (Ambrizzi & Hoskins, 1997; Karoly, 1989; Lee et al., 2009).

3 Indian Ocean-ENSO interactions

3.1 Impacts of ENSO on the Indian Ocean

ENSO can impact the Indian Ocean either through an atmospheric teleconnection, essentially by modulating the Walker circulation, or via the so-called oceanic bridge through the ITF (see Sprintall et al., 2024). ENSO exerts its most extensive impacts on the Indian Ocean basin through atmospheric teleconnection. An El Niño event shifts the Walker circulation, increasing atmospheric deep convection in the central equatorial Pacific and resulting in anomalous subsidence over the maritime continent and tropical Indian Ocean. The opposite anomalies occur for La Niña events, but they tend to be typically weaker than El Niño events, both in terms of the observed SST amplitude and the magnitude of atmospheric teleconnections to the Indian Ocean. About two-thirds of the IOD variance in boreal fall is independent of ENSO ($r^2 \approx 0.3$) and a series of IOD events have occurred independently of ENSO (e.g., 1961, 1963, 2006–08, and 2019), as highlighted in recent reviews (Behera et al., 2021; Kucharski et al., 2021; Saji, 2018). However, the interconnectedness of the Indian and Pacific Ocean basins is well-recognized (cf. Cai et al., 2019; Kosaka et al., 2021). Note that this relation undergoes decadal variations, possibly just because of noise (e.g., the null hypothesis of Gershunov et al., 2001; Yun & Timmermann, 2018) and/or possibly also because the atmospheric bridge could be stronger when the El Niño event is strong enough (Ashok et al., 2003b). These ENSO-induced circulation changes can trigger or reinforce an emergent IOD event in the Indian Ocean during boreal summer and fall: anomalous easterlies along the equatorial Indian Ocean associated with the shift in the Walker circulation during El Niño favoring the onset of a positive IOD that tends to peak in boreal fall (Fig. 5) through regional positive feedbacks (Annamalai et al., 2003; Fischer et al., 2005; Gualdi et al., 2003; Murtugudde et al., 2000; Reverdin et al., 1986; Saji et al., 1999; Webster et al., 1999). Without ENSO's impact, the IOD would be of higher frequency with a strong biennial periodicity (Behera et al., 2006). The IOD temporarily counteracts the Indian Ocean Basin Mode development in the east, causing the Indian Ocean Basin Mode to peak later in winter-spring (Fig. 5b; Annamalai et al., 2005b; see Yamagata et al., 2024). During an El Niño event, the anomalous subsidence over the maritime continent and the tropical Indian Ocean reduces deep convection, cloud cover, and wind speed. This increases

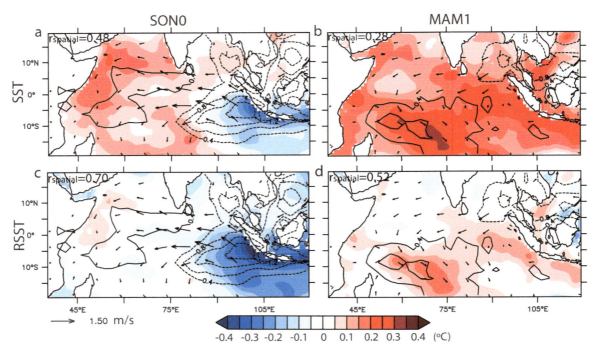

FIG. 5 Relative SST (RSST) anomalies explain ENSO remote influences on Indian Ocean precipitation better than SST anomalies. Seasonally stratified regression of September–November (SON0) averaged (a) SST anomalies and (c) relative SST anomalies (°C, color) on the following November to January (NDJ0) Niño3.4 index (normalized by its standard deviation, so that the regression slope illustrates typical amplitudes in physical units). Precipitation (contours; mm day^{-1}) and surface wind (vectors, when significant at the 90% level; at 10m; m s^{-1}) anomalies are overlaid. (b and d): same as a and c, but for March to May (MAM1) averaged anomalies regressed onto the preceding November to January (NDJ0) Niño3.4 index. The values of the uncentered spatial pattern correlations ($r_{spatial}$) between precipitation and SST patterns (top), or between precipitation and relative SST (bottom), are indicated on each panel, confirming that relative SST anomalies explain the precipitation anomalies better than SST anomalies. When relative SST is weighted by climatological precipitation, the match is even better (not shown; cf. Izumo et al., 2020 for details).

incoming solar radiation and weakens evaporative cooling, thus favoring warming across the basin as Indian Ocean Basin Mode (e.g., Klein et al., 1999; Lau & Nath, 2003; Ohba & Ueda, 2005, 2009b; Shinoda et al., 2004). The Indian Ocean Basin Mode can then last until the following summer, because of reduced evaporation (Du et al., 2009; Kawamura et al., 2001) and Somalia upwelling (Izumo et al., 2008), thus causing a delayed feedback onto the atmospheric circulation outside the Indian Ocean, as a capacitor effect (Kosaka et al., 2021; Xie et al., 2009).

Regarding ENSO-forced deep convection and precipitation in the Indian Ocean, it is well known that in the tropics SST strongly control local precipitation (Section 2) through gross moist stability (defined as the difference of moist static energy between the surface and the upper troposphere; Neelin & Held, 1987). However, large discrepancies between the spatial patterns of SST and precipitation occur (Fig. 5a and b), with reduced rainfall over positive SST anomalies. The upper troposphere actually warms during El Niño events, stabilizing the atmosphere. Relative SST (a concept originally developed to understand the effect of climate change on tropical cyclones; Vecchi & Soden, 2007; Vecchi et al., 2008), defined here as local SST minus its 20°N–20°S tropical mean, accounts for this upper tropospheric influence on gross moist stability. Relative SST hence explains ENSO-forced precipitation significantly better than absolute SST (Fig. 5c and d; see details in Izumo et al., 2020). This relation further improves when relative SST anomalies are weighted by the precipitation climatology to account for low- and high-precipitation regimes (not shown; another concept developed in parallel for the Pacific in Okumura (2019) is relevant and complementary here: The SST threshold for atmospheric convection is $\Delta T = \text{RSST} \times H(\text{RSST})$, where H is the Heaviside function.). In terms of relative SST, relative SST weighted by climatological precipitation, and precipitation, the IOD is actually almost a monopole in the eastern Indian Ocean, coupled with zonal wind stress anomalies along the equator and a subsurface zonal dipole in terms of thermocline depth (not shown) because of IOD's fully coupled nature (e.g., Saji, 2018). In contrast, the Indian Ocean Basin Mode is only a warming patch in the southwestern Indian Ocean, with a maximum over the SCTR region.

3.2 Impacts of the Indian Ocean on ENSO

3.2.1 Indian Ocean Basin Mode impact

The Indian Ocean Basin Mode impact on ENSO has been extensively studied, contrary to the IOD. The Indian Ocean Basin Mode forces anomalous easterlies in the equatorial western Pacific through a Matsuno-Gill response (i.e., atmospheric equatorial Kelvin waves east of the Indian Ocean; Annamalai et al., 2005b, 2010; Gill, 1980), thereby favoring the phase reversal from El Niño to La Niña (e.g., Kug & Kang, 2006; Kug et al., 2006; Ohba & Ueda, 2007; Santoso et al., 2012; Terray et al., 2016; Wieners et al., 2017; Yamanaka et al., 2009). Since El Niño events are usually stronger than La Niña cases, the Indian Ocean Basin Mode is asymmetrical, with larger warming than cooling events, and hence its impact on the ENSO phase reversal and duration also exhibits asymmetry (e.g., Ohba, 2013; Ohba et al., 2010; Ohba & Ueda, 2009a; Okumura & Deser, 2010; Okumura et al., 2011). However, as ENSO forces the Indian Ocean Basin Mode, this reciprocal effect should be seen as a systematic negative feedback (correlation of ENSO with Indian Ocean Basin Mode of about 0.8; e.g., Frauen & Dommenget, 2012; Jansen et al., 2009), rather than a proper external influence. Its added value in terms of ENSO predictability skill is therefore rather weak (Dayan et al., 2014, 2015; Frauen & Dommenget, 2012; Izumo et al., 2014, 2016) and asymmetric (Ohba & Watanabe, 2012).

3.2.2 IOD impact

3.2.2.1 IOD simultaneous impact

The IOD peak tends to coincide with the build-up of ENSO (Fig. 1), making it a challenge statistically to disentangle the impact of IOD onto ENSO. However, numerical sensitivity experiments have shown that, for instance, a negative IOD (Fig. 6a, vice versa for a positive IOD) forces easterly anomalies in the western equatorial Pacific during boreal summer-fall, thereby forcing initially upwelling oceanic equatorial Kelvin waves, favoring La Niña conditions in boreal fall-winter or possibly strengthening a co-occurring La Niña (Annamalai et al., 2010; Izumo et al., 2010a, 2016; Lim & Hendon, 2017; Luo et al., 2010; Wieners et al., 2018). The warm eastern IOD pole induces easterly anomalies in the western Pacific. As such, the warm eastern pole exceeds the opposing forcing by the cold western pole that results in westerly anomalies. Similarly, anomalous forcing from the cold eastern pole dominates over the warm western pole in the positive IOD case. The forcing from the eastern pole dominates over that from the western Indian Ocean for several reasons: (1) The eastern pole of the IOD has much larger relative SST anomalies than its western pole and thus forces larger deep convection diabatic heating and related wind anomalies (see also Fig. 5c; Izumo et al., 2020). (2) The maritime continent's deep convection is highly sensitive to SST anomalies in the eastern Indian Ocean since its mean state SST is close to the convective threshold (Annamalai et al., 2010). (3) Convective feedback can strengthen the influence of the eastern pole (Wieners et al., 2018).

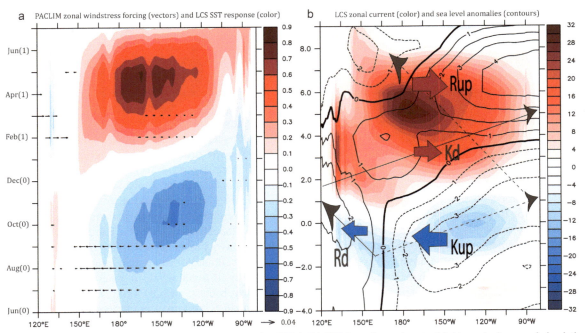

FIG. 6 Equatorial Pacific (2°N–2°S) uncoupled response to a negative IOD in PACLIM set of experiments. (a) Atmospheric general circulation model zonal wind stress τ_x (vectors shown only when significant at the 90% level, $Nm^{-2}/(2\sigma_{IOD})$) forcing the linear continuously stratified (LCS, McCreary, 1981) ocean model and linear continuously stratified SST response (*color shading*, °C/($2\sigma_{IOD}$)). (b) Linear continuously stratified zonal current (*color shading*, 0–100 m depth average, cm/s/2σIOD) and sea-level anomaly (*black contours*, cm) as a proxy of thermocline depth. *Arrows* schematically indicate equatorial Kelvin and Rossby wave propagations and reflections (for the first baroclinic mode). This analysis is inferred from the partial regression on negative IOD (without Indian Ocean Basin Mode influence; see Izumo et al., 2016, for details on the linear continuously stratified ocean model; for PACLIM atmospheric general circulation model experiments were forced by interannually varying SST from 1979 to 2010 except in the tropical Pacific where SST climatology is applied; see Dayan et al., 2015). Note that, in this former study, SST instead of relative SST was used to force the atmospheric general circulation model; hence this experimental setup likely underestimates the IOD influence. Also, in the real world's coupled context, the Bjerknes feedback would amplify the initial SST response, which would then peak in the following boreal winter (e.g., Izumo et al., 2010a). A 3-month running mean is applied to all variables to reduce intraseasonal noise.

Therefore, analysis of relative SST, rather than absolute SST, could help reconcile apparent diverging results from previous studies: While predictability and partial correlation/regression analyses suggest the dominance of the eastern pole (Dayan et al., 2014; Izumo et al., 2010a, 2016; Jourdain et al., 2016; Le et al., 2020; Luo et al., 2010; Wieners et al., 2018; Yoo et al., 2020), atmospheric general circulation model studies forced with typical IOD SST anomalies suggest a larger effect of the western pole relative to the eastern pole (e.g., Annamalai et al., 2005a; Dayan et al., 2015; Ohba & Ueda, 2007; Santoso et al., 2012; Wieners et al., 2018). When future numerical studies will use relative SST anomalies rather than absolute SST to force atmospheric general circulation model sensitivity experiments, it will likely become clearer that the influence of the eastern IOD pole on the western equatorial Pacific zonal wind is larger than that from the IOD western pole, consistent with predictability studies.

3.2.2.2 IOD delayed impact

Since the equatorial Pacific is bounded by western and eastern meridional boundaries, and because of equatorial oceanic Kelvin and Rossby wave propagation and reflection characteristics, any zonal wind forcing along the equator will force a wave adjustment causing a delayed negative feedback to the initial response, about 6 months later in the case of a western-central Pacific wind stress forcing (e.g., Battisti & Hirst, 1989; Cane & Moore, 1981; Izumo et al., 2016; Schopf & Suarez, 1988). In boreal fall, the western Pacific easterlies (for the negative IOD case) that had been induced by the simultaneous IOD rapidly collapsing in winter force a delayed negative advective-reflective feedback (Picaut et al., 1997) as follows: A significant zonal current reversal pushes the eastern edge of the warm pool eastward (Picaut et al., 1996), favoring positive SST about 6 months later in the western-central Pacific (Fig. 6) where the SST climatology is close to the ~28°C SST threshold for atmospheric deep convection to occur. This mechanism hence increases the likelihood of an El Niño onset in the following year (vice versa for a positive IOD favoring a La Niña). This IOD external influence can be combined with the tropical Pacific heat content recharge state to improve ENSO predictability at ~1-year lead (Dayan et al., 2014; Izumo et al., 2010a, 2014, 2016, 2018; Izumo & Colin, 2022; Jourdain et al., 2016). It can also favor successive El Niño events as observed in 2014–16 (Kim & Yu, 2020; Mayer & Balmaseda, 2021).

However, some aspects of these relationships are still an active area of research in the community: Questions about this delayed impact of the IOD (Jiang et al., 2021a; Stuecker, 2018) could possibly be aided by focusing on relative SST anomalies rather than absolute SST (Izumo et al., 2020). Furthermore, uncertainties remain about the influence of anthropogenic global warming on the Indian Ocean-ENSO relationships (Han & Wang, 2021; Marathe et al., 2021). Concerning the possible additional role of the oceanic pathway through the ITF, oceanic numerical experiments quantifying this effect show that it is on average much weaker than through the atmospheric bridge (Izumo et al., 2016). Yet seasonal forecasting experiments suggested that it could still play a significant role in a more specific case for a second successive El Niño event (Mayer & Balmaseda, 2021).

3.3 Impacts of Indian Ocean-ENSO interactions on climate predictability

The intrinsic interbasin coupling is important to climate variations in both the Pacific and Indian Oceans and their predictability at long-lead times (Izumo et al., 2010a; Luo et al., 2010, 2016). Improved forecasts of the Indian Ocean climate variations favor more skillful ENSO predictions and vice versa. In particular, successful prediction of the second cooling during double-peaked La Niñas often requires the correct forecast of the SST anomalies in the Indian Ocean as well as in the Atlantic (e.g., Luo et al., 2017; Zhang et al., 2019b). The interbasin coupling generally appears to have the largest impact on the predictability of the onsets of ENSO and IOD, since their subsequent growths are mainly determined by local unstable air-sea interactions (e.g., Luo et al., 2010). While a strong IOD and/or Indian Ocean Basin Mode signal may have a substantial influence on ENSO's evolution and hence its predictability, the interbasin influences are not symmetric. That is, ENSO has a much stronger influence on the predictability of the Indian Ocean climate, compared to the influence of the Indian Ocean signals on ENSO predictability. This is highlighted in sensitivity hindcast experiments suppressing Indo-Pacific coupling (Fig. 7): In the absence of Indo-Pacific interbasin coupling, observed monthly climatological SST during 1983–2006 were prescribed in the individual ocean basins between 25°S and 25°N. By doing this over the basin without air-sea coupling, the atmosphere there will respond to the climatological observed SST, rather than predicted SST, during the

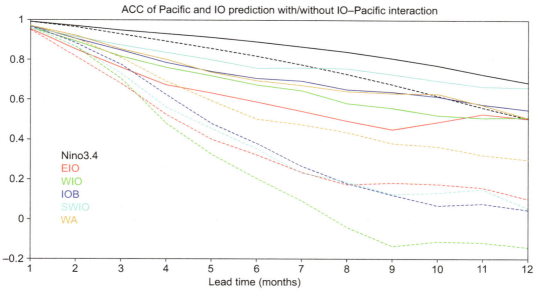

FIG. 7 Impacts of Indian Ocean-Pacific interactions on the tropical climate predictability based on SINTEX-F 9-member sensitivity hindcast experiment during 1982–2006 that prescribed monthly climatological SST in a tropical band spanning 25°S to 25°N. *Solid (dashed) lines* indicate the prediction skills of the SST anomalies in Nino3.4 *(black lines)*, eastern Indian Ocean (EIO; 90°E–110°E, 10°S–0, *red lines*), western Indian Ocean (WIO; 10°S–10°N, 50°E–70°E, *green lines*), Indian Ocean basin (IOB; 5°S–5°N, 40°E–110°E; *blue lines*), southwestern Indian Ocean (SWIO; 20°S–10°S, 50°E–70°E; *light blue lines*), and west of Australia (WA; 12°S–22°S, 95°E–115°E; *yellow lines*) in the presence (absence) of Indian Ocean-Pacific interactions. *(Reproduced with permission from Luo et al. (2016).)*

156 The Indian Ocean and its role in the global climate system

12-month forecast period. That is, oceanic feedback to the atmosphere there is suppressed, and thereby the interacting feedback between the Indian Ocean and the Pacific via the atmospheric bridge (particularly the Walker circulation) is suppressed. Readers are referred to Luo et al. (2010) for more details of the sensitivity prediction experiments.

4 The effect of long-term warming of the Indian Ocean on regional and global climate

4.1 Indian Ocean long-term warming

The Indian Ocean has experienced robust warming in the 20th century, warming at a faster rate since the 1950s than other tropical ocean basins (Fig. 8; Alory et al., 2007; Beal et al., 2020; Du & Xie, 2008; Ihara et al., 2008; Lee et al., 2015; Roxy

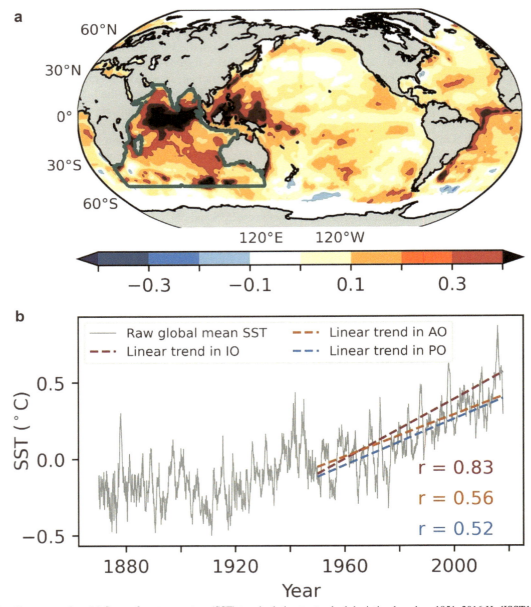

FIG. 8 Indian Ocean warming. (a) Sea surface temperature (SST) trend relative to standard deviation based on 1951–2016 HadISST1 data. (b) Time series of tropical SST anomalies averaged between 35°S and 35°N from 1861 to 2017. Linear trends of 1950–2017 SST anomalies averaged over the tropical Indian Ocean (IO, *red dashed line*), Atlantic Ocean (AO, *orange dashed line*), and Pacific Ocean (PO, *blue dashed line*). *(Modified from Dhame et al. (2020).)*

et al., 2014, 2020; Trenary & Han, 2008). While warming occurred throughout the basin, most likely due to anthropogenic greenhouse gases (Du & Xie, 2008; Zhang et al., 2018), a faster rate was recorded in the western basin compared to the east (Ihara et al., 2008; Roxy et al., 2014), and predominantly during boreal summer. This heterogeneous trend signal results in a zonal gradient in the warming that resembles the SST pattern seen during a positive IOD event. The heterogeneous warming trend has been suggested to contribute to projections of more frequent or intense positive IOD events in the future (e.g., Cai et al., 2013, 2014; Roxy et al., 2024; Yamagata et al., 2024), though care must be taken given enduring biases in model skill in representing Indian Ocean mean state and IOD variability (e.g., McKenna et al., 2020; Shinoda et al., 2024). Care is also warranted in this assessment, as internal variability can blur trends, especially in spatial patterns (e.g., Gopika et al., 2020), if the period is too short, the signal-to-noise ratio is too low, or given a short or incomplete observational record (Tozuka et al., 2024). Despite that, the long-term basin-wide warming of the Indian Ocean is a robust signal in the instrumental record and is most likely caused by a combination of anthropogenic and natural processes (Lee et al., 2015; Roxy et al., 2014; Ummenhofer et al., 2021; Zhang et al., 2018). Natural variability that has been suggested to play a role includes decadal variability in the Pacific and enhanced oceanic heat transport to the Indian Ocean (Lee et al., 2015; Nieves et al., 2015), changes in the subtropical gyre circulation (Alory et al., 2007), as well as asymmetric response to ENSO teleconnections (Roxy et al., 2014).

The Indian Ocean SST trend has important implications for interbasin interactions with the Pacific and Atlantic mean state and variability, and the potential to offset the effects of anthropogenic greenhouse gas warming in the Pacific Ocean via a thermostat mechanism (Zhang et al., 2019a). It was also shown to influence precipitation trends in Indian Ocean rim countries and modulate extratropical teleconnections to remote regions (Dhame et al., 2020). With Indian Ocean SST likely continuing to warm in the future, this has important implications for the local ocean, such as a projected slow-down of the Indian Ocean circulation (Stellema et al., 2019), likely increased marine heatwave occurrences with impacts on fisheries, coral reefs, and ecosystems (Frölicher et al., 2018; Oliver et al., 2018; for details see Feng et al., 2024), and implications for regional hydroclimate. The effects of long-term Indian Ocean warming on regional and global climate are discussed next.

4.2 Effects on Indian Ocean regional climate

Using atmospheric general circulation model experiments, Giannini et al. (2003) demonstrated that tropical Indian Ocean warming can shift the net convective activity from the African summer monsoon to the ocean and attributed this mechanism to the observed rainfall decline over the semiarid Sahel region and drought during the 1970s and 1980s (Bader & Latif, 2003; Giannini, 2010; Lu, 2009). East African long rains have declined since the 1980s (Lyon & Dewitt, 2012; Nicholson, 2017), while the short rains experienced an increase (Liebmann et al., 2014). East African short rains are closely related to SST conditions in the Indian and Pacific Oceans, and as such the increase in October–December precipitation has been attributed to local Indian Ocean SST warming (Liebmann et al., 2014). On the other hand, East African long rains are only weakly constrained by SST anomalies. Instead, the intensity of the local Indian Ocean Walker cell appears to play a more important role in modulating the low-level winds (Hastenrath et al., 2011; Mutai et al., 2012) and moisture transport to land. SST changes in the Indo-Pacific sector modulate the Somali Jet and moisture transport into East Africa (Nicholson, 2017). The increased Indian Ocean temperature and expansion of the Indo-Pacific warm pool have been suggested as possible factors for the weakening of the Somali Jet, increased precipitation over the warm ocean, and disruption of the moisture transport to East Africa (Funk et al., 2008; Williams et al., 2012).

Land-sea-atmosphere interaction over adjacent landmasses can also be affected directly by the rise of Indian Ocean temperatures. For example, the Indian Ocean warming reduces the land-sea thermal contrast and moisture flux convergence to South Asia and has been suggested to explain the observed 20th-century weakening of the Indian subcontinent monsoon (Mishra et al., 2012; Roxy et al., 2015), a trend that is, however, debated because of large observational uncertainties and large internal variability (e.g., Cohen & Boos, 2014). There is also a projected weakening of the boreal winter monsoon (Parvathi et al., 2017). In addition, local Indian Ocean SST is known to contribute to significant decadal variability in the Indian monsoon (Kucharski et al., 2006). The heterogeneous warming, with a larger warming trend in the western Indian Ocean and Arabian Sea than in the eastern part, has also been suggested to decrease (increase) rainfall during the Indian monsoon onset (withdrawal) (Chakravorty et al., 2016). In Australia, atmospheric general circulation model experiments simulated anomalous compensatory subsidence over the northern regions driven by ascending motion caused by the long-term Indian Ocean warming (Luffman et al., 2010). To what extent the Indian Ocean warming alone has influenced rainfall trends in Australia is unclear, as the Australian northern tropics have experienced an increasing trend in precipitation during the 20th century (e.g., Taschetto & England, 2009). In fact, the Indian Ocean warming also contributed to the expansion of the Indo-Pacific warm pool, which changed the residence time of the MJO since ~1980, such that it is shorter over the

Indian Ocean and longer over the western Pacific (Roxy et al., 2019). The changes in MJO propagation are consistent with trends in rainfall patterns not only locally but also in the global monsoon regions, including the rainfall intensification in the Philippines, northern Australia, and the Amazon, and the weakening trend over southern and eastern Asia.

4.3 Remote effects on global climate

The Indian Ocean's long-term warming also affects regions further afield. In numerical experiments forced with observed and idealized Indian Ocean SST warming, a positive trend of geopotential height was induced in the basin's troposphere and a warming of the tropics extended toward the Pacific and Atlantic Oceans (Dhame et al., 2020). To a certain degree, the mechanisms of remote teleconnections to long-term Indian Ocean warming are akin to those on the interannual scale (e.g., Herold & Santoso, 2018; Xie et al., 2009): A Matsuno-Gill-type pattern is generated in response to the SST forcing, giving rise to a warming trend to the east along the equator and off-equatorial centers to the west, thus resembling an equatorial Kelvin and off-equatorial Rossby waves, respectively (Dhame et al., 2020). The increased diabatic heating and interactions with the background flow can also trigger extratropical Rossby wave trains, thus affecting the circulation in both hemispheres.

4.3.1 Effect on the tropical Pacific

The Indian Ocean warming, together with the warming in the Atlantic (Cai et al., 2019; McGregor et al., 2014), has played an important role in modulating interdecadal Pacific variability. In the lower troposphere, the differential warming between the Indian and Pacific Oceans creates an anomalous zonal pressure gradient across the Indo-Pacific region, triggering easterlies and a trend toward stronger trade winds in the Pacific. This interbasin warming contrast, thus, acts to enhance the Pacific Walker circulation promoting a La Niña-like cooling trend (e.g., Dhame et al., 2020; Luo et al., 2012, 2018; Zhang et al., 2019a; Zhang & Karnauskas, 2017; Zhang et al., 2019b). This mechanism has been used to explain the cooler mean state of the Pacific Ocean in recent decades that opposes the expected weakening of the Walker circulation and the El Niño-like warming trend related to atmospheric energetic constraints (e.g., Vecchi & Soden, 2007). The increased trend in the Pacific easterlies leads to evaporative cooling, Ekman divergence, and upwelling in the central and eastern equatorial Pacific that counteracts the effect of anthropogenic warming, an ocean dynamical thermostat mechanism (Zhang et al., 2019a). Thus, by inducing a La Niña-like cooling, the Indian Ocean warming has contributed to the warming slowdown of global mean surface air temperature (or so-called hiatus) in recent decades (Arora et al., 2016; Kosaka & Xie, 2013; Yao et al., 2017). The interbasin warming contrast is underestimated in climate models, and it has been suggested to contribute to the simulated El Niño-like SST warming pattern in recent decades (Luo et al., 2018).

In addition to changes in the Pacific mean state, the long-term Indian Ocean warming can amplify the Pacific ocean-atmosphere positive feedback between the zonal SST gradient and trade winds. This is an emergent topic of research, since the historical record is relatively short to quantify changes in ENSO characteristics driven by Indian Ocean warming. However, other studies have shown that the strengthened Pacific Walker circulation in the early 21st century is dominated by multidecadal internal variability (Chung et al., 2019). Regardless of whether the observed early-21st century Walker circulation strengthening and Pacific cooling are related to multidecadal variability or a steady rise of Indian Ocean SST, the fact remains that the Indian Ocean warming can affect climate outside its own basin.

4.3.2 Effect on the tropical Atlantic

The effect of Indian Ocean warming on the Atlantic is less studied compared to its influence on the Pacific. However, studies suggest the Indian-Atlantic connection can occur via tropical and extratropical pathways. In the tropics, teleconnections can occur via changes in the Walker circulation similar to the Matsuno-Gill model. Atmospheric general circulation model experiments forced with Indian Ocean SST demonstrate that a warming Indian Ocean results in wind convergence and ascending motion over the basin with subsequent anomalous large-scale subsidence over the tropical Atlantic extending to South America (Taschetto & Ambrizzi, 2012). It is also possible that the teleconnections in the tropics occur via Kelvin waves propagating eastward through the Pacific and towards the tropical Atlantic and via increased tropospheric temperature driven by the Indian Ocean surface warming that act to enhance atmospheric stability in the tropics similarly to what occurs during strong El Niño events (e.g., Chiang & Sobel, 2002). The net effect from (thermo-)dynamical changes over the tropical Atlantic is a tendency to suppress convection and reduce precipitation, particularly along the ITCZ.

4.3.3 Effect on the extratropics

The Indian Ocean warming can project onto the Northern Hemisphere extratropical climate via a strengthening of the northern Hadley cell and enhancement of the mid-latitude atmospheric eddy feedback and wintertime storm track, leading to a positive phase of the NAO (Baker et al., 2019; Hoerling et al., 2004). These changes are consistent with the observed decline of precipitation in the Mediterranean region and wettening over northern continental Europe (Hoerling & Kumar, 2003). The Northern Hemisphere response to Indian Ocean warming is visible in atmospheric general circulation model experiments via a Rossby wave train pattern (Dhame et al., 2020). An anticyclonic anomaly is formed in the subtropical western Pacific that strengthens onshore winds in East Asia, leading to increasing summer rainfall. The downstream centers of geopotential height anomalies across the northern mid-latitudes drive a poleward shift of the jet in the Pacific sector and interfere constructively with the climatological stationary wave to favor a positive NAO phase (Fletcher & Kushner, 2011). The Indian Ocean warming is also thought to have contributed to Northern Annular Mode trends in the late 20th century, with interactions between the forced stationary wave anomalies and transient eddies being key for the changes seen in the annular mode structure (Li et al., 2010).

It has also been suggested that Indian Ocean warming can induce changes in precipitation over the North Atlantic that alter salinity and Atlantic Meridional Overturning Circulation (AMOC) strength (Ferster et al., 2021; Hu & Fedorov, 2019). Two different mechanisms were identified: (1) a fast response to tropical Indian Ocean warming acting on decadal timescales affected by surface cooling in the Labrador Sea due to a persistent positive NAO pattern, and (2) a slow response on multidecadal to centennial timescales driven by enhanced advection from the tropical Atlantic of positive salinity anomalies into the Nordic Seas that lead to a strengthening of AMOC due to increased Labrador Sea deep water formation; in contrast, Indian Ocean cooling leads to a slowdown of AMOC due to sea ice expansion in the Nordic Seas, making the response nonlinear between warming and cooling of the Indian Ocean (Ferster et al., 2021).

In the Southern Hemisphere, the Indian Ocean warming has been suggested to strengthen the mid-latitude SST gradient, thus enhancing baroclinicity and storm track activity over the Indian Ocean sector (Luffman et al., 2010). Diabatic heating in the Indian Ocean tropics has been shown to trigger an extratropical Rossby wave train in the southern mid-latitudes (see Section 2.2). Thus, it is not surprising that model experiments simulate increased Rossby wave train activity in response to a steady Indian Ocean warming trend. The wave train alternating geopotential height anomalies extends across Australia, strengthens the South Pacific subtropical high, weakens the Amundsen Sea Low, and can reach the subtropical latitudes of South America (e.g., Dhame et al., 2020; Drumond & Ambrizzi, 2008; Taschetto & Ambrizzi, 2012; Taschetto et al., 2011). Whether this pattern can effectively drive a long-term change in circulation and explain observed trends in precipitation over the regions affected by the wave train remains an open question.

5 Conclusions

The impact of Indian Ocean variability on regional and global weather and climate across a range of timescales is increasingly recognized. While disentangling the effects of Indian Ocean modes of climate variability from ENSO is challenging (Fig. 1), large strides have been made in recent decades in understanding and predicting the impacts of Indian Ocean variability on the region. The Indian Ocean exerts considerable influence and hence has societal relevance, especially in the densely populated Indian Ocean rim with agrarian societies that are vulnerable to changes in weather or climate. The interest is explained by the extensive impacts in surrounding countries of key Indian Ocean climate modes, such as the IOD (Fig. 2), Indian Ocean Basin Mode (Fig. 3), and Indian Ocean subtropical dipole: During the IOD's positive phase with anomalously cold (warm) SST in the eastern (western) Indian Ocean, this includes heavy rainfall and flooding in East Africa; droughts, wildfires, and streamflow reductions in Indonesia; Australian droughts and bushfires; and a strengthened Asian monsoon. The Indian Ocean Basin Mode pattern, as a basin-wide warming pattern in a forced response to ENSO, affects hydroclimate in East Africa and northern Australia, the South and East Asian monsoon, as well as further afield in South America. A characteristic dipolar SST pattern in the subtropical Indian Ocean affects rainfall in southern Africa and southwestern Australia, the Indian summer monsoon, and South Indian tropical cyclone trajectories. The Indian Ocean also appears to be particularly influential for regional hydroclimate on (multi)decadal timescales, such as for prolonged droughts and pluvial periods in Australia and East Africa.

Variations in Indian Ocean SST associated with key Indian Ocean climate modes can also perturb the troposphere and generate changes in circulation and precipitation in the remote tropics and extratropics, mostly transmitted via Rossby wave trains that teleconnect the tropical heating anomalies to the extratropics of both hemispheres (Fig. 4). Furthermore, the recurring patterns of Indian Ocean variability also actively affect other ocean basins, including the tropical Pacific and ENSO, through atmospheric teleconnections (Figs. 5 and 6). Understanding the feedback and interactions among the tropical ocean basins can aid seasonal prediction efforts (Fig. 7). Using relative SST is essential to understand the ENSO teleconnections and interactions between modes of tropical climate variability. Hence, considering relative SST rather than SST anomalies when designing atmospheric general circulation model sensitivity experiments to understand the Indian Ocean impacts on atmospheric deep convection, precipitation, and wind could help in disentangling some remaining debates.

On longer timescales, the Indian Ocean appears particularly vulnerable to anthropogenic climate change, with warming rates since the 1950s exceeding those of other tropical ocean basins (Fig. 8), likely due to a combination of natural and anthropogenic causes. The strong centennial warming trend in Indian Ocean SST has important implications for interbasin interactions with the Pacific and Atlantic mean state and variability, and the potential to offset the effects of anthropogenic greenhouse gas warming in the Pacific Ocean via a thermostat mechanism. It has also been shown to influence regional climate in Indian Ocean rim countries and modulate extratropical teleconnections to remote regions. To a certain degree, the mechanisms of remote teleconnections to long-term Indian Ocean warming are akin to those on the interannual scale: i.e., a Matsuno-Gill-type pattern is generated in response to the SST forcing, giving rise to a warming trend to the east along the equator and off-equatorial centers to the west, thus resembling an equatorial Kelvin and off-equatorial Rossby waves, respectively. The increased diabatic heating and interactions with the background flow can also trigger extratropical Rossby wave trains, thus affecting the circulation in both hemispheres.

6 Educational resources

Weblinks for the IOD are available at:

- http://www.bom.gov.au/climate/iod/
- https://www.jamstec.go.jp/aplinfo/sintexf/e/iod/about_iod.html
- https://www.climate.gov/news-features/blogs/enso/meet-enso%E2%80%99s-neighbor-indian-ocean-dipole

Educational movie on the IOD available at:

- https://www.youtube.com/watch?v=J6hOVatamYs

Links on the current state of Indian Ocean climate indices available at:

- https://stateoftheocean.osmc.noaa.gov/sur/ind/

Seasonal climate forecasts, including for Indian Ocean rim countries:

- https://iri.columbia.edu/our-expertise/climate/forecasts/seasonal-climate-forecasts/
- https://mol.tropmet.res.in/

Acknowledgments

Use of the following data sets is gratefully acknowledged: NCEP/NCAR reanalysis data provided by NOAA/OAR/ESRL PSD, Boulder, Colorado, the United States, through http://www.cdc.noaa.gov; Hadley Centre HadISST by the UK Met Office. We gratefully acknowledge comments by Tomoki Tozuka, MK Roxy, and two anonymous reviewers on an earlier version of the chapter. CCU acknowledges support from the National Science Foundation under AGS-2002083, the *Andrew W. Mellon Foundation Award for Innovative Research,* and the *James E. and Barbara V. Moltz Fellowship for Climate-Related Research.* AST is supported by the Australian Research Council under FT160100495 and CE110001028.

Author contributions

CCU conceived and led the chapter overall, with contributions in particular in Sections 1 and 2. AST wrote Section 4, contributed to Section 2, and drafted Figs. 1–4 and 8. TI wrote Sections 3.1 and 3.2 and drafted Figs. 5 and 6. JJL wrote Section 3.3 and drafted Fig. 7. All authors contributed to the discussion of content and overall chapter structure and provided feedback on the entire chapter.

References

Abram, N. J., Gagan, M. K., McCulloch, M. T., Chappell, J., & Hantoro, W. S. (2003). Coral reef death during the 1997 Indian Ocean Dipole linked to Indonesian wildfires. *Science, 301,* 952–955.

Abram, N. J., et al. (2021). Connections of climate change and variability to large and extreme forest fires in southeast Australia. *Communications Earth & Environment, 2.* https://doi.org/10.1038/s43247-020-00065-8.

Aldrian, E. (2008). Dominant factors of Jakarta's three largest floods. *Jurnal Hidrosfir Indones, 3,* 105–112.

Alory, G., Wijffels, S., & Meyers, G. (2007). Observed temperature trends in the Indian Ocean over 1960-1999 and associated mechanisms. *Geophysical Research Letters*, *34*, L02606. https://doi.org/10.1029/2006GL028044.

Alvarez, M. S., Vera, C. S., Kiladis, G. N., & Liebmann, B. (2016). Influence of the Madden-Julian Oscillation on precipitation and surface air temperature in South America. *Climate Dynamics*, *46*, 245–262.

Ambrizzi, T., & Hoskins, B. J. (1997). Stationary Rossby-wave propagation in a baroclinic atmosphere. *Quarterly Journal of the Royal Meteorological Society*, *123*, 919–928.

Annamalai, H., Kida, S., & Hafner, J. (2010). Potential impact of the tropical Indian Ocean-Indonesian Seas on El Niño characteristics. *Journal of Climate*, *23*, 3933–3952.

Annamalai, H., Liu, P., & Xie, S.-P. (2005a). Southwest Indian Ocean SST variability: Its local effect and remote influence on Asian monsoons. *Journal of Climate*, *18*, 4150–4167.

Annamalai, H., Murtugudde, R., Potemra, J., Xie, S. P., Liu, P., & Wang, B. (2003). Coupled dynamics over the Indian Ocean: Spring initiation of the zonal mode. *Deep Sea Research Part II: Topical Studies in Oceanography*, *50*, 2305–2330.

Annamalai, H., Okajima, H., & Watanabe, M. (2007). Possible impact of the Indian Ocean SST on the Northern Hemisphere circulation during El Niño. *Journal of Climate*, *20*, 3164–3189.

Annamalai, H., & Sperber, K. R. (2005). Regional heat sources and the active and break phases of boreal summer intraseasonal (30-50 day) variability. *Journal of Atmospheric Sciences*, *62*, 2726–2748.

Annamalai, H., Xie, S.-P., McCreary, J. P., & Murtugudde, R. (2005b). Impact of Indian Ocean sea surface temperature on developing El Niño. *Journal of Climate*, *18*, 302–319.

Ansell, T., Reason, C. J. C., & Meyers, G. (2000a). Variability in the tropical southeast Indian Ocean and links with southeast Australian winter rainfall. *Geophysical Research Letters*, *27*, 3977–3980.

Ansell, T. J., Reason, C. J. C., Smith, I. N., & Keay, K. (2000b). Evidence for decadal variability in southern Australian rainfall and relationships with regional pressure and sea surface temperature. *International Journal of Climatology*, *20*, 1113–1129.

Arora, A., Rao, S. A., Chattopadhyay, R., Goswami, T., George, G., & Sabeerali, C. T. (2016). Role of Indian Ocean SST variability on the recent global warming hiatus. *Global and Planetary Change*, *143*, 21–30.

Ash, K. D., & Matyas, C. J. (2012). The influences of ENSO and the subtropical Indian Ocean Dipole on tropical cyclone trajectories in the southwestern Indian Ocean. *International Journal of Climatology*, *32*, 41–56.

Ashok, K., et al. (2014). Decadal changes in the relationship between the Indian and Australian summer monsoons. *Climate Dynamics*, *42*, 1043–1052.

Ashok, K., Guan, Z., Saji, N. H., & Yamagata, T. (2004). Individual and combined influences of ENSO and the Indian Ocean Dipole on the Indian Summer Monsoon. *Journal of Climate*, *17*, 3141–3155.

Ashok, K., Guan, Z., & Yamagata, T. (2001). Impact of the Indian Ocean Dipole on the relationship between the Indian Monsoon rainfall and ENSO. *Geophysical Research Letters*, *28*, 4499–4502.

Ashok, K., Guan, Z. Y., & Yamagata, T. (2003a). Influence of the Indian Ocean Dipole on the Australian winter rainfall. *Geophysical Research Letters*, *30*. https://doi.org/10.1029/2003GL017926.

Ashok, K., Guan, Z., & Yamagata, T. (2003b). A look at the relationship between the ENSO and the Indian Ocean dipole. *Journal of the Meteorological Society of Japan. Ser. II*, *81*, 41–56.

Ashok, K., Nakamura, H., & Yamagata, T. (2007). Impacts of ENSO and Indian Ocean Dipole events on the Southern Hemisphere storm-track activity during austral winter. *Journal of Climate*, *20*, 3147–3163.

Bader, J., & Latif, M. (2003). The impact of decadal-scale Indian Ocean sea surface temperature anomalies on Sahelian rainfall and the North Atlantic Oscillation. *Geophysical Research Letters*, *30*. https://doi.org/10.1029/2003GL018426.

Baker, H. S., Woollings, T., Forest, C. E., & Allen, M. R. (2019). The linear sensitivity of the North Atlantic Oscillation and eddy-driven jet to SSTs. *Journal of Climate*. https://doi.org/10.1175/JCLI-D-19-0038.1.

Barreiro, M., Sitz, L., de Mello, S., Franco, R. F., Renom, M., & Farneti, R. (2019). Modelling the role of Atlantic air–sea interaction in the impact of Madden–Julian Oscillation on South American climate. *International Journal of Climatology*, *39*, 1104–1116.

Battisti, D. S., & Hirst, A. C. (1989). Interannual variability in the tropical atmosphere-ocean model: Influence of the basic state, ocean geometry and nonlinearity. *Journal of Atmospheric Science*, *45*, 1687–1712.

Beal, L. M., et al. (2020). A road map to IndOOS-2: Better observations of the rapidly warming Indian Ocean. *Bulletin of the American Meteorological Society*, *101*(11), E1891–E1913.

Behera, S. K., Doi, T., & Ratnam, J. V. (2021). Air-sea interactions in tropical Indian Ocean: The Indian Ocean Dipole. In *Tropical and extratropical air-sea interactions* (pp. 115–139). Elsevier.

Behera, S. K., Krishnan, R., & Yamagata, T. (1999). Unusual ocean-atmosphere conditions in the tropical Indian Ocean during 1994. *Geophysical Research Letters*, *26*, 3001–3004.

Behera, S. K., Luo, J.-J., Masson, S., Delecluse, P., Gualdi, S., Navarra, A., & Yamagata, T. (2005). Paramount impact of the Indian Ocean Dipole on the East African short rains: A CGCM study. *Journal of Climate*, *18*, 4514–4530.

Behera, S. K., Luo, J.-J., Masson, S., Rao, S. A., Sakuma, H., & Yamagata, T. (2006). A CGCM study on the interaction between IOD and ENSO. *Journal of Climate*, *19*, 1688–1705.

Behera, S. K., & Yamagata, T. (2001). Subtropical SST dipole events in the southern Indian Ocean. *Geophysical Research Letters*, *28*, 327–330.

Birkett, C., Murtugudde, R., & Allan, R. (1999). Indian Ocean climate event brings floods to East Africa's lakes and the Sudd Marsh. *Geophysical Research Letters*, *26*, 1031–1034.

Black, E., Slingo, J., & Sperber, K. R. (2003). An observational study of the relationship between excessively strong short rains in coastal East Africa and Indian Ocean SST. *Monthly Weather Review*, *131*, 74–94.

Cai, W., Cowan, T., & Raupach, M. (2009a). Positive Indian Ocean Dipole events precondition Southeast Australia bushfires. *Geophysical Research Letters*, *36*, L19710. https://doi.org/10.1029/2009GL039902.

Cai, W., Cowan, T., & Sullivan, A. (2009b). Recent unprecedented skewness towards positive Indian Ocean Dipole occurrences and its impact on Australian rainfall. *Geophysical Research Letters*, *36*. https://doi.org/10.1029/2009GL037604.

Cai, W., van Rensch, P., Cowan, T., & Hendon, H. H. (2011). Teleconnection pathways of ENSO and the IOD and the mechanisms for impacts on Australian rainfall. *Journal of Climate*, *24*, 3910–3923.

Cai, W., et al. (2013). Projected response of the Indian Ocean Dipole to greenhouse warming. *Nature Geoscience*, *6*, 999–1007.

Cai, W., et al. (2014). Increased frequency of extreme Indian Ocean Dipole events due to greenhouse warming. *Nature*, *510*, 254–258.

Cai, W., et al. (2019). Pan-tropical climate interactions. *Science*, *363*, eaav4236. https://doi.org/10.1126/science.aav4236.

Cai, W., et al. (2020). Climate impacts of the El Niño–Southern Oscillation on South America. *Nature Reviews Earth & Environment*, *1*, 215–231.

Cane, M. A., & Moore, D. W. (1981). A note on low-frequency equatorial basin modes. *Journal of Physical Oceanography*, *11*, 1578–1584.

Carvalho, L. M. V., Jones, C., & Liebmann, B. (2004). The South Atlantic convergence zone: Intensity, form, persistence, and relationships with intraseasonal to interannual activity and extreme rainfall. *Journal of Climate*, *17*, 88–108.

Chakravorty, S., Gnanaseelan, C., & Pillai, P. A. (2016). Combined influence of remote and local SST forcing on Indian Summer Monsoon Rainfall variability. *Climate Dynamics*, *47*, 2817–2831.

Chambers, D. P., Tapley, B. D., & Stewart, R. H. (1999). Anomalous warming in the Indian Ocean coincident with El Niño. *Journal of Geophysical Research*, *104*, 3035–3047.

Chan, S. C., Behera, S. K., & Yamagata, T. (2008). Indian Ocean Dipole influence on South American rainfall. *Geophysical Research Letters*, *35*, L14S12. https://doi.org/10.1029/2008GL034204.

Chiang, J. C. H., & Sobel, A. H. (2002). Tropical tropospheric temperature variations caused by ENSO and their influence on the remote tropical climate. *Journal of Climate*, *15*, 2616–2631.

Chowdary, J. S., et al. (2011). Predictability of Northwest Pacific climate during summer and the role of the tropical Indian Ocean. *Climate Dynamics*, *36*, 607–621.

Chung, E.-S., Timmermann, A., Soden, B. J., Ha, K.-J., Shi, L., & John, V. O. (2019). Reconciling opposing Walker circulation trends in observations and model projections. *Nature Climate Change*. https://doi.org/10.1038/s41558-019-0446-4.

Clark, C. O., Cole, J. E., & Webster, P. J. (2000). Indian Ocean SST and Indian summer rainfall: Predictive relationships and their decadal variability. *Journal of Climate*, *13*, 2503–2519.

Clark, C. O., Webster, P. J., & Cole, J. E. (2003). Interdecadal variability of the relationship between the Indian Ocean zonal mode and East African coastal rainfall anomalies. *Journal of Climate*, *16*, 548–554.

Cohen, N. Y., & Boos, W. R. (2014). Has the number of Indian summer monsoon depressions decreased over the last 30 years? *Geophysical Research Letters*, *41*, 7846–7853.

D'Arrigo, R., Abram, N., Ummenhofer, C., Palmer, J., & Mudelsee, M. (2011). Reconstructed streamflow for Citarum River, Java, Indonesia: Linkages to tropical climate dynamics. *Climate Dynamics*, *36*, 451–462.

D'Arrigo, R., & Wilson, R. (2008). El Nino and Indian Ocean influences on Indonesian drought: Implications for forecasting rainfall and crop productivity. *International Journal of Climatology*, *28*, 611–616.

Dasari, H. P., et al. (2021). Long-term changes in the Arabian Peninsula rainfall and their relationship with the ENSO signals in the tropical Indo-Pacific. *Climate Dynamics*. https://doi.org/10.1007/s00382-021-06062-7.

Dayan, H., Izumo, T., Vialard, J., Lengaigne, M., & Masson, S. (2015). Do regions outside the tropical Pacific influence ENSO through atmospheric teleconnections? *Climate Dynamics*, *45*, 583–601.

Dayan, H., Vialard, J., Izumo, T., & Lengaigne, M. (2014). Does sea surface temperature outside the tropical Pacific contribute to enhanced ENSO predictability? *Climate Dynamics*, *43*, 1311–1325.

De Souza, E. B., & Ambrizzi, T. (2006). Modulation of the intraseasonal rainfall over tropical Brazil by the Madden-Julian Oscillation. *International Journal of Climatology*, *26*, 1759–1776.

DeMott, C. A., Ruppert, J. H., Jr., & Rydbeck, A. (2024). Chapter 4: Intraseasonal variability in the Indian Ocean region. In C. C. Ummenhofer, & R. R. Hood (Eds.), *The Indian Ocean and its role in the global climate system* (pp. 79–101). Amsterdam: Elsevier. https://doi.org/10.1016/B978-0-12-822698-8.00006-8.

Deser, C., Phillips, A. S., & Alexander, M. A. (2010). Twentieth century tropical sea surface temperature trends revisited. *Geophysical Research Letters*, *37*. https://doi.org/10.1029/2010GL043321.

Dhame, S., Taschetto, A. S., Santoso, A., & Meissner, K. J. (2020). Indian Ocean warming modulates global atmospheric circulation trends. *Climate Dynamics*, *55*, 2053–2073.

Doi, T., Behera, S. K., & Yamagata, T. (2019). Merits of a 108-member ensemble system in ENSO and IOD predictions. *Journal of Climate*, *32*, 957–972.

Drumond, A. R. M., & Ambrizzi, T. (2008). The role of the South Indian and Pacific Oceans in South American monsoon variability. *Theoretical and Applied Climatology*, *94*, 125–137.

Du, Y., & Xie, S.-P. (2008). Role of atmospheric adjustments in the tropical Indian Ocean warming during the 20th century in climate models. *Geophysical Research Letters*, *35*, L08712.

Du, Y., Xie, S.-P., Huang, G., & Hu, K. (2009). Role of air–sea interaction in the long persistence of El Niño-induced North Indian Ocean Warming. *Journal of Climate*, *22*, 2023–2038.

England, M. H., Ummenhofer, C. C., & Santoso, A. (2006). Interannual rainfall extremes over Southwest Western Australia linked to Indian Ocean climate variability. *Journal of Climate*, *19*, 1948–1969.

Feba, F., Ashok, K., Collins, M., & Shetye, S. (2021). Emerging skill in multi-year prediction of the Indian Ocean Dipole. *Frontiers in Climate, 3*. https://doi.org/10.3389/fclim.2021.736759.

Feng, M., Lengaigne, M., Manneela, S., Gupta, A. S., & Vialard, J. (2024). Chapter 6: Extreme events in the Indian Ocean: Marine heatwaves, cyclones, and tsunamis. In C. C. Ummenhofer, & R. R. Hood (Eds.), *The Indian Ocean and its role in the global climate system* (pp. 121–144). Amsterdam: Elsevier. https://doi.org/10.1016/B978-0-12-822698-8.00011-1.

Ferster, B. S., et al. (2021). Sensitivity of the Atlantic meridional overturning circulation and climate to tropical Indian Ocean warming. *Climate Dynamics, 57*, 2433–2451.

Fischer, A. S., Terray, P., Guilyardi, E., Gualdi, S., & Delecluse, P. (2005). Two independent triggers for the Indian Ocean Dipole/Zonal Mode in a coupled GCM. *Journal of Climate, 18*, 3428–3449.

Fletcher, C. G., & Kushner, P. J. (2011). The role of linear interference in the annular mode response to tropical SST forcing. *Journal of Climate, 24*, 778–794.

Frauen, C., & Dommenget, D. (2012). Influences of the tropical Indian and Atlantic Oceans on the predictability of ENSO. *Geophysical Research Letters, 39*, L02706. https://doi.org/10.1029/2011GL050520.

Frölicher, T. L., Fischer, E. M., & Gruber, N. (2018). Marine heatwaves under global warming. *Nature, 560*, 360–364.

Funk, C., Dettinger, M. D., Michaelsen, J. C., Verdin, J. P., Brown, M. E., Barlow, M., & Hoell, A. (2008). Warming of the Indian Ocean threatens eastern and southern African food security but could be mitigated by agricultural development. *Proceedings of the National Academy of Sciences, 105*, 11081–11086.

Gadgil, S., Vinayachandran, P. N., Francis, P. A., & Gadgil, S. (2004). Extremes of the Indian summer monsoon rainfall, ENSO and equatorial Indian Ocean oscillation. *Geophysical Research Letters, 31*, L12213.

Gershunov, A., Schneider, N., & Barnett, T. (2001). Low-frequency modulation of the ENSO–Indian monsoon rainfall relationship: Signal or noise? *Journal of Climate, 14*, 2486–2492.

Giannini, A. (2010). Mechanisms of climate change in the semi-arid African Sahel: The local view. *Journal of Climate, 23*, 743–756.

Giannini, A., Saravanan, R., & Chang, P. (2003). Oceanic forcing of Sahel rainfall on interannual to interdecadal time scales. *Science, 302*, 1027–1030.

Gill, A. E. (1980). Some simple solutions for heat-induced tropical circulation. *Quarterly Journal of the Royal Meteorological Society, 106*, 447–462.

Goddard, L., & Graham, N. E. (1999). Importance of the Indian Ocean for simulating rainfall anomalies over eastern and southern Africa. *Journal of Geophysical Research, 104*(D16), 19099–19116.

Gopika, S., et al. (2020). Aliasing of the Indian Ocean externally-forced warming spatial pattern by internal climate variability. *Climate Dynamics, 54*, 1093–1111.

Gualdi, S., Guilyardi, E., Navarra, A., Masina, S., & Delecluse, P. (2003). The interannual variability in the tropical Indian Ocean as simulated by a CGCM. *Climate Dynamics, 20*, 567–582.

Guan, Z., Ashok, K., & Yamagata, T. (2003). Summertime response of the tropical atmosphere to the Indian Ocean sea surface temperature anomalies. *Journal of the Meteorological Society of Japan, 81*, 533–561.

Han, W., Vialard, J., McPhaden, M. J., Lee, T., Masumoto, T., Feng, M., & de Ruijter, W. P. M. (2014). Indian Ocean decadal variability: A review. *Bulletin of the American Meteorological Society, 95*, 1679–1703.

Han, X., & Wang, C. (2021). Weakened feedback of the Indian Ocean on El Niño since the early 1990s. *Climate Dynamics*. https://doi.org/10.1007/s00382-021-05745-55.

Hardiman, S. C., et al. (2020). Predictability of European winter 2019/20: Indian Ocean dipole impacts on the NAO. *Atmospheric Science Letters*. https://doi.org/10.1002/asl.1005.

Hastenrath, S., Nicklis, A., & Greischar, L. (1993). Atmospheric-hydrospheric mechanisms of climate anomalies in the western equatorial Indian Ocean. *Journal of Geophysical Research, 98*(C11), 20219–20235.

Hastenrath, S., Polzin, D., & Mutai, C. (2011). Circulation mechanisms of Kenya rainfall anomalies. *Journal of Climate, 24*, 404–412.

Hendon, H. H. (2003). Indonesian rainfall variability: Impacts of ENSO and local air-sea interaction. *Journal of Climate, 16*, 1775–1790.

Hermes, J. C., & Reason, C. J. C. (2005). Ocean model diagnosis of interannual coevolving SST variability in the South Indian and South Atlantic Oceans. *Journal of Climate*, 2864–2882.

Herold, N., & Santoso, A. (2018). Indian Ocean warming during peak El Niño cools surrounding land masses. *Climate Dynamics, 51*, 2097–2112.

Hoell, A., Gaughan, A. E., Shukla, S., & Magadzire, T. (2017). The hydrologic effects of synchronous El Niño-Southern Oscillation and subtropical Indian Ocean Dipole events over Southern Africa. *Journal of Hydrometeorology, 18*, 2407–2424.

Hoerling, M. P., Hurrell, J. W., Xu, T., Bates, G. T., & Phillips, A. S. (2004). Twentieth century North Atlantic climate change. Part II: Understanding the effect of Indian Ocean warming. *Climate Dynamics, 23*, 391–405.

Hoerling, M., & Kumar, A. (2003). The perfect ocean for drought. *Science, 299*, 691–694.

Hong, C., Li, T., & Chen, Y. (2010). Asymmetry of the Indian Ocean basinwide SST anomalies: Roles of ENSO and IOD. *Journal of Climate, 23*, 3563–3576.

Hu, S., & Fedorov, A. V. (2019). Indian Ocean warming can strengthen the Atlantic Meridional Overturning Circulation. *Nature Climate Change, 9*, 747–751.

Ihara, C., Kushnir, Y., & Cane, M. A. (2008). Warming trend of the Indian Ocean SST and Indian Ocean Dipole from 1880 to 2004. *Journal of Climate, 21*, 2035–2046.

Izumo, T., & Colin, M. (2022). Improving and harmonizing El Niño recharge indices. *Geophysical Research Letters*. https://doi.org/10.1029/2022GL101003.

Izumo, T., de Boyer Montegut, C., Luo, J.-J., Behera, S. K., Masson, S., & Yamagata, T. (2008). The role of the western Arabian Sea upwelling in Indian monsoon rainfall variability. *Journal of Climate, 21*, 5603–5623.

Izumo, T., Khodri, M., Lengaigne, M., & Suresh, I. (2018). A subsurface Indian Ocean dipole response to tropical volcanic eruptions. *Geophysical Research Letters, 45*, 9150–9159.

Izumo, T., Lengaigne, M., Vialard, J., Luo, J. J., Yamagata, T., & Madec, G. (2014). Influence of Indian Ocean Dipole and Pacific recharge on following year's El Niño: Interdecadal robustness. *Climate Dynamics, 42*, 291–310.

Izumo, T., Vialard, J., Dayan, H., Lengaigne, M., & Suresh, I. (2016). A simple estimation of equatorial Pacific response from windstress to untangle Indian Ocean Dipole and Basin influences on El Niño. *Climate Dynamics, 46*, 2247–2268.

Izumo, T., Vialard, J., Lengaigne, M., & Suresh, I. (2020). Relevance of relative sea surface temperature for tropical rainfall interannual variability. *Geophysical Research Letters, 47*. https://doi.org/10.1029/2019GL086182.

Izumo, T., et al. (2010a). Influence of the state of the Indian Ocean Dipole on the following year's El Niño. *Nature Geoscience, 3*, 168.

Izumo, T., et al. (2010b). Low and high frequency Madden-Julian oscillations in austral summer: Interannual variations. *Climate Dynamics, 35*, 669–683.

Jansen, M. F., Dommenget, D., & Keenlyside, N. (2009). Tropical atmosphere–ocean interactions in a conceptual framework. *Journal of Climate, 22*, 550–567.

Jiang, F., Zhang, W., Jin, F.-F., Stuecker, M. F., & Allan, R. (2021a). El Niño pacing orchestrates inter-basin Pacific-Indian Ocean interannual connections. *Geophysical Research Letters, 48*, e2021GL095242. https://doi.org/10.1029/2021GL095242.

Jiang, Y., Zhou, L., Roundy, P. E., Hua, W., & Raghavendra, A. (2021b). Increasing influence of Indian Ocean Dipole on precipitation over Central Equatorial Africa. *Geophysical Research Letters, 48*, e2020GL092370. https://doi.org/10.1029/2020GL092370.

Jongaramrungruang, S., Seo, H., & Ummenhofer, C. C. (2017). Intraseasonal rainfall variability in the Indian Ocean during the summer monsoon: Coupling with the ocean and modulation by the Indian Ocean Dipole. *Atmospheric Science Letters, 18*, 88–95.

Jourdain, N. C., Lengaigne, M., Vialard, J., Izumo, T., & Gupta, A. S. (2016). Further insights on the influence of the Indian Ocean dipole on the following year's ENSO from observations and CMIP5 models. *Journal of Climate, 29*, 637–658.

Kalnay, E., et al. (1996). The NCEP/NCAR 40-year reanalysis project. *Bulletin of the American Meteorological Society, 77*, 437–472.

Karoly, D. J. (1989). Southern Hemisphere circulation features associated with El Niño-Southern Oscillation events. *Journal of Climate, 2*, 1239–1252.

Kasahara, A., & da Silva Dias, P. L. (1986). Response of planetary waves to stationary tropical heating in a global atmosphere with meridional and vertical shear. *Journal of Atmospheric Sciences, 43*, 1893–1912.

Kawamura, R., Matsuura, T., & Iizuka, S. (2001). Role of equatorially asymmetric sea surface temperature anomalies in the Indian Ocean in the Asian summer monsoon and El Niño-Southern Oscillation coupling. *Journal of Geophysical Research, 106*, 4681–4693.

Kim, J.-W., & Yu, J.-Y. (2020). Understanding reintensified multiyear El Niño events. *Geophysical Research Letters, 47*, e2020GL087644. https://doi.org/10.1029/2020GL087644.

King, A. D., et al. (2020). The role of climate variability in Australian drought. *Nature Climate Change, 10*, 177–179.

Klein, S. A., Soden, B. J., & Lau, N. C. (1999). Remote sea surface temperature variations during ENSO: Evidence for a tropical atmospheric bridge. *Journal of Climate, 12*, 917–932.

Kosaka, Y., Takaya, Y., & Kamae, Y. (2021). The Indo-western Pacific Ocean capacitor effect. In *Tropical and extratropical air-sea interactions*. https://doi.org/10.1016/B978-0-12-818156-0.00012-5.

Kosaka, Y., & Xie, S.-P. (2013). Recent global-warming hiatus tied to equatorial Pacific surface cooling. *Nature, 501*, 403–407.

Kucharski, F., Biastoch, A., Ashok, K., & Yuan, D. (2021). Indian Ocean variability and interactions. In C. Mechoso (Ed.), *Interacting climates of ocean basins: Observations, mechanisms, predictability, and impacts* (pp. 153–185). Cambridge University Press.

Kucharski, F., Molteni, F., & Yoo, J. H. (2006). SST forcing of decadal Indian Monsoon rainfall variability. *Geophysical Research Letters, 33*. https://doi.org/10.1029/2005GL025371.

Kug, J.-S., & Kang, I.-S. (2006). Interactive feedback between the Indian Ocean and ENSO. *Journal of Climate, 19*, 1784–1801.

Kug, J.-S., Li, T., An, S.-I., Kang, I.-S., Luo, J.-J., Masson, S., & Yamagata, T. (2006). Role of the ENSO-Indian Ocean coupling on ENSO variability in a coupled GCM. *Geophysical Research Letters, 33*, L09710. https://doi.org/10.1029/2005GL024916.

L'Heureux, M. L., Lee, S., & Lyon, B. (2013). Recent multidecadal strengthening of the Walker circulation across the tropical Pacific. *Nature Climate Change, 3*, 571–576.

Landman, W. A., & Mason, S. J. (1999). Changes in the association between Indian Ocean sea-surface temperatures and summer rainfall over South Africa and Namibia. *International Journal of Climatology, 19*, 1477–1492.

Latif, M., Dommenget, D., Dima, M., & Grötzner, A. (1999). The role of Indian Ocean sea surface temperature in forcing East African rainfall anomalies during December-January 1997/98. *Journal of Climate, 12*, 3497–3504.

Lau, N. C., & Nath, M. J. (2003). Atmosphere-ocean variations in the Indo-Pacific sector during ENSO episodes. *Journal of Climate, 16*, 3–20.

Le, T., et al. (2020). Causal effects of Indian Ocean Dipole on El Niño–Southern Oscillation during 1950-2014 based on high-resolution models and reanalysis data. *Environmental Research Letters, 15*, 1040b6.

Lee, S.-K., Park, W., Baringer, M. O., Gordon, A. L., Huber, B., & Liu, Y. (2015). Pacific origin of the abrupt increase in Indian Ocean heat content during the warming hiatus. *Nature Geoscience*. https://doi.org/10.1038/ngeo2438.

Lee, S.-K., Wang, C., & Mapes, B. E. (2009). A simple atmospheric model of the local and teleconnection responses to tropical heating anomalies. *Journal of Climate, 22*, 272–284.

Li, S., Perlwitz, J., Hoerling, M. P., & Chen, X. (2010). Opposite annular responses of the Northern and Southern Hemispheres to Indian Ocean warming. *Journal of Climate, 23*, 3720–3738.

Liebmann, B., Hoerling, M. P., Funk, C., Bladé, I., Dole, R. M., Allured, D., et al. (2014). Understanding recent eastern Horn of Africa rainfall variability and change. *Journal of Climate, 27*, 8630–8645.

Lim, E.-P., & Hendon, H. H. (2017). Causes and predictability of the negative Indian Ocean Dipole and its impact on La Niña during 2016. *Scientific Reports, 7*. https://doi.org/10.1038/s41598-017-12674-z.

Lu, J. (2009). The dynamics of the Indian Ocean sea surface temperature forcing of Sahel drought. *Climate Dynamics, 33*, 445–460.

Luffman, J. J., Taschetto, A. S., & England, M. H. (2010). Global and regional climate response to late twentieth-century warming over the Indian Ocean. *Journal of Climate, 23*, 1660–1674.

Luo, J.-J., Behera, S., Masumoto, Y., Sakuma, H., & Yamagata, T. (2008). Successful prediction of the consecutive IOD in 2006 and 2007. *Geophysical Research Letters, 35*, L14S02. https://doi.org/10.1029/2007GL032793.

Luo, J.-J., Lee, J.-Y., Yuan, C., Sasaki, W., Masson, S., Behera, S., Masumoto, Y., & Yamagata, T. (2016). Current status of intraseasonal-seasonal-to-interannual prediction of the Indo-Pacific climate. In T. Yamagata, & S. Behera (Eds.), *World scientific series on Asia-Pacific weather and climate: Vol. 7. Indo-Pacific climate variability and predictability* (pp. 63–107). (chapter 3).

Luo, J.-J., Liu, G., Hendon, H., Alves, O., & Yamagata, T. (2017). Inter-basin sources for two-year predictability of the multi-year La Niña event in 2010-2012. *Scientific Reports, 7*, 2276. https://doi.org/10.1038/s41598-017-01479-9.

Luo, J. J., Sasaki, W., & Masumoto, Y. (2012). Indian Ocean warming modulates Pacific climate change. *Proceedings of the National Academy of Sciences, 109*, 18701–18706.

Luo, J.-J., Wang, G., & Dommenget, D. (2018). May common model biases reduce CMIP5's ability to simulate the recent Pacific La Niña-like cooling? *Climate Dynamics, 50*, 1335–1351.

Luo, J.-J., Zhang, R., Behera, S., Masumoto, Y., Jin, F.-F., Lukas, R., & Yamagata, T. (2010). Interaction between El Niño and extreme Indian Ocean Dipole. *Journal of Climate, 23*, 726–742.

Lyon, B., & Dewitt, D. G. (2012). A recent and abrupt decline in the East African long rains. *Geophysical Research Letters, 39*. https://doi.org/10.1029/2011GL050337.

Manatsa, D., & Behera, S. K. (2013). On the epochal strengthening in the relationship between rainfall of East Africa and IOD. *Journal of Climate, 26*, 5655–5673.

Manatsa, D., Chipindu, B., & Behera, S. K. (2012). Shifts in IOD and their impacts on association with East African rainfall. *Theoretical and Applied Climatology, 110*, 115–128.

Marathe, S., Terray, P., & Karumuri, A. (2021). Tropical Indian Ocean and ENSO relationships in a changed climate. *Climate Dynamics*. https://doi.org/10.1007/s00382-021-05641-y.

Marshall, A. G., Gregory, P. A., de Burgh-Day, C. O., et al. (2021). Subseasonal drivers of extreme fire weather in Australia and its prediction in ACCESS-S1 during spring and summer. *Climate Dynamics*. https://doi.org/10.1007/s00382-021-05920-8.

Mason, S. J. (1995). Sea-surface temperature—South African rainfall associations, 1910–1989. *International Journal of Climatology, 15*, 119–132.

Matsuno, T. (1966). Quasi-geostrophic motions in the equatorial area. *Journal of the Meteorological Society of Japan, 44*, 25–43.

Mayer, M., & Balmaseda, M. A. (2021). Indian Ocean impact on ENSO evolution 2014–2016 in a set of seasonal forecasting experiments. *Climate Dynamics*. https://doi.org/10.1007/s00382-020-05607-6.

McCreary, J. P. (1981). A linear stratified ocean model of the equatorial undercurrent. *Philosophical Transactions of the Royal Society of London A, 298*, 603–635.

McGregor, S., Timmermann, A., Stuecker, M. F., England, M. H., Merrifield, M., Jin, F. F., & Chikamoto, Y. (2014). Recent Walker circulation strengthening and Pacific cooling amplified by Atlantic warming. *Nature Climate Change, 4*, 888–892.

McIntosh, P. C., & Hendon, H. H. (2018). Understanding Rossby wave trains forced by the Indian Ocean Dipole. *Climate Dynamics, 50*, 2783–2798.

McKenna, S., Santoso, A., Sen Gupta, A., Taschetto, A. S., & Cai, W. (2020). Indian Ocean Dipole in CMIP5 and CMIP6: Characteristics, biases, and links to ENSO. *Scientific Reports, 10*. https://doi.org/10.1038/s41598-020-68268-9.

Mishra, V., Smoliak, B. V., Lettenmaier, D. P., & Wallace, J. M. (2012). A prominent pattern of year-to-year variability in Indian Summer Monsoon Rainfall. *Proceedings of the National Academy of Sciences, 109*, 7213–7217.

Murtugudde, R., McCreary, J. P., & Busalacchi, A. J. (2000). Oceanic processes associated with anomalous events in the Indian Ocean with relevance to 1997-1998. *Journal of Geophysical Research, 105*, 3295–3306.

Mutai, C., Polzin, D., & Hastenrath, S. (2012). Diagnosing Kenya rainfall in boreal autumn: Further exploration. *Journal of Climate, 25*, 4323–4329.

Neelin, J. D., & Held, I. M. (1987). Modeling tropical convergence based on the moist static energy budget. *Monthly Weather Review, 115*, 3–12.

Nicholson, S. E. (2015). Long-term variability of the East African "short rains" and its links to large-scale factors. *International Journal of Climatology*. https://doi.org/10.1002/joc.4259.

Nicholson, S. E. (2017). Climate and climatic variability of rainfall over eastern Africa. *Reviews of Geophysics, 55*, 590–635.

Nieves, V., Willis, J. K., & Patzert, W. C. (2015). Recent hiatus caused by decadal shift in Indo-Pacific heating. *Science, 349*, 532–535.

Nuncio, M., & Yuan, X. (2015). The influence of the Indian Ocean Dipole on Antarctic sea ice. *Journal of Climate, 28*, 2682–2690.

Ohba, M. (2013). Important factors for long-term change in ENSO transitivity. *International Journal of Climatology, 33*, 1495–1509.

Ohba, M., Nohara, D., & Ueda, H. (2010). Simulation of asymmetric ENSO transition in WCRP CMIP3 multimodel experiments. *Journal of Climate, 23*, 6051–6067.

Ohba, M., & Ueda, H. (2005). Basin-wide warming in the equatorial Indian Ocean associated with El Niño. *Scientific Online Letters on the Atmosphere*. https://doi.org/10.2151/sola.

Ohba, M., & Ueda, H. (2007). An impact of SST anomalies in the Indian Ocean in acceleration of the El Niño to La Nina transition. *Journal of the Meteorological Society of Japan, 85*, 335–348.

Ohba, M., & Ueda, H. (2009a). Role of nonlinear atmospheric response to SST on the asymmetric transition process of ENSO. *Journal of Climate, 22*, 177–192.

Ohba, M., & Ueda, H. (2009b). Seasonally different response of the Indian Ocean to the remote forcing of El Niño: Linking the dynamics and thermodynamics. *Scientific Online Letters on the Atmosphere, 5*, 176–179.

Ohba, M., & Watanabe, M. (2012). Role of the Indo-Pacific interbasin coupling in predicting asymmetric ENSO transition and duration. *Journal of Climate, 25*, 3321–3335.

Okumura, Y. M. (2019). ENSO diversity from an atmospheric perspective. *Current Climate Change Reports, 5*, 245–257.

Okumura, Y. M., & Deser, C. (2010). Asymmetry in the duration of El Niño and La Niña. *Journal of Climate, 23*, 5826–5843.

Okumura, Y. M., Ohba, M., Deser, C., & Ueda, H. (2011). A proposed mechanism for the asymmetric duration of El Niño and La Niña. *Journal of Climate, 24*, 3822–3829.

Oliver, E. C. J., et al. (2018). Longer and more frequent marine heatwaves over the past century. *Nature Communications, 9*. https://doi.org/10.1038/s41467-018-03732-9.

Parvathi, V., Suresh, I., Lengaigne, M., Izumo, T., & Vialard, J. (2017). Robust projected weakening of winter monsoon winds over the Arabian Sea under climate change. *Geophysical Research Letters, 44*, 9833–9843.

Picaut, J., Ioualalen, M., Menkes, C., Delcroix, T., & McPhaden, M. J. (1996). Mechanism of the zonal displacements of the Pacific warm pool: Implications for ENSO. *Science, 274*, 1486–1489.

Picaut, J., Masia, F., & du Penhoat, Y. (1997). An advective-reflective conceptual model for the oscillatory nature of the ENSO. *Science, 277*, 663–666.

Pradhan, P. K., Ashok, K., Preethi, B., Krishnan, R., & Sahai, A. K. (2011). Modoki, IOD and Western North Pacific typhoons: Possible implications for extreme events. *Journal of Geophysical Research, 116*, D18108. https://doi.org/10.1029/2011JD015666.

Preethi, B., Sabin, T. P., Adedoyin, J. A., & Ashok, K. (2015). Impacts of the ENSO Modoki and other tropical Indo-Pacific climate-drivers on African rainfall. *Scientific Reports, 5*. https://doi.org/10.1038/srep16653.

Ramsay, H. A., Richman, M. B., & Leslie, L. M. (2017). The modulating influence of Indian Ocean sea surface temperatures on Australian region seasonal tropical cyclone counts. *Journal of Climate, 30*, 4843–4856.

Rayner, N. A., et al. (2003). Global analyses of sea surface temperature, sea ice, and night marine air temperature since the late nineteenth century. *Journal of Geophysical Research, 108*. https://doi.org/10.1029/2002JD002670.

Reason, C. J. C. (2001). Subtropical Indian Ocean SST dipole events and southern African rainfall. *Geophysical Research Letters, 28*, 2225–2227.

Reason, C. J. C. (2002). Sensitivity of the southern African circulation to dipole sea-surface temperature patterns in the south Indian Ocean. *International Journal of Climatology, 22*, 377–393.

Reason, C. J. C., & Lutjeharms, J. R. E. (1998). Variability of the South Indian Ocean and implications for southern African rainfall. *South African Journal of Science, 94*, 115–123.

Reason, C. J. C., & Mulenga, H. (1999). Relationships between South African rainfall and SST anomalies in the Southwest Indian Ocean. *International Journal of Climatology, 19*, 1651–1673.

Reverdin, G., Cadet, D., & Gutzler, D. (1986). Interannual displacements of convection and surface circulation over the equatorial Indian Ocean. *Quarterly Journal of the Royal Meteorological Society, 112*, 43–46.

Risbey, J. S., Pook, M. J., McIntosh, P. C., Ummenhofer, C. C., & Meyers, G. (2009). Variability of synoptic features associated with cool season rainfall in southeastern Australia. *International Journal of Climatology, 29*, 1595–1613.

Rodrigues, R. R., Taschetto, A. S., Sen Gupta, A., & Foltz, G. R. (2019). Common cause for severe droughts in South America and marine heatwaves in the South Atlantic. *Nature Geoscience, 12*, 620–626.

Rodríguez-Fonseca, B., et al. (2015). Variability and predictability of West African droughts: A review on the role of sea surface temperature anomalies. *Journal of Climate, 28*, 4034–4060.

Roxy, M. K., Ritika, K., Terray, P., & Masson, S. (2014). The curious case of Indian Ocean warming. *Journal of Climate, 27*, 8501–8509.

Roxy, M. K., Ritika, K., Terray, P., Murtugudde, R., Ashok, K., & Goswami, B. N. (2015). Drying of Indian subcontinent by rapid Indian Ocean warming and a weakening land-sea thermal gradient. *Nature Communications, 6*. https://doi.org/10.1038/ncomms8423.

Roxy, M. K., Saranya, J. S., Modi, A., Anusree, A., Cai, W., Resplandy, L., ... Frölicher, T. L. (2024). Chapter 20: Future projections for the tropical Indian Ocean. In C. C. Ummenhofer, & R. R. Hood (Eds.), *The Indian Ocean and its role in the global climate system* (pp. 469–482). Amsterdam: Elsevier. https://doi.org/10.1016/B978-0-12-822698-8.00004-4.

Roxy, M. K., et al. (2019). Twofold expansion of the Indo-Pacific warm pool warps the MJO life cycle. *Nature, 575*, 647–651.

Roxy, M. K., et al. (2020). Indian Ocean warming. In *Assessment of climate change over the Indian region* (pp. 191–206). Singapore: Springer.

Saji, N. H. (2018). The Indian Ocean Dipole. In *Oxford research encyclopedia of climate science*. https://doi.org/10.1093/acrefore/9780190228620.013.619.

Saji, N., Ambrizzi, T., & Ferraz, S. T. (2005). Indian Ocean Dipole Mode events and austral surface air temperature anomalies. *Dynamics of Atmospheres and Oceans, 39*, 87–101.

Saji, N. H., Goswami, B. N., Vinayachandran, P. N., & Yamagata, T. (1999). A dipole mode in the tropical Indian Ocean. *Nature, 401*, 360–363.

Saji, N. H., & Yamagata, T. (2003). Structure of SST and surface wind variability during Indian Ocean dipole mode events: COADS observations. *Journal of Climate, 16*, 2735–2751.

Santoso, A., England, M. H., & Cai, W. (2012). Impact of Indo-Pacific feedback interactions on ENSO dynamics diagnosed using ensemble climate simulations. *Journal of Climate, 25*, 7743–7763.

Schopf, P. S., & Suarez, M. J. (1988). Vacillations in a coupled ocean-atmosphere model. *Journal of Atmospheric Sciences, 45*, 549–566.

Schott, F. A., Xie, S.-P., & McCreary, J. P. (2009). Indian Ocean circulation and climate variability. *Reviews of Geophysics, 47*, RG1002. https://doi.org/10.1029/2007RG000245.

Sena, A. C. T., & Magnusdottir, G. (2021). Influence of the Indian Ocean Dipole on the large-scale circulation in South America. *Journal of Climate, 34*, 6057–6068.

Seo, K.-H., & Lee, H.-J. (2017). Mechanisms for a PNA-like teleconnection pattern in response to the MJO. *Journal of the Atmospheric Sciences, 74*, 1767–1781.

Shinoda, T., Alexander, M. A., & Hendon, H. H. (2004). Remote response of the Indian Ocean to interannual SST variations in the tropical Pacific. *Journal of Climate, 17*, 362–372.

Shinoda, T., Jenson, T. G., Lachkar, Z., Masumoto, Y., & Seo, H. (2024). Chapter 18: Modeling the Indian Ocean. In C. C. Ummenhofer, & R. R. Hood (Eds.), *The Indian Ocean and its role in the global climate system* (pp. 421–443). Amsterdam: Elsevier. https://doi.org/10.1016/B978-0-12-822698-8.00001-9.

Smith, I. N., McIntosh, P., Ansell, T. J., Reason, C. J. C., & McInnes, K. (2000). Southwest western Australian winter rainfall and its association with Indian Ocean climate variability. *International Journal of Climatology, 20*, 1913–1930.

Sprintall, J., Biastoch, A., Gruenburg, L. K., & Phillips, H. E. (2024). Chapter 9: Oceanic basin connections. In C. C. Ummenhofer, & R. R. Hood (Eds.), *The Indian Ocean and its role in the global climate system* (pp. 205–227). Amsterdam: Elsevier. https://doi.org/10.1016/B978-0-12-822698-8.00003-2.

Stellema, A., Sen Gupta, A., & Taschetto, A. S. (2019). Projected slow down of the South Indian ocean circulation. *Scientific Reports, 9*, 17705. https://doi.org/10.1038/s41598-019-54092-3.

Stuecker, M. F., et al. (2018). Revisiting ENSO/Indian Ocean Dipole phase relationships. *Geophysical Research Letters, 44*, 2481–2492.

Suzuki, R., Behera, S. K., Iizuka, S., & Yamagata, T. (2004). Indian Ocean subtropical dipole simulated using a coupled general circulation model. *Journal of Geophysical Research, 109*. https://doi.org/10.1029/2003JC001974.

Tangang, F. T., et al. (2008). On the roles of the northeast cold surge, the Borneo vortex, the Madden-Julian Oscillation, and the Indian Ocean dipole during the extreme 2006/2007 flood in southern peninsular Malaysia. *Geophysical Research Letters, 35*, L14S07. https://doi.org/10.1029/2008GL033429.

Taschetto, A. S., & Ambrizzi, T. (2012). Can Indian Ocean SST anomalies influence South American rainfall? *Climate Dynamics, 38*, 1615–1628.

Taschetto, A. S., & England, M. H. (2009). El Niño Modoki impacts on Australian rainfall. *Journal of Climate, 22*, 3167–3174.

Taschetto, A. S., Sen Gupta, A., Hendon, H. H., Ummenhofer, C. C., & England, M. H. (2011). The contribution of Indian Ocean sea surface temperature anomalies on Australian summer rainfall during El Niño events. *Journal of Climate, 24*, 3734–3747.

Taschetto, A. S., Ummenhofer, C. C., Stuecker, M., Dommenget, D., Ashok, K., Rodrigues, R., & Yeh, S.-W. (2020). El Niño-Southern Oscillation atmospheric teleconnections. In M. J. McPhaden, A. Santoso, & W. Cai (Eds.), *El Niño Southern Oscillation in a changing climate*. https://doi.org/10.1002/9781119548164.ch14.

Terray, P., et al. (2016). Impacts of Indian and Atlantic oceans on ENSO in a comprehensive modeling framework. *Climate Dynamics, 46*, 2507–2533.

Terray, P., Delecluse, P., Labattu, S., & Terray, L. (2003). Sea surface temperature associations with the late Indian summer monsoon. *Climate Dynamics, 21*, 593–618.

Tozuka, T., Dong, L., Han, W., Lengainge, M., & Zhang, L. (2024). Chapter 10: Decadal variability of the Indian Ocean and its predictability. In C. C. Ummenhofer, & R. R. Hood (Eds.), *The Indian Ocean and its role in the global climate system* (pp. 229–244). Amsterdam: Elsevier. https://doi.org/10.1016/B978-0-12-822698-8.00014-7.

Trenary, L. L., & Han, W. (2008). Causes of decadal subsurface cooling in the tropical Indian Ocean during 1961-2000. *Geophysical Research Letters, 35*. https://doi.org/10.1029/2008GL034687.

Trewin, B. (2020). Seasonal climate summary for the Southern Hemisphere (spring 2018): Positive Indian Ocean Dipole and Australia's driest September on record. *Journal of Southern Hemisphere Earth Systems Science*. https://doi.org/10.1071/ES20007.

Tseng, K.-C., Maloney, E., & Barnes, E. (2019). The consistency of MJO teleconnection patterns: An explanation using linear Rossby Wave theory. *Journal of Climate, 32*, 531–548.

Ummenhofer, C. C., England, M. H., McIntosh, P. C., Meyers, G. A., Pook, M. J., Risbey, J. S., ... Taschetto, A. S. (2009a). What causes Southeast Australia's worst droughts? *Geophysical Research Letters, 36*, L04706. https://doi.org/10.1029/2008GL036801.

Ummenhofer, C. C., Kulüke, M., & Tierney, J. E. (2018). Extremes in East African hydroclimate and links to Indo-Pacific variability on interannual to decadal timescales. *Climate Dynamics, 50*, 2971–2991.

Ummenhofer, C. C., Murty, S. A., Sprintall, J., Lee, T., & Abram, N. J. (2021). Heat and freshwater changes in the Indian Ocean region. *Nature Reviews Earth & Environment, 2*. https://doi.org/10.1038/s43017-021-00192-6.

Ummenhofer, C. C., Sen Gupta, A., Briggs, P. R., England, M. H., McIntosh, P. C., Meyers, G. A., ... Risbey, J. S. (2011a). Indian and Pacific Ocean influences on Southeast Australian drought and soil moisture. *Journal of Climate, 24*, 1313–1336.

Ummenhofer, C. C., Sen Gupta, A., England, M. H., & Reason, C. J. C. (2009b). Contributions of Indian Ocean sea surface temperatures to enhanced East African rainfall. *Journal of Climate, 22*, 993–1013.

Ummenhofer, C. C., Sen Gupta, A., Li, Y., Taschetto, A. S., & England, M. H. (2011b). Multi-decadal modulation of the El Nino–Indian monsoon relationship by Indian Ocean variability. *Environmental Research Letters, 6*, 034006.

Ummenhofer, C. C., Sen Gupta, A., Pook, M. J., & England, M. H. (2008). Anomalous rainfall over southwest Western Australia forced by Indian Ocean sea surface temperatures. *Journal of Climate, 21*, 5113–5134.

Ummenhofer, C. C., Sen Gupta, A., Taschetto, A. S., & England, M. H. (2009c). Modulation of Australian precipitation by meridional gradients in East Indian Ocean sea surface temperature. *Journal of Climate, 22*, 5597–5610.

Vecchi, G., Harrison, D., & Bond, D. (2004). Interannual Indian rainfall variability and Indian Ocean sea surface temperature anomalies. In C. Wang, S.-P. Xie, & J. A. Carton (Eds.), *Earth climate: The Ocean-atmosphere interaction* (pp. 247–260). Washington, DC: American Geophysical Union.

Vecchi, G. A., & Soden, B. J. (2007). Effect of remote sea surface temperature change on tropical cyclone potential intensity. *Nature, 450*, 1066–1070.

Vecchi, G. A., Swanson, K. L., & Soden, B. J. (2008). Whither hurricane activity? *Science, 322*, 687–689.

Vialard, J., Foltz, G. R., McPhaden, M. J., Duvel, J. P., & de Boyer Montégut, C. (2008). Strong Indian Ocean sea surface temperature signals associated with the Madden-Julian Oscillation in late 2007 and early 2008. *Geophysical Research Letters, 35*, L19608. https://doi.org/10.1029/2008GL035238.

Wainwright, C. M., Finney, D. L., Kilavi, M., Black, E., & Marsham, J. H. (2021). Extreme rainfall in East Africa, October 2019–January 2020 and context under future climate change. *Weather, 76*. https://doi.org/10.1002/wea.3824.

Wang, G., & Cai, W. (2020). Two-year consecutive concurrences of positive Indian Ocean Dipole and Central Pacific El Niño preconditioned the 2019/2020 Australian "black summer" bushfires. *Geoscience Letters, 7*. https://doi.org/10.1186/s40562-020-00168-2.

Watanabe, M., & Jin, F.-F. (2002). Role of Indian Ocean warming in the development of Philippine Sea anticyclone during ENSO. *Geophysical Research Letters, 29*. https://doi.org/10.1029/2001GL014318.

Webster, P. J., Moore, A. M., Loschnigg, J. P., & Leben, R. R. (1999). Coupled ocean-atmosphere dynamics in the Indian Ocean during 1997-98. *Nature, 401*, 356–360.

Wieners, C. E., Dijkstra, H. A., & De Ruijter, W. P. (2017). The influence of the Indian Ocean on ENSO stability and flavor. *Journal of Climate, 30*, 2601–2620.

Wieners, C., Dijkstra, H., & de Ruijter, W. (2018). The influence of atmospheric convection on the interaction between the Indian Ocean and ENSO. *Journal of Climate.* https://doi.org/10.1175/JCLI-D-17-0081.1.

Williams, A. P., et al. (2012). Recent summer precipitation trends in the Greater Horn of Africa and the emerging role of Indian Ocean sea surface temperature. *Climate Dynamics, 39*, 2307–2328.

Xie, S.-P., Hu, K., Hafner, J., Tokinaga, H., Du, Y., Huang, G., & Sampe, T. (2009). Indian Ocean capacitor effect on Indo-Western Pacific climate during the summer following El Niño. *Journal of Climate, 22*, 730–747.

Yamagata, T., Behera, S., Doi, T., Luo, J.-J., Morioka, Y., & Tozuka, T. (2024). Chapter 5: Climate phenomena of the Indian Ocean. In C. C. Ummenhofer, & R. R. Hood (Eds.), *The Indian Ocean and its role in the global climate system* (pp. 103–119). Amsterdam: Elsevier. https://doi.org/10.1016/B978-0-12-822698-8.00009-3.

Yamanaka, G., Yasuda, T., Fujii, Y., & Matsumoto, S. (2009). Rapid termination of the 2006 El Niño and its relation to the Indian Ocean. *Geophysical Research Letters, 36*, L07702. https://doi.org/10.1029/2009GL037298.

Yang, J., Liu, Q., Liu, Z., Wu, L., & Huang, F. (2009). Basin mode of Indian Ocean sea surface temperature and Northern Hemisphere circumglobal teleconnection. *Geophysical Research Letters, 36*, L19705. https://doi.org/10.1029/2009GL039559.

Yao, S.-L., Luo, J.-J., Huang, G., & Wang, P. (2017). Distinct global warming rates tied to multiple ocean surface temperature changes. *Nature Climate Change, 7*, 486–491.

Yoo, J. H., et al. (2020). Cases for the sole effect of the Indian Ocean Dipole in the rapid phase transition of the El Niño-Southern Oscillation. *Theoretical and Applied Climatology, 141*, 999–1007.

Yun, K. S., & Timmermann, A. (2018). Decadal monsoon-ENSO relationships reexamined. *Geophysical Research Letters, 45*, 2014–2021.

Zhang, C. (2005). Madden-Julian Oscillation. *Reviews of Geophysics, 43*, RG2003. https://doi.org/10.1029/2004RG000158.

Zhang, C. (2013). Madden-Julian Oscillation: Bridging weather and climate. *Bulletin of the American Meteorological Society.* https://doi.org/10.1175/BAMS-D-12-00026.1.

Zhang, C., Adames, Á. F., Khouider, B., Wang, B., & Yang, D. (2020). Four theories of the Madden-Julian Oscillation. *Reviews of Geophysics, 58*, e2019RG000685. https://doi.org/10.1029/2019RG000685.

Zhang, L., & Han, W. (2021). Indian Ocean Dipole leads to Atlantic Niño. *Nature Communications, 12*. https://doi.org/10.1038/s41467-021-26223-w.

Zhang, L., Han, W., Karnauskas, K. B., Meehl, G. A., Hu, A., Rosenbloom, N., & Shinoda, T. (2019a). Indian Ocean warming trend reduces Pacific warming response to anthropogenic greenhouse gases: An interbasin thermostat mechanism. *Geophysical Research Letters, 46*. https://doi.org/10.1029/2019GL084088.

Zhang, L., Han, W., & Sienz, F. (2018). Unraveling causes for the changing behavior of the tropical Indian Ocean in the past few decades. *Journal of Climate, 31*, 2377–2388.

Zhang, L., & Karnauskas, K. B. (2017). The role of tropical interbasin SST gradients in forcing Walker circulation trends. *Journal of Climate, 30*, 499–508.

Zhang, C., Luo, J.-J., & Li, S. (2019b). Impacts of tropical Indian and Atlantic Ocean warming on the occurrence of the 2017/18 La Niña. *Geophysical Research Letters.* https://doi.org/10.1029/2019GL082280.

Zheng, C., Chang, E. K.-M., Kim, H.-M., Zhang, M., & Wang, W. (2018). Impacts of the Madden-Julian Oscillation on storm-track activity, surface air temperature, and precipitation over North America. *Journal of Climate, 31*, 6113–6134.

Zhou, L., Murtugudde, R., Chen, D., & Tang, Y. (2017). A central Indian Ocean mode and heavy precipitation during the Indian summer monsoon. *Journal of Climate, 30*, 2055–2067.

Chapter 8

Indian Ocean circulation[*]

Helen E. Phillips[a], Viviane V. Menezes[b], Motoki Nagura[c], Michael J. McPhaden[d], P.N. Vinayachandran[e], and Lisa M. Beal[f]

[a]*Institute for Marine and Antarctic Studies and Australian Antarctic Program Partnership, University of Tasmania, Hobart, TAS, Australia,* [b]*Woods Hole Oceanographic Institution, Woods Hole, MA, United States,* [c]*Japan Agency for Marine-Earth Science and Technology, Yokosuka, Kanagawa, Japan,* [d]*NOAA/Pacific Marine Environmental Laboratory, Seattle, WA, United States,* [e]*Centre for Atmospheric and Oceanic Sciences, Indian Institute of Science, Bangalore, India,* [f]*Rosenstiel School of Marine, Atmospheric, and Earth Science, University of Miami, Coral Gables, Miami, FL, United States*

1 Introduction

The near-surface circulation of the Indian Ocean divides naturally into the monsoon-dominated, seasonally reversing currents north of around 10°S and the currents to the south. This seasonally reversing circulation is driven by the seasonally reversing winds that are in turn driven by the contrasts in temperature between the Asian landmass and the Indian Ocean, as the ocean surface temperature varies much less than that of the land (Ummenhofer et al., 2024). In this chapter, we refer to the boreal summer (June–September) monsoon as the Southwest Monsoon, when the direction of the winds is from the southwest, driven by rising air over the hot Indian continent. The boreal winter (December–February) monsoon is referred to as the Northeast Monsoon, when the winds flow from the northeast, driven by subsidence of cold air over the frozen Tibetan plateau. Fig. 1 illustrates the dramatic changes in the Northern Hemisphere currents, which switch direction between the southwest monsoon (Fig. 1a, orange) and the northeast monsoon (Fig. 1b, green). In the Southern Hemisphere, the current pathways (blue) do not change direction although their strength and position can vary seasonally.

The near-surface ocean pathways in Fig. 1 show how the Indian Ocean circulation links the Pacific Ocean via the Indonesian Throughflow (ITF) to the Atlantic and Southern Oceans along its western and eastern boundaries (Sprintall et al., 2024). The northern Indian Ocean also receives Pacific-origin waters via the South Equatorial Current and northward along the western boundary. Below this surface circulation is an intricate network of deeper pathways that carry heat and freshwater throughout the Indian Ocean as part of the global meridional overturning circulation. The deeper pathways were inferred from water property distributions as early as the 1960s, and more direct observations have gradually built up since through research voyages, volunteer observing ships, the tropical moorings of the Research Moored Array for African-Asian-Australian Monsoon Analysis and Prediction (RAMA) and, since the early 2000s, the global Argo profiling float array (McPhaden et al., 2024).

While a poleward flowing western boundary current, the Agulhas Current in the Indian Ocean, is a common feature of all ocean basins, a poleward eastern boundary current is not. The Leeuwin Current flows poleward, bringing tropical-origin water and a nutrient supply to the otherwise oligotrophic eastern boundary along western Australia. The presence of the Leeuwin Current is due to the warm, fresh ITF waters that set up a strong north-south density gradient with the colder, saltier subtropical waters to the south. The pressure gradient also drives unique eastward-flowing near-surface currents right across the Indian Ocean south of the South Equatorial Current. Beneath the eastward surface currents, the flow is westward, revealing a more familiar wind-driven gyre circulation, as found in the Atlantic and Pacific subtropical oceans (Section 4).

The equatorial circulation of the Indian Ocean is unlike that of the Pacific and Atlantic Oceans because annual mean winds are westerly along the equator rather than easterly. These winds favor downwelling at the equator rather than the upwelling that is observed in the Pacific and Atlantic equatorial waters. Moreover, seasonal reversals of the winds along the equator lead to strong semi-annual variability, with strong eastward currents, known as Wyrtki jets, dominating the upper 100m during the transitions between the Northeast and Southwest Monsoons (Section 3).

---

[*]This book has a companion website hosting complementary materials. Visit this URL to access it: https://www.elsevier.com/books-and-journals/book-companion/9780128226988.

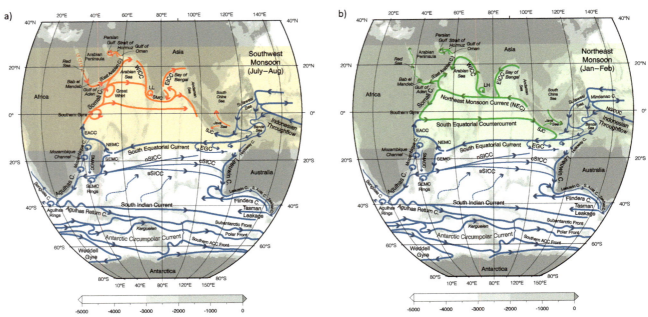

FIG. 1 Schematic near-surface circulation during (a) July–August (southwest monsoon) and (b) December–February (northeast monsoon). *Arrows* indicate year-round mean flows with no seasonal reversals (*blue*) and the monsoonally reversing circulation (*orange* and *green*). Acronyms: *NGCUC*, New Guinea Coastal Undercurrent; *SJC*, South Java Current; *EGC*, Eastern Gyral Current; *SICC*, South Indian Countercurrent (south, central, and southern branches); *NEMC*, Northeast Madagascar Current; *SEMC*, Southeast Madagascar Current; *SMACC*, Southwest MAdagascar Coastal Current; *EACC*, East Africa Coastal Current; *WICC*, West India Coastal Current; *EICC*, East India Coastal Current; *LH and LL*, Lakshadweep high and low; *SMC*, Southwest Monson Current; *NMC*, Northeast Monsoon Current. The *light gray* shading shows seafloor bathymetry. *(Figures and caption modified from Phillips et al. (2021); based on the update from Talley et al. (2011); of the original diagram from Schott and McCreary (2001).)*

In addition to these largely horizontal circulations, the Indian Ocean hosts shallow and deep overturning cells (Fig. 2). High evaporation and wintertime cooling in the southern subtropics create salty, dense surface waters that sink below the surface (blue shading in Fig. 2), aided by the downwelling-favorable winds. These subtropical mode waters supply the shallow overturning cells of the Indian Ocean that ventilate the upper part of the subtropical gyre and return to the sea surface in the northern Indian Ocean upwelling zones (Section 2). Below the shallow overturning cells, at depths around 500–2000 m, mode and intermediate waters enter the Indian Ocean from the Southern Ocean (Section 5). As they move northward, they mix with lighter waters above and progressively upwell to join the shallow overturning cells that return southward at the surface (Schott et al., 2009). The lower part of these waters mix with denser waters below to eventually return southward as deep waters (below 2000 m). This deep southward flow also receives contributions from abyssal waters that follow sea floor canyons into the northern Indian Ocean, mixing with lighter waters above on its slow northward journey (Section 5; Talley, 2013).

Our foundation knowledge of the Indian Ocean's mean circulation and its seasonal variability rests largely on the synthesis work of Schott and McCreary (2001) and Tomczak and Godfrey (2003). Since then, sustained observational efforts and modeling have provided new insight into the variability of these circulations at timescales from subseasonal to decadal. Observations also now reach more broadly across the Indian Ocean and deeper into its interior. These advances in the observing system and companion modeling work have contributed to a new understanding of the water masses, velocity structure, and driving processes of the Indian Ocean's circulation and its exchanges with the atmosphere. Recent reviews of advances in understanding the Indian Ocean circulation, air-sea exchange, and boundary current processes may be found in Beal et al. (2019, 2020), Vinayachandran et al. (2020), and Phillips et al. (2021). McCreary and Shetye (2023) provide a detailed examination of the dynamics of the northern Indian Ocean circulation.

This chapter begins with the unique monsoon circulation of the Indian Ocean in Section 2. From there, the primarily zonal currents of the equatorial regime are described, as well as their role in facilitating exchange across the basin in Section 3. The Southern Hemisphere gyre and boundary circulations are described in Section 4. Here, we focus on the interior pathways of the circulation to complement the discussion of interbasin exchanges in Sprintall et al. (2024) which describes the role of the boundaries in moving water into and out of the Indian Ocean. In Section 5, the vertical structure of the Indian Ocean circulation is presented with a discussion of the vertical exchanges that enable the Indian Ocean's contribution to the meridional overturning circulation. We conclude with a summary of the main points in Section 6.

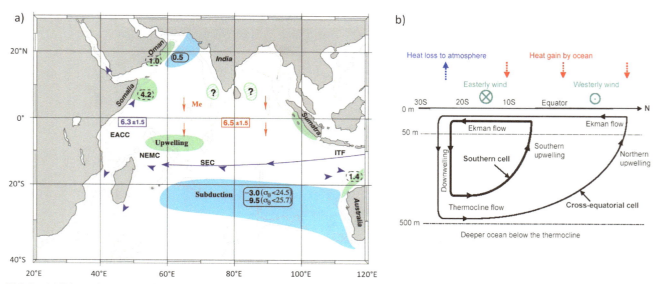

FIG. 2 (a) Schematic representation of circulation involved in connecting subduction regions (*blue shading*) with Indian Ocean upwelling regions (*green shading*). Volume of subduction (Sv = 10^6 m^3 s^{-1}) in each region is noted. Subduction in the Southern Hemisphere is for densities that can upwell off Somalia during weaker and stronger upwelling cases (Karstensen & Quadfasel, 2002). (b) Conceptual illustration of the time-mean meridional overturning circulation of the upper Indian Ocean that consists of a southern and a cross-equatorial cell. The time-mean zonal wind and surface heat flux are also shown schematically. Acronyms: *ITF*, Indonesian Throughflow; *SEC*, South Equatorial Current; *NEMC*, Northeast Madagascar Current; *EACC*, East Africa Coastal Current; *Me*, mass transport by wind-driven Ekman currents. *(Figure from (a) Schott et al. (2002) and (b) Lee (2004).)*

2 Monsoon circulation

2.1 Introduction

Fig. 1 shows the timing and broad-scale structure of the currents that represent the remarkable seasonal average monsoon circulations (Phillips et al., 2021; Schott & McCreary, 2001; Talley et al., 2011). McCreary and Shetye (2023) review the dynamics of this circulation. Sustained observations in the Indian Ocean (McPhaden et al., 2024) and associated ocean modeling advances (Shinoda et al., 2024) over the last decade or two have improved our understanding of the spatial and temporal variability of these circulations.

Circulation during the southwest monsoon (Fig. 1a) is clockwise around the continental boundaries, with northward flow across the equator in the Somali Current along the east coast of Africa feeding into the Southern Gyre and Great Whirl recirculations and on to the northern Arabian Sea via the East Arabian Current. The West India Countercurrent continues the flow southward along the west coast of India, encountering a low-pressure center with anticlockwise circulation, known as the Lakshadweep Low, that sits inshore of the West India Countercurrent (Shankar et al., 2002; Zachariah et al., 2019). The Southwest Monsoon Current and East India Coastal Current direct the flow around the southern tip of India and northward into the Bay of Bengal. To the east of Sri Lanka, an anticlockwise circulation during the southwest monsoon marks the Sri Lanka Dome (Vinayachandran & Yamagata, 1998).

During the northeast monsoon (Fig. 1b), the reversed circulation follows the East India Coastal Current southward down the western boundary of the Bay of Bengal toward Sri Lanka. It then feeds into the westward Northeast Monsoon Current, passing the southern tip of India and splitting into two branches. One branch flows strongly westward toward Africa. The second branch is steered northward by the clockwise rotation of the Lakshadweep High (Ernst et al., 2022; Shankar et al., 2002) to supply the anticlockwise circulation around the Arabian Sea via the West India Countercurrent. The East Arabian Current and Somali Current now flow southward. The Somali Current crosses the equator into the Southern Hemisphere where it meets the northward East Africa Coastal Current. This convergence deflects the flow offshore to supply the eastward South Equatorial Countercurrent that crosses the basin.

The dramatic difference in salinity between the two main basins of the northern Indian Ocean, the Bay of Bengal and the Arabian Sea, is another consequence of the monsoon forcing. Seasonal average sea surface salinity and temperature for the Southwest and Northeast Monsoons are illustrated in Fig. 3 from Rainville et al. (2022). The corresponding dynamic height

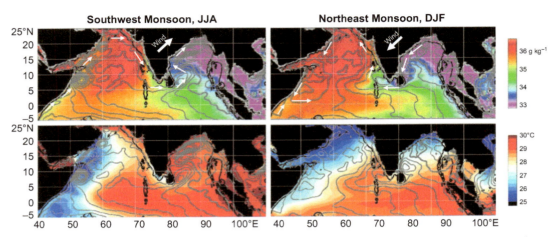

FIG. 3 Maps of surface absolute salinity (top, from Soil Moisture Active Passive satellite; Meissner et al., 2018), and temperature (bottom, from optimally interpolated microwave and infrared satellites; Wentz et al., 2000), overlaid with dynamic height contours (Ducet et al., 2000) for the southwest (left) and northeast (right) monsoons. Maps are the seasonal averages for 4 years from 2016 to 2019. The wind direction and approximate direction of boundary circulation along dynamic height contours are indicated with *white arrows* (top panels). *(Adapted from Rainville et al. (2022).)*

of the sea surface is overlaid, providing approximate streamlines of the flow. The Arabian Sea is saltier by as much as three practical salinity units throughout the year (e.g., Chatterjee et al., 2012; Gordon et al., 2016; Hormann et al., 2019). The fresh surface layer of the Bay of Bengal is maintained by large freshwater input from direct rainfall over the ocean and river runoff, especially during the southwest monsoon. The saltier sea surface of the Arabian Sea is due to strong evaporation that exceeds rainfall and river runoff over the region (e.g., Rao & Sivakumar, 2003; Sengupta et al., 2006).

The reversal of the monsoon circulation is critical for the exchanges of heat and salt between the Bay of Bengal and the Arabian Sea that maintain the climatological balances of temperature and salinity (Hormann et al., 2019; McCreary et al., 1993; Vinayachandran et al., 2013). White arrows in Fig. 3 (top panels) approximate the boundary circulation that drives these exchanges near the surface, transporting saltier water from the Arabian Sea into the Bay of Bengal during the southwest monsoon and transporting fresher, cooler water from the Bay of Bengal to the Arabian Sea during the Northeast Monsoon (Rainville et al., 2022). Subsurface exchanges of heat and salt are also influenced by the monsoons. An example of this is the passage of Arabian Sea high salinity water into the Bay of Bengal via the Indian Ocean equatorial undercurrent (Section 3.4) and Southwest Monsoon Current (Sanchez-Franks et al., 2019).

2.2 Cross-equatorial gyre circulations

While the seasonal averages presented in Figs. 1 and 3 highlight the main features of the monsoon circulation, they cannot account for the true pathways created by the constantly evolving flow. The complexity of the time-varying circulation, including propagating Rossby and Kelvin waves, means that the seasonal average maps are an entry point to understanding the monsoon circulation. Tracing the flow through the global drifter climatology using Lagrangian particle tracking reveals a more complex circulation (L'Hégaret et al., 2018; link to animation in Section 7).

L'Hégaret et al. (2018) describe the monsoon circulation using real surface drifter trajectories and simulated trajectories created by advecting hundreds of thousands of particles in the monthly drifter climatology of Laurindo et al. (2017). L'Hégaret et al. (2018) identified three cross-equatorial gyres by using probabilistic pathlines (time-dependent streamlines) defined as the most common trajectories that cross the equator within the Somali Current (Fig. 4, left column). These monsoon "gyres" are shallow and represent a combination of geostrophic flow and the wind-driven Ekman flow (Shankar et al., 2002). Fig. 4 shows the probabilistic pathlines with color coding to indicate the time of year that the path is occupied.

L'Hégaret et al. (2018) found two types of surface gyre leading up to and during the Southwest Monsoon. The cross-equatorial gyre (Fig. 4a) forms a figure of eight that takes about 2 years to complete and incorporates eastward flow along the equator in the Wyrtki Jet from August to December (Section 3.3). The overshooting equatorial gyre (Fig. 4b) is much shorter, is contained west of the Maldives, and takes only one year to complete. Southward pathlines (Fig. 4c) through the Somali Current during the northeast monsoon participate in an equatorial gyre with Somali Current flow fed by the Great Whirl. Flow around this gyre takes about 2 years to complete. In the two gyres with northward pathlines through the Somali

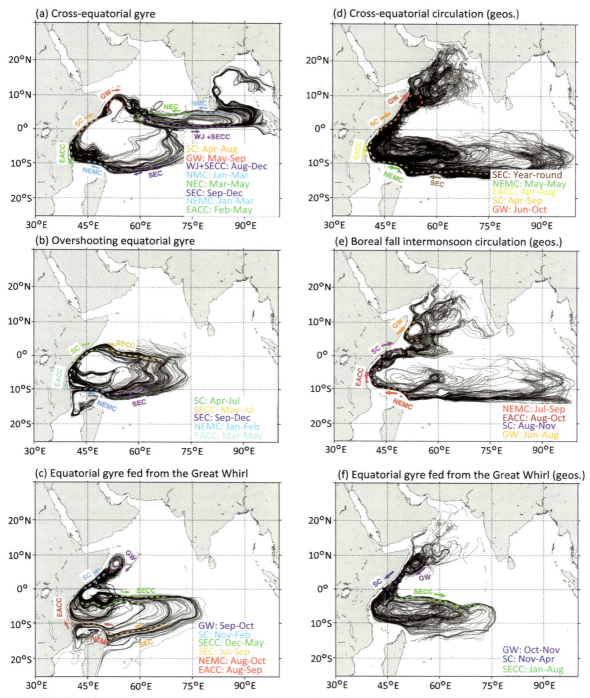

FIG. 4 Probabilistic pathlines of surface currents calculated from the monthly drifter climatology of Laurindo et al. (2017, left column) and from altimetry-derived geostrophic currents (right column). Drifter pathlines separate into three monsoon gyres: (a) cross-equatorial gyre defined by pathlines of the northward Somali Current that pass beyond 4°N, (b) an overshooting equatorial gyre defined by pathlines of the northward Somali Current that remain south of 4°N, and (c) an equatorial gyre fed from the Great Whirl, defined by pathlines of the southward Somali Current. Geostrophic flow pathlines separate into three monsoon-driven circulations: (d) cross-equatorial circulation, defined by geostrophic pathlines of the northward Somali Current, (e) boreal fall intermonsoon circulation, defined by pathlines that leave the Somali coast in an offshore direction, and (f) the equatorial gyre fed by the Great Whirl, defined by geostrophic pathlines of the southward Somali Current. Colors represent the sustained current periods as shown in the legend. Current acronyms: *SEC*, South Equatorial Current; *NEMC*, Northeast Madagascar Current; *EACC*, East Africa Coastal Current; *SC*, Somali Current; *GW*, Great Whirl; *SECC*, South Equatorial Countercurrent; *WJ*, Wyrtki Jet; *NMC*, Northeast Monsoon Current. *(Figure modified from L'Hégaret et al. (2018).)*

Current, the return to the Southern Hemisphere is primarily by southward Ekman transport in the interior. For the equatorial gyre with southward pathlines, the northward return across the equator also relies on the northward-flowing Somali Current, and East Africa Coastal Current in their reversed states. Some cross-equatorial flow was also attributed to the annual migration of the South Equatorial Countercurrent.

A surprising result of L'Hégaret et al. (2018) was that none of the pathlines entered the Arabian Sea. In contrast, probabilistic pathlines of the geostrophic flow (Fig. 4, right column), constructed from particles advected by sea surface altimetry-derived geostrophic currents did reach the northern Arabian Sea. Connectivity from the Somali Current to the Arabian Sea was found to be slow but active for most of the year. A further notable result was that the geostrophic pathlines entering the Somali Current from the south equatorial current originate from much further east (Fig. 4d and e) than for the real drifters and Ekman contribution to the pathlines. Both indicate the importance of the geostrophic flow to the south equatorial current and identifies basin-wide connectivity not evident in the drifter currents.

2.3 Somali current system and western Arabian Sea

The Somali Current is the strongest boundary current in the North Indian Ocean. During the Southwest Monsoon, current speeds in the Somali Current can reach $3\,\mathrm{m\,s^{-1}}$ with volume transport up to 37 Sverdrups (Sv, $1\,\mathrm{Sv} = 10^6\,\mathrm{m^3\,s^{-1}}$; Fischer et al., 1996; Beal & Chereskin, 2003). During the Northeast Monsoon, transports are weaker, around 5 Sv (Schott & McCreary, 2001). Strong variability of the flows on multiple timescales makes it difficult to estimate the velocity structure and transport accurately from observations (Beal & Chereskin, 2003; Zang et al., 2021).

The Somali Current and interactions with the Great Whirl and Southern Gyre are linked to important coastal upwelling cells (Fig. 2; Beal et al., 2013; Schott & McCreary, 2001). Upwelling wedges are formed leading up to and during the southwest monsoon, in response to the local southwesterly alongshore winds known as the Findlater Jet (Findlater, 1971) that drives offshore Ekman transport all along the Somali coast (Schott & McCreary, 2001). These transports are strongest in the south and weaken northward (Chatterjee et al., 2019).

Schott and McCreary (2001) describe the seasonal evolution of Somali Current upwelling zones in relation to the upper-layer flow patterns (their Fig. 32, adapted from Schott et al., 1990). Around May, during the transition period before the southwest monsoon, the northward Somali Current is an extension of the East Africa Coastal Current, turning offshore near 4°N with a cold wedge of upwelling on its inshore shoulder. The arrival of the southwest monsoon initiates the Southern Gyre, turning some of the offshore flow back toward the equator (Fig. 1a). Further north, the Great Whirl develops, with an additional cold wedge, suggested to be driven by the interaction of westward propagating Rossby waves excited by the monsoon winds and the strong anticyclonic wind stress curl off the Somali coast (Beal & Donohue, 2013; Schott & McCreary, 2001; Schott & Quadfasel, 1982). Once the northeast monsoon begins, the Somali Current turns southward and flows across the equator where it converges with the northward East Africa Coastal Current and diverts offshore into the South Equatorial Countercurrent near 4°S (Fig. 1b) (see also Section 2.2).

Upwelling in the Somali Current system during the Southwest Monsoon is nearly as strong as for the eastern boundary upwelling regimes of the Pacific and Atlantic Oceans (Schott & McCreary, 2001). This upwelling provides the primary route for the upward limb of the cross-equatorial cell (Fig. 2) bringing thermocline waters from the southern hemisphere to the sea surface in the northern Indian Ocean. The upwelling supplies nutrients to the sea surface promoting high productivity that supports the coastal fisheries of rim countries (Hermes et al., 2019; Marsac et al., 2024). This productivity is vulnerable to climate change (e.g., Kumar et al., 2018; Roxy et al., 2016, 2024) and variability in upwelling rates related to variability in monsoon winds (Hood et al., 2017, 2024a, 2024b; Vinayachandran et al., 2020).

2.4 Eastern Arabian Sea and Bay of Bengal

The large-scale wind field in the eastern Arabian Sea and the Bay of Bengal is southwesterly during the summer monsoon and northeasterly during the winter monsoon. The winds along the west coast of India, however, are directed toward the equator throughout the year (Shetye et al., 1985). The current along the west coast of India is the West India Coastal Current. During the summer monsoon, the West India Coastal Current flows equatorward in the direction of the alongshore wind (Fig. 1a, Shenoi et al., 1999). During the winter monsoon, it flows poleward against the wind (Fig. 1b). The equatorward summer West India Coastal Current is roughly 150 km wide and extends to a depth range of 75–100 m. It is stronger in the south with the volume transport increasing from about 0.5 Sv in the northeastern Arabian Sea to about 4 Sv off the southwest coast of India (Shetye et al., 1990). The poleward transport of the winter West India Coastal Current is higher at about 7 Sv (Shetye et al., 1991a).

Fig. 5 shows an almost four-year record, 2009–2012, of currents measured by four current meter moorings placed at the 1000 m depth contour along the path of the West India Coastal Current from Amol et al. (2014). More recent measurements extend the record to 2018 with interruptions (Chaudhuri et al., 2020). The West India Coastal Current has a seasonal cycle that is stronger in the north, has considerable interannual variability, and is characterized by upward phase (downward energy) propagation (Amol et al., 2014). Beneath the surface flow, there is an undercurrent that is stronger during the winter. In addition to the local winds, the seasonal cycle of the West India Coastal Current is remotely forced (McCreary et al., 1996). In fact, Kelvin waves originating from the east coast of India play a greater role than the local winds in driving the seasonal cycle (Shankar et al., 2002; Shankar & Shetye, 1997). The existence of coastally trapped waves inferred from mooring data provides observational evidence for the remote forcing (Amol et al., 2012). The West India Coastal Current is also characterized by strong intraseasonal variability in the 30- to 90-day band (Amol et al., 2014; Chaudhuri et al., 2020; Vialard et al., 2009). Thus, although the winds along the west coast of India lack a clear seasonal cycle, the ocean currents have a clear seasonal cycle and they have a significant impact on upwelling and ecosystem processes (Hood et al., 2017; Vinayachandran et al., 2020).

The western boundary current in the Bay of Bengal is known as the East India Coastal Current (Durand et al., 2009; McCreary et al., 1996; Shankar et al., 1996). Fig. 6 shows measurements of currents along the East coast of India during 2009–2012, adapted from Mukherjee et al. (2014). More recent measurements extend the record to 2018 with interruptions (Mukhopadhyay et al., 2020). The East India Coastal Current flows poleward during February–May (Shetye et al., 1993) with a transport of about 10 Sv, and equatorward during October–January with a transport of about 7 Sv (Shetye et al., 1996). This pre-monsoon East India Coastal Current is analogous to the classical western boundary current characterized by warmer waters, upwelling in its inshore edges (Shetye et al., 1993), and a basin scale anticyclonic wind stress curl that drives a Sverdrup mechanism. During the summer monsoon, the East India Coastal Current is poleward along the southern part of the Indian coast (Fig. 1b) with a transport of about 1 Sv. Along the northern coast, the flow is southward and consists of a low-salinity plume separated from the coast (Shetye et al., 1991b).

During the summer monsoon, the East India Coastal Current is characterized by upwelling in the southern part of the coast and a freshwater plume in the north. The winter East India Coastal Current is marked by a narrow band of very low salinity water that it carries into the eastern Arabian Sea (Fig. 1b; Behara & Vinayachandran, 2016; Behara et al., 2019; Chaitanya et al., 2014; Shetye et al., 1996). The seasonal cycle of the East India Coastal Current is much stronger than that of the West India Coastal Current (Mukherjee et al., 2014). Like the West India Coastal Current, the East India Coastal Current also shows upward phase propagation and occasional downward phase propagation.

The East India Coastal Current is forced by four different types of mechanisms: local alongshore winds, Ekman pumping over the bay, remote forcing from the equatorial Indian Ocean associated with Wyrtki Jets and remote forcing from the eastern boundary of the Bay of Bengal in the form of Kelvin waves excited by along shore winds at the eastern boundary (McCreary et al., 1996; Shankar et al., 1996; Vinayachandran et al., 1996). The magnitude and direction of the East India Coastal Current along the east coast of India and Sri Lanka are determined by the interplay between these four mechanisms. The East India Coastal Current is peculiar in terms of its spatial structure—the current itself is made up of several eddies; it is a discontinuous flow that is trapped to the continental shelf with the offshore flow in the opposite direction (Durand et al., 2009). The temporal variability peaks at about 120 days (Mukherjee et al., 2014; Mukhopadhyay et al., 2020). Despite its high spatiotemporal variability, the East India Coastal Current is a major pathway for the export of low-salinity water from the Bay of Bengal into the Arabian Sea (Behara & Vinayachandran, 2016; Jensen, 2001; Thompson et al., 2006).

The most organized feature of the open Bay of Bengal circulation is an anticyclonic gyre that can be seen during February–April, which is primarily driven by the weak but anticyclonic winds (Shetye et al., 1993). During the southwest monsoon, there is a general southeastward Ekman drift in the open bay and there exists a south-eastward current across the northern sector of the bay, called the North Bay Monsoon Current (Vinayachandran & Kurian, 2007), which separates the lowest salinity waters of the northern bay from the relatively higher salinity water further south. The general southward flow in the eastern bay is important for exporting the rainwater that falls in the northeastern bay to maintain the salinity balance of the Bay of Bengal (Behara & Vinayachandran, 2016).

2.5 Southwest/Northeast Monsoon Currents

The seasonally reversing open ocean currents in the Indian Ocean to the north of the equator are generally referred to as the monsoon currents. The Southwest Monsoon Current flows in a general eastward direction from May to September (Fig. 1a) and the Northeast Monsoon Current flows westward from November to February (Fig. 1b; Schott & McCreary, 2001; Shankar et al., 2002; Shenoi et al., 1999). The typical speeds of these open ocean currents are $<50\,\text{cm s}^{-1}$. The Southwest Monsoon

176 The Indian Ocean and its role in the global climate system

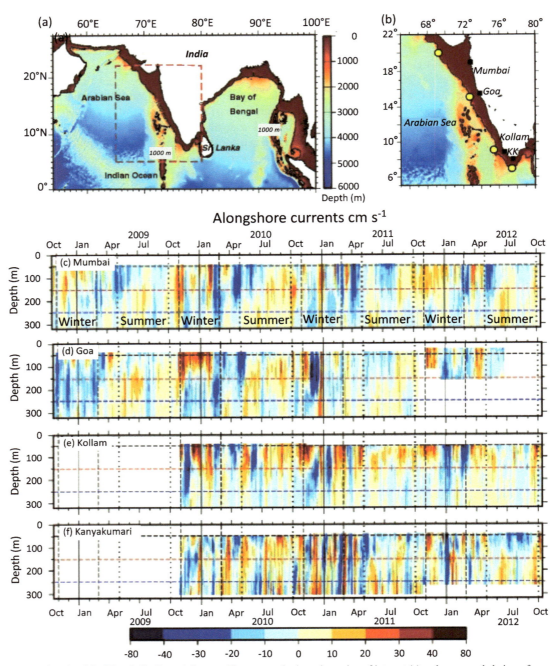

FIG. 5 The seasonal cycle of the West India Coastal Current. The top panels show the region of interest (a) and an expanded view of acoustic Doppler current profiler mooring locations off the west coast of India at 1000 m water column depth (b). KK stands for Kanyakumari. Lower panels show the 5-day low-passed alongshore currents as a function of depth from north to south: (c) Mumbai, (d) Goa, (e) Kollam, and (f) Kanyakumari. Negative currents (*blue*) indicate equatorward flow. The *dotted and dashed vertical lines* delineate summer and winter monsoons, respectively, as noted in (c). Red and blue horizontal lines are used for further analysis in Amol et al. (2014). *(Figure adapted from Amol et al. (2014).)*

Current is dominated by Ekman drift whereas the geostrophic component dominates the Northeast Monsoon Current. The monsoon currents are driven by a combination of local and remote forcing, and therefore, at any given time and location, the strength of the monsoon currents depends on local winds and remotely generated Kelvin and Rossby waves (Rath et al., 2019; Shankar et al., 2002), including signals generated along the equator (Section 3.5). Consequently, different parts of the monsoon currents come into being at different stages, and in their mature stage, they exist as trans-basin flows.

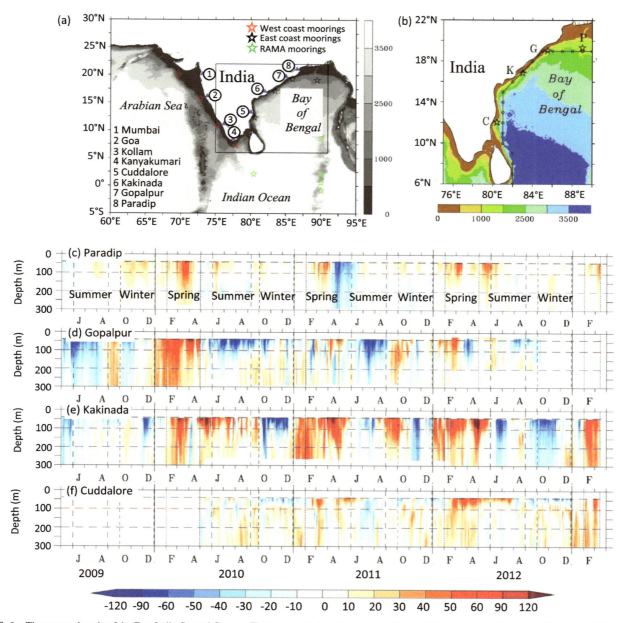

FIG. 6 The seasonal cycle of the East India Coastal Current. The top panels show the region of interest (a) and an expanded view of the acoustic Doppler current profiler mooring locations off the east coast of India at Cuddalore, Kakinada, Gopalpur, and Paradip (b). RAMA moorings (*green star* in (a)) are part of the Indian Ocean observing system described in McPhaden et al. (2024). (c–f) Lower panels show the 2.5-day low-passed, de-tided alongshore currents as a function of depth from north (Paradip—offshore) to south (Cuddalore). At Paradip, the alongshore current is zonal and an eastward current is positive. At the other three locations, a poleward current is positive. The *solid vertical lines* mark the years. The *dashed vertical lines* are used to denote the periods spring (February–May), summer monsoon (June–September), and winter (October–December), respectively. *(Figure adapted from Mukherjee et al. (2014).)*

The Southwest Monsoon Current carries high salinity water from the Arabian Sea into the fresher Bay of Bengal (Jensen, 2001; Jensen et al., 2016; Vinayachandran et al., 1999a, 2013; Webber et al., 2018; Wijesekera et al., 2015, 2016a, 2016b). Sanchez-Franks et al. (2019) found that the origins of the Arabian Sea's high salinity water are specifically from the western Arabian Sea and western equatorial Indian Ocean, and they reach the Bay of Bengal via a combination of the Indian Ocean equatorial undercurrent (Section 3.4) and the Southwest Monsoon Current. During the winter monsoon, the westward Northeast Monsoon Current links the East and West India Coastal Currents and serves as a crucial link for transporting fresh water from the Bay of Bengal into the Arabian Sea.

The Sri Lanka Dome (Vinayachandran & Yamagata, 1998) is located between the Southwest Monsoon Current and Sri Lanka and is a cyclonic (anticlockwise) circulation (Cullen & Shroyer, 2019; Rainville et al., 2022; Schott & McCreary, 2001; Vinayachandran et al., 2018, 1999a; Wijesekera et al., 2016b). It is visible as closed contours of dynamic height in Fig. 3 (left panels). It lives through the summer monsoon and is forced by thermocline upwelling due to Ekman pumping. The recurring upwelling dome and circulation enhances the Southwest Monsoon Current exchange from the Arabian Sea to the Bay of Bengal (Anutaliya et al., 2017). Upwelling associated with the Sri Lanka Dome influences the vertical exchange of water properties, enhances biological productivity, and cools sea surface temperature (SST), which affects local atmospheric convection (de Vos et al., 2014; Vinayachandran et al., 2004).

2.6 Marginal Seas

In the annual mean, Arabian Sea water enters the Persian Gulf near the surface, loses buoyancy owing to evaporation and cooling, and sinks to the bottom at about 35 m depth to form Persian Gulf Water. This water mass exits from the Gulf with a volume transport of 0.1–0.3 Sv (Johns et al., 2003; Kämpf & Sadrinasab, 2006; Thoppil & Hogan, 2010; Xue & Eltahir, 2015; Yao, 2008; Yao & Johns, 2010). Persian Gulf Water is characterized by its high salinity and enters the Arabian Sea with core salinity and temperature of ~36.5 psu and 18–19°C, respectively, and a potential density relative to the sea surface (σ_θ) of ~26.8 kg m^{-3}. It sinks to depths of 200–300 m and spreads either southward along the Omani coast during winter or first eastward along the northern boundary and then southward into the interior of the northern Arabian Sea during summer (Morrison, 1997; Premchand et al., 1986; Shenoi et al., 1993). Advection of Persian Gulf Water is both by the mean current (Prasad et al., 2001) and mesoscale eddies (L'Hégaret et al., 2013, 2015, 2016; Vic et al., 2015). Persian Gulf Water is gradually diluted while spreading due to mixing (L'Hégaret et al., 2021). Accurate simulation of this water mass is essential to represent the mean field of the Arabian Sea in climate models (Nagura et al., 2018).

Similar to the Persian Gulf, the relatively fresh and warm Gulf of Aden Surface Water enters the Red Sea at the surface predominantly during the winter monsoon through the Strait of Bab el-Mandeb (~25 km wide and 137 m deep), which connects the Red Sea and the Gulf of Aden (Bower & Farrar, 2015; Sofianos & Johns, 2015). In the Red Sea, the Gulf of Aden Surface Water is transported northward by a western boundary current to about 19°N where the current veers to the east to form the Eastern Boundary Current, which flows against the prevailing winds (Bower & Farrar, 2015; Sofianos & Johns, 2003, 2015; Yao et al., 2014a, 2014b; Zhai et al., 2015a). The poleward Eastern Boundary Current (~0.4 Sv) extends from 22°N to 27°N along the Saudi Arabia coast as a warm tongue at the surface (Bower & Farrar, 2015; Sofianos & Johns, 2003, 2015; Yao et al., 2014a, 2014b). On its way to the north, the Eastern Boundary Current water becomes less buoyant, saltier, and colder as the result of strong air-sea interactions dominated by evaporation (e.g., Bower & Farrar, 2015; Jiang et al., 2009; Menezes et al., 2018, 2019; Papadopoulos et al., 2013, 2015). Evaporation in this marginal sea is one of the highest on our planet, with an annual mean of 2.06 ± 0.22 m per year (Bower & Farrar, 2015; Eshel & Heavens, 2007; Sofianos et al., 2002; Tragou et al., 1999; Zolina et al., 2017) and average daily rates of about 5–7 m per year associated with wintertime dry-air outbreaks (Menezes et al., 2019).

In the extreme north, the Eastern Boundary Current is believed to become part of a cyclonic (anti-clockwise) gyre centered at 26°N. This is the most likely formation site for the salty Red Sea Overflow Water, which has a salinity of around 40 (Bower & Farrar, 2015; Papadopoulos et al., 2015; Zhai et al., 2015b). Once formed, the Red Sea Overflow Water is transported southward at mid-depths (100–250 m) to the Strait of Bab el-Mandeb (Zhai et al., 2015b), where it is exported to the Gulf of Aden. In situ measurements by Murray and Johns (1997) show that the Red Sea Overflow Water transport has a pronounced seasonal cycle, with a maximum (~0.7 Sv) in winter and a minimum (~0.05 Sv) in summer.

Before reaching the Arabian Sea, the Red Sea Overflow Water is transformed in the Gulf of Aden by mixing with local waters, becoming diluted and sinking to depths of 400–800 m where it is neutrally buoyant (Bower et al., 2000, 2005, 2002; Bower & Furey, 2012; Matt & Johns, 2007). The diluted Red Sea Overflow Water (salinity around 36) then spreads in the Arabian Sea and the open Indian Ocean at intermediate depths (27–27.6 kg m^{-3}; 500–1500 m), progressively deepening, mixing, and becoming fresher as it moves southward (Beal et al., 2000a; Gründlingh, 1985; L'Hégaret et al., 2021; Premchand et al., 1986; Rochford, 1964; Wyrtki et al., 1971; You, 1998). Most of the Red Sea Overflow Water transport to the south occurs along the western boundary within the Somali Undercurrent (Bruce & Volkmann, 1969; Esenkov & Olson, 2002; Quadfasel & Schott, 1983; Schott & Fischer, 2000; Schott & McCreary, 2001; Warren et al., 1966). Ultimately, the Red Sea Overflow Water is exported out of the Indian Ocean via the Agulhas Current (Beal et al., 2000a; Durgadoo et al., 2017; Roman & Lutjeharms, 2007, 2009).

The Red Sea Overflow Water has both local and remote impacts. It drives the shallow cell of the Red Sea's overturning circulation (Sofianos & Johns, 2003; Yao et al., 2014b) and is a critical source of salt for the Indian Ocean intermediate layer. The salty (and warm) Red Sea Overflow Water is in sharp contrast with both the fresher and cooler Antarctic intermediate water and the Indonesian Intermediate Water that occur at similar density levels (Section 5.2, Beal et al., 2000b; Talley & Sprintall, 2005).

3 Equatorial regime

3.1 Introduction

The region between about 10°N and 10°S is dynamically unique because the vertical component of the Coriolis parameter changes sign across the equator. This property, combined with the seasonally reversing monsoon winds, leads to distinctive features of ocean circulation found nowhere else in the world ocean. In this section, we highlight recent advances in the description and understanding of circulation in the equatorial band, many of which have been enabled in the past 20 years by the advent of the Indian Ocean Observing System (IndOOS; McPhaden et al., 2024).

3.2 Mean circulation

Along the equator, winds are strong and westerly during the transitions from the Northeast to the Southwest Monsoon in April–May (boreal spring transition) and again during the transition from the Southwest to the Northeast Monsoon in October–November (boreal fall). This results in annual mean westerly winds with a pronounced semi-annual cycle along the equator. The annual mean westerlies are downwelling favorable, in contrast to mean easterly wind regimes that lead to upwelling along the equator in the eastern Pacific and Atlantic Oceans. This downwelling circulation is characterized by meridionally convergent Ekman flow in the surface (Lumpkin & Johnson, 2013) and divergent geostrophic flow in the thermocline (Wang & McPhaden, 2017). As a consequence, the ascending branches of the shallow subtropical and cross-equatorial circulation cells are located in off-equatorial areas, e.g., along the Seychelles Chagos Thermocline Ridge and the Somali coast upwelling zones (Fig. 2). Another consequence of the mean westerly winds along the equator is that, in contrast to the Pacific and Atlantic Oceans, mean surface currents are to the east rather than to the west, and there is no permanent equatorial undercurrent, i.e., no distinctive core of eastward flow in the thermocline. An undercurrent-like structure does appear in February–March near the end of the northeast monsoon following a two-month period of sustained easterly winds along the equator, but this feature is transient and quickly disappears with the onset of westerly winds during the following monsoon transition season (McPhaden et al., 2015; Fig. 7). A weaker and more variable undercurrent appears in August–September (Iskandar et al., 2009).

3.3 Wyrtki jets

Zonal currents along the equator are characterized by strong semiannual variability, with a surface eastward jet in the upper 100 m during the boreal spring and fall transitions between the northeast and southwest monsoons. These currents are referred to as the Wyrtki jets (Fig. 7), named after Klaus Wyrtki who originally discovered them through analysis of ship drift data (Wyrtki, 1973). They are forced by zonal winds during the monsoon transition seasons and intimately linked to the large-scale variations in upper ocean mass and heat along the equator. Anomalous changes in jet transports (particularly the fall jet) are a key element in the development of Indian Ocean Dipole (IOD) events (McPhaden et al., 2015; Murtugudde et al., 2000; Thompson et al., 2006; Vinayachandran et al., 2007, 1999b). The jets are also affected by El Niño-Southern Oscillation forcing from the Pacific (Gnanaseelan et al., 2012).

The structure and dynamics of the Wyrtki jets on intraseasonal to interannual timescales have been described in the eastern and central equatorial Indian Ocean using acoustic Doppler current profiler (ADCP) data from RAMA moorings (Masumoto et al., 2005; McPhaden et al., 2009; Nagura & McPhaden, 2008, 2010a, 2010b, 2012) and surface drifter data (Qiu et al., 2009). Flow in the jets is predominantly zonal, confined to the upper 100 m within ±2° of the equator, but the jets undergo stationary wavelike meridional meanders to the east of the Maldives Islands dynamically consistent with an island wake effect (Nagura & Masumoto, 2015). McPhaden et al. (2015), using moored ADCP data from 2004 to 2013, found that the boreal spring jet volume transports peak in May at $\sim$15 Sv (1 Sv = 10^6 m^3 s^{-1}), and fall jet transports peak in November at $\sim$20 Sv, with year-to-year variations of 5–10 Sv around these seasonal means. The jets undergo decadal variations as well. Joseph et al. (2012), for example, found that the spring jet went through a systematic multiyear reduction in amplitude relative to the fall jet between 2006 and 2011 based on analyses of dynamical model output and satellite data products. These decadal changes may have affected the relative amplitudes of the spring and fall jets reported in McPhaden

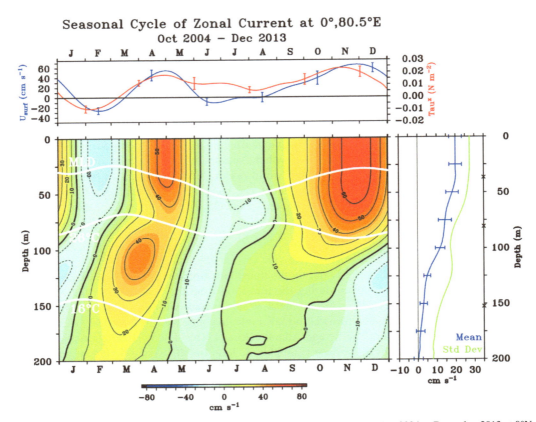

FIG. 7 Mean seasonal cycle of zonal velocity as a function of depth based on time series from October 2004 to December 2013 at 0°N, 80.5°E. Overplotted on the velocity contours are mixed layer depth (MLD) and also the 26°C and 16°C isotherms, which demarcate the depth range of upper thermocline. Plotted in the top panel are surface currents at 0°N, 80.5°E and surface wind stress averaged across the basin between 2°N and 2°S, 40° and 100°E. Right panel shows the mean and standard deviation as a function of depth. The three "x" symbols on the right axis at 37 m, 81 m, and 152 m represent the mean depths of the mixed layer and the 26°C and 16°C isotherms, respectively. *(From McPhaden et al. (2015).)*

et al. (2015). Accurate representation of the Wyrtki jets in climate models is necessary for proper simulation of monsoon rainfall because of how they affect the upper ocean heat balance and ocean-atmosphere interactions in the Indian Ocean region (Annamalai et al., 2017).

3.4 Equatorial undercurrents

Seasonal to interannual timescale variations in thermocline flow along the equator below the Wyrtki jets indicate that they too exhibit a significant semi-annual variability (Fig. 7), with eastward equatorial undercurrents during both the winter and summer monsoon seasons (Chen et al., 2015; Iskandar et al., 2009; Nagura & McPhaden, 2016; Nyadjro & McPhaden, 2014; Zhang et al., 2014). The winter season undercurrent is stronger because of the sustained easterly winds associated with the Northeast Monsoon that precedes it, while the summer season undercurrent is more temporally variable.

Interannual variations in the equatorial undercurrents are dynamically linked to wind variations associated with IOD events (Iskandar et al., 2009; Yamagata et al., 2024). Undercurrent transports in turn can affect the development of IOD SST anomalies through the supply of cold thermocline water to feed upwelling circulation in the eastern pole of the IOD (Gnanaseelan & Deshpande, 2018; Nyadjro & McPhaden, 2014; Zhang et al., 2014). A strong undercurrent supplies more cold water to the surface layer, which leads to colder SSTs and, through Bjerknes feedbacks, stronger easterly wind anomalies along the equator and thus a stronger undercurrent. Opposite tendencies occur for unusually weak undercurrents. These dynamical linkages represent positive feedbacks involving undercurrent flows in the thermocline that are an essential component in the evolution of IOD events.

3.5 Equatorial waves

The change in the sign of the Coriolis parameter across the equator allows for a unique class of planetary-scale waves of significant consequence for describing and understanding geophysical flows. The most prominent wave modes at periods longer than about 10 days that are relevant to climate scale phenomena are eastward propagating equatorial Kelvin waves and westward propagating long equatorial Rossby waves. These and other equatorial waves are described in more detail in Yamagata et al. (2024). Here we note that equatorial Kelvin and Rossby waves play a major role in seasonal timescale ocean current adjustments to wind forcing along the equator associated with Wyrtki jets and subsurface undercurrents. These waves are also crucial in the dynamical adjustment of the currents, thermocline depth, and sea surface height to changing winds associated with the IOD (McPhaden et al., 2015; McPhaden & Nagura, 2014; Murtugudde & Busalacchi, 1999; Nagura & McPhaden, 2010b), in the ocean's response to intraseasonal atmospheric forcing associated with Madden Julian Oscillation (Pujiana & McPhaden, 2020; Ummenhofer et al., 2024) and in the ocean response to convectively coupled atmospheric Kelvin waves (Pujiana & McPhaden, 2018). Equatorial waves remotely affect coastal regions and the interior ocean at higher latitudes through reflections at both eastern and western boundaries (e.g., Girishkumar et al., 2011, 2013; Iskandar & McPhaden, 2011; Sreenivas et al., 2012).

Equatorial waves propagating vertically into the deep ocean below the main thermocline were first discovered by Luyten and Roemmich (1982). They affect velocity, pressure, temperature, and the density structure of the deep ocean (Chen et al., 2020; Huang et al., 2018a, 2018b, 2019; Li et al., 2021; Zanowski & Johnson, 2019) and are particularly prominent at semi-annual periods since the semi-annual surface wind stress forcing is so pronounced. The phase of these waves propagates upward indicating downward energy propagation, consistent with an energy source at the surface. Energy in many cases can be traced along ray paths expected from WKB theory, a tool for obtaining an approximate solution to a linear differential equation (Bender & Orszag, 1999) that is named after the physicists who developed it, Wentzel-Kramers-Brillouin. At higher southern tropical latitudes where annual wind forcing is more pronounced, the signature of vertically propagating Rossby waves at annual periods is evident in the deep ocean (Johnson, 2011; Nagura, 2018).

3.6 Equatorial deep jets

Equatorial deep jets were first discovered by Luyten and Swallow (1976) in the Indian Ocean and have since been observed in both the Pacific (Firing, 1987) and the Atlantic (Gouriou et al., 1999). The jets are stacked zonal currents of alternating direction with speeds of $O(10\,\mathrm{cm\,s^{-1}})$ over vertical scales of a few hundred meters. They extend over much of the water column, are confined to within 1–2 degree of the equator, and are coherent zonally across much of the basin. The timescale for these jets is ~5 years in the Indian Ocean, and they exhibit slow downward phase propagation (Youngs & Johnson, 2015). The dynamics of the jets are poorly understood, and many theories have been proposed to describe them (Bastin et al., 2020). They are important in the generation of turbulent mixing (Dengler & Quadfasel, 2002) and for understanding tracer distributions (Bastin et al., 2020). However, they are generally not simulated accurately or at all in climate models, and how they are maintained against dissipation is an open question.

4 Southern hemisphere circulation

4.1 Introduction

The circulation in the southern Indian Ocean is relatively steady compared to the dramatic monsoonal variations to the north. The easterly trade winds and Southern Ocean westerlies drive the anticyclonic (anticlockwise) gyre circulation. The intense western boundary flow of the Agulhas Current system is supplied by the South Equatorial Current, the northern boundary of the gyre, and feeds the weak return flow in the interior of the basin to close the loop. On its eastern side, the Indian Ocean does not conform to the model of equatorward flow and eastern boundary upwelling that is well understood in the Pacific and Atlantic Oceans. Instead, the presence of the ITF drives the Leeuwin Current poleward, carrying tropical-origin waters southward along the coast of Western Australia. These waters support valuable coastal fisheries in an otherwise oligotrophic, downwelling-dominated environment (Hood et al., 2017; Thompson et al., 2011; Waite et al., 2007).

The gyre circulation reflects a depth-average flow that belies the complexity of the multilayered current pathways. The near-surface to intermediate level circulation provides a connection from the Pacific Ocean via the ITF, through the Indian Ocean and on to the Atlantic via the Agulhas Leakage, and along the western and southern coastlines of Australia to the Southern Ocean (see Sprintall et al., 2024). Intermediate level flow from the Southern Ocean follows the southern and western shelves of Australia to tropical latitudes in the Indian Ocean and also enters along the broader gyre pathway. Dense waters generated in the Red Sea flow into the interior, mixing to form Indian Intermediate Water that meets Banda Sea Intermediate Water and Antarctic intermediate water in the Southern Hemisphere tropics around 1000m depth (Rochford, 1961). Deeper still, North Atlantic Deep Water flows northward on the western side of the basin while Circumpolar Deep Water finds its way northward as boundary current flows along steep ridges in the central and eastern Indian Ocean (Hernández-Guerra & Talley, 2016). In the abyss, Antarctic Bottom Water enters the Indian Ocean constrained and steered by the sea floor topography (Purkey & Johnson, 2010).

4.2 Subtropical gyre circulation

The subtropical gyre in the South Indian Ocean is an anticlockwise horizontal gyre driven by positive wind stress curl and downward Ekman pumping velocity, as in the Pacific and Atlantic Oceans (Nagura & McPhaden, 2018; Reid, 2003; Schott et al., 2009; Schott & McCreary, 2001; Stramma & Lutjeharms, 1997). It is evident in Fig. 8a and b as an anticlockwise circulation between 15 and 30°S to the east of Madagascar. The panels in Fig. 8 represent the ocean circulation as streamlines and velocity vectors on four density levels (isopycnals) in the upper 1000m of the water column, with shading indicating the spatially varying depth of each isopycnal. The shallowest (Fig. 8a) is within the thermocline at a potential density of $1024\,\mathrm{kg\,m^{-3}}$, or $24\sigma_\theta$, and the others are at the levels of key water masses. In the southern Indian Ocean, the depth of the $24\sigma_\theta$ isopycnal varies between 60 and 150m, coming to the surface south of 20°S. The $25.5\sigma_\theta$ isopycnal (Fig. 8b) corresponds to the level of the high salinity of South Indian Subtropical Water that subducts into the subtropical gyre in the

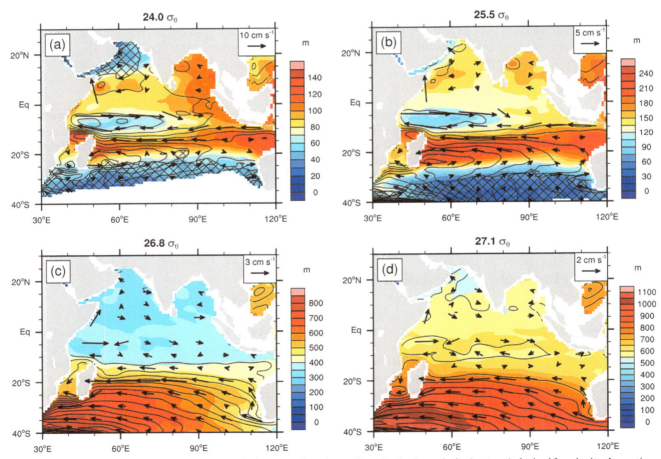

FIG. 8 Mean depth of isopycnals (colors), absolute velocity (vectors), and streamlines for absolute velocity (contours) obtained from in situ observations. (a) 24 σ_θ, (b) 25.5 σ_θ, (c) 26.8 σ_θ, and (d) 27.1 σ_θ. Contour intervals for streamlines are (a) 0.5, (b) 0.5, (c) 0.3, and (d) $0.3 \times 10^4\,\mathrm{m^2\,s^{-2}}$. Hatching shows outcropping regions, where the annual mean depth of the isopycnal is shallower than the wintertime mixed layer depth. The mixed layer depth is defined from potential density as the depth where density is higher than the 10-m value by $0.03\,\mathrm{kg\,m^{-3}}$ following de Boyer Montégut et al. (2004). *(Figure from Nagura and McPhaden (2018).)*

southeastern Indian Ocean (Fig. 2; Schott & McCreary, 2001). The two deepest levels are followed by Southern Ocean water masses in the south: $26.8\sigma_\theta$ (Fig. 8c) corresponds to the Subantarctic Mode Water and $27.1\sigma_\theta$ (Fig. 8d) to the Antarctic Intermediate Water (see Nagura & McPhaden (2018) for more detail).

The center of the subtropical gyre is located near 20°S to 40°S, retreating poleward with increasing depth. The gyre consists of interior equatorward flow, the westward South Equatorial Current near 12°S and a poleward western boundary current—the Agulhas Current (Fig. 1). At the eastern boundary of the basin near Indonesia, the ITF (Sprintall et al., 2009) carries mass into the basin, a part of which flows poleward down the west coast of Australia as the Leeuwin Current (Smith et al., 1991) and rounding the southwest corner of Australia to exit the Indian Ocean (Sprintall et al., 2024). The bulk of the ITF joins the westward South Equatorial Current, which is also fed by the interior equatorward flow. On the western side of the basin, the South Equatorial Current encounters the east coast of Madagascar, bifurcates (Chen et al., 2014), and feeds into the Northeast Madagascar Current flowing equatorward and Southeast Madagascar Current flowing poleward (Nagura, 2020; Schott et al., 1988, 2009; Swallow et al., 1988) along the Madagascan coast. The Southeast Madagascar Current joins the Agulhas Current and the Northeast Madagascar Current feeds into the East Africa Coastal Current that flows northward along the African coast (Swallow et al., 1991). The two poleward boundary currents, the Agulhas Current and the Leeuwin Current, are further described in Sections 4.3 and 4.4 and in Sprintall et al. (2024).

Whereas zonal flow is dominated by the westward South Equatorial Current north of the center of the gyre, there is a series of eastward counter currents. On the equatorward side of the gyre near 5°–10°S in the western and central basin, the thermocline is shallow and the circulation is clockwise around the low-pressure feature known as the Seychelles-Chagos Thermocline Ridge (Hermes & Reason, 2008; Yokoi et al., 2008). Flow is eastward in the upper few hundred meters north of the thermocline ridge in the South Equatorial Counter Current (Schott et al., 2009). Currents in the upper few hundred meters are also eastward on the eastern side of the basin at about 15°S and east of 105°E in the Eastern Gyral Current (Bray et al., 1997; Menezes et al., 2013; Qu & Meyers, 2005) and from Madagascar to Australia between 20° and 30°S in the South Indian Counter Current (Fig. 1; Menezes et al., 2014; Palastanga et al., 2007; Siedler et al., 2006). The South Indian Counter Current consists of a series of eastward currents interspersed by meridionally narrow westward currents (Divakaran & Brassington, 2011). The westward currents are too narrow to be seen in the smoothed fields in Fig. 8 (Nagura & McPhaden, 2018). Between 10°S and the equator, currents are eastward below the thermocline, from 200 to 1000 m depth (Nagura & McPhaden, 2018; Reid, 2003). There is also an isolated eastward current at about 15°S east of 70°E between 500 and 1000 m depths (Nagura & McPhaden, 2018), whose presence can be confirmed by the spreading pattern of salinity (Wijffels et al., 2002).

Interior equatorward flow is in geostrophic balance, and its transport can be monitored by the difference in sea surface height between the western and eastern parts of the basin (Nagura, 2020; Zhuang et al., 2013). The zonal sea surface height difference obtained from satellite altimetry showed that the gyre weakened from 1993 to 2000, strengthened from 2000 to 2004, weakened again from 2006 to 2012, and strengthened from 2012 to 2017 (Lee, 2004; Lee & McPhaden, 2008; Nagura, 2020; Zhuang et al., 2013), which indicates interannual to decadal variability in the South Indian Ocean subtropical gyre transport. A strong connectivity exists between sea surface height variability in the South Indian Ocean and that in the Pacific Ocean, due to the propagation of sea surface height signals from the Pacific to the Indian Ocean through the Indonesian Archipelago (Clarke, 1991; Clarke & Liu, 1994; Feng et al., 2010, 2011; Menezes & Vianna, 2019; Meyers, 1996; Potemra, 2001; Wijffels & Meyers, 2004). Nagura and McPhaden (2021) pointed out the importance of influence from the Pacific Ocean in generating subtropical gyre transport variability in the South Indian Ocean. In addition, there is variability in meridional geostrophic transport at intermediate levels of the South Indian Ocean, which is not in phase with the zonal sea surface height difference (Nagura, 2020).

4.3 Western boundary

Variability of the Agulhas Current at 34°S has been successfully captured by two major observing arrays over the past decade (Beal et al., 2015; Beal & Elipot, 2016; Gunn et al., 2020; McMonigal et al., 2020). In the mean, the Agulhas transports 76 Sv of warm and salty water, carrying 3.8 PW of heat, and 22,650 Sv psu of salt at transport-weighted values of 12.38°C and 34.87 psu. The well-known solitary meanders of the Agulhas Current have little imprint on these downstream fluxes, even though they dominate variability in the local temperature and salinity fields (Elipot & Beal, 2015; Leber & Beal, 2014; McMonigal et al., 2020). The Agulhas Current is strongest during austral summer (Hutchinson et al., 2018; Krug & Tournadre, 2012), but the flow deepens and cools as it strengthens and this may be the reason that a summer maximum in heat transport is not found (Beal & Elipot, 2016; McMonigal et al., 2020). The large variability of transport-weighted temperature in the Agulhas Current stands in contrast to the Gulf Stream within Florida Straits

(Johns et al., 2011) and implies that direct measurements of temperature in the Agulhas Current are necessary to constrain the meridional heat transport of the Indian Ocean to better than 0.24 PW (McMonigal et al., 2020).

In terms of recent decadal warming of the Indian Ocean (Cheng et al., 2017; Desbruyères et al., 2017; Li et al., 2017), a reduction in the meridional heat flux across the southern boundary is responsible for about 60% of the warming, with the other 40% linked to an increase in the ITF (McMonigal et al., 2022). This reduction in heat transport is likely associated with the broadening of the Agulhas Current suggested by the altimeter proxy developed by Beal and Elipot (2016) and is related to an increase in eddy kinetic energy (Backeberg et al., 2012; Elipot & Beal, 2015). Importantly, there is a disconnect between recent changes in the Agulhas Current and its extension, the Agulhas Return Current, with the latter appearing to strengthen while the Agulhas remains steady (Beal & Elipot, 2016; Sprintall et al., 2024; Yang et al., 2016). This disconnect is probably because the westerly winds have strengthened, but the subtropical wind stress curl has not, although there remain large uncertainties in these trends (Beal & Elipot, 2016; Elipot & Beal, 2018). Stronger westerlies also appear to be driving an increase in Agulhas leakage (Biastoch & Böning, 2013; Durgadoo et al., 2013; Sprintall et al., 2024). A recent estimate of the Agulhas leakage volume transport above 2000 m depth is 21.3 Sv with an error of 4.7 Sv (Daher et al., 2020).

Intermediate waters are uplifted and squeezed during meandering of the Agulhas Current, as well as during pulsing (strengthening), resulting in cooling and freshening of the near-shore thermocline (Gunn et al., 2020; Leber et al., 2017; McMonigal et al., 2020). Submesoscale variability of the inshore front occurs almost three times more often than meandering and cannot be discounted for its cryptic (invisible to satellite) impacts on upwelling and on shelf waters (Gunn et al., 2020; Krug et al., 2017). As a result, an increase in the eddy kinetic energy of the Agulhas can be expected to lead to a cooling of East African shelf waters with possible impacts on local productivity, although this remains to be substantiated.

4.4 Eastern boundary

The Leeuwin Current is the surface expression of the Leeuwin Current System that is composed of: the poleward-flowing Leeuwin Current; the equatorward-flowing Leeuwin Undercurrent that follows the continental slope beneath and just offshore of the Leeuwin Current; and the zonal flows in the interior ocean that interact with the Leeuwin Current and Leeuwin Undercurrent (Furue et al., 2017). The Leeuwin Current is approximately 200–300 m deep, extends from 22°S (North West Cape) to 34°S (Cape Leeuwin), and exists throughout the year despite significant seasonality (Feng et al., 2003; Furue et al., 2017; Ridgway & Godfrey, 2015). It owes its existence to the strong meridional pressure gradient created by the presence of the warm, fresh ITF waters in the South Equatorial Current near 10°S and denser subtropical waters to the south (e.g., Godfrey & Ridgway, 1985; Godfrey & Weaver, 1989, 1991; Wyrtki, 1973). This poleward decrease in pressure (and sea surface height) in the upper 200–300 m drives eastward geostrophic currents from as far west as Madagascar toward the eastern boundary (Divakaran & Brassington, 2011; Domingues et al., 2007; Menezes et al., 2014; Palastanga et al., 2007). As the bands of flow approach the coast of Australia, they veer southward, inducing downwelling and a surface poleward current in opposition to southerly winds (Fig. 1; Godfrey & Ridgway, 1985; McCreary et al., 1986; Thompson, 1987; Weaver & Middleton, 1989, 1990; Furue et al., 2013; Benthuysen et al., 2014). The southerly winds drive an offshore Ekman transport that is overwhelmed by the stronger onshore, eastward flows.

The primary source waters to the Leeuwin Current are thus the near-surface eastward flows (Section 4.2), including the South Indian Countercurrent jets (Menezes et al., 2014), re-circulations from the ITF via the Eastern Gyral Current (Meyers et al., 1995), with a weaker source from the Holloway Current, which flows southwestward on the Australian North West Shelf (Fig. 1; Bahmanpour et al., 2016; D'Adamo et al., 2009). The Eastern Gyral Current is an eastward flowing jet around 15°S that provides a rapid route to recirculate ITF waters back toward Australia and into the Leeuwin Current (Domingues et al., 2007; Menezes et al., 2013, and references therein). In the north, the Leeuwin Current waters are relatively fresh and warm from their tropical origins (Andrews, 1977; Domingues et al., 2007; Legeckis & Cresswell, 1981; Rochford, 1969; Wijffels & Meyers, 2004; Woo & Pattiaratchi, 2008). As the Leeuwin Current flows poleward, the saltier surface water of the subtropical South Indian Ocean joins the Leeuwin Current via the near-surface eastward flows. The density of the Leeuwin Current increases in response and is also increased due to surface cooling as it flows poleward (Furue, 2019; Woo & Pattiaratchi, 2008). As well as supplying source waters, the South Indian Countercurrent and Eastern Gyral Current, and possibly the Holloway Current, are essential components of the Leeuwin Current dynamics (Benthuysen et al., 2014; Domingues et al., 2007; Furue et al., 2013, 2017; Lambert et al., 2016; Yit Sen Bull & van Sebille, 2016).

The Leeuwin Current turns around the southwestern corner of Australia and continues eastward as the Leeuwin Current extension that supplies the shelf-break currents of the South Australia Current System, reaching as far east as the southern tip of Tasmania near 42°S, 140°E (Fig. 1; Duran et al., 2020; Oke et al., 2018; Oliver et al., 2016; Sprintall et al., 2024). This 5500-km-long boundary current was first documented as a continuous flow by Ridgway and Condie (2004). When the

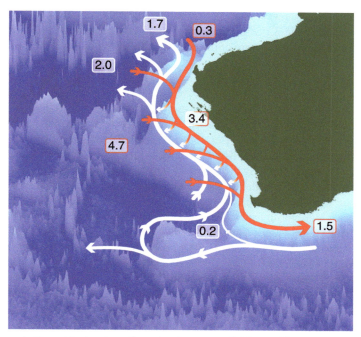

FIG. 9 Schematic summary of Leeuwin Current System three-dimensional transports (Sv) from a 1/8-degree observational climatology. The *red arrows* and *red-outline numbers* represent the upper-layer (0–200 m) meridional transport of the poleward Leeuwin Current and meridionally integrated zonal transport of the shallow eastward flows. The *white arrows* represent the lower-layer (200–900 m) flows of the Leeuwin Undercurrent. The *arrows* and *numbers* that are both *red* and *white* represent downwelling from the upper layer to the lower layer. *(Taken from Furue et al. (2017).)*

longshore current is weak, however, it tends to be somewhat fragmentary (Duran et al., 2020; Oke et al., 2018) and sometimes even reverses in places (Duran et al., 2020).

The Leeuwin Current is now understood to be part of a three-dimensional circulation that includes exchanges with the interior zonal flows in the upper 1000 m (Domingues et al., 2007; Furue et al., 2017). In this circulation, near-surface eastward flow enters the Leeuwin Current, flows poleward and at the same time sinks beneath the less dense tropical-origin Leeuwin Current waters to Leeuwin Undercurrent depths (200–900 m). In the Leeuwin Undercurrent, the flow is equatorward and turns offshore to move into the Indian Ocean interior around 22–28°S (Furue et al., 2017). Furue (2019) found that this mean downwelling in-depth coordinates is mostly along isopycnals and is likely a result of temporal variability.

The volume transport of the Leeuwin Current system estimated from a 1/8° observational climatology and corroborated by a 1/10 degree ocean general circulation model (Furue et al., 2017) is shown in Fig. 9. The annual mean Leeuwin Current carries 0.3 Sv southward at 22°S, gains 4.7 Sv from the near-surface eastward flows, loses 3.5 Sv through downwelling into the Leeuwin Undercurrent, and carries 1.5 Sv at its southern limit. A poleward transport of 1.5 Sv is lower than earlier estimates (e.g., Feng et al., 2003; Smith et al., 1991). Furue et al. (2017) found that the earlier estimates were too high primarily as a result of including energetic eddy transports and offshore flows that were not dynamically part of the Leeuwin Current. The choice of level of no motion for geostrophic calculations also contributed to the difference. While the 1.5 Sv meridional transport of the Leeuwin Current itself is relatively weak, it hosts a zonal overturning circulation of around 5 Sv in the upper 1000 m and generates a rich eddy field that carries heat, freshwater, and momentum into the interior Indian Ocean (Cyriac et al., 2019, 2021, 2022; Domingues et al., 2006; Feng et al., 2005), contributing to meridional exchanges of heat and other properties across the southern boundary.

5 Overturning circulations

5.1 Introduction

The Indian Ocean hosts intricately layered horizontal pathways that require vertical exchanges between layers to conserve volume and close the global meridional overturning circulation. The Indian Ocean upwelling is estimated to represent 40% of the upwelling required to close the global overturning, which is remarkable given the Indian Ocean only represents 20%

of the area of the world's oceans (Huussen et al., 2012). Sprintall et al. (2024) describe the overturning circulation from the perspective of water masses crossing the 32°S repeat hydrographic line, nominally the southern boundary of the Indian Ocean. In this section, we attempt to build a picture of the pathways followed by water masses within the overturning cells.

5.2 Shallow overturning cells

The shallow overturning circulation consists of subduction at mid-latitudes and upwelling at lower latitudes (Fig. 2) and is generally explained by ventilated thermocline theory, which describes the response of the ocean to wind stress curl, tropical heating, and mid-latitude cooling at the sea surface (Liu, 1994; Lu & McCreary, 1995; Luyten et al., 1983; McCreary & Lu, 1994; Pedlosky, 1996). The effect of surface heat flux is usually expressed by the meridional gradient of surface density. The shallow overturning resides in approximately the upper 1000 m of the ocean and significantly contributes to meridional heat transport (Klinger & Marotzke, 2000). In the Indian Ocean, the primary region of subduction is at mid-latitudes south of 20°S including the Antarctic Circumpolar Current region (Karstensen & Quadfasel, 2002; Liu & Huang, 2012; Sallée et al., 2010). Three water masses, high salinity subtropical water (Rochford, 1964; Talley & Baringer, 1997; Wijffels et al., 2002), Indian Ocean subtropical mode water (Tsubouchi et al., 2010), and Subantarctic Mode Water (Cerovečki et al., 2013; Herraiz-Borreguero & Rintoul, 2011; Sallée et al., 2006) subduct from the surface mixed layer into the ocean interior at mid-latitudes. A part of them flows into the Southern Hemisphere western boundary and turns poleward in the subtropical gyre (Jones et al., 2016; Nagura & McPhaden, 2018). The remaining water flows equatorward via the western boundary (Nagura & McPhaden, 2018) and eventually upwells to the surface in the Northern Hemisphere (Miyama et al., 2003). Substantial modification of water properties occurs along the interior pathways as a result of horizontal and vertical mixing with surrounding water masses (Cyriac et al., 2019, 2021; Talley, 2013).

There are two main upwelling regions in the Indian Ocean, i.e., the Seychelles-Chagos Thermocline Ridge region (near 10°S, 60°E) and the coastal region of Somalia (Fig. 2; Han, 2021; Schott et al., 2002). Both are located off the equator. This is a unique characteristic of the Indian Ocean circulation (Schott et al., 2004). In the Pacific and Atlantic Oceans, water subducts at mid-latitudes and upwells near the equator, driven by mid-latitude westerly and tropical easterly winds. Owing to the dominance of monsoonal winds and the absence of easterlies near the equator, there is no equatorial upwelling in the Indian Ocean resulting in an overturning cell that is meridionally asymmetric about the equator. A portion of the subducted water is transported across the equator by the Somali current, upwells in the Northern Hemisphere along the coast of Somalia, and returns to the Southern Hemisphere being advected by the surface flow (Han, 2021; Horii et al., 2013; Miyama et al., 2003; Pérez-Hernández et al., 2012; Wang & McPhaden, 2017; Zhang et al., 2021). This circulation is called the cross-equatorial cell (Fig. 2), which drives heat transport across the equator seasonally directed from the summer to the winter hemisphere to moderate the climate of the region (Loschnigg & Webster, 2000). The strength of the cross-equatorial cell is reported to show marked variability on interannual to decadal timescales (Meng et al., 2020; Schoenefeldt & Schott, 2006). This likely causes variability in the heat content of the northern Indian Ocean, as it is blocked to the north by land and the balance between cross-equatorial heat transport and sea surface heat flux determines the heat budget.

In addition to these cells, there is a vertical cell confined to the Arabian Sea, which consists of the subduction of Arabian Sea high salinity water (Kumar & Prasad, 1999; Prasad & Ikeda, 2002) and Persian Gulf water (Morrison, 1997; Prasad et al., 2001; Rochford, 1964) in the northern Arabian Sea and upwelling off Somalia.

5.3 Deep circulation

The deep circulation consists primarily of the North Atlantic Deep Water inflow in the western Indian Ocean and export of Indian Deep Water, which is a blend of North Atlantic Deep Water, Pacific Deep Water entering via the ITF, and dense waters formed in the Persian Gulf and Arabian Sea. Fig. 10 illustrates the vertical overturning structure and the adiabatic and diffusive upwelling processes that drive the upwelling needed to close the overturning (Talley, 2013). Inverse calculations using hydrographic observations indicate strong diffusive, cross-isopycnal upwelling in the Indian Ocean (e.g., Bryden & Beal, 2001; Ferron & Marotzke, 2003; Ganachaud & Wunsch, 2000; McDonagh et al., 2008; Sloyan & Rintoul, 2001). However, estimates from fine-scale measurements of the diapycnal (across-isopycnal) diffusivity required to accomplish this upwelling are found to be too weak at deep levels (Huussen et al., 2012; Katsumata et al., 2021). Huussen et al. (2012) found that the dissipation of internal wave energy at mid-ocean density levels (1000–3000 m) was a factor of 5–10 smaller than the dissipation needed to sustain the meridional overturning circulation at 32°S in the Indian Ocean at strengths indicated by the inverse calculations. They discounted the possibility that under-sampling of breaking internal wave hot spots could account for the difference and concluded that mixing processes other than wave breaking may be significant at deep levels.

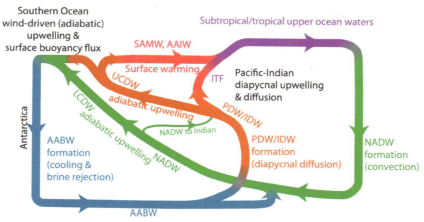

FIG. 10 Schematic of the overturning circulation in a two-dimensional view, with important physical processes listed. Colors indicate depth levels: *purple* = upper ocean and thermocline; *red* = denser thermocline and intermediate water; *orange* = Indian Deep Water and Pacific Deep Water; *green* = North Atlantic Deep Water; *blue* = Antarctic Bottom Water. Most complete version of the overturning (from Talley, 2013, revised from Talley et al., 2011), including North Atlantic Deep Water (NADW) and Antarctic Bottom Water (AABW) cells, and upwelling in the Southern, Indian, and Pacific Oceans. *SAMW*, Subantarctic Mode Water; *AAIW*, Antarctic Intermediate Water; *ITF*, Indonesian Throughflow; *PDW*, Pacific Deep Water; *IDW*, Indian Deep Water; *LCDW*, Lower Circumpolar Deep Water.

Katsumata et al. (2021) found a similar too-low estimate of the dissipation of internal wave energy in the deep layer. However, accounting for unobserved mixing events that are 10 times stronger than the mean, and enhanced mixing below 2000m across a large fraction of the Indian Ocean not sampled by the hydrographic transects, elevates their estimates of dissipation to within the error bars of many of the inverse estimates. Fig. 11a shows the spatial pattern of dissipation rate from the empirical model of Katsumata et al. (2021) that accounts for unobserved mixing due to intense wind events and unobserved topographic enhancement of deep mixing. Estimates of dissipation from research vessel profiles of temperature and salinity along hydrographic lines are overlaid as circles with matching colorbar. There is broad agreement with clear enhanced dissipation over significant topographic features and low diffusivity over deep abyssal plains.

The seasonal wind forcing of the northern Indian Ocean, driving seasonal variations in Ekman currents has also been proposed as a mechanism that influences the full-depth Indian Ocean meridional overturning circulation (e.g., Garternicht & Schott, 1997; Han & Huang, 2020; Lee & Marotzke, 1997; Schott & McCreary, 2001; Wang et al., 2014). Schott and McCreary (2001) suggested that this seasonal variability represents a sloshing back and forth of water masses along isopycnals that reflects a change in ocean heat storage rather than a change in the strength of the overturning circulation. Han (2021) investigated this hypothesis using the ECCO state estimate and found that the strength of the seasonal meridional overturning circulation in the Indian Ocean is driven by the zonally integrated Ekman pumping anomaly rather than the Ekman transport concluded in earlier studies. Han (2021) decomposed the overturning circulation into an across-isopycnal overturning and a seasonal along-isopycnal sloshing mode. Han (2021) proposed that the larger diapycnal volume fluxes across 32°S from inverse calculations are an overestimate caused by the seasonal sloshing mode that is not a diapycnal process and so not measured by fine-scale observations of dissipation.

5.4 Abyssal circulation

The abyssal circulation, by which we mean deeper than 3500m, is physically constrained by the seafloor topography. In the Indian Ocean, the complex distribution of ridges and plateaus creates several semi-isolated deep basins (Fig. 11b). The communication between them is by narrow, deep fracture zones and passages, which are predominantly north-south oriented. Because of topographic constraints, there is no strong exchange in the abyss between the western and eastern Indian Ocean (Mantyla & Reid, 1995; Reid, 2003).

The Antarctic Bottom Water, the Lower Circumpolar Deep Water, and the abyssal circulation are intrinsically connected. The cold signature of these water masses functions as a tracer for the abyssal currents (Fig. 11b). Bottom water

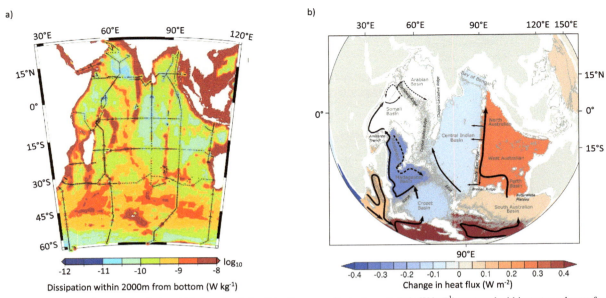

FIG. 11 Abyssal dissipation and circulation. (a) Inferred turbulent kinetic energy dissipation rate (ϵ in W kg^{-1}) averaged within water columns from the bottom to 2000 m above the bottom shown in a logarithmic scale on a 1° × 1° grid. The *black crosses with colored circles* show the observed dissipation rate plotted over the modeled estimates. (b) Schematic of abyssal circulation. *Dashed curves* highlight southward currents. *Grey contours* are the 3000 and 3500 m isobath from ETOPO-2 data. *Colors* indicate the mean local heat fluxes (over the area of each basin) at the 3000 m level required to account for the rate of warming/cooling of the water below it. SWIR stands for Southwest Indian Ridge and SEIR for Southeast Indian Ridge. *((a) Figure modified from Katsumata et al. (2021). (b) Data are from Purkey and Johnson (2010) and available at https://floats.pmel.noaa.gov/gregory-c-johnson-home-page.)*

enters the Indian Ocean by the open southern boundaries of the Mozambique, Crozet, and Perth Basins, but only the last two pathways allow further northward transport as the shallow Mozambique Channel is a physical barrier.

Unlike the Atlantic and Pacific Oceans, bottom water spreads northward on both the western and eastern sides of the Indian Ocean (Mantyla & Reid, 1995). On the western side, it reaches the subtropical Madagascar Basin (32°S–20°S) through the deep fracture zones of the Southwest Indian Ridge that separates the Crozet and Madagascar Basins. On the eastern side, it reaches the Perth Basin (35°S–25°S) through the broad gap between the Broken Ridge and the Naturaliste Plateau in the South Australian Basin.

A noteworthy difference between east and west is that the southwestern Indian Ocean is supplied by (a mixed version of) Antarctic Bottom Water formed primarily in the Weddell Sea in the South Atlantic portion of the Southern Ocean, while the southeast Indian Ocean is supplied by (a mixed version of) Antarctic Bottom Water formed in the Ross Sea and Adélie Coast regions in the Southern Ocean's Pacific sector (Mantyla & Reid, 1995). There are large uncertainties in the magnitude of bottom water transport crossing the southwestern and southeastern Indian Ocean (Sloyan, 2006; Srinivasan et al., 2009), as in situ observations of the abyss are scarce. However, it is believed that at least half of all northward deep and bottom water transport in the Indian Ocean is through the Southwest Indian Ridge fracture zones in the western Indian Ocean (MacKinnon et al., 2008).

5.4.1 Western Indian Ocean

In the western Indian Ocean, the bottom water is transported northward following the Crozet, Madagascar, Mascarene, Somali, and Arabian Basins. As a result of diapycnal mixing, the bottom water becomes progressively less dense as it moves northward. From the Madagascar Basin to the Arabian Basin, density decreases by up to ∼0.15 kg m^{-3} such that the bottom water in the Arabian Sea has the same density as water at about 2500–3000 m in the Crozet Basin (Mantyla & Reid, 1995). To date, we have only a rudimentary view of the abyssal circulation patterns within these basins.

From the four possible routes through the Southwest Indian Ridge via deep fracture zones (Indomed, Atlantis II, Novara, and Melville), the main gateway for the deep and bottom water is likely the Atlantis II at about 57°E, between 31°S and 33°S, with sill depths of 4200–4300 m. MacKinnon et al. (2008) used in situ temperature, salinity, and current measurements to estimate that this fracture zone would account for 20–30% of the deep Indian Ocean meridional overturning circulation transport. These measurements are the first and only direct evidence of an intense northward jet-like current that carries deep and bottom water (2500–4500 m) all the way along the Atlantis II fracture zone. This deep northward jet has meridional velocities above $10\,\mathrm{cm\,s^{-1}}$ with a maximum of about $30\,\mathrm{cm\,s^{-1}}$ at $3500\,\mathrm{m}$. Vertical mixing rates are about $10–100 \times 10^{-4}\,\mathrm{m^2\,s^{-1}}$ (Huussen et al., 2012; MacKinnon et al., 2008). No current measurements exist for the other deep fracture zones in the Southwest Indian Ridge to date.

Evidence suggests that nearly all bottom water inflow through the Atlantis II is carried by a northward-flowing deep western boundary current along the Madagascar Plateau and the Madagascar continental slope toward the Mascarene Basin (Donohue & Toole, 2003; Fieux et al., 1986; Warren, 1974, 1978, 1981). The long-term mean deep western boundary current velocity is largely unknown because of large bimonthly fluctuations ($O(10\,\mathrm{cm\,s^{-1}})$) associated with barotropic Rossby waves as described by Warren et al. (2002). The deep western boundary current transport seems to slightly increase from 2 Sv in the Madagascar Basin to 3 Sv in the Mascarene Basin (Donohue & Toole, 2003).

Steric height maps from Reid (2003) suggest that part of the northward flow turns eastward near 10°S and flows southward along the eastern boundary of these basins, forming a clockwise circulation. Srinivasan et al. (2009) described several permanent mesoscale eddy features in the Madagascar-Mascarene abyssal circulation based on an ocean general circulation model. One of these was a stationary clockwise circulation supplying the deep western boundary current with bottom water entering through the Southwest Indian Ridge fracture zones, consistent with the steric height maps of Reid (2003).

From the Mascarene Basin, bottom water flows to the Somali Basin through the Amirante Trench (4600–5200 m deep; 7.8–9.1°S ~ 52.5°W) between Madagascar and the boomerang-shaped Mascarene Plateau (150 m deep). Estimates of Johnson et al. (1998) indicate a northward transport of 1–1.7 Sv of bottom water within the trench.

Dengler et al. (2002) inferred that the bulk of the bottom water inflow entering the Somali Basin spreads northward across the Equator. North of 7°N, it would leave the Somali Basin through the Owen fracture zone (depths of 3800 m) in the Carlsberg Ridge (~56°E) to ventilate the abyss of the Arabian Basin. The northern portion of the Somali Basin should be dominated by an anticlockwise circulation with a southwestward deep western boundary current along the Somali slope. This cyclonic feature would be consistent with the dynamical framework of the Stommel and Arons abyssal mean circulation, as shown by Johnson et al. (1991a). It is possible that the Somali deep western boundary current also reverses direction seasonally (Beal et al., 2000b) in phase with surface winds, but this suggestion has been controversial (see, e.g., Dengler et al., 2002). The numerical model simulation of Srinivasan et al. (2009) suggests that the Somali deep western boundary current has a pronounced seasonal cycle: In boreal winter (January), the southward deep western boundary current is well developed and extends all the way to the equator; in summer (July), the deep western boundary current is undetectable in the model. Both modeling (Srinivasan et al., 2009) and observations (Dengler et al., 2002) indicate that the Somali deep western boundary current is dominated by intraseasonal oscillations.

In the Arabian Basin, a southeastern deep western boundary current along the Carlsberg Ridge has been inferred from observations (Johnson et al., 1991b).

5.4.2 Eastern Indian Ocean

The eastern Indian Ocean is supplied by Antarctic Bottom Water and Lower Circumpolar Deep Water through the broad gap between the Broken Ridge and the Naturaliste Plateau near 33°S,105°E (Mantyla & Reid, 1995; Reid, 2003). Sloyan (2006) estimated a net northward transport of 4.4–5.8 Sv below 3200 m based on in-situ observations (2.4–3.3 Sv for Lower Circumpolar Deep Water and 2–2.5 Sv for Antarctic Bottom Water).

On entering the Perth Basin, bottom water flows westward to the Ninety East Ridge (~88°E). There, a northward deep western boundary current along the eastern flank of the ridge carries bottom water to the West Australian and North Australian Basins and across the Equator (Reid, 2003; Srinivasan et al., 2009). Small westward overflows at depths below 3000 m are observed through the gaps in the Ninety East Ridge at 28°S (0.1 Sv), 10°S (1.1 Sv), and 5°S (1.0 Sv) (Toole & Warren, 1993; Warren & Johnson, 2002). These overflows supply the bottom water of the Central Indian Basin, where a deep western boundary current along the eastern flank of the Southeast Indian and Central Indian Ridges carries bottom water toward the Equator (Reid, 2003). Bottom water is lighter in the Central Indian Basin than in the most eastern basins of the South Indian Ocean (West and North Australian). However, bottom water has similar density and temperature-salinity properties in the extreme north in the Bay of Bengal and the Arabian Basin (Huussen et al., 2012; Mantyla & Reid, 1995).

5.4.3 Abyssal warming and cooling

Repeated hydrographic observations from the last three decades have revealed compelling evidence of temperature, salinity, density, and tracer changes in the deep oceanic layers (e.g., Purkey & Johnson, 2010). These changes can potentially affect circulation, oceanic heat and carbon uptake, and sea level.

In the Indian Ocean, over the past three decades, abyssal warming has been widespread on the eastern side (Fig. 11b, shading). On the western side, anomalous cooling seems to occur in the Crozet, Madagascar, and Mascarene Basins. Given the few hydrographic measurements and perhaps high-frequency natural variability in the southwest Indian Ocean abyssal layer (e.g., Warren et al., 2002), there is considerable uncertainty about the abyssal trends there. For instance, in the Crozet and Madagascar basins, uncertainties are ± 0.25 and $\pm 0.3\,\mathrm{W\,m^{-2}}$, respectively (Purkey & Johnson, 2010). Using a different approach based on the assimilation of the repeat hydrographic sections from 1989 to 2008 into a numerical model, Kouketsu et al. (2011) also found large uncertainties in the abyssal cooling of the Crozet and Madagascar Basins ($-9\pm 8\times 10^{-3}\,°C$ and $-23\pm 57\times 10^{-3}\,°C$, respectively).

If the abyssal cooling estimated from the repeat hydrographic sections in the western Indian Ocean is not due to sampling aliasing, possible causes are changes in the source Antarctic Bottom Water of the Weddell Sea, variability in the Crozet-Madagascar northward transport, and/or the Indian Ocean meridional overturning circulation strength. Following this train of thought, Kouketsu et al. (2011) describe the (model-derived) bottom transport at 35°S in the western Indian Ocean as increasing from 1970 to 1985, decreasing from 1985 to 1995, and increasing again after 1995 (their simulation ended around 2005). They speculated that the recent multidecadal abyssal cooling in the southwestern Indian Ocean might be due to the increase in the northward bottom water transport. However, they caution that the relationship between temperature and bottom water transport in the western Indian Ocean is not well constrained by observations because the variability of northward transport is relatively large and the topography is complicated.

6 Conclusions

The seasonally reversing monsoon winds drive seasonally reversing near-surface currents in the northern Indian Ocean. This complex network of currents provides pathways for the exchange of heat, freshwater, nutrients, and marine species between the Arabian Sea and the Bay of Bengal. Excess freshwater from monsoon precipitation in the Bay of Bengal and extreme evaporation in the western Arabian Sea set the conditions for sharp zonal gradients in properties that are kept in climatological balance by these exchanges, near the sea surface and below. Instabilities and mesoscale eddies modify these exchanges through vertical mixing and air-sea interactions. Communication between the Northern and Southern Hemispheres is primarily through Ekman transport in the interior and transport along the western boundary in the Somali Current system. The flow there is southward during the northeast monsoon and northward during the southwest monsoon when intense upwelling occurs along the coasts of Somalia and Oman. This upwelling returns waters subducted in the Southern Hemisphere back to the surface to communicate stored climate signals to the tropical atmosphere.

Along the equator in the Indian Ocean, the annual mean winds are westerly, downwelling-favorable, with a pronounced semi-annual cycle. This is in contrast to mean upwelling conditions along the equator in the eastern Pacific and Atlantic Oceans. Consequently, mean surface currents flow to the east rather than to the west and there is no permanent equatorial undercurrent. Wyrtki jets, strong eastward currents forced by zonal winds during the monsoon transition seasons, affect the large-scale variations in ocean mass and heat along the equator and are a key element of the development of IOD events. Equatorial waves propagating vertically into the deep ocean below the main thermocline affect velocity, pressure, temperature, and the density structure of the deep ocean. They are particularly prominent at semi-annual periods because of the monsoonal variations in wind stress. Equatorial deep jets, stacked zonal currents of alternating direction, extend over much of the water column, are confined to within 1–2° of the equator, and are coherent zonally across much of the basin. The timescale for these jets is ~5 years with slow downward phase propagation. They are important in the generation of turbulent mixing and for understanding tracer distributions. They are remarkably persistent against dissipation but are not simulated well in climate models.

The presence of the low-latitude connection to the Pacific Ocean, the ITF, creates an ocean circulation quite distinct from that found in the southern basins of the Atlantic and Pacific Oceans. As a result, poleward boundary transport is accomplished along the eastern boundary as well as the western boundary. Ekman transport at the eastern boundary by the upwelling favorable winds is overwhelmed by an onshore geostrophic flow driven by the meridional density gradient established by the ITF. Consequently, eastern boundary upwelling is present only at very small scales, and sporadically, and a downwelling environment dominates at the coast. The associated oligotrophic conditions are relieved to some degree by the more productive Leeuwin Current flow and westward propagation of anticyclonic eddies into the interior. The Leeuwin Current only transports 1.5 Sv poleward in the mean. However, it hosts a zonal overturning circulation of around 5 Sv in the upper 1000 m, and the generation of a rich eddy field that carries heat, freshwater, and momentum into the interior Indian Ocean, contributing to meridional heat transport across the southern boundary.

The depth-integrated circulation in the southern Indian Ocean reveals a familiar gyre circulation with the south equatorial current and Agulhas Current as its northern and western boundaries, respectively. The Agulhas transports 76 Sv of warm, salty water poleward at 34°S, contributing 3.8 PW of heat to the Southern Ocean, some of which is carried into the South Atlantic. A reduction in the meridional heat flux across 34°S is suggested to account for 60% of the recent warming of the Indian Ocean, with the other 40% linked to an increase in the ITF. The reduced poleward transport is likely due to the broadening of the Agulhas Current, related to an increase in eddy kinetic energy. This could lead to a cooling of East African shelf waters with possible impacts on local productivity. The Agulhas Return Current, the eastward extension of the Agulhas Current, appears to have strengthened while the Agulhas Current remains steady. This is probably because the westerly winds have strengthened, but the subtropical wind stress curl has not. Stronger westerlies also appear to be driving an increase in Agulhas leakage.

The shallow meridional overturning circulation, consisting of subduction at southern midlatitudes and upwelling at lower latitudes, resides in approximately the upper 1000 m of the ocean and includes high salinity subtropical water, Indian Ocean subtropical water, and Subantarctic Mode Water. Part of these subducted waters flows into the western boundary and turns poleward in the subtropical gyre and the remaining water flows equatorward. Upwelling in the Indian Ocean is located off the equator in the Seychelles-Chagos Thermocline Ridge region near 10°S and along the Somali coast. This is a unique characteristic of the Indian Ocean circulation, unlike in the Pacific and Atlantic Oceans, where water subducts at midlatitudes and upwells along the equator. The portion that is transported across the equator by the Somali current, upwells off Somalia, and returns to the Southern Hemisphere in the surface flow is called the cross-equatorial cell. It drives heat transport across the equator from the summer to the winter hemisphere and varies in strength on interannual to decadal timescales, contributing to variability in the heat content of the northern Indian Ocean. The portion that upwells at the Seychelles-Chagos thermocline ridge is called the Southern Cell.

The Indian Ocean is thought to play a substantial role in the upwelling of deep and abyssal waters to close the global meridional overturning circulation, based on inverse calculations from hydrographic lines. However, fine-scale observational estimates of the turbulent dissipation of kinetic energy are much weaker at deep levels than the inverse solutions imply. Accounting for unobserved strong mixing events and the enhancement of deep dissipation around rough topography goes someway to closing the gap between observations of dissipation and that required to close the Indian Ocean overturning circulation. A recent investigation of a seasonal sloshing of water masses, in response to Ekman pumping by the reversing monsoon winds, may also account for some of the upwelling, suggesting that the diapycnal diffusivity from inverse solutions has been overestimated.

7 Educational resources

The Global Drifter Program of the National Oceanic and Atmospheric Administration has produced animations of sea surface currents around the world, including the tropical Indian Ocean https://www.aoml.noaa.gov/phod/gdp/animations.php . A visualization of the speed and direction of surface currents from a daily climatology may be found here https://www.aoml.noaa.gov/phod/graphics/dacdata/seasonal_tropind.gif.

NASA's Scientific Visualization Studio's animation Perpetual Ocean provides a visualization of ocean surface currents around the world during the period from June 2005 through December 2007. This visualization was produced using model output from the joint MIT/Jet Propulsion Laboratory project: Estimating the Circulation and Climate of the Ocean, Phase II or ECCO2 https://svs.gsfc.nasa.gov/3827.

Acknowledgments

HP acknowledges support from the Australian Government under the National Environmental Science Program, the Antarctic Science Collaboration Initiative (ASCI000002), and the Australian Research Council Discovery Projects (DP210100643). This research was supported by the Australian Research Council Special Research Initiative, Australian Centre for Excellence in Antarctic Science (Project Number SR200100008). VVM acknowledges support from the National Science Foundation (OCE1736823, OCE1924431, and OCE2122964), and National Aeronautics and Space Administration (80NSSC18K1333). MN is supported by the Japan Agency for Marine-Earth Science and Technology and JSPS Grants-in-Aid for Scientific Research (KAKENHI) grant (JP18K03750). MJM is supported by NOAA. PNV acknowledges partial financial support from J C Bose National Fellowship, SERB, DST, Govt. India, and the BoBBLE program funded by the Ministry of Earth Sciences, Govt. of India under its Monsoon Mission program. We thank Lynne Talley for comments on an earlier version of the chapter.

Author contributions

HP led the writing, editing, and organization of the manuscript. Author HP wrote Sections 1 and 6; authors MN, PNV, VM, and HP wrote Section 2; author MJM wrote Section 3; authors MN, LB, and HP wrote Section 4; authors VM, MN, and HP wrote Section 5. VM contributed Fig. 11b. All authors contributed to the discussion of content and overall structure and editing of the chapter.

References

Amol, P., Shankar, D., Aparna, G., Shenoi, S. S. C., Fernando, Vijayan, F., Shetye, S. R., Mukherjee, A., Agarvadekar, Y., Khalap, S., & Satelkar, N. (2012). Observational evidence from direct current measurements for propagation of remotely forced waves on the shelf off the west coast of India. *Journal of Geophysical Research, Oceans, 117*. https://doi.org/10.1029/2011JC007606.

Amol, P., Shankar, D., Fernando, V., Mukherjee, A., Aparna, S. G., Fernandes, R., Michael, G. S., Khalap, S. T., Satelkar, N. P., Agarvadekar, Y., Gaonkar, M. G., Tari, A. P., Kankonkar, A., & Vernekar, S. P. (2014). Observed intraseasonal and seasonal variability of the West India Coastal Current on the continental slope. *Journal of Earth System Science, 123*, 1045–1074. https://doi.org/10.1007/s12040-014-0449-5.

Andrews, J. C. (1977). Eddy structure and the West Australian current. *Deep Sea Research, 24*(12), 1133–1148. https://doi.org/10.1016/0146-6291(77)90517-3.

Annamalai, H., Taguchi, B., McCreary, J. P., Nagura, M., & Miyama, T. (2017). Systematic errors in South Asian Monsoon simulation: Importance of equatorial indian ocean processes. *Journal of Climate, 30*(20), 8159–8178.

Anutaliya, A., Send, U., Mcclean, J., Sprintall, J., Rainville, L., Lee, M., Jinadasa, S., Wallcraft, A. J., & Metzger, E. (2017). An undercurrent off the east coast of Sri Lanka. *Ocean Science Discussions, 13*, 1–15.

Backeberg, B. C., Penven, P., & Rouault, M. (2012). Impact of intensified Indian Ocean winds on mesoscale variability in the Agulhas system. *Nature Climate Change, 2*(8), 608–612.

Bahmanpour, M. H., Pattiaratchi, C., Wijeratne, E. M. S., Steinberg, C., & D'Adamo, N. (2016). Multi-year observation of Holloway Current along the shelf edge of North Western Australia. *Journal of Coastal Research, 517–521*. https://doi.org/10.2112/SI75-104.1.

Bastin, S., Claus, M., Brandt, P., & Greatbatch, R. J. (2020). Equatorial deep jets and their influence on the mean equatorial circulation in an idealized ocean model forced by intraseasonal momentum flux convergence. *Geophysical Research Letters, 47*. https://doi.org/10.1029/2020GL087808.

Beal, L. M., & Chereskin, T. K. (2003). The volume transport of the Somali Current during the 1995 southwest monsoon. *Deep Sea 350 Research Part II: Topical Studies in Oceanography, 50*, 2077–2089.

Beal, L. M., & Donohue, K. A. (2013). The Great Whirl: Observations of its seasonal development and interannual variability. *Journal of Geophysical Research: Oceans, 118*(1), 1–13. https://doi.org/10.1029/2012jc008198.

Beal, L. M., & Elipot, S. (2016). Broadening not strengthening of the Agulhas Current since the early 1990s. *Nature, 540*(7634), 570–573.

Beal, L. M., Elipot, S., Houk, A., & Leber, G. M. (2015). Capturing the transport variability of a western boundary jet: Results from the Agulhas Current Time-Series Experiment (ACT). *Journal of Physical Oceanography, 45*(5), 1302–1324.

Beal, L. M., Ffield, A., & Gordon, A. L. (2000a). Spreading of Red Sea overflow waters in the Indian Ocean. *Journal of Geophysical Research, 105*, 8549–8564. https://doi.org/10.1029/1999JC900306.

Beal, L. M., Hormann, V., Lumpkin, R., & Foltz, G. R. (2013). The response of the surface circulation of the Arabian Sea to Monsoonal Forcing. *Journal of Physical Oceanography, 43*(9), 2008–2022. https://journals.ametsoc.org/view/journals/phoc/43/9/jpo-d-13-033.1.xml.

Beal, L. M., Vialard, J., Roxy, M. K., Li, J., Andres, M., Annamalai, H., & Parvathi, V. (2020). A road map to IndOOS-2: Better observations of the rapidly warming Indian Ocean. *Bulletin of the American Meteorological Society, 101*(11), E1891–E1913.

Beal, L. M., Molinari, R. L., Chereskin, T. K., & Robbins, P. E. (2000b). Reversing bottom circulation in the Somali Basin. *Geophysical Research Letters, 27*, 2565–2568. https://doi.org/10.1029/1999GL011316.

Beal, L. M., Vialard, J., Roxy, M. K., et al. (2019). *IndOOS-2: A roadmap to sustained observations of the Indian Ocean for 2020–2030*. CLIVAR-4/2019, GOOS-237, 204 pp. https://doi.org/10.36071/clivar.rp.4.2019.

Behara, A., & Vinayachandran, P. N. (2016). An OGCM study of the impact of rain and river water forcing on the Bay of Bengal. *Journal of Geophysical Research, 121*, 2425–2446. https://doi.org/10.1002/2015JC011325.

Behara, A., Vinayachandran, P. N., & Shankar, D. (2019). Influence of rainfall over eastern Arabian Sea on its salinity. *Journal of Geophysical Research, Oceans, 124*. https://doi.org/10.1029/2019JC014999.

Bender, C. M., & Orszag, S. A. (1999). WKB theory. In *Advanced mathematical methods for scientists and engineers I*. New York, NY: Springer. https://doi.org/10.1007/978-1-4757-3069-2_10.

Benthuysen, J., Furue, R., McCreary, J. P., Bindoff, N. L., & Phillips, H. E. (2014). Dynamics of the Leeuwin Current: Part 2. Impacts of mixing, friction, and advection on a buoyancy-driven eastern boundary current over a shelf. *Dynamics of Atmospheres and Oceans, 65*, 39–63. https://doi.org/10.1016/j.dynatmoce.2013.10.004.

Biastoch, A., & Böning, C. W. (2013). Anthropogenic impact on Agulhas leakage. *Geophysical Research Letters, 40*(6), 1138–1143.

Bower, A. S., & Farrar, J. T. (2015). Air-sea interaction and horizontal circulation in the Red Sea. In N. M. A. Rasul, & I. C. F. Stewart (Eds.), *The Red Sea* (pp. 329–342). Berlin, Heidelberg: Springer. https://doi.org/10.1007/978-3-662-45201-1_19.

Bower, A. S., Fratantoni, D. M., Johns, W. E., & Peters, H. (2002). Gulf of Aden eddies and their impact on Red Sea Water. *Geophysical Research Letters, 29*, 2025. https://doi.org/10.1029/2002GL015342.

Bower, A. S., & Furey, H. H. (2012). Mesoscale eddies in the Gulf of Aden and their impact on the spreading of Red Sea Outflow Water. *Progress in Oceanography, 96*, 14–39. https://doi.org/10.1016/j.pocean.2011.09.003.

Bower, A. S., Hunt, H. D., & Price, J. F. (2000). Character and dynamics of the Red Sea and Persian Gulf outflows. *Journal of Geophysical Research, 105*, 6387–6414. https://doi.org/10.1029/1999JC900297.

Bower, A. S., Johns, W. E., Fratantoni, D. M., & Peters, H. (2005). Equilibration and circulation of Red Sea outflow water in the western Gulf of Aden. *Journal of Physical Oceanography, 35*, 1963–1985. https://doi.org/10.1175/JPO2787.1.

Bray, N. A., Wijffels, S. E., Chong, J. C., Fieux, M., Hautala, S., Meyers, G., & Morawitz, W. M. L. (1997). Characteristics of the Indo-Pacific throughflow in the eastern Indian Ocean. *Geophysical Research Letters, 24*, 2569–2572. https://doi.org/10.1029/97GL51793.

Bruce, J. G., & Volkmann, G. H. (1969). Some measurements of current off the Somali Coast during the Northeast Monsoon. *Journal of Geophysical Research, 74*(8), 1958–1967. https://doi.org/10.1029/jb074i008p01958.

Bryden, H. L., & Beal, L. M. (2001). Role of the Agulhas Current in Indian Ocean circulation and associated heat and freshwater fluxes. *Deep Sea Research Part I: Oceanographic Research Papers, 48*, 1821–1845.

Cerovečki, I., Talley, L. D., Mazloff, M. R., & Maze, G. (2013). Subantarctic Mode Water formation, destruction, and export in the eddy-permitting Southern Ocean state estate. *Journal of Physical Oceanography, 43*, 1485–1511. https://doi.org/10.1175/JPO-D-12-0121.1.

Chaitanya, A., Lengaigne, M., Vialard, J., Gopalakrishna, V., Durand, F., Kranthikumar, C., Amritash, S., Suneel, V., Papa, F., & Ravichandran, M. (2014). Salinity measurements collected by fishermen reveal a river in the sea flowing along the eastern coast of India. *Bulletin of the American Meteorological Society, 95*(12), 1897–1908. https://doi.org/10.1175/BAMS-D-12-00243.1.

Chatterjee, A., Kumar, B. P., Prakash, S., & Singh, P. (2019). Annihilation of the Somali upwelling system during summer monsoon. *Scientific Reports, 9*(1), 7598. https://doi.org/10.1038/s41598-019-44099-1.

Chatterjee, A., Shankar, D., Shenoi, S. S. C., Reddy, G. V., Michael, G. S., Ravichandran, M., Gopalkrishna, V. V., Rao, E. P. R., Bhaskar, T. V. S. U., & Sanjeevan, V. N. (2012). A new atlas of temperature and salinity for the North Indian Ocean. *Journal of Earth System Science, 121*(3), 559–593. https://doi.org/10.1007/s12040-012-0191-9.

Chaudhuri, A., Shankar, D., Aparna, S. G., Amol, P., Fernando, V., Kankonkar, A., Michael, G. S., Satelkar, N. P., Khalap, S. T., Tari, A. P., Gaonkar, M. G., Ghatkar, S., & Khedekar, R. R. (2020). Observed variability of the West India Coastal Current on the continental slope from 2009–2018. *Journal of Earth System Science, 129*, 57. https://doi.org/10.1007/s12040-019-1322-3.

Chen, G., Han, W., Li, Y., Wang, D., & McPhaden, M. J. (2015). Seasonal-to-interannual time-scale dynamics of the equatorial undercurrent in the Indian Ocean. *Journal of Physical Oceanography, 45*, 1532–1553. https://doi.org/10.1175/JPO-D-14-0225.1.

Chen, G., Han, W., Zhang, X., Liang, L., Xue, H., Huang, K., He, Y., & Li, W. D. (2020). Determination of spatiotemporal variability of the Indian equatorial intermediate current. *Journal of Physical Oceanography, 50*, 3095–3108. https://doi.org/10.1175/JPO-D-20-0042.1.

Chen, Z., Wu, L., Qiu, B., Sun, S., & Jia, F. (2014). Seasonal variation of the south equatorial current bifurcation off Madagascar. *Journal of Physical Oceanography, 44*, 618–631. https://doi.org/10.1175/JPO-D-13-0147.1.

Cheng, L., Trenberth, K. E., Fasullo, J., Boyer, T., Abraham, J., & Zhu, J. (2017). Improved estimates of ocean heat content from 1960 to 2015. *Science Advances, 3*, e1601545. https://doi.org/10.1126/sciadv.1601545.

Clarke, A. J. (1991). On the reflection and transmission of low- frequency energy at the irregular western Pacific Ocean boundary. *Journal of Geophysical Research, 96*, 3289–3305. https://doi.org/10.1029/90JC00985.

Clarke, A. J., & Liu, X. (1994). Interannual sea level in the northern and eastern Indian Ocean. *Journal of Physical Oceanography, 24*, 1224–1235. https://doi.org/10.1175/1520-0485(1994)024,1224:ISLITN.2.0.CO;2.

Cullen, K. E., & Shroyer, E. L. (2019). Seasonality and interannual variability of the Sri Lanka dome. *Deep Sea Research Part II: Topical Studies in Oceanography, 168*, 104642.

Cyriac, A., McPhaden, M., Phillips, H., Bindoff, N., & Feng, M. (2019). Surface layer heat balance in the subtropical Indian Ocean. *Journal of Geophysical Research, Oceans, 124*, 6459–6477. https://doi.org/10.1029/2018JC014559.

Cyriac, A., Phillips, H. E., Bindoff, N. L., & Feng, M. (2022). Characteristics of wind-generated near-inertial waves in the Southeast Indian Ocean. *Journal of Physical Oceanography, 52*, 557–578.

Cyriac, A., Phillips, H. E., Bindoff, N. L., Mao, H., & Feng, M. (2021). Observational estimates of turbulent mixing in the southeast Indian Ocean. *Journal of Physical Oceanography*. https://doi.org/10.1175/jpo-d-20-0036.1.

D'Adamo, N., Fandry, C., Buchan, S., & Domingues, C. (2009). Northern sources of the Leeuwin Current and the "Holloway Current" on the North West Shelf. *Journal of the Royal Society of Western Australia, 92*(2), 53–66. http://nora.nerc.ac.uk/id/eprint/526029.

Daher, H., Beal, L. M., & Schwarzkopf, F. U. (2020). A new improved estimation of Agulhas Leakage using observations and simulations of Lagrangian floats and drifters. *Journal of Geophysical Research, Oceans, 125*. https://doi.org/10.1029/2019JC015753.

de Boyer Montégut, C., Madec, G., Fischer, A. S., Lazar, A., & Iudicone, D. (2004). Mixed layer depth over the global ocean: An examination of profile data and a profile-based climatology. *Journal of Geophysical Research, 109*, C12003. https://doi.org/10.1029/2004JC002378.

de Vos, A., Pattiaratchi, C. B., & Wijeratne, E. M. S. (2014). Surface circulation and upwelling patterns around Sri Lanka. *Biogeosciences*. https://doi.org/10.5194/bg-11-5909-2014.

Dengler, M., & Quadfasel, D. (2002). Equatorial deep jets and abyssal mixing in the Indian Ocean. *Journal of Physical Oceanography, 32*, 1165–1180.

Dengler, M., Quadfasel, D., Schott, F., & Fischer, J. (2002). Abyssal circulation in the Somali Basin. *Deep Sea Research, Part II, 49*, 1297–1322. https://doi.org/10.1016/S0967-0645(01)00167-9.

Desbruyères, D., McDonagh, E. L., King, B. A., & Thierry, V. (2017). Global and full-depth ocean temperature trends during the early twenty-first century from Argo and repeat hydrography. *Journal of Climate, 30*, 1985–1997. https://doi.org/10.1175/JCLI-D-16-0396.1.

Divakaran, P., & Brassington, G. B. (2011). Arterial ocean circulation of the southeast Indian Ocean. *Geophysical Research Letters, 38*, L01802. https://doi.org/10.1029/2010GL045574.

Domingues, C. M., Maltrud, M. E., Wijffels, S. E., Church, J. A., & Tomczak, M. (2007). Simulated Lagrangian pathways between the Leeuwin Current System and the upper-ocean circulation of the southeast Indian Ocean. *Deep-Sea Research Part II, 54*, 797–817. https://doi.org/10.1016/j.dsr2.2006.10.003.

Domingues, C. M., Wijffels, S. E., Maltrud, M. E., Church, J. A., & Tomczak, M. (2006). Role of eddies in cooling the Leeuwin Current. *Geophysical Research Letters, 33*.

Donohue, K. A., & Toole, J. M. (2003). A near-synoptic survey of the Southwest Indian Ocean. *Deep-Sea Research Part II, 50*, 1893–1931. https://doi.org/10.1016/S0967-0645(03)00039-0.

Ducet, N., Traon, P. L., & Reverdin, G. (2000). Global high-resolution mapping of ocean circulation from TOPEX/Poseidonand ERS-1 and -2. *Journal of Geophysical Research, 105*, 19477–19498. https://doi.org/10.1029/2000JC900063.

Duran, E. R., Phillips, H. E., Furue, R., Spence, P., & Bindoff, N. L. (2020). Southern Australia Current System based on a gridded hydrography and a high-resolution model. *Progress in Oceanography, 181*, 102254. https://doi.org/10.1016/j.pocean.2019.102254.

Durand, F., Shankar, D., Birol, F., & Shenoi, S. S. C. (2009). Spatiotemporal structure of the East India Coastal Current from satellite altimetry. *Journal of Geophysical Research, 114*, CO2013. https://doi.org/10.1029/2008JC004807.

Durgadoo, J. V., Loveday, B. R., Reason, C. J., Penven, P., & Biastoch, A. (2013). Agulhas leakage predominantly responds to the Southern Hemisphere westerlies. *Journal of Physical Oceanography, 43*(10), 2113–2131.

Durgadoo, J. V., Rühs, S., Biastoch, A., & Böning, C. W. B. (2017). Indian Ocean sources of Agulhas leakage. *Journal of Geophysical Research, Oceans, 122*, 3481–3499.

Elipot, S., & Beal, L. M. (2015). Characteristics, energetics, and origins of Agulhas Current meanders and their limited influence on ring shedding. *Journal of Physical Oceanography, 45*(9), 2294–2314.

Elipot, S., & Beal, L. M. (2018). Observed Agulhas Current sensitivity to interannual and long-term trend atmospheric forcings. *Journal of Climate, 31*(8), 3077–3098.

Ernst, P. A., Subrahmanyam, B., & Trott, C. B. (2022). Lakshadweep high propagation and impacts on the Somali current and Eddies during the Southwest Monsoon. *Journal of Geophysical Research: Oceans, 127*(3). https://doi.org/10.1029/2021jc018089.

Esenkov, O. E., & Olson, D. B. (2002). A numerical study of the Somali coastal undercurrents. *Deep Sea Research Part II: Topical Studies in Oceanography, 49*(7–8), 1253–1277. https://doi.org/10.1016/s0967-0645(01)00152-7.

Eshel, G., & Heavens, N. (2007). Climatological evaporation seasonality in the northern Red Sea. *Paleoceanography, 22*, PA4201. https://doi.org/10.1029/2006PA001365.

Feng, M., Böning, C., Biastoch, A., Behrens, E., Weller, E., & Masumoto, Y. (2011). The reversal of the multidecadal trends of the equatorial Pacific easterly winds, and the Indonesian Throughflow and Leeuwin Current transports. *Geophysical Research Letters, 38*, L11604. https://doi.org/10.1029/2011GL047291.

Feng, M., McPhaden, M. J., & Lee, T. (2010). Decadal variability of the Pacific subtropical cells and their influence on the south- east Indian Ocean. *Geophysical Research Letters, 37*, L09606. https://doi.org/10.1029/2010GL042796.

Feng, M., Meyers, G., Pearce, A., & Wijffels, S. (2003). Annual and interannual variations of the Leeuwin Current at 32°S. *Journal of Geophysical Research, 108*(C11), 3355. https://doi.org/10.1029/2002JC001763.

Feng, M., Wijffels, S., Godfrey, S., & Meyers, G. (2005). Do Eddies play a role in the momentum balance of the Leeuwin current? *Journal of Physical Oceanography, 35*, 964–975.

Ferron, B., & Marotzke, J. (2003). Impact of 4D-variational assimilation of WOCE hydrography on the meridional circulation of the Indian Ocean. *Deep Sea Research Part II: Topical Studies in Oceanography, 50*, 2005–2021.

Fieux, M., Schott, F., & Swallow, J. C. (1986). Deep boundary currents in the western Indian Ocean revisited. *Deep Sea Research Part A, 33*, 415–426. https://doi.org/10.1016/0198-0149(86)90124-X.

Findlater, J. (1971). Mean monthly airflow at low levels over the western Indian Ocean. *Geophysical Memoirs, 115*, 55.

Firing, E. (1987). Deep zonal currents in the central equatorial Pacific. *Journal of Marine Research, 45*, 791–812. https://doi.org/10.1357/002224087788327163.

Fischer, J., Schott, F., & Stramma, L. (1996). Current and transports of the Great Whirl-Socotra Gyre system during the summer monsoon, August 1993. *Journal of Geophysical Research, 101*(1996), 3573–3687.

Furue, R. (2019). The three-dimensional structure of the Leeuwin Current System in density coordinates in an eddy-resolving OGCM. *Ocean Modelling, 138*, 36–50. https://doi.org/10.1016/j.ocemod.2019.03.001.

Furue, R., Guerreiro, K., Phillips, H. E., McCreary, J. P., & Bindoff, N. L. (2017). On the Leeuwin Current System and its linkage to zonal flows in the South Indian Ocean as inferred from a gridded hydrography. *Journal of Physical Oceanography, 47*, 583–602. https://doi.org/10.1175/JPO-D-16-0170.1.

Furue, R., McCreary, J. P., Benthuysen, J., Phillips, H. E., & Bindoff, N. L. (2013). Dynamics of the Leeuwin Current: Part 1. Coastal flows in an inviscid, variable-density, layer model. *Dynamics of Atmospheres and Oceans, 63*, 24–59. https://doi.org/10.1016/j.dynatmoce.2013.03.003.

Ganachaud, A. C., & Wunsch, C. (2000). Improved estimates of global ocean circulation, heat transport and mixing from hydrographic data. *Nature, 408*, 453–457.

Garternicht, U., & Schott, F. (1997). Heat fluxes of the Indian Ocean from a global eddy-resolving model. *Journal of Geophysical Research, 102*, 21147–21159. https://doi.org/10.1029/97JC01585.

Girishkumar, M. S., Ravichandran, M., & Han, W. (2013). Observed intraseasonal thermocline variability in the Bay of Bengal. *Journal of Geophysical Research, Oceans, 118*. https://doi.org/10.1002/jgrc.20245.

Girishkumar, M. S., Ravichandran, M., McPhaden, M. J., & Rao, R. R. (2011). Intraseasonal variability in barrier layer thickness in the south central Bay of Bengal. *Journal of Geophysical Research, 116*, C03009. https://doi.org/10.1029/2010JC006657.

Gnanaseelan, C., & Deshpande, A. (2018). Equatorial Indian Ocean subsurface current variability in an ocean general circulation model. *Climate Dynamics, 50*, 1705–1717. https://doi.org/10.1007/s00382-017-3716-8.

Gnanaseelan, C., Deshpande, A., & McPhaden, M. J. (2012). Impact of Indian Ocean Dipole and El Niño/Southern Oscillation forcing on the Wyrtki jets. *Journal of Geophysical Research, 117*, C08005. https://doi.org/10.1029/2012JC007918.

Godfrey, J. S., & Ridgway, K. R. (1985). The large-scale environment of the poleward-flowing Leeuwin Current, Western Australia: longshore steric height gradients, wind stresses and geostrophic flow. *Journal of Physical Oceanography, 15*(5), 481–495. https://doi.org/10.1175/1520-0485 (1985)015<0481:TLSEOT>2.0.CO;2.

Godfrey, J. S., & Weaver, A. (1989). Why are there such strong steric height gradients off Western Australia? In *Proceedings of the Western Pacific international meeting and workshop on TOGA COARE, May 24–30, 1989, Noumea, New Caledonia* (pp. 215–222). http://hdl.handle.net/102.100.100/262338.

Godfrey, J. S., & Weaver, A. J. (1991). Is the Leeuwin Current driven by Pacific heating and winds? *Progress in Oceanography, 27*(3–4), 225–272. https://doi.org/10.1016/0079-6611(91)90026-I.

Gordon, A. L., Shroyer, E. L., Mahadevan, A., Sengupta, D., & Freilich, M. (2016). Bay of Bengal: 2013 Northeast monsoon upper-ocean circulation. *Oceanography*. https://doi.org/10.5670/oceanog.2016.41.

Gouriou, Y., Bourlès, B., Mercier, H., & Chuchla, R. (1999). Deep jets in the equatorial Atlantic Ocean. *Journal of Geophysical Research, 104*, 21217–21226. https://doi.org/10.1029/1999JC900057.

Gründlingh, M. L. (1985). Occurrence of Red Sea Water in the Southwestern Indian Ocean, 1981. *Journal of Physical Oceanography, 15*, 207–212. https://doi.org/10.1175/1520-0485(1985)015<0207:OORSWI>2.0.CO;2.

Gunn, K. L., Beal, L. M., Elipot, S., McMonigal, K., & Houk, A. (2020). Mixing of subtropical, central, and intermediate waters driven by shifting and pulsing of the Agulhas current. *Journal of Physical Oceanography, 50*, 3545–3560.

Han, L. (2021). The sloshing and diapycnal meridional overturning circulations in the Indian Ocean. *Journal of Physical Oceanography, 51*, 701–725. https://doi.org/10.1175/JPO-D-20-0211.1.

Han, L., & Huang, R. X. (2020). Using the Helmholtz decomposition to define the Indian Ocean meridional overturning streamfunction. *Journal of Physical Oceanography, 50*, 679–694. https://doi.org/10.1175/JPO-D-19-0218.1.

Hermes, J. C., Masumoto, Y., Beal, L. M., Roxy, M. K., Vialard, J., Andres, M., Annamalai, H., Behera, S., D'Adamo, N., Doi, T., Feng, M., Han, W., Hardman-Mountford, N., Hendon, H., Hood, R., Kido, S., Lee, C., Lee, T., Lengaigne, M., & Yu, W. (2019). A sustained ocean observing system in the Indian Ocean for climate related scientific knowledge and societal needs. *Frontiers in Marine Science, 6*, 355. https://doi.org/10.3389/fmars.2019.00355.

Hermes, J. C., & Reason, C. J. C. (2008). Annual cycle of the south Indian Ocean (Seychelles-Chagos) thermocline ridge in a regional ocean model. *Journal of Geophysical Research, 113*, C04035. https://doi.org/10.1029/2007JC004363.

Hernández-Guerra, A., & Talley, L. D. (2016). Meridional overturning transports at 30°S in the Indian and Pacific Oceans in 2002–2003 and 2009. *Progress in Oceanography, 146*, 89–120.

Herraiz-Borreguero, L., & Rintoul, S. R. (2011). Subantarctic mode water: Distribution and circulation. *Ocean Dynamics, 61*(1), 103–126. https://doi.org/10.1007/s10236-010-0352-9.

Hood, R. R., Beckley, L. E., & Wiggert, J. D. (2017). Biogeochemical and ecological impacts of boundary currents in the Indian Ocean. *Progress in Oceanography, 156*, 290–325. https://doi.org/10.1016/j.pocean.2017.04.011.

Hood, R. R., Coles, V. J., Huggett, J. A., Landry, M. R., Levy, M., Moffett, J. W., & Rixen, T. (2024a). Chapter 13: Nutrient, phytoplankton, and zooplankton variability in the Indian Ocean. In C. C. Ummenhofer, & R. R. Hood (Eds.), *The Indian Ocean and its role in the global climate system* (pp. 293–327). Amsterdam: Elsevier. https://doi.org/10.1016/B978-0-12-822698-8.00020-2.

Hood, R. R., Rixen, T., Levy, M., Hansell, D. A., Coles, V. J., & Lachkar, Z. (2024b). Chapter 12: Oxygen, carbon, and pH variability in the Indian Ocean. In C. C. Ummenhofer, & R. R. Hood (Eds.), *The Indian Ocean and its role in the global climate system* (pp. 265–291). Amsterdam: Elsevier. https://doi.org/10.1016/B978-0-12-822698-8.00017-2.

Horii, T., Mizuno, K., Nagura, M., Miyama, T., & Ando, K. (2013). Seasonal and interannual variation in the cross-equatorial meridional currents observed in the eastern Indian Ocean. *Journal of Geophysical Research, Oceans, 118*. https://doi.org/10.1002/2013JC009291.

Hormann, V., Centurioni, L. R., & Gordon, A. L. (2019). Freshwater export pathways from the Bay of Bengal. *Deep Sea Research Part II: Topical Studies in Oceanography, 168*, 104645. https://doi.org/10.1016/j.dsr2.2019.104645.

Huang, K., Han, W., Wang, D., Wang, W., Xie, Q., Chen, J., & Chen, G. (2018a). Features of the equatorial intermediate current associated with basin resonance in the Indian Ocean. *Journal of Physical Oceanography, 48*, 1333–1347.

Huang, K., et al. (2018b). Vertical propagation of middepth zonal currents associated with surface wind forcing in the equatorial Indian Ocean. *Journal of Geophysical Research: Oceans, 123*. https://doi.org/10.1029/2018JC013977.

Huang, K., Wang, D., Han, W., Feng, M., Chen, G., Wang, W., Chen, J., & Li, J. (2019). Semiannual variability of middepth zonal currents along 5°N in the eastern Indian Ocean: Characteristics and causes. *Journal of Physical Oceanography, 49*, 2715–2729. https://doi.org/10.1175/JPO-D-19-0089.1.

Hutchinson, K., Beal, L. M., Penven, P., Ansorge, I., & Hermes, J. (2018). Seasonal phasing of Agulhas current transport tied to a baroclinic adjustment of near-field winds. *Journal of Geophysical Research: Oceans, 123*(10), 7067–7083.

Huussen, T. N., Naveira-Garabato, A. C., Bryden, H. L., & McDonagh, E. L. (2012). Is the deep Indian Ocean MOC sustained by breaking internal waves? *Journal of Geophysical Research, 117*, C08024. https://doi.org/10.1029/2012JC008236.

Iskandar, I., Masumoto, Y., & Mizuno, K. (2009). Subsurface equatorial zonal current in the eastern Indian Ocean. *Journal of Geophysical Research, 114*, C06005. https://doi.org/10.1029/2008JC005188.

Iskandar, I., & McPhaden, M. J. (2011). Dynamics of wind-forced intraseasonal zonal current variations in the equatorial Indian Ocean. *Journal of Geophysical Research, 116*, C06019. https://doi.org/10.1029/2010JC006864.

Jensen, T. G. (2001). Arabian Sea and Bay of Bengal exchange of salt and tracers in an ocean model. *Geophysical Research Letters, 28*(20), 3967–3970. https://doi.org/10.1029/2001GL013422.

Jensen, T. G., Wijesekera, H. W., Nyadjro, E. S., Thoppil, P. G., Shriver, J., Sandeep, K. K., & Pant, V. (2016). Modeling salinity exchanges between the equatorial Indian Ocean and the Bay of Bengal. *Oceanography, 29*(2), 92–101. https://doi.org/10.5670/oceanog.2016.42.

Jiang, H., Farrar, J. T., Beardsley, R. C., Chen, R., & Chen, C. (2009). Zonal surface wind jets across the Red Sea due to mountain gap forcing along both sides of the Red Sea. *Geophysical Research Letters, 36*, L19605. https://doi.org/10.1029/2009GL040008.

Johns, W. E., Baringer, M. O., Beal, L. M., Cunningham, S., Kanzow, T., Bryden, H. L., ... Curry, R. (2011). Continuous, array-based estimates of Atlantic Ocean heat transport at 26.5°N. *Journal of Climate, 24*, 2429–2449.

Johns, W. E., Yao, F., Olson, D. B., Josey, S. A., Grist, J. P., & Smeed, D. A. (2003). Observations of seasonal exchange through the Straits of Hormuz and the inferred heat and freshwater budgets of the Persian Gulf. *Journal of Geophysical Research, 108*(C12), 3391. https://doi.org/10.1029/2003JC001881.

Johnson, G. C. (2011). Deep signatures of southern tropical Indian Ocean annual Rossby waves. *Journal of Physical Oceanography, 41*(10), 1958–1964. https://doi.org/10.1175/JPO-D-11-029.1.

Johnson, G. C., Musgrave, D. L., Warren, B. A., Ffield, A., & Olson, D. B. (1998). Flow of bottom and deep water in the Amirante Passage and Mascarene Basin. *Journal of Geophysical Research, 103*, 30973–30984. https://doi.org/10.1029/1998JC900027.

Johnson, G. C., Warren, B. A., & Olson, D. B. (1991a). Flow of bottom water in the Somali Basin. *Deep-Sea Research, 38*, 637–652. https://doi.org/10.1016/0198-0149(91)90003-.

Johnson, G. C., Warren, B. A., & Olson, D. B. (1991b). A deep boundary current in the Arabian Basin. *Deep Sea Research Part A, 38*, 653–661. https://doi.org/10.1016/0198-0149(91)90004-Y.

Jones, D. C., Meijers, A. J. S., Shuckburgh, E., Sallée, J.-B., Haynes, P., McAufield, E. K., & Mazloff, M. R. (2016). How does subantarctic mode water ventilate the southern hemisphere subtropics? *Journal of Geophysical Research, Oceans, 121*, 6558–6582. https://doi.org/10.1002/2016JC011680.

Joseph, S., Wallcraft, A. J., Jensen, T. G., Ravichandran, M., Shenoi, S. S. C., & Nayak, S. (2012). Weakening of spring Wyrtki jets in the Indian Ocean during 2006–2011. *Journal of Geophysical Research, 117*, C04012. https://doi.org/10.1029/2011JC007581.

Kämpf, J., & Sadrinasab, M. (2006). The circulation of the Persian Gulf: A numerical study. *Ocean Science, 2*, 27–41. https://doi.org/10.5194/os-2-27-2006.

Karstensen, J., & Quadfasel, D. (2002). Water subducted into the Indian Ocean subtropical gyre. *Deep-Sea Research Part II, 49*, 1441–1457. https://doi.org/10.1016/S0967-0645(01)00160-6.

Katsumata, K., Talley, L. D., Capuano, T. A., & Whalen, C. B. (2021). Spatial and temporal variability of diapycnal mixing in the Indian Ocean. *Journal of Geophysical Research: Oceans, 126*(7). https://doi.org/10.1029/2021jc017257.

Klinger, B. A., & Marotzke, J. (2000). Meridional heat transport by the subtropical cell. *Journal of Physical Oceanography, 30*, 696–705. https://doi.org/10.1175/1520-0485(2000)030,0696: MHTBTS.2.0.CO;2.

Kouketsu, S., Doi, T., Kawano, T., Masuda, N., Sasaki, Y., et al. (2011). Deep ocean heat content changes estimated from observation and reanalysis product and their influence on sea level change. *Journal of Geophysical Research, 116*, C03012. https://doi.org/10.1029/2010JC006464.

Krug, M., Swart, S., & Gula, J. (2017). Submesoscale cyclones in the Agulhas current. *Geophysical Research Letters, 44*(1), 346–354.

Krug, M., & Tournadre, J. (2012). Satellite observations of an annual cycle in the Agulhas Current. *Geophysical Research Letters, 39*(15).

Kumar, S., Bhavya, P. S., Ramesh, R., Gupta, G. V. M., Chiriboga, F., Singh, A., Karunasagar, I., Rai, A., Rehnstam-Holm, A.-S., Edler, L., & Godhe, A. (2018). Nitrogen uptake potential under different temperature-salinity conditions: Implications for nitrogen cycling under climate change scenarios. *Marine Environmental Research, 141*, 196–204. https://doi.org/10.1016/j.marenvres.2018.09.001.

Kumar, S. P., & Prasad, T. G. (1999). Formation and spreading of Arabian Sea high-salinity water mass. *Journal of Geophysical Research, 104*, 1455–1464. https://doi.org/10.1029/1998JC900022.

L'Hégaret, P., Beal, L. M., Elipot, S., & Laurindo, L. (2018). Shallow cross-equatorial gyres of the Indian Ocean driven by seasonally reversing Monsoon winds. *Journal of Geophysical Research, Oceans, 123*, 8902–8920. https://doi.org/10.1029/2018JC014553.

L'Hégaret, Carton, X., Louazel, S., & Boutin, G. (2016). Mesoscale eddies and submesoscale structures of Persian Gulf Water off the Omani coast in spring 2011. *Ocean Science, 12*, 687–701. https://doi.org/10.5194/os-12-687-2016.

L'Hégaret, P., de Marez, C., Morvan, M., Meunier, T., & Carton, X. (2021). Spreading and vertical structure of the Persian Gulf and Red Sea outflows in the northwestern Indian Ocean. *Journal of Geophysical Research: Oceans, 126*. https://doi.org/10.1029/2019JC015983.

L'Hégaret, P., Duarte, R., Carton, X., Vic, C., Ciani, D., Baraille, R., & Corréard, S. (2015). Mesoscale variability in the Arabian Sea from HYCOM model results and observations: Impact on the Persian Gulf Water path. *Ocean Science, 11*, 667–693. https://doi.org/10.5194/os-11-667-2015.

L'Hégaret, P., Lacour, L., Carton, X., Roullet, G., Baraille, R., & Corréard, S. (2013). A seasonal dipolar eddy near Ras Al Hamra (Sea of Oman). *Ocean Dynamics, 63*, 633–659. https://doi.org/10.1007/s10236-013-0616-2.

Lambert, E., Le Bars, W., & de Ruijter, W. P. M. (2016). The connection of the Indonesian throughflow, South Indian Ocean Countercurrent and the Leeuwin Current. *Ocean Science, 12*(3), 771–780. https://doi.org/10.5194/os-12-771-2016.

Laurindo, L. C., Mariano, A. J., & Lumpkin, R. (2017). An improved near-surface velocity climatology for the global ocean from drifter observations. *Deep-Sea Research Part I: Oceanographic Research Papers*. https://doi.org/10.1016/j.dsr.2017.04.009.

Leber, G. M., & Beal, L. M. (2014). Evidence that Agulhas Current transport is maintained during a meander. *Journal of Geophysical Research, Oceans, 119*, 3806–3817. https://doi.org/10.1002/2014JC009802.

Leber, G. M., Beal, L. M., & Elipot, S. (2017). Wind and current forcing combine to drive strong upwelling in the Agulhas Current. *Journal of Physical Oceanography, 47*, 123–134.

Lee, T. (2004). Decadal weakening of the shallow overturning circulation in the south Indian Ocean. *Geophysical Research Letters, 31*, L18305. https://doi.org/10.1029/2004GL020884.

Lee, T., & Marotzke, J. (1997). Inferring meridional mass and heat transports of the Indian Ocean by fitting a general circulation model to climatological data. *Journal of Geophysical Research, 102*, 10585–10602. https://doi.org/10.1029/97JC00464.

Lee, T., & McPhaden, M. J. (2008). Decadal phase change in large-scale sea level and winds in the Indo-Pacific region at the end of the 20th century. *Geophysical Research Letters, 35*, L01605. https://doi.org/10.1029/2007GL032419.

Legeckis, R., & Cresswell, G. (1981). Satellite observations of sea-surface temperature fronts off the coast of western and southern Australia. *Deep Sea Research, Part I, 28*, 297–306. https://doi.org/10.1016/0198-0149(81)90069-8.

Li, Z., Aiki, H., Nagura, M., & Ogata, T. (2021). The vertical structure of annual wave energy flux in the tropical Indian Ocean. *Progress in Earth and Planetary Science, 8*(43). https://doi.org/10.1186/s40645-021-00432-9.

Li, Y., Han, W., & Zhang, L. (2017). Enhanced decadal warming of the southeast Indian Ocean during the recent global surface warming slowdown. *Geophysical Research Letters, 44*, 9876–9884. https://doi.org/10.1002/2017GL075050.

Liu, Z. (1994). A simple model of the mass exchange between the subtropical and tropical ocean. *Journal of Physical Oceanography, 24*, 1153–1165. https://doi.org/10.1175/1520-0485(1994)024,1153:ASMOTM.2.0.CO;2.

Liu, L. L., & Huang, R. X. (2012). The global subduction/obduction rates: Their interannual and decadal variability. *Journal of Climate, 25*, 1096–1115. https://doi.org/10.1175/2011JCLI4228.1.

Loschnigg, J., & Webster, P. J. (2000). A coupled ocean–atmosphere system of SST modulation for the Indian Ocean. *Journal of Climate, 13*(19), 3342–3360.

Lu, P., & McCreary, J. P. (1995). Influence of the ITCZ on the flow of thermocline water from the subtropical to the equatorial Pacific Ocean. *Journal of Physical Oceanography, 25*, 3076–3088. https://doi.org/10.1175/1520-0485(1995)025,3076:IOTIOT.2.0.CO;2.

Lumpkin, R., & Johnson, G. C. (2013). Global ocean surface velocities from drifters: Mean, variance, El Nino–Southern Oscillation response, and seasonal cycle. *Journal of Geophysical Research, Oceans, 118*, 2992–3006. https://doi.org/10.1002/jgrc.20210.

Luyten, J., Pedlosky, J., & Stommel, H. (1983). The ventilated thermocline. *Journal of Physical Oceanography, 13*, 292–309.

Luyten, J. R., & Roemmich, D. H. (1982). Equatorial currents at semiannual period in the Indian Ocean. *Journal of Physical Oceanography, 12*(5), 406–413. https://doi.org/10.1175/1520-0485(1982)012<0406:Ecasap>2.0.Co;2.

Luyten, J. R., & Swallow, J. C. (1976). Equatorial undercurrents. *Deep Sea Research and Oceanographic Abstracts, 23*(10), 999–1001. https://doi.org/10.1016/0011-7471(76)90830-5.

MacKinnon, J. A., Johnston, T. M. S., & Pinkel, R. (2008). Strong transport and mixing of deep water through the Southwest Indian Ridge. *Nature Geoscience, 1*, 755–758. https://doi.org/10.1038/ngeo340.

Mantyla, A. W., & Reid, J. L. (1995). On the origins of deep and bottom waters of the Indian Ocean. *Journal of Geophysical Research, 100*, 2417–2439. https://doi.org/10.1029/94JC02564.

Marsac, F., Everett, B., Shahid, U., & Strutton, P. G. (2024). Chapter 11: Indian Ocean primary productivity and fisheries variability. In C. C. Ummenhofer, & R. R. Hood (Eds.), *The Indian Ocean and its role in the global climate system* (pp. 245–264). Amsterdam: Elsevier. https://doi.org/10.1016/B978-0-12-822698-8.00019-6.

Masumoto, Y., Hase, H., Kuroda, Y., Matsuura, H., & Takeuchi, K. (2005). Intraseasonal variability in the upper layer currents observed in the eastern equatorial Indian Ocean. *Geophysical Research Letters, 32*, L02607. https://doi.org/10.1029/2004GL021896.

Matt, S., & Johns, W. E. (2007). Transport and entrainment in the Red Sea outflow plume. *Journal of Physical Oceanography, 37*, 819–836.

McCreary, J. P., Han, W., Shankar, D., & Shetye, S. R. (1996). Dynamics of the East India coastal current 2. Numerical solutions. *Journal of Geophysical Research, 101*, 13993–14010. https://doi.org/10.1029/96jc00560.

McCreary, J. P., Jr., & Lu, P. (1994). Interaction between the subtropical and equatorial ocean circulations: The subtropical cell. *Journal of Physical Oceanography, 24*, 466–497. https://doi.org/10.1175/1520-0485(1994)024,0466:IBTSAE.2.0.CO;2.

McCreary, J. P., Kundu, P. K., & Molinari, R. L. (1993). A numerical investigation of dynamics, thermodynamics, and mixed layer processes in the Indian Ocean. *Progress in Oceanography, 31*, 181–224.

McCreary, J. P., & Shetye, S. R. (2023). Observations and dynamics of circulations in the North Indian Ocean. *Atmosphere, Earth, Ocean & Space*. https://doi.org/10.1007/978-981-19-5864-9.

McCreary, J. P., Shetye, S. R., & Kundu, P. K. (1986). Thermohaline forcing of eastern boundary currents: With application to the circulation off the west coast of Australia. *Journal of Marine Research, 44*, 71–92. https://doi.org/10.1357/002224086788460184.

McDonagh, E. L., Bryden, H. L., King, B. A., & Sanders, R. J. (2008). The circulation of the Indian Ocean at 32S. *Progress in Oceanography, 79*, 20–36.

McMonigal, K., Beal, L. M., Elipot, S., Gunn, K. L., Hermes, J., Morris, T., & Houk, A. (2020). The impact of meanders, deepening and broadening, and seasonality on Agulhas current temperature variability. *Journal of Physical Oceanography, 50*, 1–51.

McMonigal, K., Gunn, K. L., Beal, L. M., Elipot, S., & Willis, J. K. (2022). Reduction in meridional heat export contributes to recent Indian Ocean warming. *Journal of Physical Oceanography, 52*(3), 329–345. https://doi.org/10.1175/jpo-d-21-0085.1.

McPhaden, M. J., Beal, L. M., Bhaskar, T. V. S. U., Lee, T., Nagura, M., Strutton, P. G., & Yu, L. (2024). Chapter 17: The Indian Ocean Observing System (IndOOS). In C. C. Ummenhofer, & R. R. Hood (Eds.), *The Indian Ocean and its role in the global climate system* (pp. 393–419). Amsterdam: Elsevier. https://doi.org/10.1016/B978-0-12-822698-8.00002-0.

McPhaden, M. J., Meyers, G., Ando, K., Masumoto, Y., Murty, V. S. N., Ravichandran, M., et al. (2009). RAMA the research moored array for African-Asian-Australian monsoon analysis and prediction. *Bulletin of the American Meteorological Society, 90*, 459–480. https://doi.org/10.1175/2008BAMS2608.1.

McPhaden, M. J., & Nagura, M. (2014). Indian Ocean dipole interpreted in terms of recharge oscillator theory. *Climate Dynamics, 42*, 1569–1586.

McPhaden, M. J., Wang, Y., & Ravichandran, M. (2015). Volume transports of the Wyrtki jets and their relationship to the Indian Ocean Dipole. *Journal of Geophysical Research, Oceans, 120*(8), 5302–5317. https://doi.org/10.1002/2015JC010901.

Meissner, T., Wentz, F. J., & Le Vine, D. M. (2018). The salinity retrieval algorithms for the NASA Aquarius version 5 and SMAP version 3 releases. *Remote Sensing, 10*, 1121. https://doi.org/10.3390/rs10071121.

Menezes, V. V., Farrar, J. T., & Bower, A. S. (2018). Westward mountain-gap wind jets of the northern Red Sea as seen by QuikSCAT. *Remote Sensing of Environment, 209*, 677–699. https://doi.org/10.1016/j.rse.2018.02.075.

Menezes, V. V., Farrar, J. T., & Bower, A. S. (2019). Evaporative implications of dry-air outbreaks over the northern Red Sea. *Journal of Geophysical Research: Atmospheres, 124*. https://doi.org/10.1029/2018JD028853.

Menezes, V. V., Phillips, H. E., Schiller, A., Bindoff, N. L., Domingues, C. M., & Vianna, M. L. (2014). South Indian Countercurrent and associated fronts. *Journal of Geophysical Research, Oceans, 119*, 6763–6791. https://doi.org/10.1002/2014JC010076.

Menezes, V. V., Phillips, H. E., Schiller, A., Domingues, C. M., & Bindoff, N. L. (2013). Salinity dominance on the Indian Ocean Eastern Gyral Current. *Geophysical Research Letters, 40*, 5716–5721. https://doi.org/10.1002/2013GL057887.

Menezes, V. V., & Vianna, M. L. (2019). Quasi-biennial Rossby and Kelvin waves in the south Indian ocean: Tropical and subtropical modes and the Indian Ocean Dipole. *Deep-Sea Research Part II, 166*, 43–63. https://doi.org/10.1016/j.dsr2.2019.05.002.

Meng, L., Zhuang, W., Zhang, W., Yan, C., & Yan, X.-. H. (2020). Variability of the shallow overturning circulation in the Indian Ocean. *Journal of Geophysical Research: Oceans, 125*. https://doi.org/10.1029/2019JC015651.

Meyers, G. (1996). Variation of Indonesian throughflow and El Niño-Southern oscillation. *Journal of Geophysical Research, 101*, 12255–12263. https://doi.org/10.1029/95JC03729.

Meyers, G., Bailey, R. J., & Worby, A. P. (1995). Geostrophic transport of Indonesian throughflow. *Deep-Sea Research Part I, 42*, 1163–1174.

Miyama, T., McCreary, J. P., Jensen, T. G., Loschnigg, J. L., Godfrey, S., & Ishida, A. (2003). Structure and dynamics of the Indian-Ocean cross-equatorial cell. *Deep Sea Research, Part II, 50*, 2023–2047. https://doi.org/10.1016/S0967-0645(03)00044-4.

Morrison, J. M. (1997). Inter-monsoonal changes in the T-S properties of the near-surface waters of the northern Arabian Sea. *Geophysical Research Letters, 24*, 2553–2556. https://doi.org/10.1029/97GL01876.

Mukherjee, A., Shankar, D., Fernando, V., Amol, P., Aparna, S. G., Fernandes, R., Michael, G. S., Khalap, S. T., Satelkar, N. P., Agarvadekar, Y., Gaonkar, M. G., Tari, A. P., Kankonkar, A., & Vernekar, S. (2014). Observed seasonal and intraseasonal variability of the East India Coastal Current on the continental slope. *Journal of Earth System Science, 123*, 1197–1232. https://doi.org/10.1007/s12040-014-0471-7.

Mukhopadhyay, S., Shankar, D., Aparna, S. G., Mukherjee, A., Fernando, V., Kankonkar, A., Khalap, S., Satelkar, N., Gaonkar, M. G., Tari, A. P., Khedekar, R., & Ghatkar, S. (2020). Observed variability of the East India Coastal Current on the continental slope during 2009–2018. *Journal of Earth System Science, 129*, 77. https://doi.org/10.1007/s12040-020-1346-8.

Murray, S. P., & Johns, W. E. (1997). Direct observations of seasonal exchange through the Bab el Mandab Strait. *Geophysical Research Letters, 24*, 2557–2560.

Murtugudde, R., & Busalacchi, A. J. (1999). Interannual variability of the dynamics and thermodynamics of the tropical Indian Ocean. *Journal of Climate, 12*(8), 2300–2326.

Murtugudde, R., McCreary, J. P., & Busalacchi, A. J. (2000). Oceanic processes associated with anomalous events in the Indian Ocean with relevance to 1997–1998. *Journal of Geophysical Research: Oceans, 105*(C2), 3295–3306.

Nagura, M. (2018). Annual Rossby waves below the pycnocline in the Indian Ocean. *Journal of Geophysical Research: Oceans, 123*, 9405–9415. https://doi.org/10.1029/2018JC014362.

Nagura, M. (2020). Variability in meridional transport of the subtropical circulation in the south Indian Ocean for the period from 2006 to 2017. *Journal of Geophysical Research: Oceans, 124*. https://doi.org/10.1029/2019JC015874.

Nagura, M., & Masumoto, Y. (2015). A wake due to the Maldives in the eastward Wyrtki jet. *Journal of Physical Oceanography, 45*, 1858–1876.

Nagura, M., McCreary, J. P., & Annamalai, H. (2018). Origins of coupled model biases in the Arabian Sea climatological state. *Journal of Climate, 31*, 2005–2029. https://doi.org/10.1175/JCLI-D-17-0417.1.

Nagura, M., & McPhaden, M. J. (2008). The dynamics of zonal current variations in the central equatorial Indian Ocean. *Geophysical Research Letters, 35*, L23603. https://doi.org/10.1029/2008GL035961.

Nagura, M., & McPhaden, M. J. (2010a). Wyrtki Jet dynamics: Seasonal variability. *Journal of Geophysical Research, 115*, C07009. https://doi.org/10.1029/2009JC005922.

Nagura, M., & McPhaden, M. J. (2010b). Dynamics of zonal current variations associated with the Indian Ocean dipole. *Journal of Geophysical Research, 115*, C11026. https://doi.org/10.1029/2010JC006423.

Nagura, M., & McPhaden, M. J. (2012). The dynamics of wind-driven intraseasonal variability in the equatorial Indian Ocean. *Journal of Geophysical Research, 117*, C02001. https://doi.org/10.1029/2011JC007405.

Nagura, M., & McPhaden, M. J. (2016). Zonal propagation of near-surface zonal currents in relation to surface wind forcing in the equatorial Indian Ocean. *Journal of Physical Oceanography, 46*, 3623–3638.

Nagura, M., & McPhaden, M. J. (2018). The shallow overturning circulation in the Indian Ocean. *Journal of Physical Oceanography, 48*, 413–434.

Nagura, M., & McPhaden, M. J. (2021). Interannual variability in sea surface height at southern midlatitudes of the Indian Ocean. *Journal of Physical Oceanography, 51*, 1595–1609.

Nyadjro, E., & McPhaden, M. J. (2014). Variability of zonal currents in the eastern equatorial Indian Ocean on seasonal to interannual time scales. *Journal of Geophysical Research, Oceans, 119*, 7969–7986. https://doi.org/10.1002/2014JC010380.

Oke, P. R., Griffin, D. A., Rykova, T., & de Oliveira, H. B. (2018). Ocean circulation in the Great Australian Bight in an eddy-resolving ocean reanalysis: The eddy field, seasonal and interannual variability. *Deep Sea Research, Part II, 157–158*, 11–26. https://doi.org/10.1016/j.dsr2.2018.09.012.

Oliver, E. C. J., Herzfeld, M., & Holbrook, N. J. (2016). Modelling the shelf circulation off eastern Tasmania. *Continental Shelf Research, 130*, 14–33.

Palastanga, V., van Leeuwen, P. J., Schouten, M. W., & de Ruijter, W. P. M. (2007). Flow structure and variability in the subtropical Indian Ocean: Instability of the South Indian Ocean Countercurrent. *Journal of Geophysical Research, 112*, C01001. https://doi.org/10.1029/2005JC003395.

Papadopoulos, V. P., Abualnaja, Y., Josey, S. A., Bower, A., Raitsos, D. E., Kontoyiannis, H., & Hoteit, I. (2013). Atmospheric forcing of the winter air-sea heat fluxes over the northern Red Sea. *Journal of Climate, 593*, 1685–1701. https://doi.org/10.1175/JCLI-D-12-00267.1.

Papadopoulos, V. P., Zhan, P., Sofianos, S. S., Raitsos, D. E., Qurban, M., Abualnaja, Y., et al. (2015). Factors governing the deep ventilation of the Red Sea. *Journal of Geophysical Research: Oceans, 120*, 7493–7505. https://doi.org/10.1002/2015JC010996.

Pedlosky, J. (1996). *Ocean circulation theory*. Berlin, Heidelberg: Springer-Verlag.

Pérez-Hernández, M. D., Hernández-Guerra, A., Joyce, T. M., & Vélez-Belchí, P. (2012). Wind-driven cross-equatorial flow in the Indian Ocean. *Journal of Physical Oceanography, 42*, 2234–2253. https://doi.org/10.1175/JPO-D-12-033.1.

Phillips, H. E., Tandon, A., Furue, R., Hood, R., Ummenhofer, C., Benthuysen, J., Menezes, V., Hu, S., Webber, B., Sanchez-Franks, A., Cherian, D., Shroyer, E., Feng, M., Wijesekera, H., Chatterjee, A., Yu, L., Hermes, J., Murtugudde, R., Tozuka, T., ... Wiggert, J. (2021). Progress in understanding of Indian Ocean circulation, variability, air-sea exchange and impacts on biogeochemistry. *Ocean Science*. https://doi.org/10.5194/os-2021-1.

Potemra, J. T. (2001). Contribution of equatorial Pacific winds to southern tropical Indian Ocean Rossby waves. *Journal of Geophysical Research, 106*, 2407–2422. https://doi.org/10.1029/1999JC000031.

Prasad, T. G., & Ikeda, M. (2002). The wintertime water mass formation in the northern Arabian Sea: A model study. *Journal of Physical Oceanography, 32*, 1028–1040. https://doi.org/10.1175/1520-0485(2002)032,1028:TWWMFI.2.0.CO;2.

Prasad, T. G., Ikeda, M., & Kumar, S. P. (2001). Seasonal spreading of the Persian Gulf water mass in the Arabian Sea. *Journal of Geophysical Research, 106*, 17059–17071. https://doi.org/10.1029/2000JC000480.

Premchand, K., Sastry, J. S., & Murty, C. S. (1986). Watermass structure in the western Indian Ocean—Part II: The spreading and transformation of the Persian Gulf water. *Mausam, 37*, 179–186.

Pujiana, K., & McPhaden, M. J. (2018). Ocean's response to the convectively coupled Kelvin waves in the eastern equatorial Indian Ocean. *Journal of Geophysical Research, 123*, 5727–5741. https://doi.org/10.1029/2018JC013858.

Pujiana, K., & McPhaden, M. J. (2020). Intraseasonal Kelvin waves in the equatorial Indian Ocean and their propagation into the Indonesian Seas. *Journal of Geophysical Research, 25*. https://doi.org/10.1029/2019JC015839.

Purkey, S. G., & Johnson, G. C. (2010). Warming of global abyssal and deep Southern Ocean waters between the 1990s and 2000s: Contributions to global heat and sea level rise budgets. *Journal of Climate, 23*, 6336–6351. https://doi.org/10.1175/2010JCLI3682.1.

Qiu, Y., Li, L., & Yu, W. (2009). Behavior of the Wyrtki Jet observed with surface drifting buoys and satellite altimeter. *Geophysical Research Letters, 36*, L18607. https://doi.org/10.1029/2009GL039120.

Qu, T., & Meyers, G. (2005). Seasonal characteristics of circulation in the southeastern tropical Indian Ocean. *Journal of Physical Oceanography, 35*, 255–267. https://doi.org/10.1175/JPO-2682.1.

Quadfasel, D. R., & Schott, F. (1983). Southward subsurface flow below the Somali Current. *Journal of Geophysical Research: Oceans, 88*(C10), 5973–5979. https://doi.org/10.1029/jc088ic10p05973.

Rainville, L., Lee, C. M., Arulananthan, K., Jinadasa, S. U. P., Fernando, H. J. S., Priyadarshani, W. N. C., & Wijesekera, H. (2022). Water mass exchanges between the Bay of Bengal and Arabian Sea from multiyear sampling with autonomous gliders. *Journal of Physical Oceanography, 52*, 2377–2396. https://doi.org/10.1175/JPO-D-21-0279.1.

Rao, R. R., & Sivakumar, R. (2003). Seasonal variability of sea surface salinity and salt budget of the mixed layer of the north Indian Ocean. *Journal of Geophysical Research, 108*(C1), 3009. https://doi.org/10.1029/2001JC000907.

Rath, S., Vinayachandran, P. N., Behara, A., & Neema, C. P. (2019). Dynamics of summer monsoon current around Sri Lanka. *Ocean Dynamics, 69*(10), 1133–1154. https://doi.org/10.1007/s10236-019-01295-x.

Reid, J. L. (2003). On the total geostrophic circulation of the Indian Ocean: Flow patterns, tracers, and transports. *Progress in Oceanography, 56*, 137–186. https://doi.org/10.1016/S0079-6611(02)00141-6.

Ridgway, K. R., & Condie, S. A. (2004). The 5500-km–long boundary flow off western and southern Australia. *Journal of Geophysical Research, 109*, C04017. https://doi.org/10.1029/2003JC001921.

Ridgway, K. R., & Godfrey, J. (2015). The source of the Leeuwin Current seasonality. *Journal of Geophysical Research, 120*(10), 6843–6864. https://doi.org/10.1002/2015JC011049.

Rochford, D. J. (1961). Hydrology of the Indian Ocean. I. The water masses in intermediate depths of the South-East Indian Ocean. *Marine and Freshwater Research, 12*, 129–149.

Rochford, D. J. (1964). Salinity maxima in the upper 1000 metres of the north Indian Ocean. *Australian Journal of Marine & Freshwater Research, 15*, 1–24. https://doi.org/10.1071/MF9640001.

Rochford, D. J. (1969). *Seasonal interchange of high and low salinity surface waters off south-west Australia*. Technical Paper Australia: Division of Fisheries and Oceanography, CSIRO. http://hdl.handle.net/102.100.100/321788?index=1.

Roman, R. E., & Lutjeharms, J. R. E. (2007). Red Sea intermediate water at the agulhas current termination. *Deep Sea Research Part I, 54*, 1329–1340. https://doi.org/10.1016/j.dsr.2007.04.009.

Roman, R. E., & Lutjeharms, J. R. E. (2009). Red Sea intermediate water in the source regions of the Agulhas Current. *Deep Sea Research Part I: Oceanographic Research Papers, 56*, 939–962. https://doi.org/10.1016/j.dsr.2009.01.003.

Roxy, M. K., Modi, A., Murtugudde, R., Valsala, V., Panickal, S., Kumar, S. P., Ravichandran, M., Vichi, M., & Levy, M. (2016). A reduction in marine primary productivity driven by rapid warming over the tropical Indian Ocean. *Geophysical Research Letters, 43*, 826–833.

Roxy, M. K., Saranya, J. S., Modi, A., Anusree, A., Cai, W., Resplandy, L., ... Frölicher, T. L. (2024). Chapter 20: Future projections for the tropical Indian Ocean. In C. C. Ummenhofer, & R. R. Hood (Eds.), *The Indian Ocean and its role in the global climate system* (pp. 469–482). Amsterdam: Elsevier. https://doi.org/10.1016/B978-0-12-822698-8.00004-4.

Sallée, J.-B., Speer, K., Rintoul, S., & Wijffels, S. (2010). Southern Ocean thermocline ventilation. *Journal of Physical Oceanography, 40*, 509–529. https://doi.org/10.1175/2009JPO4291.1.

Sallée, J., Wienders, N., Morrow, R., & Speer, K. (2006). Formation of subantarctic mode water in the southeastern Indian Ocean. *Ocean Dynamics, 56*, 525–542. https://doi.org/10.1007/s10236-005-0054-x.

Sanchez-Franks, A., Webber, B. G. M., King, B. A., Vinayachandran, P. N., Matthews, A. J., Sheehan, P. M. F., Behara, A., & Neema, C. P. (2019). The railroad switch effect of seasonally reversing currents on the Bay of Bengal high salinity core. *Geophysical Research Letters*. https://doi.org/10.1029/2019gl082208.

Schoenefeldt, R., & Schott, F. A. (2006). Decadal variability of the Indian Ocean cross-equatorial exchange in SODA. *Geophysical Research Letters, 33*, L08602. https://doi.org/10.1029/2006GL025891.

Schott, F. A., Dengler, M., & Schoenefeldt, R. (2002). The shallow overturning circulation of the Indian Ocean. *Progress in Oceanography, 53*, 57–103.

Schott, F. A., Fieux, M., Kindle, J., Swallow, J., & Zantopp, R. (1988). The boundary currents east and north of Madagascar: 2. Direct measurements and model comparisons. *Journal of Geophysical Research, 93*, 4963–4974. https://doi.org/10.1029/JC093iC05p04963.

Schott, F., & Fischer, J. (2000). The winter monsoon circulation of the northern Arabian Sea and Somali Current. *Journal of Geophysical Research, 105*, 6359–6376.

Schott, F. A., & McCreary, J. P. (2001). The monsoon circulation of the Indian ocean. *Progress in Oceanography, 51*, 1–123.

Schott, F. A., McCreary, J. P., & Johnson, G. C. (2004). Shallow overturning circulations of the tropical-subtropical oceans. In C. Wang, S. P. Xie, & J. A. Carton (Eds.), *Earth's climate*. Washington, DC: American Geophysical Union. https://doi.org/10.1029/147GM15.

Schott, F., & Quadfasel, D. R. (1982). Variability of the Somali Current system during the onset of the southwest monsoon. *Journal of Physical Oceanography, 12*, 1343–1357.

Schott, F., Swallow, J. C., & Fieux, M. (1990). The Somali Current at the equator: Annual cycle of currents and transports in the upper 1000 m and connection to neighboring latitudes. *Deep-Sea Research, 37*, 1825–1848.

Schott, F. A., Xie, S.-P., & McCreary, J. P. (2009). Indian Ocean circulation and climate variability. *Reviews of Geophysics, 47*, RG1002. https://doi.org/10.1029/2007RG000245.

Sengupta, D., Bharath Raj, G. N., & Shenoi, S. S. C. (2006). Surface freshwater from Bay of Bengal runoff and Indonesian throughflow in the tropical Indian Ocean. *Geophysical Research Letters*. https://doi.org/10.1029/2006GL027573.

Shankar, D., McCreary, J. P., Han, W., & Shetye, S. R. (1996). Dynamics of the East India Coastal Current 1. Analytic solutions forced by interior Ekman pumping and local alongshore Winds. *Journal of Geophysical Research: Oceans, 101*. https://doi.org/10.1029/96JC00559.

Shankar, D., & Shetye, S. R. (1997). On the dynamics of the Lakshadweep high and low in the southeastern Arabian Sea. *Journal of Geophysical Research, 102*, 12551–12562. https://doi.org/10.1029/97JC00465.

Shankar, D., Vinayachandran, P. N., & Unnikrishnan, A. S. (2002). The monsoon currents in the north Indian Ocean. *Progress in Oceanography, 52*, 63–120. https://doi.org/10.1016/s0079-6611(02)00024-1.

Shenoi, S. S. C., Saji, P., & Almeida, A. (1999). Near-surface circulation and kinetic energy in the tropical Indian Ocean derived from Lagrangian drifters. *Journal of Marine Research, 57*, 885–907. https://doi.org/10.1357/002224099321514088.

Shenoi, S. S. C., Shetye, S. R., Gouveia, A. D., & Michael, G. S. (1993). Salinity extrema in the Arabian Sea. In V. Ittekkot, & R. R. Nair (Eds.), *Monsoon biogeochemistry* (pp. 37–49). Hamburg: Geologisch-Paläontologischen Institutes der Universität.

Shetye, S. R., Chandra Shenoi, S. S., Antony, M. K., & Kumar, V. K. (1985). Monthly-mean wind stress along the coast of the north Indian Ocean. *Proceedings of the Indian Academy of Sciences-Earth and Planetary Sciences, 94*, 129–137. https://doi.org/10.1007/BF02871945.

Shetye, S. R., Gouveia, A. D., Shankar, D., Shenoi, S. S. C., Vinayachandran, P. N., Sundar, D., Michael, G. S., & Nampoothiri, G. (1996). Hydrography and circulation in the western Bay of Bengal during the northeast monsoon. *Journal of Geophysical Research, 101*, 14011–14025. https://doi.org/10.1029/95jc03307.

Shetye, S. R., Gouveia, A. D., Shenoi, S. S. C., Michael, G. S., Sundar, D., Almeida, A. M., & Santanam, K. (1991a). The coastal current off western India during the northeast monsoon. *Deep Sea Research, 38*, 1517–1529. https://doi.org/10.1016/0198-0149(91)90087-V.

Shetye, S. R., Gouveia, A. D., Shenoi, S. S. C., Sundar, D., Michael, G. S., Almeida, A. M., & Santanam, K. (1990). Hydrography and circulation off the west coast of India during the southwest monsoon 1987. *Journal of Marine Research, 48*, 359–378. https://doi.org/10.1357/002224090784988809.

Shetye, S. R., Gouveia, A. D., Shenoi, S. S. C., Sundar, D., Michael, G. S., & Nampoothiri, G. (1993). The western boundary current of the seasonal subtropical gyre in the Bay of Bengal. *Journal of Geophysical Research, 98*, 945–954. https://doi.org/10.1029/92jc02070.

Shetye, S. R., Shenoi, S. S. C., Gouveia, A. D., Michael, G. S., Sundar, D., & Nampoothiri, G. (1991b). Wind-driven coastal upwelling along the western boundary of the Bay of Bengal during the southwest monsoon. *Continental Shelf Research, 11*, 1397–1408. https://doi.org/10.1016/0278-4343(91)90042-5.

Shinoda, T., Jenson, T. G., Lachkar, Z., Masumoto, Y., & Seo, H. (2024). Chapter 18: Modeling the Indian Ocean. In C. C. Ummenhofer, & R. R. Hood (Eds.), *The Indian Ocean and its role in the global climate system* (pp. 421–443). Amsterdam: Elsevier. https://doi.org/10.1016/B978-0-12-822698-8.00001-9.

Siedler, G., Rouault, M., & Lutjeharms, J. (2006). Structure and origin of the subtropical South Indian Ocean Countercurrent. *Geophysical Research Letters, 33*, L24609. https://doi.org/10.1029/2006GL027399.

Sloyan, B. M. (2006). Antarctic bottom and lower circumpolar deep water circulation in the eastern Indian Ocean. *Journal of Geophysical Research, 111*, C02006. https://doi.org/10.1029/2005JC003011.

Sloyan, B. M., & Rintoul, S. R. (2001). The Southern Ocean limb of the global deep overturning circulation. *Journal of Physical Oceanography, 31*, 143–173.

Smith, R. L., Huyer, A., Godfrey, J. S., & Church, J. A. (1991). The Leeuwin Current off Western Australia, 1986–1987. *Journal of Physical Oceanography, 21*, 323–345.

Sofianos, S. S., & Johns, W. E. (2003). An Oceanic General Circulation Model (OGCM) investigation of the Red Sea circulation: 2. Three-dimensional circulation in the Red Sea. *Journal of Geophysical Research, 108*(C3), 3066. https://doi.org/10.1029/2001JC001185.

Sofianos, S. S., & Johns, W. E. (2015). Water mass formation, overturning circulation, and the exchange of the Red Sea with the adjacent basins. In N. M. A. Rasul, & I. C. F. Stewart (Eds.), *The Red Sea* (pp. 343–353). Berlin, Heidelberg: Springer. https://doi.org/10.1007/978-3-662-45201-1_20.

Sofianos, S. S., Johns, W. E., & Murray, S. P. (2002). Heat and freshwater budgets in the Red Sea from direct observations at Bab elMandeb. *Deep-Sea Research Part II: Topical Studies in Oceanography, 49*, 1323–1340. https://doi.org/10.1016/S0967-0645(01)00164-3.

Sprintall, J., Biastoch, A., Gruenburg, L. K., & Phillips, H. E. (2024). Chapter 9: Oceanic basin connections. In C. C. Ummenhofer, & R. R. Hood (Eds.), *The Indian Ocean and its role in the global climate system* (pp. 205–227). Amsterdam: Elsevier. https://doi.org/10.1016/B978-0-12-822698-8.00003-2.

Sprintall, J., Wijffels, S. E., Molcard, R., & Jaya, I. (2009). Direct estimates of the Indonesian Throughflow entering the Indian Ocean: 2004–2006. *Journal of Geophysical Research, 114*, C07001. https://doi.org/10.1029/2008JC005257.

Sreenivas, P., Gnanaseelan, C., & Prasad, K. V. S. R. (2012). Influence of El Niño and Indian Ocean Dipole on sea level variability in the Bay of Bengal. *Global and Planetary Change, 80-81*, 215–225. https://doi.org/10.1016/j.gloplacha.2011.11.001.

Srinivasan, A., Garraffo, Z., & Iskandarani, M. (2009). Abyssal circulation in the Indian Ocean from a 1/12 resolution global hindcast. *Deep-Sea Research Part I, 56*, 1907–1926. https://doi.org/10.1016/j.dsr.2009.07.001.

Stramma, L., & Lutjeharms, J. R. E. (1997). The flow field of the subtropical gyre of the south Indian Ocean. *Journal of Geophysical Research, 102*, 5513–5530. https://doi.org/10.1029/96JC03455.

Swallow, J., Fieux, M., & Schott, F. (1988). The boundary currents east and north of Madagascar: 1. Geostrophic currents and transports. *Journal of Geophysical Research, 93*, 4951–4962. https://doi.org/10.1029/JC093iC05p04951.

Swallow, J. C., Schott, F., & Fieux, M. (1991). Structure and transport of the east African coastal current. *Journal of Geophysical Research, 96*(C12), 22245–22257.

Talley, L. (2013). Closure of the global overturning circulation through the Indian, Pacific, and Southern Oceans: Schematics and transports. *Oceanography, 26*, 80–97.

Talley, L. D., & Baringer, M. O. (1997). Preliminary results from WOCE hydrographic sections at 80°E and 32°S in the central Indian Ocean. *Geophysical Research Letters, 24*, 2789–2792. https://doi.org/10.1029/97GL02657.

Talley, L., Pickard, G., Emery, W., & Swift, J. (2011). *Descriptive physical oceanography: An introduction* (6th ed.). Boston, MA: Elsevier.

Talley, L. D., & Sprintall, J. (2005). Deep expression of the Indonesian throughflow: Indonesian intermediate water in the south equatorial current. *Journal of Geophysical Research: Oceans (1978–2012), 110*(C10). https://doi.org/10.1029/2004jc002826.

Thompson, R. O. R. Y. (1987). Continental-shelf-scale model of the Leeuwin Current. *Journal of Marine Research, 45*(4), 813–827. https://doi.org/10.1357/002224087788327190.

Thompson, B., Gnanaseelan, C., & Salvekar, P. S. (2006). Variability in the Indian Ocean circulation and salinity and its impact on SST anomalies during dipole events. *Journal of Marine Research, 64*, 853–880.

Thompson, P. A., Wild-Allen, K., Lourey, M., Rousseaux, C., Waite, A. M., Feng, M., & Beckley, L. E. (2011). Nutrients in an oligotrophic boundary current: Evidence of a new role for the Leeuwin Current. *Progress in Oceanography, 91*(4), 345–359. https://doi.org/10.1016/j.pocean.2011.02.011.

Thoppil, P., & Hogan, P. J. (2010). A modeling study of circulation and eddies in the Persian Gulf. *Journal of Physical Oceanography, 40*, 2122–2134. https://doi.org/10.1175/2010JPO4227.1.

Tomczak, M., & Godfrey, J. (2003). *Regional oceanography: An introduction* (2nd ed.). Daya Publishing House.

Toole, J. M., & Warren, B. A. (1993). A hydrographic section across the subtropical South Indian Ocean. *Deep-Sea Research Part I, 40*, 1973–2019. https://doi.org/10.1016/0967-0637(93)90042-2.

Tragou, E., Garrett, C., Outerbridge, R., & Gilman, C. (1999). The heat and freshwater budgets of the Red Sea. *Journal of Physical Oceanography, 29*, 2504–2522. https://doi.org/10.1175/1520-0485(1999)029<2504:THAFBO>2.0.CO;2.

Tsubouchi, T., Suga, T., & Hanawa, K. (2010). Indian Ocean subtropical mode water: Its water characteristics and spatial distribution. *Ocean Science, 6*, 41–50. https://doi.org/10.5194/os-6-41-2010.

Ummenhofer, C. C., Geen, R., Denniston, R. F., & Rao, M. P. (2024). Chapter 3: Past, present, and future of the South Asian monsoon. In C. C. Ummenhofer, & R. R. Hood (Eds.), *The Indian Ocean and its role in the global climate system* (pp. 49–78). Amsterdam: Elsevier. https://doi.org/10.1016/B978-0-12-822698-8.00013-5.

Vialard, J., Shenoi, S. S. C., Mc Creary, J., Shankar, D., Durand, F., Fernando, V., & Shetye, S. R. (2009). Intraseasonal response of the Northern Indian Ocean coastal waveguide to the Madden–Julian Oscillation. *Geophysical Research Letters, 36*. https://doi.org/10.1029/2009GL038450.

Vic, C., Roullet, G., Capet, X., Carton, X., Molemaker, M. J., & Gula, J. (2015). Eddy-topography interactions and the fate of the Persian Gulf Outflow. *Journal of Geophysical Research, Oceans, 120*, 6700–6717. https://doi.org/10.1002/2015JC011033.

Vinayachandran, P. N., Chauhan, P., Mohan, M., & Nayak, S. (2004). Biological response of the sea around Sri Lanka to summer monsoon. *Geophysical Research Letters, 31*. https://doi.org/10.1029/2003GL018533.

Vinayachandran, P., & Kurian, J. (2007). Hydrographic observations and model simulation of the Bay of Bengal freshwater plume. *Deep Sea Research Part I: Oceanographic Research Papers, 54*(4), 471–486. https://doi.org/10.1016/j.dsr.2007.01.007.

Vinayachandran, P. N., Kurian, J., & Neema, C. P. (2007). Indian Ocean response to anomalous conditions in 2006. *Geophysical Research Letters, 34*, L15602. https://doi.org/10.1029/2007GL030194.

Vinayachandran, P. N. M., Masumoto, Y., Roberts, M. J., Huggett, J. A., Halo, I., Chatterjee, A., Amol, P., Gupta, G. V. M., Singh, A., Mukherjee, A., Prakash, S., Beckley, L. E., Raes, E. J., & Hood, R. (2020). Reviews and syntheses: Physical and biogeochemical processes associated with upwelling in the Indian Ocean. *Biogeosciences, 18*(22), 5967–6029. https://doi.org/10.5194/bg-18-5967-2021.

Vinayachandran, P. N., Masumoto, Y., Mikawa, T., & Yamagata, T. (1999a). Intrusion of the Southwest Monsoon Current into the Bay of Bengal. *Journal of Geophysical Research: Oceans, 104*(C5), 11077–11085. https://doi.org/10.1029/1999JC900035.

Vinayachandran, P. N., Matthews, A. J., Kumar, K. V., Sanchez-Franks, A., Thushara, V., George, J., Vijith, V., Webber, B. G. M., Queste, B. Y., Roy, R., Sarkar, A., Baranowski, D. B., Bhat, G. S., Klingaman, N. P., Peatman, S. C., Parida, C., Heywood, K. J., Hall, R., King, B., & Joshi, M. (2018). BoBBLE: Ocean–atmosphere interaction and its impact on the South Asian Monsoon. *Bulletin of the American Meteorological Society, 99*(8), 1569–1587. https://doi.org/10.1175/BAMS-D-16-0230.1.

Vinayachandran, P. N., Saji, N. H., & Yamagata, T. (1999b). Response of the equatorial Indian Ocean to an anomalous wind event during 1994. *Geophysical Research Letters, 26*(11), 1613–1615.

Vinayachandran, P. N., Shankar, D., Vernekar, S., Sandeep, K. K., Amol, P., Neema, C. P., & Chatterjee, A. (2013). A summer monsoon pump to keep the Bay of Bengal salty. *Geophysical Research Letters, 40*, 1777–1782. https://doi.org/10.1002/grl.50274.

Vinayachandran, P. N., Shetye, S. R., Sengupta, D., & Gadgil, S. (1996). Forcing mechanisms of the Bay of Bengal circulation. *Current Science, 71*.

Vinayachandran, P. N., & Yamagata, T. (1998). Monsoon Response of the Sea around Sri Lanka: Generation of thermal domes and anticyclonic vortices. *Journal of Physical Oceanography, 28*(10), 1946–1960. https://doi.org/10.1175/1520-0485.

Waite, A. M., Thompson, P. A., Pesant, S., Feng, M., Beckley, L. E., Domingues, C. M., Gaughan, D., Hanson, C. E., Holl, C. M., Koslow, T., Meuleners, M., Montoya, J. P., Moore, T., Muhling, B. A., Paterson, H., Rennie, S., Strzelecki, J., & Twomey, L. (2007). The Leeuwin Current and its eddies: An introductory overview. *Deep Sea Research Part II: Topical Studies in Oceanography, 54*(8–10), 789–796. https://doi.org/10.1016/j.dsr2.2006.12.008.

Wang, Y., & McPhaden, M. J. (2017). Seasonal cycle of cross-equatorial flow in the central Indian Ocean. *Journal of Geophysical Research, 122*. https://doi.org/10.1002/2016JC012537.

Wang, W., Zhu, X., Wang, C., & Köhl, A. (2014). Deep meridional overturning circulation in the Indian Ocean and its relation to Indian Ocean Dipole. *Journal of Climate, 27*, 4508–4520. https://doi.org/10.1175/JCLI-D-13-00472.1.

Warren, B. A. (1974). Deep flow in the Madagascar and Mascarene basins. *Deep Sea Research, 21*, 1–21. https://doi.org/10.1016/0011-7471(74)90015-1.

Warren, B. A. (1978). Bottom water transport through the Southwest Indian Ridge. *Deep Sea Research, 25*, 315–321. https://doi.org/10.1016/0146-6291(78)90596-9.

Warren, B. A. (1981). Transindian hydrographic section at Lat. 18°S: Property distributions and circulation in the South Indian Ocean. *Deep-Sea Research Part A, 28*, 759–788. https://doi.org/10.1016/S0198-0149(81)80001-5.

Warren, B. A., & Johnson, G. C. (2002). The overflows across the Ninetyeast Ridge. *Deep-Sea Research Part I, 49*, 1423–1439. https://doi.org/10.1016/S0967-0645(01)00156-4.

Warren, B. A., Whitworth, T., & LaCasce, J. H. (2002). Forced resonant undulation in the deep Mascarene Basin. *Deep-Sea Research Part II, 49*, 1513–1526. https://doi.org/10.1016/S0967-0645(01)00151-5.

Warren, B., et al. (1966). Water mass and patterns of flow in the Somali basin during the southwest monsoon of 1964. *Deep Sea Research, 13*(5), 825–860.

Weaver, A. J., & Middleton, J. H. (1989). On the dynamics of the Leeuwin Current. *Journal of Physical Oceanography, 19*, 626–648. https://doi.org/10.1175/1520-0485(1989)019<0626:OTDOTL>2.0.CO;2.

Weaver, A. J., & Middleton, J. H. (1990). An analytic model for the Leeuwin Current off western Australia. *Continental Shelf Research, 10*(2), 105–122. https://doi.org/10.1016/0278-4343(90)90025-H.

Webber, B. G. M., Matthews, A. J., Vinayachandran, P. N., Neema, C. P., Sanchez-Franks, A., Vijith, V., Amol, P., & Baranowski, D. B. (2018). The dynamics of the Southwest Monsoon Current in 2016 from high-resolution in situ observations and models. *Journal of Physical Oceanography, 48*(10), 2259–2282. https://doi.org/10.1175/JPO-D-17-0215.1.

Wentz, F. J., Gentemann, C., Smith, D., & Chelton, D. (2000). Satellite measurements of sea surface temperature through clouds. *Science, 288*, 847–850. https://doi.org/10.1126/science.288.5467.847.

Wijesekera, H. W., Jensen, T. G., Jarosz, E., Teague, W. J., Metzger, E. J., Wang, D. W., Jinadasa, S. U. P., Arulananthan, K., Centurioni, L. R., & Fernando, H. J. S. (2015). Southern Bay of Bengal currents and salinity intrusions during the northeast monsoon. *Journal of Geophysical Research, Oceans*. https://doi.org/10.1002/2015JC010744.

Wijesekera, H. W., Shroyer, E., Tandon, A., Ravichandran, M., Sengupta, D., Jinadasa, S. U. P., … Whalen, C. B. (2016a). ASIRI: An ocean-atmosphere initiative for Bay of Bengal. *Bulletin of the American Meteorological Society*. https://doi.org/10.1175/BAMS-D-14-00197.1.

Wijesekera, H., Teague, W., Jarosz, E., Wang, D., Jensen, T., Jinadasa, S. U. P., … Moum, J. (2016b). Observations of currents over the deep Southern Bay of Bengal—With a little luck. *Oceanography, 29*(2), 112–123. https://doi.org/10.5670/oceanog.2016.44.

Wijffels, S., & Meyers, G. (2004). An intersection of oceanic waveguides: Variability in the Indonesian throughflow region. *Journal of Physical Oceanography, 34*, 1232–1253. https://doi.org/10.1175/1520-0485(2004)034,1232:AIOOWV.2.0.CO;2.

Wijffels, S., Sprintall, J., Fieux, M., & Bray, N. (2002). The JADE and WOCE I10/IR6 Throughflow sections in the southeast Indian Ocean. Part I: Water mass distribution and variability. *Deep-Sea Research Part II, 49*, 1341–1362. https://doi.org/10.1016/S0967-0645(01)00155-2.

Woo, L. M., & Pattiaratchi, C. B. (2008). Hydrography and water masses off the western Australian coast. *Deep Sea Research, Part I, 55*(9), 1090–1104. https://doi.org/10.1016/j.dsr.2008.05.005.

Wyrtki, K. (1973). An equatorial jet in the Indian Ocean. *Science, 181*(4096), 262–264. https://doi.org/10.1126/science.181.4096.262.

Wyrtki, K., Bennett, E. B., & Rochford, D. J. (1971). *Oceanographic Atlas of the International Indian Ocean expeditions*. National Science Foundation. OCE/NSF 86-00-001.

Xue, P., & Eltahir, E. A. B. (2015). Estimation of the heat and water budgets of the Persian (Arabian) Gulf using a regional climate model. *Journal of Climate, 28*, 5041–5062.

Yamagata, T., Behera, S., Doi, T., Luo, J.-J., Morioka, Y., & Tozuka, T. (2024). Chapter 5: Climate phenomena of the Indian Ocean. In C. C. Ummenhofer, & R. R. Hood (Eds.), *The Indian Ocean and its role in the global climate system* (pp. 103–119). Amsterdam: Elsevier. https://doi.org/10.1016/B978-0-12-822698-8.00009-3.

Yang, H., et al. (2016). Intensification and poleward shift of subtropical western boundary currents in a warming climate. *Journal of Geophysical Research, Oceans, 121*, 4928–4945.

Yao, F. (2008). *Water mass formation and circulation in the Persian Gulf and water exchange with the Indian Ocean*. Ph.D. dissertation University of Miami. 144 pp.

Yao, F., Hoteit, I., Pratt, L. J., Bower, A. S., Kohl, A., Gopalakrishnan, G., & Rivas, D. (2014a). Seasonal overturning circulation in the Red Sea: 2. Winter circulation. *Journal of Geophysical Research: Oceans, 119*, 2263–2289. https://doi.org/10.1002/2013JC009331.

Yao, F., Hoteit, I., Pratt, L. J., Bower, A. S., Zhai, P., Kohl, A., & Gopalakrishnan, G. (2014b). Seasonal overturning circulation in the Red Sea: 1. Model validation and summer circulation. *Journal of Geophysical Research: Oceans, 119*, 2238–2262. https://doi.org/10.1002/2013JC009004.

Yao, F., & Johns, W. E. (2010). A HYCOM modeling study of the Persian Gulf: 1. Model configurations and surface circulation. *Journal of Geophysical Research, 115*, C11017. https://doi.org/10.1029/2009JC005781.

Yit Sen Bull, C., & van Sebille, E. (2016). Sources, fate, and pathways of Leeuwin Current water in the Indian Ocean and Great Australian Bight: A Lagrangian study in an eddy-resolving ocean model. *Journal of Geophysical Research, Oceans, 121*, 1626–1639. https://doi.org/10.1002/2015JC011486.

Yokoi, T., Tozuka, T., & Yamagata, T. (2008). Seasonal variation of the Seychelles Dome. *Journal of Climate, 21*, 3740–3754. https://doi.org/10.1175/2008JCLI1957.1.

You, Y. (1998). Intermediate water circulation and ventilation of the Indian Ocean derived from water-mass contributions. *Journal of Marine Research, 56*, 1029–1067. https://doi.org/10.1357/002224098765173455.

Youngs, M. K., & Johnson, G. C. (2015). Basin-wavelength equatorial deep jet signals across three oceans. *Journal of Physical Oceanography, 45*, 2134–2148. https://doi.org/10.1175/JPO-D-14-0181.1.

Zachariah, J., Babu, C. A., & Varikoden, H. (2019). Dynamics of westward propagation and intensification of Lakshadweep low in the southern Arabian Sea. *Ocean Dynamics, 69*(5), 519–528. https://doi.org/10.1007/s10236-019-01263-5.

Zang, N., Sprintall, J., Ienny, R., & Wang, F. (2021). Seasonality of the Somali current/undercurrent system. *Deep-Sea Research Part II*. https://doi.org/10.1016/j.dsr2.2021.104953.

Zanowski, H., & Johnson, G. C. (2019). Semiannual variations in 1,000-dbar equatorial Indian Ocean velocity and isotherm displacements from Argo data. *Journal of Geophysical Research: Oceans, 124*. https://doi.org/10.1029/2019JC015342.

Zhai, P., Bower, A. S., Smethie, W., & Pratt, L. (2015a). Formation and spreading of Red Sea outflow water in the Red Sea. *Journal of Geophysical Research: Oceans, 120*, 6542–6563. https://doi.org/10.1002/2015JC010751.

Zhai, P., Pratt, L. J., & Bower, A. (2015b). On the crossover of boundary currents in an idealized model of the Red Sea. *Journal of Physical Oceanography, 45*, 1410–1425. https://doi.org/10.1175/JPO-D-14-0192.1.

Zhang, D., McPhaden, M. J., & Lee, T. (2014). Observed interannual variability of zonal currents in the equatorial Indian ocean thermocline and their relation to Indian ocean dipole. *Geophysicahl Research Letters, 41*, 7933–7941. https://doi.org/10.1002/2014GL061449.

Zhang, T., Wang, W., Xie, Q., Liu, K., Wang, D., Wu, X., Zhang, X., Xu, K., & Yuan, W. (2021). A new presentation of the Indian Ocean shallow overturning circulation from a vertical perspective. *Atmospheric and Oceanic Science Letters, 100061*. https://doi.org/10.1016/j.aosl.2021.100061.

Zhuang, W., Feng, M., Du, Y., Schiller, A., & Wang, D. (2013). Low-frequency sea level variability in the southern Indian Ocean and its impacts on the oceanic meridional transports. *Journal of Geophysical Research: Oceans, 118*, 1302–1315. https://doi.org/10.1002/jgrc.20129.

Zolina, O., Dufour, A., Gulev, S., & Stenchikov, G. (2017). Regional hydrological cycle over the Red Sea in ERA-Interim. *Journal of Hydrometeorology, 18*, 65–83. https://doi.org/10.1175/JHM-D-16-0048.1.

Chapter 9

Oceanic basin connections[*]

Janet Sprintall[a], Arne Biastoch[b], Laura K. Gruenburg[c], and Helen E. Phillips[d]

[a]Climate, Atmospheric Sciences and Physical Oceanography, Scripps Institution of Oceanography, University of California San Diego, La Jolla, CA, United States, [b]GEOMAR Helmholtz Centre for Ocean Research and Kiel University, Kiel, Germany, [c]School of Marine and Atmospheric Sciences, Stony Brook University, Stony Brook, NY, United States, [d]Institute for Marine and Antarctic Studies and Australian Antarctic Program Partnership, University of Tasmania, Hobart, TAS, Australia

1 Introduction

The global thermohaline circulation driven by density variations in response to variable surface heat and freshwater fluxes acts to connect the different ocean basins. Such connectivity is maintained via open boundaries around ocean borders that permit interbasin exchanges of the different water masses and their biogeochemical properties. The boundaries of the Indian Ocean basin are characterized by some unique features that distinctively set its connectivity apart from the other oceans affecting the transport of heat, freshwater, and other properties (Fig. 1).

Unlike other ocean basins, the Indian Ocean is closed off in the north by the Asian land mass. This not only limits the northern extent of the currents within the ocean basin but also results in the extreme land-sea surface temperature contrasts that control the Asian monsoon system, which is of fundamental importance for the Indian Ocean circulation (Ummenhofer et al., 2024). The Indian Ocean uniquely has a low-latitude ocean connection via the various channels of the Indonesian archipelago. Critically, this conduit provides a linkage between the warm and fresh pools of the tropical Pacific and Indian Oceans. The coastal subtropical boundary currents within the Indian Ocean also have noticeable differences compared to their global counterparts. The Leeuwin Current is the only poleward flowing eastern boundary current in the world, flowing along the Western Australian coastline to feed into a complex series of currents south of Australia. This current system influences regional climate and provides a key conveyance of many important tropical marine species and their larvae that are vital to sustaining one of Australia's largest fisheries (Caputi et al., 1996). On the other side of the basin, the Agulhas Current is the strongest western boundary current in the Southern Hemisphere, flowing poleward along the east coast of the southern African continent. Most of the Agulhas Current retroflects south of the Agulhas Bank and returns eastward, while a smaller part leaks westward into the Atlantic Ocean primarily through large eddies known as Agulhas Rings (Lutjeharms, 2006). These eddies return warm and salty Indian Ocean waters into the Atlantic and play a significant role in the upper branch of the global meridional overturning circulation (Gordon, 1986). The eastward-flowing Antarctic Circumpolar Current in the Southern Ocean marks the southern boundary of the Indian Ocean and provides direct sources for both the shallow and deep subtropical cells in the Indian Ocean circulation. South of Australia has a remarkable series of westward-flowing streams that provide inflow into the Indian Ocean north of the Antarctic Circumpolar Current. These streams include the Pacific-sourced Tasman leakage and the Flinders Current that extend into the southeastern Indian Ocean subtropical gyre. There is also some evidence that the Indian Ocean subtropical gyre is embedded in a Southern Hemisphere supergyre that broadly connects the southern Atlantic, Indian, and Pacific Ocean subtropical gyres (Speich et al., 2002, 2007; Ridgway & Dunn, 2007).

The interbasin exchanges thus play a critical role in the transportation of oceanic properties that sustain our unifying view of a single, vast, interconnected global ocean circulation. This chapter examines our knowledge of interocean exchanges with the Indian Ocean from observational evidence and numerical simulations. Section 2 highlights the tropical interbasin connection from the Pacific, addressing the pathways and variability of the Indonesian Throughflow (ITF) and its relative influence on the Indian Ocean circulation, water mass, and biogeochemical properties. Section 3 focuses on the exports and imports that stem from the south of Australia. Section 4 features the exchanges that occur within the greater

---

[*]This book has a companion website hosting complementary materials. Visit this URL to access it: https://www.elsevier.com/books-and-journals/book-companion/9780128226988.

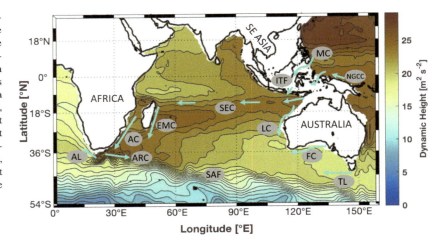

FIG. 1 Schematic of the boundary current circulation around the Indian Ocean. Color contours are dynamic height (m² s⁻²) at the sea surface relative to 1975m averaged between 2004 and 2015, estimated using the gridded Argo climatology data (Roemmich & Gilson, 2009). The solid teal arrows show the surface circulation in the New Guinea Coastal Current (NGCC), Mindanao Current (MC), Indonesian Throughflow (ITF), Leeuwin Current (LC), Tasman leakage (TL), Flinders Current (FC), South Equatorial Current (SEC), Eastern Madagascar Current (EMC), Agulhas Current (AC), Agulhas leakage (AL), Agulhas Return Current (ARC), and the SubAntarctic Front (SAF) of the Antarctic Circumpolar Current.

Agulhas system, and Section 5 details the Southern Ocean exchanges. Projected changes in all these interbasin connections in a warming world are briefly considered in Section 6, and conclusions are given in Section 7.

2 The Indonesian Throughflow

2.1 Introduction

The Indonesian seas provide the low-latitude pathway between the Pacific and Indian Oceans and play a critical role in regional climate, air–sea interaction, and the heat and freshwater budgets of the Indian Ocean. This unique choke point in the global ocean circulation allows for the transport of relatively warm low salinity tropical waters via multiple circuitous pathways collectively called the ITF (Fig. 2).

Strong easterly trade winds create high sea level in the tropical western Pacific, generating the Pacific–Indian interbasin pressure gradient that drives the ITF on annual and longer timescales (Wyrtki, 1987). Hence, any phenomena that influence

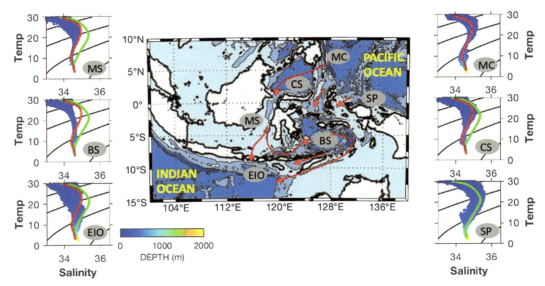

FIG. 2 Schematic of pathways of flow through the Indonesian seas. The side diagrams show the temperature (°C; y-axis) versus salinity (PSS-78; x-axis) diagrams color-coded by depth (m) in the inflow regions from the Mindanao Current (MC) and the South Pacific (SP) through the Celebes Sea (CS) into Makassar Strait (MS), then the Banda Sea (BS) and exiting into the Eastern Indian Ocean (EIO). In each panel, the *red* curve represents the isopycnal mean temperature and salinity in the MC of the North Pacific and the *green* curve represents the isopycnal mean temperature and salinity in the South Pacific. Note the erosion of both the salinity maxima in the thermocline layer and the salinity minimum at intermediate depth as the Pacific inflow waters traverse the regional Indonesia seas into the Indian Ocean. The temperature and salinity data are from the Argo climatology.

the strength or timing of this pressure gradient will, in turn, impact the ITF. One such phenomenon is the El Niño-Southern Oscillation (ENSO). During the La Niña phase, the Pacific trade winds are stronger and thereby increase the pressure gradient and enhance the ITF transport (e.g., Gordon et al., 2008, 2012). The opposite occurs during El Niño. ENSO can also affect temperature stratification, with a deepening of the thermocline during La Niña (Ffield et al., 2000; Susanto et al., 2012). These combined effects on the volume transport and temperature act to increase the ITF heat flux into the Indian Ocean during La Niña (Gruenburg & Gordon, 2018), especially during boreal summer months (Gordon et al., 2019). Similarly, on interdecadal timescales, the large-scale wind changes in the Pacific Walker Circulation associated with the Interdecadal Pacific Oscillation (IPO) can also impact the Indo-Pacific pressure gradient, changing the ITF transport and its subsequent influence on the Indian Ocean heat and freshwater (Hu & Sprintall, 2016; Ummenhofer et al., 2021). More details on the impact of the ITF on the Indian Ocean basin on decadal and longer time scales can be found in Tozuka et al. (2024).

On seasonal timescales, the annual reversal of winds associated with the large-scale East Asian Monsoon system causes a significant annual and semi-annual (during monsoon transitions) variation in the ITF transport with maxima typically occurring during the boreal summer season (Gordon et al., 2008, 2012; Sprintall et al., 2009; Susanto et al., 2012). Paleo coral records have shown that the strength of the monsoon can also change over time and this in turn modulates the ITF (Murty et al., 2017). This nonstationarity can challenge interpretation of longer-term trends related to both natural and anthropogenic forcing in the region (Ummenhofer et al., 2021).

2.2 Pathways through the Indonesian seas

The inflow of ITF waters is primarily drawn from the Mindanao (North Pacific) and Halmahera (South Pacific) Retroflections. Still, the ratio of the relatively fresher North Pacific to the saltier South Pacific contributions to the ITF and how this might change over time remains poorly understood. Most of the ITF is thought to originate from the North Pacific taking a western route through the Celebes Sea by way of Makassar Strait (Gordon & Fine, 1996). The North Pacific water is characterized by a salinity maximum in the upper thermocline layer and a salinity minimum in the lower thermocline intermediate layer (Fig. 2). Secondary portals of fresher North Pacific surface waters can also enter the Indonesian seas via the South China Sea to the Sulu Sea through the shallow Sibutu Passage as well as via the Karimata Strait between Sumatra and Kalimantan. Saltier lower thermocline and fresher intermediate South Pacific waters (Fig. 2) primarily enter via an eastern route through the Maluku and Halmahera Seas, as well as through density-driven overflows in the deep Lifamatola Passage that dominate the deeper layers of the Banda Sea (Van Aken et al., 1988). However, the various inflows via these northeastern passages remain inadequately observed and have poorly known pathways and transit times, which may vary over many timescales (Gordon et al., 2010).

As the North and South Pacific Ocean water masses make their way through into the internal Indonesian seas they undergo significant alteration. The salinity extrema of the Pacific source waters erode to create a unique Indonesian sea water mass with an intense thermocline but a more isohaline profile (Fig. 2). The water mass transformation seems to occur almost as soon as the Pacific waters enter the Indonesian seas (Koch-Larrouy et al., 2008; Sprintall et al., 2014). Processes causing the modification of the Pacific water masses include diapycnal mixing related to topography that induces strong internal tides (Koch-Larrouy et al., 2008, 2010, 2015), as well as contributions from enhanced air-sea heat and freshwater fluxes (Wijffels et al., 2008), and Ekman pumping induced by the monsoon winds (Gordon & Susanto, 2001).

The ITF water masses that are exported into the Indian Ocean are thus not the same water masses that entered the Pacific Ocean (Fig. 2). Flow into the Indian Ocean occurs primarily via the shallow Lombok Strait and the deeper exit channels of Ombai Strait and Timor Passage (Sprintall et al., 2009). However, because the vertical structure of the transport through each exit passage is quite different, each passage also admits distinct variants of the ITF water masses into the Indian Ocean. The transport profile is surface intensified in the wider Timor Passage but subsurface velocity maxima are evident in Ombai and Lombok Straits. A mean total volume transport of ~15 Sv enters the Indian Ocean via these combined ITF exit channels with a mean transport weighted temperature of 17.9°C (Sprintall et al., 2009). However, the wide-ranging temporal variability of the wind and buoyancy forcings and the contrasting vertical profiles causes variability in the volume transport that can be as large as the mean, and this significantly impacts the subsequent amount of heat and freshwater transport that is partitioned through each channel into the Indian Ocean.

Local and remote wind forcing drives planetary waves that radiate and scatter into the Indonesian seas and affect the currents and water mass transformation. Pacific Ocean wind anomalies force low-frequency off-equatorial Rossby waves that interact with the western Pacific maritime boundary to excite coastally trapped waves that propagate through the Indonesian seas, along the northwest coast of Australia, subsequently generating Rossby waves that move offshore into the Indian Ocean (England & Huang, 2005; McClean et al., 2005; Wijffels & Meyers, 2004). In the Indian Ocean, equatorial wind anomalies on timescales ranging from intraseasonal to interannual spawn eastward propagating equatorial Kelvin waves. The Kelvin waves strongly impact the timing, location, and strength of the ITF exit flow

(Drushka et al., 2010; Sprintall et al., 2009) and have been shown to radiate up into Makassar Strait and beyond (Hu et al., 2019; Napitu et al., 2019; Pujiana et al., 2013, 2019). Nonetheless, the impacts that planetary waves have on the water masses and residence times within the internal seas are not well known (Wijffels & Meyers, 2004).

2.3 Biogeochemistry within the Indonesian seas

The Indonesian seas are an integral part of the "Coral Triangle," a vast network of abundant marine ecosystems that contains ~80% of all the marine species in the Indo-Pacific region, making it a region of global significance for marine biodiversity (Roberts et al., 2002; Tun et al., 2004). The ITF is thought to be an important source of nutrients to the Indian Ocean (Ayers et al., 2014) although the North Pacific Mindanao Current that sources the ITF is not very rich in nutrients (Xie et al., 2019). Hence, the Indonesian seas themselves must act as a significant nutrient source. Strong mixing and upwelling within the Indonesian seas draw the nutrient-rich deeper waters upward to become available to primary producers (e.g., Broecker et al., 1986; Coatanoan et al., 1999; van Bennekom & Muchtar, 1988). River runoff into the Indonesian coastal regions provides large amounts of terrestrial organic matter that may also result in the delivery of nutrients (Asanuma et al., 2003; Hendiarti et al., 2004). The high percentage of diatoms in the euphotic zone within the Indonesian seas accumulates dissolved silica in the surface water, especially during the Southeast Monsoon in July–September (van Bennekom, 1988). The sedimentation of the diatom frustules and their subsequent dissolution then enriches the silica input throughout the water column (Talley & Sprintall, 2005). Silica concentrations are also observed to be relatively high in shallow waters, such as the Java Sea, which could be caused by a terrestrial silica supply via river runoff (Kartadikaraia et al., 2015).

Internal waves generated in response to the strong tidal flows within the internal Indonesian seas and in the vicinity of straits (Koch-Larrouy et al., 2007) induce changes in upper ocean stratification that impact chlorophyll blooms and primary productivity that are evident in remotely sensed imagery (Moore & Marra, 2002; Nugroho et al., 2018; Susanto et al., 2006; Xu et al., 2018). The seasonal reversal of monsoon winds appears to act like a "switch," turning on much higher primary production during the Southeast Monsoon and leaving an oligotrophic regime for the remainder of the year. The ENSO cycle acts in much the same way: the El Niño phase strengthens and lengthens the biological response to the Southeast Monsoon but has little apparent effect during the rest of the year (Moore et al., 2003).

Little attention has been given to the carbon budget within the Indonesian seas. The Indonesian seas are situated between appreciable CO_2 sink regions in the western Pacific ITF entrance and the Indian Ocean exit. However, overall the Indonesian seas appear to act as a CO_2 source to the atmosphere with an air–sea flux of $3.7 \pm 2.2\,\mathrm{mol\,m^{-2}\,year^{-1}}$ (Kartadikaraia et al., 2015), consistent with that of other low-latitude coastal regions (Gruber, 2015). Nonetheless, significant regional differences are evident. The regional carbon cycle is influenced by the carbonate system, water mass characteristics, nutrients, organic matter, upwelling, river inflow, and atmospheric conditions. The shallow Karimata Strait and Java Sea release more CO_2 to the atmosphere compared to the deeper Flores and Banda Seas, while considerable CO_2 sinks have been observed in the northern Makassar Strait and the northern Banda Sea (Hamzah et al., 2020; Kartadikaraia et al., 2015). Physical mixing appears responsible for the carbon sink in the eastern seas. Somewhat surprisingly, temperature had only limited seasonal influence on the carbon cycle but becomes more significant during ENSO events (Kartadikaraia et al., 2015).

2.4 The ITF influence on the properties and currents within the Indian Ocean

From the outflow passages, the signature of the Indonesian water mass is readily identifiable across nearly the entire Indian Ocean basin (e.g., Wyrtki, 1971) (Fig. 1). Observations show a separate low salinity surface core (Gordon et al., 1997) from a low salinity, high silicate intermediate depth core (Talley & Sprintall, 2005) within the South Equatorial Current between around 10° and 15°S. Somewhat paradoxically, because the ITF is relatively fresh compared to Indian Ocean water masses, the spreading of the ITF along isopycnal surfaces acts to cool and freshen the Indian Ocean (Song & Gordon, 2004; Talley & Sprintall, 2005).

Sharp biogeochemical fronts coincide with the hydrological fronts associated with the pathway of the ITF waters in the tropical Indian Ocean. The biogeochemical characteristics of the water masses that originate from the Indonesian seas have tracers that clearly separate the waters to the north and south in the Indian Ocean with distinct cores of elevated tritium (Fine, 1985) and high phosphate, nitrate, and silicate at mid-depth (Ayers et al., 2014; Coatanoan et al., 1999; Fieux et al., 1996) and a separate intermediate depth high silica core (Reid, 2003; Talley & Sprintall, 2005). The mid-depth signal occurs because the nutricline is elevated in the well-mixed Indonesian seas relative to that of the Indian Ocean subtropical gyre, and so appears as a larger effective nutrient flux within lighter density classes in the Indian Ocean (Ayers et al., 2014).

The nutrient enrichment by the ITF thermocline waters can support a substantial amount of new production and significantly impact the basin-wide biogeochemical cycles in the Indian Ocean (Ayers et al., 2014).

The ITF flow from the South Equatorial Current also contributes to the Agulhas Current system (Section 4). Le Bars et al. (2013) showed that ~10 Sv (two-thirds) of a modeled ITF transport arrives in the Agulhas Current, with 3 Sv of that exiting via the Agulhas leakage into the Atlantic while Durgadoo et al. (2017) found 6 Sv of ITF water contributes to the Agulhas leakage. These models suggest a 10- to 30-year timescale for the ITF water to reach the Agulhas region, and during the journey, the ITF waters are subject to significant cooling and salinification through air-sea interaction and mixing with the ambient saltier Indian Ocean water masses. Hence, upon entering the Atlantic Ocean, there is little resemblance of the water masses to the characteristic isohaline ITF profile that exited from the Indonesian seas into the southeast Indian Ocean. The ITF also feeds into the Leeuwin Current (Section 3.2.1), where changes in the ITF contribution can affect marine ecosystems along the West Australian coast (Feng et al., 2013).

In recent decades, the role of the ITF in interbasin heat exchange has been striking. During the period of global surface warming slowdown or "hiatus" in the early 21st century, numerous studies suggested that the excess atmospheric heat uptake within the Pacific Ocean was transferred westwards via the ITF resulting in an increase in upper ocean heat content within the Indian Ocean (Lee et al., 2015; Nieves et al., 2015; Zhang et al., 2018). This increase in upper ocean heat content contributed to a reduction in the meridional pressure gradient within the Indian basin acting to reduce the strength of the monsoonal winds (Vidya et al., 2020) and even influenced the Atlantic meridional overturning circulation (Hu & Fedorov, 2019, 2020). Further details on longer-term heat content related to decadal Indian Ocean variability are found in Tozuka et al. (2024).

3 Southeastern boundary exchanges

3.1 Introduction

A network of boundary currents along Australia's western and southern coastlines allows the exchange of waters between the Indian Ocean and the waters south of Australia. A schematic view (Fig. 3) illustrates the export from the Indian Ocean via the near-surface, poleward-flowing Leeuwin Current, and a recently identified deep eastern boundary current that both continue south of Australia (Tamsitt et al., 2019). Import into the Indian Ocean is accomplished by the Leeuwin Undercurrent at thermocline and intermediate depths and the Flinders Current and Tasman Outflow (or Tasman leakage) between the surface and intermediate depths. The general circulation within the Indian Ocean is discussed further in Phillips et al. (2024).

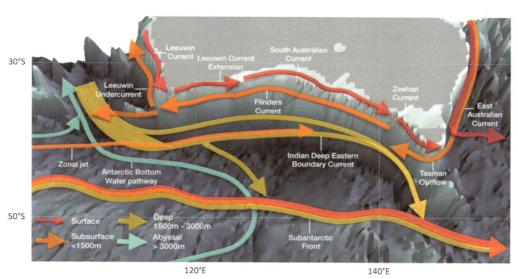

FIG. 3 Schematic of the circulation pathways of the southeastern Indian Ocean boundary exchanges south of Australia. Upper ocean current pathway on the continental shelf/slope is in *red*; subsurface (<1500m) pathway in *orange*; deep (1500–3000m) circulation in *yellow*; abyssal (>3000m) circulation of Antarctic Bottom Water is shown in teal. The major currents and features of the region are labeled. *(Figure from Tamsitt et al. (2019).)*

3.2 Exports from the Indian Ocean

3.2.1 The Leeuwin Current

The Leeuwin Current is an eastern boundary current that uniquely flows poleward along the shelf break of Western Australia (Smith et al., 1991), unlike the equatorward flowing eastern boundary currents found elsewhere. It owes its existence to the strong meridional pressure gradient created by the presence of the warm, fresh ITF waters carried westward by the South Equatorial Current and denser subtropical waters to the south. This poleward decrease in pressure in the upper 200–300 m drives eastward geostrophic currents from as far west as Madagascar toward the eastern boundary (Divakaran & Brassington, 2011; Domingues et al., 2007; Menezes et al., 2014; Palastanga et al., 2007; Phillips et al., 2024). As the bands of flow approach Australia, they veer southward because of the pressure gradient set up by the Australian landmass, inducing downwelling and a surface poleward current in opposition to the southerly winds along the coast (Furue et al., 2013; Godfrey & Ridgway, 1985; McCreary et al., 1986; Thompson, 1987).

The primary source waters to the Leeuwin Current are the near-surface eastward flow, the ITF, and a weaker source from the Holloway Current on the Australian Northwest Shelf. In the north, the Leeuwin Current waters are relatively fresh and warm due to their tropical origins (Andrews, 1977; Domingues et al., 2007; Legeckis & Cresswell, 1981; Phillips et al., 2005; Rochford, 1969; Woo & Pattiaratchi, 2008). As the Leeuwin Current flows poleward, the saltier surface waters of the subtropical South Indian Ocean and additional surface cooling together contribute to the Leeuwin Current densification (Domingues et al., 2006; Furue, 2019; Woo & Pattiaratchi, 2008).

The Leeuwin Current creates mixed barotropic and baroclinic instabilities that generate mesoscale eddies (Feng et al., 2005; Meuleners et al., 2007, 2008; Pearce & Griffiths, 1991). The eddies and energetic meandering drive the strongest eddy kinetic energy level found in any subtropical eastern boundary system (Fang & Morrow, 2003; Feng et al., 2005). The eddies thus play a key role in the momentum balance of the Leeuwin Current. In an eddy-permitting model, Domingues et al. (2006) found that 70% of the heat advected by the Leeuwin Current was exported by eddy heat fluxes to the ocean interior, facilitating a transfer of over $40\,\mathrm{W\,m^{-2}}$ of heat to the atmosphere in the southeast Indian Ocean. Anticyclonic eddies generated within the Leeuwin Current carry coastal water with elevated chlorophyll into the oligotrophic offshore waters, with enhanced productivity that can persist for months (Dufois et al., 2014; Feng et al., 2007; Moore et al., 2007; Phillips et al., 2021).

3.2.2 The southern Australia current system

The Leeuwin Current pivots around the southwestern corner of Australia and continues to flow eastward in the Leeuwin Current Extension along the shelf break of the south coast of Australia (Duran et al., 2020b; Ridgway & Condie, 2004) (Fig. 3). It merges with a series of shelf-break currents that extend to the southern tip of Tasmania near 140°E. Starting from the North West Cape (21.8°S) off western Australia and extending to South East Cape, Tasmania (43.6°S), the combined Leeuwin Current and shelf-break currents represent the longest boundary circulation in the world (Middleton & Bye, 2007; Ridgway & Condie, 2004). The unique factors that contribute to this remarkably long boundary current are the ITF, which maintains the longshore pressure gradient off the west coast, and the uniquely broad zonal extent of the south coast of Australia (Ridgway & Condie, 2004). Monthly sea surface height anomalies (Fig. 4) show the gradual strengthening of the cross-shore pressure gradient, driving poleward flow along western Australia and continuing as eastward flow along southern Australia peaks in May–June (Ridgway & Condie, 2004).

The shelf-break currents are the surface expression of coupled surface and deep currents collectively known as the Southern Australia Current System (Duran et al., 2020b). The system contains the eastward shelf-break currents including the eastward Leeuwin Current Extension, the predominantly eastward South Australian Current, and the poleward Zeehan Current; the counter-flowing Flinders Current is an undercurrent to the shelf-break current along the shelf break but extends to the sea surface further offshore (Fig. 3). When the longshore surface current is weak, the shelf-break currents tend to be somewhat fragmentary (Oke et al., 2018) and sometimes even reverse (Duran et al., 2020b). For this reason, and also because of a lack of observations, the SBCs have not traditionally been viewed as a single current.

The forcing mechanisms of the flow on the western and southern Australian shelves are distinct. While the Leeuwin Current is driven by the longshore pressure gradient that overwhelms the local opposing winds (Section 3.2.1), the southern shelf-break currents are directly forced by the winds, with high coastal sea level established by onshore Ekman flow driven by the winter westerly wind. The seasonal timing of these two different forcing mechanisms means that the west coast pressure gradient, strongest in austral winter (Feng et al., 2003), delivers the Leeuwin Current to the south coast just as the winds reverse and are thus able to maintain the eastward passage of the shelf-break currents (Ridgway & Condie, 2004) (Fig. 4).

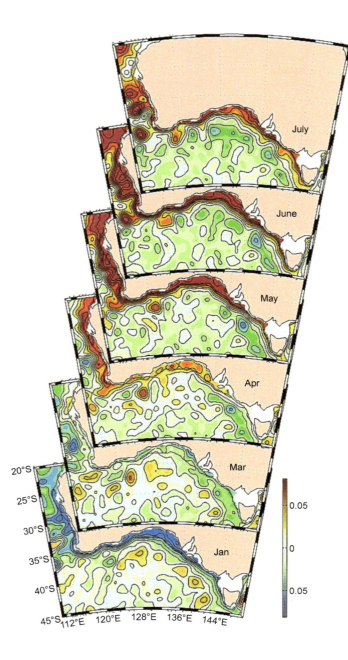

FIG. 4 Monthly composite maps (July through January) of the sea surface height along the western and southern Australian coastline illustrate the 5500 km continuity of the southeastern boundary current of the Indian Ocean. The monthly sea surface height (m) maps are constructed from the merged remotely sensed altimetric data, month is indicated in each map, and contour interval is 0.02 m. *(Figure from Ridgway and Condie (2004).)*

The impact of the eastward shelf-break current on the ecology of the southern shelf region is extensive (Edyvane, 2000). The long distances traveled by biological organisms within the current system were in fact what led to the discovery of the 5500 km long boundary current (Maxwell & Cresswell, 1981; Saville-Kent, 1897). A range of observations provide evidence that demonstrates an "Indo-Pacific" character of waters in the Great Australian Bight (Maxwell & Cresswell, 1981). The presence of a subtropical eastward current flowing from Cape Leeuwin to western Bass Strait was surmised based on the distribution of dinoflagellates that are excellent indicators of water mass origins (Wood, 1954). Wood (1954) further suggested that the discovery of warm water turtles on the west coast of Tasmania confirmed the presence of the warm water current.

3.2.3 A deep eastern boundary current

Poleward-flowing deep-water pathways exist along the eastern boundary of each Southern Hemisphere basin although their dynamics are not well understood (Faure & Speer, 2012; Schulze Chretien & Speer, 2018; Tamsitt et al., 2019; Van Sebille et al., 2012). The little-studied deep pathway in the eastern Indian Ocean was first suggested from oxygen, chlorofluorocarbon, and carbon sections (Hufford et al., 1997; Schodlok et al., 1997; Talley, 2013a, 2013b) and supported by some limited direct velocity measurements (McCartney & Donohue, 2007). The Deep Eastern Boundary Current spanning 1200–3000m depth along the southern boundary of Australia carries the low oxygen, high dissolved inorganic carbon signature characteristic of Indian Deep Water (Tamsitt et al., 2019), that is delivered to the upwelling regions of the Southern Ocean (Tamsitt et al., 2017, 2018, 2019).

The Deep Eastern Boundary Current is composed of two branches: eastward flow along the continental slope beneath the westward flowing Flinders Current and a second core of eastward flow offshore (Tamsitt et al., 2019) (Fig. 3). The offshore flow is part of a vertically coherent full-depth jet south of the Flinders Current (Duran et al., 2020b; Tamsitt et al., 2019). Models suggest the variability in the part of the Deep Eastern Boundary Current against the slope—the true boundary current—is strongly correlated to the variability in the Flinders Current above (Tamsitt et al., 2019). Part of the offshore Deep Eastern Boundary Current deviates southward between 120° and 130°E and appears to follow the abyssal path of the Antarctic Bottom Water inflow to the Indian Ocean. While the dynamics of the Deep Eastern Boundary Current are still unclear, eddy variability is thought to play a key role (Drake et al., 2018; Tamsitt et al., 2019).

3.3 Imports to the Indian Ocean

3.3.1 Leeuwin Undercurrent

Beneath and just offshore of the Leeuwin Current is the equatorward-flowing Leeuwin Undercurrent that extends from 200 to 900m (Church et al., 1989; Furue et al., 2017; Smith et al., 1991; Thompson, 1984). The Leeuwin Undercurrent begins at Cape Leeuwin (34°S, 114°E) and is fed by a northward bend of a small fraction of the Flinders Current (Furue et al., 2017) (Fig. 3). Near 22°S, most of the Leeuwin Undercurrent leaves the continental slope and flows offshore (Duran, 2015; Furue et al., 2017; Zheng, 2018), apparently following the southern flank of the Exmouth Plateau although its vertical extent of 900m is much shallower than the Plateau.

The long-term average volume transport of the Leeuwin Undercurrent, based on the CARS 1/8-degree climatology (Ridgway et al., 2002), reveals a complex three-dimensional overturning circulation. The Leeuwin Undercurrent carries 0.2 Sv northward at its southern end and receives an additional inflow of 1.5 Sv from offshore between 32°S and 28°S that appears to be fed by a cyclonic recirculation of Flinders Current water around the Naturaliste Plateau. In addition to these imports, the Leeuwin Undercurrent also receives 3.5 Sv by a vertical exchange along isopycnals from the layer above that is supplied by shallow eastward flow into the Leeuwin Current from the Indian Ocean interior (Furue, 2019; Furue et al., 2017). At its northern end near 22°S, the Leeuwin Undercurrent loses 3.6 Sv to westward flow into the Indian Ocean interior and carries 1.7 Sv northward (Furue et al., 2017). No systematic seasonal variability of the Leeuwin Undercurrent was evident in the high-resolution CARS climatology nor in an eddy-permitting ocean model (Furue et al., 2017).

The upper part of the Leeuwin Undercurrent carries high salinity, low oxygen South Indian Central Water (Duran, 2015; Woo & Pattiaratchi, 2008) supplied by the downwelling along isopycnals from the layer above (Furue, 2019; Furue et al., 2017). The middle section of the Leeuwin Undercurrent carries cool, salty, high-oxygen Subantarctic Mode Water northward, centered near 400m depth (Thompson, 1984; Woo & Pattiaratchi, 2008) and is also evident as a vertical minimum in potential vorticity (Duran, 2015). Antarctic Intermediate Water salinity minimum occupies the lower part of the Leeuwin Undercurrent in the depth range 600–900m, shoaling northward (Duran, 2015; Woo & Pattiaratchi, 2008). The Subantarctic Mode Water and Antarctic Intermediate Water water masses are most likely supplied to the Leeuwin Undercurrent by the 0.2 Sv at its southern end and the additional 1.7 Sv from the recirculation of the Flinders Current, based on the CARS climatology and an eddy-permitting ocean model (Furue et al., 2017). These Southern Ocean water mass signatures are still evident in the waters that outflow to the ocean interior at the northern end of the Leeuwin Undercurrent near 22°S (Duran, 2015; Zheng, 2018), forming a direct route from the Southern Ocean into the tropical Indian Ocean.

3.3.2 Tasman leakage

A portion of the western boundary of East Australian Current that does not retroflect back into the Pacific Ocean finds its way into the Indian Ocean. This Tasman leakage (Ridgway & Dunn, 2003) is an important interocean contribution that, together with the ITF, builds the upper branch of the global thermohaline circulation passing through the Indian Ocean toward the Atlantic (Speich et al., 2002), thus connecting the subtropical gyres in all three oceans to form a "supergyre" (Speich et al., 2007).

High-resolution ocean models suggest that the Tasman leakage has a transport of about 4 Sv with variability of about the same amplitude (Van Sebille et al., 2012, 2014). The main portion of Tasman leakage is found at intermediate depths that observational estimates report as 3.8 ± 1.3 Sv (Rosell-Fieschi et al., 2013). Owing to a southward shift and increase in the Southern Hemisphere westerlies (Swart & Fyfe, 2012), the supergyre has spun up in recent decades (Cai, 2006; Duran et al., 2020a). However, it remains unclear and unconfirmed from oceanic hindcasts (i.e., ocean models forced by atmospheric reanalyzes) (Van Sebille et al., 2012, 2014) if the Tasman leakage has undergone a subsequent strengthening.

3.3.3 Flinders current

The Flinders Current, found between the surface and 2000 m depth, follows the southern Australian shelf break from the southern tip of Tasmania to Cape Leeuwin (Duran et al., 2020b; Middleton & Cirano, 2002) and continues westward into the Indian Ocean (Hufford et al., 1997; McCartney & Donohue, 2007). The Flinders Current is located beneath the shelf-break currents along the continental slope but strengthens and shoals from east to west (Cresswell & Peterson, 1993; Duran et al., 2020b; Middleton & Bye, 2007).

Model experiments show the mechanism driving the Flinders Current is positive wind-stress curl in the South Australian Basin leading to a northward barotropic Sverdrup transport that is deflected westward along the southern Australia continental slope (Middleton & Cirano, 2002). Where the Flinders Current has a northward orientation along the coast (east of 132°E), the Sverdrup theory may fail because it is along an eastern boundary (McCreary, 1981; McCreary et al., 1991) and offshore propagation of Rossby waves should eliminate the current (Anderson & Gill, 1975). Recirculating flows in the South Australian Basin also supply the Flinders Current. In particular, the Flinders Current is the westward-flowing arm on the northern side of a large, quasi-stationary anticyclonic eddy off 120°E, known as the Albany High (Middleton & Bye, 2007) while the southern eastward-flowing arm coincides with a full depth eastward jet (McCartney & Donohue, 2007; Tamsitt et al., 2019) (see Section 3.2.3).

The Tasman leakage is a potential source of the Flinders Current (Feng et al., 2016; Middleton & Cirano, 2002; Rosell-Fieschi et al., 2013). However, an observational climatology and high-resolution ocean circulation model suggest that the Flinders Current receives no input from the Tasman leakage (Duran et al., 2020b). Rather, the entire Tasman leakage model transport of 11 Sv flows due west into the South Australian Basin. As the Flinders Current flows westward, its transport increases slightly to 1.9 Sv in the eastern Great Australian Bight, then increases more strongly once the shelf becomes zonal due to stronger onshore Ekman flows from the south. By the western end of the Bight, the transport is 7.8 Sv increasing to 12.3 Sv by Cape Leeuwin.

4 The Agulhas leakage

4.1 Introduction to the greater Agulhas current system

The Agulhas Current is the strongest western boundary current in the Southern Hemisphere and connects the Indian and Atlantic Oceans through the Agulhas leakage. This section describes the physical circulation within the Agulhas system, and we refer the interested reader to the reviews of Hood et al. (2017) and Phillips et al. (2021) for a discussion of biogeochemistry.

Although the dynamics of the Agulhas Current depend on the strength and interplay of the large-scale wind systems, it does not form a classical closure of interior Sverdrup dynamics (Beal et al., 2015; Biastoch et al., 2009b) like other western boundary currents. The Agulhas Current gathers additional transport through the thermohaline-driven global overturning circulation (Le Bars et al., 2013). Models typically simulate a nonlinear recirculation within the southwestern Indian Ocean subgyre (Biastoch et al., 2018; Loveday et al., 2014), although inverse models do not necessarily support its existence (Casal et al., 2009).

On a basin scale, the Agulhas Current is fed from the north by the ITF and by the Tasman leakage south of Australia (Fig. 5). Based on a Lagrangian analysis in a global eddy-rich model, Durgadoo et al. (2017) estimated that 10–11 Sv of the inflow from the Pacific arrives in the Agulhas Current through the Mozambique Channel and the East Madagascar Current. Other modeling and observational studies also support these sources (Biastoch & Krauss, 1999; De Ruijter et al., 2004; Halo et al., 2014; Schouten et al., 2002a). Smaller contributions come from the Persian Gulf and the Red Sea, the latter quite prominently shaping the high salinity below the thermocline and eroding the salinity minimum of the Antarctic Intermediate Water (Beal et al., 2006). The Indian Ocean significantly modifies the inflow water masses in their thermohaline properties, for example, cooling and salinifying the upper part of the ITF during its passage across the Indian Ocean (Durgadoo et al., 2017).

The Agulhas Current changes its character along its pathway toward the southern tip of Africa. The northern part, fully constituted south of 27°S, hugs the African shelf break in a v-shaped profile with surface velocities >1.5 m s^{-1} in the apex (Lutjeharms, 2006). At 32°S, it reaches below 2000 m, transporting 78 Sv (Bryden et al., 2005, integrated over the upper

214 The Indian Ocean and its role in the global climate system

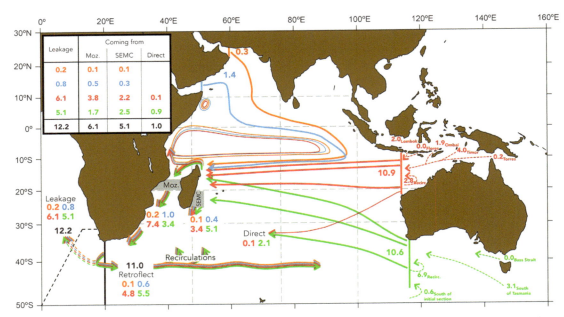

FIG. 5 Schematic of the major circulation pathways that cross longitude 20°E south of Africa. The transport (Sv) pathways are based on a Lagrangian analysis in a global eddy-rich model for a 100-year run. The 32.2 Sv total transport that crosses 20°E comes from the Persian Gulf (*orange*), the Red Sea (*blue*), the Indonesian Throughflow (ITF: *red*), and the Tasman leakage (TL: *green*), with a total transport of 12.2 Sv entering the Atlantic Ocean and 11 Sv in the Agulhas Retroflection. The table inset details the decomposition of the Agulhas leakage in Sv from the Mozambique Channel (Moz), the southeast Madagascar Current (SEMC), and directly from the interior Indian Ocean (direct). Sizes of the arrows vary only for clarity, dashed arrows itemize the upstream sources for the ITF and TL. *(Figure from Durgadoo et al. (2017).)*

2400 m). At 34°S, it has established a transport of 84 ± 24 Sv (Beal & Elipot, 2016). South of ~34°S, where the shelf begins to widen, the upper part of the Agulhas Current can meander toward the coast. As a consequence, the Agulhas Current is highly variable and subject to nonlinear recirculation.

Along its pathway, the Agulhas Current is subject to occasional meanders, abruptly shifting the boundary current by more than 100 km offshore (Fig. 6). These "Natal Pulses" occur 1.6 times per year on average (Elipot & Beal, 2015), although there are year-to-year variations (Yamagami et al., 2019). The Natal Pulses are a result of the mesoscale eddies arriving from within the Mozambique Channel and from the East Madagascar Current, interacting with the Agulhas Current through barotropic instability (Elipot & Beal, 2015; Tsugawa & Hasumi, 2010; Yamagami et al., 2019). Natal Pulses are a direct connection between the upstream sources and the variability south of Africa as described below (Schouten et al., 2002b).

At about 34°S, the African continent terminates, with the submerged Agulhas Bank forming a relatively shallow (<200 m) and wide (250 km) shelf. At 20°E, the Agulhas Current enters the Atlantic Ocean, but due to its inertia, flows for another five or more degrees of longitude into the South Atlantic. Because of the zero line of the wind stress curl positioned south of Africa, the Agulhas Current then retroflects, with the major portion returning into the Indian Ocean (Lutjeharms, 2006). This Agulhas Return Current loses water through recirculation in the subtropical gyre of the Indian Ocean, as it flows eastward and eventually completes the gyre circulation. The remainder of the Agulhas waters enter the Atlantic Ocean, quite distinctly as cyclonic and anticyclonic mesoscale features, the latter called Agulhas rings, but also in the form of a direct inflow. This Agulhas leakage connects the subtropical gyres to form the supergyre of the Southern Hemisphere (Speich et al., 2007).

Agulhas rings are among the largest and most long-lived mesoscale features in the world ocean, extending below 2000 m depth and featuring diameters of several 100 km (Chelton et al., 2011; Van Aken et al., 2003). Both the Agulhas Return Current and the ring shedding are subject to strong variability caused by internal dynamics and can be triggered by the Natal Pulses in the Agulhas Current (Elipot & Beal, 2015; Schouten et al., 2002b). Consequently, the amount of Agulhas leakage entering (and remaining in) the Atlantic Ocean is highly variable on a range of timescales (Biastoch et al., 2009b). Observational estimates of Agulhas leakage are challenged because of sparse data and the strong variability of the region. Using (at that time) all available surface drifters and surface floats, Richardson (2007) proposed a mean Agulhas leakage of 15 Sv for the upper 1000 m. A more recent update, including profiling floats, resulted in a leakage of 21.3 ± 4.7 Sv

(Daher et al., 2020). Nonetheless, it remains unclear whether the updated higher amount is because the transport is estimated over the upper 2000 m or rather if it points to an increase of Agulhas leakage over time.

The Agulhas Current system also provides an exchange from west to east. Entering the Cape Basin as a sluggish flow from the north (Arhan et al., 2003), North Atlantic Deep Water finds its way around the Cape into the Indian Ocean. Part of the North Atlantic Deep Water is concentrated in a northward flow beneath the Agulhas Current. This Agulhas Undercurrent transports 4.2 ± 5.2 Sv (Beal, 2009), albeit with high variability as a result of the interplay between the meandering Agulhas Current along the inshore that permits upwelling of deeper water masses (Biastoch et al., 2009a; Goschen et al., 2015; Leber et al., 2017).

4.2 Temporal variability of the Agulhas current and Agulhas leakage

The transport of the Agulhas Current has short-term variations (Beal et al., 2015) possibly related to the intermittent Natal Pulses that introduce cyclonic circulations and steer the Agulhas Current offshore (Elipot & Beal, 2015). Owing to the high frequency and internal variability, the seasonal cycle of the Agulhas Current can be difficult to detect. Using a combination of observations from Argo hydrography, satellite altimetry, and mooring transports, McMonigal et al. (2018) found an annual cycle with a 22 Sv peak-to-peak amplitude, with a minimum in austral winter and a maximum in austral summer. Seasonal changes in the Agulhas Current transport are related to the timescale of Rossby waves communicating the wind change from the southern Indian Ocean into the Agulhas regime (Hutchinson et al., 2018). In contrast to the volume transport, the temperature transport of the Agulhas Current and hence the amount of warm water brought from the equatorial Indian Ocean to the southern tip of Africa does not indicate a clear seasonal cycle (McMonigal et al., 2020), although it varies in response to the presence of meanders (Fig. 6). Over longer timescales, Beal and Elipot (2016) showed that enhanced eddy activity and associated meandering have caused the Agulhas Current to broaden instead of strengthen since the early 1990s.

In addition to the external drivers of variability, the nonlinearity in the system creates a significant level of internal variability. Biastoch et al. (2009b) demonstrated that both the Agulhas Current transport and the Agulhas leakage display the same level of interannual variability, independent of whether the atmospheric forcing shows year-to-year variability or not. Since the interannual variability is in large part dominated by internal variability, the influence of large-scale climate

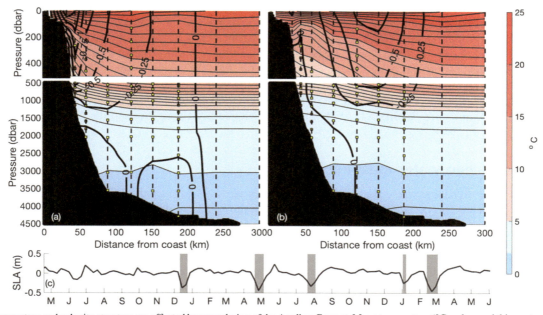

FIG. 6 Temperature and velocity structure are affected by meandering of the Agulhas Current. Mean temperature (°C; colors and thin contours) and mean cross-track velocity (m s^{-1}; thick contours) estimated from April 2016 to June 2018 moored observations in the Agulhas Current during (a) nonmeandering and (b) meandering periods. The temperature contour interval is 1°C and the velocity contour interval is 0.25 m s^{-1}. (c) Time series of sea level anomaly (SLA, m) at 33.6°S, 28°E in the Agulhas Current is used to define meanders when SLA < 0.2 m (*gray shading*). *(Figure from McMonigal et al. (2020).)*

modes on the Agulhas Current system is hard to detect. Paris et al. (2018) found that the warm Indian Ocean during El Niño becomes evident in the Agulhas leakage region as a subsurface signal 2 years after the event, communicated by baroclinic Rossby waves. Using a coupled climate model, Putrasahan et al. (2016) also linked sea surface temperature variability in the Agulhas Current to El Niño events. While ENSO explains ~11.5% of the variance of the Agulhas Current transport, the interannual Southern Annular Mode does not show a significant correlation (Elipot & Beal, 2018).

Oceanic hindcasts demonstrate a strengthening of the Southern Hemisphere westerlies that has increased the Agulhas leakage by more than 1.2 Sv per decade, or 30% from the 1960s to the 2000s (Biastoch et al., 2009b; Durgadoo et al., 2013; Schwarzkopf et al., 2019). An observational index for the Agulhas leakage, based on the SST difference between the Cape Basin and the Agulhas Return Current (Biastoch et al., 2015), suggests a similar transport increase from the 1960s with a leveling off occurring in the 1990s. This confirms that Agulhas leakage is not only subject to strong interannual variability but also suggests decadal variability and a general increasing trend.

4.3 Impact of the Agulhas leakage on the Atlantic Ocean

Agulhas leakage forms an important part of the global overturning circulation and injects salt into the Atlantic Ocean, potentially stabilizing a declining Atlantic Meridional Overturning Circulation (AMOC) (Beal et al., 2011). Agulhas leakage impacts the Atlantic Ocean in two ways: through a planetary wave response and through an advection of water masses.

Systematically isolating the mesoscale dynamics in the Agulhas Current region using a nested model approach, Biastoch et al. (2008) demonstrated that decadal anomalies of 1–2 Sv in the Agulhas leakage stem from the Agulhas dynamics and are rapidly communicated toward the North Atlantic. The signals find their way to the Northern Hemisphere after only a few years, through an interplay between westward propagating Agulhas rings and/or baroclinic Rossby waves crossing the South Atlantic that induce fast propagating topographic shelf waves along the American continental margin. The waves themselves do not transport any water masses but have the potential to lift and lower the density interfaces far away from the southern tip of Africa and hence induce additional variability within the AMOC.

More important than the wave effect is the transfer of water masses from the Indian to the Atlantic Ocean. Since the Indian is warmer and more saline than the Atlantic at the same latitude on either side of the African continent, the Agulhas leakage provides a fan of warm and saline waters into the Atlantic. The estimated Agulhas Leakage fluxes are 0.2 PW of heat and 8×10^{13} kg yr^{-1} of salt (Biastoch et al., 2015; Gunn et al., 2020; McMonigal et al., 2020). The leakage is most visible at depth (see Fig. 2 in Biastoch et al., 2008) but is also evident in SST (Biastoch et al., 2015). Part of the Agulhas leakage recirculates within the subtropical gyre of the South Atlantic, whereby more than half finds its way toward the equator, primarily transported by the North Brazil Undercurrent, and then into the North Atlantic (Rühs et al., 2013, 2019).

Agulhas leakage warms and salinifies the Atlantic Ocean (Biastoch et al., 2015; Lübbecke et al., 2015). In a series of sensitivity experiments, Lee et al. (2011) demonstrated that the increase in North Atlantic heat content since the mid-20th century can be explained by the injection of warmer water from the Agulhas Current system. Changing wind systems may also affect the flow from the Pacific toward the Atlantic, eventually compensating for part of the Agulhas leakage through fresher and colder water masses (Cessi & Jones, 2017; Rühs et al., 2019). In addition, both contributions (particularly their upper portions) are significantly modified through air–sea exchange on their way toward the North Atlantic (Rousselet et al., 2020; Rühs et al., 2019).

5 Southern Ocean water mass exchanges

The Indian Ocean also exchanges water, heat, and other properties across its large southern border with the Southern Ocean. Exchanges between these two oceans are often described by the movement of water masses across a zonal transect from southern Africa to western Australia along 32°S (Fig. 7). Multiple full-depth hydrographic sampling of this transect has been conducted since 1987, repeated roughly every 7–10 years. Estimates across 32°S of the meridional overturning circulation within the Indian Ocean are quite varied, ranging from ~10 Sv (Bryden & Beal, 2001; Ganachaud et al., 2000) to ~27 Sv (Toole & Warren, 1993). Here we focus on the water masses exchanged across the 32°S Southern Ocean boundary. A more detailed discussion of the pathways of the Indian Ocean meridional overturning circulation is found in Phillips et al. (2024).

In the upper water column at 32°S, Subantarctic Mode Water flows equatorward and is distinguishable by its relatively high oxygen in a thick thermostad layer in the upper 300–500 m (Fig. 7). Subantarctic Mode Water forms east of the Kerguelen Plateau between the Subantarctic and Subtropical Fronts in the northern Antarctic Circumpolar Current from unusually deep winter mixed layers (Sallée et al., 2006). These waters then flow northward and westward, subducting into

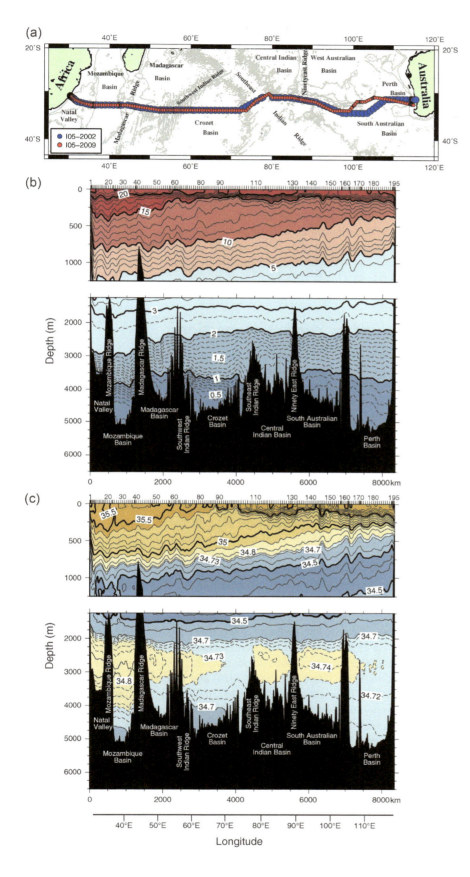

FIG. 7 Water mass properties across the 32°S Southern Ocean boundary of the Indian Ocean. (a) Station map of repeat hydrographic section along 32°S in 2002 and 2009. Vertical sections along the transect of (b) potential temperature (°C) and (c) practical salinity. *Black shading is the sea floor bathymetry with key features marked. (Figure from Hernandez-Guerra and Talley (2016).)*

the subtropical Indian Ocean (Herraiz-Borreguero & Rintoul, 2011) and ventilating the subthermocline (Hanawa & Talley, 2001) throughout the Indian Ocean. The subsurface oxygen maximum associated with Subantarctic Mode Water is observed as far north as the poleward side of the South Equatorial Current and even further north along the western boundary within the Somali Current (Talley et al., 2011). The Antarctic Intermediate Water exists as a low salinity layer beneath the Subantarctic Mode Water, also flowing equatorward across the 32°S boundary. While Subantarctic Mode Water is strongest in the eastern part of the basin, Antarctic Intermediate Water is most prominent in the western half (Fine, 1993). Both Subantarctic Mode Water and Antarctic Intermediate Water return southward across 32°S within the Agulhas Current.

No deep waters are formed within the Indian Ocean basin itself; therefore, all abyssal water masses enter from the Southern Ocean. As in all ocean basins, the deep pathways are largely constrained by the seafloor topography. Deep water masses including Circumpolar Deep Water, North Atlantic Deep Water, and Antarctic Bottom Water flow northward in deep western boundary currents along Madagascar and African coasts in the western Indian, and along the Southeast Indian and Ninety-East Ridges in the eastern Indian Ocean. Antarctic Bottom Water is the primary deep water within the Indian Basin, with the volume of North Atlantic Deep Water contributing around 10%–20% (Johnson, 2008). Since the 1990s, abyssal water masses within the Indian Ocean have experienced warming, although this signal seems mainly constrained to the deep basins east of the Ninety-East Ridge and weakens to the north (Purkey, 2010). Collectively, the deep waters are upwelled within the Indian Ocean and then return south to exit the basin at shallower depths as part of the return flow within the meridional overturning circulation.

6 Predicted changes to Interbasin boundary current connections

Climate models project robust changes in the global ocean circulation in response to greenhouse gas and other anthropogenic forcing, and the current systems that participate in the interbasin connections with the Indian Ocean are no exception. Nonetheless, there can be challenges to the confidence in the simulated predictions since these boundary currents are often poorly represented in coupled climate models partly because of bias in the forcing terms (particularly the winds), but also because the models' coarse horizontal resolutions (often 1–2 degree) are unable to realistically resolve the narrow jets and eddy variability that are characteristic of the boundary flows. Various approaches are used to somewhat alleviate these concerns such as the application of ensembles constructed from multiple climate models that enable some assessment of the statistical uncertainty. More recently, global eddy-resolving ocean model projections have been constructed and in addition, many studies have embedded regionally downscaled higher-resolution models under the premise that these eddy-resolving models have greater fidelity to the observations and provide sharper details in boundary currents. However, downscaled ocean-only models can also omit the important physics associated with air–sea interaction in boundary current regions. With these model limitations and caveats in mind, here we highlight some of the recent projections of change in the current streams that feed into and out of the Indian Ocean basin. Additional information on future projections within the Indian Ocean can be found in Roxy et al. (2024).

A growing consensus of climate models projects a decrease in the ITF mass and heat transport in a warming world by as much as 20%–30% of the modern-day transport estimates under the RCP8.5 scenario that assumes an increase of greenhouse gases reaches ~1370 ppm CO_2-equivalent concentrations by 2100 (Feng et al., 2017; Hu et al., 2015; Ma et al., 2020; Sen Gupta et al., 2016; Stellema et al., 2019; Sun et al., 2012). Nonetheless, there remains some debate as to the ultimate drivers of the ITF weakening. Future trade winds in the equatorial Pacific Ocean are projected to decrease, driving a weaker Mindanao Current that is the dominant contributor to the ITF (Hu et al., 2015; Sen Gupta et al., 2012), and hence one might expect this to subsequently produce a slowdown in the ITF. However, various studies have questioned whether the decrease in the Mindanao Current is sufficient to fully explain the ITF change (Hu et al., 2015) and indeed there appears little relationship in the Climate Model Intercomparison Project (CMIP) models between the variability in the Mindanao Current and that of the ITF (Sen Gupta et al., 2016). One alternative hypothesis suggested for the ITF transport weakening is a projected reduction in the upwelling and circulation of the deep waters entering the Pacific from the Southern Ocean that ultimately upwell to exit via the ITF (Feng et al., 2017; Sen Gupta et al., 2016). Another hypothesis suggests that the weakened ITF on centennial time scales is closely tied to a projected weakening of the AMOC (Cheng et al., 2013; Weijer et al., 2020) that initiates a transient response that propagates into the Indian Ocean via planetary waves, acting to reduce the Indo-Pacific pressure gradient responsible for the ITF (Sun & Thompson, 2020). The relative importance of local and remotely driven wind and buoyancy processes on the projected variability of the ITF within CMIP simulations requires further analysis.

Both CMIP (Stellema et al., 2019) and ocean downscaled models (Sun et al., 2012) also project a weakening of the Leeuwin Current, on average by about 5%–10% by the end of the 21st century under the RCP8.5 high emissions scenario, albeit with large intermodel differences. This transport reduction has been primarily attributed to reduced onshore flow in response to the weakening of the longshore pressure gradient that sets up the Leeuwin Current. Using regional downscaled models, Sun et al. (2012) suggested this was consistent with the projected weakening of the ITF. However, more recent CMIP5 multimodel analyses show little correspondence between the projected variability in the ITF and the Leeuwin Current (Stellema et al., 2019). Hence, it remains unclear what drives the longshore change in sea surface height on these longer time scales. Significant decreases are also projected in the Leeuwin Undercurrent due to a reduction in the westward outflow from the Flinders Current system south of Australia (Stellema et al., 2019).

Multiple climate models forecast a robust weakening in the East Madagascar Current and through the Mozambique Channel with a corresponding weakening of transport (11%–23%) into the Agulhas Current under increased greenhouse gas emission RCP8.5 scenarios (Stellema et al., 2019). The weakened Agulhas transport reduces the southward heat transport out of the Indian Ocean, which together with increased air–sea heat flux input, is projected to amplify the already significant Indian Ocean warming trend (Cheng et al., 2017; Ma et al., 2020; Roxy et al., 2014). CMIP5 analysis suggests the reduction in the greater Agulhas system appears to be partially related to the reduced ITF and partially related to an overall slow-down of the Indian Ocean meridional overturning circulation manifested as more sluggish equatorward deep flow and reduced upwelling that ultimately acts to reduce the basin-wide upper-ocean outflow (Stellema et al., 2019). Future coupled climate models that adequately simulate mesoscale processes will be critical for determining the key mechanisms responsible for driving projected oceanic changes in energetic regions such as the Agulhas Current system.

In contrast to the weakening Agulhas Current, the Agulhas leakage south of Africa into the Atlantic Ocean is consistently projected to strengthen with climate change (Biastoch & Böning, 2013; Stellema et al., 2019). In a warming world, the Southern Ocean westerlies are anticipated to intensify and shift poleward and consequently there is a southward expansion in the subtropical supergyre (Cai, 2006; Sen Gupta et al., 2009) and subsequent increase in the Agulhas leakage (Biastoch & Böning, 2013). As a result, the enhanced contribution of saltier Indian Ocean water causes a net gain in density within the Atlantic Ocean that may potentially counteract some of the projected slow-down in the AMOC due to greenhouse warming (Weijer & van Sebille, 2014).

7 Conclusions

This chapter has highlighted the oceanic interbasin exchanges of mass and other properties into and out of the Indian Ocean via a series of complex pathways.

The transport of Pacific waters into the Indian Ocean via the ITF affects the salinity, temperature, and nutrient budgets of the Indian Ocean and serves as a pathway for the propagation of oceanic waves. The ITF is influenced by a wide array of regional climate forcing on intraseasonal timescales, the Indian Ocean Dipole and ENSO on interannual timescales, and the IPO on decadal timescales and potentially anthropogenic activity on longer timescales. Nonetheless, a paucity of observations in the Indonesian seas, particularly of marine biogeochemistry, means that there remains a relatively poor understanding of the gating of the flow and properties through the maritime continent and their influence on the Indian Ocean basin.

The Leeuwin Current system provides an efficient route for Indian Ocean surface and deep waters to be exported poleward and for intermediate waters from the Southern Ocean to penetrate into the tropical Indian Ocean. This route is due to the unique presence of the ITF exiting into the southeast Indian Ocean and does not exist in any other basin. The recent availability of Argo hydrographic profiles invigorated the discovery of new current pathways, such as westward outflow from the Leeuwin Undercurrent (Furue, 2019; Furue et al., 2017; Zheng, 2018) and the Deep Eastern Boundary Current (Tamsitt et al., 2019) south of Australia.

Surface transport from the Pacific Ocean via both the ITF and the Tasman leakage crosses the Indian Ocean to join the Agulhas system south of Africa, albeit undergoing strong water mass changes along the way. The Agulhas Current is the strongest of the Southern Hemisphere's western boundary currents and is subject to vigorous dynamics through wind and internal variability. Export into the Atlantic Ocean via the Agulhas leakage is highly nonlinear and exhibits strong decadal variability and long-term trends.

Along the southern border of the Indian Ocean, the inflow of mode and intermediate waters from the Southern Ocean acts to ventilate the basin. The deep limb of the meridional overturning circulation is characterized by the topographically constrained inflow of deep waters that then upwell and return via the Agulhas western boundary current system and the southward flow of Indian Deep Water. Changes in the property characteristics and pathways of these waters can affect the circulation, ocean heat content, and sea level rise.

Projected changes in the oceanic teleconnection pathways due to a warming climate are expected to alter the Indian Ocean circulation. In contrast to the modern world, the Indian Ocean projections of inflow from the ITF and outflow via the Leeuwin Current and the Agulhas Current system appear to show relatively little co-variability on climate time scales. However, there remains much uncertainty given that these narrow and eddying boundary currents are poorly simulated in coarse resolution climate models. Nonetheless, future changes in the Indian Ocean boundary currents will have important implications for marine ecosystems and connectivity (Barange et al., 2014; Bopp et al., 2013; Funk et al., 2008) and hence are of significant consequence for the Indian Ocean rim countries that rely on marine resources for their sustenance and economic well-being.

8 Educational resources

Additional general references that describe sampling and data collection in the boundary currents of the Indian Ocean can be found in:

Wyrtki, K. (1971). Oceanographic Atlas of the International Indian Ocean Expedition. National Science Foundation, Washington, D.C., 531 pp.

Beal, L.M., Vialard, J., Roxy, M.K., and lead authors (2019). Executive Summary. IndOOS-2: A roadmap to sustained observations of the Indian Ocean for 2020–2030. *CLIVAR-4/2019, GOOS-237*, 204 pp. https://doi.org/10.36071/clivar.rp.4.2019.

Beal, L.M., et al. (2020). A roadmap to IndOOS-2: Better observations of the rapidly-warming Indian Ocean, *Bulletin of the American Meteorological Society*. https://doi.org/10.1175/BAMS-D-19-0209.1.

Hermes, J.C., et al. (2019). A sustained ocean observing system in the Indian Ocean for climate related scientific knowledge and societal needs, *Frontiers of Marine Sciences*. https://doi.org/10.3389/fmars.2019.00355.

Acknowledgments

JS acknowledges funding to support her effort by the National Science Foundation under Grant Number OCE-1851316. HEP acknowledges support from the Australian Government Department of the Environment and Energy National Environmental Science Programme, the Australian Government Antarctic Science Collaboration Initiative (ASCI000002), and the ARC Centre of Excellence in Climate Extremes.

Author contributions

JS led the overall writing, editing, and organization of the manuscript. JS wrote Sections 1 and 6; LKG and JS wrote Section 2; HEP and AB wrote Section 3; AB wrote Section 4; HEP and LKG wrote Section 5; all authors wrote Section 7. All authors contributed comments and feedback on the overall structure of the chapter and contributed to revisions.

References

Anderson, D. L., & Gill, A. (1975). Spin-up of a stratified ocean, with applications to upwelling. *Deep Sea Research and Oceanographic Abstracts*, 22(9), 583–596. https://doi.org/10.1016/0011-7471(75)90046-7.

Andrews, J. C. (1977). Eddy structure and the West Australian current. *Deep Sea Research*, 24(12), 1133–1148. https://doi.org/10.1016/0146-6291(77)90517-3.

Arhan, M., Mercier, H., & Park, Y.-H. (2003). On the deep water circulation of the eastern South Atlantic Ocean. *Deep Sea Research Part I: Oceanographic Research Papers*, 50, 889–916. https://doi.org/10.1016/S0967-0637(03)00072-4.

Asanuma, I., Matsumoto, K., Okano, H., Kawano, T., Hendiarti, N., & Sachoemar, S. I. (2003). Spatial distribution of phytoplankton along the Sunda Islands: The monsoon anomaly in 1998. *Journal of Geophysical Research, Oceans*. https://doi.org/10.1029/1999jc000139.

Ayers, J. M., Strutton, P. G., Coles, V. J., Hood, R. R., & Matear, R. J. (2014). Indonesian throughflow nutrient fluxes and their potential impact on Indian Ocean productivity. *Geophysical Research Letters*, 41(14), 5060–5067. https://doi.org/10.1002/2014GL060593.

Barange, M., Merino, G., Blanchard, J., et al. (2014). Impacts of climate change on marine ecosystem production in societies dependent on fisheries. *Nature Climate Change*, 4, 211–216. https://doi.org/10.1038/nclimate2119.

Beal, L. M. (2009). A time series of Agulhas undercurrent transport. *Journal of Physical Oceanography*, 39, 2436–2450. https://doi.org/10.1175/2009JPO4195.1.

Beal, L. M., Chereskin, T. K., Lenn, Y. D., & Elipot, S. (2006). The sources and mixing characteristics of the Agulhas current. *Journal of Physical Oceanography*, 36, 2060–2074. https://doi.org/10.1175/JPO2964.1.

Beal, L. M., De Ruijter, W. P. M., Biastoch, A., Zahn, R., & members of S. W. G. (2011). On the role of the Agulhas system in ocean circulation and climate. *Nature*, 472(7344), 429–436. https://doi.org/10.1038/nature09983.

Beal, L. M., & Elipot, S. (2016). Broadening not strengthening of the Agulhas current since the early 1990s. *Nature*. https://doi.org/10.1038/nature19853.

Beal, L. M., Elipot, S., Houk, A., & Leber, G. M. (2015). Capturing the transport variability of a western boundary jet: Results from the Agulhas current time-series experiment (ACT). *Journal of Physical Oceanography*, 45(5), 1302–1324. https://doi.org/10.1175/JPO-D-14-0119.1.

Biastoch, A., Beal, L. M., Lutjeharms, J. R. E., & Casal, T. G. D. (2009a). Variability and coherence of the Agulhas undercurrent in a high-resolution ocean general circulation model. *Journal of Physical Oceanography, 39*(10), 2417–2435. https://doi.org/10.1175/2009JPO4184.1.

Biastoch, A., & Böning, C. W. (2013). Anthropogenic impact on Agulhas leakage. *Geophysical Research Letters, 40*, 1138–1143. https://doi.org/10.1002/grl.50243.

Biastoch, A., Böning, C. W., & Lutjeharms, J. R. E. (2008). Agulhas leakage dynamics affects decadal variability in Atlantic overturning circulation. *Nature, 456*(7221), 489–492. https://doi.org/10.1038/nature07426.

Biastoch, A., Böning, C. W., Schwarzkopf, F. U., & Lutjeharms, J. R. E. (2009b). Increase in Agulhas leakage due to poleward shift of southern hemisphere westerlies. *Nature, 462*(7272), 495–498. https://doi.org/10.1038/nature08519.

Biastoch, A., Durgadoo, J. V., Morrison, A. K., Van Sebille, E., Weijer, W., & Griffies, S. M. (2015). Atlantic multi-decadal oscillation covaries with Agulhas leakage. *Nature Communications, 6*, 10082. https://doi.org/10.1038/ncomms10082.

Biastoch, A., & Krauss, W. (1999). The role of mesoscale eddies in the source regions of the Agulhas current. *Journal of Physical Oceanography, 29*(9), 2303–2317. https://doi.org/10.1175/1520-0485(1999)029<2303:TROMEI>2.0.CO;2.

Biastoch, A., Sein, D., Durgadoo, J. V., & Wang, Q. (2018). Simulating the Agulhas system in global ocean models—Nesting vs. multi-resolution unstructured meshes. *Ocean Modelling, 121*, 117–131. https://doi.org/10.1016/j.ocemod.2017.12.002.

Bopp, L., Resplandy, L., Orr, J. C., Doney, S. C., Dunne, J. P., Gehlen, M., Halloran, P., Heinze, C., Ilyina, T., Séférian, R., Tjiputra, J., & Vichi, M. (2013). Multiple stressors of ocean ecosystems in the 21st century: Projections with CMIP5 models. *Biogeosciences, 10*, 6225–6245. https://doi.org/10.5194/bg-10-6225-2013.

Broecker, W. S., Patzert, W. C., Toggweiler, J. R., & Stuiver, M. (1986). Hydrography, chemistry, and radioisotopes in the Southeast Asian basins. *Journal of Geophysical Research.* https://doi.org/10.1029/jc091ic12p14345.

Bryden, H. L., & Beal, L. M. (2001). Role of the Agulhas current in Indian Ocean circulation and associated heat and freshwater fluxes. *Deep Sea Research, Part I, 48*, 1821–1845.

Bryden, H. L., Beal, L. M., & Duncan, L. M. (2005). Structure and transport of the Agulhas current and its temporal variability. *Journal of Oceanography, 61*, 479–492.

Cai, W. (2006). Antarctic ozone depletion causes an intensification of the Southern Ocean super-gyre circulation. *Geophysical Research Letters, 33*, L03712. https://doi.org/10.1029/2005GL024911.

Caputi, N., Fletcher, W. J., Pearce, A., & Chubb, C. F. (1996). Effect of the Leeuwin current on the recruitment of fish and invertebrates along the Western Australian coast. *Marine and Freshwater Research.* https://doi.org/10.1071/MF9960147.

Casal, T. G. D., Beal, L. M., Lumpkin, R., & Johns, W. E. (2009). Structure and downstream evolution of the Agulhas current system during a quasi-synoptic survey in February–march 2003. *Journal of Geophysical Research, 114*, C03001. https://doi.org/10.1029/2008JC004954.

Cessi, P., & Jones, C. S. (2017). Warm-route versus cold-route interbasin exchange in the meridional overturning circulation. *Journal of Physical Oceanography, 47*(8), 1981–1997. https://doi.org/10.1175/JPO-D-16-0249.1.

Chelton, D. B., Schlax, M. G., & Samelson, R. M. (2011). Global observations of nonlinear mesoscale eddies. *Progress in Oceanography, 91*(2), 167–216. https://doi.org/10.1016/j.pocean.2011.01.002.

Cheng, W., Chiang, J. C. H., & Zhang, D. (2013). Atlantic meridional overturning circulation (AMOC) in CMIP5 models: RCP and historical simulations. *Journal of Climate, 26*, 7187–7197.

Cheng, L., Trenberth, K. E., Fasullo, J., Boyer, T., Abraham, J., & Zhu, J. (2017). Improved estimates of ocean heat content from 1960 to 2015. *Science Advances, 3*(3), 1–10. https://doi.org/10.1126/sciadv.1601545.

Church, J. A., Cresswell, G. R., & Godfrey, J. S. (1989). The Leeuwin current. In S. J. Neshyba, C. N. K. Mooers, R. L. Smith, & R. T. Barber (Eds.), *Vol. 34. Poleward flows along eastern ocean boundaries, coastal and estuarine studies (formerly lecture notes on coastal and estuarine studies)* (pp. 230–254). New York: Springer. https://doi.org/10.1007/978-1-4613-8963-7_16.

Coatanoan, C., Metzl, N., Fieux, M., & Coste, B. (1999). Seasonal water mass distribution in the Indonesian throughflow entering the Indian Ocean. *Journal of Geophysical Research: Oceans, 104*(C9), 20801–20826. https://doi.org/10.1029/1999jc900129.

Cresswell, G. R., & Peterson, J. L. (1993). The Leeuwin current south of Western Australia. *Marine and Freshwater Research.* https://doi.org/10.1071/MF9930285.

Daher, H., Beal, L. M., & Schwarzkopf, F. U. (2020). A new improved estimation of Agulhas leakage using observations and simulations of Lagrangian floats and drifters. *Journal of Geophysical Research, Oceans.* https://doi.org/10.1029/2019JC015753.

De Ruijter, W. P. M., Aken, H. M., Beier, E. J., Lutjeharms, J. R. E., Matano, R. P., & Schouten, M. W. (2004). Eddies and dipoles around South Madagascar: Formation, pathways and large-scale impact. *Deep Sea Research, Part I, 51*(3), 383–400.

Divakaran, P., & Brassington, G. B. (2011). Arterial Ocean circulation of the Southeast Indian Ocean. *Geophysical Research Letters, 38*, L01802. https://doi.org/10.1029/2010GL045574.

Domingues, C. M., Maltrud, M. E., Wijffels, S. E., Church, J. A., & Tomczak, M. (2007). Simulated Lagrangian pathways between the Leeuwin current system and the upper-ocean circulation of the Southeast Indian Ocean. *Deep Sea Research, Part II, 54*, 797–817. https://doi.org/10.1016/j.dsr2.2006.10.003.

Domingues, C. M., Wijffels, S. E., Maltrud, M. E., Church, J. A., & Tomczak, M. (2006). Role of eddies in cooling the Leeuwin current. *Geophysical Research Letters, 33*.

Drake, H., Morrison, A. K., Griffies, S. M., Sarmiento, J. L., Weijer, W., & Gray, A. R. (2018). Lagrangian timescales of Southern Ocean upwelling in a hierarchy of model resolutions. *Geophysical Research Letters, 45*, 891–898. https://doi.org/10.1002/2017GL076045.

Drushka, K., Sprintall, J., & Gille, S. T. (2010). Vertical structure of Kelvin waves in the Indonesian Throughflow exit passages. *Journal of Physical Oceanography, 40*, 1965–1987.

Dufois, F., Hardman-Mountford, N. J., Greenwood, J., Richardson, A. J., Feng, M., Herbette, S., & Matear, R. (2014). Impact of eddies on surface chlorophyll in the South Indian Ocean. *Journal of Geophysical Research, Oceans, 119*, 8061–8077. https://doi.org/10.1002/2014JC010164.

Duran, E. R. (2015). *An investigation of the Leeuwin undercurrent source waters and pathways*. (Honours thesis). University of Tasmania.

Duran, E. R., England, M. H., & Spence, P. (2020a). Surface ocean warming around Australia driven by interannual variability and long-term trends in Southern Hemisphere westerlies. *Geophysical Research Letters, 47*, e2019GL086605. https://doi.org/10.1029/2019GL086605.

Duran, E. R., Phillips, H. E., Furue, R., Spence, P., & Bindoff, N. L. (2020b). Southern Australia current system based on a gridded hydrography and a high-resolution model. *Progress in Oceanography*. https://doi.org/10.1016/j.pocean.2019.102254.

Durgadoo, J. V., Loveday, B. R., Reason, C. J. C., Penven, P., & Biastoch, A. (2013). Agulhas leakage predominantly responds to the southern hemisphere westerlies. *Journal of Physical Oceanography, 43*(10).

Durgadoo, J. V., Rühs, S., Biastoch, A., & Böning, C. W. B. (2017). Indian Ocean sources of Agulhas leakage. *Journal of Geophysical Research: Oceans, 122*(4), 3481–3499. https://doi.org/10.1002/2016JC012676.

Edyvane, K. (2000). The great Australian bight. In C. R. C. Sheppard (Ed.), *Vol. II. Seas at the millennium: An environmental evaluation* (pp. 673–690). New York: Pergamon.

Elipot, S., & Beal, L. M. (2015). Characteristics, energetics, and origins of Agulhas current meanders and their limited influence on ring shedding. *Journal of Physical Oceanography*, 150626133140004. https://doi.org/10.1175/JPO-D-14-0254.1.

Elipot, S., & Beal, L. M. (2018). Observed Agulhas current sensitivity to interannual and long-term trend atmospheric forcings. *Journal of Climate, 31*(8), 3077–3098. https://doi.org/10.1175/JCLI-D-17-0597.1.

England, M., & Huang, F. (2005). On the interannual variability of the Indonesian Throughflow and its linkage with ENSO. *Journal of Climate, 18*(9), 1435–1444. https://doi.org/10.1175/JCLI3322.1.

Fang, F., & Morrow, R. (2003). Evolution, movement and decay of warm-core Leeuwin current eddies. *Deep Sea Research, Part II, 50*(12–13), 2245–2261. https://doi.org/10.1016/S0967-0645(03)00055-9.

Faure, V., & Speer, K. (2012). Deep circulation in the eastern South Pacific Ocean. *Journal of Marine Research, 70*(5), 748–778.

Feng, M., Majewski, L. J., Fandry, C. B., & Waite, A. M. (2007). Characteristics of two counter-rotating eddies in the Leeuwin current system off the Western Australian coast. *Deep Sea Research, Part II, 54*(8–10), 961–980. https://doi.org/10.1016/j.dsr2.2006.11.022.

Feng, M., McPhaden, M. J., Xie, S., & Hafner, J. (2013). La Niña forces unprecedented Leeuwin current warming in 2011. *Scientific Reports, 3*.

Feng, M., Meyers, G., Pearce, A., & Wijffels, S. (2003). Annual and interannual variations of the Leeuwin current at 32 degrees S. *Journal of Geophysical Research, Oceans, 108*(C11).

Feng, M., Wijffels, S., Godfrey, S., & Meyers, G. (2005). Do eddies play a role in the momentum balance of the Leeuwin current? *Journal of Physical Oceanography, 35*, 964–975.

Feng, M., Zhang, X., Oke, P., Monselesan, D., Chamberlain, M., Matear, R., & Schiller, A. (2016). Invigorating ocean boundary current systems around Australia during 1979–2014: As simulated in a near-global eddy-resolving ocean model. *Journal of Geophysical Research, Oceans, 121*, 3395–3408.

Feng, M., Zhang, X., Sloyan, B., & Chamberlain, M. (2017). Contribution of the deep ocean to the centennial changes of the Indonesian Throughflow. *Geophysical Research Letters, 44*(6), 2859–2867. https://doi.org/10.1002/2017GL072577.

Ffield, A., Vranes, K., Gordon, A. L., Susanto, R. D., & Garzoli, S. L. (2000). Temperature variability within the Makassar Strait. *Geophysical Research Letters, 27*, 237–240.

Fieux, M., Andrié, C., Charriaud, E., Ilahude, A. G., Metzl, N., Molcard, R., & Swallow, J. C. (1996). Hydrological and chlorofluoromethane measurements of the Indonesian throughflow entering the Indian Ocean. *Journal of Geophysical Research, C: Oceans*. https://doi.org/10.1029/96JC00207.

Fine, R. A. (1985). Direct evidence using tritium data for throughflow from the Pacific into the Indian Ocean. *Nature*. https://doi.org/10.1038/315478a0.

Fine, R. A. (1993). Circulation of Antarctic intermediate water in the South Indian Ocean. *Deep Sea Research, Part I, 40*, 2021–2042.

Funk, C., Dettinger, M. D., Michaelsen, J. C., Verdin, J. P., Brown, M. E., Barlow, M., & Hoell, A. (2008). Warming of the Indian Ocean threatens eastern and southern African food security but could be mitigated by agricultural development. *Proceedings of the National Academy of Sciences, 105*(32), 11081–11086.

Furue, R. (2019). The three-dimensional structure of the Leeuwin current system in density coordinates in an eddy-resolving OGCM. *Ocean Modelling, 138*, 36–50. https://doi.org/10.1016/j.ocemod.2019.03.001.

Furue, R., Guerreiro, K., Phillips, H. E., McCreary, J. P., Jr., & Bindoff, N. L. (2017). On the Leeuwin current system and its linkage to zonal flows in the South Indian Ocean as inferred from a gridded hydrography. *Journal of Physical Oceanography, 47*, 583–602.

Furue, R., McCreary, J. P., Benthuysen, J., Phillips, H., & Bindoff, N. (2013). Dynamics of the Leeuwin current: Part 1. Coastal flows in an inviscid, variable-density, layer model. *Dynamics of Atmospheres and Oceans, 63*, 24–59.

Ganachaud, A., Wunsch, C., Marotzke, J., & Toole, J. (2000). Meridional overturning and large-scale circulation of the Indian Ocean. *Journal of Geophysical Research, Oceans*. https://doi.org/10.1029/2000JC900122.

Godfrey, J. S., & Ridgway, K. R. (1985). The large-scale environment of the poleward-flowing Leeuwin current, Western Australia: Longshore steric height gradients, wind stresses and geostrophic flow. *Journal of Physical Oceanography*. https://doi.org/10.1175/1520-0485(1985)015<0481:tlseot>2.0.co;2.

Gordon, A. L. (1986). Interocean exchange of thermocline water. *Journal of Geophysical Research*. https://doi.org/10.1029/jc091ic04p05037.

Gordon, A. L., & Fine, R. (1996). Pathways of water between the Pacific and Indian oceans in the Indonesian seas. *Nature, 379*, 146–149.

Gordon, A. L., Huber, B. A., Metzger, J., Susanto, R. D., Hulbert, H. E., & Adi, T. R. (2012). South China Sea throughflow impact on the Indonesian throughflow. *Geophysical Research Letters, 39*, LI1602.

Gordon, A. L., Ma, S., Olson, D. B., Hacker, P., Ffield, A., Talley, L. D., Wilson, D., & Baringer, M. (1997). Advection and diffusion of Indonesian throughflow water within the Indian Ocean south equatorial current. *Geophysical Research Letters, 24*, 2573–2576.

Gordon, A. L., Napitu, A., Huber, B. A., Gruenburg, L. K., Pujiana, K., Agustiadi, T., Kuswardani, A., Mbay, N., & Setiawan, A. (2019). Makassar Strait throughflow seasonal and interannual variability: An overview. *Journal of Geophysical Research, Oceans, 124*. https://doi.org/10.1029/2018JC014502.

Gordon, A. L., Sprintall, J., Van Aken, H. M., Susanto, D., Wijffels, S., Molcard, R., Ffield, A., Pranowo, W., & Wirasantosa, S. (2010). The Indonesian Throughflow during 2004-2006 as observed by the INSTANT program in modeling and observing the Indonesian Throughflow. In A. L. Gordon, & V. M. Kamenkovich (Eds.), *Vol. 50. Dynamics of atmosphere and oceans* (pp. 115–128). https://doi.org/10.1016/j.dynatmoce.2009.12.002.

Gordon, A. L., & Susanto, R. D. (2001). Banda Sea surface-layer divergence. *Ocean Dynamics, 52*(1), 2–10.

Gordon, A. L., Susanto, R. D., Ffield, A., Huber, B. A., Pranowo, W., & Wirasantosa, S. (2008). Makassar Strait throughflow, 2004 to 2006. *Geophysical Research Letters, 35*, L24605.

Goschen, W. S., Bornman, T. G., Deyzel, S. H. P., & Schumann, E. H. (2015). Coastal upwelling on the far eastern Agulhas Bank associated with large meanders in the Agulhas current. *Continental Shelf Research, 101*, 34–46. https://doi.org/10.1016/j.csr.2015.04.004.

Gruber, N. (2015). Ocean biogeochemistry: Carbon at the coastal interface. *Nature*. https://doi.org/10.1038/nature14082.

Gruenburg, L. K., & Gordon, A. L. (2018). Variability in Makassar Strait heat flux and its effect on the eastern tropical Indian Ocean. *Oceanography, 31*(2), 80–87.

Gunn, K. L., Beal, L. M., Elipot, S., McMonigal, K., & Houk, A. (2020). Mixing of subtropical, central, and intermediate waters driven by shifting and pulsing of the Agulhas current. *Journal of Physical Oceanography, 50*, 3545–3560. https://doi.org/10.1175/JPO-D-20-0093.1.

Halo, I., Penven, P., Backeberg, B., Ansorge, I., Shillington, F., & Roman, R. (2014). Mesoscale eddy variability in the southern extension of the East Madagascar Current: Seasonal cycle, energy conversion terms, and eddy mean properties. *Journal of Geophysical Research: Oceans, 119*(10), 7324–7356. https://doi.org/10.1002/2014JC009820.

Hamzah, F., Agustiadi, T., Susanto, R. D., Wei, Z., Guo, L., Cao, Z., & Dai, M. (2020). Dynamics of the carbonate system in the western Indonesian seas during the southeast monsoon. *Journal of Geophysical Research: Oceans, 125*(1), 1–18. https://doi.org/10.1029/2018JC014912.

Hanawa, K., & Talley, L. (2001). In G. Siedler, & J. Church (Eds.), *International Geophysics Series. Ocean circulation and climate, chapter mode waters* (pp. 373–386). New York: Academic.

Hendiarti, N., Siegel, H., & Ohde, T. (2004). Investigation of different coastal processes in Indonesian waters using SeaWiFS data. *Deep Sea Research Part II: Topical Studies in Oceanography*. https://doi.org/10.1016/j.dsr2.2003.10.003.

Hernandez-Guerra, A., & Talley, L. D. (2016). Meridional overturning transports at 30°S in the Indian and Pacific oceans in 2002-2003 and 2009. *Progress in Oceanography, 146*, 89–120. https://doi.org/10.1016/j.pocean.2016.06.005.

Herraiz-Borreguero, L., & Rintoul, S. R. (2011). Subantarctic mode water: Distribution and circulation. *Ocean Dynamics, 61*, 103–126.

Hood, R. R., Beckley, L. E., & Wiggert, J. D. (2017). Biogeochemical and ecological impacts of boundary currents in the Indian Ocean. *Progress in Oceanography, 156*, 290–325.

Hu, S., & Fedorov, A. V. (2019). Indian Ocean warming can strengthen the Atlantic meridional overturning circulation. *Nature Climate Change, 9*, 747–751. https://doi.org/10.1038/s41558-019-0566-x.

Hu, S., & Fedorov, A. V. (2020). Indian Ocean warming as a driver of the North Atlantic warming hole. *Nature Communications, 11*, 4785. https://doi.org/10.1038/s41467-020-18522-5.

Hu, S., & Sprintall, J. (2016). Interannual variability of the Indonesian Throughflow: The salinity effect. *Journal of Geophysical Research, 121*, 2596–2615. https://doi.org/10.1002/2015JC011495.

Hu, X., Sprintall, J., Yuan, D., Tranchant, B., Gaspar, P., Koch-Larrouy, A., Reffray, G., Li, X., Wang, Z., Li, Y., Nugroho, D., Corvianawatie, C., & Surinati, D. (2019). Interannual variability of the Sulawesi Sea circulation forced by indo-Pacific planetary waves. *Journal of Geophysical Research, Oceans*. https://doi.org/10.1029/2018JC014356.

Hu, D., Wu, L., Cai, W., Gupta, A. S., Ganachaud, A., Qiu, B., et al. (2015). Pacific western boundary currents and their roles in climate. *Nature, 522*(7556). https://doi.org/10.1038/nature14504.

Hufford, G. E., McCartney, M. S., & Donohue, K. A. (1997). Northern boundary currents and adjacent recirculations off southwestern Australia. *Geophysical Research Letters, 24*(22), 2797–2800.

Hutchinson, K., Beal, L. M., Penven, P., Ansorge, I., & Hermes, J. (2018). Seasonal phasing of Agulhas current transport tied to a baroclinic adjustment of near-field winds. *Journal of Geophysical Research: Oceans, 123*(10), 7067–7083. https://doi.org/10.1029/2018JC014319.

Johnson, G. C. (2008). Quantifying Antarctic bottom water and North Atlantic deep water volumes. *Journal of Geophysical Research, Oceans*. https://doi.org/10.1029/2007JC004477.

Kartadikaraia, A. R., Watanabe, A., Nadaoka, K., Adi, N. S., Prayitno, H. B., Soemorumekso, S., Muchtar, M., Triyulianti, I., Setiawan, A., Suratno, S., & Khasanah, E. N. (2015). CO2 sink/source characteristics in the tropical Indonesian seas. *Journal of Geophysical Research: Oceans*, 7842–7856. https://doi.org/10.1002/2015JC010925.Received.

Koch-Larrouy, A., Atmadipoera, A., van Beek, P., Madec, G., Aucan, J., Lyard, F., Grelet, J., & Souhaut, M. (2015). Estimates of tidal mixing in the Indonesian archipelago from multidisciplinary INDOMIX in-situ data. *Deep Sea Research, Part I, 106*, 136–153. DSR1-D-14-00245R1.

Koch-Larrouy, A., Lengaigne, M., Terray, P., Madec, G., & Masson, S. (2010). Tidal mixing in the Indonesian seas and its effect on the tropical climate system. *Climate Dynamics, 34*, 891–904.

Koch-Larrouy, A., Madec, G., Bouruet-Aubertot, P., Gerkema, T., Bessières, L., & Molcard, R. (2007). On the transformation of Pacific water into Indonesian Throughflow water by internal tidal mixing. *Geophysical Research Letters*. https://doi.org/10.1029/2006GL028405.

Koch-Larrouy, A., Madec, G., Iudicone, D., Molcard, R., & Atmadipoera, A. (2008). Physical processes contributing in the water mass transformation of the Indonesian ThroughFlow. *Ocean Dynamics, 58*, 275–288.

Le Bars, D., Dijkstra, H. A., & De Ruijter, W. P. M. (2013). Impact of the Indonesian Throughflow on Agulhas leakage. *Ocean Science, 9*(5), 773–785. https://doi.org/10.5194/os-9-773-2013.

Leber, G. M., Beal, L. M., & Elipot, S. (2017). Wind and current forcing combine to drive strong upwelling in the Agulhas current. *Journal of Physical Oceanography, 47*(1), 123–134. https://doi.org/10.1175/JPO-D-16-0079.1.

Lee, S.-K., Park, W., Baringer, M. O., Gordon, A. L., Huber, B., & Liu, Y. (2015). Pacific origin of the abrupt increase in Indian Ocean heat content during the warming hiatus. *Nature Geoscience, 8*, 445–449. https://doi.org/10.1038/ngeo2438.

Lee, S.-K., Park, W., van Sebille, E., Baringer, M. O., Wang, C., Enfield, D. B., et al. (2011). What caused the significant increase in Atlantic Ocean heat content since the mid-20th century? *Geophysical Research Letters, 38*(17), L17607. https://doi.org/10.1029/2011GL048856.

Legeckis, R., & Cresswell, G. (1981). Satellite observations of sea-surface temperature fronts off the coast of western and southern Australia. *Deep Sea Research, Part I, 28*, 297–306. https://doi.org/10.1016/0198-0149(81)90069-8.

Loveday, B. R., Durgadoo, J. V., Reason, C. J. C., Biastoch, A., & Penven, P. (2014). Decoupling the Agulhas current and Agulhas leakage. *Journal of Physical Oceanography, 44*, 1776–1797. https://doi.org/10.1175/JPO-D-13-093.1.

Lübbecke, J. F., Durgadoo, J. V., & Biastoch, A. (2015). Contribution of increased agulhas leakage to tropical Atlantic warming. *Journal of Climate, 28*(24). https://doi.org/10.1175/JCLI-D-15-0258.1.

Lutjeharms, J. R. E. (2006). The Agulhas. *Current*. https://doi.org/10.1007/3-540-37212-1.

Ma, J., Feng, M., Lan, J., & Hu, D. (2020). Projected future changes of meridional heat transport and heat balance of the Indian Ocean. *Geophysical Research Letters, 47*(4), e2019GL086803. https://doi.org/10.1029/2019GL086803.

Maxwell, J. G. H., & Cresswell, G. R. (1981). Dispersal of tropical marine fauna to the great Australian bight by the Leeuwin current. *Australian Journal of Marine and Freshwater Research, 32*, 493–500.

McCartney, M. S., & Donohue, K. A. (2007). A deep cyclonic gyre in the Australian-Antarctic Basin. *Progress in Oceanography, 75*(4), 675–750.

McClean, J. L., Ivanova, D. P., & Sprintall, J. (2005). Remote origins of interannual variability in the Indonesian Throughflow region from data and a global parallel ocean program simulation. *Journal of Geophysical Research, 110*, C10013. https://doi.org/10.1029/2004JC002477.

McCreary, J. P. (1981). A linear stratified ocean model of the equatorial undercurrent. *Philosophical Transactions of the Royal Society A - Mathematical Physical and Engineering Sciences, 298*(1444), 603–635. https://doi.org/10.1098/rsta.1981.0002.

McCreary, J. P., Fukamachi, Y., & Kundu, P. K. (1991). A numerical investigation of jets and eddies near an eastern ocean boundary. *Journal of Geophysical Research, Oceans, 96*(C2), 2515–2534. https://doi.org/10.1029/90JC02195.

McCreary, J. P., Shetye, S. R., & Kundu, P. K. (1986). Thermohaline forcing of eastern boundary currents: With application to the circulation off the west coast of Australia. *Journal of Marine Research, 44*, 71–92. https://doi.org/10.1357/002224086788460184.

McMonigal, K., Beal, L. M., Elipot, S., Gunn, K. L., Hermes, J., Morris, T., & Houk, A. (2020). The impact of meanders, deepening and broadening, and seasonality on Agulhas current temperature variability. *Journal of Physical Oceanography*. https://doi.org/10.1175/jpo-d-20-0018.1.

McMonigal, K., Beal, L. M., & Willis, J. K. (2018). The seasonal cycle of the South Indian Ocean subtropical gyre circulation as revealed by Argo and satellite data. *Geophysical Research Letters, 45*(17), 9034–9041. https://doi.org/10.1029/2018GL078420.

Menezes, V. V., Phillips, H. E., Schiller, A., Bindoff, N. L., Domingues, C. M., & Vianna, M. L. (2014). South Indian countercurrent and associated fronts. *Journal of Geophysical Research, Oceans, 119*, 6763–6791. https://doi.org/10.1002/2014JC010076.

Meuleners, M. J., Ivey, G. N., & Pattiaratchi, C. B. (2008). A numerical study of the eddying characteristics of the Leeuwin current system. *Deep Sea Research, Part I, 55*(3), 261–276. https://doi.org/10.1016/j.dsr.2007.12.004.

Meuleners, M. J., Pattiaratchi, C. B., & Ivey, G. N. (2007). Numerical modelling of the mean flow characteristics of the Leeuwin current system. *Deep Sea Research, Part II, 54*(8–10), 837–858. https://doi.org/10.1016/j.dsr2.2007.02.003.

Middleton, J. F., & Bye, J. A. T. (2007). A review of the shelf-slope circulation along Australia's southern shelves: Cape Leeuwin to Portland. *Progress in Oceanography, 75*(1), 1–41. https://doi.org/10.1016/j.pocean.2007.07.001.

Middleton, J. F., & Cirano, M. (2002). A northern boundary current along Australia's southern shelves: The Flinders current. *Journal of Geophysical Research, Oceans*. https://doi.org/10.1029/2000jc000701.

Moore, T. S., & Marra, J. (2002). Satellite observations of bloom events in the strait of Ombai: Relationships to monsoons and ENSO. *Geochemistry, Geophysics, Geosystems*. https://doi.org/10.1029/2001gc000174.

Moore, T. S., Marra, J., & Alkatiri, A. (2003). Response of the Banda Sea to the southeast monsoon. *Marine Ecology Progress Series*. https://doi.org/10.3354/meps261041.

Moore, T. S., Matear, R. J., Marra, J., & Clementson, L. (2007). Phytoplankton variability off the Western Australian coast: Mesoscale eddies and their role in cross-shelf exchange. *Deep Sea Research Part II: Topical Studies in Oceanography*. https://doi.org/10.1016/j.dsr2.2007.02.006.

Murty, S. A., Goodkin, N. F., Halide, H., Natawidjaja, D., Suwargadi, B., Suprihasnto, I., … Gordon, A. L. (2017). Climatic influences on southern Makassar Strait salinity over the past century. *Geophysical Research Letters, 44*, 11967–11975. https://doi.org/10.1002/2017GL075504.

Napitu, A. M., Pujiana, K., & Gordon, A. L. (2019). The madden-Julian Oscillation's impact on the Makassar Strait surface layer transport. *Journal of Geophysical Research*. https://doi.org/10.1029/2018JC014729.

Nieves, V., Willis, J. K., & Patzert, W. C. (2015). Recent hiatus caused by decadal shift in Indo-Pacific heating. *Science, 349*, 532–535. https://doi.org/10.1126/science.aaa4521.

Nugroho, D., Koch-Larrouy, A., Gaspar, P., Lyard, F., Reffray, G., & Tranchant, B. (2018). Modelling explicit tides in the Indonesian seas: An important process for surface sea water properties. *Marine Pollution Bulletin*. https://doi.org/10.1016/j.marpolbul.2017.06.033.

Oke, P. R., Griffin, D. A., Rykova, T., & de Oliveira, H. B. (2018). Ocean circulation in the great Australian bight in an eddy-resolving ocean reanalysis: The eddy field, seasonal and interannual variability. *Deep Sea Research, Part II, 157–158*, 11–26. https://doi.org/10.1016/j.dsr2.2018.09.012.

Palastanga, V., van Leeuwen, P. J., Schouten, M. W., & de Ruijter, W. P. M. (2007). Flow structure and variability in the subtropical Indian Ocean: Instability of the South Indian Ocean countercurrent. *Journal of Geophysical Research, 112*, C01001. https://doi.org/10.1029/2005JC003395.

Paris, M. L., Subrahmanyam, B., Trott, C. B., & Murty, V. S. N. (2018). Influence of ENSO events on the Agulhas leakage region. *Remote Sensing in Earth Systems Sciences, 1*(3–4), 79–88. https://doi.org/10.1007/s41976-018-0007-z.

Pearce, A. F., & Griffiths, R. W. (1991). The mesoscale structure of the Leeuwin current: A comparison of laboratory model and satellite images. *Journal of Geophysical Research, 96*, 16730–16757. https://doi.org/10.1029/91JC01712.

Phillips, H. E., Menezes, V. V., Nagura, M., McPhaden, M. J., Vinayachandran, P. N., & Beal, L. M. (2024). Chapter 8: Indian Ocean circulation. In C. C. Ummenhofer, & R. R. Hood (Eds.), The Indian Ocean and its role in the global climate system (pp. 169–203). Amsterdam: Elsevier. https://doi.org/10.1016/B978-0-12-822698-8.00012-3.

Phillips, H. E., Tandon, A., Furue, R., Hood, R., Ummenhofer, C., Benthuysen, J., Menezes, V., Hu, S., Webber, B., Sanchez-Franks, A., Cherian, D., Shroyer, E., Feng, M., Wijeskera, H., Chatterjee, A., Yu, L., Hermes, J., Murtugudde, R., Tozuka, T., … Wiggert, J. (2021). Progress in understanding of Indian Ocean circulation, variability, air-sea exchange and impacts on biogeochemistry. *Ocean Science, 17*, 1677–1751. https://doi.org/10.5194/os-17-1677-2021.

Phillips, H. E., Wijffels, S. E., & Feng, M. (2005). Interannual variability in the freshwater content of the Indonesian-Australian Basin. *Geophysical Research Letters, 32*, L03603. https://doi.org/10.1029/2004GL021755.

Pujiana, K., Gordon, A. L., & Sprintall, J. (2013). Intraseasonal Kelvin waves in Makassar Strait. *Journal of Geophysical Research, 118*, 2023–2034.

Pujiana, K., McPhaden, M. J., Gordon, A. L., & Napitu, A. M. (2019). Unprecedented response of Indonesian throughflow to anomalous Indo-Pacific climatic forcing in 2016. *Journal of Geophysical Research*. https://doi.org/10.1029/2018JC014574.

Purkey, S. G., & Johnson, G. C. (2010). Warming of global abyssal and deep Southern Ocean waters between the 1990s and 2000s: Contributions to global heat and sea level rise budgets. *Journal of Climate, 23*, 6336–6351.

Putrasahan, D., Kirtman, B. P., & Beal, L. M. (2016). Modulation of SST interannual variability in the Agulhas leakage region associated with ENSO. *Journal of Climate, 29*(19), 7089–7102. https://doi.org/10.1175/JCLI-D-15-0172.1.

Reid, J. L. (2003). On the total geostrophic circulation of the Indian Ocean: Flow patterns, tracers, and transports. *Progress in Oceanography*. https://doi.org/10.1016/S0079-6611(02)00141-6.

Richardson, P. L. (2007). Agulhas leakage into the Atlantic estimated with subsurface floats and surface drifters. *Deep-Sea Research Part I: Oceanographic Research Papers*. https://doi.org/10.1016/j.dsr.2007.04.010.

Ridgway, K. R., & Condie, S. A. (2004). The 5500-km–long boundary flow off western and southern Australia. *Journal of Geophysical Research, 109*, C04017. https://doi.org/10.1029/2003JC001921.

Ridgway, K. R., & Dunn, J. R. (2003). Mesoscale structure of the mean East Australian current system and its relationship with topography. *Progress in Oceanography, 56*, 189–222.

Ridgway, K. R., & Dunn, J. R. (2007). Observational evidence for a southern hemisphere oceanic 'Supergyre'. *Geophysical Research Letters, 34*, L13612. https://doi.org/10.1029/2007GL030392.

Ridgway, K. R., Dunn, J. R., & Wilkin, J. L. (2002). Ocean interpolation by four-dimensional weighted least squares—Application to the waters around Australasia. *Journal of Atmospheric and Oceanic Technology, 19*(9), 1357–1375.

Roberts, C. M., McClean, C. J., Veron, J. E. N., Hawkins, J. P., Allen, G. R., McAllister, D. E., et al. (2002). Marine biodiversity hotspots and conservation priorities for tropical reefs. *Science*. https://doi.org/10.1126/science.1067728.

Rochford, D. J. (1969). *Seasonal interchange of high and low salinity surface waters off south-West Australia*. Technical Paper Australia: Division of Fisheries and Oceanography, CSIRO. http://hdl.handle.net/102.100.100/321788?index=1.

Roemmich, D., & Gilson, J. (2009). The 2004–2008 mean and annual cycle of temperature, salinity, and steric height in the global ocean from the Argo Program. *Progress in Oceanography, 82*(2), 81–100. https://doi.org/10.1016/j.pocean.2009.03.004.

Rosell-Fieschi, M., Rintoul, S. R., Gourrion, J., & Pelegrí, J. L. (2013). Tasman leakage of intermediate waters as inferred from Argo floats. *Geophysical Research Letters, 40*(20), 5456–5460. https://doi.org/10.1002/2013GL057797.

Rousselet, L., Cessi, P., & Forget, G. (2020). Routes of the upper branch of the Atlantic meridional overturning circulation according to an ocean state estimate. *Geophysical Research Letters, 47*(18). https://doi.org/10.1029/2020GL089137.

Roxy, M. K., Ritika, K., Terray, P., & Masson, S. (2014). The curious case of Indian Ocean warming. *Journal of Climate, 27*(22), 8501–8509. https://doi.org/10.1175/JCLI-D-14-00471.1.

Roxy, M. K., Saranya, J. S., Modi, A., Anusree, A., Cai, W., Resplandy, L., … Frölicher, T. L. (2024). Chapter 20: Future projections for the tropical Indian Ocean. In C. C. Ummenhofer, & R. R. Hood (Eds.), *The Indian Ocean and its role in the global climate system* (pp. 469–482). Amsterdam: Elsevier. https://doi.org/10.1016/B978-0-12-822698-8.00004-4.

Rühs, S., Durgadoo, J. V., Behrens, E., & Biastoch, A. (2013). Advective timescales and pathways of Agulhas leakage. *Geophysical Research Letters, 40*(15), 3997–4000. https://doi.org/10.1002/grl.50782.

Rühs, S., Schwarzkopf, F. U., Speich, S., & Biastoch, A. (2019). Cold vs. warm water route-sources for the upper limb of the Atlantic meridional overturning circulation revisited in a high-resolution ocean model. *Ocean Science, 15*(3), 489–512. https://doi.org/10.5194/os-15-489-2019.

Sallée, J.-B., Wienders, N., Speer, K., & Morrow, R. (2006). Formation of subantarctic mode water in the southeastern Indian Ocean. *Ocean Dynamics*, *56*, 525–542. https://doi.org/10.1007/s10236-005-0054-x.

Saville-Kent, W. (1897). *The Naturaliste in Australia*. New York: Chapman and Hall. 302 pp.

Schodlok, M. P., Tomczak, M., & White, N. (1997). Deep sections through the South Australian Basin and across the Australian-Antarctic discordance. *Geophysical Research Letters*, *24*(22), 2785–2788.

Schouten, M. W., de Ruijter, W. P. M., & van Leeuwen, P. J. (2002a). Upstream control of Agulhas ring shedding. *Journal of Geophysical Research*, *107* (10.1029).

Schouten, M. W., de Ruijter, W. P. M., van Leeuwen, P. J., & Dijkstra, H. (2002b). An oceanic teleconnection between the equatorial and southern Indian Ocean. *Geophysical Research Letters*, *29*(16), 51–59.

Schulze Chretien, L. M., & Speer, K. (2018). A deep eastern boundary current in the Chile Basin. *Journal of Geophysical Research: Oceans*, *124*, 27–40. https://doi.org/10.1029/2018JC014400.

Schwarzkopf, F. U., Biastoch, A., Böning, C. W., Chanut, J., Durgadoo, J. V., Getzlaff, K., et al. (2019). The INALT family—A set of high-resolution nests for the Agulhas current system within global NEMO Ocean/sea-ice configurations. *Geoscientific Model Development*, *12*(7), 3329–3355. https://doi.org/10.5194/gmd-12-3329-2019.

Sen Gupta, A. R., Ganachaud, A., McGregor, S., Brown, J. N., & Muir, L. (2012). Drivers of the projected changes to the Pacific Ocean equatorial circulation. *Geophysical Research Letters*, *39*. https://doi.org/10.1029/2012GL051447.

Sen Gupta, A., McGregor, S., Van Sebille, E., Ganachaud, A., Brown, J. N., & Santoso, A. (2016). Future changes to the Indonesian Throughflow and Pacific circulation: The differing role of wind and deep circulation changes. *Geophysical Research Letters*, *43*(4), 1669–1678. https://doi.org/10.1002/2016GL067757.

Sen Gupta, A., Santoso, A., Taschetto, A. S., Ummenhofer, C. C., Trevena, J., & England, M. H. (2009). Projected changes to the Southern Hemisphere Ocean and sea ice in the IPCC AR4 climate models. *Journal of Climate*, *22*(11), 3047–3078. https://doi.org/10.1175/2008JCLI2827.1.

Smith, R. L., Huyer, A., Godfrey, J. S., & Church, J. A. (1991). The Leeuwin current off Western Australia, 1986–1987. *Journal of Physical Oceanography*, *21*, 323–345. https://doi.org/10.1175/1520-0485(1991)021<0323:TLCOWA>2.0.CO;2.

Song, Q., & Gordon, A. L. (2004). Significance of the vertical profile of Indonesian Throughflow transport on the Indian Ocean. *Geophysical Research Letters*, *31*, L16307. https://doi.org/10.1029/2004GL020360.

Speich, S., Blanke, B., & Cai, W. (2007). Atlantic meridional overturning circulation and the southern hemisphere supergyre. *Geophysical Research Letters*, *34*, 1–5. https://doi.org/10.1029/2007GL031583.

Speich, S., Blanke, B., de Vries, P., Drijfhout, S., Doos, K., Ganachaud, A., & Marsh, R. (2002). Tasman leakage- A new route in the global ocean conveyor belt. *Geophysical Research Letters*, *29*(10), 51–55.

Sprintall, J., Gordon, A. L., Koch-Larrouy, A., Lee, T., Potemra, J. T., Pujiana, K., & Wijffels, S. J. (2014). The Indonesian seas and their role in the coupled ocean-climate system. *Nature Geoscience*. https://doi.org/10.1038/NGEO2188.

Sprintall, J., Wijffels, S. E., Molcard, R., & Jaya, I. (2009). Direct estimates of the Indonesian Throughflow entering the Indian Ocean: 2004-2006. *Journal of Geophysical Research, Oceans*, *114*. https://doi.org/10.1029/2008JC005257.

Stellema, A., Sen Gupta, A., & Taschetto, A. S. (2019). Projected slow down of South Indian Ocean circulation. *Scientific Reports*, *9*(1), 1–15. https://doi.org/10.1038/s41598-019-54092-3.

Sun, C., Feng, M., Matear, R. J., Chamberlain, M. A., Craig, P., Ridgway, K. R., & Schiller, A. (2012). Marine downscaling of a future climate scenario for Australian boundary currents. *Journal of Climate*, *25*(8), 2947–2962. https://doi.org/10.1175/JCLI-D-11-00159.1.

Sun, S., & Thompson, A. F. (2020). Centennial changes in the Indonesian Throughflow connected to the Atlantic meridional overturning circulation : The ocean's transient conveyor belt. *Geophysical Research Letters*, *47*, 1–9. https://doi.org/10.1029/2020GL090615.

Susanto, R. D., Ffield, A., Gordon, A. L., & Adi, T. A. (2012). Variability of the Indonesian throughflow within Makassar Strait 2004-2009. *Journal of Geophysical Research*, *117*, C09013.

Susanto, R. D., Moore, T. S., & Marra, J. (2006). Ocean color variability in the Indonesian seas during the SeaWiFS era. *Geochemistry, Geophysics, Geosystems*. https://doi.org/10.1029/2005GC001009.

Swart, N. C., & Fyfe, J. C. (2012). Observed and simulated changes in the Southern Hemisphere surface westerly wind-stress. *Geophysical Research Letters*, *39*, L16711. https://doi.org/10.1029/2012GL052810.

Talley, L. D. (2013a). Closure of the global overturning circulation through the Indian, Pacific, and southern oceans: Schematics and transports. *Oceanography*, *26*, 80–97. https://doi.org/10.5670/oceanog.2011.65.

Talley, L. D. (2013b). In I. N. M. Sparrow, P. Chapman, & J. Gould (Eds.), *Hydrographic atlas of the World Ocean circulation experiment (WOCE). Volume 4: Indian Ocean*. Southampton, UK: International WOCE Project Office.

Talley, L. D., Pickard, G. L., Emery, W. J., & Swift, J. H. (2011). *Descriptive physical oceanography: An introduction* (6th ed.). Academic Press, Elsevier Ltd. 983pp.

Talley, L. D., & Sprintall, J. (2005). Deep expression of the Indonesian Throughflow: Indonesian intermediate water in the south equatorial current. *Journal of Geophysical Research, C: Oceans*, *110*(10). https://doi.org/10.1029/2004JC002826.

Tamsitt, V., Abernathey, R. P., Mazloff, M. R., Wang, J., & Talley, L. D. (2018). Transformation of deep water masses along Lagrangian upwelling pathways in the Southern Ocean. *Journal of Geophysical Research: Oceans*, *123*, 1994–2017. https://doi.org/10.1002/2017JC013409.

Tamsitt, V., Drake, H., Morrison, A., Talley, L. D., Dufour, C., Gray, A., et al. (2017). Spiraling up: Pathways of global deep water from the deep ocean to the surface of the Southern Ocean. *Nature Communications*, *8*, 172.

Tamsitt, V., Talley, L. D., & Mazloff, M. R. (2019). A deep eastern boundary current carrying Indian deep water south of Australia. *Journal of Geophysical Research: Oceans, 124*(3), 2218–2238. https://doi.org/10.1029/2018JC014569.

Thompson, R. O. R. Y. (1984). Observations of the Leeuwin current off Western Australia. *Journal of Physical Oceanography, 14*, 623–628. https://doi.org/10.1175/1520-0485(1984)014<0623:OOTLCO>2.0.CO;2.

Thompson, R. O. R. Y. (1987). Continental-shelf-scale model of the Leeuwin current. *Journal of Marine Research, 45*(4), 813–827. https://doi.org/10.1357/002224087788327190.

Toole, J. M., & Warren, B. A. (1993). A hydrographic section across the subtropical South Indian Ocean. *Deep Sea Research, Part I, 40*, 1973–2019.

Tozuka, T., Dong, L., Han, W., Lengainge, M., & Zhang, L. (2024). Chapter 10: Decadal variability of the Indian Ocean and its predictability. In C. C. Ummenhofer, & R. R. Hood (Eds.), The Indian Ocean and its role in the global climate system (pp. 229–244). Amsterdam: Elsevier. https://doi.org/10.1016/B978-0-12-822698-8.00014-7.

Tsugawa, M., & Hasumi, H. (2010). Generation and growth mechanism of the Natal pulse. *Journal of Physical Oceanography, 40*, 1597–1612.

Tun, K., Chou, L. M., Cabanban, A., Tuan, V. S., Yeemin, T., Suharsono, et al. (2004). Status of coral reefs, coral reef monitoring and management in Southeast Asia, 2004. In *Status of Coral Reefs of the World: 2004—Volume 1.*

Ummenhofer, C. C., Geen, R., Denniston, R. F., & Rao, M. P. (2024). Chapter 3: Past, present, and future of the South Asian monsoon. In C. C. Ummenhofer, & R. R. Hood (Eds.), The Indian Ocean and its role in the global climate system (pp. 49–78). Amsterdam: Elsevier. https://doi.org/10.1016/B978-0-12-822698-8.00013-5.

Ummenhofer, C. C., Murty, S. A., Sprintall, J., Lee, T., & Abram, N. (2021). Decadal to centennial Indian Ocean heat and freshwater changes in a regional context. *Nature Reviews Earth and Environment, 2*, 525–541. https://doi.org/10.1038/s43017-021-00192-6.

Van Aken, H. M., Punjanan, J., & Saimima, S. (1988). Physical aspects of the flushing of the east Indonesian basins. *Netherlands Journal of Sea Research, 22*(4), 315–339. https://doi.org/10.1016/0077-7579(88)90003-8.

Van Aken, H. M., Van Veldhoven, A. K., Veth, C., De Ruijter, W. P. M., Van Leeuwen, P. J., Drijfhout, S. S., et al. (2003). Observations of a young Agulhas ring, Astrid, during MARE in March 2000. *Deep Sea Research Part II: Topical Studies in Oceanography, 50*(1), 167–195. https://doi.org/10.1016/S0967-0645(02)00383-1.

van Bennekom, A. J. (1988). Deep-water transit times in the eastern Indonesian basins, calculated from dissolved silica in deep and interstitial waters. *Netherlands Journal of Sea Research.* https://doi.org/10.1016/0077-7579(88)90004-X.

van Bennekom, A. J., & Muchtar, M. (1988). Reactive phosphate in the eastern Indonesian seas. *Netherlands Journal of Sea Research.* https://doi.org/10.1016/0077-7579(88)90006-3.

Van Sebille, E., England, M. H., Zika, J. D., & Sloyan, B. M. (2012). Tasman leakage in a fine-resolution ocean model. *Geophysical Research Letters, 39*, L06601. https://doi.org/10.1029/2012GL051004.

Van Sebille, E., Sprintall, J., Schwarzkopf, F. U., Sen Gupta, A., Santoso, A., England, M. H., et al. (2014). Pacific-to-Indian Ocean connectivity: Tasman leakage, Indonesian Throughflow, and the role of ENSO. *Journal of Geophysical Research: Oceans, 119*(2). https://doi.org/10.1002/2013JC009525.

Vidya, P. J., Ravichandran, M., Subeesh, M. P., Chatterjee, S., & N.M. (2020). Global warming hiatus contributed weakening of the Mascarene high in the Southern Indian Ocean. *Scientific Reports, 10*, 3255. https://doi.org/10.1038/s41598-020-59964-7.

Weijer, W., Cheng, W., Garuba, O. A., Hu, A., & Nadiga, B. T. (2020). CMIP6 models predict significant 21st century decline of the Atlantic meridional overturning circulation. *Geophysical Research Letters, 47*, e2019GL086075. https://doi.org/10.1029/2019GL086075.

Weijer, W., & van Sebille, E. (2014). Impact of Agulhas leakage on the Atlantic overturning circulation in the CCSM4. *Journal of Climate, 27*, 101–110.

Wijffels, S., & Meyers, G. (2004). An intersection of oceanic waveguides: Variability in the Indonesian Throughflow region. *Journal of Physical Oceanography.* https://doi.org/10.1175/1520-0485(2004)034<1232:AIOOWV>2.0.CO.

Wijffels, S. E., Meyers, G. M., & Godfrey, J. S. (2008). A 20-Yr average of the Indonesian Throughflow: Regional currents and the interbasin exchange. *Journal of Physical Oceanography, 38*, 1965–1978.

Woo, M., & Pattiaratchi, C. (2008). Hydrography and water masses of the western Australian coast. *Deep Sea Research, 55*, 1090–1104.

Wood, E. J. F. (1954). Dinoflagellates in Australian waters. *Australian Journal of Marine & Freshwater Research, 5*, 171–351.

Wyrtki, K. (1971). *Oceanographic atlas of the international Indian Ocean expedition.* Washington, D.C.: National Science Foundation. 531 pp.

Wyrtki, K. (1987). Indonesian Throughflow and the associated pressure gradient. *Journal of Geophysical Research, 92*, 12941–12946.

Xie, T., Newton, R., Schlosser, P., Du, C., & Dai, M. (2019). Long-term mean mass, heat and nutrient flux through the Indonesian seas, based on the tritium inventory in the Pacific and Indian oceans. *Journal of Geophysical Research: Oceans, 124*(6), 3859–3875. https://doi.org/10.1029/2018JC014863.

Xu, T., Li, S., Hamzah, F., Setiawan, A., Susanto, R. D., Cao, G., & Wei, Z. (2018). Intraseasonal flow and its impact on the chlorophyll-a concentration in the Sunda Strait and its vicinity. *Deep-Sea Research Part I: Oceanographic Research Papers.* https://doi.org/10.1016/j.dsr.2018.04.003.

Yamagami, Y., Tozuka, T., & Qiu, B. (2019). Interannual variability of the Natal pulse. *Journal of Geophysical Research: Oceans, 124*(12), 9258–9276. https://doi.org/10.1029/2019JC015525.

Zhang, Y., Feng, M., Du, Y., Phillips, H. E., Bindoff, N. L., & McPhaden, M. J. (2018). Strengthened Indonesian Throughflow drives decadal warming in the Southern Indian Ocean. *Geophysical Research Letters, 45*(6167), 6175. https://doi.org/10.1029/2018GL078265.

Zheng, Q. (2018). *Water mass and velocity structure of the subsurface westward flows in the South Indian Ocean.* (Honours thesis). University of Tasmania. 50 pp.

Chapter 10

Decadal variability of the Indian Ocean and its predictability

Tomoki Tozuka[a], Lu Dong[b], Weiqing Han[c], Matthieu Lengaigne[d], and Lei Zhang[e,f]

[a]Department of Earth and Planetary Science, Graduate School of Science, The University of Tokyo, Tokyo, Japan, [b]Frontier Science Centre for Deep Ocean Multispheres and Earth System and Physical Oceanography Laboratory, Ocean University of China, Qingdao, People's Republic of China, [c]Department of Atmospheric and Oceanic Sciences, University of Colorado, Boulder, CO, United States, [d]MARBEC, University of Montpellier, CNRS, IFREMER, IRD, Sète, France, [e]State Key Laboratory of Tropical Oceanography, South China Sea Institute of Oceanology, Chinese Academy of Sciences, Guangzhou, Guangdong, People's Republic of China, [f]Global Ocean and Climate Research Center, South China Sea Institute of Oceanology, Guangzhou, Guangdong, People's Republic of China

1 Introduction

The Indian Ocean exhibits variations across different timescales, with decadal variations (encompassing periods ranging from approximately 7 years to several decades) being the least understood. Unlike the Pacific Ocean, which has been extensively studied for its Pacific Decadal Oscillation (PDO) and/or Interdecadal Pacific Oscillation (IPO) (Newman, 2016; Power, 2021), and the Atlantic Ocean, which has received significant attention for its Atlantic multidecadal variability (Sutton et al., 2018), the modes of decadal internal climate variability in the Indian Ocean remain a "gray area" due to limited data and dedicated research (Han et al., 2014). However, this does not imply that the Indian Ocean is devoid of decadal variations. On the contrary, it poses a serious challenge when attributing climate change signals specific to the region. To overcome this challenge, there is a pressing need to thoroughly characterize the natural decadal variability of the Indian Ocean, enabling us to attribute regional climate change signals accurately and enhance our global climate prediction capabilities for the coming decades.

There has been renewed interest in Indian Ocean decadal variability recently due to its potential involvement in the recent global surface warming "hiatus" (England, 2014; Kosaka & Xie, 2013). It is believed that the negative IPO phase played a crucial role by causing increased heat absorption in the tropical Pacific, but also a rapid surface warming in the Indian Ocean, which in turn, likely contributed to this negative IPO phase by enhancing the Walker circulation (Dong & McPhaden, 2017a; Hamlington et al., 2014; Han et al., 2017a; Li & Han, 2015; Luo et al., 2012; Zhang et al., 2019). Additionally, a substantial fraction of the Pacific heat uptake has been transported to the Indian Ocean through the Indonesian Throughflow (ITF; Lee et al., 2015; Nieves et al., 2015; Zhang et al., 2018b). Consequently, the Indian Ocean has been responsible for more than 70% of the increase in global oceanic heat content within the upper 700 m during the past decade (Lee et al., 2015). The fate of this extra heat in the Indian Ocean—whether it re-surfaces or remains at depth in the upcoming decades—has the potential to influence the evolution of global surface temperature (Vialard, 2015). As a result, the Indian Ocean emerges as a region of significant importance for decadal climate predictions.

This chapter aims to present an up-to-date review of progress made in our understanding of Indian Ocean decadal variability since a seminal review paper by Han et al. (2014). Addressing outstanding questions raised in that review paper is a key focus, particularly exploring the relative importance of external forcing and internal variability within the Indian Ocean in generating decadal fluctuations. This chapter is organized as follows: the next section delves into what instrumental records reveal about decadal variability in the Indian Ocean. Sections 3 and 4 then offer comprehensive reviews of recent advancements in understanding internal and externally forced decadal climate variability, respectively. Section 5 presents the latest findings on decadal climate predictions. Conclusions and noteworthy unresolved questions are discussed in the final section. It is important to note that this chapter only provides brief mentions of decadal variability in biogeochemical variables

---

[⋆]This book has a companion website hosting complementary materials. Visit this URL to access it: https://www.elsevier.com/books-and-journals/book-companion/9780128226988.

(for more details, see Al-Yamani et al., 2024; Hood et al., 2024a), decadal variations recorded in paleoclimate proxies (Mohtadi et al., 2024), future climate projections (Roxy et al., 2024), and impact of Indian Ocean decadal/multidecadal signals on regional and global climate (Ummenhofer et al., 2024) as these aspects are reviewed more extensively in other chapters.

2 Observational datasets

Data sparsity before 1980. To investigate the decadal variability of the Indian Ocean and differentiate it from externally forced long-term climate change signals, comprehensive and consistent climate observations spanning the entire basin are essential. However, the availability of high-resolution instrumental climate datasets with global coverage has been limited to the introduction of satellites in the late 1970s. Before that, the instrumental records of climate variables were characterized by significant uncertainties due to the scarcity of observations and the presence of time-varying biases in the collected data (Parker et al., 2000). As a consequence, studying Indian Ocean decadal variability before the satellite era poses considerable challenges, primarily due to the lack of reliable and homogeneous data.

SST and tide gauges. Sea surface temperature (SST) stands as a vital variable in studying decadal climate variability, owing to its key role in large-scale atmosphere-ocean interactions. The instrumental record of SST dates back to the mid-19th century. However, the spatial coverage during earlier periods was generally inadequate, with SST measurements mostly restricted to shipping lanes (Fig. 1a). Although the coverage expanded to more areas in the 20th century, certain areas, such as the tropics and the Southern Ocean, still exhibit poor data coverage (Fig. 1b and c). Another pertinent example is the global sea level record derived from tide gauges (Fig. 1d), which offers essential information for studying both regional and global scale sea level changes (Han, 2010; Unnikrishnan & Shankar, 2007). While some tide gauge data provide more than a century of continuous sea level records—surpassing the duration of satellite-derived products—their utility is limited for analyzing basin-scale climate variations due to coarse spatial sampling. Consequently, data sparsity

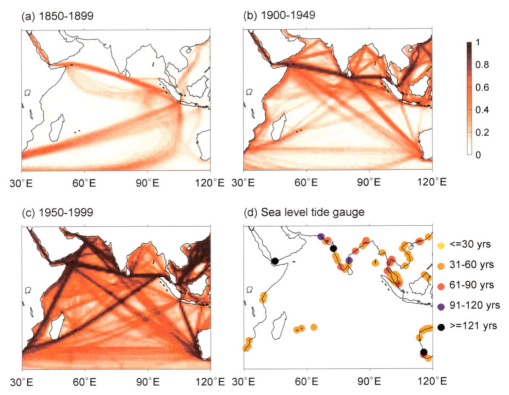

FIG. 1 (a)–(c) Distributions of sea surface temperature (SST) observations from HadSST2. Shown are the percentage of months with SST observations over the 50-year period. (d) Distribution of sea level tide gauges. Different colors represent different time length of available sea level data.

may introduce sampling errors, leading to significant uncertainties in studying decadal climate variability and long-term climate change signals based on instrumental climate datasets.

Instrumental biases. Instrumental measurements are not without biases or errors, which may even evolve over different time periods due to changing observational practices. In the 19th and early 20th centuries, SST was gauged by collecting water samples using uninsulated or partially insulated buckets from ships. From the middle of the 20th century, SST measurements transitioned to utilizing water intake from the engine room. This change in measurement techniques could lead to abrupt shifts in the SST record obtained from the same ships (Barnett, 1984), necessitating "correction" of the instrumental SST data. Since the late 1970s, global coverage satellite-derived SST data have become available. However, satellite measurements also carry biases, particularly due to the influence of aerosols and clouds. Additionally, unlike ships that measure bulk temperature, satellite observations capture skin temperature (Reynolds & Smith, 1994). These inconsistencies in SST measurements throughout the instrumental period add another layer of uncertainties when using the instrumental record for climate research, alongside the scarcity of observations in earlier periods.

Reconstructions. The presence of significant gaps and time-varying biases in the instrumental climate record introduce considerable uncertainties, restricting our ability to analyze Indian Ocean decadal climate variability. To address this, modern SST reconstructions employ "interpolation" techniques to fill gaps, especially in earlier periods, and "adjustment" or "correction" methods to mitigate the effects of the time-varying biases in the SST measurements (e.g., Nidheesh et al., 2017; Rayner et al., 2003). Similarly, sea level reconstructions combine large-scale sea level anomaly patterns associated with climate modes derived from satellite observations and tide gauge data to extend the observational record back in time (Church et al., 2004; Kumar et al., 2020). These products yield homogeneous gridded global SST and sea level data dating back to the mid-19th century. However, due to variations in the methods employed in different SST reconstructions and the scarcity of the raw data from the instrumental record for both SST and sea level datasets, significant cross-data differences persist (e.g., Nidheesh et al., 2017; Zhang et al., 2018b). Despite these efforts, uncertainties in the historical climate records continue to pose challenges in comprehensively understanding Indian Ocean decadal climate variability.

Uncertainties in Indian Ocean decadal variability and trends. The considerable uncertainties in the SST reconstructions present challenges in examining the decadal variability and long-term climate change in the Indian Ocean. In Fig. 2, we explore the decadal variability of the tropical Indian Ocean SST anomalies using empirical orthogonal function analysis of 8-year low-pass filtered monthly SST data spanning 1890 to 2015 from six different products. Overall, the filtered decadal variability explains 8% to 18% of the total variance in the six data sets. Before conducting the empirical orthogonal function analysis, the SST data were linearly interpolated onto uniform 2.5×2.5 degree grids, and the linear trend was removed to partly eliminate the influence of anthropogenic global warming. The first empirical orthogonal function mode (EOF1) reveals the main mode of decadal SST variability in the tropical Indian Ocean, characterized mainly by basin-scale SST anomalies of the same sign (Fig. 2a–f). As noted in prior studies, this behavior is primarily driven by the IPO through atmospheric teleconnections (Han et al., 2014). However, the six SST data sets clearly exhibit distinct SST anomaly patterns. Cobev2, HadISST, and Hurrell SST demonstrate maximum SST anomaly signals located in the central tropical Indian Ocean. In contrast, the 20CRv2c and ERSSTv4 datasets are associated with the strongest SST anomalies in the South Indian Ocean, especially near the so-called Seychelles-Chagos thermocline ridge (SCTR) or Seychelles Dome region, where the mean thermocline depth is shallow. ERA-20C, on the other hand, shows relatively weaker signals compared to other SST products, resulting in an explained variance of EOF1 of only 45%, smaller than others, which range roughly between 60% and 70%. These variations in SST anomaly patterns highlight the importance of accounting for data source differences when studying Indian Ocean decadal climate variability.

While the time evolution of the corresponding principal components (PC1) generally shows agreement across the six SST products, noticeable discrepancies arise during specific periods (Fig. 2g), particularly in the early 20th century, which is likely due to limited observations during that time. In recent decades, results from different SST products tend to converge, thanks to the advances in satellite observations that provide high-resolution global-scale SST measurements. In addition to uncertainties in internal decadal climate variability, long-term climate change signals also exhibit uncertainties to some extent (Fig. 2h). For instance, although all SST datasets indicate a relatively steady warming trend in the tropical Indian Ocean, there was a slowdown in the basin warming between the 1940s and 1960s. However, the magnitudes of the warming trend differ among the datasets. ERSSTv4 shows a warming trend of 0.74°C/century, whereas ERA-20C reflects a lower warming trend of 0.49°C/century. It is worth noting that such differences in warming trends may partly stem from uncertainties in internal climate variability. Consequently, both externally forced climate change signals (assessed by the linear trends) and internal decadal climate variability exhibit uncertainties across different SST reconstructions, posing challenges when investigating their associated mechanisms.

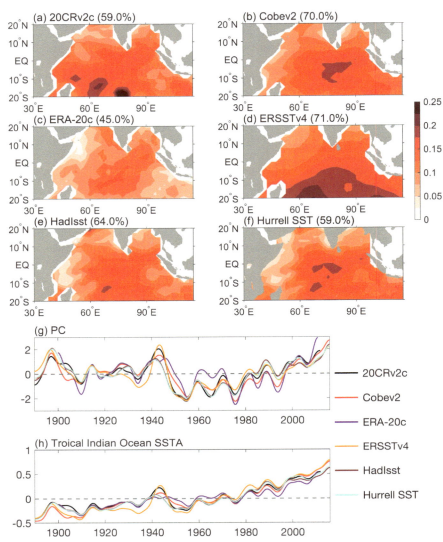

FIG. 2 (a)–(f) First Empirical Orthogonal Function (EOF1) mode of 8-year low-pass filtered SST anomalies in the tropical Indian Ocean from various SST data sets. Linear trend has been removed prior to the empirical orthogonal function analysis. Unit is °C. (g) Corresponding Principal Component (PC1). (h) 8-year lowpass filtered tropical Indian Ocean SST anomalies averaged in 30°E–120°E and 20°S–20°N. All SST data sets used here were interpolated into the same 2.5° by 2.5° grids.

3 Internal decadal climate variability

In comparison with the Pacific (e.g., Newman, 2016) and Atlantic (e.g., Liu, 2012), the exploration of decadal climate variations in the Indian Ocean has been relatively restricted. This section provides an overview of the existing literature that examines Indian Ocean decadal climate variability, addressing certain pertinent questions raised by a seminal review paper on this topic by Han et al. (2014).

3.1 Remote forcing from other regions

The decadal-to-interdecadal variability of SST in the tropical Indian Ocean has been observed to be in sync with the IPO as evidenced by instrumental records (Zhang et al., 1997) and proxy data (Cole et al., 2000). Until the 1960s, the SST variability of the tropical Indian Ocean was in phase with the IPO but a significant shift occurred after 1980, causing it to become out of phase with the Pacific SST variability (Zhang et al., 2018b). Observations and coupled climate model experiments suggest that this altered relationship can be primarily attributed to enhanced external forcing (Dong & McPhaden, 2017a; Zhang et al., 2018b). Zhang et al. (2018b) attributed this change to an earlier emergence of anthropogenic warming in the tropical Indian Ocean and volcanic eruptions in the 1980s and 1990s, leading to cooling over the region. In line with these findings, Dong and McPhaden (2017b) demonstrated that greenhouse gas forcing has contributed to warming the Indian Ocean, overpowering changes that would have arisen from the IPO, thereby weakening the dynamical links between the two ocean basins. Through coupled climate model experiments, it has been shown that in the absence of external forcing, the evolution of the Indian Ocean decadal variability would be considerably influenced by the IPO.

Numerous studies have emphasized the significance of the atmospheric bridge in shaping decadal variability. Copsey et al. (2006) attempted to simulate decadal sea level pressure (SLP) anomalies over the Indian Ocean using an atmospheric general circulation model but encountered difficulties in doing so. They proposed that remote forcing impacts atmospheric decadal variations in the Indian Ocean. To test this hypothesis, a recent study (Dong et al., 2016) used a coupled general circulation model and prescribed observed SST over the eastern tropical Pacific, predominantly influenced by the IPO on the decadal timescale This approach successfully reproduced observed decadal variations in the tropical Indian Ocean. Specifically, the IPO exerts a ~50% amplifying (or weakening) effect on the externally forced warming effect by ~50% during its warm (cold) phases. This indicates that the IPO influence via an atmospheric bridge plays an important role in the decadal SST variability of the tropical Indian Ocean. During a positive IPO phase, the Walker circulation weakens, resulting in anomalous descending motion over the tropical Indian Ocean. Consequently, there is reduced cloud cover and increased shortwave radiation reaching the ocean surface. Moreover, easterly wind stress anomalies over the equatorial Indian Ocean, associated with the weakened Walker circulation, trigger anomalous Ekman downwelling in off-equatorial regions. The combined effect of these processes leads to anomalous warming of the tropical Indian Ocean during a positive IPO. Importantly, these processes bear resemblance to the influence of the El Niño-Southern Oscillation (ENSO) on the Indian Ocean via the atmospheric bridge at interannual timescales (Alexander et al., 2002; Klein et al., 1999).

However, the above-mentioned decadal atmospheric bridge exhibits seasonal dependence. Han et al. (2017a) showed that the Walker circulations over the Pacific and Indian Oceans co-vary on the decadal timescale during boreal winter, primarily through fluctuations of atmospheric convection over the warm pool. In contrast, during boreal summer, the warm-pool convection only co-varies with the Pacific Walker circulation and not with the Indian Ocean Walker circulation. The latter is more strongly influenced by convective activity associated with the Indian summer monsoon, especially after the 1990s.

The role of the oceanic tunnel has also been under scrutiny. Dong and McPhaden (2016) found that an increase in heat transport by the ITF due to stronger trade winds during the recent hiatus period, associated with the negative IPO, led to enhanced warming in the Indian Ocean south of 10°S. A more comprehensive review of decadal sea level variability in the Indian Ocean due to the oceanic bridge from the Pacific can be found in Sprintall et al. (2024).

Both atmospheric bridge and oceanic tunnel from the IPO are suggested to influence interdecadal variations in the Ningaloo Niño, a climate mode characterized by positive SST anomalies off the west coast of Australia (Feng et al., 2013; Kataoka et al., 2014, 2017; Zhang et al., 2018a). Feng et al. (2015) and Li et al. (2017) showed that the occurrence of the Ningaloo Niño has increased since the late 1990s owing to the recent shift of the IPO to its negative phase. The enhanced ITF during the negative IPO is suggested to play a key role in this phenomenon. Furthermore, Tanuma and Tozuka (2020) revealed that the local positive feedback is intensified during the negative IPO phase, facilitated by the anomalously warm conditions to the northwest of Australia induced by the negative IPO phase through the atmospheric bridge (Li et al., 2019).

Decadal variability of sea level has also been observed over the Indian Ocean and shown to be enhanced during the 20th century along the coasts of global oceans (Little, 2023). Decadal variabilities of ENSO, Indian Ocean dipole (IOD), and monsoon largely account for the observed decadal sea level variability in different regions of the Indian Ocean (Han et al., 2018). Surface winds associated with these climate modes over the Indian Ocean are the main drivers for decadal sea level variability, while the Pacific influence via the ITF is strong mainly in the southeast basin.

3.2 Intrinsic Indian Ocean decadal variability

Ashok et al. (2004) proposed the existence of an intrinsic decadal climate mode known as the "decadal IOD," which is not correlated with decadal variability in the tropical Pacific and whose spatial pattern bears resemblance with the interannual IOD (Saji et al., 1999). Conversely, Tozuka et al. (2007) pointed out that decadal air-sea interactions associated with the "decadal IOD" may be a statistical artifact resulting from low-pass filtering. They suggested that it might emerge as an outcome of the decadal modulation of interannual IOD events or as a consequence of asymmetry in the occurrence of positive and negative IOD events. Additionally, they inferred that the decadal modulations in the interannual IOD are attributable to both intrinsic and remotely forced variability.

The Indian Ocean subtropical dipole is a dominant climate mode in the southern Indian Ocean (Behera & Yamagata, 2001; Reason, 2001). Positive (negative) SST anomalies are generated when the mixed layer becomes anomalously shallow (deep), with sensitivity to shortwave radiation being more (less) pronounced (Morioka et al., 2010). Yamagami and Tozuka (2015) conducted mixed-layer heat budget analyses, revealing that the thinner mixed layer in the recent decade amplifies this effect, making it possible for even weak atmospheric anomalies to trigger the Indian Ocean subtropical dipole. An examination of the Monin-Obukhov depth revealed an increasing trend in surface heat flux, attributed to decreased wind speeds (and increased specific humidity near the sea surface) linked to the poleward shift of the westerly jet in January (strengthening of the Mascarene High in February), resulting in a decreasing trend of the mixed layer depth. Regarding amplitude, Yan et al. (2013) demonstrated a weakening of the Indian Ocean subtropical dipole after 1979/80, although the exact mechanism behind this remains unknown. More recently, Zhang et al. (2022) highlighted a decadal eastward shift in the center of action for interannual climate modes, contributing to the recent weakening of the Indian Ocean subtropical dipole, while the amplitude of Ningaloo Niño has recently increased. Furthermore, based on an analysis of coupled model experiments, they proposed that internal climate variability plays a dominant role in these changes.

Li and Han (2015) conducted sensitivity experiments using an ocean general circulation model to investigate the mechanisms underlying decadal sea level variations in the Indian Ocean (Han et al., 2017b). This study revealed that the dominant mechanisms vary depending on the regions within the Indian Ocean. In most of the tropics, decadal modulations of wind stress anomalies associated with climate modes make a dominant contribution to sea level variations. However, off the Somali coast and the western Bay of Bengal, oceanic internal variability emerges as the primary driver. In the subtropical southern Indian Ocean, both surface heat flux and oceanic internal variability contribute, while stochastic winds become important in the southwest Indian Ocean, especially south of 30°S.

To assess the contribution of climate modes to decadal sea level variability in the Indian Ocean, Han et al. (2018) used a Bayesian dynamic linear model. Their study revealed that both decadal IOD and Indian monsoon surface wind variability play significant roles in regional decadal sea level variability, with some seasonal dependence. During boreal winter, both factors contribute to decadal variability to the northeast of Madagascar, while the Indian monsoon induces decadal variability off the Sumatra coast and along the Bay of Bengal coast. Conversely, in boreal summer, both decadal IOD and Indian monsoon influence decadal variability in the three aforementioned regions. For decadal sea level variability north of 5°S, the primary explanation may lie in the thermosteric sea level of the upper 700m (Srinivasu et al., 2017) induced by decadal variations in surface winds over the Indian Ocean (Srinivasu et al., 2017; Thompson et al., 2016). Thompson et al. (2016) emphasized the importance of the cross-equatorial cell and deep equatorial upwelling, both influencing the upper ocean thermal distribution. The former is forced by zonal wind stress curl anomalies at the equator, and its decadal variability is associated with changes in the strength and position of the Mascarene High associated with the Indian Ocean subtropical dipole. However, Srinivasu et al. (2017) noted qualitative differences among different ocean reanalyzes in their finding.

Owing to the limited availability of sea level observational data over an extended period and inconsistencies among different ocean reanalysis products, Nidheesh et al. (2019) turned to control runs of phases 3 and 5 of the coupled model intercomparison project (CMIP) (Meehl et al., 2007; Taylor et al., 2012) to investigate decadal variations in sea level over the Indian Ocean. The leading mode of variability, explaining about 38% of the total variance on the decadal timescale, is associated with the decadal modulation of the IOD, which shows only a weak correlation with ENSO. During its positive phase, negative sea level anomalies occur in the eastern tropical Indian Ocean, with minima found along the Indonesian coast and west coast of Australia. Simultaneously, positive sea level anomalies are found in the western tropical Indian Ocean with a maximum located along 10°S. Conversely, the mode of variability, explaining around 14% of the variance, is characterized by positive sea level anomalies east of Madagascar. The mechanism driving this mode is model dependent. In some models, downwelling Rossby waves induced by anticyclonic wind stress anomalies in the southeastern tropical Indian Ocean (Nidheesh et al., 2013) associated with the meridional shift of the Mascarene High contribute to this mode. In other models, positive sea level anomalies induced by alongshore wind stress anomalies off the west coast of Australia are radiated away from the coast as westward propagating Rossby waves and also play a role. Zhuang et al. (2013) highlighted the importance of wind stress curl anomalies in the region of 70°E–95°E for triggering the Rossby waves, using a baroclinic Rossby wave model. Additionally, Zhang et al. (2019) demonstrated that the meridional shift in the Mascarene High, which is uncorrelated with ENSO, contributes to the generation of sea level anomalies east of Madagascar, emphasizing the importance of intrinsic Indian Ocean variability.

4 Externally forced signals

4.1 Detection and attribution of Indian Ocean warming

Over recent decades, the Indian Ocean has undergone a rapid basin-wide warming trend, exceeding that of other tropical oceans (Han et al., 2014; Luo et al., 2012). Despite being detectable amidst background "climate noise" (Gopika et al., 2020), the underlying cause of this warming remains a topic of debate. The Indian Ocean warming in observations is likely the result of external forcing, particularly greenhouse gas forcing (Alory et al., 2007; Barnett et al., 2005; Dong et al., 2014; Dong & Zhou, 2014; Knutson, 2006; Pierce et al., 2006; Roxy et al., 2020). To evaluate the exact contributions of greenhouse gases and anthropogenic aerosols, including direct and indirect aerosol effects, to the Indian Ocean warming, the CMIP protocol encouraged simulations encompassing all forcing, greenhouse gas-only forcing, natural-only forcing simulations, and anthropogenic aerosol-only forcing.

Fig. 3 compares the time evolution from the basin-averaged warming in observations and the multimodel ensemble averages of different forcings. A significant warming trend during 1870–2005 is evident in the observations, particularly rapid in the latter half of the 20th century (Fig. 3a). This warming trend is well reproduced by all forcing runs (Fig. 3b). Greenhouse gas forcing exerts the dominant role on the Indian Ocean warming trend, with a persistent warming effect throughout the 20th century (Fig. 3c). In contrast, the total anthropogenic aerosols effect (including both direct and indirect effects) cools the basin-averaged SST and slows down the warming trend (Fig. 3d). The direct aerosol effect reduces shortwave radiation reaching the Earth's surface by absorbing or scattering the sunlight, while the indirect effect of aerosols affects surface radiative flux through its influence on clouds. Notably, the cooling was more pronounced after the 1950s due to the increased emission of anthropogenic aerosols with postwar economic growth (Xie et al., 2013). However, when considering only the direct effect of aerosols, the cooling effect is weakened (Fig. 3e), although the SST anomaly signals may still contain natural decadal variability due to the limited number of ensemble members. The observed trend of 0.40 K century^{-1} closely aligns with all forcing runs of 0.41 ± 0.16 K century^{-1} (standard deviation of intermodel variations). Under greenhouse gas-only forcing, the warming trend is stronger than in observations and the all forcing run (0.66 ± 0.14 K century^{-1}). For each of the 17 CMIP5 models, the warming trend under greenhouse gas forcing is positive

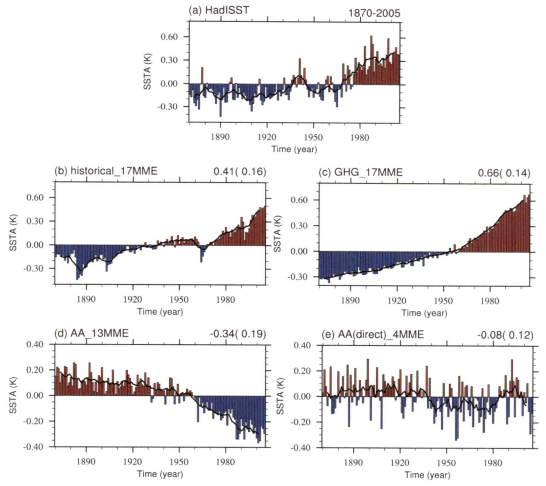

FIG. 3 Time series of Indian Ocean (40°S–15°N, 40°–100°E) annual mean (bar) and 8-yr running average (line) sea surface temperature anomaly (SSTA) from (a) observations, (b) all forcing runs of 17 models from the multimodel ensemble (MME), (c) Greenhouse gas (GHG)-only forcing runs of 17 models of the MME, (d) Anthropogenic aerosols (AA)-only forcing of 13 models from the MME that include both direct and indirect effects, and (e) AA-only forcing of four models from the MME that only include direct effects. Only one member is used for each model. SSTA is relative to the period of 1870–2005 mean. Units: K. The numbers on the top right of (b)–(e) denote the trend values and standard deviation of intermodel variations (in K century^{-1}), respectively. *(From Dong and Zhou (2014). © American Meteorological Society. Used with permission.)*

with the multimodel ensemble mean accounting for about 163.6%. In contrast, the emission of anthropogenic aerosols has mitigated the warming trend, with negative contributions to the warming, except for CCSM4, which exhibits weak warming. Based on the result of the multimodel ensemble mean, anthropogenic aerosols reduce the warming trend of all forcing runs by 72.7%. If only the direct effect of aerosols is considered, there is a weak cooling trend of -0.08 ± 0.12 K century^{-1} (Fig. 3e). However, when including both direct and indirect effects, the cooling trend is enhanced to -0.34 ± 0.19 K century^{-1} (Fig. 3d), offsetting roughly half of greenhouse gas warming. Consequently, the warming rate of the Indian Ocean, though faster than the global mean, reflects the competing influences of greenhouse gases and anthropogenic aerosols on the Indian Ocean SST changes, akin to their effects on global mean SST (Han et al., 2014).

Apart from the centennial and multidecadal warming trends, the Indian Ocean SST also exhibits decadal-to-interdecadal variability (Fig. 3) arising from both internal climate variability (Section 3) and external forcing. As discussed in Section 3, strong external forcing weakens the dynamical connection between the Indian and Pacific Oceans, resulting in the inability of the negative phase of the IPO after 2000 to induce a cold SST anomaly over the Indian Ocean in observations (Dong & McPhaden, 2017a). This enhanced external forcing after the 1980s can be primarily attributed to greenhouse gases, with anthropogenic aerosol forcing and natural forcing exhibiting relatively weak impacts (Dong & McPhaden, 2017b).

4.2 Nonuniform warming patterns

The Indian Ocean basin-averaged SST displays a robust warming trend, yet this warming is spatially heterogeneous, exhibiting an uneven warming pattern. Understanding this pattern is crucial for comprehending the global climate response to increasing greenhouse gases. For instance, this SST warming pattern strongly constrains the regional distribution of precipitation changes over the tropical ocean, following a warmer-get-wetter mechanism (Grose et al., 2014; Kent et al., 2015; Kosaka & Xie, 2016; Xie et al., 2010). Additionally, the pattern of tropical SST change can influence large-scale circulation in the tropical atmosphere, exerting a notable impact on extratropical climates (Lee et al., 2022; Trenberth et al., 1998) such as in the North Atlantic and Arctic regions (Hoerling et al., 2004; Hu & Fedorov, 2020; Xu et al., 2022). Consequently, detecting the spatial warming pattern in the Indian Ocean and exploring the associated mechanisms are of utmost importance.

Attribution of spatial patterns. Fig. 4 illustrates the pattern of SST trend in the Indian Ocean during 1870–2005 based on observations and multimodel ensemble simulations with various external forcings. Observations display basin-wide warming trends over the past century, with the most substantial warming observed in the western tropical Indian Ocean and relatively weaker warming in the tropical south Indian Ocean, resembling the positive IOD-like SST pattern (Cai et al., 2013; Du & Xie, 2008; Roxy et al., 2014; Xie et al., 2010; Zheng et al., 2010; see Fig. 4a). Externally forced enhanced warming in the Arabian Sea and western tropical Indian Ocean relative to the rest of the basin simulated by CMIP models is already detectable, which aligns with observations. The multimodel ensemble mean of all forcing runs captures a significant portion of the observed east-west dipole warming pattern (Fig. 4a and b), predominantly due to greenhouse gas forcing (Fig. 4c). This pattern weakens the Walker cell, leading to easterly wind anomalies in the equatorial Indian Ocean and consequently the positive IOD-like warming pattern (Cai et al., 2013). However, observational analyses present conflicting views on the Walker cell changes in recent decades (Han et al., 2014). In contrast, direct anthropogenic aerosol forcing results in a weak but opposite SST change pattern, counteracting the effect of greenhouse gas warming (Fig. 4e). The total (direct and indirect) aerosol forcing yields stronger cooling in the north Indian Ocean and comparatively less cooling in the subtropical south Indian Ocean (Fig. 4c and d). Moreover, the warming (cooling) rate in response to greenhouse gas (anthropogenic aerosol) forcing is more pronounced in the northern basin than in the southern basin (Fig. 4c and d). The stronger cooling effect from anthropogenic aerosol in the northern basin may be attributed to the presence of anthropogenic aerosols originating from Asia.

Mechanisms. The mechanism underlying the Indian Ocean response to the greenhouse gas forcing remains a subject of investigation in climate models (Du & Xie, 2008). Some studies suggested that basin-wide SST warming is directly triggered by atmospheric processes, initiated by the increase in downward longwave radiation due to greenhouse gases, and then amplified by the water vapor feedback and atmospheric adjustment in climate models (Du & Xie, 2008). On the other hand, other studies suggest a link between this SST change and changes in ocean heat transport. For example, Indian Ocean basin-wide warming is associated with ocean wave-induced thermocline changes (Li et al., 2003) or a decrease in upwelling related to a slowdown of the wind-driven Ekman pumping (Alory & Meyers, 2009). Few modeling studies have indicated contributions from both atmospheric and oceanic processes. Ocean temperature advection is regarded as the dominant process for the increase of the northern Indian Ocean heat content, whereas control by surface heat fluxes prevails in other areas of the Indian Ocean basin (Barnett et al., 2005).

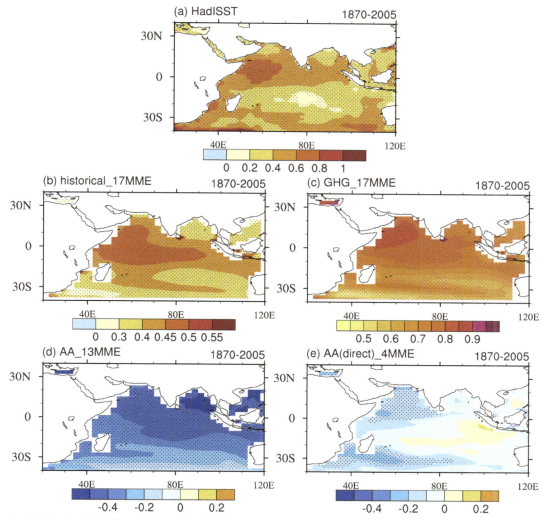

FIG. 4 Trends of SST in the Indian Ocean during 1870–2005 from (a) observations, (b) all forcing runs of a 17-model multimodel ensemble (MME), (c) Greenhouse gas (GHG)-only forcing runs of a 17-model MME, (d) Anthropogenic aerosol (AA)-only forcing of a 13-model MME, and (e) direct effects of AA-only forcing of a 4-model MME. The dotted areas are statistically significant at the 1% level by Student's t-test. Note that the color scale intervals are not the same for each figure. Units: K $(100\,year)^{-1}$. *(From Dong and Zhou (2014). © American Meteorological Society. Used with permission.)*

Water cycle. As for SST, the observed sea surface salinity (SSS) trend patterns from 1950 to 2019 are reasonably captured by the multimodel ensemble mean of CMIP6 simulations, albeit an underestimated amplitude (Fig. 5). Both the observed and simulated SSS changes generally enhance the mean SSS patterns, resulting in fresher conditions in the Bay of Bengal and equatorial basin, where mean salinity is low and precipitation (P) exceeds evaporation (E) ($E-P<0$), and saltier conditions in the subtropical south Indian Ocean, where mean salinity is high and $E-P>0$. These findings suggest an intensified water cycle driven by greenhouse gas forcing (Durack, 2015; Eyring et al., 2021), with an 8% pattern amplification for E-P change, which is close to the 7% increase in tropospheric relative humidity, corresponding to each degree of global surface warming (Douville et al., 2022; Durack et al., 2012). Projections from climate models indicate that this pattern of amplification of SSS is expected to increase by the end of the 21st century.

Sea level trends. Distinct spatial patterns of sea level trend since the 1950s have been identified after removing the global mean sea level rise, showing regional sea level fall in the southwest tropical Indian Ocean and rise in the eastern Indian Ocean (Fig. 6a–d; Dunne et al., 2012; Han, 2010; Han et al., 2018; Schwarzkopf & Böning, 2011; Timmermann et al., 2010). While thermosteric sea level dominates the upper 2000 m total steric sea level trends for the 1950–2008 period in most regions of the Indian Ocean, halosteric changes associated with the water cycle can be the leading

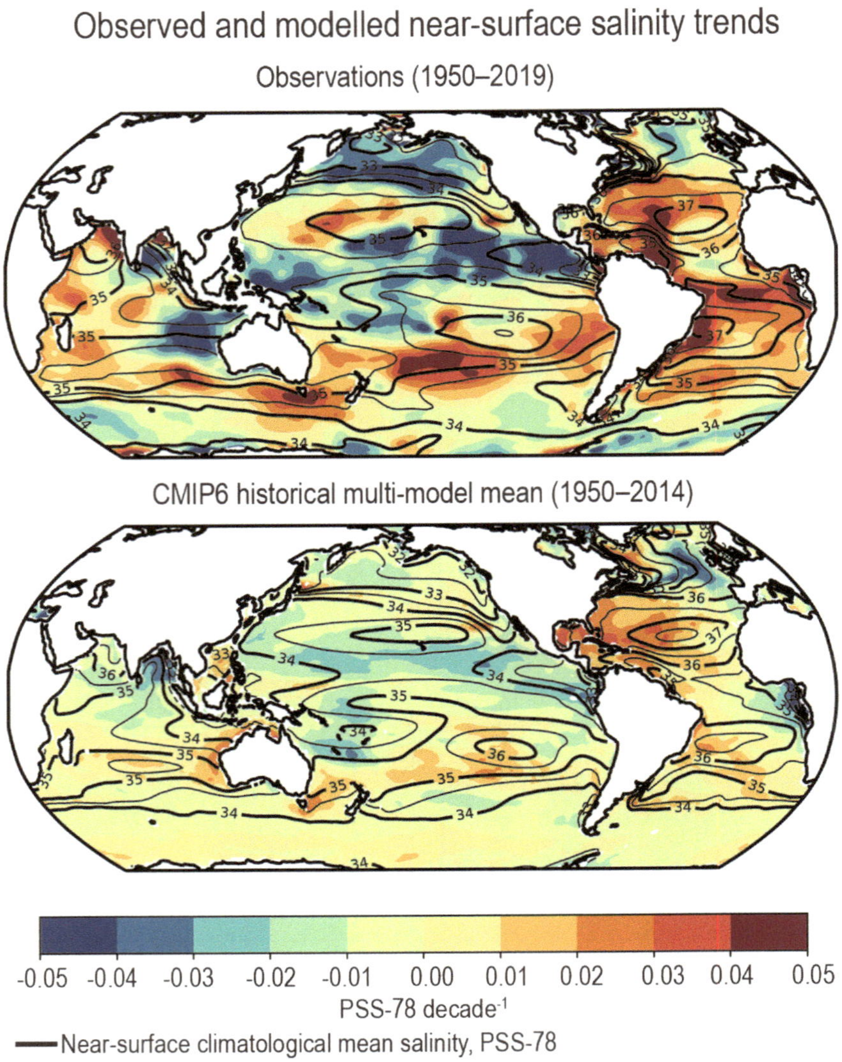

FIG. 5 Multidecadal salinity trends for the near-surface ocean (in Practical Salinity Scale 1978 [PSS-78] per decade). *(Top)* Observed trend. *(Bottom)* Simulated trend from the CMIP6 historical experiment multimodel mean (1950–2014). Black contours show the climatological mean salinity in increments of 0.5 PSS-78 (thick lines 1 PSS-78). *(From Eyring et al. (2021).)*

contributor to total steric changes in the tropical southeast Indian Ocean and subtropical South Indian Ocean (Durack et al., 2012). Indeed, the importance of salinity in causing positive sea level trends in the tropical southeast Indian Ocean for the 2005–2013 decade has also been shown, due to enhanced precipitation in the maritime continent region and strengthened ITF (Llovel & Lee, 2015). In addition, halosteric sea level shows large-scale decreasing trends due to increased salinity in the North Indian Ocean (10°S–20°N) and increasing trends due to decreased salinity in the South Indian Ocean (30°S–10°S) for the 2005–2015 period. Trend patterns of dynamical sea level (global mean removed) from large ensemble experiments using two global climate models bear some resemblance with the observed patterns, particularly the sea level fall in the southwest basin (Fig. 6e–h). This relative decline was primarily attributed to internal climate variability with external forcing contributing $\sim 19\% \pm 2.4\%$ (Han et al., 2018). However, significant qualitative and quantitative differences exist between the two climate model simulations in most regions of the Indian Ocean, introducing considerable uncertainty in the climate model's ability to simulate the forced signals of regional sea level changes.

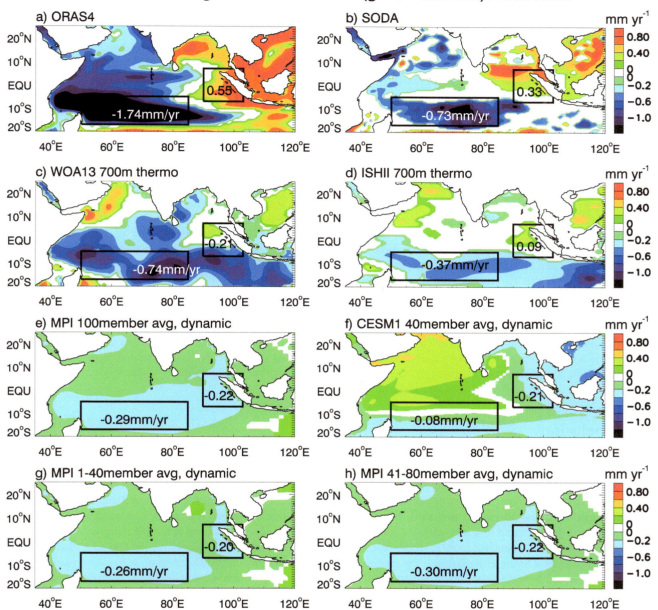

FIG. 6 The linear trends of regional sea level over the tropical Indian Ocean from 1958 to 2005. Monthly data are used, with the 1958–2005 monthly climatology and global mean sea level rise time series removed from each dataset before the trend calculation. (a) The ECMWF Ocean Reanalysis System 4 (ORAS4) reanalysis; (b) the Simple Ocean Data Assimilation (SODA) reanalysis; (c) the world ocean atlas 2013 (WOA13) upper 700 m thermosteric sea level; (d) Ishii upper 700 m thermosteric sea level; (e) the Max-Planck-Institute of Meteorology (MPI) 100-member ensemble mean dynamic sea level; (f) the National Center for Atmospheric Research (NCAR) community earth system model version 1 (CESM1) 40-member ensemble mean dynamic sea level; (g) MPI ensemble mean for members 1–40; (h) MPI ensemble mean for members 41–80. The two boxes show the maximum sea level fall and rise areas in both the ORAS4 and SODA dynamic sea level, with Reg. 1 located in the thermocline ridge of the southwest tropical basin (50° E–85° E, 17° S–5° S) and Reg. 2, eastern equatorial Indian Ocean (90° E–103° E, 7° S–7° N). The number in each box shows the trend value of SLA averaged in the box. Trend values exceeding 95% significance are shown in color contours, and those below 95% are shown in white. *(From Han et al. (2018).)*

5 Predictability

Earlier CMIP (CMIP3 and before) have demonstrated that near-term climate predictions, ranging between 2 and 30 years, were influenced by a combination of external forcing and internal variability (Doblas-Reyes et al., 2013). However, to bridge the gap between subseasonal-to-seasonal predictions and climate projections, the focus has shifted to decadal predictions, which have become a prominent area of research since CMIP5 (Doblas-Reyes et al., 2013; Kirtman et al., 2013). Coordinated experiments, as conducted in CMIP5, have explored the potential of initializing coupled models with observational data. This initialization allows for the prediction of internal variability in addition to changes associated with external forcing, offering valuable insights into decadal climate trends (Doblas-Reyes et al., 2013; Kirtman et al., 2013; Smith et al., 2007).

The Indian Ocean stands out as the region with the highest decadal prediction skill for SST over the global ocean in 2–9 years lead forecasts. Guemas et al. (2013) demonstrated that anomaly correlation coefficients of SST anomalies averaged over the Indian Ocean (40°S–30°N, 20°–120°E) remain consistently high at 0.7 for 2–9 years lead forecasts, surpassing other basins. This enhanced prediction skill can be attributed to the predominance of varying radiative forcings linked to increasing anthropogenic greenhouse gases and volcanic aerosols, which exert a stronger influence than relatively weak internal variability. Interestingly, initializing coupled models with observed climate states did not lead to significant skill improvements in the Indian Ocean due to the dominance of external forces. This is in stark contrast with other basins, where natural decadal climate variability plays a more significant role. However, in the western Indian Ocean, multimodel decadal SST predictions from the CMIP5 models indicate some improvements with initialization, even though SLP forecasts deteriorate (Smith et al., 2019). This contrasting impact of initialization on SST and SLP suggests potential issues in the representation of ocean-atmosphere interaction over the western Indian Ocean or the involvement of different mechanisms governing their predictability. Morioka et al. (2018) conducted a study on decadal climate variability in the southern Indian Ocean. They found that observed decadal SST variations were particularly large east of Madagascar, off the west coast of Australia, and in the Agulhas Return Current region. Furthermore, their coupled model demonstrated relatively high skill in the Agulhas Return Current region for a lead time of up to 9 years. They suggested that this enhanced predictability may be attributed to the model's successful representation of eastward propagation of anomalously warm SST coupled with positive SLP anomalies from the Atlantic and local air-sea interactions.

6 Conclusions

In Section 3, our review has highlighted recent advancements in understanding decadal climate variability in the Indian Ocean, arising from both remote forcing from other basins and intrinsic processes. Remote forcing, particularly from the IPO, exerts its influence through both the atmospheric bridge and oceanic tunnel, with the Walker circulation and the ITF playing key roles. However, the relative contributions of these effects still require further quantification, and improved coupled general circulation models are needed to realistically simulate interbasin interactions. Regarding intrinsic variability, research has explored how interannual climate modes like IOD and Indian Ocean subtropical dipole vary on decadal timescales. Yet, there is a need for more in-depth investigations to fully uncover the underlying mechanisms. One intriguing question that warrants future exploration is the potential role of reemergence in decadal climate variability, particularly in the subtropical southern Indian Ocean. Reemergence is a process by which wintertime SST anomalies that extend throughout a deep mixed layer are preserved beneath the thin summer mixed layer after the mixed layer shoals, but when the mixed layer becomes deeper again in the subsequent autumn and winter, subsurface water with temperature anomalies is re-entrained into the mixed layer and SST anomalies with the same sign recur. While reemergence has been implicated in the PDO (Newman, 2016), its significance in the Indian Ocean remains relatively unexplored.

In summary of Section 4, our understanding of externally forced signals in the Indian Ocean heavily relies on multimodel ensemble analyses. We have quantitatively compared the relative contributions of greenhouse gases and anthropogenic aerosols, revealing greenhouse gases as the dominant factor driving rapid warming and anthropogenic aerosols mitigating the warming rate. The mechanisms governing the basin-wide warming pattern have been attributed to atmospheric processes involving radiative and turbulent fluxes, while the positive IOD-like warming pattern is closely related to the surface wind anomalies. Nonetheless, uncertainties persist in climate model simulations, warranting further investigation into the relative impact of model biases and internal variability in reproducing and projecting the multidecadal warming, sea level, and SSS spatial patterns. Quantifying and understanding these uncertainties are vital for providing informed and reliable projections of future changes.

In Section 5, we delved into decadal predictions in the Indian Ocean, where SST prediction skill is notably high, primarily driven by varying radiative forcings linked to anthropogenic greenhouse gases and volcanic aerosols. However, our ability to predict internal variability on decadal time scales remains limited and requires further advancements.

As stated in the introduction, a crucial motivation for understanding decadal variations in the Indian Ocean is its potential role in the global warming hiatus. While the extra heat absorbed by the tropical Pacific is largely transported to the Indian Ocean via the ITF (Lee et al., 2015; Nieves et al., 2015; Zhang et al., 2018b), the ultimate fate of this heat is still uncertain. Unraveling whether the extra heat is released back into the atmosphere within the Indian Ocean or transported southward to the Southern Ocean (Vialard, 2015) remains an outstanding question. Both observational data analyses and model studies are necessary to address this important issue in the coming decade.

Although Section 2 has reviewed the continued efforts in observing the Indian Ocean (Beal et al., 2020; Hermes et al., 2019), historical observational datasets still exhibit large uncertainties, particularly in earlier periods. Hence, establishing a sustained observational system in the Indian Ocean is essential to extend data coverage with improved quality and enhance our understanding of its decadal climate variability. The IndOOS initiative, as detailed in McPhaden et al. (2024), plays a critical role in this endeavor, and incorporating subsurface observations can further enhance our understanding of decadal variations in the Indian Ocean, including the fate of extra heat stored in the Indian Ocean during the recent hiatus in global warming.

7 Educational resources

- The Decadal Climate Prediction Project of the World Climate Research Programme (WCRP) (https://www.wcrp-climate.org/dcp-overview).
- WCRP Grand Challenge on Near-Term Climate Prediction (https://www.wcrp-climate.org/gc-near-term-climate-prediction).

Acknowledgments

We sincerely thank two reviewers, Mike McPhaden and Tangdong Qu, for their constructive comments and suggestions that helped us greatly to improve our book chapter. WH is supported by NSF AGS 1935279 and NSF OCE 2242193.

Author contributions

For the first draft, TT wrote Sections 1, 3, 5, and 6, LZ wrote Section 2, and LD and WH wrote Section 4. ML contributed to the writing of all sections. All authors read and approved the final manuscript.

References

Alexander, M. A., Bladé, I., Newman, M., Lanzante, J. R., Lau, N.-C., & Scott, J. D. (2002). The atmospheric bridge: The influence of ENSO teleconnections on air–sea interaction over the global oceans. *Journal of Climate, 15*, 2205–2231.

Alory, G., & Meyers, G. (2009). Warming of the upper equatorial Indian Ocean and changes in the heat budget (1960–99). *Journal of Climate, 22*, 93–113.

Alory, G., Wijffels, S., & Meyers, G. (2007). Observed temperature trends in the Indian Ocean over 1960–1999 and associated mechanisms. *Geophysical Research Letters, 34*, L02606.

Al-Yamani, F. Y., Burt, J. A., Goes, J. I., Jones, B., Nagappa, R., Murty, V. S. N., ... Yamamoto, T. (2024). Chapter 16: Physical and biogeochemical characteristics of the Indian Ocean marginal seas. In C. C. Ummenhofer, & R. R. Hood (Eds.), *The Indian Ocean and its role in the global climate system* (pp. 365–391). Amsterdam: Elsevier. https://doi.org/10.1016/B978-0-12-822698-8.00008-1.

Ashok, K., Chan, W.-L., Motoi, T., & Yamagata, T. (2004). Decadal variability of the Indian Ocean dipole. *Geophysical Research Letters, 31*, L24207.

Barnett, T. P. (1984). Long-term trends in surface temperature over the oceans. *Monthly Weather Review, 112*, 303–312.

Barnett, T. P., Pierce, D. W., AchutaRao, K. M., Gleckler, P. J., Santer, B. D., Gregory, J. M., & Washington, W. M. (2005). Penetration of human-induced warming into the world's oceans. *Science, 309*, 284–287.

Beal, L., et al. (2020). A roadmap to IndOOS-2: Better observations of the rapidly-warming Indian Ocean. *Bulletin of the American Meteorological Society, 101*, E1891–E1913.

Behera, S. K., & Yamagata, T. (2001). Subtropical SST dipole events in the southern Indian Ocean. *Geophysical Research Letters, 28*, 327–330.

Cai, W., Zheng, X., Weller, E., Collins, M., Cowan, T., Lengaigne, M., Yu, W., & Yamagata, T. (2013). Projected response of the Indian Ocean dipole to greenhouse warming. *Nature Geoscience, 6*, 999–1007.

Church, J. A., White, N. J., Coleman, R., Layback, K., & Mitrovica, J. X. (2004). Estimates of the regional distribution of sea level rise over the 1950–2000 period. *Journal of Climate, 17*, 2609–2625.

Cole, J. E., Dunbar, R. B., McClanahan, T. R., & Muthiga, N. A. (2000). Tropical Pacific forcing of decadal SST variability in the western Indian Ocean over the past two centuries. *Science, 287*, 617–619.

Copsey, D., Sutton, R., & Knight, J. R. (2006). Recent trends in sea level pressure in the Indian Ocean region. *Geophysical Research Letters, 33*, L19712.

Doblas-Reyes, F. J., Andreu-Burillo, I., Chikamoto, Y., García-Sarrano, J., Guemas, V., Kimoto, M., Mochizuki, T., Rodrigues, L. R. L., & van Oldenborgh, G. J. (2013). Initialized near-term regional climate change prediction. *Nature Communications, 4*, 1715.

Dong, L., & McPhaden, M. J. (2016). Interhemispheric SST gradient trends in the Indian Ocean prior to and during the recent global warming hiatus. *Journal of Climate, 29*, 9077–9095.

Dong, L., & McPhaden, M. J. (2017a). Why has the relationship between Indian and Pacific Ocean decadal variability changed in recent decades? *Journal of Climate, 30*, 1971–1983.

Dong, L., & McPhaden, M. J. (2017b). The effects of external forcing and internal variability of the formation of interhemispheric sea surface temperature gradient trends in the Indian Ocean. *Journal of Climate, 30*, 9077–9095.

Dong, L., & Zhou, T. J. (2014). The Indian Ocean Sea surface temperature warming simulated by CMIP5 models during the twentieth century: Competing forcing roles of GHGs and anthropogenic aerosols. *Journal of Climate, 27*, 3348–3362.

Dong, L., Zhou, T. J., Dai, A., Song, F. F., Wu, B., & Chen, X. L. (2016). The footprint of the inter-decadal Pacific oscillation in Indian Ocean Sea surface temperatures. *Scientific Reports, 6*, 21251.

Dong, L., Zhou, T. J., & Wu, B. (2014). Indian Ocean warming during 1958–2004 simulated by a climate system model and its mechanism. *Climate Dynamics, 42*, 203–217.

Douville, H., Qasmi, S., Ribes, A., & Bock, O. (2022). Global warming at near-constant tropospheric relative humidity is supported by observations. *Communications Earth & Environment, 3*, 237.

Du, Y., & Xie, S.-P. (2008). Role of atmospheric adjustments in the tropical Indian Ocean warming during the 20th century in climate models. *Geophysical Research Letters, 35*, L08712.

Dunne, R. P., Barbosa, S. M., & Woodworth, P. L. (2012). Contemporary sea level in the Chagos Archipelago, Central Indian Ocean. *Global and Planetary Change, 82–83*, 25–37.

Durack, P. J. (2015). Ocean salinity and the global water cycle. *Oceanography, 28*, 20–31.

Durack, P. J., Wijffels, S. E., & Matear, R. J. (2012). Ocean salinities reveal strong global water cycle intensification during 1950–2000. *Science, 336*, 455–458.

England, M. H., et al. (2014). Recent intensification of wind-driven circulation in the Pacific and the ongoing warming hiatus. *Nature Climate Change, 4*, 222–227.

Eyring, V., Gillett, N. P., Achuta Rao, K. M., Barimalala, R., Barreiro Parrillo, M., Bellouin, N., Cassou, C., Durack, P. J., Kosaka, Y., McGregor, S., Min, S., Morgenstern, O., & Sun, Y. (2021). Human influence on the climate system. In V. Masson-Delmotte, P. Zhai, A. Pirani, S. L. Connors, C. Péan, S. Berger, N. Caud, Y. Chen, L. Goldfarb, M. I. Gomis, M. Huang, K. Leitzell, E. Lonnoy, J. B. R. Matthews, T. K. Maycock, T. Waterfield, O. Yelekçi, R. Yu, & B. Zhou (Eds.), Climate change 2021: The physical science basis. Contribution of working group I to the sixth assessment report of the intergovernmental panel on climate change. Cambridge, United Kingdom and New York, NY: Cambridge University Press, pp. 423–552. https://doi.org/10.1017/9781009157896.005. Chapter 3.

Feng, M., Hendon, H. H., Xie, S.-P., Marshall, A. G., Schiller, A., Kosaka, Y., Caputi, N., & Pearce, A. (2015). Decadal increase in Ningaloo Niño since the late 1990s. *Geophysical Research Letters, 42*, 104–112.

Feng, M., McPhaden, M. J., Xie, S.-. P., & Hafner, J. (2013). La Niña forces unprecedented Leeuwin current warming in 2011. *Scientific Reports, 3*, 1277.

Gopika, S., Izumo, T., Vialard, J., Lengaigne, M., Suresh, I., & Kumar, M. R. R. (2020). Aliasing of the Indian Ocean externally-forced warming spatial pattern by internal climate variability. *Climate Dynamics, 54*, 1093–1111.

Grose, M. R., Bhend, J., Narsey, S., Gupta, A. S., & Brown, J. R. (2014). Can we constrain CMIP5 rainfall projections in the tropical Pacific based on surface warming patterns? *Journal of Climate, 27*, 9123–9138.

Guemas, V., Corti, S., García-Serrano, J., Doblas-Reyes, F. J., Balmaseda, M., & Magnusson, L. (2013). The Indian Ocean: The region of highest skill worldwide in decadal climate prediction. *Journal of Climate, 26*, 726–739.

Hamlington, B. D., Strassburg, M. W., Leben, R. R., Han, W., Nerem, R. S., & Kim, K.-Y. (2014). Uncovering an anthropogenic sea-level rise signal in the Pacific Ocean. *Nature Climate Change, 4*, 782–785.

Han, W., Meehl, G. A., Hu, A., Zheng, J., Kenigson, J., Vialard, J., … Yanto. (2017a). Decadal variability of Indian and Pacific Walker cells: Do they co-vary on decadal timescales? *Journal of Climate, 30*, 8447–8468.

Han, W., Meehl, G., Stammer, D., Hu, A., Hamlington, B., Kenigson, J., … Thompson, P. (2017b). Spatial patterns of sea level variability associated with natural internal climate modes. *Surveys in Geophysics, 38*, 217–250.

Han, W., Stammer, D., Meehl, G. A., Hu, A., Sienz, F., & Zhang, L. (2018). Multi-decadal trend and decadal variability of the regional sea level over the Indian Ocean since the 1960s: Roles of climate modes and external forcing. *Climate, 6*, 51.

Han, W., Vialard, J., McPhaden, M. J., Lee, T., Masumoto, Y., Feng, M., & de Ruijter, W. P. M. (2014). Indian Ocean decadal variability: A review. *Bulletin of the American Meteorological Society, 95*, 1679–1703.

Han, W., et al. (2010). Patterns of Indian Ocean sea-level change in a warming climate. *Nature Geoscience, 3*, 546–550.

Hermes, J. C., et al. (2019). A sustained ocean observing system in the Indian Ocean for climate related scientific knowledge and societal needs. *Frontiers in Marine Science, 6*, 355.

Hoerling, M., Hurrell, J. W., Xu, T., Bates, G., & Phillips, A. (2004). Twentieth century North Atlantic climate change. Part II: Understanding the effect of Indian Ocean warming. *Climate Dynamics, 23*, 391–405.

Hood, R. R., Coles, V. J., Huggett, J. A., Landry, M. R., Levy, M., Moffett, J. W., & Rixen, T. (2024a). Chapter 13: Nutrient, phytoplankton, and zooplankton variability in the Indian Ocean. In C. C. Ummenhofer, & R. R. Hood (Eds.), *The Indian Ocean and its role in the global climate system* (pp. 293–327). Amsterdam: Elsevier. https://doi.org/10.1016/B978-0-12-822698-8.00020-2.

Hood, R. R., Rixen, T., Levy, M., Hansell, D. A., Coles, V. J., & Lachkar, Z. (2024b). Chapter 12: Oxygen, carbon, and pH variability in the Indian Ocean. In C. C. Ummenhofer, & R. R. Hood (Eds.), *The Indian Ocean and its role in the global climate system* (pp. 265–291). Amsterdam: Elsevier. https://doi.org/10.1016/B978-0-12-822698-8.00017-2.

Hu, S., & Fedorov, A. (2020). Indian Ocean warming as a driver of the North Atlantic warming hole. *Nature Communications, 11*, 4785.

Kataoka, T., Tozuka, T., Behera, S., & Yamagata, T. (2014). On the Ningaloo Niño/Niña. *Climate Dynamics, 43*, 1463–1482.

Kataoka, T., Tozuka, T., & Yamagata, T. (2017). Generation and decay mechanisms of Ningaloo Niño/Niña. *Journal of Geophysical Research, Oceans, 122*, 8913–8932.

Kent, C., Chadwick, R., & Rowell, D. P. (2015). Understanding uncertainties in future projections of seasonal tropical precipitation. *Journal of Climate, 28*, 4390–4413.

Kirtman, B., et al. (2013). Near-term climate change: Projections and predictability. In *Climate change 2013 the Physical Science Basis: Working group I contribution to the fifth assessment report of the Inter- Governmental Panel on Climate Change*.

Klein, S. A., Soden, B. J., & Lau, N. C. (1999). Remote sea surface temperature variations during ENSO: Evidence for a tropical atmospheric bridge. *Journal of Climate, 12*, 917–932.

Knutson, T. R., et al. (2006). Assessment of twentieth-century regional surface temperature trends using the GFDL CM2 coupled models. *Journal of Climate, 19*, 1624–1651.

Kosaka, Y., & Xie, S.-P. (2013). Recent global-warming hiatus tied to equatorial Pacific surface cooling. *Nature, 501*, 403–407.

Kosaka, Y., & Xie, S. P. (2016). The tropical Pacific as a key pacemaker of the variable rates of global warming. *Nature Geoscience, 9*, 669–673.

Kumar, P., Hamlington, B., Cheon, S.-H., Han, W., & Thompson, P. (2020). 20th century multivariate Indian Ocean regional sea level reconstruction. *Journal of Geophysical Research, Oceans, 125*. e2020JC016270.

Lee, S., L'Heureux, M., Wittenberg, A. T., Seager, R., O'Gorman, P. A., & Johnson, N. C. (2022). On the future zonal contrasts of equatorial Pacific climate: Perspectives from observations, simulations, and theories. *npj Climate and Atmospheric Science, 5*, 82.

Lee, S. K., Park, W., Baringer, M., Gordon, A. L., Huber, B., & Liu, Y. (2015). Pacific origin of the abrupt increase in Indian Ocean heat content during the warming hiatus. *Nature Geoscience, 8*, 445–449.

Li, Y., & Han, W. (2015). Decadal Sea level variations in the Indian Ocean investigated with HYCOM: Roles of climate modes, ocean internal variability, and stochastic wind forcing. *Journal of Climate, 28*, 9143–9165.

Li, Y., Han, W., & Zhang, L. (2017). Enhanced decadal warming of the Southeast Indian Ocean during the recent global surface warming slowdown. *Geophysical Research Letters, 44*, 9876–9884.

Li, Y., Han, W., Zhang, L., & Wang, F. (2019). Decadal SST variability in the Southeast Indian Ocean and its impact on regional climate. *Journal of Climate, 32*, 6299–6318.

Li, T., Wang, B., Chang, C. P., & Zhang, Y. (2003). A theory for the Indian Ocean dipole–zonal mode. *Journal of the Atmospheric Sciences, 60*, 2119–2135.

Little, C. (2023). Coastal Sea level observations record the twentieth-century enhancement of decadal climate variability. *Journal of Climate, 36*, 243–260.

Liu, Z. (2012). Dynamics of interdecadal climate variability: A historical perspective. *Journal of Climate, 25*, 1963–1995.

Llovel, W., & Lee, T. (2015). Importance and origin of halosteric contribution to sea level change in the southeast Indian Ocean during 2005–2013. *Geophysical Research Letters, 42*, 1148–1157.

Luo, J.-J., Sasakia, W., & Masumoto, Y. (2012). Indian Ocean warming modulates Pacific climate change. *Proceedings of the National Academy of Sciences of the United States of America, 109*, 18701–18706.

McPhaden, M. J., Beal, L. M., Bhaskar, T.V.S. U., Lee, T., Nagura, M., Strutton, P. G., & Yu, L. (2024). Chapter 17: The Indian Ocean Observing System (IndOOS). In C. C. Ummenhofer, & R. R. Hood (Eds.), *The Indian Ocean and its role in the global climate system* (pp. 393–419). Amsterdam: Elsevier. https://doi.org/10.1016/B978-0-12-822698-8.00002-0.

Meehl, G. A., Covey, C., Delworth, T., Latif, M., McAvaney, B., Mitchell, J. F. B., Stouffer, R. J., & Taylor, K. E. (2007). The WCRP CMIP3 multi-model dataset: A new era in climate change research. *Bulletin of the American Meteorological Society, 88*, 1383–1394.

Mohtadi, M., Abram, N. J., Clemens, S. C., Pfeiffer, M., Russell, J. M., Steinke, S., & Zinke, J. (2024). Chapter 19: Paleoclimate evidence of Indian Ocean variability across a range of timescales. In C. C. Ummenhofer, & R. R. Hood (Eds.), *The Indian Ocean and its role in the global climate system* (pp. 445–467). Amsterdam: Elsevier. https://doi.org/10.1016/B978-0-12-822698-8.00007-X.

Morioka, Y., Doi, T., & Behera, S. K. (2018). Decadal climate predictability in the southern Indian Ocean captured by SINTEX-F using a simple SST-nudging scheme. *Scientific Reports, 8*, 1029.

Morioka, Y., Tozuka, T., & Yamagata, T. (2010). Climate variability in the southern Indian Ocean as revealed by self-organizing maps. *Climate Dynamics, 35*, 1059–1072.

Newman, M., et al. (2016). The Pacific decadal oscillation, revisited. *Journal of Climate, 29*, 4399–4427.

Nidheesh, A. G., Lengaigne, M., Vialard, J., Izumo, T., Unnikrishnan, A. S., & Krishnan, R. (2019). Natural decadal sea-level variability in the Indian Ocean: Lessons from CMIP models. *Climate Dynamics, 53*, 5653–5673.

Nidheesh, A. G., Lengaigne, M., Vialard, J., Izumo, T., Unnikrishnan, A. S., Meyssignac, B., Hamlington, B., & de Boyer Montegut, C. (2017). Robustness of observation-based decadal sea level variability in the Indo-Pacific Ocean. *Geophysical Research Letters, 44*, 7391–7400.

Nidheesh, A. G., Lengaigne, M., Vialard, J., Unnikrishnan, A. S., & Dayan, H. (2013). Decadal and long-term sea level variability in the tropical Indo-Pacific Ocean. *Climate Dynamics, 41*, 381–402.

Nieves, V., Willis, J. K., & Patzert, W. C. (2015). Recent hiatus caused by decadal shift in Indo-Pacific heating. *Science, 349*, 532–535.

Parker, D. E., Basnett, T. A., Brown, S. J., Gordon, M., Horton, E. B., & Rayner, N. A. (2000). Climate observations—The instrumental record. *Space Science Reviews, 94*, 309–320.

Pierce, D. W., Barnett, T. P., AchutaRao, K. M., Glecker, P. J., Gregory, J. M., & Washington, W. M. (2006). Anthropogenic warming of the oceans: Observations and model results. *Journal of Climate, 19*, 1873–1900.

Power, S., et al. (2021). Decadal climate variability in the tropical Pacific: Characteristics, causes, predictability and prospects. *Science, 374*, eaay9165.

Rayner, N. A., Parker, D. E., Horton, E. B., Folland, C. K., Alexander, L. V., Rowell, D. P., Kent, E. C., & Kaplan, A. (2003). Global analyses of sea surface temperature, sea ice, and night marine air temperature since the late nineteenth century. *Journal of Geophysical Research, 108*, 4407.

Reason, C. J. C. (2001). Subtropical Indian Ocean SST dipole events and southern African rainfall. *Geophysical Research Letters, 28*, 2225–2227.

Reynolds, R. W., & Smith, T. M. (1994). Improved global sea surface temperature analyses using optimum interpolation. *Journal of Climate, 7*, 929–948.

Roxy, M. K., Saranya, J. S., Modi, A., Anusree, A., Cai, W., Resplandy, L., ... Frölicher, T. L. (2024). Chapter 20: Future projections for the tropical Indian Ocean. In C. C. Ummenhofer, & R. R. Hood (Eds.), *The Indian Ocean and its role in the global climate system* (pp. 469–482). Amsterdam: Elsevier. https://doi.org/10.1016/B978-0-12-822698-8.00004-4.

Roxy, M. K., Gnanaseelan, C., Parekh, A., Chowdary, J. S., Singh, S., Modi, A., Kakatkar, R., Mohapatra, S., Dhara, C., Shenoi, S. C., & Rajeevan, M. (2020). Indian Ocean warming. In R. Krishnan, J. Sanjay, C. Gnanaseelan, M. Mujumdar, A. Kulkarni, & S. Chakraborty (Eds.), *Assessment of climate change over the Indian region*. Singapore: Springer.

Roxy, M. K., Ritika, K., Terray, P., & Masson, S. (2014). The curious case of Indian Ocean warming. *Journal of Climate, 27*, 8501–8509.

Saji, N. H., Goswami, B. N., Vinayachandran, P. N., & Yamagata, T. (1999). A dipole mode in the tropical Indian Ocean. *Nature, 401*, 360–363.

Schwarzkopf, F., & Böning, C. (2011). Contribution of Pacific wind stress to multi-decadal variations in upper-ocean heat content and sea level in the tropical South Indian Ocean. *Geophysical Research Letters, 38*, L12602.

Smith, D. M., Cusack, S., Colman, A. W., Folland, C. K., Harris, G. R., & Murphy, J. M. (2007). Improved surface temperature prediction for the coming decade from a global climate model. *Science, 317*, 796–799.

Smith, D. M., et al. (2019). Robust skill of decadal climate predictions. *npj Climate and Atmospheric Science, 2*, 13.

Sprintall, J., Biastoch, A., Gruenburg, L. K., & Phillips, H. E. (2024). Chapter 9: Oceanic basin connections. In C. C. Ummenhofer, & R. R. Hood (Eds.), *The Indian Ocean and its role in the global climate system* (pp. 205–227). Amsterdam: Elsevier. https://doi.org/10.1016/B978-0-12-822698-8.00003-2.

Srinivasu, U., Ravichandran, M., Han, W., Sivareddy, S., Rahman, H., Li, Y., & Nayak, S. (2017). Causes for the reversal of North Indian Ocean decadal sea level trend in recent two decades. *Climate Dynamics, 49*, 3887–3904.

Sutton, R., McCarthy, G. D., Robson, J., Sinha, B., Archibald, A., & Gray, L. (2018). Atlantic multidecadal variability and the UK ACSIS program. *Bulletin of the American Meteorological Society, 99*, 415–425.

Tanuma, N., & Tozuka, T. (2020). Influences of the interdecadal Pacific oscillation on the locally amplified Ningaloo Nino. *Geophysical Research Letters, 47*. e2020GL088712.

Taylor, K. E., Stouffer, R. J., & Meehl, G. A. (2012). An overview of CMIP5 and the experiment design. *Bulletin of the American Meteorological Society, 93*, 485–498.

Thompson, P. R., Piechch, C. G., Merrifield, M. A., McCreary, J. P., & Firing, E. (2016). Forcing of recent decadal variability in the equatorial and North Indian Ocean. *Journal of Geophysical Research, Oceans, 121*, 6762–6778.

Timmermann, A., McGregor, S., & Jin, F.-F. (2010). Wind effects on past and future regional sea level trends in the southern Indo-Pacific. *Journal of Climate, 23*, 4429–4437.

Tozuka, T., Luo, J. J., Masson, S., & Yamagata, T. (2007). Decadal modulations of the Indian Ocean dipole in the SINTEX-F1 coupled GCM. *Journal of Climate, 20*, 2881–2894.

Trenberth, K. E., Branstator, G. W., Karoly, D., Kumar, A., Lau, N. C., & Ropelewski, C. (1998). Progress during TOGA in understanding and modeling global teleconnections associated with tropical sea surface temperatures. *Journal of Geophysical Research, 103*, 14291–14324.

Ummenhofer, C. C., Taschetto, A. S., Izumo, T., & Luo, J.-J. (2024). Chapter 7: Impacts of the Indian Ocean on regional and global climate. In C. C. Ummenhofer, & R. R. Hood (Eds.), *The Indian Ocean and its role in the global climate system* (pp. 145–168). Amsterdam: Elsevier. https://doi.org/10.1016/B978-0-12-822698-8.00018-4.

Unnikrishnan, A. S., & Shankar, D. (2007). Are sea-level-rise trends along the coasts of the North Indian Ocean consistent with global estimates? *Global and Planetary Change, 57*, 301–307.

Vialard, J. (2015). Hiatus heat in the Indian Ocean. *Nature Geoscience, 8*, 423–424.

Xie, S. P., Deser, C., Vecchi, G. A., Ma, J., Teng, H., & Wittenberg, A. T. (2010). Global warming pattern formation: Sea surface temperature and rainfall. *Journal of Climate, 23*, 966–986.

Xie, S.-P., Lu, B., & Xiang, B. (2013). Similar spatial patterns of climate responses to aerosol and greenhouse gas changes. *Nature Geoscience, 6*, 828–832.

Xu, J., Luo, J.-J., & Yuan, C. (2022). Tropical Indian Ocean warming contributes to Arctic warming. *Geophysical Research Letters, 49*. e2022GL101339.

Yamagami, Y., & Tozuka, T. (2015). Interdecadal changes of the Indian Ocean subtropical dipole mode. *Climate Dynamics, 44*, 3057–3066.

Yan, L., Du, Y., & Zhang, L. (2013). Southern Ocean SST variability and its relationship with ENSO on inter-decadal time scales. *Journal of Ocean University of China, 12*, 287–294.

Zhang, L., Han, W., Karnauskas, K. B., Li, Y., & Tozuka, T. (2022). Eastward shift of interannual climate variability in the South Indian Ocean since 1950. *Journal of Climate, 35*, 561–575.

Zhang, L., Han, W., & Li, Y. (2018a). Mechanisms for generation and development of the Ningaloo Niño. *Journal of Climate, 31*, 9239–9259.

Zhang, L., Han, W., Li, Y., & Lovenduski, N. (2019). Variability of sea level and upper-ocean heat content in the Indian Ocean: Effects of subtropical Indian Ocean dipole and ENSO. *Journal of Climate, 32*, 7227–7245.

Zhang, Y., Feng, M., Du, Y., Philips, H., Bindoff, N., & McPhaden, M. J. (2018b). Strengthened Indonesian throughflow drives decadal warming in the southern Indian Ocean. *Geophysical Research Letters, 45*, 6167–6175.

Zhang, Y., Wallace, J. M., & Battisti, D. S. (1997). ENSO-like interdecadal variability. *Journal of Climate, 10*, 1004–1020.

Zheng, X.-T., Xie, S.-P., Vecchi, G. A., Liu, Q., & Hafner, J. (2010). Indian Ocean dipole response to global warming: Analysis of ocean–atmospheric feedbacks in a coupled model. *Journal of Climate, 23*, 1240–1253.

Zhuang, W., Feng, M., Du, Y., Schiller, A., & Wang, D. (2013). Low-frequency sea level variability in the southern Indian Ocean and its impacts on the oceanic meridional transports. *Journal of Geophysical Research, Oceans, 118*, 1302–1315.

Chapter 11

Indian Ocean primary productivity and fisheries variability[*]

Francis Marsac[a,b], Bernadine Everett[c], Umair Shahid[d], and Peter G. Strutton[e,f]

[a]MARBEC, Univ Montpellier, CNRS, Ifremer, IRD, Sète, France, [b]IRD, Seychelles Fishing Authority, Victoria, Seychelles, [c]Oceanographic Research Institute, South African Association for Marine Biological Research, Durban, South Africa, [d]WWF, Karachi, Pakistan, [e]Institute for Marine and Antarctic Studies, University of Tasmania, Hobart, TAS, Australia, [f]Australian Research Council Centre of Excellence for Climate Extremes, University of Tasmania, Hobart, TAS, Australia

1 Introduction

The Indian Ocean is a highly complex and dynamic system under monsoonal influence (Schott & McCreary, 2001). The semi-annual wind reversal north of 10°S drives changes in the ocean circulation and nutrient enrichment in the upper layer, reshaping the biological productivity landscape over the seasons. Although the semi-annual monsoon signal is the primary physical driver in the Indian Ocean, there is also a well-developed intraseasonal variability (30–90-day bands) in air-sea fluxes, rainfall, surface winds, and mixed layer depth in the tropical region, leading to fluctuations in chlorophyll concentration (Keerthi et al., 2016; Resplandy et al., 2009). At interannual timescales, the basin-wide dynamic variability is under the influence of an inherent climate mode, the Indian Ocean Dipole, or IOD (Saji et al., 1999; Webster et al., 1999). The most striking anomalies in surface winds, circulation, mixed layer depth, and spatial extent of plankton blooms occur during the positive IOD phase (Murtugudde et al., 2000; Wiggert et al., 2009).

Fisheries play a vital role in the Indian Ocean by ensuring food security and providing jobs to a large proportion of the population of 2.5 billion inhabiting its bordering countries. The Indian Ocean is the only oceanic region where marine capture fisheries have been on a steadily ascending curve (FAO, 2020a). The year 1995 was a tipping point for world marine production, as its multidecadal progression halted, and catches started to plateau around 80 Million tons. In the meantime, Indian Ocean catches increased by 50% between 1995 and 2018 (8 to 12 Mt, FAO, 2020a). These estimates are somewhat uncertain, as the monitoring of small-scale and multispecies fisheries is particularly problematic. Such data limitations are responsible for the high degree of uncertainty in the assessment of several stocks. It is estimated that 32% of the assessed stocks in the Indian Ocean are fished at biologically unsustainable levels (FAO, 2020a). Indeed, the demand for fish and marine products, in general, is the main driver in regions where the rate of increase of the population is among the highest in the world, such as in southern and eastern Africa (2.3% increase annually) and the Middle East (1.8%, World Odometer, https://www.worldometers.info/world-population).

Fisheries mostly exploit species groups that are in the intermediate and high trophic levels. Consequently, any significant climate-induced changes at the base of the food web will propagate through the ecosystem, altering the abundance of marine resources, their accessibility to the fleets, and the efficacy of the fishing gear. This is how primary productivity and fisheries are connected, thus requiring special attention to account for climate variability in fisheries management.

In this chapter, we first describe the main patterns of productivity, focusing on the interannual mode of variability in several selected areas of the Indian Ocean. We review the main characteristics, the socio-economic importance, and vulnerability to climate change of the coastal fisheries in three large regions, the Western, Northern, and Eastern Indian Ocean. Finally, we focus on the well-documented offshore and high-seas tuna fisheries.

---

[*]This book has a companion website hosting complementary materials. Visit this URL to access it: https://www.elsevier.com/books-and-journals/book-companion/9780128226988.

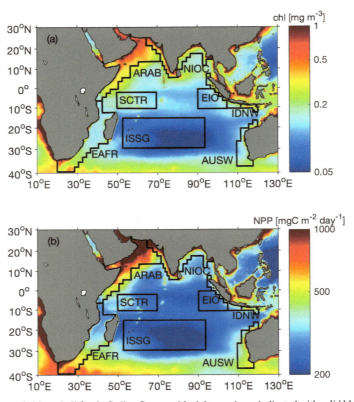

FIG. 1 The climatological mean annual chlorophyll for the Indian Ocean, with eight provinces indicated with solid black line boundaries and their labels. The main text explains the abbreviations. (a) The chlorophyll data are averaged from January 1998 to December 2019, from Ocean Color Climate Change Initiative project (https://www.oceancolour.org/). (b) Similar to panel (a) but net primary productivity (NPP) calculated using the Vertically Integrated Productivity Model (Behrenfeld & Falkowski, 1997), averaged from 2002 to 2020 and downloaded from http://sites.science.oregonstate.edu/ocean.productivity/index.php.

2 Indian Ocean productivity: Variability and trends

The Indian Ocean is home to areas of extreme oligotrophy with some of the lowest chlorophyll concentrations anywhere in the global ocean observed in the southern Indian Ocean gyre. However, the Indian Ocean boundaries are home to very productive systems. The Arabian Sea, Somali Coast, and the west coast of Indonesia are upwelling-influenced, and the coast of India is heavily impacted by river runoff. Here, with reference to previous work and new analyses, we describe the mean state, variability, and inter-annual trends in the productivity of eight provinces in the Indian Ocean Basin (Fig. 1). We selected five mostly coastal provinces that are very similar to those proposed in the Longhurst (2007) biogeography. They are

(1) East Africa (labeled EAFR in Fig. 1), based on Longhurst's province of the same name, but excluding the region east of Madagascar.
(2) The upwelling-dominated Arabian Sea (ARAB)
(3) The North Indian Ocean Coast (NIOC), which pools two Longhurst provinces: INDW and INDE.
(4) The west coast of Indonesia (IDNW), which is the Indonesian part of Longhurst's AUSW province.
(5) The Australian West Coast (AUSW), a subset of the original Longhurst AUSW.

We also added three open ocean, less productive, regions:

(6) A widened Seychelles-Chagos Thermocline Ridge (SCTR)
(7) The Eastern Indian Ocean (EIO), and
(8) The central oligotrophic gyre (ISSG).

SCTR and EIO were created to permit investigation of the perturbations to chlorophyll caused by the IOD.

Fig. 1 shows the mean annual climatology of satellite-derived chlorophyll and net primary productivity (NPP). We use chlorophyll as a proxy for phytoplankton biomass and NPP as a measure of the rate of photosynthesis and carbon fixation. The chlorophyll data are a merged product developed by the Ocean Color Climate Change Initiative (https://www.

oceancolour.org/), and the NPP data are from the Vertically Integrated Productivity Model (Behrenfeld & Falkowski, 1997). The satellite chlorophyll data are reasonably well-validated in the Indian Ocean (Sathyendranath et al., 2019), but the representation of in situ observations is still poor compared to other basins. This may influence some of the behavior observed in the time series and noted later. The validation of the NPP data is much weaker. As noted in Strutton and Hood (2019), there were no data from the temperate or tropical Indian Ocean in the database that developed the Vertically Integrated Productivity Model. But for the time being, it is probably still the best satellite NPP data we have.

To the naked eye, the distributions look very similar. If we take the ratio of the two (not shown), the very northern part of the Arabian Sea and the East African region between the southern tip of Madagascar and Cape Agulhas support higher productivity relative to chlorophyll. The same is true of the southern Indian Ocean gyre, but that is likely an artifact of the very low denominator (low chlorophyll) in that region. Areas that show up as lower in productivity relative to chlorophyll include the northern Leeuwin Current, flowing southwards along the Western Australian coast, the west coast of Indonesia, and the North Indian Ocean coastal areas. Because of the similarity in spatial patterns described by both satellite chlorophyll concentration and NPP, and as NPP validation is still a matter of concern, we shall only use satellite chlorophyll when referring to ocean productivity in the rest of the chapter.

The interannual variability is depicted in Fig. 2, with some statistics summarized in Table 1. Obviously, the central subtropical southern Indian Ocean oligotrophic gyre has the lowest productivity, as quantified by chlorophyll concentration, and it also has the lowest seasonal and interannual variability. The SCTR and eastern Indian Ocean regions are similar to each other in both their mean chlorophyll and variability, as are the Australian West coast and East African regions. The Arabian Sea region exhibits consistent and large seasonal variability driven by seasonal upwelling. The interannual variability in the magnitude of the seasonal chlorophyll peak is very large for the west coast of Indonesia, presumably due to interannual variability in upwelling strength. The North Indian Ocean Coast region is curious. It is the highest productivity system with a large variability in the magnitude of the seasonal cycle (compare 2002–2012 vs 2012–2016 in Fig. 2). These periods of contrasting magnitude of chlorophyll variability also correspond to what appears to be a step change in mean chlorophyll. It is higher from 2002 to 2012 compared to the rest of the record. It is possible that this is due to the challenging optical environment (river sediments) and the mix of satellites in the Ocean Color Climate Change Initiative algorithm; MODIS-Aqua commenced in July 2002 and MERIS died in 2012. This deserves further investigation.

The time series in Fig. 2 represents the most recent and longest compilation of satellite chlorophyll data for the Indian Ocean. As discussed in Hood et al. (2024a), Goes et al. (2005) used the first seven years of SeaWiFS satellite ocean color (1997 to 2004) to document an increase in Arabian Sea chlorophyll (47°E–55°E, 5°N–10°N). This trend was attributed to warming and reduced snow cover over Eurasia, which created a thermal gradient conducive to intensified upwelling

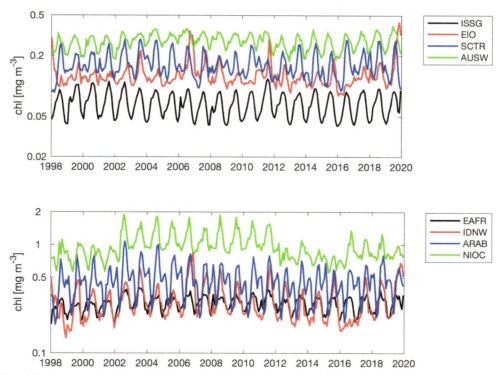

FIG. 2 Time series of the mean monthly chlorophyll from each of the eight regions identified in Fig. 1.

TABLE 1 A summary of statistics for the eight regions, based on the time series of the spatial monthly mean in Fig. 2.

Region	Mean chl (mg m^{-3})	Max chl (mg m^{-3})	Min chl (mg m^{-3})	Range (mg m^{-3})	chl trend (mg m^{-3} year^{-1})
EAFR	0.29	0.40	0.18	0.22	0.00051
ARAB	0.47	1.07	0.18	0.89	−0.00208
NIOC	0.96	1.88	0.56	1.32	−0.00585[a]
IDNW	0.30	0.86	0.14	0.72	0.00061
AUSW	0.28	0.40	0.18	0.22	0.00018
EIO	0.13	0.43	0.08	0.34	0.00045
SCTR	0.17	0.29	0.09	0.20	−0.00032
ISSG	0.07	0.12	0.04	0.08	−0.00027

[a] *For the trend in NIOC indicates that it is significant at the 95% level.*

favorable winds in the western Arabian Sea during the Southwest Monsoon (boreal summer). A decade later, Roxy et al. (2016) looked at a broader area of the Indian Ocean (50°E–65°E, 5°N–25°N, still arguably the Arabian Sea) and a longer time series (1997 to 2013) to draw the opposite conclusion—that productivity was decreasing due to a warming ocean and greater stratification. More recently, Goes et al. (2020) showed that increased stratification and decreased dissolved nitrate concentrations are likely driving a change in winter blooms, from diatoms to a mixotrophic dinoflagellate (*Noctiluca scintillans*). These recent findings do not necessarily resolve the contrasting conclusions, but they do point to consequences for fisheries that we will return to later.

Using the time series compiled here, trends can be assessed in at least two different ways. First, linear trends in the Fig. 2 time series were calculated and tested for significance. The only region with a significant trend was the North Indian Ocean Coast, where the mean chlorophyll seems to be decreasing, but this may be due to potential issues with the satellite product mentioned above. Based on our 23-year time series, there is no evidence of any trends in the mean chlorophyll concentration in any of the eight regions identified here. Second, it is possible to look at trends in the expanse of extremely oligotrophic waters, similar to Polovina et al. (2008). Using an upper threshold of 0.1 mg m^{-3} chlorophyll, linear trends in the proportion of each region that could be considered "low productivity" were calculated, but none of these trends were significant. Therefore, similar to the conclusions for trends in the mean, there has been no change over the last two decades in the size of low-productivity regions in the Indian Ocean. However, the analysis of linear trends can mask other variability that may or may not be periodic. Marsac and Demarcq (2019) analyzed chlorophyll variability in four large ecoregions that covered almost all of the Indian Ocean between 10°N and 30°S. They found a general uniform trend of decreasing chlorophyll from about 2000 to 2011 in the central subtropical southern Indian Ocean oligotrophic gyre and Mozambique Channel, while the western Indian Ocean (west of 77°E) remained rich in chlorophyll from 2000 to 2005. A common feature is a regime with depleted chlorophyll in all ecoregions from 2007 to 2014, increasing thereafter to the present. Interestingly, this low chlorophyll regime, possibly triggered by the 2006–2007 positive dipole, covers several years with sustained positive loadings of the dipole mode index (Fig. 3).

The Indian Ocean is also strongly influenced by both the IOD and the El Niño-Southern Oscillation (ENSO). Several studies have investigated the impact of these climate modes on the primary productivity of the basin (Currie et al., 2013; Keerthi et al., 2017; McCreary et al., 2009; Wiggert et al., 2009). Using the time series from the SCTR and eastern Indian Ocean regions, the zonal oscillations in productivity in response to the IOD can be quantified. We calculated, at monthly resolution, the difference between SCTR and eastern Indian Ocean chlorophyll, $SCTR_{chl}$-EIO_{chl}, and correlated that with the dipole mode index (DMI) downloaded from https://psl.noaa.gov/gcos_wgsp/Timeseries/DMI/. There was a significant negative correlation (P value <0.01), which indicates that, consistent with previous work, during a positive IOD, chlorophyll is enhanced in the eastern Indian Ocean compared to the SCTR and vice versa. Aspects of seasonal and interannual variability at higher trophic levels are taken up elsewhere in this chapter, and Strutton and Hood (2019) describe seasonal and regional variability in productivity in more detail than here. Compared to the Pacific and Atlantic, the Indian Ocean is very poorly sampled with respect to biogeochemistry and ocean productivity in particular. Recent efforts (Beal et al., 2020; Hermes et al., 2019) have described priorities and opportunities for improved ocean observations in the future.

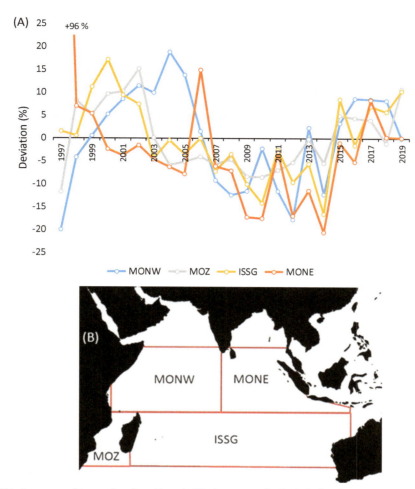

FIG. 3 (a) Deviation (in %) to the mean multi-annual surface chlorophyll by large ecoregion in the Indian Ocean. Two datasets were used: SeaWiFS from 1997 to 2002 and MODIS from 2003 to present. Climatological means used to compute the deviation cover the period 2000–2008 for SeaWiFS and 2003–2011 for MODIS. (b) Study ecoregions: Monsoon West (MONW: 40°E–77°E, 10°N–12°S); Monsoon East (MONE: 77°E–120°E, 10°N–12°S, excluding non-Indian Ocean regions); the Mozambique Channel (MOZ: 30°E–47°E, 12°S–30°S); and the subtropical gyre (ISSG: 47°E–120°E, 12°S–30°S). *(From Marsac and Demarcq (2019).)*

While the Longhurst marine provinces were used to outline the distribution and variability of phytoplankton biomass, the fisheries description hereafter will be based on large marine ecosystems (Sherman & Alexander, 1986). Large Marine Ecosystems are expert-derived systems taking into account bathymetry, hydrography, productivity, and trophically dependent populations. Hosting about 90% of the world's fish catch, Large Marine Ecosystems are tools to assist with transboundary distributions and management issues, and thus form an appropriate classification in a fisheries context.

3 Trends in coastal fisheries

3.1 The western Indian Ocean

3.1.1 Description of the fisheries

The western Indian Ocean supports a large diversity of fisheries along the coasts of East Africa, Madagascar, Seychelles, Mauritius, Comoros, and Reunion (Groeneveld, 2015). The catches comprise a mix of species including invertebrates and fishes inhabiting coral reefs, mangroves, seagrass beds, and shallow inshore waters (WIOFish, 2020). Fish contribute the most to catches in the region (Fig. 4a–d) with invertebrates providing much smaller contributions. For instance, crustaceans represent no more than 10% of the fisheries production in the western Indian Ocean region (FAO, 2021). Western Indian

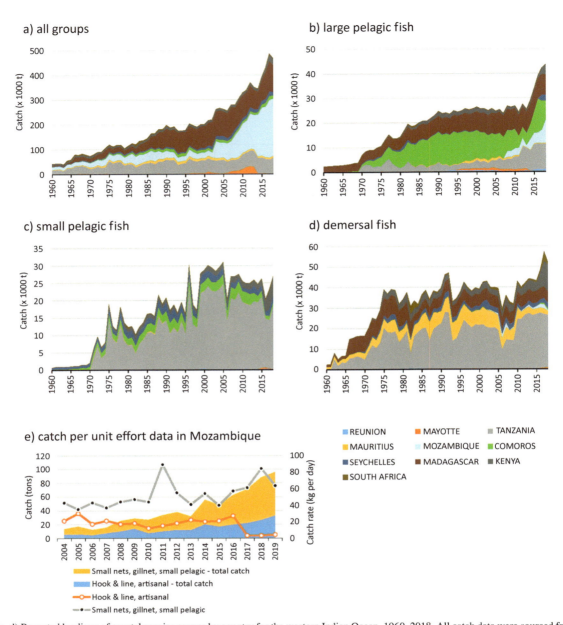

FIG. 4 (a–d) Reported landings of coastal species groups by country for the western Indian Ocean, 1960–2018. All catch data were sourced from FAO's FIGIS database (FAO, 2021). The Indian Ocean Tuna Commission database (IOTC, 2021) was used to remove the industrial component of large pelagic fish catches from the FAO-derived catch data (the Indian Ocean Tuna Commission currently categorizes as industrial fisheries those operated by vessels larger than 24 m in length, or by smaller vessels fishing outside the exclusive economic zones). The "all groups" category includes marine fish, crustaceans, and mollusks. (e) Catch per unit effort by gear group for Mozambique, 2004–2019, WIOFISH database (WIOFish, 2020).

Ocean total catches, as reported to the FAO, have been increasing over time and they peaked in 2017 at approximately 494,000 t (FAO, 2021). Catches of large pelagic fish have dramatically increased since 2014, surpassing small pelagic fish catches, as more coastal operators ventured into the more lucrative fishing activity for large fish. Artisanal fisheries, comprising fishing households with limited access to capital and utilizing simple gears operated on the shore or from small boats, dominate in the region (van der Elst et al., 2005). The boats vary from traditional dugouts to small vessels made from planks or fiberglass and propelled by sail and/or outboard engines. Larger vessels (e.g., dhows) have triangular sails and inboard engines (Fulanda et al., 2011; Munga et al., 2014). Fishing gears include sticks, harpoons, nets of various designs, handlines with hooks, and numerous types of traps made from locally sourced materials (Fulanda et al., 2011; Jiddawi & Öhman, 2002; WIOFish, 2020). Of particular importance to artisanal fisheries are the small pelagic species that have become more dominant in urban consumption in place of the reef-associated and several demersal species that have declined in abundance over time (Jiddawi & Öhman, 2002; Van Hoof & Kraan, 2017). These trends become obvious when observing the catch rates (gross proxies for abundance) of specific fisheries that target these species groups (WIOFish, 2020). Fig. 4e shows the catch rates for two Mozambique fisheries, the artisanal hook and line fishery for reef-associated and demersal fish and the gillnet fishery that targets small pelagic species. Catch rates for the hook and line fishery show a dramatic drop in catch rates in 2016 indicating that much more fishing effort is required to increase the catches. Around the same time, after 2015, the catch rates of the small pelagic species started to increase along with the total catches indicating that less fishing effort was expended to catch more fish.

3.1.2 Socio-economic importance

In the early 2000s, it was estimated that approximately 60 million people were living within 100 km of the coast in the western Indian Ocean (World Resources Institute, 2002). This translates to a high proportion of people that are likely to be dependent on fisheries in one form or another for economic, social, and cultural security (Cox, 2012). The western Indian Ocean includes some of the poorest countries in the world, based on per capita gross domestic product (GDP) (World Bank, 2020) and Human Development Index (UNDP, 2020). These include Mozambique (GDP: $504 per capita, Human Development Index: 0.456), Madagascar GDP: ($523 per capita, Human Development Index: 0.528) and the United Republic of Tanzania (GDP: $1122 per capita, Human Development Index: 0.529). Most of these countries face severe socio-economic challenges in terms of a massive and expanding need for employment opportunities and food security through the provision of fish protein (van der Elst et al., 2005). While coastal fisheries' contributions to these countries' GDP may be quite low, between 0.5% for Kenya and 5.5% for Madagascar (World Bank, 2012), at the provincial or district level, they may provide the main source of employment and income. An example is in the Rufiji Delta in Tanzania where 61% of households are involved in fishing (Turpie, 2000). Small-scale and, in particular, artisanal fisheries play the biggest role in contributing toward the livelihoods of some of the poorest sectors of these communities through providing employment, daily protein for consumption, and financial revenues from sales of catches (Walmesley et al., 2006). However, because these artisanal fishers have very little access to opportunities to modify their fishing vessels, equipment, and areas where to operate, they are highly vulnerable to changes within the ecosystems due to the impacts of factors such as climate change.

3.1.3 Vulnerability to climate change

The western Indian Ocean has been warming for more than a century at a faster rate than any other tropical region (Roxy et al., 2014). This warming of the generally cooler western Indian Ocean against the rest of the tropical region has the potential to alter marine ecosystems (Taylor et al., 2019). Adding to this warming trend is the influence of IOD events. During a positive IOD event, the western Indian Ocean becomes warmer than average (Fan et al., 2016). These positive IOD events are projected to become more frequent (Cai et al., 2014; Ihara et al., 2008) placing a heavier burden on the marine ecosystems to try to adapt to the changing situation. Warming has led to increased ocean stratification, suppression of nutrient mixing, decreased phytoplankton (Roxy et al., 2015), or a change from a winter bloom dominated by diatoms, to a dinoflagellate with the potential to cause nontoxic but otherwise harmful blooms (Goes et al., 2020). The changes in phytoplankton, on which the small pelagic fish rely, consequently lead to a decline in the availability of these fishes to coastal fisheries (Sekadende et al., 2020). This was evident during the strong 2006–2007 positive IOD event when small pelagic catches in Tanzania declined from 22,000 t in 2005 to 17,000 t in 2006 and 2007 (Breuil & Bodiguel, 2015). Another impact of a warmer ocean is the increased sensitivity and die-off of corals and the decline of reef-associated fishes (McClanahan, 2017; See also Hood et al., 2024b). Variability of catches does not, however, show a true reflection of resource changes as these are masked by increasing effort applied to the fishery to overcome catch shortages. In Tanzania, between 2007 and 2016, the number of fishers active in the small pelagic fishery rose by 15% with an associated increase in small fishing vessels of 100% and dhows by 69% (Department of Fisheries Development, 2016). Given the high dependence of western Indian Ocean coastal communities on the ocean for their livelihoods and food security, the continued warming of the western Indian Ocean will threaten food security and will most likely push more people into poverty (Taylor et al., 2019).

3.2 North Indian Ocean

3.2.1 Description of the fisheries

The North Indian Ocean fisheries are embedded in two major ecoregions, the Arabian Sea Large Marine Ecosystem and the Bay of Bengal Large Marine Ecosystem (Monolisha et al., 2018). The Large Marine Ecosystems are characterized by freshwater discharge from the Indus River in the Arabian Sea and seven major rivers within the Bay of Bengal influencing coastal productivity. Positive IOD events generate higher river outflow through anomalously high precipitation during the southwest monsoon (Hussain et al., 2017). The coastal fisheries in both Large Marine Ecosystems are supported by rich habitats and nearshore ecosystems, which provide spawning grounds for several commercial species. Coastal fisheries are small-scale in nature and generally multigear and multispecies, including subsistence and artisanal vessels with major exploited groups including Indian oil sardine (*Sardinella longiceps*), Indian mackerel (*Rastrelliger kanagurta*), and ribbonfish (*Lepturacanthus savala* among others) (Bhathal, 2005). In both regions, fisheries are open-access resulting in high fishing effort (Townsley, 2012; World Bank, 2018). Coastal fisheries harvest marine catadromous/anadromous fish species such as "Palla" (*Tenualosa ilisha*), small and large pelagics (such as Scianidae, Sparidae, Mullidae, Latidae), demersal species such as elasmobranchs (mainly Pristidae, and several species of Rajidae), and Perciforms that are the largest dominating group in coastal areas. A considerable quantity of meso-pelagic fish, in particular lantern fish (Myctophidae), are exploited in the offshore areas (Butler et al., 2001; Gartner, 1993; Gjøsæter & Tilseth, 1983; Johannesson & Valinassab, 1994; Nafpaktitis & Nafpaktitis, 1969).

In the Arabian Sea Large Marine Ecosystem, the bulk of fishing occurs within the exclusive economic zones (EEZs; Palomares et al., 2021). The coastal catches of fishes, mollusks, and crustaceans were above 2.5 Mt in 2018 (Fig. 5a), with

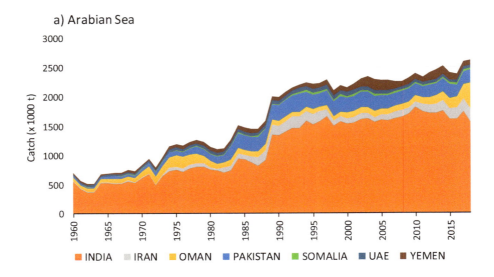

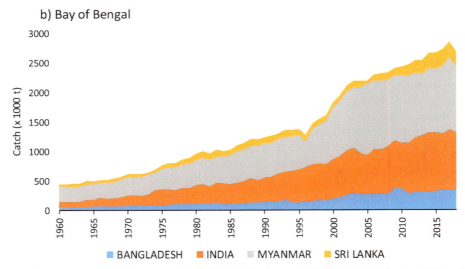

FIG. 5 Reported landings of coastal species groups by country for the North Indian Ocean, 1960–2018: (a) Arabian Sea, (b) Bay of Bengal. *All catch data were sourced from FAO's FIGIS database (FAO (2021).)*

India as the primary contributor (58% of the production). Fishes represent 79% of the landings. The heavy fishing pressure on several groups such as the demersal fish and sharks and the subsequent decline in catches has caused a shift to small pelagic fishes, e.g., Indian oil sardine (mean annual catch of 472,000 t during 2000–2018) and Indian mackerel (114,000 t/year during 2000–2013, 303,000 t/year for 2014–2018). Over the years, the fisheries sector has grown gradually with vessels equipped with advanced technology, resulting in massive overcapacity in the region (Palomares et al., 2021). In the Bay of Bengal Large Marine Ecosystem, the coastal catches reached a peak at 2.9 Mt in 2017 and decreased by 6% in 2018 (Fig. 5b). Fishes represent 88% of the landings. Among fish, small pelagics such as clupeoids, including sardines and anchovies, are the dominant species groups in the catch (250,000 t per annum over 2000–2018). Several species are fished at their full potential (>50% of the stocks) or overfished, with high-value demersal fish facing the highest pressure. A trend in both Large Marine Ecosystems is the shift to unsustainable practices to target low-value "trash" fish that contribute to feed for poultry and aquaculture (APFIC, 2012).

3.2.2 Socio-economic importance

The Bay of Bengal Large Marine Ecosystem hosts about a quarter of the world's population, including 187 million people who live in 405 coastal cities directly dependent on coastal and marine fisheries (Emerton, 2014). The Arabian Sea Large Marine Ecosystem region collectively with India had a population of 1.2 billion in 2002 (Heileman et al., 2008). This indicates a very high population engaged in fishing with a large variety of gears. However, countries in both Arabian Sea and Bay of Bengal Large Marine Ecosystems are at different levels of management and their contribution to fisheries varies significantly. The Arabian Sea Large Marine Ecosystem exhibits low to medium levels of economic development, whereas economic development is stagnating at a low level in the Bay of Bengal Large Marine Ecosystem. Both Large Marine Eecosystems are assessed by a combined measure of Human Development Index and marine activities taking place in the area. The Arabian Sea and the Bay of Bengal Large Marine Ecosystems face an increased risk of unsustainable fisheries due to poor management regimes, which if not addressed through adequate and robust regulations, may lead to a fisheries collapse with subsequent devastating impacts on the socio-economic development of coastal communities (Elayaperumal et al., 2019; Palomares et al., 2021).

The reported catch in the Arabian Sea Large Marine Ecosystem is worth ~1.6 billion US$ but it exhibited a dramatic decline in preferred species within a decade (Dwivedi & Choubey, 1998; FAO, 2005; Sherman, 2003). An estimate of reported landings value in Arabian Sea Large Marine Ecosystem peaked at 5.5 billion US$ in 1992 and then declined to around 3.8 billion US$ in 2008 without change in total catch. In the Bay of Bengal Large Marine Ecosystem, the reported landing value was estimated at 5.7 billion US$ for the period 2005–2010; however, this figure is not considered accurate, and no recent update was made available.

3.2.3 Vulnerability to climate change

The Arabian Sea Large Marine Ecosystem region is a highly productive ecosystem (up to $1 gC\ m^{-2}\ day^{-1}$ in the northern and western areas) strongly influenced by the monsoons, which enhances fish productivity (Baars et al., 1998; Desai & Bhargava, 1998). The subsurface waters of the Arabian Sea are oxygen-depleted and, in spite of elevated primary productivity, the abundance of coastal pelagic fish is low (Bakun & Csirke, 1998). Moreover, climate change and variability along the coastlines of Arabian Sea countries will increase the risk to coastal communities and natural resources in their adaptation toward severe climatic events (Khan et al., 2016). Changing climate, erratic weather patterns, intense monsoon winds, and strong current flows in the Arabian Sea will induce persistent unfavorable feeding conditions for coastal fish larvae, which supply fisheries stocks. These environmental conditions have often given rise to blooms, for instance, jellyfish outbreaks in recent years, widely reported in both Arabian Sea and Bay of Bengal Large Marine Ecosystem regions (Baliarsingh et al., 2020). It is reported that the prevalence of large numbers of jellyfish in coastal waters of the northwest Bay of Bengal has also negatively influenced plankton ecology (Baliarsingh et al., 2020).

In the Arabian Sea Large Marine Ecosystem, climate change is projected to cause large impacts on marine biodiversity with the Arabian Gulf countries ranking high on the vulnerability list (Wabnitz et al., 2018). There, some degree of species loss (5 to 9 species) is predicted for the second half of the 21st century, along with a drastic reduction in the total habitat biodiversity suitability for all species inhabiting the Arabian Gulf and a projected drop of 30% of the future fisheries catch potential (Wabnitz et al., 2018). The impact of climate change on sea turtle biology was documented in another area, at Karachi Beach (Pakistan). Phillott et al. (2018) concluded that, due to continuously rising temperatures, male hatching would become dominant over female hatching, altering the sex ratio, and the viability of turtle populations. Both Large

Marine Ecosystems are mostly affected by anthropogenic factors, and the majority of impacts due to climate change may be latent. As such, climate impacts on marine resources need to be assessed, as their depletion would threaten the livelihoods of the coastal populations, increasing food security issues and poverty.

3.3 The Eastern Indian Ocean

3.3.1 Description of fisheries

The countries included in this review are Thailand, Malaysia, Indonesia, and Australia. These countries lie between the Indian and Pacific Ocean, but we only consider Indian Ocean fisheries here. The largest player in marine capture fisheries in this region is Indonesia (56%), followed by Malaysia, Thailand, and Australia with 26%, 14%, and 4%, respectively, during 2010–2018 (Fig. 6a). The Indonesian catch of marine fishes has been on a steadily increasing trend since the mid-1970s, from around 230,000 t to a production of 1.7 million tons in 2018 (FAO, 2021). Thailand catches plateaued at an average of about 800,000 t during 1993–2006 then started to decline, to stabilize around 320,000 t from 2015 onwards. In Malaysia, catches increased gradually from 77,000 t in 1960 to 700,000 t in 2009 and they have remained stable since (FAO, 2021). The Strait of Malacca, located between Malaysia and the east coast of Sumatra, yields 48% of the total marine production for Malaysia (Jagerroos, 2016). Indonesia cumulates the largest fractions for all the major groups of fishes, 89% of large pelagic catches of the whole region, 55% of small pelagics, and 60% of demersal catches (Fig. 6b–d). The coastal fisheries are primarily catching small pelagic fishes (35% of total catches) before demersal (20%) and large pelagic fishes

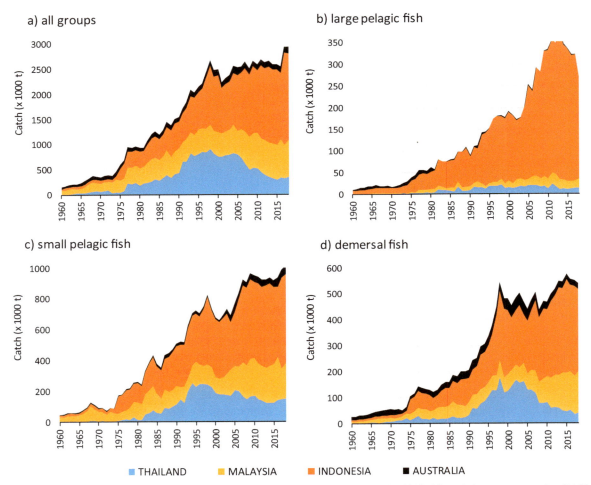

FIG. 6 Reported landings of coastal species groups by country for the eastern Indian Ocean, 1960–2018. All catch data were sourced on FAO's FIGIS database (FAO, 2021). The Indian Ocean Tuna Commission database was used to remove the industrial component of large pelagic fish catches from the FAO-derived catch data. The "all groups" category includes marine fish, crustaceans and mollusks.

(12%). The artisanal fleets are composed of thousands of vessels, largely dominated by small-sized vessels of less than 5 GT (<12 m). A wide range of fishing gears are used: purse seine, trawls, drift and bottom longline, handline, gillnet, and other nets (www.seafdec.org).

3.3.2 Socio-economic importance

The fisheries sector in the region plays a key role in supporting national food security. In Indonesia, for instance, 54% of all animal protein consumed comes from fishery products. The fish consumption in 2016 was 33 kg/capita/year in Thailand, 41 kg in Indonesia, and 56 kg in Malaysia (www.seafdec.org). These figures contrast with Australia where the fish consumption is barely 14 kg/capita/year (ABARES, 2018) The livelihoods of small-scale fishermen in developing coastal states are facing multiple complex issues resulting from the decline of fish resources, degradation of mangroves and coastal habitats, fuel crises, urbanization, and pollution in densely populated areas. In Indonesia, 95% of the fishers are working in the artisanal sector (Murdiyanto, 2011). They operate in nearshore areas, and they are considered having low welfare standards of living. The artisanal fisheries sector in Indonesia has the highest rate of poverty compared to any other sector. The contribution of the fisheries sector to the GDP is 0.7% in Thailand, 1.1% in Malaysia, 2.5% in Indonesia, and 0.3% in Australia (BDO EconSearch, 2019; Choongan, 2018; KKP, 2015; Rozhan, 2019).

3.3.3 Vulnerability to climate change

The productivity of the coastal waters over the region is high during the boreal summer, when upwelling develops along the south coast of Java and west coast of Sumatra, in association with the local southeast monsoon over the region (Susanto et al., 2001; Wyrtki, 1962). The coasts of Thailand and the Malacca Strait also exhibit seasonal productivity, however, with peak values occurring during the northeast monsoon in association with lower sea surface temperature (Siswanto & Tanaka, 2014). Another player in climate variability at interannual timescales is the IOD that is phase-locked to the seasonal upwelling. There, in addition to the local wind forcing, the eastern cooling is forced remotely by equatorial and coastal Kelvin waves shoaling the thermocline (Murtugudde et al., 1999). The positive phase of the IOD sustains high oceanic productivity beyond the sole upwelling season, until the decay of the IOD event, i.e., January–February of the following year (Annamalai & Murtugudde, 2004). In Thailand and along the Strait of Malacca, atmospheric deposition during El Niño and positive IOD events elevates surface nutrients and promotes chlorophyll blooms (Siswanto & Tanaka, 2014). Fundamentally, an elevated biological productivity, measured in terms of chlorophyll concentration, boosts the energy transfer in the food web, inducing prosperous fisheries. For instance, the catches of oil sardine (*Sardinella lemuru*) south of Java respond positively to upwelling activity within a lag of three months, and there is a clear interannual signal associated with IOD variability (Sartimbul et al., 2018). Sardine production increased by 200%–300% in 1997–98 and 2006–07, which were strong positive IOD years. As the recurrence of extremely positive IOD events is projected to increase by almost a factor of three from current, with one event every 6.3 years over the 21st century (Cai et al., 2014), enhanced carrying capacity can be hypothesized over the eastern Indian Ocean region. Cheung et al. (2018) project a different outcome in a global study focusing on the EEZs. This study suggests a decrease of the catch potential by 25%–30% by the end of the century in the eastern Indian Ocean EEZs. However, the authors point out the key uncertainties of their projected changes in the coastal regions, due to the low spatial resolution (100–200 km) of the underlying Earth System models used in the study, and the considerable range of ecosystem response at the EEZ scale (Bopp et al., 2013).

4 Trends in tuna fisheries

4.1 Outlook for tuna fisheries in the Indian Ocean

Tuna fishing has existed for centuries along the coasts of the Indian Ocean, as reported by the 14th-century explorer, Ibn Battuta (Dunn, 2012; Walker, 2024). Fisheries expanded offshore after the Second World War; however, it was only in the 1980s that the Indian Ocean gradually became a prominent tuna-fishing area. In 2019, tuna catches in the Indian Ocean were 1.8 million tons, representing 23% of the world's tuna and billfish catch (FAO, 2020b). A range of gears exploits the tuna and billfish resources: hand lines, troll lines, pole and lines, gillnets (including driftnets), mostly operated by the coastal fisheries, and, longlines and purse seines operated offshore by industrial and semi-industrial fisheries (Fig. 7). A major characteristic of the Indian Ocean fisheries is that coastal and artisanal gears catch almost as much tropical tunas and billfish than do the industrial fisheries (490,000 t, 48% vs 530,000 t, 52% respectively, for 2015–2019). Moreover, most of the neritic tunas are caught by coastal gears (353,000 t, 90% for 2015–2019) and reported with high uncertainties.

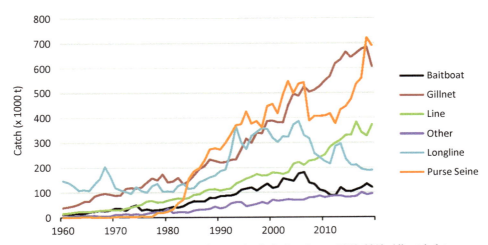

FIG. 7 Reported landings of tuna, tuna-like, and billfish species by gear for the Indian Ocean, 1960–2019. All catch data were sourced on the Indian Ocean Tuna Commission database (IOTC, 2021). The gears indicated as baitboat, gillnet, line, and other are mostly operated by artisanal and coastal fleets, whereas the longline is operated by industrial and semi-industrial fleets and the purse seine by industrial fleets only.

Tuna coastal fisheries have great socio-economic importance, ensuring food security and providing jobs along the Indian Ocean rim. Therefore, efficient management of fisheries to keep the stocks at safe and sustainable levels is the major challenge faced by the fishing nations forming the Indian Ocean Tuna Commission.

4.2 Status and management of tuna and billfish stocks

Composed of 31 member countries, the Indian Ocean Tuna Commission has a mandate to monitor catch, to carry out research and stock assessments for 16 species of large pelagic fishes (11 tuna and five billfish species), and to manage the fisheries exploiting these resources in the Indian Ocean, while developing mechanisms to better integrate ecosystem considerations into the scientific advice. The assessment of the stocks is updated on a 3-year cycle. The three species of tropical tunas, yellowfin (*Thunnus albacares*), bigeye (*Thunnus obesus*) and skipjack (*Katsuwonus pelamis*) have different status (IOTC, 2021). The Indian Ocean Tuna Commission has determined the yellowfin stock as overexploited and subject to overfishing. A restoration plan has been in place since 2014 to reduce catches and rebuild the stock. However, the management measure is not effective in reducing fishing mortality due to poor monitoring of catches by several fishing countries, as well as weaknesses in the plan itself, requiring several adjustments. The catch reductions achieved by the fleets bound to the management measure were offset by increases in the catch from national fleets not bound to the measure, resulting in an overall increase of yellowfin catches of ~5% from 2014 to 2019. Facing this undesirable trend, leading international retailers and supply chain companies have threatened to stop sourcing yellowfin tuna from the Indian Ocean. Such a ban would dramatically affect the economy of developing countries exporting tuna to the world market. In 2021, a new yellowfin stock rebuilding plan was adopted to straighten out the situation despite the fact that six fishing countries objected to the measure, therefore not being bound to it. In 2018, the bigeye tuna stock was not overexploited, however it was subject to excessive fishing mortality (i.e., overfishing). The 17% catch reduction recorded in 2019 was a positive response to this situation. The skipjack stock assessment carried out in 2020 indicated that neither overexploitation nor overfishing is occurring and that the stock is exploited at the target reference point. Albacore tuna (*Thunnus alalunga*) is the only temperate tuna directly assessed by the Indian Ocean Tuna Commission. Its stock status determination is quite uncertain; however, the fishing mortality might be too high, and scientists recommend catch reductions as catches exceeded the maximum sustainable yield by 7% in 2019. As for billfish, only the swordfish (*Xiphias gladius*) stock is exploited optimally, while the blue (*Makaira nigricans*) and striped marlins (*Kajikia audax*) are overexploited. The black marlin (*Istiompax indica*) stock is considered not overfished; however, this stock status is largely uncertain (IOTC, 2021).

4.3 Vulnerability to climate change

The large tunas are among the most highly evolved groups of fish in the oceans. In their evolution, they have developed a vascular counter-current heat-exchange system keeping their internal bodies at metabolically optimal temperatures (Graham & Dickson, 2001) Thus, they can perform deep dives in cold waters to escape predation, or to forage on components of the deep scattering layer. However, thermoregulation can also be a cause of body overheating in areas of elevated ambient temperature. Therefore, warmer oceans may lead tuna to spend more time in the subsurface, around the thermocline, to balance the positive and negative effects of temperature on their metabolism. As tunas have high metabolic requirements, their oxygen demand is high, which explains why the optimal habitat for tropical tunas is the well-oxygenated mixed layer. Skipjack, the less tolerant species relative to dissolved oxygen content, increases their swimming speed to enhance oxygen uptake when dissolved oxygen falls below $4\,mg\,L^{-1}$. Lethal levels for yellowfin and skipjack are within a range of dissolved oxygen saturation levels of 23%–43%. By contrast, bigeye has a greater tolerance to low dissolved oxygen (and to low temperature), which allows this species to extend its habitat to deeper layers of the ocean (Korsmeyer & Dewar, 2001; Sharp & Dizon, 1978). Tuna larvae are also sensitive to dissolved oxygen levels. Lethal conditions for post-hatched and first-feeding yellowfin larvae occur at dissolved oxygen levels below $2.2\,mg\,L^{-1}$.

Tunas have developed a generalized and opportunistic feeding strategy to survive in an oligotrophic environment. The spatial distribution of actively feeding tuna schools is highly patchy in the tropical ocean (Fonteneau, 1986; Ravier et al., 2000). This reflects discontinuities in the distribution of prey, as depicted by satellite imagery of chlorophyll blooms in the open ocean, which condition the development of the food chain. The largest tuna concentration ever observed in the Indian Ocean was clearly associated with a chlorophyll bloom, which developed 2–3 weeks before the catch by the tuna vessels, at the edge of an anticyclonic eddy (Fonteneau et al., 2008). In many instances, such a contrasted "foraging" landscape is related to mesoscale eddies, although the impact of eddies on the distribution of marine top predators is stronger for seabirds than for tunas (Tew Kai & Marsac, 2010). Still, mesoscale features tend to promote tuna aggregations in the water column, and this applies also to large cetaceans (Digby et al., 1999). Several authors have identified links between changes in mesoscale activity and large-scale climate signals such as the ENSO and IOD, but also the Pacific decadal oscillation (PDO) and the southern annular mode (not discussed here). In the Mozambique Channel, the eddy kinetic energy (EKE) weakens during La Niña and negative IOD events (Palastanga et al., 2006; Tew Kai & Marsac, 2009). An opposite response (i.e., lesser EKE during El Niño) seems to prevail in the southeast Indian Ocean due to a weaker Leeuwin Current occurring during El Niño events off western Australia (Zheng et al., 2018). Current coupled climate models fail to resolve mesoscale dynamics rendering predictions on the future state of the ocean mesoscale turbulence highly uncertain. In the event of a reduction of the mesoscale activity in the low latitudes under climate change, tropical tunas would endure challenging conditions to find appropriate foraging grounds in the open ocean.

At interannual timescales, spatial shifts in the distribution of tuna fleets were observed in the Pacific and Indian Oceans. In the Pacific, large-scale movements of tuna correlate with the position of the convergence on the eastern edge of the western Pacific warm pool (Lehodey et al., 1997). This convergence, induced by westward advection of the cold and productive waters of the adjacent central and eastern Pacific equatorial upwelling (Chavez & Barber, 1987), accumulates plankton and micronekton (tuna prey). The mechanism linking favorable physical and biological habitat (prey enhancement) was successfully modeled to relate the response of the skipjack tuna population to ENSO-related interannual changes (Lehodey et al., 2003). In the equatorial Indian Ocean, surface tuna fleets displaced from west to east during the 1997–1998 massive IOD as a response to mitigate the adverse effect of an anomalously deep mixed layer and reduced biological productivity in the regular fishing grounds (Marsac, 2017; Marsac & Le Blanc, 1998, 1999). These physical and biogeochemical processes, which are magnified during intense positive IODs, have been well described and modeled by several authors (Murtugudde et al., 1999; Murtugudde & Busalacchi, 1999; Webster et al., 1999; Wiggert et al., 2009). Due to wind anomalies, a positive IOD disrupts the typical seasonal evolution of phytoplankton bloom dynamics over the whole ocean basin. Because of a lesser occurrence of tuna schooling at the surface in the western Indian Ocean during the 1997–1998 positive IOD, the purse seiner fleets could not operate efficiently and had to relocate their activities to areas with higher biological productivity, notably in the east equatorial Indian Ocean (Fig. 8) (Marsac, 2017). During a weaker positive IOD event, in January 2007, the catch rates of the purse seiners were again very low (three times less than normal), but the fleets remained in the west Indian Ocean as several tuna schools could be fished over a shallow thermocline ridge associated with a pocket of higher productivity that formed to the North of Madagascar (Fig. 16.14 in Marsac, 2017). It appears that, for both Pacific and Indian Oceans, the prey-enhancement mechanism is a major driver of tuna fleet relocation. However, the variability in the thermocline depth in the western Indian Ocean plays an additional role in lowering catch rates with regard to the lesser vulnerability of tuna schools to the purse seine gear.

Multidecadal regimes are observed in tuna fisheries. However, such investigations require a very long time series of catch data, which is a rare situation in fisheries. Actually, this is the case in the Mediterranean and eastern Atlantic bluefin tuna fisheries, where substantial fluctuations in abundance were brought out from four centuries of catch records by the fixed traps set along the coast of the Mediterranean (Ravier & Fromentin, 2001). The annual fluctuations in the recruitment of juveniles appear to be amplified during alternating multidecadal regimes, with abundance negatively linked to temperature changes, and oceanographic conditions potentially affecting the spawning migration patterns (Ravier & Fromentin,

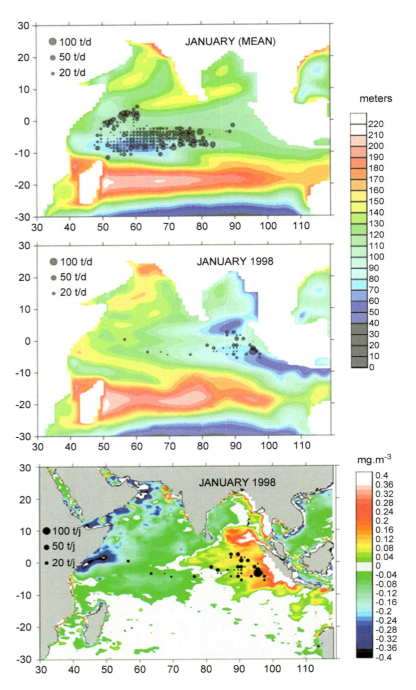

FIG. 8 Distribution of purse seine tuna Catch Per Unit Effort (circles, t day^{-1}) on free-swimming schools overlaid on environmental factors. *Top panel*: January mean, Catch Per Unit Effort 1991–2002 and 20°C isothermal depth (m, color shaded), 1980–2005. *Middle panel*: Catch Per Unit Effort and 20°C depth in January 1998 (positive dipole). These panels show that purse seine catches are associated with relatively shallow thermocline, located in the western Indian Ocean in normal conditions and in the eastern Indian Ocean during the positive 1997–98 IOD. *Bottom panel*: Catch Per Unit Effort and surface chlorophyll anomalies (mg m^{-3}, color shaded) in January 1998. The shallow thermocline in the eastern Indian Ocean triggered a chlorophyll bloom with which tuna catches were associated. *(Data sources: Tuna data from the Indian Ocean Tuna Commission web site (http://www.iotc.org/data/datasets); 20°C isothermal depth from the GODAS model outputs of the National Centers for Environmental Prediction https://cpc.ncep.noaa.gov/products/GODAS/; chlorophyll from SeaWiFs data. Reproduced from Marsac (2017).)*

2004). As for tunas in the Pacific, a study suggested that the high-frequency of ENSO-related recruitment signals can convert into low-frequency decadal fluctuations of the population biomass (Lehodey et al., 2006). The authors identified decadal regimes promoting high or low recruitment levels, as an accumulation over time of positive or negative recruitment anomalies. La Niña-dominant regimes, associated with a negative phase of the PDO, would promote the recruitment of the south Pacific albacore, a temperate tuna. By contrast, El Niño-dominated regimes, associated with positive PDO, would push up the recruitment of the tropical tunas, yellowfin, and skipjack.

Based on these studies, tuna fisheries may undergo substantial changes along with climate change. For the Indian Ocean, the western region is projected to be the most impacted with large positive changes in sea surface temperature and dramatic declines in NPP (Roxy et al., 2014). Unsuitable habitats are likely to expand over the 21st century, with

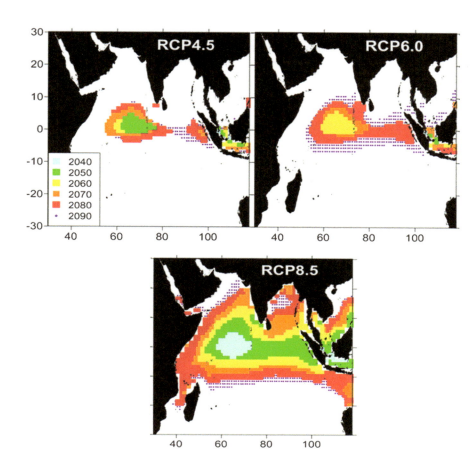

FIG. 9 Expansion of unsuitable thermal habitats for tropical tunas (SST >31°C) from years 2040 to 2099, through a range of Representative Concentration Pathways (RCP) scenarios (left to right: from reduced to high radiative forcing scenarios). *(Adapted from Fig. 6 in Dueri et al. (2014).)*

a magnitude varying across various representative concentration pathways scenarios (Fig. 9), as simulated by the APECOSM (Maury et al., 2007) coupled ecosystem model (Dueri et al., 2014). Tunas, thanks to their wide habitat, large-scale movements, and capabilities to adapt to changing conditions, would survive unless unregulated or poorly managed fisheries drag the tuna stocks down to minimal levels. On the other hand, the fleets will have to adapt their strategies and gears to a profound reshaping of the tuna habitat, both horizontally and in depth. The mobile industrial fleets could easily relocate outside the western Indian Ocean, which is currently the most productive region in terms of tuna catches, while locally operated domestic fleets will need to innovate to mitigate the impact of evolving habitat (Moustahfid et al., 2018). Obviously, these changes would also affect the socio-economic landscape of the coastal countries that are highly reliant on tuna resources.

There are good reasons to incorporate climate change considerations in the management of tuna fisheries. Firstly, climate change can affect the setting of targets and limit reference points used to inform the status of stocks. Secondly, changing environmental conditions will impair the reliability of forward projections of stock status from management strategy evaluation models (Hobday et al., 2018). Therefore, building species resilience and ensuring sustainable fisheries is based on proper monitoring of fishing activities and robust projections of ocean conditions in order to implement adaptive strategies.

5 Conclusion

The oceanographic and climate data provide contexts to which changes in resource availability and vulnerability to the various fishing methods and gears can be related (Sharp, 2001). Scales do matter, as they play on distinct but connected phases of the life cycle of the species exploited by the fisheries. Short-term responses, such as collective behavior and school formation, occur on the scale of days to weeks. Seasons shape the biological enrichment processes that induce movements, as well as trophic and spawning migrations. This timescale is well understood by the fishers as it often determines their area of operations, the fishing methods, and the gears. The interannual timescale can induce dramatic changes in the spatial distribution of the resources and has long been an unknown for the fishers. During the past four decades, the improved knowledge of ocean-atmosphere interactions now provides insight into the processes and to some extent, predictions on the future plausible states of the ocean. This story started with the onset of the satellite era, devoted to synoptic observations of land and oceans, especially by the polar-orbiting satellites since the late 1970s. The satellite-based sea surface temperature and ocean color products have been instrumental in identifying fronts and discontinuities leading to aggregations of pelagic fishes. The development of satellite altimetry in 1992 has set another milestone to unveil processes occurring at mesoscale (100–200 km), which play a structuring role through the ecosystem, from plankton to megafauna. The coupled ocean-atmosphere models took off in the 1990s thanks to the advent of supercomputers (Meehl, 1990). They started to provide an integrated 3D understanding of the ocean processes at basin scales, in relation to large-scale climate modes. The in situ ocean observations, which developed largely through international programs since the 1980s, now forming part of the Global Ocean Observing System, are assimilated into models to provide the community with higher quality ocean-state products (Zhang et al., 2007). The coupling of biogeochemical models and ecosystem models, in the 2000s, achieved the connection between the "blue" and the "green" ocean. This provided a range of tools that are key for fisheries oceanographers to better appraise the potential biological responses of fish resources from a climate change perspective. The new challenge is now to produce realistic high-resolution simulations to scales that are relevant to fine biological processes, which are on the order of a kilometer. Ecosystem variability at intraseasonal frequencies also remains largely unexplored (Wiggert et al., 2009).

Finally, the gradual improvement in understanding the biophysical interactions should not be a means to increase the fishing power and the catch rates of the fishing fleets, but it should deliver a robust scientific background to manage fisheries better and to delineate sensitive areas. It is noteworthy that nonclimate stressors such as overfishing and overcapacity, illegal, unreported, and unregulated fishing, and insufficient fisheries management can worsen the impacts of climate change alone.

6 Educational resources

NASA ocean color web site: https://oceancolor.gsfc.nasa.gov/
NASA Giovanni for data visualization: https://giovanni.gsfc.nasa.gov/giovanni/
The Bay of Bengal large marine ecosystem project: https://www.boblme.org/documentRepository/BOBLME-2015-Brochure-05.pdf and https://www.bobpigo.org/webroot/img/Climate%20Exchange.pdf
The WIOFish Database: https://www.wiofish.org
Marine resources of the Pemba Channel and climate change: https://www.youtube.com/watch?v=aUrsAI79iJc
The IOTC web site: www.iotc.org
The Economics of Adapting Fisheries to Climate Change | READ online (oecd-ilibrary.org).

Acknowledgments

The authors thank MM Shoaib Abdul Razzaq (WWF Pakistan), Mohammad Moazzam Khan (WWF Pakistan), Mohammad Shoaib Kiani (Centre of Excellence Marine Biology, University of Karachi, Pakistan), and Fabio Fiorellato (IOTC) for their contribution to data and various information on the North Indian Ocean fisheries.

Authors contribution

FM: Conceptualization, writing (original draft, review, and editing), and supervision. BE: Writing (original draft, review, and editing). US: Writing (original draft, review, and editing). PGS: Writing (original draft, review, and editing).

References

ABARES. (2018). *Australian fisheries and aquaculture statistics 2018*. Canberra, Australia: Australian Bureau of Agricultural and Resource Economics and Sciences.
Annamalai, H., & Murtugudde, R. (2004). Role of the Indian Ocean in regional climate variability. In Wang, et al. (Eds.), *Geophysical monograph series. Earth's climate: The ocean atmosphere interaction* (p. 147).
APFIC. (2012). Fourth APFIC regional consultative forum meeting. In *Improving management and governance of fisheries and aquaculture in the Asia-Pacific region, Da Nang, Vietnam, 17–19 September 2012*.
Baars, M., Schalk, P., & Veldhuis, M. (1998). Seasonal fluctuations in plankton biomass and productivity in the ecosystems of the Somali current, Gulf of Aden and Southern Red Sea. In K. Sherman, E. Okemwa, & M. Ntiba (Eds.), *Large marine ecosystems of the Indian Ocean: Assessment, sustainability and management* (pp. 143–174). Cambridge, U.S: Blackwell Science.
Bakun, A., & Csirke, J. (1998). Environmental processes and recruitment variability (Chapt. 6). *FAO Fisheries Technical Paper, 376*, 105–124.

Baliarsingh, S. K., Lotliker, A. A., Srichandan, S., Samanta, A., Kumar, N., & Nair, T. B. (2020). A review of jellyfish aggregations, focusing on India's coastal waters. *Ecological Processes, 9*(1), 1–9.

BDO EconSearch. (2019). *Australian fisheries and aquaculture industry 2017/2018: Economic contributions estimates report.* 119 p.

Beal, L. M., Vialard, J., Roxy, M. K., Li, J., Andres, M., Annamalai, H., Feng, M., Han, W., Hood, R., Lee, T., Lengaigne, M., Lumpkin, R., Masumoto, Y., McPhaden, M. J., Ravichandran, M., Shinoda, T., Sloyan, B. M., Strutton, P. G., Subramanian, A. C., ... Parvathi, V. (2020). A road map to IndOOS-2: Better observations of the rapidly warming Indian Ocean. *Bulletin of the American Meteorological Society, 101*(11), E1891–E1913. https://doi.org/10.1175/BAMS-D-19-0209.1.

Behrenfeld, M. J., & Falkowski, P. G. (1997). Photosynthetic rates derived from satellite-based chlorophyll concentration. *Limnology and Oceanography, 42*(1), 1–20.

Bhathal, B. (2005). Historical reconstruction of Indian marine fisheries catches, 1950-2000, as a basis for testing the marine trophic index. *Fisheries Centre Research Reports, 13*(5), 122.

Bopp, L., Resplandy, L., Orr, J. C., Doney, S. C., Dunne, J. P., Gehlen, M., Halloran, P., Heinze, C., Ilyina, T., Séférian, R., Tjiputra, J., & Vichi, M. (2013). Multiple stressors of ocean ecosystems in the 21st century: Projections with CMIP5 models. *Biogeosciences, 10*, 6225–6245.

Breuil, C., & Bodiguel, C. (2015). *Report of the meeting on marine small pelagic fishery in the United Republic of Tanzania SFFAO/2015/34, IOC-SmartFish Programme.* FAO. 96 pp http://www.fao.org/3/a-bl755e.pdf. (Accessed 6 January 2021).

Butler, M., Bollens, S. M., Burkhalter, B., Madin, L. P., & Horgan, E. (2001). Mesopelagic fishes of the Arabian Sea: Distribution, abundance and diet of Chauliodus pammelas, Chauliodus sloani, Stomias affinis, and Stomias nebulosus. *Deep Sea Research, Part II, 48*(6), 1369–1383.

Cai, W., Santoso, A., Wang, G., Weller, E., Wu, L., Ashok, K., Masumoto, Y., & Yamagata, T. (2014). Increased frequency of extreme Indian Ocean dipole events due to greenhouse warming. *Nature, 510*, 254–258.

Chavez, F. P., & Barber, R. T. (1987). An estimate of new production in the equatorial Pacific. *Deep Sea Research, 34*, 1229–1243.

Cheung, W. W. L., Bruggemen, J., & Butenschön, M. (2018). Projected changes in global and national potential marine fisheries catch under climate change scenarios in the twenty-first century. In M. Barange, T. Bahri, M. C. M. Beveridge, K. L. Cochrane, & S. Funge-Smith (Eds.), *Impacts of climate change on fisheries and aquaculture: Synthesis of current knowledge, adaptation and mitigation options (chapter 4)* (pp. 63–85). FAO fish Aquac tech paper 627.

Choongan, C. (2018). *Thailand's fisheries situation.* Bangkok: Department of Fisheries of Thailand. Bangkok, Thailand.

Cox, J. (2012). *Assessment report on small-scale fisheries in Africa.* Prepared by Masifundise development trust for AU-IBAR. 9 p.

Currie, J. C., Lengaigne, M., Vialard, J., Kaplan, D. M., Aumont, O., Naqvi, S. W. A., & Maury, O. (2013). Indian Ocean dipole and El Niño/southern oscillation impacts on regional chlorophyll anomalies in the Indian Ocean. *Biogeosciences, 10*(10), 6677–6698. https://doi.org/10.5194/bg-10-6677-2013.

Department of Fisheries Development. (2016). *Marine fisheries frame survey 2016, Zanzibar. Department of Fisheries Development Ministry of agriculture, natural resources, livestock and fisheries, Zanzibar. SWIOFish Project/World Bank* (p. 54).

Desai, B. N., & Bhargava, R. M. S. (1998). Biologic production and fishery potential of the exclusive economic zone of India. In K. Sherman, E. Okemwa, & M. Ntiba (Eds.), *Large marine ecosystems of the Indian Ocean: Assessment, sustainability and management* (pp. 322–333). Cambridge, U.S: Blackwell Science.

Digby, S., Antczak, T., Leben, R., Born, G., Barth, S., Cheney, R., Foley, D., Goni, G. J., Jacobs, G., & Shay, N. (1999). Altimeter data for operational use in the marine environment. *Oceans, 9*, 605–613. MTS/IEEE, riding the crest into the 21st century 2.

Dueri, S., Bopp, M., & Maury, O. (2014). Projecting the impacts of climate change on skipjack tuna abundance and spatial distribution. *Global Change Biology, 20*, 742–753.

Dunn, R. E. (2012). *The adventures of Ibn Battuta: A Muslim traveler of the fourteenth century.* University of California press. 384 p.

Dwivedi, S. N., & Choubey, A. K. (1998). Indian Ocean large marine ecosystems: Need for national and regional framework for conservation and sustainable development. In K. Sherman, E. Okemwa, & M. Ntiba (Eds.), *Large marine ecosystems of the Indian Ocean: Assessment, sustainability and management* (pp. 361–367). Cambridge, U.S: Blackwell Science.

Elayaperumal, V., Hermes, R., & Brown, D. (2019). An ecosystem based approach to the assessment and governance of the Bay of Bengal Large Marine Ecosystem. *Deep Sea Research II, 163*, 87–95. https://doi.org/10.1016/j.dsr2.2019.01.001.

Emerton, L. (2014). *Assessing, demonstrating and capturing the economic value of marine and coastal ecosystem services in the Bay of Bengal large marine ecosystem.* BOBLME-2014-Socioec-02. 86 pp.

Fan, L., Liu, Q., Wang, C., & Guo, F. (2016). Indian Ocean dipole modes associated with different types of ENSO development. *Journal of Climate, 30*(6), 2233–2249.

FAO. (2005). *Fishery Country Profiles.* www.fao.org/countryprofiles/selectiso.asp?lang=en.

FAO. (2020a). *The state of world fisheries and aquaculture 2020.* Rome: Sustainability in action. 206 pp.

FAO. (2020b). FishStatJ—Software for fishery and aquaculture statistical time series. In *FAO fisheries division [online].* Rome. (Updated 14 September 2020).

FAO. (2021). *Global capture production 1950–2018.* http://www.fao.org/fishery/statistics/global-capture-production/query/en. (Accessed 6 January 2021).

Fonteneau, A. (1986). Analysis of exploitation of some yellowfin tuna concentrations by purse seiners during the period 1980–1983 in the East Atlantic. In *25. ICCAT Coll. Vol. Sci. Pap.* (pp. 81–98).

Fonteneau, A., Lucas, V., Tew Kai, E., Delgado, A., & Demarcq, H. (2008). Mesoscale exploitation of a major tuna concentration in the Indian Ocean. *Aquatic Living Resources, 21*, 109–121.

Fulanda, B., Ohtomi, J., Mueni, E., & Kimani, E. (2011). Fishery trends, resource-use and management system in the Ungwana Bay fishery Kenya. *Ocean and Coastal Management, 54*(5), 401–414.

Gartner, J. V. (1993). Patterns of reproduction in the dominant lanternfish species (Pisces: Myctophidae) of the eastern Gulf of Mexico, with a review of reproduction among tropical-subtropical Myctophidae. *Bulletin of Marine Science, 52*(2), 721–750.

Gjøsæter, J., & Tilseth, S. (1983). *Survey on mesopelagic fish resources in the Gulf of Oman*. Reports on surveys with RV Dr. Fridtjof Nansen Bergen: Institute of Marine Research. NORAD/FAO/UNDP project GLO/82/001, 33 p.

Goes, J. I., Thoppil, P. G., Gomes, H.d. R., & Fasullo, J. T. (2005). Warming of the Eurasian landmass is making the Arabian Sea more productive. *Science, 308*(5721), 545–547. https://doi.org/10.1126/science.1106610.

Goes, J. I., Tian, H., Gomes, H.d. R., Anderson, O. R., Al-Hashmi, K., de Rada, S., Luo, H., Al-Kharusi, L., Al-Azri, A., & Martinson, D. G. (2020). Ecosystem state change in the Arabian Sea fuelled by the recent loss of snow over the Himalayan-Tibetan plateau region. *Scientific Reports, 10*(1), 7422. https://doi.org/10.1038/s41598-020-64360-2.

Graham, B. G., & Dickson, K. A. (2001). Anatomical and physiological specializations for endothermy. In B. A. Block, & E. D. Stevens (Eds.), *Tuna: Physiology, ecology, and evolution* (pp. 121–165). Academic Press. Chapter 4.

Groeneveld, J. C. (2015). Capture fisheries. In J. Paula (Ed.), *The regional state of the coast report: Western Indian Ocean. UNEP-Nairobi convention and WIOMSA*. (Chapter 21), Nairobi, Kenya, 546 pp.

Heileman, S., Eghtesadi-Araghi, P., & Mistafa, N. (2008). VI-9 Arabian Sea large marine ecosystem (LME 32). In K. Sherman, & G. Hempel (Eds.), *The UNEP large marine ecosystem report. A perspective on changing conditions in LMEs of the world's regional seas* (pp. 221–234). Nairobi, Kenya: United Nations Environment Programme. UNEP regional seas report and studies no. 182.

Hermes, J. C., Masumoto, Y., Beal, L. M., Roxy, M. K., Vialard, J., Andres, M., Annamalai, H., Behera, S., D'Adamo, N., Doi, T., Feng, M., Han, W., Hardman-Mountford, N., Hendon, H., Hood, R., Kido, S., Lee, C., Lee, T., Lengaigne, M., … Yu, W. (2019). A sustained ocean observing system in the Indian Ocean for climate related scientific knowledge and societal needs. *Frontiers in Marine Science, 6*, 355. https://doi.org/10.3389/fmars.2019.00355.

Hobday, A. J., Pecl, G. T., Fulton, B., Pethybridge, H., Bulman, C., & Villanueva, C. (2018). Climate change impacts, vulnerabilities and adaptation: Australian marine fisheries. In M. Barange, T. Bahri, M. C. M. Beveridge, K. L. Cochrane, & S. Funge-Smith (Eds.), *Impacts of climate change on fisheries and aquaculture: Synthesis of current knowledge, adaptation and mitigation options* (pp. 347–362). Rome: FAO. Chapter 16. FAO fish Aquac tech paper 627.

Hood, R. R., Coles, V. J., Huggett, J. A., Landry, M. R., Levy, M., Moffett, J. W., & Rixen, T. (2024a). Chapter 13: Nutrient, phytoplankton, and zooplankton variability in the Indian Ocean. In C. C. Ummenhofer, & R. R. Hood (Eds.), *The Indian Ocean and its role in the global climate system* (pp. 293–327). Amsterdam: Elsevier. https://doi.org/10.1016/B978-0-12-822698-8.00020-2.

Hood, R. R., Rixen, T., Levy, M., Hansell, D. A., Coles, V. J., & Lachkar, Z. (2024b). Chapter 12: Oxygen, carbon, and pH variability in the Indian Ocean. In C. C. Ummenhofer, & R. R. Hood (Eds.), *The Indian Ocean and its role in the global climate system* (pp. 265–291). Amsterdam: Elsevier. https://doi.org/10.1016/B978-0-12-822698-8.00017-2.

Hussain, M. S., Kim, S., & Lee, S. (2017). On the relationship between Indian Ocean dipole and the precipitation of Pakistan. *Theoretical and Applied Climatology*. https://doi.org/10.1007/s00704-016-1902-y.

Ihara, C., Kushnir, Y., & Cane, M. A. (2008). Warming trend of the Indian Ocean SST and Indian Ocean dipole from 1880 to 2004. *Journal of Climate, 21*(10), 2035–2046.

IOTC. (2021). *Indian Ocean tuna commission online data querying service*. https://iotc.org/oqs. (Accessed 12 January 2021).

Jagerroos, S. (2016). Assessment of living resources in the straits of Malacca, Malaysia: Case study. *Journal of Aquaculture & Marine Biology, 4*(1), 00070. https://doi.org/10.15406/jamb.2016.04.00070.

Jiddawi, N. S., & Öhman, M. C. (2002). Marine fisheries in Tanzania. *Ambio, 31*, 518–527.

Johannesson, K., & Valinassab, T. (1994). Survey of mesopelagic fish resources within the Iranian exclusive economic zone of the Oman Sea. In *Final report Govt/FAO project, UTF-IRA-020/IRA*. 85 p.

Keerthi, M. G., Lengaigne, M., Drushka, K., Vialard, J., de Boyer Montegut, C., Pous, S., Levy, M., & Muraleedharan, P. M. (2016). Intraseasonal variability of mixed layer depth in the tropical Indian Ocean. *Climate Dynamics, 46*, 2633–2655.

Keerthi, M. G., Lengaigne, M., Levy, M., Vialard, J., Parvathi, V., de Boyer Montégut, C., Ethé, C., Aumont, O., Suresh, I., Akhil, V. P., & Muraleedharan, P. M. (2017). Physical control of interannual variations of the winter chlorophyll bloom in the northern Arabian Sea. *Biogeosciences, 14*(15), 3615–3632.

Khan, M. A., Khan, J. A., Ali, Z., et al. (2016). The challenge of climate change and policy response in Pakistan. *Environment and Earth Science, 75*, 412.

KKP. (2015). *Analisis Data Pokok. Pusat Data dan Statistik. Kementerian Kelautan dan Perikanan*. Available at: http://statistik.kkp.go.id/sidatik-dev/Publikasi/src/analisisdatakkp2015.pdf.

Korsmeyer, K. E., & Dewar, H. (2001). Tuna metabolism and energetics. In B. A. Block, & E. D. Stevens (Eds.), *Tuna: Physiology, ecology, and evolution* (pp. 35–78). Academic Press. Chapter 2.

Lehodey, P., Alheit, J., Barange, M., Baumgartner, T., et al. (2006). Climate variability, fish and fisheries. *Journal of Climate, 19*, 5009–5030.

Lehodey, P., Bertignac, M., Hampton, J., Lewis, A., & Picaut, J. (1997). El Niño southern oscillation and tuna in the western Pacific. *Nature, 389*, 715–718.

Lehodey, P., Chai, F., & Hampton, J. (2003). Modelling climate-related variability of tuna populations from a coupled ocean biogeochemical-populations dynamics model. *Fisheries Oceanography, 12*, 483–494.

Longhurst, A. R. (2007). In A. R. Longhurst (Ed.), *The Indian Ocean. Ecological Geography of the Sea (2nd edition)* (pp. 275–325). Burlington: Academic Press (Chapter 10).

Marsac, F. (2017). The Seychelles tuna fishery and climate change. In B. Phillips, & M. Perez-Ramirez (Eds.), *Climate change impacts on fisheries and aquaculture—Vol II* (pp. 523–568). Wiley Blackwell.

Marsac, F., & Demarcq, H. (2019). *Outline of climate and oceanic conditions in the Indian Ocean: An update to mid-2019*. IOTC-2019-WPTT21-24_Rev1. 19 p.

Marsac, F., & Le Blanc, J. L. (1998). Interannual and ENSO-associated variability of the coupled ocean-atmosphere system with possible impacts on the yellowfin tuna fisheries of the Indian and Atlantic oceans. In J. S. Beckett (Ed.), *ICCAT tuna symposium. ICCAT collective volume of scientific papers, L (1)* (pp. 345–377).

Marsac, F., & Le Blanc, J. L. (1999). Oceanographic changes during the 1997–1998 El Niño in the Indian ocean and their impact on the purse seine fishery. First session of the IOTC working party on tropical tunas, Mahé, Seychelles. *IOTC Proceedings, 2*, 147–157.

Maury, O., Faugeras, B., Shin, Y.-J., Poggiale, J.-C., Ben Ari, T., & Marsac, F. (2007). Modelling environmental effects of size-structured energy flow through marine ecosystems. Part 1: The model. *Progress in Oceanography, 74*, 479–499.

McClanahan, T. R. (2017). Changes in coral sensitivity to thermal anomalies. *Marine Ecology Progress Series, 570*(71–85), 2017. https://doi.org/10.3354/meps12150.

McCreary, J. P., Murtugudde, R., Vialard, J., Vinayachandran, P. N., Wiggert, J. D., Hood, R. R., Shankar, D., & Shetye, S. (2009). Biophysical processes in the Indian Ocean. In J. D. Wiggert, R. R. Hood, S. W. A. Naqvi, S. L. Smith, & K. H. Brink (Eds.), *Geophysical Monograph Series 185. Indian Ocean biogeochemical processes and ecological variability* (pp. 9–32). Washington, D. C: American Geophysical Union.

Meehl, G. A. (1990). Development of global coupled ocean-atmosphere general circulation models. *Climate Dynamics, 5*, 19–33.

Monolisha, S., Grinson, G., & Platt, T. (2018). Biogeography of the Northern Indian Ocean (chapter 25). In *Winter School on structure and function of the marine ecosystem: Fisheries* (pp. 221–227). Fishery Resources Assessment Division ICAR-Central Marine Fisheries Research Institute.

Moustahfid, H., Marsac, F., & Gangopadhyay, A. (2018). Climate change impacts, vulnerabilities and adaptations: Western Indian Ocean marine fisheries. In M. Barange, T. Bahri, M. C. M. Beveridge, K. L. Cochrane, & S. Funge-Smith (Eds.), *Impacts of climate change on fisheries and aquaculture: Synthesis of current knowledge, adaptation and mitigation options* (pp. 251–279). Rome: FAO. Chapter 12. FAO fish Aquac tech paper 627.

Munga, C. N., Omukoto, J. O., Kimani, E. N., & Vanreusel, A. (2014). Propulsion-gear-based characterisation of artisanal fisheries in the Malindi-Ungwana Bay, Kenya and its use for fisheries management. *Ocean and Coastal Management, 98*, 130–139.

Murdiyanto, B. (2011). In T. W. Nurani, D. Simbolon, A. Solihin, & S. Yuniarta (Eds.), *"Perikanan Tangkap: Dulu and Sekarang," new paradigm in marine fisheries: Pemanfaatan and Pengelolaan Sumberdaya Perikanan Laut Berkelanjutan* (pp. 33–44). Bogor: Departemen Pemanfaatan dan Sumberdaya Perikanan. IPB.

Murtugudde, R., & Busalacchi, A. J. (1999). Interannual variability of the dynamics and thermodynamics of the tropical Indian Ocean. *Journal of Climate, 12*, 2300–2326.

Murtugudde, R., McCreary, J. P., Jr., & Busalacchi, A. J. (2000). Oceanic processes associated with anomalous events in the Indian Ocean with relevance to 1997–1998. *Journal of Geophysical Research, 105*, 3295–3306.

Murtugudde, R., Signorini, S. R., Christian, J. R., Busalacchi, A. J., McClain, C. R., & Picaut, J. (1999). Ocean color variability of the tropical Indo-Pacific basin observed by SeaWiFS during 1997–1998. *Journal of Geophysical Research, 104*(C8), 18351–18366.

Nafpaktitis, B., & Nafpaktitis, M. (1969). *Lanternfishes (family Myctophidae) collected during cruises 3 and 6 of the R/V Anton Bruun in the Indian Ocean.* Los Angeles County Museum of Natural History.

Palastanga, V., van Leeuwen, P. J., & de Ruijter, W. P. M. (2006). A link between low frequency mesoscale eddy variability around Madagascar and the large scale Indian Ocean variability. *Journal of Geophysical Research, Oceans, 111*(C9), C09029.

Palomares, M. L. D., Khalfallah, M., Zeller, D., & Pauly, D. (2021). The fisheries of the Arabian Sea large marine ecosystem. In L. A. Jawad (Ed.), *The Arabian Sea: Biodiversity, environmental challenges and conservation measures* (pp. 883–897). Springer. https://doi.org/10.1007/978-3-030-51506-5_38.

Phillott, A. D., Firdous, F., & Shahid, U. (2018). Sea turtle hatchery practices and hatchling production in Karachi, Pakistan, from 1979-1997. *Indian Ocean Turtle Newsletter, 27*, 2–8.

Polovina, J. J., Howell, E. A., & Abecassis, M. (2008). Ocean's least productive waters are expanding. *Geophysical Research Letters, 35*(3). https://doi.org/10.1029/2007GL031745.

Ravier, C., & Fromentin, J. M. (2001). Long-term fluctuations in the eastern Atlantic and Mediterranean bluefin tuna population. *ICES Journal of Marine Science, 58*, 1299–1317.

Ravier, C., & Fromentin, J. M. (2004). Are the long-term fluctuations in Atlantic bluefin tuna (Thunnus thynnus) population related to environmental changes? *Fisheries Oceanography, 13*, 145–160.

Ravier, C., Marsac, F., Fonteneau, A., & Pallares, P. (2000). Contribution to the study of tuna concentrations in the eastern tropical Atlantic. *Collective Volume of Scientific Papers ICCAT, 51*(2), 679–712.

Resplandy, L., Vialard, J., Lévy, M., Aumont, O., & Dandonneau, Y. (2009). Seasonal and intraseasonal biogeochemical variability in the thermocline ridge of the southern tropical Indian Ocean. *Journal of Geophysical Research, 114*, C07024.

Roxy, M. K., Modi, A., Murtugudde, R., Valsala, V., Panickal, S., Prasanna Kumar, S., Ravichandran, M., Vichi, & Lévy, M. (2016). A reduction in marine primary productivity driven by rapid warming over the tropical Indian Ocean. *Geophysical Research Letters, 43*, 826–833. https://doi.org/10.1002/2015GL066979.

Roxy, M. K., Ritika, K., Terray, P., & Masson, S. (2014). The curious case of the Indian Ocean warming. *Journal of Climate, American Meteorological Society, 27*(22), 8501–8509.

Roxy, M. K., Ritika, K., Terray, P., Murtugudde, R., Ashok, K., & Goswami, B. N. (2015). Drying of Indian subcontinent by rapid Indian Ocean warming and a weakening land-sea thermal gradient. *Nature Communications, 6*, 7423.

Rozhan, A. D. (2019). Malaysia's Agrofood policy (2011−2020)—Performance and new direction. *FFTC Agricultural Policy Platform*, 1368.

Saji, N. H., Goswami, B. N., Vinayachandran, P. N., & Yamagata, T. (1999). A dipole mode in the tropical Indian Ocean. *Nature, 401*, 360–363.

Sartimbul, A., Rohadi, E., Yona, D., Yuli, H. E., Sambah, A. B., & Arleston, J. (2018). Cange in species composition and its implication on climate variation in Bali Strait: Case study in 2006 and 2010. *Journal of Survey in Fisheries Science, 4*(2), 38–46.

Sathyendranath, S., Brewin, R. J. W., Brockmann, C., Brotas, V., Calton, B., Chuprin, A., Cipollini, P., Couto, A. B., Dingle, J., Doerffer, R., Donlon, C., Dowell, M., Farman, A., Grant, M., Groom, S., Horseman, A., Jackson, T., Krasemann, H., Lavender, S., … Platt, T. (2019). An ocean-colour time series for use in climate studies: The experience of the ocean-colour climate change initiative (OC-CCI). *Sensors, 19*(19). https://doi.org/10.3390/s19194285.

Schott, F. A., & McCreary, J. P. (2001). The monsoon circulation of the Indian Ocean. *Progress in Oceanography*, *51*, 1–123.

Sekadende, B., Scott, L., Anderson, J., Aswani, S., Francis, J., Jacobs, Z., Jebri, F., Jiddawi, N., Kamukuru, A. T., Kelly, S., Kizenga, H., Kuguru, B., Kyewalyanga, M., Noyon, M., Nyandwi, N., Painter, S. C., Palmer, M., Raitsos, D. E., Roberts, M., ... Popova, E. (2020). The small pelagic fishery of the Pemba Channel, Tanzania: What we know and what we need to know for management under climate change. *Ocean and Coastal Management*, *197*. https://doi.org/10.1016/j.ocecoaman.2020.105322.

Sharp, G. D. (2001). Tuna oceanography—An applied science. In B. A. Block, & E. D. Stevens (Eds.), *Tuna: Physiology, ecology, and evolution* (pp. 345–389). Academic Press. Chapter 9.

Sharp, G. D., & Dizon, A. E. (1978). *The physiological ecology of tunas*. San Francisco/New York: Academic Press. 485 pp.

Sherman, K. (2003). *Physical, biological and human forcing of biomass yields in large marine ecosystems*. ICES CM 2003/P:12, 26 pp.

Sherman, K., & Alexander, L. M. (1986). *Variability and management of large marine ecosystems. American Association for the Advancement of Science (AAAS) Selected symposium 99*. Boulder, CO: Westview Press.

Siswanto, E., & Tanaka, K. (2014). Phytoplankton biomass dynamics in the strait of Malacca within the period of the SeaWiFS full mission: Seasonal cycles, interannual variations and decadal-scale trends. *Remote Sensing*, *6*, 2718–2742. https://doi.org/10.3390/rs6042718.

Strutton, P. G., & Hood, R. (2019). Ocean primary productivity: Variability and change. lead authors In L. M. Beal, J. Vialard, & M. K. Roxy (Eds.), *IndOOS-2: A roadmap to sustained observations of the Indian Ocean for 2020-2030* (pp. 83–88). Chapter 16. CLIVAR-4/2019, GOOS-237, 206 pp.

Susanto, R. D., Gordon, A. L., & Zheng, Q. (2001). Upwelling along the coasts of Java and Sumatra and its relation to ENSO. *Geophysical Research Letters*, *28*, 1599–1602. https://doi.org/10.1029/2000GL011844.

Taylor, S. F. W., Roberts, M. J., Milligan, B., & Ncwadi, R. (2019). Measurement and implications of marine food security in the Western Indian Ocean: An impending crisis? *Food Security*, *11*, 1395–1415. https://doi.org/10.1007/s12571-019-00971-6.

Tew Kai, E., & Marsac, F. (2009). Patterns of variability of sea surface chlorophyll in the Mozambique Channel: A quantitative approach. *Journal of Marine Systems*, *77*, 77–88.

Tew Kai, E., & Marsac, F. (2010). Influence of mesoscale eddies on spatial structuring of top predators' communities in the Mozambique Channel. *Progress in Oceanography*, *86*, 214–223.

Townsley, P. (2012). *Review of coastal and marine livelihoods and food security in the Bay of Bengal large marine ecosystem region*. Bay of Bengal Large Marine Ecosystem Programme. 116 pp.

Turpie, J. (2000). *The use and value of natural resources of the Rufiji floodplain and delta, Tanzania*. REMP/IUCN Technical Report 17 (p. 108).

UNDP. (2020). *Planetary pressures-adjusted Human Development Index*. Human Development Report 2020 http://hdr.undp.org/sites/default/files/hdr2020.pdf.

van der Elst, R., Everett, B., Jiddawi, N., Mwatha, G., Afonso, P. S., & Boulle, D. (2005). Fish, fishers and fisheries of the Western Indian Ocean: Their diversity and status. A preliminary assessment. *Philosophical Transactions of the Royal Society A - Mathematical Physical and Engineering Sciences*, *363*, 263–284.

Van Hoof, L., & Kraan, M. (2017). Mission report Tanzania; scoping Mission marine fisheries Tanzania. In *Wageningen, Wageningen Marine Research (University & Research centre), Wageningen Marine Research report*. 66pp.

Wabnitz, C. C., Lam, V. W., Reygondeau, G., Teh, L. C., Al-Abdulrazzak, D., Khalfallah, M., et al. (2018). Climate change impacts on marine biodiversity, fisheries and society in the Arabian Gulf. *PLoS One*, *13*(5), e0194537.

Walker, T. D. (2024). Chapter 2: A brief historical overview of the maritime Indian Ocean World (ancient times to 1950). In C. C. Ummenhofer, & R. R. Hood (Eds.), *The Indian Ocean and its role in the global climate system* (pp. 33–48). Amsterdam: Elsevier. https://doi.org/10.1016/B978-0-12-822698-8.00005-6.

Walmesley, S., Purvis, J., & Ninnes, C. (2006). The role of small-scale fisheries management in the poverty reduction strategies in the Western Indian Ocean region. *Ocean and Coastal Management*, *49*, 812–833.

Webster, P. J., Moore, A. M., Loschnigg, J. P., & Leben, R. R. (1999). Coupled ocean-temperature dynamics in the Indian Ocean during 1997–98. *Nature*, *401*, 356–360.

Wiggert, J. D., Vialard, J., & Behrenfeld, M. J. (2009). Basin-wide modification of dynamical and biogeochemical processes by the positive phase of the Indian Ocean dipole during the SeaWiFS era. In J. D. Wiggert, R. R. Hood, S. W. A. Naqvi, S. L. Smith, & K. H. Brink (Eds.), *Indian Ocean biogeochemical processes and ecological variability* (pp. 385–407). Washington, D. C: American Geophysical Union. Geophysical Monograph Series 185.

WIOFish. (2020). *Western Indian Ocean Fisheries Database*. Retrieved from www.wiofish.org. Accessed December 2020.

World Bank. (2012). *The hidden harvest: The global contribution of capture fisheries*. Washington, DC: World Bank. https://openknowledge.worldbank.org/handle/10986/11873. (Accessed 8 January 2021).

World Bank. (2018). *Revitalizing Pakistan's fisheries: Options for sustainable development*. World Bank.

World Bank. (2020). *World development indicators*. https://databank.worldbank.org/reports.aspx?source=world-development-indicators.

World Resources Institute. (2002). *People and ecosystems*. Washington, DC: World Resources Institute.

Wyrtki, K. (1962). The upwelling in the region between Java and Australia during the South-East monsoon. *Australian Journal of Marine & Freshwater Research*, *13*, 217–225. https://doi.org/10.1071/MF9620217.

Zhang, S., Harrison, M. J., Rosati, A., & Wittenberg, A. (2007). System design and evaluation of coupled ensemble data assimilation for global oceanic climate studies. *Monthly Weather Review*, *135*(10), 3541–3564.

Zheng, S., Feng, M., Du, Y., Meng, X., & Yu, W. (2018). Interannual variability of eddy kinetic energy in the subtropical Southeast Indian Ocean associated with the El Niño-Southern oscillation. *Journal of Geophysical Research, Oceans*, *123*, 1048–1061.

Chapter 12

Oxygen, carbon, and pH variability in the Indian Ocean☆

Raleigh R. Hood[a], Timothy Rixen[b], Marina Levy[c], Dennis A. Hansell[d], Victoria J. Coles[a], and Zouhair Lachkar[e]

[a]Horn Point Laboratory, University of Maryland Center for Environmental Science, Cambridge, MD, United States, [b]Leibniz Centre for Tropical Marine Research, University of Bremen, Bremen, Germany, [c]LOCEAN-IPSL, Sorbonne Universités (UPMC, Univ Paris 06)-CNRS-IRD-MNHN, Paris, France, [d]Rosenstiel School of Marine and Atmospheric Science, University of Miami, Miami, FL, United States, [e]Center for Prototype Climate Modeling, New York University Abu Dhabi, Abu Dhabi, United Arab Emirates

1 Introduction

1.1 The northern Indian Ocean oxygen minimum zones (OMZs)

Open ocean oxygen minimum zones (OMZs) occur below the mixed layer of the ocean and are characterized by oxygen concentrations that are low in comparison with surface and deep waters (Dietrich, 1936; Seiwell, 1937; Sverdrup, 1938). Even though mid-water oxygen minimum zones are common features in the ocean, they are typically classified as OMZs only when the lowest minimum oxygen concentrations in the vertical profile drop below certain thresholds (Paulmier & Ruiz-Pino, 2009). However, due to the varying oxygen tolerances of marine organisms, there are multiple definitions for threshold concentrations. Table 1 outlines common definitions, thresholds, and impacts of varying definitions of oxygen limitations. A threshold of about 20 μmol/kg is often used to define OMZs because concentrations below this have profound implications for microbes and marine biogeochemical cycles. These oxygen levels are therefore referred to as "microbial hypoxia" (Table 1). The total volume of water characterized by microbial hypoxia in the global ocean is approximately $15 \times 10^{15} m^3$, of which 21% ($3.13 \times 10^{15} m^3$) is located in the northern Indian Ocean (Acharya & Panigrahi, 2016; Garcia et al., 2010). Even though the Indian Ocean OMZ constitutes only 0.23% of the ocean's volume ($1355 \times 10^{15} m^3$), its influence on the cycles of nitrogen, carbon, and associated elements is globally significant.

The thickest OMZ in the world is found in the Arabian Sea (e.g., Morrison et al., 1999) where functional anoxia (concentrations <0.05 μmol/kg, Table 1) occurs in intermediate water (~200–800 m), with profound impacts on the nitrogen cycle as a result of denitrification. Very low oxygen concentrations are also found in intermediate water in the Bay of Bengal, but there are important physical and biogeochemical differences between the Arabian Sea and the Bay of Bengal, which have, so far, prevented the development of persistent functional anoxia in the Bay of Bengal OMZ (Figs. 1 and 2). Hence, in contrast to the Bay of Bengal, the Arabian Sea OMZ is a globally important zone of denitrification (Naqvi et al., 2005), where NO_3^- and NO_2^- are converted to N_2O and N_2 gas, which is then released to the atmosphere. This process of denitrification removes nitrogen-containing compounds from the ocean (Figs. 1 and 2) and generates N_2O, a prominent greenhouse gas (Bange et al., 2001; Ramaswamy et al., 2001). The Arabian Sea OMZ is most intense in the eastern part of the basin (Figs. 1 and 2), with the water column contributing ~10%–20% of global open-ocean mid-water column denitrification (Anju et al., 2022; Codispoti et al., 2001; Rixen et al., 2020).

Questions remain regarding the relative roles of biological oxygen demand derived from surface organic matter export versus circulation and ventilation timescales in maintaining subtle differences in the deep oxygen fields in the Arabian Sea and the Bay of Bengal (Bopp et al., 2017; McCreary et al., 2013; Rixen et al., 2020; Valsala, 2009). Recent observational and modeling studies in the Indian Ocean suggest that the OMZs are expanding in response to global warming (Lachkar et al., 2020; Rixen et al., 2020). Perhaps the low nitrogen-to-phosphorus ratios in the western Bay of Bengal are a harbinger of what is to come there (Fig. 1). This OMZ expansion is consistent with global modeling studies (Doney, 2010; Stramma

---

☆This book has a companion website hosting complementary materials. Visit this URL to access it: https://www.elsevier.com/books-and-journals/book-companion/9780128226988.

TABLE 1 Definitions, thresholds, and impacts of different hypoxia thresholds from Hofmann et al. (2011).

Terminology	Threshold for impacts to occur may be in different units	Impacts	Indian Ocean Regions with oxygen minimums below threshold	References
Mild hypoxia	107 μmol/kg, 109 μM, 3.5 mgO$_2$/L, ~53% saturation,	Sensitive species show avoidance reactions	Eastern and Central Indian Ocean south of 20S, Western Indian Ocean south of 10S	Cenr (2010)
Hypoxia	61 μmol/kg, 63 μM, 2 mgO$_2$/L, ~30% saturation	Fishes and the majority of higher organisms suffer from oxygen deficiency, ecosystem adapted to low oxygen conditions	Arabian Sea, Bay of Bengal	Ekau et al. (2010) and Vaquer-Sunyer and Duarte (2008)
Microbial hypoxia, suboxic, severe hypoxia, typical OMZ definition	22 μmol/kg, 22 μM, 0.71 mgO$_2$/L, ~11% saturation	Microbes start to experience the toxic effects of oxygen on anaerobic processes, and only highly specialized species survive	Northern tropical Indian Ocean	Rixen et al. (2020) and Diaz and Rosenberg (2008)
Functional anoxia	0.05 μM	Anaerobic microbial processes dominate, nitrite accumulation	Central and Eastern Arabian Sea	Rixen et al. (2020), Morrison et al. (1999), and Ulloa et al. (2012)

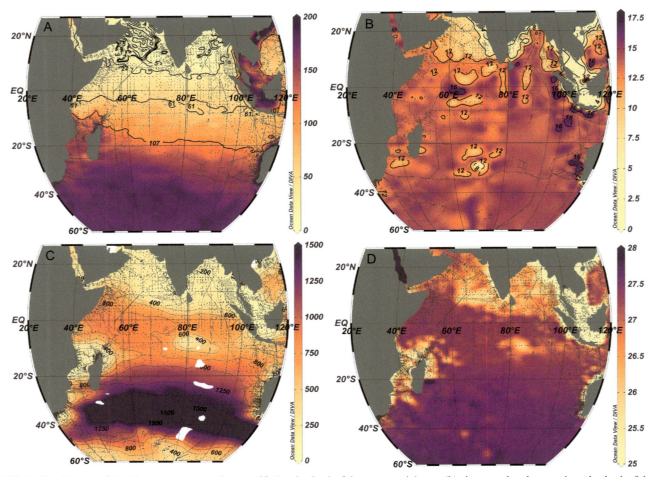

FIG. 1 Spatial maps of the (a) oxygen concentration (umol/kg) at the depth of the oxygen minimum, (b) nitrate to phosphorus ratio at the depth of the oxygen minimum, (c) the depth (m) of the oxygen minimum, (d) the potential density anomaly (kg/m^3) at the depth of the oxygen minimum. *(Data from Garcia et al., 2018).*

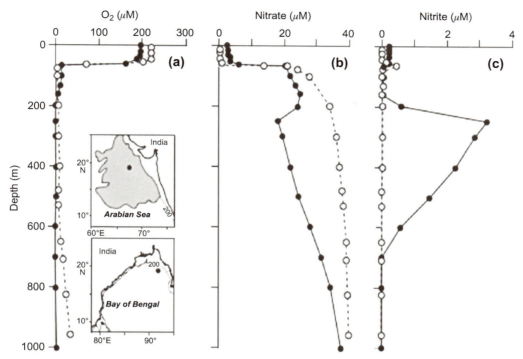

FIG. 2 Comparison of vertical profiles of (a) oxygen, (b) nitrate, and (c) nitrite in the Arabian Sea (filled circles) and Bay of Bengal (open circles). Station locations are shown in insets. The Arabian Sea inset also shows the limit of the denitrification zone to the eastern-central basin. *(Reproduced from Naqvi et al. (2006).)*

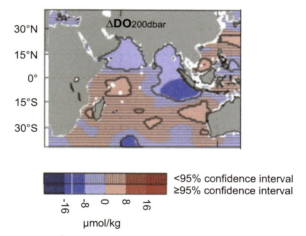

FIG. 3 Changes in dissolved oxygen (DO) (μmol kg^{-1}) at 200 dbar between 1960 and 1974 and 1990–2008 in the Indian Ocean. Increases (decreases) in dissolved oxygen are indicated in *red* (*blue*). Areas with differences below the 95% confidence interval are shaded by black horizontal lines. *(Modified from Stramma et al. (2010).)*

et al., 2008, 2010; Fig. 3), but uncertainties in global model predictions are large (Bopp et al., 2017; Kwiatkowski et al., 2020; McCreary et al., 2013; Rixen et al., 2020; Schmidt et al., 2020, 2021).

Given the importance of the Indian Ocean OMZs in the global carbon and nitrogen cycles, including the production of radiatively active greenhouse gas, it is essential to understand the biogeochemical variability associated with these regions, their rates of change, and the potential implications for global warming.

1.2 Role of the Indian Ocean in the global carbon cycle

The Indian Ocean plays an important role in the global carbon cycle, yet it remains one of the most poorly sampled ocean regions with respect to inorganic and organic carbon pools and air-sea carbon fluxes (Figs. 4–6). Estimates suggest that the

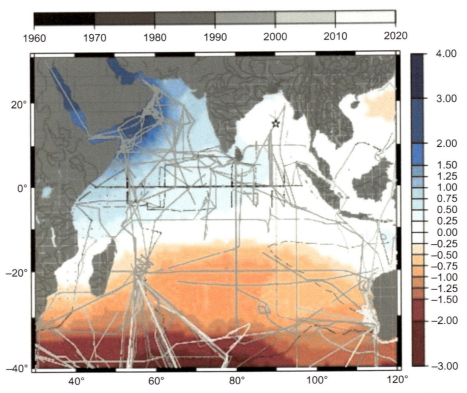

FIG. 4 Annual CO_2 flux ($mol\,C\,m^{-2}\,yr^{-1}$) referenced to the year 2000 (Takahashi et al., 2009) over the Indian Ocean. Data points colored by year of collection are overlaid as points. Major rivers are also delineated over the continents. The star in the central Bay of Bengal shows the location of a mooring that has been collecting continuous CO_2 and pH measurements since November 2013.

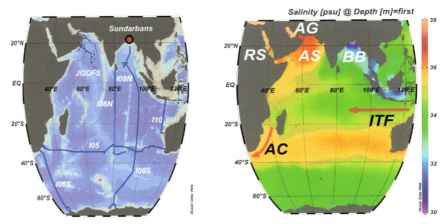

FIG. 5 Map of the Indian Ocean showing oceanographic sections discussed (left) and gridded surface salinity (right) during austral summer (based on Garcia et al. (2018)), with transports associated with the Indonesian Throughflow (ITF) and Agulhas Current (AC) indicated. Relevant basins are labeled: Arabian Sea (AS); Bay of Bengal (BB); Red Sea (RS); and Arabian Gulf (AG), also known as the Persian Gulf.

Indian Ocean accounts for a significant fraction of the global oceanic uptake of atmospheric CO_2 (de Verneil et al., 2021; Sarma et al., 2013; Sreeush et al., 2018, 2020; Takahashi et al., 2002; Valsala & Maksyutov, 2013). This uptake of anthropogenic CO_2 drives a decrease in pH, and indeed, surface pH values in the Indian Ocean are among the lowest of the major ocean basins (Feely et al., 2009; Sreeush et al., 2019). Projected increases in oceanic CO_2 concentrations will lead to further acidification of the Indian Ocean over the coming decades, with potentially severe negative impacts on coral reefs and other calcifying organisms (Doney, 2010; Hoegh-Guldberg et al., 2007).

Dissolved organic carbon (DOC) is a significant carbon pool in the ocean. As observed elsewhere in the global ocean, DOC concentrations in the Indian Ocean tend to be high in stratified near-surface tropical and subtropical waters where DOC is produced and accumulates, and DOC concentrations are lowest in the deep ocean where heterotrophic consumption of DOC exceeds autotrophic production (Hansell, 2009; Hansell et al., 2009; Figs. 6 and 7). Unsurprisingly, the highest

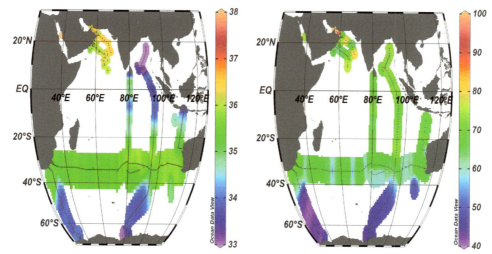

FIG. 6 Surface distributions of salinity (left) and DOC (μmolC/kg; right) observed during cruises. *(Data from Hansell and Orellana (2021).)*

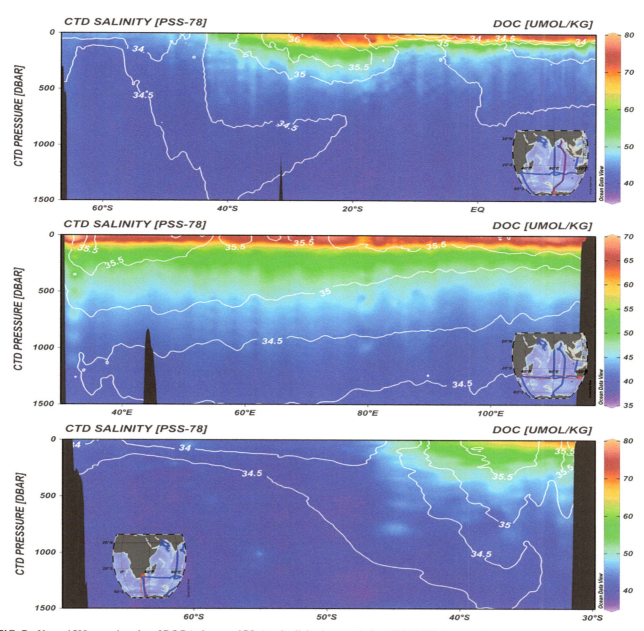

FIG. 7 Upper 1500 m section plot of DOC (color; μmolC/kg) and salinity (contours) along I09N/I08S (top panel, occupied in 2016), I05 (middle panel, occupied in 2009) and I06S (bottom panel, occupied in 2019). *(Data from Hansell and Orellana (2021).)*

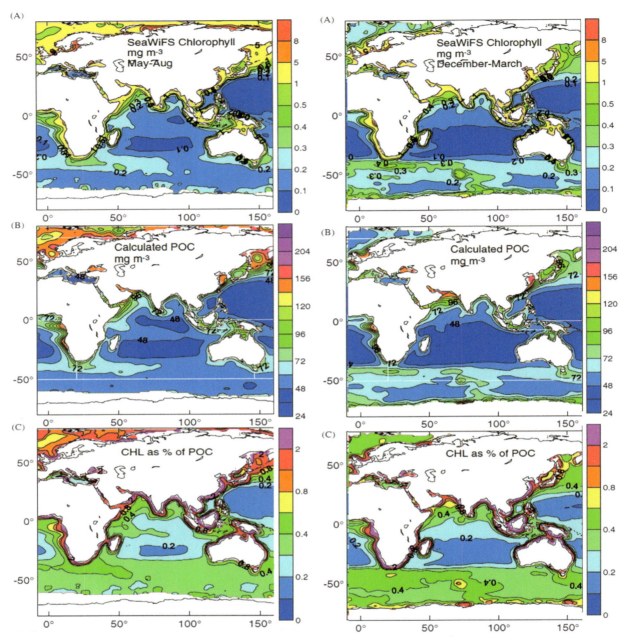

FIG. 8 Distribution of: (a) SeaWiFS CHL (mg m^{-3}, level 3, reprocessing 4 data); (b) average POC (mgC/m^3) over one attenuation depth calculated from K_{490}:c_p:POC; (c) CHL as a % of POC for (left panels) summer season (1997–2002, May–August, 20 months); (right panels) winter season (1997–2002, December–March, 20 months). *White lines* in (b) mark boundaries separating ocean basins as used by Behrenfeld and Falkowski (1997). *(Modified from Gardner et al. (2006).)*

concentrations of DOC are found in the near-surface waters of the central Arabian Sea due to high autotrophic production, and in the northern Bay of Bengal due to the large inputs of fresh water and associated terrigenous DOC flux from rivers (Hansell, 2009; Shah et al., 2018; Fig. 6). In contrast, particulate organic carbon (POC) concentrations tend to be high in coastal regions of the Indian Ocean where autotrophic production is high, and lowest in the oligotrophic southern subtropical gyre and equatorial waters where autotrophic production is low (Gardner et al., 2006; Fig. 8). The highest POC concentrations are found in the western and northern Arabian Sea during the Southwest Monsoon and Northeast Monsoon, respectively (Gardner et al., 2006). The spatial and temporal variability in POC export flux in the Indian Ocean is similar to the productivity and POC concentration patterns, consistent with the idea that primary productivity is the main control of the spatial and temporal variability of organic carbon fluxes (Rixen et al., 2019a). However, in regions strongly influenced by river inputs, like the Bay of Bengal, the spatial variability of organic carbon export flux is also strongly influenced by lithogenic matter content, which provides ballast that increases POC sinking rates (Rixen et al., 2019b).

Understanding and predicting these disparate elements of the carbon cycle and its role in basin acidification in the Indian Ocean is critical for understanding the biogeochemical and ecological evolution of the Indian Ocean under the impact of human activities.

2 Oxygen concentrations and the biogeochemical impacts of the OMZs

The first large ocean-going oceanographic expeditions discovered OMZs in the Indian Ocean between the end of the 19th and the first third of the 20th century (Sewell, 1934; Sewell & Fage, 1948 and references therein). During one of these cruises, the John Murray expedition of 1933–34, Gilson (1933) discovered the secondary nitrite maximum of the Arabian Sea, seen as an accumulation of nitrite within the upper part of the OMZ between 200 and 300 m depth (Fig. 2). Its existence indicates active denitrification (Naqvi, 1991), which in addition to anammox (Dalsgaard et al., 2003; Kuypers et al., 2003) are the main sinks of fixed nitrogen (NH_4^+, NO_2^-, NO_3^-) in the ocean (Gruber, 2004). Denitrifying microbes use nitrate (NO_3^-) and nitrite (NO_2^-) to fuel heterotrophic respiration in microbially hypoxic waters, in turn reducing these compounds, ultimately, to nitrogen gas (N_2). In contrast, the anammox reaction is carried out by chemoautotrophic microbes that derive energy from the oxidation of NH_4^+ with NO_2^- to produce N_2 in microbially hypoxic waters.

In contrast to fixed nitrogen, N_2 is inaccessible to eukaryotic phytoplankton. Even though there are specific bacterial phytoplankton clades capable of fixing or breaking apart and oxidizing N_2, the surface ocean is often depleted in fixed nitrogen in comparison to phosphate (Gruber & Sarmiento, 1997). Since phosphate and fixed nitrogen are the primary macronutrients required for phytoplankton growth, the intensification of OMZs and the expansion of functional anoxia is assumed to lower marine productivity by favoring denitrification and anammox and thereby the loss of fixed nitrogen in the ocean (Altabet et al., 1995; McElroy, 1983). An interesting feedback is that, over time, reduced productivity and carbon export due to fixed nitrogen loss in turn reduces biological oxygen consumption at depth, such that the expansion of the volume of anoxic waters in the OMZ core can be accompanied by a reduction of the volume of hypoxic waters (Deutsch et al., 2007; Lachkar et al., 2016).

Changes in the coupled marine nitrogen and oxygen cycles also affect the role of the ocean as a sink and/or source of greenhouse gases such as CO_2 and N_2O. The response of the biologically mediated CO_2 uptake of the ocean, referred to as the biological carbon pump (Boyd et al., 2019; Volk & Hoffert, 1985), to environmental changes is difficult to predict (Bopp et al., 2013; de la Rocha & Passow, 2014) because of the multiple processes involved in its functioning (Boyd et al., 2019; Chakraborty et al., 2018). In contrast, the N_2O source function of the ocean is assumed to be directly linked to the expansion and intensity of OMZs, since N_2O is an intermediate product formed during denitrification (Bange et al., 2001; Fuhrman & Capone, 1991).

Even though denitrification and anammox occur at microbially hypoxic levels of oxygen (concentrations <20 μmol/kg; Table 1), their impact on the nitrogen cycle becomes significant only with the occurrence of functional anoxia (concentrations <0.05 μmol/kg; Table 1). At such low oxygen concentrations, the reduction of nitrite to N_2 outcompetes the re-oxidation of nitrite to nitrate, leading to a net loss of fixed nitrogen (Bristow et al., 2017; Gaye et al., 2013). The re-oxidation of nitrite to nitrate prevents, in turn, the loss of fixed nitrogen and the formation of the secondary nitrite maximum at higher oxygen concentrations within the OMZ. In contrast to the Bay of Bengal OMZ, which is on the verge of becoming functionally anoxic (Bristow et al., 2017), functional anoxia is wide spread in the Arabian Sea OMZ as indicated by the extent of the secondary nitrite maximum (Naqvi, 1991; Rixen et al., 2014; Fig. 2).

2.1 Oxygen distributions, sources, and sinks

Low oxygen conditions in the water column are usually associated with slow ventilation and high biological oxygen demand. OMZs are hence found in productive but poorly ventilated shadow zones of the subtropical oceans, where subsurface water masses recirculate beneath a shallow mixed-layer with minimal direct advective or mixing connection to the surface ocean (Luyten et al., 1983). In these domains, the residence time below the thermocline is at a maximum, and oxygen is slowly supplied mainly through isopycnal (horizontal) mixing processes (Gnanadesikan et al., 2012; Lévy et al., 2021). Moreover, the strong and localized subsurface biological oxygen consumption in OMZs generates strong spatial gradients in oxygen at their edges, making mixing central to the oxygen balance. The Arabian Sea OMZ is not a classical shadow zone due to the northern boundary of the Asian subcontinent, but diapycnal and isopycnal mixing are likely the main oxygen sources balancing strong biological depletion in the core of the OMZ, in roughly equal proportions (Resplandy et al., 2012). Model simulations suggest that without oxygen supplied by eddy mixing, the volume of the Arabian Sea OMZ would double (Lachkar et al., 2016).

In addition to mixing processes in the core of the OMZ, the inflow of oxygen-enriched Indian Ocean Central Water and Persian Gulf Water are assumed to be the main physical oxygen supply mechanism to the edges of the OMZ in the Arabian Sea (Gupta & Naqvi, 1984; Lachkar et al., 2019; McCreary et al., 2013; Resplandy et al., 2012; Swallow, 1984). Indian Ocean Central Water forms through convective mixing as Subantarctic Mode Water in the southern Indian Ocean and is advected northward into the OMZs in both the Arabian Sea and the Bay of Bengal (Fine, 1993; McCartney, 1979; Sverdrup et al., 1942). Oxygen-enriched Persian Gulf Water is introduced into the Arabian Sea OMZ after its outflow from the Persian Gulf (Lachkar et al., 2019; Rixen & Ittekkot, 2005; Schmidt et al., 2020, 2021; Tchernia, 1980). The negative water balance of the Persian Gulf drives this localized deep-water formation by increasing the salinity and hence the density of Persian Gulf Water.

The oxygen-poor Arabian Sea intermediate water, in turn, flows into the Bay of Bengal OMZ (Fig. 1), where high freshwater inputs from rivers and monsoon rainfall (Fig. 5) reduce the vertical oxygen supply by increasing stratification in the surface layers (Rixen et al., 2020). However, in comparison to the Arabian Sea, lower biomass (Fig. 8), productivity, and a stronger ballast effect prevent the development of functional anoxia in the Bay of Bengal OMZ by keeping biological oxygen consumption lower (Al Azhar et al., 2017; Rao et al., 1994; Rixen et al., 2019b). Ballast minerals, which are supplied from land via rivers or as dust, accelerate the sinking of particles, lowering the residence time of exported organic matter in the water column and thereby its decomposition in the OMZ (Haake & Ittekkot, 1990; Ramaswamy et al., 1991). It should be noted, however, that the ballast effect appears to be important primarily in the northern Bay of Bengal (Ittekkot et al., 1991), which suggests that the primary cause of the higher oxygen concentrations further south in the Bay of Bengal is the lower biomass and productivity that results in reduced export and a lower biological oxygen demand. In contrast to the Bay of Bengal, a weak ballast effect and higher biological production (Fig. 8) sustain a higher biological oxygen consumption at intermediate depths in the Arabian Sea. This enhanced biological oxygen consumption balances the higher physical oxygen supply in the Arabian Sea and explains the more intense OMZ (Rixen et al., 2019a).

Contrary to expectations, the Arabian Sea OMZ is more intense in the eastern part of the basin and less intense in the western parts of the basin where productivity is highest (Antoine et al., 1996; McCreary et al., 2013; Naqvi, 1991). The thickest part of the OMZ is in the northeastern Arabian Sea (Sarma et al., 2020). Multiple factors are presumed to cause this asymmetry. The inflow of Indian Ocean Central Water and Persian Gulf Water ventilates the western Arabian Sea preferentially and is assumed to cause this eastward displacement (McCreary et al., 2013; Resplandy et al., 2012; Rixen et al., 2014). Additionally, the seasonal monsoon-driven reversal of the surface ocean circulation leads to intense vertical eddy mixing in the western Arabian Sea causing an eastward shift of the upper OMZ relative to the region of highest productivity (Lachkar et al., 2016; McCreary et al., 2013; Resplandy et al., 2012). Similarly, the eastward shift of the OMZ has been attributed to weak mixing and high penetration time of intermediate water masses in the northeastern Arabian Sea, with the additional influence of organic matter transport from the shelf region that enhances biological oxygen demand (Sarma et al., 2020).

2.2 Biogeochemical impacts of the northern Indian Ocean OMZs

The Arabian Sea and the Bay of Bengal collectively contain ~59% of the Earth's marine sediments exposed to hypoxia (Helly & Levin, 2004). Denitrification within sediments (benthic denitrification) underlying functionally anoxic conditions in the water column is the largest sink for fixed nitrogen in the ocean (DeVries et al., 2013; Gruber, 2004). However, estimates of the relative rates of benthic and water column denitrification are still fraught with large uncertainties on global as well as regional scales. Hence, the role of the Arabian Sea as a sink of fixed nitrogen is difficult to quantify. Although denitrification in the Arabian Sea has been much more intensively studied than in the Bay of Bengal, estimates of benthic and water column denitrification in the Arabian Sea still encompass a wide range with means of 3.9 ± 2.9 and $17 \pm 16\,Tg\,N\,yr^{-1}$, respectively (Bange et al., 2000; Bristow et al., 2017; Dueser et al., 1978; Gaye et al., 2013; Howell et al., 1997; Naqvi et al., 1982; Somasundar et al., 1990). Considering global mean estimates of benthic and water-column denitrification of 183 ± 118 and $155 \pm 116\,Tg\,N\,yr^{-1}$ (Eugster & Gruber, 2012; Gruber, 2004; Somes et al., 2013), on average the Arabian Sea comprises approximately 2% and 11% of the global mean benthic and water column denitrification, respectively. It should be noted, however, that more recent data indicate that benthic denitrification at the Pakistan continental margin alone could be up to $10.5\,Tg\,N\,yr^{-1}$ (Schwartz et al., 2009; Somes et al., 2013), greatly exceeding the previously estimated benthic denitrification of $3.9 \pm 2.9\,Tg\,N\,yr^{-1}$ for the entire Arabian Sea. Moreover, the most recent observations suggest the northeastern Arabian Sea has a water column denitrification rate of $25.3 \pm 7.0\,\mathrm{Tg\,N\,yr^{-1}}$ (Anju et al., 2022). These estimates suggest that the Arabian Sea comprises at least 6% and 16% of the global mean benthic and water column denitrification, respectively.

The volume of hypoxic waters and rate of denitrification in the Arabian Sea strongly increase in response to enhanced monsoon winds (Lachkar et al., 2018). Stronger winds intensify the upwelling, increasing biological productivity and respiration in excess of the increase in wind-driven ventilation and are also responsible for the deepening of the OMZ. However, as discussed above, increases in denitrification have the potential to reduce biological productivity through nitrogen removal, and hence the efficiency of the biological pump of carbon, at the basin scale (and beyond) on timescales of decades to centuries (Canfield et al., 2019; Lachkar et al., 2018; McElroy, 1983). Therefore, an expansion of Indian Ocean OMZs can affect the large-scale biogeochemical cycles of nitrogen and carbon and contribute to climate variations over long timescales (Altabet et al., 2002; Gaye et al., 2018). The extent of this control depends on the magnitude of the nitrogen removal and the importance of concurrent stabilizing negative feedbacks, including the feedback of reduced productivity on biological oxygen demand and the tight coupling between denitrification and N_2 fixation downstream of OMZs (Deutsch et al., 2007; Gruber & Galloway, 2008). If the excess in phosphorus relative to nitrogen supports local nitrogen fixation within the Indian Ocean basin, the timescales for feedback with denitrification could be much shorter than the timescale of the global ocean circulation.

2.3 Recent (decadal) changes in oxygen concentrations

Early reports on the occurrence of coastal hydrogen sulfide in the north-eastern Arabian Sea and off Oman at Ras-al-Hadd (Carruthers et al., 1959; Ivanenkov & Rozanov, 1966) indicate that the Arabian Sea OMZ was more intense in the past because the emergence of hydrogen sulfide indicates the transition from functional anoxia (oxygen <0.05 µmol/kg) toward anoxic (zero oxygen) conditions (Table 1). These are the only reports of the occurrence of hydrogen sulfide in the Arabian Sea except for Naqvi et al. (2000), who discovered an anoxic event that developed along the western Indian coast off Mumbai in the late summer of 1999. Such strong events do not develop every year (Gupta et al., 2016; Sudheesh et al., 2016), but their appearance shows that the spreading of oxygen-depleted zones in coastal regions is a global phenomenon that does not spare the Indian shelf (Altieri et al., 2017; Diaz et al., 2019; Diaz & Rosenberg, 2008). However, contrary to this prevailing understanding, Gupta et al. (2021) argue that the hypoxic-anoxic zone along the west coast of India is formed through a natural process, i.e., upwelling of deoxygenated waters during the summer monsoon and that the persistence and extent of this coastal oxygen deficiency depend on the degree of deoxygenation of source waters for the upwelling. Moreover, the volume of anoxic waters is strongly modulated by the Indian Ocean Dipole (IOD), with positive IOD events preventing anoxia due to wind-forced downwelling coastal Kelvin waves propagating along the west coast of India (Vallivattathillam et al., 2017).

In contrast to these shelf processes, there is only a weak decadal decline in dissolved oxygen concentrations in the OMZs of the Arabian Sea and the Bay of Bengal in comparison to OMZs of the South Atlantic Ocean and the Pacific Ocean (Ito et al., 2017; Naqvi, 2019; Schmidtko et al., 2017; Stramma et al., 2008). The analysis of all oxygen data available from the Arabian Sea between 1959 and 2004 by Banse et al. (2014) ascribes this modest change to opposing regional trends within the Arabian Sea, where oxygen concentrations increased in the southern part of the Arabian Sea and declined in the central Arabian Sea. In contrast, Sarma et al. (2018) ascribe the modest change in oxygen levels in the Bay of Bengal to the influence of anticyclonic eddies that precluded the OMZ from intensifying. Follow-up studies report decreasing oxygen concentrations in the western and northern Arabian Sea (Piontkovski & Al-Oufi, 2015; Queste et al., 2018). In the northern Arabian Sea, dissolved oxygen concentrations in the surface mixed layer largely reflect the decreasing trend seen in the OMZ, as indicated by a compilation of dissolved oxygen data covering the period from the 1960s to 2010 (Gomes et al., 2014). In response to this deoxygenation, the secondary nitrite maximum expanded southward and westward in the early 1990s (Rixen et al., 2014). Perhaps in response to changing oxygen concentrations, the planktonic community structure has changed as shown by the development and now regular occurrence of large *Noctiluca* winter blooms since the 2000s in the Arabian Sea (Goes & Gomes, 2016; Goes et al., 2020; Gomes et al., 2014; Hood et al., 2024). It should be noted, however, that it has been shown that *Noctiluca* blooms occur in oxic conditions and that both natural and anthropogenic processes appear to have contributed to the development of the massive blooms in the northeastern Arabian Sea (Sarma et al., 2019; Sredivi & Sarma, 2022). Numerical model results suggest that reduced ventilation caused by a reduced inflow of Persian Gulf Water in response to warming, combined with reduced solubility of oxygen in the surface water of the Persian Gulf, may have contributed to these developments (Lachkar et al., 2019). In addition to changes in the Gulf outflow, recent numerical model results also attribute decreasing oxygen concentrations in the northern Arabian Sea to reduced local ventilation due to global warming-induced increases in stratification and weakening of winter convective mixing (Lachkar et al., 2020).

Nonetheless, widespread and or more frequent outbreaks of hydrogen sulfide as seen in the upwelling systems off Peru (Schunck et al., 2013) and Namibia (Weeks et al., 2002) have not so far been reported in the northern Indian Ocean during the last 50 years. This finding implies that the interplay between physical oxygen supply and biological oxygen consumption has prevented the development of persistent anoxia (zero oxygen) in the Arabian Sea and functional anoxia (oxygen <0.05 μM) in the Bay of Bengal OMZ (Rixen et al., 2020).

2.4 Future changes in oxygen concentrations

There is high uncertainty in the future evolution of the Arabian Sea OMZ, and of all major OMZs, projected by Earth System models by the year 2100 (Bopp et al., 2013; Cabré et al., 2015; Kwiatkowski et al., 2020; Rixen et al., 2020). The uncertainty largely arises from differences among models in the magnitude and timing of changes in ventilation and biological oxygen demand and how strongly they offset each other (Resplandy, 2018). Both processes are strongly influenced by model resolution and are thus challenging to simulate at the centennial scale. It is projected that OMZs may either shrink or expand under projected climate change, depending on the efficiency of mixing (Bahl et al., 2019; Duteil & Oschlies, 2011; Lévy et al., 2021). As discussed above, this uncertainty is particularly large in the northern Indian Ocean where global forecast models generally fail to reproduce the current oxygen concentrations and distributions (McCreary et al., 2013; Rixen et al., 2020; Schmidt et al., 2021). This failure, and the uncertainty in future projections, may also be related to the lack of consideration of the role of cross-shelf transport of organic matter in Earth System models, which is a process that is known to contribute to the development of the OMZs in both the Arabian Sea and the Bay of Bengal (Sarma et al., 2020; Udaya Bhaskar et al., 2021).

3 Carbon concentrations and fluxes

The earliest CO_2, pH, and POC measurements in the Indian Ocean date back to the International Indian Ocean Expedition in the early 1960s (Fig. 4; e.g., Behrman, 1981; Newell, 1969). Carbonate system measurements were also made in the Indian Ocean under the Geochemical Ocean Section Study in the 1970s (Moore, 1984), the Joint Global Ocean Flux Study (JGOFS; Fasham, 2003; see also https://en.wikipedia.org/wiki/Joint_Global_Ocean_Flux_Study) and the World Ocean Circulation Experiment (WOCE) in the 1990s (Woods, 1985). Subsequently, in the first decade of the 21st-century Climate and Ocean: Variability, Predictability, and Change (CLIVAR) Repeat Hydrography project started (Gould et al., 2013), and carbonate system measurements continue through the present under the ongoing GO-SHIP program (Talley et al., 2017; see also https://www.go-ship.org/About.html) and the Second International Indian Ocean Expedition (IIOE-2; Hood et al., 2015). In addition to these major international efforts, there have been numerous national expeditions and programs that have contributed to the current inventory of carbonate system measurements in the Indian Ocean. For example, time-series observations of pH and pCO_2 are being made in the coastal waters of India revealing recent trends (Sarma et al., 2021). There is also now a mooring in the central Bay of Bengal (Fig. 4) that has been collecting continuous CO_2 and pH measurements since November 2013 (see https://www.pmel.noxsaa.gov/co2/story/BOBOA).

In contrast, the history of DOC measurements in the Indian Ocean dates back only to the early 1990s when it was discovered (thanks to advances in measurement methods that achieved ∼1 μmol/kg precision sufficient to reveal DOC variability over space and time) that DOC concentrations are more dynamic than previously thought (Hansell, 2009). The timing of this discovery coincided with the onset of the JGOFS Arabian Sea Process Study when some of the earliest "modern" Indian Ocean DOC measurements were made. Subsequently, DOC concentration measurements were made in the Indian Ocean under the US CLIVAR Repeat Hydrography project and they continue under the ongoing GO-SHIP program and through national expeditions. DOC isotope measurements are also now being made in the northern Indian Ocean to provide insight into DOC sources (Rao & Sarma, 2022).

3.1 Inorganic carbon distributions and fluxes

The Indian Ocean accounts for ∼20% of the global oceanic uptake of atmospheric CO_2 (Takahashi et al., 2002). The Arabian Sea is a source of CO_2 to the atmosphere due to elevated pCO_2 within the Southwest Monsoon-driven upwelling (Fig. 4; see also de Verneil et al., 2021; Takahashi et al., 2009, 2014; Valsala & Maksyutov, 2010, 2013). North of 14°S the Indian Ocean loses CO_2 at a rate of 0.12–0.16 PgC/yr (de Verneil et al., 2021; Takahashi et al., 2002, 2009, 2014). The most intense air-sea CO_2 exchange occurs during the Southwest Monsoon where outgassing rates reach ∼6 molC/m^2/yr in the upwelling regions off Oman and Somalia, but the entire Arabian Sea contributes CO_2 to the atmosphere (de Verneil et al., 2021; Sreeush et al., 2018, 2020; Valsala & Murtugudde, 2015).

Time series measurements have shown that the rates of increase in dissolved inorganic carbon (DIC) and surface partial pressure of CO_2 (pCO_2) per decade are consistent with global trends in the southwestern coastal Bay of Bengal, whereas rates in the northwestern coastal Bay of Bengal have been observed to be 3–5 times higher than the global trends (Sarma et al., 2015a). Thus, the northwestern Bay of Bengal, which was previously considered to be a significant sink for atmospheric CO_2, now seems to have become a source of CO_2 to the atmosphere. Variability in CO_2 fluxes in the northwestern shelf region of the Bay of Bengal depends on the river discharge characteristics and the East India Coastal Current that distributes this water along the coast (Sarma et al., 2012). Nonetheless, it is still uncertain whether the entire Bay of Bengal is a net CO_2 source or sink due to the high levels of variability combined with sparse spatial and temporal sampling (Fig. 4; Bates et al., 2006). South of 14°S the Indian Ocean appears to be a strong net CO_2 sink (−0.44 PgC/yr in the band 14°S–50°S; Fig. 4). The solubility pump (CO_2 dissolution and its physical mixing and transport) and the biological pump (biologically mediated processes that export carbon) contribute equally to the CO_2 sink in the south Indian Ocean region (Valsala et al., 2012). Cold temperatures increase CO_2 solubility at higher latitudes and subduction in the subtropical front can transport this CO_2 into the ocean interior, but there is also evidence that chemical and biological factors are important, e.g., potential iron fertilization that facilitates particulate carbon export in the southern hemisphere of the Indian Ocean (Piketh et al., 2000).

In addition, the IOD leads to a substantial sea-to-air CO_2 flux variability in the southeastern tropical Indian Ocean over a broad region (70–105°E, 0–20°S), with the most intense effects manifesting near the coast of Java-Sumatra due to the impacts on upwelling dynamics and associated westward propagating anomalies. The sea-to-air CO_2 fluxes, pCO_2, DIC, and ocean alkalinity range as much as ± 1.0 mol m^{-2} yr^{-1}, ± 20 μatm, ± 35 μmole kg^{-1}, and ± 22 μmole kg^{-1}, respectively, within 80–105°E, 0–10°S due to the IOD. The DIC and alkalinity are significant drivers of pCO_2 variability associated with IOD (Valsala et al., 2020).

Synthesis of the seasonal, annual, and interannual air-sea CO_2 fluxes based on both models (ocean, atmospheric inversions) and observations (Takahashi et al., 2009) reveals that the net sea-air CO_2 uptake estimated from observations (−0.24 PgC/yr) is consistent with uptake derived from models (−0.37 PgC/yr), given the uncertainties (Sarma et al., 2013). However, some models overestimate flux in the Bay of Bengal and underestimate flux in the Arabian Sea/northwestern Indian Ocean (Sarma et al., 2013). These offsets are likely due to the models being inadequately constrained by CO_2 observations. There are fewer observations in the Indian Ocean (especially north of 20°S) compared to other oceans in recent years (Fig. 4; Bakker et al., 2014, see also www.socat.info). More observations of ocean DIC concentrations and carbon flux are needed in the Indian Ocean to constrain the models and reduce uncertainties in the fluxes. Toward this goal, Valsala et al. (2021) have recently done observing system simulation experiments for Indian Ocean pCO_2 arrays and recommended the deployment of additional moorings at suitable locations to monitor and better constrain the Indian Ocean air-sea CO_2 fluxes. A few key ship-of-opportunity routes for underway pCO_2 sampling in the Indian Ocean are also recommended in their study.

3.2 Spatial and temporal variability in pH

The uptake of anthropogenic CO_2 by the ocean results in fundamental changes in seawater chemistry that can have significant impacts on ocean ecosystems (Doney, 2010; Gattuso & Hansson, 2011). Intergovernmental Panel on Climate Change (IPCC) business-as-usual emission scenarios indicate that atmospheric CO_2 levels will reach 800 ppm near the end of this century (Feely et al., 2009). The associated increase in oceanic CO_2 concentrations will lead to acidification (lower pH) of the Indian Ocean over the coming decades, with potentially severe negative impacts on coral reefs and other calcifying organisms (Doney, 2010; IPCC, 2019). In 1995, the surface pH values for the northern (20°E–120°E, 0°–24.5°N) and southern (20°E–120°E, 0°–40°S) Indian Oceans were 8.068 ± 0.03 and 8.092 ± 0.03, respectively, which is the lowest of the major ocean basins (Feely et al., 2009). It is not entirely clear why surface pH values are so low in the Indian Ocean (Takahashi et al., 2014). Increases in sulfate and nitrogen aerosol loadings over the Bay of Bengal from the Indo-Gangetic Plain and Southeast Asia may be mainly responsible for the increased acidity in the northwestern Bay of Bengal in recent years (Sarma et al., 2015a). Reduced Godavari River discharge together with a positive IOD event have also been shown to contribute to enhanced acidification and pCO_2 levels in the coastal waters in the western Bay of Bengal (Sarma et al., 2015b). A study from the eastern Bay of Bengal indicates a decline in pH of 0.2 from 1994 to 2012 (Rashid et al., 2013), which is considerably faster than the global decline of 0.1 over the last century (IPCC, 2007). Upwelling of low pH subsurface waters in the Arabian Sea results in low surface pH (<7.9) during the Southwest Monsoon (Takahashi et al., 2014). Sreeush et al. (2019) estimated that in addition to the anthropogenic causes, ocean warming exacerbates acidification in the western Arabian Sea by an additional 16%. Chakraborty et al. (2021) studied the seasonal drivers of surface ocean pH in the Arabian Sea and Bay of Bengal and found that DIC and SST make complementary contributions to the seasonal cycle in pH. Tarique et al. (2021) reported the first time series of Arabian Sea pH from proxy records of Boron Isotopes from 1990 to 2013. Their investigation reveals that physical oceanographic processes, for example, upwelling, downwelling, and convective mixing modulated by El Niño-Southern Oscillation (ENSO), largely control surface pH variability and mask expected long-term ocean acidification trends resulting from anthropogenic CO_2 rise. An increase in pH has been observed in the eastern and southern Bay of Bengal during all seasons associated with warming and a decrease in salinity (Sridevi & Sarma, 2021). In contrast, a decrease in pH (-0.001 yr^{-1}) and a pCO_2 increase ($+0.1$ to $+0.7$ µatm yr^{-1}) has been observed in the western and northern Bay of Bengal during winter and spring seasons due to the deposition of atmospheric pollutants (Sridevi & Sarma, 2021). These studies suggest that increases in freshwater input due to the melting of Himalayan ice cover and deposition of atmospheric pollutants are dominant controlling factors on surface ocean pH and pCO_2 in the Bay of Bengal between 1998 and 2015, and this region is acting as a stronger sink for the atmospheric CO_2 in the present than in the past two decades (Sarma et al., 2021; Sridevi & Sarma, 2021).

The susceptibility of the Indian Ocean coral communities to warming has been revealed by the large-scale coral bleaching events of 1998, 2005, 2011 as well as between 2014 and 2017 caused by high SST (Cerutti et al., 2020; McClanahan et al., 2007; Moore et al., 2012; Obura et al., 2017). Ocean acidification has the potential to exacerbate these negative impacts on coral reef ecosystems. For example, the 1998 bleaching event altered the age distribution of commercially harvested fish (Graham et al., 2007). Because of the combined effects of acidification, human development, and global warming, coral reef ecosystems may be at greater risk than previously thought (Hoegh-Guldberg et al., 2007; IPCC, 2019). Moreover, some commercially fished species (e.g., shelled mollusks) are directly vulnerable to ocean acidification (Hoegh-Guldberg et al., 2014). A study in Somalia suggests that human-induced ocean acidification reduced the rate at which foraminifera calcify, resulting in lighter shells (de Moel et al., 2009). Due to the effects of acidification on calcifying pteropods, which are preyed on by many higher trophic level organisms, the Southern Ocean sector of the Indian Ocean could experience major disruptions in pelagic food webs (Bednarsek et al., 2012).

In addition to the direct impacts of acidification, increasing CO_2 in the upper ocean could lead to increased primary productivity for some species (e.g., diazotrophic cyanobacteria; Hutchins et al., 2007), altering rates of nitrogen fixation and therefore the biogeochemistry of particulate organic matter formation and remineralization. Declines in pH also shift the chemical equilibrium from ammonia (NH_3) to ammonium (NH_4^+), which could alter key biogeochemical processes such as nitrogen assimilation by phytoplankton and microbial nitrification (Gattuso & Hansson, 2011).

3.3 Large-scale DOC and POC distribution and fluxes

Near-surface (<50 m) DOC concentrations in the Indian Ocean vary between ~50 and 80 µmolC/kg with the lowest values occurring south of 40°S (Hansell, 2009; Hansell et al., 2009; Figs. 6 and 7). In general, near-surface concentrations of DOC are positively correlated with temperature and, to a lesser extent, they are negatively correlated with nutrient concentrations. The positive correlation with temperature happens because warm, low-latitude waters tend to be more strongly stratified, and there is reduced vertical mixing. As a result, DOC concentrations increase in these waters even though there are low rates of net DOC production because the DOC is not mixed away (Hansell, 2009). As a result, near-surface DOC concentrations in the Indian Ocean tend to be highest (>70 µmolC/kg) in tropical and subtropical near-surface waters (Hansell, 2009; Hansell et al., 2009; Fig. 7). These patterns are also found in the Atlantic and Pacific (Hansell, 2009; Hansell et al., 2009).

Near-surface DOC concentrations tend to be inversely correlated with nitrate, silicate, and phosphate concentrations because, when these nutrients are consumed, DOC is produced (Hansell, 2009). Thus, a negative correlation is also observed between DOC and nitrate and phosphate concentrations in the deep waters (>500 m) of the Indian Ocean where DOC concentrations decline to <50 µmolC/kg (Fig. 7) and nitrate and phosphate concentrations are substantially enriched. However, these correlations are not observed between DOC and silicate concentrations because in deep waters of the Indian Ocean there are large gradients in silicate that are associated with very little change in DOC (Hansell, 2009).

As in the Pacific and Atlantic, the lowest DOC concentrations in the Indian Ocean (40–42 µmol/kg at 3000 m) are found below 1000 m depth (Fig. 7; Fig. 2 in Hansell et al., 2009). Global observations show that DOC concentrations below 1000 m depth in the Indian Ocean are comparable to those in the Atlantic and significantly higher than deep DOC concentrations in the Pacific (<40 µmolC/kg at 3000 m; Hansell et al., 2009). DOC concentrations decline below the euphotic zone because heterotrophic consumption exceeds autotrophic production (Hansell et al., 2009). Presumably, a significant fraction of the DOC below 1000 m in the Indian Ocean is very old (>4000 years) and refractory (resistant to bacterial consumption) as has been shown to be the case in the Atlantic and Pacific (Bercovici et al., 2018).

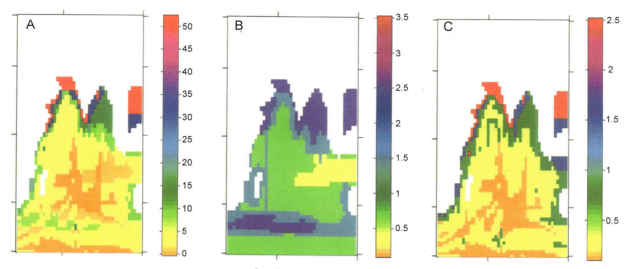

FIG. 9 Estimated (a) organic carbon burial rate (mmolC m^{-2} yr^{-1}); (b) sediment accumulation rate (grams per square centimeter per thousand years); and (c) seafloor distribution of organic carbon (weight percent). *(Modified from Jahnke (1996).)*

As observed elsewhere in the global ocean, satellite-estimated near-surface POC concentrations are elevated in coastal regions of the Indian Ocean with values often exceeding 120 mgC/m^3 (=10 μmolC/kg; Fig. 8; Gardner et al., 2006). The highest POC concentrations in the Indian Ocean are observed in the northwestern part of the basin in the Arabian Sea and off of the coast of Somalia with values estimated to be greater than 96 mgC/m^3 (=8 μmolC/kg) extending well into the open ocean during the Northeast Monsoon (Fig. 8). These high POC values are associated with high chlorophyll concentrations, although chlorophyll itself only accounts for <2% of the POC and presumably the remainder is cellular and detrital material (Gardner et al., 2006). Note that C:Chl ratios range from ~15 to 158 by weight (Sathyendranath et al., 2009). Elevated POC concentrations (>72 mgC/m^3 = 6 μmolC/kg) are also observed in the Indonesian Throughflow region (particularly during the Southeast Monsoon) and off of southern Africa over the Agulhas Bank, and in the Agulhas Current and its retroflection, again associated with high chlorophyll concentrations (~0.4%–0.8% of POC; Fig. 8, Gardner et al., 2006). The lowest POC concentrations in the Indian Ocean are observed between 40°S and 10°N in the oligotrophic southern subtropical gyre and equatorial waters where primary production is very low (Gardner et al., 2006; Fig. 8; see below).

Unsurprisingly, the spatial and temporal variability in POC export flux in the Indian Ocean is similar to the abovementioned patterns in POC and chlorophyll concentration, consistent with primary production as the main control on the spatial and temporal variability of organic carbon fluxes (Rixen et al., 2019a). However, as discussed above, in river-influenced regions like the Bay of Bengal, the spatial variability of organic carbon flux is also strongly influenced by lithogenic matter content. Rixen et al., (2019b) estimate that lithogenic matter content enhances organic carbon flux rates on average by 45% and by up to 62% in river-influenced regions of the Indian Ocean. This strong ballast effect explains why organic carbon fluxes are lower in the highly productive western Arabian Sea than they are in the relatively unproductive southern Java Sea (Rixen et al., 2019b). This explanation appears to be broadly consistent with the earlier global estimates of organic carbon burial and sediment accumulation rates calculated by Jahnke (1996; Fig. 9), indicating that both are elevated in the coastal and northern regions of the Arabian Sea where rates of primary production are very high, but river influence is small. In contrast, both organic carbon burial and sediment accumulation rates are elevated over a much larger area in the Bay of Bengal where primary production is lower, but river influence is large (Jahnke, 1996; Fig. 9).

3.4 Regional DOC and POC distributions and fluxes

3.4.1 Arabian Sea

DOC in the surface ocean is a short-term reservoir for carbon, expanding and contracting seasonally. In the Arabian Sea the highest near-surface (<50 m) DOC concentrations (80–100 μmolC/kg) are observed near the coast when upwelling is not active. During the Southwest Monsoon, upwelling results in lowered near-surface DOC concentrations along the western side of the basin, i.e., DOC concentrations increase seaward from <70 μmolC/kg near the coast to >80 μmolC/kg in the central Arabian Sea. In the open ocean, the highest surface DOC concentrations (80–95 μmolC/kg) are observed during the Northeast Monsoon and they remain high through mid-Southwest Monsoon. The lowest open ocean near-surface DOC concentrations (65–75 μmolC/kg) are observed during late Southwest Monsoon and during the Fall Intermonsoon (Hansell, 2009; Hansell & Peltzer, 1998).

The seasonal accumulation of DOC north of 15°N in the Arabian Sea happens mostly during the Notheast Monsoon, and it has been estimated to be equivalent to 6%–8% of annual primary production and 80% of net community production. In contrast, net DOC production is very small during the Southwest Monsoon (Hansell & Peltzer, 1998).

In the vertical, DOC concentrations decrease to values of <55 μmolC/kg at the top of the OMZ (~100 m). Vertical layering is also observed in DOC concentrations in the Arabian Sea and is associated with vertical layering of the water masses, with Persian Gulf Water producing a clear signal of elevated DOC. DOC concentrations below 500 m are low (<50 μmolC/kg) and relatively uniform across the basin and there appears to be little impact of the OMZ on DOC concentrations (Hansell, 2009; Hansell & Peltzer, 1998).

Consistent with satellite estimates (Gardner et al., 2006), in situ measurements of near-surface (0–150 m) POC concentrations in the Arabian Sea are generally high near the coast of Oman and decrease offshore throughout the year with concentrations often exceeding 12.5 μmolC/kg (Gundersen et al., 1998; Fig. 10). The spatial distributions are always patchy due to the influence of mesoscale eddies and jets. Seasonal variations in POC concentrations in the upper 150 m can be the same order of magnitude as basin-wide spatial variations, and these concentrations are strongly influenced by nutrient availability and mixing (Gundersen et al., 1998). POC and chlorophyll concentrations are clearly correlated (Figs. 10 and 11), indicating that variability in phytoplankton production drives much of the observed POC variations (Gardner et al., 2006; Gundersen et al., 1998). Subsurface POC and chlorophyll maxima are also observed, particularly offshore and during the intermonsoon periods with the latter generally deeper and more pronounced, as observed elsewhere

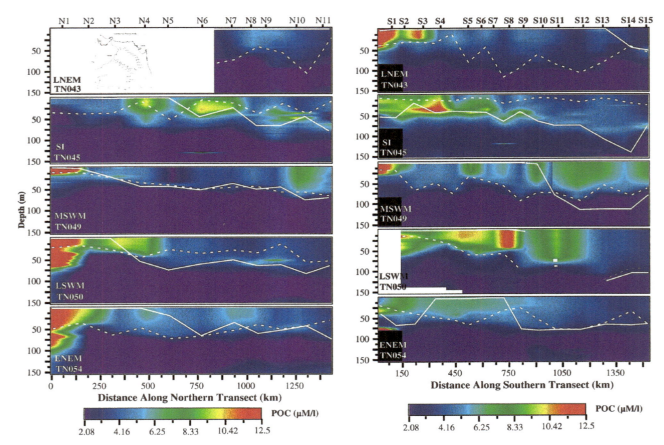

FIG. 10 POC sections of the northern transect (left panels) and southern transect (right panels) for each cruise. Mean mixed layer depth [delta sigma=0.03] (dotted line) and 0.5 μM/L nitrate isopleth (solid line) are shown for each cruise. The station locations are shown in the map inset in the upper left panel. *(Modified from Gundersen et al. (1998).)*

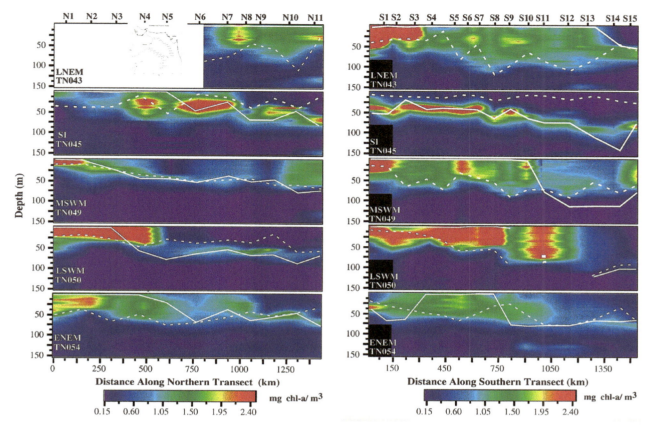

FIG. 11 Chlorophyll sections of the northern transect (left panels) and southern transect (right panels) for each cruise. Mean mixed layer depth [delta sigma = 0.03] (dotted line) and 0.5 µM/L nitrate isopleth (solid line) are shown for each cruise. The station locations are shown in the map inset in the upper left panel. *(Modified from Gundersen et al. (1998).)*

(e.g., Fennel & Boss, 2003; see also review by Cullen, 2015). Interestingly, when POC and chlorophyll concentrations are vertically integrated, the values can be higher during the spring intermonsoon than during the Southwest Monsoon because of the strong subsurface POC and chlorophyll maxima that develop during the spring intermonsoon (Gundersen et al., 1998).

In the Arabian Sea, POC export normalized to 2000 m ranges from <6 to >22 mgC m^{-2} d^{-1} (equal to <0.5 to >1.8 mmolC m^{-2} d^{-1}) depending upon location and season (Rixenet et al., 2019b; Honjo et al., 1999). In contrast, POC export from the base of the euphotic zone in the western Arabian Sea estimated from the satellite-derived net primary production data and export fluxes calculated with a model give 258.3 and 335.7 mgC m^{-2} d^{-1} (equal to 21.5 and 27.9 mmolC m^{-2} d^{-1}), respectively (Sreeush et al., 2018), suggesting dramatic declines in export at depth compared to the surface. The POC fluxes are generally highest in the western Arabian Sea during the Southwest Monsoon (>20 mgC m^{-2} d^{-1} equal to >1.7 mmolC m^{-2} d^{-1}) where large seasonal variations in flux (7–24 mgC m^{-2} d^{-1} = 0.6–2 mmolC m^{-2} d^{-1}) are also observed (Honjo et al., 1999; Rixen et al., 2019b). In contrast, POC fluxes at 2000 m are generally much lower in the central Arabian Sea (4–12 mgC m^{-2} d^{-1} = 0.25–0.5 mmolC m^{-2} d^{-1}) with the highest fluxes (>10 mgC m^{-2} d^{-1} equal to >0.8 mmolC m^{-2} d^{-1}) also occurring during Southwest Monsoon, though with much less pronounced seasonal variations (Rixen et al., 2019b; Honjo et al., 1999). These patterns of POC flux in the Arabian Sea are broadly consistent with organic carbon burial and sediment accumulation rates calculated by Jahnke (1996), though with the latter an order of magnitude lower (Fig. 9; <5 to >50 mmolC m^{-2} yr^{-1} which is equal to <0.01 to >0.14 mmolC m^{-2} d^{-1}). Major flux events are also observed in the western Arabian Sea during both the Southwest Monsoon and Northeast Monsoon in association with passing eddies and wind-curl events; these events can dominate the annual mass flux (Honjo et al., 1999). ^{234}Th-estimated POC export fluxes in the Arabian Sea reveal similar spatial and temporal patterns with POC export efficiencies varying from <2% to 5% (Subha Anand et al., 2018a, 2018b).

3.4.2 Bay of Bengal

Some of the highest concentrations of DOC in the Bay of Bengal (75–100 μmolC/kg) are observed in the near-surface (<50 m) waters of the northern Bay of Bengal, primarily due to the large inputs of fresh water and terrigenous DOC from rivers (Hansell, 2009; Shah et al., 2018; Fig. 7). In addition, this fresh water enhances near-surface stratification which helps to maintain the elevated DOC concentrations because it inhibits vertical mixing, as discussed above. The concentration of DOC in the upper ocean in the Bay of Bengal also exhibits a significant relationship with Chlorophyll-a, POC, and DOC exudation rates, suggesting possible sources through in situ biological processes (Rao & Sarma, 2022). Elevated DOC concentrations are also observed at >200 m depth in the Bay of Bengal due to the remineralization of sinking particulate organic matter from the surface waters (Shah et al., 2018; Fig. 7). Near-surface DOC concentrations tend to decline at lower latitudes (<14°N, Hansell, 2009; Shah et al., 2018). Rao et al. (2021) report that about half of the primary production in the Bay of Bengal is released as DOC during the Southwest Monsoon due to the existence of oligotrophic conditions, warm waters, and dominance of pico-phytanktion biomass, and Shah et al. (2018) estimate that DOC remineralization fuels ~18% of the apparent oxygen utilization. As observed in the Arabian Sea, DOC concentrations below 500 m in the Bay of Bengal are low (<50 μmolC/kg) and relatively uniform across the basin, and there appears to be little impact of the OMZ on DOC concentrations (Fig. 7).

In the Bay of Bengal, the near-surface concentrations of POC can vary from <4 to >10 μmol/kg depending on location and season (Fernandes et al., 2009; Fig. 12). Spatial and temporal variations are smaller compared to the Arabian Sea with POC concentrations varying from 4.3 to 11.1 μmolC/kg, 3.1 to 10.9 μmolC/kg, and 4.3 to 9.0 μmolC/kg during Southwest Monsoon, fall intermonsoon, and spring intermonsoon, respectively, in one study (Fernandes et al., 2009). In contrast to the Arabian Sea, POC and chlorophyll concentrations can be higher offshore compared to coastal stations in the Bay of Bengal, especially during Southwest Monsoon (Fernandes et al., 2009; Thushara et al., 2019; Vinayachandran, 2009). These elevated offshore POC and chlorophyll concentrations are not, however, revealed by remote sensing. Rather, the satellite estimates generally indicate that the highest concentrations occur near the coast (Gardner et al., 2006; Fig. 8). As in the Arabian Sea, POC and chlorophyll concentrations are correlated in the Bay of Bengal with subsurface chlorophyll maxima generally deeper and more pronounced, as discussed above (e.g., Fernandes et al., 2009). In general, POC concentrations and percent POC contribution to the total suspended particulate matter tend to decrease with increasing water column depth indicating heterotrophic remineralization of sinking particles (Fernandes et al., 2009). High chlorophyll concentrations and high Chl*a*/POC ratios observed during the Southwest Monsoon and fall intermonsoon indicate the presence of relatively fresh particulate organic matter in the Bay Bengal during these two seasons (Fernandes et al., 2009). POC concentrations measured during the spring intermonsoon were lowest in the southern Bay of Bengal (0.51–0.65 μmolC/kg, Subha Anand et al., 2017). In general, seasonal and spatial differences in river influence, combined with physical forces such as eddies and

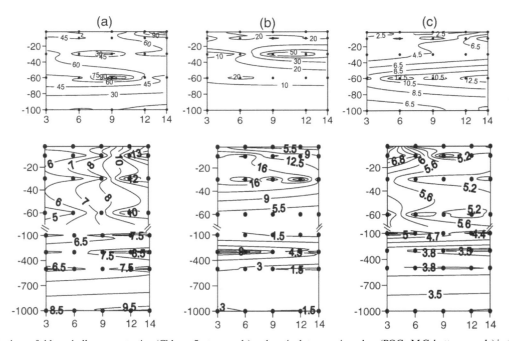

FIG. 12 Distributions of chlorophyll *a* concentration (Chl a ng/L, top panels) and particulate organic carbon (POC μM C, bottom panels) in the central Bay of Bengal during (a) Southwest Monsoon, (b) Fall intermonoon, and (c) Spring intermonsoon. *(Modified from Fernandes et al. (2009).)*

mixing/entrainment that pump nutrients into the euphotic zone, drive seasonal and spatial variations in the quantity and quality of particulate organic matter in the Bay of Bengal (Fernandes et al., 2009; Subha Anand et al., 2017; Thushara et al., 2019; Vinayachandran, 2009).

POC export normalized to 2000 m in the Bay of Bengal is generally lower and less strongly seasonal than in the Arabian Sea, with fluxes varying between 2 and 14 mgC m^{-2} d^{-1} (=0.17–1.2 mmolC m^{-2} d^{-1}; Rixen et al., 2019b). In contrast, POC export from the euphotic zone in the Sri Lanka Dome region estimated from the satellite-derived net primary production data and export fluxes calculated with a model give 118.5 and 427.4 mgC m^{-2} d^{-1} (equal to 9.8 and 35.6 mmolC m^{-2} d^{-1}), respectively (Sreeush et al., 2018), again suggesting dramatic declines in export at depth compared to the surface. In general, riverine freshwater and nutrient inputs to the Bay of Bengal increase POC export near the coast. This pattern is revealed by ^{234}Th-estimated POC export flux in the Bay of Bengal during the Northeast Monsoon, which varies from 0.1 to 1.6 mmolC m^{-2} d^{-1} with the highest flux observed in the east Indian coastal zone (Subha Anand et al., 2017). These patterns of POC flux are also broadly consistent with older organic carbon burial and sediment accumulation rates calculated by Jahnke (1996), where the former are highest along the east coast of India (Fig. 9). However, as in the Arabian Sea, the organic carbon burial rates estimated by Jahnke (1996) are an order of magnitude lower than the POC fluxes normalized to 2000 m from Rixen et al. (2019b).

3.4.3 Equatorial waters, Indonesian Throughflow, and the Leeuwin current

A modeling study reported by Hansell (2009) and Hansell et al. (2009) suggests a zonal decline in near-surface (30 m) DOC concentrations along the equator in the Indian Ocean with values dropping from ~75 μmolC/kg in the east off of northern Sumatra to <70 μmolC/kg in the west off of central east Africa. This pattern appears to be linked to the aforementioned declines in near-surface DOC concentrations to <70 μmolC/kg that are associated with upwelling along the western side of the basin (Hansell, 2009; Hansell & Peltzer, 1998). The modeling study reported by Hansell (2009) and Hansell et al. (2009) also suggests that near-surface DOC concentrations are elevated to >75 μmolC/kg in an open ocean region between 10° and 20°S and 80° and 100°E coincident with shallow OMZ (Fig. 1), minimum sedimentation rate (Fig. 9), and lower salinity (Fig. 5) and that concentrations decline eastward to <70 μmolC/kg in the Indonesian Throughflow region and in the Leeuwin Current, but these features are not readily apparent in meridional or zonal sections (Fig. 7). DOC concentrations below 500 m in these waters are also low (<50 μmolC/kg) and uniform (Fig. 7).

Satellite-estimated near-surface POC concentrations in open ocean equatorial waters of the Indian Ocean are generally low (<60 mgC/m^3 which is equal to <5 μmolC/kg) with concentrations increasing during the Southwest Monsoon, especially in the west due to the aforementioned influence of coastal upwelling (Gardner et al., 2006; Fig. 8). Elevated POC concentrations (>72 mgC/m^3 which is equal to >6 μmolC/kg) are also observed in the eastern equatorial Indian Ocean along the coast off Sumatra and Java and in the Indonesian Throughflow region, particularly during the Southeast Monsoon. All of these regions of elevated POC are associated with elevated chlorophyll concentrations (>0.4% of POC; Fig. 8, Gardner et al., 2006). Unsurprisingly, POC and chlorophyll concentrations are generally low along the coast of western Australia where the downwelling Leeuwin Current flows southward (Fig. 8). Estimated organic carbon burial rates show similar patterns, i.e., low rates (<5 mmolC m^{-2} yr^{-1}) in open ocean equatorial waters and higher rates in the west and east in association with upwelling and the influence of the Indonesian Throughflow (Jahnke, 1996; Fig. 9). Surprisingly, estimated organic carbon burial rates are distinctly elevated (5–10 mmolC m^{-2} yr^{-1}) along the coast of Western Australia (Jahnke, 1996; Fig. 9), likely due to coastal mixing processes, intermittent localized upwelling, and eddy generation in the otherwise oligotrophic downwelling Leeuwin Current (Hood et al., 2017).

^{234}Th-estimates of POC export flux have revealed elevated values in eastern-central equatorial waters of the Indian Ocean during spring intermonsoon (up to 7.7 mmolC m^{-2} d^{-1}) in association with elevated rates of primary production (Subha Anand et al., 2017). These elevated rates are somewhat surprising given the downwelling circulation that is associated with the eastward-flowing equatorial currents (Strutton et al., 2015), but they are consistent with eastward increases in organic carbon burial and sedimentation rates along the equator (Jahnke, 1996; Fig. 9). They are also consistent with eastward increases POC export from the euphotic zone estimated from the satellite-derived net primary production data and export fluxes calculated with a model that gives 149.4 and 590.5 mgC m^{-2} d^{-1} (equal to 12.54 and 49.2 mmolC m^{-2} d^{-1}), respectively, off Sumatra, compared to similarly estimated values of 110.8 and 220.1 mgC m^{-2} d^{-1} (equal to 9.2 and 18.3 mmolC m^{-2} d^{-1}), respectively, in the Seychelles-Chagos Thermocline Ridge region (Sreeush et al., 2018).

There are few direct measurements of POC concentrations and export flux off of western Australia. Waite et al. (2016) have shown that mesoscale eddies generated by the Leeuwin Current can strongly impact carbon export fluxes. Specifically, the subsurface distribution of particles in these eddies funnel into a wineglass shape down to 1000 m (Fig. 13), leading to a sevenfold increase of vertical carbon flux in the eddy center versus the eddy flanks. This is consistent with the

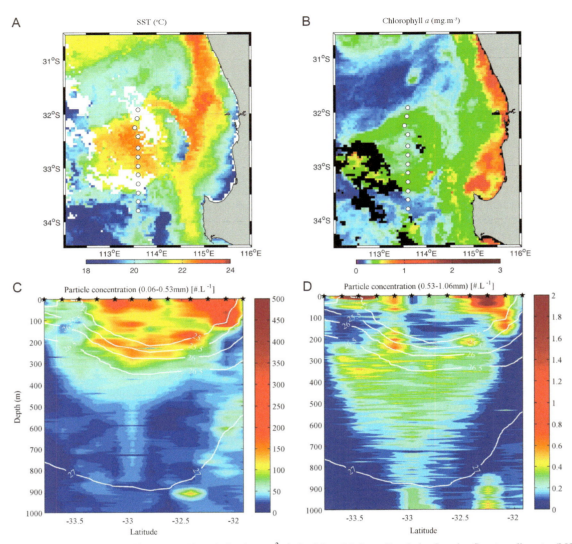

FIG. 13 (a) Sea-surface temperature (°C) and (b) chlorophyll a (mg m^{-3}) derived from Moderate Resolution Imaging Spectroradiometer (MODIS) satellite, showing the mesoscale eddy in the Leeuwin Current off Australia. *White circles* indicate stations sampled. Spatial distributions of particles across the eddy: (c) Small (0.06–0.5 mm) and (d) large (0.5–1 mm) particles shown in color, with isopycnals contoured in *white*. *(Modified from Waite et al. (2016).)*

aforementioned idea that eddy generation in the Leeuwin Current gives rise to elevated organic carbon burial rates along the coast of Western Australia (Jahnke, 1996; Fig. 9).

3.4.4 Southwestern Indian Ocean

Although the modeling study reported by Hansell (2009) and Hansell et al. (2009) suggests that near-surface (30 m) DOC concentrations are lower (<65 μmolC/kg) in the southwestern Indian Ocean and in the Mozambique Channel, these declines are not apparent in zonal sections (e.g., Fig. 7). Rather, a meridional section along ~30°E (IO6N from Arctica to western South Africa) shows a very clear increase in near-surface (<50 m) DOC concentrations north of 45° S from <50 to >70 μmolC/kg (Hansell, 2009; Fig. 7). Moreover, these elevated concentrations extend downward to >300 m depth (Fig. 7). These data suggest that the Agulhas Current advects relatively high tropical and subtropical DOC concentrations southwestward along the coast of South Africa and also eastward in the Agulhas Retroflection, and that these elevated concentrations are being mixed downward in the current. This speculation is consistent with the fact that the Agulhas Current extends to >1000 m in depth and is derived from oligotrophic tropical and subtropical sources waters

(Hood et al., 2017) that have relatively high DOC concentrations (Fig. 7). DOC concentrations below 1000 m in these waters are low (<50 μmolC/kg) and relatively uniform (Fig. 7).

Satellite estimates suggest that POC concentrations are generally low (<72 mgC/m^3 which is equal to <6 μmolC/kg) in the southwestern Indian Ocean and in the Agulhas Current during both the austral summer and winter seasons (Gardner et al., 2006; Fig. 8). However, some elevated POC concentrations (>72 mgC/m^3, >6 μmolC/kg) are observed in the austral winter in the Mozambique Channel, along the southwestern coast of South Africa, over the Agulhas Bank and in the Agulhas retroflection (Gardner et al., 2006; Fig. 8). All of these regions are associated with elevated chlorophyll concentrations (>0.4% of POC; Fig. 8, Gardner et al., 2006). In the Mozambique Channel, the elevated concentrations are associated with entrainment and offshore advection of coastal particulate organic matter in eddies (Kolasinski et al., 2012), whereas the elevated particulate organic matter concentrations along the southwestern coast of South Africa and over the Agulhas Bank are associated with wind-induced and topographically controlled coastal upwelling (Hood et al., 2017, 2024). The tongue of elevated POC and chlorophyll concentration that is associated with the Agulhas retroflection is particularly striking (Fig. 8). This is a region where strong eastward flows and persistent eddies are observed (Pazan & Niiler, 2004). It is possible that these eddies mix or upwell nutrients into the euphotic zone, fueling a low level of primary production. This region of elevated POC and chlorophyll extends all the way to the Kerguelen Islands during the austral summer (Fig. 8; Gardner et al., 2006).

Estimated organic carbon burial rates in the southwestern Indian Ocean range from <5 to ~10 mmolC m^{-2} yr^{-1} with the highest rates observed along the coastal zone of southeastern Africa (Jahnke, 1996; Fig. 9). Interestingly, these estimates also clearly show elevated organic carbon burial (and sedimentation) rates in the same region where there is a tongue of elevated near-surface POC and chlorophyll concentration associated with the Agulhas retroflection (Gardner et al., 2006; compare Figs. 8 and 9). These elevated carbon burial rates also extend all the way to the Kerguelen Islands. This is in contrast to the low carbon burial rates (<5 mmolC m^{-2} yr^{-1}; Jahnke, 1996; Fig. 9) and low ^{234}Th-estimated POC export fluxes (from 0.10 to 2.53 mmolC m^{-2} d^{-1} below 100 m) that are observed slightly further south (Coppola et al., 2005).

3.4.5 Southern subtropical gyre

Near-surface (<50 m) DOC concentrations in the southern subtropical gyre of the Indian Ocean are significantly elevated (>75 μmolC/kg) between 10° and 35°S (Fig. 7), consistent with the positive correlation between DOC concentrations and temperature, i.e., the southern subtropical gyre is warm and stratified and DOC concentrations become elevated because there is reduced vertical mixing. Near-surface DOC concentrations decline precipitously to <50 μmolC/kg south of 40°S due to increased vertical mixing in the higher latitude waters (Fig. 7). There is also evidence of subduction and northward transport of DOC in the subtropical front between 10° and 35°S where elevated DOC concentrations can be seen extending to >250 m depth (Fig. 7). The source of this subducted water has DOC concentrations between 65 and 70 μmolC/kg, and as this water moves downward and equatorward the DOC is remineralized, dropping to ~55 μmolC/kg at 250 m depth (Hansell, 2009; Fig. 7). DOC concentrations below 500 m in the southern subtropical Indian Ocean are also low (<50 μmolC/kg) and relatively uniform (Fig. 7).

The lowest POC concentrations (<48 mgC/m^3 equal to <4 μmolC/kg) in the Indian Ocean are observed in the oligotrophic southern subtropical gyre. These low POC values are associated with low chlorophyll concentrations (<0.3% of POC; Fig. 8, Gardner et al., 2006). These low POC and low percent chlorophyll waters dramatically increase in area during the Southern Hemisphere summer due to increases in summer stratification and lower nutrient availability (Gardner et al., 2006; Fig. 8). Nonetheless, subsurface chlorophyll maxima exist below these oligotrophic regions (see Hood et al., 2024). It is likely that the subsurface chlorophyll maxima are associated with somewhat shallower subsurface POC maxima, with the latter weaker as discussed above.

Sediment trap-measured POC fluxes in the southern subtropical gyre of the Indian Ocean are some of the lowest recorded worldwide (~0.50 mgC m^{-2} day^{-1} = 0.04 mmolC m^{-2} d^{-1}, measured at 500–600 and 2600–3500 m; Harms et al., 2021). Low POC fluxes are consistent with the extremely low organic carbon burial (<2 mmolC m^{-2} yr^{-1}) and sediment accumulation rates calculated by Jahnke (1996; Fig. 9). These low POC fluxes are the result of the strongly stratified, nutrient-depleted, and low productivity near-surface waters in the gyre (Harms et al., 2021; Hood et al., 2024). These continuously oligotrophic conditions result in almost constant POC fluxes in space and time. The lack of seasonality in the POC fluxes can also be attributed to intense organic matter degradation in the water column (Harms et al., 2021). The small amount of variability that is observed is related to variations wind-induced physical mixing events and the passage of eddies. Preliminary estimates indicate that the average POC export efficiency is extremely low (~0.03%) in these waters (Harms et al., 2021).

4 Summary and conclusions

The thickest OMZ in the world is found in the northern Indian Ocean in the Arabian Sea where intermediate water (~200–800 m) oxygen concentrations decline to nearly zero, with consequent impacts on nitrogen cycling. These impacts include denitrification-driven reductions in deep NO_3^- concentrations, the appearance of NO_2^- maxima, the generation of greenhouse gases (N_2O), and globally significant losses of fixed nitrogen (N_2) from the ocean. In contrast, these biogeochemical impacts are not observed in the Bay of Bengal where intermediate water dissolved oxygen concentrations are poised just above the threshold below which denitrification becomes significant. The low oxygen conditions in the water column in both the Arabian Sea and the Bay of Bengal are associated with slow ventilation and high biological oxygen demand. It appears that a weaker ballast effect and higher biological productivity sustain higher biological oxygen consumption in the Arabian Sea compared to the Bay of Bengal, which balances the higher physical oxygen supply in the Arabian Sea and explains its more intense OMZ.

The volume of hypoxic waters and denitrification in the Arabian Sea strongly increase in response to increases in monsoon winds. Such increases in Arabian Sea denitrification are expected to cause increases in N_2 and N_2O production. However, global syntheses of long-term OMZ variability reveal only a weak decrease of dissolved oxygen concentrations in the OMZs of the Arabian Sea and the Bay of Bengal in comparison to OMZs of the South Atlantic Ocean and the Pacific Ocean. Moreover, outbreaks of hydrogen sulfide have so far not been reported in the northern Indian Ocean during the last 50 years, other than in bottom waters on the Indian shelf. The absence of H_2S implies that the interplay between physical oxygen supply and biological oxygen consumption has largely maintained the current hypoxic/anoxic conditions in the Arabian Sea and the Bay of Bengal OMZs. However, recent observational and modeling studies indicate that oxygen concentrations are now decreasing in the Arabian Sea and that these decreases are having significant biogeochemical and ecological impacts. The future evolution of the northern Indian Ocean OMZs projected by Earth System models is highly uncertain.

The Indian Ocean remains one of the most poorly sampled ocean regions with respect to inorganic and organic carbon pools and air-sea carbon fluxes. The system accounts for ~20% of the global oceanic uptake of atmospheric CO_2. The Arabian Sea is a source of CO_2 to the atmosphere due to elevated pCO_2 within the Southwest Monsoon-driven upwelling whereas it is still uncertain whether the Bay of Bengal is a CO_2 source or sink due to the sparse spatial and temporal sampling. South of 14°S the Indian Ocean appears to be a strong net CO_2 sink due to the combined effects of both the solubility pump and the biological pump. Surface pH values in the Indian Ocean are among the lowest of the major ocean basins, and the reasons for this are poorly understood. Increases in sulfate and nitrogen aerosol loadings over the Bay of Bengal may be mainly responsible for the increased acidity in the northwestern Bay of Bengal in recent years and reduced river discharge together with a positive IOD event have also been shown to contribute to enhanced acidification and pCO_2 levels in the coastal waters in the western Bay of Bengal. Projected increases in oceanic CO_2 concentrations will lead to further acidification of the Indian Ocean over the coming decades, with potentially severe negative impacts on coral reefs and other calcifying organisms.

DOC concentrations in the Indian Ocean vary between ~40 and 80 μmolC/kg. DOC concentrations in the Indian Ocean tend to be high in stratified near-surface tropical and subtropical waters where DOC is produced and accumulates, and DOC concentrations are lowest in the deep ocean where heterotrophic consumption of DOC exceeds autotrophic production. As observed elsewhere in the global ocean, satellite-estimated near-surface POC concentrations are elevated in coastal regions of the Indian Ocean with values often exceeding 120 mgC/m^3 (=10 μmolC/kg). The highest POC concentrations in the Indian Ocean are observed in the northwestern part of the basin in the Arabian Sea and off the coast of Somalia with values estimated to be greater than 96 mgC/m^3 (=8 μmolC/kg). These high POC values are associated with high chlorophyll concentrations. The lowest POC concentrations (<48 mgC/m^3 =4 μmolC/kg) are observed in the oligotrophic southern subtropical gyre. These low POC values are associated with low chlorophyll concentrations.

The spatial and temporal variability in POC export flux in the Indian Ocean is similar to the above-mentioned patterns in POC and chlorophyll concentration, consistent with primary production as the main control on the spatial and temporal variability of organic carbon fluxes. However, in river-influence regions, like the Bay of Bengal, the spatial variability of organic carbon flux is also strongly influenced by lithogenic matter content. In the Arabian Sea POC export at 2000 m ranges from <6 to >22 mgC m^{-2} d^{-1} (<0.5 to >1.8 mmolC m^{-2} d^{-1}) with the highest fluxes observed during the Soutwest Monsoon. In contrast, POC export at 2000 m in the Bay of Bengal is generally lower and less strongly seasonal than in the Arabian Sea with fluxes varying between 2 and 14 mgC m^{-2} d^{-1} (0.17–1.1 mmolC m^{-2} d^{-1}). ^{234}Th-estimates of POC export flux have revealed elevated values in eastern-central equatorial waters of the Indian Ocean (up to 7.7 mmolC m^{-2} d^{-1}) in association with elevated rates of primary production. These elevated rates are surprising given the downwelling equatorial circulation, but they are consistent with zonal variations in satellite and model-based estimates of near-surface POC export flux and organic carbon burial rates. There are few direct measurements of POC concentrations and export flux off of Western Australia where estimated organic carbon burial rates are distinctly elevated, perhaps due to the influence of Leeuwin Current eddies. Estimated organic carbon burial rates in the southwestern Indian Ocean range from $<5-\sim10\,mmol\,C\,m^{-2}\,yr^{-1}$, with the highest rates observed along the coastal zone of southeastern Africa and in the Agulhas retroflection. POC export fluxes in the southern subtropical gyre of the Indian Ocean are some of the lowest recorded worldwide ($\sim 0.50\,mg\,C\,m^{-2}\,d^{-1} = 0.04\,mmol\,C\,m^{-2}\,d^{-1}$).

As in the other ocean basins, it is clear that there is a strong connection in the Indian Ocean between the physics that drives (or suppresses) nutrient delivery to the photic zone and responses of oxygen, CO_2 flux, pH, DOC, POC, and POC export. Monsoonal wind forcing is a major driver of biogeochemical variability throughout the northern Indian Ocean and in equatorial waters. In addition, there are regionally specific processes that significantly modulate oxygen, CO_2 flux, pH, DOC, POC, and POC export. For example, upwelling and strong advective impacts in the Arabian Sea; freshwater and stratification in the Bay of Bengal; the influence of ITF, poleward transport, downwelling, and eddies in the southeastern Indian Ocean; and poleward transport of tropical waters, combined with localized upwelling in the southwestern Indian Ocean. The southern subtropical gyre is extremely oligotrophic.

Observational and modeling research should be aimed at improving the understanding of northern Indian Ocean OMZ variability at seasonal and decadal time scales. Further, uncertainty in future projections needs to be reduced. Additional observations are also needed to better constrain air-sea CO_2 fluxes and the carbon cycle in general. Measurements of CO_2 and DOC concentrations, capturing spatial and temporal variability, are particularly sparse. This improved understanding is needed to predict the impacts of anthropogenic influence and global warming on Indian Ocean biogeochemistry and ecosystems and also for understanding the role of the Indian Ocean in global ocean biogeochemical cycles both now and in the future.

5 Educational resources

- Ocean Data View, free software for plotting oceanographic data. Available at: https://odv.awi.de
- World Ocean Atlas, a collection of objectively analyzed, quality-controlled temperature, salinity, oxygen, phosphate, silicate, and nitrate means based on profile data from the World Ocean Database. Available at: https://www.ncei.noaa.gov/products/world-ocean-atlas
- Surface Ocean CO_2 Atlas (SOCAT) is a synthesis of quality-controlled, surface ocean fCO_2 (fugacity of carbon dioxide) observations by the international marine carbon research community. Available at: https://www.socat.info
- Satellite ocean color data. Available at: https://oceancolor.gsfc.nasa.gov

Acknowledgments

The development of this article was supported by the Scientific Committee for Oceanic Research via direct funding to the Second International Indian Ocean Expedition and indirect funding through the Integrated Marine Biosphere Research regional program SIBER (Sustained Indian Ocean Biogeochemistry and Ecosystem Research). Additional support was provided by NASA grant no. 80NSSC17K0258 49A37A, NOAA grant no. NA15NMF4570252 NCRS-17, and NSF grant no. 2009248 to R. Hood. The Ocean Data View software package (https://odv.awi.de) was used in developing this article's original graphics. The article also benefitted from extensive comments provided by Dr. V.V.S.S. Sarma and one anonymous reviewer. This is UMCES contribution 6346.

Author contributions

All authors contributed to the writing of the text, discussion of content, and structure of the chapter.

References

Acharya, S. S., & Panigrahi, M. K. (2016). Eastward shift and maintenance of Arabian Sea oxygen minimum zone: Understanding the paradox. *Deep Sea Research Part I: Oceanographic Research Papers, 115*, 240–252.

Al Azhar, M., Lachkar, Z., Lévy, M., & Smith, S. (2017). Oxygen minimum zone contrasts between the Arabian Sea and the Bay of Bengal implied by differences in remineralization depth. *Geophysical Research Letters, 44*, 11106–11114.

Altabet, M. A., Francois, R., Murray, D. W., & Prell, W. L. (1995). Climate-related variations in denitrification in the Arabian Sea from sediment N-15/N-14 ratios. *Nature, 373*, 506–509.

Altabet, M. A., Higginson, M. J., & Murray, D. W. (2002). The effect of millennial-scale changes in Arabian Sea denitrification on atmospheric CO 2. *Nature, 415*, 159–162.

Altieri, A. H., Harrison, S. B., Seemann, J., Collin, R., Diaz, R. J., & Knowlton, N. (2017). Tropical dead zones and mass mortalities on coral reefs. *Proceedings of the National Academy of Sciences, 114*, 3660–3665.

Anju, M., Valsala, V., Smitha, B. R., Bharathi, G., & Naidu, C. V. (2022). Observed denitrification in the northeast Arabian Sea during the winter-spring transition of 2009. *Journal of Marine Systems, 227*, 103680.

Antoine, D., André, J.-. M., & Morel, A. (1996). Oceanic primary production: 2. Estimation at global scale from satellite (coastal zone color scanner) chlorophyll. *Global Biogeochemical Cycles, 10*, 57–69.

Bahl, A., Gnanadesikan, A., & Pradal, M.-. A. (2019). Variations in ocean deoxygenation across earth system models: Isolating the role of parameterized lateral mixing. *Global Biogeochemical Cycles, 33*, 703–724.

Bakker, D. C. E., et al. (2014). An update to the Surface Ocean CO_2 atlas (SOCAT version 2). *Earth Systems Science Data, 6*, 69–90.

Bange, H. W., Andreae, M. O., Lal, S., Law, C. S., Naqvi, S. W. A., Patra, P. K., Rixen, T., & Upstill-Goddard, R. C. (2001). Nitrous oxide emissions from the Arabian Sea: A synthesis. *Atmospheric Chemistry and Physics, 1*, 61–71.

Bange, H. W., et al. (2000). A revised nitrogen budget for the Arabian Sea. *Global Biogeochemical Cycles, 14*, 1283–1297.

Banse, K., Naqvi, S. W. A., Narvekar, P. V., Postel, J. R., & Jayakumar, D. A. (2014). Oxygen minimum zone of the open Arabian Sea: Variability of oxygen and nitrite from daily to decadal timescales. *Biogeosciences, 11*, 2237–2261.

Bates, N. R., Pequignet, A. C., & Sabine, C. L. (2006). Ocean carbon cycling in the Indian Ocean: I. Spatio-temporal variability of inorganic carbon and air-sea CO_2 gas exchange. *Global Biogeochemical Cycles, 20*, GB3020.

Bednarsek, N., Tarling, G. A., Bakker, D. C. E., Fielding, S., Jones, E. M., Venables, H. J., Ward, P., Kuzirian, A., Leze, B., Feely, R. A., & Murphy, E. J. (2012). Extensive dissolution of live pteropods in the Southern Ocean. *Nature Geoscience, 5*, 881–885.

Behrenfeld, M. J., & Falkowski, P. G. (1997). Photosynthetic rates derived from satellite-based chlorophyll concentration. *Limnology and Oceanography, 42*(1), 1–20. https://doi-org.proxy-um.researchport.umd.edu/10.4319/lo.1997.42.1.0001.

Behrman, D. (1981). *Assault on the largest unknown: The international Indian Ocean expedition, 1959–1965*. UNESCO Press: Paris.

Bercovici, S. K., McNichol, A. P., Xu, L., & Hansell, D. A. (2018). Radiocarbon content of dissolved organic carbon in the South Indian Ocean. *Geophysical Research Letters, 45*, 872–879.

Bopp, L., Resplandy, L., Orr, J. C., Doney, S. C., Dunne, J. P., Gehlen, M., Halloran, P., Heinze, C., Ilyina, T., & Seferian, R. (2013). Multiple stressors of ocean ecosystems in the 21st century: Projections with CMIP5 models. *Biogeosciences, 10*, 6225–6245.

Bopp, L., Resplandy, L., Untersee, A., Le Mezo, P., & Kageyama, M. (2017). Ocean (de) oxygenation from the last glacial maximum to the twenty-first century: Insights from earth system models. *Philosophical transactions of the Royal Society A: Mathematical, Physical and Engineering Sciences, 375*, 20160323.

Boyd, P. W., Claustre, H., Levy, M., Siegel, D. A., & Weber, T. (2019). Multi-faceted particle pumps drive carbon sequestration in the ocean. *Nature, 568*, 327–335.

Bristow, L. A., Callbeck, C. M., Larsen, M., Altabet, M. A., Dekaezemacker, J., Forth, M., Gauns, M., Glud, R. N., Kuypers, M. M. M., & Lavik, G. (2017). N_2 production rates limited by nitrite availability in the Bay of Bengal oxygen minimum zone. *Nature Geoscience, 10*, 24–29.

Cabré, A., Marinov, I., Bernardello, R., & Bianchi, D. (2015). Oxygen minimum zones in the tropical Pacific across CMIP5 models: Mean state differences and climate change trends. *Biogeosciences, 12*, 5429–5454.

Canfield, D. E., Kraft, B., Löscher, C. R., Boyle, R. A., Thamdrup, B., & Stewart, F. J. (2019). The regulation of oxygen to low concentrations in marine oxygen-minimum zones. *Journal of Marine Research, 77*, 297–324.

Carruthers, J. N., Gogate, S. S., Naidu, J. R., & Laevastu, T. (1959). Shorewards upslope of the layer of minimum oxygen off Bombay: Its influence on marine biology, especially fisheries. *Nature, 183*, 1084–1087.

Cenr. (2010). Scientific Assessment of Hypoxia in U.S. Coastal Waters. Technical Report, Committee on Environment and Natural Resources, Interagency Working Group on Harmful Algal Blooms, Hypoxia, and Human Health of the Joint Subcommittee on Ocean Science and Technology. *Washington, DC*.

Cerutti, J., Burt, A. J., Haupt, P., Bunbury, N., Mumby, P. J., & Schaepman-Strub, G. (2020). Impacts of the 2014–2017 global bleaching event on a protected remote atoll in the Western Indian Ocean. *Coral Reefs, 39*, 15–26.

Chakraborty, K., Valsala, V., Bhattacharya, T., & Ghosh, J. (2021). Seasonal cycle of surface ocean pCO2 and pH in the northern Indian Ocean and their controlling factors. *Progress in Oceanography, 198*, 102683.

Chakraborty, K., Valsala, V., Gupta, G. V. M., & Sarma, V. V. S. S. (2018). Dominant biological control over upwelling on pCO2 in sea east of Sri Lanka. *Journal of Geophysical Research: Biogeosciences, 123*, 3250–3261.

Codispoti, L., Brandes, J. A., Christensen, J., Devol, A., Naqvi, S., Paerl, H. W., & Yoshinari, T. (2001). The oceanic fixed nitrogen and nitrous oxide budgets: Moving targets as we enter the anthropocene? *Scientia Marina (Barcelona), 65*(Suppl. 2), 85–105.

Coppola, L., Roy-Barman, M., Mulsow, S., Povinec, P., & Jeandel, C. (2005). Low particulate organic carbon export in the frontal zone of the Southern Ocean (Indian sector) revealed by 234Th. *Deep Sea Research Part I: Oceanographic Research Papers, 52*, 51–68.

Cullen, J. J. (2015). Subsurface chlorophyll maximum layers: Enduring enigma or mystery solved? *Annual Review of Marine Science, 7*, 207–239.

Dalsgaard, T., Canfield, D. E., Petersen, J., Thamdrup, B., & Acuna-Gonzalez, J. (2003). N_2 production by the anammox reaction in the anoxic water column of Golfo Dulce, Costa Rica. *Nature, 422*, 606–608.

de la Rocha, C., & Passow, U. (2014). *The biological pump*. Elsevier Science.

de Moel, H., Ganssen, G. M., Peeters, F. J. C., Jung, S. J. A., Kroon, D., Brummer, G. J. A., & Zeebe, R. E. (2009). Planktonic foraminiferal shell thinniing in the Arabian Sea due to anthropogenic ocean acidification? *Biogeosciences, 6*, 1917–1925.

de Verneil, A., Lachkar, Z., Smith, S., & Lévy, M. (2021). Evaluating the Arabian Sea as a regional source of atmospheric CO2: Seasonal variability and drivers. *Biogeosciences Discussions*, 1–38.

Deutsch, C., Sarmeinto, J. L., Sigman, D. M., Gruber, N., & Dunne, J. P. (2007). Spatial coupling of nitrogen inputs and losses in the ocean. *Nature*, *445*, 163–167.

DeVries, T., Deutsch, C., Rafter, P. A., & Primeau, F. (2013). Marine denitrification rates determined from a global 3-D inverse model. *Biogeosciences*, *10*, 2481–2496.

Diaz, R. J., & Rosenberg, R. (2008). Spreading dead zones and consequences for marine ecosystems. *Science*, *321*, 926–929.

Diaz, R. J., Rosenberg, R., & Sturdivant, K. (2019). Hypoxia in estuaries and semi-enclosed seas. In D. Laffoley, & J. M. Baxter (Eds.), *Ocean deoxygenation: Everyone's problem*. Gland, Switzerland: IUCN.

Dietrich, G. (1936). Aufbau und Bewegung von Golfstrom und Agulhasstrom, eine vergleichende Betrachtung. *Naturwissenschaften*, *24*, 225–230.

Doney, S. C. (2010). The growing human footprint on coastal and open-ocean biogeochemistry. *Science*, *328*, 1512–1516.

Dueser, W. G., Ross, E. H., & Mlodzinska, Z. J. (1978). Evidence for and rate of denitrification in the Arabian Sea. *Deep Sea Research*, *25*(5), 431–445. https://doi.org/10.1016/0146-6291(78)90551-9.

Duteil, O., & Oschlies, A. (2011). Sensitivity of simulated extent and future evolution of marine suboxia to mixing intensity. *Geophysical Research Letters*, *38*.

Ekau, W., Auel, H., Portner, H.-O., & Gilbert, D. (2010). Impacts of hypoxia on the structure and processes in pelagic communities (zooplankton, macro-invertebrates and fish). *Biogeosciences*, *7*(5), 1669–1699. https://doi.org/10.5194/bg-7-1669-2010.

Eugster, O., & Gruber, N. (2012). A probabilistic estimate of global marine N-fixation and denitrification. *Global Biogeochemical Cycles*, *26*.

Fasham, M. J. R. (2003). JGOFS: A retrospective view. In *Ocean Biogeochemistry* Springer.

Feely, R. A., Doney, S. C., & Cooley, S. R. (2009). Ocean acidification: Present conditions and future changes in a high-CO_2 world. *Oceanography*, *22*, 36–47.

Fennel, K., & Boss, E. (2003). Subsurface maxima of phytoplankton and chlorophyll: Steady-state solutions from a simple model. *Limnology and Oceanography*, *48*, 1521–1534.

Fernandes, L., Bhosle, N. B., Prabhu Matondkar, S. G., & Bhushan, R. (2009). Seasonal and spatial distribution of particulate organic matter in the Bay of Bengal. *Journal of Marine Systems*, *77*, 137–147.

Fine, R. A. (1993). Circulation of Antarctic intermediate water in the South Indian Ocean. *Deep Sea Research Part I: Oceanographic Research Papers*, *40*, 2021–2042.

Fuhrman, J. A., & Capone, D. G. (1991). Possible biogeochemical consequences of ocean fertilization. *Limnology and Oceanography*, *36*, 1951–1959.

Garcia, H. E., Locarnini, R. A., Boyer, T. P., Antonov, J. I., Zweng, M. M., Baranova, O. K., & Johnson, D. R. (2010). World Ocean atlas 2009. In S. Levitus (Ed.), *Vol. 4. NOAA atlas NESDIS 71*. Washington, DC: U.S. Government Printing Office.

Garcia, H. E., Weathers, K. W., Paver, C. R., Smolyar, I., Boyer, T. P., Locarnini, M. M., … Reagan, J. R. (2018). World Ocean Atlas 2018. Vol. 4: Dissolved inorganic nutrients (phosphate, nitrate and nitrate+nitrite, silicate). *NOAA Atlas NESDIS*, *84*, 35 pp. https://archimer.ifremer.fr/doc/00651/76336/.

Gardner, W. D., Mishonov, A. V., & Richardson, M. J. (2006). Global POC concentrations from in-situ and satellite data. *Deep Sea Research Part II: Topical Studies in Oceanography*, *53*, 718–740.

Gattuso, J.-P., & Hansson, L. (2011). *Ocean acidification*. Oxford University Press: Oxford, UK.

Gaye, B., Böll, A., Segschneider, J., Burdanowitz, N., Emeis, K.-C., Ramaswamy, V., Lahajnar, N., Lückge, A., & Rixen, T. (2018). Glacial–interglacial changes and Holocene variations in Arabian Sea denitrification. *Biogeosciences*, *15*, 507–527.

Gaye, B., Nagel, B., Dähnke, K., Rixen, T., & Emeis, K. C. (2013). Evidence of parallel denitrification and nitrite oxidation in the ODZ of the Arabian Sea from paired stable isotopes of nitrate and nitrite. *Global Biogeochemical Cycles*, *27*, 1059–1071.

Gilson, H. C. (1933). The nitrogen cycle. *The John Murray Expedition*, *1934*, 21–81.

Gnanadesikan, A., Dunne, J. P., & John, J. (2012). Understanding why the volume of suboxic waters does not increase over centuries of global warming in an Earth System Model. *Biogeosciences*, *9*, 1159–1172.

Goes, J. I., & Gomes, H.d. R. (2016). An ecosystem in transition: the emergence of mixotrophy in the Arabian Sea. In *Aquatic Microbial Ecology and Biogeochemistry: A Dual Perspective* Springer.

Goes, J. I., Tian, H., do Rosario Gomes, H., Anderson, O. R., Al-Hashmi, K., de Rada, S., Luo, H., Al-Kharusi, L., Al-Azri, A., & Martinson, D. G. (2020). Ecosystem state change in the Arabian Sea fuelled by the recent loss of snow over the Himalayan-tibetan plateau region. *Scientific Reports*, *10*, 1–8.

Gomes, H. D., Goes, J. I., Motondkar, S. G. P., Buskey, E. J., Basu, S., Parab, S., & Thoppil, P. G. (2014). Massive outbreaks of Notiluca scintillans blooms in teh Arabian Sea due to spread of hypoxia. *Nature Communications*, *5*. https://doi.org/10.1038/ncomms5862.

Gould, J., Sloyan, B., & Visbeck, M. (2013). In situ ocean observations: A brief history, present status, and future directions. In *International Geophysics* Elsevier.

Graham, N. A. J., Wilson, S. K., Jennings, S., Polunin, N. V. C., Robinson, J., Bijoux, J. P., & Daw, T. M. (2007). Lag effects in the impacts of mass coral bleaching on coral reef fish, fisheries, and ecosystems. *Conservation Biology*, *21*, 1291–1300.

Gruber, N. (2004). The dynamics of the marine nitrogen cycle and its influence on atmospheric CO2 variations. In *The ocean carbon cycle and climate* Springer.

Gruber, N., & Galloway, J. N. (2008). An Earth-system perspective of the global nitrogen cycle. *Nature*, *451*, 293–296.

Gruber, N., & Sarmiento, J. L. (1997). Global patterns of marine nitrogen fixation and denitrification. *Global Biogeochemical Cycles*, *11*, 235–266.

Gundersen, J. S., Gardner, W. D., Richardson, M. J., & Walsh, I. D. (1998). Effects of monsoons on the seasonal and spatial distributions of POC and chlorophyll in the Arabian Sea. *Deep Sea Research Part II*, *45*, 2103–2132.

Gupta, G. V. M., Jyothibabu, R., Ramu, C. V., Yudhistir Reddy, A., Balachandran, K. K., Sudheesh, V., Kumar, S., Chari, N. V. H. K., Bepari, K. F., & Marathe, P. H. (2021). The world's largest coastal deoxygenation zone is not anthropogenically driven. *Environmental Research Letters*, *16*, 054009.

Gupta, R. S., & Naqvi, S. W. A. (1984). Chemical oceanography of the Indian Ocean, north of the equator. *Deep Sea Research Part A. Oceanographic Research Papers*, *31*, 671–706.

Gupta, G. V. M., Sudheesh, V., Sudharma, K. V., Saravanane, N., Dhanya, V., Dhanya, K. R., Lakshmi, G., Sudhakar, M., & Naqvi, S. W. A. (2016). Evolution to decay of upwelling and associated biogeochemistry over the southeastern Arabian Sea shelf. *Journal of Geophysical Research: Biogeosciences, 121*, 159–175.

Haake, B., & Ittekkot, V. (1990). Die Wind-getriebene "biologische Pumpe" und der Kohlenstoffentzug im Ozean. *Naturwissenschaften, 77*, 75–79.

Hansell, D. A. (2009). Dissolved organic carbon in the carbon cycle of the Indian Ocean. *Indian Ocean Biogeochemical Processes and Ecological Variability, Geophysical Monograph Series, 185*, 217–230.

Hansell, D. A., Carlson, C. A., Repeta, D. J., & Schlitzer, R. (2009). Dissolved organic matter in the ocean: A controversy stimulates new insights. *Oceanography, 22*, 202–211.

Hansell, D. A., & Orellana, M. V. (2021). Dissolved organic matter in the Global Ocean: A primer. *Gels, 7*, 128.

Hansell, D. A., & Peltzer, E. T. (1998). Spatial and temporal variations of total organic carbon in the Arabian Sea. *Deep Sea Research Part II: Topical Studies in Oceanography, 45*, 2171–2193.

Harms, N. C., Lahajnar, N., Gaye, B., Rixen, T., Schwarz-Schampera, U., & Emeis, K.-C. (2021). Sediment trap-derived particulate matter fluxes in the oligotrophic subtropical gyre of the South Indian Ocean. *Deep Sea Research Part II: Topical Studies in Oceanography*, 104924.

Helly, J. J., & Levin, L. A. (2004). Global distribution of naturally occurring marine hypoxia on continental margins. *Deep Sea Research Part I, 51*, 1159–1168.

Hoegh-Guldberg, O., Mumby, P. J., Hooten, A. J., Steneck, R. S., Greenfield, P., Gomez, E., Harvell, C. D., Sale, P. F., Edwards, A. J., Caldeira, K., Knowlton, N., Eakin, C. M., Iglesias-Prieto, R., Muthiga, N., Bradbury, R. H., Dubi, A., & Hatziolos, M. E. (2007). Coral reefs under rapid climate change and oceand acidification. *Science, 318*, 1737–1742.

Hoegh-Guldberg, O., et al. (2014). Chapter 30: The ocean. In *IPCC WGII AR5 (Ed.), Climate change 2014: Impacts, adaptation, and vulnerability. Contribution of working group II to the fifth assessment report of the intergovernmental panel on climate change*. Cambridge, UK, and New York, NY, USA: Cambridge University Press.

Hofmann, A. F., Peltzer, E. T., Walz, P. M., & Brewer, P. G. (2011). Hypoxia by degrees: Establishing definitions for a changing ocean. *Deep Sea Research Part I: Oceanographic Research Papers, 58*(12), 1212–1226. https://doi-org.proxy-um.researchport.umd.edu/10.1016/j.dsr.2011.09.004.

Honjo, S., Dymond, J., Prell, W., & Ittekkot, V. (1999). Monsoon-controlled export fluxes to the interior of the Arabian Sea. *Deep-Sea Research (Part II, Topical Studies in Oceanography), 46*, 1859–1902.

Hood, R. R., Bange, H. W., Beal, L., Beckley, L. E., Burkill, P., Cowie, G. L., D'Adamo, N., Ganssen, G., Hendon, H., Hermes, J., Honda, M., McPhaden, M., Roberts, M., Singh, S., Urban, E., & Yu, W. (2015). *Science plan of the second international Indian Ocean expedition (IIOE-2): A basin-wide research program*. Newark, Delaware, USA: Scientific Committee on Oceanic Research.

Hood, R. R., Beckley, L. E. B., & Wiggert, J. D. (2017). Biogeochemical and ecological impacts of boundary currents in the Indian Ocean. *Progress in Oceanography, 156*, 290–325.

Hood, R. R., Coles, V. J., Huggett, J. A., Landry, M. R., Levy, M., Moffett, J. W., & Rixen, T. (2024). Chapter 13: Nutrient, phytoplankton, and zooplankton variability in the Indian Ocean. In C. C. Ummenhofer, & R. R. Hood (Eds.), *The Indian Ocean and its role in the global climate system* (pp. 293–327). Amsterdam: Elsevier. https://doi.org/10.1016/B978-0-12-822698-8.00020-2.

Howell, E. A., Doney, S. C., Fine, R. A., & Olson, D. B. (1997). Geochemical estimates of denitrification in the Arabian Sea and the Bay of Bengal during WOCE. *Geophysical Research Letters, 24*, 2549–2552.

Hutchins, D. A., Fu, F.-X., Zhang, Y., Warner, M. E., Feng, Y., Portune, K., Bernhardt, P. W., & Mulholland, M. R. (2007). CO_2 control of Trichodesmium N_2 fixation, photosynthesis, growth rates, and elemental ratios: Implications for past, present, and future ocean biogeography. *Limnology and Oceanography, 52*, 1293–1304.

IPCC. (2007). In Core_Writing_Team, R. K. Pachauri, & A. Reisinger (Eds.), *Contribution of working groups I, II and III to the fourth assessment report of the intergovernmental panel on climate change*. Geneva, Switzerland: IPCC.

IPCC. (2019). *IPCC special report on the ocean and cryosphere in a changing climate*.

Ito, T., Minobe, S., Long, M. C., & Deutsch, C. (2017). Upper ocean O2 trends: 1958–2015. *Geophysical Research Letters, 44*, 4214–4223.

Ittekkot, V., Nair, R. R., Honjo, S., Ramaswamy, V., Bartsch, M., Manganini, S., & Desai, B. N. (1991). Enhanced particle fluxes in Bay of Bengal induced by injection of fresh water. *Nature, 351*, 385–387.

Ivanenkov, V. N., & Rozanov, A. G. (1966). *Hydrogen sulphide contamination of the intermediate layers of the Arabian Sea and the Bay of Bengal*. National Institute of Oceanography.

Jahnke, R. A. (1996). The global ocean flux of particulate organic carbon: Areal distribution and magnitude. *Global Biogeochemical Cycles, 10*, 71–88.

Kolasinski, J., Kaehler, S., & Jaquemet, S. (2012). Distribution and sources of particulate organic matter in a mesoscale eddy dipole in the Mozambique Channel (South-Western Indian Ocean): Insight from C and N stable isotopes. *Journal of Marine Systems, 96*, 122–131.

Kuypers, M. M. M., Sliekers, A. O., Lavik, G., Schmid, M., Joergensen, B. B., Kuenen, J. G., Damste, J. S. S., Strous, M., & Jetten, M. S. M. (2003). Anaerobic ammonium oxidation by anammox bacteria in the Black Sea. *Nature, 422*, 608–611.

Kwiatkowski, L., Torres, O., Bopp, L., Aumont, O., Chamberlain, M., Christian, J. R., Dunne, J. P., Gehlen, M., Ilyina, T., & John, J. G. (2020). Twenty-first century ocean warming, acidification, deoxygenation, and upper-ocean nutrient and primary production decline from CMIP6 model projections. *Biogeosciences, 17*, 3439–3470.

Lachkar, Z., Lévy, M., & Shafer Smith, K. (2019). Strong intensification of the Arabian Sea oxygen minimum zone in response to Arabian Gulf warming. *Geophysical Research Letters, 46*, 5420–5429.

Lachkar, Z., Lévy, M., & Smith, S. (2018). Intensification and deepening of the Arabian Sea oxygen minimum zone in response to increase in Indian monsoon wind intensity. *Biogeosciences, 15*, 159–186.

Lachkar, Z., Mehari, M., Al Azhar, M., Lévy, M., & Smith, S. (2020). Fast local warming of sea-surface is the main factor of recent deoxygenation in the Arabian Sea. *Biogeosciences Discussions*, 1–27.

Lachkar, Z., Smith, S., Lévy, M., & Pauluis, O. (2016). Eddies reduce denitrification and compress habitats in the Arabian Sea. *Geophysical Research Letters, 43*, 9148–9156.

Lévy, M., Resplandy, L., Palter, J. B., Couespel, D., & Lachkar, Z. (2021). *The crucial contribution of mixing to present and future ocean oxygen distribution*. Elsevier.

Luyten, J. R., Pedlosky, J., & Stommel, H. (1983). The ventilated thermocline. *Journal of Physical Oceanography, 13*, 292–309.

McCartney, M. S. (1979). Subantarctic mode water. *Woods Hole Oceanographic Institution Contribution, 3773*, 103–119.

McClanahan, T. R., Ateweberhan, M., Graham, N. A. J., Wilson, S. K., Sebastian, C. R., Guillaume, M. M. M., & Bruggemann, J. H. (2007). Western Indian Ocean coral communities: Bleaching responses and susceptibility to extinction. *Marine Ecology Progress Series, 337*, 1–13.

McCreary, J. P., Yu, Z., Hood, R. R., Vinayachandran, P. N., Furue, R., Ishida, A., & Richards, K. J. (2013). Dynamcis of the Indian-Ocean oxygen minumum zones. *Progress in Oceanography, 112–113*, 15–37.

McElroy, M. B. (1983). Marine biological controls on atmospheric CO2 and climate. *Nature, 302*, 328–329.

Moore, W. S. (1984). Review of the geosecs project. *Nuclear Instruments and Methods in Physics Research, 223*, 459–465.

Moore, J. A., Bellchambers, L. M., Depczynski, M. R., Evans, R. D., Evans, S. N., Field, S. N., Friedman, K. J., Gilmour, J. P., Holmes, T. H., Middlebrook, R., Radford, B. T., Ridgway, T., Shedrawi, G., Taylor, H., Thompson, D. P., & Wilson, S. K. (2012). Unprecedented mass bleaching and loss of coral across 12 degrees of latitude in Western Australia in 2010-11. *PLoS ONE, 7*, e51087. https://doi.org/10.1317/journal.pone.0051807.

Morrison, J. M., Codispoti, L. A., Smith, S. L., Wishner, K., Flagg, C., Gardner, W. D., Gaurin, S., Naqvi, S. W. A., Manghnani, V., Prosperie, L., & Gundersen, J. S. (1999). The oxygen minimum zone in the Arabian Sea during 1995. *Deep-Sea Research (Part II, Topical Studies in Oceanography), 46*, 1903–1931.

Naqvi, S. W. A. (1991). Geographical extent of denitrification in the Arabian Sea in relation to some physical processes. *Oceanologica Acta, 14*, 281–290.

Naqvi, S. W. A. (2019). Evidence for ocean deoxygenation and its patterns: Indian Ocean. In D. Laffoley, & J. M. Baxter (Eds.), *Ocean deoxygenation: Everyone's problem*. Gland, Switzerland: IUCN.

Naqvi, S. W. A., Bange, H. W., Gibb, S. W., Goyet, C., Hatton, A. D., & Upstill-Goddard, R. C. (2005). Biogeochemical Ocean-atmosphere transfers in the Arabian Sea. *Progress in Oceanography, 65*, 116–144.

Naqvi, S. W. A., Jayakumar, D. A., Narvekar, P. V., Naik, H., Sarma, V. V. S. S., D'Souza, W., Joseph, S., & George, M. D. (2000). Increased marine production of N_2O due to intensifying anoxia on the Indian continental shelf. *Nature, 408*, 346–349.

Naqvi, S. W. A., Narvekar, P. V., & Desa, E. (2006). Coastal biogeochemical processes in the North Indian Ocean. In Robinson, A., & Brink, K. (Eds.), *The Sea* (Vol. 14, pp. 723–780). Harvard University Press.

Naqvi, S. W. A., Noronha, R. J., & Gangadhara Reddy, C. V. (1982). Denitrification in the Arabian Sea. *Deep Sea Research Part A. Oceanographic Research Papers, 29*, 459–469.

Newell, B. S. (1969). Seasonal variations in the Indian Ocean along 110° E. II. Particulate carbon. *Marine and Freshwater Research, 20*, 51–54.

Obura, D., Gudka, M., Rabi, F. A., Gian, S. B., Bigot, L., Bijoux, J., Freed, S., Maharavo, J., Munbodhe, V., & Mwaura, J. (2017). Coral reef status report for the Western Indian Ocean. Global Coral Reef Monitoring Network (GCRMN). *International Coral Reef Initiative (ICRI)*, 144.

Paulmier, A., & Ruiz-Pino, D. (2009). Oxygen minimum zones (OMZs) in the modern ocean. *Progress in Oceanography, 80*, 113–128.

Pazan, S. E., & Niiler, P. (2004). *New global drifter data set available*. Wiley Online Library.

Piketh, S. J., Tyson, P. D., & Steffen, W. (2000). Aeolian transport from southern Africa and iron fertilization of marine biota in the South Indian Ocean. *South African Journal of Science, 96*, 244–246.

Piontkovski, S. A., & Al-Oufi, H. S. (2015). The Omani shelf hypoxia and the warming Arabian Sea. *International Journal of Environmental Studies, 72*, 256–264.

Queste, B. Y., Vic, C., Heywood, K. J., & Piontkovski, S. A. (2018). Physical controls on oxygen distribution and denitrification potential in the north west Arabian Sea. *Geophysical Research Letters, 45*, 4143–4152.

Ramaswamy, V., Boucher, O., Haigh, J., Hauglustaine, D., Haywood, J., Myhre, G., Nakajima, T., Shi, G. Y., & Solomon, S. (2001). Radiative forcing of climate change. In J. T. Houghton, Y. Ding, D. J. Griggs, M. Noguer, P. J. Van der Linden, X. Dai, K. Maskell, & C. A. Johnson (Eds.), *Climate change 2001: The scientific basis: Contribution of working group I to the third assessment report of the intergovernmental panel on climate change*. Cambridge, UK: Cambridge University Press.

Ramaswamy, V., Nair, R. R., Manganini, S., Haake, B., & Ittekkot, V. (1991). Lithogenic fluxes to the deep Arabian Sea measured by sediment traps. *Deep Sea Research Part A. Oceanographic Research Papers, 38*, 169–184.

Rao, D. N., Chopra, M., Rajula, G. R., Durgadevi, D. S. L., & Sarma, V. V. S. S. (2021). Release of significant fraction of primary production as dissolved organic carbon in the Bay of Bengal. *Deep Sea Research Part I: Oceanographic Research Papers, 168*, 103445.

Rao, C. K., Naqvi, S. W. A., Dileep Kumar, M., Varaprasad, S. J. D., Jayakumar, D. A., George, M. D., & Singbal, S. Y. S. (1994). Hydrochemistry of the Bay of Bengal: Possible reasons for a different water-column cycling of carbon and nitrogen from the Arabian Sea. *Marine Chemistry, 47*, 279–290.

Rao, D. N., & Sarma, V. V. S. S. (2022). Accumulation of dissolved organic carbon and nitrogen in the photic zone in the nitrogen-depleted waters of the Bay of Bengal. *Marine Chemistry, 239*, 104074.

Rashid, T., Hoque, S., & Akter, F. (2013). Ocean acidification in the Bay of Bengal. *Scientific Reports, 2*, 699.

Resplandy, L. (2018). Climate change and oxygen in the ocean. *Nature, 557*(7705), 314–315. https://doi.org/10.1038/d41586-018-05034-y.

Resplandy, L., Lévy, M., Bopp, L., Echevin, V., Pous, S., Sarma, V. V. S. S., & Kumar, D. (2012). Controlling factors of the oxygen balance in the Arabian Sea's OMZ. *Biogeosciences, 9*, 5095–5109.

Rixen, T., Baum, A., Gaye, B., & Nagel, B. (2014). Seasonal and interannual variations in the nitrogen cycle in the Arabian Sea. *Biogeosciences, 11*, 5733–5747.

Rixen, T., Cowie, G., Gaye, B., Goes, J., do Rosário Gomes, H., Hood, R. R., Lachkar, Z., Schmidt, H., Segschneider, J., & Singh, A. (2020). Reviews and syntheses: Present, past, and future of the oxygen minimum zone in the northern Indian Ocean. *Biogeosciences, 17*, 6051–6080.

Rixen, T., Gaye, B., & Emeis, K.-C. (2019a). The monsoon, carbon fluxes, and the organic carbon pump in the northern Indian Ocean. *Progress in Oceanography, 175*, 24–39.

Rixen, T., Gaye, B., Emeis, K.-C., & Ramaswamy, V. (2019b). The ballast effect of lithogenic matter and its influences on the carbon fluxes in the Indian Ocean. *Biogeosciences, 16*, 485–503.

Rixen, T., & Ittekkot, V. (2005). Nitrogen deficits in the Arabian Sea, implications from a three component mixing analysis. *Deep Sea Research Part II: Topical Studies in Oceanography, 52*, 1879–1891.

Sarma, V. V. S. S., Jagadeesan, L., Dalabehera, H. B., Rao, D. N., Kumar, G. S., Durgadevi, D. S., Yadav, K., Behera, S., & Priya, M. M. R. (2018). Role of eddies on intensity of oxygen minimum zone in the Bay of Bengal. *Continental Shelf Research, 168*, 48–53.

Sarma, V. V. S. S., Krishna, M. S., Paul, Y. S., & Murty, V. S. N. (2015a). Observed changes in ocean acidity and carbon dioxide exchange in the coastal Bay of Bengal—A link to air pollution. *Tellus B: Chemical and Physical Meteorology, 67*, 24638.

Sarma, V. V. S. S., Krishna, M. S., Rao, V. D., Viswanadham, R., Kumar, N. A., Kumari, T. R., Gawade, L., Ghatkar, S., & Tari, A. (2012). Sources and sinks of CO2 in the west coast of Bay of Bengal. *Tellus B: Chemical and Physical Meteorology, 64*, 10961.

Sarma, V. V. S. S., Krishna, M. S., Srinivas, T. N. R., Kumari, V. R., Yadav, K., & Kumar, M. D. (2021). Elevated acidification rates due to deposition of atmospheric pollutants in the coastal Bay of Bengal. *Geophysical Research Letters, 48*, e2021GL095159.

Sarma, V. V. S. S., Lenton, A., Law, R. M., Metzl, N., Patra, P. K., Doney, S., Lima, I. D., Dlugokencky, E., Ramonet, M., & Valsala, V. (2013). Sea-air CO_2 fluxes in the Indian Ocean between 1990 and 2009. *Biogeosciences, 10*, 7035–7052.

Sarma, V. V. S. S., Patil, J. S., Shankar, D., & Anil, A. C. (2019). Shallow convective mixing promotes massive Noctiluca scintillans bloom in the northeastern Arabian Sea. *Marine Pollution Bulletin, 138*, 428–436.

Sarma, V. V. S. S., Paul, Y. S., Vani, D. G., & Murty, V. S. N. (2015b). Impact of river discharge on the coastal water pH and pCO2 levels during the Indian Ocean dipole (IOD) years in the western Bay of Bengal. *Continental Shelf Research, 107*, 132–140.

Sarma, V. V. S. S., Udaya Bhaskar, T. V. S., Pavan Kumar, J., & Chakraborty, K. (2020). Potential mechanisms responsible for occurrence of core oxygen minimum zone in the north-eastern Arabian Sea. *Deep Sea Research Part I: Oceanographic Research Papers, 165*, 103393.

Sathyendranath, S., Stuart, V., Nair, A., Oka, K., Nakane, T., Bouman, H., Forget, M.-H., Maass, H., & Platt, T. (2009). Carbon-to-chlorophyll ratio and growth rate of phytoplankton in the sea. *Marine Ecology Progress Series, 383*, 73–84.

Schmidt, H., Czeschel, R., & Visbeck, M. (2020). Seasonal variability of the circulation in the Arabian Sea at intermediate depth and its link to the Oxygen Minimum Zone. *Ocean Science Discussions*. https://doi.org/10.5194/os-2020-9.

Schmidt, H., Getzlaff, J., Löptien, U., & Oschlies, A. (2021). Causes of uncertainties in the representation of the Arabian Sea oxygen minimum zone in CMIP5 models. *Ocean Science Discussions*, 1–32.

Schmidtko, S., Stramma, L., & Visbeck, M. (2017). Decline in global oceanic oxygen content during the past five decades. *Nature, 542*, 335–339.

Schunck, H., Lavik, G., Desai, D. K., Großkopf, T., Kalvelage, T., Löscher, C. R., Paulmier, A., Contreras, S., Siegel, H., & Holtappels, M. (2013). Giant hydrogen sulfide plume in the oxygen minimum zone off Peru supports chemolithoautotrophy. *PLoS ONE, 8*, e68661.

Schwartz, M. C., Woulds, C., & Cowie, G. L. (2009). Sedimentary denitrification rates across the Arabian Sea oxygen minimum zone. *Deep Sea Research Part II: Topical Studies in Oceanography, 56*, 324–332.

Seiwell, H. R. (1937). *The minimum oxygen concentration in the western basin of the North Atlantic*. Massachusetts Institute of Technology.

Sewell, R. B. S. (1934). The John Murray expedition to the Arabian Sea. *Nature, 133*, 669–672.

Sewell, R. B. S., & Fage, L. (1948). Minimum oxygen layer in the ocean. *Nature, 162*, 949–951.

Shah, C., Sudheer, A. K., & Bhushan, R. (2018). Distribution of dissolved organic carbon in the Bay of Bengal: Influence of sediment discharge, fresh water flux, and productivity. *Marine Chemistry, 203*, 91–101.

Somasundar, K., Rajendran, A., Dileep Kumar, M., & Sen Gupta, R. (1990). Carbon and nitrogen budgets of the Arabian Sea. *Marine Chemistry, 30*, 363–377.

Somes, C. J., Oschlies, A., & Schmittner, A. (2013). Isotopic constraints on the pre-industrial oceanic nitrogen budget. *Biogeosciences, 10*, 5889–5910.

Sredivi, B., & Sarma, V. V. S. S. (2022). Enhanced atmospheric pollutants strengthened winter convective mixing and phytoplankton blooms in the northern Arabian Sea. *Journal of Geophysical Research, Biogeosciences, 127*, e2021JG006527.

Sreeush, M. G., Rajendran, S., Valsala, V., Pentakota, S., Prasad, K. V. S. R., & Murtugudde, R. (2019). Variability, trend and controlling factors of Ocean acidification over Western Arabian Sea upwelling region. *Marine Chemistry, 209*, 14–24.

Sreeush, M. G., Valsala, V., Pentakota, S., Siva Rama Prasad, K. V., & Murtugudde, R. (2018). Biological production in the Indian Ocean upwelling zones—Part 1: Refined estimation via the use of a variable compensation depth in ocean carbon models. *Biogeosciences, 15*, 1895–1918.

Sreeush, M. G., Valsala, V., Santanu, H., Pentakota, S., Prasad, K. V. S. R., Naidu, C. V., & Murtugudde, R. (2020). Biological production in the Indian Ocean upwelling zones—Part 2: Data based estimates of variable compensation depth for ocean carbon models via cyclo-stationary Bayesian inversion. *Deep Sea Research Part II: Topical Studies in Oceanography, 179*, 104619.

Sridevi, B., & Sarma, V. V. S. S. (2021). Role of river discharge and warming on ocean acidification and pCO2 levels in the Bay of Bengal. *Tellus B: Chemical and Physical Meteorology, 73*, 1–20.

Stramma, L., Johnson, G. C., Sprintall, J., & Mohrholz, V. (2008). Expanding oxygen-minimum zones in the tropical oceans. *Science, 320*, 655–658.

Stramma, L., Schmidtko, S., Levin, L. A., & Johnson, G. C. (2010). Ocean oxygen minima expansions and their biological impacts. *Deep-Sea Research, Part I, 57*, 587–595.

Strutton, P. G., Coles, V. J., Hood, R. R., Matear, R. J., McPhaden, M. J., & Phillips, H. E. (2015). Biogeochemical variability in the central equatorial Indian Ocean during the monsoon transition. *Biogeosciences, 12*, 2367–2382.

Subha Anand, S., Rengarajan, R., & Sarma, V. V. S. S. (2018a). 234Th-based carbon export flux along the Indian GEOTRACES GI02 section in the Arabian Sea and the Indian Ocean. *Global Biogeochemical Cycles, 32*, 417–436.

Subha Anand, S., Rengarajan, R., Sarma, V. V. S. S., Sudheer, A. K., Bhushan, R., & Singh, S. K. (2017). Spatial variability of upper ocean POC export in the B ay of B engal and the I ndian O cean determined using particle-reactive 234 T h. *Journal of Geophysical Research: Oceans, 122*, 3753–3770.

Subha Anand, S., Rengarajan, R., Shenoy, D., Gauns, M., & Naqvi, S. W. A. (2018b). POC export fluxes in the Arabian Sea and the Bay of Bengal: A simultaneous 234Th/238U and 210Po/210Pb study. *Marine Chemistry, 198*, 70–87.

Sudheesh, V., Gupta, G. V. M., Sudharma, K. V., Naik, H., Shenoy, D. M., Sudhakar, M., & Naqvi, S. W. A. (2016). Upwelling intensity modulates N2O concentrations over the western I ndian shelf. *Journal of Geophysical Research: Oceans, 121*, 8551–8565.

Sverdrup, H. U. (1938). On the explanation of the oxygen minima and maxima in the oceans. *ICES Journal of Marine Science, 13*, 163–172.

Sverdrup, H. U., Johnson, M. W., & Fleming, R. H. (1942). *The oceans: Their physics, chemistry and general biology*. Prentice-Hall: Englewood Cliffs, NJ.

Swallow, J. (1984). Some aspects of the physical oceanography of the Indian Ocean. *Deep Sea Research, 31*, 639–650.

Takahashi, T., Sutherland, S. C., Chipman, D., Goddard, J. G., Ho, C., Newberger, T., Sweeney, C., & Munro, D. R. (2014). Climatological distributions of pH, pCO2, total CO2, alkalinity, and CaCO3 saturationh in the global surface ocean, and temporal changes at selected locations. *Marine Chemistry, 164*, 95–125.

Takahashi, T., Sutherland, S. C., Sweeney, C., Poisson, A., Metzl, N., Tilbrook, B., Bates, N., Wanninkhof, R., Feely, R. A., Sabine, C., Olafsson, J., & Nojiri, Y. (2002). Global Sea-air CO_2 flux based on climatological surface ocean pCO_2, and seasonal biological and temperature effects. *Deep-Sea Research Part II, 49*, 1601–1623.

Takahashi, T., et al. (2009). Climatological mean and decadal change in surface ocean pCO2, and net sea-air CO2 flux over the global ocean. *Deep-Sea Rearch, Part II, 56*, 554–577.

Talley, L. D., Johnson, G. C., Purkey, S., Feely, R. A., & Wanninkhof, R. (2017). *Global Ocean Ship-based Hydrographic Investigations Program (GO-SHIP) provides key climate-relevant deep ocean observations* (p. 15). US CLIVAR Variations.

Tarique, M., Rahaman, W., Fousiya, A. A., Lathika, N., Thamban, M., Achyuthan, H., & Misra, S. (2021). Surface pH record (1990–2013) of the Arabian Sea from boron isotopes of Lakshadweep corals—Trend, variability, and control. *Journal of Geophysical Research: Biogeosciences, 126*, e2020JG006122.

Tchernia, P. (1980). *Descriptive regional oceanography* (p. 253). Pergamon Press.

Thushara, V., Vinayachandran, P. N. M., Matthews, A. J., Webber, B. G. M., & Queste, B. Y. (2019). Vertical distribution of chlorophyll in dynamically distinct regions of the southern Bay of Bengal. *Biogeosciences, 16*, 1447–1468.

Udaya Bhaskar, T. V. S., Sarma, V. V. S. S., & Pavan Kumar, J. (2021). Potential mechanisms responsible for spatial variability in intensity and thickness of oxygen minimum zone in the Bay of Bengal. *Journal of Geophysical Research: Biogeosciences, 126*, e2021JG006341.

Ulloa, O., Canfield, D. E., DeLong, E. F., & Stewart, F. J. (2012). Microbial oceanography of anoxic oxygen minimum zones. *Proceedings of the National Academy of Sciences of the United States of America, 109*(40), 15996–16003. https://doi-org.proxy-um.researchport.umd.edu/10.1073/pnas.1205009109.

Vallivattathillam, P., Iyyappan, S., Lengaigne, M., Ethé, C., Vialard, J., Levy, M., Suresh, N., Aumont, O., Resplandy, L., & Naik, H. (2017). Positive Indian Ocean dipole events prevent anoxia off the west coast of India. *Biogeosciences, 14*, 1541–1559.

Valsala, V. (2009). Different spreading of Somali and Arabian coastal upwelled waters in the northern Indian Ocean: A case study. *Journal of Oceanography, 65*, 803–816.

Valsala, V., & Maksyutov, S. (2010). Simulation and assimilation of global ocean pCO2 and air–sea CO2 fluxes using ship observations of surface ocean pCO2 in a simplified biogeochemical offline model. *Tellus B: Chemical and Physical Meteorology, 62*, 821–840.

Valsala, V., & Maksyutov, S. (2013). Interannual variability of the air–sea CO2 flux in the North Indian Ocean. *Ocean Dynamics, 63*, 165–178.

Valsala, V., Maksyutov, S., & Murtugudde, R. (2012). A window for carbon uptake in the southern subtropical Indian Ocean. *Geophysical Research Letters, 39*, L17605.

Valsala, V., & Murtugudde, R. (2015). Mesoscale and intraseasonal air–sea CO2 exchanges in the western Arabian Sea during boreal summer. *Deep Sea Research Part I: Oceanographic Research Papers, 103*, 101–113.

Valsala, V., Sreeush, M. G., Anju, M., Sreenivas, P., Tiwari, Y. K., Chakraborty, K., & Sijikumar, S. (2021). An observing system simulation experiment for Indian Ocean surface pCO2 measurements. *Progress in Oceanography, 194*, 102570.

Valsala, V., Sreeush, M. G., & Chakraborty, K. (2020). The IOD impacts on the Indian Ocean carbon cycle. *Journal of Geophysical Research: Oceans, 125*, e2020JC016485.

Vaquer-Sunyer, R., & Duarte, C. M. (2008). Thresholds of hypoxia for marine biodiversity. *Proceedings of the National Academy of Sciences, 105*, 15452–15457.

Vinayachandran, P. N. (2009). Impact of physical processes on chlorophyll distribution in the Bay of Bengal. In J. D. Wiggert, R. R. Hood, S. W. A. Naqvi, K. H. Brink, & S. L. Smith (Eds.), *Indian Ocean biogeochemical processes and ecological variability*. Washington DC: American Geophysical Union.

Volk, T., & Hoffert, M. I. (1985). The carbon cycle and atmospheric CO2, natural variation archean to present. In E. T. Sundquist, & W. S. Broecker (Eds.), *The carbon cycle and atmospheric CO2: Natural variations Archean to present* (pp. 99–110). Washington: AGU. Geophysical Monograph 32.

Waite, A. M., Stemmann, L., Guidi, L., Calil, P. H. R., Hogg, A. M. C., Feng, M., Thompson, P. A., Picheral, M., & Gorsky, G. (2016). The wineglass effect shapes particle export to the deep ocean in mesoscale eddies. *Geophysical Research Letters, 43*, 9791–9800.

Weeks, S. J., Currie, B., & Bakun, A. (2002). Massive emissions of toxic gas in the Atlantic. *Nature, 415*, 493–494.

Woods, J. D. (1985). The world ocean circulation experiment. *Nature, 314*, 501–511.

Chapter 13

Nutrient, phytoplankton, and zooplankton variability in the Indian Ocean[*]

Raleigh R. Hood[a], Victoria J. Coles[a], Jenny A. Huggett[b], Michael R. Landry[c], Marina Levy[d], James W. Moffett[e], and Timothy Rixen[f]

[a]Horn Point Laboratory, University of Maryland Center for Environmental Science, Cambridge, MD, United States, [b]Oceans and Coasts, Department of Forestry, Fisheries and the Environment, Cape Town, South Africa, [c]Scripps Institution of Oceanography, University of California San Diego, La Jolla, CA, United States, [d]LOCEAN-IPSL, Sorbonne Universités (UPMC, Univ Paris 06)-CNRS-IRD-MNHN, Paris, France, [e]Department of Biological Sciences, Earth Sciences and Civil and Environmental Engineering, University of Southern California, Los Angeles, CA, United States, [f]Leibniz Centre for Tropical Marine Research, University of Bremen, Bremen, Germany

1 Introduction

The northern Indian Ocean experiences strong monsoonal wind forcing that reverses seasonally due to seasonal heating and cooling over the Eurasian landmass (Schott et al., 2009; Schott & McCreary, 2001; Shankar et al., 2002). As a result, all of the boundary currents in the northern Indian Ocean also reverse seasonally, which is unique among ocean basins. The combination of winds, boundary currents, and remote forcing from Kelvin waves drives localized regions of upwelling during the Southwest Monsoon and downwelling (as well as buoyancy-driven convection) during the Northeast Monsoon. In addition, the partitioning of the northern basin of the Indian Ocean by the Indian subcontinent, combined with substantial differences in evaporation, precipitation, river runoff, and connectivity to marginal seas, gives rise to large differences in salinity and stratification between the Arabian Sea and the Bay of Bengal. The surface waters of the Bay of Bengal are much fresher and more stratified than in the Arabian Sea. The northern Indian Ocean is also unique in having one of three major open-ocean oxygen minimum zones (OMZs) as well as large dust and aerosol inputs that occur all year round from both natural (Guieu et al., 2019; Leon & Legrand, 2003; Pease et al., 1998) and anthropogenic sources (Lelieveld et al., 2001; McGowan & Clark, 2008).

Unlike in the easterly trade-wind-forced Pacific and Atlantic Oceans, upwelling in the equatorial Indian Ocean is not persistent because the mean winds along the equator are toward the east, driving convergent rather than divergent Ekman transport (Schott et al., 2009; Wang & McPhaden, 2017). However, off-equatorial upwelling in the Seychelles-Chagos Thermocline Ridge (SCTR) and the Sri Lankan Dome is geographically extensive and persistent. The equatorial Indian Ocean is strongly influenced by oscillations and perturbations unique to it, such as the Wyrtki Jets and the Indian Ocean Dipole (IOD), as well as by the Madden-Julian Oscillation (MJO), which originates in the Indian Ocean (Madden & Julian, 1972). The equatorial region further influences the Indian Ocean more broadly through coastal Kelvin waves generated along the equator by the Wyrtki Jets that affect the eastern equatorial Indian Ocean and the northern boundary currents (Hood et al., 2017; Schott et al., 2009; Schott & McCreary, 2001; Shankar et al., 2002).

In the southeastern tropical Indian Ocean, the Indonesian Throughflow (ITF) connects the Pacific and Indian Ocean basins, altering water mass properties through exchanges of heat and freshwater (Schott & McCreary, 2001) and biogeochemical water mass properties through exchanges of nutrients (Ayers et al., 2014; Talley & Sprintall, 2005). Presumably, the ITF also influences Indian Ocean ecosystems directly through passive and active transport of plankton and other higher trophic levels from the Pacific and the Maritime Continent, but this has not been investigated.

The southern Indian Ocean subtropical gyre is an oligotrophic ocean habitat. To the north, the South Equatorial Current transports warm, nutrient-enriched freshwater from the ITF across the basin (Schott & McCreary, 2001). To the east, the Leeuwin Current is a small downwelling-favorable current that is anomalous in being the global ocean's only

---

[*]This book has a companion website hosting complementary materials. Visit this URL to access it: https://www.elsevier.com/books-and-journals/book-companion/9780128226988.

poleward-flowing eastern boundary current and in having the largest eddy kinetic energy among all mid-latitude eastern boundary current systems (Hood et al., 2017). Combined with the topographic influence of the Ninety East Ridge, the South Equatorial Current and Leeuwin Current generate numerous westward-propagating eddies (Gaube et al., 2013; Waite et al., 2007b) and unusual circulation patterns. To the west, the Agulhas Current is an unusually large, poleward-flowing, "upwelling-favorable" western boundary current (Bryden et al., 2005; Hood et al., 2017). To the south, the gyre is bounded by the Antarctic Circumpolar Current.

2 Nutrient, phytoplankton and zooplankton variability

Variability in nutrients and plankton communities in the Indian Ocean is strongly influenced by the physical characteristics and by land-ocean coupling in different regions, which are presented separately as: (1) the Arabian Sea and the northwestern Indian Ocean; (2) the Bay of Bengal; (3) the equatorial Indian Ocean including the Java Upwelling and SCTR; (4) the eastern Indian Ocean and Leeuwin Current; (5) the southwestern Indian Ocean and Agulhas Current; and (6) the southern subtropical gyre.

2.1 Arabian Sea and the Northwestern Indian Ocean

2.1.1 Macro nutrients

During the Southwest Monsoon, strong upwelling occurs in the western Arabian Sea off Somalia, Yemen, and Oman (Fig. 1), where near-surface nitrate and silicate concentrations can increase to $>15\,\mu M$ and phosphate to $>1\,\mu M$ (Morrison et al., 1998). However, environmental conditions vary dramatically over the coastal to open-ocean gradient, with mixed-layer nitrate, silicate, and phosphate all declining to very low concentrations offshore (Gupta & Naqvi, 1984; Mantoura et al., 1993; Moffett et al., 2015; Moffett & Landry, 2020; Morrison et al., 1998; Rixen et al., 2009). Inorganic nitrogen-to-phosphate ratios tend to be lower than the Redfield ratio of 16:1, suggesting nitrogen is a more limiting nutrient than phosphorus (Morrison et al., 1998; Fig. 1). Nitrogen-to-silicate ratios tend to be higher ($\sim$2:1) than the uptake ratio (1:1) for diatom growth, suggesting the potential for silicate limitation of diatoms during the Southwest Monsoon (Morrison et al., 1998; Rixen et al., 2009; Fig. 1). Modeling studies further indicate that phytoplankton growth in Southwest Monsoon upwelled waters in the western Arabian Sea is prone to limitation by silicate (Kone et al., 2009; Resplandy et al., 2011) and/or iron (see Section 2.2.2 below).

Southwest Monsoon winds and the southward-flowing West India Coastal Current in the eastern Arabian Sea also promote upwelling that outcrops at the surface along the west coast of India. This upwelling is most pronounced between $\sim$8°N and 14°N (Shetye et al., 1990; Smitha et al., 2014) and gives rise to elevated nitrate, phosphate, and silicate concentrations in the near-surface layers (Hood et al., 2017), though this is not very evident in Fig. 1. The Laccadive Low that develops off the southwest coast of India during the Southwest Monsoon is an anticyclonic upwelling circulation that is evident in sea surface height (SSH) (Bruce et al., 1998; Shankar & Shetye, 1997; Subrahmanyam & Robinson, 2000) but does not appear to have a strong influence on surface nutrient concentrations (Lierheimer & Banse, 2002; Fig. 1).

Winds and circulation shift to downwelling favorable during the Northeast Monsoon. However, cooling and deep buoyancy-driven convection entrain nutrients that promote modest euphotic zone increases in nitrate, phosphate, and silicate over much of the northern Arabian Sea (Kone et al., 2009; McCreary et al., 2009; Resplandy et al., 2012; Wiggert et al., 2000; Fig. 2).

Between 200 and 400 m depth, denitrification in the central and western Arabian Sea OMZ reduces nitrate concentration to $\sim$20 μM (Figs. 3 and 4), and nitrite concentration increases to $\sim$5 μM (Fig. 2 in Moffett et al., 2015; Rixen et al., 2020), with nitrification also contributing to the latter (Li et al., 2006). The signature of denitrification is relatively low (<10) N:P ratios in these waters (Figs. 3 and 4). Estimates of denitrification rates in the OMZ are still fraught with large uncertainties and amount to $\sim$7–16 Tg N yr^{-1}, according to recent estimates (Gaye et al., 2013; Rixen et al., 2020). Moreover, the relative contributions and regulating mechanisms regarding N_2 gas production and release by either anammox or denitrification are highly variable and still debatable in different oceans/regions (Choi et al., 2016; Dähnke & Thamdrup, 2016; Ward et al., 2009), and the anammox reaction has been shown to be an important sink for fixed nitrogen in other OMZs (Arrigo, 2005; Dalsgaard et al., 2005; Dalsgaard & Thamdrup, 2002). However, denitrification is the dominant loss process in the Arabian Sea (Ward et al., 2009). Therefore, the low nitrate and high nitrite concentrations emphasize the role of the OMZ as a significant sink for fixed nitrogen in the Arabian Sea due to denitrification.

Nitrogen fixation has been intensively studied in the eastern Arabian Sea, and according to the few estimates for the entire Arabian Sea, it contributes a modest 3.3 Tg N yr^{-1} (Ahmed et al., 2017; Bange et al., 2000; Parab & Matondkar, 2012), which constitutes a small fraction (less than 1%) of primary production.

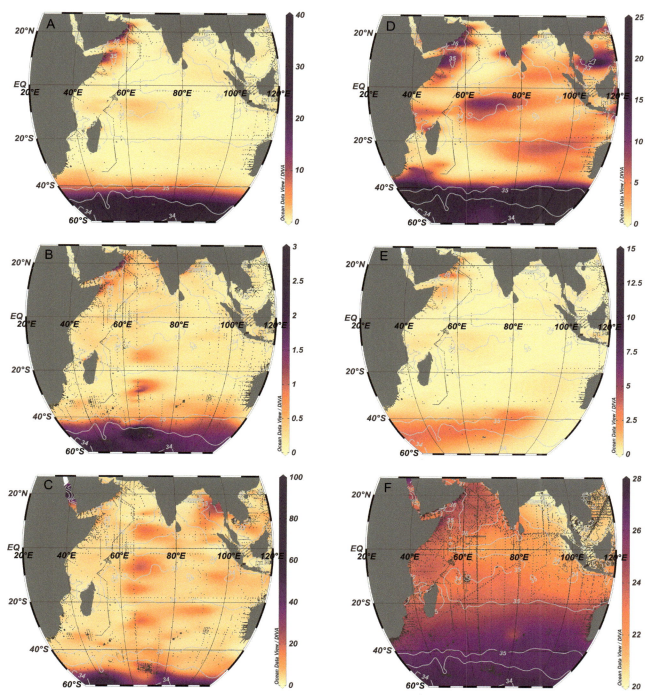

FIG. 1 Maps of the Southwest Monsoon period (July–August) for: (a) 10 m nitrate (mmol m^{-3}); (b) 10 m phosphate (mmol m^{-3}); (c) 10 m silicate (mmol m^{-3}); (d) 10 m nitrate to phosphate ratio; (e) 10 m nitrate to silicate ratio; (f) 10 m potential density anomaly. The *gray* contours are surface salinity. The station locations are shown as dots. Near surface nutrient ratios may have large errors due to the small surface nutrient concentrations. *(Data from Garcia et al. (2018).)*

2.1.2 Iron

Iron concentrations in the Arabian Sea are influenced by dust from the Arabian Peninsula and India, redox processes associated with the OMZ, biological scavenging, and possibly hydrothermal inputs from the Gulf of Aden (Guieu et al., 2019; Moffett & German, 2020). Seasonal variability arises because biological scavenging and dust inputs are strongly influenced

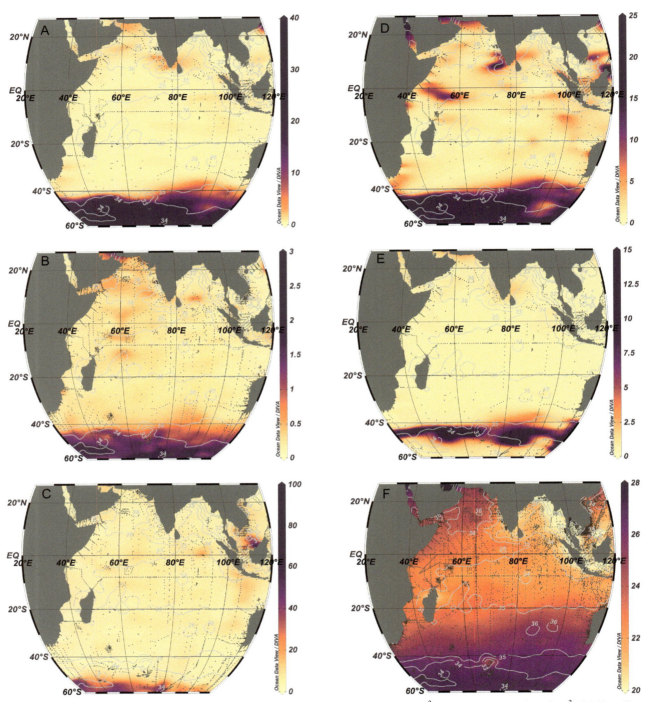

FIG. 2 Maps of the Northeast Monsoon period (December–January) for: (a) 10 m nitrate (mmol m^{-3}); (b) 10 m phosphate (mmol m^{-3}); (c) 10 m silicate (mmol m^{-3}); (d) 10 m nitrate to phosphate ratio; (e) 10 m nitrate to silicate ratio; (f) 10 m potential density anomaly. The *gray* contours are surface salinity. The station locations are shown as dots. Near-surface nutrient ratios may have large errors due to the small surface nutrient concentrations. *(Data from Garcia et al. (2018).)*

by the Northeast Monsoon and Southwest Monsoon. During the Northeast Monsoon, dust inputs (Pease et al., 1998; Tindale & Pease, 1999) elevate surface Fe and Al throughout the Arabian Sea (Measures & Vink, 1999). During the spring and fall inter-monsoon, dust deposition is lower than in winter (Pease et al., 1998; Tindale & Pease, 1999), but there is little biological scavenging of Fe due to prevailing oligotrophic conditions. Surprisingly, the Southwest Monsoon is the period of lowest Fe deposition (Tindale & Pease, 1999).

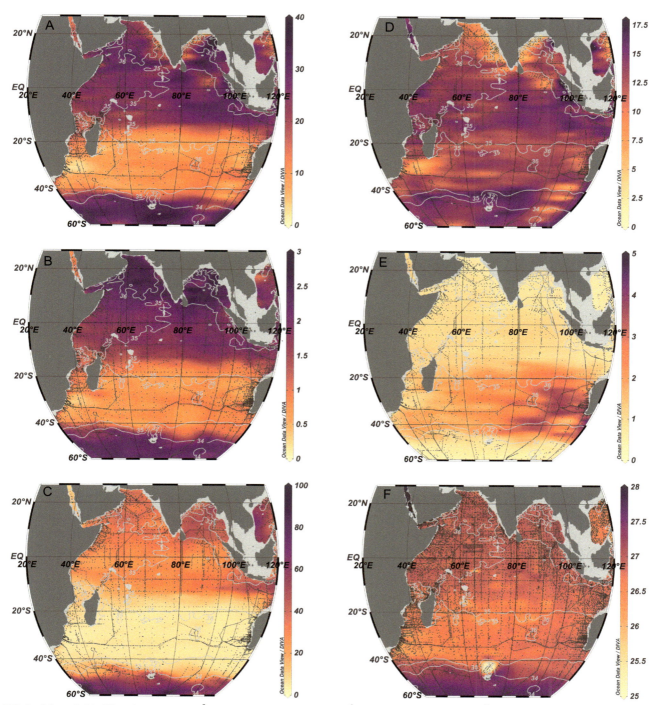

FIG. 3 Maps of: (a) 400m nitrate (mmol m^{-3}); (b) 400m phosphate (mmol m^{-3}); (c) 400m silicate (mmol m^{-3}); (d) 400m nitrate to phosphate ratio; (e) 400m nitrate to silicate ratio; (f) 400m potential density anomaly. The *gray* contours are surface salinity. The station locations are shown as dots. Data from Soviet cruises were removed for nutrient biases. *(Data from Garcia et al. (2018).)*

In the suboxic core of the OMZ, high iron levels accumulate (Fig. 5) as reduced Fe(II). Local Fe maxima coincide with the secondary nitrite maxima (Kondo & Moffett, 2013; Moffett et al., 2007, 2015; Fig. 5), suggesting a relationship between Fe redox cycling and denitrification that has yet to be identified. Fe(II) and total dissolved Fe decrease outside the OMZ. Reductive mobilization of iron within the OMZ is probably an important mechanism for Fe supply from the Indian continental margin to the Arabian Sea interior (Moffett et al., 2015).

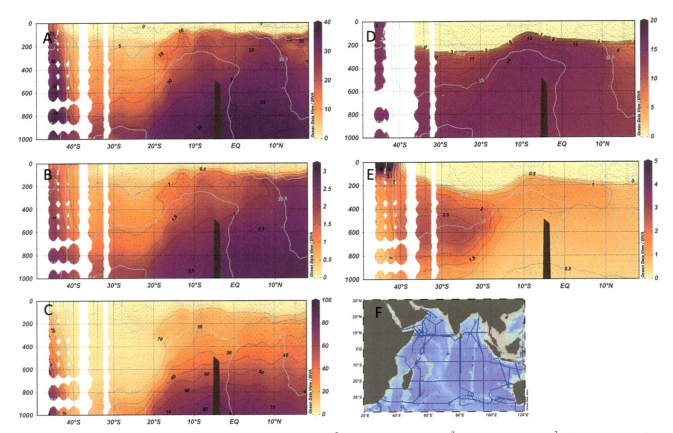

FIG. 4 Vertical sections nominally along 55–65°E of (a) nitrate (mmol m^{-3}); (b) phosphate (mmol m^{-3}); (c) silicate (mmol m^{-3}); (d) nitrate to phosphate ratio; (e) nitrate to silicate ratio; (f) map of the section. The *gray* contours are salinity, and the *black* contours are the colored property. The station locations for each property are shown as dots. *(Data from Garcia et al. (2018), comprising mainly data after 2005 and the 2018 IO7 GoShip line.)*

Naqvi et al. (2010) observed that chlorophyll-*a* (Chl*a*) and Fe:N conditions in the Omani upwelling region in 2004 were consistent with Fe limitation of primary production. Most importantly, the Arabian Sea becomes strongly depleted in Si relative to N during the Southwest Monsoon (Fig. 1). Preferential depletion of Si relative to N is a powerful diagnostic of Fe limitation in upwelling systems like Central California and southern Peru, which overlie narrow shelves and resemble the Oman upwelling regime (Hutchins et al., 1998, 2002; Johnson et al., 1999). Moffett et al. (2015) showed that Chl*a* responded strongly to added Fe compared to controls in the central Arabian Sea during the 2007 Southwest Monsoon. The strongest response was at stations with high nutrients and low sea surface temperature (SST) associated with filaments of upwelled coastal water from Oman. Nitrate was high ($\geq$6 mmole m^{-3}), but Si was depleted in most samples. Results were consistent with the modeling study of Wiggert et al. (2006), which predicted an intensifying region of Fe limitation in the western Arabian Sea through the Southwest Monsoon and into October (Fig. 6). The modeling results of Guieu et al. (2019) suggest that without atmospheric Fe inputs through dust deposition during the Southwest Monsoon, annual net primary production would be reduced by half.

2.1.3 Phytoplankton

Ocean color images over the Arabian Sea show a seasonal cycle of phytoplankton characterized by two distinct growth periods, one in summer during the Southwest Monsoon and the other in winter during the Northeast Monsoon. The bloom areas during the Southwest and the Northeast Monsoons show different regional patterns, with lags in bloom timing (Lévy et al., 2006). Due to upwelling, the western margin of the Arabian Sea transitions to a eutrophic coastal upwelling system during the Southwest Monsoon (Hood et al., 2017; Lévy et al., 2007; Wiggert et al., 2005; Figs. 7 and 8). Satellite ocean color reveals dramatic Chl*a* increases along the coasts of Somalia, Yemen, and Oman (Banse & English, 2000; Brock & McClain, 1992; George et al., 2013; Hood et al., 2017; Kumar et al., 2000; Lierheimer & Banse, 2002; Wiggert et al., 2005;

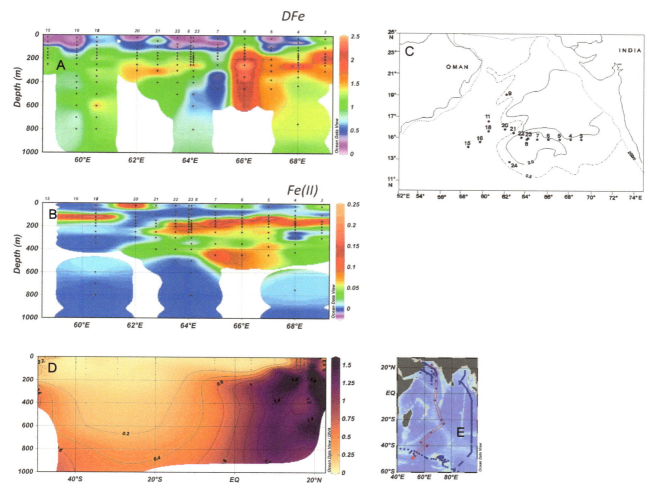

FIG. 5 Panels a and b are east to west sections of dissolved iron (Dfe) and Fe^{2+} (Fe(ii)) in the Arabian Sea down to 1000 m depth. The units are $\mu mole\,m^{-3}$. The station locations are shown in panel c. Figures reproduced with permission from Moffett et al. (2015). Panel d is dissolved Iron ($\mu mole\,m^{-3}$) along approximately 60–70°E in the Indian Ocean. The station locations are shown in panel e. *(Data from Tagliabue et al. (2012) database.)*

Fig. 7). They also show that the coastal Southwest Monsoon bloom onset occurs as early as March–April, peaks during the Southwest Monsoon, and coincides spatially with wind-driven upwelling. Further offshore, away from the direct influence of the upwelling, the bloom initiates ~1 month later and is co-located with a pattern of deep-mixed layers (Lévy et al., 2007). The modeling study of Resplandy et al. (2011) shows that the central Arabian Sea bloom is largely explained by lateral and vertical eddy supplies of nitrate associated with filaments originating from the coastal upwelling region, which is consistent with observations (Rixen et al., 2006a).

Vertically integrated Chl*a* in the western Arabian Sea can exceed $40\,mg\,m^{-2}$ during the Southwest Monsoon, with production rates of $2.5\,g\,C\,m^{-2}\,d^{-1}$ (Hood et al., 2017; Marra et al., 1998; Wiggert et al., 2005). However, in the oligotrophic offshore waters, Chl*a* and primary production decline dramatically (Brown et al., 1999; Hood et al., 2017; Wiggert et al., 2005; Figs. 7 and 8). In the western Arabian Sea, larger cells (diatoms) dominate phytoplankton community structure during the Southwest Monsoon (Brown et al., 1999; Shalapyonok et al., 2001; Tarran et al., 1999), while picophytoplankton (e.g., *Prochlorococcus sp.*) predominate during the oligotrophic spring and fall intermonsoon periods (Garrison et al., 2000; Roy et al., 2015). During all seasons, subsurface Chl*a* maxima occur between 40 and 140 m in the central, southeastern Arabian Sea (Gundersen et al., 1998; Ravichandran et al., 2012; Fig. 9a), in some places below the oxyclines of the OMZ (Goericke et al., 2000; Fig. 9b).

During the Southwest Monsoon, topographically locked eddies off Oman, Yemen, and Somalia advect high Chl*a* concentrations and coastal phytoplankton communities hundreds of kilometers offshore into more oligotrophic waters

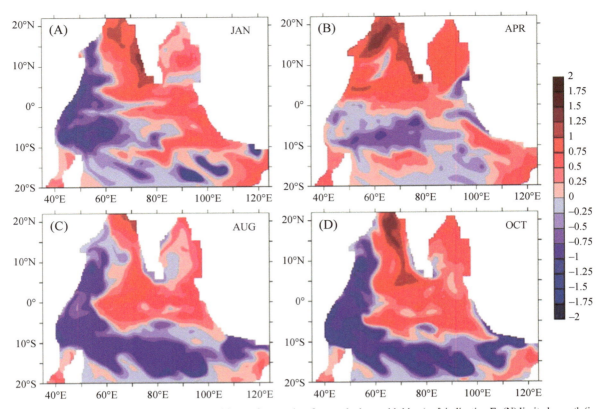

FIG. 6 Model-simulated seasonal evolution of most limiting surface nutrient for net plankton with *blue* (*red*) indicating Fe (N) limited growth (i.e., *red* is iron replete). The four seasons consist of: (a) January (Northeast Monsoon); (b) April (Spring Intermonsoon); (c) August (Southwest Monsoon); and (d) October (Fall Intermonsoon). *(Figure and caption modified from Wiggert et al. (2006).)*

(Gundersen et al., 1998; Hitchcock et al., 2000; Keen et al., 1997; Kim et al., 2001; Latasa & Bidigare, 1998; Lee et al., 2000; Manghnani et al., 1998; Roy et al., 2015; Fig. 7), which is also revealed by ocean-color images and models (Resplandy et al., 2011).

During the Northeast Monsoon, offshore waters are subject to shear and buoyancy-driven convection associated with cold, dry northeasterly winds, driving mixing that entrains nutrients and increases Chl*a* and primary production over the northern Arabian Sea (Hood et al., 2017; Lévy et al., 2007; Resplandy et al., 2011; Wiggert et al., 2000, 2005; Figs. 7 and 8). The amplitude of this winter bloom varies between years and is significantly tied to the interannual winter mixed-layer depth anomaly (Keerthi et al., 2017). Convective mixing during the Northeast Monsoon in the northern Arabian Sea inhibits picoplankton (specifically *Prochlorococcus*) growth (Roy & Anil, 2015). Resplandy et al. (2011) suggest that, in addition to convective mixing, nearly 50% of Northeast Monsoon productivity could be sustained by eddy-driven vertical nutrient supplies. The Northeast Monsoon increases in Chl*a* and primary production were associated with diatoms in early studies (Banse & McClain, 1986; Sawant & Madhupratap, 1996). However, *Phaeocystis* was first reported in Si-deficient upwelled waters in the central Arabian Sea during the Joint Global Ocean Flux Study (JGOFS) Arabian Sea Study (Garrison et al., 1998). Recent observations suggest a broader climate change-induced regime shift in the composition of winter phytoplankton blooms from diatoms to *Noctiluca scintillans*, a large mixotrophic dinoflagellate (Goes et al., 2020; Gomes et al., 2014). Indeed, *Noctiluca* and *Phaeocystis* blooms have become more common and widespread (Gomes et al., 2009). A succession from diatoms to *Phaeocystis* associated with Si removal may be common during the late Southwest Monsoon (Madhupratap et al., 2000).

Southwest Monsoon upwelling along the west coast of India gives rise to modest increases in Chl*a* and primary production compared to the central Arabian Sea (Hood et al., 2017; Kumar et al., 2000; Lévy et al., 2007; Luis & Kawamura, 2004; Naqvi et al., 2000; Figs. 7 and 8). In this region, upwelling is primarily forced by remotely generated planetary waves and secondarily by local winds (Shankar et al., 2002; Shankar & Shetye, 1997). This upwelling is associated with increased diatom abundance (Sawant & Madhupratap, 1996).

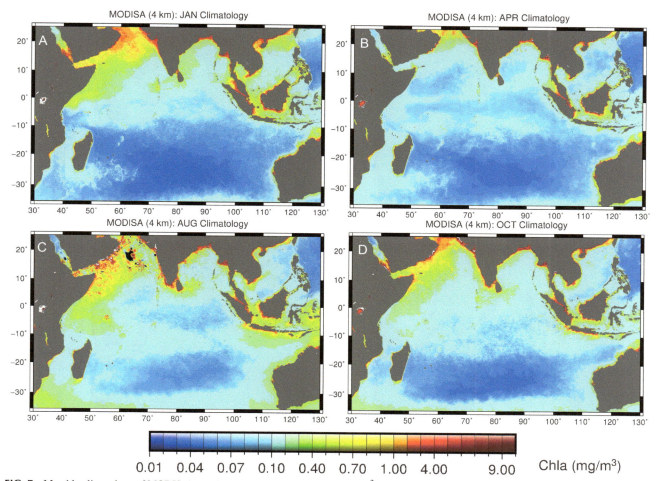

FIG. 7 Monthly climatology of MODIS-Aqua (4 km resolution) chlorophyll (mg m^{-3}): (a) January; (b) April; (c) August; (d) October. The climatology fields were obtained from the Goddard DAAC (http://daac.gsfc.nasa.gov). *(Figure and caption from Hood et al. (2017).)*

During the Northeast Monsoon, winds and currents along western India are downwelling-favorable, suppressing primary production, most notably along the southwestern coast. Nutrient depletion in this region coincides with blooms of *Trichodesmium* and dinoflagellates during the Northeast Monsoon (Matondkar et al., 2007; Parab et al., 2006). Vertically integrated Chl*a* and primary production off western India has been observed to decrease from ~24 to ~9 mg Chl*a* m^{-2} and from ~2.25 to ~1 g C m^{-2} d^{-1}, respectively, from the Southwest to the Northeast Monsoon (Hood et al., 2017; Luis & Kawamura, 2004; Figs. 7 and 8).

A comparison of the phytoplankton bloom cases reported before and after the 1950s (data from 1908 to 2009; D'Silva et al., 2012) reveals an increase in the number of algal bloom occurrences along the west coast of India, especially the southern part. The majority of these were caused by dinoflagellates. There have been 39 causative species responsible for the blooms, of which *N. scintillans* is one of the most common.

Satellite estimates of Chl*a* and primary production are low year-round off southwestern India in the Laccadive Sea despite the transition from Northeast Monsoon downwelling (cyclonic) to Southwest Monsoon upwelling (anticyclonic) circulations (Hood et al., 2017; Lierheimer & Banse, 2002; Figs. 7 and 8).

2.1.4 Zooplankton

Mesozooplankton sampling in the Arabian Sea has highlighted the relative seasonal constancy of biomass over much of the region, known as the "Arabian Sea Paradox" because it runs counter to expectations for a system with dramatic seasonal contrasts from intensive upwelling to subtropical oligotrophy (Baars, 1999; Madhupratap et al., 1992). During JGOFS, for example, biomass estimates in the northeastern Arabian Sea were 1.1 g C m^{-2} (upper 1000 m) in both coastal and oceanic

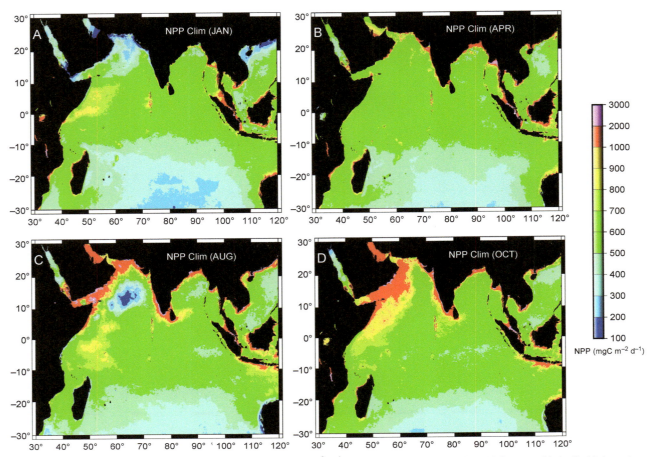

FIG. 8 Monthly climatology of net primary production (NPP; $mgC\,m^{-2}\,d^{-1}$) estimated from SeaWiFs data: (a) January; (b) April; (c) August; and (d) October. The climatology fields were obtained from the Goddard DAAC (http://daac.gsfc.nasa.gov). *(Figure and caption from Hood et al. (2017).)*

waters during the Southwest and Northeast Monsoons and slightly higher during the spring intermonsoon (1.3–1.9 g C m^{-2}; Smith & Madhapratap, 2005). Distributions of biomass and species at mid-water depths vary with the presence or absence of the OMZ (Wishner et al., 2008). In the western Arabian Sea, coastal stations averaged only 60% higher biomass in the upper 200 m compared to offshore (1.4 vs 0.9 g C m^{-2}, respectively) over four seasonal cruises, with ~twofold higher biomass during the Southwest Monsoon (1.5 vs 0.7 g C m^{-2}) and the peak shifting offshore during the Northeast Monsoon and spring intermonsoon (Wishner et al., 1998). Along the same sampling transect, protistan microzooplankton (heterotrophic nanoflagellates, dinoflagellates, and ciliates) averaged 0.52 ± 0.25 g C m^{-2} (upper 100 m), with similar ~twofold seasonal variability at individual stations but with biomass peaks shifting (opposite to mesozooplankton) to coastal stations during the late Northeast Monsoon and spring intermonsoon (Garrison et al., 2000). In the eastern central sector, higher protistan biomass occurs both in coastal and open-ocean waters during the spring intermonsoon compared to the Southwest and Northeast Monsoons (Gauns et al., 1996). However, in the southeastern Arabian Sea (15°N, east of 65°E), both microzooplankton abundances and mesozooplankton biomass are highest during the Soutwest Monsoon (Jyothibabu et al., 2010, 2008a).

To first order, the observed Arabian Sea zooplankton biomass patterns are consequences of physical drivers and circulation features that connect primary production to zooplankton consumers. In the western Arabian Sea, fast currents, eddies, and filaments disperse nutrients, production, and zooplankton widely over the central and northern regions (Baars, 1999; Kumar et al., 2001; Manghnani et al., 1998; Resplandy et al., 2011; Smith & Madhapratap, 2005). Nutrient entrainment by deep convective mixing during the Northeast Monsoon stimulates a second productivity peak that sustains high zooplankton stocks over a broad region during winter (Banse & McClain, 1986; Madhupratap et al., 1992, 1996a, 1996b).

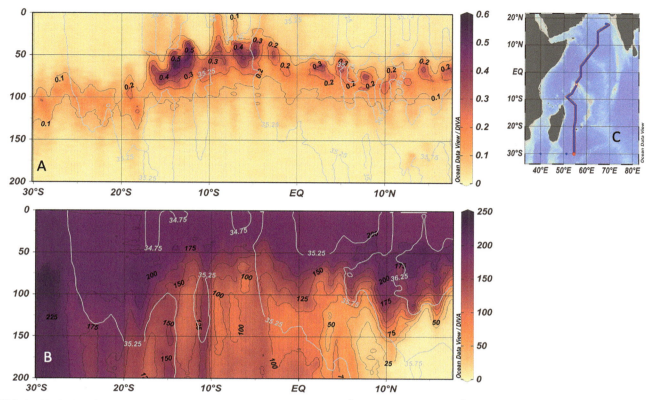

FIG. 9 Vertical sections nominally along 55–65°E of (a) Chlorophyll-a (mg m^{-3}); (b) oxygen (mmole m^{-3}); and (c) map of the section. The *gray* contours are salinity, and the *black* contours are the colored property. The station locations for each property are shown as dots. *(Data from the 2018 IO7 GoShip line.)*

While productivity shows some seasonal variation between these peaks, it is notable that primary production during JGOFS exhibited only a threefold total range (0.8–2.4 g C m^{-2} d^{-1}) among the 10 stations visited on 5 seasonal cruises, with only a 37% difference between the spring intermonsoon and higher Northeast and Southwest Monsoon averages (1.0 vs 1.6 g C m^{-2} d^{-1}; Barber et al., 2001). South of 15°N, however, the southeastern Arabian Sea differs from the seasonally and spatially distributed productivity of the northern and western areas in being strongly stratified by the inflow of low salinity Bay of Bengal water during the Northeast Monsoon, which suppresses convective mixing and results in oligotrophic conditions in offshore waters for much of the year. Thus, zooplankton variability aligns with the seasonal productivity peak driven by upwelling and river runoff during the Southwest Monsoon (Devi et al., 2010; Jyothibabu et al., 2010).

Microzooplankton dominates Arabian Sea grazing processes, consuming an average of 71% of primary production for all coastal to offshore measurements during JGOFS (Caron & Dennett, 1999; Edwards et al., 1999; Landry et al., 1998; Marra & Barber, 2005; Reckermann & Veldhuis, 1997; as summarized in Landry, 2009). High microzooplankton grazing (58%–97% of production) was also estimated for a eutrophic estuary during the Southwest Monsoon (Gauns et al., 2015). In a JGOFS synthesis of food web fluxes (Landry, 2009), carbon consumption by microzooplankton was 8 times higher than their grazing on heterotrophic bacteria (the "microbial loop," Azam et al., 1983). Mesozooplankton, in turn, derived ~40% of their nutritional requirements (Roman et al., 2000) from predation on microzooplankton. Importantly, however, the relative grazing contributions of micro- and mesozooplankton are not fixed but vary seasonally and spatially in ways that modulate total impact in a manner consistent with hypothesized grazing control of phytoplankton biomass in the Arabian Sea (Marra & Barber, 2005; Smith, 2001). Where mesozooplankton biomass is highest in the coastal upwelling area during the Southwest Monsoon (coincident with large phytoplankton), their predatory pressure significantly diminishes the relative grazing impact of microzooplankton. When the mesozooplankton grazing peak shifts offshore during the Northeast Monsoon, microzooplankton grazing increases closer to the coast, and the balance of production and grazing is maintained. Such dynamical flexibility is key to the idea of spatially separated co-regulation by iron and grazing in the Arabian Sea, as suggested by Moffett and Landry et al. (2020).

2.2 Bay of Bengal

2.2.1 Nutrients

The effects of the monsoon winds and associated currents are more cryptic in the Bay of Bengal due to weaker winds and large freshwater inputs from northern rivers (i.e., the Ganges-Brahmaputra and Irrawaddy). Most (95%) of the inorganic nitrogen and phosphorus loads are removed before the rivers reach the coast, with the remaining nutrients utilized within 10 km from the coast (De Sousa et al., 1981; Krishna et al., 2015; Sarma et al., 2013a, 2013b. Freshwater in the Bay of Bengal surface layers leads to strong salinity stratification (Kumar et al., 2007; Vinayachandran, 2009; Wijesekera et al., 2016) that inhibits vertical mixing and upwelling processes (Gomes et al., 2000; Thushara et al., 2019; Vinayachandran, 2009; Vinayachandran et al., 2002, 2005). Coastal upwelling occurs along eastern India during summer (June to August) due to Southwest Monsoon winds, but strong stratification prevents surface outcropping of nutrients in the northern Bay of Bengal (Sarma et al., 2012, 2015; Shetye et al., 1991, 1993).

In general, nutrient concentrations are much lower in the Bay of Bengal than in the Arabian Sea, resulting in strong nutrient limitation of phytoplankton growth (Kone et al., 2009). In contrast to the Arabian Sea, there is little or no enhancement of climatological surface nitrate, phosphate, or silicate concentrations in the Bay of Bengal coastal zones during the SWM (compare Figs. 1 and 2). Nonetheless, nutrient ratios in east Indian coastal waters (11–15°N) over all seasons (molar ratios N:P < 16; Si:P > 7 and N:Si < 1) indicate that nitrogen and silicate are the primary limiting nutrients (Thangaradjou et al., 2014). In the central Bay of Bengal, climatological nitrate, phosphate, and silicate concentrations are all undetectable down to ∼75 m (Fig. 10, Baer et al., 2019).

Nutrient limitation experiments in the Bay of Bengal indicate that N is the primary limiting nutrient for autotrophic growth (Twining et al., 2019). However, experiments aimed at specific phytoplankton groups suggest a more complex, species-specific, nutrient-limitation mosaic, i.e., different phytoplankton groups can be limited by different nutrients (Twining et al., 2019). In addition, eddy-mediated processes have a strong influence on Bay of Bengal circulation

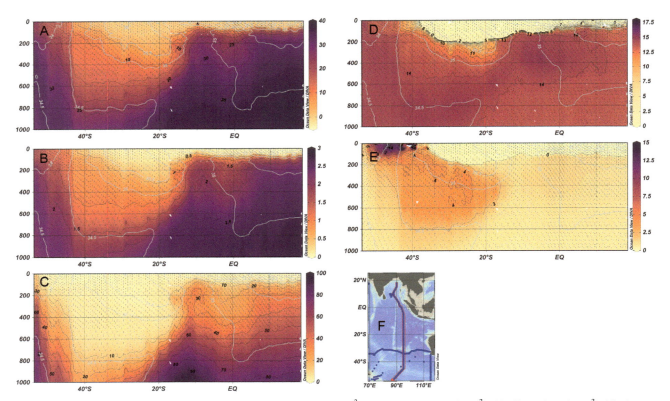

FIG. 10 Vertical sections nominally along 90–95°E of (a) nitrate (mmole m^{-3}); (b) phosphate (mmole m^{-3}); (c) silicate (mmole m^{-3}); (d) nitrate to phosphate ratio; (e) nitrate to silicate ratio; and (f) map of the section. The *gray* contours are salinity, and the *black* contours are the colored property. The station locations for each property are shown as dots. *(Data from Garcia et al. (2018), comprising mainly data from the IO9N and IO8S repeat hydrography lines.)*

(Sarma et al., 2020a) and euphotic zone nutrient concentrations and ratios (Jyothibabu et al., 2015; Kumar et al., 2007; Mahadevan, 2016; Narvekar & Kumar, 2006; Nuncio & Prasanna Kumar, 2013; Vidya & Kumar, 2013). High N:P ratios (~25) observed in offshore upwelling eddies indicate the potential for phosphate limitation of phytoplankton growth, though these high ratios appear to have been observed in only one study (Sarma et al., 2020b).

In the southern Bay of Bengal, vertical stratification is weaker because of reduced riverine influence. As a result, monsoon-forced mixing and upwelling are strong enough to bring nutrients into the euphotic zone (Jyothibabu et al., 2015; Lévy et al., 2006, 2007; Vinayachandran, 2004; Vinayachandran et al., 2005; Vinayachandran & Mathew, 2003). The southern coasts of India and Sri Lanka are characterized by intense upwelling of nutrients during the Southwest Monsoon (de Vos et al., 2014; Lévy et al., 2007; Vinayachandran, 2004). In addition, Southwest Monsoon winds drive open-ocean upwelling of nutrients in the Sri Lanka Dome east of Sri Lanka (Thushara et al., 2019; Vinayachandran, 2004; Vinayachandran & Yamagata, 1998). However, the impacts of these upwelling features are not readily apparent in climatological Southwest Monsoon surface nitrate, phosphate, and silicate concentrations (Fig. 1), suggesting rapid consumption by phytoplankton in the deep Chl$a$ maximum (Fig. 11c). Indeed, direct measurements of southern Sri Lanka from August to October 2015 revealed only a small enhancement of surface nutrients (Thushara et al., 2019). Similarly, model simulations of the Sri Lanka Dome during the Southwest Monsoon reveal a distinct shoaling of the nitracline (2 mmol m^{-3} isoline) to 20 m, with little surface nutrient enhancement (Thushara et al., 2019).

While oxygen is exceedingly low in the Bay of Bengal, surface oligotrophic conditions result in low export of carbon for bacterial consumption at depth, which may limit the formation of a suboxic OMZ (for details on this topic, see Hood et al., 2024). There is no significant water-column denitrification (Bristow et al., 2017; Naqvi et al., 2006), but there is Fe enrichment in the OMZ (Fig. 11a) as in the Arabian Sea (Fig. 5), which suggests that Fe can be mobilized and transported even in the absence of water-column denitrification. Recent surveys (Chinni et al., 2019; Grand et al., 2015a, 2015b, 2015c; Vu & Sohrin, 2013) show a strong depletion of Fe in surface waters, consistent with Fig. 11a. However, prevailing oligotrophic conditions result in production limitation by macronutrients rather than Fe (Twining et al., 2019).

2.2.2 Phytoplankton

While strong salinity stratification generally depresses primary production in the Bay of Bengal, there are regions of elevated Chl$a$ and production associated with river plumes, upwelling eddies and where the monsoonal forcing can overcome the stratification (Lévy et al., 2006, 2007; Thushara et al., 2019; Vinayachandran, 2009). In general, Chl$a$ and primary production rates are elevated close to the coast, particularly off river mouths, with the highest values occurring during the season of peak discharge (Vinayachandran, 2009; Figs. 7 and 8). Along the Indian coast, the flow of Chl$a$ rich water is determined by the East Indian Coastal Current, which flows northward during the spring intermonsoon and Southwest Monsoon and southward during the fall intermonsoon and Northeast Monsoon. When the East Indian Coastal Current meanders seaward from the Indian coast, it leads to offshore Chl$a$ increases (Vinayachandran, 2009). Eddies in the East Indian Coastal Current and the open ocean sometimes have relatively high Chl$a$, but those nearer to the coast are typically more enriched by offshore entrainment of high Chl$a$ coastal water (Vinayachandran, 2009). As in the Arabian Sea, dinoflagellate blooms are common in east Indian coastal waters in the Bay of Bengal, with variations in species abundance attributed to seasonal variations in the stratification (Naik et al., 2011b). Pigment measurements also reveal the importance of diatom blooms in east Indian coastal waters, whereas prokaryotes and flagellates tend to dominate phytoplankton species composition offshore (Naik et al., 2011a).

A second region of elevated Chl$a$ and primary production occurs offshore in the southwestern Bay of Bengal during the Northeast Monsoon (Lévy et al., 2006, 2007; Vinayachandran, 2009; Vinayachandran & Mathew, 2003). Modeling studies suggest that elevated surface Chl$a$ is caused by wind-driven entrainment of both subsurface nutrients and phytoplankton from the subsurface Chl$a$ maximum that is present during the fall intermonsoon (Vinayachandran, 2009; Vinayachandran et al., 2005). In contrast, coastal productivity is suppressed during the Northeast Monsoon when the winds and southward flowing East Indian Coastal Current are downwelling favorable. Subsurface Chl$a$ maxima (0.3–1.2 mg m^{-3}) occur throughout the Bay of Bengal wherever wind forcing and currents are insufficient to entrain them into the surface layer (Kumar et al., 2007; Murty et al., 2000; Sarma & Aswanikumar, 1991; Sarojini & Sarma, 2001; Thushara et al., 2019; Fig. 11c). The depth and intensity of these maxima are strongly influenced by eddies (Kumar et al., 2007; Fig. 11c) that tend to shoal near the coast (Murty et al., 2000; Sarma & Aswanikumar, 1991).

Elevated Chl$a$ and primary production are also observed along southern Sri Lanka during the Southwest Monsoon, along with Chl$a$-rich waters advected from southwestern India by the Southwest Monsoon Current (de Vos et al., 2014; Hood et al., 2017; Lévy et al., 2006, 2007; Strutton et al., 2015; Thushara et al., 2019; Vinayachandran, 2004,

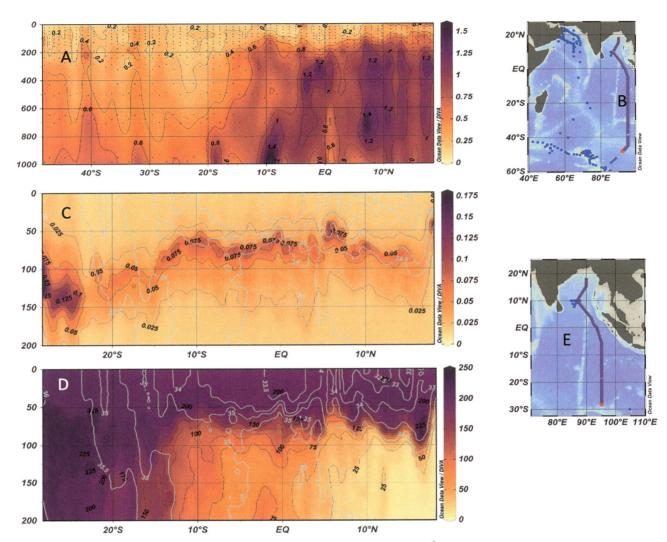

FIG. 11 Meridional vertical sections in the eastern Indian Ocean of (a) dissolved Iron (μmole m^{-3}) along approximately 90–95°E. The station locations are shown in panel b. The *black* contours are the colored property. (c) CTD Fluorescence (volts) and (d) oxygen (mmole m^{-3}) along approximately 90–95°E. The station locations are shown in panel e. In panels c and d, the *gray* contours are salinity, and the *black* contours are the colored property. The data in panels c and d are from the 2017 IO9 GoShip line. *(The data in panel a are from Tagliabue et al. (2012) database.)*

2009; Figs. 7 and 8). Surface Chl*a* and primary production can exceed $10\,\text{mg m}^{-3}$ and $1\,\text{g C m}^{-2}\,\text{d}^{-1}$, respectively, compared to much lower values during the Northeast Monsoon, when the westward-flowing Northeast Monsoon Current and winds favor downwelling (de Vos et al., 2014; Hood et al., 2017; Figs. 7 and 8). Diatom blooms are likely responsible.

2.2.3 Zooplankton

Density stratification and low oxygen content of waters beneath the pycnocline strongly affect the depth distributions of most mesozooplankton species such that 70%–80% of the Bay of Bengal biomass to 1000m generally resides above the thermocline and mainly in a thin mixed layer (Fernandes, 2008; Fernandes & Ramaiah, 2014). Within this upper layer, zooplankton abundances and trophic interactions are likely more vertically concentrated in the Bay of Bengal than in other regions, even if integrated biomasses are similar or less. Moreover, low oxygen concentrations and reduced salinity in the Bay of Bengal are associated with lower species richness (Sutton & Beckley, 2017). Depth-integrated biomass estimates have a large range (<0.2 to $6.7\,\text{g C m}^{-2}$), with 3-20× variability within seasons (Fernandes, 2008; Fernandes & Ramaiah, 2009, 2013, 2014). For most seasons, zooplankton biomass variability reflects the productivity. Along the western (coastal) bay, Ramaiah et al. (2010) indicated a >10-fold seasonal difference in mean mesozooplankton biomass from the Southwest Monsoon minimum to spring intermonsoon maximum (0.22 and 2.43 g C m^{-2}), with the fall intermonsoon and Northwest Monsoon being intermediate (0.91 and 0.99 g C m^{-2}). However, in the southwestern coastal region, Muraleedharan et al. (2007) found that wind-driven upwelling and a cyclonic eddy during the Southwest Monsoon enhanced local biomass estimates by sevenfold and fourfold, respectively, compared to stratified waters. Seasonal averages were more similar for a transect along 88°E, with an Southwest Monsoon minimum and fall intermonsoon maximum (0.59 and 1.04 g C m^{-2}) and 0.81–0.82 g C m^{-2} during the Norhtwest Monsoon and spring intermonsoon. Both western coastal and central stations show strong south-to-north biomass increases during the spring intermonsoon (Fernandes & Ramaiah, 2019). On the eastern side of the basin (Andaman Sea), seasonal variability in zooplankton standing stocks and production indices are small despite pronounced differences in environmental conditions (Nielsen et al., 2004; Pillai et al., 2014; Satapoomin et al., 2004).

Gauns et al. (2005) documented substantially lower microzooplankton biomass in the western and central Bay of Bengal than in the eastern and central Arabian Sea during all monsoon seasons. Mean biomasses are highest during the spring intermonsoon in both the central and western bay (0.33 and 0.23 g C m^{-2}, respectively), decline by ~one-third during the Southwest Monsoon (0.22 and 0.14 g C m^{-2}) and drop precipitously during the fall intermonsoon (0.02 and 0.03 g C m^{-2}) before increasing again during the Northwest Monsoon (0.09 g C m^{-2}) (Jyothibabu et al., 2008b; Ramaiah et al., 2010).

Zooplankton grazing studies are extremely rare in the Bay of Bengal. Fernandes and Ramaiah (2014) estimated that herbivorous feeding by copepods accounted for ~25% of primary production during the Northwest Monsoon. In the sole study of microzooplankton grazing, consumption of phytoplankton production in shallow coastal waters off Southwest India (Kochi) varied from 80% during the spring intermonsoon to 26% during an Southwest Monsoon upwelling bloom (Anjusha et al., 2018). Among key issues unexplored in regional food-web investigations are the nutritional sources and grazing pathways that sustain zooplankton stocks during different seasons. Isotope measurements have suggested, for example, that zooplankton in river-influenced areas during the Southwest Monsoon might derive more food value from suspended detritus than phytoplankton (Mukherjee et al., 2018). High zooplankton biomass during the spring intermonsoon has also been attributed to enhanced fluxes via the microbial loop (Anjusha et al., 2013; Jyothibabu et al., 2008b; Madhupratap et al., 1996a), but the magnitudes of grazing flows from heterotrophic bacteria and dominant picophytoplankton (e.g., Brown et al., 1999, 2002) have not been measured for any season.

2.3 Equatorial Indian Ocean, including the Java upwelling and the SCTR

2.3.1 Nutrients

Unlike the Atlantic and Pacific, mean winds along the equator in the monsoon-dominated Indian Ocean are westerly and, therefore, predominantly downwelling favorable (Schott et al., 2009; Wang & McPhaden, 2017). As a result, near-surface waters along the equator tend to be oligotrophic, i.e., typically undetectable nitrate, phosphate and silicate down to ~100 m (Figs. 4 and 10), and upwelling centers in the Indian Ocean are generally found in off-equatorial regions. One example is the SCTR, an upwelling region in the southern tropical Indian Ocean between ~5–15°S and ~50–80°E characterized by a thin (~30 m) mixed layer and relatively shallow thermocline (Vialard et al., 2009; Figs. 4 and 12a). The thermocline ridge and associated upwelling are driven by local wind stress curl (Hermes & Reason, 2008; McPhaden & Nagura, 2014; Nyadjro et al., 2017; Xie et al., 2002; Yokoi et al., 2008). The SCTR coincides with the southernmost latitudes of monsoon-driven circulation in the Indian Ocean, south of which a steadier Trade Wind regime prevails. The SCTR influence on near-surface nutrients can be seen in Fig. 4, where the nitricline, phosphocline, and silicline shoal sharply at 20°S by 700 m over about 5° and remain near 100 m to the equator and beyond. This shoaling can also be seen in Fig. 12b (Resplandy et al., 2009), where the average nitrate concentration between the surface and 80 m depth exceeds 5 mmole m^{-3} in a bullseye centered at about 62°E, 8°S. The SCTR is the largest and most persistent upwelling region in the Indian Ocean.

The Java upwelling region of Indonesia is the only example of eastern boundary upwelling in the Indian Ocean. In contrast to the large eastern boundary upwelling in the Pacific and Atlantic Oceans, it occurs only seasonally during the Southeast Monsoon in association with the reversing South Java Current (Sprintall et al., 1999; Susanto et al., 2001) and, like the Bay of Bengal, is strongly influenced by freshwater inputs from the maritime Indonesian continent (Rixen et al., 2006b). Upwelling effects are strong along the coasts of the islands of Sumatra, Java, and Bali between ~5° and 10°S, and also influence the southern coasts of Lombok, Flores, and Alor (Fig. 10 in Hood et al., 2017). Nutrient measurements along a transect from the northwestern coast of Australia to Lombok Strait clearly show the upwelling signature off Java, where elevated nitrate and phosphate concentrations (8 and 0.8 mmole m^{-3}, respectively) can extend to the surface near the coast (Fig. 14 in Hood et al., 2017).

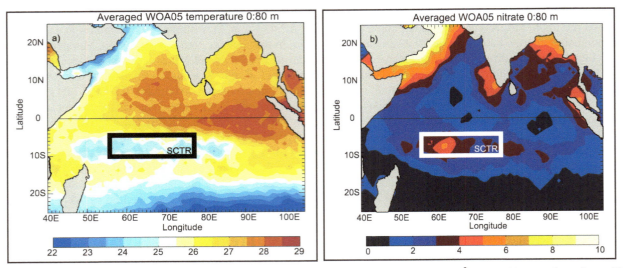

FIG. 12 Annual World Ocean Atlas (2005) (a) temperature (degrees C) and (b) nitrate concentration (mmol m^{-3}) averaged between the surface and 80 m in the Indian Ocean. *(Figure and caption modified and reproduced with permission from Resplandy et al. (2009).)*

2.3.2 Phytoplankton

Satellite observations and model results reveal an annual cycle in surface Chl*a* and primary production in the SCTR region, with the highest values in austral winter (June–August; >0.20 mg m^{-3} and >0.6 g C m^{-2} d^{-1}, respectively, Figs. 7 and 8) due to the strong southeasterly winds that increase wind stirring and induce upwelling (Dilmahamod, 2014; Resplandy et al., 2009). Vertical sections through the SCTR region also reveal a deep chl*a* maximum that shoals and intensifies from >100 m at 18°S, 55°E to ~50 m at 10°S, 55°E due to upwelling (George et al., 2013; Fig. 9a).

There is significant seasonal variability in near-surface Chl*a* and primary production in equatorial waters (10°N–10°S) associated with monsoon forcing, with the lowest concentrations (<0.1 mg m^{-3}) and rates (<0.8 g C m^{-2} d^{-1}) occurring during the spring intermonsoon (Figs. 7 and 8, Strutton et al., 2015; Wiggert et al., 2006). During the Southwest Monsoon, Chl*a* and production increase with wind mixing and upwelling in far western equatorial waters but stay relatively low (<0.5 mg m^{-3}, <0.8 g C m^{-2} d^{-1}, respectively) in the central and eastern equatorial waters. Island wake effects give rise to high Chl*a* water (>0.5 mg m^{-3}) advected eastward along the equator from the Chagos-Laccadive ridge at 73°E during the fall intermonsoon and westward during the spring intermonsoon by Wyrtki Jets (Fig. 1 in Strutton et al., 2015).

Well-developed deep chl*a* maximum is observed in the western, central, and eastern equatorial Indian Ocean centered between 50 and 100 m (George et al., 2013; Sorokin et al., 1985; Figs. 9a and 11c). It is unknown whether this subsurface maximum exists along the equator throughout the year, but it is probably present whenever the water column is stratified. Models predict the presence of a deep chl*a* maximum at ~60 m along 87°E in eastern Indian Ocean equatorial waters throughout the year except when high Chl*a* surface water is advected into the region from the west during the Soutwest Monsoon (Wiggert et al., 2006). Pigment measurements indicate that prokaryotic cyanobacteria and flagellates dominate the flora in the central equatorial Indian Ocean (Naik et al., 2020).

Enhanced Chl*a* and primary production are also observed along southern Indonesia during the Southeast Monsoon (Figs. 7 and 8). Satellites reveal that elevated Chl*a* (>2 mg m^{-3}) first appears off Java in June and persists into November (Hood et al., 2017; Lévy et al., 2006, 2007), with primary production estimates in August exceeding 1 g C m^{-2} d^{-1} (Fig. 8). Relaxation of Southeast Monsoon winds and downwelling Kelvin waves (Sprintall et al., 1999) suppress productivity in the fall. Satellite observations also reveal that upwelling-enhanced Chl*a* progresses northwestward during the Southeast Monsoon and extends to southwestern Sumatra in September (Hood et al., 2017). These Chl*a* and production responses are associated with diatom blooms (Hood et al., 2017; Romero et al., 2009; Yu et al., 2015).

The biogeochemistry of the equatorial Indian Ocean is strongly influenced by physical processes at intraseasonal to interannual time scales (i.e., Wyrtki Jets, MJO, and IOD). For example, IOD events can significantly increase Chl*a* and production in eastern equatorial waters, and it has been shown that relaxation of an IOD event can decrease biological productivity (Currie et al., 2013; Kumar et al., 2012; Marsac et al., 2024; Wiggert et al., 2009). In the SCTR region, wind-induced mixing during MJO episodes (typically between January and March) can also lead to enhanced Chl*a* at intra-seasonal time scales; their efficacy is strongly related to basin-scale interannual variability of the thermocline depth (Dilmahamod et al., 2016; Resplandy et al., 2009).

2.3.3 Zooplankton

Although the central equatorial Indian Ocean and the off-equatorial SCTR and Java upwelling regions lack modern systematic plankton studies comparable to the Arabian Sea and Bay of Bengal, historical net sampling in these areas provides some basis for comparing their seasonality and relative magnitudes of mesozooplankton biomass (Fig. 13). The narrow sampling band close to the equator (1°S–2°N) shows the lowest biomass during and following the Northeast Monsoon into March, ascending to ~three times higher level for the remainder of the year, with no additional increase during the Southwest Monsoon. Given the general downwelling characteristics of this region, such a seasonal pattern may be more consistent with advective transport of zooplankton into the central equatorial area than local secondary production. Zooplankton communities dominated by calanoid copepods of the family Pontellidae (bright blue-pigmented calanoid copepods) have been observed at the ocean surface in the open equatorial Indian Ocean during the Southwest Monsoon (Venkataramana et al., 2017), which could also be related to advective transport.

In the SCTR upwelling area directly to the south (2–10°S), biomass is relatively low for most of the year, with a pronounced four- to fivefold increase during the Southeast Monsoon upwelling in August (austral winter) (Fig. 13). West of 60°E, in the SCTR area renowned for its tuna fishery, Gallienne et al. (2004) reported a 10-fold increase in optically sensed zooplankton biovolume in April–May 2001 relative to subtropical waters to the south, which they attributed to upwelling production driven by the collision of the South Equatorial Current with the shallow (20–200 m) ridge crest of the Mascarene Plateau. In addition to zooplankton biomass being elevated downstream (west) compared to upstream (east) of the Mascarene Plateau in June–July 2002 (3–6 fold over 12–19°S), it was also elevated sixfold or more over the northern plateau

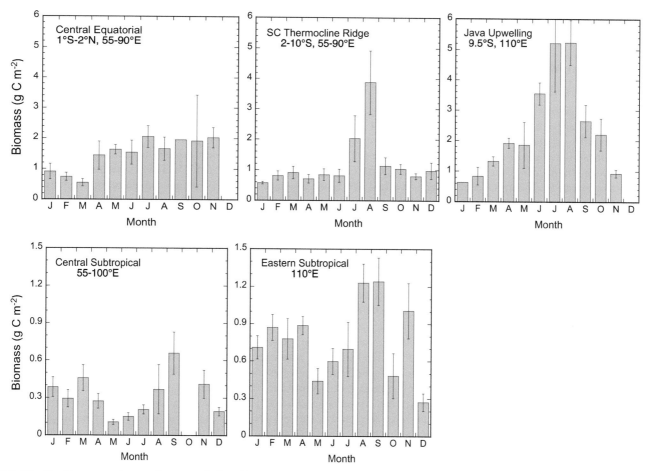

FIG. 13 Monthly mean estimates of mesozooplankton biomass in the equatorial and off-equatorial upwelling systems of the Indian Ocean (*top panels*) and in the central and eastern subtropical regions of the Indian Ocean (*bottom panels*). Data are from historical collections in the Coastal and Oceanic Plankton Ecology, Production, and Observation Database (COPEPOD, http://www.st.nmfs.noaa.gov/copepod) were measured as wet weight and displacement volume (upper ~200 m) and corrected for different mesh sizes to carbon equivalents m^{-2} for a 200-μm mesh net, according to Moriarty and O'Brien (2013). Uncertainties are standard errors of mean estimates. Plots are meant to compare and contrast seasonal patterns in different regions. Carbon may be overestimated where gelatinous animals are abundant.

compared to the southern part, increasing northwards from the gap between Saya de Malha and Nazareth Banks (13–14°S; Gallienne & Smythe-Wright, 2005). A similar north-south contrast in zooplankton biomass (as well as phytoplankton biomass and particulate organic matter) was observed in October–November 2008, with mean values of 0.2 g C m^{-2} (max 1.3) over Seychelles Bank (4–6°S), 0.1 g C m^{-2} (max 0.4) over Nazareth Bank (14–17°S), and 0.05 g C m^{-2} (max 0.3) north of Mauritius (19–20°S; Huggett & Kyewalyanga, 2017). Sampling with a very coarse net (1-mm mesh that would have missed smaller size fractions), Roger (1994) found little seasonal difference in zooplankton biomass on the Mascarene Plateau from March (spring intermonsoon) to September (Southwest Monsoon). However, total large zooplankton biomass was fourfold higher in this area than in southern subtropical waters, and the biomass of specific zooplankton that contribute most to the tuna food chain was enriched sevenfold, with a maximum at 4–6°S (west of the Seychelles).

The Java upwelling region is at the northern end (9.5°S) of the 110°E transect line sampled ~bi-monthly from August 1962 to August 1963 during the International Indian Ocean Expedition (IIOE, Tranter & Kerr, 1969, 1977), providing most of the seasonal biomass data and consistent methods for that area (Fig. 13). Following standard conversions from mostly wet weight measurements to carbon and from 333 to 200-μm nets (Moriarty & O'Brien, 2013), zooplankton carbon estimates increase by about an order of magnitude seasonally in the Java upwelling (Fig. 13). The maximum of ~5 g C m^{-2} occurs during the Southeast Monsoon when upwelling prevails along southern Indonesia. The biomass minimum of ~0.6 g C m^{-2} occurs during the Northwest Monsoon (December–January, Austral summer) when downwelling prevails. At least for the 1962–1963 year from which most data were derived, the seasonal changes were smooth, with a broad peak of >2 g C m^{-2} persisting from June to October (Fig. 13). The 110°E transect was revisited in May/June of 2019 but did not go far enough north to sample the Java upwelling region (the northernmost station was located at 11.5°S). The 2019 data reveal increases in mesozooplankton biomass (up to ~0.5 g C m^{-2}) at the northernmost stations in early June, consistent with observations from the 1960s (Landry et al., 2022).

2.4 Eastern Indian Ocean and the Leeuwin Current

2.4.1 Nutrients

The South Equatorial Current carries ITF waters into the Indian Ocean, flowing westward between 10 and 20°S (George et al., 2013; Talley & Sprintall, 2005; Van Sebille et al., 2014). The ITF influences Indian Ocean biogeochemistry by supplying a net flux of nutrients primarily to thermocline waters (Ayers et al., 2014). A comparison of N:P ratios in the ITF and Indian Ocean by Ayers et al. (2014) shows that the ITF water is enriched in nitrate, which increases the N:P ratio of the Indian Ocean thermocline, though it remains below the classical Redfield ratio of 16:1. The ITF also represents a significant silicate source to the Indian Ocean, which is evident as a broad tongue of elevated silicate concentration between 10° and 20°S extending from the ITF region on the 31.96 σ_1 density surface (~1000 m) across the basin to 60°E and beyond (Fig. 9b in Talley & Sprintall, 2005). This silicate feature, associated with the westward advection of Indonesian Intermediate Water, can also be seen along 95°E between 5° and 20°S in association with less pronounced increases in nitrate and phosphate extending from 1000 to 100 m depth (Fig. 10). Elevated silicate concentrations derived from ITF sources have also been observed in the Leeuwin Current (Thompson et al., 2011).

The tropical origins of the Leeuwin Current combine with its southward-flowing downwelling tendency to create a warm oligotrophic current. However, local wind forcing during the austral summer can override this general tendency and drive localized upwelling (Gersbach et al., 1999; Hanson et al., 2005a, 2005b; Pearce & Pattiaratchi, 1999; Rossi et al., 2013a, 2013b). Seasonal nutrient climatologies from the southwestern Australian shelf, the Leeuwin Current, and offshore show low surface nitrate (<0.5 μM) throughout the year, suggesting N-limitation of primary production (Lourey et al., 2006). Phosphate concentrations in the Leeuwin Current are also low, but its relatively high levels of silicate (up to 4 mmole m^{-3}) may be a source of silicate for surrounding waters (Lourey et al., 2006). Shelf waters inshore of the current also generally have low levels of nitrate (<0.5 mmole m^{-3}), phosphate (<0.25 mmole m^{-3}), and silicate (<2 mmole m^{-3}) year-round, but deepening of the mixed layer can lead to nutrient entrainment in offshore waters during austral winter (Lourey et al., 2006). In addition, the Leeuwin Current generates numerous warm (downwelling) and cold core (upwelling) eddies between 20° and 35°S (Gaube et al., 2013; Waite et al., 2007b, 2015). These eddies propagate directly westward, and some appear to be very long-lived (Dufois et al., 2014; Feng et al., 2005, 2007; Gaube et al., 2013; Moore et al., 2007; Waite et al., 2007b), although their persistence and potential impacts on open-ocean biogeochemistry have not been fully investigated.

Low oxygen, high-nitrate layers have been observed in the eastern Indian Ocean off northwest Australia (Thompson et al., 2011; Waite et al., 2013). These layers are formed from multiple subduction events of the Eastern Gyral Current beneath the Leeuwin Current and have been shown to directly impact the ecological function of western Australian coastal waters. Isotopic measurements suggest that 40%–100% of the nitrate in these layers is derived from nitrogen fixation (Waite et al., 2013). Slightly elevated N:P ratios at 400 m in Fig. 3 are consistent with these data. Based on measurements from several cruises in the region bounded by 28°S–34°S and 110°E and 114°E, Raes et al. (2015) similarly concluded that nitrogen fixation is an important source of fixed nitrogen in the eastern Indian Ocean, with most of the nitrate in the photic zone derived from oxidized ammonium with a nitrogen-fixation origin.

2.4.2 Phytoplankton

Satellite estimates of Chl$a$ and primary production are generally elevated in the South Equatorial Current in the eastern Indian Ocean, with Chl$a$ varying from ~0.10 to 1.0 mg m^{-3} and production from ~0.4 to 1 g C m^{-2} d^{-1} (Figs. 7 and 8). The highest values for the eastern South Equatorial Current occur in July–August during the Southeast Monsoon (austral winter) associated with ITF nutrient sources and Java upwelling. The lowest values are observed in January (austral summer, Figs. 7 and 8).

The downwelling Leeuwin Current is a warm, oligotrophic current with low Chl$a$ (<30 mg m^{-2}) and production (<0.5 g C m^{-2} d^{-1}; Koslow et al., 2008; Lourey et al., 2006, 2013). Productivity is lowest during austral summer when the water column is stratified, and the Chl$a$ maximum is between 50 and 120 m (Hanson et al., 2007). However, primary production in nearshore regions can attain very high levels of 3–8 g C m^{-2} d^{-1} in localized upwelling centers (Furnas, 2007). For some examples of local upwelling mechanisms along the Western Australia coastline, see a review by Vinayachandran et al. (2021).

Meanders in the Leeuwin Current give rise to the aforementioned warm-core anticyclonic eddies that carry high-chlorophyll water and coastal diatom communities offshore into cooler oligotrophic waters dominated by open-ocean picophytoplankton (Paterson et al., 2008a; Waite et al., 2015, 2007b). These eddies, which can extend to more than 250 m depth, are unusual because they are downwelling (anticyclonic) circulations with high chl$a$.

2.4.3 Zooplankton

Zooplankton distributions in the eastern Indian Ocean off western and northwestern Australia reflect the influences of distinct tropical and subtropical/temperate habitats (Buchanan & Beckley, 2016; Tranter & Kerr, 1977), transport by the Leeuwin Current, and the ecotones and mesoscale features where they mix and interact. In the oceanic realm along 110°E, tropical waters north of ~17°S have the highest total biomass of mesozooplankton as well as the highest abundances of copepods, chaetognaths, appendicularians, euphausiids and foraminiferans (Landry et al., 2022; Tranter & Kerr, 1969, 1977). Seasonality in this tropical area is driven by upwelling off of Java and northwest Australia during the Southeast Monsoon, leading to a broad mesozooplankton maximum across the ITF region in August–September and minimum biomass during the Northwest Monsoon downwelling season in December–January (Tranter & Kerr, 1977), with the former also observed in June of 2019 (Landry et al., 2022). Larger predators follow the seasonal development of smaller prey, with larger size classes of zooplankton peaking in September and the micronekton maximum occurring later in October–November (Legand, 1969).

Compared to the tropical region, biomass seasonality in subtropical/temperate waters south of 27–28°S displays a more bimodal pattern, with peaks during the winter Southeast Monsoon and summer seasons (Tranter & Kerr, 1969, 1977). Total mesozooplankton biomass is generally 2–5 fold lower in subtropical than tropical waters (Landry et al., 2020, 2022; Säwström et al., 2014; Tranter & Kerr, 1969), but mean biomasses of chaetognaths and pteropods are reported to be higher in the subtropics (Legand, 1969; Tranter & Kerr, 1969). Gelatinous predators appear to have a biomass maximum at intermediate latitudes where tropical and subtropical waters mix (Legand, 1969).

From its low-latitude origin, the Leeuwin Current transports the enhanced zooplankton biomass and species assemblages of tropical waters southward along the continental margin of Western Australia (Buchanan & Beckley, 2016; Sutton & Beckley, 2017). The current contribution to mesoscale zooplankton variability in the oceanic region is well established. The center areas of the anticyclonic warm-core eddies that form off of western Australia contain Leeuwin Current water, with a tropical signature of lower salinity and higher zooplankton biomass, as well as the aforementioned coastal phytoplankton (diatom) influences (Feng et al., 2007; Paterson et al., 2007). The center areas of cyclonic cold-core eddies are waters of subtropical origin with higher salinity, lower chlorophyll, and ~twofold lower zooplankton stocks than the warm-core eddies (Feng et al., 2007; Strzelecki et al., 2007). Microzooplankton standing stocks are correspondingly higher in warm-core eddies (Paterson et al., 2007; Strzelecki et al., 2007), and Waite et al. (2007a) have argued, based on δ^{15}N and δ^{13}C isotopic indices, that zooplankton are also healthier in warm-core eddies. Paradoxically, however, the oceanic phyllosoma larvae of western rock lobster, the most valuable fishery of western Australia, accumulate more lipid and fatty acids as energy stores in the cold-core eddies (Wang et al., 2014), which Waite et al. (2019) attributed to more efficient energy transfer in the flagellate-dominated food webs and shallower mixed layers of cold-core features relative to the deeper diatom-dominated food webs of warm-core eddies. The flow fields between warm and cold eddy pairs that direct loops of the Leeuwin Current shoreward are also believed to be an important mechanism in returning phyllosoma larvae from oceanic subtropical waters to their recruitment into the benthic habitat and fishery on the continental margin (Säwström et al., 2014).

Zooplankton respiration, grazing, and production have been systematically investigated based on biogeochemical indices and size-rate relationships in shallow tropical coastal habitats of northwestern Australia (McKinnon et al., 2015a, 2015b; Moritz et al., 2006), but rate measurements are rare for deeper waters. During a time (May) when zooplankton biomass along 110°E is seasonally minimal, Landry et al. (2020) determined that mesozooplankton consumed 3% of phytoplankton (Chl$a$) standing stock d^{-1} in subtropical waters and 21% in the tropical region. In addition, Landry et al. (2020) showed a strong and positive correlation between the mesozooplankton grazing rates and temperature ($r^2 = 0.85$), which exceeded the temperature effects on gut turnover and metabolic rates. Paterson et al. (2007) found that microzooplankton consumed all or most phytoplankton production in both warm- and cold-core eddies, with large dinoflagellates as major grazers in the diatom-rich warm-core eddies while ciliates dominated in the flagellate-rich cold-core eddies. In seasonal studies from the continental slope to a coastal lagoon, microzooplankton grazing also generally balanced the cell growth of picoplankton populations, while the total phytoplankton community exhibited positive net growth (Paterson et al., 2008b).

2.5 Southwestern Indian Ocean and the Agulhas Current

2.5.1 Nutrients

Surface waters in the Mozambique Channel are generally warm and nutrient-depleted, having been derived from oligotrophic surface waters from the southwestern tropical Indian Ocean. However, areas of nutrient enrichment (nitrate >0.2 mmole m^{-3}) are sometimes found in the core of cyclonic eddies, as well as on the continental shelf (Jose et al., 2014, 2016; Marsac et al., 2014). Counter to conventional wisdom, modeling studies of the Mozambique Channel indicate that anticyclonic downwelling eddies sometimes have high nutrient concentrations at their cores, while cyclonic upwelling eddies sometimes have low nutrients (Jose et al., 2014, 2016). These eddies mediate the lateral transport of nutrients from the coasts of Africa and Madagascar (Jose et al., 2014, 2016; Lamont et al., 2014; Roberts et al., 2014). In contrast, along eastern Madagascar, topographically forced coastal upwelling in the East Madagascar Current brings cold nutrient-rich water to the surface (Ho et al., 2004; Lutjeharms & Machu, 2000; Quartly & Srokosz, 2004; Ramanantsoa et al., 2018). This upwelling and its impacts, which are most pronounced along the southern and southeastern coasts of Madagascar, are enhanced in both austral winter and summer (Ho et al., 2004).

The Agulhas Current is also derived from oligotrophic surface waters from the southwestern tropical Indian Ocean (Lutjeharms, 2006). Its northern source waters are from the Mozambique Channel eddies and the East Madagascar Current, which coalesce off southern Mozambique (Lutjeharms, 2006). The Agulhas Current is additionally supplied from the east via recirculation in the southwest Indian Ocean sub-gyre (Stramma & Lutjeharms, 1997). Even at 400 m depth, climatological concentrations of nitrate, phosphate, and silicate are all relatively low (<10, <1, and <10 mmole m^{-3}, respectively) off southeastern Africa (Fig. 3). N:P ratios in these waters appear low. However, data from only a single cruise drives this signature, so it should be further investigated (Fig. 3). Nutrient concentrations over the southeast African shelf can be strongly influenced by the Agulhas Current via topographically forced upwelling at specific locations (e.g., the St. Lucia upwelling cell in the KwaZulu-Natal Bight; Meyer et al., 2002). In general, inorganic nutrient concentrations (nitrate, silicate, and phosphate) off southeastern Africa decline in surface waters from the inner to the outer shelf and into the Agulhas Current, where they drop to very low (oligotrophic) levels (Barlow et al., 2015; Carter & D'Aubery, 1988). However, the interaction of upwelling with horizontal advection and mixing can give rise to complex nutrient distribution patterns in the coastal waters of southeastern Africa (Barlow et al., 2015; Meyer et al., 2002).

2.5.2 Phytoplankton

Chl$a$ and production in Mozambique Channel surface waters are generally low (<0.4 mg m^{-3} and <0.7 g C m^{-2} d^{-1}) and not significantly different in cyclonic and anticyclonic eddies (Barlow et al., 2014; Hood et al., 2017; Lamont et al., 2014; Figs. 7 and 8). Deep chl$a$ maxima occur between 25 and 125 m depending on eddy influences and shelf proximity (Barlow et al., 2014; Lamont et al., 2014). Enhanced production in both cyclonic and anticyclonic eddies in the Mozambique Channel often occurs in response to nutrient inputs by lateral advection from the coasts of Madagascar and Africa rather than eddy-induced upwelling (Jose et al., 2014; Kyewalyanga et al., 2007; Lamont et al., 2014; Roberts et al., 2014).

In contrast, the topographically induced coastal upwelling in the East Madagascar Current gives rise to high Chl*a* and production (Ho et al., 2004; Hood et al., 2017; Lutjeharms & Machu, 2000; Quartly & Srokosz, 2004).

The Agulhas Current itself is oligotrophic (Lutjeharms, 2006). Chl*a* and production in surface waters are particularly low during austral summer (<0.2 mg m^{-3} and <0.5 g C m^{-2} d^{-1}) and higher in austral winter (Hood et al., 2017; Machu & Garçon, 2001; Figs. 7 and 8). However, the Agulhas Current can elevate primary production in the coastal zone through meandering and topographic interactions that drive upwelling, and it can suppress primary production when it impinges onto the shelf (Hood et al., 2017; Schumann et al., 2005).

In general, Chl*a* and production are elevated in southeast coastal Africa along the inshore side of the Agulhas Current (Goschen et al., 2012; Hood et al., 2017; Machu & Garçon, 2001). This effect is most pronounced in austral summer and further southward, associated with upwelling favorable (southeasterly) winds and topographically induced upwelling. For example, Chl*a* concentrations from the Kwa-Zulu Natal Bight range from ~0.1 to 1.5 mg m^{-3}, and primary production ranges from ~0.3 to 2.6 g C m^{-2} d^{-1} (Lamont & Barlow, 2015). Elevated values are also observed over the Agulhas Bank, where near-surface Chl*a* can exceed 3 mg m^{-3}, and production rates often exceed 0.5 g C m^{-2} d^{-1} (Boyd & Shillington, 1994; Burchall, 1968; Carter & Schleyer, 1988; Jackson et al., 2012; Lutjeharms et al., 1996; Roberts, 2005).

2.5.3 Zooplankton

Surveys of Mozambique between 1977 and 2014 (Nehring et al., 1987; reviewed by Huggett & Kyewalyanga, 2017) have generally shown lower mesozooplankton biomass in the monsoon-influenced region north of 18°S compared to the southern sector. For example, lower biomass was measured in the north during winter 2009 (mean 0.3, maximum 1.4 g C m^{-2}; Huggett, unpublished), while values up to 5.7 g C m^{-2} (mean 1.2–1.7 g C m^{-2}) were recorded in the south in early summer 2014 (Huggett & Kyewalyanga, 2017). Macrozooplankton biomass (>500 μm, upper 100 m) averaged 0.5 g C m^{-2} on the Mozambique shelf during seasonal sampling in 1977–1978, with maximum values of 2.1 g C m^{-2} during summer (October–December), but unlike mesozooplankton showed no clear longitudinal patterns (Huggett & Kyewalyanga, 2017). Monthly zooplankton density has been shown to track the Chl*a* seasonal cycle on the Mozambique shelf, with a primary peak in March/April following the summer rains and high nutrient inputs from rivers and a lesser peak in September (Paula et al., 1998).

Mesozooplankton biomass measured over four surveys from 2007 to 2010 in the Mozambique Channel averaged 0.4 g C m^{-2} in the upper 200 m (range 0.1–1.8 g C m^{-2}), concentrated in the upper 100 m (Huggett, 2014). Biomass was significantly higher in cyclonic eddies (and divergence areas) compared to anticyclonic eddies (and convergence areas) but highest overall in the shelf region. Off the southern Madagascar shelf, a young cyclonic eddy entrained material from the productive shelf region in its periphery during July 2013 (Noyon et al., 2019), demonstrating the potential to transport meroplankton across the Mozambique Channel (Ockhuis et al., 2017). Zooplankton data are scarce for the rest of the Madagascan perimeter (Huggett & Kyewalyanga, 2017).

The biogeographical distributions of zooplankton species off South Africa clearly reflect the influence of the Agulhas Current, with specific assemblages associated with Agulhas water (De Decker, 1984; Schleyer, 1985). It has also been shown that the current transports Indo-Pacific zooplankton species southward (Sutton & Beckley, 2017) and into waters over the Agulhas Bank (De Decker, 1973). In general, zooplankton biomass is greater inshore of the current compared to within, but species diversity is higher in the current (Carter & Schleyer, 1988; Pretorius et al., 2016), consistent with its tropical origins. Measurements for the northeastern continental shelf of South Africa indicate a relatively low mean biomass of 0.4 g C m^{-2} (max 1.1; Huggett, unpublished). Further downstream in the KwaZulu-Natal Bight, sampling during 2010 indicated high spatial variability with greater zooplankton biomass during the dry winter compared to the wet summer (~0.6 vs 0.2 g C m^{-2}; from data in Pretorius et al., 2016), as observed for phytoplankton biomass. In contrast, biomass on the southeastern shelf (between Port Edward and Cape St Francis) during 2017 was greater in summer (mean 0.3, max 1.0 g C m^{-2}) compared to winter (mean 0.2, max 0.5 g C m^{-2}; Huggett, unpublished).

Off the south coast of South Africa, the shelf broadens into the extensive Agulhas Bank, where the zooplankton community is dominated by the large calanoid copepod *Calanus agulhensis* (De Decker & Marska, 1991; Huggett & Richardson, 2000). Hutchings et al. (1991) estimated mesozooplankton biomass there to average 0.9 g C m^{-2} (range 0.2–2.0 g C m^{-2}) and macroplankton biomass 0.1 g C m^{-2}. During annual surveys in late spring to early summer 1988–1990, mean copepod biomass over the Agulhas Bank ranged from 1.1 to 1.6 g C m^{-2} (Verheye et al., 1994), with the highest biomass east of Cape Agulhas (mean 1.7–1.8 g C m^{-2}), dominated (85%) by *Caffrogobius agulhensis*. These years likely represent a short-term peak in biomass, with a long-term (24 years, 1988–2011) mean of 1 g C m^{-2} for the whole bank and 1.1 g C m^{-2} for the area east of Cape Agulhas (Huggett, unpublished). Seasonal studies are limited, but De Decker and Marska (1991) noted peak abundance of *C. agulhensis* from August to November 1964 (austral spring) and extremely high settled volumes in late spring or summer due to vast swarms of Thaliaceans over the southern and central Bank (De Decker, 1973). High biomass of *C. agulhensis* on the central and eastern Agulhas Bank is associated with a quasi-permanent shallow but mainly subsurface "ridge" of cool upwelled water over the mid-shelf (Boyd & Shillington, 1994). This feature, which is most prominent in summer (Roberts, 2005), is thought to stimulate primary and secondary production and enhance local retention of copepods through cyclonic circulation around the ridge, although there is a net westward advection of copepods across the Bank (Huggett & Richardson, 2000).

Copepod feeding studies on the Agulhas Bank are few but suggest that younger stages of *C. agulhensis* graze preferentially on small cells prevalent in the upper mixed layer, while older stages prefer larger cells more commonly found at the subsurface maximum (Verheye et al., 1994), and are thus more likely to be food-limited when chlorophyll concentrations are low (Huggett & Richardson, 2000). Grazing estimates vary, but copepod grazing impact during spring and summer, based on daily consumption of $0.6-0.9\,g\,C\,m^{-2}$, was estimated as 30%–50% of daily primary production for a whole phytoplankton diet and 15%–25% for a diet with equal proportions of phytoplankton and microzooplankton (Verheye et al., 1994). Gibbons (1997) estimated that grazing impact by high densities of *Thalia democratica* on the Agulhas Bank could exceed 100% of phytoplankton production.

2.6 The southern subtropical gyre

2.6.1 Nutrients

The core of the southern Indian Ocean subtropical gyre between 20° and 40°S is oligotrophic, with undetectable nitrate, phosphate, and silicate concentrations in near-surface waters (Figs. 1–4; Harms et al., 2019). N:P ratios are generally below 7.5, suggesting nitrogen limitation (Figs. 4 and 10) consistent with nutrient addition experiments conducted in the southern subtropical gyre along 90–95°E (Garcia et al., 2018; Twining et al., 2019). Climatological nutrient concentrations reveal greater depression of the nitracline and phosphacline (between ~25° and 35°S) in the western gyre (Figs. 3, 4, and 10). The vertical section along 90–95°E (Fig. 10) reveals a sharp increase in N:Si (>10) in the surface waters of the subtropical front between 35° and 50°S and subduction of this high N:Si water northward beneath the subtropical gyre. These features are visible, though less pronounced, in the vertical section along 55–65°E (Fig. 4). High N:Si water in the Subtropical Front is a circum-global feature where silicate can limit diatom growth (e.g., Bishop et al., 2004; Coale et al., 2004; Martin et al., 2013).

2.6.2 Phytoplankton

In the southern Indian Ocean subtropical gyre, near-surface Chl*a*, and primary production are low (<$0.1\,mg\,m^{-3}$ and <$0.5\,g\,C\,m^{-2}\,d^{-1}$, respectively; Figs. 7 and 8). A well-defined deep chl*a* maximum generally occurs between 50 and 150 m (Figs. 9a and 11c). Fig. 11c reveals a particularly pronounced deep chl*a* maximum along 90–95°E, with concentrations exceeding $0.125\,mg\,m^{-3}$ between 100 and 150 m at the southern end of the transect. This feature may be associated with the aforementioned subduction of high N:Si water between 35° and 50°S. Pigment measurements indicate that prokaryotic cyanobacteria and flagellates dominate the phytoplankton community composition in the gyre, transitioning to flagellate dominance in the Subtropical Front (Naik et al., 2020).

The Southeast Madagascar bloom is a prominent exception to the general oligotrophy of the southern Indian Ocean subtropical gyre. This bloom occurs in the late austral summer/fall (Jan-April) and extends eastward into the subtropical gyre. It can cover a $2500\,km^2$ area with near-surface Chl*a* of $2-3\,mg\,m^3$ (Longhurst, 2001). While the mechanism that generates the bloom is not entirely clear, it occurs within a warm (>26.5°C), shallow mixed layer (~30 m) overlying a strong pycnocline (Dilmahamod et al., 2019; Srokosz & Quartly, 2013; Uz, 2007). It has been suggested that new nitrogen from N_2 fixation fuels the bloom, with eddies and/or the South Indian Counter Current advecting and dispersing it eastward while potentially also contributing additional nutrients and/or phytoplankton biomass (Dilmahamod et al., 2019; Huhn et al., 2012; Poulton et al., 2009; Raj et al., 2010; Srokosz et al., 2004, 2015; Srokosz & Quartly, 2013; Uz, 2007). This hypothesis is supported by observations of nitrogen-fixing species southeast of Madagascar in the Southeast Madagascar bloom region (Poulton et al., 2009) and elevated nitrogen fixation rate measurements in the western part of the southern subtropical gyre (Hörstmann et al., 2021) during austral summer/fall. An alternative explanation is that the Southeast Madagascar bloom initiates off Madagascar from coastal processes that bring limiting nutrients into the photic zone (Srokosz et al., 2004).

2.6.3 Zooplankton

A recent update of calanoid copepod biogeography in the subtropical region south of Madagascar by Cedras et al. (2020) revealed three main species clusters corresponding to water masses associated with the Agulhas Return Current, the Subtropical Front and the Sub-Antarctic Front, all with distinct Chl*a* concentrations. Mean mesozooplankton biomass in the western subtropical zone, including both open-ocean and seamount stations, increases latitudinally from $0.1-0.2\,g\,C\,m^{-2}$

in the north (~27–33°S) to 0.3–0.4 g C m^{-2} at seamounts farther south and in the Subtropical Front (~36–42°S) (Sonnekus et al., 2017). Zooplankton communities differ among seamounts, which has been attributed to latitudinal or mesoscale (e.g., eddy-related) oceanographic variability and connectivity with the Madagascar shelf rather than topographic effects on mesozooplankton biomass or size composition (Noyon et al., 2020). These findings are consistent with the observations of Devi et al. (2020) from the Subtropical Front (40–43°S, 52–60°E), which reveal a strong impact of eddies on productivity, food web structure, and zooplankton community composition. Venkataramana et al. (2019) observed a high-standing stock of small copepods in the Subtropical Front (~42°S, 57°E), transitioning to larger copepods further south in the Polar Front.

Seasonal variability of zooplankton biomass in the central southern subtropical Indian Ocean shows a primary peak during the late Southeast Monsoon (August–September, winter), a broad secondary peak during and following the Northeast Monsoon (January–April, summer), and minima during May and December (Fig. 13). Biomass magnitudes differ by a factor of two, however, between the western-central region (55–100°E) and the eastern area along 110°E, which retains water mass characteristics of South Indian Central Water (Rochford, 1969, 1977). The primary maximum in the central southern subtropical Indian Ocean is probably related to increased wind mixing and nutrient entrainment during winter. In contrast, the summertime secondary maximum might be related to increased primary production during the most stratified summer period due to nitrogen fixation, as observed at station ALOHA in the subtropical North Pacific (Landry et al., 2001; Valencia et al., 2016). The higher biomass values in all seasons along 110°E compared to the central-eastern subtropics may reflect the intensification of winds and increased eddy activity due to proximity to the Australian land mass and Leeuwin Current.

Based on gut pigment analyses of mesozooplankton in eastern subtropical waters along 110°E, Landry et al. (2020) determined grazing rates of $3.0 \pm 0.4\%$ of euphotic zone Chl$a$ d^{-1}, equivalent to $14 \pm 2\%$ of zooplankton C standing stock d^{-1}. For two stations in the central subtropical region, Jaspers et al. (2009) used measured rates of fecal pellet production, biomass scaling, and literature relationships to estimate total consumption by copepods and larvaceans of 7%–18% of zooplankton C d^{-1}. These estimates are indicative of highly oligotrophic waters where the dominant pico-sized phytoplankton largely escape direct feeding by mesozooplankton, and microzooplankton are the major grazers and trophic intermediaries.

3 Summary and conclusions

The Indian Ocean has many special attributes that impact its biogeochemical and ecological dynamics. The Arabian Sea and the northwestern Indian Ocean are strongly influenced by monsoon winds, dust deposition, upwelling/downwelling circulations, and the OMZ. The strongest upwelling occurs during the Southwest Monsoon, driving nutrient enrichment and elevated phytoplankton productivity in coastal waters. However, multiple lines of evidence suggest that maximum production is capped by Si and Fe limitations. In contrast, the Northeast Monsoon drives downwelling except in the northern central Arabian Sea, where wind-driven nutrient entrainment increases primary production. There is surprisingly weak seasonal and spatial variability in mesozooplankton biomass in the Arabian Sea. Nonetheless, the relative grazing contributions of micro- and mesozooplankton vary seasonally and spatially in a manner consistent with spatially separated co-regulation by iron and grazing.

The Bay of Bengal is influenced by the same factors that impact the Arabian Sea, but seasonal wind effects are more muted due to weaker winds and strong freshwater stratification. Although the Bay of Bengal generally exhibits surface oligotrophy, there are regions of high Chl$a$ and production associated with river plumes, upwelling eddies, and wind-induced nutrient entrainment, which drive significant variability in mesozooplankton biomass. The low oxygen content of waters beneath the pycnocline force the Bay of Bengal mesozooplankton to reside in a thin mixed layer where trophic interactions are likely more concentrated than in other Indian Ocean regions. The lack of zooplankton biomass and grazing rate measurements in the Bay of Bengal leads to uncertainty in zooplankton grazing impacts and trophic transfer in the region.

The equatorial Indian Ocean encompasses important open-ocean and coastal upwelling systems driven by monsoon winds. South of the equator, observations and models of the SCTR show strong nutrient, Chl$a$, production, and zooplankton responses to Southeast Monsoon wind forcing in austral winter. Along the southern coasts of the Indonesian island chains, Southeast Monsoon winds and the upwelling-favorable South Java Current also drive nutrient inputs that give rise to enhanced productivity and seasonal zooplankton biomass an order of magnitude higher than in open-ocean waters further south.

The eastern Indian Ocean and the Leeuwin Current are strongly influenced by the ITF, downwelling circulations, and seaward-propagating eddies. In the Leeuwin Current, a unique poleward-flowing eastern boundary current, tropical zooplankton are transported southward, and warm-core anticyclonic eddies carry moderately high Chl$a$ coastal water offshore. In the oceanic realm, tropical waters to the north have the highest total biomass of mesozooplankton, with seasonality driven by upwelling off of Java and northwest Australia during the Southeast Monsoon.

The southwestern Indian Ocean is dominated by alongshore-propagating eddies in the Mozambique Channel and Agulhas Current that elevate (suppress) nutrients and productivity by meanders and topographic interactions that drive upwelling (downwelling) and lateral transport. In general, zooplankton biomass is also low in these waters, except in coastal/shelf regions. The Agulhas Current transports Indo-Pacific zooplankton southward and over the Agulhas Bank, where primary production is somewhat elevated, and the zooplankton community is dominated by the large calanoid copepod *Calanus agulhensis* in austral spring.

The southern Indian Ocean subtropical gyre is extremely oligotrophic except when the Southeast Madagascar bloom occurs in the late austral summer/fall. It has been hypothesized that this bloom is driven by nitrogen fixation, but the causes remain unclear. Low N:P ratios in the gyre suggest perennial nitrogen limitation. Phytoplankton production and mesozooplankton abundance are particularly low in the western subtropical gyre, with variability linked to different water masses and chlorophyll concentrations. Zooplankton biomass is highest in the eastern subtropical gyre, which likely reflects the intensification of winds and increased eddy activity associated with proximity to the Australian land mass and Leeuwin Current. Grazing rate estimates from the southern subtropical gyre of the Indian Ocean are indicative of highly oligotrophic waters where microzooplankton are the major grazers.

In general, it is clear that in all subregions of the Indian Ocean, there is a strong connection between the physics that drives (or suppresses) nutrient delivery to the photic zone and responses of phytoplankton (Chl$a$), primary production, and zooplankton. Wind forcing during the Southwest Monsoon/Southeast Monsoon (boreal summer, austral winter) is a major biogeochemical and ecological driver with broad stimulatory effects throughout many regions (Arabian Sea, Bay of Bengal, SCTR, Java, central/eastern subtropical gyre). In contrast, substantial stimulatory response to the Northwest Monsoon/Northeast Monsoon is less pronounced (Arabian Sea and Bay of Bengal). In addition, there are regionally specific processes that significantly modulate the biogeochemical and ecological responses. For example, strong advective impacts and Fe/Si limitation in the Arabian Sea; freshwater and stratification in the Bay of Bengal; the influence of ITF nutrient inputs, poleward transport, downwelling, and seaward-propagating eddies in the southeastern Indian Ocean; and alongshore-propagating eddies, meanders, upwelling and poleward transport in the southwestern Indian Ocean. The southern subtropical gyre is just extremely oligotrophic except when/where the Southeast Madagascar bloom occurs.

It should be emphasized that other than basic patterns, the details of mechanisms and biogeochemical and ecological responses are poorly explored in many of these Indian Ocean sub-regions, despite evidence that a main driver, the monsoon winds, is being impacted by climate change. There is clearly a need for more studies aimed at improved understanding of planktonic food web dynamics in the Indian Ocean, especially in regions other than the Arabian Sea.

4 Educational resources

Ocean Data View, free software for plotting oceanographic data. Available at: https://odv.awi.de.

World Ocean Atlas, a collection of objectively analyzed, quality-controlled temperature, salinity, oxygen, phosphate, silicate, and nitrate means based on profile data from the World Ocean Database. Available at: https://www.ncei.noaa.gov/products/world-ocean-atlas.

Satellite ocean color data. Available at: https://oceancolor.gsfc.nasa.gov.

GFDL Sea Surface Temperature Simulation for the Indian Ocean. Available at: https://www.youtube.com/watch?v=ZVssbK0K4wc.

Acknowledgments

The development of this article was supported by the Scientific Committee for Oceanic Research via direct funding to the Second International Indian Ocean Expedition and indirect funding through the Integrated Marine Biosphere Research regional program SIBER (Sustained Indian Ocean Biogeochemistry and Ecosystem Research). Additional support was provided by NASA grant no. 80NSSC17K0258 49A37A, NOAA grant no. NA15NMF4570252 NCRS-17 and NSF grant no. 2009248 to R. Hood. The Ocean Data View software package (https://odv.awi.de) was used in developing this article's original graphics. The article also benefitted from extensive comments provided by three anonymous reviewers. This is UMCES contribution 6347.

Author contributions

RRH conceived and led the chapter overall with contributions in particular to all sections on nutrients and phytoplankton. JWM wrote Sections 2.1.1 and 2.1.2 on Fe limitation. MRL wrote Sections 2.1.4, 2.2.3, 2.3.3, 2.4.3, and 2.6.3 on zooplankton. JAH wrote Section 2.5.3 on zooplankton. VJC drafted Figs. 1–4, 5d, and 9–11 and edited the entire chapter. ML, TR, and all authors contributed to the discussion of content, overall chapter structure, and provided feedback on the entire chapter.

References

Ahmed, A., Gauns, M., Kurian, S., Bardhan, P., Pratihary, A., Naik, H., Shenoy, D. M., & Naqvi, S. (2017). Nitrogen fixation rates in the eastern Arabian Sea. *Estuarine, Coastal and Shelf Science, 191*, 74–83.

Anjusha, A., Jyothibabu, R., Jagadeesan, L., Mohan, A. P., Sudheesh, K., Krishna, K., Ullas, N., & Deepak, M. (2013). Trophic efficiency of plankton food webs: Observations from the Gulf of Mannar and the Palk Bay, southeast coast of India. *Journal of Marine Systems, 115*, 40–61.

Anjusha, A., Jyothibabu, R., Jagadeesan, L., Savitha, K., & Albin, K. (2018). Seasonal variation of phytoplankton growth and microzooplankton grazing in a tropical coastal water (off Kochi), southwest coast of India. *Continental Shelf Research, 171*, 12–20.

Arrigo, K. R. (2005). Marine microorganisms and global nutrient cycles. *Nature, 437*(7057), 349–355.

Ayers, J. M., Strutton, P. G., Coles, V. J., Hood, R. R., & Matear, R. J. (2014). Indonesian throughflow nutrient fluxes and their potential impact on Indian Ocean productivity. *Geophysical Research Letters, 41*(14), 5060–5067.

Azam, F., Fenchel, T., Field, J. G., Gray, J. S., Meyer-Reil, L.-A., & Thingstad, F. (1983). The ecological role of water-column microbes in the sea. *Marine Ecology Progress Series*, 257–263.

Baars, M. A. (1999). On the paradox of high mesozooplankton biomass, throughout the year in the western Arabian Sea: Re-analysis of IIOE data and comparison with newer data. *International Journal of Marine Science, 28*, 125–127.

Baer, S. E., Rauschenberg, S., Garcia, C. A., Garcia, N. S., Martiny, A. C., Twining, B. S., & Lomas, M. W. (2019). Carbon and nitrogen productivity during spring in the oligotrophic Indian Ocean along the GO-SHIP IO9N transect. *Deep Sea Research Part II: Topical Studies in Oceanography, 161*, 81–91.

Bange, H. W., Rixen, T., Johansen, A. M., Siefert, R. L., Ramesh, R., Ittekkot, V., Hoffmann, M. R., & Andreae, M. O. (2000). A revised nitrogen budget for the Arabian Sea. *Global Biogeochemical Cycles, 14*(4), 1283–1297.

Banse, K., & English, D. C. (2000). Geographical differences in seasonality of CZCS-derived phytoplankton pigment in the Arabian Sea for 1978-1986. *Deep Sea Research, Part II, 47*(7–8), 1623–1677.

Banse, K., & McClain, C. R. (1986). Winter blooms of phytoplankton in the Arabian Sea as observed by the coastal zone color scanner. *Marine Ecology Progress Series*, 201–211.

Barber, R. T., Marra, J., Bidigare, R. R., Codispoti, L. A., Halpern, D., Johnson, Z., Latasa, M., Goericke, R., & Smith, S. L. (2001). Primary productivity and its regulation in the Arabian Sea during 1995. *Deep Sea Research, Part II, 48*, 1127–1172.

Barlow, R. G., Lamont, T., Gibberd, M. J., van den Berg, M., & Britz, K. (2015). Chemotaxonomic investigation of phytoplankton in the shelf ecosystem of the KwaZulu-Natal bight, South Africa. *African Journal of Marine Science, 37*(4), 467–484.

Barlow, R., Lamont, T., Morris, T., Sessions, H., & Van Den Berg, M. (2014). Adaptation of phytoplankton communities to mesoscale eddies in the Mozambique Channel. *Deep Sea Research Part II: Topical Studies in Oceanography, 100*, 106–118.

Bishop, J. K. B., Wood, T. J., Davis, R. E., & Sherman, J. T. (2004). Robotic observations of enhanced carbon biomass and export at 55 degrees S during SOFeX. *Science, 304*(5669), 417–420.

Boyd, A. J., & Shillington, F. A. (1994). Physical forcing and circulation patterns on the Agulhas Bank. *South African Journal of Marine Science, v. 90, no. 114-122*.

Bristow, L. A., Callbeck, C. M., Larsen, M., Altabet, M. A., Dekaezemacker, J., Forth, M., Gauns, M., Glud, R. N., Kuypers, M. M., & Lavik, G. (2017). N_2 production rates limited by nitrite availability in the bay of Bengal oxygen minimum zone. *Nature Geoscience, 10*(1), 24–29.

Brock, J. C., & McClain, C. R. (1992). Interannual variability in phytoplankton blooms observed in the northwestern Arabian Sea during the southwest monsoon. *Journal of Geophysical Research, Oceans, 97*, 733–750.

Brown, S. L., Landry, M. R., Barber, R. T., Campbell, L., Garrison, D. L., & Gowing, M. M. (1999). Picophytoplankton dynamics and production in the Arabian Sea during the 1995 southwest monsoon. *Deep Sea Research Part II: Topical Studies in Oceanography*, (8-9), 1745–1768.

Brown, S., Landry, M., Christensen, S., Garrison, D., Gowing, M., Bidigare, R., & Campbell, L. (2002). Microbial community dynamics and taxon-specific phytoplankton production in the Arabian Sea during the 1995 monsoon seasons. *Deep Sea Research Part II: Topical Studies in Oceanography, 49*(12), 2345–2376.

Bruce, J. G., Kindle, J. C., Kantha, L. H., Kerling, J. L., & Bailey, J. F. (1998). Recent observations and modeling in the Arabian Sea Laccadive high region. *Journal of Geophysical Research, 102*(C4), 7593–7600.

Bryden, H. L., Beal, L. M., & Duncan, L. M. (2005). Structure and transport of the Agulhas current and its temporal variability. *Journal of Oceanography, 61*, 479–492.

Buchanan, P., & Beckley, L. (2016). Chaetognaths of the Leeuwin current system: Oceanographic conditions drive epi-pelagic zoogeography in the south-East Indian Ocean. *Hydrobiologia, 763*(1), 81–96.

Burchall, J. (1968). *An evaluation of primary productivity studies in the continental shelf region of the Agulhas current near Durban (1961-1966)*. Investigational Report Vol. 20. Oceanographic Research Institute. 16 pp.

Caron, D. A., & Dennett, M. R. (1999). Phytoplankton growth and mortality during the 1995 northeast monsoon and spring Intermonsoon in the Arabian Sea. *Deep Sea Research Part II: Topical Studies in Oceanography, 46*(8–9), 1665–1690.

Carter, R. A., & D'Aubery, J. (1988). Inorganic nutrients in Natal continental shelf waters. In E. H. Schumann (Ed.), *Lecture notes on coastal and estuarine studies, volume 26: Berlin. Coastal Ocean studies off Natal* (pp. 131–151). Springer.

Carter, R. A., & Schleyer, M. H. (1988). Plankton distributions in Natal coastal waters. In E. H. Schumann (Ed.), *Coastal Ocean studies off Natal, South Africa: New York* (pp. 152–177). Springer-Verlag.

Cedras, R., Halo, I., & Gibbons, M. (2020). Biogeography of pelagic calanoid copepods in the Western Indian Ocean. *Deep Sea Research Part II: Topical Studies in Oceanography, 179*, 104740.

Chinni, V., Singh, S. K., Bhushan, R., Rengarajan, R., & Sarma, V. (2019). Spatial variability in dissolved iron concentrations in the marginal and open waters of the Indian Ocean. *Marine Chemistry, 208*, 11–28.

Choi, A., Cho, H., Kim, S.-H., Thamdrup, B., Lee, S., & Hyun, J.-H. (2016). Rates of N2 production and diversity and abundance of functional genes associated with denitrification and anaerobic ammonium oxidation in the sediment of the Amundsen Sea polynya, Antarctica. *Deep Sea Research Part II: Topical Studies in Oceanography, 123*, 113–125.

Coale, K. H., Johnson, K. S., Chavez, F. P., Buesseler, K. O., Barber, R. T., Brzezinski, M. A., Cochlan, W. P., Millero, F. J., Falkowski, P. G., & Bauer, J. E. (2004). Southern Ocean iron enrichment experiment: Carbon cycling in high-and low-Si waters. *Science, 304*(5669), 408–414.

Currie, J. C., Lengaigne, M., Vialard, J., Kaplan, D. M., Aumont, O., Naqvi, S. W. A., & Maury, O. (2013). Indian Ocean dipole and El Niño/southern oscillation impacts on regional chlorophyll anomalies in the Indian Ocean. *Biogeosciences, 10*(10), 6677–6698.

D'Silva, M. S., Anil, A. C., Naik, R. K., & D'Costa, P. M. (2012). Algal blooms: A perspective from the coasts of India. *Natural Hazards, 63*(2), 1225–1253.

Dähnke, K., & Thamdrup, B. (2016). Isotope fractionation and isotope decoupling during anammox and denitrification in marine sediments. *Limnology and Oceanography, 61*(2), 610–624.

Dalsgaard, T., & Thamdrup, B. (2002). Factors controlling anaerobic ammonium oxidation with nitrite in marine sediments. *Applied and Environmental Microbiology, 68*(8), 3802–3808.

Dalsgaard, T., Thamdrup, B., & Canfield, D. E. (2005). Anaerobic ammonium oxidation (anammox) in the marine environment. *Research in Microbiology, 156*(4), 457–464.

De Decker, A. H. B. (1973). Agulhas Bank plankton. In B. Zeitzschel (Ed.), *The biology of the Indian Ocean* (pp. 189–219). Berlin: Springer-Verlag.

De Decker, A. H. B. (1984). Near-surface copepod distribution in the southwestern Indian and southeastern Atlantic Ocean. *Annals. South African Museum, 93*(5), 303–370.

De Decker, A., & Marska, G. (1991). *A new species of Calanus (Copepoda, Calanoida) from south African waters*. South African Museum.

De Sousa, S., Naqvi, S., & Reddy, C. (1981). *Distribution of nutrients in the western Bay of Bengal*.

de Vos, A., Pattiaratchi, C. B., & Wijeratne, E. M. S. (2014). Surface circulation and upwelling patterns around Sri Lanka. *Biogeosciences, 11*, 5909–5930.

Devi, C. A., Jyothibabu, R., Sabu, P., Jacob, J., Habeebrehman, H., Prabhakaran, M., Jayalakshmi, K., & Achuthankutty, C. (2010). Seasonal variations and trophic ecology of microzooplankton in the southeastern Arabian Sea. *Continental Shelf Research, 30*(9), 1070–1084.

Devi, C. A., Sabu, P., Naik, R., Bhaskar, P., Achuthankutty, C., Soares, M., Anilkumar, N., & Sudhakar, M. (2020). Microzooplankton and the plankton food web in the subtropical frontal region of the Indian Ocean sector of the Southern Ocean during austral summer 2012. *Deep Sea Research Part II: Topical Studies in Oceanography, 178*, 104849.

Dilmahamod, A. F. (2014). *Links between the Seychelles-Chagos thermocline ridge and large scale climate modes and primary productivity; and the annual cycle of chlorophyll-a*. [Ph.D.: University of Cape Town.

Dilmahamod, A., Hermes, J., & Reason, C. (2016). Chlorophyll-a variability in the Seychelles–Chagos Thermocline Ridge: Analysis of a coupled bio-physical model. *Journal of Marine Systems, 154*, 220–232.

Dilmahamod, A. F., Penven, P., Aguiar-González, B., Reason, C., & Hermes, J. (2019). A new definition of the South-East Madagascar bloom and analysis of its variability. *Journal of Geophysical Research: Oceans, 124*(3), 1717–1735.

Dufois, F., Hardman-Mountford, N. J., Greenwood, J., Richardson, A. J., Feng, M., Herbette, S., & Matear, R. (2014). Impact of eddies on surface chlorophyll in the South Indian Ocean. *Journal of Geophysical Research: Oceans, 119*(11), 8061–8077.

Edwards, E. S., Burkill, P. H., & Stelfox, C. E. (1999). Zooplankton herbivory in the Arabian Sea during and after the SW monsoon. *Deep Sea Research, Part II, 46*, 843–863.

Feng, M., Majewski, L., Fandry, C., & Waite, A. (2007). Characteristics of two counter-rotating eddies in the Leeuwin current system off the Western Australian coast. *Deep Sea Research, Part II, 54*, 961–980.

Feng, M., Wijffels, S., Godfrey, S., & Meyers, G. (2005). Do eddies play a role in the momentum balance of the Leeuwin current? *Journal of Physical Oceanography, 35*, 964–975.

Fernandes, V. (2008). The effect of semi-permanent eddies on the distribution of mesozooplankton in the central Bay of Bengal. *Journal of Marine Research, 66*(4), 465–488.

Fernandes, V., & Ramaiah, N. (2009). Mesozooplankton community in the Bay of Bengal (India): Spatial variability during the summer monsoon. *Aquatic Ecology, 43*(4), 951–963.

Fernandes, V., & Ramaiah, N. (2013). Mesozooplankton community structure in the upper 1,000 m along the western Bay of Bengal during the 2002 fall intermonsoon. *Zoological Studies, 52*(1), 1–16.

Fernandes, V., & Ramaiah, N. (2014). *Distributional characteristics of surface-layer mesozooplankton in the Bay of Bengal during the 2005 winter monsoon*.

Fernandes, V., & Ramaiah, N. (2019). Spatial structuring of zooplankton communities through partitioning of habitat and resources in the Bay of Bengal during spring intermonsoon. *Turkish Journal of Zoology, 43*(1), 68–93.

Furnas, M. (2007). Intra-seasonal and inter-annual variations in phytoplankton biomass, primary production and bacterial production at North West Cape, Western Australia: Links to the 1997–1998 El Niño event. *Continental Shelf Research, 27*(7), 958–980.

Gallienne, C., Conway, D., Robinson, J., Naya, N., Williams, J., Lynch, T., & Meunier, S. (2004). Epipelagic mesozooplankton distribution and abundance over the Mascarene plateau and basin, South-Western Indian Ocean. *Journal of the Marine Biological Association of the United Kingdom, 84*, 1–8.

Gallienne, C. P., & Smythe-Wright, D. (2005). Epipelagic mesozooplankton dynamics around the Mascarene plateau and basin, Southwestern Indian Ocean. *Philosophical Transactions of the Royal Society A: Mathematical, Physical and Engineering Sciences, 363*(1826), 191–202.

Garcia, C. A., Baer, S. E., Garcia, N. S., Rauschenberg, S., Twining, B. S., Lomas, M. W., & Martiny, A. C. (2018). Nutrient supply controls particulate elemental concentrations and ratios in the low latitude eastern Indian Ocean. *Nature Communications*, *9*(1), 1–10.

Garcia, H. E., Weathers, K. W., Paver, C. R., Smolyar, I., Boyer, T. P., Locarnini, M. M., ... Reagan, J. R. (2018). World Ocean Atlas 2018. Vol. 4: Dissolved inorganic nutrients (phosphate, nitrate and nitrate+nitrite, silicate). NOAA Atlas NESDIS 84, 35 pp. https://archimer.ifremer.fr/doc/00651/76336/.

Garrison, D. L., Gowing, M. M., & Hughes, M. P. (1998). Nano- and microplankton in the northern Arabian Sea during the Southwest Monsoon, August–September 1995 A US–JGOFS study. *Deep Sea Research Part II*, *45*(10–11), 2269–2300.

Garrison, D. L., Gowing, M. M., Hughes, M. P., Campbell, L., Caron, D. A., Dennett, M. R., Shalapyonok, A., Olson, R. J., Landry, M. R., Brown, S. L., Liu, H. B., Azam, F., Steward, G. F., Ducklow, H. W., & Smith, D. C. (2000). Microbial food web structure in the Arabian Sea: A US JGOFS study. *Deep Sea Research Part II: Topical Studies in Oceanography*, (7-8), 1387–1422.

Gaube, P., Chelton, D. B., Strutton, P. G., & Behrenfeld, M. J. (2013). Satellite observations of chlorophyll, phytoplankton biomass, and Ekman pumping in nonlinear mesoscale eddies. *Journal of Geophysical Research, Oceans*, *118*(12), 6349–6370.

Gauns, M., Madhupratap, M., Ramaiah, N., Jyothibabu, R., Fernandes, V., Paul, J. T., & Kumar, S. P. (2005). Comparative accounts of biological productivity characteristics and estimates of carbon fluxes in the Arabian Sea and the Bay of Bengal. *Deep Sea Research Part II: Topical Studies in Oceanography*, *52*(14–15), 2003–2017.

Gauns, M., Mochemadkar, S., Patil, S., Pratihary, A., Naqvi, S., & Madhupratap, M. (2015). Seasonal variations in abundance, biomass and grazing rates of microzooplankton in a tropical monsoonal estuary. *Journal of Oceanography*, *71*(4), 345–359.

Gauns, M., Mohanraju, R., & Madhupratap, M. (1996). Studies on the microzooplankton from the central and eastern Arabian Sea. *Current Science*, *71*(11).

Gaye, B., Nagel, B., Dähnke, K., Rixen, T., & Emeis, K. C. (2013). Evidence of parallel denitrification and nitrite oxidation in the ODZ of the Arabian Sea from paired stable isotopes of nitrate and nitrite. *Global Biogeochemical Cycles*, *27*(4), 1059–1071.

George, J. V., Nuncio, M., Chacko, R., Anilkumar, N., Noronha, S. B., Patil, S. M., Pavithran, S., Alappattu, D. P., Krishnan, K., & Achuthankutty, C. (2013). Role of physical processes in chlorophyll distribution in the western tropical Indian Ocean. *Journal of Marine Systems*, *113*, 1–12.

Gersbach, G. H., Pattiaratchi, C., Ivey, G. N., & Cresswell, G. R. (1999). Upwelling on the south-west coast of Australia—Source of the capes current? *Continental Shelf Research*, *19*, 363–400.

Gibbons, M. (1997). Vertical distribution and feeding of Thalia democratica on the Agulhas Bank during March 1994. *Journal of the Marine Biological Association of the United Kingdom*, *77*(2), 493–505.

Goericke, R., Olson, R. J., & Shalapyonok, A. (2000). A novel niche for Prochlorococcus sp. in low-light suboxic environments in the Arabian Sea and the Eastern Tropical North Pacific. *Deep Sea Research Part I: Oceanographic Research Papers*, *47*(7), 1183–1205.

Goes, J. I., Tian, H., do Rosario Gomes, H., Anderson, O. R., Al-Hashmi, K., deRada, S., Luo, H., Al-Kharusi, L., Al-Azri, A., & Martinson, D. G. (2020). Ecosystem state change in the Arabian Sea fuelled by the recent loss of snow over the Himalayan-tibetan plateau region. *Scientific Reports*, *10*(1), 1–8.

Gomes, H. D., Goes, J. I., Motondkar, S. G. P., Buskey, E. J., Basu, S., Parab, S., & Thoppil, P. G. (2014). Massive outbreaks of Notiluca scintillans blooms in teh Arabian Sea due to spread of hypoxia. *Nature Communications*, *5*(4863). https://doi.org/10.1038/ncomms5862.

Gomes, H. R., Goes, J. I., & Saino, T. (2000). Influence of physical processes and freshwater discharge on the seasonality of phytplankton regime in the Bay of Bengal. *Continental Shelf Research*, *20*, 313–330.

Gomes, H. R., Prabhu Matondkar, S. G., Parab, S. G., Goes, J. I., Pednekar, S., Al-Azri, A. R. N., & Thoppil, P. G. (2009). Unusual blooms of green *Noctiluca miliaris* (Dinophyceae) in the Arabian Sea. In J. D. Wiggert, R. R. Hood, S. W. A. Naqvi, K. H. Brink, & S. L. Smith (Eds.), *Indian Ocean biogeochemical processes and ecological variability*. Washington, D.C., USA: American Geophysical Union.

Goschen, W. S., Schumann, E., Bernard, K. S., Bailey, S., & Deyzel, S. (2012). Upwelling and ocean structures off Algoa Bay and the south-east coast of South Africa. *African Journal of Marine Science*, *34*(4), 525–536.

Grand, M. M., Measures, C. I., Hatta, M., Hiscock, W. T., Buck, C. S., & Landing, W. M. (2015a). Dust deposition in the eastern Indian Ocean: The ocean perspective from Antarctica to the Bay of Bengal. *Global Biogeochemical Cycles*, *29*(3), 357–374.

Grand, M. M., Measures, C. I., Hatta, M., Hiscock, W. T., Landing, W. M., Morton, P. L., Buck, C. S., Barrett, P. M., & Resing, J. A. (2015b). Dissolved Fe and Al in the upper 1000 m of the eastern Indian Ocean: A high-resolution transect along 95° E from the Antarctic margin to the Bay of Bengal. *Global Biogeochemical Cycles*, *29*(3), 375–396.

Grand, M. M., Measures, C. I., Hatta, M., Morton, P. L., Barrett, P., Milne, A., ... Landing, W. M. (2015c). The impact of circulation and dust deposition in controlling the distributions of dissolved Fe and Al in the south Indian subtropical gyre. *Marine Chemistry*, *176*, 110–125.

Guieu, C., Al Azhar, M., Aumont, O., Mahowald, N., Lévy, M., Éthé, C., & Lachkar, Z. (2019). Major impact of dust deposition on the productivity of the Arabian Sea. *Geophysical Research Letters*, *46*(12), 6736–6744.

Gundersen, J. S., Gardner, W. D., Richardson, M. J., & Walsh, I. D. (1998). Effects of monsoons on the seasonal and spatial distributions of POC and chlorophyll in the Arabian Sea. *Deep Sea Research, Part II*, *45*, 2103–2132.

Gupta, R. S., & Naqvi, S. (1984). Chemical oceanography of the Indian Ocean, north of the equator. *Deep Sea Research Part A: Oceanographic Research Papers*, *31*(6–8), 671–706.

Hanson, C. E., Pattiaratchi, C. B., & Waite, A. M. (2005a). Sporadic upwelling on a downwelling coast: Phytoplankton responses to spatially variable nutrient dynamics off the Gascoyne region of Western Australia. *Continental Shelf Research*, *25*, 1561–1582.

Hanson, C. E., Pattiaratchi, C. B., & Waite, A. M. (2005b). Seasonal production regimes off South-Western Australia: Influence of the capes and Leeuwin currents on phytoplankton dynamics. *Marine and Freshwater Research*, *56*, 1011–1026.

Hanson, C. E., Pesant, S., Waite, A. M., & Pattiaratchi, C. B. (2007). Assessing the magnitude and significance of deep chlorophyll maxima of the coastal eastern Indian Ocean. *Deep Sea Research, Part II, 54*, 884–901.

Harms, N. C., Lahajnar, N., Gaye, B., Rixen, T., Dähnke, K., Ankele, M., Schwarz-Schampera, U., & Emeis, K.-C. (2019). Nutrient distribution and nitrogen and oxygen isotopic composition of nitrate in water masses of the subtropical southern Indian Ocean. *Biogeosciences, 16*(13), 2715–2732.

Hermes, J. C., & Reason, C. J. C. (2008). Annual cycle of the South Indian Ocean (Seychelles-Chagos) thermocline ridge in a regional ocean model. *Journal of Geophysical Research, 113*, C04035. https://doi.org/10.01029/02007JC004363.

Hitchcock, G. L., Key, E., & Masters, J. (2000). The fate of upwelled waters in the Great Whirl, August 1995. *Deep Sea Research, Part II, 47*, 1605–1621.

Ho, C.-R., Zheng, Q., & Kuo, N.-J. (2004). SeaWiFS observations of upwelling south of Madagascar: Long-term variability and interaction with the East Madagascar current. *Deep Sea Research, Part II, 51*(1), 59–67.

Hood, R. R., Beckley, L. E. B., & Wiggert, J. D. (2017). Biogeochemical and ecological impacts of boundary currents in the Indian Ocean. *Progress in Oceanography, 156*, 290–325.

Hood, R. R., Rixen, T., Levy, M., Hansell, D. A., Coles, V. J., & Lachkar, Z. (2024). Chapter 12: Oxygen, carbon, and pH variability in the Indian Ocean. In C. C. Ummenhofer, & R. R. Hood (Eds.), *The Indian Ocean and its role in the global climate system* (pp. 265–291). Amsterdam: Elsevier. https://doi.org/10.1016/B978-0-12-822698-8.00017-2.

Hörstmann, C., Raes, E. J., Buttigieg, P. L., Lo Monaco, C., John, U., & Waite, A. M. (2021). Hydrographic fronts shape productivity, nitrogen fixation, and microbial community composition in the southern Indian Ocean and the Southern Ocean. *Biogeosciences, 18*(12), 3733–3749.

Huggett, J. A. (2014). Mesoscale distribution and community composition of zooplankton in the Mozambique Channel. *Deep Sea Research Part II: Topical Studies in Oceanography, 100*, 119–135.

Huggett, J., & Kyewalyanga, M. (2017). Ocean productivity: The RV Dr Fridtjof Nansen in the Western Indian Ocean. In J. C. Groeneveld, & K. A. Koranteng (Eds.), *The RV Dr Fridtjof Nansen in the Western Indian Ocean: Voyages of marine research and capacity development (pp. 55-80), Appendix (pp. 189-216)* (p. 55). Rome: FAO. ISBN 978-92-5-109872-1 http://www.fao.org/3/a-i7652e.pdf.

Huggett, J., & Richardson, A. (2000). A review of the biology and ecology of Calanus agulhensis off South Africa. *ICES Journal of Marine Science, 57*(6), 1834–1849.

Huhn, F., Von Kameke, A., Pérez-Muñuzuri, V., Olascoaga, M. J., & Beron-Vera, F. J. (2012). The impact of advective transport by the South Indian Ocean countercurrent on the Madagascar plankton bloom. *Geophysical Research Letters, 39*(6).

Hutchings, L., Pillar, S., & Verheye, H. (1991). Estimates of standing stock, production and consumption of meso-and macrozooplankton in the Benguela ecosystem. *South African Journal of Marine Science, 11*(1), 499–512.

Hutchins, D. A., DiTullio, G. R., Zhang, Y., & Bruland, K. W. (1998). An iron limitation mosaic in the California upwelling regime. *Limnology and Oceanography, 43*(6), 1037–1054.

Hutchins, D., Hare, C., Weaver, R., Zhang, Y., Firme, G., DiTullio, G., Alm, M., Riseman, S., Maucher, J., & Geesey, M. (2002). Phytoplankton iron limitation in the Humboldt current and Peru upwelling. *Limnology and Oceanography, 47*(4), 997–1011.

Jackson, J. M., Rainville, L., Roberts, M. J., McQuaid, C. D., & Lutjeharms, J. R. E. (2012). Mesoscale bio-physical interactions between the Agulhas current and the Agulhas Bank, South Africa. *Continental Shelf Research, 49*, 10–24.

Jaspers, C., Nielsen, T. G., Carstensen, J., Hopcroft, R. R., & Møller, E. F. (2009). Metazooplankton distribution across the Southern Indian Ocean with emphasis on the role of Larvaceans. *Journal of Plankton Research, 31*(5), 525–540.

Johnson, K. S., Chavez, F. P., & Friederich, G. E. (1999). Continental-shelf sediment as a primary source of iron for coastal phytoplankton. *Nature, 398*(6729), 697–700.

Jose, Y. S., Aumont, O., Machu, E., Penven, P., Moloney, C. L., & Maury, O. (2014). Influence of mesoscale eddies on biological production in the Mozambique Channel: Severl contrasted examples from a coupled ocean-biogeochemistry model. *Deep Sea Research, Part II, 100*, 79–93.

Jose, Y. S., Penven, P., Aumont, O., Machu, E., Moloney, C., Shillington, F., & Maury, O. (2016). Suppressing and enhancing effects of mesoscale dynamics on biological production in the Mozambique Channel. *Journal of Marine Systems, 158*, 129–139.

Jyothibabu, R., Devi, C. A., Madhu, N., Sabu, P., Jayalakshmy, K., Jacob, J., … Nair, K. (2008a). The response of microzooplankton (20–200 μm) to coastal upwelling and summer stratification in the southeastern Arabian Sea. *Continental Shelf Research, 28*(4–5), 653–671.

Jyothibabu, R., Madhu, N., Habeebrehman, H., Jayalakshmy, K., Nair, K., & Achuthankutty, C. (2010). Re-evaluation of 'paradox of mesozooplankton' in the eastern Arabian Sea based on ship and satellite observations. *Journal of Marine Systems, 81*(3), 235–251.

Jyothibabu, R., Madhu, N., Maheswaran, P., Jayalakshmy, K., Nair, K., & Achuthankutty, C. (2008b). Seasonal variation of microzooplankton (20–200 μm) and its possible implications on the vertical carbon flux in the western Bay of Bengal. *Continental Shelf Research, 28*(6), 737–755.

Jyothibabu, R., Vinayachandran, P. N., Madhu, N. V., Robin, R. S., Karnan, C., Jagadeesan, L., & Anjusha, A. (2015). Phytoplankton size structure in the southern Bay of Bengal modified by the summer monsoon current and associated eddies: Implications on the vertical biogenic flux. *Journal of Marine Systems, 143*, 98–119.

Keen, T. R., Kindle, J. C., & Young, D. K. (1997). The interaction of southwest monsoon upwelling, advection and primary production in the northwest Arabian Sea. *Journal of Marine Systems, 13*, 61–82.

Keerthi, M. G., Lengaigne, M., Levy, M., Vialard, J., Parvathi, V., Boyer Montégut, C.d., Ethé, C., Aumont, O., Suresh, I., & Akhil, V. P. (2017). Physical control of interannual variations of the winter chlorophyll bloom in the northern Arabian Sea. *Biogeosciences, 14*(15), 3615–3632.

Kim, H. S., Flagg, C., & Howden, S. D. (2001). Northern Arabian Sea variability from TOPEX/Poseidon altimetry data: An extension of the US JGOFS/ONR shipboard ADCP study. *Deep Sea Research, Part II, 48*, 1069–1096.

Kondo, Y., & Moffett, J. W. (2013). Dissolved Fe (II) in the Arabian Sea oxygen minimum zone and western tropical Indian Ocean during the inter-monsoon period. *Deep Sea Research Part I: Oceanographic Research Papers, 73*, 73–83.

Kone, V., Aumont, O., Levy, M., & Resplandy, L. (2009). Physical and biogeochemical controls of the phytoplankton seasonal cycle in the Indian Ocean: A modeling study. In J. D. Wiggert, R. R. Hood, S. W. A. Naqvi, K. H. Brink, & S. L. Smith (Eds.), *Indian Ocean biogeochemical processes and ecological variability* (pp. 147–166). Washington, DC: American Geophysical Union.

Koslow, J. A., Pesant, S., Feng, M., Pearce, A. F., Fearns, P., Moore, T., Matear, R. J., & Waite, A. (2008). The effect of the Leeuwin current on phytoplankton biomass and production off southwestern Australia. *Journal of Geophysical Research-Oceans, 113*(C07050).

Krishna, M., Prasad, V., Sarma, V., Reddy, N., Hemalatha, K., & Rao, Y. (2015). Fluxes of dissolved organic carbon and nitrogen to the northern Indian Ocean from the Indian monsoonal rivers. *Journal of Geophysical Research: Biogeosciences, 120*(10), 2067–2080.

Kumar, S. P., David, T. D., Byju, P., Narvekar, J., Yoneyama, K., Nakatani, N., Ishida, A., Horii, T., Masumoto, Y., & Mizuno, K. (2012). Bio-physical coupling and ocean dynamics in the central equatorial Indian Ocean during 2006 Indian Ocean Dipole. *Geophysical Research Letters, 39*(14). https://doi.org/10.1029/2012GL052609.

Kumar, S. P., Madhupratap, M., Kumar, M. D., Gauns, M., Muraleedharan, P., Sarma, V., & De Souza, S. (2000). Physical control of primary productivity on a seasonal scale in central and eastern Arabian Sea. *Journal of Earth System Science, 109*(4), 433–441.

Kumar, S. P., Madhupratap, M., Kumar, M. D., Muraleedharan, P., De Souza, S., Gauns, M., & Sarma, V. (2001). High biological productivity in the central Arabian Sea during the summer monsoon driven by Ekman pumping and lateral advection. *Current Science*, 1633–1638.

Kumar, S. P., Nuncio, M., Ramaiah, N., Sardesai, S., Narvekar, J., Fernandes, V., & Paul, J. T. (2007). Eddy-mediated biological productivity in the Bay of Bengal during fall and spring intermonsoons. *Deep Sea Research Part I: Oceanographic Research Papers, 54*(9), 1619–1640.

Kyewalyanga, M. S., Naik, R., Hegde, S., Raman, M., Barlow, R., & Roberts, M. (2007). Phytoplankton biomass and primary production in Delagoa bight Mozambique: Application of remote sensing. *Estuarine, Coastal and Shelf Science, 74*(3), 429–436.

Lamont, T., & Barlow, R. G. (2015). Environmental influence on phytoplankton production during summer on the KwaZulu-Natal shelf of the Agulhas ecosystem. *African Journal of Marine Science, 37*(4), 485–501.

Lamont, T., Barlow, R., Morris, T., & Van Den Berg, M. (2014). Characterisation of mesoscale features and phytoplankton variability in the Mozambique Channel. *Deep Sea Research Part II: Topical Studies in Oceanography, 100*, 94–105.

Landry, M. R. (2009). Grazing processes and secondary production in the Arabian Sea: A simple food web synthesis with measurement constraints. In J. D. Wiggert, R. R. Hood, S. W. A. Naqvi, K. H. Brink, & S. L. Smith (Eds.), *Indian Ocean biogeochemical processes and ecological variability* (pp. 133–146). Washington D.C.: American Geophysical Union.

Landry, M. R., Al-Mutairi, H., Selph, K. E., Christensen, S., & Nunnery, S. (2001). Seasonal patterns of mesozooplankton abundance and biomass at station ALOHA. *Deep Sea Research Part II: Topical Studies in Oceanography, 48*(8–9), 2037–2061.

Landry, M. R., Brown, S. L., Campbell, L., Constantinou, J., & Liu, H. (1998). Spatial patterns in phytoplankton growth and microzooplankton grazing in the Arabian Sea during monsoon forcing. *Deep Sea Research Part II: Topical Studies in Oceanography, 45*(10–11), 2353–2368.

Landry, M. R., Hood, R. R., & Davies, C. H. (2020). Mesozooplankton biomass and temperature-enhanced grazing along a 110 E transect in the eastern Indian Ocean. *Marine Ecology Progress Series, 649*, 1–19.

Landry, M. R., Hood, R. R., Davies, C. H., Selph, K. E., Antoine, D., Carl, M. C., & Beckley, L. E. (2022). Microbial community biomass, production and grazing along 110°E in the eastern Indian Ocean. *Deep Sea Research, Part II, 202*, 105134. https://doi.org/10.1016/j.dsr2.2022.105134.

Latasa, M., & Bidigare, R. R. (1998). A comparison of phytoplankton populations of the Arabian Sea during the spring Intermonsoon and southwest monsoon of 1995 as described by HPLC-analyzed pigments. *Deep Sea Research Part II: Topical Studies in Oceanography, 45*(10–11), 2133–2170.

Lee, C. M., Jones, B. H., Brink, K. H., & Fischer, A. S. (2000). The upper-ocean response to monsoonal forcing in the Arabian Sea: Seasonal and spatial variability. *Deep Sea Research, Part II, 47*, 1177–1226.

Legand, M. (1969). Seasonal variations in the Indian Ocean along 110° E. IV. *Macropankton and Micronekton Biomass: Marine and Freshwater Research, 20*(1), 85–104.

Lelieveld, J.o., Crutzen, P., Ramanathan, V., Andreae, M., Brenninkmeijer, C., Campos, T., Cass, G., Dickerson, R., Fischer, H., & De Gouw, J. (2001). The Indian Ocean experiment: Widespread air pollution from South and Southeast Asia. *Science, 291*(5506), 1031–1036.

Leon, J.-F., & Legrand, M. (2003). Mineral dust sources in the surroundings of the North Indian Ocean. *Geophysical Research Letters, 30*(6). https://doi.org/10.1029/2002GL016690.

Lévy, M., André, J.-M., Shankar, D., Durand, F., & Shenoi, S. S. C. (2006). A quantitative method for describing the seasonal cycles of surface chlorophyll in the Indian Ocean. In *Vol. 6406. Proceedings Remote Sensing of the Marine Environment* (p. 640611). International Society for Optics and Photonics.

Lévy, M., Shankar, D., André, J. M., Shenoi, S. S. C., Durand, F., & Montégut, C. D. B. (2007). Basin-wide seasonal evolution of the Indian Ocean's phytoplankton blooms. *Journal of Geophysical Research, 112*. https://doi.org/10.1029/2007JC004090.

Li, Y.-H., Menviel, L., & Peng, T.-H. (2006). Nitrate deficits by nitrification and denitrification processes in the Indian Ocean. *Deep Sea Research Part I: Oceanographic Research Papers, 53*(1), 94–110.

Lierheimer, L. J., & Banse, K. (2002). Seasonal and interannual variability of phytoplankton pigment in the Laccadive (Lakshadweep) sea as observed by the coastal zone color scanner. *Proceedings of the Indian Academy of Sciences-Earth and Planetary Sciences, 111*(2), 163–185.

Longhurst, A. (2001). A major seasonal phytoplankton bloom in Madagascar. *Deep Sea Research, Part I, 48*(11), 2413–2422.

Lourey, M. J., Dunn, J. R., & Waring, J. (2006). A mixed layer nutrient climatology of Leeuwin current and Western Australian shelf waters: Seasonal nutrient dynamics and biomass. *Journal of Marine Systems, 59*, 25–51.

Lourey, M. J., Thompson, P. A., McLaughlin, M. J., Bonham, P., & Feng, M. (2013). Primary production and phytoplankton community structure during a winter shelf-scale phytoplankton bloom off Western Australia. *Marine Biology, 160*, 355–369.

Luis, A. J., & Kawamura, H. (2004). Air-sea interaction, coastal circulation and primary production in the eastern Arabian Sea: A review. *Journal of Oceanography, 60*, 205–218.

Lutjeharms, J. R. E. (2006). *The Agulhas current*. Berlin, Heidelbert, New York: Springer. 329 p.

Lutjeharms, J. R. E., & Machu, E. (2000). An upwelling cell inshore of the East Madagascar current. *Deep Sea Research, Part I, 47*, 2405–2411.

Lutjeharms, J. R. E., Meyer, A. A., Ansorge, I. J., Eagle, G. A., & Orren, M. J. (1996). The nutrient characteristics of the Agulhas Bank. *South African Journal of Marine Science, 17*, 253–274.

Machu, E., & Garçon, V. (2001). Phytoplankton seasonal distribution from SeaWiFS data in the Agulhas current system. *Journal of Marine Research, 59*(5), 795–812.

Madden, R. A., & Julian, P. R. (1972). Description of global-scale circulation cells in the tropics with a 40-50 day period. *Journal of Atmospheric Sciences, 29*(6), 1109–1123.

Madhupratap, M., Gopalakrishnan, T., Haridas, P., Nair, K., Aravindakshan, P., Padmavati, G., & Paul, S. (1996a). Lack of seasonal and geographic variation in mesozooplankton biomass in the Arabian Sea and its structure in the mixed layer. *Current Science (Bangalore), 71*(11), 863–868.

Madhupratap, M., Haridas, P., Ramaiah, N., & Achuthankutty, C. (1992). *Zooplankton of the southwest coast of India: abundance, composition, temporal and spatial variability in 1987*.

Madhupratap, M., Kumar, S. P., Bhattathiri, P., Kumar, M. D., Raghukumar, S., Nair, K., & Ramaiah, N. (1996b). Mechanism of the biological response to winter cooling in the northeastern Arabian Sea. *Nature, 384*(6609), 549–552.

Madhupratap, M., Sawant, S., & Gauns, M. (2000). A first report on a bloom of the marine prymnesiophycean, *Phaeocystis globosa* from the Arabian Sea. *Oceanologica Acta, 23*(1), 83–90.

Mahadevan, A. (2016). The impact of submesoscale physics on primary productivity of plankton. *Annual Review of Marine Science, 8*, 161–184.

Manghnani, V., Morrison, J. M., Hopkins, T. S., & Bohm, E. (1998). Advection of upwelled waters in the form of plumes off Oman during the southwest monsoon. *Deep Sea Research, Part II, 45*, 2027–2052.

Mantoura, R. F. C., Law, C. S., Owens, N. P. J., Burkill, P. H., Woodward, E. M. S., Howland, R. J. M., & Llewellyn, C. A. (1993). *Nitrogen biogeochemical cycling in the northwestern Indian Ocean Deep Sea research part II. 40* (pp. 651–671).

Marra, J., & Barber, R. T. (2005). Primary productivity in the Arabian Sea: A synthesis of JGOFS data. *Progress in Oceanography, 65*, 159–175.

Marra, J., Dickey, T. D., Ho, C., Kinkade, C. S., Sigurdson, D. E., Weller, R. A., & Barber, R. T. (1998). Variability in primary production as observed from moored sensors in the central Arabian Sea in 1995. *Deep Sea Research, Part II, 45*, 2253–2267.

Marsac, F., Barlow, R., Ternon, J. F., Menard, F., & Roberts, M. (2014). Ecosystem functioning in the Mozambique Channel: Synthesis and future research. *Deep Sea Research, Part II, 100*, 212–220.

Marsac, F., Everett, B., Shahid, U., & Strutton, P. G. (2024). Chapter 11: Indian Ocean primary productivity and fisheries variability. In C. C. Ummenhofer, & R. R. Hood (Eds.), *The Indian Ocean and its role in the global climate system* (pp. 245–264). Amsterdam: Elsevier. https://doi.org/10.1016/B978-0-12-822698-8.00019-6.

Martin, P., van der Loeff, M. R., Cassar, N., Vandromme, P., d'Ovidio, F., Stemmann, L., Rengarajan, R., Soares, M., González, H. E., & Ebersbach, F. (2013). Iron fertilization enhanced net community production but not downward particle flux during the Southern Ocean iron fertilization experiment LOHAFEX. *Global Biogeochemical Cycles, 27*(3), 871–881.

Matondkar, S. G. P., Dwivedi, R. M., Parab, S., Pednekar, S., Mascarenhas, A., Raman, M., & Singh, S. (2007). *Phytoplankton in the northeastern Arabian Sea exhibit seasonality* (pp. 1–4). SPIE Newsroom.

McCreary, J. P., Murtugudde, R., Vialard, J., Vinayachandran, P. N., Wiggert, J. D., Hood, R. R., Shankar, D., & Shetye, S. R. (2009). Biophysical processes in the Indian Ocean. In J. Wiggert, R. R. Hood, S. W. A. Naqvi, K. H. Brink, & S. L. Smith (Eds.), *Vol. 185. Indian Ocean biogeochemical processes and ecological variability* (pp. 9–32). Washington D.C.: American Geophysical Union.

McGowan, H., & Clark, A. (2008). Identification of dust transport pathways from Lake Eyre, Australia using Hysplit. *Progress in Oceanography, 86*, 302–315.

McKinnon, A. D., Doyle, J., Duggan, S., Logan, M., Lønborg, C., & Brinkman, R. (2015a). Zooplankton growth, respiration and grazing on the Australian margins of the tropical Indian and Pacific oceans. *PLoS One, 10*(10), e0140012.

McKinnon, A., Duggan, S., Holliday, D., & Brinkman, R. (2015b). Plankton community structure and connectivity in the Kimberley-browse region of NW Australia. *Estuarine, Coastal and Shelf Science, 153*, 156–167.

McPhaden, M. J., & Nagura, M. (2014). Indian Ocean dipole interpreted in terms of recharge oscillator theory. *Climate Dynamics, 42*, 1569–1586.

Measures, C., & Vink, S. (1999). Seasonal variations in the distribution of Fe and Al in the surface waters of the Arabian Sea. *Deep Sea Research, Part II, 46*, 1597–1622.

Meyer, A. A., Lutjeharms, J. R. E., & de Villiers, S. (2002). The nutrient characteristics of the Natal bight, South Africa. *Journal of Marine Systems, 35*, 11–37.

Moffett, J. W., & German, C. R. (2020). Distribution of iron in the Western Indian Ocean and the Eastern tropical South pacific: An inter-basin comparison. *Chemical Geology, 532*, 119334.

Moffett, J. W., Goepfert, T. J., & Naqvi, S. W. A. (2007). Reduced iron associated with secondary nitrite maxima in the Arabian Sea. *Deep Sea Research Part I: Oceanographic Research Papers, 54*(8), 1341–1349.

Moffett, J. W., & Landry, M. R. (2020). Grazing control and iron limitation of primary production in the Arabian Sea: Implications for anticipated shifts in Southwest Monsoon intensity. *Deep Sea Research Part II: Topical Studies in Oceanography, 179*, 104687.

Moffett, J. W., Vedamati, J., Goepfert, T. J., Pratihary, A., Gauns, M., & Naqvi, S. W. A. (2015). Biogeochemistry of iron in the Arabian Sea. *Limnology and Oceanography, 60*(5), 1671–1688.

Moore, T. S., Matear, R. J., Marra, J., & Clementson, L. (2007). Phytoplankton variability off the Western Australian coast: Mesoscale eddies and their role in cross-shelf exchange. *Deep Sea Research Part II: Topical Studies in Oceanography, 54*(8–10), 943–960.

Moriarty, R., & O'Brien, T. (2013). Distribution of mesozooplankton biomass in the global ocean. *Earth System Science Data, 5*(1), 45–55.

Moritz, C., Montagnes, D., Carleton, J., Wilson, D., & McKinnon, A. (2006). The potential role of microzooplankton in a northwestern Australian pelagic food web. *Marine Biology Research, 2*(01), 1–13.

Morrison, J. M., Codispoti, L. A., Gaurin, S., Jones, B., Manghnani, V., & Zheng, Z. (1998). Seasonal variation of hydrographic and nutrient fields during the US JGOFS Arabian Sea process study. *Deep Sea Research, Part II, 45*, 2053–2101.

Mukherjee, J., Naidu, S. A., Sarma, V., & Ghosh, T. (2018). Influence of river discharge on zooplankton diet in the Godavari estuary (Bay of Bengal, Indian Ocean). *Advances in Oceanography and Limnology, 9*(1).

Muraleedharan, K., Jasmine, P., Achuthankutty, C., Revichandran, C., Kumar, P. D., Anand, P., & Rejomon, G. (2007). Influence of basin-scale and mesoscale physical processes on biological productivity in the Bay of Bengal during the summer monsoon. *Progress in Oceanography, 72*(4), 364–383.

Murty, V., Gupta, G., Sarma, V., Rao, B., Jyothi, D., Shastri, P., & Supraveena, Y. (2000). Effect of vertical stability and circulation on the depth of the chlorophyll maximum in the Bay of Bengal during May–June, 1996. *Deep Sea Research Part I: Oceanographic Research Papers, 47*(5), 859–873.

Naik, R., Anil, A., Narale, D., Chitari, R., & Kulkarni, V. (2011a). Primary description of surface water phytoplankton pigment patterns in the Bay of Bengal. *Journal of Sea Research, 65*(4), 435–441.

Naik, R. K., George, J., Soares, M., Anilkumar, N., Mishra, R., Roy, R., Bhaskar, P., Sivadas, S., Murukesh, N., & Chacko, R. (2020). Observations of surface water phytoplankton community in the Indian Ocean: A transect from tropics to polar latitudes. *Deep Sea Research Part II: Topical Studies in Oceanography, 178*, 104848.

Naik, R. K., Hegde, S., & Anil, A. C. (2011b). Dinoflagellate community structure from the stratified environment of the Bay of Bengal, with special emphasis on harmful algal bloom species. *Environmental Monitoring and Assessment, 182*(1), 15–30.

Naqvi, S. W. A., Jayakumar, D. A., Narvekar, P. V., Naik, H., Sarma, V., D'Souza, W., Joseph, S., & George, M. D. (2000). Increased marine production of N_2O due to intensifying anoxia on the Indian continental shelf. *Nature, 408*(6810), 346–349.

Naqvi, S. W. A., Moffett, J. W., Gauns, M. U., Narvekar, P. V., Pratihary, A. K., Naik, H., Shenoy, D. M., Jayakumar, D. A., Goepfert, T. J., Patra, P. K., Al-Azri, A., & Ahmed, S. I. (2010). The Arabian Sea as a high-nutrient, low-chlorophyll region during the late southwest monsoon. *Biogeosciences, 7*, 2091–2100.

Naqvi, S. W. A., Narvekar, P. V., & Desa, E. (2006). Coastal biogeochemical processes in the North Indian Ocean. In A. Robinson, & K. Brink (Eds.), *Vol. 14. The sea* (pp. 723–780). Harvard University Press.

Narvekar, J., & Kumar, S. P. (2006). Seasonal variability of the mixed layer in the central Bay of Bengal and associated changes in nutrients and chlorophyll. *Deep Sea Research Part I: Oceanographic Research Papers, 53*(5), 820–835.

Nehring, D., Hagen, E., Jorge da Silva, A., Schemainda, R., Wolf, G., Michelchen, N., Kaiser, W., Postel, L., Gosselck, F., Brenning, U., Kühner, E., Arlt, G., Siegel, H., Gohs, L., & Bublitz, G. (1987). Results of oceanological studies in the Mozambique Channel in February—March 1980. *Beiträge zur Meereskunde, 56*, 51–63.

Nielsen, T. G., Bjørnsen, P. K., Boonruang, P., Fryd, M., Hansen, P. J., Janekarn, V., Limtrakulvong, V., Munk, P., Hansen, O. S., & Satapoomin, S. (2004). Hydrography, bacteria and protist communities across the continental shelf and shelf slope of the Andaman Sea (NE Indian Ocean). *Marine Ecology Progress Series, 274*, 69–86.

Noyon, M., Morris, T., Walker, D., & Huggett, J. (2019). Plankton distribution within a young cyclonic eddy off South-Western Madagascar. *Deep Sea Research Part II: Topical Studies in Oceanography, 166*, 141–150.

Noyon, M., Rasoloarijao, Z., Huggett, J., Ternon, J.-F., & Roberts, M. (2020). Comparison of mesozooplankton communities at three shallow seamounts in the South West Indian Ocean. *Deep Sea Research Part II: Topical Studies in Oceanography, 176*, 104759.

Nuncio, M., & Prasanna Kumar, S. (2013). Evolution of cyclonic eddies and biogenic fluxes in the northern Bay of Bengal. *Biogeosciences Discussions, 10*(10), 16213–16236.

Nyadjro, E. S., Jensen, T. G., Richman, J. G., & Shriver, J. F. (2017). On the Relationship Between Wind, SST, and the Thermocline in the Seychelles–Chagos Thermocline Ridge. *IEEE Geoscience and Remote Sensing Letters, 14*(12), 2315–2319.

Ockhuis, S., Huggett, J., Gouws, G., & Sparks, C. (2017). The 'suitcase hypothesis': Can entrainment of meroplankton by eddies provide a pathway for gene flow between Madagascar and KwaZulu-Natal, South Africa? *African Journal of Marine Science, 39*(4), 435–451.

Parab, S. G., & Matondkar, S. (2012). Primary productivity and nitrogen fixation by Trichodesmium spp. in the Arabian Sea. *Journal of Marine Systems, 105*, 82–95.

Parab, S. G., Matondkar, S. G. P., Gomes, H. D., & Goes, J. I. (2006). Monsoon driven changes in phytoplankton populations in the eastern Arabian Sea as revealed by microscopy and HPLC pigment analysis. *Continental Shelf Research, 26*, 2538–2558.

Paterson, H. L., Feng, M., Waite, A. M., Gomis, D., Beckley, L. E., Holliday, D., & Thompson, P. A. (2008a). Physical and chemical signatures of a developing anticyclonic eddy in the Leeuwin current, eastern Indian Ocean. *Journal of Geophysical Research, Oceans, 113*. https://doi.org/10.1029/2007JC004707.

Paterson, H., Knott, B., Koslow, A., & Waite, A. (2008b). The grazing impact of microzooplankton off south West Western Australia: As measured by the dilution technique. *Journal of Plankton Research, 30*(4), 379–392.

Paterson, H. L., Knott, B., & Waite, A. M. (2007). Microzooplankton community structure and grazing on phytoplankton, in an eddy pair in the Indian Ocean off Western Australia. *Deep Sea Research Part II: Topical Studies in Oceanography, 54*(8–10), 1076–1093.

Paula, J., Pinto, I., Guambe, I., Monteiro, S., Gove, D., & Guerreiro, J. (1998). Seasonal cycle of planktonic communities at Inhaca Island, southern Mozambique. *Journal of Plankton Research, 20*(11), 2165–2178.

Pearce, A., & Pattiaratchi, C. (1999). The capes current: A summer countercurrent flowing past cape Leeuwin and cape Naturaliste, Western Australia. *Continental Shelf Research, 19*(3), 401–420.

Pease, P. P., Tchakerian, V. P., & Tindale, N. W. (1998). Aerosols over the Arabian Sea: Geochemistry and source areas for aeolian desert dust. *Journal of Arid Environments, 39*, 477–496.

Pillai, H. U., Jayalakshmy, K., Biju, A., Jayalakshmi, K., Paulinose, V., Devi, C., Nair, V., Revichandran, C., Menon, N., & Achuthankutty, C. (2014). A comparative study on mesozooplankton abundance and diversity between a protected and an unprotected coastal area of Andaman Islands. *Environmental Monitoring and Assessment, 186*(6), 3305–3319.

Poulton, A. J., Stinchcombe, M. C., & Quartly, G. D. (2009). High numbers of Trichodesmium and diazotrophic diatoms in the Southwest Indian Ocean. *Geophysical Research Letters, 36*(15). https://doi.org/10.1029/2009GL039719.

Pretorius, M., Huggett, J., & Bibbons, M. (2016). Summer and winter diferences in zooplankton biomass, distribution and size composition in the KwaZulu-Natal Bight. In M. J. Roberts, S. T. Fennesy, & R. G. Barlow (Eds.), *v. 38. Ecosystem processes in the KwaZulu-Natal bight*. African journal of marine science. Supplement, in press).

Quartly, G. D., & Srokosz, M. A. (2004). Eddies in the southern Mozambique Channel. *Deep Sea Research, Part II, 51*(1), 69–83.

Raes, E. J., Thompson, P. A., McInnes, A. S., Nguyen, H. M., Hardman-Mountford, N., & Waite, A. M. (2015). Sources of new nitrogen in the Indian Ocean. *Global Biogeochemical Cycles, 29*(8), 1283–1297.

Raj, R. P., Peter, B. N., & Pushpadas, D. (2010). Oceanic and atmospheric influences on the variability of phytoplankton bloom in the Southwestern Indian Ocean. *Journal of Marine Systems, 82*(4), 217–229.

Ramaiah, N., Fernandes, V., Paul, J., Jyothibabu, R., Gauns, M., & Jayraj, E. (2010). *Seasonal variability in biological carbon biomass standing stocks and production in the surface layers of the Bay of Bengal*.

Ramanantsoa, J. D., Krug, M., Penven, P., Rouault, M., & Gula, J. (2018). Coastal upwelling south of Madagascar: Temporal and spatial variability. *Journal of Marine Systems, 178*, 29–37.

Ravichandran, M., Girishkumar, M. S., & Riser, S. (2012). Observed variability of chlorophyll-a using Argo profiling floats in the southeastern Arabian Sea. *Deep Sea Research Part I: Oceanographic Research Papers, 65*, 15–25.

Reckermann, M., & Veldhuis, M. J. (1997). Trophic interactions between picophytoplankton and micro-and nanozooplankton in the western Arabian Sea during the NE monsoon 1993. *Aquatic Microbial Ecology, 12*(3), 263–273.

Resplandy, L., Lévy, M., Bopp, L., Echevin, V., Pous, S., Sarma, V., & Kumar, D. (2012). Controlling factors of the oxygen balance in the Arabian Sea's OMZ. *Biogeosciences, 9*(12), 5095–5109.

Resplandy, L., Levy, M., Madec, G., Pous, S., Aumont, O., & Kumar, D. (2011). Contribution of mesoscale processes to nutrient budgets in the Arabian Sea. *Journal of Geophysical Research, 116*, C11007.

Resplandy, L., Vialard, J., Lévy, M., Aumont, O., & Dandoneau, Y. (2009). Seasonal and intraseasonal biogeochemical variability in the thermocline ridge of the Indian Ocean. *Journal of Geophysical Research, 114*, C07024. https://doi.org/10.1029/2008JC005246.

Rixen, T., Cowie, G., Gaye, B., Goes, J., do Rosário Gomes, H., Hood, R. R., Lachkar, Z., Schmidt, H., Segschneider, J., & Singh, A. (2020). Reviews and syntheses: Present, past, and future of the oxygen minimum zone in the northern Indian Ocean. *Biogeosciences, 17*(23), 6051–6080.

Rixen, T., Goyet, C., & Ittekkot, V. (2006a). Diatoms and their influence on the biologically mediated uptake of atmospheric CO2 in the Arabian Sea upwelling system. *Biogeosciences, 3*(1), 1–13.

Rixen, T., Ittekkot, V., Herunadi, B., Wetzel, P., Maier-Reimer, E., & Gaye-Haake, B. (2006b). ENSO-driven carbon see saw in the Indo-Pacific. *Geophysical Research Letters, 33*(7).

Rixen, T., Ramaswamy, V., Gaye, B., Herunadi, B., Maier-Reimer, E., Bange, H. W., & Ittekkot, V. (2009). Monsoonal and ENSO impacts on export fluxes and the biological pump in the Indian Ocean. In R. R. Hood, J. D. Wiggert, S. W. A. Naqvi, S. Smith, & K. Brink (Eds.), *Indian Ocean biogeochemical processes and ecological variability* AGU.

Roberts, M. J. (2005). Chokka squid Loglio vulgaris reynaudii abundance may be linked to changes in the Agulhas Bank (South Africa) ecosystem during spawning and the early life cycle. *ICES Journal of Marine Science, 62*, 33–55.

Roberts, M. J., Ternon, J.-F., & Morris, T. (2014). Interaction of dipole eddies with the western continental slope of the Mozambique Channel. *Deep Sea Research, Part II, 100*, 54–67.

Rochford, D. (1969). Seasonal variations in the Indian Ocean along 110 El hydrological structure of the upper 500 m. *Marine and Freshwater Research, 20*(1), 1–50.

Rochford, D. (1977). Further studies of plankton ecosystems in the eastern Indian Ocean. II. Seasonal variations in water mass distribution (upper 150 m) along 110° E. (August 1962-August 1963). *Marine and Freshwater Research, 28*(5), 541–555.

Roger, C. (1994). The plankton of the tropical western Indian ocean as a biomass indirectly supporting surface tunas (yellowfin, Thunnus albacares and skipjack, Katsuwonus pelamis). *Environmental Biology of Fishes, 39*(2), 161–172.

Roman, M., Smith, S., Wishner, K., Zhang, X., & Gowing, M. (2000). Mesozooplankton production and grazing in the Arabian Sea. *Deep Sea Research, Part II, 47*(7–8), 1423–1450.

Romero, O. E., Rixen, T., & Herunadi, B. (2009). Effects of hydrographic and climatic forcing on diatom production and export in the tropical southeastern Indian Ocean. *Marine Ecology Progress Series, 384*, 69–82.

Rossi, V., Feng, M., Pattiaratchi, C., Roughan, M., & Waite, A. M. (2013a). On the factors influencing the development of sporadic upwelling in the Leeuwin current system. *Journal of Geophysical Research, 118*(7), 3608–3621.

Rossi, V., Feng, M., Pattiaratchi, C., Roughan, M., & Waite, A. M. (2013b). Linking synoptic forcing and local mesoscale processes with biological dynamics off Ningaloo reef. *Journal of Geophysical Research, 118*(3), 1211–1225.

Roy, R., & Anil, A. (2015). Complex interplay of physical forcing and Prochlorococcus population in ocean. *Progress in Oceanography, 137*, 250–260.

Roy, R., Chitari, R., Kulkarni, V., Krishna, M., Sarma, V., & Anil, A. (2015). CHEMTAX-derived phytoplankton community structure associated with temperature fronts in the northeastern Arabian Sea. *Journal of Marine Systems, 144*, 81–91.

Sarma, V., & Aswanikumar, V. (1991). Subsurface chlorophyll maxima in the north-western Bay of Bengal. *Journal of Plankton Research, 13*(2), 339–352.

Sarma, V., Chopra, M., Rao, D., Priya, M., Rajula, G., Lakshmi, D., & Rao, V. (2020a). Role of eddies on controlling total and size-fractionated primary production in the Bay of Bengal. *Continental Shelf Research, 204*, 104186.

Sarma, V., Krishna, M., Rao, V., Viswanadham, R., Kumar, N., Kumari, T., Gawade, L., Ghatkar, S., & Tari, A. (2012). Sources and sinks of CO2 in the west coast of Bay of Bengal. *Tellus B: Chemical and Physical Meteorology, 64*(1), 10961.

Sarma, V., Krishna, M., Viswanadham, R., Rao, G., Rao, V., Sridevi, B., Kumar, B., Prasad, V., Subbaiah, C. V., & Acharyya, T. (2013a). Intensified oxygen minimum zone on the western shelf of Bay of Bengal during summer monsoon: Influence of river discharge. *Journal of Oceanography, 69*(1), 45–55.

Sarma, V., Paul, Y., Vani, D., & Murty, V. (2015). Impact of river discharge on the coastal water pH and pCO2 levels during the Indian Ocean Dipole (IOD) years in the western Bay of Bengal. *Continental Shelf Research, 107*, 132–140.

Sarma, V., Sridevi, B., Maneesha, K., Sridevi, T., Naidu, S., Prasad, V., Venkataramana, V., Acharya, T., Bharati, M., & Subbaiah, C. V. (2013b). Impact of atmospheric and physical forcings on biogeochemical cycling of dissolved oxygen and nutrients in the coastal Bay of Bengal. *Journal of Oceanography, 69*(2), 229–243.

Sarma, V., Vivek, R., Rao, D., & Ghosh, V. (2020b). Severe phosphate limitation on nitrogen fixation in the Bay of Bengal. *Continental Shelf Research, 205*, 104199.

Sarojini, Y., & Sarma, N. S. (2001). Vertical distribution of phytoplankton around Andaman and Nicobar Islands, Bay of Bengal. *Indian Journal of Marine Sciences, 30*, 65–69.

Satapoomin, S., Nielsen, T. G., & Hansen, P. J. (2004). Andaman Sea copepods: Spatio-temporal variations in biomass and production, and role in the pelagic food web. *Marine Ecology Progress Series, 274*, 99–122.

Sawant, S., & Madhupratap, M. (1996). Seasonality and composition of phytoplankton in the Arabian Sea. *Current Science, 71*(11), 869–873.

Säwström, C., Beckley, L. E., Saunders, M. I., Thompson, P. A., & Waite, A. M. (2014). The zooplankton prey field for rock lobster phyllosoma larvae in relation to oceanographic features of the South-Eastern Indian Ocean. *Journal of Plankton Research, 36*(4), 1003–1016.

Schleyer, M. H. (1985). *Chaetognaths as indicators of water masses in the Agulhas current system.* Investigations Reports Vol. 61 (pp. 1–20). Durban: Oceanographic Research Institute.

Schott, F. A., & McCreary, J. P. (2001). The monsoon circulation in the Indian Ocean. *Progress in Oceanography, 51*(1), 1–123.

Schott, F. A., Xie, S. P., & McCreary, J. P. (2009). Indian ocean circulation and climate variability. *Reviews of Geophysics, 47*, RG1002.

Schumann, E. H., Churchill, J. R. S., & Zaayman, H. J. (2005). Oceanic variability in the western sector of Algoa Bay, South Africa. *African Journal of Marine Science, 27*(1), 65–80.

Shalapyonok, A., Olson, R. J., & Shalapyonok, L. S. (2001). Arabian Sea phytoplankton during southwest and northeast monsoons 1995: Composition, size structure and biomass from individual cell properties measured by flow cytometry. *Deep Sea Research, Part II, 48*, 1231–1261.

Shankar, D., & Shetye, S. R. (1997). On the dynamics of the Lakshadweep high and low in the southeastern Arabian Sea. *Journal of Geophysical Research, 102*, 12551–12562.

Shankar, D., Vinayachandran, P. N., & Unnikrishnan, A. S. (2002). The monsoon currents in the North Indian Ocean. *Progress in Oceanography, 52*, 63–120.

Shetye, S. R., Gouveia, A. D., Shenoi, S. S. C., Sundar, D., Michael, G. S., Almeida, A. M., & Santanam, K. (1990). Hydrography and circulation off the west coast of India during the southwest monsoon 1987. *Journal of Marine Research, 48*, 359–378.

Shetye, S. R., Gouveia, A. D., Shenoi, S. S., Sundar, D., Michael, G. S., & Nampoothiri, G. (1993). The western boundary current of the seasonal subtropical gyre in the Bay of Bengal. *Journal of Geophysical Research, Oceans, 98*, 945–954.

Shetye, S., Shenoi, S., Gouveia, A., Michael, G., Sundar, D., & Nampoothiri, G. (1991). Wind-driven coastal upwelling along the western boundary of the Bay of Bengal during the southwest monsoon. *Continental Shelf Research, 11*(11), 1397–1408.

Smith, S. L. (2001). Understanding the Arabian Sea: Reflections on the 1994-1996 Arabian Sea expedition. *Deep Sea Research, Part II, 48*(6–7), 1385–1402.

Smith, S. L., & Madhapratap, M. (2005). Mesozooplankton of the Arabian Sea: Patterns influenced by seasons, upwelling and oxygen concentrations. *Progress in Oceanography, 65*(2–4), 214–239.

Smitha, A., Joseph, K. A., Jayaram, C., & Balchand, A. N. (2014). Upwelling in the southeastern Arabian Sea as evidenced by Ekman mass transport using wind observatoins from OCEANSAT-II Scatterometer. *Indian Journal of Geo-Marine Sciences, 43*(1), 111–116.

Sonnekus, M. J., Bornman, T. G., & Campbell, E. E. (2017). Phytoplankton and nutrient dynamics of six South West Indian Ocean seamounts. *Deep Sea Research Part II: Topical Studies in Oceanography, 136*, 59–72.

Sorokin, Y. I., Kopylov, A., & Mamaeva, N. (1985). Abundance and dynamics of microplankton in the central tropical Indian Ocean. *Marine Ecology Progress Series. Oldendorf, 24*(1), 27–41.

Sprintall, J., Chong, J., Syamsudin, F., Morawitz, W., Hautala, S., Bray, N., & Wijffels, S. (1999). Dynamics of the South Java current in the indo-Australian Basin. *Geophysical Research Letters, 26*(16), 2493–2496.

Srokosz, M. A., & Quartly, G. D. (2013). The Madagascar bloom: A serendipitous study. *Journal of Geophysical Research, 118*(1), 14–25.

Srokosz, M. A., Quartly, G. D., & Buck, J. J. (2004). A possible plankton wave in the Indian Ocean. *Geophysical Research Letters, 31*(13).

Srokosz, M., Robinson, J., McGrain, H., Popova, E., & Yool, A. (2015). Could the Madagascar bloom be fertilized by Madagascan iron? *Journal of Geophysical Research: Oceans*, *120*(8), 5790–5803.

Stramma, L., & Lutjeharms, J. R. E. (1997). The flow field of the subtropical gyre of the South Indian Ocean. *Journal of Geophysical Research, Oceans*, *102*(C3), 5513–5530.

Strutton, P. G., Coles, V. J., Hood, R. R., Matear, R. J., McPhaden, M. J., & Phillips, H. E. (2015). Biogeochemical variability in the central equatorial Indian Ocean during the monsoon transition. *Biogeosciences*, *12*(8), 2367–2382.

Strzelecki, J., Koslow, J., & Waite, A. (2007). Comparison of mesozooplankton communities from a pair of warm-and cold-core eddies off the coast of Western Australia. *Deep Sea Research Part II: Topical Studies in Oceanography*, *54*(8–10), 1103–1112.

Subrahmanyam, B., & Robinson, I. S. (2000). Sea surface height variability in the Indian Ocean from TOPEX/POSEIDON altimetry and model simulations. *Marine Geodesy*, *23*(3), 167–195.

Susanto, R. D., Gordon, A. L., & Zheng, Q. (2001). Upwelling along the coasts of Java and Sumatra and its relation to ENSO. *Geophysical Research Letters*, *28*(8), 1599–1602.

Sutton, A. L., & Beckley, L. E. (2017). Species richness, taxonomic distinctness and environmental influences on euphausiid zoogeography in the Indian Ocean. *Diversity*, *9*(2), 23.

Tagliabue, A., Mtshali, T., Aumont, O., Bowie, A. R., Klunder, M. B., Roychoudhury, A. N., & Swart, S. (2012). A global compilation of dissolved iron measurements: Focus on distributions and processes in the Southern Ocean. *Biogeosciences*, *9*, 2333–2349. https://doi.org/10.5194/bg-9-2333-2012.

Talley, L. D., & Sprintall, J. (2005). Deep expression fo the Indonesian Throughflow: Indonesian intermediate water in the south equatorial current. *Journal of Geophysical Research, Oceans*, *110*(C10009). https://doi.org/10.1029/2004JC002826.

Tarran, G. A., Burkill, P. H., Edwards, E. S., & Woodward, E. M. S. (1999). Phytoplankton community structure in the Arabian Sea during and after the SW monsoon, 1994. *Deep Sea Research, Part II*, *46*, 655–676.

Thangaradjou, T., Sarangi, R., Shanthi, R., Poornima, D., Raja, K., Saravanakumar, A., & Balasubramanian, S. (2014). Changes in nutrients ratio along the central Bay of Bengal coast and its influence on chlorophyll distribution. *Journal of Environmental Biology*, *35*(3), 467.

Thompson, P. A., Wild-Allen, K., Lourey, M., Rousseaux, C., Waite, A. M., Feng, M., & Beckley, L. E. (2011). Nutrients in an oligotrophic boundary current: Evidence of a new role for the Leeuwin current. *Progress in Oceanography*, *91*, 345–359.

Thushara, V., Vinayachandran, P. N. M., Matthews, A. J., Webber, B. G., & Queste, B. Y. (2019). Vertical distribution of chlorophyll in dynamically distinct regions of the southern Bay of Bengal. *Biogeosciences*, *16*(7), 1447–1468.

Tindale, N. W., & Pease, P. P. (1999). Aerosols over the Arabian Sea: Atmospheric transport pathways and concentrations of dust and sea salt. *Deep Sea Research Part II: Topical Studies in Oceanography*, *46*(8–9), 1577–1595.

Tranter, D., & Kerr, J. (1969). Seasonal variations in the Indian Ocean along 110 EV zooplankton biomass. *Marine and Freshwater Research*, *20*(1), 77–84.

Tranter, D. J., & Kerr, J. D. (1977). Further studies of plankton ecosystems in the eastern Indian Ocean. III. Numerical abundance and biomass. *Australian Journal of Marine and Freshwater Research*, *28*, 557–583.

Twining, B. S., Rauschenberg, S., Baer, S. E., Lomas, M. W., Martiny, A. C., & Antipova, O. (2019). A nutrient limitation mosaic in the eastern tropical Indian Ocean. *Deep Sea Research Part II: Topical Studies in Oceanography*, *166*, 125–140.

Uz, B. M. (2007). What causes the sporadic phytoplankton bloom southeast of Madagascar? *Journal of Geophysical Research*, *112*(C9). https://doi.org/10.1029/2006JC003685.

Valencia, B., Landry, M. R., Décima, M., & Hannides, C. C. (2016). Environmental drivers of mesozooplankton biomass variability in the North Pacific subtropical gyre. *Journal of Geophysical Research: Biogeosciences*, *121*(12), 3131–3143.

Van Sebille, E., Sprintall, J., Schwarzkopf, F. U., Sen Gupta, A., Santoso, A., England, M. H., Biastoch, A., & Böning, C. W. (2014). Pacific-to-Indian Ocean connectivity: Tasman leakage, Indonesian Throughflow, and the role of ENSO. *Journal of Geophysical Research: Oceans*, *119*(2), 1365–1382.

Venkataramana, V., Anilkumar, N., Naik, R., Mishra, R., & Sabu, P. (2019). Temperature and phytoplankton size class biomass drives the zooplankton food web dynamics in the Indian Ocean sector of the Southern Ocean. *Polar Biology*, *42*(4), 823–829.

Venkataramana, V., Tripathy, S. C., & Anilkumar, N. P. (2017). The occurrence of blue-pigmented Pontella valida Dana, 1852 (Copepoda: Calanoida: Pontellidae) in the equatorial Indian Ocean. *The Journal of Crustacean Biology*, *37*(4), 512–515.

Verheye, H., Hutchings, L., Huggett, J., Painting, S., Carter, R., & Peterson, W. (1994). Community structure, distribution and trophic ecology of zooplankton on the Agulhas rank with special reference to copepods. *South African Journal of Science*, *90*(3), 154–166.

Vialard, J., Duvel, J. P., McPhaden, M., Bouruet-Aubertot, P., Ward, B., Key, E., Bourras, D., Weller, R., Minnett, P., Weill, A., Cassou, C., Eymard, L., Fristedt, T., Basdevant, C., Dandoneau, Y., Duteil, O., Izumo, T., Montégut, C.d. B., Masson, S., ... Kennan, S. (2009). Cirene: Air-sea interactions in the Seychelles-Chagos thermocline ridge region. *Bulletin of the American Meteorological Society*, *90*(1), 45–61.

Vidya, P., & Kumar, S. P. (2013). Role of mesoscale eddies on the variability of biogenic flux in the northern and central Bay of Bengal. *Journal of Geophysical Research: Oceans*, *118*(10), 5760–5771.

Vinayachandran, P. N. (2004). Biological response of the sea around Sri Lanka to summer monsoon. *Geophysical Research Letters*, *31*(1).

Vinayachandran, P. N. (2009). Impact of physical processes on chlorophyll distribution in the Bay of Bengal. In J. D. Wiggert, R. R. Hood, S. W. A. Naqvi, K. H. Brink, & S. L. Smith (Eds.), *Vol. 185*. *Indian Ocean biogeochemical processes and ecological variability*. Washington D.C.: American Geophysical Union.

Vinayachandran, P. N. M., Masumoto, Y., Roberts, M. J., Huggett, J. A., Halo, I., Chatterjee, A., Amol, P., Gupta, G. V., Singh, A., & Mukherjee, A. (2021). Reviews and syntheses: Physical and biogeochemical processes associated with upwelling in the Indian Ocean. *Biogeosciences*, *18*(22), 5967–6029.

Vinayachandran, P., & Mathew, S. (2003). Phytoplankton bloom in the Bay of Bengal during the northeast monsoon and its intensification by cyclones. *Geophysical Research Letters, 30*(11).

Vinayachandran, P. N., McCreary, J. P., Hood, R. R., & Kohler, K. E. (2005). A numerical investigation of phytoplankton blooms in the Bay of Bengal during the northeast monsoon. *Journal of Geophysical Research, 110*(C12001). https://doi.org/10.1029/2005JC002966.

Vinayachandran, P. N., Murty, V. S. N., & Ramesh Babu, V. (2002). Observations of barrier layer formation in the Bay of Bengal during summer monsoon. *Journal of Geophysical Research, 107*.

Vinayachandran, P., & Yamagata, T. (1998). Monsoon response of the sea around Sri Lanka: Generation of thermal domesand anticyclonic vortices. *Journal of Physical Oceanography, 28*(10), 1946–1960.

Vu, H. T. D., & Sohrin, Y. (2013). Diverse stoichiometry of dissolved trace metals in the Indian Ocean. *Scientific Reports, 3*.

Waite, A. M., Beckley, L. E., Guidi, L., Landrum, J., Holliday, D., Montoya, J., Paterson, H., Feng, M., Thompson, P. A., & Raes, E. J. (2015). Cross-shelf transport, oxygen depletion and nitrate release within a forming mesoscale eddy in the eastern Indian Ocean. *Limnology and Oceanography, 61*(1), 103–121.

Waite, A. M., Muhling, B. A., Holl, C. M., Beckley, L. E., Montoya, J. P., Strzelecki, J., ... Pesant, S. (2007a). Foodweb structure in two counter-rotating eddies based on delta ^{15}N and delta ^{13}C isotopic analyses. *Deep Sea Research, 54*, 1055–1075.

Waite, A. M., Raes, E., Beckley, L. E., Thompson, P. A., Griffin, D., Saunders, M., Säwström, C., O'Rorke, R., Wang, M., & Landrum, J. P. (2019). Production and ecosystem structure in cold-core vs. warm-core eddies: Implications for the zooplankton isoscape and rock lobster larvae. *Limnology and Oceanography, 64*(6), 2405–2423.

Waite, A. M., Rossi, V., Roughan, M., Tilbrook, B., Thompson, P. A., Feng, M., Wyatt, A. S., & Raes, E. J. (2013). Formation and maintenance of high-nitrate, low pH layers in the eastern Indian Ocean and the role of nitrogen fixation. *Biogeosciences, 10*(8), 5691–5702.

Waite, A. M., Thompson, P. A., Pesant, S., Feng, M., Beckley, L. E., Domingues, C. M., ... Twomey, L. (2007b). The Leeuwin current and its eddies: An introductory overview. *Deep Sea Research, Part II, 54*, 789–796.

Wang, Y., & McPhaden, M. J. (2017). Seasonal cycle of cross-equatorial flow in the central Indian Ocean. *Journal of Geophysical Research: Oceans, 122*(5), 3817–3827.

Wang, M., O'Rorke, R., Waite, A., Beckley, L., Thompson, P., & Jeffs, A. (2014). Fatty acid profiles of phyllosoma larvae of western rock lobster (Panulirus cygnus) in cyclonic and anticyclonic eddies of the Leeuwin current off Western Australia. *Progress in Oceanography, 122*, 153–162.

Ward, B., Devol, A., Rich, J., Chang, B., Bulow, S., Naik, H., Pratihary, A., & Jayakumar, A. (2009). Denitrification as the dominant nitrogen loss process in the Arabian Sea. *Nature, 461*(7260), 78–81.

Wiggert, J. D., Hood, R. R., Banse, K., & Kindle, J. C. (2005). Monsoon-driven biogeochemical processes in the Arabian Sea. *Progress in Oceanography, 65*(2–4), 176–213.

Wiggert, J. D., Jones, B. H., Dickey, T. D., Weller, R. A., Brink, K. H., Marra, J., & Codispoti, L. A. (2000). The northeast monsoon's impact on mixing, phytoplankton biomass and nutrient cycling in the Arabian Sea. *Deep-Sea Research Part II, 47*, 1353–1385.

Wiggert, J. D., Murtugudde, R. G., & Christian, J. R. (2006). Annual ecosystem variability in the tropical Indian Ocean: Results of a coupled bio-physical ocean general circulation model. *Deep Sea Research, Part II, 53*, 644–676.

Wiggert, J. D., Vialard, J., & Behrenfeld, M. (2009). Basinwide modification of dynamical and biogeochemical processes by the positive phase of the Indian Ocean dipole during the SeaWiFS era. In J. D. Wiggert, R. R. Hood, S. W. A. Naqvi, S. L. Smith, & K. H. Brink (Eds.), *Indian Ocean biogeochemical processes and ecological variability* (pp. 385–407). Washington, D. C: American Geophysical Union.

Wijesekera, H. W., Shroyer, E., Tandon, A., Ravichandran, M., Sengupta, D., Jinadasa, S., Fernando, H. J., Agrawal, N., Arulananthan, K., & Bhat, G. (2016). ASIRI: An ocean–atmosphere initiative for Bay of Bengal. *Bulletin of the American Meteorological Society, 97*(10), 1859–1884.

Wishner, K. F., Gelfman, C., Gowing, M. M., Outram, D. M., Rapien, M., & Williams, R. L. (2008). Vertical zonation and distributions of calanoid copepods through the lower oxycline of the Arabian Sea oxygen minimum zone. *Progress in Oceanography, 78*(2), 163–191.

Wishner, K. F., Gowing, M. M., & Gelfman, C. (1998). Mesozooplankton biomass in the upper 1000 m in the Arabian Sea: Overall seasonal and geographic patterns, and relationship to oxygen gradients. *Deep Sea Research Part II: Topical Studies in Oceanography, 45*(10–11), 2405–2432.

Xie, S. P., Annamalai, H., Schott, F. A., & McCreary, J. P. (2002). Structure and mechanisms of South Indian Ocean climate variability. *Journal of Climate, 15*, 874–878.

Yokoi, T., Tozuka, T., & Yamagata, T. (2008). Seasonal variation of the Seychelles Dome. *Journal of Climate, 21*(15), 3740–3754.

Yu, W., Masumoto, Y., Hood, R. R., D'Adamo, N., McPhaden, M. J., Adi, R., Tisiana, R., Kuswardani, D., Feng, M., Ivey, G., Lee, T., Meyers, G., Ueki, I., Landry, M., Ji, R., Davis, C., & Pranowo, W. (2015). *The eastern Indian Ocean upwelling research initiative science plan and implementation strategy*. First Institution of Oceanography.

Chapter 14

Air-sea exchange and its impacts on biogeochemistry in the Indian Ocean[*]

Hermann W. Bange[a,*], Damian L. Arévalo-Martínez[a], Srinivas Bikkina[b,c], Christa A. Marandino[a], Manmohan Sarin[d], Susann Tegtmeier[e], and Vinu Valsala[f]

[a]*GEOMAR Helmholtz Centre for Ocean Research Kiel, Kiel, Germany,* [b]*Chubu University, Kasugai, Japan,* [c]*National Institute of Oceanography, Dona Paula, India,* [d]*Physical Research Laboratory, Ahmedabad, India,* [e]*University of Saskatchewan, Saskatoon, SK, Canada,* [f]*Indian Institute of Tropical Meteorology, Pune, India*

1 Introduction

The interactions between the atmosphere and the ocean are driven by the exchange of momentum, energy (Phillips et al., 2021), and material (i.e., gases and particles) across the ocean/atmosphere interface (Liss & Johnson, 2014). The ocean is both an important source and sink of atmospheric trace gases and aerosols, which, in turn, play significant roles in the climate and chemistry of the Earth's atmosphere. Moreover, the surface ocean can act as a sink for atmospheric pollutants.

Carbon dioxide (CO_2), methane (CH_4), and nitrous oxide (N_2O) are long-lived atmospheric trace gases. They act as strong greenhouse gases in the troposphere (IPCC, 2021). Additionally, CH_4 and N_2O are involved in the depletion of ozone in the stratosphere (WMO, 2018). On the global scale, the ocean is a major sink of atmospheric CO_2, a minor source of atmospheric CH_4 (Saunois et al., 2020; Weber et al., 2019), and a major source of atmospheric N_2O (Tian et al., 2020a; Yang et al., 2020). About 25% of the atmospheric CO_2 emitted due to anthropogenic activities since the beginning of industrialization is taken up by the global ocean (Friedlingstein et al., 2020). The increasing dissolved CO_2 concentrations in seawater, in turn, lead to increasing hydrogen ion (H^+) concentrations and thus decreasing ocean pH (i.e., ocean acidification). The excess H^+ easily combines with the natural carbonate ions (CO_3^{2-}) in seawater to form more bicarbonate and reduce the carbonate concentration in seawater, which is important for shell formation (pH is also discussed in Chapter 12 (Hood et al., 2024b)).

Dimethylsulfide (DMS), isoprene, halocarbons (e.g., bromoform, dibromomethane, methyliodide), and carbon monoxide (CO) are short-lived atmospheric trace gases. They play important roles in the oxidative capacity of the atmosphere (CO, isoprene), are precursors of aerosols (e.g., DMS, isoprene), and are involved in atmospheric ozone cycles (halocarbons). All these gases are released from the ocean to the atmosphere in significant amounts (Liss et al., 2014).

For many atmospheric pollutants (e.g., nitrogen oxides, NO_x, sulfur dioxide, SO_2, ozone, O_3, and mercury, Hg) as well as aerosols, deposition to the sea surface is a major atmospheric loss pathway. The deposition, and thus the impact of the pollutants and aerosols on oceanic biogeochemical processes in the surface ocean, depends strongly on the proximity of their sources and their atmospheric transport patterns and lifetimes. In the case of aerosols, their elemental composition determines their effects on biogeochemical processes in the ocean, such as counteracting the nutrient limitation of biological productivity in the nutrient-poor (oligotrophic) open ocean (Kanakidou et al., 2018). The Indian Ocean is known to experience periods of strong atmospheric pollution and plumes of aerosols due to the large-scale monsoon circulation in combination with anthropogenic emissions from the surrounding countries (Lelieveld et al., 2001).

Here, we summarize the knowledge about the air-sea exchange of selected long- and short-lived trace gases as well as the deposition of atmospheric pollutants and aerosols in the Indian Ocean.

---

[*]This book has a companion website hosting complementary materials. Visit this URL to access it: https://www.elsevier.com/books-and-journals/book-companion/9780128226988.
*Current affiliation: Radboud University, Nijmegen, The Netherlands.

2 Greenhouse gases

2.1 Carbon dioxide

The air-sea exchange of CO_2 in the Arabian Sea is characterized by a perennial source (i.e., release) of CO_2 to the atmosphere due to the prominence of the coastal upwelling of CO_2-enriched subsurface waters along the coasts of the Arabian Peninsula and Somalia (Chakraborty et al., 2021; de Verneil et al., 2022) (see also Chapter 12 (Hood et al., 2024b)). This is also a venue for enhanced biological productivity. The combined role of biological and solubility pumps governs the CO_2 fluxes of the western Arabian Sea. The export production estimated from the satellite-derived net primary production data and export fluxes calculated with a model across the euphotic zone range from 94 to $123 \pm 25\,g\,C\,m^{-2}\,yr^{-1}$ in the western Arabian Sea (Sreeush et al., 2018). The biological pump and solubility pump were estimated to be 127 ± 24 and $13 \pm 3\,g\,C\,m^{-2}\,yr^{-1}$, respectively, in the western Arabian Sea (Sreeush et al., 2018). This net source of atmospheric CO_2 drives the basin-scale seasonal mean CO_2 fluxes of the Indian Ocean, and, on a regional scale, it also affects the acidification of the western Arabian Sea (Sarma, 2002; Sreeush et al., 2019). In contrast, the Bay of Bengal appears to be a small annual sink of atmospheric CO_2, albeit considerable uncertainty in the estimates exists due to sparse observations (Chakraborty et al., 2018; Sreeush et al., 2020), which does not allow to quantify the significance of the CO_2 sink or source (see also Chapter 12 (Hood et al., 2024b)). The oligotrophic gyre of the subtropical southern Indian Ocean is a sink of atmospheric CO_2 due to the dominance of the solubility pump and shows a marked decadal variability. Between 30°S and 45°S, both the solubility and the biological pumps contribute equally to the CO_2 sink (Valsala et al., 2012). Fig. 1 shows the spatial distribution of the seasonal fluxes of CO_2 in the Indian Ocean, and Table 1 presents estimates for the net annual CO_2 fluxes across the basin. Overall, the tropical Indian Ocean stores 14–22 Gt (10^9 t) of anthropogenic carbon, accounting for up to 21% of the global ocean carbon inventory, which is about half of the CO_2 inventory of the Atlantic (38%) or Pacific (41%) Oceans (Hall et al., 2004; Sabine et al., 2004).

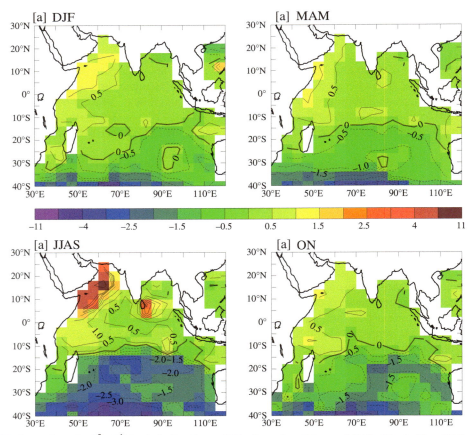

FIG. 1 Seasonal CO_2 fluxes (in mol m^{-2} yr^{-1}) in the Indian Ocean. DJF = December–February, MAM = March–May, JJAS = June–September and ON = October–November (based on Takahashi et al., 2009). Negative fluxes indicate uptake of CO_2 by the ocean.

TABLE 1 Estimates of the mean annual CO_2 flux in the Indian Ocean (north of 40°S).

Methodology	CO_2 flux (Gt C yr^{-1})	Reference
Variational assimilation of surface ocean pCO_2 in a biogeochemical model	-0.28 ± 0.18	Valsala and Maksyutov (2010)
Synthesis of top-down and bottom-up approaches	-0.37 ± 0.06	Sarma et al. (2013)
Climatology of surface ocean pCO_2 based on direct measurements	-0.24 ± 0.12	Takahashi et al. (2014)
Neural network-self organizing map-based methods	-0.17 ± 0.12	Landschützer et al. (2016)

Negative values indicate CO_2 uptake by the ocean. pCO_2 stands for the partial pressure of dissolved CO_2, C stands for carbon, and $Gt = 10^9 t = 10^{15} g$.

Apart from the regional variability described above, fluxes of CO_2 in the Indian Ocean exhibit significant temporal variability on intraseasonal, interannual, and interdecadal timescales. The western Arabian Sea features intense intraseasonal variability of the CO_2 fluxes due to the interactions with eddies such as the Great Whirl and the Southern Gyre (Valsala & Murtugudde, 2015), while the prominence of eddies in the Bay of Bengal affects the regional carbon biogeochemistry (Sarma et al., 2020). The oceanic pCO_2 variability in the intraseasonal band over the Somali region is also remarkably consistent with the other variables and is found to be driven by sea surface temperatures (SST), albeit with a counteracting but relatively minor influence from the dynamics of dissolved inorganic carbon (DIC). The 20-to-90-day band in pCO_2 accounts for about 40% of the monthly mean variability of the sea-to-air CO_2 fluxes of this region in boreal summer (Valsala & Murtugudde, 2015). The dominant interannual variability of CO_2 fluxes is located in the southeastern Indian Ocean and is related to the Indian Ocean Dipole mode (IOD; Valsala et al., 2020). The IOD leads to a substantial sea-to-air CO_2 flux variability in the southeastern tropical Indian Ocean over a broad region (70–105°E, 0–20°S), with a focus near the coast of Java/Sumatra due to the prevailing upwelling dynamics and associated westward propagating anomalies. The sea-to-air CO_2 fluxes, surface ocean partial pressure of CO_2 (pCO_2), the concentration of DIC, and ocean alkalinity range as much as $\pm 1.0\,mol\,m^{-2}\,yr^{-1}$, $\pm 20\,\mu atm$, $\pm 35\,\mu mol\,kg^{-1}$, and $\pm 22\,\mu mol\,kg^{-1}$ between 80°E and 105°E, and 0–10°S due to the IOD. The DIC and alkalinity are significant drivers of pCO_2 variability associated with IOD (Valsala et al., 2020). The counteracting influences of IOD and El Niño-Southern Oscillation (ENSO) affect the variability of the CO_2 fluxes of the western Arabian Sea and the region of south Sri Lanka (Valsala & Maksyutov, 2013).

The Indian Ocean is rapidly warming with profound implications for the upper ocean carbon cycle and biogeochemistry (Hermes et al., 2019; Valsala et al., 2021). This warming exacerbates the Arabian Sea acidification by 16% (Sreeush et al., 2019). However, large parts of the Indian Ocean are still significantly under-sampled for the major components of the carbonate system, and thus, estimates of the CO_2 fluxes are associated with a high degree of uncertainty (Olsen et al., 2020; see also Chapter 12 (Hood et al., 2024b)).

2.2 Methane

Since the early measurements of methane (CH_4) in the Red Sea/Indian Ocean (Lamontagne et al., 1973), the majority of studies were conducted in the northern Indian Ocean. There are only a few studies of CH_4 in the southern Indian Ocean (Farías et al., 2015) and the Indian Ocean sector of the Southern Ocean (Bui et al., 2018). CH_4 in coastal systems such as estuaries, river deltas and mangrove and seagrass ecosystems have been studied intensively along the west and east coasts of India, in coastal areas of the Andaman and Red Seas, and the Zambezi River delta (Barnes et al., 2006; Dutta et al., 2017; Garcias-Bonet & Duarte, 2017; Purvaja & Ramesh, 2001; Rao & Sarma, 2016; Teodoru et al., 2015). Seeps of CH_4 of geological origin have been identified in the Arabian Sea, the Bay of Bengal, the Nicobar-Simeulue Basin, and the Kerguelen Plateau (Dewangan et al., 2021; Siegert et al., 2011; Spain et al., 2020; Von Rad et al., 1996), and CH_4 measurements from brines in the Red Sea have been reported (Faber et al., 1998).

Dissolved CH_4 concentrations in coastal areas are significantly higher compared to the open ocean. This is illustrated by the remarkably high CH_4 concentrations measured in the mangrove-dominated Sundarbans region in the northern Bay of Bengal and the Zuari estuary on the west coast of India (max. $129\,nmol\,L^{-1}$ and $1022\,nmol\,L^{-1}$, respectively; Araujo et al., 2018; Biswas et al., 2007). In contrast, dissolved CH_4 surface concentrations in the open Indian Ocean are usually close to the equilibrium concentrations of $2-3\,nmol\,L^{-1}$ (Bange et al., 1998; Berner et al., 2003; Bui et al., 2018; Patra et al., 1998). Enhanced CH_4 concentrations in shallow coastal areas result from the interaction between in situ production from sedimentary methanogenesis, discharge of CH_4-enriched river water, and physical processes such as coastal upwelling, which occurs along the west and east coasts of the Arabian Sea.

While early estimates suggested a significant contribution of the Arabian Sea to the global CH_4 budget in the atmosphere (Owens et al., 1991), follow-up measurements showed that at the basin scale, the Arabian Sea is not a significant hotspot of CH_4 emissions (Bange et al., 1998). However, recent observations evidenced that, at a regional scale, shelf areas in the southeastern Arabian Seas can be significant sources of CH_4 (Patra et al., 1998; Sudheesh et al., 2020). Overall, atmospheric CH_4 over the Indian Ocean is thus mainly affected by land sources such as oil and gas production around the Arabian Gulf or emissions from rice paddies in Asia (Tegtmeier et al., 2020).

CH_4 depth profiles usually show enhanced concentrations associated with either permanent suboxic/anoxic oxygen minimum zones (OMZs) (e.g., northern Indian Ocean) or anoxic events at the shelf (Fig. 2). The CH_4 subsurface maxima observed in open Arabian Sea and Bay of Bengal have been attributed to in situ production by methanogenesis in sinking organic particles and/or advection from the continental margins (Berner et al., 2003; Jayakumar et al., 2001; Owens et al., 1991; Patra et al., 1998). CH_4 concentrations resulting from anoxic events at the shelf (up to $104\,nmol\,L^{-1}$) might be fuelled by sedimentary release (Shirodkar et al., 2018).

2.3 Nitrous oxide

The majority of nitrous oxide (N_2O) studies in the Indian Ocean have been conducted in the Arabian Sea (Bange, 2009). Additional measurements from the Bay of Bengal, Red Sea, southern Indian Ocean, and the Indian Ocean sector of the Southern Ocean are available (Bange et al., 2019; Butler et al., 1989; Farías et al., 2015; Hashimoto et al., 1998; Ma et al., 2020; Naqvi et al., 1994; Zhan et al., 2015). Moreover, N_2O measurements have been conducted in shelf regions, estuaries, and mangrove ecosystems along the west and east coasts of India, the Andaman Sea, and the Zambezi River delta (Barnes et al., 2006; Hershey et al., 2019; Naqvi et al., 2000; Rao et al., 2013; Teodoru et al., 2015).

N_2O surface concentrations in the open Indian Ocean are generally close to the equilibrium concentration of $5-17\,nmol\,L^{-1}$ (Bange et al., 1996; Lal & Patra, 1998; Ma et al., 2020; Zhan et al., 2015), whereas coastal areas feature

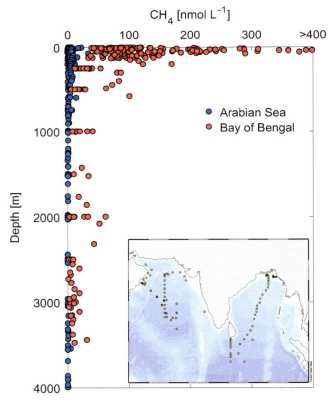

FIG. 2 Water column distribution of CH_4 concentrations in the Arabian Sea (*blue*) and Bay of Bengal (*red*). The insert shows sampling locations of the depth profiles as found in the Marine Methane and Nitrous Oxide Database (MEMENTO; access March 2021; Kock & Bange, 2015).

enhanced surface concentrations. For instance, a maximum N_2O concentration of $436\,nmol\,L^{-1}$ was measured during an anoxic event on a shelf off the west coast of India (Naqvi et al., 2000). Moreover, coastal upwelling of N_2O-enriched subsurface waters can result in surface concentrations up to $\sim 40\,nmol\,L^{-1}$ in the western Arabian Sea (De Wilde & Helder, 1997) and up to $\sim 165\,nmol\,L^{-1}$ in the eastern Arabian Sea (Sudheesh et al., 2016). N_2O concentrations of up to $46\,nmol\,L^{-1}$ have been measured in the open ocean upwelling region of the southwestern Indian Ocean (5–10°S; Ma et al. (2020)).

N_2O production generally increases with decreasing O_2 concentrations. Hence, N_2O depth distribution mirrors that of O_2 in large parts of the Indian Ocean (e.g., the southern Indian Ocean and the Bay of Bengal; see Fig. 3). Maximal N_2O concentrations result from nitrification in the low-O_2 subsurface layer where organic matter remineralization takes place. In contrast, N_2O depth profiles in the central Arabian Sea show a pronounced two-peak structure with enhanced N_2O concentrations (up to $\sim 110\,nmol\,L^{-1}$ (Law & Owens, 1990)) at the upper and lower boundaries of the OMZ (Fig. 3). The extreme N_2O accumulation in the Arabian Sea OMZ is caused by nitrification, denitrification, or coupling of both (Bange et al., 2005). In the core of the Arabian Sea OMZ, denitrification results in significant N_2O consumption (Bange et al., 2005). Recent model studies indicate that increased anthropogenic inputs of nitrogen (nitrogenous aerosol deposition) to the Arabian Sea might have augmented N_2O production by 5%–30% since the pre-industrial era (Suntharalingam et al., 2019).

Early studies (e.g., Law & Owens, 1990; Naqvi & Noronha, 1991) showed that coastal upwelling sites off the Arabian Peninsula, Somalia, and the west coast of India represent hotspots for N_2O emissions to the atmosphere. Currently, N_2O

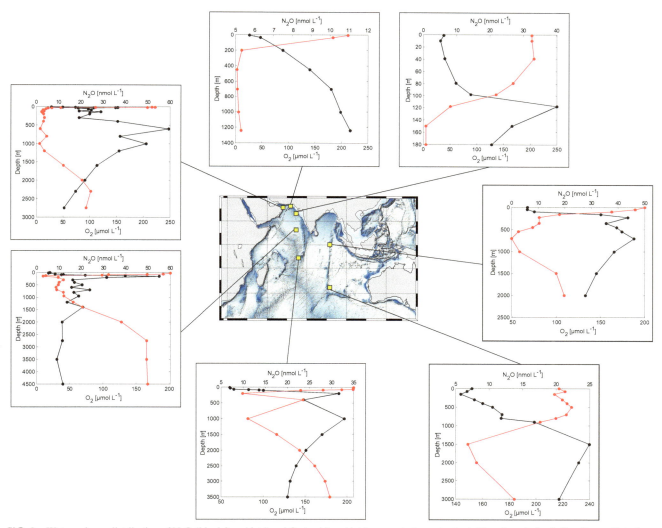

FIG. 3 Water column distribution of N_2O (*black lines/dots*) and O_2 (*red lines/dots*) concentrations at selected locations in the Indian Ocean. Data from MEMENTO database (access March 2021; Kock & Bange, 2015).

emissions from the Arabian Sea contribute significantly (2%–31%) to the global budget of atmospheric N_2O (Suntharalingam et al., 2019). N_2O emissions from the Bay of Bengal are considerably lower than those of the Arabian Sea and are, therefore, of minor importance for the global oceanic emissions. Emission estimates from the other parts of the Indian Ocean are still hampered by the scarce data coverage.

3 Reactive gases

3.1 Biogenic gases: Dimethlysulfide, isoprene, and very short-lived halocarbons

Marine DMS is the largest source of biogenic sulfur in the atmosphere. The precursor to DMS, dimethylsulfoniopropionate, is produced by algae and cleaved by marine microbes to form DMS. The CLAW hypothesis proposed a feedback loop between DMS production, emissions, and climate via aerosol and cloud formation (Charlson et al., 1987) and led to decades of DMS research (Quinn & Bates, 2011). Isoprene (2-methyl-1,3-butadiene) is a biogenic volatile organic compound and affects the cycles of atmospheric ozone, hydroxyl radicals, and secondary organic aerosols (Carlton et al., 2009). Most are emitted from terrestrial vegetation (e.g., Arneth et al., 2008), while the phytoplankton-derived oceanic source is much lower and uncertain (e.g., Booge et al., 2016). The oceanic halocarbons, such as bromoform ($CHBr_3$), dibromomethane (CH_2Br_2), and methyliodide (CH_3I), influence atmospheric halogen budgets, the NO_x and HO_x cycles, as well as ozone abundances (e.g., Read et al., 2008; Saiz-Lopez & von Glasow, 2012) that, in turn, impact inorganic iodine emissions (see Section 3.2) (Carpenter et al., 2021). Brominated halocarbons stem from macroalgae and phytoplankton production (e.g., Gschwend et al., 1985; Tokarczyk & Moore, 1994) as well as anthropogenic sources along the coastlines (e.g., Maas et al., 2020). Iodinated halocarbons also have biotic sources (e.g., phytoplankton and cyanobacteria) and show elevated oceanic abundances in the subtropical gyre regions indicative of photochemical production (Richter & Wallace, 2004).

Biogenic trace gases are undersampled in the Indian Ocean. Lana et al. (2011) is the most comprehensive monthly DMS concentration and flux climatology, but only 6271 non-uniformly spaced data points over 40 years are available in the Indian Ocean. Since 2010, significant improvements have been made to address the lack of data for DMS (NOAA-PMEL, 2020), isoprene (Booge et al., 2016, 2018; Rodríguez-Ros et al., 2020; Tripathi et al., 2020) and halocarbons (e.g., Fiehn et al., 2018). Interestingly, Galí et al. (2018) compared satellite-based proxies to the Lana et al. (2011) climatology and found that the climatology overestimates the DMS in the Indian Ocean, in agreement with direct measurements (Zavarsky et al., 2018b). Nonetheless, these new proxy-based estimations suggest an increase in oceanic DMS concentrations in the global oceans, including the Indian Ocean (Galí et al., 2018). Generally, the surface concentrations of all three gases are spatially and temporally highly variable, depending mostly on biological activity and by-products. Maximum DMS, isoprene, and halocarbon values are linked to enhanced phytoplanktonic activity during the boreal summer and in upwelling water (Tegtmeier et al., 2020), with the latter being linked to the monsoon seasonality. Isoprene production rates are affected further by light, SST, salinity, and nutrients (Booge et al., 2018). In the Arabian Sea, isoprene peaks in the OMZ (Tripathi et al., 2020).

The biogenic trace gas emissions exhibit clear seasonality. DMS shows the highest fluxes during the summer monsoon period in the Arabian Sea (Fig. 4) due to a combination of high seawater concentrations and strong, steady winds. The lowest fluxes are computed during the spring transition period, likely associated with low productivity and low wind speeds. DMS direct flux measurements using the eddy covariance technique were found to be 60% lower than the Lana climatology in the western tropical Indian Ocean (Zavarsky et al., 2018b). Nonetheless, the Arabian Sea appears to be a hotspot for DMS emissions during the summer monsoon (Edtbauer et al., 2020). Computed isoprene fluxes have the highest values during the summer monsoon over the entire Indian Ocean extent (Fig. 4; Booge et al., 2016, 2018), which is a hotspot of emissions to the atmosphere. Fluxes during the winter monsoon are high in the northern region of the Indian Ocean, especially in the Arabian Sea. Booge et al. (2016) used a top-down approach to calculate isoprene emissions and found the top-down fluxes to be one order of magnitude greater than those computed using the bottom-up estimate. One possible explanation could be that production in the surface microlayer is not considered with the bottom-up approach (Ciuraru et al., 2015). The most recent bottom-up $CHBr_3$ emission inventory suggests the tropical Indian Ocean as a productive source region (Fiehn et al., 2018; Fig. 4). Emissions peak along the coastlines due to macroalgae production and anthropogenic sources, agreeing with other emission estimates. High emissions along the coasts of Somalia and Oman due to coastal upwelling detected in biogeochemical modeling studies (Stemmler et al., 2015) are not captured here due to missing $CHBr_3$ measurements in this biogeochemical regime. The emissions show a pronounced seasonal cycle with a peak during the summer monsoon period due to higher wind speeds over the whole Indian Ocean during this time of year.

Biogenic trace gas emissions have a clear impact on the atmosphere. Over the Indian Ocean, DMS was found among the 10 most important OH sinks (Pfannerstill et al., 2019). Zavarsky et al. (2018a) correlated directly measured DMS fluxes, as

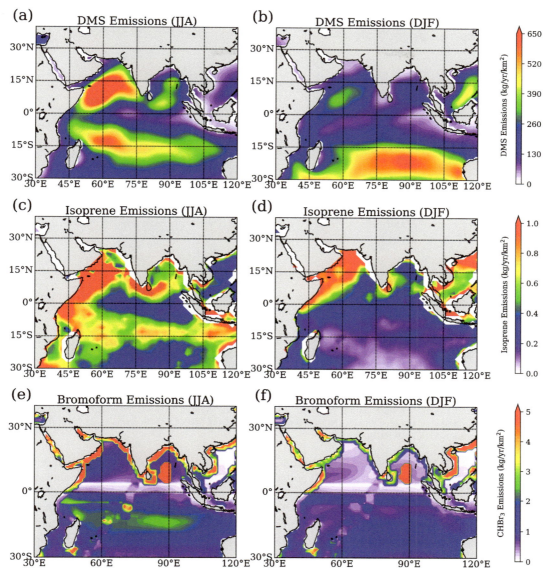

FIG. 4 Seasonal (monsoonal) emissions for DMS (a, b), isoprene (c, d), and bromoform (e, f). JJA stands for June–July–August (summer monsoon) and DJF stands for December–January–February (winter monsoon). The maps are based on data from Lana et al. (2011) for DMS, Booge et al. (2016) for isoprene and Fiehn et al. (2018) for bromoform. *Color bars* indicating trace gas emission quantity is shown to the right of the panels.

well as calculated isoprene and sea spray fluxes, with satellite-derived aerosol products over the region. The aerosol distribution more closely resembled the trace gas fluxes than the sea spray flux distributions, which is supported by tropical observations by Quinn et al. (2017). In addition, iodine chemistry by itself can be responsible for as much as 25% of the ozone destruction in the northern Indian Ocean marine boundary layer (Mahajan et al., 2021). Observations have confirmed the presence of iodine oxide in the marine atmosphere, although at low levels (e.g., Oetjen, 2009). These results suggest that, despite their low levels, the reactive species could contribute to local chemistry (Mahajan et al., 2010).

3.2 Pollutants: Carbon monoxide (CO), sulfur dioxide (SO_2), nitrogen oxides (NO_x), ozone (O_3), and mercury (Hg0)

Carbon monoxide (CO), nitrogen oxides ($NO_x = NO + NO_2$), and sulfur dioxide (SO_2) emissions from the continents bordering the Indian Ocean generally correspond to population densities and economic activities with emission centers in the Indo-Gangetic Plain, northern China and Java (Emissions Database for Global Atmospheric Research; Crippa et al., 2020).

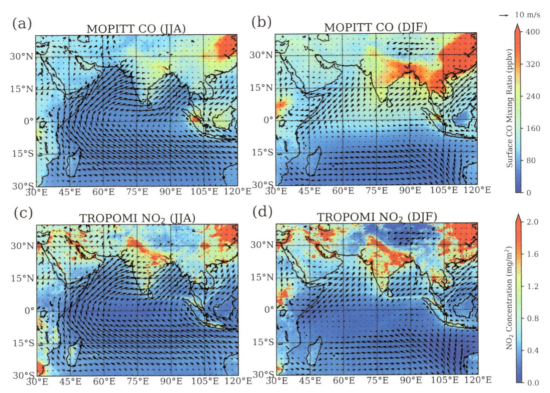

FIG. 5 Surface carbon monoxide (CO) mixing ratios from MOPITT (Deeter et al., 2019; colored shading) and surface wind from the ERA-Interim reanalyzes (Dee et al., 2011; *black arrows*) for (a) summer monsoon 2014–2018 (June–August, JJA) and (b) winter monsoon 2014/2015–2018/2019 (December–February, DJF). Tropospheric nitrogen dioxide (NO_2) column abundance from TROPOMI (Boersma et al., 2018) for (c) summer monsoon 2019 (JJA) and (d) winter monsoon 2018/2019 (DJF).

In addition, ship traffic leads to anthropogenic NO_x and SO_2 discharge directly over the open ocean (e.g., Franke et al., 2009). Over the last two decades, the emissions of all pollutants increased in almost all regions around the Indian Ocean, with the exception of decreasing Chinese emissions after 2013 as a consequence of clean air policies (Zheng et al., 2018).

The distributions of atmospheric CO, NO_x, and ozone (O_3) show pronounced differences between the landmass source regions and the Indian Ocean, with strong gradients over the coasts (Fig. 5). Satellite measurements and ship campaigns have revealed that the highest surface pollution occurs during winter over the Bay of Bengal due to advection from the Indo-Gangetic Plain and Southeast Asia (e.g., David et al., 2011) and the Arabian Sea coastal waters due to South Asian outflow (Girach et al., 2020). During the winter monsoon, the north-south pollution gradient continues over the open Indian Ocean, driven by southward transport along with chemical processing, dilution, and surface deposition. The contrast between polluted Northern Hemisphere and pristine Southern Hemisphere air results in a sharp gradient across the Intertropical Convergence Zone (ITCZ). During the summer monsoon, clean air dominates the atmospheric composition over the Indian Ocean, leading to a different chemical regime with low atmospheric pollutant levels.

Among the pollutants discussed here, SO_2 and O_3 are deposited from the atmosphere to the ocean surface. For SO_2, the few available field measurements suggest that air-sea transfer is controlled by the atmospheric side due to high effective solubility (Porter et al., 2020). For the Indian Ocean, this implies higher SO_2 deposition in the coastal regions. However, no current measurements are available to confirm this assumption. The deposition of O_3 at the sea surface plays an important role in the emissions of inorganic iodine compounds such as I_2 and HOI to the atmosphere (Carpenter et al., 2013). Given the higher ozone abundances over the northern Indian Ocean, the interplay of O_3 uptake and iodine emissions can be assumed to occur here but currently requires observational quantification.

For NO_2, the ocean acts as a sink via the reaction of N_2O_5 (which is formed from the reaction of NO_2 with NO_3 during the night) with seawater (McCaslin et al., 2019; Tian et al., 2021). In addition, there exists an oceanic source of NO via its photochemical production in the surface ocean (Tian et al., 2019, 2020a, 2020b; Zafiriou & McFarland, 1981). For the Indian Ocean, NO_x air-sea exchange and the respective contributions of the different pathways, as well as their impact on atmospheric O_3 and OH levels, are currently unknown. Based on atmospheric observations, the largest gaseous NO_x deposition can be expected over the polluted coastal regions and shipping lanes.

CO and mercury are different from the other pollutants as the ocean can also act as a significant source for these compounds. CO is produced in the surface ocean from organic material photochemistry and biological processes. Indian Ocean CO emission measurements are sparse, and net fluxes range from $\sim$0.1 to $\sim$1.4 $\times$ 10^6 g km^{-2} yr^{-1} (Conte et al., 2019 and references therein). Such values are similar to regional ship emissions but considerably smaller than continental emissions. In the western Indian Ocean, CO emissions peak in the upwelling band between 5°S and 15°S and decrease to zero outside this region.

The air-sea exchange of mercury is a key process in global mercury cycling and occurs in the form of gaseous elemental mercury (Hg0) or via the deposition of reactive mercury to the sea surface. Studies from the Pacific and Atlantic suggest elevated oceanic Hg0 emissions along the ITCZ due to enhanced precipitation in this region scavenging reactive mercury from the upper atmosphere and driving elevated Hg0 in seawater (Soerensen et al., 2014). Thus, Hg0 emissions in the Indian Ocean can be expected to peak in the ITCZ and other regions of enhanced precipitation.

Anthropogenic reactive pollutants discussed in Section 3 can impact oceanic processes in manifold ways. A well-known example is the air-sea exchange of Hg0 mediating the amount of oceanic mercury available for the formation of methylmercury (MeHg), the most bioaccumulative and toxic form of mercury in seafood. Deposition of pollutants, such as ozone and nitrogen, leads to chemical reactions at the sea surface and subsequent emissions of chemically active gases into the atmosphere. As these processes can be expected to occur specifically in polluted coastal regions of the Indian Ocean, they may lead to chemical feedback mechanisms. In addition, pollutants can affect oceanic production and biogeochemistry by modifying oceanic nutrient levels. The interactions of atmospheric pollution, biogeochemistry, trace gas cycling, and climate feedback in the Indian Ocean are severely understudied.

4 Aerosols

4.1 Sources

Aerosols, the suspension of tiny particulates and/or liquid droplets in the ambient air over the Indian Ocean and surrounding marginal seas, typically comprise chemical constituents derived from both natural and anthropogenic source emissions (Lelieveld et al., 2001; Mahowald, 2011; Patra et al., 2007; Ramanathan et al., 2001) (Fig. 6). While sea salt and mineral dust are the two major classes of naturally occurring aerosols over the Indian Ocean, anthropogenic sources such as biomass burning (agricultural waste and wood/dung-cake combustion) and fossil-fuel combustion (vehicular and coal-fired thermal power plant emissions) also significantly contribute to particulate load (Aswini et al., 2020; Bikkina et al., 2017a, 2019a, 2020c; Kumar et al., 2008a). Apart from the source emissions, the condensable semi-volatile organic vapors involved in gas-to-particle formation (e.g., non-sea-salt SO_4^{2-} aerosol formed by the oceanic DMS; see Section 3.1), an oxidation product of phytoplankton exudate also contribute and, hence, affect the composition of atmospheric aerosols during transport (Bikkina et al., 2020a, 2020b; Sarin et al., 2010). Because of the high surface-area-to-volume ratio, ambient

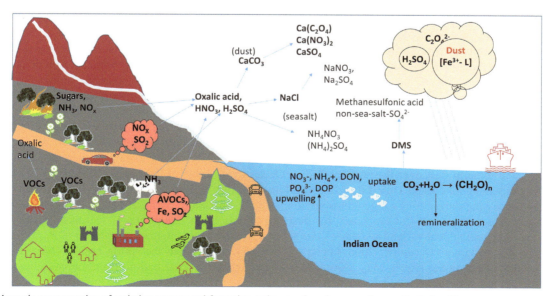

FIG. 6 Schematic representation of emission sources and formation pathways of marine aerosols over the Indian Ocean.

aerosols are conducive for various heterogeneous chemical reactions to occur during long-range transport (Bikkina et al., 2011; Rengarajan et al., 2007). Aerosol chemical composition is modified during transit from the source regions to the marine atmosphere (e.g., adsorption of SO_2 and NO_x on dust/sea-salt particles and their subsequent oxidation to form SO_4^{2-} and NO_3^- aerosols) despite their short residence time in the atmosphere ($\sim$ a week) (Bikkina et al., 2017b, 2019a).

4.2 Transport

The two marine basins of the northern Indian Ocean, Bay of Bengal on the eastern side and Arabian Sea on the western side receive compositionally different aerosol regimes during the early to late Northeast Monsoon, November–April (Bikkina et al., 2019b; Bosch et al., 2014; Budhavant et al., 2020; Dasari et al., 2019), a time frame coinciding with the development of brown haze in the lower troposphere of the Indo-Gangetic Plain. The Indo-Gangetic Plainis a highly fertile land and home to $\sim$40% of the Indian population. Every year, open burning of the post-harvest agricultural waste in the early winter (October–November: rice-paddy) and late spring (April–May: wheat crop) seasons contribute a variety of airborne pollutants (NH_4^+, NO_3^-, K^+, organic carbon, black carbon) over the Indo-Gangetic Plain (Bikkina et al., 2014, 2019b; Rajput et al., 2011; Rastogi et al., 2016). Wood burning and dung-cake combustion over the Indo-Gangetic Plain and other rural regions in India are the perennial sources of these pollutants (Mallik et al., 2013a; Venkataraman et al., 2005). In addition, the widespread SO_2 emissions from coal-fired thermal power plants and vehicular emissions of NO_x (NO + NO_2) (see Section 3.2) contribute to high concentrations of SO_4^{2-} and NO_3^- aerosols over the Indo-Gangetic Plain (Bikkina et al., 2019b; Mallik et al., 2013b). In winter, the stagnant weather and low temperatures caused by the solar insolation facilitate the entrapment of airborne pollutants (e.g., fine alluvium, black carbon, brown carbon, SO_4^{2-}, and NO_3^- aerosols) in the lower atmosphere (Bikkina et al., 2017b; Ram et al., 2010). As a result, a thick brown haze develops (Atmospheric Brown Clouds) over the Indian subcontinent during this period of the year and moves eastward along with NE monsoonal winds (Bikkina et al., 2019b; Jayaraman et al., 1998; Ramanathan et al., 2005).

The large-scale advection of pollution plumes from the Indo-Gangetic Plain and Southeast Asia during the winter season (December–February) influences the chemical composition of marine aerosols over the northern Indian Ocean (Agnihotri et al., 2011; Bikkina & Sarin, 2012, 2013c; Kumar et al., 2010; Sudheer & Sarin, 2008). In contrast, mineral dust from the desert regions in the Middle East (e.g., the Arab and Iranian Desert) and India contribute significantly to aerosol composition over the northern Indian Ocean during spring seasons (March–May; Bikkina & Sarin, 2012; Johansen & Hoffmann, 2004; Sarin et al., 1999; Tindale & Pease, 1999). Besides the source emissions, the extent of heterogeneous reactive uptake of atmospheric acidic species (H_2SO_4, HNO_3) on mineral aerosol surface profoundly influences the aerosol composition over both marine basins. However, this inference is based on the observed differences in the concentration ratios of crustal elements (Ca/Al and Mg/Al; Bikkina & Sarin, 2013b) and the percentage fractional solubility of aerosol iron (Fe_{ws} (%) = $Fe_{ws}/Fe_{Tot} \times 100$; Fe_{ws} = soluble iron; Fe_{Tot} = total aerosol Fe) over the Bay of Bengal and Arabian Sea between winter and spring seasons (Bikkina et al., 2012).

Analogous to these ratios, other proxies of aerosol water-soluble inorganic phosphorus (P_{Inorg} and P_{Inorg}/nss-Ca^{2+}, P_{Inorg}/nss-K^+, and P_{Inorg}/nss-SO_4^{2-}) in marine aerosols have exhibited marked differences between both these two basins (Bay of Bengal and Arabian Sea). Until recently, global modeling studies have assumed that the dust deposition from the desert regions mostly dictates the eolian supply of P_{Inorg} to the seawater because of the large spatial extent of dust sources in the Northern Hemisphere (Nenes et al., 2011; Patra et al., 2007). However, this contention is questionable because of the ongoing rise in anthropogenic emissions (biomass burning, fertilizers) from South and Southeast Asia and, hence, their relevance to air-to-sea deposition. Bikkina and Sarin (2015) observed significant linear relationships between the concentrations of P_{Inorg} and mineral dust, assessed based on the assumption that Al comprises 8.04% of upper continental crust (McLennan, 2001; source of dust) and using the Al concentration in marine aerosols sampled during winter and spring-intermonsoon cruises in the northern Indian Ocean. These preliminary observations reveal that there is a firm association between P_{Irnog} and mineral dust over the northern Indian Ocean irrespective of their origin (fine alluvium of the Indo-Gangetic Plain in winter vs. coarser Desert dust from Middle-East/Thar in spring-intermonsoon) and extent of influence from the acid-processing during long-range atmospheric transport (Bikkina & Sarin, 2012). In addition, the molar mass ratio of P_{Inorg} to the non-sea-salt-Ca^{2+} (a proxy for dust; Rastogi & Sarin, 2006) shows seasonal (winter and spring-intermonsoon) and geographical variability (Bay of Bengal and Arabian Sea). Higher molar ratios of P_{Inorg}/nss-Ca^{2+} over the Bay of Bengal in winter (December 2008–January 2009: 0.54 ± 0.54) than the spring cruises (March–April 2006: 0.09 ± 0.05) and the Arabian Sea (April–May 2006: 0.03 ± 0.02), suggest the contributions from sources other than mineral dust (fertilizers, biomass burning: P_{Anth}; Bikkina & Sarin, 2012) (Eqs. 1 and 2).

$$P_{Dust} = [Al]_{aerosol} \times [P/Al]_{dust} \qquad (1)$$

$$P_{Anth} = P_{Inorg} - P_{dust} \qquad (2)$$

Thus, P_{Anth}/P_{Inorg} over the Bay of Bengal during the winter cruise (0.8 ± 0.2) is comparable to those observed during spring-intermonsoon (0.8 ± 0.1) but somewhat higher than those from the Arabian Sea (April–May 2006: 0.3 ± 0.3; Bikkina & Sarin, 2012). Overall, all these preliminary investigations reveal that Bay of Bengal is relatively more influenced by anthropogenic source emissions than the Arabian Sea for the P_{Inorg}.

4.3 Air-sea deposition

Air-to-sea deposition (mostly via dry-fallout) of nutrients (and trace metals) to the Indian Ocean has immense potential to influence the biogeochemistry of the surface waters on a regional scale (Krishnamurthy et al., 2009; Mahowald, 2011; Schulz et al., 2012). This is because the frequency of rain events over the two marine basins of the northern Indian Ocean is rather sparse during the continental outflow (late November-early April; Bikkina & Sarin, 2013b; Bikkina et al., 2011). Therefore, most studies have focused only on the dry-deposition fluxes of aerosol-N, P, and Fe to the northern Indian Ocean (Bikkina & Sarin, 2013a, 2015). Nevertheless, the cyclonic storms and associated rain events over the surface waters of the Bay of Bengal during winter/spring seasons could lead to precipitation scavenging of continentally derived nutrients/trace metals and cause enhanced primary productivity (Kumar et al., 2004). Likewise, the dust deposition to the Arabian Sea during the spring-intermonsoon season (March–May) also triggers phytoplankton blooms (Banerjee & Prasanna Kumar, 2014; Guieu et al., 2019; Kessler et al., 2020). Due to the episodic and high spatial heterogeneity of rain events, it is rather difficult to quantify the wet-deposition fluxes of nutrients (Bange et al., 2000; Duce et al., 1991). To circumvent this, Singh et al. (2012) adopted the use of scavenging ratios for some reactive nitrogen species (NO_3^- and NH_4^+) to the northern Indian Ocean times the literature documented aerosol concentrations (Johansen & Hoffmann, 2004; Kumar et al., 2008b; Sarin et al., 1999). Subsequently, Bikkina and Sarin (2013a) have first compiled the concentrations of NO_3^-, NH_4^+, water-soluble organic nitrogen, inorganic phosphorus (P_{Inorg}) and Fe_{ws} in aerosols collected over the Bay of Bengal and the Arabian Sea from literature data and then estimated the total atmospheric deposition fluxes (both dry and wet) based on the approach of Duce et al. (1991).

The dry-deposition fluxes are estimated as the product of the concentration of nutrients in aerosols (C_{aero}) and dry-deposition velocity (V_{dry}) (Eq. 3).

$$\text{Dry-deposition flux}, f_{dry}\ (\mu mol\ m^{-2}\ d^{-1}) = C_{aero} \times V_{dry} \qquad (3)$$

Bikkina and Sarin (2013a) choose a V_{dry} value of $0.1\,cm\,s^{-1}$ for the chemical species existing mostly in fine mode (NH_4^+, water-soluble organic nitrogen, and Fe_{ws}) and originate mostly from the anthropogenic combustion sources/secondary formation processes. Likewise, Bikkina and Sarin (2013a) used a V_{dry} value of $2.0\,cm\,s^{-1}$ for those chemical species typically occurring in coarse mode (NO_3^- and P_{Inorg}) due to their association with mineral dust. A comparison of NH_4^+, water-soluble organic nitrogen, and Fe_{ws} concentrations in $PM_{2.5}$ samples with those in PM_{10} aerosols has clearly revealed their predominant fine mode occurrence (Bikkina et al., 2011). Likewise, NO_3^- and P_{Inorg} concentrations show higher loading in PM_{10} than $PM_{2.5}$, emphasizing their coarse mode association with mineral dust (inferred based on the prevailing strong correlations with non-sea-salt Ca^{2+}, which is a proxy for mineral dust (Bikkina & Sarin, 2012; Bikkina et al., 2011)). The wet-deposition fluxes are constrained by the precipitation rate (P, in $mm\,d^{-1}$; Eq. 4), scavenging ratio (S, the ratio of concentration of target chemical species in water to that in ambient air), C_{aero}, and the densities of water ($\rho_{water} = 1000\,kg\,m^{-3}$) and air ($\rho_{air} = 1.2\,kg\,m^{-3}$). S is a dimensionless quantity, usually expressed in the units of "kg of air per kg of water," (Eq. 5), and depends mostly on the particle size, chemical/physical form, rain droplet size, ambient temperature, and cloud type (Bange et al., 2000, 2005).

$$\text{Wet-deposition flux}, f_{wet}\ (\mu mol\ m^{-2}\ d^{-1}) = P \times S \times C_{aero} \times \rho_{air}^{-1} \times \rho_{water} \qquad (4)$$

$$\text{Scavenging ratio}, S\ (\text{dimensionless quantity}) = \frac{C_{rain} \times \rho_{air}}{C_{aerosol} \times \rho_{water}} \qquad (5)$$

The Tropical Rainfall Measuring Mission (TRMM) data for the last 10 years further suggest that Arabian Sea and Bay of Bengal receive $\sim 750\,mm\,yr^{-1}$ (i.e., $<150\,mm/month$) and $2000\,mm\,yr^{-1}$ ($<400\,mm/month$) of rainfall, respectively, during the continental outflow (for 5 months, December–April). The literature-based estimate of S values ranges between 200 and 2000 for the typically measured chemical composition of ambient aerosols (Singh et al., 2012). We adopt the previously used S values from Duce et al. (1991) and Bange et al. (2000) for estimating the wet deposition fluxes of NO_3^- (330) and NH_4^+ (200) to the Arabian Sea. Here, the relatively low S value for NH_4^+ in marine aerosols is because of its preferential existence in the fine mode as NH_4HSO_4 or $(NH_4)_2SO_4$ aerosols (Eqs. 6 and 7) (Bikkina et al., 2011). In contrast, higher S value for NO_3^- than NH_4^+ in marine aerosols is mainly due to the coarse mode association with either mineral dust (Eq. 8) or sea salts (Eq. 9) (Bikkina et al., 2011, 2020b; Kumar et al., 2012).

$$2NH_3 \text{ (l)} + H_2SO_4 \text{ (l)} \rightarrow (NH_4)_2SO_4 \text{ (s)} \tag{6}$$

$$NH_3 \text{ (l)} + H_2SO_4 \text{ (l)} \rightarrow NH_4HSO_4 \text{ (s)} \tag{7}$$

$$CaCO_3 \text{ (surrogate for dust)} + 2HNO_3 \text{ (l)} \rightarrow Ca(NO_3)_2 + CO_2 + H_2O \tag{8}$$

$$NaCl \text{ (surrogate for sea salt)} + HNO_3 \text{ (l)} \rightarrow NaNO_3 + HCl \tag{9}$$

The total atmospheric deposition of aerosol-N to the Arabian Sea (0.51–1.63 Tg yr^{-1}) and Bay of Bengal (0.06–2.63 Tg yr^{-1}) are estimated based on the concentration data from Bikkina et al. (2020b) and Bikkina et al. (2011), respectively. It is also an important point here that the real-time atmospheric dry-deposition fluxes of aerosol-soluble N over the Bay of Bengal for the year 2009 (10–853 mg m^{-2} yr^{-1}; Bikkina et al., 2011) are much lower than the projection made for the year 2000 (800–1200 mg m^{-2} yr^{-1}) using the state-of-the-art global chemical transport models (Duce et al., 2008).

4.4 Biogeochemical significance

Dust deposition to the surface waters of the Indian Ocean has a profound influence on the biogeochemical cycles of carbon and nitrogen through the phytoplankton primary production and nitrogen fixation (Fig. 6; Chase et al., 2011; Jickells et al., 2016; Okin et al., 2011; Schulz et al., 2012). Because of the regionally varying provenance of mineral aerosols in the continental outflow as well as inadvertent and disproportionate contributions from anthropogenic sources (e.g., biomass burning, coal fly ash, and industrial emissions), both the marine basins (Bay of Bengal and Arabian Sea) receive variable nutrient load (Bikkina & Sarin, 2013a; Kumar et al., 2008a). In addition, heterogeneous reactive uptake of acid species on mineral dust during atmospheric transport promotes the dissolution of aerosol-Fe and P_{Inorg} (Bikkina & Sarin, 2012; Bikkina et al., 2012). Though the dust load over the Arabian Sea and the Bay of Bengal are comparable during the continental outflows, there are many elemental proxies showing differences (e.g., Fe/Al, Ca/Al, Mg/Al, Fe-solubility, and P_{Dust}/P_{Inorg}) (Bikkina & Sarin, 2014; Bikkina et al., 2015, 2014). In contrast to persisting continental outflows in winter, the summertime tropical wildfires in Southeast Asia have a profound influence on the water-column primary productivity in the Indian Ocean. In particular, the wildfires in Southeast Asia (e.g., Sumatra Island in Indonesia) significantly contribute (i.e., almost by a factor of 3–8 during hazy days) to the airborne nutrients (Sundarambal et al., 2010). Accordingly, air-to-sea deposition of nutrients from the wildfires could trigger massive phytoplankton blooms in the surface waters of the eastern tropical Indian Ocean right after, for example, the famous 1997 Indonesian wildfires during the positive phase of IOD (Abram et al., 2003). Although lesser in magnitude, such an influence of atmospheric deposition of wildfires on the surface waters has been documented in subsequent modeling studies in the summertime from the 2006 positive phase of the IOD (Wiggert et al., 2009). Based on the coupled ocean-atmosphere models and satellite observations, Siswanto (2015) has argued that atmospheric deposition of nutrients from major wildfires in Southeast Asia significantly contribute to enhancing primary productivity even in the upwelling zones.

The residence time of trace metals (τ_{Me}; Eq. 10) serves as another useful parameter to understand the relevance of atmospheric input of trace metals on the biogeochemistry of the surface ocean (Bikkina & Sarin, 2013b).

$$\text{Residence time}, \tau_{Me} = \frac{[Me]_{seawater} \times Z_{mix}}{f_{dep-Me} \times Me_{solubility}(\%)} \tag{10}$$

Here, $[Me]_{seawater}$ is the concentration of dissolved trace metal in the surface waters, and Z_{mix} is the surface mixed layer depth. Likewise, f_{dep-Me} is the dry-deposition flux of desired trace metal at the air-sea interface, and $Me_{solubility}$ (%) is the fractional solubility of trace metal in marine aerosols. We thus constrained the τ_{Fe} in the northern Indian Ocean with respect to eolian input (Bay of Bengal: ∼27 days; Arabian Sea: ∼8.1 years) using the concentrations in seawater (Bay of Bengal: 0.44 ± 0.11 nM and Arabian Sea: 0.68 ± 0.03 nM; Chinni et al., 2019), the f_{dep-Me} (Bay of Bengal: 7.1 ± 6.9 μmol m^{-2} d^{-1} and Arabian Sea: 6.9 ± 4.1 μmol m^{-2} d^{-1}) and the $Fe_{solubility}$ data (Bay of Bengal: 6.9 ± 4.3% and Arabian Sea: 0.09 ± 0.10%; Bikkina & Sarin, 2013a). Despite having similar dry-deposition fluxes of Fe_{Tot} over the northern Indian Ocean, relatively short τ_{Fe} for the Bay of Bengal is a result of inorganic acid processing (Bikkina et al., 2012; Kumar et al., 2010) and/or organic ligand-promoted dissolution of aeolian dust (Bikkina et al., 2020c).

Diatoms or sometimes dinoflagellates are the most abundant phytoplankton species in the surface waters of the Bay of Bengal (80%–90%; Madhupratap et al., 2003) and Arabian Sea (∼86%; Sambrotto, 2001; Sawant & Madhupratap, 1996) during the winter season (see also Chapter 13 (Hood et al., 2024a)). However, the central Arabian Sea is characterized by the abundance of *Trichodesmium* Spp. (N-fixers that grow under oligotrophic conditions) during spring-intermonsoon and fall-inter-monsoon seasons (Ahmed et al., 2017; do Rosario Gomes et al., 2008, 2014; Gandhi et al., 2011; Prakash et al., 2017; Rixen et al., 2020) (see also Chapter 13 (Hood et al., 2024a)). Considering the prevalence of these phytoplankton assemblages and by assuming the Fe-limitation in the surface waters (see also Chapter 13 (Hood et al., 2024a)), we can, thus, constrain the potential C- and N-fixation rates in the northern Indian Ocean (Eqs. 11 and 12). The idea behind this simple calculation is to deduce a maximum supportable estimate of C- and N-fixation by the atmospheric deposition to the northern Indian Ocean.

$$C - \text{fixation}, C^{Fe}_{\text{diatoms}} = f_{\text{dep-Fe}} \times Fe - \text{solubility}(\%) \times \left(\frac{C}{Fe}\right)_{\text{diatoms}} \quad (11)$$

$$N - \text{fixation}, N^{Fe}_{\text{Trichodesmium Spp.}} = f_{\text{dep-Fe}} \times Fe - \text{solubility}(\%) \times \left(\frac{N}{Fe}\right)_{\text{Trichodesmium Spp.}} \quad (12)$$

Using the basin-wide estimates of Fe deposition (Bay of Bengal: 0.13 Tg yr^{-1}; Arabian Sea: 0.28 Tg yr^{-1}), Fe-solubility (Bay of Bengal: ~6.9%; AS: 0.09%) and the proposed ratio of C:Fe cellular quota for diatoms by Moore et al. (2001), Bikkina and Sarin (2013a) have estimated potential C-fixation rate of ~1.1 and 0.03 Pg-C yr^{-1}, respectively, in the surface waters. Here, the Fe$_{\text{dep}}$ rates are constrained by using the mean concentrations of total aerosol-Fe over the Bay of Bengal (460 ng m^{-3}) and Arabian Sea (444 ng m^{-3}) in the flux calculations for the continental outflow period (150 days; V$_{\text{dry}}$ = 0.01 m s^{-1}). These dry-deposition fluxes of Fe$_{\text{Tot}}$ were further integrated for ~2.2 × 10^{12} m^2 for the Bay of Bengal and 4.93 × 10^{12} m^2 for the Arabian Sea (Bange et al., 2000) to obtain the basin-wide deposition fluxes. Likewise, Bikkina and Sarin (2013a) also estimated the potential N-fixation rates according to Eq. (8), where a range of values (50–1050 g/g) were used as the cellular quota of N:Fe for the typical diazotrophs (Okin et al., 2011). This preliminary calculation based on the average concentration of total aerosol soluble iron yields a potential N-fixation rate of 0.01–0.26 Tg yr^{-1} over the Arabian Sea and 0.45–9.2 Tg yr^{-1} over the Bay of Bengal supported by the atmospheric input of aerosol soluble iron (Table 2). The estimated maximum N-fixation fuelled by the atmospheric deposition is comparable to that measured in situ by Ahmed et al. (2017) but appears to be lower than other direct measurements (Table 3). However, it is worth pointing out that the Fe$_{\text{dep}}$ and cellular quota-based C-fixation and N-fixation estimates cannot apply to the same system since Fe taken up by diatoms would not then be available to diazotrophs and diazotroph C-fixation has a dramatically different C:Fe ratio than diatoms.

5 Conclusions and outlook

In general, dissolved trace gas concentrations and the resulting air-sea fluxes show a pronounced temporal and spatial variability in the Indian Ocean which is caused by the variability of both physical processes and biological productivity. In the Arabian Sea, for example, the seasonally occurring coastal upwelling events result in temporarily high surface concentrations of trace gases. Trace gas concentrations in the (oligotrophic) open Indian Ocean are usually close to the equilibrium concentrations, whereas in the (eutrophic) coastal areas, such as estuaries, river deltas, mangrove and seagrass ecosystems, high (and in some cases extremely high) concentrations have been measured. Therefore, the emissions from upwelling and coastal areas contribute significantly to the overall emissions from the Indian Ocean. In a global context, trace gas emissions from the Indian Ocean are not exceptionally high or low, except for the Arabian Sea, a globally significant hotspot of N$_2$O, DMS, and isoprene emissions.

TABLE 2 Estimates of C- and N-fixation rates (C-fix., N-fix; based on Eqs. 11 and 12) in the northern Indian Ocean based on the atmospheric concentrations of Fe$_{\text{Tot}}$ and the associated fractional solubility.

Basin	Fe$_{\text{Tot}}$ (ng m^{-3})	Area (10^{12} m^2)	Fe$_{\text{dep}}$ (Tg yr^{-1})	Fe-sol. (%)	C-fix. (Pg. yr^{-1})	(N:Fe)$_{\text{Trichodesmium}}$	N-fix. (Tg yr^{-1})
Bay of Bengal	460	2.20	0.13	0.069	1.12	50–1050	0.45–9.42
Arabian Sea	444	4.93	0.28	0.001	0.03	50–1050	0.01–0.26

TABLE 3 Comparison of atmospheric soluble-Fe based estimates of nitrogen fixation rates (i.e., based on Eq. 12) by diazotrophs with the literature-based direct estimates.

Basin	N-fixation rates (μmol-N m^{-2} d^{-1})	Reference
Arabian Sea	1–29	Bikkina and Sarin (2013a, 2013b, 2013c)
Arabian Sea	25–35	Shiozaki et al. (2014)
Arabian Sea	<1739	Ahmed et al. (2017)
Arabian Sea	25	Capone et al. (1998)
Arabian Sea	2–540	Gandhi et al. (2011)
Arabian Sea	1108	Parab and Matondkar (2012)
Arabian Sea	1300–2500	Singh et al. (2019)
Bay of Bengal	98–2057	Bikkina and Sarin (2013a, 2013b, 2013c)
Bay of Bengal	0–<2088	Löscher et al. (2020)
Bay of Bengal	4–75	Saxena et al. (2020)

The Indian Ocean is a net sink for anthropogenic CO_2. However, the overall uptake of CO_2 (i.e., the storage capacity for CO_2 in the entire water column) in the Indian Ocean is low compared to the other major ocean basins because of its comparably small area and its special geographic conditions, resulting in the absence of deep-water formation areas in the northern Indian Ocean.

The coastal areas of the Indian Ocean act as a sink for atmospheric pollutants such as CO, NO_x and SO_2 which mainly originate from land sources. Ship traffic leads to additional emissions of NO_x and SO_2 directly over the open ocean and thus to an additional uptake in the surface layer of the open ocean.

Comparable to the atmospheric pollutants, the aerosols in the atmosphere over the Indian Ocean originate mainly from natural and anthropogenic sources on land. Their deposition to the surface waters of the Indian Ocean has a profound influence on the biogeochemical cycles of carbon and nitrogen as the input of nitrogen and Fe-containing aerosols stimulates primary production by phytoplankton and nitrogen fixation in the surface layer of the open Indian Ocean.

The Indian Ocean is under-sampled for trace gases, atmospheric pollutants, and aerosols in both space and time. Therefore, estimates of trace gas emissions and aerosol/pollutants depositions are associated with a high degree of uncertainty, and any quantitative assessment of their significance for biogeochemical processes remains uncertain at the moment.

The major environmental changes affecting the Indian Ocean system are warming, acidification, eutrophication, increasing atmospheric pollution, deoxygenation, and changing wind patterns. Each of the ongoing individual changes or a combination of them has a great potential to affect future trace gas fluxes and aerosol deposition in the Indian Ocean. To this end, joint measurement campaigns and the deployment of novel autonomous (mobile) measurement platforms and moored observatories, in combination with modeling efforts, will help to improve our knowledge about air-sea exchange and its impacts on biogeochemistry in the Indian Ocean.

6 Educational resources

- Ocean-Atmosphere Interactions of Gases and Particles (collection of various review chapters, edited by P.S. Liss and M.T. Johnson, Springer, 2014; open access): https://link.springer.com/book/10.1007%2F978-3-642-25643-1
- 2nd International Indian Ocean Expedition: https://iioe-2.incois.gov.in
- Surface Ocean – Lower Atmosphere Study: www.solas-int.org
- MarinE MethanE and NiTtrous Oxide database: https://memento.geomar.de
- Surface Ocean CO_2 ATtlas: www.socat.info
- Emission Database for Global Atmospheric Research: https://edgar.jrc.ec.europa.eu
- Global surface seawater dimethylsulfide (DMS) database: https://saga.pmel.noaa.gov/dms
- FluxEngine (open source atmosphere–ocean gas flux data processing toolbox): https://github.com/oceanflux-ghg/FluxEngine

Acknowledgments

SB and MS acknowledge the funding support received through the Geosphere-Biosphere Program of the Indian Space Research Organization (Bengaluru, India) for aerosol studies conducted over the Indian Ocean. This chapter is a contribution to the 2nd International Indian Ocean Expedition (IIOE-2) and the Indian Ocean Integrated Study of the Surface Ocean—Lower Atmosphere Study (SOLAS) programs.

Author contributions

HWB drafted the chapter outline and coordinated writing. VV wrote the section on CO_2, DLAM/HWB wrote the section on CH_4 and N_2O, CAM/ST wrote the section on reactive gases, and SB/MS wrote the section on aerosols. HWB wrote the Abstract, Introduction, Conclusions, and Outlook sections with contributions from the co-authors.

References

Abram, N. J., Gagan, M. K., McCulloch, M. T., Chappell, J., & Hantoro, W. S. (2003). Coral reef death during the 1997 Indian Ocean dipole linked to Indonesian wildfires. *Science, 301*(5635), 952–955.

Agnihotri, R., et al. (2011). Stable carbon and nitrogen isotopic composition of bulk aerosols over India and northern Indian Ocean. *Atmospheric Environment, 45*(17), 2828–2835.

Ahmed, A., et al. (2017). Nitrogen fixation rates in the eastern Arabian Sea. *Estuarine, Coastal and Shelf Science, 191*, 74–83.

Araujo, J., Naqvi, S. W. A., Naik, H., & Naik, R. (2018). Biogeochemistry of methane in a tropical monsoonal estuarine system along the west coast of India. *Estuarine, Coastal and Shelf Science, 207*, 435–443. https://doi.org/10.1016/j.ecss.2017.07.016.

Arneth, A., Monson, R. K., Schurgers, G., Niinemets, Ü., & Palmer, P. I. (2008). Why are estimates of global terrestrial isoprene emissions so similar (and why is this not so for monoterpenes)? *Atmospheric Chemistry and Physics, 8*, 4605–4620. https://doi.org/10.5194/acp-8-4605-2008.

Aswini, A. R., Hegde, P., Aryasree, S., Girach, I. A., & Nair, P. R. (2020). Continental outflow of anthropogenic aerosols over Arabian Sea and Indian Ocean during wintertime: ICARB-2018 campaign. *Science of the Total Environment, 712*, 135–214.

Banerjee, P., & Prasanna Kumar, S. (2014). Dust-induced episodic phytoplankton blooms in the Arabian Sea during winter monsoon. *Journal of Geophysical Research: Oceans, 119*(10), 7123–7138.

Bange, H. W. (2009). Nitrous oxide in the Indian Ocean. In J. D. Wiggert, R. R. Hood, S. W. A. Naqvi, K. H. Brink, & S. L. Smith (Eds.), *Indian Ocean biogeochemical processes and ecological variability* (pp. 205–216). Washington, DC: American Geophysical Union.

Bange, H. W., Kock, A., Pelz, N., Schmidt, M., Schütte, F., Walter, S., Jones, B., & Kürten, B. (2019). Nitrous oxide in the northern Gulf of Aqaba and the Central Red Sea. *Deep-Sea Research Part II, 166*, 90–103. https://doi.org/10.1016/j.dsr2.2019.06.015.

Bange, H. W., Naqvi, S. W. A., & Codispoti, L. A. (2005). The nitrogen cycle in the Arabian Sea. *Progress in Oceanography, 65*, 145–158.

Bange, H. W., Ramesh, R., Rapsomanikis, S., & Andreae, M. O. (1998). Methane in the surface waters of the Arabian Sea. *Geophysical Research Letters, 25*(19), 3547–3550.

Bange, H. W., Rapsomanikis, S., & Andreae, M. O. (1996). Nitrous oxide emissions from the Arabian Sea. *Geophysical Research Letters, 23*(22), 3175–3178.

Bange, H. W., et al. (2000). A revised nitrogen budget for the Arabian Sea. *Global Biogeochemical Cycles, 14*(4), 1283–1297.

Barnes, J., Ramesh, R., Purvaja, R., Nirmal Rajkumar, A., Senthil Kumar, B., Krithika, K., … Upstill-Goddard, R. (2006). Tidal dynamics and rainfall control N_2O and CH_4 emissions from a pristine mangrove creek. *Geophysical Research Letters, 33*, L15405. https://doi.org/10.1029/2006GL026829.

Berner, U., Poggenburg, J., Faber, E., Quadfasel, D., & Frische, A. (2003). Methane in the ocean waters of the Bay of Bengal: Its sources and exchange with the atmosphere. *Deep-Sea Research Part II, 50*, 925–950.

Bikkina, S., Haque, M. M., Sarin, M., & Kawamura, K. (2019a). Tracing the relative significance of primary versus secondary organic aerosols from biomass burning plumes over Coastal Ocean using sugar compounds and stable carbon isotopes. *ACS Earth and Space Chemistry, 3*, 1471–1484.

Bikkina, S., Kawamura, K., & Sarin, M. (2017a). Secondary organic aerosol formation over Coastal Ocean: Inferences from atmospheric water-soluble low molecular weight organic compounds. *Environmental Science & Technology, 51*(8), 4347–4357.

Bikkina, S., Kawamura, K., Sarin, M., & Tachibana, E. (2020c). 13C probing of ambient photo-Fenton reactions involving iron and oxalic acid: Implications for oceanic biogeochemistry. *ACS Earth and Space Chemistry, 4*, 964–976.

Bikkina, S., & Sarin, M. (2012). Atmospheric pathways of phosphorous to the Bay of Bengal: Contribution from anthropogenic sources and mineral dust. *Tellus B, 64*.

Bikkina, S., & Sarin, M. M. (2013a). Atmospheric deposition of N, P and Fe to the Northern Indian Ocean: Implications to C- and N-fixation. *Science of the Total Environment, 456–457*, 104–114.

Bikkina, S., & Sarin, M. M. (2013b). Atmospheric dry-deposition of mineral dust and anthropogenic trace metals to the Bay of Bengal. *Journal of Marine Systems, 126*, 56–68.

Bikkina, S., & Sarin, M. M. (2013c). Light absorbing organic aerosols (brown carbon) over the tropical Indian Ocean: Impact of biomass burning emissions. *Environmental Research Letters, 8*(4), 044042.

Bikkina, S., & Sarin, M. M. (2014). PM2.5, EC and OC in atmospheric outflow from the Indo-Gangetic Plain: Temporal variability and aerosol organic carbon-to-organic mass conversion factor. *Science of the Total Environment, 487*, 196–205.

Bikkina, S., & Sarin, M. M. (2015). Atmospheric deposition of phosphorus to the Northern Indian Ocean. *Current Science, 108*(7), 1300–1305.

Bikkina, S., Sarin, M. M., & Chinni, V. (2015). Atmospheric 210Pb and anthropogenic trace metals in the continental outflow to the Bay of Bengal. *Atmospheric Environment, 122*, 737–747.

Bikkina, S., Sarin, M., & Kumar, A. (2012). Impact of anthropogenic sources on aerosol iron solubility over the Bay of Bengal and the Arabian Sea. *Biogeochemistry, 110*, 257–268.

Bikkina, S., Sarin, M. M., & Rengarajan, R. (2014). Atmospheric transport of mineral dust from the Indo-Gangetic Plain: Temporal variability, acid processing, and iron solubility. *Geochemistry, Geophysics, Geosystems, 15*(8), 3226–3243.

Bikkina, S., Sarin, M. M., & Sarma, V. V. S. S. (2011). Atmospheric dry deposition of inorganic and organic nitrogen to the Bay of Bengal: Impact of continental outflow. *Marine Chemistry, 127*(1–4), 170–179.

Bikkina, S., et al. (2017b). Carbon isotope-constrained seasonality of carbonaceous aerosol sources from an urban location (Kanpur) in the Indo-Gangetic Plain. *Journal of Geophysical Research: Atmospheres, 122*(9), 4903–4923.

Bikkina, S., et al. (2019b). Air quality in megacity Delhi affected by countryside biomass burning. *Nature Sustainability, 2*(3), 200–205.

Bikkina, P., et al. (2020a). Evidence for brown carbon absorption over the Bay of Bengal during the southwest monsoon season: A possible oceanic source. *Environmental Science: Processes & Impacts, 22*, 1743–1758.

Bikkina, P., et al. (2020b). Chemical characterization of wintertime marine aerosols over the Arabian Sea: Impact of marine sources and long-range transport. *Atmospheric Environment, 117749*.

Biswas, H., Mukhopadhyay, S. K., Sen, S., & Jana, T. K. (2007). Spatial and temporal patters of methane dynamics in the tropical mangrove dominated estuary, NE coast of Bay of Bengal, India. *Journal of Marine Systems, 68*, 55–64.

Boersma, K. F., Eskes, H. J., Richter, A., De Smedt, I., Lorente, A., Beirle, S., van Geffen, J. H. G. M., Zara, M., Peters, E., Van Roozendael, M., Wagner, T., Maasakkers, J. D., van der, A. R. J., Nightingale, J., De Rudder, A., Irie, H., Pinardi, G., Lambert, J.-C., & Compernolle, S. C. (2018). Improving algorithms and uncertainty estimates for satellite NO2 retrievals: Results from the quality assurance for the essential climate variables (QA4ECV) project. *Atmospheric Measurement Techniques, 11*, 6651–6678. https://doi.org/10.5194/amt-11-6651-2018.

Booge, D., Marandino, C. A., Schlundt, C., Palmer, P. I., Schlundt, M., Atlas, E. L., Bracher, A., Saltzman, E. S., & Wallace, D. W. R. (2016). Can simple models predict large-scale surface ocean isoprene concentrations? *Atmospheric Chemistry and Physics, 16*, 11807–11821. https://doi.org/10.5194/acp-16-11807-2016.

Booge, D., Schlundt, C., Bracher, A., Endres, S., Zäncker, B., & Marandino, C. A. (2018). Marine isoprene production and consumption in the mixed layer of the surface ocean—A field study over two oceanic regions. *Biogeosciences, 15*, 649–667. https://doi.org/10.5194/bg-15-649-2018.

Bosch, C., et al. (2014). Source-diagnostic dual-isotope composition and optical properties of water-soluble organic carbon and elemental carbon in the South Asian outflow intercepted over the Indian Ocean. *Journal of Geophysical Research: Atmospheres, 119*(20), 11743–11759.

Budhavant, K., et al. (2020). Enhanced light-absorption of black carbon in rainwater compared with aerosols over the Northern Indian Ocean. *Journal of Geophysical Research: Atmospheres, 125*(2). e2019JD031246.

Bui, O. T. N., Kameyama, S., Yoshikawa-Inoue, H., Ishii, M., Sasano, D., Uchida, H., & Tsunogai, U. (2018). Estimates of methane emissions from the Southern Ocean from quasi-continuous underway measurements of the partial pressure of methane in surface seawater during the 2012/13 austral summer. *Tellus B: Chemical and Physical Meteorology, 70*(1), 1–15. https://doi.org/10.1080/16000889.2018.1478594.

Butler, J. H., Elkins, J. W., Thompson, T. M., & Egan, K. B. (1989). Tropospheric and dissolved N_2O of the West Pacific and east Indian Oceans during the El Nino southern oscillation event of 1987. *Journal of Geophysical Research, 94*(D12), 14865–14877.

Capone, D. G., Subramaniam, A., Montoya, J. P., Voss, M., Humborg, C., Johansen, A. M., ... Carpenter, E. J. (1998). An extensive bloom of the N2-fixing cyanobacterium Trichodesmium erythraeum in the central Arabian Sea. *Marine Ecology Progress Series, 172*, 281–292.

Carlton, A. G., Wiedinmyer, C., & Kroll, J. H. (2009). A review of secondary organic aerosol (SOA) formation from isoprene. *Atmospheric Chemistry and Physics, 9*, 4987–5005. https://doi.org/10.5194/acp-9-4987-2009.

Carpenter, L. J., Chance, R. J., Sherwen, T., Adams, T. J., Ball, S. M., Evans, M. J., Hepach, H., Hollis, L. D. J., Hughes, C., Jickells, T. D., Mahajan, A., Stevens, D. P., Tinel, L., & Wadley, M. R. (2021). Marine iodine emissions in a changing world. *Proceedings of the Royal Society A, 477*, 20200824. https://doi.org/10.1098/rspa.2020.0824.

Carpenter, L. J., MacDonald, S. M., Shaw, M. D., Kumar, R., Saunders, R. W., Parthipan, R., Wilson, J., & Plane, J. M. C. (2013). Atmospheric iodine levels influenced by sea surface emissions of inorganic iodine. *Nature Geoscience, 6*(2), 108–111. https://doi.org/10.1038/ngeo1687.

Chakraborty, K., Valsala, V., Bhattacharya, T., & Ghosh, J. (2021). Seasonal cycle of surface ocean pCO2 and pH in the northern Indian Ocean and their controlling factors. *Progress in Oceanography, 198*, 102683.

Chakraborty, K., Valsala, V., Gupta, G. V. M., & Sarma, V. V. S. S. (2018). Dominant biological control over upwelling on pCO_2 in the sea east of Sri Lanka. *Journal of Geophysical Research, 13*, 3250–3261.

Charlson, R., Lovelock, J., Andreae, M., & Warren, S. G. (1987). Oceanic phytoplankton, atmospheric sulphur, cloud albedo and climate. *Nature, 326*, 655–661. https://doi.org/10.1038/326655a0.

Chase, Z., et al. (2011). Evaluating the impact of atmospheric deposition on dissolved trace-metals in the Gulf of Aqaba, Red Sea. *Marine Chemistry, 126*(1), 256–268.

Chinni, V., Singh, S. K., Bhushan, R., Rengarajan, R., & Sarma, V. V. S. S. (2019). Spatial variability in dissolved iron concentrations in the marginal and open waters of the Indian Ocean. *Marine Chemistry, 208*, 11–28.

Ciuraru, R., Fine, L., Pinxteren, M. V., D'Anna, B., Herrmann, H., & George, C. (2015). Unravelling new processes at interfaces: Photochemical isoprene production at the sea surface. *Environmental Science & Technology, 49*, 13199–13205. https://doi.org/10.1021/acs.est.5b02388.

Conte, L., Szopa, S., Séférian, R., & Bopp, L. (2019). The oceanic cycle of carbon monoxide and its emissions to the atmosphere. *Biogeosciences, 16*, 881–902. https://doi.org/10.5194/bg-16-881-2019.

Crippa, M., Solazzo, E., Huang, G., Guizzardi, D., Koffi, E., Muntean, M., Schieberle, C., Friedrich, R., & Janssens-Maenhout, G. (2020). High resolution temporal profiles in the emissions database for global atmospheric research. *Scientific Data, 7*, 121. https://doi.org/10.1038/s41597-020-0462-2.

Dasari, S., et al. (2019). Photochemical degradation affects the light absorption of water-soluble brown carbon in the south Asian outflow. *Science Advances, 5*(1), eaau8066. https://doi.org/10.1126/sciadv.aau8066.

David, L. M., Girach, I. A., & Nair, P. R. (2011). Distribution of ozone and its precursors over Bay of Bengal during winter 2009: Role of meteorology. *Annales de Geophysique, 29*, 1613–1627. https://doi.org/10.5194/angeo-29-1613-2011.

de Verneil, A., Lachkar, Z., Smith, S., & Lévy, M. (2022). Evaluating the Arabian Sea as a regional source of atmospheric CO2: Seasonal variability and drivers. *Biogeosciences Discussions, 19*, 907–929. https://doi.org/10.5194/bg-19-907-2022.

De Wilde, H. P. J., & Helder, W. (1997). Nitrous oxide in the Somali Basin: The role of upwelling. *Deep-Sea Research Part II, 44*(6–7), 1319–1340.

Dee, D. P., Uppala, S. M., Simmons, A. J., Berrisford, P., Poli, P., Kobayashi, S., Andrae, U., Balmaseda, M. A., Balsamo, G., Bauer, P., Bechtold, P., Beljaars, A. C. M., van de Berg, L., Bidlot, J., Bormann, N., Delsol, C., Dragani, R., Fuentes, M., Geer, A. J., ... Vitart, F. (2011). The ERA-interim reanalysis: Configuration and performance of the data assimilation system. *Quarterly Journal of the Royal Meteorological Society, 137*, 553–597. https://doi.org/10.1002/qj.828.

Deeter, M. N., Edwards, D. P., Francis, G. L., Gille, J. C., Mao, D., Martínez-Alonso, S., Worden, H. M., Ziskin, D., & Andreae, M. O. (2019). Radiance-based retrieval bias mitigation for the MOPITT instrument: The version 8 product. *Atmospheric Measurement Techniques, 12*, 4561–4580. https://doi.org/10.5194/amt-12-4561-2019.

Dewangan, P., Sriram, G., Kumar, A., Mazumdar, A., Peketi, A., Mahale, V., ... Babu, A. (2021). Widespread occurrence of methane seeps in deep-water regions of Krishna-Godavari basin, Bay of Bengal. *Marine and Petroleum Geology, 124*, 104783. https://doi.org/10.1016/j.marpetgeo.2020.104783.

do Rosário Gomes, H., Goes, J. I., Matondkar, S. P., Buskey, E. J., Basu, S., Parab, S., & Thoppil, P. (2014). Massive outbreaks of Noctiluca scintillans blooms in the Arabian Sea due to spread of hypoxia. *Nature Communications, 5*, 4862.

do Rosario Gomes, H., Goes, J. I., Matondkar, S. P., Parab, S. G., Al-Azri, A. R., & Thoppil, P. G. (2008). Blooms of Noctiluca miliaris in the Arabian Sea—An in situ and satellite study. *Deep Sea Research Part I: Oceanograhic Research Papers, 55*, 751–765.

Duce, R. A., et al. (1991). The atmospheric input of trace species to the world ocean. *Global Biogeochemical Cycles, 5*(3), 193–259.

Duce, R., et al. (2008). Impacts of atmospheric anthropogenic nitrogen on the open ocean. *Science, 320*(5878), 893–897.

Dutta, M. K., Bianchi, T. S., & Mukhopadhyay, S. K. (2017). Mangrove methane biogeochemistry in the Indian Sundarbans: A proposed budget. *Frontiers in Marine Science, 4*(187). https://doi.org/10.3389/fmars.2017.00187.

Edtbauer, A., Stönner, C., Pfannerstill, E. Y., Berasategui, M., Walter, D., Crowley, J. N., Lelieveld, J., & Williams, J. (2020). A new marine biogenic emission: Methane sulfonamide (MSAM), DMS and $DMSO_2$ measured in air over the Arabian Sea. *Atmospheric Chemistry and Physics, 20*, 6091–6094. https://doi.org/10.5194/acp-20-6081-2020.

Faber, E., Botz, R., Poggenburg, J., Schmidt, M., Stoffers, P., & Hartmann, M. (1998). Methane in Rea Sea brines. *Organic Geochemistry, 29*(1–3), 363–379.

Farías, L., Florez-Leiva, L., Besoain, V., Sarthou, G., & Fernández, C. (2015). Dissolved greenhouse gases (nitrous oxide and methane) associated with the naturally iron-fertilized Kerguelen region (KEOPS 2 cruise) in the Southern Ocean. *Biogeosciences, 12*(6), 1925–1940. https://doi.org/10.5194/bg-12-1925-2015.

Fiehn, A., Quack, B., Stemmler, I., Ziska, F., & Krüger, K. (2018). Importance of seasonally resolved oceanic emissions for bromoform delivery from the tropical Indian Ocean and West Pacific to the stratosphere. *Atmospheric Chemistry and Physics, 18*(16), 11973–11990. https://doi.org/10.5194/acp-18-11973-2018.

Franke, K., Richter, A., Bovensmann, H., Eyring, V., Jöckel, P., Hoor, P., & Burrows, J. P. (2009). Ship emitted NO_2 in the Indian Ocean: Comparison of model results with satellite data. *Atmospheric Chemistry and Physics, 9*, 7289–7301. https://doi.org/10.5194/acp-9-7289-2009.

Friedlingstein, P., O'Sullivan, M., Jones, M. W., Andrew, R. M., Hauck, J., Olsen, A., & Zaehle, S. (2020). Global carbon budget 2020. *Earth System Science Data, 12*(4), 3269–3340. https://doi.org/10.5194/essd-12-3269-2020.

Galí, M., Levasseur, M., Devred, E., Simó, R., & Babin, M. (2018). Sea-surface dimethylsulfide (DMS) concentration from satellite data at global and regional scales. *Biogeosciences, 15*(11), 3497–3519. https://doi.org/10.5194/bg-15-3497-2018.

Gandhi, N., et al. (2011). First direct measurements of N2 fixation during a Trichodesmium bloom in the eastern Arabian Sea. *Global Biogeochemical Cycles, 25*(4).

Garcias-Bonet, N., & Duarte, C. M. (2017). Methane production by seagrass ecosystems in the Red Sea. *Frontiers in Marine Science, 4*(340). https://doi.org/10.3389/fmars.2017.00340.

Girach, I. A., Nair, P. R., Ojha, N., & Sahu, S. K. (2020). Tropospheric carbon monoxide over the northern Indian Ocean during winter: Influence of intercontinental transport. *Climate Dynamics, 54*, 5049–5064. https://doi.org/10.1007/s00382-020-05269-4.

Gschwend, P. M., MacFarlane, J. K., & Newman, K. A. (1985). Volatile halogenated organic compounds released to seawater from temperate marine macroalgae. *Science, 227*, 1033–1035. https://doi.org/10.1126/science.227.4690.1033.

Guieu, C., et al. (2019). Major impact of dust deposition on the productivity of the Arabian Sea. *Geophysical Research Letters, 46*(12), 6736–6744.

Hall, T. M., Waugh, D. W., Haine, T. W. N., Robbins, P. E., & Khatiwala, S. (2004). Estimates of anthropogenic carbon in the Indian Ocean with allowance for mixing and time-varying air-sea CO2 disequilibrium. *Global Biogeochemical Cycles, 18*(1). https://doi.org/10.1029/2003GB002120.

Hashimoto, S., Kurita, Y., Takasu, Y., & Otsuki, A. (1998). Significant difference in the vertical distribution of nitrous oxide in the central Bay of Bengal from that in the western area. *Deep-Sea Research Part I, 45*, 301–316.

Hermes, J. C., Masumoto, Y., Beal, L., Roxy, M., Vialard, J., Andres, M., Annamalai, H., Behera, S., D'Adamo, N., Doi, T., Feng, M., Han, W., Hardman-Mountford, N., Hendon, H., Hood, R. R., Kido, S., Lee, C., Lee, T., Lengainge, M., … Yu, W. (2019). A sustained ocean observing system in the Indian Ocean for climate related scientific knowledge and societal needs. *Frontiers in Marine Science, 6*, 355.

Hershey, R. N., Bijoy Nandan, S., Jayachandran, P. R., Vijay, A., Neelima Vasu, K., & Sudheesh, V. (2019). Nitrous oxide flux from a tropical estuarine system (Cochin estuary, India). *Regional Studies in Marine Science, 30*, 100725. https://doi.org/10.1016/j.rsma.2019.100725.

Hood, R. R., Coles, V. J., Huggett, J. A., Landry, M. R., Levy, M., Moffett, J. W., & Rixen, T. (2024a). Chapter 13: Nutrient, phytoplankton, and zooplankton variability in the Indian Ocean. In C. C. Ummenhofer, & R. R. Hood (Eds.), *The Indian Ocean and its role in the global climate system* (pp. 293–327). Amsterdam: Elsevier. https://doi.org/10.1016/B978-0-12-822698-8.00020-2.

Hood, R. R., Rixen, T., Levy, M., Hansell, D. A., Coles, V. J., & Lachkar, Z. (2024b). Chapter 12: Oxygen, carbon, and pH variability in the Indian Ocean. In C. C. Ummenhofer, & R. R. Hood (Eds.), *The Indian Ocean and its role in the global climate system* (pp. 265–291). Amsterdam: Elsevier. https://doi.org/10.1016/B978-0-12-822698-8.00017-2.

IPCC. (2021). *Climate change 2021: The physical science basis. Contribution of working group I to the sixth assessment report of the intergovernmental panel on climate change.* Cambridge, UK: Cambridge University Press (in press).

Jayakumar, D. A., Naqvi, S. W. A., Narveka, P. V., & George, M. D. (2001). Methane in the coastal and offshore waters of the Arabian Sea. *Marine Chemistry, 74*, 1–13.

Jayaraman, A., et al. (1998). Direct observations of aerosol radiative forcing over the tropical Indian Ocean during the January–February 1996 pre-INDOEX cruise. *Journal of Geophysical Research: Atmospheres, 103*(D12), 13827–13836.

Jickells, T. D., Baker, A. R., & Chance, R. (2016). Atmospheric transport of trace elements and nutrients to the oceans. *Philosophical Transactions of the Royal Society A: Mathematical, Physical and Engineering Sciences, 374*(2081), 20150286.

Johansen, A. M., & Hoffmann, M. R. (2004). Chemical characterization of ambient aerosol collected during the northeast monsoon season over the Arabian Sea: Anions and cations. *Journal of Geophysical Research: Atmospheres, 109*(D5).

Kanakidou, M., Myriokefalitakis, S., & Tsigaridis, K. (2018). Aerosols in atmospheric chemistry and biogeochemical cycles of nutrients. *Environmental Research Letters, 13*(6), 063004. https://doi.org/10.1088/1748-9326/aabcdb.

Kessler, N., et al. (2020). Selective collection of iron-rich dust particles by natural Trichodesmium colonies. *The ISME Journal, 14*(1), 91–103.

Kock, A., & Bange, H. W. (2015). Counting the ocean's greenhouse gas emissions. *Eos (Washington. DC), 96*(3), 10–13. https://doi.org/10.1029/2015EO023665.

Krishnamurthy, A., et al. (2009). Impacts of increasing anthropogenic soluble iron and nitrogen deposition on ocean biogeochemistry. *Global Biogeochemical Cycles, 23*(3).

Kumar, A., Sarin, M. M., & Srinivas, B. (2010). Aerosol iron solubility over Bay of Bengal: Role of anthropogenic sources and chemical processing. *Marine Chemistry, 121*(1), 167–175.

Kumar, A., Sarin, M. M., & Sudheer, A. K. (2008a). Mineral and anthropogenic aerosols in Arabian Sea–atmospheric boundary layer: Sources and spatial variability. *Atmospheric Environment, 42*(21), 5169–5181.

Kumar, A., Sudheer, A. K., Goswami, V., & Bhushan, R. (2012). Influence of continental outflow on aerosol chemical characteristics over the Arabian Sea during winter. *Atmospheric Environment, 50*, 182–191.

Kumar, A., Sudheer, A., & Sarin, M. (2008b). Chemical characteristics of aerosols in MABL of Bay of Bengal and Arabian Sea during spring inter-monsoon: A comparative study. *Journal of Earth System Science, 117*(1), 325–332.

Kumar, S., Ramesh, R., Sardesai, S., & Sheshshayee, M. S. (2004). High new production in the Bay of Bengal: Possible causes and implications. *Geophysical Research Letters, 31*(18).

Lal, S., & Patra, P. K. (1998). Variabilities in the fluxes and annual emissions of nitrous oxide from the Arabian Sea. *Global Biogeochemical Cycles, 12*, 321–327.

Lamontagne, R. A., Swinnerton, J. W., Linnenbom, V. J., & Smith, W. D. (1973). Methane concentrations in various marine environments. *Journal of Geophysical Research, 78*, 5317–5325.

Lana, A., Bell, T. G., Simó, R., Vallina, S. M., Ballabrera-Poy, J., Kettle, A. J., Dachs, J., Bopp, L., Saltzman, E. S., Stefels, J., Johnson, J. E., & Liss, P. S. (2011). An updated climatology of surface dimethylsulfide concentrations and emission fluxes in the global ocean. *Global Biogeochemical Cycles, 25*. https://doi.org/10.1029/2010gb003850.

Landschützer, P., Gruber, N., & Bakker, D. C. E. (2016). Decadal variations and trends of the global ocean carbon sink. *Global Biogeochemical Cycles, 30*(10), 1396–1417.

Law, C. S., & Owens, N. J. P. (1990). Significant flux of atmospheric nitrous oxide from the Northwest Indian Ocean. *Nature, 346*, 826–828.

Lelieveld, J., Crutzen, P. J., Ramanathan, V., Andreae, M. O., Brenninkmeijer, C. A. M., Campos, T., … Williams, J. (2001). The Indian Ocean experiments: Widespread air pollution from south and southeast Asia. *Science, 291*, 1031–1036.

Liss, P. S., & Johnson, M. T. (Eds.). (2014). *Ocean-atmosphere interactions of gases and particles.* Heidelberg: Springer. https://doi.org/10.1007/978-3-642-25643-1 (315 pp.).

Liss, P. S., Marandino, C. A., Dahl, E. E., Helmig, D., Hintsa, E. J., Hughes, C., … Williams, J. (2014). Short-lived trace gases in the surface ocean and atmosphere. In P. S. Liss, & M. T. Johnson (Eds.), *Ocean-atmosphere interactions of gases and particles* (pp. 1–54). Heidelberg: Springer.

Löscher, C. R., Mohr, W., Bange, H. W., & Canfield, D. E. (2020). No nitrogen fixation in the Bay of Bengal? *Biogeosciences, 17*(4), 851–864.

Ma, X., Bange, H. W., Eirund, G. K., & Arévalo-Martínez, D. L. (2020). Nitrous oxide and hydroxylamine measurements in the Southwest Indian Ocean. *Journal of Marine Systems, 209*, 103062. https://doi.org/10.1016/j.jmarsys.2018.03.003.

Maas, J., Jia, Y., Quack, B., Durgadoo, J. V., Biastoch, A., & Tegtmeier, S. (2020). Simulations of anthropogenic bromoform indicate high emissions at the coast of East Asia. *Atmospheric Chemistry and Physics Discussions*, 1–31. https://doi.org/10.5194/acp-2019-1004.

Madhupratap, M., et al. (2003). Biogeochemistry of the Bay of Bengal: Physical, chemical and primary productivity characteristics of the central and western Bay of Bengal during summer monsoon 2001. *Deep Sea Research Part II: Topical Studies in Oceanography*, *50*(5), 881–896.

Mahajan, A. S., Li, Q., Inamdar, S., Ram, K., Badia, A., & Saiz-Lopez, A. (2021). Modelling the impacts of iodine chemistry on the northern Indian Ocean marine boundary layer. *Atmospheric Chemistry and Physics*, 8437–8454. https://doi.org/10.5194/acp-21-8437-2021.

Mahajan, A. S., Plane, J. M. C., Oetjen, H., Mendes, L., Saunders, R. W., Saiz-Lopez, A., Jones, C. E., Carpenter, L. J., & McFiggans, G. B. (2010). Measurement and modelling of tropospheric reactive halogen species over the tropical Atlantic Ocean. *Atmospheric Chemistry and Physics*, *10*, 4611–4624. https://doi.org/10.5194/acp-10-4611-2010.

Mahowald, N. (2011). Aerosol indirect effect on biogeochemical cycles and climate. *Science*, *334*(6057), 794–796.

Mallik, C., Lal, S., Venkataramani, S., Naja, M., & Ojha, N. (2013a). Variability in ozone and its precursors over the Bay of Bengal during post monsoon: Transport and emission effects. *Journal of Geophysical Research: Atmospheres*, *118*(17), 10190–10209.

Mallik, C., et al. (2013b). Enhanced SO2 concentrations observed over northern India: Role of long-range transport. *International Journal of Remote Sensing*, *34*(8), 2749–2762.

McCaslin, L. M., Johnson, M. A., & Gerber, R. B. (2019). Mechanisms and competition of halide substitution and hydrolysis in reactions of N2O5 with seawater. *Science Advances*, *5*(6), eaav6503. https://doi.org/10.1126/sciadv.aav6503.

McLennan, S. M. (2001). Relationships between the trace element composition of sedimentary rocks and upper continental crust. *Geochemistry, Geophysics, Geosystems*, *2*(4). https://doi.org/10.1029/2000GC000109.

Moore, J. K., Doney, S. C., Glover, D. M., & Fung, I. Y. (2001). Iron cycling and nutrient-limitation patterns in surface waters of the World Ocean. *Deep Sea Research Part II: Topical Studies in Oceanography*, *49*(1), 463–507.

Naqvi, S. W. A., Jayakumar, D. A., Nair, M., Dileep Kumar, M., & George, M. D. (1994). Nitrous oxide in the western Bay of Bengal. *Marine Chemistry*, *47*, 269–278.

Naqvi, S. W. A., Jayakumar, D. A., Narveka, P. V., Naik, H., Sarma, V. V. S. S., D'Souza, W., … George, M. D. (2000). Increased marine production of N_2O due to intensifying anoxia on the Indian continental shelf. *Nature*, *408*, 346–349.

Naqvi, S. W. A., & Noronha, R. J. (1991). Nitrous oxide in the Arabian Sea. *Deep Sea Research*, *38*, 871–890.

Nenes, A., Michael, D. K., Nikos, M., Philippe, V. C., Zongbo, S., Aikaterini, B., … Barak, H. (2011). Atmospheric acidification of mineral aerosols: A source of bioavailable phosphorus for the oceans. *Atmospheric Chemistry and Physics*, *11*(13), 6265–6272.

Oetjen, H. (2009). *Measurements of halogen oxides by scattered sunlight differential optical absorption spectroscopy*. University of Bremen.

Okin, G. S., et al. (2011). Impacts of atmospheric nutrient deposition on marine productivity: Roles of nitrogen, phosphorus, and iron. *Global Biogeochemical Cycles*, *25*(2), n/a-n/a.

Olsen, A., Lange, N., Key, R. M., Tanhua, T., Bittig, H. C., & Kozyr, A. (2020). An updated version of the global interior ocean biogeochemical data product, GLODAPv2.2020. *Earth System Science Data*, *12*, 3653–3678. https://doi.org/10.5194/essd-12-3653-2020.

Owens, N. J. P., Law, C. S., Mantoura, R. F. C., Burkill, P. H., & Llewellyn, C. A. (1991). Methane flux to the atmosphere from the Arabian Sea. *Nature*, *354*, 293–296.

Parab, S. G., & Matondkar, S. G. P. (2012). Primary productivity and nitrogen fixation by Trichodesmium spp. in the Arabian Sea. *Journal of Marine Systems*, *105*, 82–95.

Patra, P. K., Kumar, M. D., Mahowald, N., & Sarma, V. V. S. S. (2007). Atmospheric deposition and surface stratification as controls of contrasting chlorophyll abundance in the North Indian Ocean. *Journal of Geophysical Research: Oceans*, *112*.

Patra, P. K., Lal, S., Venkataramani, S., Gauns, M., & Sarma, V. V. S. S. (1998). Seasonal variability in distribution and fluxes of methane in the Arabian Sea. *Journal of Geophysical Research*, *103*, 1167–1176.

Pfannerstill, E. Y., Wang, N., Edtbauer, A., Bourtsoukidis, E., Crowley, J. N., Dienhart, D., Eger, P. G., Ernle, L., Fischer, H., Hottmann, B., Paris, J.-D., Stönner, C., Tadic, I., Walter, D., Lelieveld, J., & Williams, J. (2019). Shipborne measurements of total OH reactivity around the Arabian Peninsula and its role in ozone chemistry. *Atmospheric Chemistry and Physics*, *19*, 11501–11523. https://doi.org/10.5194/acp-19-11501-2019.

Phillips, H. E., Tandon, A., Furue, R., Hood, R., Ummenhofer, C. C., Benthuysen, J. A., Menezes, V., Hu, S., Webber, B., Sanchez-Franks, A., Cherian, D., Shroyer, E., Feng, M., Wijesekera, H., Chatterjee, A., Yu, L., Hermes, J., Murtugudde, R., Tozuka, T., … Wiggert, J. (2021). Progress in understanding of Indian Ocean circulation, variability, air–sea exchange, and impacts on biogeochemistry. *Ocean Science*, *17*, 1677–1751.

Porter, J. G., de Bruyn, W. J., Miller, S. D., & Saltzman, E. S. (2020). Air/sea transfer of highly soluble gases over coastal waters. *Geophysical Research Letters*, *47*. https://doi.org/10.1029/2019GL085286. e2019GL085286.

Prakash, S., Roy, R., & Lotliker, A. (2017). Revisiting the Noctiluca scintillans paradox in northern Arabian Sea. *Current Science*, *113*, 1429.

Purvaja, R., & Ramesh, R. (2001). Natural and anthropogenic methane emission from coastal wetlands of South India. *Environmental Management*, *27*(4), 547–557.

Quinn, P., & Bates, T. (2011). The case against climate regulation via oceanic phytoplankton sulphur emissions. *Nature*, *480*, 51–56. https://doi.org/10.1038/nature10580.

Quinn, P. K., Coffman, D. J., Johnson, J. E., Upchurch, L. M., & Bates, T. S. (2017). Small fraction of marine cloud condensation nuclei made up of sea spray aerosol. *Nature Geoscience*, *10*(9), 674–679. https://doi.org/10.1038/ngeo3003.

Rajput, P., Sarin, M. M., Rengarajan, R., & Singh, D. (2011). Atmospheric polycyclic aromatic hydrocarbons (PAHs) from post-harvest biomass burning emissions in the Indo-Gangetic Plain: Isomer ratios and temporal trends. *Atmospheric Environment*, *45*(37), 6732–6740.

Ram, K., Sarin, M., & Tripathi, S. (2010). A 1 year record of carbonaceous aerosols from an urban site in the Indo-Gangetic Plain: Characterization, sources, and temporal variability. *Journal of Geophysical Research, 115*(D24). https://doi.org/10.1029/2010JD014188.

Ramanathan, V., Crutzen, P., Kiehl, J., & Rosenfeld, D. (2001). Aerosols, climate, and the hydrological cycle. *Science, 294*(5549), 2119–2124.

Ramanathan, V., et al. (2005). Atmospheric brown clouds: Impacts on South Asian climate and hydrological cycle. *Proceedings of the National Academy of Sciences of the United States of America, 102*(15), 5326–5333.

Rao, G. D., Rao, V. D., & Sarma, V. V. S. S. (2013). Distribution and air-sea exchange of nitrous oxide in the coastal Bay of Bengal during peak discharge (southwest monsoon). *Marine Chemistry, 155*, 1–9. https://doi.org/10.1016/j.marchem.2013.04.014.

Rao, G. D., & Sarma, V. V. S. S. (2016). Variability in concentrations and fluxes of methane in the Indian estuaries. *Estuaries and Coasts, 39*(6), 1639–1650. https://doi.org/10.1007/s12237-016-0112-2.

Rastogi, N., & Sarin, M. M. (2006). Chemistry of aerosols over a semi-arid region: Evidence for acid neutralization by mineral dust. *Geophysical Research Letters, 33*(23).

Rastogi, N., Singh, A., Sarin, M. M., & Singh, D. (2016). Temporal variability of primary and secondary aerosols over northern India: Impact of biomass burning emissions. *Atmospheric Environment, 125*, 396–403. https://doi.org/10.1016/j.atmosenv.2015.06.010.

Read, K. A., et al. (2008). Extensive halogen-mediated ozone destruction over the tropical Atlantic Ocean. *Nature, 453*, 1232–1235.

Rengarajan, R., Sarin, M., & Sudheer, A. (2007). Carbonaceous and inorganic species in atmospheric aerosols during wintertime over urban and high-altitude sites in North India. *Journal of Geophysical Research: Atmospheres (1984–2012), 112*(D21). https://doi.org/10.1029/2006JD008150.

Richter, U., & Wallace, D. W. R. (2004). Production of methyl iodide in the tropical Atlantic Ocean. *Geophysical Research Letters, 31*(23). https://doi.org/10.1029/2004GL020779.

Rixen, T., Cowie, G., Gaye, B., Goes, J. I., Gomes, H., Hood, R. R., Lachkar, Z., Schmidt, H., Segschneider, J., & Singh, A. (2020). Reviews and syntheses: Present, past, and future of the oxygen minimum zone in the northern Indian Ocean. *Biogeosciences, 17*, 1–30. https://doi.org/10.5194/bg-17-1-2020.

Rodríguez-Ros, P., Cortés, P., Robinson, C. M., Nunes, S., Hassler, C., Royer, S. J., Estrada, M., Sala, M. M., & Simó, R. (2020). Distribution and drivers of marine isoprene concentration across the Southern Ocean. *Atmosphere (Basel), 11*(6), 1–19. https://doi.org/10.3390/atmos11060556.

Sabine, C. L., Feely, R. A., Gruber, N., Key, R. M., Lee, K., Bullister, J. L., Wanninkhof, R., Wong, C. S., Wallace, D. W. R., Tilbrook, B., Millero, F. J., Peng, T., Kozhyr, A., Ono, T., & Rios, A. F. (2004). The oceanic sink for anthropogenic CO_2. *Science, 305*, 367–371.

Saiz-Lopez, A., & von Glasow, R. (2012). Reactive halogen chemistry in the troposphere. *Chemical Society Reviews, 41*, 6448–6472.

Sambrotto, R. N. (2001). Nitrogen production in the northern Arabian Sea during the Spring Intermonsoon and Southwest Monsoon seasons. *Deep Sea Research Part II: Topical Studies in Oceanography, 48*(6–7), 1173–1198.

Sarin, M., Kumar, A., Srinivas, B., Sudheer, A., & Rastogi, N. (2010). Anthropogenic sulphate aerosols and large Cl-deficit in marine atmospheric boundary layer of tropical Bay of Bengal. *Journal of Atmospheric Chemistry, 66*(1–2), 1–10.

Sarin, M. M., Rengarajan, R., & Krishnaswami, S. (1999). Aerosol NO-3 and 210Pb distribution over the central-eastern Arabian Sea and their air-sea deposition fluxes. *Tellus B: Chemical and Physical Meteorology, 51*(4), 749–758.

Sarma, V. V. S. S. (2002). An evaluation of physical and biogeochemical processes regulating perennial suboxic conditions in the water column of the Arabian Sea. *Global Biogeochemical Cycles, 16*(4), 29-1–29-11.

Sarma, V. V. S. S., Lenton, A., Law, R. M., Metzl, N., Patra, P. K., Doney, S., Lima, I. D., Dlugokencky, E., Ramonet, M., & Valsala, V. (2013). Sea–air CO_2 fluxes in the Indian Ocean between 1990 and 2009. *Biogeosciences, 10*, 7035–7052.

Sarma, V. V. S. S., Rajula, G. R., Durgadevi, D. S. L., Sampath Kumar, G., & Loganathan, J. (2020). Influence of eddies on phytoplankton composition in the Bay of Bengal. *Continental Shelf Research, 208*, 104241.

Saunois, M., Stavert, A. R., Poulter, B., Bousquet, P., Canadell, J. G., Jackson, R. B., ... Zhuang, Q. (2020). The global methane budget 2000–2017. *Earth System Science Data, 12*(3), 1561–1623. https://doi.org/10.5194/essd-12-1561-2020.

Sawant, S., & Madhupratap, M. (1996). Seasonality and composition of phytoplankton in the Arabian Sea. *Current Science, 71*, 869–873.

Saxena, H., Sahoo, D., Khan, M. A., Kumar, S., Sudheer, A., & Singh, A. (2020). Dinitrogen fixation rates in the Bay of Bengal during summer monsoon. *Environmental Research Communications, 2*, 051007. https://doi.org/10.1088/2515-7620/ab89fa.

Schulz, M., et al. (2012). Atmospheric transport and deposition of mineral dust to the ocean: Implications for research needs. *Environmental Science & Technology, 46*(19), 10390–10404.

Shiozaki, T., Minoru, L., Kodama, T., Takeda, S., & Furuya, K. (2014). Heterotrophic bacteria as major nitrogen fixers in the euphotic zone of the Indian Ocean. *Global Biogeochemical Cycles, 28*, 1096–1110.

Shirodkar, G., Naqvi, S. W. A., Naik, H., Pratihary, A. K., Kurian, S., & Shenoy, D. M. (2018). Methane dynamics in the shelf waters of the west coast of India during seasonal anoxia. *Marine Chemistry, 203*, 55–63. https://doi.org/10.1016/j.marchem.2018.05.001.

Siegert, M., Krüger, M., Teichert, B., Wiedicke, M., & Schippers, A. (2011). Anaerobic oxidation of methane at a marine methane seep in a Forearc Sediment Basin off Sumatra, Indian Ocean. *Frontiers in Microbiology, 2*(249). https://doi.org/10.3389/fmicb.2011.00249.

Singh, A., Gandhi, N., & Ramesh, R. (2012). Contribution of atmospheric nitrogen deposition to new production in the nitrogen limited photic zone of the northern Indian Ocean. *Journal of Geophysical Research: Oceans, 117*(C6).

Singh, A., Gandhi, N., & Ramesh, R. (2019). Surplus supply of bioavailable nitrogen through N2 fixation to primary producers in the eastern Arabian Sea during autumn. *Continental Shelf Research, 181*, 103–110.

Siswanto, E. (2015). Atmospheric deposition—Another source of nutrients enhancing primary productivity in the eastern tropical Indian Ocean during positive Indian Ocean dipole phases. *Geophysical Research Letters, 42*(13), 5378–5386.

Soerensen, A. L., Mason, R. P., Balcom, P. H., Jacob, D. J., Zhang, Y., Kuss, J., & Sunderland, E. M. (2014). Elemental mercury concentrations and fluxes in the tropical atmosphere and ocean. *Environmental Science & Technology, 48*, 11312–11319. https://doi.org/10.1021/es503109p.

Spain, E. A., Johnson, S. C., Hutton, B., Whittaker, J. M., Lucieer, V., Watson, S. J., ... Coffin, M. F. (2020). Shallow seafloor gas emissions near Heard and McDonald Islands on the Kerguelen Plateau, Southern Indian Ocean. *Earth and Space Science*, *7*(3). https://doi.org/10.1029/2019EA000695. e2019EA000695.

Sreeush, M. G., Saran, R., Valsala, V., Pentakota, S., Prasad, K. V. S. R., & Murtugudde, R. (2019). Variability, trend and controlling factors of ocean acidification over Western Arabian Sea upwelling region. *Marine Chemistry*, *209*, 14–24.

Sreeush, M. G., Valsala, V., Pentakota, S., Prasad, K. V. S. R., & Murtugudde, R. (2018). Biological production in the Indian Ocean upwelling zones—Part 1: Refined estimation via the use of a variable compensation depth in ocean carbon models. *Biogeosciences*, *15*, 1895–1918. https://doi.org/10.5194/bg-15-1895-2018.

Sreeush, M. G. V., Valsala, H., Santanu, S., Pentakota, K. V. S. R., Prasad, C. V. N., & Murtugudde, R. (2020). Biological production in the Indian Ocean upwelling zones—Part 2: Data based estimates of variable compensation depth for ocean carbon models via cyclo-stationary Bayesian inversion. *Deep Sea Research Part II*, *179*, 104619.

Stemmler, I., Hense, I., & Quack, B. (2015). Marine sources of bromoform in the global open ocean—Global patterns and emissions. *Biogeosciences*, *12*, 1967–1981. https://doi.org/10.5194/bg-12-1967-2015.

Sudheer, A. K., & Sarin, M. M. (2008). Carbonaceous aerosols in MABL of Bay of Bengal: Influence of continental outflow. *Atmospheric Environment*, *42*(18), 4089–4100.

Sudheesh, V., Gupta, G. V. M., & Naqvi, S. W. A. (2020). Massive methane loss during seasonal hypoxia/anoxia in the nearshore waters of Southeastern Arabian Sea. *Frontiers in Marine Science*, *7*, 324. https://doi.org/10.3389/fmars.2020.00324.

Sudheesh, V., Gupta, G., Sudharma, K., Naik, H., Shenoy, D., Sudhakar, M., & Naqvi, S. (2016). Upwelling intensity modulates N_2O concentrations over the western Indian shelf. *Journal of Geophysical Research: Oceans*, *121*, 8551–8565.

Sundarambal, P., Balasubramanian, R., Tkalich, P., & He, J. (2010). Impact of biomass burning on ocean water quality in Southeast Asia through atmospheric deposition: Field observations. *Atmospheric Chemistry and Physics*, *10*(23), 11323–11336.

Suntharalingam, P., Zamora, L. M., Bange, H. W., Bikkina, S., Buitenhuis, E., Kanakidou, M., ... Singh, A. (2019). Anthropogenic nitrogen inputs and impacts on oceanic N2O fluxes in the northern Indian Ocean: The need for an integrated observation and modelling approach. *Deep-Sea Research Part II*, *166*, 104–113. https://doi.org/10.1016/j.dsr2.2019.03.007.

Takahashi, T., Sutherland, S. C., Chipman, D. W., Goddard, J. G., Ho, C., Newberger, T., ... Munro, D. R. (2014). Climatological distributions of pH, pCO2, total CO2, alkalinity, and CaCO3 saturation in the global surface ocean, and temporal changes at selected locations. *Marine Chemistry*, *164*, 95–125. https://doi.org/10.1016/j.marchem.2014.06.004.

Takahashi, T., Sutherland, S. C., Wanninkhof, R., Sweeney, C., Feely, R. A., et al. (2009). Climatological mean and decadal changes in surface ocean pCO$_2$ and net sea-air CO$_2$ flux over the global oceans. *Deep Sea Research Part II*, *56*, 554–577.

Tegtmeier, S., Marandino, C., Jia, Y., Quack, B., & Mahajan, A. S. (2020). Atmospheric gas-phase composition over the Indian Ocean. *Atmospheric Chemistry and Physics Discussions*, *2020*, 1–84. https://doi.org/10.5194/acp-2020-718.

Teodoru, C. R., Nyoni, F. C., Borges, A. V., Darchambeau, F., Nyambe, I., & Bouillon, S. (2015). Dynamics of greenhouse gases (CO_2, CH_4, N_2O) along the Zambezi River and major tributaries, and their importance in the riverine carbon budget. *Biogeosciences*, *12*(8), 2431–2453. https://doi.org/10.5194/bg-12-2431-2015.

Tian, Y., Wang, K.-K., Yang, G.-P., Li, P.-F., Liu, C.-Y., Ingeniero, R. C. O., & Bange, H. W. (2021). Continuous chemiluminescence measurements of dissolved nitric oxide (NO) and nitrogen dioxide (NO2) in the ocean surface layer of the East China Sea. *Environmental Science & Technology*, *55*(6), 3668–3675. https://doi.org/10.1021/acs.est.0c06799.

Tian, H., Xu, R., Canadell, J. G., Thompson, R. L., Winiwarter, W., Suntharalingam, P., ... Yao, Y. (2020a). A comprehensive quantification of global nitrous oxide sources and sinks. *Nature*, *586*(7828), 248–256. https://doi.org/10.1038/s41586-020-2780-0.

Tian, Y., Xue, C., Liu, C. Y., Yang, G. P., Li, P. F., Feng, W. H., & Bange, H. W. (2019). Nitric oxide (NO) in the Bohai Sea and the Yellow Sea. *Biogeosciences*, *16*(22), 4485–4496. https://doi.org/10.5194/bg-16-4485-2019.

Tian, Y., Yang, G. P., Liu, C. Y., Li, P. F., Chen, H. T., & Bange, H. W. (2020b). Photoproduction of nitric oxide in seawater. *Ocean Science*, *16*(1), 135–148. https://doi.org/10.5194/os-16-135-2020.

Tindale, N. W., & Pease, P. P. (1999). Aerosols over the Arabian Sea: Atmospheric transport pathways and concentrations of dust and sea salt. *Deep Sea Research Part II: Topical Studies in Oceanography*, *46*(8), 1577–1595.

Tokarczyk, R., & Moore, R. M. (1994). Production of volatile organohalogens by phytoplankton cultures. *Geophysical Research Letters*, *21*, 285–288. https://doi.org/10.1029/94GL00009.

Tripathi, N., Sahu, L. K., Singh, A., Yadav, R., & Karati, K. K. (2020). High levels of isoprene in the marine boundary layer of the Arabian Sea during spring inter-monsoon: Role of phytoplankton blooms. *ACS Earth and Space Chemistry*, *4*(4), 583–590. https://doi.org/10.1021/acsearthspacechem.9b00325.

Valsala, V., & Maksyutov, S. (2010). Simulation and assimilation of global ocean pCO2 and sea-to-air CO2 fluxes using ship observations of surface ocean pCO$_2$ in a simplified biogeochemical offline model. *Tellus Series B: Chemical and Physical Meteorology*, *62*, 821–840.

Valsala, V., & Maksyutov, S. (2013). Interannual variability of the air–sea CO2 flux in the North Indian Ocean. *Ocean Dynamics*, *63*, 165–178.

Valsala, V., Maksyutov, S., & Murtugudde, R. G. (2012). A window for carbon uptake in the southern subtropical Indian Ocean. *Geophysical Research Letters*, *39*, L17605.

Valsala, V., & Murtugudde, R. (2015). Mesoscale and intraseasonal air-sea CO$_2$ exchanges in the Western Arabian Sea during boreal summer. *Deep Sea Research, Part I*, *103*, 101–113.

Valsala, V., Sreeush, M. G., Anju, M., Sreenivas, P., Tiwari, Y. K., Chakraborty, K., & Sijikumar, S. (2021). An observing system simulation experiment for Indian Ocean surface pCO2 measurements. *Progress in Oceanography, 194*, 102570. https://doi.org/10.1016/j.pocean.2021.102570, 1-14.

Valsala, V., Sreeush, M. G., & Chakraborty, K. (2020). IOD impacts on Indian the ocean carbon cycle. *Journal of Geophysical Research, 125*. e2020JC016485.

Venkataraman, C., Habib, G., Eiguren-Fernandez, A., Miguel, A., & Friedlander, S. (2005). Residential biofuels in South Asia: Carbonaceous aerosol emissions and climate impacts. *Science, 307*(5714), 1454–1456.

Von Rad, U., Rösch, H., Berner, U., Geyh, M., Marchig, V., & Schulz, H. (1996). Authigenic carbonates derived from oxidized methane vented from the Makran accretionary prism off Pakistan. *Marine Geology, 136*, 55–77.

Weber, T., Wiseman, N. A., & Kock, A. (2019). Global ocean methane emissions dominated by shallow coastal waters. *Nature Communications, 10*(1), 4584. https://doi.org/10.1038/s41467-019-12541-7.

Wiggert, J. D., Vialard, J., & Behrenfeld, M. J. (2009). Basin-wide modification of dynamical and biogeochemical processes by the positive phase of the Indian Ocean dipole during the SeaWiFS era. In *Indian Ocean biogeochemical processes and ecological variability* (pp. 385–407). Washington, DC: American Geophysical Union. https://doi.org/10.1029/2008GM000776.

WMO. (2018). *Scientific assessment of ozone depletion: 2018*. (Global Ozone Research and Monitoring Project–Report No. 58). Retrieved from Geneva, Switzerland.

Yang, S., Chang, B. X., Warner, M. J., Weber, T. S., Bourbonnais, A. M., Santoro, A. E., ... Bianchi, D. (2020). Global reconstruction reduces the uncertainty of oceanic nitrous oxide emissions and reveals a vigorous seasonal cycle. *Proceedings of the National Academy of Sciences of the United States of America, 201921914*. https://doi.org/10.1073/pnas.1921914117.

Zafiriou, O. C., & McFarland, M. (1981). Nitric oxide from nitrite photolysis in the central equatorial Pacific. *Journal of Geophysical Research, 86*(C4), 3173–3182. https://doi.org/10.1029/JC086iC04p03173.

Zavarsky, A., Booge, D., Fiehn, A., Krüger, K., Atlas, E., & Marandino, C. (2018a). The influence of air-sea fluxes on atmospheric aerosols during the summer monsoon over the tropical Indian Ocean. *Geophysical Research Letters, 45*, 418–426. https://doi.org/10.1002/2017gl076410.

Zavarsky, A., Goddijn-Murphy, L., Steinhoff, T., & Marandino, C. A. (2018b). Bubble-mediated gas transfer and gas transfer suppression of DMS and CO_2. *Journal of Geophysical Research, 123*, 6624–6647. https://doi.org/10.1029/2017jd028071.

Zhan, L., Chen, L., Zhang, J., Yan, J., Li, Y., Wu, M., ... Zhao, J. (2015). Austral summer N_2O sink and source characteristics and their impact factors in Prydz Bay, Antarctica. *Journal of Geophysical Research: Oceans, 120*(8), 5836–5849. https://doi.org/10.1002/2015jc010944.

Zheng, B., Tong, D., Li, M., Liu, F., Hong, C., Geng, G., Li, H., Li, X., Peng, L., Qi, J., Yan, L., Zhang, Y., Zhao, H., Zheng, Y., He, K., & Zhang, Q. (2018). Trends in China's anthropogenic emissions since 2010 as the consequence of clean air actions. *Atmospheric Chemistry and Physics, 18*, 14095–14111. https://doi.org/10.5194/acp-18-14095-2018.

Chapter 15

Microbial ecology of the Indian Ocean

Carolin Regina Löscher[a] and Christian Furbo Reeder[a,b]
[a]University of Southern Denmark, Odense, Denmark, [b]Marseille Institute for Ocean Science, Marseille, France

1 Microbial diversity in the Indian Ocean

1.1 Introduction

Microbes, including bacteria, archaea, and microbial eukaryotes, are the key drivers of marine biogeochemical cycles. They moderate Earth's climate and are at the base of the marine food web. To explore the role of microbes in an environment, we can study their diversity using classic methods such as metabolic assays, modern sequencing-based methods (Jasmin et al., 2017), or functional studies, including isotope labeling or inhibitor experiments. The warm water temperatures of the Indian Ocean, as well as its specific properties, such as the monsoon, large and intense oxygen minimum zones (OMZs), and mesoscale activity, strongly influence the diversity and functionality of microbes (Díez et al., 2016). Especially the comparison between the Arabian Sea with its intense OMZ, higher nutrient levels and primary productivity (Gregg et al., 2017; Gregg & Rousseaux, 2019; Marra & Barber, 2005) and the oligotrophic, river-influenced Bay of Bengal with an OMZ with oxygen concentrations just above zero (see also Hood et al., 2024a, 2024b) makes the Indian Ocean an excellent natural laboratory to study microbial responses to climate variability and change and to explore the evolution of OMZs. Still, the Indian Ocean is one of the least studied oceans regarding microbial diversity, activity, and feedback with climate change when compared to other oceans (Li et al., 2012; Venter et al., 2004; Wang et al., 2021).

1.2 Microbial research in the Indian Ocean: Early expeditions and methods

First, reports on microbial diversity and the function of the Indian Ocean are available from early expeditions. The Galathea and RV Anton Bruun expeditions in the early 1950s, followed by the International Indian Ocean Expedition (IIOE) from 1959 to 1965 (Snider, 1961) provide, for instance, the first record of carbon fixation, however, without providing details on the microbes producing those rates. To investigate the microbes fixing carbon, earlier environmental microbial studies used culture-dependent methods to study microbial identity and function. As modern sequencing technology developed, more recent studies typically use metagenomic deep-sequencing methods, thus providing a new angle to our understanding of microbial diversity in those waters. However, the number of studies on the benthic and pelagic microbiomes in the Indian Ocean is very limited (Fernandes et al., 2008; Lincy & Manohar, 2020; Orsi et al., 2017; Wu et al., 2019). Another way of understanding the Indian Ocean's microbes is to derive their presence, abundance, and activity from other biogeochemical parameters, including chlorophyll distribution, nutrient, carbon, and oxygen budgets, isotope records and rate measurements of elemental turnover carried out by microbes (e.g., Bristow et al., 2017; Jayakumar et al., 2012; Jensen et al., 2011; Menezes et al., 2020; Ward et al., 2009). A general pattern, which can be deducted from, e.g., satellite observations, is the overall higher primary production in the Arabian Sea compared to the Bay of Bengal (Fig. 1), leading to a higher organic matter production and export (see also Hood et al., 2024a, 2024b). This fundamental difference is largely responsible and important for shaping the biogeochemistry of both basins, as it allows for respiration of organic matter, nitrogen, and sulfur cycling to take place in different intensities.

---

[*]This book has a companion website hosting complementary materials. Visit this URL to access it: https://www.elsevier.com/books-and-journals/book-companion/9780128226988.

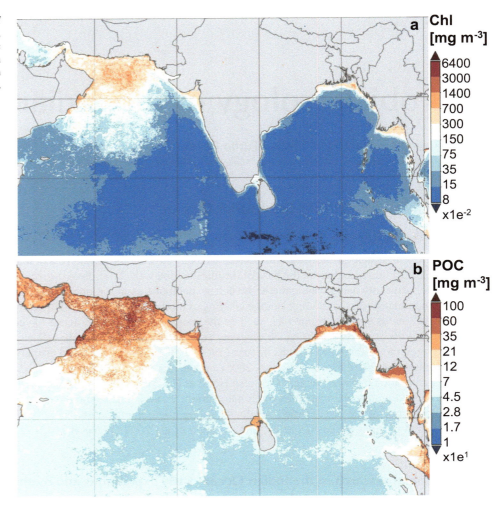

FIG. 1 (a) Time-averaged map of chlorophyll concentrations (Chl, mg m^{-3}) and (b) particulate organic carbon (POC, mg m^{-3}) distribution in the Indian Ocean, obtained from MODIS-Aqua (4 km resolution, monthly measurements).

2 Microbial diversity in the Indian Ocean

2.1 Prokaryotic microbes in the Arabian Sea and the bay of Bengal—Who is out there?

Available studies are derived from coastal and open waters and from sediments from within or outside the OMZs. As mentioned above, these studies are based on a variety of sampling and measurement techniques. This results in a certain challenge to compare different datasets. However, while relative abundances and quantities may vary, a general pattern of key taxa and functional groups of microbes can be derived. By comparing the microbial diversity of Arabian Sea and Bay of Bengal waters, a generally high diversity among microbes was found. A key study describing and comparing the microbial diversity of Arabian Sea and Bay of Bengal waters found a generally high diversity among microbes with, however, lower diversity in surface and OMZ core waters (Fernandes et al., 2020). The highest diversity was detected around the upper oxycline in both basins. In this picture, as is the case for many ocean regions, Proteobacteria are most dominant, with γ- and δ-Proteobacteria being key players (Bandekar et al., 2018; Bristow et al., 2017; Fernandes et al., 2020; Wang et al., 2021). One study demonstrated the particular importance of γ- Proteobacteria in the South Indian Ocean, with the metabolic potential of this diverse group largely governed by the depth-related environmental parameters of the Agulhas Current (Phoma et al., 2018). Other important taxa in the Arabian Sea and Bay of Bengal OMZ include β- Proteobacteria, Actinobacteria, Bacteroidetes, Chloroflexi, Marinomicrobia, Nitrospinia, Planctomycetes and the SAR406 cluster (Bandekar et al., 2018; Fernandes et al., 2020; Wang et al., 2021).

Representatives among the Proteobacteria, Actinobacteria, and Bacteroidetes are known to contribute to an active heterotrophic community degrading organic material and thus removing oxygen from the OMZ (Glöckner et al., 1999). Because we see higher organic carbon loads in the Arabian Sea, higher respiration is possible, providing one explanation for the more intense OMZ (see also Hood et al., 2024b; Fig. 1b). Modeled functional predictions suggested involvement of this community in nitrogen and sulfur cycling, which is in line with previous studies on isolated representatives of γ- and δ-Proteobacteria from both basins (Fernandes et al., 2020). Metagenomic studies identified denitrifiers and anammox bacteria in the Arabian Sea OMZ (Bandekar et al., 2018), with *Candidatus Scalindua* being the dominant organism among the latter functional group (Pitcher et al., 2011). In the Bay of Bengal, γ-proteobacterial sulfur oxidizers of the SUP05 clade were detected harboring nitrate reductases, thus also suggesting a potential of a (coupled) sulfur-nitrogen cycle (Bristow et al., 2017; Wang et al., 2021). Recent studies have emphasized the quantitative importance of cyanobacteria in the eastern Indian Ocean upper water column, while α-Proteobacteria were shown to dominate deeper waters. In the equatorial waters of the Indian Ocean, Prochlorococcus and Synechococcus can account for up to 90% of the cyanobacterial reads (Díez et al., 2016; Wu et al., 2022a). Prochlorococcus populations are defined according to their light adaptation, separating them into high light-adapted populations and low light-adapted populations. In the Indian Ocean, the high-light adapted ecotype is abundant in the Prochlorococcus community and plays an important role in microbial primary production but also in nutrient removal from the upper water column (Farrant et al., 2016).

To understand which environmental parameters shape the microbial community structure and distribution throughout the Arabian Sea and Bay of Bengal basins, a statistical approach was applied (Fernandes et al., 2020). Here, six environmental variables (temperature, salinity, dissolved oxygen (DO), pH, nitrate (NO_3^-), and nitrite (NO_2^-)) were explored regarding their potential to describe patterns of microbial community distribution. This study demonstrated that in the Arabian Sea, dissolved oxygen, nitrate, and pH are most important in shaping the community. For instance, the genera Nitrospina, Nitrospira, Woeseia, SUP05, Pelagibacteraceae, and *Candidatus Scalindua* were negatively correlated with DO and positively correlated with NO_2^-. In the Bay of Bengal, Pelagibacteraceae, SUP05, Bacillus, and Nitrospina, were found at low DO, NO_2^- and pH conditions. In Bay of Bengal surface waters, however, Synechococcus and Prochlorococcus strains were associated with high DO, NO_2^-, and temperature. While this picture is not entirely conclusive, it speaks to the importance of the differences in the biogeochemical character of the two OMZs for shaping the microbial community and vice versa.

While there is a growing body of work available on Arabian Sea and the Bay of Bengal microbial ecology, the southern Indian Ocean is largely unexplored in this regard. One of the few studies is based on the Tara Ocean dataset, which provides 16S rDNA sequence data for the southern Indian Ocean (Castillo et al., 2022). Here, a pattern with generally low proportions of cyanobacteria and Bacteroidota and high relative abundances of Chloroflexi and Verrumicrobiota has been observed (Castillo et al., 2022; Delmont et al., 2018). A study exploring the microbial biogeography of the Indian Ocean (Jeffries et al., 2015) delivered a description of microbial diversity in relation to the Longhurst provinces (Longhurst, 1998, 2006; Longhurst et al., 1995), dividing the Indian Ocean into the Indian Monsoon Gyres province containing the Bay of Bengal and Arabian Sea and extending to about 10°S, and the Indian Southern Subtropical Gyre Province expanding from about 10°S to about 40°S. A key outcome of this study was the identification of a highly distinct diversity of microbes in connection with the Chagos archipelago, a pristine lagoon of the Solomon Islands in the Indian Monsoon Gyres province, as compared to their mid-latitude Southern Ocean, Southern Ocean, and Bay of Bengal samples. An overall high degree of dissimilarity between the microbial diversity identified in the Bay of Bengal versus the southern Indian Ocean resulted from increased abundances of SAR 86, Alteromonas clades, Synechococcus, and also archaeal sequences of the Marine group II in the Bay of Bengal compared to the southern Indian Ocean samples, and an opposite trend for SAR11 clades (Jeffries et al., 2015).

2.2 Primary producers—Who is fixing the carbon?

Historically, phytoplankton diversity records were derived mostly from direct or microscopic phytoplankton counts; therefore, small-sized phytoplankton and cyanobacteria were likely underrepresented. In both the Arabian Sea and the Bay of Bengal, diatoms were found to be the dominant group of phytoplankton (Devassy et al., 1983; Gauns et al., 2005; Madhupratap et al., 2003; Sawant & Madhupratap, 1996; see also Hood et al., 2024a) with descriptions down to the species level available from the Bay of Bengal showing a diversity of diatoms including *Thalassiothrix, Nitzschia, Thalassionema, Skeletonema, Chaetoceros* and *Coscinodiscus* (Devassy et al., 1983; Nair & Gopinathan, 1983; Ramaiah et al., 2010). Other groups were later identified through sequencing-based approaches, including *Haptophyceae, Chrysophyceae, Eustigamatophyceae, Xanthophyceae, Cryptophyceae, Dictyochophyceae, Pelagophyceae,* and *Pinguiophyceae* adding as well small cyanobacteria (Bemal et al., 2019; Larkin et al., 2020; Li et al., 2012; Löscher et al., 2020; Pujari et al., 2019; Yuqiu et al., 2020). The identified cyanobacteria accounted for up to 60% of primary producers in the central Bay of Bengal and included *Synechococcus* and *Prochlorococcus*. The former has been detected from the surface down to the chlorophyll maximum, while the latter has been found abundant in the lower margin of the chlorophyll maximum at around 50–80 m water depth, slightly deeper than the maximum of eukaryotic primary producers (Löscher et al., 2020; Yuqiu et al., 2020). The *Prochlorococcus* population has been described to consist of several different ecotypes of the HLII clade, with their respective abundances being governed by macro- and micronutrient distribution and by temperature (Larkin et al., 2020; Pujari et al., 2019). These distribution patterns have also been found in other OMZ regions (Beman & Carolan, 2013; Franz et al., 2012; Meyer et al., 2015). A change in Prochlorococcus ecotypes in response to changes in temperature, nutrient availability, and iron stress has been observed by Larkin et al. (2020). It remains unclear, however, how far this change in ecotypes would impact cyanobacterial primary production. In addition to picocyanobacteria, the nitrogen-fixing cyanobacteria *Trichodesmium* has been observed in both basins, the Arabian Sea and the Bay of Bengal (Devassy et al., 1983; Hegde et al., 2008; Jyothibabu et al., 2017; Sahu et al., 2017; Shetye et al., 2013; Wu et al., 2019). In the southern Indian Ocean Prochlorococcus has also been identified to dominate the cyanobacterial pool of primary producers with the exception of the Chagos archipelago, where a pronounced shift to Synechococcus was observed (Jeffries et al., 2015), along with an increase in chlorophyll concentrations most pronounced during the time of the southwest monsoon (Dalpadado et al., 2021). *Trichodesmium* abundances have been shown to decrease from the BoB into the southern Indian Ocean (Wu et al., 2019).

In the Arabian Sea, *Noctiluca scintillans*, a mixotrophic dinoflagellate, is taking on a special role, as it has been observed to increasingly contribute to algal blooms in the northern Arabian Sea from the early 2000s on (e.g., Gomes et al., 2008; Goes et al., 2020; Lotliker et al., 2020; see also Hood et al., 2024a). This phenomenon has been controversially discussed, with one line of reasoning suggesting an impact of influx of OMZ water into surface waters (do Rosário Gomes et al., 2014; Gomes et al., 2008) and another line proposing varying intensities of winter convective mixing changing silicate:nitrate ratios (Lotliker et al., 2020; Prakash et al., 2017; Sarma et al., 2019). A community shift from diatoms to *Noctiluca* in the water column can severely impact the total primary production in the northern Arabian Sea (Sawant & Madhupratap, 1996) with potentially severe consequences also for organic matter export and remineralization in the OMZ.

3 Functional diversity of microbes in the Indian Ocean

3.1 Microbes of the nitrogen (N) cycle

The N cycle (Fig. 2), consisting of N_2 fixation as the only external biological source of new N to the ocean, of N loss processes, and oxidative processes, is chiefly important when aiming at understanding the difference between the two basins, the Arabian Sea and Bay of Bengal, and the different character of their OMZs. The fixation of dinitrogen (N_2) gas, hereby, is highly relevant as it directly and quantitatively impacts primary production and, with that, the character and amount of organic matter exported to OMZ waters. In the OMZ, denitrification and the anaerobic oxidation of ammonia with nitrate or nitrite (anammox) are key processes not only regarding organic matter and nutrient remineralization but also because they respire nitrate, making it unavailable for primary production and reduce the amount of electron acceptors available in the water column. Other processes in the N cycle, which will briefly be discussed here, are nitrification, the oxidation of ammonia to nitrite and further to nitrate. Nitrification, as well as denitrification, also plays an important role in the production of the strongest natural greenhouse gas, nitrous oxide (N_2O). The quantity of N_2O production is strongly dependent on the microbial community present in the different Indian Ocean waters, with the potential to either produce or consume N_2O and therefore providing a switch between the Indian Ocean becoming a greenhouse gas sink or source (Bange, 2009). For additional review and discussion of the impacts of the OMZs on the N cycle and the biogeochemistry of the northern Indian Ocean, see Hood et al., 2024b.

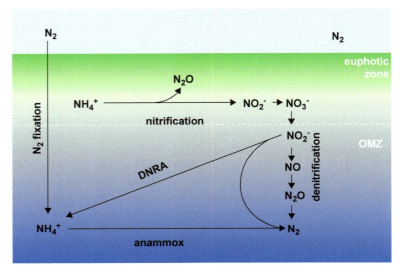

FIG. 2 Systematic depiction of the nitrogen cycle with N_2 fixation providing new nitrogen in the form of ammonia (NH_4^+) for nitrification and anammox. During nitrification, ammonia is oxidized to nitrite (NO_2^-) and further to nitrate (NO_3^-); nitrous oxide (N_2O) is a side product. Denitrification can make use of nitrate, which is respired in a 4-step reaction to N_2. N_2 is also the end-product of the anammox reaction using NH_4^+ and NO_2^-.

3.1.1 N₂ fixation

The microbial fixation of N_2 is essential to life in the ocean, making the otherwise unavailable N_2 gas bioavailable to primary producers. This process is limited to a diverse group of microbes also referred to as diazotrophs. Classically, marine diazotrophs were considered to mostly belong to cyanobacteria and, due to the light need of those photosynthetic bacteria, be limited to the photic zone of the water column. Over the last decade, however, a tremendous diversity of diazotrophs not associated with cyanobacteria has been found (Fernandez et al., 2011; Halm et al., 2011; Hamersley et al., 2011; Jayakumar et al., 2012, 2017; Löscher et al., 2014; Wang et al., 2021), which seems to consistently be present throughout the oceans, especially in low oxygen environments (Jayakumar & Ward, 2020; Reeder & Löscher, 2022). The discovery of those novel clades can be considered a paradigm shift, especially because those non-cyanobacterial diazotrophs do not seem to be governed by the same environmental controls as cyanobacterial diazotrophs, making it extremely difficult to estimate their contribution to N budgets and include their rates in models.

While N_2 fixation has been found to play a minor role in large parts of the Bay of Bengal (Löscher et al., 2020; Saxena et al., 2020; Singh et al., 2012; Singh & Ramesh, 2011), estimates for N_2 fixation in the Arabian Sea vary widely but are commonly found in the upper range with rates of $3.3-15.4\,Tg\,N\,yr^{-1}$ (Bange et al., 2005; Gandhi et al., 2011; Mulholland & Capone, 2009). One reason for this extreme difference can be found in the difference in general biogeochemistry and topography between the two basins, translating into two different communities of N_2 fixers (Fig. 3). The most obvious difference

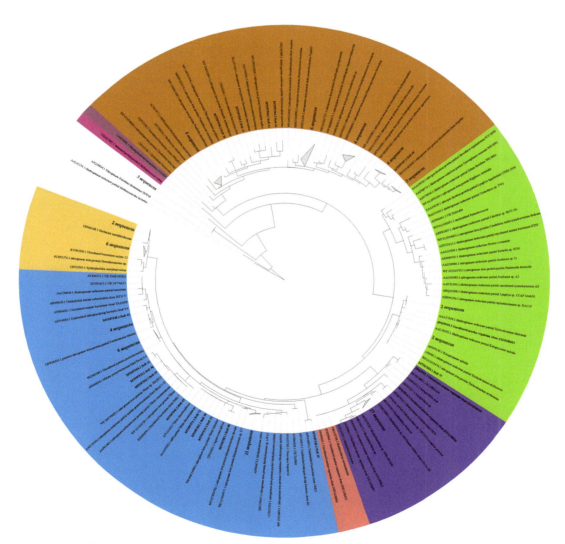

FIG. 3 Phylogenetic tree of diazotrophs present in the Arabian Sea and the Bay of Bengal, datasets from Jayakumar and Ward (2020) and Löscher et al. (2020). The highest diversity is present among heterotroph bacteria of Cluster I and III.

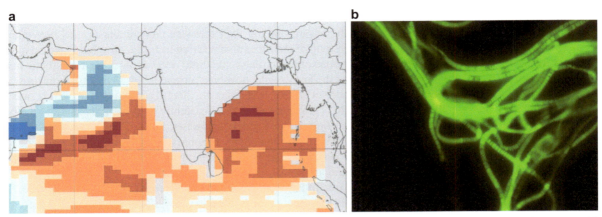

FIG. 4 (a) Satellite-based distribution of cyanobacteria in the Indian Ocean in spring 2015, when a bloom of *Trichodesmium* has been reported from the Bay of Bengal. (b) The photograph shows a *Trichodesmium* culture. *Photo C.F. Reeder.*

is the presence of *Trichodesmium*, a cyanobacterial diazotroph, which has also been described to form blooms producing high N_2 fixation rates in the Arabian Sea (Fig. 4; Gandhi et al., 2011). Other studies, however, also emphasize the potential contribution of non-cyanobacterial diazotrophs in suboxic waters to N_2 fixation in the Arabian Sea (Bird & Wyman, 2013; Jayakumar et al., 2012).

In the Bay of Bengal, the N_2 fixer community consists largely of heterotrophic N_2 fixers, which do not necessarily fix N_2 if there is a deficit of nitrogen compared to phosphorus but respond to yet unresolved factors (Löscher et al., 2020; Sahu et al., 2017; Saxena et al., 2020). There are studies available demonstrating the presence of *Trichodesmium* as well as diatom-diazotroph associations playing a role in nitrogen fixation (Bhaskar et al., 2007; Sahu et al., 2017), but data sets indicate high spatial and temporal variability.

Previous reports suggested that N_2 fixers in the Bay of Bengal might be limited by iron, other micronutrients, or organic matter (Benavides et al., 2018; Löscher et al., 2020; Saxena et al., 2020; Shetye et al., 2013), with organic matter limitation possibly imposing a feedback regulation for primary production, thus providing one explanation of the low productivity of the Bay of Bengal as a whole.

In the Southeast Indian Ocean, a decrease of UCYN-B and *Trichodesmium*-related diazotrophs has been demonstrated along with an increase in heterotrophic diazotrophs, particularly the Alphaproteobacterium *Sagittula castanea* (Wu et al., 2019, 2022b).

3.1.2 Nitrification and N loss processes

Nitrification has been described to take place in both basins, with the respective community of ammonia-oxidizing archaea being the key microbes catalyzing the first step of nitrification, the oxidation of ammonia, and *Nitrospina* and *Nitrospira*, as well as another to date unidentified nitrite oxidizer in the Arabian Sea catalyzing the second step of nitrite oxidation to nitrite (Bristow et al., 2017; Lüke et al., 2016; Newell et al., 2011; Schouten et al., 2012; Villanueva et al., 2014). A detailed comparison of the nitrifier communities in both basins is challenged due to the use of different methods. A comparison of available sequence data presented in earlier studies (Löscher et al., 2020; Lüke et al., 2016) shows, however, similar ammonia-oxidizing archaea communities based on the functional marker gene *amoA* (Fig. 5). Quantitatively, it also appears from the few available datasets that ammonia-oxidizing archaea occur in similar abundances in both the Arabian Sea, and the Bay of Bengal (Bristow et al., 2017; Newell et al., 2011). Given the autotrophic nature of ammonia-oxidizing archaea and their need to actively oxidize ammonia to survive, this would speak for a similar potential for nitrification in both basins.

In the Arabian Sea, genes for ammonia-oxidizing archaea nitrifier denitrification, a process in which the produced nitrite is further reduced to N_2O and potentially to N_2, have been detected (Lüke et al., 2016). Ammonia-oxidizing archaea are known to harbor the genetic potential for a copper-containing nitrite oxidase (Stahl & Torre, 2012). Ammonia-oxidizing archaea nitrifier denitrification might play a role in N_2O production in the Arabian Sea and other ocean regions (Babbin et al., 2015; Kozlowski et al., 2016).

N loss processes, namely anammox and denitrification, mark a key difference between the Arabian Sea and the Bay of Bengal, both qualitatively and quantitatively, with the Bay of Bengal having been described to be a site of low to absent N

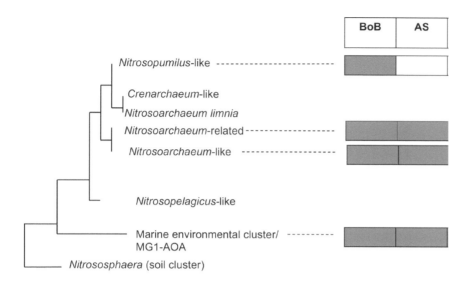

FIG. 5 Phylogenetic tree of ammonia-oxidizing archaea *amo*A, based on data from Löscher et al. (2020) and Lüke et al. (2016). Except for some Nitrosopumilus-like sequences, the community of ammonia-oxidizing archaea seems to be similar in both basins.

loss (Bristow et al., 2017), however with some potential for it if oxygen is fully depleted (Johnson et al., 2019), while the Arabian Sea being a classic site of intense N loss by both processes, denitrification and anammox (Bandekar et al., 2018; Jayakumar et al., 2004; Jensen et al., 2011; Naqvi, 2008; Naqvi et al., 1998, 2000, 2006; Pitcher et al., 2011; Villanueva et al., 2014; Ward et al., 2009; see also Hood et al., 2024b). Still, it is debated and not fully explored whether denitrification or anammox dominates N loss in the Arabian Sea. There are two reports demonstrating the quantitative importance of denitrification in the Arabian Sea (Bulow et al., 2010; Ward et al., 2009), and one study showing that anammox coupled to dissimilatory nitrate reduction to ammonia (Fig. 2) is the key N loss process in this system (Jensen et al., 2011). Other studies described a high abundance of anammox bacteria at the core of the Arabian Sea OMZ (Pitcher et al., 2011; Villanueva et al., 2014), an observation in line with the omnipresence of anammox bacteria and the quantitative importance of anammox in many OMZ regions (Hamersley et al., 2007; Kuypers et al., 2005; Lam et al., 2009; Thamdrup et al., 2006).

Consistent with results from other OMZs (Lam et al., 2009; Ulloa et al., 2012), the enzyme catalyzing the first step of both anammox and denitrification, the reduction of nitrate to nitrite has been found to be the most dominant nitrogen cycle enzyme encoded in the Arabian Sea OMZ core (78% of normalized *rpoB* gene abundance; Lüke et al., 2016). This genetic potential is likely a result of many different microbes being able to use nitrate often to remineralize organic matter (Kalvelage et al., 2015).

The same study, however, found only very limited potential for other denitrification steps, for instance, nitrite oxidation in the Arabian Sea OMZ. While many reads associated with the nitrite reductase encoded by either *nirS or nirK* were found, 50% of all *nirS* reads were classified as related to the anammox bacterium *Scalindua* with a more detailed genetic analysis indicating the presence of one dominant uncultivated anammox ecotype and thus suggestive of a low diversity but a high abundance of anammox bacteria at the time of sampling. Consistent with the dominance of nitrite reduction genes, downstream denitrification genes (nitric oxide and N_2O reductase genes) were also affiliated with *Scalindua*.

The low genetic potential for denitrification stands in contrast to the findings of Ward et al. (2009), but this study demonstrated denitrification to be faster and quantitatively more important than anammox at the time of sampling and documented denitrification to be directly dependent on organic matter availability and to be sustained by one dominant denitrifier type.

3.1.3 Niches in the N cycle

Both nitrification and N loss processes are connected via nitrite and nitrate produced by nitrification, which can potentially be taken up by primary producers or be used as a substrate for anammox and denitrification. However, niche segregation between ammonia oxidizers mainly present in the upper oxycline at oxygen concentrations of around 5 μM and anammox bacteria peaking in the OMZ core has been described (Fig. 6; Pitcher et al., 2011). This niche segregation has been found in other OMZ regions, too (Kalvelage et al., 2013), and would classically be explained by the oxygen requirement of nitrification. However, in other OMZ waters, ammonia-oxidizing archaea have been found to actively oxidize ammonia at anoxic conditions (Löscher et al., 2012). A very recent study provided an explanation for this enigma and showed that ammonia-oxidizing archaea are able to produce oxygen and are thus able to sustain aerobic ammonia oxidation even within

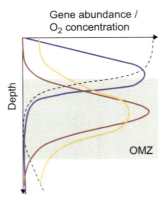

FIG. 6 Systematic depiction of niches taken by ammonia oxidizers (*purple*), denitrifiers (*yellow*), and anammox bacteria in both the Arabian Sea and the Bay of Bengal along the natural oxygen gradient (*dashed black line*). The OMZ is indicated with a *gray* box. Following the gene quantification presented by Ward et al. (2009) and Bristow et al. (2017), a difference of two to three orders of magnitude in gene abundance can be found, with higher abundances in the Arabian Sea compared to the Bay of Bengal.

OMZ core waters (Kraft et al., 2022). Therefore, other factors, including substrate affinity or trace metal requirements, might provide additional explanations for the different niches of aerobic and anaerobic ammonia oxidizers in this OMZ.

3.2 The microbial sulfur cycle

Besides the nitrogen cycle, the sulfur cycle plays an important role in OMZ regions. Sulfur cycling is most important in coastal OMZ regions with high organic matter load, such as the coastal Arabian Sea and possibly parts of the coastal Bay of Bengal (Fig. 1b). In such regions, sediments can become sulfide-rich in response to high surface productivity and organic matter export, and eutrophic and anoxic conditions can result in flux from sediments to the water column that can form the basis of a coupled benthic-pelagic sulfur cycle (Callbeck et al., 2018, 2021). Sulfide oxidizing microbes couple the sulfur and nitrogen cycle by detoxifying sulfide and reducing nitrate to N_2. They further generate nitrite and ammonia, which can promote anammox or denitrification. In the coastal upwelling region of the Arabian Sea, high surface productivity would possibly facilitate sulfur cycling, but constant ventilation of the West Indian shelf typically contributes to low but persistent oxygen concentrations (Jensen et al., 2011; Naqvi et al., 2000, 2006). During pre-monsoon, however, a freshwater lens in surface waters increased stratification, and high surface productivity can cause oxygen and nitrate depletion (Naqvi et al., 2006) and facilitate the release of sulfide from sediments (Naqvi, 2008; Naqvi et al., 2006; Shenoy et al., 2012). Large sulfide plumes can develop over the West Indian shelf (Naqvi & Jayakumar, 2000; Shirodkar et al., 2018), similar to other OMZ regions (Dugdale et al., 1977; Schunck et al., 2013). Sulfur oxidizing and reducing microbes have been identified in the Arabian Sea and the Bay of Bengal; an active sulfur cycle has, however, only been observed in the Arabian Sea (Fernandes et al., 2020; Menezes et al., 2020).

4 Microbial feedback loops impacting OMZ intensity and expansion

In both the Arabian Sea and the Bay of Bengal, a nitrite loop (Fig. 7) has been proposed with the nitrite formed by nitrate reduction to be re-oxidized to nitrate, ultimately resulting in the removal of additional organic matter and release of more ammonium (Bristow et al., 2017; Lüke et al., 2016). It has been suggested that this loop results from the low but permanently present trace concentrations of oxygen in the Bay of Bengal OMZ, and it has also been proposed to provide a substrate for anammox bacteria in the Arabian Sea OMZ (Jensen et al., 2011; Pitcher et al., 2011).

An N_2 fixation feedback loop connected to the OMZ in the Bay of Bengal has been proposed, which is based mainly on the low observable N_2 fixation and the specific community of diazotrophs lacking active N_2 fixing cyanobacteria in surface waters. Here, low primary production and organic matter export directly hinder heterotrophic N_2 fixation in the OMZ. A stratified water column further hinders the export of fixed N into surface waters, thus, again, reducing primary production (Fig. 8). This feedback cycle could, however, at any time change, as soon as upwelling increases (Löscher et al., 2020). In this case, the Bay of Bengal could become more productive, increased organic matter load would allow for increased respiration in the OMZ, and the Bay of Bengal could become fully anoxic. Enhanced upwelling could be a response to global warming; therefore, such a scenario might be possible in the future (Bakun, 1990).

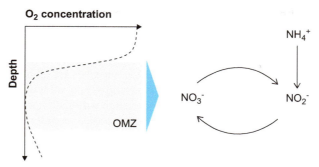

FIG. 7 A nitrite loop with nitrite from either ammonia oxidation or from nitrate reduction is suggested to limit nitrite accumulation by nitrite being re-oxidized to nitrate. This feedback loop can, for instance, occur at the upper oxycline of the Arabian Sea but also in OMZ core waters of the Bay of Bengal, where trace oxygen concentrations are still left.

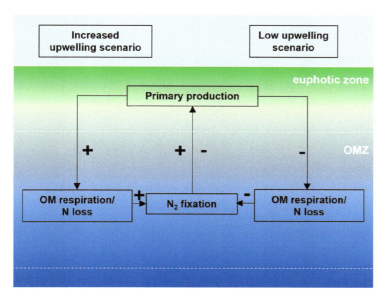

FIG. 8 N_2 fixation-based feedback loop as proposed for the Bay of Bengal. Low primary production leads to organic matter export to the OMZ. The result is low respiration, maintaining traces of oxygen even in the OMZ core. Heterotrophic N_2 fixation, which has a requirement for organic matter and is highly oxygen-sensitive is hindered; no new N_2 is fixed. Sluggish upwelling circulation further hinders the flux of fixed nitrogen to surface waters, where production, therefore, remains low. If upwelling would increase, the external supply of nutrients from the atmosphere, rivers, or land would increase primary production, higher organic matter loads would reach the OMZ, and the system could develop anoxia.

5 Conclusions and outlook

Microbes largely shape the biogeochemistry of the Indian Ocean, starting with primary production, where different communities determine the productivity of the two opposing systems of the Arabian Sea and the Bay of Bengal. Trends in primary production include the occurrence of Noctiluca blooms in the Arabian Sea, which changes the community composition of phytoplankton, with the potential to directly impact the amount and character of organic matter exported to deeper waters. The character of organic material has been shown to be of key importance for respiration processes, including those of the nitrogen cycle (Babbin et al., 2014). Remineralization of organic matter, as well as the quantity of organic matter exported in coastal regions such as the shelf in the western Arabian Sea, can further impact sulfur cycling in areas where sulfide fluxes between sediment and water column evolve. The presence of a specific microbial community capable of denitrification, anammox, and sulfide oxidation, along with heterotrophic degraders, harbors the potential to promote and increase anoxia in those regions. In open waters of the Arabian Sea, N_2 fixation, especially if carried out by efficiently fixing cyanobacterial diazotrophs, including *Trichodesmium*, can further promote offshore productivity, organic matter export to the OMZ, and enhance anoxia and OMZ expansion beyond shelf regions. While we see different results on the dominant N loss process, we certainly can conclude that the Arabian Sea is a strong source of N loss, with denitrification, anammox, and dissimilatory nitrate reduction to ammonia contributing to N turnover, removal, and respiration in the OMZ (Jensen et al., 2011; Lüke et al., 2016; Villanueva et al., 2014; Ward et al., 2009). The low diversity of anammox bacteria is a remarkable feature of this region, which is not fully understood yet but might have relevance for predicting the future of N loss in the Arabian Sea.

In the Bay of Bengal, lower and possibly decreasing productivity might result from a specific community of primary producers and N_2 fixers, the latter dominated by heterotrophic bacteria. Those "atypical" N_2 fixers are, to date, not fully understood, and future surveys will have to explore what factors impact and control those organisms in greater detail. The Bay of Bengal is an unusual setting with the enigma of the not-fully anoxic OMZ. This feature will allow us to study under which circumstances an OMZ can become anoxic and what specific microbes are required to allow for anoxia or prevent it. So far, we can see that the genetic potential of the microbial community is not all too different from the Arabian Sea, with the difference being quantitative rather than qualitative. Future research will likely look into describing the microbial communities in both regions, as in combined metagenomic-functionality studies, in order to understand who the key drivers of biogeochemical cycles in this region are and under which circumstances they are active. This information will be critically needed to understand how microbes in the Indian Ocean will respond to climate change and how far the present microbes will be able to stabilize each system in microbial feedback loops.

A largely underexplored region is the southern Indian Ocean, with only a handful of studies describing the present microbiology available, and thus demonstrating the need for future studies exploring microbial diversity and function, including primary productivity and nitrogen turnover.

6 Educational resources

- Galathea expedition: https://en.natmus.dk/historical-knowledge/historical-knowledge-the-world/asia/india/tranquebar/danish-era-1620-1845/publications-on-the-danish-era/the-galathea-expedition-1845-1847/
- IIOE-2 (2nd International Indian Ocean Expedition) program: https://iioe-2.incois.gov.in
- Tara Ocean expeditions: https://fondationtaraocean.org/en/expedition/tara-oceans/

Acknowledgments

The authors thank the Danish Research Council and the Villum Fonden for financial support.

Author contributions

CRL and CFR analyzed data, and CRL wrote the chapter with input from CFR.

References

Babbin, A. R., Bianchi, D., Jayakumar, A., & Ward, B. B. (2015). Rapid nitrous oxide cycling in the suboxic ocean. *Science, 348*, 1127–1129.

Babbin, A., Keil, R., Devol, A., & Ward, B. (2014). Organic matter stoichiometry, flux, and oxygen control nitrogen loss in the ocean. *Science, 344*, 406–408.

Bakun, A. (1990). Global climate change and intensification of coastal ocean upwelling. *Science, 247*, 198–201.

Bandekar, M., Ramaiah, N., Jain, A., & Meena, R. M. (2018). Seasonal and depth-wise variations in bacterial and archaeal groups in the Arabian Sea oxygen minimum zone. *Deep Sea Research Part II: Topical Studies in Oceanography, 156*, 4–18.

Bange, H. W. (2009). Nitrous oxide in the Indian Ocean. In J. D. Wiggert, R. R. Hood, S. W. A. Naqvi, K. H. Brink, & S. L. Smith (Eds.), *Indian Ocean biogeochemical processes and ecological variability* (pp. 205–216). Washington, DC: American Geophysical Union.

Bange, H. W., Naqvi, S. W. A., & Codispoti, L. A. (2005). The nitrogen cycle in the Arabian Sea. *Progress in Oceanography, 65*, 145–158.

Bemal, S., Anil, A. C., & Amol, P. (2019). Picophytoplankton variability: Influence of Rossby wave propagation in the southeastern Arabian Sea. *Journal of Marine Systems, 199*, 103221.

Beman, J. M., & Carolan, M. T. (2013). Deoxygenation alters bacterial diversity and community composition in the oceans largest oxygen minimum zone. *Nature Communications, 4*.

Benavides, M., Martias, C., Elifantz, H., Berman-Frank, I., Dupouy, C., & Bonnet, S. (2018). Dissolved organic matter influences N2 fixation in the new Caledonian lagoon (Western Tropical South Pacific). *Frontiers in Marine Science, 5*.

Bhaskar, J. T., Ramaiah, N., Gauns, M., & Fernandes, V. (2007). Preponderance of a few diatom species among the highly diverse microphytoplankton assemblages in the Bay of Bengal. *Journal of Marine Biology, 152*, 63–75.

Bird, C., & Wyman, M. (2013). Transcriptionally active heterotrophic diazotrophs are widespread in the upper water column of the Arabian Sea. *FEMS Microbiology Ecology, 84*, 189–200.

Bristow, L. A., Callbeck, C. M., Larsen, M., et al. (2017). N2 production rates limited by nitrite availability in the bay of Bengal oxygen minimum zone. *Nature Geoscience, 10*, 24–29.

Bulow, S. E., Rich, J. J., Naik, H. S., Pratihary, A. K., & Ward, B. B. (2010). Denitrification exceeds anammox as a nitrogen loss pathway in the Arabian Sea oxygen minimum zone. *Deep Sea Research Part I: Oceanographic Research Papers, 57*, 384–393.

Callbeck, C. M., Canfield, D. E., Kuypers, M. M. M., Yilmaz, P., Lavik, G., Thamdrup, B., Schubert, C. J., & Bristow, L. A. (2021). Sulfur cycling in oceanic oxygen minimum zones. *Limnology and Oceanography, 66*, 2360–2392.

Callbeck, C. M., Lavik, G., Ferdelman, T. G., et al. (2018). Oxygen minimum zone cryptic sulfur cycling sustained by offshore transport of key sulfur oxidizing bacteria. *Nature Communications, 9*, 1729.

Castillo, D. J., Dithugoe, S. D., Bezuidt, O. K., & Makhalanyane, T. P. (2022). Microbial ecology of the Southern Ocean. *FEMS Microbiology Ecology, 98*, 11. https://doi.org/10.1093/femsec/fiac123.

Dalpadado, P., Arrigo, K. R., van Dijken, G. L., Gunasekara, S. S., Ostrowski, M., Bianchi, G., & Sperfeld, E. (2021). Warming of the Indian Ocean and its impact on temporal and spatial dynamics of primary production. *Progress in Oceanography, 198*. https://doi.org/10.1016/j.pocean.2021.102688.

Delmont, T. O., Quince, C., Shaiber, A., et al. (2018). Nitrogen-fixing populations of planctomycetes and proteobacteria are abundant in surface ocean metagenomes. *Nature Microbiology, 3*, 804–813.

Devassy, V. P., Bhattathiri, P. M. A., & Radhakrishna, K. (1983). Primary production in the Bay of Bengal during august, 1977. *Mahasagar-Bulletin of National Institute of Oceanography, 16*, 443–447.

Díez, B., Nylander, J. A. A., Ininbergs, K., Dupont, C. L., Allen, A. E., Yooseph, S., Rusch, D. B., & Bergman, B. (2016). Metagenomic analysis of the Indian Ocean picocyanobacterial community: Structure, potential function and evolution. *PLoS One, 11*, e0155757.

do Rosário Gomes, H., Goes, J. I., SGP, M., Buskey, E. J., Basu, S., Parab, S., & Thoppil, P. (2014). Massive outbreaks of Noctiluca scintillans blooms in the Arabian Sea due to spread of hypoxia. *Nature Communications, 5*, 4862.

Dugdale, R. C., Goering, J. J., Barber, R. T., Smith, R. L., & Packard, T. T. (1977). Denitrification and hydrogen sulfide in the Peru upwelling region during 1976. *Deep-Sea Research, 24*.

Farrant, G. K., Doré, H., Cornejo-Castillo, F. M., et al. (2016). Delineating ecologically significant taxonomic units from global patterns of marine picocyanobacteria. *Proceedings of the National Academy of Sciences, 113*, E3365–E3374.

Fernandes, V., Nagappa, R., Bhaskar, J., Sardessai, S., JyotiBabu, R., & Gauns, M. (2008). Strong variability in bacterioplankton abundance and production in central and western Bay of Bengal. *Marine Biology, 153*.

Fernandes, G., Shenoy, B., & Damare, S. (2020). Diversity of bacterial community in the oxygen minimum zones of Arabian Sea and Bay of Bengal as deduced by Illumina sequencing. *Frontiers in Microbiology, 10*.

Fernandez, C., Farias, L., & Ulloa, O. (2011). Nitrogen fixation in denitrified marine waters. *PLoS One, 6*, 9.

Franz, J., Krahmann, G., Lavik, G., Grasse, P., Dittmar, T., & Riebesell, U. (2012). Dynamics and stoichiometry of nutrients and phytoplankton in waters influenced by the oxygen minimum zone in the eastern tropical Pacific. *Deep-Sea Research Part I: Oceanographic Research Papers, 62*, 20–31.

Gandhi, N., Singh, A., Prakash, S., Ramesh, R., Raman, M., Sheshshayee, M. S., & Shetye, S. (2011). First direct measurements of N2 fixation during a Trichodesmium bloom in the eastern Arabian Sea. *Global Biogeochem Cycles, 25*, 10.

Gauns, M., Madhupratap, M., Nagappa, R., Retnamma, J., Fernandes, V., Bhaskar, J., & PrasannaKumar, S. (2005). Comparative accounts of biological productivity characteristics and estimates of carbon fluxes in the Arabian Sea and the Bay of Bengal. *Deep Sea Research Part II Topical Studies in Oceanography, 52*.

Glöckner, F. O., Fuchs, B. M., & Amann, R. (1999). Bacterioplankton compositions of lakes and oceans: A first comparison based on fluorescence in situ hybridization. *Environmental Microbiology, 65*, 3721–3726.

Goes, J. I., Tian, H., Gomes, H. R., Anderson, O. R., Al-Hashmi, K., deRada, S., Luo, H., Al-Kharusi, L., Al-Azri, A., & Martinson, D. G. (2020). Ecosystem state change in the Arabian Sea fuelled by the recent loss of snow over the Himalayan-Tibetan Plateau region. *Scientific Reports, 10*, 7422.

Gomes, H. R., Goes, J. I., Matondkar, S. G. P., Parab, S. G., Al-Azri, A. R., & Thoppil, P. G. (2008). *Blooms of Noctiluca miliaris in the Arabian Sea—An in situ and satellite study*.

Gregg, W. W., & Rousseaux, C. S. (2019). Global Ocean primary production trends in the modern ocean color satellite record (1998–2015). *Environmental Research Letters, 14*, 124011.

Gregg, W. W., Rousseaux, C. S., & Franz, B. A. (2017). Global trends in ocean phytoplankton: A new assessment using revised ocean colour data. *Remote Sensing Letters (Print), 8*, 1102–1111.

Halm, H., Lam, P., Ferdelman, T. G., Lavik, G., Dittmar, T., LaRoche, J., D'Hondt, S., & Kuypers, M. M. M. (2011). *Heterotrophic organisms dominate nitrogen fixation in the South Pacific Gyre*. Submitted.

Hamersley, M. R., Lavik, G., Woebken, D., et al. (2007). Anaerobic ammonium oxidation in the Peruvian oxygen minimum zone. *Limnology and Oceanography, 52*, 923–933.

Hamersley, M., Turk-Kubo, K., Leinweber, A., Gruber, N., Zehr, J., Gunderson, T., & Capone, D. (2011). Nitrogen fixation within the water column associated with two hypoxic basins in the Southern California Bight. *Aquatic Microbial Ecology, 63*.

Hegde, S., Anil, A., Patil, J., Mitbavkar, S., Krishnamurthy, V., & Gopalakrishna, V. (2008). Influence of environmental settings on the prevalence of Trichodesmium spp. in the Bay of Bengal. *Marine Ecology Progress Series, 356*.

Hood, R. R., Coles, V. J., Huggett, J. A., Landry, M. R., Levy, M., Moffett, J. W., & Rixen, T. (2024a). Chapter 13: Nutrient, phytoplankton, and zooplankton variability in the Indian Ocean. In C. C. Ummenhofer, & R. R. Hood (Eds.), *The Indian Ocean and its role in the global climate system* (pp. 293–327). Amsterdam: Elsevier. https://doi.org/10.1016/B978-0-12-822698-8.00020-2.

Hood, R. R., Rixen, T., Levy, M., Hansell, D. A., Coles, V. J., & Lachkar, Z. (2024b). Chapter 12: Oxygen, carbon, and pH variability in the Indian Ocean. In C. C. Ummenhofer, & R. R. Hood (Eds.), *The Indian Ocean and its role in the global climate system* (pp. 265–291). Amsterdam: Elsevier. https://doi.org/10.1016/B978-0-12-822698-8.00017-2.

Jasmin, C., Anas, A., Tharakan, B., Jaleel, A., Puthiyaveettil, V., Narayanane, S., Lincy, J., & Nair, S. (2017). Diversity of sediment-associated planctomycetes in the Arabian Sea oxygen minimum zone. *Journal of Basic Microbiology, 57*, 1010–1017.

Jayakumar, A., Al-Rshaidat, M. M. D., Ward, B. B., & Mulholland, M. R. (2012). Diversity, distribution, and expression of diazotroph *nifH* genes in oxygen-deficient waters of the Arabian Sea. *FEMS Microbiology Ecology, 82*, 597–606.

Jayakumar, A., Chang, B. X., Widner, B., Bernhardt, P., Mulholland, M. R., & Ward, B. B. (2017). Biological nitrogen fixation in the oxygen-minimum region of the eastern tropical North Pacific Ocean. *The ISME Journal, 11*, 2356–2367.

Jayakumar, D. A., Chris, A. F., Naqvi, S. W. A., & Bess, B. W. (2004). Diversity of nitrite reductase genes (nirS) in the denitrifying water column of the coastal Arabian Sea. *Aquatic Microbial Ecology, 34*, 69–78.

Jayakumar, A., & Ward, B. B. (2020). Diversity and distribution of nitrogen fixation genes in the oxygen minimum zones of the world oceans. *Biogeosciences, 17*, 5953–5966.

Jeffries, T., Ostrowski, M., Williams, R., et al. (2015). Spatially extensive microbial biogeography of the Indian Ocean provides insights into the unique community structure of a pristine coral atoll. *Scientific Reports, 5*, 15383. https://doi.org/10.1038/srep15383.

Jensen, M. M., Lam, P., Revsbech, N. P., Nagel, B., Gaye, B., Jetten, M. S. M., & Kuypers, M. M. M. (2011). Intensive nitrogen loss over the Omani Shelf due to anammox coupled with dissimilatory nitrite reduction to ammonium. *The ISME Journal, 5*.

Johnson, K. S., Riser, S. C., & Ravichandran, M. (2019). Oxygen variability controls denitrification in the Bay of Bengal oxygen minimum zone. *Geophysical Research Letters, 46*, 804–811.

Jyothibabu, R., Karnan, C., Jagadeesan, L., Arunpandi, N., Pandiarajan, R. S., Muraleedharan, K. R., & Balachandran, K. K. (2017). Trichodesmium blooms and warm-core ocean surface features in the Arabian Sea and the Bay of Bengal. *Marine Pollution Bulletin, 121*, 201–215.

Kalvelage, T., Lavik, G., Jensen, M. M., et al. (2015). Aerobic microbial respiration in oceanic oxygen minimum zone. *PLoS One, 10*.

Kalvelage, T., Lavik, G., Lam, P., Contreras, S., Arteaga, L., Loscher, C. R., Oschlies, A., Paulmier, A., Stramma, L., & Kuypers, M. M. M. (2013). Nitrogen cycling driven by organic matter export in the South Pacific oxygen minimum zone. *Nature Geoscience, 6*, 228–234.

Kozlowski, J. A., Stieglmeier, M., Schleper, C., Klotz, M. G., & Stein, L. Y. (2016). Pathways and key intermediates required for obligate aerobic ammonia-dependent chemolithotrophy in bacteria and Thaumarchaeota. *The ISME Journal, 10*, 1836–1845.

Kraft, B., Jehmlich, N., Larsen, M., Bristow, L. A., Könneke, M., Thamdrup, B., & Canfield, D. E. (2022). Oxygen and nitrogen production by an ammonia-oxidizing archaeon. *Science, 375*, 97–100.

Kuypers, M. M. M., Lavik, G., Woebken, D., Schmid, M., Fuchs, B. M., Amann, R., Jorgensen, B. B., & Jetten, M. S. M. (2005). Massive nitrogen loss from the Benguela upwelling system through anaerobic ammonium oxidation. *Proceedings of the National Academy of Sciences of the United States of America, 102*, 6478–6483.

Lam, P., Lavik, G., Jensen, M. M., van de Vossenberg, J., Schmid, M., Woebken, D., Dimitri, G., Amann, R., Jetten, M. S. M., & Kuypers, M. M. M. (2009). Revising the nitrogen cycle in the Peruvian oxygen minimum zone. *Proceedings of the National Academy of Sciences of the United States of America, 106*, 4752–4757.

Larkin, A. A., Garcia, C. A., Ingoglia, K. A., Garcia, N. S., Baer, S. E., Twining, B. S., Lomas, M. W., & Martiny, A. C. (2020). Subtle biogeochemical regimes in the Indian Ocean revealed by spatial and diel frequency of Prochlorococcus haplotypes. *Limnology and Oceanography, 65*, S220–S232.

Li, G., Lin, Q., Ni, G., Shen, P., Fan, Y., Huang, L., & Tan, Y. (2012). Vertical patterns of early summer chlorophyll a concentration in the Indian Ocean with special reference to the variation of deep chlorophyll maximum. *Journal of Marine Biology, 2012*, 801248.

Lincy, J., & Manohar, C. S. (2020). A comparison of bacterial communities from OMZ sediments in the Arabian Sea and the Bay of Bengal reveals major differences in nitrogen turnover and carbon recycling potential. *Marine Biology Research, 16*, 656–673.

Longhurst, A. R. (1998). *Ecological geography of the sea*. San Diego: Academic Press. 397 p. (IMIS).

Longhurst, A. R. (2006). *Ecological geography of the sea* (2nd ed.). San Diego: Academic Press. 560 p.

Longhurst, A. R., Sathyendranath, S., Platt, T., & Caverhill, C. (1995). An estimate of global primary production in the ocean from satellite radiometer data. *Journal of Plankton Research, 17*, 1245–1271.

Löscher, C. R., Großkopf, T., Desai, F., et al. (2014). Facets of diazotrophy in the oxygen minimum zone off Peru. *The ISME Journal, 8*, 2180–2192.

Löscher, C. R., Kock, A., Könneke, M., LaRoche, J., Bange, H. W., & Schmitz, R. A. (2012). Production of oceanic nitrous oxide by ammonia-oxidizing archaea. *Biogeosciences, 9*, 2419–2429.

Löscher, C. R., Mohr, W., Bange, H. W., & Canfield, D. E. (2020). No nitrogen fixation in the Bay of Bengal? *Biogeosciences, 17*, 851–864.

Lotliker, A. A., Baliarsingh, S. K., Samanta, A., & Varaprasad, V. (2020). Growth and decay of high-biomass algal bloom in the northern Arabian Sea. *Journal of the Indian Society of Remote Sensing, 48*, 465–471.

Lüke, C., Speth, D. R., Kox, M. A. R., Villanueva, L., & Jetten, M. S. M. (2016). Metagenomic analysis of nitrogen and methane cycling in the Arabian Sea oxygen minimum zone. *PeerJ, 4*, e1924.

Madhupratap, M., Gauns, M., Nagappa, R., PrasannaKumar, S., Muraleedharan, P. M., DeSousa, S. N., Sardessai, S., & Muraleedharan, D. U. (2003). Biogeochemistry of the Bay of Bengal: Physical, chemical and primary productivity characteristics of the central and western Bay of Bengal during summer monsoon 2001. *Deep Sea Research Part II Topical Studies in Oceanography, 50*.

Marra, J., & Barber, R. T. (2005). Primary productivity in the Arabian Sea: A synthesis of JGOFS data. *Progress in Oceanography, 65*, 159–175.

Menezes, L. D., Fernandes, G. L., Mulla, A. B., Meena, R. M., & Damare, S. R. (2020). Diversity of culturable Sulphur-oxidising bacteria in the oxygen minimum zones of the northern Indian Ocean. *Journal of Marine Systems, 209*, 103085.

Meyer, J., Löscher, C. R., Neulinger, S. C., Reichel, A. F., Loginova, A., Borchard, C., Schmitz, R. A., Hauss, H., Kiko, R., & Riebesell, U. (2015). Changing nutrient stoichiometry affects phytoplankton production, DOP build up and dinitrogen fixation—A mesocosm experiment in the eastern tropical North Atlantic. *Biogeosciences Discussions, 12*, 9991–10029.

Mulholland, M. R., & Capone, D. G. (2009). Dinitrogen fixation in the Indian Ocean. *Geophysical Monograph Series, 185*, 167–186. https://doi.org/10.1029/2009GM000850.

Nair, P. V. R., & Gopinathan, C. P. (1983). Primary production in coastal waters. *CMFRI Bulletin, 34*, 29–32.

Naqvi, S. W. A. (2008). The Indian Ocean. In D. G. Capone (Ed.), *Nitrogen in the marine environment* (pp. 631–681). Burlington, Mass., USA: Elsevier.

Naqvi, S. W. A., & Jayakumar, D. A. (2000). Ocean biogeochemistry and atmospheric composition: Significance of the Arabian Sea. *Current Science, 78*, 289–299.

Naqvi, S. W. A., Jayakumar, D. A., Narveka, P. V., Naik, H., Sarma, V. V. S. S., D'Souza, W., Joseph, S., & George, M. D. (2000). Increased marine production of N_2O due to intensifying anoxia on the Indian continental shelf. *Nature, 408*, 346–349.

Naqvi, S. W. A., Naik, H., Pratihary, A., D'Souza, W., Narvekar, P. V., Jayakumar, D. A., Devol, A. H., Yoshinari, T., & Saino, T. (2006). Coastal versus open-ocean denitrification in the Arabian Sea. *Biogeosciences, 3*, 621–633.

Naqvi, S. W. A., Yoshinari, T., Brandes, J. A., Devol, A. H., Jayakumar, D. A., Narvekar, P. V., Altabet, M. A., & Codispoti, L. A. (1998). Nitrogen isotopic studies in the suboxic Arabian Sea. *Proceedings of the Indian Academy of Sciences. Earth and Planetary Sciences, 107*, 367–378.

Newell, S. E., Babbin, A. R., Jayakumar, A., & Ward, B. B. (2011). Ammonia oxidation rates and nitrification in the Arabian Sea. *Global Biogeochemical Cycles, 25*.

Orsi, W. D., Coolen, M. J. L., Wuchter, C., et al. (2017). Climate oscillations reflected within the microbiome of Arabian Sea sediments. *Scientific Reports, 7*, 6040.

Phoma, S., Vikram, S., Jansson, J. K., Ansorge, I. J., Cowan, D. A., Van de Peer, Y., & Makhalanyane, T. P. (2018). Agulhas current properties shape microbial community diversity and potential functionality. *Scientific Reports, 8*, 10542.

Pitcher, A., Villanueva, L., Hopmans, E. C., Schouten, S., Reichart, G.-J., & Sinninghe Damsté, J. S. (2011). Niche segregation of ammonia-oxidizing archaea and anammox bacteria in the Arabian Sea oxygen minimum zone. *The ISME Journal, 5*, 1896–1904.

Prakash, S., Roy, R., & Lotliker, A. A. (2017). Revisiting the *Noctiluca scintillans* paradox in northern Arabian Sea. *Current Science, 113*, 1429.

Pujari, L., Wu, C., Kan, J., Li, N., Wang, X., Zhang, G., Shang, X., Wang, M., Zhou, C., & Sun, J. (2019). Diversity and spatial distribution of chromophytic phytoplankton in the Bay of Bengal revealed by RuBisCO genes (rbcL). *Frontiers in Microbiology, 10*.

Ramaiah, N., Fernandes, V., Bhaskar, J., Retnamma, J., Gauns, M., & Jayraj, E. A. (2010). Seasonal variability in biological carbon biomass standing stocks and production in the surface layers of the Bay of Bengal. *Indian Journal of Marine Sciences, 39*.

Reeder, C. F., & Löscher, C. R. (2022). Nitrogenases in oxygen minimum zone waters. *Frontiers in Marine Science, 9*.

Sahu, B. K., Baliarsingh, S. K., Lotliker, A. A., Parida, C., Srichandan, S., & Sahu, K. C. (2017). Winter thermal inversion and Trichodesmium dominance in north-western Bay of Bengal. *Ocean Science Journal, 52*, 301–306.

Sarma, V., Patil, J., Shankar, D., & Anil, A. (2019). Shallow convective mixing promotes massive Noctiluca scintillans bloom in the northeastern Arabian Sea. *Marine Pollution Bulletin, 138*, 428–436.

Sawant, S., & Madhupratap, M. (1996). Seasonality and composition of phytoplankton in the Arabian Sea. *Current Science, 17*.

Saxena, H., Sahoo, D., Khan, M. A., Kumar, S., Sudheer, A. K., & Singh, A. (2020). Dinitrogen fixation rates in the Bay of Bengal during summer monsoon. *Environmental Research Communications, 2*, 051007.

Schouten, S., Pitcher, A., Hopmans, E. C., Villanueva, L., van Bleijswijk, J., & Sinninghe Damsté, J. S. (2012). Intact polar and core glycerol dibiphytanyl glycerol tetraether lipids in the Arabian Sea oxygen minimum zone: I. Selective preservation and degradation in the water column and consequences for the TEX86. *Geochimica et Cosmochimica Acta, 98*, 228–243.

Schunck, H., Lavik, G., Desai, D. K., et al. (2013). Giant hydrogen sulfide plume in the oxygen minimum zone off Peru supports chemolithoautotrophy. *PLoS One, 8*.

Shenoy, D. M., Sujith, K. B., Gauns, M. U., Patil, S., Sarkar, A., Naik, H., Narvekar, P. V., & Naqvi, S. W. A. (2012). Production of dimethylsulphide during the seasonal anoxia off Goa. *Biogeochemistry, 110*, 47–55.

Shetye, S., Sudhakar, M., Jena, B., & Mohan, R. (2013). Occurrence of nitrogen fixing cyanobacterium Trichodesmium under elevated CO_2 conditions in the Western Bay of Bengal. *International Journal of Oceanography, 2013*, 350465.

Shirodkar, G., Naqvi, S. W. A., Naik, H., Pratihary, A., Kurian, S., & Shenoy, D. M. (2018). Methane dynamics in the shelf waters of the West coast of India during seasonal anoxia. *Marine Chemistry, 203*.

Singh, A., Gandhi, N., & Ramesh, R. (2012). Contribution of atmospheric nitrogen deposition to new production in the nitrogen limited photic zone of the northern Indian Ocean. *Journal of Geophysical Research, 117*.

Singh, A., & Ramesh, R. (2011). Contribution of riverine dissolved inorganic nitrogen flux to new production in the coastal northern Indian Ocean: An assessment. *International Journal of Oceanography, 2011*, 7p.

Snider, R. G. (1961). The Indian Ocean expedition—An international venture. *Transactions. American Geophysical Union, 42*, 289–294.

Stahl, D. A., & Torre, J. (2012). Physiology and diversity of ammonia-oxidizing archaea. *Annual Review of Microbiology, 66*, 83–101.

Thamdrup, B., Dalsgaard, T., Jensen, M. M., Ulloa, O., Farias, L., & Escribano, R. (2006). Anaerobic ammonium oxidation in the oxygen-deficient waters off northern Chile. *Limnology and Oceanography, 51*, 2145–2156.

Ulloa, O., Canfield, D. E., DeLong, E. F., Letelier, R. M., & Stewart, F. J. (2012). Microbial oceanography of anoxic oxygen minimum zones. *Proceedings of the National Academy of Sciences of the United States of America, 109*, 15996–16003.

Venter, J. C., Remington, K., Heidelberg, J. F., et al. (2004). Environmental genome shotgun sequencing of the Sargasso Sea. *Science, 304*, 66–74.

Villanueva, L., Speth, D. R., van Alen, T., Hoischen, A., & Jetten, M. S. (2014). Shotgun metagenomic data reveals significant abundance but low diversity of "Candidatus Scalindua" marine anammox bacteria in the Arabian Sea oxygen minimum zone. *Frontiers in Microbiology, 5*, 31.

Wang, Y., Liao, S., Gai, Y., et al. (2021). Metagenomic analysis reveals microbial community structure and metabolic potential for nitrogen acquisition in the oligotrophic surface water of the Indian Ocean. *Frontiers in Microbiology, 12*.

Ward, B. B., Devol, A. H., Rich, J. J., Chang, B. X., Bulow, S. E., Naik, H., Pratihary, A., & Jayakumar, A. (2009). Denitrification as the dominant nitrogen loss process in the Arabian Sea. *Nature*, *461*, 78–U77.

Wu, C., Kan, J., Liu, H., Pujari, L., Guo, C., Wang, X., & Sun, J. (2019). Heterotrophic bacteria dominate the diazotrophic Community in the Eastern Indian Ocean (EIO) during pre-southwest monsoon. *Microbial Ecology*, *78*, 804–819.

Wu, C., Narale, D., Cui, Z., Wang, X., Liu, H., Xu, W., Guicheng, Z., & Sun, J. (2022a). Diversity, structure, and distribution of bacterioplankton and diazotroph communities in the bay of Bengal during the winter monsoon. *Frontiers in Microbiology*, *13*. https://doi.org/10.3389/fmicb.2022.987462.

Wu, C., Sun, J., Liu, H., Xu, W., Guicheng, Z., Lu, H., & Guo, Y. (2022b). Evidence of the significant contribution of heterotrophic diazotrophs to nitrogen fixation in the eastern Indian Ocean during pre-southwest monsoon period. *Ecosystems*, *24*. https://doi.org/10.1007/s10021-021-00702-z.

Yuqiu, W., Xiangwei, Z., Zhao, Y., Huang, D., & Sun, J. (2020). Biogeographic variations of picophytoplankton in three contrasting seas: The Bay of Bengal, South China Sea and Western Pacific Ocean. *Aquatic Microbial Ecology*, *84*.

Chapter 16

Physical and biogeochemical characteristics of the Indian Ocean marginal seas[✦]

Faiza Y. Al-Yamani[a], John A. Burt[b], Joaquim I. Goes[c], Burton Jones[d], Ramaiah Nagappa[e], V.S.N. Murty[e], Igor Polikarpov[a], Maria Saburova[a], Mohammed Alsaafani[f], Alkiviadis Kalampokis[d], Helga do R. Gomes[c], Sergio de Rada[g], Dale Kiefer[h], Turki Al-Said[a], Manal Al-Kandari[a], Khalid Al-Hashmi[i], and Takahiro Yamamoto[a]

[a]Kuwait Institute for Scientific Research, Kuwait City, Kuwait, [b]Mubadala ACCESS Center, New York University Abu Dhabi, Abu Dhabi, United Arab Emirates, [c]Lamont-Doherty Earth Observatory at Columbia University, Palisades, NY, United States, [d]King Abdullah University of Science and Technology, Thuwal, Saudi Arabia, [e]CSIR – National Institute of Oceanography, Dona Paula, Goa, India, [f]King Abdulaziz University, Jeddah, Saudi Arabia, [g]US Naval Research Laboratory, Stennis Space Centre, MS, United States, [h]University of Southern California, Los Angeles, CA, United States, [i]Sultan Qaboos University, Al-Khod, Muscat, Oman

1 Introduction

A marginal sea, as the nomenclature implies, is a partially enclosed sea region bordering continents adjacent to an open vast ocean, occupying only a small fraction of the ocean surface, yet remaining distinguished from the latter by several traits. The hydrography and circulation of marginal seas tend to become local in character, and their chemistry is highly influenced by proximity to land and, at times, by large river run-off, advecting great quantities of nutrients. In terms of biological attributes, the marginal seas harbor a variety of ecosystems, ranging from sandy, muddy, and rocky shores to saltmarshes, sea grass beds, mangroves, and coral reefs. The diversity of habitats also leads to a great diversity of fauna and flora, with some of them being endemic.

Proximity to land also results in the development of coastal fisheries, which sustain both artisanal and industrial fishing practices. Some of the marginal sea areas are also home to large urban settlements or industrial assemblies, resulting in eutrophication and/or pollution of coastal waters as well as degradation of habitats. Several marine protected areas were established to conserve biodiversity and restore degraded habitats (Lavieren & Klaus, 2013; PERSGA (The Regional Organization for the Conservation of the Environment of the Red Sea and Gulf of Aden), 2006). The marginal seas are also associated with a variety of hinterlands, ranging from dry deserts to tropical rainforests. The influence and impacts of all these properties render each marginal sea to be a unique environment.

The marginal seas of the Indian Ocean are the Andaman Sea, the Arabian (also known as Persian) Gulf, the Gulf of Aden, the Red Sea, and the Sea of Oman, all lying in the northern part of the Indian Ocean (Fig. 1a). The characteristics of each of these are elaborated in detail in this chapter, and their unique attributes are highlighted. What is remarkable in these narratives is the fact that though these marginal seas are few in number (Britannica, 2020), they share, among themselves, almost all the properties described above. Two of them, the Red Sea and the Arabian/Persian Gulf, even rise above these in that they qualify as natural laboratories where the impacts of climate change can be gainfully studied. The synthetic accounts of these seas are arranged in what follows below, in alphabetical order.

2 The Andaman Sea

2.1 Physical setting

The Andaman Sea (4–20°N; 92–100°E), a marginal sea of the Bay of Bengal, extends over 600,000 km^2 (average depth 1096 m) between the Malayan Peninsula and Andaman and Nicobar Islands (Fig. 1b).

[✦]This book has a companion website hosting complementary materials. Visit this URL to access it: https://www.elsevier.com/books-and-journals/book-companion/9780128226988.

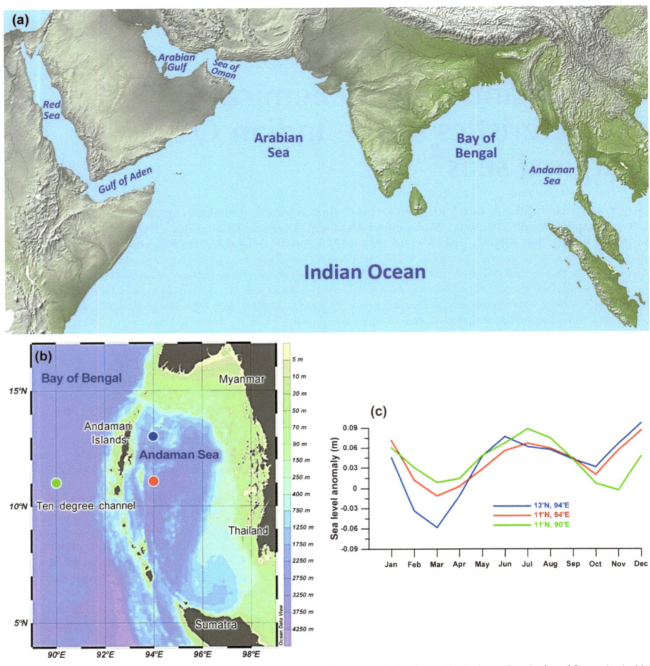

FIG. 1 Marginal seas, including the Andaman Sea. (a) Map of the marginal seas of the Indian Ocean (the Andaman Sea, the Sea of Oman, the Arabian Gulf, the Gulf of Aden, and the Red Sea), (b) the Andaman Sea with detailed bathymetry *(color shading)* and depths are in meters, and (c) mean annual variation of satellite altimetric sea level anomaly (m) at the locations in the northern (13°N, 94°E, *blue curve*) and central (11°N, 94°E, *red curve*) Andaman Sea and in the eastern Bay of Bengal (11°N, 90°E, *green curve*) using the data for the period 1993–2018. *((a) Modified from https://www.freeworldmaps.net/ocean/indian/.)*

2.2 Oceanographic setting

The Northeast Monsoon during November–April, the Southwest Monsoon during June–September, and the enormous freshwater discharges of about $13.6 \times 10^3 \, m^3/s$ from the Irrawaddy River (Tomczak & Godfrey, 2013) govern climate and hydrography of the Andaman Sea, the notable consequence of which is a reduction in surface salinity to 28 during

October–November (Wyrtki, 1961) in the northern region and to <33 in the western region (Rameshbabu & Sastry, 1976), with the meridional gradients coinciding with periods of highest and lowest freshwater influx (Sewell, 1932). Seasonal variations of temperature in the depth interval of 549–1829 m demonstrate identical patterns at all depths (Sewell, 1932), with the deep (>900 m) waters of the Andaman Sea being warmer by 1.7°C at 1829 m than those of the Bay of Bengal. Annually, precipitation exceeds evaporation from May to November but reverses during January–April, with net heat gain during warmer periods of March–April and October–November (Wyrtki, 1961). Strong air-sea interactions cause cyclogenesis of heightened importance in the context of global warming.

Cyclonic surface currents prevail during Southwest Monsoon (March–September) and anti-cyclonic currents during Northeast Monsoon (December–February). Clockwise circulation gyres on various horizontal surfaces during Northeast Monsoon are normal features in the Andaman Sea (Varkey et al., 1996), and the Hybrid Coordinated Ocean Model simulations of sea surface elevation and surface currents have shown a well-defined large clockwise gyre circulation in August 2014 with inflows from the Preparis and Ten Degree Channels and outflow from the Six Degree Channel (Ramya, 2015). Other notable features of circulation in the Andaman Sea are dominant semidiurnal tides (M2) of period 12 h 25 min (Rizal et al., 2012; Mandal et al., 2020) and large-amplitude (80 m) internal waves of wavelength 2 km and period of M2 tides (Osborne & Burch, 1980; Perry & Schimke, 1965) which have been validated by in situ observations in the central Andaman Sea (Mohanty et al., 2018). These internal waves play an important role in coral survival and metabolism (Roder et al., 2010).

Many contemporary in situ studies on surface currents and circulation, satellite-derived sea surface height (SSH) anomaly variations (Mandal et al., 2020), propagation of coastally trapped Kelvin waves (Chatterjee et al., 2017), and westward propagating radiated Rossby waves (Nienhaus et al., 2012; Rao et al., 2010) have been useful in understanding the dynamics of circulation and impacts of climatic events. Compared with the adjacent Bay of Bengal, bimodality is absent in the annual variation of SSH in the Andaman Sea at the two selected locations (northern and central Andaman Sea) and one location in the eastern Bay of Bengal (Fig. 1c), which could be due to the impact of the Indian Ocean Dipole (IOD) and El Niño-Southern Oscillation (ENSO) events and the associated dynamics of propagating Kelvin and Rossby waves (Mandal et al., 2020; Nienhaus et al., 2012; Ramya, 2015).

2.3 Biogeochemical features and biological characteristics

The single most important biogeochemical feature is the annual advection of >600 Mt sediment through the Irrawaddy, Salween, and Sittang rivers into a largely macro-tidal northern Andaman Sea. The Irrawaddy–Salween River system may be the second largest point source of organic carbon to the global ocean, with an estimated delivery of 5.7–8.8 Mt of organic carbon (Ramaswamy et al., 2008). The dispersal of these sediments is governed by seasonally reversing monsoonal currents, river influxes, and tidal currents (Ramaswamy & Rao, 2014; Ramaswamy et al., 2004). This renders the northern Andaman Sea a highly turbid region, with the turbidity-front oscillating ∼150 km in tune with spring-neap tidal cycles. The swath of this turbid zone with >15 mg/L suspended sediment concentration increases from <15,000 km^2 during neap tide to >45,000 km^2 during spring tide, transporting some of the sediments into the deep Andaman Sea via the Martaban canyon, where seaward deposition of sediments is up to 60 m thick in the Gulf of Martaban (Liu et al., 2020).

The low salinities and swift incorporation of $CaCO_3$ in the euphotic zone by foraminifera, pteropods, and coccolithophores, giving rise to a carbonate pump (Sabine et al., 2002), reduce total alkalinities to 2120–2160 μmol/kg in the surface waters. Both calcite and aragonite remain at super-saturation down to 300 m, but aragonite gets undersaturated from 300 to 1500 m (Sarma & Narvekar, 2001), presumably due to warmer temperatures in the deep layers. Concentrations of dissolved organic carbon (DOC) in the western Andaman Sea range from sparse to 106.19 μM (Mohan et al., 2016) and, along 88°E, they decrease from 170 μM at 10 m to 123 μM at 120 m (Ramaiah et al., 2010).

Perennial capping with low saline waters (31.8–33.4) and warmer temperatures (∼29°C) have their own effects on oxygen and nutrient histories of the top (50 m) layers (Sen Gupta & Naqvi, 1984). Concentrations of dissolved oxygen at the surface have been reported in the range of 210–230 μmol/kg (Sarma & Narvekar, 2001), decreasing to 90 μmol/kg at 2000 m, which could be due to the inflow of oxygen-poor intermediate water masses from the Bay of Bengal. Nutrient concentrations have been reported to be low/below detection limits in the top 50 m (Table 1) and have been attributed to poor vertical mixing caused by permanent stratification (Sen Gupta & Naqvi, 1984). Concentrations of silicate in nearshore waters have also been reported to be low, presumably due to utilization by abundant diatoms (Egge & Aksnes, 1992), but concentrations have been observed to increase in the far offshore surface waters (Table 1), brought in with surface waters of Irrawaddy-Salween River system (Sarma & Narvekar, 2001).

TABLE 1 Ecological parameters in spatially separated sampling locations in the Andaman Sea.

Off Myanmar (Data from 60 sampled stations; Jyothibabu et al., 2014; Sampling year 2002)

Clusters (# of sampled locations)	Rakhine Coast (11)	Off Gulf of Mottama (6)	Northern Andaman Sea (17)	Off Thanintharyi Coast (18)	Gulf of Mottama (8)
MLD (m)	19.3 ± 4.29	14.3 ± 3.60	20.8 ± 8.1	31.4 ± 11.83	10.5 ± 4.6
Salinity	33.36 ± 0.32	32.20 ± 0.26	32.63 ± 0.21	33.09 ± 0.29	31.80 ± 0.22
DO (μM)	207 ± 3.7	195 ± 12.0	205 ± 11.6	205 ± 5.0	185 ± 14.8
NO_3 (μM)	0.12 ± 0.05	0.56 ± 0.46	0.05 ± 0.04	0.09 ± 0.05	1.53 ± 1.04
PO_4 (μM)	0.07 ± 0.05	0.16 ± 0.12	0.03 ± 0.0	0.05 ± 0.0	0.48 ± 0.15
SiO_4 (μM)	0.81 ± 0.40	1.61 ± 1.32	0.29 ± 0.23	0.49 ± 0.35	6.24 ± 1.42
Chl-a (mg m^{-3})	0.34 ± 0.12	0.28 ± 0.14	0.19 ± 0.17	0.39 ± 0.24	0.89 ± 0.72

Off Thailand (Data from 38 stations along three transects pooled; Nielsen et al., 2004)

Sampling month	March 1996	August 1996	Notes
MLD (m)	34 ± 2	62 ± 2	Density based pycnocline
Salinity	32.8–35	33–35	Data range: Surface to 150 m
NO_3 (μM)	<0.1–24 ± 0.1	<0.1–23.3 ± 0.5	Data range: Surface to 150 m
PO_4 (μM)	0.13 ± 0.02–2.18 ± 0.05	0.05 ± 0.004–1.79 ± 0.13	Data range: Surface to 150 m
Chl-a (mg m^{-3})	0.05–0.1	0.05–0.1	0.5–1.5 in pycnocline; in both seasons

Southern Andaman Sea (Data from Sarma & Narvekar, 2001; 23 stations sampled October–November 1996)

MLD (m)	50	Notes: Sampling depth surface to 2000 m
Salinity	31.8–33.4 (MLD)	33.4–34.90 below MLD to 2000 m
DO (μM)	210 at surface; 20 at 200 m	increase from 20 μM at 200 m to 60 μM at 2000 m
NO_3 (μM)	<0.01 at surface; 25 at 200 m	6–8 (50–100 m column); 8–25 (100–2000 m)
SiO_4 (μM)	0.1 at surface; 60 μM at 50 m	Increase from 60 μM at 50 m to ~100 μM at 2000 m
Chl-a (mg m^{-3})	0.1 at surface; 5 at 35 m	0.24–1.80 at 50 m; secondary maxima (Gomes et al., 1992)

Brief characteristics of sampling locations in each cluster. See Jyothibabu et al. (2014) for details. Cluster 1, Rakhine coast: Highest salinity, lowest temperature, nutrient concentration, and suspended sediments low; Cluster 2, Off Gulf of Mottama: Low salinity, warmer temperatures, nutrients, and suspended sediments high; Cluster 3, Offshore waters of the northern Andaman Sea: High salinity, highest temperature, lowest nutrients, and suspended sediments; Cluster 4, Off Irrawaddy delta and south off Thanintharyi coast: High salinity, low temperature, low nutrients, and suspended sediments; Cluster 5, Gulf of Mottama: Lowest salinity, high temperature, highest nutrient, and suspended sediments.

Table 2 presents a cross-section of biological properties across various trophic levels gathered from published sources (Bhattathiri & Devassy, 1981; Jyothibabu et al., 2014; Kiørboe et al., 1991; Madhupratap et al., 2003, 1981a, 1981b; Nielsen et al., 2004; Paul et al., 2007; Ramaiah et al., 2010; Ramakrishna & Sivaperuman, 2010). The annual harvest of fish and shellfish by eight countries bordering this marginal sea is of the order of >2 million tonnes (FAO (Food and Agriculture Organization), 2003).

The Andaman Sea region is a region of great biological diversity. The western coast of the Malay Peninsula, the Myanmar coast, the Andaman and Nicobar Islands, and many other islands are rich in coral reefs with over 200 species of corals and 1200 reef-associated fishes (Ramakrishna & Sivaperuman, 2010). This region is also home to 61 mangrove plant species in 39 genera and 30 families (Ramakrishna & Sivaperuman, 2010). The occurrence of over 16 species of seagrasses was reported on the coasts of different countries/islands surrounding the Andaman Sea (Ragavan et al., 2016).

Systematic studies to elucidate physical and biogeochemical processes, measurements of biological productivity characteristics, and diversity analyses are essential to recognize the climate change impacts on the Andaman Sea.

TABLE 2 Plankton carbon biomass and production rates during different seasons within the Andaman Sea (Jyothibabu et al., 2014; Nielsen et al., 2004) and from the central Bay of Bengal along 88°E (Ramaiah et al., 2010).

Off Thailand coast in waters of <300 m; Data pooled by Nielsen et al. (2004) from three transects covering 38 stations

Sampling month	March 1996	August 1996	
Carbon standing stocks (mg C m^{-2})			
Phytoplankton (chlorophyll-a)	~5–24 (top 100 m)	5–23 (top 100 m)	Note: Chl range 0.5–1.5 mg m^{-3}; both seasons
Heterotrophic bacteria	652 ± 65	867 ± 72	
Microzooplankton	38 ± 2	33 ± 4	Average biomass: 0.5–1.0 mg m^{-3}
Production rate (mg C m^{-2} d^{-1})			
Primary production	567 ± 76	684 ± 79	Within top 100 m; annual av. 229 mg C m^{-2} d^{-1}
Bacterial production	140 ± 47	139 ± 13	Cell vol range: 12–17 µm^3; larger near bottom of ~100 m

Off Myanmar coast data modified from Jyothibabu et al. (2014); 60 stations sampled from depth <100 m

Sampling month April–May 2002		
Chlorophyll-a (mg C m^{-2})	4–16; top 20 m	Surf Chl range 0.6–1 mg m^{-3} in situ; 1.5–3 mg m^{-3} satellite derived
Mesozooplankton (vol; mL)	5.8–99.4	Wide variation between stations and, within each of the five clusters

S-120 m column-integrated data from 8 depths each at 5 stations along 88°E; modified from Ramaiah et al. (2010)

Sampling season[a]	SWM	FIM	NEM	SIM
Carbon standing stocks (mg C m^{-2})				
Phytoplankton (chlorophyll-a)	518	904	1023	789
Heterotrophic bacteria	677	227	333	104
Microzooplankton	215	22	92	327
Mesozooplankton	587	1036	808	821
Production rate (mg C m^{-2} d^{-1})				
Primary production	144	306	375	241
Bacterial production	248	85	196	70
Mesozooplankton production	127	133	94	225

[a]Southwest monsoon (SWM): June–September; Fall intermonsoon (FIM): October–November; Northeast monsoon (NEM): December–February; Spring intermonsoon (SPIM): March–May.

3 The Arabian/Persian Gulf

3.1 Physical setting

The Arabian/Persian Gulf is a young sea that formed following the Holocene glacial retreat, with modern coastlines forming just 6000 years before the present (Kennett & Kennett, 2006). The Arabian/Persian Gulf is a subtropical biogeographic province of the northern Indian Ocean, occurring between 24–30°N and 48–57°E and covering 250,000 km^2. It is bounded by the Zagros and Hajjar mountains to the east and by the sedimentary Arabian coastline to the south and west (Vaughan et al., 2019).

The Gulf is characterized by extreme conditions (Sheppard et al., 1992). Surrounding terrestrial systems are arid to hyperarid due to their location in the subtropical high-pressure zone, and air temperatures vary from below-freezing to >50°C seasonally (Vaughan et al., 2019). The Gulf is shallow (mean 36 m) and has limited oceanic exchange through the narrow (42 km wide) Strait of Hormuz (Reynolds, 1993). As a result, the Gulf is hypersaline (>42) due to high evaporation rates and limited freshwater input (Reynolds, 1993); it is also characterized by extreme sea temperatures, ranging from <12°C in winter to >36°C each summer (Fig. 2), making it the world's hottest sea annually (Riegl & Purkis, 2012; Sheppard et al., 1992).

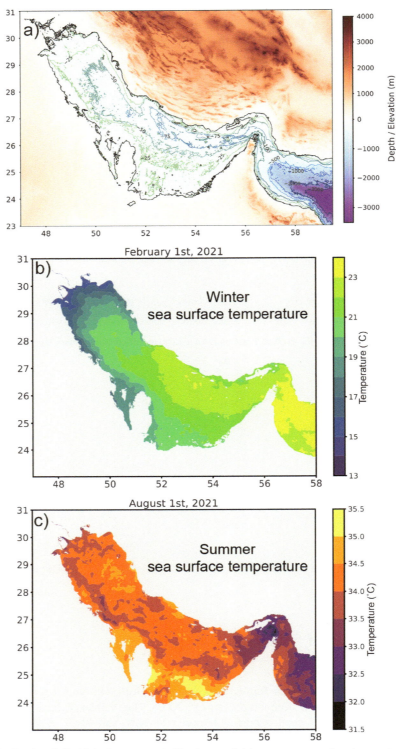

FIG. 2 Map of the Arabian/Persian Gulf, (a) bathymetry, and (b) winter and (c) summer sea surface temperatures. *(Modified from Burt and Paparella (2023).)*

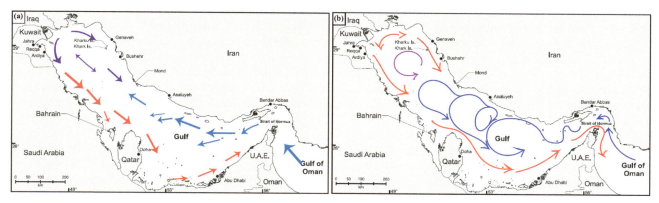

FIG. 3 Illustration of (a) general Arabian/Persian Gulf circulation and (b) shamal-driven seasonal eddies. *Blue arrows*: lower salinity surface flows; *red arrows*: outflows of saline bottom water. *Purple arrows*: transitional areas in the northern Gulf. *(Modified from Vaughan et al. (2019).)*

3.2 Oceanographic setting

Gulf circulation patterns are largely driven by halocline forces structured by basin-scale evaporation (Reynolds, 1993; Swift & Bower, 2003; Thoppil & Hogan, 2010a). Oceanic water enters at the surface of the Strait of Hormuz, with predominant currents carrying this water along the Iranian coast toward the northern Gulf, where it turns and begins a southward progression near Kuwait (Fig. 3a). High evaporation causes increasing salinity, peaking in the southern Gulf. As density increases, the highly saline water sinks, flowing along the seabed to discharge as bottom outflow into the Sea of Oman (Fig. 3a) (Kämpf & Sadrinasab, 2006; Reynolds, 1993; Yao & Johns, 2010b). Winds also play important roles in mesoscale circulation (Pous et al., 2015; Thoppil & Hogan, 2010a; Yao & Johns, 2010a). Seasonally persistent northerly (*shamal*) winds drive formation of several large-scale (>100 km) eddies in the central Gulf (Thoppil & Hogan, 2010a, 2010b; Yao & Johns, 2010a), and coastal currents in the south (Cavalcante et al., 2016; Elshorbagy et al., 2006) (Fig. 3b). Shamals peak between November to March (Al Senafi & Anis, 2015), but they are also ecologically important in summer when they dramatically reduce sea temperatures (Burt et al., 2019; Paparella et al., 2019) and can disrupt recurrent nighttime hypoxia on coral reefs (de Verneil et al., 2022). Except for areas near the Hormuz Strait, tidal circulation is negligible (Swift & Bower, 2003; Vaughan et al., 2019).

3.3 Biogeochemical features and biological characteristics

Since the 1970s, basic research on chemical processes has been available (Brewer & Dyrssen, 1985; Grasshoff, 1976; Hashimoto et al., 1998). However, a recent review indicates that large-scale biogeochemical processes in the Gulf (e.g., nitrogen budget and carbon cycling) remain understudied (Al-Yamani & Naqvi, 2019).

Inflowing Gulf surface waters are enriched with phosphate (Brewer & Dyrssen, 1985; Grasshoff, 1976) and high levels of nitrate (maximum 3.9 μM) (Al-Yamani & Naqvi, 2019). In the northern Gulf, marine biochemistry is influenced by the Shatt Al-Arab River (Al-Yamani et al., 2004; Al-Yamani & Naqvi, 2019), where maximum values of nitrate (10.0 μM), phosphate (3.5 μM) and silicate (10.8 μM) have been reported (Al-Said et al., 2018c). Silicate, in particular, is strongly influenced by this river discharge, having a low concentration in the Gulf (≤1–5 μM) except in Kuwait (mean of 5.9 μM) (Al-Yamani et al., 2004; Al-Yamani & Naqvi, 2019). Southerly returning currents are nutrient-depleted (Brewer & Dyrssen, 1985; Hashimoto et al., 1998), being low in nitrate relative to phosphate and negligible silicate (Brewer et al., 1978). Sedimentary processes likely play an important role in controlling pelagic biogeochemistry as surface layers directly exchange materials with the seafloor in the shallow Gulf.

Dust storms are also important for Gulf geochemistry, bringing trace metals, especially iron (Ezzati, 2012; Maher et al., 2010), and are a source of nitrogen and phosphorus (Herut et al., 2002). Total dissolved iron can exhibit wide spatiotemporal variability (ranging from 0.44 to 12.80 nM in summer and from 1.36 to 28.15 nM during winter) (Al-Said et al., 2018a).

Al-Said et al. (2018b) reported on total organic carbon in the NW Gulf, with maximal concentrations occurring within the polluted Kuwait Bay and decreasing offshore (101.0–318.4 μM, mean 161.2 μM), indicating substantial anthropogenic influence.

Systematic observations are vital for improving the current understanding of the biogeochemical processes in the Gulf, especially given increasing anthropogenic and climate change impacts. Future studies should focus on processes controlling micro- and macronutrient concentrations and distributions and the effect of dust on biochemistry.

The Gulf environment acts as a biological filter for many marine organisms. The diversity of fish, corals, and other Gulf marine fauna is negatively correlated with environmental extremes (Bauman et al., 2013; Burt et al., 2011; Sheppard et al., 1992), resulting in fish and coral diversity in the Arabian/Persian Gulf that is one-third of the Sea of Oman (Claereboudt, 2019), and a tenth of the Indo-Pacific (Coles, 2003); this pattern is reflected in many taxa (Sheppard et al., 1992; Vaughan et al., 2019).

Despite the extremes, Gulf ecosystems are far more diverse and productive than terrestrial habitats (Sheppard et al., 1992). Important subtidal ecosystems include coral reefs, oyster reefs, seagrass beds, and macroalgal beds, as well as coastal mangrove forests, saltmarshes, sabkha (salt flats), and mudflats (Vaughan et al., 2019). Several guidebooks were recently produced on the Gulf biodiversity, such as on plankton (Al-Yamani & Saburova, 2019; Al-Yamani et al., 2011; Richards, 2008), benthos (Al-Yamani & Saburova, 2010, 2011; Al-Yamani et al., 2012, 2019), and macroalgae (Al-Yamani et al., 2014).

As the world's hottest sea, the Gulf has become a focus for climate change research (Burt, 2013; Burt et al., 2020; Feary et al., 2013; Vaughan & Burt, 2016). Gulf corals tolerate temperatures that would kill corals on most reefs (Riegl et al., 2011). This thermal tolerance is, in part, the result of genetic adaptations (Howells et al., 2016; Kirk et al., 2018; Smith et al., 2022, 2017a), and heritable epigenetic modifications (Liew et al., 2020). In addition, corals in extreme environments associate with a highly thermo-tolerant symbiotic algae *Cladocopium thermophilum* (Hume et al., 2015, 2016; Smith et al., 2017b).

But Gulf ecosystems are under pressure. Coastal development has degraded numerous coastal ecosystems (Burt, 2014). Eutrophication has enhanced harmful algal blooms that have caused mass fish kills and likely contribute to the large central Gulf hypoxic zone (Al-Ansari et al., 2015; Lachkar et al., 2022). Parts of the Gulf are warming at nearly twice the global average (Al-Rashidi et al., 2009; Lachkar et al., 2020), leading to widespread reef degradation (Burt et al., 2013, 2016, 2019). Robust monitoring, management, and conservation programs are needed but lacking across much of the Gulf (Sale et al., 2011; Sheppard et al., 2010).

4 The Gulf of Aden

4.1 Physical setting

The Gulf of Aden lies on the western side of the Arabian Sea between Somalia and Yemen. It extends approximately 1200 km from the western end near Djibouti to the eastern tip of Socotra Island (Fig. 4). It is about 300 km wide and tapers

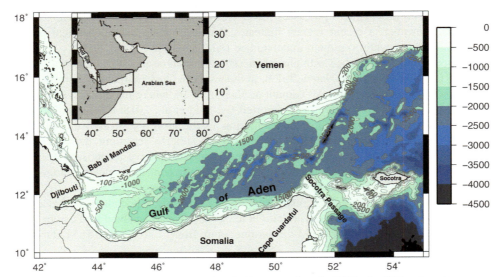

FIG. 4 Map of the Gulf of Aden region. The overall dimensions of the Gulf are approximately 1250 km by 300 km, from the western end to the eastern tip of Socotra Island. The placement of this area within the northern Arabian Sea is shown in the inset.

slightly toward the west. Most of the Gulf is less than 2500 m in depth, but some deep holes exceed 3000 m, particularly within the rift zone.

The Gulf of Aden, along with the Red Sea and the Gulf of Aqaba, was formed by the spreading between the Arabian and the Nubian plates (Rasul et al., 2015). The rifting began about 17 Ma, and the Strait of Bab al Mandab ("Gate of Tears") opened about 5 Ma, filling the Red Sea rift with seawater (Bonatti et al., 2015).

The Gulf of Aden hosts a range of ecological environments. The basin's eastern end is linked to the productive coastal environments along the northwestern Arabian Sea responding to monsoonal forcing. In contrast, the western end contains coral reefs and biodiversity similar to the southern Red Sea.

4.2 Oceanographic setting

4.2.1 Monsoonal forcing

As with the Arabian Sea, physical dynamics within the Gulf are tightly coupled with the monsoonal forcing. Schott and McCreary (2001) extensively covered the monsoonal forcing and the Indian Ocean response. One consequence is the reversal of coastal currents along the African and Arabian coasts with the reversing monsoonal winds. During the Southwest Monsoon, currents along the Somali, Yemen, and Oman coasts are toward the northeast, reversing toward the southwest during the Northeast Monsoon. During the Southwest Monsoon, orographic effects along the Somali coast and the transfer of energy through the Socotra Passage contribute to the complex physical circulation inside the Gulf. During the Northeast Monsoon, winds from the northeast funnel into the Gulf and through Bab al Mandab into the southern Red Sea. Water exchange through Bab al Mandab is controlled by the reversing wind patterns between the monsoons (Murray & Johns, 1997; Smeed, 1997, 2000; Sofianos & Johns, 2002; Sofianos et al., 2002).

4.2.2 Red Sea Outflow

A major feature in the Gulf is the export of warm, salty Red Sea Outflow Water, especially during the Northeast Monsoon (Aiki et al., 2006; Bower et al., 2000, 2005; Matt & Johns, 2007; Murray & Johns, 1997; Naqvi & Fairbanks, 1996; Peters & Johns, 2006). The Gulf provides cooler, fresher water to the Red Sea, which then modifies the water through heating and evaporation, returning it to the Gulf as warm, salty, dense water (Bower et al., 2000; Sofianos et al., 2002). RSOW enters the Gulf through Bab al Mandab and descends from the shallow depth of Hanish sill, ~150 m, until it reaches its equilibrium density, 27–27.5 σ_θ as it mixes with ambient Gulf water (Bower et al., 2005). Red Sea Water is then dispersed throughout the Indian Ocean, a portion observed entering the Agulhas Current (Beal et al., 2000), and sometimes observed in the Bay of Bengal (Jain et al., 2017).

4.2.3 Water masses

Four major water masses have been identified within the Gulf of Aden (Alsaafani & Shenoi, 2007). Ship-based observations in the Gulf (Bower & Furey, 2012; Bower et al., 2005; Peters et al., 2005) and, most recently, Argo observations (Fig. 5a) provide further clarification of these water masses. Gulf of Aden Surface Water is seasonally variable with densities less than 25 σ_θ. Its properties resemble those of Arabian Sea High Salinity Water. Gulf of Aden Intermediate Water lies between 26 and 27 σ_θ, with salinity centered near 35.5 and temperature 10–15°C. Gulf of Aden Intermediate Water inflow into the Red Sea during the Southwest Monsoon contributes nutrients and supports increased primary productivity in the southern Red Sea during this period. Red Sea Outflow Water, with its uniquely high salinity, >37, warm temperature, and density between 26.5 and 27.5 σ_θ is the most prominent water mass feature. Gulf of Aden Bottom Water in the deep basin has a density range of 27.5–27.8 σ_θ, salinity of 34.7–35.6, and temperature of 3–9°C.

4.2.4 Eddies

Eddies dominate the Gulf's circulation (Bower et al., 2002; Fratantoni et al., 2006). Their diameters are constrained by the basin width, they can easily penetrate to depths of 1000 m, and have azimuthal velocities on the order of 0.4–0.5 m/s. Eddy sources include energy transfer through the Socotra Passage (Fratantoni et al., 2006; Fig. 5b), westward propagation across the Indian Ocean (Alsaafani et al., 2007), and wind stress shear due to the local orography (Morvan et al., 2020; Yao & Hoteit, 2015). The eddies stir and mix Red Sea Outflow Water with Arabian Sea water before finally exiting into the Arabian Sea. Submesoscale eddies and filaments are also important in the Gulf, generated by upwelling, coastal instabilities, and flow around capes (Morvan et al., 2020). Chlorophyll imagery reveals the complexity of the submesoscale circulation in the Gulf (Lévy et al., 2018).

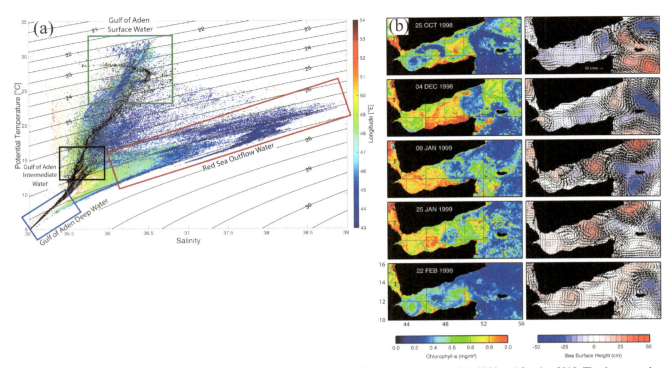

FIG. 5 Gulf of Aden. (a) θ-S plot from 9 Argo profiling floats deployed in the Gulf of Aden between July 2008 and October 2019. The data are color-coded according to longitude (see *color bar*), with *blue* being the most westward and *red* the most eastward. For background reference, observations from a 2-year time-series of biogeochemical Argo Float #2902199 in the central Arabian Sea are shown in *black*. (b) Westward propagation of an anticyclonic eddy originating from flow through the Socotra Passage from the coast of Somalia into the Gulf. The anticyclonic eddy is indicated by the region of low chlorophyll (left panels), elevated SSH, and anticyclonic circulation. *(Panel b from fig. 3 in Fratantoni et al. (2006).)*

4.3 Biogeochemical features and biological characteristics

There is a dearth of literature on the biogeochemical and biological characteristics of the Gulf of Aden. Although few publications are available, historical ocean observations dating back to the first half of the 20th century that include hydrographic characteristics (temperature, salinity, oxygen, pH, nutrients, etc.) are available in the World Ocean Database (Boyer et al., 2013). A recent biogeochemical Argo float spent 400 days west of the northeastern tip of Somalia within the Gulf of Aden. Although the Gulf is supplied with water predominantly from the Arabian Sea and its biogeochemical characteristics are thus similar, depths between 300 and more than 1000 m are affected by the inflow of Red Sea Outflow Water. This region of the water column is warmer, saltier, and thus more oxygenated than water found at similar depths in the Arabian Sea. An oxygen minimum layer (nominally $\sim$20 μmol/kg) extends from about 125 m to 1200–1300 m depth. Because the water column is not suboxic, the denitrification found in many areas of the Arabian Sea is not evident in the Gulf (Bange et al., 2005). Nutrient patterns within the Gulf are similar to those of the Arabian Sea. The primary nutrients generally increase with depth. At a depth of 500 m, nitrate concentration is typically 20–30 μM, and phosphate concentration is about 2–2.5 μM. Near the surface, nitrate is usually less than 10 μM, and phosphate is less than 1 μM.

A few papers and reports provide an overview of the biology of the Gulf. The most detailed information comes from remote sensing and modeling studies. Chlorophyll distributions often correlate with the eddy structures—anticyclonic eddies containing low chlorophyll concentrations and cyclonic eddies with higher concentrations (Fratantoni et al., 2006). Similarly, wind forcing and orography significantly affect the distribution of chlorophyll within the basin (e.g., Yao & Hoteit, 2015). A phenological analysis of chlorophyll patterns within the Gulf demonstrated the differential response of chlorophyll in different regions of the Gulf (Gittings et al., 2017). The Gulf provides both nutrients and chlorophyll to the Red Sea via the flux of Gulf of Aden Intermediate Water (Dreano et al., 2016). In contrast, because the Red Sea water entering the Gulf sinks rapidly, it has little to no effect on the upper-layer productivity of the Gulf. As in the Arabian Sea, the Gulf of Aden can experience significant *Noctiluca* blooms, as demonstrated in the NASA MODIS image from February 2018 (https://earthobservatory.nasa.gov/images/91937/bloom-in-the-gulf-of-aden).

The Gulf contains high biological diversity with more than 100 species of coral and 600 reef-associated fish species and is known for its copious fishery production due to the influence of nutrient-rich upwelling waters (PERSGA (The Regional Organization for the Conservation of the Environment of the Red Sea and Gulf of Aden), 2006; Sakaff & Esseen, 1999). Coral reefs, located primarily in the western Gulf, are estimated to have an area of more than 900 km^2 and are similar to reefs in the southern Red Sea (ICRI et al., 2021). Along the Yemeni coastline, mangroves, sea grasses, and coral reefs are absent along the open coast due to the combination of a high-energy environment and cold water temperatures due to upwelling (Wilson et al., 2003). However, mangroves are present along the Yemeni coast in sheltered embayments. Mangroves are also widespread in Djibouti, although the stands are patchy. Three species of sea turtles (green, hawksbill, and loggerhead) occur and nest in the Gulf region (Wilson et al., 2003).

More research is needed to understand the biogeochemical regime and the dynamics of the Gulf of Aden. Documentation of the biodiversity of marine fauna and flora of the Gulf, including semiclosed ports and isolated shores, as well as areas that experience intensive evaporation, is needed.

5 The Red Sea

5.1 Physical setting

The Red Sea is located between the Arabian Peninsula and the African mainland (Tomczak & Godfrey, 2003). It extends over 2000 km between 12°N and 30°N latitude and reaches its maximum width of 350 km at the Eritrean coast. Roughly 40% of the Red Sea is shallow, with a depth of less than 100 m (Rasul et al., 2015). In the central deep-water part, about 25 bathymetric depressions occur at depths of 1500–2800 m, filled with hot, hypersaline, acidic, and anoxic water (Rasul & Stewart, 2019; Rasul et al., 2015).

The northern end of the Red Sea is bifurcated by the Sinai Peninsula into the Gulf of Aqaba to the east and the Gulf of Suez to the west; the latter is connected to the Mediterranean Sea through the artificial Suez Canal. To the south, the Red Sea is connected to the Arabian Sea and the Indian Ocean via the Gulf of Aden through the Bab al Mandab strait, where the exchange is restricted by the Hanish Sill with a maximum depth of 137 m (Rasul & Stewart, 2019).

During the summer (May–September), the Southwest Monsoon winds blow southward over the entire Red Sea. In the winter monsoon (October–April), the Northeast Monsoon winds in the northern and southern basins blow in opposite directions, forming a convergence zone in the central part, marking the boundary between the monsoon-dominated weather system in the south and the continental eastern Mediterranean weather system in the north (Chiffings, 1995; Johns et al., 1999; Menezes et al., 2018; Sofianos & Johns, 2002). The rainfall varies from a few mm per year along the northern part of the western shore, gradually increasing to 180 mm at Suakin in Sudan, averaging 6 mm per year (Edwards, 1987).

5.2 Oceanographic setting

The surface waters of the Red Sea are characterized by a broad latitudinal temperature gradient with 2–4°C seasonal variability. The lowest temperatures occur in the north, gradually increasing southward (Fig. 6a). Water temperatures range from a winter minimum of 22°C in the Gulf of Aqaba to a summer maximum >31°C in the southern basin, locally up to 34°C (at the Dahlak Archipelago). Salinity generally follows an opposite pattern in that the highest values of 40–41 occur in the north around the Gulf of Suez and decrease southward down to 36 in the waters connected to the Gulf of Aden (Berumen et al., 2019; Edwards, 1987; Rasul et al., 2015).

Water temperature decreases gradually within the upper 400-m layer but does not fall below 21°C even deeper than 2000 m (Roder et al., 2013). Temperatures around 68°C, salinity up to 257, and pH of 5.3 have been recorded within hydrothermal Atlantis II Deep at 1900–2200 m depth (Kaartvedt et al., 2016).

Intense air-sea interactions over the Red Sea coupled with along-basin change in salinity, scarcity of rainfall, and no major source of fresh water to the sea result in one of the highest evaporation rates globally, with an annual mean of 2.06 ± 0.22 m per year with peaks exceeding 5 m/year in the central part (Bower & Farrar, 2015; Eshel & Heavens, 2007; Menezes et al., 2018; Sofianos et al., 2002). Evaporation rates increase northward and are stronger in winter (Eshel & Heavens, 2007; Zolina et al., 2017).

Because of the excessive evaporation, the hydrologic budget of the Red Sea is negative, causing seawater influx from the Indian Ocean through a two-layer system from October to May and a three-layer system from June to September via Bab al Mandab strait (Sofianos & Johns, 2002). The high salinity Red Sea water mass is the main source for the intermediate layer, which enters the Indian Ocean and can be traced at intermediate depths as far as about 40°S (Beal et al., 2000; Roman & Lutjeharms, 2009).

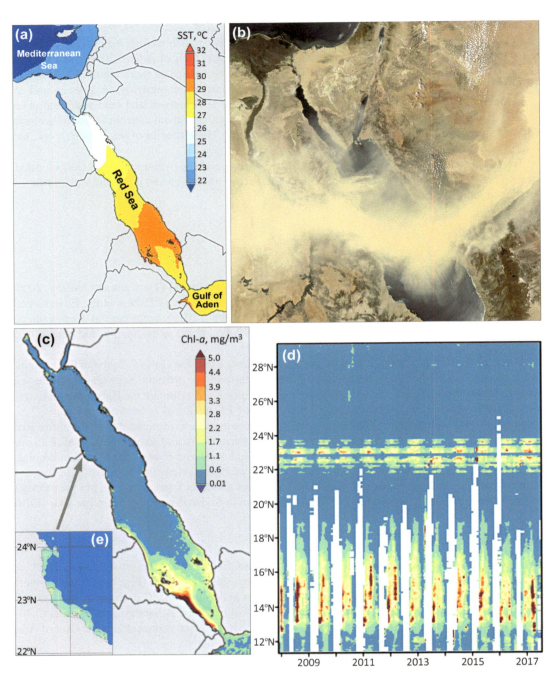

FIG. 6 Remote sensing of the Red Sea area. (a) Time-averaged map of the night sea surface temperature (SST) showing a latitudinal gradient of SST along the Red Sea compared to adjacent waters with monthly temporal and 4 km spatial resolution (data derived from the 11 μm infrared band of MODIS-Aqua spectroradiometer for period of October 2002 to September 2020), (b) Dust storm across the Red Sea (data acquired March 13, 2005, MODIS-Aqua), (c) Time-averaged map of the surface chlorophyll-*a* concentrations along the Red Sea, with monthly temporal and 4 km spatial resolution (data derived from the MODIS Aqua dataset processed using Level-3 Chl for the period of August 2002 to August 2020), (d) Hovmöller longitude vs time plot of the same data delimited by 2008–2018. Noticeable vertical white strips correspond to missing chlorophyll data due to extensive atmospheric aerosol events, such as dust storms. Visible elevated chlorophyll-*a* signal seen between 22°N and 24°N is associated with the Mirear Island reef complex (see Acker et al., 2008) located at 23°10′N and 35°40′E (e). *(Analyses and visualizations used for (a), (c), and (d) were produced with the Giovanni online data system, developed and maintained by the NASA GES DISC, see Acker and Leptoukh (2007); (b)—Image credit to NASA Visible Earth; Collection: Rapid Response Gallery, https://visibleearth.nasa.gov/collection/1653/rapid-response-gallery.)*

The evaporation-driven thermohaline circulation results in northward water transport within the upper layer of the basin (Chiffings, 1995; Johns et al., 1999; Sofianos & Johns, 2003). The basic circulation pattern consists of a gyre in the northern Red Sea, which reverses from an anticyclonic rotation during winter to a cyclonic rotation during the summer, and a northward-flowing eastern boundary current (Zhai et al., 2015). This current initiates along the western boundary in the south and crosses over to the eastern coast mid-basin influenced by topographically-steered winds, but disappears in the summer when the direction of the winds over the southern Red Sea and the Gulf of Aden is reversed (Menezes, 2018; Zhai et al., 2015). Mesoscale eddies are mostly associated with the central and northern regions, where they actively redistribute water masses horizontally and contribute to vertical convection (Zhan et al., 2019).

5.3 Biogeochemical features and biological characteristics

The Red Sea is an oligotrophic system. Its biological productivity is largely governed by nutrient balance throughout the entire basin (Qurban et al., 2019; Raitsos et al., 2015; Wafar et al., 2016). The primary production increases from north to south and where maximum values in the south are due to the influx of elevated nutrient concentrations in water from the upwelling areas of the Arabian Sea through the Bab al Mandab strait, particularly during winter monsoons (Table 3). This pattern is apparent in remotely sensed chlorophyll-*a* concentrations (Fig. 6c and d). The latitudinal gradient in phytoplankton biomass and productivity through the Red Sea translates into similar distribution patterns of productivity and biomass in other components of the pelagic food web (Kürten et al., 2014).

The primary production in the Red Sea is essentially (up to 80%–90%) contributed by nano- and picoplankton (e.g., cyanobacteria *Prochlorococcus* and *Synechococcus*, and diverse picoeukaryotes), which further the nutrient retention through the recycling of organic matter within the microbial loop with a small proportion being transferred into higher trophic levels (Pearman et al., 2017; Qurban et al., 2019). Nitrogen deficiency over most parts of the basin may facilitate sporadic blooms of diazotroph cyanobacterium *Trichodesmium* (Gokul et al., 2019), which contribute substantially to the input of nitrogen in near-surface waters through biogenic fixation of atmospheric nitrogen. The high water transparency in the Red Sea allows enough light penetration to sustain benthic primary production up to great depths and enhances the contribution of micro- and macrophytes, seagrasses, and photosynthetic symbionts of corals to the total carbon fixation and overall productivity (Cardini et al., 2016; Ziegler et al., 2019b).

With negligible riverine runoff, frequent dust storms over the Red Sea (15–20 events annually; Fig. 6b and d) are a very important component of the nutrient cycles through the fertilization of its oligotrophic waters. Annually, estimated desert dust deposition to the Red Sea (6–7.5 Mt; Prakash et al., 2015, 2016) corresponds to 76 kt of iron oxides and 6 kt of phosphorus per year (Anisimov et al., 2017).

The high biodiversity of the Red Sea is supported by a wide range of habitats along its length, foremost by coral reefs, as well as mangroves, seagrass beds, soft-bottom sediment flats, islands, and lagoons. The evolutionary history of the Red Sea, combined with its unique environmental properties and restricted connection to the Indian Ocean, has likely played a role in promoting a high level of endemism (Berumen et al., 2019; DiBattista et al., 2016).

The extensive reefs support a great diversity of coral-associated organisms (Carvalho et al., 2019; Churchill et al., 2019). In total, 2,710 reef-associated species have been identified in the Red Sea, with levels of endemism ranging from <10% (scleractinian corals, mollusks, and echinoderms) to 10%–12.9% (polychaetes, crustaceans, and fish), 16.5% for ascidians (DiBattista et al., 2016), and up to 50% in butterflyfishes (Roberts et al., 1992). Scleractinian corals are associated with a phylogenetically diverse microbiome comprised of endosymbiotic photosynthetic dinoflagellates (65 Symbiodiniaceae types encountered in 57 host genera) and a multitude of other eukaryotic and prokaryotic microorganisms (Ziegler et al., 2019a).

The marine ecosystems of the Red Sea are threatened by global warming and ocean acidification. The coral reefs in the southern Red Sea live near or above their maximum temperature tolerance and have experienced bleaching events in the recent past. However, corals in the northern part withstand water temperature anomalies and show far above-average physiological thermal tolerance. In the northern Red Sea, including the Gulf of Aqaba, the average summer maximum temperature of ~26°C are still far under the predicted bleaching threshold of 32°C for corals, despite a rapid warming rate. These empirical data suggest the Gulf of Aqaba could potentially be one of the planet's largest coral refuges from global warming and acidification (e.g., Fine et al., 2013, 2019; Krueger et al., 2017; Osman et al., 2018; Voolstra et al., 2015).

6 The Sea of Oman

6.1 Physical setting

The Sea of Oman, considered the northwest extension of the Arabian Sea, is a semienclosed ocean basin bound by Iran and Pakistan in the north and the Sultanate of Oman to the south. Roughly 340 km wide and over 3000 m deep at the entrance, it tapers northward to about 34 km and <100 m in depth at the Strait of Hormuz.

TABLE 3 Selected key parameters related to biogeochemical cycles in the Red Sea.

Parameter	NRS	CRS	SRS	Red Sea	Reference
Water column:					
Total alkalinity, µmol/kg		2391–2492		2360–2475	[1–3]
Macronutrients, µM:					[4–9]
Nitrite (NO_2)	0.03 ± 0.007	0.45–1.47			
Nitrate (NO_3)	0.45 ± 0.88	9.62–18.64	15.7–23.4		
Ammonium (NH_4)		0.014–0.22			
Phosphate (PO_4)	0.02 ± 0.001	<0.05–0.50	1.4–1.9		
Silicate (SiO_3)	0.51 ± 2.00	0.9–5.51	9.3–17.9		
Dissolved inorganic nitrogen (DIN)		bdl–1.21			
Dissolved inorganic phosphorus (DIP)		bdl–0.17			
Chlorophyll-a concentrations:					
Chl-a, mg m^{-3}	0.24	0.1–0.83	3.0		[5–7, 9, 10]
Deep Chl-a maximum, mg m^{-3}	0.01–0.6	0.28–2.08			[9, 11, 12]
Average column-integrated Chl-a, mg m^{-2}	0.1–24.3	9.1–38.8	59.2		[6, 13–15]
Primary production:					
mg C m^{-2} h^{-1}	17.5	33–96.3	133		[16, 17]
mg C m^{-2} d^{-1}	250–650	300–420	360–3150	<100–346	[18–21]
N_2 fixation rate, nmol N L^{-1} d^{-1}	0.01–0.8				[22–24]
Coral reefs:					
Carbonate budget, kg CaCO$_3$ m^{-2} year^{-1}		0.66 ± 2.01			[3]
Corals-Symbiodiniaceae symbiotic complex:					
N_2 fixation rate, nmol N cm^{-2} d^{-1}		0.04–0.23			[25]
Net photosynthesis, µmol O_2 cm^{-2} h^{-1}		0.21–0.66			
Gross photosynthesis, µmol O_2 cm^{-2} h^{-1}		0.42–0.90			

Average ± standard deviation; *bdl*, below detection level; *Chl-a*, chlorophyll-a; *CRS*, Central Red Sea; *NRS*, Northern Red Sea; *Red Sea*, entire basin; *SRS*, Southern Red Sea.
Published reports are cited as follows: [1]—Metzl et al. (1989); [2]—Elsheikh (2008); [3]—Roik et al. (2018); [4]—Souvermezoglou et al. (1989); [5]—Zarokanellos et al. (2017); [6]—Kürten et al. (2019); [7]—Al-Mur (2020); [8]—Qurban et al. (2020); [9]—Calleja et al. (2019); [10]—Raitsos et al. (2013); [11]—Fahmy (2003); [12]—Levanon-Spanier et al. (1979); [13]—Petzold (1986); [14]—Dorgham et al. (2012); [15]—Qurban et al. (2014); [16]—Khmeleva (1970); [17]—Shaikh et al. (1986); [18]—Yentsch and Wood (1961); [19]—Koblentz-Mishke (1970); [20]—Veldhuis et al. (1997); [21]—Kheireddine et al. (2017); [22]—Foster et al. (2009); [23]—Rahav et al. (2013); [24]—Rahav et al. (2015); [25]—Tilstra et al. (2019).

6.2 Oceanographic setting

Surface circulation and mixing in the Sea of Oman are largely driven by highly dynamic mesoscale eddies that form from a combination of local winds, outflow of high salinity, high-density Persian Gulf Outflow Water through the Strait of Hormuz and monsoonal wind-driven circulation of the Arabian Sea.

Typically, Persian Gulf Outflow Water is colder and denser in winter. As these waters traverse through the Strait of Hormuz, they are rapidly advected by gravitational flow toward Ra's Al Hamra, where they become part of the southeastward flowing Oman Coastal Current (Bower et al., 2000; Pous et al., 2004a, 2004b; Senjyu et al., 1998).

Along the way, small eddies and recirculating features are shed from the Oman Coastal Current (Vic et al., 2015), mixing and stirring the denser Persian Gulf Outflow Water throughout the Sea of Oman (Ezam et al., 2010; L'Hégaret et al., 2013; Pous et al., 2004a, 2004b) producing a three-layered vertical water column, made up of low salinity western

Arabian Sea inflow water (200–350 m), denser Persian Gulf Outflow Water (~150 and 350 m), and a low salinity, suboxic water mass below the Persian Gulf Outflow Water layer. The depth to which the Persian Gulf Outflow Water equilibrates varies seasonally. It is usually deepest (>350 m) in winter and about 200 m in summer (Ezam et al., 2010).

At the Ra's al Hadd cape, the summer Oman Coastal Current becomes part of the Ra's al Hadd jet with the formation of a dipole of intense horizontal velocity. The instability of the Oman Coastal Current generates meanders and mesoscale eddies around the Arabian Peninsula, inducing horizontal transports into the Sea of Oman. As the Southwest Monsoon ends and the Ra's Al Hadd jet completely weakens, circulation is dominated by a large, annually recurring, cyclonic eddy (Morvan & Carton, 2020) that stalls southwest of the mouth of the Sea of Oman transporting subsurface, suboxic, nutrient-rich Arabian Sea waters into the Sea of Oman (Goes et al., unpublished data).

6.3 Biogeochemical features and biological characteristics

6.3.1 Phytoplankton communities

Although the biological cycle of the Sea of Oman mirrors that of the adjacent monsoon-driven Arabian Sea, it is unclear how oceanographic features that develop over seasonal, annual, and interannual time scales impact biogeochemistry and productivity. A major reason for this lack of clarity is the paucity of multidisciplinary studies relating phytoplankton biomass and species distribution to large-scale oceanographic processes. One of the most extensive studies is that of Polikarpov et al. (2016), who showed that coastal phytoplankton communities in the Sea of Oman differ substantially from those in the Strait of Hormuz, Arabian Gulf, and the central Sea of Oman.

Dedicated monthly sampling from 2001 at a coastal station in the Bandar Khayran Bay, and from 2005 at an offshore location (Al-Azri et al., 2010; Al-Hashmi et al., 2014, 2015), have provided evidence, that like the Arabian Sea (Garrison et al., 2000), the Sea of Oman also has a summer (July) and winter (February) peak in phytoplankton biomass and diversity. These time series studies have established that dinoflagellates and diatoms predominate in the coastal waters of Oman, but green *Noctiluca scintillans* also forms extensive blooms intermittently, especially in winter. *Noctiluca* is a large (~1 mm in diameter) mixotrophic dinoflagellate (Gomes et al., 2020 and references within), that sustains itself via photosynthesis from a green free-swimming endosymbiont, *Pedinomonas noctilucae* (Wang et al., 2016) and/or by ingestion of exogenous prey. While blooms of the exclusively heterotrophic red *Noctiluca* strain, which is devoid of endosymbionts (Gomes et al., 2020) have been observed prior to 2000 (Thangaraja et al., 2007), the emergence of the mixotrophic green *Noctiluca* as the dominant winter bloom is recent (Al-Azri et al., 2007, 2010; Al-Hashmi et al., 2015).

6.3.2 Noctiluca blooms

Gomes et al. (2014) were the first to suggest that winter green *Noctiluca* blooms in the Arabian Sea are facilitated by the shoaling of hypoxic waters caused by the expansion of the permanent oxygen minimum zone (OMZ) (Bopp et al., 2013). Prior to appearing as surface blooms in the Sea of Oman, *Noctiluca* is seen close to the oxycline (Goes et al., unpublished data; Piontkovski et al., 2017), where the actively photosynthesizing endosymbionts are advantaged by higher nutrient, hypoxic waters. Surface shoaling of these hypoxic waters is facilitated by mesoscale eddies, which thus play a significant role in the development and dispersal of *Noctiluca* blooms (Gomes et al., 2009). Outputs from a coupled, physical-biogeochemical model, based on the Navy Coastal Ocean Model (Gomes et al., 2016; Huang et al., 2015) for the Sea of Oman have revealed that the onset of *Noctiluca* blooms (mid-December) occurs in the Strait of Hormuz aided by the transport of cold, nutrient-rich, hypoxic waters from the western Arabian Sea into the Sea of Oman, by a large, annually recurrent (~350 km) cyclonic eddy that forms at the mouth of the Sea of Oman (Fig. 7a–c).

Within a week after *Noctiluca* blooms are established in the Strait Hormuz near the coast of Iran, they are incorporated into the Oman Coastal Current and rapidly advected toward the Omani coast, and subsequently injected into the central Sea of Oman by recirculating currents (Fig. 7d). In contrast, in summer, transport of Arabian Sea waters into the Sea of Oman is disrupted by the strong front of the Ra's Al Hadd jet (Fig. 7e–g) (Böhm et al., 1999; Queste et al., 2018). Nutrient-rich waters that shoal on the western side of the jet are recirculated in the Sea of Oman, fueling small coastal dinoflagellate and diatoms blooms (Fig. 7h). *Noctiluca* blooms have also recently begun appearing in summer but are largely confined to coastal embayments (Al-Hashmi et al., 2015) where their growth and proliferation appears to be favored by water column stability (Goes et al., 2020).

In an Arabian Sea-focused study (Goes et al., 2020), water column stability was shown to be essential for the formation of surface *Noctiluca* blooms. They noted that enhanced water column stratification in the Arabian Sea caused by the recent warming trend was fueling the range expansion of *Noctiluca*. Consistent with observations in the Arabian Sea, the Sea of Oman appears to be experiencing a similar range expansion (Fig. 8a–c). The almost twofold increase in annual average

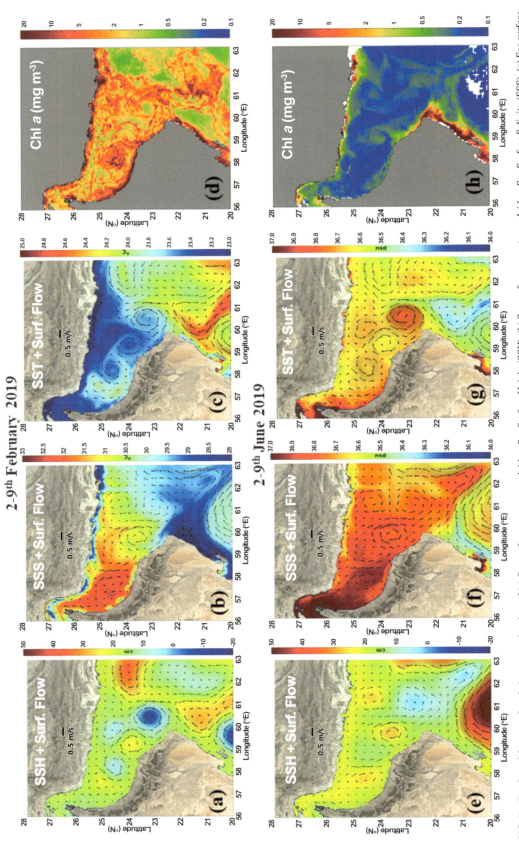

FIG. 7 The Sea of Oman surface 8-day composite plots of (a) Sea surface currents overlaid on Sea Surface Height (SSH), (b) Sea surface currents overlaid on Sea Surface Salinity (SSS), (c) Sea surface currents overlaid on Sea Surface Temperature (SST), (d) Sea surface chlorophyll-*a* (Chl-*a*) concentrations for a representative period during the winter monsoon, (e) Sea surface currents overlaid on Sea Surface Height (SSH), (f) Sea surface currents overlaid on Sea Surface Salinity (SSS), (g) Sea Surface Currents overlaid on Sea Surface Temperature (SST), and (h) Sea surface chlorophyll-*a* (Chl-*a*) concentrations for a representative period during the summer monsoon. (*Data from Navy Coastal Ocean Model-COSiNE.*)

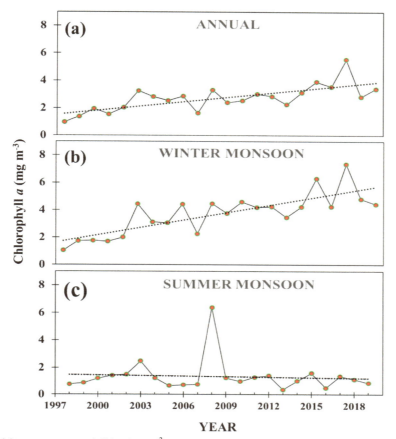

FIG. 8 Time-series of Sea of Oman area averaged Chl-a (mg m^{-3}) (a) Annual, (b) Winter monsoon (December–February), and (c) Summer monsoon (June–August).

chlorophyll-a (Chl-a) in the Sea of Oman in recent years (Fig. 8a) has been largely driven by winter monsoon *Noctiluca* bloom outbreaks (Fig. 8b). No such trend is observed in summer, except for the large Chl-a peak in 2008 (Fig. 8c) from an exceptionally large and unprecedented bloom of the dinoflagellate *Cochlodinium polykrikoides* (Al-Azri et al., 2013).

Ocean color data suggest that the Sea of Oman is becoming greener, largely because of the increasing intensity of green *Noctiluca* blooms, but the role of eddies as modulators of their appearance and growth needs better investigation. *Noctiluca*'s major consumers are gelatinous tunicates (Gomes et al., 2014), and the recent unprecedented swarms of salps and jellyfish (Moazzam, 2020; Sahu & Panigrahy, 2013) portend long-term impacts on Omani fisheries.

7 Conclusions

The descriptions of the marginal seas not only summarize what we know of these seas but also highlight the gaps in our knowledge and suggest directions for future research. These gaps and recommendations are summarized as follows:

Understanding the responses of these marginal seas to global changes needs to be prioritized, as the ecosystems of these marginal seas are clearly changing. The need for this is particularly well-illustrated by the changing composition of phytoplankton and increasing incidences of *Noctiluca* blooms in the Sea of Oman, which could affect fish yield at local and regional levels. The impacts of climate change on the Andaman Sea ecosystem and systematic studies on the physical and biogeochemical processes are needed, and this also illustrates the potential for increased instances of cyclones in the Bay of Bengal. Even though they are marginal, they have the potential to affect adjacent large marine ecosystems of which they are constituents, and there can be reciprocal impacts. Biological diversity is generally given a secondary importance in most of these studies. However, they are first responders to large oceanographic changes, and all five sections have been able to provide concise accounts of the current status of the habitats and their biological constituents, including endemism.

At least two of these marginal seas (Arabian/Persian Gulf and the Red Sea), thanks to the extremes of temperature and salinity, have the potential to serve as natural laboratories to predict ecosystem and population level changes likely to happen elsewhere in other seas of the world. A concerted effort focused on understanding these changes, not only by laboratories in the region but also by those in developed countries, would help advance our knowledge considerably in this domain.

The Gulf of Aden and the Sea of Oman provide excellent laboratories for studying the physical, biogeochemical, and ecological impacts of the forcing reversal from the monsoons and their formation dynamics.

All these marginal seas are bordered by multiple countries with varying degrees of political and economic differences. Coordinated international efforts need to be motivated to further investigate the dynamics and changes of the observed oceanographic variables and understand the linkage across physical, biogeochemical, and ecological processes. If basin-scale level research is needed, then regional and international scientific bodies need to become more proactive in organizing scientific cruises, capacity-building programs, and enabling the cross-border deployment of data-collecting devices such as gliders and Argo floats. This could be facilitated by concerted efforts among regional countries and organizations, such as PERSGA (the Regional Organization for the Conservation of the Environment of the Red Sea and the Gulf of Aden) and ROPME (the Regional Organization for the Protection of the Marine Environment).

8 Educational resources

The Editors of Encyclopaedia Britannica, Andaman Sea; Encyclopaedia Britannica April 14, 2020 URL https://www.britannica.com/place/Andaman-Sea.

Al-Yamani, F. Y. (2021). *Fathoming the Oceanography and Marine Biology of the Northwestern Arabian Gulf*. Kuwait Institute for Scientific Research (Publisher), 450 pp. ISBN 978-99966-37-40-7.

Geist, E. L.; Titov, V. V.; Arcas, D.; Pollitz, F. F.; Bilek, S. L. (2007). Implications of the 26 December 2004 Sumatra–Andaman Earthquake on tsunami forecast and assessment models for Great Subduction-Zone Earthquakes. *Bulletin of the Seismological Society of America*, 97(1A), S249–S270. https://doi.org/10.1785/0120050619.

Kiran, S. R. (2017). General circulation and principal wave modes in Andaman Sea from observations. *Indian Journal of Science and Technology*, 10(24), 1–11. 10.17485/ijst/2017/v10i24/115764.

Burt, J. A., Killilea, M. E., & Ciprut, S. (2019). Coastal urbanization and environmental change: Opportunities for collaborative education across a global network university. *Regional Studies in Marine Science*, 26, 1–10. https://doi.org/10.1016/j.rsma.2019.100501.

Sale, P. F., Feary, D., Burt, J. A., Bauman, A., Cavalcante, G., Drouillard, K., Kjerfve, B., Marquis, E., Trick, C., Usseglio, P., & van Lavieren, H. (2011). The growing need for sustainable ecological management of marine communities of the Persian Gulf. *Ambio*, 40, 4–17. https://doi.org/10.1007/s13280-010-0092-6.

Vaughan, G. O., Al-Mansoori, N., & Burt, J. (2019). The Arabian Gulf. In: Sheppard, C. (Ed.) *World Seas: An Environmental Evaluation*. 2nd edition. Elsevier Science, Amsterdam, NL, 1–23. https://doi.org/10.1016/B978-0-08-100,853-9.00001-4.

Bower, A. S., & Furey, H. H. (2012). Mesoscale eddies in the Gulf of Aden and their impact on the spreading of Red Sea Outflow Water. *Progress in Oceanography*, 96(1), 14–39. https://doi.org/10.1016/j.pocean.2011.09.003. *(Provides a good summary of eddies in the Gulf of Aden)*.

Rasul, N. M. A., & Stewart I. C. F. (Eds.) (2015). *The Red Sea: The Formation, Morphology, Oceanography and Environment of a Young Ocean Basin*. Springer Berlin Heidelberg, Berlin, Heidelberg. https://doi.org/10.1007/978-3-662-45,201-1. *(Book contains significant content also on Gulf of Aden. Geological papers by Bonatti et al., and Bosworth et al., provide significant geological background on the gulf)*.

Calleja, M., & Morán, X. (2020). Red Sea fishes that travel into the deep ocean daily. *Frontiers for Young Minds*, 8, 85. https://doi.org/10.3389/frym.2020.00085.

Fine, M., Gildor, H., & Genin, A. (2013). A coral reef refuge in the Red Sea. *Global Change Biology*, 19, 3640–3647. https://doi.org/10.1111/gcb.12356.

Menezes, V. (2018). Mysteries of the Red Sea. *Oceanus*, https://www.whoi.edu/oceanus/feature/mysteries-of-the-red-sea/.

Triantafyllou, G., Yao, F., Petihakis, G., Tsiaras, K. P., Raitsos, D. E., & Hoteit, I. (2014). Exploring the Red Sea seasonal ecosystem functioning using a three-dimensional biophysical model. *Journal of Geophysical Research: Oceans*, 119, 1791–1811. https://doi.org/10.1002/2013JC009641.

Acknowledgments

John Burt was supported by the NYUAD Water Research Center, funded by Tamkeen under the NYUAD Research Institute Award (Project CG007), and by the NYUAD Center for Genomics and Systems Biology research program in Sustainability (Project CGSB5).

Faiza Al-Yamani, Igor Polikarpov, Maria Saburova, Turki Al-Said, Manal Al-Kandari, and Takahiro Yamamoto are supported by Kuwait Institute for Scientific Research.

Ramaiah Nagappa and V.S.N. Murty are grateful to the CSIR-NIO for giving them the opportunity of researching on the seas around India.

Joaquim I. Goes and Helga do R. Gomes are supported by grants from the National Science Foundation (NSF 2019983), the National Aeronautics and Space Administration (NASA NNX18AG66G-ECO4CAST, NASA NNH18ZDA001-CMP, NOAA-STAR-GST, the Vetlesen Foundation at Lamont Doherty Earth Observatory, and the Gordon Betty Moore Foundation). Sergio de Rada and Dale Kiefer acknowledge funding from NASA NNX18AG66G-ECO4CAST. Khalid Al-Hashmi acknowledges support from Sultan Qaboos University.

Burton Jones and Alkiviadis Kalampokis are supported through baseline funding from King Abdullah University of Science and Technology. Mohammed Alsaafani is supported by the Department of Marine Physics, Faculty of Marine Science, King Abdulaziz University.

Authors contributions

Author FYY wrote Section 1, RN and VSNM wrote Section 2, and Section 3 was led by JAB, who wrote Section 3.1, Section 3.2 was written by JAB with contribution from TY, and FAY, TAS, MAK, and JAB contributed to Section 3.3, while all authors contributed to the structure, reviewing, and editing of Section 3. Authors BJ, AK, and MA wrote collaboratively Section 4. Authors FYY, IP, and MS wrote Section 5. JIG and HRG led the writing of Section 6 with contributions from SD, DK, and KH. All authors contributed to the overall structure of the chapter and contributed to Section 7. FYY compiled and edited the full chapter. MS drafted Fig. 1a, RN and VSNM provided Fig. 1b and c, JAB created Figs. 2 and 3, MA drafted Fig. 4, and BJ provided Fig. 5 and drafted Fig. 5a. IP and MS contributed to the conceptual design for Fig. 6, and MS drafted Fig. 6, JIG and HRG developed Figs. 7 and 8 with contributions from SD, DK, and KHRN and VSNM provided the summary in Tables 1 and 2, and MS conducted the analyses summarized in Table 3. FYY, JAB, RN and VSNM, BJ, and JIG provided the educational material.

References

Acker, J., Leptoukh, G., Shen, S., Zhu, T., & Kempler, S. (2008). Remotely-sensed chlorophyll a observations of the northern Red Sea indicate seasonal variability and influence of coastal reefs. *Journal of Marine Systems, 69*, 191–204. https://doi.org/10.1016/j.jmarsys.2005.12.006.

Acker, J. G., & Leptoukh, G. (2007). Online analysis enhances use of NASA Earth science data. *Eos, 88*(2), 14–17. https://doi.org/10.1029/2007EO020003.

Aiki, H., Takahashi, K., & Yamagata, T. (2006). The Red Sea outflow regulated by the Indian monsoon. *Continental Shelf Research, 26*(12–13), 1448–1468. https://doi.org/10.1016/J.Csr.2006.02.017.

Al Senafi, F., & Anis, A. (2015). Shamals and climate variability in the Northern Arabian/Persian Gulf from 1973 to 2012. *International Journal of Climatology, 35*, 4509–4528. https://doi.org/10.1002/joc.4302.

Al-Ansari, E. M. A. S., Rowe, G., Abdel-Moati, M. A. R., Yigiterhan, O., Al-Maslamani, I., Al-Yafei, M. A., Al-Shaikh, I., & Upstill-Goddard, R. (2015). Hypoxia in the central Arabian Gulf Exclusive Economic Zone (EEZ) of Qatar during summer season. *Estuarine, Coastal and Shelf Science, 159*, 60–68. https://doi.org/10.1016/j.ecss.2015.03.022.

Al-Azri, A., Al-Hashmi, K., Goes, J. I., Gomes, H.d. R., Rushdi, A., Al-Habsi, H., Al-Khusaibi, S., Al-Kindi, R., & Al-Azri, N. (2007). Seasonality of the bloom-forming heterotrophic dinoflagellate *Noctiluca scintillans* in the Gulf of Oman in relation to environmental conditions. *International Journal of Oceans and Oceanography, 2*, 51–60.

Al-Azri, A. R., Piontkovski, S. A., Al-Hashmi, K. A., Goes, J. I., & Gomes, H.d. R. (2010). Chlorophyll *a* as a measure of seasonal coupling between phytoplankton and the monsoon periods in the Gulf of Oman. *Aquatic Ecology, 44*, 449–461. https://doi.org/10.1007/s10452-009-9303-2.

Al-Azri, A. R., Piontkovski, S. A., Al-Hashmi, K. A., Goes, J. I., Gomes, H.d. R., & Glibert, P. M. (2013). Mesoscale and nutrient conditions associated with the massive 2008 *Cochlodinium polykrikoides* bloom in the Sea of Oman/Arabian Gulf. *Estuaries and Coasts, 37*, 325–338. https://doi.org/10.1007/s12237-013-9693-1.

Al-Hashmi, K. A., Goes, J., Claereboudt, M., Piontkovski, S. A., Al-Azri, A., & Smith, S. L. (2014). Variability of dinoflagellates and diatoms in the surface waters of Muscat, Sea of Oman: Comparison between enclosed and open ecosystem. *International Journal of Oceans and Oceanography, 8*, 137–152.

Al-Hashmi, K. A., Smith, S. L., Claereboudt, M., Piontkovski, S. A., & Al-Azri, A. (2015). Dynamics of potentially harmful phytoplankton in a semi-enclosed bay in the Sea of Oman. *Bulletin of Marine Science, 91*, 141–166. https://doi.org/10.5343/bms.2014.1041.

Al-Mur, B. (2020). Assessing nutrient salts and trace metals distributions in the coastal water of Jeddah, Red Sea. *Saudi Journal of Biological Sciences, 27*(11), 3087–3098. https://doi.org/10.1016/j.sjbs.2020.07.012.

Al-Rashidi, T., El-Gamily, H., Amos, C., & Rakha, K. (2009). Sea surface temperature trends in Kuwait Bay, Arabian Gulf. *Natural Hazards, 50*, 73–82. https://doi.org/10.1007/s11069-008-9320-9.

Alsaafani, M. A., & Shenoi, S. S. C. (2007). Water masses in the Gulf of Aden. *Journal of Oceanography, 63*(1), 1–14. https://doi.org/10.1007/s10872-007-0001-1.

Alsaafani, M. A., Shenoi, S. S. C., Shankar, D., Aparna, M., Kurian, J., Durand, F., & Vinayachandran, P. N. (2007). Westward movement of eddies into the Gulf of Aden from the Arabian Sea. *Journal of Geophysical Research-Oceans, 112*(C11). https://doi.org/10.1029/2006jc004020.

Al-Said, T., Madhusoodhanan, R., Pokavanich, T., Al-Yamani, F., Kedila, R., Al-Ghunaim, A., & Al-Hashem, A. (2018a). Environmental characterization of a semiarid hyper saline system based on dissolved trace metal-macronutrient synergy: A multivariate spatio-temporal approach. *Marine Pollution Bulletin, 129*, 846–858. https://doi.org/10.1016/j.marpolbul.2017.10.009.

Al-Said, T., Naqvi, S. W. A., Al-Yamani, F., Goncharov, A., & Fernandes, L. (2018b). High total organic carbon in surface waters of the northern Arabian Gulf: Implications for the oxygen minimum zone of the Arabian Sea. *Marine Pollution Bulletin, 129*, 35–42. https://doi.org/10.1016/j.marpolbul.2018.02.013.

Al-Said, T., Yamamoto, T., Madhusoodhanan, R., Al-Yamani, F., & Pokavanich, T. (2018c). Summer hydrographic characteristics in the northern ROPME Sea Area: Role of ocean circulation and water masses. *Estuarine, Coastal and Shelf Science, 213*, 18–27. https://doi.org/10.1016/j.ecss.2018.07.026.

Al-Yamani, F. Y., Al-Kandari, M., Polikarpov, I., & Grinstov, V. (2019). *Field guide of order Amphipoda (Malacostraca, Crustacea) of Kuwait*. Kuwait: Kuwait Institute for Scientific Research.

Al-Yamani, F. Y., Bishop, J., Ramadhan, E., Al-Husaini, M., & Al-Ghadban, A. (2004). *Oceanographic atlas of Kuwait's waters*. Kuwait: Kuwait Institute for Scientific Research.

Al-Yamani, F., & Naqvi, S. W. A. (2019). Chemical oceanography of the Arabian Gulf. *Deep Sea Research Part II: Topical Studies in Oceanography, 161*, 72–80. https://doi.org/10.1016/j.dsr2.2018.10.003.

Al-Yamani, F. Y., Polikarpov, I., Al-Ghunaim, A., & Mikhailova, T. (2014). *Field guide to the marine macroalgae (Chlorophyta, Rhodophyta, Phaeophyceae) of Kuwait*. Kuwait: Kuwait Institute for Scientific Research.

Al-Yamani, F. Y., & Saburova, M. A. (2010). *Illustrated guide on the flagellates of Kuwait's intertidal soft sediments*. Kuwait: Kuwait Institute for Scientific Research.

Al-Yamani, F. Y., & Saburova, M. A. (2011). *Illustrated guide on the benthic diatoms in Kuwait's marine environment*. Kuwait: Kuwait Institute for Scientific Research.

Al-Yamani, F. Y., & Saburova, M. A. (2019). *Marine phytoplankton of Kuwait's waters. Vol. 1: Cyanobacteria, dinoflagellates, flagellates; Vol. 2: Diatoms*. Kuwait: Kuwait Institute for Scientific Research.

Al-Yamani, F. Y., Skryabin, V., Boltachova, N., Revkov, N., Makarov, M., Grinstov, V., & Kolesnikova, E. (2012). *Illustrated atlas on the Zoobenthos of Kuwait*. Kuwait: Kuwait Institute for Scientific Research.

Al-Yamani, F. Y., Skryabin, V., Gubanova, A., Khvorov, S., & Prusova, I. (2011). *Marine zooplankton practical guide (Volumes 1 and 2) for the Northwestern Arabian Gulf*. Kuwait: Kuwait Institute for Scientific Research.

Anisimov, A., Tao, W., Stenchikov, G., Kalenderski, S., Prakash, P. J., Yang, Z. L., & Shi, M. (2017). Quantifying local-scale dust emission from the Arabian Red Sea coastal plain. *Atmospheric Chemistry and Physics, 17*, 993–1015. https://doi.org/10.5194/acp-17-993-2017.

Bange, H. W., Naqvi, S. W. A., & Codispoti, L. A. (2005). The nitrogen cycle in the Arabian Sea. *Progress in Oceanography, 65*(2–4), 145–158. https://doi.org/10.1016/J.Pocean.2005.03.002.

Bauman, A., Feary, D., Heron, S., Pratchett, M. S., & Burt, J. (2013). Multiple environmental factors influence the spatial distribution and structure of reef communities in the northeastern Arabian Peninsula. *Marine Pollution Bulletin, 72*, 302–312. https://doi.org/10.1016/j.marpolbul.2012.10.013.

Beal, L. M., Ffield, A., & Gordon, A. L. (2000). Spreading of red sea overflow waters in the Indian ocean. *Journal of Geophysical Research-Oceans, 105*(C4), 8549–8564. https://doi.org/10.1029/1999jc900306.

Berumen, M. L., Voolstra, C. R., Daffonchio, D., Agusti, S., Aranda, M., Irigoien, X., Jones, B. H., Morán, X. A. G., & Duarte, C. M. (2019). The Red Sea: Environmental gradients shape a natural laboratory in a nascent ocean. In M. L. Berumen, & C. R. Voolstra (Eds.), *Coral reefs of the Red Sea* (pp. 1–10). Cham: Springer International Publishing.

Bhattathiri, P. M. A., & Devassy, V. P. (1981). Primary productivity of the Andaman Sea. *Indian Journal of Marine Sciences, 10*, 248–251.

Böhm, E., Morrison, J. M., Manghnani, V., Kim, H., & Flagg, C. (1999). Remotely sensed and acoustic doppler current profiler observations of the Ras Al Hadd Jet. *Deep Sea Research Part II: Topical Studies in Oceanography, 46*, 1531–1549.

Bonatti, E., Cipriani, A., & Lupi, L. (2015). The Red Sea: Birth of an ocean. In N. M. A. Rasul, & I. C. F. Stewart (Eds.), *Springer earth system sciences. The Red Sea* (pp. 29–44). Berlin/Heidelberg: Springer-Verlag. https://doi.org/10.1007/978-3-662-45201-1_2.

Bopp, L., Resplandy, L., Orr, J. C., Doney, S. C., Dunne, J. P., Gehlen, M., Halloran, P., Heinze, C., Ilyina, T., Seferian, R., Tjiputra, J., & Vichi, M. (2013). Multiple stressors of ocean ecosystems in the 21st century: Projections with CMIP5 models. *Biogeosciences, 10*, 6225–6245. https://doi.org/10.5194/bg-10-6225-2013.

Bower, A. S., & Farrar, J. T. (2015). Air-sea interaction and horizontal circulation in the Red Sea. In N. M. A. Rasul, & I. C. F. Stewart (Eds.), *Springer earth system sciences. The Red Sea* (pp. 329–342). Berlin/Heidelberg: Springer-Verlag. https://doi.org/10.1007/978-3-662-45201-1_19.

Bower, A. S., Fratantoni, D. M., Johns, W. E., & Peters, H. (2002). Gulf of Aden eddies and their impact on Red Sea Water. *Geophysical Research Letters, 29*(21), 21-1–21-4. https://doi.org/10.1029/2002gl015342.

Bower, A. S., & Furey, H. H. (2012). Mesoscale eddies in the Gulf of Aden and their impact on the spreading of Red Sea Outflow Water. *Progress in Oceanography, 96*(1), 14–39. https://doi.org/10.1016/j.pocean.2011.09.003.

Bower, A. S., Hunt, H. D., & Price, J. F. (2000). Character and dynamics of the Red Sea and Persian Gulf outflows. *Journal of Geophysical Research-Oceans, 105*(C3), 6387–6414. https://doi.org/10.1029/1999jc900297.

Bower, A. S., Johns, W. E., Fratantoni, D. M., & Peters, H. (2005). Equilibration and circulation of Red Sea Outflow water in the western Gulf of Aden. *Journal of Physical Oceanography, 35*(11), 1963–1985. https://doi.org/10.1175/Jpo2787.1.

Boyer, T. P., Antonov, J. I., Baranova, O. K., Coleman, C., Garcia, H. E., Grodsky, A., Johnson, D. R., Locarnini, R. A., Mishonov, A. V., O'Brien, T. D., Paver, C. R., Reagan, J. R., Seidov, D., Smolyar, I. V., & Zweng, M. M. (2013). World ocean database 2013. In *NOAA atlas NESDIS 72*. Silver Spring, MD: NOAA NESDIS. https://doi.org/10.7289/V5NZ85MT.

Brewer, P. G., & Dyrssen, D. (1985). Chemical oceanography of the Persian Gulf. *Progress in Oceanography, 14*, 41–55.

Brewer, P., Fleer, A., Kadar, S., Shafer, D., & Smith, C. (1978). *Chemical oceanographic data from the Persian Gulf and Gulf of Oman*. Woods Hole: Woods Hole Oceanographic Institution Technical Report, 78-37.

Britannica. (2020). *Indian Ocean*. https://www.britannica.com/place/Indian-Ocean (Accessed 16 October 2020).
Burt, J. (2013). The growth of coral reef science in the Gulf: A historical perspective. *Marine Pollution Bulletin, 72*, 289–301. https://doi.org/10.1016/j.marpolbul.2013.05.016.
Burt, J. (2014). The environmental costs of coastal urbanization in the Arabian Gulf. *City: Analysis of Urban Trends, Culture, Theory, Policy, Action, 18*, 760–770. https://doi.org/10.1080/13604813.2014.962889.
Burt, J., Al-Khalifa, K., Khalaf, E., AlShuwaik, B., & Abdulwahab, A. (2013). The continuing decline of coral reefs in Bahrain. *Marine Pollution Bulletin, 72*, 357–363. https://doi.org/10.1016/j.marpolbul.2012.08.022.
Burt, J., Camp, E., Enochs, I. C., Johansen, J. L., Morgan, K. M., Riegl, B., & Hoey, A. S. (2020). Insights from extreme coral reefs in a changing world. *Coral Reefs, 39*(3), 495–507. https://doi.org/10.1007/s00338-020-01966-y.
Burt, J., Feary, D., Bauman, A., Usseglio, P., Cavalcante, G., & Sale, P. (2011). Biogeographic patterns of reef fish community structure in the northeastern Arabian Peninsula. *ICES Journal of Marine Science, 68*, 1875–1883. https://doi.org/10.1093/icesjms/fsr129.
Burt, J. A., Paparella, F., Al-Mansoori, N., Al-Mansoori, A., & Al-Jailani, H. (2019). Causes and consequences of the 2017 coral bleaching event in the southern Persian/Arabian Gulf. *Coral Reefs, 38*, 567–589. https://doi.org/10.1007/s00338-019-01767-y.
Burt, J. A., & Paparella, F. (2023). The marine environment of the Emirates. In J. A. Burt (Ed.), *A natural history of the Emirates* (pp. 95–117). Switzerland: Springer Nature. https://doi.org/10.1007/978-3-031-37397-8_4 under CC-BY 4.0.
Burt, J. A., Smith, E. G., Warren, C., & Dupont, J. (2016). An assessment of Qatar's coral communities in a regional context. *Marine Pollution Bulletin, 105*, 473–479. https://doi.org/10.1016/j.marpolbul.2015.09.025.
Calleja, M. L., Al-Otaibi, N., & Moran, X. A. G. (2019). Dissolved organic carbon contribution to oxygen respiration in the central Red Sea. *Scientific Reports, 9*, 4690. https://doi.org/10.1038/s41598-019-40753-w.
Cardini, U., Bednarz, V. N., van Hoytema, N., Rovere, A., Naumann, M. S., Al-Rshaidat, M. M. D., & Wild, C. (2016). Budget of primary production and dinitrogen fixation in a highly seasonal Red Sea coral reef. *Ecosystems, 19*, 771–785. https://doi.org/10.1007/s10021-016-9966-.
Carvalho, S., Kürten, B., Krokos, G., Hoteit, I., & Ellis, J. (2019). The Red Sea. In C. Sheppard (Ed.), *The Indian Ocean to the Pacific: Vol. II. World seas: An environmental evaluation* (2nd ed., pp. 49–74). Amsterdam: Elsevier.
Cavalcante, G. H., Feary, D. A., & Burt, J. A. (2016). The influence of extreme winds on coastal oceanography and its implications for coral population connectivity in the southern Arabian Gulf. *Marine Pollution Bulletin, 105*, 489–497. https://doi.org/10.1016/j.marpolbul.2015.10.031.
Chatterjee, A., Shankar, D., McCreary, J. P., Vinayachandran, P. N., & Mukherjee, A. (2017). Dynamics of Andaman Sea circulation and its role in connecting the equatorial Indian Ocean to the Bay of Bengal. *Journal of Geophysical Research: Oceans, 122*, 3200–3218. https://doi.org/10.1002/2016JC012300.
Chiffings, A. W. (1995). Marine region 11: Arabian seas. In G. Kelleher, C. Bleakley, & S. Wells (Eds.), *Vol. III. A global representative system of marine protected areas* (pp. 39–70). Washington, DC: The Great Barrier Reef Marine Park Authority, The World Bank and IUCN.
Churchill, J., Davis, K., Wurgaft, E., & Shaked, Y. (2019). Environmental setting for reef building in the Red Sea. In M. L. Berumen, & C. R. Voolstram (Eds.), *Coral reefs of the Red Sea* (pp. 11–32). Cham: Springer International Publishing.
Claereboudt, M. R. (2019). Oman. In C. Sheppard (Ed.), *World seas: An environmental evaluation* (2nd ed., pp. 25–47). Academic Press (chapter 2).
Coles, S. (2003). Coral species diversity and environmental factors in the Arabian Gulf and the Gulf of Oman: A comparison to the Indo-Pacific region. *Atoll Research Bulletin, 507*, 1–19. https://doi.org/10.5479/si.00775630.507.1.
de Verneil, A., Lachkar, Z., Smith, S., & Lévy, M. (2022). Evaluating the Arabian Sea as a regional source of atmospheric CO_2: Seasonal variability and drivers. *Biogeosciences, 19*, 907–929. https://doi.org/10.5194/bg-19-907-2022.
DiBattista, J. D., Roberts, M. B., Bouwmeester, J., Bowen, B. W., Coker, D. J., Lozano-Cortés, D. F., Choat, J. H., Gaither, M. R., Hobbs, J. P. A., Khalil, M. T., Kochzius, M., Myers, R. F., Paulay, G., Robitzch, V. S. N., Saenz-Agudelo, P., Salas, E., Sinclair-Taylor, T. H., Toonen, R. J., Westneat, M. W., ... Berumen, M. L. (2016). A review of contemporary patterns of endemism for shallow water reef fauna in the Red Sea. *Journal of Biogeography, 43*, 423–439. https://doi.org/10.1111/jbi.12649.
Dorgham, M. M., El-Sherbiny, M. M., & Hanifi, M. H. (2012). Environmental properties of the southern Gulf of Aqaba, Red Sea, Egypt. *Mediterranean Marine Science, 13*, 179–186. https://doi.org/10.12681/mms.297.
Dreano, D., Raitsos, D. E., Gittings, J., Krokos, G., & Hoteit, I. (2016). The Gulf of Aden intermediate water intrusion regulates the southern Red Sea summer phytoplankton blooms. *PLoS One, 11*(12), e0168440. https://doi.org/10.1371/journal.pone.0168440.
Edwards, F. J. (1987). Climate and oceanography. In A. J. Edwards, & S. M. Head (Eds.), *Key environments: Red Sea* (pp. 45–70). Oxford: Pergamon Press.
Egge, J. K., & Aksnes, D. L. (1992). Silicate as regulating nutrient in phytoplankton competition. *Marine Ecology Progress Series, 83*, 281–289.
Elsheikh, A. B. (2008). *The inorganic carbon cycle in the Red Sea*. Master thesis in chemical oceanography University of Bergen Geophysical Institute.
Elshorbagy, W., Azam, M. H., & Taguchi, K. (2006). Hydrodynamic characterization and modeling of the Arabian Gulf. *Journal of Waterway, Port, Coastal, and Ocean Engineering, 132*, 47–56. https://doi.org/10.1061/(ASCE)0733-950X(2006)132:1(47).
Eshel, G., & Heavens, N. (2007). Climatological evaporation seasonality in the northern Red Sea. *Paleoceanography, 22*, PA4201. https://doi.org/10.1029/2006PA001365.
Ezam, M., Bidokhti, A. A., & Javid, A. H. (2010). Numerical simulation of spreading of the Persian Gulf outflow in the Oman Sea. *Ocean Science, 6*, 887–900. https://doi.org/10.5194/os-6-887-2010.
Ezzati, R. (2012). An investigation on atmospheric dust which transports to the Persian Gulf. *Journal of Oceanography, 2*, 21–30.
Fahmy, M. (2003). Water quality in the Red Sea coastal waters (Egypt): Analysis of spatial and temporal variability. *Journal of Chemical Ecology, 19*, 67–77. https://doi.org/10.1080/0275754031000087074.

FAO (Food and Agriculture Organization). (2003). *A Paper on Thailand's marine capture fisheries management in the Andaman Sea*. http://www.fao.org/3/a0477e/a0477e0f.htm.

Feary, D. A., Burt, J., Bauman, A., Al-Hazeem, S., Abdel-Moati, M., Al-Khalifa, K., Anderson, D., Amos, C., Baker, A., Bartholomew, A., Bento, R., Cavalcante, G., Chen, A., Coles, S., Dabo, K., Fowler, A., George, D., Grandcourt, E., Hill, R., ... Wiedenmann, J. (2013). Future changes in the Gulf marine ecosystem: Identifying critical research needs. *Marine Pollution Bulletin, 72*, 406–416.

Fine, M., Cinar, M., Voolstra, C. R., Safa, A., Rinkevich, B., Laffoley, D., Hilmi, N., & Allemand, D. (2019). Coral reefs of the Red Sea—Challenges and potential solutions. *Regional Studies in Marine Science, 25*, 100498. https://doi.org/10.1016/j.rsma.2018.100498.

Fine, M., Gildor, H., & Genin, A. (2013). A coral reef refuge in the Red Sea. *Global Change Biology, 19*, 3640–3647. https://doi.org/10.1111/gcb.12356.

Foster, R. A., Paytan, A., & Zehr, J. P. (2009). Seasonality of N_2 fixation and nifH gene diversity in the Gulf of Aqaba (Red Sea). *Limnology and Oceanography, 54*, 219–233. https://doi.org/10.4319/lo.2009.54.1.0219.

Fratantoni, D. M., Bower, A. S., Johns, W. E., & Peters, H. (2006). Somali Current rings in the eastern Gulf of Aden. *Journal of Geophysical Research-Oceans, 111*(C9), C09039. https://doi.org/10.1029/2005jc003338.

Garrison, D. L., Gowing, M. M., Hughes, M. P., Campbell, L., Caron, D. A., Dennett, M. R., Shalapyonok, A., Olson, R. J., Landry, M. R., Brown, S. L., Liu, H. B., Azam, F., Steward, G. F., Ducklow, H. W., & Smith, D. C. (2000). Microbial food web structure in the Arabian Sea: A US JGOFS study. *Deep Sea Research Part II: Topical Studies in Oceanography, 47*, 1387–1422. https://doi.org/10.1016/S0967-0645(99)00148-4.

Gittings, J. A., Raitsos, D. E., Racault, M. F., Brewin, R. J. W., Pradhan, Y., Sathyendranath, S., & Platt, T. (2017). Seasonal phytoplankton blooms in the Gulf of Aden revealed by remote sensing. *Remote Sensing of Environment, 189*, 56–66. https://doi.org/10.1016/j.rse.2016.10.043.

Goes, J. I., Tian, H., Gomes, H.d. R., Anderson, O. R., Al-Hashmi, K., de Rada, S., Luo, H., Al-Kharusi, L., Al-Azri, A., & Martinson, D. G. (2020). Ecosystem state change in the Arabian Sea fuelled by the recent loss of snow over the Himalayan-Tibetan Plateau region. *Scientific Reports, 10*, 7422. https://doi.org/10.1038/s41598-020-64360-2.

Gokul, E. A., Raitsos, D. E., Gittings, J. A., Alkawri, A., & Hoteit, I. (2019). Remotely sensing harmful algal blooms in the Red Sea. *PLoS One, 14*(4), e0215463. https://doi.org/10.1371/journal.pone.0215463.

Gomes, H.d. R., de Rada, S., Goes, J. I., & Chai, F. (2016). Examining features of enhanced phytoplankton biomass in the Bay of Bengal using a coupled physical-biological model. *Journal of Geophysical Research: Oceans, 121*, 5112–5133. https://doi.org/10.1002/2015JC011508.

Gomes, H.d. R., Goes, J. I., Al-Hashmi, K., & Al-Kharusi, L. (2020). Global distribution and range expansion of green vs red *Noctiluca scintillans*. In D. S. Rao (Ed.), *Dinoflagellates: Morphology, life history, and ecological significance* (pp. 499–525). New York: Nova Publishers.

Gomes, H.d. R., Goes, J. I., Matondkar, S. G. P., Buskey, E. J., Basu, S., Parab, S., & Thoppil, P. (2014). Massive outbreaks of *Noctiluca scintillans* blooms in the Arabian Sea due to spread of hypoxia. *Nature Communications, 5*, 4862. https://doi.org/10.1038/ncomms5862.

Gomes, H., Goes, J. I., Matondkar, S. G. P., Parab, S. G., Al-Azri, A., & Thoppil, P. G. (2009). Unusual blooms of the green *Noctiluca miliaris* (Dinophyceae) in the Arabian Sea during the winter monsoon. In J. D. Wiggert, R. R. Hood, S. W. A. Naqvi, S. L. Smith, & K. H. Brink (Eds.), *Geophysical Monograph Series: Vol. 185. Indian Ocean: Biogeochemical processes and ecological variability* (pp. 347–363). American Geophysical Union.

Gomes, H.d. R., Goes, J. I., & Parulekar, A. H. (1992). Size-fractionated biomass, photosynthesis and dark CO_2 fixation in a tropical oceanic environment. *Journal of Plankton Research, 14*(9), 1307–1329.

Grasshoff, K. (1976). *Review of hydrographic and productivity conditions in the Gulf region*. Paris: UNESCO.

Hashimoto, S., Tsujimoto, R., Maeda, M., Ishimaru, T., Yoshida, J., Takasu, Y., Koike, Y., Mine, Y., Kamatani, A., & Otsuki, A. (1998). Distribution of nutrient, nitrous oxide and chlorophyll a of RSA: Extremely high ratios of nitrite to nitrate in whole water column. In A. Otsuki, M. Abdulraheem, & M. Reynolds (Eds.), *Offshore environment of the ROPME Sea Area after the war-related oil spill: Results of the 1993–94 Umitaka-Maru cruises* (pp. 99–124). Tokyo: Terra Scientific.

Herut, B., Collier, R., & Krom, M. D. (2002). The role of dust in supplying nitrogen and phosphorus to the Southeast Mediterranean. *Limnology and Oceanography, 47*, 870–878. https://doi.org/10.4319/lo.2002.47.3.0870.

Howells, E. J., Abrego, D., Meyer, E., Kirk, N. L., & Burt, J. A. (2016). Host adaptation and unexpected symbiont partners enable reef-building corals to tolerate extreme temperatures. *Global Change Biology, 22*, 2702–2714. https://doi.org/10.1111/gcb.13250.

Huang, K., Derada, S., Xue, H., Xiu, P., Chai, F., Xie, Q., & Wang, D. (2015). A 1/8 coupled biochemical-physical Indian Ocean Regional Model: Physical results and validation. *Ocean Dynamics, 65*, 1121–1142. https://doi.org/10.1007/s10236-015-0860-8.

Hume, B., D'Angelo, C., Smith, E., Stevens, J., Burt, J., & Wiedenmann, J. (2015). *Symbiodinium thermophilum* sp. nov., a thermotolerant symbiotic alga prevalent in corals of the world's hottest sea, the Persian/Arabian Gulf. *Scientific Reports, 5*, 1–8. https://doi.org/10.1038/srep08562.

Hume, B. C. C., Voolstra, C. R., Arif, C., D'Angelo, C., Burt, J. A., Eyal, G., Loya, Y., & Wiedenmann, J. (2016). Ancestral genetic diversity associated with the rapid spread of stress-tolerant coral symbionts in response to Holocene climate change. *Proceedings of the National Academy of Sciences, 113*, 4416–4421. https://doi.org/10.1073/pnas.1601910113.

ICRI, GCRMN, AIMS, & UNEP. (2021). Status and trends of coral reefs of the Red Sea and Gulf of Aden. In D. Souter, S. Planes, J. Wicquart, M. Logan, D. Obura, & R. Staub (Eds.), *Status of coral reefs of the world: 2020* GCRMN (chapter 3).

Jain, V., Shankar, D., Vinayachandran, P. N., Kankonkar, A., Chatterjee, A., Amol, P., Almeida, A. M., Michael, G. S., Mukherjee, A., Chatterjee, M., Fernandes, R., Luis, R., Kamble, A., Hegde, A. K., Chatterjee, S., Das, U., & Neema, C. P. (2017). Evidence for the existence of Persian Gulf Water and Red Sea Water in the Bay of Bengal. *Climate Dynamics, 48*, 3207–3226. https://doi.org/10.1007/s00382-016-3259-4.

Johns, W. E., Jacobs, G. A., Kindle, J. C., Murray, S. P., & Carron, M. (1999). *Arabian marginal seas and gulfs: Report of a workshop held at Stennis Space Center, Miss. 11-13 May, 1999*. Tech. Rep. University of Miami RSMAS Technical Report 2000-01.

Jyothibabu, R., Win, N. N., Shenoy, D. M., Tint Swe, U., Pratik, M., Thwin, S., & Jagadeesan, L. (2014). Interplay of diverse environmental settings and their influence on the plankton community off Myanmar during the Spring Intermonsoon. *Journal of Marine Systems, 139*, 446–459. https://doi.org/10.1016/j.jmarsys.2014.08.003.

Kaartvedt, S., Antunes, A., Røstad, A., Klevjer, T. A., & Vestheim, H. (2016). Zooplankton at deep Red Sea brine pools. *Journal of Plankton Research, 38*, 679–684. https://doi.org/10.1093/plankt/fbw013.

Kämpf, J., & Sadrinasab, M. (2006). The circulation of the Persian Gulf: A numerical study. *Ocean Science, 2*, 27–41. https://doi.org/10.5194/os-2-27-2006.

Kennett, D. J., & Kennett, J. P. (2006). Early state formation in Southern Mesopotamia: Sea levels, shorelines, and climate change. *The Journal of Island and Coastal Archaeology, 1*, 67–99. https://doi.org/10.1080/15564890600586283.

Kheireddine, M., Ouhssain, M., Claustre, H., Uitz, J., Gentili, B., & Jones, B. H. (2017). Assessing pigment-based phytoplankton community distributions in the Red Sea. *Frontiers in Marine Science, 4*, 2296–7745. https://doi.org/10.3389/fmars.2017.00132.

Khmeleva, N. N. (1970). On the primary production in the Red Sea and the Gulf of Aden. *Biologia Morja, Kiew, 21*, 107–133.

Kiørboe, T., Janekarn, V., Poung-in, S., Sawangareeruks, S., & Piukhao, P. (1991). New fisheries resources in the Andaman Shelf Sea? Indirect oceanographical evidence. *Thai Fisheries Gazette, 44*, 261–270.

Kirk, N., Howells, E., Abrego, D., Burt, J., & Meyer, E. (2018). Genomic and transcriptomic signals of thermal tolerance in heat-tolerant corals (*Platygyra daedalea*) of the Arabian/Persian Gulf. *Molecular Ecology, 27*, 5180–5194. https://doi.org/10.1111/mec.14934.

Koblentz-Mishke, O. J. (1970). Plankton primary production of the world ocean. In W. S. Wooster (Ed.), *Scientific exploration of the South Pacific* (pp. 189–193). Washington, DC: National Academy of Science.

Krueger, T., Horwitz, N., Bodin, J., Giovani, M. E., Escrig, S., Meibom, A., & Fine, M. (2017). Common reef-building coral in the Northern Red Sea resistant to elevated temperature and acidification. *Royal Society Open Science, 4*, 170038. https://doi.org/10.1098/rsos.170038.

Kürten, B., Khomayis, H. S., Devassy, R., Audritz, S., Sommer, U., Struck, U., El-Sherbiny, M. M., & Al-Aidaroos, A. M. (2014). Ecohydrographic constraints on biodiversity and distribution of phytoplankton and zooplankton in coral reefs of the Red Sea, Saudi Arabia. *Marine Ecology, 36*, 1195–1214. https://doi.org/10.1111/maec.12224.

Kürten, B., Zarokanellos, N. D., Devassy, R. P., El-Sherbiny, M. M., Struck, U., Capone, D. G., Schulz, I. K., Al-Aidaroos, A. M., Irigoien, X., & Jones, B. H. (2019). Seasonal modulation of mesoscale processes alters nutrient availability and plankton communities in the Red Sea. *Progress in Oceanography, 173*, 238–255. https://doi.org/10.1016/j.pocean.2019.02.007.

Lachkar, Z., Mehari, M., Al Azhar, M., Lévy, M., & Smith, S. (2020). Fast local warming of sea-surface is the main factor of recent deoxygenation in the Arabian Sea. *Biogeosciences Discussions*, 1-27. https://doi.org/10.5194/bg-2020-325.

Lachkar, Z., Mehari, M., Lévy, M., Paparella, F., & Burt, J. A. (2022). Recent expansion and intensification of hypoxia in the Arabian Gulf and its drivers. *Frontiers in Marine Science, 9*. https://doi.org/10.3389/fmars.2022.891378.

Lavieren, H. V., & Klaus, R. (2013). An effective regional Marine Protected Area network for the ROPME Sea Area: Unrealistic vision or realistic possibility? *Marine Pollution Bulletin, 72*, 389–404. https://doi.org/10.1016/j.marpolbul.2012.09.004.

Levanon-Spanier, I., Padan, E., & Reiss, Z. (1979). Primary production in a desert-enclosed sea—The Gulf of Elat (Aqaba), Red Sea. *Deep Sea Research Part A: Oceanographic Research Papers, 26*, 673–685. https://doi.org/10.1016/0198-0149%2879%2990040-2.

Lévy, M., Franks, P. J. S., & Smith, K. S. (2018). The role of submesoscale currents in structuring marine ecosystems. *Nature Communications, 9*(1), 4758. https://doi.org/10.1038/s41467-018-07059-3.

L'Hégaret, P., Lecour, L., Carton, X., Roullet, G., Baraille, R., & Corréard, S. (2013). A seasonal dipolar eddy near Ras Al Hamra (Sea of Oman). *Ocean Dynamics, 63*, 633–659. https://doi.org/10.1007/s10236-013-0616-2.

Liew, Y. J., Howells, E. J., Wang, X., Michell, C. T., Burt, J. A., Idaghdour, Y., & Aranda, M. (2020). Intergenerational epigenetic inheritance in reef-building corals. *Nature Climate Change, 10*, 254–259. https://doi.org/10.1038/s41558-019-0687-2.

Liu, J. P., Kuehl, S. A., Pierce, A. C., Williams, J., Blair, N. E., Harris, C., Aung, D. W., & Aye, Y. Y. (2020). Fate of Ayeyarwady and Thanlwin Rivers sediments in the Andaman Sea and Bay of Bengal. *Marine Geology*, 106137. https://doi.org/10.1016/j.margeo.2020.106137.

Madhupratap, M., Gauns, M., Ramaiah, N., Prasanna Kumar, S., Muraleedharan, P. M., de Sousa, S. N., Sardessai, S., & Muraleedharan, U. (2003). Biogeochemistry of the Bay of Bengal: Physical, chemical and primary productivity characteristics of the central and western Bay of Bengal during summer monsoon 2001. *Deep Sea Research Part II: Topical Studies in Oceanography, 50*, 881–896. https://doi.org/10.1016/S0967-0645(02)00611-2.

Madhupratap, M., Nair, S. R. S., Achuthankutty, C. T., & Nair, V. R. (1981a). Major crustacean groups & zooplankton diversity around Andaman-Nicobar Islands. *Indian Journal of Marine Sciences, 10*, 266–269.

Madhupratap, M., Nair, V. R., Nair, S. R. S., & Achuthankutty, C. T. (1981b). Thermocline and zooplankton distribution. *Indian Journal of Marine Sciences, 10*, 262–265.

Maher, B. A., Prospero, J. M., Mackie, D., Gaiero, D., Hesse, P. P., & Balkanski, Y. (2010). Global connections between aeolian dust, climate and ocean biogeochemistry at the present day and at the last glacial maximum. *Earth-Science Reviews, 99*, 61–97. https://doi.org/10.1016/j.earscirev.2009.12.001.

Mandal, S., Sil, S., Gangopadhyay, A., Jena, B. K., Venkatesan, R., & Glen, G. (2020). Seasonal and tidal variability of surface currents in the western Andaman Sea using HF Radars and Buoy Observations during 2016-2017. *IEEE Transactions on Geoscience and Remote Sensing*, 7235–7244. https://doi.org/10.1109/TGRS.2020.3032885.

Matt, S., & Johns, W. E. (2007). Transport and entrainment in the Red Sea outflow plume. *Journal of Physical Oceanography, 37*(4), 819–836. https://doi.org/10.1175/Jpo2993.1.

Menezes, V. (2018). Mysteries of the Red Sea. *Oceanus*. https://www.whoi.edu/oceanus/feature/mysteries-of-the-red-sea/ (Accessed 24 October 2020).

Menezes, V. V., Farrar, J. T., & Bower, A. S. (2018). Westward mountain-gap wind jets of the northern Red Sea as seen by QuikSCAT. *Remote Sensing of Environment, 209*, 677–699. https://doi.org/10.1016/j.rse.2018.02.075.

Metzl, N., Moore, B., III, Papaud, A., & Poisson, A. (1989). Transport and carbon exchange in Red Sea inverse methodology. *Global Biogeochemical Cycles, 3*, 1–26.

Moazzam, M. (2020). Jellyfish *Crambionella orsini*: A menace for fishing in the Arabian Sea. *Wildlife and Environment, 26*, 15.

Mohan, P. M., Kumari, R. K., Muruganantham, M., Ubare, V. V., Jeeva, C., & Chakraborty, S. (2016). Dissolved organic (DOC) and inorganic carbon (DIC) distribution in the nearshore waters off Port Blair, Andaman Islands. *India International Journal of Oceans and Oceanography, 10*, 29–48.

Mohanty, S., Rao, A. D., & Latha, G. (2018). Energetics of semidiurnal internal tides in the Andaman Sea. *Journal of Geophysical Research: Oceans, 123*. https://doi.org/10.1029/2018JC013852.

Morvan, M., & Carton, X. (2020). Sub-mesoscale frontal instabilities in the Omani Coastal Current. *Mathematics, 8*, 562. https://doi.org/10.3390/math8040562.

Morvan, M., Carton, X., Correard, S., & Baraille, R. (2020). Submesoscale dynamics in the Gulf of Aden and the Gulf of Oman. *Fluids, 5*, 146. https://doi.org/10.3390/fluids5030146.

Murray, S. P., & Johns, W. (1997). Direct observations of seasonal exchange through the Bab el Mandab Strait. *Geophysical Research Letters, 24*(21), 2557–2560. https://doi.org/10.1029/97GL02741.

Naqvi, W. A., & Fairbanks, R. G. (1996). A 27,000 year record of Red Sea Outflow: Implication for timing of post-glacial monsoon intensification. *Geophysical Research Letters, 23*(12), 1501–1504. https://doi.org/10.1029/96gl01030.

Nielsen, T. G., Bjørnsen, P. K., Boonruang, P., Fryd, M., Hansen, P. J., Janekarn, V., Limtrakulvong, V., Munk, P., Hansen, O. S., Satapoomin, S., Sawangarreruks, S., Thomsen, H. A., & Østergaard, J. B. (2004). Hydrography, bacteria and protist communities across the continental shelf and shelf slope of the Andaman Sea (NE Indian Ocean). *Marine Ecology Progress Series, 274*, 69–86. https://doi.org/10.3354/meps274069.

Nienhaus, M. J., Subrahmanyam, B., & Murty, V. S. N. (2012). Altimetric observations and model simulations of coastal Kelvin Waves in the Bay of Bengal. *Marine Geodesy, 35*(1), 190–216. https://doi.org/10.1080/01490419.2012.718607.

Osborne, A. R., & Burch, T. L. (1980). Internal solitons in the Andaman Sea. *Science, 208*, 451–460. https://doi.org/10.1126/science.208.4443.451.

Osman, E. O., Smith, D. J., Zielger, M., Kürten, B., Conrad, C., El-Haddad, K. M., & Voolstra, C. R. (2018). Thermal regufia against coral bleaching throughout the northern Red Sea. *Global Change Biology, 24*, e474–e484. https://doi.org/10.1111/gcb.13895.

Paparella, F., Xu, C., Vaughan, G. O., & Burt, J. A. (2019). Coral bleaching in the Persian/Arabian Gulf is modulated by summer winds. *Frontiers in Marine Science, 6*, 1–15. https://doi.org/10.3389/fmars.2019.00205.

Paul, J. T., Ramaiah, N., Gauns, M., & Fernandes, V. (2007). Preponderance of a few diatom species among the highly diverse microphytoplankton assemblages in the Bay of Bengal. *Marine Biology, 152*, 63–75. https://doi.org/10.1007/s00227-007-0657-5.

Pearman, J. K., Ellis, J., Irigoien, X., Sarma, Y. V. B., Jones, B. H., & Carvalho, S. (2017). Microbial planktonic communities in the Red Sea: High levels of spatial and temporal variability shaped by nutrient availability and turbulence. *Scientific Reports, 7*, 6611. https://doi.org/10.1038/s41598-017-06928-z.

Perry, R. B., & Schimke, G. R. (1965). Large-amplitude internal waves observed off the northwest coast of Sumatra. *Journal of Geophysical Research: Oceans, 70*(10), 2319–2324. https://doi.org/10.1029/JZ070i010p02319.

PERSGA (The Regional Organization for the Conservation of the Environment of the Red Sea and Gulf of Aden). (2006). *State of the marine environment: Report for the Red Sea and Gulf of Aden—2006*. https://www.cbd.int/doc/meetings/mar/ebsaws-2015-02/other/ebsaws-2015-02-persga-submission1-en.pdf.

Peters, H., & Johns, W. F. (2006). Bottom layer turbulence in the Red Sea outflow plume. *Journal of Physical Oceanography, 36*(9), 1763–1785. https://doi.org/10.1175/Jpo2939.1.

Peters, H., Johns, W. E., Bower, A. S., & Fratantoni, D. M. (2005). Mixing and entrainment in the Red Sea outflow plume. Part I: Plume structure. *Journal of Physical Oceanography, 35*(5), 569–583. https://doi.org/10.1175/Jpo2679.1.

Petzold, M. (1986). *Untersuchungen zur horizontalen und vertikalen Verteilung des Phytoplanktons in Roten Meer*. Diplomarbeit Institut fur Hydrobiologie und Fischereiwissenschaft, Hamburg University.

Piontkovski, S. A., Queste, B. Y., Al-Hashmi, K. A., Al-Shaaibi, A., Bryantseva, Y. V., & Popova, E. A. (2017). Subsurface algal blooms of the northwestern Arabian Sea. *Marine Ecology Progress Series, 566*, 67–78. https://doi.org/10.3354/meps11990.

Polikarpov, I., Saburova, M., & Al-Yamani, F. (2016). Diversity and distribution of winter phytoplankton in the Arabian Gulf and the Sea of Oman. *Continental Shelf Research, 119*, 85–99. https://doi.org/10.1016/j.csr.2016.03.009.

Pous, S., Carton, X., & Lazure, P. (2004a). Hydrology and circulation in the Strait of Hormuz and the Gulf of Oman—Results from the GOGP99 Experiment. Part I. Strait of Hormuz. *Journal of Geophysical Research, 109*(C12037), 1–15. https://doi.org/10.1029/2003JC002145.

Pous, S., Carton, X., & Lazure, P. (2004b). Hydrology and circulation in the Strait of Hormuz and the Gulf of Oman—Results from theGOGP99 Experiment. Part II. Gulf of Oman. *Journal of Geophysical Research, Oceans, 109*(C12038), 1–26. https://doi.org/10.1029/2003JC002146.

Pous, S., Lazure, P., & Carton, X. (2015). A model of the general circulation in the Persian Gulf and in the Strait of Hormuz: Intraseasonal to interannual variability. *Continental Shelf Research, 94*, 55–70. https://doi.org/10.1016/j.csr.2014.12.008.

Prakash, P. J., Stenchikov, G., Kalenderski, S., Osipov, S., & Bangalath, H. (2015). The impact of dust storms on the Arabian Peninsula and the Red Sea. *Atmospheric Chemistry and Physics, 15*, 199–222. https://doi.org/10.5194/acp-15-199-2015.

Prakash, P. J., Stenchikov, G., Tao, W., Yapici, T., Warsama, B., & Engelbrecht, J. P. (2016). Arabian Red Sea coastal soils as potential mineral dust sources. *Atmospheric Chemistry and Physics, 16*, 11991–12004. https://doi.org/10.5194/acp-16-11991-2016.

Queste, B. Y., Vic, C., Heywood, K. J., & Piontkovski, S. A. (2018). Physical controls on oxygen distribution and denitrification potential in the North West Arabian Sea. *Geophysical Research Letters, 45*, 4143–4152. https://doi.org/10.1029/2017GL076666.

Qurban, M. A., Balala, A. C., Kumar, S., Bhavya, P. S., & Wafar, M. (2014). Primary production in the northern Red Sea. *Journal of Marine Systems, 132*, 75–82. https://doi.org/10.1016/j.jmarsys.2014.01.006.

Qurban, M. A., Krishnakumar, P. K., Joydas, T. V., Manikandan, K., Ashraf, T., Sampath, G., Thiyagarajan, D., He, S., & Cairns, S. (2020). Discovery of deep-water coral frameworks in the northern Red Sea waters of Saudi Arabia. *Scientific Reports, 10*, 15356. https://doi.org/10.1038/s41598-020-72344-5.

Qurban, M., Wafar, M., & Heinle, M. (2019). Phytoplankton and primary production in the Red Sea. In N. Rasul, & I. Stewart (Eds.), *Oceanographic and biological aspects of the Red Sea* (pp. 491–506). Cham: Springer International Publishing. https://doi.org/10.1007/978-3-319-99417-8_27.

Ragavan, P., Jayaraj, R. S. C., Muruganantham, M., Jeeva, C., Ubare, V. V., Saxena, A., & Mohan, P. M. (2016). Species composition and distribution of sea grasses of the Andaman and Nicobar Islands. *Vegetos*, *29*, 76–87.

Rahav, E., Bar Zeev, E., Ohayon, S., Elifantz, H., Belkin, N., Herut, B., Mulholland, M., & Berman-Frank, I. (2013). Dinitrogen fixation in aphotic oxygenated marine environments. *Frontiers in Microbiology*, *4*, 227. https://doi.org/10.3389/fmicb.2013.00227.

Rahav, E., Herut, B., Mulholland, M., Belkin, N., Elifantz, H., & Berman-Frank, I. (2015). Heterotrophic and autotrophic contribution to dinitrogen fixation in the Gulf of Aqaba. *Marine Ecology Progress Series*, *522*, 67–77. https://doi.org/10.3354/meps11143.

Raitsos, D. E., Pradhan, Y., Brewin, R. J. W., Stenchikov, G., & Hoteit, I. (2013). Remote sensing the phytoplankton seasonal succession of the Red Sea. *PLoS One*, *8*, e64909. https://doi.org/10.1371/journal.pone.0064909.

Raitsos, D. E., Yi, X., Platt, T., Racault, M. F., Brewin, R. J., Pradhan, Y., Papadopoulos, V. P., Sathyendranath, S., & Hoteit, I. (2015). Monsoon oscillations regulate fertility of the Red Sea. *Geophysical Research Letters*, *42*, 855–862. https://doi.org/10.1002/2014GL062882.

Ramaiah, N., Fernandes, V., Paul, J. T., Jyothibabu, R., Mangesh, G., & Jayraj, E. A. (2010). Seasonal variability in biological carbon biomass standing stocks and production in the surface layers of the Bay of Bengal. *Indian Journal of Marine Sciences*, *39*(3), 369–379.

Ramakrishna, R. C., & Sivaperuman, C. (2010). *Recent trends in biodiversity of Andaman and Nicobar Islands* (pp. 1–542). Kolkata: The Director, Zool. Surv. India.

Ramaswamy, V., Gaye, B., Shirodkar, P. V., Rao, P. S., Chivas, A. R., Wheeler, D., & Thwin, S. (2008). Distribution and sources of organic carbon, nitrogen and their isotopic signatures in sediments from the Ayeyarwady (Irrawaddy) continental shelf, northern Andaman Sea. *Marine Chemistry*, *111*, 137–150. https://doi.org/10.1016/j.marchem.2008.04.006.

Ramaswamy, V., & Rao, P. S. (2014). The Myanmar continental shelf. In F. L. Chiocci, & A. R. Chivas (Eds.), *Vol. 41. Continental shelves of the world: Their evolution during the last Glacio-Eustatic cycle* (pp. 231–240). London: Geological Society, Memoirs. https://doi.org/10.1144/M41.17.

Ramaswamy, V., Rao, P. S., Rao, K. H., Thwin, S., Srinivasa Rao, N., & Raikera, V. (2004). Tidal influence on suspended sediment distribution and dispersal in the northern Andaman Sea and Gulf of Martaban. *Marine Geology*, *208*, 33–42. https://doi.org/10.1016/j.margeo.2004.04.019.

Rameshbabu, V., & Sastry, J. S. (1976). Hydrography of the Andaman Sea during late winter. *Indian Journal of Marine Sciences*, *5*, 179–189.

Ramya, S. P. (2015). *Analyses of satellite altimeter derived sea surface height variability and upper ocean circulation variability in the Andaman Sea during 2004-2015*. Dissertation Project Report for M.Tech Degree in Geomatics, Work carried out at the CSIR-National Institute of Oceanography, Dona Paula, Goa.

Rao, R. R., Girish Kumar, M. S., Ravichandran, M., Rao, A. S., Gopalakrishna, V. V., & Thadathil, P. (2010). Interannual variability of Kelvin wave propagation in the wave guides of the equatorial Indian Ocean, the coastal Bay of Bengal and the southeastern Arabian Sea during 1993–2006. *Deep Sea Research Part I: Oceanographic Research Papers*, *57*, 1–13. https://doi.org/10.1016/j.dsr.2009.10.008.

Rasul, N., & Stewart, I. (Eds.). (2019). *Oceanographic and biological aspects of the Red Sea*. Cham: Springer International Publishing. https://doi.org/10.1007/978-3-319-99417-8.

Rasul, N. M. A., Stewart, I. C. F., & Nawab, Z. A. (2015). Introduction to the Red Sea: Its origin, structure, and environment. In N. M. A. Rasul, & I. C. F. Stewart (Eds.), *The Red Sea: The formation, morphology, oceanography and environment of a Young Ocean basin*. Berlin/Heidelberg: Springer. https://doi.org/10.1007/978-3-662-45201-11.

Reynolds, M. (1993). Physical oceanography of the Gulf, Strait of Hormuz, and the Gulf of Oman: Results from the Mt Mitchell expedition. *Marine Pollution Bulletin*, *27*, 35–39. https://doi.org/10.1016/0025-326X(93)90007-7.

Richards, W. J. (2008). In F. Al-Yamani (Ed.), *Identification guide of the early life history stages of fishes from the waters of Kuwait in the Arabian Gulf, Indian Ocean*. Kuwait: Kuwait Institute for Scientific Research.

Riegl, B., & Purkis, S. (2012). Environmental constraints for reef building in the Gulf. In B. M. Riegl, & S. J. Purkis (Eds.), *Coral reefs of the world: Vol. 3. Coral reefs of the Gulf* (pp. 5–32). Dordrecht: Springer. https://doi.org/10.1007/978-94-007-3008-3_2.

Riegl, B. M., Purkis, S. J., Al-Cibahy, A. S., Abdel-Moati, M. A., & Hoegh-Guldberg, O. (2011). Present limits to heat-adaptability in corals and population-level responses to climate extremes. *PLoS One*, *6*, e24802. https://doi.org/10.1371/journal.pone.0024802.

Rizal, S., Damm, P., Wahid, M. A., Sundermann, J., Ilhamsyah, Y., Iskandar, T., & Muhammad. (2012). General circulation in the Malacca Strait and Andaman Sea: A numerical model study. *American Journal of Environmental Sciences*, *8*(5), 479–488. https://doi.org/10.3844/ajessp.2012.479.488.

Roberts, C. M., Shepherd, A. R. D., & Ormond, R. F. G. (1992). Large-scale variation in assemblage structure of Red Sea butterflyfishes and angelfishes. *Journal of Biogeography*, *19*, 239–250. https://doi.org/10.2307/2845449.

Roder, C., Berumen, M. L., Bouwmeester, J., Papathanassiou, E., Al-Suwailem, A., & Voolstra, C. R. (2013). First biological measurements of deep-sea corals from the Red Sea. *Scientific Reports*, *3*, 2802. https://doi.org/10.1038/srep02802.

Roder, C., Fillinger, L., Jantzen, C., Schmidt, G. M., Khokiattiwong, S., & Richte, C. (2010). Trophic response of corals to large amplitude internal waves. *Marine Ecology Progress Series*, *412*, 113–128. https://doi.org/10.3354/meps08707.

Roik, A. K., Röthig, T., Pogoreutz, C., Saderne, V., & Voolstra, C. R. (2018). Coral reef carbonate budgets and ecological drivers in the central Red Sea—A naturally high temperature and high total alkalinity environment. *Biogeosciences*, *15*, 6277–6296. https://doi.org/10.5194/bg-15-6277-2018.

Roman, R. E., & Lutjeharms, J. R. E. (2009). Red Sea Intermediate Water in the source regions of the Agulhas Current. *Deep Sea Research Part I: Oceanographic Research Papers*, *56*, 939–962. https://doi.org/10.1016/j.dsr.2009.01.003.

Sabine, C. L., Key, R. M., Feely, R. A., & Greeley, D. (2002). Inorganic carbon in the Indian Ocean: Distribution and dissolution processes. *Global Biogeochemical Cycles*, *16*. https://doi.org/10.1029/2002GB001869.

Sahu, B. K., & Panigrahy, R. C. (2013). Jellyfish bloom along the south Odisha coast, Bay of Bengal. *Current Science*, *104*, 410–411.

Sakaff, H. A., & Esseen, M. (1999). Occurrence and distribution of fish species off Yemen (Gulf of Aden and Arabian Sea). *NAGA*, *22*(1), 43–47.

Sale, P. F., Feary, D., Burt, J. A., Bauman, A., Cavalcante, G., Drouillard, K., Kjerfve, B., Marquis, E., Trick, C., Usseglio, P., & van Lavieren, H. (2011). The growing need for sustainable ecological management of marine communities of the Persian Gulf. *Ambio, 40,* 4–17. https://doi.org/10.1007/s13280-010-0092-6.

Sarma, V. V. S. S., & Narvekar, P. V. (2001). A study on inorganic carbon components in the Andaman Sea during the post monsoon season. *Oceanologica Acta, 24,* 125–134. https://doi.org/10.1016/S0399-1784(00)01133-6.

Schott, F. A., & McCreary, J. P. (2001). The monsoon circulation of the Indian Ocean. *Progress in Oceanography, 51*(1), 1–123. https://doi.org/10.1016/S0079-6611(01)00083-0.

Sen Gupta, R., & Naqvi, S. W. A. (1984). Chemical oceanography of the Indian Ocean, north of the equator. *Deep Sea Research Part A: Oceanographic Research Papers, 31,* 671–706. https://doi.org/10.1016/0198-0149(84)90035-9.

Senjyu, T., Ishimaru, T., Matsuyama, M., & Koike, Y. (1998). High salinity lens from the Strait of Hormuz. In A. Otsuki, M. Y. Abdulraheem, & R. M. Reynolds (Eds.), *Offshore environment cruises of the ROPME Sea Area after the war-related oil spill* (pp. 35–48). Tokyo: Terra SciencePub.

Sewell, R. B. S. (1932). Geographic and oceanographic research in Indian waters: Temperature and salinity of the deeper waters of the Bay of Bengal and Andaman Sea. *Memoirs of the Asiatic Society of Bengal, 9,* 357–423.

Shaikh, E. A., Roff, J. C., & Dowidar, N. M. (1986). Phytoplankton ecology and production in the Red Sea off Jiddah, Saudi Arabia. *Marine Biology, 92,* 405–416.

Sheppard, C., Al-Husiani, M., Al-Jamali, F., Al-Yamani, F., Baldwin, R., Bishop, J., Benzoni, F., Dutrieux, E., Dulvy, N., Durvasula, S., Jones, D., Loughland, R., Medio, D., Nithyanandan, M., Pilling, G., Polikarpov, I., Price, A., Purkis, S., Riegl, B., ... Zainal, K. (2010). The Gulf: A young sea in decline. *Marine Pollution Bulletin, 60,* 13–38. https://doi.org/10.1016/j.marpolbul.2009.10.017.

Sheppard, C., Price, A., & Roberts, C. (1992). *Marine ecology of the Arabian region: Patterns and processes in extreme tropical environments.* Toronto: Academic Press.

Smeed, D. (1997). Seasonal variation of the flow in the strait of Bab al Mandab. *Oceanologica Acta, 20*(6), 773–781.

Smeed, D. A. (2000). Hydraulic control of three-layer exchange flows: Application to the Bab al Mandab. *Journal of Physical Oceanography, 30*(10), 2574–2588. https://doi.org/10.1175/1520-0485(2000)030<2574:Hcotle>2.0.Co;2.

Smith, E. G., Hazzouri, K. M., Choi, J. Y., Delaney, P., Al-Kharafi, M., Howells, E. J., Aranda, M., & Burt, J. A. (2022). Signatures of selection underpinning rapid coral adaptation to the world's warmest reefs. *Science Advances, 8,* eabl7287. https://doi.org/10.1126/sciadv.abl7287.

Smith, E., Hume, B., Delaney, P., Wiedenmann, J., & Burt, J. (2017a). Genetic structure of coral-*Symbiodinium* symbioses on the world's warmest reefs. *PLoS One, 12,* 1–12. https://doi.org/10.1371/journal.pone.0180169.

Smith, E. G., Vaughan, G. O., Ketchum, R. N., McParland, D., & Burt, J. A. (2017b). Symbiont community stability through severe coral bleaching in a thermally extreme lagoon. *Scientific Reports, 7,* 1–9. https://doi.org/10.1038/s41598-017-01569-8.

Sofianos, S. S., & Johns, W. E. (2002). An Oceanic General Circulation Model (OGCM) investigation of the Red Sea circulation, 1. Exchange between the Red Sea and the Indian Ocean. *Journal of Geophysical Research-Oceans, 107*(C11), 3196. https://doi.org/10.1029/2001jc001184.

Sofianos, S. S., & Johns, W. E. (2003). An Oceanic General Circulation Model (OGCM) investigation of the Red Sea circulation: 2. Three-dimensional circulation in the Red Sea. *Journal of Geophysical Research: Oceans, 108,* 3066. https://doi.org/10.1029/2001JC001185.

Sofianos, S. S., Johns, W. E., & Murray, S. P. (2002). Heat and fresh-water budgets in the Red Sea from direct observations at Bab el Mandab. *Deep Sea Research Part II: Topical Studies in Oceanography, 49,* 1323–1340. https://doi.org/10.1016/S0967-0645(01)00164-3.

Souvermezoglou, E., Metzl, N., & Poisson, A. (1989). Red Sea budgets of salinity, nutrients and carbon calculated in the Strait of Bab-El-Mandab during the summer and winter seasons. *Journal of Marine Research, 47,* 441–456. https://doi.org/10.1357/002224089785076244.

Swift, S. A., & Bower, A. S. (2003). Formation and circulation of dense water in the Persian/Arabian Gulf. *Journal of Geophysical Research: Oceans, 108,* 4-1-4-21. https://doi.org/10.1029/2002JC001360.

Thangaraja, M., Al-Aisery, A., & Al-Kharusi, L. (2007). Harmful algal blooms and their impacts in the middle and outer ROPME Sea area. *International Journal of Oceans and Oceanography, 2,* 85–98.

Thoppil, P., & Hogan, P. (2010a). A modeling study of circulation and eddies in the Persian Gulf. *Journal of Physical Oceanography, 40,* 2122–2134. https://doi.org/10.1175/2010JPO4227.1.

Thoppil, P. G., & Hogan, P. J. (2010b). Persian Gulf response to a wintertime shamal wind event. *Deep Sea Research Part I: Oceanographic Research Papers, 57,* 946–955. https://doi.org/10.1016/j.dsr.2010.03.002.

Tilstra, A., El-Khaled, Y. C., Roth, F., Rädecker, N., Pogoreutz, C., Voolstra, C., & Wild, C. (2019). Denitrification aligns with N_2 fixation in Red Sea corals. *Scientific Reports, 9,* 19460. https://doi.org/10.1038/s41598-019-55408-z.

Tomczak, M., & Godfrey, J. S. (2003). *Regional oceanography: An introduction* (2nd ed.). New Delhi: Daya Publishing House.

Tomczak, M., & Godfrey, J. S. (2013). *Regional oceanography: An introduction.* Amsterdam: Pergamon.

Varkey, M. J., Murty, V. S. N., & Suryanarayana, A. (1996). Physical oceanography of the Bay of Bengal and Andaman Sea. *Oceanography and Marine Biology: An Annual Review, 34,* 1–70.

Vaughan, G. O., Al-Mansoori, N., & Burt, J. (2019). The Arabian Gulf. In C. Sheppard (Ed.), *World seas: An environmental evaluation* (2nd ed., pp. 1–23). Amsterdam, NL: Elsevier Science.

Vaughan, G. O., & Burt, J. A. (2016). The changing dynamics of coral reef science in Arabia. *Marine Pollution Bulletin, 105,* 441–458. https://doi.org/10.1016/j.marpolbul.2015.10.052.

Veldhuis, M. J., Kraay, G. W., Van Bleijswijk, J. D., & Baars, M. A. (1997). Seasonal and spatial variability in phytoplankton biomass, productivity and growth in the northwestern Indian Ocean: The southwest and northeast monsoon, 1992-1993. *Deep Sea Research Part I: Oceanographic Research Papers, 44,* 425–449. https://doi.org/10.1016/S0967-0637(96)00116-1.

Vic, C., Roullet, G., Capet, X., Carton, X., Molemaker, M. J., & Gula, J. (2015). Eddy-topography interactions and the fate of the Persian Gulf Outflow. *Journal of Geophysical Research: Oceans, 120,* 6700–6717. https://doi.org/10.1002/2015JC011033.

Voolstra, C. R., Miller, D. J., Ragan, M. A., Hoffmann, A. A., Hoegh-Guldberg, O., Bourne, D. G., Ball, E. E., Ying, H., Forêt, S., Takahashi, S., Weynberg, K. D., van Oppen, M. J. H., Morrow, K., Chan, C. X., Rosic, N., Leggat, W., Sprungala, S., Imelfort, M., Tyson, G. W., ... Fyffe, T. (2015). The ReFuGe 2020 consortium-using 'omics' approaches to explore the adaptability and resilience of coral holobionts to environmental change. *Frontiers in Marine Science, 2*, 68. https://doi.org/10.3389/fmars.2015.00068.

Wafar, M., Ashraf, M., Manikandan, K. P., Qurban, M. A., & Kattan, Y. (2016). Propagation of Gulf of Aden Intermediate Water (GAIW) in the Red Sea during autumn and its importance to biological production. *Journal of Marine Systems, 154*, 243–251. https://doi.org/10.1016/j.jmarsys.2015.10.016.

Wang, L., Lin, X., Goes, J. I., & Lin, S. (2016). Phylogenetic analyses of three genes of *Pedinomonas noctilucae*, the green endosymbiont of the marine dinoflagellate *Noctiluca scintillans*, reveal its affiliation to the order Marsupiomonadales (Chlorophyta, Pedinophyceae) under the reinstated name *Protoeuglena noctilucae*. *Protist, 167*, 205–216. https://doi.org/10.1016/j.protis.2016.02.005.

Wilson, G., Price, A., Huntington, T., & Wilson, S. (2003). Environmental status of Yemen's Gulf of Aden coast determined from rapid field assessment and satellite imagery. *Aquatic Ecosystem Health & Management, 6*, 119–129. https://doi.org/10.1080/14634980301465.

Wyrtki, K. (1961). *Physical oceanography of the southeast Asian waters*. NAGA Report *Vol. 2*. California: Scripps Institution of Oceanography.

Yao, F. C., & Hoteit, I. (2015). Thermocline regulated seasonal evolution of surface chlorophyll in the Gulf of Aden. *PLoS One, 10*(3), e0119951. https://doi.org/10.1371/journal.pone.0119951.

Yao, F., & Johns, W. E. (2010a). A HYCOM modeling study of the Persian Gulf: 1. Model configurations and surface circulation. *Journal of Geophysical Research: Oceans, 115*, C11017. https://doi.org/10.1029/2009JC005781.

Yao, F., & Johns, W. E. (2010b). A HYCOM modeling study of the Persian Gulf: 2. Formation and export of Persian Gulf Water. *Journal of Geophysical Research: Oceans, 115*, C11018. https://doi.org/10.1029/2009JC005788.

Yentsch, C., & Wood, L. (1961). *Measurements of primary productivity in the Red Sea, Gulf of Aden and Indian Ocean* (pp. 61–66). Woods Hole Oceanographic Institution. Ref 61-6 (Appendix 8:6).

Zarokanellos, N. D., Kürten, B., Churchill, J. H., Roder, C., Voolstra, C. R., Abualnaja, Y., & Jones, B. (2017). Physical mechanisms routing nutrients in the central Red Sea. *Journal of Geophysical Research: Oceans, 122*, 9032–9046. https://doi.org/10.1002/2017JC013017.

Zhai, P., Pratt, L. J., & Bower, A. (2015). On the crossover of boundary currents in an idealized model of the Red Sea. *Journal of Physical Oceanography, 45*, 1410–1425. https://doi.org/10.1175/JPO-D-14-0192.1.

Zhan, P., Krokos, G., Guo, D., & Hoteit, I. (2019). Three-dimensional signature of the Red Sea eddies and eddy-induced transport. *Geophysical Research Letters, 46*(4), 2167–2177. https://doi.org/10.1029/2018GL081387.

Ziegler, M., Arif, C., & Voolstra, C. R. (2019a). Symbiodiniaceae diversity in Red Sea coral reefs & coral bleaching. In M. L. Berumen, & C. R. Voolstra (Eds.), *Coral reefs of the Red Sea* (pp. 69–90). Cham: Springer International Publishing. https://doi.org/10.1007/978-3-030-05802-9_5.

Ziegler, M., Roik, A., Röthig, T., Wild, C., Rädecker, N., Bouwmeester, J., & Voolstra, C. R. (2019b). Ecophysiology of reef-building in the Red Sea. In M. L. Berumen, & C. R. Voolstra (Eds.), *Coral reefs of the Red Sea* (pp. 33–52). Cham: Springer International Publishing. https://doi.org/10.1007/978-3-030-05802-9_3.

Zolina, O., Dufour, A., Gulev, S., & Stenchikov, G. (2017). Regional hydrological cycle over the Red Sea in ERA-interim. *Journal of Hydrometeorology, 18*, 65–83. https://doi.org/10.1175/JHM-D-16-0048.1.

Chapter 17

The Indian Ocean Observing System (IndOOS)[*]

Michael J. McPhaden[a], Lisa M. Beal[b], T.V.S. Udaya Bhaskar[c], Tong Lee[d], Motoki Nagura[e], Peter G. Strutton[f,g], and Lisan Yu[h]

[a]NOAA/Pacific Marine Environmental Laboratory, Seattle, WA, United States, [b]Rosenstiel School of Marine, Atmospheric, and Earth Science, University of Miami, Coral Gables, Miami, FL, United States, [c]ESSO-Indian National Centre for Ocean Information Services, Hyderabad, India, [d]Jet Propulsion Laboratory, California Institute of Technology, Pasadena, CA, United States, [e]Japan Agency for Marine-Earth Science and Technology, Yokosuka, Kanagawa, Japan, [f]Institute for Marine and Antarctic Studies, University of Tasmania, Hobart, TAS, Australia, [g]Australian Research Council Centre of Excellence for Climate Extremes, University of Tasmania, Hobart, TAS, Australia, [h]Woods Hole Oceanographic Institution, Woods Hole, MA, United States

1 Introduction

The Indian Ocean Observing System (IndOOS) was inaugurated in the mid-2000s as a multi-national effort to provide basin-scale measurements for improved description, understanding, and prediction of the Indian Ocean's role in weather and climate both regionally and globally (Beal et al., 2020; International CLIVAR Project Office, 2006). It comprises a collection of in situ observing system elements designed to complement the constellation of Earth-observing satellites. Before the advent of IndOOS, observational efforts in the Indian Ocean were either regional in scope, fixed duration in time, or targeted at a specific set of processes. The goal of IndOOS, in contrast, was to establish a systematic, sustained, and internationally-coordinated observational effort that could support both scientific research in ocean-atmospheric interactions and ocean dynamics as well as provide a real-time capability for operational oceanography that would contribute to weather and climate forecasting.

The Indian Ocean is unique geographically in that it is blocked by the Asian land mass to the north. Land-sea heating contrasts energize a vigorous seasonally reversing monsoon wind circulation over the Indian Ocean in contrast to the relatively steady trade wind regimes of the tropical Atlantic and Pacific Oceans (Ummenhofer et al., 2024). The associated monsoon rainfall patterns in the countries within and surrounding the Indian Ocean are indelibly linked to the cultural identity and economic vitality of the region, particularly as it relates to rainfed agriculture (Gadgil & Gadgil, 2006).

Monsoon rainfalls are modulated by active-break periods associated with the 30–90 day Madden Julian Oscillation (MJO; Zhang, 2005; Yamagata et al., 2024) and the 10–20 day and 3–7 day monsoon intraseasonal oscillations (Goswami & Mohan, 2001; Goswami et al., 2016; Roman-Stork et al., 2020; Subrahmanyam et al., 2020). Year-to-year variations in the monsoons and climate across the basin are affected by the Indian Ocean Dipole (IOD; Gadgil et al., 2004; Saji et al., 1999; Webster et al., 1999; Yamagata et al., 2024) and influenced by remote forcing from the tropical Pacific from El Niño and the Southern Oscillation (ENSO) (Cai et al., 2019; Yamagata et al., 2024). Likewise, decadal variations are linked to the Atlantic Multidecadal Oscillation (Feng & Hu, 2008) and the Pacific Decadal Oscillation (PDO; Han et al., 2014). Sea surface temperatures (SSTs) in the Indian Ocean are the warmest on average compared to other ocean basins, and they have continued to warm (by >1°C over the past 70 years) as a consequence of anthropogenic greenhouse gas-forced climate change (Dong et al., 2014; Du & Xie, 2008; Roxy et al., 2024; Ummenhofer et al., 2021). Increasing precipitation extremes over the Indian subcontinent have been linked to rising SSTs in the Arabian Sea (Roxy et al., 2017). There is a concern that tropical cyclones, which regularly afflict the region's most vulnerable populations, may increase in intensity as oceanic and atmospheric temperatures continue to rise (Vellore et al., 2020). Rising sea levels also represent a major threat to low-lying coastal regions and island nations within and around the basin (Oppenheimer et al., 2019).

---

[*]This book has a companion website hosting complementary materials. Visit this URL to access it: https://www.elsevier.com/books-and-journals/book-companion/9780128226988.

SST is an essential ocean variable affecting weather and climate in the Indian Ocean region. The spatial patterns and temporal evolution of SST are determined by air-sea heat, moisture, and momentum exchanges, ocean current variations, and turbulent ocean mixing processes. The complexity of these drivers in the Indian Ocean is remarkable. In the Bay of Bengal, for example, heavy seasonal river runoff and rainfall create thin freshwater-mixed layers overlying salt-stratified "barrier layers" that buffer the mixing of surface water with colder thermocline water below (Shenoi et al., 2002). The tendency for SSTs to warm as a result is limited by competing processes (Krishnamohan et al., 2019). However, it is nonetheless noteworthy that SSTs in the Bay of Bengal are some of the highest observed in the world ocean.

Seasonally reversing monsoon winds north of 10°S drive dramatic reversals in major ocean currents along coastal boundaries, near and across the equator, and in upwelling zones (Beal et al., 2013; L'Hegaret et al., 2018; Phillips et al., 2024; Schott et al., 2009). These current reversals contribute to a cross-equatorial circulation cell unique to the Indian Ocean that moderates regional climate by transporting heat from the Northern to the Southern Hemisphere in boreal summer and from the Southern to the Northern Hemisphere in boreal winter (Schott et al., 2009). Interbasin exchanges associated with the Indonesian Throughflow (ITF) from the Pacific to the Indian Ocean and the Agulhas leakage to the Atlantic Ocean make the Indian Ocean a major player in the global "conveyor belt" circulation. Increased heat transport from the Pacific via the ITF and a broadening of the Agulhas Current related to an increase in eddy kinetic energy were key mechanisms accounting for the rapid storage of heat in the southern Indian Ocean during the first decade and half of the 21st century (Lee et al., 2015; McMonigal et al., 2022; Meng et al., 2020). Interannually, increased ITF transports during La Niña events contribute to the development of marine heatwaves off the west coast of Australia, referred to as Ningaloo Niños (Feng et al., 2013, 2024; Yamagata et al., 2024).

How ocean circulation and mixed layer dynamics affect the overlying atmosphere in the Indian Ocean is an issue not just for the region but also for the globe. The MJO, for example, which is spawned over the Indian Ocean, has a global reach that affects the generation of atmospheric rivers in the Pacific, west coast U.S. rainfall, tropical storm formation in all three ocean basins, and the onset and evolution of El Niño events (Zhang, 2013). ENSO affects the contemporaneous development of the IOD through changes in the atmospheric Walker Circulation, but the reverse is also true: IOD-induced changes in the Walker Circulation affect the development of ENSO at lead times of a year or more (Cai et al., 2019; Izumo et al., 2010; Yamagata et al., 2024). Similarly, anomalous Indian Ocean SSTs can affect the prevalence of drought in the Sahel region of West Africa (Giannini et al., 2003), the occurrence of Atlantic Niños (Liao & Wang, 2021; Zhang & Han, 2021), and the strength of the Atlantic Meridional Overturning Circulation (AMOC; Hu & Fedorov, 2019).

IndOOS was designed by the Indian Ocean Regional Panel of the Climate and Ocean: Variability, Predictability, and Change (CLIVAR) program and the Global Ocean Observing System to address the broad range of scientific issues described above. Many recent advances in our knowledge of variability across the Indian Ocean have come from physical IndOOS measurements targeted at ocean circulation, ocean–atmosphere interactions, weather, and climate. However, many outstanding questions of fundamental importance related to marine biogeochemistry and ecosystems are of great societal consequence because of their impact on fisheries and living marine resources (Hood et al., 2024a; Marsac et al., 2024; Roxy et al., 2024). As will be described below, IndOOS is evolving to address these questions as well.

Most IndOOS observations are transmitted to shore in real-time several times a day via satellite relay and distributed worldwide on the Global Telecommunications System. Hence, they are also routinely available for operational model forecast systems and the development of various data and reanalysis products. However, our emphasis in this chapter will be to highlight research progress and directions rather than more practical applications of IndOOS data.

We also make a distinction between sustained, basin-scale IndOOS observations and those from process-oriented field studies designed to explore a particular set of phenomena in great detail for a limited duration in a particular subregion, oftentimes using novel and/or experimental measurement technologies. IndOOS provides a basin-scale, long-term context in which to interpret the results of these shorter-term, regionally focused field programs. Process studies, on the other hand, provide a test bed for new technologies that may later be incorporated into the sustained observing system, as well as a better understanding of phenomena that may need to be more systematically measured as part of the sustained observing system. There is an obvious synergy between process studies and sustained observations, but our focus in this chapter will be the latter.

In the following sections, we describe the historical development of IndOOS (Section 2) and its overall design (Section 3). We then highlight the scientific advances enabled by IndOOS in Section 4. The chapter concludes with a brief perspective on current challenges and future directions in Section 5.

2 Historical background

The World Weather Watch, established in the early 1960s at the dawn of the space age, provided a framework for the collection and distribution of sustained meteorological observations for atmospheric research and weather forecasting. No such framework for the ocean existed, however, until the advent of the international Tropical Ocean Global Atmosphere

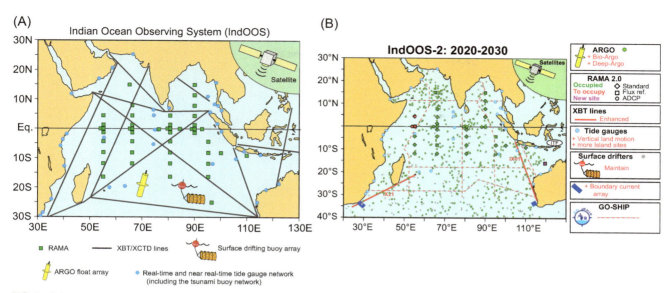

FIG. 1 Schematic of the Indian Ocean Observing System (a) in its original configuration (after McPhaden et al., 2009b) and (b) as proposed following the decadal review (Beal et al., 2020). (© *American Meteorological Society. Used with permission.*)

(TOGA) program (1985–1994). TOGA was the first major initiative of the World Climate Research Program (WCRP), which was established in 1980 to advance international cooperation in climate studies. TOGA and later another WCRP program, the World Ocean Circulation Experiment (WOCE; 1990–1997), laid the groundwork for developing a comprehensive and sustained ocean observing system to support climate research and forecasting (McPhaden et al., 2010).

Much of the emphasis on ocean observing system development in the early stages of the WCRP was on the Pacific because of the strength and global impact of ENSO. The Indian Ocean, in contrast, was underdeveloped for a variety of reasons (McPhaden, 2020). Two advances in the late 1990s helped to crystallize the motivations for IndOOS. One was the discovery of the IOD, in which ocean dynamics played a critical role as for El Niño in the Pacific, and the other was the growing appreciation for how ocean mixed layer processes contributed to the initiation and organization of the MJO.

In the early 2000s, the Indian Ocean Regional Panel (known then as the Indian Ocean Panel) was formed with sponsorship from CLIVAR and Global Ocean Observing System to design a sustained Indian Ocean observing system. That observing system (International CLIVAR Project Office, 2006; Fig. 1a) consists of a variety of in situ components designed to complement the existing constellation of earth-observing satellites. Satellite observations remain the foundation for the observing system, providing broad spatial coverage of ocean surface variables such as winds, SST, sea surface salinity (SSS), sea surface height (SSH), and ocean color. In situ measurements are essential, however, for calibration and validation of satellite retrievals as well as for observing below the air-sea interface at depths not observable from space.

After more than a decade of successful operation, the International CLIVAR program and Global Ocean Observing System sponsored a review of the system, which resulted in several design modifications as well as a broadened set of scientific priorities (Beal et al., 2020; Fig. 1b). These new priorities include a greater emphasis on biogeochemical measurements, on better resolving upper-ocean processes for monsoon forecasting, and on constraining the basin-wide energy budget. The components of the observing system are described next.

3 IndOOS design

3.1 In situ components

In configuring the in situ components of IndOOS, priority is given to proven technologies that can be cost-effectively deployed and maintained for long periods and to systems that can deliver data in real-time to support weather and climate forecasting. It is built around a moored buoy array, the Argo float program, a network of surface drifters, underway measurements from ships-of-opportunity, island and coastal tide gauge stations, and decadal research cruises (Fig. 1). The abundance and variety of IndOOS in situ data (Fig. 2) have grown in time, providing an enormously valuable resource for describing, modeling, understanding, and predicting variability and change in the Indian Ocean region.

The moored buoy component, referred to as the Research Moored Array for African-Asian-Australian Monsoon Analysis and Prediction (RAMA), was designed to sample several key regions across the basin: the Arabian Sea, Bay

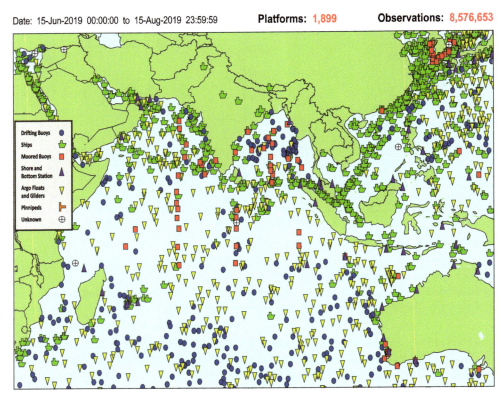

FIG. 2 An example of IndOOS data distribution for one 3-month period (June–August 2019) chosen at random. Display generated from NOAA's Observing System Monitoring Center (http://www.osmc.noaa.gov).

of Bengal, equatorial waveguide, Seychelles Chagos Thermocline Ridge (SCTR), the eastern pole of the IOD, and the southeastern subtropical subduction zone of the southeastern basin (McPhaden et al., 2009b). Emphasis is on providing marine meteorological and upper ocean measurements to a depth of 500 m, with data transmitted to shore in real-time every day via satellite relay. Rapid and regular temporal sampling ensures that mooring data resolve higher frequency variations on time scales of hours to weeks that might otherwise be aliased into the lower frequency climate signals of primary interest. One key outcome of the IndOOS decadal review (Beal et al., 2020) was to reduce the number of moorings in the array for practical purposes but to expand coverage to the Timor Sea, where the MJO signature in SST is very strong (Feng et al., 2020; Vialard et al., 2008). (The updated version of the array is dubbed RAMA-2.0. For the purposes of this review, we will refer generically to RAMA, except where the distinction between RAMA and RAMA-2.0 is necessary for clarity.) Another outcome was a plan to more closely coordinate the maintenance and data sharing between RAMA and the Indian National Moored Buoy Network for the Northern Indian Ocean (Venkatesan et al., 2016a, 2016b).

Argo is a global array of free-drifting floats, roughly one float per $3° \times 3°$ area, each profiling the ocean at a nominal 10-day interval measuring temperature and salinity at preset pressure levels from the surface down to 2000 m (Gould et al., 2004). The design goal of 450 floats to cover the Indian Ocean north of 40°S was achieved by 2008. In late 2022 there were 551 active floats providing more than 20,000 profiles per year. An increasing number of floats (63) are equipped with biogeochemical sensors that measure dissolved oxygen, chlorophyll, backscatter, and nitrate. While the data from core Argo are used to observe changes in ocean circulation, temperature, and salinity, data from the biogeochemical sensors can track processes related to phytoplankton blooms, marine ecology, and Oxygen Minimum Zones (OMZs) in the Arabian Sea and Bay of Bengal.

The Global Drifter Program consists of an array of satellite-tracked surface drifting buoys drogue to follow horizontal ocean currents at a depth of 15 m with an accuracy better than $1\,\mathrm{cm\,s^{-1}}$ per $10\,\mathrm{m\,s^{-1}}$ of wind (Niiler et al., 1995). Each drifter measures surface currents (via displacement) and SST, and a growing number measure sea-level atmospheric pressure (Centurioni, 2018; Centurioni et al., 2017a), surface wave height variations (Centurioni et al., 2017b), and other parameters. These data have been relayed to a satellite at an average interval of 1.2 h since 2005 (Elipot et al., 2016). The Global Drifter

Program design of one drifter per 5° by 5° area reached 95% density in 2015, with undersampled regions remaining in the western equatorial region, the central and northern Somali Current system, and the region of the inflowing Indonesian Throughflow.

The XBT (eXpendable BaythyThermograph) network consists of ship-of-opportunity measurements of temperature over the upper 800 m of the water column. The Indian Ocean network began in the 1980s and, at its peak, comprised 11 transects crisscrossing the basin. XBTs have now largely been superseded by the Argo program, but ship-of-opportunity transects are invaluable in regions of strong flows and variability where eddy-resolving sampling is needed. Two transects remain important components of IndOOS: IX01 across the ITF began in 1983 and is currently occupied monthly at 20–30 km resolution (Wijffels & Meyers, 2004), and IX21 across the Agulhas Current is occupied seasonally at 30 km resolution. XBT data are publicly available through the National Centers for Environmental Information (https://www.ncei.noaa.gov). The relatively long time history of XBT measurements in the Indian Ocean provides one of the very few baselines for decadal variability and trends.

The Global Sea Level Observing System (https://www.gloss-sealevel.org/) tide-gauge network provides sea-level measurements near the coasts or islands around the Indian Ocean. Many tide-gauge stations provide multi-decadal or longer records that are important for sea-level research and historical sea-level reconstruction on the basin and global scales. Tide-gauge data—when the effects of vertical land motion, global isostatic adjustment, and surface pressure are properly accounted for—are an important validation of satellite altimetry measurements of sea level. The need for storm surge and tsunami monitoring and early warning has spurred an enhancement of the tide gauge network in the Indian Ocean.

The GO-SHIP (https://www.go-ship.org/) program uses research vessels to conduct decadal surveys of large-scale ocean water property distributions, collecting contemporaneous measures of physical, biogeochemical, and biological parameters throughout the entire water column. Among the comprehensive suite of measurements are dissolved inorganic carbon, total alkalinity, pH, dissolved oxygen, nutrients, chlorofluorocarbons, pCO_2, N_2O, ^{14}C, $\delta^{13}C$ of dissolved inorganic carbon (DIC), dissolved organic carbon, dissolved organic nitrogen, helium, and trace metals. GO-SHIP also provides deployment opportunities and calibration data for autonomous observing platforms and sensors, such as Argo floats, as well as underway meteorological data, surface ocean properties, and acoustic Doppler current observations that can extend to 1000 m depth.

To constrain the basin scale heat balance, the IndOOS decadal review (Beal et al., 2020) recommended a continuous observing system near 36°S, similar to the successful system in the North Atlantic (McCarthy et al., 2015) for the Atlantic Meridional Overturning Circulation (AMOC), comprising a western boundary array to resolve the Agulhas Current, plus hydrographic end-point moorings in the west and east to capture the large-scale interior geostrophic shear (Beal et al., 2020). Gliders can be useful for capturing the inshore front of the Agulhas jet, where scales cascade down to less than 10 km (Krug et al., 2017). Because the Indian Ocean heat budget is dominated by the Agulhas and a shallow overturning cell (Bryden & Beal, 2001; McDonagh et al., 2005; Sprintall et al., 2024a), circulation below 2000 m in the interior may be sufficiently captured through enhanced Deep-Argo coverage. Setting up this observing system close to latitude 36°S leverages a long-term altimetry crossing of the Agulhas (Beal & Elipot, 2016), is a region of weak meridional eddy fluxes close to the subtropical gyre maximum, and avoids the shallow topography of the southeast Indian ridge (McMonigal et al., 2018).

3.2 Satellites

The satellite component of IndOOS, which is also part of the global Earth-observing satellite system, provides basin-scale measurements of various oceanic and surface meteorological variables at higher spatial resolution than routinely available from in situ measurement systems (Fu et al., 2019). The oceanic variables include SST, SSS, SSH, significant wave height, ocean mass from gravity measurements, and ocean color (e.g., chlorophyll concentration). The marine surface meteorological variables include ocean-surface winds (wind stress, vector wind velocity, or wind speed) relative to the moving ocean surface, precipitation, ocean-surface shortwave radiation, and outgoing longwave radiation. These observations are typically retrieved as electromagnetic signals using either active sensors that send radar pulses and measure the return signals or passive sensors that measure radiation emitted from the ocean surface. Various technologies have been used, including radiometry for SST, SSS, wind speed, and ocean color, radar altimetry for SSH, radar scatterometry for vector wind, accelerometry, and laser ranging interferometry for ocean mass. The measurements using higher frequencies of the electromagnetic spectrum (visible ocean color and infrared SST) can have spatial resolutions down to km scale. The lower-frequency passive microwave measurements (e.g., of SST and SSS) have typical resolutions of 40–55 km. Active microwave measurements of precipitation and vector wind have typical resolutions of 10–25 km. Ocean mass measurements have a resolution of hundreds of kilometers. Gridded satellite products are provided at various temporal intervals, for example, down to 30 min for precipitation, daily for SST and ocean color, weekly for SSS, 5–10 days for SSH, and monthly for ocean mass.

4 Scientific highlights

IndOOS has sparked an avalanche of scientific progress since its inception. Our intention here is to be illustrative rather than comprehensive in highlighting recent advances enabled by IndOOS. We also try to avoid overlap with other chapters in this book by focusing primarily on studies that have depended critically on the availability of IndOOS data. The chapter is organized along several topical themes: Ocean circulation, equatorial waves, air-sea interaction and mixed layer processes, IOD, sea level rise, and biogeochemical cycles. There is no unique way to parse themes and organize the results, but we hope that most of the key highlights are captured in this section regardless of their specific location.

4.1 Ocean circulation

Prior to IndOOS, our view of the large-scale circulation of the Indian Ocean was comprehensively summarized by Schott and McCreary (2001) and Schott et al. (2009). There is a shallow, time-averaged meridional overturning circulation—the subtropical cell—involving mid-latitude subduction in the southeastern Indian Ocean where subtropical mode water is formed and upwelling in the SCTR south of the equator. A portion of the mode waters crosses the equator to upwelling off the coasts of Somalia and Oman during the summer monsoon, eventually returning southward near the surface in the cross-equatorial circulation cell (Lee, 2004). North of 10°S, there are dramatic reversals of surface currents like the Somali Current associated with the seasonally reversing monsoon winds (Beal et al., 2013; Düing et al., 1980; Schott, 1983). Currents along the equator are characterized by strong semiannual Wyrtki jets, with a surface eastward jet in boreal spring and fall during the transitions between the Northeast and Southwest Monsoons (Wyrtki, 1973). The eastward subsurface equatorial undercurrent, a permanent feature of the circulation in the Pacific and Atlantic Oceans, is most prominent in boreal winter in the Indian Ocean during the Northeast Monsoon (Knauss & Taft, 1964).

With the advent of IndOOS, our understanding of the mean circulation and its variations has sharpened (see also Phillips et al., 2024, for a more comprehensive treatment of this topic). Using Argo and GO-SHIP observations, for example, Nagura and McPhaden (2018) describe the climatological mean circulation along isopycnal surfaces over the upper 2000 m, updating the seminal study of Reid (2003) with two orders of magnitude more observations. The mean interior ocean circulation is weak in the Northern Hemisphere because the currents reverse in direction seasonally. But in the Southern Hemisphere, a very clear definition emerges of the anticyclonic gyre that resides between 10°S and 30°S and the cyclonic gyre is located between the equator and 10°S west of 80°E (Fig. 3a), the latter of which is related to the SCTR. The thermocline deepens at the center of the anticyclonic gyre and shoals in the cyclonic gyre (Fig. 3b and c), forced by wind-driven downwelling and upwelling, respectively (Fig. 3d). A high salinity tongue—the south Indian subtropical water (Rochford, 1964; Talley & Baringer, 1997; Wijffels et al., 2002)—is observed south of 15°S along 25–26 σ_θ surfaces (Fig. 3c). This water mass is generated near the surface by high evaporation, then subducts into the thermocline due to wind-driven downwelling (Fig. 3e–g) and is advected by equatorward flow in the interior (Fig. 3a). This interior flow has no direct path to the equator but instead is deflected to the western boundary between 10°S and 20°S (Fig. 3a) due to a zonal band of high potential vorticity water along 15°S (Nagura & McPhaden, 2018). Thus, the only pathway for water subducted in the Southern Hemisphere thermocline to cross the equator is through the western boundary.

Surface evaporation is also high in the Arabian Sea and the nearby marginal seas. The high salinity water generated there is subducted into the thermocline, advected southward by interior flow, and eventually crosses the equator. Its signature is observed in Fig. 3c north of 6°S in the upper 200 m. The convergence of that water mass with northward flowing water from the south leads to upwelling in the SCTR. Upwelling is evident in the vertical velocity estimates in Fig. 3e–g in this region and downwelling to the south of the thermocline ridge in the southern subtropical cell. The volume transports estimated from the three representations of vertical velocity in Fig. 3e–g are 7.2 Sv based on wind stress (Fig. 3e), 3.2 Sv based on Argo and GO-SHIP data (Fig. 3f), and 2.4 Sv based on the ocean reanalysis (Fig. 3g) over the region 5°–10°S, 50°–90°E. Schott et al. (2002), using the Sverdrup relation and two different wind stress products, estimated about 8 Sv of upwelling transport in the SCTR. However, those estimates were over 2°–12°S, 50°–90°E, an area approximately twice the size of ours. So, though our estimates are roughly the same order of magnitude, the areal mismatch complicates direct comparison.

IndOOS surface drifters (Beal et al., 2013; Lumpkin & Johnson, 2013), RAMA moored acoustic Doppler current profiler data (Horii et al., 2013a, 2013b; Wang & McPhaden, 2017), Argo float data and XBT observations (Pérez-Hernández et al., 2012; Zang et al., 2021) have revealed the full complexity of seasonally reversing currents in the Indian Ocean. For instance, L'Hegaret et al. (2018) recently found three shallow, time-dependent monsoon gyres that link the reversing Somali Current and monsoon currents in the northern hemisphere with the steadier South Equatorial Current and East

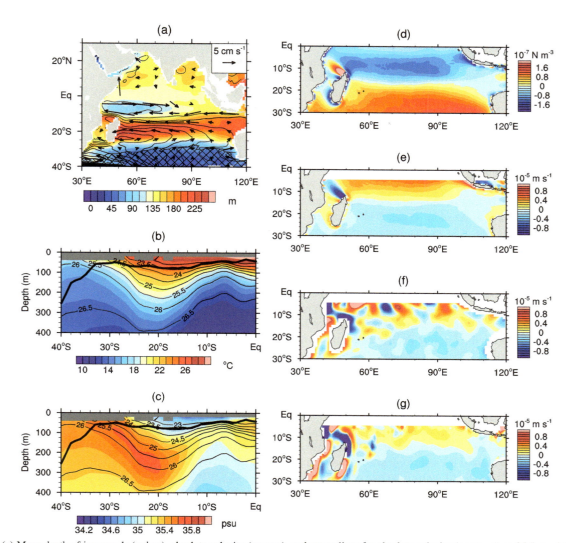

FIG. 3 (a) Mean depth of isopycnals (*colors*), absolute velocity (*vectors*), and streamlines for absolute velocity (*contours*) on 25.5 σ_θ obtained from in-situ observations. Contour intervals are $0.5 \times 10^4 \, m^2 \, s^{-2}$. Hatching shows outcropping regions where the annual mean depth of the isopycnal is shallower than the wintertime mixed layer depth. Vertical sections along 60°E for (b) temperature and (c) salinity were obtained from in-situ observations. Thin lines are for σ_θ. Thick lines are for wintertime mixed layer depth along the same longitude. *Gray* shading indicates regions with no available estimates. (d) Wind stress curl and (e) Ekman pumping velocity calculated from surface wind stress obtained from ERA-Interim (Dee et al., 2011). (f) Vertical velocity at 24.5 σ_θ estimated from meridional transport obtained from in situ observations assuming the Sverdrup relation. (g) As in (f), but obtained from meridional transport in an ocean reanalysis (ORAS4; Balmaseda et al., 2013). *(From Nagura and McPhaden (2018). Used with permission.)*

African Coastal Current of the southern hemisphere. These monsoon gyres also efficiently link the Arabian Basin with the Bay of Bengal via the monsoon currents and the intra-monsoon Wyrtki jets at the equator.

Cross-equatorial flow in the interior ocean is directed southward during the Summer Monsoon and northward during the Winter Monsoon to regulate the heat balance of the Indian Ocean basin (Schott et al., 2009). The dynamics of these reversing flows is consistent with monsoon wind-forced Ekman and Sverdrup theory (Miyama et al., 2003; Wang & McPhaden, 2017). To close the mass balance on mean seasonal time scales, 13 Sv ($1 \, Sv = 10^6 \, m^3 \, s^{-1}$) of Somali Current transport is directed southward during the Northeast Monsoon, and 20–30 Sv of transport is directed northward during the Southwest Monsoon based on the analysis of IndOOS data (Pérez-Hernández et al., 2012; Zang et al., 2021).

The prevalence of annual mean westerly winds along the equator in the Indian Ocean implies there is a mean downwelling circulation, unlike the easterly tradewind-driven upwelling in the Atlantic and Pacific. Wang and McPhaden (2017) described this mean downwelling circulation near the equator from direct velocity observations, showing convergent Ekman flow at the surface and divergent geostrophic flow below driven by the mean westerly winds. Around these mean westerlies is a very energetic semi-annual zonal wind variation with strong westerlies during the transition seasons between the Southwest and Northeast monsoons. The winds drive the Wyrtki jets, whose structure and dynamics have been detailed using 5–10 years of acoustic Doppler current profiler data from RAMA moorings (Masumoto et al., 2005; McPhaden et al., 2015; Nagura & McPhaden, 2010a, 2010b, 2012) and surface drifters (Qiu et al., 2009). The jets are confined to the upper 100 m of the water column and to ±2° latitude around the equator. They accelerate as the zonal winds pick up in spring and fall until an opposing zonal pressure gradient develops (via equatorial wave-induced zonal wind setup) to limit further growth of the jets (Nagura & McPhaden, 2008). Spring jet volume transports peak in May at ~15 Sv, and fall jet transports peak in November at ~20 Sv, around which there are year-to-year variations of 5–10 Sv (McPhaden et al., 2015). The Wyrtki jet develops a standing meander to the east of the Maldives Islands, dynamically consistent with an island wake effect (Nagura & Masumoto, 2015).

Like in the Pacific and Atlantic Oceans, undercurrents are evident in the equatorial thermocline of the Indian Ocean. Unlike in the other two oceans, though, the Indian Ocean undercurrents are highly variable on seasonal to interannual time scales (Chen et al., 2015; Iskandar et al., 2009; Nagura & McPhaden, 2016; Nyadjro & McPhaden, 2014; Zhang et al., 2014). As with the surface Wyrtki jets, zonal thermocline flows exhibit a significant semi-annual variability, with eastward equatorial undercurrents during both the winter and summer monsoon seasons separated by weak intervening westward flows. The winter season undercurrent is stronger because of the sustained easterly winds associated with the Northeast Monsoon that precedes it, while the summer season undercurrent is more variable from year to year. Interannual variations in both the Wyrtki jets and the undercurrents are dynamically linked to IOD events (Iskandar et al., 2009; Nyadjro & McPhaden, 2014) and, to a lesser extent, ENSO (Gnanaseelan et al., 2012).

Several studies using Argo, satellite altimetry, and/or XBT data have examined the role of both local Indian Ocean wind forcing and forcing from the tropical Pacific associated with ENSO and the PDO on the tropical and subtropical circulation of the Indian Ocean (e.g., Lee & McPhaden, 2008; Nagura & McPhaden, 2021; Trenary & Han, 2013; Volkov et al., 2020; Zhuang et al., 2013). There are two routes for remote inter-basin communication with the Pacific, an oceanic route through the Indonesian seas (e.g., Lee & McPhaden, 2008; Liu et al., 2015; Nagura & McPhaden, 2021; Sprintall et al., 2024a; Volkov et al., 2020) and an atmospheric route—"the atmospheric bridge"—involving changes in the Walker Circulation (Dong & McPhaden, 2016; Han et al., 2014; Xie et al., 2002; Yu et al., 2005). Forcing from the Pacific is evident at all southern tropical and subtropical latitudes and is especially pronounced in the subtropics where local wind forcing is weak (Nagura & McPhaden, 2021). Volkov et al. (2020), for example, found a decade-long (2004–2013) increase in basin-wide sea level and heat content in the subtropical southern Indian Ocean forced by strengthened Pacific trade winds associated with the recent hiatus in global warming (Lee et al., 2015; Nieves et al., 2015). This increase, mediated by increased ITF transport, abruptly ended with an unprecedented drop in sea level and heat content that was driven by the 2014–2016 El Niño in the Pacific via the atmospheric bridge. SSH differences between the eastern and western boundaries of the Indian Ocean have been used to infer decadal variations in the lower branch (i.e., meridional pycnocline flow) of the shallow overturning circulation in the Indian Ocean (e.g., Lee & McPhaden, 2008), though Nagura (2020) showed using Argo data that not all meridional mass transport variations in the subtropical cell are reflected in satellite altimetry alone.

The structure and variability of the Agulhas Current have been successfully captured by two major observing arrays during the past decade (Fig. 4; Beal et al., 2015; McMonigal et al., 2020), and these observations have led to a long-term, altimeter-based proxy for Agulhas Current volume transport (Beal & Elipot, 2016). In the mean, the Agulhas transports 76 Sv of warm and salty water southward, carrying 3.8 PW of heat and 2650 Sv pss (pss = practical salinity scale) of salt. The well-known solitary meanders of the Agulhas Current—locally known as Natal Pulses (Lutjeharms, 2006)—have little imprint on these downstream fluxes, even though they dominate variability in the local temperature and salinity fields (Elipot & Beal, 2015; Gunn et al., 2020; Leber & Beal, 2014; McMonigal et al., 2020). The Agulhas is strongest during austral summer (Hutchinson et al., 2018; Krug & Tournadre, 2012), but the flow deepens and cools as it strengthens, and this may be the reason there is no summer maximum in heat transport (Beal & Elipot, 2016; McMonigal et al., 2020). The large variability of transport-weighted temperature in the Agulhas Current stands in contrast to the Gulf Stream within the Florida Straits (Johns et al., 2011) and implies that sustained measurements of temperature in the Agulhas Current, together with a transport proxy, are necessary to constrain the meridional heat transport of the Indian Ocean to better than 0.24 PW (McMonigal et al., 2020). A broadening of the Agulhas Current since the early 1990s, due to a small increase in meandering (Beal & Elipot, 2016) led to a reduction in the heat exported from the basin via the Agulhas Current, contributing about one-third of the rapid heat gain in the Indian Ocean during the global warming hiatus (McMonigal et al., 2022). These changes in the Agulhas Current are related to a strengthening and poleward shift of the Southern Hemisphere westerlies, with negligible change in wind curl (Beal & Elipot, 2016).

The remaining two-thirds of the heat gain in the Indian Ocean is attributable to a strengthening of the ITF (e.g., Dong & McPhaden, 2016; Feng et al., 2011; Lee et al., 2015; Nieves et al., 2015). Liu et al. (2015) used XBT data to quantify a trend

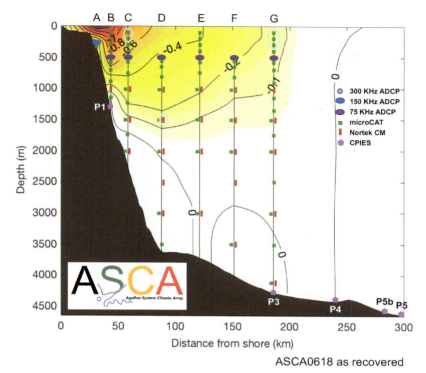

FIG. 4 The Agulhas System Climate Array (ASCA) was recovered in June 2018. The background colors show the 2-year mean current velocity as measured by the ASCA (in m s^{-1}, with red through *yellow shades* indicating southward flow). Data were recovered from all instruments except those circled in *cyan*. *(From McMonigal et al. (2020). Used with permission.)*

of ~1 Sv per decade in the ITF over 1984–2013, associated with a strengthening of the Pacific trade winds (England et al., 2014; Kosaka & Xie, 2013). The rapid warming during this period was also reflected in decadal changes in SSH both along the west coast of Australia and in the interior of the Indian Ocean, as inferred from a century-long tide gauge record at Fremantle (Feng et al., 2010) and satellite altimetry SSH data (Volkov et al., 2020). Decadal warming of the Indian Ocean, mediated by an ITF that was further strengthened by the strong 2010–2011 La Niña, set the stage for development of the Ningaloo Niño, an unprecedented marine heatwave discovered off the west coast of Australia in February–March 2011 during which anomalous SST warming of 5°C led to major fish kills and widespread coral bleaching (Feng et al., 2013, 2024; Yamagata et al., 2024).

4.2 Equatorial waves

At low latitudes, the ocean adjusts to changes in wind forcing via the excitation and propagation of equatorial waves, which are a unique class of planetary-scale waves that result from a change in the sign of the Coriolis parameter across the equator (Moore & Philander, 1977). Given the energetic surface wind forcing in the Indian Ocean on intraseasonal to interannual timescales, these waves are very prominent in ocean currents, temperature, and SSH. The most prominent wave modes at periods longer than about 10 days are eastward propagating equatorial Kelvin waves, westward propagating long equatorial Rossby waves, and mixed Rossby-gravity waves. We will refer to these waves as "low frequency" waves to distinguish them from higher frequency inertia-gravity waves that are evident at periods shorter than 10 days.

Wind-forced Kelvin and Rossby waves play a major role in seasonal timescale ocean current adjustments to wind forcing along the equator associated with semiannual equatorial surface and subsurface jets, i.e., the Wyrtki jet and undercurrents in the thermocline (Chen et al., 2015; Iskandar et al., 2009; Nagura & McPhaden, 2010a, 2016). They are also important in the ocean's response to intraseasonal atmospheric forcing (Fu, 2007; Masumoto et al., 2005; Nagura & McPhaden, 2012; Pujiana & McPhaden, 2018, 2020) and in the generation of interannual climate variations associated with the IOD (Chen et al., 2015; Joseph et al., 2012; Nagura & McPhaden, 2010b; Zhang et al., 2014). Wind-forced equatorial waves also remotely affect coastal regions and the interior ocean at higher latitudes through reflections at both eastern and western boundaries (e.g., Girishkumar et al., 2011, 2013a; Iskandar & McPhaden, 2011; McPhaden & Nagura, 2014; Suresh et al., 2013, 2016, 2018).

Luyten and Roemmich (1982) were the first to document equatorial waves propagating into the deep ocean below the main thermocline in the Indian Ocean. Since their seminal discovery, IndOOS data, especially from RAMA and Argo, have greatly expanded our ability to describe and understand the properties of these waves (Chen et al., 2020; Huang et al., 2019, 2018a, 2018b; Zanowski & Johnson, 2019). Equatorial waves in the deep ocean are particularly prominent at semi-annual periods because surface wind stress forcing at semi-annual periods is so pronounced. The phase of the waves propagates upward, indicative of downward energy propagation, consistent with an energy source at the surface. At higher southern tropical latitudes where semi-annual period wind forcing is weaker than at annual periods, the signature of vertically propagating annual period Rossby waves is evident in geostrophic velocities obtained from Argo float observations in the deep ocean (Johnson, 2011; Nagura, 2018).

Ocean waves exhibiting characteristics consistent with mixed Rossby-gravity waves are prominent in all tropical oceans. They have periods between roughly 10–30 days and are often referred to as "biweekly" waves. Sengupta et al., 2004 were the first to observe these waves from RAMA deep ocean moorings in the Indian Ocean. Later studies confirmed that these ocean waves were forced by atmospheric wind variations at similar periods (Miyama et al., 2006; Ogata et al., 2008; Nagura & McPhaden, 2023; Pujiana & McPhaden, 2021) with phase propagating westward and a zonal wavelength of approximately 3000 km (Arzeno et al., 2020; Pujiana and McPhaden, 2020). Pujiana and McPhaden (2020) showed using RAMA data that while phase propagates westward, energy propagates eastward and downward. The biweekly waves converge heat onto the equator (Nagura et al., 2014; Smyth et al., 2015), generating a time-averaged upwelling below the thermocline via nonlinear effects (Ogata et al., 2017). However, they have little impact on SST (Horii et al., 2011).

4.3 Air-sea fluxes and mixed layer processes

Much progress has been made in improving the quantification of surface radiative and turbulent heat fluxes in the tropical Indian Ocean over the past 15 years. The progress was a result of a variety of efforts, including the RAMA air-sea measurements (McPhaden et al., 2009b), improved retrievals of flux-relevant variables from satellite remote sensing, advanced applications of satellite observations as constraints in the latest versions of atmospheric reanalyses (Gelaro et al., 2017; Hersbach et al., 2020; Saha et al., 2010), and the increased number of flux products that are sourced from atmospheric reanalyses, satellite, or a combination of both (Praveen Kumar et al., 2012, 2013; Yu et al., 2007). Current flux products are capable of producing regional subseasonal-seasonal heat flux variability with significant fidelity when compared to RAMA observations (Cyriac et al., 2019; Pokhrel et al., 2020; Sanchez-Franks et al., 2018). However, the uncertainty of net surface heat flux, Q_{net}—the sum of surface radiative and turbulent heat fluxes—is still large (Pokhrel et al., 2020). As happens in other ocean basins (Yu, 2019), the spread in the Q_{net} products (Fig. 5) increases sharply in convective regimes where SST exceeds 26°C and in frontal regions associated with western boundary currents. This spread results mainly because of differences in evaporative heat fluxes and surface radiation in atmospheric reanalyses. Atmospheric moist physics and parameterization of cloud microphysics remain challenging problems but are key to constraining surface heat flux estimates for improved atmospheric reanalyses.

Surface heat fluxes are critical in determining changes in SST and upper ocean heat content across a broad range of time and space scales in the Indian Ocean (Godfrey, 1996; Webster et al., 2002). For example, analysis of RAMA, Argo, and satellite observations have established air-sea fluxes as a primary driver of the seasonal cycle of SST (e.g., Foltz et al., 2010; Thangaprakash et al., 2016) and SST variability on intraseasonal timescales associated with the eastward propagating MJO and the northward propagating monsoon intraseasonal oscillations (Drushka et al., 2012; Girishkumar et al., 2017; Parampil et al., 2016; Vialard et al., 2008). However, vertical mixing, entrainment, and horizontal advection are also important at times depending on the season, location, and phase of the MJO and monsoon intraseasonal oscillations (e.g., McPhaden & Foltz, 2013; Girishkumar et al., 2013b, 2020; Han et al., 2007; Warner et al., 2016).

Near-surface salinity can dominate upper-ocean stratification in parts of the Indian Ocean and in particular, the Bay of Bengal, where there is a large volume of freshwater input from river runoff and the excess in precipitation over evaporation. This large freshwater flux at the surface introduces strong haline stratification, leading to the formation of thick "barrier layers" between the base of the mixed layer and the top of the thermocline (Vinayachandran et al., 2002). Barrier layers inhibit the turbulent entrainment of cooler thermocline waters into the mixed layer and thereby play an important role in the ocean surface layer heat budget that controls SST (McPhaden & Foltz, 2013; Moum et al., 2014; Warner et al., 2016). Thin mixed layers defined by shallow salinity stratification also have lower heat capacity, so for the same amount of surface heat flux, SSTs tend to be warmer and air-sea interactions more vigorous (Krishnamohan et al., 2019).

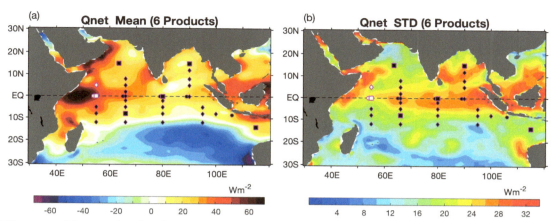

FIG. 5 (a) Time-mean of the net surface flux (Q_{net}, positive for oceanic heat gain) at the ocean surface from the ensemble mean of six different flux products for the 2001–2015 period. (b) Standard deviations (STDs) around the mean of the six flux products over that period, giving an idea of the area where flux estimates are most uncertain. The STDs in climatological Qnet are up to 25 Wm^{-2} in a large part of the Indian Ocean north of 10°S, of the same order of magnitude as the mean Qnet itself. Buoy locations of RAMA-2.0 are superimposed, with diamonds denoting RAMA surface mooring sites and squares corresponding to "flux reference sites" that provide the essential benchmark time series for validating and improving air-sea parameterizations in models and for improving uncertainty quantification in air-sea flux products. *(Adapted from McPhaden et al. (2009b). Used with permission.)*

Barrier layers in the Indian Ocean vary in thickness on intraseasonal (Girishkumar et al., 2011; Li et al., 2017a, 2017b), seasonal (Thadathil et al., 2007), and interannual (de Boyer Montégut et al., 2007) timescales. They can cause temperature inversions by trapping a significant part of the penetrating solar radiation below shallow mixed layers. In the Bay of Bengal, these thermal inversions play a significant role in warming the mixed layer by entrainment of warm subsurface water during the winter monsoon (de Boyer Montégut et al., 2007; Girishkumar et al., 2013b; Nagura et al., 2015; Sengupta et al., 2008; Thangaprakash et al., 2016), during post summer monsoon cyclones (Neetu et al., 2012; Jarugula & McPhaden, 2022) and on interannual timescales (Nagura et al., 2015). The combination of satellite SSS, Argo data, and RAMA mooring data have greatly increased our ability to observe and understand barrier layers and their effects in the Bay of Bengal and other regions of the Indian Ocean where they are prominent (e.g., Felton et al., 2014).

Tropical cyclones regularly occur every year in both the northern and southern Indian Ocean, often resulting in devastating property damage and loss of life. IndOOS data are widely distributed in real-time to support forecasting of extreme weather events; they are also used in research to understand the processes involved in cyclone development. Investigations have been undertaken using IndOOS and related data to examine specific storms over the past 15 years, such as Dora in 2007 in the southwest Indian Ocean (Cuypers et al., 2013); Nargis in 2009 (Lin et al., 2009; Maneesha et al., 2012; McPhaden et al., 2009a; Yu & McPhaden, 2011), Phailin in 2013 (Qiu et al., 2019), Hudhud in 2014 (Warner et al., 2016) in the Bay of Bengal; Riley, Veronica, and Wallace in 2019 off northwest Australia (Song et al., 2021), and many others (Venkatesan et al., 2016a, 2016b). These storms extract heat from the ocean at a rate of many hundreds of Watts per square meter at their height (McPhaden et al., 2009a; Song et al., 2021), extreme fluxes that are not well captured by atmospheric reanalysis products (Song et al., 2021). Tropical cyclones in the Indian Ocean also mix the upper ocean via generation of intense turbulence (Warner et al., 2016) mediated by the excitation of resonant inertial oscillations (Cuypers et al., 2013; Maneesha et al., 2012; Warner et al., 2016). This mixing up typically generates a cold wake in SST after the storm passage (Kuttippurath et al., 2022), but post-monsoon storms in the Bay of Bengal can initially increase SST because of the mixing upward of temperature inversions trapped in the barrier layer (Neetu et al., 2012, 2017; Warner et al., 2016).

The frequency and intensity of tropical storms in the Indian Ocean varies with the phase of ENSO and the IOD (Burns et al., 2016; Girishkumar & Ravichandran, 2012; Mahala et al., 2015; Xie et al., 2002). There has been a downward trend in the number of tropical cyclones over the past 70 years in the northern Indian Ocean, though whether this trend is due to climate change is unresolved (Vellore et al., 2020). It has been hypothesized that the larger long-term warming over the Arabian Sea than in other parts of the tropics (e.g., Gopika et al., 2020) promotes more intense cyclones in this region (e.g., Murakami et al., 2017).

Air-sea interactions on sub-daily timescales are pronounced in the Indian Ocean and have consequences for longer-term weather and climate variations. The diurnal cycle of SST, for example, plays a fundamental role in regulating ocean–atmosphere interactions because of the elevated mean (typically >25°C) SSTs on which diurnal variations occur. Turbulent exchanges of heat, moisture, and carbon dioxide across the air-sea interface are sensitive to diurnal SST variations such that neglect of these variations degrades the accuracy of turbulent flux estimates on daily mean and longer time scales (Praveen Kumar et al., 2012). Not accounting for the diurnal cycle of SST also affects the fidelity of models used in climate studies and forecasting of, for example, the MJO and the Indian monsoon (e.g., Terray et al., 2012; Woolnough et al., 2007). RAMA data, because it is available at 10 min to 1 h sampling intervals, have allowed for a systematic definition of the diurnal cycle of SST, its seasonal variations, and the factors responsible for it in different regions of the Indian Ocean (Yang et al., 2015). RAMA data have also allowed a definition of atmospheric "cold pools" (drops in air temperature of over 1°C in 30 min associated with convective downdrafts) in the Bay of Bengal. These cold pools occur most frequently in boreal summer and fall in association with the diurnal cycle of rainfall (Joseph et al., 2021) and can enhance turbulent air-sea fluxes by as much as 40–80 W m^{-2} over the course of individual events. During the convectively active phase of the MJO, the number of cold pools nearly doubles compared to the suppressed phase (de Szoeke et al., 2017).

The Indian Ocean has experienced some of the strongest warming trends in the world ocean over the last 50–100 years. The cause of increasing SST on decadal time scales is not yet fully explained (Roxy et al., 2014), though it is at least in part related to climate change (Dong et al., 2014; Gopika et al., 2020; Roxy et al., 2020, 2024). The role of wind-induced ocean circulation changes in modulating tropical Indian Ocean warming patterns has been suggested using models and atmospheric reanalysis products (Rahul & Gnanaseelan, 2016; Rao et al., 2011; Swapna et al., 2017), but the mechanisms have yet to be clearly articulated owing to the lack of data of sufficient length, spatial coverage, and/or quality.

Decadal variations, such as the accelerated warming of the Indian Ocean during the recent hiatus in global surface warming, appear to be driven not by Q_{net} (Alory & Meyers, 2009) but by the excess transfer of heat from the Pacific Ocean via the ITF (Dong & McPhaden, 2016; Lee et al., 2015; Vialard, 2015; Zhang et al., 2018) and a reduction in heat export from the basin via the Agulhas Current (McMonigal et al., 2022). Srinivasu et al. (2017) speculated that the ITF impacts should be limited to the Southern Indian Ocean, with the warming of the Northern Indian Ocean attributable to changes in wind-driven meridional heat transport across 5°S (see also Thompson et al., 2016). Thus, it appears that surface heat fluxes are largely responding to, rather than driving, long-term changes in Indian Ocean SST.

4.4 Oceanic processes involved in IOD development

The IOD is an interannual fluctuation of the climate system that arises from coupled ocean–atmosphere interactions and large-scale changes in ocean circulation similar to those affecting ENSO in the Pacific (Saji et al., 1999; Webster et al., 1999; Yamagata et al., 2024). Like ENSO, altered patterns of weather variability associated with IOD events can have significant socio-economic impacts, particularly in Indian Ocean rim countries (e.g., Behera et al., 2005; Cai et al., 2011; Moihamette et al., 2022). Positive IOD events are associated with easterly wind stress anomalies along the equator, unusually cold SSTs in the eastern basin off Sumatra and Java, and unusually warm SSTs in the western basin. Deep atmospheric convection and rainfall are suppressed over the cool waters in the east and enhanced over warm waters in the west. Anomalies of opposite signs characterize negative IOD events.

IndOOS data have provided fundamental insights into the oceanic processes associated with IOD development. Mixed layer heat balance studies indicate that a variety of processes are involved in the generation of SST anomalies associated with the IOD, including vertical advection and mixing, zonal advection, and surface heat flux forcing in both the eastern and western poles of the dipole (e.g., Horii et al., 2009; Murtugudde et al., 2000; Praveen Kumar et al., 2014). The precise mix of processes varies over the course of IOD development and also from event to event (Rao et al., 2009). Dynamical processes are essential in generating the anomalies, while surface heat fluxes may both favor or, particularly at later stages of development, damp SST anomalies. Salinity variations are important in affecting vertical mixing processes and generally enhance SST anomalies through variations in barrier layer thickness that are correlated with the phase of the IOD (Kido & Tozuka, 2017; Qiu et al., 2012).

Anomalous upwelling is a critical process in the eastern pole of the IOD. It is fed by the anomalous wind-forced 3-dimensional circulation in the thermocline (Zhang et al., 2014), bringing cool water to the surface during positive events as inferred from both observations (Masumoto et al., 2008) and ocean reanalyses (Nyadjro & McPhaden, 2014). At the surface, anomalous wind-driven currents during September–November drain the eastern basin of water that has been upwelled and transport it westward (Fig. 6). This east–west transfer of mass elevates the thermocline (and depresses the sea level, which is a mirror image of the thermocline) in the east while depressing the thermocline in the west (and elevating the sea level there). Anomalies of opposite sign occur for negative IOD events. An important distinction between

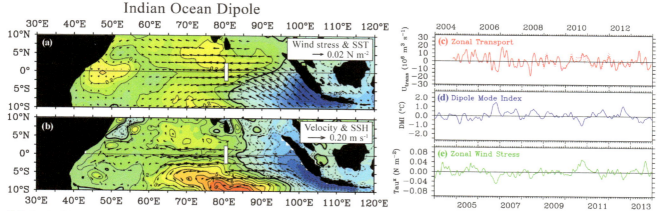

FIG. 6 (a) SST and surface wind stress and (b) SSH and surface layer zonal currents (from Bonjean & Lagerloef, 2002) regressed against SON zonal transport time series for transport equal to one standard deviation (6.3 Sv). The white strip indicates the location of the 80.5°E volume transport array used to compute the index time series. Vectors are omitted when cross-correlations between the two-time series are <0.2. The patterns are those expected for the positive phase of the Indian Ocean Dipole. Time series of (c) zonal transport anomalies, (d) Dipole Mode Index, and (e) zonal wind stress anomalies averaged over 2°N–2°S, 75°–95°E. The 10-year transport anomaly time series estimates in (c) are based on regression from 5 years of actual measured transport anomalies (*dotted line*) over the second half of the record (after McPhaden et al., 2015).

positive and negative IOD events, though, is that nonlinear zonal and vertical advection always tend to cool the eastern Indian Ocean (cf., Horii et al., 2011, 2013b), contributing to the asymmetry in magnitude of positive and negative IOD events, the former of which are generally stronger (Hong et al., 2008; Ogata et al., 2013).

Wind-forced equatorial Kelvin and Rossby waves are crucial in the dynamical adjustment of the currents, thermocline depth, and SSH to changing winds associated with the IOD (McPhaden et al., 2015; McPhaden & Nagura, 2014; Murtugudde & Busalacchi, 1999; Nagura & McPhaden, 2010b). Moreover, reflection of Kelvin waves into Rossby waves at the eastern boundary is important in the decay phase of the IOD. These reflected waves cause surface zonal velocity on the equator to decay faster and earlier than zonal wind stress anomalies by about 1 month, limiting the growth of SST anomalies and thus helping to terminate IOD events (Iskandar et al., 2013; Nagura & McPhaden, 2010b).

Using RAMA data, Horii et al. (2008) suggested that a shoaling thermocline along the equator preceded surface cooling by 3 months associated with the 2006 IOD event. McPhaden and Nagura (2014) explored this idea in the framework of the Recharge Oscillator theory (Jin, 1997) that is based on the premise that upper-ocean heat content in the tropical Pacific Ocean is a major predictor of ENSO development (e.g., Meinen & McPhaden, 2000). Using altimeter-derived SSH as a proxy for heat content and an analytical, linear equatorial wave model, they found that, just as for ENSO development, there are zonally coherent changes in SSH along the equatorial Indian Ocean prior to IOD onset, an indication that the recharge oscillator mechanism is at work for IOD development. However, upper-ocean heat content is not as strong a predictor for the IOD as it is for ENSO because IOD is strongly affected by ENSO itself through atmospheric teleconnections.

ITF transports on interannual timescales are modulated by ENSO variations in western Pacific SSH, which strongly controls the Pacific to Indian pressure difference that is the driving force for the ITF. However, this pressure difference is also affected by the Indian Ocean. In particular, during the major negative 2016 IOD event, elevated SST and SSH off the coast of Sumatra and Java overwhelmed the Pacific influence, leading to an unprecedented reduction in the ITF transport (Pujiana et al., 2019). This weakened transport may have been a factor in the unexpectedly weak 2016–2017 La Niña by keeping subsurface heat content high in the Pacific Ocean (Mayer et al., 2018).

4.5 Sea level variability and rise

Global sea level has been rising since 1900, and the pace of sea level rise has accelerated in recent decades due to the combined effects of melting mountain glaciers, ice melt from Greenland and Antarctica, and thermal expansion of the ocean (IPCC, 2019). Sea level has also risen in the Indian Ocean; however, this rise is non-uniform in space and non-stationary in time (Kumar et al., 2020). Hundreds of millions of people living near the coast in countries surrounding the Indian Ocean and many low-lying island nations are at risk from rising sea levels (Oppenheimer et al., 2019; Swapna et al., 2017). Thus, understanding the processes involved in sea level rise and its variability is an urgent societal problem.

An example of how sea level has varied in the Indian Ocean for the period of the 1960s–2000s (Fig. 7; Han et al., 2010) shows a pattern of positive and negative trends in SSH from tide gauges and simulations of a validated Hybrid Coordinate Ocean Model. Trends over much of the subtropical midlatitude South Indian Ocean are larger than in the North Indian Ocean, while sea level in the western equatorial Indian Ocean dropped, a decrease recorded at the Zanzibar tide-gauge station. The decreasing trend of SSH in this upwelling zone is consistent with an observed subsurface cooling trend that has been attributed to climate change (e.g., Alory et al., 2007). However, SSH in the Indian Ocean is modulated by natural

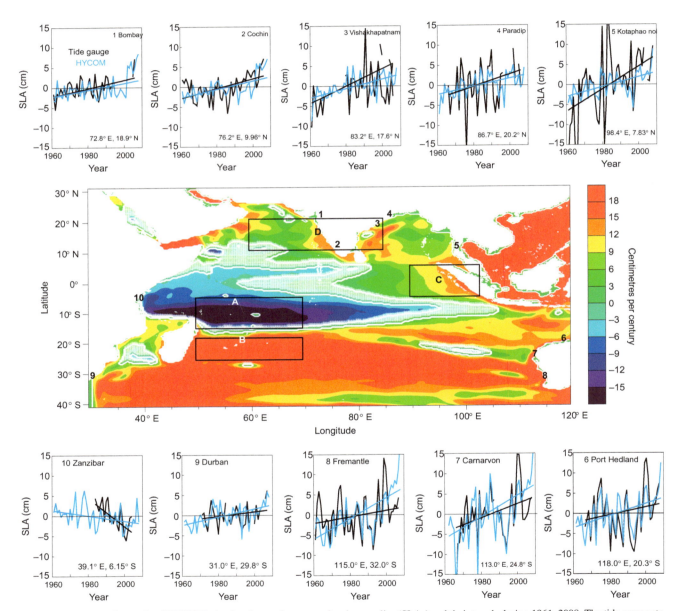

FIG. 7 Tide-gauge-observed and HYCOM-simulated annual mean sea level anomalies (SLAs) and their trends during 1961–2008. The tide-gauge stations (marked 1–10) with records longer than 30 years (20 years for Zanzibar) are shown. The effects of vertical land motion, global isostatic adjustment, and surface pressure have been removed from the tide-gauge data. All trends exceed 95% significance except for stations 6 and 9 tide-gauge data. The map in the middle shows SSH trends simulated by the HYCOM model for 1961–2008. The *light blue/green* regions are below, and the rest are above 95% significance. The *rectangles* labeled A–D mark regions of which sea level changes were further discussed in the source of this figure. *(From Han et al. (2010). Reprinted by permission of Nature Geoscience.)*

decadal variations (Han et al., 2018), with the multi-decadal decreasing SSH trend in the western basin reversing after the mid-2000s due to decadal variability in the wind forcing and the associated ocean dynamics (Thompson et al., 2016).

Variations in SSH can be related to the three-dimensional structure of salinity and temperature variations in the Indian Ocean derived from Argo. Indian Ocean sea level rise is typically dominated by ocean thermal expansion (Swapna et al., 2017), but there are regional variations. For instance, over the upper 250 m of the southeast Indian Ocean during 2005–2013 (Llovel & Lee, 2015), a freshening and warming trend accelerated the rate of sea level rise beyond the global average. The halosteric effect caused by the freshening accounted for almost 2/3 of this regional increase in steric height, more prominent even than in the subpolar North Atlantic, where the halosteric effect is also known to be important. The origin of the thermosteric effect can be linked to warming in the northwestern tropical Pacific, while the halosteric effect was caused by a freshening due to excess precipitation over the maritime continent (Hu & Sprintall, 2017). These signals were advected into the Indian Ocean via the ITF (Du et al., 2015; Feng et al., 2015; Ummenhofer et al., 2021).

4.6 Ocean biogeochemistry

IndOOS originally focused primarily on physical oceanographic and meteorological measurements, with little emphasis on ocean biogeochemistry. Only recently has biogeochemistry become a priority within IndOOS, thanks in part to the recent IndOOS review (Beal et al., 2020). This section is intended, therefore, to highlight just a few topics in ocean biogeochemistry that IndOOS has contributed to either directly or indirectly and to underscore the opportunities that more systematic and sustained IndOOS biogeochemical measurements hold for the future. A more complete description of ocean biogeochemistry measurements can be found in Hood et al. (2024a,b).

Using in situ hydrographic data collected prior to the advent of IndOOS, Prasanna Kumar et al. (2009) gave detailed insights into the factors responsible for biological productivity in the Arabian Sea and Bay of Bengal that serve as a basis for further investigation using IndOOS data. They concluded that productivity in the Arabian Sea is triggered by wind-driven mixing, Ekman pumping and advection of nutrient-rich upwelled waters in summer and winter convection, and injection of nutrients into the euphotic zone in winter. Conversely, productivity in the Bay of Bengal is inhibited by salt-stratified barrier layers induced by freshwater influx from river runoff and rainfall during the boreal summer. Thus, biological productivity in the Bay of Bengal is lower, mediated by energetic mesoscale eddies and occasional tropical cyclones whose intense winds significantly mix the upper ocean and entrain nutrients into the euphotic zone (e.g., Chowdhury et al., 2020; McPhaden et al., 2009a; Vinayachandran, 2009).

Koné et al. (2009) simulated the seasonal cycle of productivity in a biogeochemical model, and Wiggert et al. (2009) analyzed the SeaWiFS ocean color data set to quantify how the IOD influences the seasonal cycle of productivity. During a positive IOD, the eastern Indian Ocean is more productive, while the western tropics and the Arabian Sea experience reduced productivity. These changes in productivity are mediated in part by large-scale ocean dynamics that alter the depth of the thermocline and nutricline so as to either enhance or inhibit nutrient fluxes into the mixed layer. At the southern tip of India, easterly wind anomalies associated with positive IOD events drive local downwelling, which then propagates as coastal Kelvin waves to the west coast of India, where they depress the OMZ and reduce anoxia. Negative IODs have the opposite effect and may promote coastal anoxia (Vallivattathillam et al., 2017). These relationships suggest a degree of biogeochemical predictability based on physical observations.

Goes et al. (2005) used the first 7 years of SeaWiFS satellite ocean color (1997–2004) to document an increase in Arabian Sea chlorophyll (47–55°E, 5–10°N). This trend was attributed to warming and reduced snow cover over Eurasia, which created a land-sea thermal gradient conducive to intensified upwelling favorable winds in the western Arabian Sea. A decade later, Roxy et al. (2016) looked at a broader area of the northwestern Indian Ocean (50–65°E, 5–25°N) over a longer period of time (1997–2013) to draw the opposite conclusion—that productivity was decreasing due to a warming ocean and greater upper ocean thermal stratification. The resolution of these conflicting results requires further investigation.

A relatively large number of biogeochemical Argo floats have been deployed in the Indian Ocean. They have been used to validate satellite ocean color (Wojtasiewicz et al., 2018a), to investigate the link between upper ocean primary productivity and the thickness of the oxygen minimum zone in the Arabian Sea (Sarma et al., 2020; Wojtasiewicz et al., 2018b), and productivity during cyclones (Chowdhury et al., 2020; Girishkumar et al., 2019). Bio-optical sensors deployed on RAMA moorings have shown that variability in primary productivity along the equator is very episodic on intraseasonal time scales and modulated by some combination of MJO-induced wind mixing and/or mixed Rossby gravity wave dynamics (Strutton et al., 2015). Prasanna Kumar et al. (2012) also inferred a bio-physical coupling involving wave dynamics, nitrate, and chlorophyll concentrations using RAMA mooring and shipboard data in the central Indian Ocean during the 2006 IOD event.

5 Challenges and future directions

IndOOS has been a great success in expanding the current state of our knowledge about oceanic and coupled ocean–atmosphere processes that give rise to the many unique phenomena found within the Indian Ocean region. Nonetheless, IndOOS has fallen short in meeting some of the most pressing societal needs. The recent scientific assessment of IndOOS Beal et al. (2020), and the CLIVAR/GOOS review it is based on Beal et al. (2020), outline these shortfalls and provide a comprehensive roadmap for addressing them as part of the Global Ocean Observatory 2030 Strategy (http://goosocean.org/2030strategy). Of particular importance for the monsoon-dominated Indian Ocean is the urgent need for better monsoon rainfall forecasts. To improve these forecasts, more information is needed on upper-ocean conditions within the tropics. This is where diurnal and longer timescale variations in near-surface ocean stratification impact regional SST patterns, which in turn affect the development and propagation of the MJO and Monsoon Intraseasonal Oscillations (e.g., DeMott et al., 2015; Girishkumar et al., 2017; Woolnough et al., 2007). Monsoon predictability would also benefit from observations of the Somali Current and upwelling system, where strong oceanic fluxes are linked to variations in mixed layer depth and air-sea fluxes (Izumo et al., 2008).

There is an urgent need for sustained biogeochemical measurements as an integral part of the IndOOS. Aside from the satellite chlorophyll studies cited here (Goes et al., 2005; Roxy et al., 2016) modeled changes in net primary productivity, dissolved oxygen, acidification, and ecosystem structure (Kwiatkowski et al., 2020; Tagliabue et al., 2021) are largely unconstrained by sustained observations throughout the Indian Ocean. Effective management of the Indian Ocean's natural resources, including coral reefs and wild-catch fisheries, will require sustained measurements of biogeochemical parameters, particularly in regions of high variability and change, such as the Arabian Sea, Bay of Bengal, and the eastern equatorial Indian Ocean. Expanding the global Argo array to include biogeochemical measurements is underway and will mark a step-change in ocean observation over the next decade.

IndOOS must be able to support ocean state estimations and reanalyses that are used to initialize climate predictions and drive biogeochemistry models. Lack of sustained observations in energetic boundary regions leaves these products and the basin-scale heat budget poorly constrained. For instance, discrepancies among climatologies of heat exchange at the air-sea interface over boundary current and upwelling regions, such as the Agulhas Current system in the west and the Java-Sumatra upwelling cells in the east, remain an order of magnitude larger than the global energy imbalance due to climate change (e.g., Fig. 5). While we have been able to account for much of the missing heat of the recent warming hiatus within the upper 2000 m of the Indian Ocean using Argo, the ultimate fate of this excess heat and its possible contribution to a future acceleration in global warming, is unknown (Vialard, 2015). Additional observations of the Agulhas Current and of the deep ocean are needed to constrain the Indian Ocean heat budget (McMonigal et al., 2022).

We need a sustained IndOOS to improve our understanding, modeling, and prediction of the processes responsible for warming trends in the Indian Ocean. Continuation and enhancement of satellite measurements and in-situ networks are necessary for understanding the fundamental physical mechanisms, inter-basin connections, land-sea linkages, and the relationships of climate variability and change with human systems. However, even while we argue for enhancements of IndOOS, there are immediate threats to sustaining the observing system as it exists now.

The observing system and the ocean observing community have always been under-resourced. The COVID-19 pandemic turned a challenging situation into a crisis, with some elements of the observing system in disarray as a result of cruises that were delayed or canceled. International cooperation was also curtailed because of COVID-19 travel restrictions, further exacerbating the problem. RAMA was especially hard hit because of reliance on regular servicing cruises from research vessels. The full extent of COVID-19 impacts on the Global Ocean Observing System in general and IndOOS in particular will take years to completely assess (Sprintall et al., 2024b). In the meantime however, more effort should be devoted to strengthening institutional commitments and to building resilience by introducing more diversity in IndOOS sampling technologies. In particular, introduction of uncrewed measurement systems, such as ocean gliders and saildrones, should be considered for deployment where appropriate. This will require both new financial investments and capacity building with partners in the region, but such systems have been implemented with success in other oceanographic settings.

The original IndOOS design targeted measurements of basin-scale variability in the open ocean. Thus, with a few exceptions (e.g., tide gauges networks), the in situ elements of IndOOS are concentrated outside of the exclusive economic zones (EEZs) of Indian Ocean rim nations. However, a complete picture of ocean dynamics and ocean-atmosphere interactions in the region requires linking variability in the open ocean to the coastal zone. Several national, regional observing systems have been established partially or wholly within the jurisdictions of these EEZs, such as the Integrated Marine Observing System (https://imos.org.au/) of Australia and the Indian National Moored Buoy Network for the Northern Indian Ocean. Improving coordination between these efforts and IndOOS, as well as creating new partnerships with rim nations, are important ingredients to the overall success of a sustained ocean observing system in the Indian Ocean.

Indeed, one of the greatest imperatives for the future of the IndOOS is increased engagement and partnerships among Indian Ocean rim countries (Beal et al., 2020). Much of the desired expansion of the IndOOS into coastal and upwelling regions is reliant on increased capability and commitment from regional countries and agencies. Identifying shared goals with these nations and following up with collaborations, resource and data sharing, and capacity building are essential.

IndOOS has been developed within The Framework for Ocean Observing (www.goosocean.org), which defines guiding principles for the implementation of GOOS through coordinated international efforts and Global Ocean Observing System regional alliances. New commitments are needed from the World Meteorological Organization, the International Council for Science, and the Intergovernmental Oceanographic Commission of UNESCO to support and expand the Global Ocean Observing System in the future. The current UN Ocean Decade (www.oceandecade.org) is an ongoing opportunity to raise awareness about the ocean's role in our changing climate and the necessity of sustained ocean observing systems worldwide, including IndOOS.

6 Educational resources

- RAMA data (https://www.pmel.noaa.gov/gtmba/)
- Argo data (https://argodata.com)
- Biogeochemical Argo (https://biogeochemical-argo.org/)
- Global Drifter Program data (https://www.aoml.noaa.gov/phod/gdp/)
- Global Sea Level Observing System for tide gauge data (https://www.psmsl.org/gloss/)
- Satellite data (https://podaac.jpl.nasa.gov/)
- OAFlux Surface flux data (http://oaflux.whoi.edu/)
- Tropflux data (https://incois.gov.in/tropflux/)
- Global Ocean Observing System (https://www.goosocean.org/)

Acknowledgments

We greatly appreciate the thorough reviews and constructive comments provided by three anonymous reviewers. Special thanks to Kevin Obrien and Sarah Battle of PMEL for help with Fig. 2. MJM is supported by National Oceanic and Atmospheric Administration. This is PMEL contribution no. 5212. Part of this research was carried out at the Jet Propulsion Laboratory, California Institute of Technology, under a contract with the National Aeronautics and Space Administration. LMB acknowledges support of National Science Foundation award 1459543, the Agulhas System Climate Array.

Author contributions

All authors contributed to the writing of the text, discussion of content, and structure of the chapter.

References

Alory, G., & Meyers, G. (2009). Warming of the upper equatorial Indian ocean and changes in the heat budget (1960–99). *Journal of Climate, 22*, 93–113.

Alory, G., Wijffles, S., & Meyers, G. (2007). Observed temperature trends in the Indian Ocean over 1960_1999 and associated mechanisms. *Geophysical Research Letters, 34*, L02606. https://doi.org/10.1029/2006GL028044.

Arzeno, I. B., Giddings, S. N., Pawlak, G., & Pinkel, R. (2020). Generation of quasi-biweekly Yanai waves in the equatorial Indian Ocean. *Geophysical Research Letters, 47*. https://doi.org/10.1029/2020GL088915. e2020GL088915.

Balmaseda, M. A., Mogensen, K., & Weaver, A. T. (2013). Evaluation of the ECMWF ocean reanalysis system ORAS4. *Quarterly Journal of the Royal Meteorological Society, 139*, 1132–1161. https://doi.org/10.1002/qj.2063.

Beal, L. M., & Elipot, S. (2016). Broadening not strengthening of the Agulhas Current since the early 1990s. *Nature, 540*(7634), 570–573.

Beal, L. M., Elipot, S., Houk, A., & Leber, G. M. (2015). Capturing the transport variability of a western boundary jet: Results from the Agulhas Current Time-Series Experiment (ACT). *Journal of Physical Oceanography, 45*(5), 1302–1324.

Beal, L. M., Hormann, V., Lumpkin, R., & Foltz, G. R. (2013). The response of the surface circulation of the Arabian Sea to monsoonal forcing. *Journal of Physical Oceanography, 43*(9), 2008–2022. https://journals.ametsoc.org/view/journals/phoc/43/9/jpo-d-13-033.1.xml.

Beal, L. M., Vialard, J., Roxy, M. K., Li, J., Andres, M., Annamalai, H., Feng, M., Han, W., Hood, R., Lee, T., Lengaigne, M., Lumpkin, R., Masumoto, Y., McPhaden, M. J., Ravichandran, M., Shinoda, T., Sloyan, B. M., Strutton, P. G., Subramanian, A. C., ... Parvathi, V. (2020). A roadmap to IndOOs-2: Better observations of the rapidly-warming Indian Ocean. *Bulletin of the American Meteorological Society, 101*, E1891–E1913. https://doi.org/10.1175/BAMS-D-19-0209.1.

Behera, S. K., Luo, J.-J., Masson, S., Delecluse, P., Gualdi, S., Navarra, A., & Yamagata, T. (2005). Paramount impact of the Indian Ocean dipole on the east African short rains: A 2CGCM study. *Journal of Climate, 18*(21), 4514–4530.

Bonjean, F., & Lagerloef, G. S. E. (2002). Diagnostic model and analysis of the surface currents in the tropical Pacific Ocean. *Journal of Physical Oceanography*, *32*, 2938–2954.

Bryden, H. L., & Beal, L. M. (2001). Role of the Agulhas Current in Indian Ocean circulation and associated heat and freshwater fluxes. *Deep-Sea Research Part I: Oceanographic Research Papers*, *48*, 1821–1845. https://doi.org/10.1016/S0967-0637(00)00111-4.

Burns, J. M., Subrahmanyam, B., Nyadjro, E. S., & Murty, V. S. N. (2016). Tropical cyclone activity over the southwest tropical Indian Ocean. *Journal of Geophysical Research, Oceans*, *121*, 6389–6402. https://doi.org/10.1002/2016JC011992.

Cai, W., van Rensch, P., Cowan, T., & Hendon, H. H. (2011). Teleconnection pathways of ENSO and the IOD and the mechanisms for impacts on Australian rainfall. *Journal of Climate*, *24*(15), 3910–3923.

Cai, W., Wu, L., Lengaigne, M., Li, T., McGregor, S., Kug, J.-S., Yu, J.-Y., Stuecker, M. F., Santoso, A., Li, X., Ham, Y.-G., Chikamoto, Y., Ng, B., McPhaden, M. J., Du, Y., Dommenget, D., Jia, F., Kajtar, J. B., Keenlyside, N., ... Chang, P. (2019). Pan-tropical climate interactions. *Science*, *36*, eaav4236. https://doi.org/10.1126/science.aav4236.

Centurioni, L. R. (2018). Drifter technology and impacts for sea surface temperature, sea-level pressure, and ocean circulation studies. In R. Venkatesan, A. Tandon, E. D'Asaro, & M. Atmanand (Eds.), *Observing the oceans in real time* Springer Oceanography. Springer. https://doi.org/10.1007/978-3-319-66493-4_3.

Centurioni, L. R., Hormann, V., Talley, L. D., Arzeno, I., Beal, L., Caruso, M., ... Wang, H. (2017a). Northern Arabian Sea circulation-autonomous research (NASCar): A research initiative based on autonomous sensors. *Oceanography*, *30*(2), 74–87. https://doi.org/10.5670/oceanog.2017.224.

Centurioni, L. R., Hornayi, A., Cardinali, C., Charpentier, E., & Lumpkin, R. (2017b). A global observing system for measuring sea level atmospheric pressure: Effects and impacts on numerical weather prediction. *Bulletin of the American Meteorological Society*, *98*, 231–238. https://doi.org/10.1175/BAMS-D-15-00080.1.

Chen, G., Han, W., Li, Y., Wang, D., & McPhaden, M. J. (2015). Seasonal-to-interannual time-scale dynamics of the equatorial undercurrent in the Indian Ocean. *Journal of Physical Oceanography*, *45*, 1532–1553. https://doi.org/10.1175/JPO-D-14-0225.1.

Chen, G., Han, W., Zhang, X., Liang, L., Xue, H., Huang, K., He, Y., Li, J., & Wang, D. (2020). Determination of spatiotemporal variability of the Indian equatorial intermediate current. *Journal of Physical Oceanography*, *50*, 3095–3108.

Chowdhury, R., Prasanna Kumar, S., & Chakraborty, A. (2020). A study on the physical and biogeochemical responses of the Bay of Bengal due to cyclone Madi. *Journal of Operational Oceanography*. https://doi.org/10.1080/1755876X.2020.1817659.

Cuypers, Y., Le Vaillant, X., Bouruet-Aubertot, P., Vialard, J., & McPhaden, M. J. (2013). Tropical storm-induced near-inertial internal waves during the Cirene experiment: Energy fluxes and impact on vertical mixing. *Journal of Geophysical Research*, *118*(1), 358–380. https://doi.org/10.1029/2012JC007881. https://doi-org.proxy-um.researchport.umd.edu/10.1029/2012JC007881.

Cyriac, A., McPhaden, M. J., Phillips, H. E., Bindoff, N. L., & Feng, M. (2019). Seasonal evolution of the surface layer heat balance in the eastern subtropical Indian Ocean. *Journal of Geophysical Research, Oceans*, *124*, 6459–6477. https://doi.org/10.1029/2018JC014559.

de Boyer Montégut, C., Vialard, J., Shenoi, S. S. C., Shankar, D., Durand, F., Ethé, C., & Madec, G. (2007). Simulated seasonal and interannual variability of the mixed layer heat budget in the northern Indian Ocean. *Journal of Climate*, *20*, 3249–3268. https://doi.org/10.1175/JCLI4148.1.

de Szoeke, S. P., Skyllingstad, E. D., Zuidema, P., & Chandra, A. S. (2017). Cold pools and their influence on the tropical marine boundary layer. *Journal of the Atmospheric Sciences*, *74*, 1149–1168. https://journals.ametsoc.org/view/journals/atsc/74/4/jas-d-16-0264.1.xml.

Dee, D. P., et al. (2011). The ERA-interim reanalysis: Configuration and performance of the data assimilation system. *Quarterly Journal of the Royal Meteorological Society*, *137*, 553–597. https://doi.org/10.1002/qj.828.

DeMott, C. A., Klingaman, N. P., & Woolnough, S. J. (2015). Atmosphere-ocean coupled processes in the Madden-Julian oscillation. *Reviews of Geophysics*, *53*, 1099–1154. https://doi.org/10.1002/2014RG000478.

Dong, L., & McPhaden, M. J. (2016). Interhemispheric SST gradient trends in the Indian Ocean prior to and during the recent global warming hiatus. *Journal of Climate*, *29*, 9077–9095.

Dong, L., Zhou, T., & Wu, B. (2014). Indian Ocean warming during 1958–2004 simulated by a climate system model and its mechanism. *Climate Dynamics*, *42*, 203–217. https://doi.org/10.1007/s00382-013-1722-z.

Drushka, K., Sprintall, J., Gille, S. T., & Wijffels, S. (2012). In situ observations of Madden–Julian oscillation mixed layer dynamics in the Indian and Western Pacific Oceans. *Journal of Climate*, *25*, 2306–2328. https://doi.org/10.1175/JCLI-D-11-00203.1.

Du, Y., & Xie, S. P. (2008). Role of atmospheric adjustments in the TIO warming during the 20th century in climate models. *Geophysical Research Letters*, *35*, L08712.

Du, Y., Zhang, Y., Feng, M., Wang, T., Zhang, N., & Wijffels, S. (2015). Decadal trends of the upper ocean salinity in the tropical Indo-Pacific since mid-1990s. *Scientific Reports*, *5*, 16050.

Düing, W., Molinari, R. L., & Swallow, J. C. (1980). Somali Current: Evolution of surface flow. *Science*, *209*(4456), 588–590.

Elipot, S., & Beal, L. M. (2015). Characteristics, energetics, and origins of agulhas current meanders and their limited influence on ring shedding. *Journal of Physical Oceanography*, *45*, 2294–2314. https://doi.org/10.1175/JPO-D-14-0254.1.

Elipot, S., Lumpkin, R., Perez, R. C., Lilly, J. M., Early, J. J., & Sykulski, A. M. (2016). A global surface drifter data set at hourly resolution. *Journal of Geophysical Research, Oceans*, *121*, 2937–2966. https://doi.org/10.1002/2016JC011716.

England, M. H., McGregor, S., Spence, P., Meehl, G. A., Timmermann, A., Cai, W., Sen Gupta, A., McPhaden, M. J., Purich, A., & Santoso, A. (2014). Recently intensified Pacific Ocean wind–driven circulation and the ongoing warming hiatus. *Nature Climate Change*, *4*, 222–227. https://doi.org/10.1038/nclimate2106.

Felton, C. S., Subrahmanyam, B., Murty, V. S. N., & Shriver, J. F. (2014). Estimation of the barrier layer thickness in the Indian Ocean using aquarius salinity. *Journal of Geophysical Research, Oceans*, *119*(7), 4200–4213. https://doi.org/10.1002/2013JC009759.

Feng, M., Benthuysen, J., Zhang, N., & Slawinski, D. (2015). Freshening anomalies in the Indonesian throughflow and impacts on the Leeuwin Current during 2010–2011. *Geophysical Research Letters, 42*, 8555–8562.

Feng, M., Böning, C., Biastoch, A., Behrens, E., Weller, E., & Masumoto, Y. (2011). The reversal of the multidecadal trends of the equatorial Pacific easterly winds, and the Indonesian throughflow and Leeuwin Current transports. *Geophysical Research Letters, 38*, L11604. https://doi.org/10.1029/2011GL047291.

Feng, M., Duan, Y., Wijffels, S., Hsu, J., Li, C., Wang, H., Yang, Y., Shen, H., Liu, J., Ning, C., & Yu, W. (2020). Tracking air–sea exchange and upper-ocean variability in the Indonesian–Australian Basin during the onset of the 2018/19 Australian summer monsoon. *Bulletin of the American Meteorological Society, 101*(8), E1397–E1412. https://journals.ametsoc.org/view/journals/bams/101/8/bamsD190278.xml.

Feng, M., Lengaigne, M., Manneela, S., Gupta, A. S., & Vialard, J. (2024). Chapter 6: Extreme events in the Indian Ocean: Marine heatwaves, cyclones, and tsunamis. In C. C. Ummenhofer, & R. R. Hood (Eds.), *The Indian Ocean and its role in the global climate system* (pp. 121–144). Amsterdam: Elsevier. https://doi.org/10.1016/B978-0-12-822698-8.00011-1.

Feng, S., & Hu, Q. (2008). How the North Atlantic multidecadal oscillation may have influenced the Indian summer monsoon during the past two millennia. *Geophysical Research Letters, 35*, L01707. https://doi.org/10.1029/2007GL032484.

Feng, M., McPhaden, M. J., & Lee, T. (2010). Decadal variability of the Pacific subtropical cells and their influence on the Southeast Indian Ocean. *Geophysical Research Letters, L09606*. https://doi.org/10.1029/2010GL042796.

Feng, M., McPhaden, M. J., Xie, S., & Hafner, J. (2013). La Nina forces unprecedented Leeuwin Current warming in 2011. *Scientific Reports, 3*. https://doi.org/10.1038/srep01277.

Foltz, G. R., Vialard, J., Kumar, P., & McPhaden, M. J. (2010). Seasonal mixed layer heat balance of the southwestern tropical Indian Ocean. *Journal of Climate, 23*, 947–965.

Fu, L.-L. (2007). Intraseasonal variability of the equatorial Indian Ocean observed from sea surface height, wind and temperature data. *Journal of Physical Oceanography, 37*, 188–202. https://doi.org/10.1175/JPO3006.1.

Fu, L.-L., Lee, T., Liu, W. T., & Kwok, R. (2019). Fifty years of satellite remote sensing of the ocean. *Meteorological Monographs, 59*, 5.1–5.46. https://doi.org/10.1175/AMSMONOGRAPHS-D-18-0010.1.

Gadgil, S., & Gadgil, S. (2006). The Indian monsoon, GDP, and agriculture. *Economic and Political Weekly, 41*(57), 4887–4895.

Gadgil, S., Vinayachandran, P. N., Francis, P. A., & Gadgil, S. (2004). Extremes of the Indian summer monsoon rainfall, ENSO and equatorial Indian Ocean oscillation. *Geophysical Research Letters, 31*, L12213.

Gelaro, R., McCarty, W., Suárez, M. J., Todling, R., Molod, A., et al. (2017). The modern-era retrospective analysis for research and applications, version 2 (MERRA-2). *Journal of Climate, 30*, 5419–5454. https://doi.org/10.1175/JCLI-D-16-0758.1.

Giannini, A., Saravanan, R., & Chang, P. (2003). Oceanic forcing of Sahel rainfall on interannual to interdecadal time scales. *Science, 302*, 1027–1030.

Girishkumar, M. S., Ashin, K., McPhaden, M. J., Balaji, B., & Praveenkumar, B. (2020). Estimation of vertical heat diffusivity at the base of the mixed layer in the Bay of Bengal. *Journal of Geophysical Research, 125*. https://doi.org/10.1029/2019JC015402. e2019JC015402.

Girishkumar, M. S., Joseph, P., Thangaprakash, N. P., Vijay, P., & McPhaden, M. J. (2017). Mixed layer temperature budget for the northward propagating summer monsoon intraseasonal oscillation (MISO) in the central Bay of Bengal. *Journal of Geophysical Research, 122*, 8841–8854. https://doi.org/10.1002/2017JC013073.

Girishkumar, M. S., & Ravichandran, M. (2012). The influences of ENSO on tropical cyclone activity in the Bay of Bengal during October–December. *Journal of Geophysical Research, 117*(C2), C02033. https://doi.org/10.1029/2011JC007417. https://doi-org.proxy-um.researchport.umd.edu/10.1029/2011JC007417.

Girishkumar, M. S., Ravichandran, M., & Han, W. (2013a). Observed intraseasonal thermocline variability in the Bay of Bengal. *Journal of Geophysical Research, Oceans, 118*. https://doi.org/10.1002/jgrc.20245.

Girishkumar, M. S., Ravichandran, M., & McPhaden, M. J. (2013b). Temperature inversions and their influence on the mixed layer heat budget during the winters of 2006–2007 and 2007–2008 in the Bay of Bengal. *Journal of Geophysical Research, 118*, 2426–2437. https://doi.org/10.1002/jgrc.20192.

Girishkumar, M. S., Ravichandran, M., McPhaden, M. J., & Rao, R. R. (2011). Intraseasonal variability in barrier layer thickness in the south central Bay of Bengal. *Journal of Geophysical Research, 116*, C03009. https://doi.org/10.1029/2010JC006657.

Girishkumar, M. S., Thangaprakash, V. P., Udaya Bhaskar, T. V. S., Suprit, K., Sureshkumar, N., Baliarsingh, S. K., Jofia, J., Pant, V., Vishnu, S., George, G., Abhilash, K. R., & Shivaprasad, S. (2019). Quantifying tropical cyclone's effect on the biogeochemical processes using profiling float observations in the Bay of Bengal. *Journal of Geophysical Research, 124*, 1945–1963. https://doi.org/10.1029/2017JC013629.

Gnanaseelan, C., Deshpande, A., & McPhaden, M. J. (2012). Impact of Indian Ocean dipole and El Niño/southern oscillation forcing on the Wyrtki jets. *Journal of Geophysical Research, 117*, C08005. https://doi.org/10.1029/2012JC007918.

Godfrey, J. S. (1996). The effect of the Indonesian throughflow on ocean circulation and heat exchange with the atmosphere: A review. *Journal of Geophysical Research, 101*, 12217–12237.

Goes, J. I., Thoppil, P. G., Gomes, H. D. R., & Fasullo, J. T. (2005). Warming of the Eurasian landmass is making the Arabian Sea more productive. *Science, 308*(5721), 545. https://doi.org/10.1126/science.1106610.

Gopika, S., Izumo, T., Vialard, J., et al. (2020). Aliasing of the Indian Ocean externally-forced warming spatial pattern by internal climate variability. *Climate Dynamics, 54*, 1093–1111. https://doi.org/10.1007/s00382-019-05049-9.

Goswami, B. N., & Mohan, R. S. A. (2001). Intraseasonal oscillations and interannual variability of the Indian summer monsoon. *Journal of Climate, 14*(6), 1180–1198. Retrieved March 13, 2022, from https://journals.ametsoc.org/view/journals/clim/14/6/1520-0442_2001_014_1180_ioaivo_2.0.co_2.xml.

Goswami, B. N., Rao, S. A., Sengupta, D., & Chakravorty, S. (2016). Monsoons to mixing in the Bay of Bengal: Multiscale air-sea interactions and monsoon predictability. *Oceanography, 29*(2), 18–27. https://doi.org/10.5670/oceanog.2016.35.

Gould, J., Roemmich, D., Wijffels, S., Freeland, H., Ignaszewsky, M., Jianping, X., et al. (2004). Argo profiling floats bring new era of in situ ocean observations. *Eos, Transactions American Geophysical Union, 85*(19), 185–191.

Gunn, K. L., Beal, L. M., Elipot, S., McMonigal, K., & Houk, A. (2020). Mixing of subtropical, central, and intermediate waters driven by shifting and pulsing of the Agulhas Current. *Journal of Physical Oceanography, 50*(12), 3545–3560. https://journals.ametsoc.org/view/journals/phoc/50/12/jpo-d-20-0093.1.xml.

Han, W., Meehl, G. A., Rajagopalan, B., et al. (2010). Patterns of Indian Ocean sea-level change in a warming climate. *Nature Geoscience, 3*, 546–550. https://doi.org/10.1038/NGEO901.

Han, W., Stammer, D., Meehl, G. A., Hu, A., Sienz, F., & Zhang, L. (2018). Multi-decadal trend and decadal variability of the Regional Sea level over the Indian Ocean since the 1960s: Roles of climate modes and external forcing. *Climate, 6*, 51. https://doi.org/10.3390/cli6020051.

Han, W., Vialard, J., McPhaden, M. J., Lee, T., Masomoto, Y., Feng, M., & de Ruijter, W. (2014). Indian Ocean decadal variability: A review. *Bulletin of the American Meteorological Society, 95*, 1679–1703.

Han, W., Yuan, D., Liu, W. T., & Halkides, D. (2007). Intraseasonal variability of Indian Ocean sea surface temperature during boreal winter: Madden-Julian oscillation versus submonthly forcing and processes. *Journal of Geophysical Research, 112*, C04001. https://doi.org/10.1029/2006JC003791.

Hersbach, H., Bell, B., Berrisford, P., Hirahara, S., Horányi, A., et al. (2020). The ERA5 global reanalysis. *Quarterly Journal of the Royal Meteorological Society, 146*, 1999–2049. https://doi.org/10.1002/qj.3803.

Hong, C., Li, T., & Kug, J. (2008). Asymmetry of the Indian Ocean dipole. Part I: Observational analysis. *Journal of Climate, 21*(18), 4834–4848. https://journals.ametsoc.org/view/journals/clim/21/18/2008jcli2222.1.xml.

Hood, R. R., Coles, V. J., Huggett, J. A., Landry, M. R., Levy, M., Moffett, J. W., & Rixen, T. (2024a). Chapter 13: Nutrient, phytoplankton, and zooplankton variability in the Indian Ocean. In C. C. Ummenhofer, & R. R. Hood (Eds.), *The Indian Ocean and its role in the global climate system* (pp. 293–327). Amsterdam: Elsevier. https://doi.org/10.1016/B978-0-12-822698-8.00020-2.

Hood, R. R., Rixen, T., Levy, M., Hansell, D. A., Coles, V. J., & Lachkar, Z. (2024b). Chapter 12: Oxygen, carbon, and pH variability in the Indian Ocean. In C. C. Ummenhofer, & R. R. Hood (Eds.), *The Indian Ocean and its role in the global climate system* (pp. 265–291). Amsterdam: Elsevier. https://doi.org/10.1016/B978-0-12-822698-8.00017-2.

Horii, T., Hase, H., Ueki, I., & Masumoto, Y. (2008). Oceanic precondition and evolution of the 2006 Indian Ocean dipole. *Geophysical Research Letters, 35*, L03607. https://doi.org/10.1029/2007GL032464.

Horii, T., Masumoto, Y., Ueki, I., Hase, H., & Mizuno, K. (2009). Mixed layer temperature balance in the eastern Indian Ocean during the 2006 Indian Ocean dipole. *Journal of Geophysical Research, 114*, C07011. https://doi.org/10.1029/2008JC005180.

Horii, T., Masumoto, Y., Ueki, I., Kumar, S. P., & Mizuno, K. (2011). Intraseasonal vertical velocity variation caused by the equatorial wave in the central equatorial Indian Ocean. *Journal of Geophysical Research, 116*, C09005. https://doi.org/10.1029/2011JC007081.

Horii, T., Mizuno, K., Nagura, M., Miyama, T., & Ando, K. (2013a). Seasonal and interannual variation in the cross-equatorial meridional currents observed in the eastern Indian Ocean. *Journal of Geophysical Research, 118*, 6658–6671. https://doi.org/10.1002/2013JC009291.

Horii, T., Ueki, I., Ando, K., & Mizuno, K. (2013b). Eastern Indian Ocean warming associated with the negative Indian Ocean dipole: A case study of the 2010 event. *Journal of Geophysical Research, 118*, 536–549. https://doi.org/10.1002/jgrc.20071.

Hu, S., & Fedorov, A. V. (2019). Indian Ocean warming can strengthen the Atlantic meridional overturning circulation. *Nature Climate Change, 9*, 747–751. https://doi.org/10.1038/s41558-019-0566-x.

Hu, S., & Sprintall, J. (2017). Observed strengthening of interbasin exchange via the Indonesian seas due to rainfall intensification. *Geophysical Research Letters, 44*, 1448–1456. https://doi.org/10.1002/2016GL072494.

Huang, K., Han, W., Wang, D., Wang, W., Xie, Q., Chen, J., & Chen, G. (2018a). Features of the equatorial intermediate current associated with basin resonance in the Indian Ocean. *Journal of Physical Oceanography, 48*, 1333–1347.

Huang, K., McPhaden, M. J., Wang, D., Wang, W., Xie, Q., Chen, J., et al. (2018b). Vertical propagation of middepth zonal currents associated with surface wind forcing in the equatorial Indian Ocean. *Journal of Geophysical Research: Oceans, 123*. https://doi.org/10.1029/2018JC013977.

Huang, K., Wang, D., Han, W., Feng, M., Chen, G., Wang, W., Chen, J., & Li, J. (2019). Semiannual variability of middepth zonal currents along 5°N in the eastern Indian Ocean: Characteristics and causes. *Journal of Physical Oceanography, 49*, 2715–2729. https://doi.org/10.1175/JPO-D-19-0089.1.

Hutchinson, K., Beal, L. M., Penven, P., Ansorge, I., & Hermes, J. (2018). Seasonal phasing of Agulhas Current transport tied to a baroclinic adjustment of near-field winds. *Journal of Geophysical Research: Oceans, 123*(10), 7067–7083.

International CLIVAR Project Office. (2006). *Understanding the role of the Indian Ocean in the climate system—Implementation plan for sustained observations*. International CLIVAR Project Office, CLIVAR Publication Series No. 100.

IPCC. (2019). In H.-O. Pörtner, D. C. Roberts, V. Masson-Delmotte, P. Zhai, M. Tignor, E. Poloczanska, ... N. Weyer (Eds.), *IPCC Special Report on the Ocean and Cryosphere in a Changing Climate*. https://www.ipcc.ch/srocc/chapter/summary-for-policymakers/.

Iskandar, I., Irfan, M., & Saymsuddin, F. (2013). Why was the 2008 Indian Ocean Dipole a short-lived event? *Ocean Science Journal, 48*, 149–160. https://doi.org/10.1007/s12601-013-0012-3. https://doi-org.proxy-um.researchport.umd.edu/10.1007/s12601-013-0012-3.

Iskandar, I., Masumoto, Y., & Mizuno, K. (2009). Subsurface equatorial zonal current in the eastern Indian Ocean. *Journal of Geophysical Research, 114*, C06005. https://doi.org/10.1029/2008JC005188.

Iskandar, I., & McPhaden, M. J. (2011). Dynamics of wind-forced intraseasonal zonal current variations in the equatorial Indian Ocean. *Journal of Geophysical Research, 116*, C06019. https://doi.org/10.1029/2010JC006864.

Izumo, T., de Boyer Montegut, C., Luo, J. J., Behera, S. K., Masson, V., & Yamagata, T. (2008). The role of the western Arabian Sea upwelling in Indian monsoon rainfall variability. *Journal of Climate, 21*, 5603–5623. https://doi.org/10.1175/2008JCLI2158.1.

Izumo, T., Vialard, J., Lengaigne, M., de Boyer Montegut, C., Behera, S. K., Luo, J. J., Cravatte, S., Masson, S., & Yamagata, T. (2010). Influence of the Indian Ocean dipole on following year's El Niño. *Nature Geoscience, 3*, 168–172.

Jarugula, S. L., & McPhaden, M. J. (2022). Ocean mixed layer response to two post-monsoon cyclones in the Bay of Bengal in 2018. *Journal of Geophysical Research, 127*, e2022JC018874. https://doi.org/10.1029/2022JC018874.

Jin, F.-F. (1997). An equatorial ocean recharge paradigm for ENSO. Part I: Conceptual model. *Journal of the Atmospheric Sciences, 54*, 811–829.

Johns, W. E., Baringer, M. O., Beal, L. M., Cunningham, S. A., Kanzow, T., Bryden, H. L., Hirschi, J. J. M., Marotzke, J., Meinen, C. S., Shaw, B., & Curry, R. (2011). Continuous, array-based estimates of Atlantic Ocean heat transport at 26.5°N. *Journal of Climate, 24*(10), 2429–2449. https://journals.ametsoc.org/view/journals/clim/24/10/2010jcli3997.1.xml.

Johnson, G. C. (2011). Deep signatures of southern tropical Indian Ocean annual Rossby waves. *Journal of Physical Oceanography, 41*(10), 1958–1964. https://doi.org/10.1175/JPO-D-11-029.1.

Joseph, J., Girishkumar, M. S., McPhaden, M. J., & Pattabhi Rama Rao, E. (2021). Diurnal variability of atmospheric cold pool events and associated air-sea interactions in the Bay of Bengal during the summer monsoon. *Climate Dynamics, 56*, 837–853. https://doi.org/10.1007/s00382-020-05506-w.

Joseph, S., Wallcraft, A. J., Jensen, T. G., Ravichandran, M., Shenoi, S. S. C., & Nayak, S. (2012). Weakening of spring Wyrtki jets in the Indian Ocean during 2006–2011. *Journal of Geophysical Research, 117*, C04012. https://doi.org/10.1029/2011JC007581.

Kido, S., & Tozuka, T. (2017). Salinity variability associated with the positive Indian Ocean dipole and its impact on the upper ocean temperature. *Journal of Climate, 30*(19), 7885–7907. https://journals.ametsoc.org/view/journals/clim/30/19/jcli-d-17-0133.1.xml.

Knauss, J. A., & Taft, B. A. (1964). Equatorial undercurrent of the Indian Ocean. *Science, 143*, 354–356.

Koné, V., Aumont, O., Levy, M., & Resplandy, L. (2009). Physical and biogeochemical controls of the phytoplankton seasonal cycle in the Indian Ocean: A modeling study. *American Geophysical Union Geophysical Monograph Series, 185*, 147–166. https://doi.org/10.1029/2008GM000700.

Kosaka, Y., & Xie, S.-P. (2013). Recent global-warming hiatus tied to equatorial Pacific surface cooling. *Nature, 501*, 403–407. https://doi.org/10.1038/nature12534.

Krishnamohan, K. S., Vialard, J., Lengaigne, M., Masson, S., Samson, G., Pous, S., Neetu, S., Durand, F., Shenoi, S. S. C., & Madec, G. (2019). Is there an effect of Bay of Bengal salinity on the northern Indian Ocean climatological rainfall? *Deep-Sea Research Part II, 166*, 19–33. https://doi.org/10.1016/j.dsr2.2019.04.003.

Krug, M., Swart, S., & Gula, J. (2017). Submesoscale cyclones in the Agulhas Current. *Geophysical Research Letters, 44*(1), 346–354.

Krug, M., & Tournadre, J. (2012). Satellite observations of an annual cycle in the Agulhas Current. *Geophysical Research Letters, 39*(15).

Kumar, P., Hamlington, B., Cheon, S.-. H., Han, W., & Thompson, P. (2020). 20th Century multivariate Indian Ocean regional sea level reconstruction. *Journal of Geophysical Research: Oceans, 125*. https://doi.org/10.1029/2020JC016270. e2020JC016270.

Kuttippurath, J., Akhila, R., Martin, M. V., et al. (2022). Tropical cyclone–induced cold wakes in the Northeast Indian Ocean. *Environmental Science: Atmospheres.* https://doi.org/10.1039/D1EA00066G.

Kwiatkowski, L., Torres, O., Bopp, L., Aumont, O., Chamberlain, M., Christian, J. R., Dunne, J. P., Gehlen, M., Ilyina, T., John, J. G., Lenton, A., Li, H., Lovenduski, N. S., Orr, J. C., Palmieri, J., Santana-Falcón, Y., Schwinger, J., Séférian, R., Stock, C. A., ... Ziehn, T. (2020). Twenty-first century ocean warming, acidification, deoxygenation, and upper-ocean nutrient and primary production decline from CMIP6 model projections. *Biogeosciences, 17*(13), 3439–3470. https://doi.org/10.5194/bg-17-3439-2020.

L'Hegaret, P., Beal, L. M., Elipot, S., & Laurindo, L. C. (2018). Shallow cross-equatorial gyres of the Indian Ocean driven by seasonally reversing monsoon winds. *Journal of Geophysical Research, 123*, 8902–8920. https://doi.org/10.1029/2018JC014553.

Leber, G. M., & Beal, L. M. (2014). Evidence that Agulhas Current transport is maintained during a meander. *Journal of Geophysical Research, Oceans, 119*, 3806–3817. https://doi.org/10.1002/2014JC009802.

Lee, T. (2004). Decadal weakening of the shallow overturning circulation in the South Indian Ocean. *Geophysical Research Letters, 31*, L18305. https://doi.org/10.1029/2004GL020884.

Lee, T., & McPhaden, M. J. (2008). Decadal phase change in large-scale sea level and winds in the Indo-Pacific region at the end of the 20th century. *Geophysical Research Letters, 35*, L01605. https://doi.org/10.1029/2007GL032419.

Lee, S.-K., Park, W., Baringer, M. O., Gordon, A. L., Huber, B., & Liu, Y. (2015). Pacific origin of the abrupt increase in Indian Ocean heat content during the warming hiatus. *Nature Geoscience, 8*(6), 445–449. https://doi.org/10.1038/ngeo2438.

Li, Y., Han, W., Ravichandran, M., Wang, W., Shinoda, T., & Lee, T. (2017a). Bay of Bengal salinity stratification and Indian summer monsoon intraseasonal oscillation: 1. Intraseasonal variability and causes. *Journal of Geophysical Research: Oceans, 122*(5), 4291–4311. https://doi.org/10.1002/2017jc012691.

Li, Y., Han, W., Wang, W., Ravichandran, M., Lee, T., & Shinoda, T. (2017b). Bay of Bengal salinity stratification and Indian summer monsoon intraseasonal oscillation: 2. Impact on SST and convection. *Journal of Geophysical Research: Oceans, 122*(5), 4312–4328. https://doi.org/10.1002/2017jc012692.

Liao, H., & Wang, C. (2021). Sea surface temperature anomalies in the western Indian Ocean as a trigger for Atlantic Niño events. *Geophysical Research Letters, 48.* e2021GL092489.

Lin, I.-I., Chen, C.-H., Pun, I.-F., Liu, W. T., & Wu, C.-C. (2009). Warm ocean anomaly, air sea fluxes, and the rapid intensification of tropical cyclone Nargis (2008). *Geophysical Research Letters, 36*, L03817. https://doi.org/10.1029/2008GL035815.

Liu, Q.-Y., Feng, M., Wang, D., & Wijffels, S. (2015). Interannual variability of the Indonesian throughflow transport: A revisit based on 30 year expendable bathythermograph data. *Journal of Geophysical Research, Oceans, 120*, 8270–8282. https://doi.org/10.1002/2015JC011351.

Llovel, W., & Lee, T. (2015). Importance and origin of halosteric contribution to sea level change in the southeast Indian Ocean during 2005-2013. *Geophysical Research Letters, 42*, 1148–1157. https://doi.org/10.1002/2014GL062611.

Lumpkin, R., & Johnson, G. C. (2013). Global ocean surface velocities from drifters: Mean, variance, El Nino–Southern Oscillation response, and seasonal cycle. *Journal of Geophysical Research, Oceans, 118*, 2992–3006. https://doi.org/10.1002/jgrc.20210.

Lutjeharms, J. R. E. (2006). *The Agulhas Current*. Berlin: Springer-Verlag. https://doi.org/10.1007/3-540-37212-1. 329pp.

Luyten, J. R., & Roemmich, D. H. (1982). Equatorial currents at semi-annual period in the Indian Ocean. *Journal of Physical Oceanography*, *12*, 406–413.

Mahala, B. K., Nayak, B. K., & Mohanty, P. K. (2015). Impacts of ENSO and IOD on tropical cyclone activity in the Bay of Bengal. *Natural Hazards*, *75*, 1105–1125. https://doi.org/10.1007/s11069-014-1360-8.

Maneesha, K., Murty, V. S. N., Ravichandran, M., Lee, T., Yu, W., & McPhaden, M. J. (2012). Upper ocean variability in the Bay of Bengal during the tropical cyclones Nargis and Laila. *Progress in Oceanography*, *106*, 49–61. https://doi-org.proxy-um.researchport.umd.edu/10.1016/j.pocean.2012.06.006.

Marsac, F., Everett, B., Shahid, U., & Strutton, P. G. (2024). Chapter 11: Indian Ocean primary productivity and fisheries variability. In C. C. Ummenhofer, & R. R. Hood (Eds.), *The Indian Ocean and its role in the global climate system* (pp. 245–264). Amsterdam: Elsevier. https://doi.org/10.1016/B978-0-12-822698-8.00019-6.

Masumoto, Y., Hase, H., Kuroda, Y., Matsuura, H., & Takeuchi, K. (2005). Intraseasonal variability in the upper layer currents observed in the eastern equatorial Indian Ocean. *Geophysical Research Letters*, *32*, L02607. https://doi.org/10.1029/2004GL021896.

Masumoto, Y., Horii, T., Ueki, I., Hase, H., Ando, K., & Mizuno, K. (2008). Short-term upper-ocean variability in the central equatorial Indian Ocean during 2006 Indian Ocean Dipole event. *Geophysical Research Letters*, *35*(14), L14S09. https://doi.org/10.1029/2008GL033834. https://doi-org.proxy-um.researchport.umd.edu/10.1029/2008GL033834.

Mayer, M., Alonso Balmaseda, M., & Haimberger, L. (2018). Unprecedented 2015/2016 Indo-Pacific heat transfer speeds up tropical Pacific heat recharge. *Geophysical Research Letters*, *45*. https://doi.org/10.1002/2018GL077106.

McCarthy, G. D., Smeed, D. A., Johns, W. E., Frajka-Williams, E., Moat, B. I., Rayner, D., et al. (2015). Measuring the Atlantic meridional overturning circulation at 26 N. *Progress in Oceanography*, *130*, 91–111.

McDonagh, E. L., Bryden, H. L., King, B. A., Sanders, R. J., Cunningham, S. A., & Marsh, R. (2005). Decadal changes in the South Indian Ocean thermocline. *Journal of Climate*, *18*(10), 1575–1590.

McMonigal, K., Beal, L. M., Elipot, S., Gunn, K. L., Hermes, J., Morris, T., & Houk, A. (2020). The impact of meanders, deepening and broadening, and seasonality on Agulhas current temperature variability. *Journal of Physical Oceanography*, *50*(12), 3529–3544. https://journals.ametsoc.org/view/journals/phoc/50/12/jpo-d-20-0018.1.xml.

McMonigal, K., Beal, L. M., & Willis, J. K. (2018). The seasonal cycle of the South Indian Ocean subtropical gyre circulation as revealed by Argo and satellite data. *Geophysical Research Letters*, *45*, 9034–9041. https://doi.org/10.1029/2018GL078420.

McMonigal, K., Gunn, K. L., Beal, L. M., Elipot, S., & Willis, J. K. (2022). Reduction in meridional heat export contributes to recent Indian Ocean warming. *Journal of Physical Oceanography*, *52*(3), 329–345. https://journals.ametsoc.org/view/journals/phoc/52/3/JPO-D-21-0085.1.xml.

McPhaden, M. J. (2020). *Reflections on the origins of the Indian Ocean observing system (IndOOS)*. CLIVAR exchanges, no. 78 (February 2020) (pp. 64–73). https://doi.org/10.36071/clivar.78.2020.

McPhaden, M. J., Busalacchi, A. J., & Anderson, D. L. T. (2010). A TOGA retrospective. *Oceanography*, *23*, 86–103.

McPhaden, M. J., & Foltz, G. R. (2013). Intraseasonal variations in the surface layer heat balance of the central equatorial Indian Ocean: The importance of zonal advection and vertical mixing. *Geophysical Research Letters*, *40*, 1–5. https://doi.org/10.1029/GL056092.

McPhaden, M. J., Foltz, G. R., Lee, T., Murty, V. S. N., Ravichandran, M., Vecchi, G. A., … Yu, L. (2009a). Ocean-atmosphere interactions during cyclone Nargis. *Eos, Transactions of the American Geophysical Union*, *90*, 53–54.

McPhaden, M. J., Meyers, G., Ando, K., Masumoto, Y., Murty, V. S. N., Ravichandran, M., et al. (2009b). RAMA the research moored array for African-Asian-Australian monsoon analysis and prediction. *Bulletin of the American Meteorological Society*, *90*, 459–480. https://doi.org/10.1175/2008BAMS2608.1.

McPhaden, M. J., & Nagura, M. (2014). Indian Ocean dipole interpreted in terms of recharge oscillator theory. *Climate Dynamics*, *42*, 1569–1586. https://doi.org/10.1007/s00382-013-1765-1.

McPhaden, M. J., Wang, Y., & Ravichandran, M. (2015). Volume transports of the Wyrtki jets and their relationship to the Indian Ocean dipole. *Journal of Geophysical Research*, *120*. https://doi.org/10.1002/2015JC010901.

Meinen, C. S., & McPhaden, M. J. (2000). Observations of warm water volume changes in the equatorial Pacific and their relationship to El Niño and La Niña. *Journal of Climate*, *13*, 3551–3559.

Meng, L., Zhuang, W., Zhang, W., Yan, C., & Yan, X.-. H. (2020). Variability of the shallow overturning circulation in the Indian Ocean. *Journal of Geophysical Research: Oceans*, *125*. https://doi.org/10.1029/2019JC015651. e2019JC015651.

Miyama, T., McCreary, J. P., Jensen, T. G., Loschnigg, J., Godfrey, S., & Ishida, A. (2003). Structure and dynamics of the Indian-Ocean crossequatorial cell. *Deep Sea Research, Part I*, *50*(12), 2023–2047.

Miyama, T., Sengupta, D., & Senan, R. (2006). Dynamics of biweekly oscillations in the equatorial Indian Ocean. *Journal of Physical Oceanography*, *36*, 827–846.

Moihamette, F., Pokam, W. M., Diallo, I., & Washington, R. (2022). Extreme Indian Ocean dipole and rainfall variability over Central Africa. *International Journal of Climatology*, 1–18. https://doi.org/10.1002/joc.7531.

Moore, D. W., & Philander, S. G. H. (1977). Modeling the tropical ocean circulation. In *Vol. 6. The Sea* (pp. 319–361). Wiley Interscience.

Moum, J. N., de Szoeke, S. P., Smyth, W., DEdson, J. B., DeWitt, H. L., Moullin, A. J., Thompson, E. J., Zappa, C. J., Rutledge, S. A., Johnson, R. H., & Fairall, C. W. (2014). Air-sea interactions from westerly wind bursts during the November 2011 MJO in the Indian Ocean. *Bulletin of the American Meteorological Society*, *95*, 1185–1199.

Murakami, H., Vecchi, G. A., & Underwood, S. (2017). Increasing frequency of extremely severe cyclonic storms over the Arabian Sea. *Nature Climate Change*, *7*, 885–889. https://doi.org/10.1038/s41558-017-0008-6.

Murtugudde, R., & Busalacchi, A. J. (1999). Interannual variability of the dynamics and thermodynamics of the tropical Indian Ocean. *Journal of Climate*, *12*(8), 2300–2326.

Murtugudde, R., McCreary, J. P., & Busalacchi, A. J. (2000). Oceanic processes associated with anomalous events in the Indian Ocean with relevance to 1997–1998. *Journal of Geophysical Research: Oceans, 105*(C2), 3295–3306.

Nagura, M. (2018). Annual Rossby waves below the pycnocline in the Indian Ocean. *Journal of Geophysical Research: Oceans, 123*, 9405–9415. https://doi.org/10.1029/2018JC014362.

Nagura, M. (2020). Variability in meridional transport of the subtropical circulation in the South Indian Ocean for the period from 2006 to 2017. *Journal of Geophysical Research: Oceans, 124*. https://doi.org/10.1029/2019JC015874. e2019JC015874.

Nagura, M., & Masumoto, Y. (2015). A wake due to the Maldives in the eastward Wyrtki jet. *Journal of Physical Oceanography, 45*, 1858–1876.

Nagura, M., Masumoto, Y., & Horii, T. (2014). Meridional heat advection due to mixed Rossby gravity waves in the equatorial Indian Ocean. *Journal of Physical Oceanography, 44*, 343–358.

Nagura, M., & McPhaden, M. J. (2008). The dynamics of zonal current variations in the central equatorial Indian Ocean. *Geophysical Research Letters, 35*, L23603. https://doi.org/10.1029/2008GL035961.

Nagura, M., & McPhaden, M. J. (2010a). Wyrtki jet dynamics: Seasonal variability. *Journal of Geophysical Research, 115*, C07009. https://doi.org/10.1029/2009JC005922.

Nagura, M., & McPhaden, M. J. (2010b). Dynamics of zonal current variations associated with the Indian Ocean dipole. *Journal of Geophysical Research, 115*, C11026. https://doi.org/10.1029/2010JC006423.

Nagura, M., & McPhaden, M. J. (2012). The dynamics of wind-driven intraseasonal variability in the equatorial Indian Ocean. *Journal of Geophysical Research, 117*, C02001. https://doi.org/10.1029/2011JC007405.

Nagura, M., & McPhaden, M. J. (2016). Zonal propagation of near-surface zonal currents in relation to surface wind forcing in the equatorial Indian Ocean. *Journal of Physical Oceanography, 46*, 3623–3638.

Nagura, M., & McPhaden, M. J. (2018). The shallow overturning circulation in the Indian Ocean. *Journal of Physical Oceanography, 48*, 413–434. https://doi.org/10.1175/JPO-D-17-0127.1.

Nagura, M., & McPhaden, M. J. (2021). Interannual variability in sea surface height at southern mid-latitudes of the Indian Ocean. *Journal of Physical Oceanography, 51*, 1595–1609. https://doi.org/10.1175/JPO-D-20-0279.1.

Nagura, M., & McPhaden, M. J. (2023). Dual-frequency wind-driven mixed Rossby-gravity waves in the equatorial Indian Ocean. *Journal of Physical Oceanography, 53*(6), 1535–1553. https://doi.org/10.1175/JPO-D-22-0222.1.

Nagura, M., Terao, T., & Hashizume, M. (2015). The role of temperature inversions in the generation of seasonal and interannual SST variability in the far northern Bay of Bengal. *Journal of Climate, 28*, 3671–3693. https://doi.org/10.1175/JCLI-D-14-00553.1.

Neetu, S., Lengaigne, M., Menon, H. B., Vialard, J., Mangeas, M., Menkès, C. E., Ali, M. M., Suresh, I., & Knaff, J. A. (2017). Global assessment of tropical cyclone intensity statistical-dynamical hindcasts. *Quarterly Journal of the Royal Meteorological Society, 143*, 2143–2156. https://doi.org/10.1002/qj.3073.

Neetu, S., Lengaigne, M., Vincent, E. M., Vialard, J., Madec, G., Samson, G., Ramesh Kumar, M. R., & Durand, F. (2012). Influence of upper-ocean stratification on tropical cyclone-induced surface cooling in the Bay of Bengal. *Journal of Geophysical Research, 117*, C12020. https://doi.org/10.1029/2012JC008433.

Nieves, V., Willis, J. K., & Patzert, W. C. (2015). Recent hiatus caused by decadal shift in Indo-Pacific heating. *Science, 349*(6247), 532–535. https://doi.org/10.1126/science.aaa4521. https://doi-org.proxy-um.researchport.umd.edu/10.1126/science.aaa4521.

Niiler, P. P., Sybrandy, A. S., Bi, K., Poulain, P. M., & Bitterman, D. (1995). Measurements of the water-following capability of holey-sock and TRISTAR drifters. *Deep-Sea Research, Part 1, 42*, 1951–1955.

Nyadjro, E., & McPhaden, M. J. (2014). Variability of zonal currents in the eastern equatorial Indian Ocean on seasonal to interannual time scales. *Journal of Geophysical Research, Oceans, 119*, 7969–7986. https://doi.org/10.1002/2014JC010380.

Ogata, T., Nagura, M., & Masumoto, Y. (2017). Mean subsurface upwelling induced by intraseasonal variability over the equatorial Indian Ocean. *Journal of Physical Oceanography, 47*, 1347–1365. https://doi.org/10.1175/JPO-D-16-0257.1.

Ogata, T., Sasaki, H., Murty, V. S. N., Sarma, M. S. S., & Masumoto, Y. (2008). Intraseasonal meridional current variability in the eastern equatorial Indian Ocean. *Journal of Geophysical Research, 113*, C07037. https://doi.org/10.1029/2007JC004331.

Ogata, T., Xie, S.-P., Lan, J., & Zheng, X. (2013). Importance of ocean dynamics for the skewness of the Indian Ocean dipole mode. *Journal of Climate, 26*, 2145–2159. https://doi.org/10.1175/JCLI-D-11-00615.1.

Oppenheimer, M., Glavovic, B. C., Hinkel, J., van de Wal, R., Magnan, A. K., et al. (2019). Sea level rise and implications for low-lying islands, coasts and communities. In H.-O. Pörtner, D. C. Roberts, V. Masson-Delmotte, P. Zhai, M. Tignor, E. Poloczanska, ... N. M. Weyer (Eds.), *IPCC special report on the ocean and cryosphere in a changing climate*. https://www.ipcc.ch/srocc/.

Parampil, S. R., Bharathraj, G. N., Harrison, M., & Sengupta, D. (2016). Observed subseasonal variability of heat flux and the SST response of the tropical Indian Ocean. *Journal of Geophysical Research, Oceans, 121*, 7290–7307. https://doi.org/10.1002/2016JC011948.

Pérez-Hernández, M. D., Hernández-Guerra, A., Joyce, T. M., & Vélez-Belchí, P. (2012). Wind-driven cross-equatorial flow in the Indian Ocean. *Journal of Physical Oceanography, 42*, 2234–2253. https://doi.org/10.1175/JPO-D-12-033.1.

Phillips, H. E., Menezes, V. V., Nagura, M., McPhaden, M. J., Vinayachandran, P. N., & Beal, L. M. (2024). Chapter 8: Indian Ocean circulation. In C. C. Ummenhofer, & R. R. Hood (Eds.), *The Indian Ocean and its role in the global climate system* (pp. 169–203). Amsterdam: Elsevier. https://doi.org/10.1016/B978-0-12-822698-8.00012-3.

Pokhrel, S., Dutta, U., Rahaman, H., Chaudhari, H., Hazra, A., Saha, S. K., & Veeranjaneyulu, C. (2020). Evaluation of different heat flux products over the tropical Indian Ocean. *Earth and Space Science, 7*. https://doi.org/10.1029/2019EA000988. e2019EA000988.

Prasanna Kumar, S., Divya David, T., Byju, P., Narvekar, J., Yoneyama, K., Nakatani, N., Ishida, A., Horii, T., Masumoto, Y., & Mizuno, K. (2012). Bio-physical coupling and ocean dynamics in the central equatorial Indian Ocean during the 2006 Indian Ocean dipole. *Geophysical Research Letters, 39*, L14601. https://doi.org/10.1029/2012GL052609.

Prasanna Kumar, S., Narvekar, J., Nuncio, M., Gauns, M., & Sardesai, S. (2009). What drives the biological productivity of the northern Indian Ocean? *Geophysical Monograph Series, 185*, 33–56. https://doi.org/10.1029/2008GM000757.

Praveen Kumar, B., Vialard, J., Lengaigne, M., Murty, V. S. N., Foltz, G. R., McPhaden, M. J., Pous, S., & Montegut, C. D. (2014). Processes of interannual mixed layer temperature variability in the thermocline ridge of the Indian Ocean. *Climate Dynamics, 43*, 2377–2397.

Praveen Kumar, B., Vialard, J., Lengaigne, M., Murty, V. S. N., & McPhaden, M. J. (2012). Tropflux: Air-sea fluxes for the global tropical oceans-description and evaluation. *Climate Dynamics, 38*(7–8), 1521–1543. https://doi.org/10.1007/s00382-011-1115-0.

Praveen Kumar, B., Vialard, J., Lengaigne, M., Murty, V. S. N., McPhaden, M. J., Cronin, M. F., et al. (2013). TropFlux wind stresses over the tropical oceans: Evaluation and comparison with other products. *Climate Dynamics, 40*, 2049–2071. https://doi.org/10.1007/s00382-012-1455-4.

Pujiana, K., & McPhaden, M. J. (2018). Ocean's response to the convectively coupled Kelvin waves in the eastern equatorial Indian Ocean. *Journal of Geophysical Research, 123*, 5727–5741. https://doi.org/10.1029/2018JC013858.

Pujiana, K., & McPhaden, M. J. (2020). Intraseasonal Kelvin waves in the equatorial Indian Ocean and their propagation into the Indonesian seas. *Journal of Geophysical Research, 25*. https://doi.org/10.1029/2019JC015839.

Pujiana, K., & McPhaden, M. J. (2021). Biweekly mixed Rossby-gravity waves in the equatorial Indian Ocean. *Journal of Geophysical Research, 126*, e2020JO016840. https://doi.org/10.1029/2020JC016840.

Pujiana, K., McPhaden, M. J., Gordon, A. L., & Napitu, A. M. (2019). Unprecedented response of Indonesian throughflow to anomalous indo-Pacific climatic forcing in 2016. *Journal of Geophysical Research, 124*, 3737–3754. https://doi.org/10.1029/2018JC014574.

Qiu, Y., Cai, W., Li, L., & Guo, X. (2012). Argo profiles variability of barrier layer in the tropical Indian Ocean and its relationship with the Indian Ocean dipole. *Geophysical Research Letters, 39*, L08605. https://doi.org/10.1029/2012GL051441.

Qiu, Y., Han, W., Lin, X., West, B. J., Li, Y., Xing, W., Zhang, X., Arulananthan, K., & Guo, X. (2019). Upper-ocean response to the super tropical cyclone Phailin (2013) over the freshwater region of the Bay of Bengal. *Journal of Physical Oceanography, 49*, 1201–1228. https://journals.ametsoc.org/view/journals/phoc/49/5/jpo-d-18-0228.1.xml.

Qiu, Y., Li, L., & Yu, W. (2009). Behavior of the Wyrtki jet observed with surface drifting buoys and satellite altimeter. *Geophysical Research Letters, 36*, L18607. https://doi.org/10.1029/2009GL039120.

Rahul, S., & Gnanaseelan, C. (2016). Can large scale surface circulation changes modulate the sea surface warming pattern in the tropical Indian Ocean? *Climate Dynamics, 46*, 3617–3632. https://doi.org/10.1007/s00382-015-2790-z.

Rao, S. A., Dhakate, A. R., Saha, S. K., Mahapatra, S., Chaudhari, H. S., Pokhrel, S., & Sahu, S. K. (2011). Why is Indian Ocean warming consistently? *Climatic Change, 110*(3–4), 709–719.

Rao, S. A., Luo, J. J., Behera, S. K., et al. (2009). Generation and termination of Indian Ocean dipole events in 2003, 2006 and 2007. *Climate Dynamics, 33*, 751–767. https://doi.org/10.1007/s00382-008-0498-z.

Reid, J. L. (2003). On the total geostrophic circulation of the Indian Ocean: Flow patterns, tracers, and transports. *Progress in Oceanography, 56*, 137–186. https://doi.org/10.1016/S0079-6611(02)00141-6.

Rochford, D. (1964). Hydrology of the Indian Ocean. III. Water masses of the upper 500 metres of the south-East Indian Ocean. *Australian Journal of Marine & Freshwater Research, 15*, 25–55. https://doi.org/10.1071/MF9640025.

Roman-Stork, H. L., Subrahmanyam, B., & Trott, C. B. (2020). Monitoring intraseasonal oscillations in the Indian Ocean using satellite observations. *Journal of Geophysical Research: Oceans, 125*. https://doi.org/10.1029/2019JC015891. e2019JC015891.

Roxy, M. K., Ghosh, S., Pathak, A., et al. (2017). A threefold rise in widespread extreme rain events over Central India. *Nature Communications, 8*, 708. https://doi.org/10.1038/s41467-017-00744-9.

Roxy, M. K., Modi, A., Murtugudde, R., Valsala, V., Panickal, S., Prasanna Kumar, S., Ravichandran, M., Vichi, M., & Lévy, M. (2016). A reduction in marine primary productivity driven by rapid warming over the tropical Indian Ocean. *Geophysical Research Letters, 43*, 826–833. https://doi.org/10.1002/2015GL066979.

Roxy, M. K., Ritika, K., Terray, P., & Masson, S. (2014). The curious case of Indian Ocean warming. *Journal of Climate, 27*(22), 8501–8509. https://doi.org/10.1175/jcli-d-14-00471.1.

Roxy, M. K., et al. (2020). Indian Ocean warming. In R. Krishnan, J. Sanjay, C. Gnanaseelan, M. Mujumdar, A. Kulkarni, & S. Chakraborty (Eds.), *Assessment of climate change over the Indian region*. Singapore: Springer. https://doi.org/10.1007/978-981-15-4327-2_10.

Roxy, M. K., Saranya, J. S., Modi, A., Anusree, A., Cai, W., Resplandy, L.,…Frölicher, T. L. (2024). Chapter 20: Future projections for the tropical Indian Ocean. In C. C. Ummenhofer, & R. R. Hood (Eds.), *The Indian Ocean and its role in the global climate system* (pp. 469–482). Amsterdam: Elsevier. https://doi.org/10.1016/B978-0-12-822698-8.00004-4.

Saha, S., Moorthi, S., Pan, H. L., Wu, X., Wang, J., Nadiga, S., et al. (2010). The NCEP climate forecast system reanalysis. *Bulletin of the American Meteorological Society, 91*(8), 1015–1058. https://doi.org/10.1175/2010BAMS3001.1.

Saji, N. H., Goswami, B. N., Vinayachandran, P. N., & Yamagata, T. (1999). A dipole mode in the tropical Indian Ocean. *Nature, 401*, 360–363.

Sanchez-Franks, A., Kent, E. C., Matthews, A. J., Webber, B. G. M., Peatman, S. C., & Vinayachandran, P. N. (2018). Intraseasonal variability of air–sea fluxes over the Bay of Bengal during the southwest monsoon. *Journal of Climate, 31*(17), 7087–7109.

Sarma, V. V. S. S., Udaya Bhaskar, T. V. S., Pavan Kumar, J., & Chakraborty, K. (2020). Potential mechanisms responsible for occurrence of core oxygen minimum zone in the north-eastern Arabian Sea. *Deep Sea Research Part I: Oceanographic Research Papers, 165*. https://doi.org/10.1016/j.dsr.2020.103393.

Schott, F. (1983). Monsoon response of the Somali current and associated upwelling. *Progress in Oceanography, 12*(3), 357–381.

Schott, F. A., Dengler, M., & Schoenefeldt, R. (2002). The shallow overturning circulation of the Indian Ocean. *Progress in Oceanography, 53*, 57–103.

Schott, F. A., & McCreary, J. P. (2001). The monsoon circulation of the Indian Ocean. *Progress in Oceanography, 51*, 1–123.

Schott, F. A., Xie, S.-P., & McCreary, J. P., Jr. (2009). Indian Ocean circulation and climate variability. *Reviews of Geophysics, 47*, RG1002. https://doi.org/10.1029/2007RG000245.

Sengupta, D., Goddalehundi, B. R., & Anitha, D. S. (2008). Cyclone induced mixing does not cool SST in the post-monsoon Bay of Bengal. *Atmospheric Science Letters, 9*, 1–6. https://doi.org/10.1002/asl.162.

Sengupta, D., Senan, R., Murty, V. S. N., & Fernando, V. (2004). A biweekly mode in the equatorial Indian Ocean. *Journal of Geophysical Research, 109*(C10), C10003. https://doi.org/10.1029/2004JC002329. https://doi-org.proxy-um.researchport.umd.edu/10.1029/2004JC002329.

Shenoi, S. S. C., Shankar, D., & Shetye, S. R. (2002). Differences in heat budgets of the near-surface Arabian Sea and Bay of Bengal: Implications for the summer monsoon. *Journal of Geophysical Research, 107*(C6), 1–14.

Smyth, W. D., Durland, T. S., & Moum, J. N. (2015). Energy and heat fluxes due to vertically propagating Yanai waves observed in the equatorial Indian Ocean. *Journal of Geophysical Research, Oceans, 120*, 1–15. https://doi.org/10.1002/2014JC010152.

Song, X., Ning, C., Duan, Y., Wang, H., Li, C., Yang, Y., … Yu, W. (2021). Observed extreme air-sea heat flux variations during three tropical cyclones in the tropical southeastern Indian Ocean. *Journal of Climate, 34*, 3683–3705. https://doi.org/10.1175/JCLI-D-20-0170.1.

Sprintall, J., Biastoch, A., Gruenburg, L. K., & Phillips, H. E. (2024a). Chapter 9: Oceanic basin connections. In C. C. Ummenhofer, & R. R. Hood (Eds.), *The Indian Ocean and its role in the global climate system* (pp. 205–227). Amsterdam: Elsevier. https://doi.org/10.1016/B978-0-12-822698-8.00003-2.

Sprintall, J., Nagura, M., Hermes, J., Roxy, M. K., McPhaden, M. J., Pattabhi Ram Rao, E., Srinivasa Kumar, T., Thurston, S., Li, J., Belbeoch, M., & Turpin, V. (2024b). COVID Impacts cause critical gaps in the Indian Ocean observing system. *Bulletin of the American Meteorological Society*. https://doi.org/10.1175/BAMS-D-22-0270.1, in press.

Srinivasu, U., Ravichandran, M., Han, W., Sivareddy, S., Rahman, H., Li, Y., & Nayak, S. (2017). Causes for the reversal of North Indian Ocean decadal sea level trend in recent two decades. *Climate Dynamics, 49*, 3887–3904. https://doi.org/10.1007/s00382-017-3551-y.

Strutton, P. G., Coles, V. J., Hood, R. R., Matear, R. J., McPhaden, M. J., & Phillips, H. E. (2015). Biogeochemical variability in the central equatorial Indian Ocean during the monsoon transition. *Biogeosciences, 12*(8), 2367–2382. https://doi.org/10.5194/bg-12-2367-2015.

Subrahmanyam, B., Roman-Stork, H. L., & Murty, V. S. N. (2020). Response of the Bay of Bengal to 3-7-day synoptic oscillations during the southwest monsoon of 2019. *Journal of Geophysical Research: Oceans, 125*. https://doi.org/10.1029/2020JC016200. e2020JC016200.

Suresh, I., Vialard, J., Izumo, T., Lengaigne, M., Han, W., McCreary, J., & Muraleedharan, P. M. (2016). Dominant role of winds near Sri Lanka indriving seasonal sea level variations along the west coast of India. *Geophysical Research Letters, 43*, 7028–7035. https://doi.org/10.1002/2016GL069976.

Suresh, I., Vialard, J., Lengaigne, M., Han, W., McCreary, J., Durand, F., & Muraleedharan, P. M. (2013). Origins of wind-driven intraseasonalsea level variations in the North Indian Ocean coastal waveguide. *Geophysical Research Letters, 40*, 5740–5744. https://doi.org/10.1002/2013GL058312.

Suresh, I., Vialard, J., Lengaigne, M., Izumo, T., Parvathi, V., & Muraleedharan, P. M. (2018). Sea level interannualvariability along the west coast of India. *Geophysical Research Letters, 45*, 12440–12448. https://doi.org/10.1029/2018GL080972.

Swapna, P., Jyoti, J., Krishnan, R., Sandeep, N., & Griffies, S. M. (2017). Multidecadal weakening of Indian summer monsoon circulation induces an increasing northern Indian Ocean Sea level. *Geophysical Research Letters, 44*, 10560–10572. https://doi.org/10.1002/2017GL074706.

Tagliabue, A., Kwiatkowski, L., Bopp, L., Butenschön, M., Cheung, W., Lengaigne, M., & Vialard, J. (2021). Persistent uncertainties in ocean net primary production climate change projections at regional scales raise challenges for assessing impacts on ecosystem services. *Frontiers in Climate, 3*. https://doi.org/10.3389/fclim.2021.738224.

Talley, L. D., & Baringer, M. O. (1997). Preliminary results from WOCE hydrographic sections at 808E and 328S in the central Indian Ocean. *Geophysical Research Letters, 24*, 2789–2792. https://doi.org/10.1029/97GL02657.

Terray, P., Kamala, K., Masson, S., Madec, G., Sahai, A., Luo, J.-J., & Yamagata, T. (2012). The role of the intra-daily SST variability in the Indian monsoon variability and monsoon-ENSO–IOD relationships in a global coupled model. *Climate Dynamics, 39*, 729–754. https://doi.org/10.1007/s00382-011-1240-9.

Thadathil, P., Muraleedharan, P. M., Rao, R. R., Somayajulu, Y. K., Reddy, G. V., & Revichandran, C. (2007). Observed seasonal variability of barrier layer in the Bay of Bengal. *Journal of Geophysical Research, 112*, C02009. https://doi.org/10.1029/2006JC003651.

Thangaprakash, V. P., Girishkumar, M. S., Suprit, K., Suresh Kumar, N., Chaudhuri, D., Dinesh, K., Kumar, A., Shivaprasad, S., Ravichandran, M., Farrar, J. T., Sundar, R., & Weller, R. A. (2016). What controls seasonal evolution of sea surface temperature in the Bay of Bengal? Mixed layer heat budget analysis using moored buoy observations along 90°E. *Oceanography, 29*, 202–213. https://doi.org/10.5670/oceanog.2016.52.

Thompson, P. R., Piecuch, C. G., Merrifield, M. A., McCreary, J. P., & Firing, E. (2016). Forcing of recent decadal variability in the equatorial and North Indian Ocean. *Journal of Geophysical Research, Oceans, 121*, 6762–6778. https://doi.org/10.1002/2016JC012132.

Trenary, L., & Han, W. (2013). Local and remote forcing of decadal sea level and thermocline depth variability in the South Indian Ocean. *Journal of Geophysical Research, 118*, 381–398. https://doi.org/10.1029/2012JC008317.

Ummenhofer, C. C., Geen, R., Denniston, R. F., & Rao, M. P. (2024). Chapter 3: Past, present, and future of the South Asian monsoon. In C. C. Ummenhofer, & R. R. Hood (Eds.), The Indian Ocean and its role in the global climate system (pp. 49–78). Amsterdam: Elsevier. https://doi.org/10.1016/B978-0-12-822698-8.00013-5.

Ummenhofer, C. C., Murty, S. A., Sprintall, J., Lee, T., & Abram, N. J. (2021). Heat and freshwater changes in the Indian Ocean region. *Nature Reviews Earth and Environment, 2*(8), 525–541.

Vallivattathillam, P., Iyyappan, S., Lengaigne, M., et al. (2017). Positive Indian Ocean dipole events prevent anoxia off the west coast of India. *Biogeosciences, 14*, 1541–1559. https://doi.org/10.5194/bg-14-1541-2017.

Vellore, R. K., Deshpande, N., Priya, P., Singh, B. B., Bisht, J., & Ghosh, S. (2020). Extreme storms. In R. Krishnan, J. Sanjay, C. Gnanaseelan, M. Mujumdar, A. Kulkarni, & S. Chakraborty (Eds.), *Assessment of climate change over the Indian region*. Singapore: Springer. https://doi.org/10.1007/978-981-15-4327-2_8.

Venkatesan, R., Lix, J. K., Phanindra Reddy, A., Arul Muthiah, M., & Atmanand, M. A. (2016a). Two decades of operating the Indian moored buoy network: Significance and impact. *Journal of Operational Oceanography, 9*(1), 45–54. https://doi.org/10.1080/1755876X.2016.1182792.

Venkatesan, R., Sundar, R., Vedachalam, N., & Jossia Joseph, K. (2016b). India's ocean observation network: Relevance to society. *Marine Technology Society Journal, 50*(3), 34–46.

Vialard, J. (2015). Hiatus heat in the Indian Ocean. *Nature Geoscience, 8*, 423–424. https://doi.org/10.1038/ngeo2442.

Vialard, J., Foltz, G. R., McPhaden, M. J., Duvel, J. P., & de Boyer Montegut, C. (2008). Strong Indian Ocean sea surface temperature signals associated with the Madden-Julian oscillation in late 2007 and early 2008. *Geophysical Research Letters, 35*, L19608. https://doi.org/10.1029/2008gl035238.

Vinayachandran, P. N. (2009). Impact of physical processes on chlorophyll distribution in the Bay of Bengal. In J. D. Wiggert, R. R. Hood, S. W. A. Naqvi, K. H. Brink, & S. L. Smith (Eds.), *Indian Ocean biogeochemical processes and ecological variability*. Washington, DC: American Geophysical Union.

Vinayachandran, P. N., Murty, V. S. N., & Babu, V. R. (2002). Observations of barrier layer formation in the Bay of Bengal during summer monsoon. *Journal of Geophysical Research, 107*, 8018. https://doi.org/10.1029/2001JC000831.

Volkov, D. L., Lee, S.-K., Gordon, A. L., & Rudko, M. (2020). Unprecedented reduction and quick recovery of the South Indian Ocean heat content and sea level in 2014-2018. *Science Advances, 6*, eabc1151.

Wang, Y., & McPhaden, M. J. (2017). Seasonal cycle of cross-equatorial flow in the Central Indian Ocean. *Journal of Geophysical Research, 122*. https://doi.org/10.1002/2016JC012537.

Warner, S. J., Becherer, J., Pujiana, K., Shroyer, E. L., Ravichandran, M., Thangaprakash, V. P., & Moum, J. N. (2016). Monsoon mixing cycles in the Bay of Bengal: A year-long subsurface mixing record. *Oceanography, 29*(2), 158–169. https://doi.org/10.5670/oceanog.2016.48.

Webster, P. J., Clark, C., Cherikova, G., Fasulla, J., Han, W., Loschnigg, J., & Sahami, K. (2002). The monsoon as a self-regulating coupled ocean–atmosphere system. *International Geophysics, 83*, 198–219. https://doi.org/10.1016/S0074-6142(02)80168-1.

Webster, P. J., Moore, A. M., Loschnigg, J. P., & Leben, R. R. (1999). Coupled ocean-atmosphere dynamics in the Indian Ocean during 1997–98. *Nature, 401*, 356–360.

Wiggert, J., Vialard, J., & Behrenfeld, M. (2009). Basin-wide modification of dynamical and biogeochemical processes by the positive phase of the Indian Ocean dipole during the SeaWiFS era. *American Geophysical Union Geophysical Monograph Series, 185*, 385–407. https://doi.org/10.1029/2008GM000776.

Wijffels, S., & Meyers, G. (2004). An intersection of oceanic waveguides: Variability in the Indonesian Throughflow region. *Journal of Physical Oceanography, 34*, 1232–1253.

Wijffels, S., Sprintall, J., Fieux, M., & Bray, N. (2002). The JADE and WOCE I10/IR6 throughflow sections in the southeast Indian Ocean. Part 1: Water mass distribution and variability. *Deep Sea Research Part II: Topical Studies in Oceanography, 49*(7–8), 1341–1362. https://doi.org/10.1016/S0967-0645(01)00155-2.

Wojtasiewicz, B., Hardman-Mountford, N. J., Antoine, D., Dufois, F., Slawinski, D., & Trull, T. W. (2018a). Use of bio-optical profiling float data in validation of ocean colour satellite products in a remote ocean region. *Remote Sensing of Environment, 209*, 275–290. https://doi.org/10.1016/j.rse.2018.02.057.

Wojtasiewicz, B., Trull, T. W., Udaya Bhaskar, T. V. S., Gauns, M., Prakash, S., Ravichandran, M., … Hardman-Mountford, N. J. (2018b). Autonomous profiling float observations reveal the dynamics of deep biomass distributions in the denitrifying oxygen minimum zone of the Arabian Sea. *Journal of Marine Systems*. https://doi.org/10.1016/j.jmarsys.2018.07.002.

Woolnough, S. J., Vitart, F., & Balmaseda, M. A. (2007). The role of the ocean in the Madden–Julian oscillation: Implications for MJO prediction. *Quarterly Journal of the Royal Meteorological Society, 133*, 117–128. https://doi.org/10.1002/qj.4.

Wyrtki, K. (1973). An equatorial jet in the Indian Ocean. *Science, 181*, 262–264.

Xie, S.-P., Annamalai, H., Schott, F. A., & McCreary, J. P. (2002). Structure and mechanisms of south Indian Ocean climate variability. *Journal of Climate, 15*, 867–878.

Yamagata, T., Behera, S., Doi, T., Luo, J.-J., Morioka, Y., & Tozuka, T. (2024). Chapter 5: Climate phenomena of the Indian Ocean. In C. C. Ummenhofer, & R. R. Hood (Eds.), *The Indian Ocean and its role in the global climate system* (pp. 103–119). Amsterdam: Elsevier. https://doi.org/10.1016/B978-0-12-822698-8.00009-3.

Yang, Y., Li, T., Li, K., & Yu, W. (2015). What controls seasonal variations of the diurnal cycle of sea surface temperature in the eastern tropical Indian Ocean? *Journal of Climate, 28*, 8466–8485. https://doi.org/10.1175/JCLI-D-14-00826.1.

Yu, L. (2019). Global air–sea fluxes of heat, fresh water, and momentum: Energy budget closure and unanswered questions. *Annual Review of Marine Science, 11*(1), 227–248.

Yu, L., Jin, X., & Weller, R. A. (2007). Annual, seasonal, and interannual variability of air-sea heat fluxes in the Indian Ocean. *Journal of Climate, 20*, 3190–3209. https://doi.org/10.1175/JCLI4163.1.

Yu, L., & McPhaden, M. J. (2011). Ocean pre-conditioning of cyclone Nargis in the Bay of Bengal: Interaction between Rossby waves, surface fresh waters, and sea surface temperatures. *Journal of Physical Oceanography, 41*(9), 1741–1755. https://doi.org/10.1175/2011JPO4437.1.

Yu, W., Xiang, B., Liu, L., & Liu, N. (2005). Understanding the origins of interannual thermocline variations in the tropical Indian Ocean. *Geophysical Research Letters, 32*, L24706. https://doi.org/10.1029/2005GL024327.

Zang, N., Sprintall, J., Ienny, R., & Wang, F. (2021). Seasonality of the Somali current/undercurrent system. *Deep-Sea Research Part II, 191–192*. https://doi.org/10.1016/j.dsr2.2021.104953.

Zanowski, H., & Johnson, G. C. (2019). Semiannual variations in 1,000-dbar equatorial Indian Ocean velocity and isotherm displacements from Argo data. *Journal of Geophysical Research: Oceans, 124*. https://doi.org/10.1029/2019JC015342.

Zhang, C. (2005). Madden-Julian oscillation. *Reviews of Geophysics, 43*, RG2003. https://doi.org/10.1029/2004RG000158.

Zhang, C. (2013). Madden–Julian oscillation: Bridging weather and climate. *Bulletin of the American Meteorological Society, 94*(12), 1849–1870. https://journals.ametsoc.org/view/journals/bams/94/12/bams-d-12-00026.1.xml.

Zhang, Y., Feng, M., Du, Y., Phillips, H. E., Bindoff, N. L., & McPhaden, M. J. (2018). Strengthened Indonesian throughflow drives decadal warming in the southern Indian Ocean. *Geophysical Research Letters, 45*, 6167–6175. https://doi.org/10.1029/2018GL078265.

Zhang, L., & Han, W. (2021). Indian Ocean dipole leads to Atlantic Niño. *Nature Communications, 12*(1), 5952. https://doi.org/10.1038/s41467-021-26223-w.

Zhang, D., McPhaden, M. J., & Lee, T. (2014). Observed interannual variability of zonal currents in the equatorial Indian Ocean thermocline and their relation to Indian Ocean dipole. *Geophysical Research Letters, 41*, 7933–7941. https://doi.org/10.1002/2014GL061449.

Zhuang, W., Feng, M., Du, Y., Schiller, A., & Wang, D. (2013). Low-frequency sea level variability in the southern Indian Ocean and its impacts on the oceanic meridional transports. *Journal of Geophysical Research: Oceans, 118*, 1302–1315. https://doi.org/10.1002/jgrc.20129.

Chapter 18

Modeling the Indian Ocean⊛

Toshiaki Shinoda[a], Tommy G. Jensen[b], Zouhair Lachkar[c], Yukio Masumoto[d], and Hyodae Seo[e]

[a]*Department of Physical and Environmental Sciences, Texas A&M University-Corpus Christi, Corpus Christi, TX, United States,* [b]*Ocean Sciences Division, US Naval Research Laboratory, Stennis Space Center, MS, United States,* [c]*Arabian Center for Climate and Environmental Sciences, New York University Abu Dhabi, Abu Dhabi, United Arab Emirates,* [d]*Department of Earth and Planetary Science, University of Tokyo, Tokyo, Japan,* [e]*Department of Physical Oceanography, Woods Hole Oceanographic Institution, Woods Hole, MA, United States*

1 Introduction

Over the last few decades, the critical roles of the Indian Ocean in global climate variability have been recognized in climate research community (Ummenhofer et al., 2024b; Yamagata et al., 2024). Hence, the Indian Ocean is becoming one of the focus regions in the modeling community since the accurate representation of oceanic and atmospheric phenomena over the Indian Ocean in models is crucial for predicting weather and climate in many regions over the globe. Accordingly, the performance of global climate models over the Indian Ocean has begun to be evaluated, and many modeling studies that focus on examining physical processes in the Indian Ocean have been reported.

Because of the unique features of the Indian Ocean, accurate simulations of upper-ocean currents, temperature, and salinity have been a major challenge. For example, due to strong monsoons, the ocean circulation including strong western boundary currents (e.g., Somali Current) varies substantially with season. Also, a huge river discharge associated with heavy precipitation during the summer monsoon season generates very strong salinity stratification in the Bay of Bengal, resulting in a large upper ocean salinity contrast between the eastern and western portions of the basin. Unlike the eastern boundaries of other ocean basins, the boundary current along the southeast Indian Ocean (Leeuwin Current) flows poleward against the prevailing equatorward surface winds. Simulations of these unique features by large-scale ocean general circulation models have been substantially improved in recent years due primarily to the use of eddy-resolving horizontal grid scale (1/10° or finer).

Given the importance of air-sea coupled processes over the Indian Ocean for climate variability, ocean-atmosphere coupled models have been widely used recently for identifying key processes. For example, because of the improvement of global coupled general circulation models (GCMs), they have been used for examining interbasin interaction of climate variability such as the impact of the Indian Ocean Dipole (IOD) on other ocean basins. Also, regional coupled models are shown to be useful for examining a variety of air-sea feedback processes over the Indian Ocean such as the complex air-sea-land interaction over the maritime continent (Hood et al., 2024b).

Predicting marine ecosystem state over the Indian Ocean has been receiving increasing attention because of the socio-economic and environmental importance. Both regional and global ecosystem models have been used to study a variety of biogeochemical and ecological processes that are unique and highly diverse. Yet accurate simulations of various characteristics in the Indian Ocean such as large oxygen minimum zones (OMZs) have been a major challenge due to the complex interaction between physical and biogeochemical processes.

This chapter primarily covers the topics described above including the simulation of Indian Ocean circulation and upper-ocean processes by regional and global ocean models, the simulation and prediction of climate variability over the Indian Ocean, and modeling of biogeochemical processes. In addition, regional coupled modeling over the Indian Ocean, which provides a useful tool for understanding air-sea coupled processes, is discussed in a separate section. Because of the broad scope in Indian Ocean modeling, this chapter cannot fully cover all aspects of modeling. Some of the topics that are not covered in this chapter are found in other chapters, including the decadal prediction (Roxy et al., 2024; Tozuka et al., 2024), intraseasonal variability and prediction (DeMott et al., 2024), and modeling of monsoons (Ummenhofer et al., 2024a).

⊛This book has a companion website hosting complementary materials. Visit this URL to access it: https://www.elsevier.com/books-and-journals/book-companion/9780128226988.

2 Ocean circulation and upper-ocean structure

In this section, upper-ocean circulation and structure simulated by regional and global models as well as physical and dynamical processes identified based on the modeling or model/data comparisons are discussed.

2.1 Boundary currents

2.1.1 Somali Current

The annual reversal of monsoon winds causes a reversal of the currents north of the equator, including the strong western boundary current, the Somali Current (Fig. 4 in Phillips et al., 2024). The Somali Current reverses flow direction from an intense, deep northward flow during boreal summer with a transport in the range of 32–42 Sv (Beal et al., 2003) to a weaker shallow southward flow of about 14 Sv during the Northeast Monsoon. The importance of equatorial Rossby waves in establishing the strong Somali Current during the Southwest Monsoon has been demonstrated in previous studies (e.g., Beal & Donohue, 2013; Lighthill, 1969; McCreary et al., 1993). The onset of northward flow in the Somali Current starts in April nearly 2 months before the onset of southerly winds, which is attributed to the arrival of Rossby waves (Beal & Donohue, 2013).

During boreal summer, the Somali Current is associated with a large anticyclonic eddy off the coast of Somalia, known as the Great Whirl (e.g., Beal & Donohue, 2013; Melzer et al., 2019; Schott & McCreary, 2001; Schott et al., 1990). The persistence of the Great Whirl during boreal fall has been shown to contribute to the existence of northward flow in the northern part of Somail Current after the weakening of local southwesterly winds (e.g., Quadfasel, 1982; Wang et al., 2018). The generation mechanism of the Great Whirl has been investigated by regional ocean models. Proposed mechanisms of the generation include local wind stress curl (McCreary & Kundu, 1985), inertial instability due to crossing the equator (Anderson & Moore, 1979), and barotropic instability (Jensen, 1991).

During July and August, the cross-equatorial Southern Gyre migrates northward (Swallow & Fieux, 1982) and merges with the Great Whirl. Cyclonic eddies often appear on the eastern flank of the Great Whirl and south of the Socotra Eddy and are advected clockwise around the Great Whirl (Beal & Donohue, 2013; Jensen, 1991). Such northward migration has been simulated by a high-resolution (3.5 km) regional coupled model (Chen et al., 2010; Hodur, 1997) (Fig. 1). In this simulation, an intense cyclonic eddy is shredded and subsequently absorbed into the Great Whirl. This process increases the potential vorticity of the anti-cyclonic Great Whirl and permits it to migrate northward. The annual reversal of the Somali Current has been well simulated by a high-resolution (0.1°) global ocean GCM and the important role of the Great Whirl and Rossby waves in the Somali Current seasonal cycle was demonstrated by the analysis of the model output (Wang et al., 2018).

2.1.2 Agulhas Current

The Agulhas Current is part of the surface pathway of the global conveyor belt, flowing southward from 27°S to 40°S with a transport of 67–77 Sv (Beal et al., 2015; Beal & Bryden, 1999; Bryden et al., 2005; Gordon, 1985; Lutjeharms, 2006). The Agulhas Current is fed from the South Equatorial Current via the Mozambique Current and the East Madagascar Current. A notable feature of the Agulhas Current is a retroflection south of Africa near 18°E. Large cyclonic meanders, known as Natal pulses, appear on the Agulhas Current southern edge. They play a significant role in the generation of Agulhas rings that enter the South Atlantic Ocean and transport significant volumes of warm and saline water. While the Agulhas Current retroflection is difficult to represent well in most models even at high resolution (e.g., Thoppil et al., 2011), a recent study by Biastoch et al. (2018) demonstrated that a global ocean GCM with an unstructured-mesh grid (~8 km spacing in the AC region) was able to simulate the Agulhas Current, including its retroflection, reasonably well. Renault et al. (2017) pointed out the importance of including the ocean current in the momentum flux exchange between air and sea to reduce the eddy kinetic energy (EKE) in the Agulhas Current system so that it is closer to observations, and they demonstrated improvement of AC simulation in a coupled atmosphere-ocean model.

2.1.3 Leeuwin Current

The Leeuwin Current is an eastern boundary current in the southeast Indian Ocean, which flows along the west coast of Australia (Fig. 9 in Phillips et al., 2024). Among eastern boundary currents, it is unique as its near-surface water (0–250 m) flows poleward against the prevailing winds (Church et al., 1989; Cresswell & Golding, 1980). The transport is about 1.4 Sv in February and 6.8 Sv in June, while the Leeuwin Undercurrent (250–800 m at 30°S) flows equatorward with a transport of about 5 Sv (Smith et al., 1991). Variability of the Leeuwin Current strongly impacts regional climate variability such as the Ningaloo Niño (e.g., Feng et al., 2013, 2023, 2024; Yamagata et al., 2024) and local marine ecosystems (e.g., Pearce & Feng, 2013; Wernberg et al., 2013).

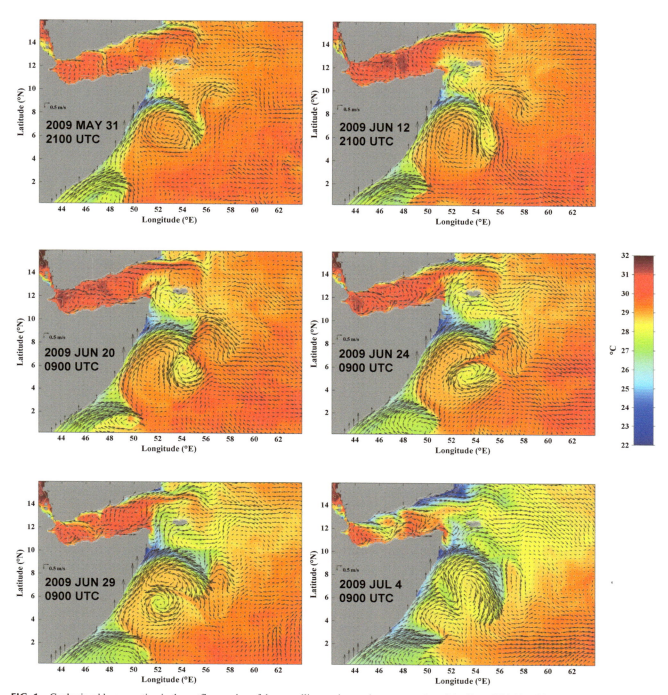

FIG. 1 Cyclonic eddy generation in the outflow region of the upwelling wedge on the western edge of the Great Whirl in a high-resolution fully coupled regional atmosphere-ocean-wave model. The color shading and arrows show sea surface temperature and surface currents, respectively. The observational data are assimilated in the atmospheric component only.

Accurate simulation of the Leeuwin Current in large-scale ocean models has been a challenge due partly to its narrow width (30–50 km). Because of the recent development of high-resolution ocean models, some of the global ocean GCMs are able to realistically simulate the Leeuwin Current. For example, a global ocean GCM with a grid spacing of 1/10° was able to reproduce the Leeuwin Current with the strength (~30–50 cm/s) and the seasonal cycle estimated from hydrographic observations (Furue, 2019). The model also generated the Leeuwin Undercurrent that flows northward around 300–500 m. However, the simulated undercurrent was weaker and almost vanishes north of 22°S where a strong Leeuwin Undercurrent is observed.

2.2 Equatorial currents

During the transitions between the monsoons in boreal spring and summer, intense eastward jets appear along the equator. These jets are known as Wyrtki Jets (Wyrtki, 1973). Numerical models (Han et al., 1999; Jensen, 1993; Nagura & McPhaden, 2010) have been used to examine the generation mechanism of Wyrtki Jets, showing that the semiannual jets are strong due to an equatorial basin-resonance of the second vertical baroclinic mode (Cane & Moore, 1981; Gent, 1981). The basin-mode is a combined oscillatory mode of eastward propagating equatorial Kelvin waves with a speed of about 1.5 m/s, and westward propagating equatorial Rossby waves with one third of that speed.

While the seasonal change of the monsoons forces the semiannual Wyrtki Jets, zonal wind bursts on the equator drive intermittent surface jets as explained by Yoshida (1959). For example, when an intense Madden-Julian Oscillation (MJO) occurred in November 2011, a rapidly intensifying Yoshida Jet was generated in the central equatorial Indian Ocean with zonal currents exceeding 1 m/s in the mixed layer. The jet was observed by RAMA (Research Moored Array for African-Asian-Australian Monsoon Analysis and Prediction) buoys and simulated well by models (e.g., Jensen et al., 2015; Shinoda et al., 2016). During this period, the Yoshida jet is superimposed on the seasonal Wyrtki Jets that has a subsurface local maximum between 50 and 150 m (Jensen et al., 2015).

Equatorial Undercurrent in the Indian Ocean, which is observed during boreal winter/spring (e.g., Knauss & Taft, 1964; Knox, 1974; Swallow, 1964) and summer/fall (e.g., Bruce, 1973; Reppin et al., 1999) seasons, is highly transient. The mechanisms of Equatorial Undercurrent generation and variability have been recently investigated using regional ocean GCMs (e.g., Chen et al., 2019a, 2019b), showing the importance of equatorial Rossby and Kelvin waves forced by seasonally varying zonal winds. In the western basin, the seasonal variation of the Equatorial Undercurrent is primarily controlled by directly forced Kelvin and Rossby waves, whereas reflected Rossby waves from the eastern boundary largely contribute to the Equatorial Undercurrent variability in the eastern basin. Although high-resolution global ocean GCMs are able to reproduce the magnitude of the semiannual cycle of the Equatorial Undercurrent reasonably well, most models are not able to accurately simulate the timing of the peak value of current (Rahman et al., 2020).

2.3 Indonesian Throughflow

The Indonesian Throughflow (ITF) is an important component of the global thermohaline and wind-driven circulations carrying upper ocean waters from the Pacific to the Indian Ocean (Fig. 2 in Sprintall et al., 2024). Since the total transport of the ITF can be determined to first order by the "island rule" theory (Godfrey, 1989), ocean GCMs with coarse resolution can simulate the mean transport in agreement with observations (e.g., Lee et al., 2002; Schiller et al., 1998). However, the distribution of transport among different straits and passages cannot be accurately simulated in coarse resolution models. Recent development of high-resolution global ocean GCMs as well as comprehensive observations in the Indonesian Seas such as the International Nusantara Stratification and Transport program (Gordon et al., 2010) allow evaluation of the ability of ocean models to simulate ITF structures. It has been demonstrated that a global eddy-resolving ($1/12°$) ocean GCM is able to simulate net ITF transports as well as its distribution among inflow/outflow passages reasonably well when compared to observations (Metzger et al., 2010). The simulation was further improved by the higher resolution ($1/25°$) model (Metzger et al., 2010; Shinoda et al., 2016).

Seasonal and intraseasonal variations of the ITF are also simulated well by global ocean GCMs where remotely forced equatorial waves from the Indian and Pacific Oceans and local wind forcing over the Indonesian Seas determine the transport (e.g., Shinoda et al., 2012, 2016). Simulating interannual/decadal variations of the ITF structure is still a major challenge. Long integrations of global ocean GCMs have begun in recent years (e.g., Sasaki et al., 2018) but long-term in situ measurements of currents are unavailable in most areas. Sustained measurements in most major straits in the Indonesian Seas are necessary to validate the simulated long-term ITF variability.

2.4 Temperature and salinity upper-ocean structure

One of the notable characteristics of the upper ocean structure in the tropical Indian Ocean is the dome-like feature of the main thermocline located around 2°S-10°S in the western portion of the basin. This area of shallow thermocline is often referred to as the Seychelles-Chagos thermocline ridge (SCTR). The SCTR is maintained primarily by strong Ekman pumping due to the northward weakening of the southeast trades (McCreary et al., 1993; Xie et al., 2002). Because of the shallow mixed layer and thermocline, subseasonal to interannual sea surface temperature (SST) signals and air-sea interaction are strong in the SCTR region (e.g., Han et al., 2007; Saji et al., 2006; Xie et al., 2002) and play an important role in the generation of atmospheric convection associated with the MJO and IOD.

Ocean GCMs with coarse resolutions can simulate SCTR variability at least qualitatively (e.g., Masumoto & Meyers, 1998; Murtugudde & Busalacchi, 1999). Models suggest that several processes impact SCTR variability: One ocean GCM showed that it is primarily controlled by local Ekman pumping (Yokoi et al., 2008) while others imply that the annual cycle is due to complex interactions between the response to local and remote forcing (Hermes & Reason, 2008; Nyadjro et al., 2017). The relative importance of surface heat fluxes and ocean dynamics in controlling subseasonal SST variability over the SCTR is still unclear and results vary substantially between different modeling studies (Duvel et al., 2004; Halkides et al., 2015; Han et al., 2007; Jayakumar et al., 2011; Li et al., 2014; Shinoda et al., 2017; Vinayachandran & Saji, 2008; Yuan et al., 2020).

Another unique feature of the upper ocean is found in the Bay of Bengal, where a complex stratification including a strong halocline and temperature inversion is caused by a substantial amount of freshwater input through river runoff and precipitation. In particular, the strong salinity stratification causes a shallow mixed layer and thick barrier layer (the isothermal layer below the mixed layer) (e.g., Howden & Murtugudde, 2001; Li et al., 2017a, 2017b; Yu & McCreary, 2004), and its accurate simulation by models is still a challenge (e.g., Rahman et al., 2020). Modeling of the upper-ocean structure of the Bay of Bengal is further discussed in Section 4.

2.5 Future perspectives

As described in this section, ocean CGM's ability to simulate major upper ocean currents over the Indian Ocean such as western and eastern boundary currents, equatorial currents, and circulations in the Indonesian Seas have been substantially improved in recent years primarily because of the development of high-resolution (eddy-resolving) global models. Yet, the performance of ocean GCMs depends on their resolution and model physics, and the details of such model dependence are not covered in this section. A recent ocean GCM intercomparison study provides a detailed assessment of simulations from multiple state-of-the-art global ocean models (Rahman et al., 2020).

While in situ data coverage in the Indian Ocean has been improving during recent years (Beal et al., 2020), the coverage of subsurface data in many locations such as the interior southern Indian Ocean is sparse, and thus quantitative model/data comparisons are difficult. Although recent studies discuss the circulations in the southern Indian Ocean using available in situ data, numerical model outputs, and reanalysis products (e.g., Menezes et al., 2014, 2016; Nagura & McPhaden, 2018), the comprehensive evaluation of model performance on different time scales is not possible at present due to the insufficient coverage of in situ observations. Hence, in addition to maintaining and enhancing the current observational network in boundary and equatorial regions, it is desirable to extend the regions of sustained observations to the data sparse areas such as the interior southern Indian Ocean for further evaluating the ocean GCM's ability to reproduce observed ocean circulations over the entire Indian Ocean (McPhaden et al., 2024).

3 Climate variability

3.1 Modeling of climate variations

The challenge of modeling climate variations in the Indian Ocean boils down to how accurately the models can reproduce the upper-ocean conditions at the intraseasonal, seasonal, interannual, and longer time scales. Early efforts to reproduce the upper-ocean variability in the Indian Ocean can be traced back to the concept of a reduced gravity model by Lighthill (1969). Due to the unique geographic nature of the Indian Ocean, with the northern boundary blocked by the landmass at rather a low latitude, this simple dynamical framework can successfully simulate the seasonal upper-ocean variability (e.g., Kindle & Thompson, 1989; Luther & O'Brien, 1985; McCreary et al., 1993; Murtugudde & Busalacchi, 1999).

Efforts to simulate the three-dimensional circulations and thermohaline structures in the ocean were pioneered by Cox (1970, 1976), with his epoch-making work using an ocean GCM. Subsequent developments of ocean GCM simulations of the Indian Ocean make it possible to reproduce realistic ocean circulations and tracer distributions (e.g., Godfrey & Weaver, 1991; Lee & Marotzke, 1998) and their variability (e.g., Anderson & Carrington, 1993).

It is natural to extend simulations to longer time-scale variations and/or incorporate air-sea coupled processes for the study of climate variations. This research direction was accelerated by the discovery and recognition of the IOD, an air-sea coupled climate variation inherent in the tropical Indian Ocean (Saji et al., 1999; Yamagata et al., 2024). It is relatively easy with a modeling study to conduct quantitative analyses of processes and mechanisms involved in simulated ocean variability, and sensitivity experiments can be carried out to tease out the roles of these processes. These studies are particularly useful for examining climate variations, for which the observations are limited in terms of spatial and temporal coverages. This section briefly introduces the present status of modeling of the climate variability in the Indian Ocean, mainly focusing on the IOD (see also Yamagata et al., 2024 for more discussion of the IOD).

3.2 Ocean dynamics

Large-scale surface wind stress variability is considered as the dominant force for the major dynamical responses associated with the climate variations. However, both local and remote responses of the upper ocean to the wind forcing need to be considered. Planetary-scale wave dynamics play a key role in this remote response.

Piece-wise investigations on processes during different phases of the IOD have been conducted, with much attention to the generation and termination periods. These studies demonstrated that equatorial Kelvin and Rossby waves play essential roles, as well as local ocean responses to the wind forcing (e.g., Effy et al., 2020; Gualdi et al., 2003; Prasad & McClean, 2004; Rao et al., 2007; Vinayachandran et al., 2002a; Wang & Yuan, 2015). A major example of the remote influence can be seen for the equatorial Kelvin waves and subsequent propagation as coastal Kelvin waves along the Sumatra/Java coasts for the eastern pole of the IOD. Arrival of the upwelling (downwelling) favorable Kelvin wave at the eastern pole is associated with shallowing (deepening) of the thermocline, resulting in cooler (warmer) SST there. Atmospheric intraseasonal oscillations may also contribute to the IOD initiation and termination through Kelvin wave propagation (e.g., Han et al., 2006). Another example is westward propagating Rossby waves generated by the wind stress curl over the eastern/central Indian Ocean for the western pole of the IOD, with a similar influence on the SST variations through the vertical movement of the thermocline. However, the degree of importance of these wave dynamics to other processes such as local upwelling and surface heat flux depends on the events and regions of interest.

Coupled atmosphere-ocean general circulation models can reasonably simulate the interannual climate variations similar to the observed IOD (e.g., Fischer et al., 2004; Gualdi et al., 2003; Iizuka et al., 2000; Lau & Nath, 2004; Song et al., 2007; Wajsowicz, 2004; Yao et al., 2016), although some biases are observed in all the models. The coupled GCMs are used for further analyses of key processes involved in the evolution of simulated IODs and for investigations on their relations to ENSO in the Pacific Ocean. For example, Iizuka et al. (2000) successfully reproduced the IOD characteristics and indicated that the variability in the Indian Ocean is independent of the Pacific variations. Lau and Nath (2004) suggested that some of the IOD-like events in their coupled GCM are associated with atmospheric teleconnection of the ENSO-related changes, while some strong IOD-like events exist with the absence of ENSO influences (see also Yang et al., 2015; Yamagata et al., 2024). Fischer et al. (2004) suggested two possible triggers for the IOD in their coupled GCM; one is anomalously stronger southeasterly trade winds penetrated over the southeastern Indian Ocean earlier than usual, and the other is a zonal shift of the convective region over the maritime continent associated with the El Niño phenomenon in the Pacific Ocean. Other coupled GCMs also reproduced observed characteristics of the IOD events, with some of them co-occurring with ENSO events in the Pacific while some are not (e.g., Song et al., 2007; Yamagata et al., 2024). These coupled GCM experiments suggest that air-sea interactions inherent in the Indian Ocean are essential for the development of the anomalous conditions, whereas there are several possibilities of the triggering processes of the IOD.

3.3 Heat budget analyses and salinity impacts for the IOD

Heat budget analysis has been utilized extensively, with a focus on processes responsible for changes in the SST or mixed-layer temperature for various regions and phenomena within the Indian Ocean. This is also the case for the two poles of the IOD. It has been shown that both the surface heat flux and local/remote oceanic processes are important to determine the temperature tendencies (e.g., Delman et al., 2018; Halkides & Lee, 2009; Iizuka et al., 2000; Thompson et al., 2009; Vinayachandran et al., 2002a, 2007). However, the degree of relative importance among the processes seems to be different for each target event, the region of budget analysis, and due to differences in model settings. Recent mixed-layer heat budget analysis of an eddy-resolving ocean GCM suggested the importance of the winds off the Sumatra coast to generate temperature cooling through an advective process during the positive IOD events (Delman et al., 2018). For the termination period of the positive IOD events, Thompson et al. (2009) showed the dominant contribution of surface heat flux in the 1997 event and ocean advective processes in 1961 and 1994 cases, whereas Ogata and Masumoto (2010) suggested a possible role of eddy heat transport to hasten the IOD termination in the strong events in 1994. These analyses suggest combined influences of different processes to determine the surface mixed-layer temperature variations during the IOD, which may result in relatively large uncertainty of the IOD prediction (Tanizaki et al., 2017).

An additional complication comes from salinity contributions to the upper-layer temperature and circulation variability in the tropical Indian Ocean, because of the large spatial contrasts in salinity distribution in the tropical Indian Ocean (e.g., Kido et al., 2019a; Masson et al., 2003; Thompson et al., 2006; Vinayachandran & Nanjundiah, 2009). A series of sensitivity experiments conducted by Kido et al. (2019a, 2019b) demonstrated that salinity anomalies associated with the positive IOD events are generated by both the surface freshwater flux anomaly and ocean dynamical processes, particularly due to advective processes including the rectified effects of the high-frequency variability. They also showed that the salinity

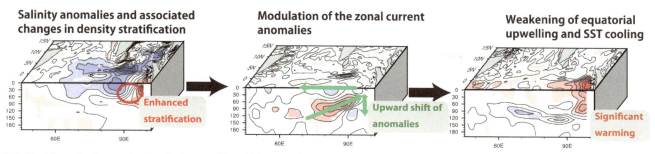

FIG. 2 Schematic diagram showing the impact of the salinity anomaly associated with the positive IOD on the upper-ocean circulations and temperature in the equatorial Indian Ocean (see Kido et al., 2019b for details).

anomalies in turn contribute to the temperature anomaly through modification of the circulations within the vertical sections along the equator (Fig. 2).

3.4 Dynamical prediction of the IOD variations

Prediction and predictability studies of the IOD are important not only for scientific understanding of the phenomena but also for socioeconomic activities in surrounding regions of the Indian Ocean. Previous studies using dynamical coupled models showed reasonable prediction skill with the lead time of a few months or a season (e.g., Doi et al., 2019; Feng et al., 2014; Luo et al., 2007, 2008; Wajsowicz, 2005, 2007; Yamagata et al., 2024). The potential predictability is different for the SST variability averaged over the eastern pole and western pole of the IOD, showing a shorter period of skillful prediction for the eastern pole (Wajsowicz, 2005). Some of the coupled GCMs demonstrated much longer prediction skills for the IOD up to two to three seasons ahead (Lu et al., 2018; Luo et al., 2007; Song et al., 2008; Yamagata et al., 2024) in the case when anomalous conditions in the ocean played a key role in the evolution of the events. This longer prediction skill seems to come from the oceanic processes before the IOD events, such as the wave propagations, or influences from other basins, including a teleconnection from the Pacific Ocean (Doi et al., 2020; Yamagata et al., 2024).

At the same time, however, these processes may provide seeds for prediction barriers. One example of the known issues for the IOD prediction is the so-called "winter prediction barrier" (Feng et al., 2014; Wajsowicz, 2007; Yamagata et al., 2024). This prediction barrier may be caused by initial errors amplified by the surface flux and/or oceanic processes (Feng et al., 2014; Feng & Duan, 2019). Doi et al. (2020) suggested that signals from the Pacific Ocean associated with the El Niño Modoki overcame the winter prediction barrier in the 2019 case. The skills for the IOD prediction may be improved by increasing model resolutions (Doi et al., 2016), initialization of subsurface ocean conditions (Doi et al., 2017), and increasing ensemble size for the prediction (Doi et al., 2019) (see Yamagata et al., 2024 for further discussion).

3.5 Future perspectives

The basin-scale to meso-scale variability necessary to simulate the climate variations in the Indian Ocean can be reproduced reasonably well in recent models. Air-sea coupled models and earth system models have the capability to reproduce IOD-like climate variations, showing successful prediction skills for one or two seasons ahead. Analyses of outputs from these models have highlighted important dynamical and physical processes responsible for initiation, evolution, and termination of IOD events. Our understanding of the IOD variability for individual events has been accumulated to discuss common features and differences among the events. However, further analyses of detailed mechanisms to build integrated and comprehensive pictures of the IOD and associated interactions with different time and space scales are needed.

High-resolution modeling for the purpose of process understanding is necessary to elucidate impacts from smaller-scale phenomena and boundary processes. Coordinated collaborative research will be required to do this. Analyses on long-term variability are also necessary to explore relations and interactions with longer time-scale phenomena, such as decadal variations and global warming trends. This can be achieved partly by analyzing CMIP (Coupled Model Intercomparison Project) type model output and partly by conducting long-term simulations of the ocean and/or coupled atmosphere-ocean models. Finally, a better understanding of subgrid-scale variability and turbulent processes is key to improve mixing parameterizations, which are unavoidable even in much higher resolution models.

4 Regional coupled climate modeling

Emphasis on regional air-sea interaction is becoming an increasingly greater focus within the climate modeling communities. However, most of the current global models use horizontal and vertical scales that are often inadequate to resolve regional and local air-sea interaction and deep convection. Such processes have begun to be investigated by regional coupled climate models, dynamically downscaling global analyses, or coarse-resolution models in the atmosphere and ocean over a limited domain. Various regional coupled climate model studies in the Indian Ocean and other basins have unraveled novel coupled processes and advanced our understanding of how the oceans and air-sea interactions shape regional weather and climate (see reviews by Giorgi, 2019; Miller et al., 2017; Schrum, 2017; Xue et al., 2020). This section offers a brief overview of the regional coupled climate model studies specifically designed for examining air-sea interaction in the Indian Ocean.

4.1 Processes and regions suitable for regional coupled climate model applications

4.1.1 Mesoscale air-sea interactions

One of the most crucial benefits of an regional coupled climate model is the improved representation of air-sea interaction mediated by ocean mesoscale eddies. Despite their small spatial scales, the mesoscale flow fields typically feature strong anomalies in SSTs and their gradients (Chelton et al., 2007; Small et al., 2008; Vecchi et al., 2004; Fig. 3). Seo et al. (2008) used a 25-km regional coupled climate model (Seo et al., 2007) in the Arabian Sea to study the influence of surface wind variations on mesoscale SST variability, suggesting that the local boundary layer coupling is a vital forcing mechanism of baroclinic instability that governs the regional ocean circulation. In the Bay of Bengal, Seo et al. (2019) used a 4-km regional coupled climate model (Seo et al., 2014) to show how the effect of a surface current in the wind stress (Renault et al., 2016; Seo et al., 2016) acts to stabilize the ocean circulation and increase the upper ocean stratification. Seo (2017) further conducted a series of 7-km regional coupled climate model simulations for the Arabian Sea to quantify relative impacts of the mesoscale SST-driven and current-driven wind stress responses (Renault et al., 2019), showing that the two couplings exert distinct influences on the ocean: the thermal (mechanical) coupling acts to shift the position (intensity) of the mesoscale flow fields.

4.1.2 Indian summer monsoons and Intraseasonal variability (ISV)

In the tropics, the local air-sea coupling is effectively communicated to the deep troposphere via their impacts on cumulus convection, leading to well-defined regional air-sea coupled effects on the Indian summer monsoon precipitation and its ISV (DeMott et al., 2015, 2024). Using regional coupled climate model simulations of Li et al. (2012), Misra et al. (2017) concluded that their 15-km model not only produces improved simulation skill in Indian summer monsoon rainfall but also accurately captured the transition of the active/break cycles of the Indian summer monsoon and remote impacts by ENSO. Theories and global modeling studies have demonstrated that local air-sea coupling is critical for realistic simulation of the northward propagating Indian summer monsoon ISV (DeMott et al., 2013; Fu et al., 2007, 2008). The subsequent studies by Misra and colleagues (Karmakar & Misra, 2020a, 2020b; Misra et al., 2018) have illustrated the importance of the regional air-sea interaction in the simulation of the Indian Ocean subseasonal, seasonal and interannual climate (Fig. 4).

4.1.3 Maritime Continent

The maritime continent is an ideal region for regional coupled climate model applications. It is the largest archipelago with complex coastlines and island geometry, surrounded by oceans of significant heat content supporting deep convection. The thermally forced diurnal convection from land migrates offshore while interacting with oceanic and land-surface processes (Yoneyama & Zhang, 2020), which are too small in scales to be reliably represented in global models and analyses (Xue et al., 2020). Aldrian et al. (2005) were the first to apply a regional coupled climate model to the maritime continent, identifying significant skill enhancement in rainfall as compared to uncoupled runs. Wei et al. (2014) demonstrated improved simulations of SST and ITF transport in their regional coupled climate model. However, they did not find the corresponding improvement in the rainfall bias, which is at odds with the recent observational and modeling studies showing significant impacts of air-sea coupling on the atmosphere. Li et al. (2017c) explored the resolution sensitivity of the diurnal precipitation and its land-sea characteristics in the Maritime Continent using the regional coupled climate model of Samson et al. (2014). They found that the mean SST and rainfall biases are reduced from $\tfrac{3}{4}°$ to $\tfrac{1}{4}°$ ocean models, but the rainfall becomes too strong in the $1/12°$ atmospheric model, in agreement with previous high-resolution atmospheric modeling studies (Vincent & Lane, 2017). While there are numerous contributors to the excessive precipitation bias in high-resolution models (e.g., radiation and cloud microphysics parameterizations) in the context of air-sea coupled modeling, improved representations of upper-ocean processes and air-sea fluxes, for example, due to tides and surface waves, are likely significant in the maritime continent. However, to date, there have been no regional coupled climate model studies that quantify the impacts of tides and surface waves on convective precipitation and MJO in the maritime continent.

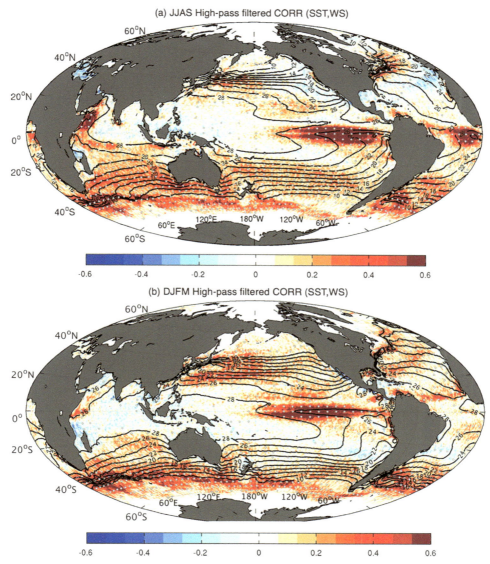

FIG. 3 Maps of correlation coefficients between daily NOAA-OI SST and surface wind speed from QuikSCAT for (a) boreal summer (June–September) and (b) boreal winter (December–March) of 2001–2009. Wind speed and SST are zonally high-pass filtered (10° longitudes) to remove large-scale air-sea coupling. The black contours denote the climatological SST [CI = 2°C] and the gray dots denote the significant correlation at the 95% level. The correlation is positive over most global oceans, including the western Indian Ocean in boreal summer and the southern Indian Ocean year around, indicating that rich mesoscale SST variability there induces local responses in wind speed. *(Adapted from Seo (2017).)*

4.1.4 Upper-ocean processes

Other processes well-suited for regional coupled climate model studies include the air-sea interaction associated with better resolved upper ocean processes, including the barrier layer and the diurnal SST variability. Here, examples of some of these processes are briefly summarized.

Shenoi et al. (2002) hypothesized that positive feedback exists over the Bay of Bengal during the Indian summer monsoon. The thick barrier layer forced by strong freshwater forcing suppresses the vertical mixing and increases the SST. This leads to enhanced Indian summer monsoon rainfall and higher river discharges, in turn reinforcing the thick barrier layer. Seo et al. (2009) used a 25-km regional coupled climate model to explore the SST and precipitation responses to the enhanced stratification in the Bay of Bengal. Their results support this positive feedback, but the identified SST and precipitation responses were generally weak and limited geographically near the major river mouths. However, their model significantly underestimated the observed upper-ocean stratification in the Bay of Bengal with cold SST bias, and hence it

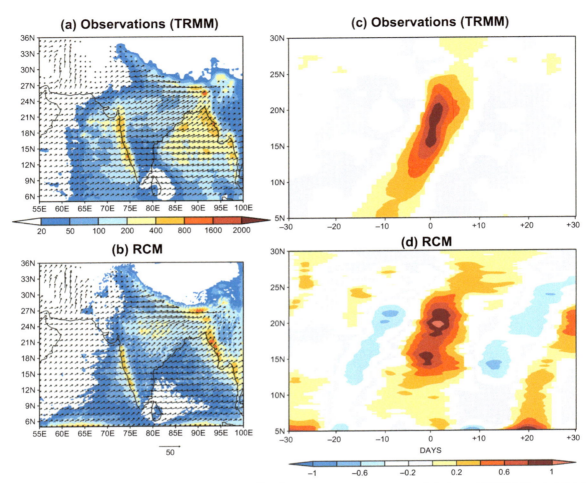

FIG. 4 (*Left*) ISV (20–90 days) variances of rainfall (shaded; mm^2/day^2) and 850-hPa winds (vectors; m^2/s^2) from (a) TRMM (Tropical Rainfall Measuring Mission) and (b) a regional coupled climate model. The vectors represent the variance of the zonal and meridional wind components. (*Right*) The regression of zonally averaged ISV rainfall between 70 and 95°E on the ISV rainfall anomalies over central India (75–85°E and 15–25°N) from (c) TRMM and (d) a regional coupled climate model. Negative (positive) lags denote that central Indian rainfall anomalies lead (lag) to the zonally averaged rainfall anomaly. *(Adapted from Misra et al. (2018).)*

remains unclear if and how such mean state biases affect the sensitivity of the Indian summer monsoon rainfall responses. Krishnamohan et al. (2019) revisited this problem using the 25-km regional coupled climate model of Samson et al. (2014) and explained that the weak SST response results from two compensating effects; while the barrier layer suppresses the vertical mixing and increases the mixed layer temperature, the associated temperature increase is quickly damped by turbulent heat flux via air-sea interaction, leaving only a small net change in SST.

Until recently, conventional CMIP-class global climate models did not consider subdaily air-sea interaction due to increased computational burdens associated with frequent model coupling and/or lack of relevant physics for subdaily air-sea coupling and turbulent vertical mixing. Yet, the importance of the SST diurnal cycle in the tropical oceans has long been recognized (Soloviev & Lukas, 1996; Weller & Anderson, 1996), especially in the context of modulation of ISV (Ruppert & Johnson, 2015; Shinoda, 2005; Shinoda et al., 2021; Sui et al., 1997). Seo et al. (2014) examined the impacts of resolving diurnal SSTs in simulations of the MJO, where the diurnal coupling effect is adjusted by varying the coupling frequency between 1 and 24 h. The resolved diurnal variability with higher coupling frequency (1 h) during the suppressed MJO phase raises the time-mean SST and moistens the low-troposphere, facilitating the subsequent convection. Zhao and Nasuno (2020) used the 7 km regional coupled climate model of Warner et al. (2010) with varied coupling frequency to quantify the impact of subdaily air-sea interaction over the maritime continent. They suggest that the higher model coupling frequency (1 h or shorter) leads to the warmer pre-convection SST aided by shallower mixed layer depth, leading to the more vigorous MJO convection.

4.2 Synthesis, issues, and outlook

Recent regional coupled climate model studies in the Indian Ocean have demonstrated significant "added values" in the context of advancing process-level understanding of air-sea interaction and upper ocean processes and their large-scale impacts. However, several challenges remain, many of which are not unique to the regional coupled climate model or the Indian Ocean, as reviewed previously (Schrum, 2017; Xue et al., 2020). But some of these generic issues can be amplified by the regional coupled climate models. One such example is the their inability to represent a realistic ocean mean state in some areas of the Indian Ocean. In regional coupled climate models, additional internal variability arises from better-resolved air-sea interaction, such that slight mean state bias can be amplified by the air-sea coupling. This is in contrast to uncoupled regional models where the observed states are prescribed as surface forcing.

The most striking ocean mean state bias may be found in the northern Indian Ocean during the Indian summer monsoon (e.g., Goswami et al., 2016), especially in the Bay of Bengal, where existing regional coupled climate models exhibit significant cold bias (Krishnamohan et al., 2019; Misra et al., 2018; Seo et al., 2009, 2019). The thermal coupling of the boreal summer intraseasonal oscillation with the oceans is likely to be influenced by the cold bias of SST, which otherwise would hover above the convective thresholds and support the moist convection. The parameterized turbulent mixing schemes in the ocean tend to overestimate the mixing given the wind stress in the BoB, failing to maintain the observed salinity-driven stratification (e.g., Goswami et al., 2016). The coastal upwelling along India's east coast also tends to be strong in high-resolution regional coupled climate models, providing for the cold bias in the interior basin. Such basin-scale bias persists even with strong salinity restoring (Seo et al., 2019), realistic wind stress forcings (Krishnamohan et al., 2019), or at exceptionally high-model resolutions (Seo et al., 2019). Recent field experiments in the Bay of Bengal (e.g., Vinayachandran et al., 2018; Wijesekara et al., 2016) revealed the intricate patterns of upper-ocean variability and vertical fluxes driven by fine-scale ocean dynamics, leading to the formation of interleaving layers and submesoscale restratification (e.g., Jaeger et al., 2020; Ramachandran et al., 2018). These observed phenomena are not well parameterized in the current ocean models (Chowdary et al., 2016). The excessive vertical mixing and the cooler SST bias are reinforced by the too strong wind bias typical of high-resolution atmospheric modeling.

Other notable regions of significant ocean mean state biases can be found in the Arabian Sea (Weller et al., 2002), where wind-driven mixing and ocean mesoscale dynamics dominate the mixed layer temperature balance, and the maritime continent, where vertical mixing due to tides and the diurnal cycle rectify the ocean mean state and atmosphere-ocean coupling. These regions deserve further regional coupled climate model studies with special attention to the physical representation of upper ocean physics and oceanic mesoscale processes. These include, but are not limited to, vertical mixing affected by the symmetric instability (Bachman & Taylor, 2014; Dong et al., 2021), Langmuir turbulence (Kantha & Clayson, 2004; Li et al., 2016), diurnal cycle (Danabasoglu et al., 2006; Takaya et al., 2010), and wave impacts on ocean mixing and wind stress (Pianezze et al., 2018; Shi & Bourassa, 2019), and mesoscale eddy impacts on air-sea heat fluxes (Bishop et al., 2020).

5 Biogeochemistry

5.1 Modeling phytoplankton bloom dynamics in the Indian Ocean

5.1.1 Arabian Sea

The Arabian Sea is an exceptionally unique and productive marine ecosystem. Summer monsoon southwesterly winds drive upwelling along the coasts of Oman and Somalia, whereas winter northeasterly winds cause convective mixing in its northern part; each bringing nutrients from the deep into the well-lit upper ocean, driving two major phytoplankton blooms per year (Wiggert et al., 2005). Early coupled physical-biogeochemical modeling efforts of the Arabian Sea have focused on the study of phytoplankton bloom dynamics and its seasonal variability (Hood et al., 2003; McCreary et al., 1996; Ryabchenko et al., 1998). Using intermediate complexity physical models coupled to simple nutrient-phytoplankton-zooplankton-detritus ecosystem models, these pioneering works highlighted the critical role of wind-driven mixing in the bloom dynamics and identified distinct physical processes such as upwelling, entrainment, and detrainment that control Arabian Sea phytoplankton blooms. Despite relative success in reproducing several observed bloom characteristics, these models showed some fundamental flaws associated with deficiencies in the representation of circulation in the Arabian Sea region (Friedrichs et al., 2006; Hood et al., 2003). Many of these discrepancies may be traceable to the inaccurate climatological forcing employed in these studies as well as the inaccurate representation of mixed layer and lateral export of nutrients associated with eddies and filaments in coarse resolution models (Hood et al., 2003; Wiggert et al., 2005, 2006). Indeed, mesoscale and submesoscale eddies and filaments have been suggested to play a dominant role in advecting nutrients from the coastal upwelling region into the central and northern Arabian Sea (e.g., Koné et al., 2009). Using an eddy-resolving (1/12°) horizontal resolution model of the northern Indian Ocean, Resplandy et al. (2011) found that mesoscale eddies substantially increase nutrient supply to the surface and the oligotrophic open sea through both vertical eddy-mixing and lateral stirring. This in turn results in enhanced biological productivity and an improved agreement of the model with observations relative to coarse resolution models.

While modeling studies suggest that nitrogen is generally the most limiting nutrient of biological productivity in the Arabian Sea, silicate and iron have also been suggested to limit productivity locally and intermittently (see also Hood et al., 2024a). For instance, Koné et al. (2009) have shown that silicate limitation becomes dominant in local patches along the west coast of India. In another modeling study of the northern Indian Ocean, Resplandy et al. (2011) found that silicate limitation can also be important in the northern Arabian Sea during the northeast monsoon season. Finally, while the modeling studies of Koné et al. (2009) and Resplandy et al. (2011) suggest that iron limitation is rather marginal in the Arabian Sea, previous works by Wiggert et al. (2006) and Wiggert and Murtugudde (2007) indicate that the region off the coasts of Somalia and Oman is prone to iron limitation during both summer and winter monsoon seasons (see also Hood et al., 2024a). A recent modeling study by Guieu et al. (2019) confirms the critical role that iron input through dust deposition plays in fueling biological production in the Arabian Sea. Indeed, this study suggests that nearly half of summer primary production in the Arabian Sea is dependent on iron supply through eolian dust deposition (Fig. 5).

5.1.2 Bay of Bengal

Biological productivity in the Bay of Bengal is weaker than in the Arabian Sea, especially during the summer monsoon season (Prasanna Kumar et al., 2002) despite the presence of coastal upwelling in the northwestern area driven by the Findlater jet during the summer monsoon season (Gomes et al., 2000; Lévy et al., 2007). This is thought to be caused by the large freshwater input through rivers and rainfall that leads to the formation of a barrier layer near the surface limiting vertical mixing and nutrient transfer from the subsurface (e.g., Gomes et al., 2000; Vinayachandran et al., 2002b). Previous modeling work suggests that the bloom in the northwestern sector is essentially driven by upwelling and vertical mixing (Koné et al., 2009). During the northeast monsoon, observations reveal higher surface chlorophyll concentrations in the

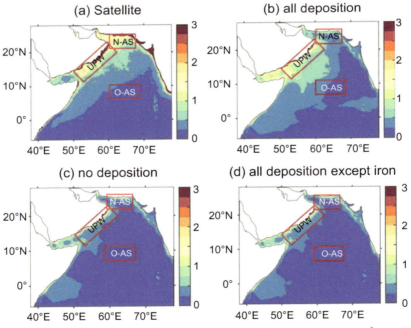

FIG. 5 Impact of nutrient atmospheric deposition on Arabian Sea productivity. Annual patterns of Chl-a (mg m^{-3}) in satellite data (a) and model simulation with (b) atmospheric deposition of soluble iron, nitrogen, and phosphate (all deposition), (c) no atmospheric deposition of nutrients (zero deposition), and (d) atmospheric deposition of nitrogen and phosphate only (all deposition except iron). Note that the "no deposition" and the "all deposition except iron" simulations fail to properly reproduce the observed Chl-a patterns, in contrast to the simulation where the deposition of all nutrients is taken into account. *(Adapted from Guieu et al. (2019).)*

western and northwestern Bay of Bengal (Gomes et al., 2000). The dynamics of the phytoplankton bloom during winter in the Bay of Bengal was first investigated using an intermediate complexity coupled physical-biogeochemical model by Vinayachandran et al. (2005). The model was similar to the one used to study the dynamics of blooms in the Arabian Sea by McCreary et al. (1996) and Hood et al. (2003). The model reproduces several characteristics of the observed bloom from the remotely sensed data. Their main finding was that the entrainment of not only subsurface nutrients but also phytoplankton contributes to the observed surface bloom. This entrainment is driven by both Ekman pumping and the deepening of the mixed layer (Vinayachandran et al., 2005). Finally, previous modeling studies suggest that nitrogen is generally the most limiting nutrient during bloom onset in most of the Bay of Bengal, except in limited areas along the east coast of India and in the eastern part of the bay where diatom growth tends to be silicate limited (Koné et al., 2009).

5.1.3 Southern tropical Indian Ocean

In the southern tropical Indian Ocean, observations and model simulations reveal the presence of a single bloom peaking in summer (Koné et al., 2009). Due to an upwelling-favorable wind stress curl, the nutricline is relatively shallow between 10°S and 15°S during the summer season. Episodic strong wind events leading to convective mixing cause a deepening of the mixed layer and entrainment of nutrients to the surface there (Koné et al., 2009). Resplandy et al. (2009) explored the biogeochemical response to variability in atmospheric forcing at intraseasonal scales such as that associated with the MJO in the SCTR. Using a coupled physical-biogeochemical eddy-permitting model of the Indian Ocean, they found that vertical mixing associated with MJO events fertilizes the mixed layer through entrainment, thus enhancing surface biological productivity. They also found that this response is highly sensitive to interannual variability of thermocline depth in the SCTR region.

5.2 Modeling Indian Ocean oxygen minimum zones (OMZs)

The OMZs of the Arabian Sea and the Bay of Bengal are among the most intense in the world ocean (Hood et al., 2024b; Paulmier & Ruiz-Pino, 2009). While both OMZs display low oxygen concentrations, important contrasts in their intensity are observed. While active denitrification and oxygen levels in the suboxic range ($< 5\,\mathrm{mmol\,m^{-3}}$) are observed over large swaths of the Arabian Sea, oxygen concentrations in the Bay of Bengal remain above the suboxic thresholds, preventing large-scale denitrification from occurring (Bristow et al., 2017; Hood et al., 2024b). Using a series of model simulations, Al Azhar et al. (2017) have demonstrated that deeper remineralization in the Bay of Bengal driven by faster particle sinking speeds contributes to weakening the OMZ there (see also Hood et al., 2024b). Indeed, ballast minerals associated with the large riverine input in the Bay of Bengal increase the particle sinking speed, thus reducing the organic matter residence time in the OMZ layer, and hence weakening the oxygen consumption there. Additional factors have been identified as potentially contributing to the varying OMZ intensity between the two seas (Fig. 6). For instance, McCreary et al. (2013) investigated the role of lateral advection and transport of organic matter in controlling the oxygen distribution in the Arabian Sea and Bay of Bengal. Their modeling work highlights the importance of lateral transport of organic matter from the western Arabian Sea into the central and eastern sectors of the Arabian Sea, where it enhances remineralization and contributes to lowering O_2 levels to below suboxic thresholds. According to these authors, the lack of such a source of organic matter in the Bay of Bengal may contribute to the weakness of its OMZ. In addition, differences in the intensity of eddy activity between the two OMZs have also been suggested to contribute to their contrasting intensities. Indeed, as eddies were shown to dominate both the vertical and lateral supply of oxygen in the Arabian Sea (Lachkar et al., 2016; Resplandy et al., 2012), the higher eddy activity in the Bay of Bengal may contribute to weakening its OMZ intensity (Sarma & Udaya Bhaskar, 2018).

5.3 Modeling Indian Ocean ecosystems under a warmer climate

The northern and western tropical Indian Ocean have experienced strong warming throughout most of the twentieth century, accelerating since the early 1990s (Kumar et al., 2009; Roxy et al., 2014, 2024). This warming has the potential to alter the marine ecosystems and the biogeochemistry of the region. For instance, using satellite observations as well as hindcast simulations, Roxy et al. (2016) found a significant drop in biological productivity in the western Indian Ocean that they attributed to enhanced vertical stratification suppressing vertical mixing and limiting nutrient supply from the ocean interior to the surface (see also Hood et al., 2024a). Their analysis has also suggested that the projected future warming in the region will likely cause a further decline in its biological productivity. Recent model projections confirm a general future decline of summer productivity in the western Indian Ocean and the Arabian Sea with important implications for the biological pump of carbon and fisheries (Bindoff et al., 2019; Kwiatkowski et al., 2020). In addition to increased stratification and dropping productivity, the ongoing surface warming in the northern and western Indian Ocean is also predicted to

FIG. 6 The contrasts in oxygen minimum zone (OMZ) intensity between the Arabian Sea and the Bay of Bengal and their potential drivers. (a) Despite stronger ventilation from the south (Somali Current) and in the north (associated with Persian Gulf water subduction and winter convective mixing), a very large flux of organic matter remineralized at shallow depth in the Arabian Sea leads to an intense OMZ with an important shallow denitrifying suboxic core. (b) A stronger stratification, a deeper remineralization due to ballasting by riverine particles, and stronger ventilation by eddies limit the intensity of the OMZ and prevent the occurrence of large-scale denitrification.

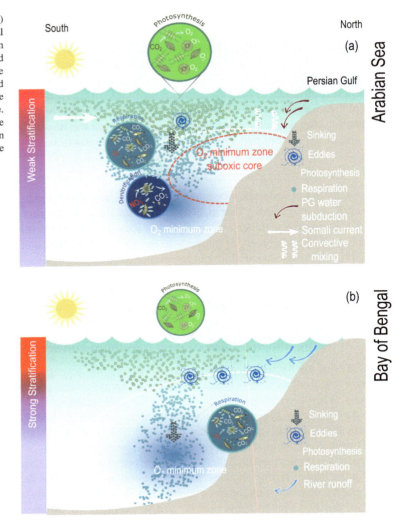

affect the Indian Ocean OMZs (Kwiatkowski et al., 2020). Yet, these projections come with high uncertainties that stem from multiple factors among which the representation of mixing in earth system models (Lévy et al., 2021), the strong sensitivity of OMZs to subtle changes in the balance between ventilation-driven O_2 replenishment and biology-induced O_2 depletion (Resplandy, 2018), and the uncertainty around future regional changes in Indian monsoon winds (e.g., Lachkar et al., 2018).

6 Conclusions and discussion

Ocean circulation and upper-ocean structure over the Indian Ocean are unique and complex. For example, a large seasonal cycle of upper-ocean currents is found almost everywhere in the tropics such as the seasonal reversal of low-latitude western boundary currents due to the strong influence of the monsoon on the ocean. Because of the improvement of large-scale high-resolution ocean GCMs in recent years, the mean and seasonal cycle of major ocean currents and upper-ocean structure, including narrow boundary currents, are now simulated by ocean models reasonably well. Yet quantitative comparisons with observations are still difficult in many locations because of the sparse coverage of in situ measurements. While recent satellite observations can be used to estimate surface current variability, it is still difficult to characterize the detailed structure of narrow boundary currents. Also, subsurface data are necessary to estimate the transport of major currents. Although the coverage of in situ measurements has been substantially improved in recent years due to Argo float measurements since the early 2000s (McPhaden et al., 2024; Phillips et al., 2024), it may not be sufficient for the comprehensive evaluation of model performance on different time scales in many locations such as the interior southern Indian Ocean.

Ocean GCMs' ability to reproduce SSTs over the Indian Ocean is crucial for climate model simulation and prediction, and comparisons with satellite-derived SSTs suggest that many models can simulate the mean and seasonal cycle of SST reasonably well. However, there are still a number of issues for evaluating the model performance quantitatively. For example, SST simulations of ocean GCMs critically depend on the quality of surface fluxes of heat, momentum, and freshwater. Yet a significant uncertainty exists in these surface fluxes (e.g., Yu, 2019), and thus model errors in SSTs are partly due to the errors in surface fluxes. The uncertainty of surface fluxes, especially surface heat fluxes, needs to be reduced for further quantitative evaluation of the model performance.

Upper ocean salinity stratification is shown to be important to control SSTs in the tropical Indian Ocean. In particular, accurate simulations of strong salinity stratification and thick barriers in the Bay of Bengal may largely influence SSTs. However, evaluating a model's capability of simulating sea surface salinity and subsurface salinity stratification is still challenging because of the uncertainty of surface freshwater fluxes and river runoff. Significant errors in the upper ocean salinity over the Bay of Bengal are evident in most ocean GCMs. These errors are partly caused by the uncertainty of freshwater input and model deficiencies, such as mixing parameterization, and the relative magnitude of these errors is still unknown.

Model simulation and prediction of climate variability over the Indian Ocean (such as IOD) using coupled ocean-atmosphere models have been improving in recent years. Some IOD events have been predicted with a lead time of one or two seasons, but the prediction skill is often reduced due to various prediction challenges such as the winter prediction barrier (Yamagata et al., 2024). To further improve current climate model simulations, it is at least necessary to advance our understanding of oceanic, atmospheric, and regional air-sea coupled processes associated with climate variability not only over the Indian Ocean but also in other ocean basins. For example, IOD is influenced by climate variability in other ocean basins including ENSO in the Pacific. Hence, simulation and prediction of Indian Ocean climate variability critically depend on simulations in other regions including ENSO in the Pacific. Dynamical and physical processes for the generation and development of climate variability such as IOD and Ningaloo Niño are not well understood (see also Yamagata et al., 2024). A better understanding of interbasin interaction of climate variability may help to improve climate prediction in the Indian Ocean by climate models.

Since biogeochemical processes are largely influenced by physical processes such as ocean currents, upwelling/downwelling, mixing, and upper ocean temperature changes, coupled biological/physical models that include adequate representations of physical processes are necessary to examine the unique biogeochemical processes in the Indian Ocean. Biogeochemical modeling in the Indian Ocean has become an active research area in recent years. For example, simulating the generation and maintenance of large OMZs in the Arabian Sea is one of the focus areas in recent research efforts. Identifying significant changes in the ecosystem in the Indian Ocean under a warming climate using coupled biological/physical modeling and observations has also been a focus of recent studies.

7 Educational resources

The source codes of some of the state-of-the-art ocean models, atmospheric models, and coupled models are publicly available. For example, major OGCMs widely used in climate research are mostly community models. Web sites of these community models are listed in the following:

Ocean GCMs

Modular Ocean Model version 6 (MOM6). https://github.com/NOAA-GFDL/MOM6-examples/wiki
Hybrid Coordinate Ocean Model (HYCOM). https://www.hycom.org
Regional Ocean Modeling System (ROMS). https://www.myroms.org
MITgcm. http://mitgcm.org

Atmosphere GCMs

The Community Atmosphere Model. http://www.cesm.ucar.edu/models/cesm2/atmosphere/
FV3GFS (Finite Volume 3 Global Forecast System). https://github.com/NOAA-EMC/fv3gfs

Coupled Atmosphere-Ocean GCMs

Community Earth System Model. http://www.cesm.ucar.edu
Unified Forecast System (UFS). https://ufscommunity.org

Regional coupled model

Scripps Coupled Ocean-Atmosphere Regional (SCOAR) Model. https://github.com/hyodae-seo/SCOAR

Acknowledgments

TS acknowledges support from National Oceanic and Atmospheric Administration Grant NA17OAR4310256 and DOD Grant W911NF-20-1-0309. TGJ is sponsored by the U.S. Office of Naval Directed Research Initiative PISTON (73-4347-27-5). HS is grateful for support from National Oceanic and Atmospheric Administration NA17OAR4310255 and ONR N00014-17-1-2398. ZL is supported by Tamkeen through Research Institute grant CG009 to the NYUAD Arabian Center for Climate and Environmental Sciences (ACCESS). YM is supported by the Japan Society for the Promotion of Science KAKENHI Grant Number JP17H01663.

Author contributions

TS led the overall writing, editing and organization of the manuscript. TS wrote Sections 1 and 6; TGJ and TS wrote Section 2; YM wrote Section 3; HS wrote Section 4; ZL wrote Section 5; all authors wrote Section 7. All authors contributed comments and feedback on overall structure of the chapter and contributed to revisions.

References

Al Azhar, M., Lachkar, Z., Lévy, M., & Smith, S. (2017). Oxygen minimum zone contrasts between the Arabian Sea and the Bay of Bengal implied by differences in remineralization depth. *Geophysical Research Letters, 44*(21), 11–106.

Aldrian, E., Sein, D., Jacob, D., Gates, L. D., & Podzun, R. (2005). Modelling Indonesian rainfall with a coupled regional model. *Climate Dynamics, 25*(1), 1–17.

Anderson, D. L. T., & Carrington, D. (1993). Modeling interannual variability in the Indian Ocean using momentum fluxes from the operational weather analyses of the United Kingdom Meteorological Office and European Centre for Medium Range Weather Forecasts. *Journal of Geophysical Research, 98*, 12483–12499.

Anderson, D. L. T., & Moore, D. W. (1979). Cross-equatorial inertial jets with special relevance to very remote forcing of the Somali current. *Deep Sea Research, 26*, 1–22.

Bachman, S. D., & Taylor, J. R. (2014). Modeling of partially-resolved oceanic symmetric instability. *Ocean Modelling, 82*, 15–27.

Beal, L. M., & Bryden, H. L. (1999). The velocity and vorticity structure of the Agulhas Current at 32°S. *Journal of Geophysical Research, 104*(C3), 5151–5176.

Beal, L. M., Chereskin, T. K., Bryden, H. L., & Ffield, A. (2003). Variability of water properties, heat and salt fluxes in the Arabian Sea, between the onset and wane of the 1995 Southwest monsoon. *Deep Sea Research, Part II, 50*, 2049–2075.

Beal, L. M., & Donohue, K. A. (2013). The Great Whirl: Observations of its seasonal development and interannual variability. *Journal of Geophysical Research-Oceans, 118*, 1–13. https://doi.org/10.1029/2012JC008198.

Beal, L. M., Elipot, S., Houk, A., & Leber, G. M. (2015). Capturing the transport variability of a western boundary jet: Results from the Agulhas Current Time-Series Experiment (ACT). *Journal of Physical Oceanography, 45*, 1302–1324. https://doi.org/10.1175/JPO-D-14-0119.1.

Beal, L., et al. (2020). A roadmap to IndOOS-2: Better observations of the rapidly-warming Indian Ocean. *Bulletin of the American Meteorological Society*, 1–50.

Biastoch, A., Sein, D., Durgadoo, J. V., Wang, Q., & Danilov, S. (2018). Simulating the Agulhas system in global ocean models—Nesting vs. multi-resolution unstructured meshes. *Ocean Modelling, 121*(2018), 117–131. https://doi.org/10.1016/j.ocemod.2017.12.002.

Bindoff, N. L., Cheung, W. W., Kairo, J. G., Arístegui, J., Guinder, V. A., Hallberg, R., ... Williamson, P. (2019). *Changing ocean, marine ecosystems, and dependent communities*. IPCC special report on the ocean and cryosphere in a changing climate (pp. 477–587).

Bishop, S. P., Small, R. J., & Bryan, F. O. (2020). The global sink of available potential energy by mesoscale air-sea interaction. *Journal of Advances in Modeling Earth Systems, 12*, e2020MS002118.

Bristow, L. A., Callbeck, C. M., Larsen, M., Altabet, M. A., Dekaezemacker, J., Forth, M., & Canfield, D. E. (2017). N2 production rates limited by nitrite availability in the bay of Bengal oxygen minimum zone. *Nature Geoscience, 10*(1), 24–29.

Bruce, J. (1973). Equatorial undercurrent in the western IndianOcean during the southwest monsoon. *Journal of Geophysical Research, 78*, 6386–6394. https://doi.org/10.1029/JC078i027p06386.

Bryden, H. L., Beal, L. M., & Duncan, L. (2005). Structure and transport of the Agulhas current and its temporal variability. *Journal of Oceanography, 61*, 479–492. https://doi.org/10.1007/s10872-005-0057-8.

Cane, M., & Moore, D. W. (1981). A note on low-frequency equatorial basin modes. *Journal of Physical Oceanography, 11*, 1578–1584.

Chelton, D. B., Schlax, M. G., Samelson, R. M., & de Szoeke, R. A. (2007). Global observations of large oceanic eddies. *Geophysical Research Letters, 34*(L15606). https://doi.org/10.1029/2007GL030812.

Chen, S., Campbell, T. J., Jin, H., Gabersek, S., Hodur, R. M., & Martin, P. (2010). Effect of two-way air-sea coupling in high and low wind speed regimes. *Monthly Weather Review, 138*, 3579–3602.

Chen, G. X., Han, W. Q., Li, Y. L., Yao, J., & Wang, D. (2019a). Intraseasonal variability of the equatorial undercurrent in the Indian Ocean. *Journal of Physical Oceanography, 45*, 85–101.

Chen, G. X., Han, W. Q., Li, Y. L., Wang, D. X., & McPhaden, M. J. (2019b). Seasonal-to-interannual time-scale dynamics of the equatorial undercurrent in the Indian Ocean. *Journal of Physical Oceanography, 45*, 1532–1553.

Chowdary, J. S., Srinivas, G., Fousiya, T. S., Parekh, A., Gnanaseelan, C., Seo, H., & MacKinnon, J. A. (2016). Representation of Bay of Bengal upper-ocean salinity in general circulation models. *Oceanography, 29*(2), 38–49. https://doi.org/10.5670/oceanog.2016.37.

Church, J. A., Cresswell, G. R., & Godfrey, J. S. (1989). The Leeuwin current. In S. J. Neshyba, C. N. K. Moors, R. L. Smith, & R. T. Barber (Eds.), *34. Polewardflows along eastern ocean boundaries. Coastal and estuarine studies* (pp. 230–254). New York: Springer.

Cox, M. D. (1970). A mathematical model of the Indian Ocean. *Deep Sea Research, 17*, 47–75.

Cox, M. D. (1976). Equatorially trapped waves and the generation of the Somali Current. *Deep Sea Research, 23*, 1139–1152.

Cresswell, G. R., & Golding, T. J. (1980). Observations of a south-flowing current in the southeastern Indian Ocean. *Deep Sea Research, 27*, 449–466.

Danabasoglu, G., Large, W. G., Tribbia, J. J., Gent, P. R., Briegleb, B. P., & McWilliams, J. C. (2006). Diurnal coupling in the tropical oceans of CCSM3. *Journal of Climate, 19*(11), 2347–2365.

Delman, A. S., McClean, J. L., Sprintall, J., Talley, L. D., & Bryan, F. O. (2018). Process-specific contributions to anomalous Java mixed layer cooling during positive IOD events. *Journal of Geophysical Research: Oceans, 123*, 4153–4176. https://doi.org/10.1029/2017JC013749.

DeMott, C. A., Klingaman, N. P., & Woolnough, S. J. (2015). Atmosphere-ocean coupled processes in the Madden-Julian oscillation. *Reviews of Geophysics, 53*, 1099–1154. https://doi.org/10.1002/2014RG000478.

DeMott, C. A., Ruppert, J. H., Jr., & Rydbeck, A. (2024). Chapter 4: Intraseasonal variability in the Indian Ocean region. In C. C. Ummenhofer, & R. R. Hood (Eds.), *The Indian Ocean and its role in the global climate system* (pp. 79–101). Amsterdam: Elsevier. https://doi.org/10.1016/B978-0-12-822698-8.00006-8.

DeMott, C. A., Stan, C., & Randall, D. A. (2013). Northward propagation mechanisms of the boreal summer intraseasonal oscillation in the ERA-Interim and SP-CCSM. *Journal of Climate, 26*, 1973–1992.

Doi, T., Behera, S. K., & Yamagata, T. (2016). Improved seasonal prediction using the SINTEX-F2 coupled model. *Journal of Advances in Modeling Earth Systems, 8*(4), 1847–1867. https://doi.org/10.1002/2016MS000744.

Doi, T., Behera, S. K., & Yamagata, T. (2019). Merits of a 108-member ensemble system in ENSO and IOD predictions. *Journal of Climate, 32*, 957–972.

Doi, T., Behera, S. K., & Yamagata, T. (2020). Predictability of the super IOD event in 2019 and its link with El Niño Modoki. *Geophysical Research Letters, 47*. https://doi.org/10.1029/2019GL086713.

Doi, T., Storto, A., Behera, S. K., Navarra, A., & Yamagata, T. (2017). Improved prediction of the Indian Ocean dipole mode by use of subsurface ocean observations. *Journal of Climate, 30*, 7953–7970. https://doi.org/10.1175/JCLI-D-16-0915.1.

Dong, J., Fox-Kemper, B., Zhu, J., & Dong, C. (2021). Application of symmetric instability parameterization in the coastal and Regional Ocean community model (CROCO). *Journal of Advances in Modeling Earth Systems, 13*.

Duvel, J. P., Roca, R., & Vialard, J. (2004). Ocean mixed layer temperature variations induced by intraseasonal convective perturbations over the Indian Ocean. *Journal of the Atmospheric Sciences, 61*, 1004–1022.

Effy, J. B., Francis, P. A., Ramakrishna, S. S. V. S., & Mukherjee, A. (2020). Anomalous warming of the western equatorial Indian Ocean in 2007: Role of ocean dynamics. *Ocean Modelling, 147*, 101542. https://doi.org/10.1016/j.ocemod.2019.101542.

Feng, R., & Duan, W. S. (2019). Indian Ocean Dipole–related predictability barriers induced by initial errors in the tropical Indian Ocean in a CGCM. *Advances in Atmospheric Sciences, 36*(6), 658–668. https://doi.org/10.1007/s00376-019-8224-9.

Feng, R., Duan, W., & Mu, M. (2014). The "winter predictability barrier" for IOD events and its error growth dynamics: Results from a fully coupled GCM. *Journal of Geophysical Research, Oceans, 119*, 8688–8708. https://doi.org/10.1002/2014JC010473.

Feng, M., Lengaigne, M., Manneela, S., Gupta, A. S., & Vialard, J. (2024). Chapter 6: Extreme events in the Indian Ocean: Marine heatwaves, cyclones, and tsunamis. In C. C. Ummenhofer, & R. R. Hood (Eds.), *The Indian Ocean and its role in the global climate system* (pp. 121–144). Amsterdam: Elsevier. https://doi.org/10.1016/B978-0-12-822698-8.00011-1.

Feng, M., McPhaden, M. J., Xie, S.-P., & Hafner, J. (2013). La Niña forces unprecedented Leeuwin Current warming in 2011. *Scientific Reports, 3*, 1–9. https://doi.org/10.1038/srep01277.

Feng, X., Shinoda, T., & Han, W. (2023). Topographic trapping of the Leeuwin Current and its impact on the 2010/11 Ningaloo Niño. *Journal of Climate, 36*(6), 1587–1603. https://doi.org/10.1175/JCLI-D-22-0218.1.

Fischer, A. S., Terray, P., Guilyardi, E., Gualdi, S., & Delecluse, P. (2004). Two independent triggers for the Indian Ocean dipole/zonal mode in a coupled GCM. *Journal of Climate, 18*, 3428–3449.

Friedrichs, M. A., Hood, R. R., & Wiggert, J. D. (2006). Ecosystem model complexity versus physical forcing: Quantification of their relative impact with assimilated Arabian Sea data. *Deep Sea Research Part II: Topical Studies in Oceanography, 53*(5–7), 576–600.

Fu, X., Wang, B., Waliser, D. E., & Tao, L. (2007). Impact of atmosphere—Ocean coupling on the predictability of monsoon intraseasonal oscillations. *Journal of the Atmospheric Sciences, 64*, 157–174.

Fu, X., Yang, B., Bao, Q., & Wang, B. (2008). Sea surface temperature feedback extends the predictability of tropical intraseasonal oscillation. *Monthly Weather Review, 136*, 577–597.

Furue, R. (2019). The three-dimensional structure of the Leeuwin Current System in density coordinates in an eddy-resolving OGCM. *Ocean Modelling, 138*, 36–50. https://doi.org/10.1016/j.ocemod.2019.03.001.

Gent, P. R. (1981). Forced standing equatorial ocean wave modes. *Journal of Marine Research, 39*, 695–709.

Giorgi, F. (2019). Thirty years of regional climate modeling: Where are we and where are we going next? *Journal of Geophysical Research-Atmospheres, 124*, 5696–5723.

Godfrey, J. S. (1989). A Sverdrup model of the depth-integrated flow for the world ocean allowing for island circulations. *Geophysical and Astrophysical Fluid Dynamics, 45*, 89 112. https://doi.org/10.1080/03091928908208894.

Godfrey, J. S., & Weaver, A. J. (1991). Is the Leeuwin Current driven by Pacific heating and winds? *Progress in Oceanography, 27*, 225–272.

Gomes, H. R., Goes, J. I., & Saino, T. (2000). Influence of physical processes and freshwater discharge on the seasonality of phytoplankton regime in the Bay of Bengal. *Continental Shelf Research, 20*(3), 313–330.

Gordon, A. L. (1985). Indian-Atlantic transfer of thermocline water at the Agulhas retroflection. *Science, 227*, 1030–1033.

Gordon, A. L., Sprintall, J., Van Aken, H. M., Susanto, D., Wijffels, S., Molcard, R., Ffield, A., Pranowo, W., & Wirasantosa, S. (2010). The Indonesian Throughflow during 2004-2006 as observed by the INSTANT program. *Dynamics of Atmosphere and Oceans, 50*, 115–128. https://doi.org/10.1016/j.dynatmoce.2009.12.002.

Goswami, B. N., Rao, S. A., Sengupta, D., & Chakravorty, S. (2016). Monsoons to mixing in the Bay of Bengal: Multiscale air-sea interactions and monsoon predictability. *Oceanography, 29*(2), 18–27.

Gualdi, S., Guilyardi, E., Navarra, A., Masina, S., & Delecluse, P. (2003). The interannual variability in the tropical Indian Ocean as simulated by a CGCM. *Climate Dynamics, 20*, 567–582. https://doi.org/10.1007/s00382-002-0295-z.

Guieu, C., Al Azhar, M., Aumont, O., Mahowald, N., Lévy, M., Éthé, C., & Lachkar, Z. (2019). Major impact of dust deposition on the productivity of the Arabian Sea. *Geophysical Research Letters, 46*(12), 6736–6744.

Halkides, D. J., & Lee, T. (2009). Mechanisms controlling seasonal-to-interannual mixed layer temperature variability in the southeastern tropical Indian Ocean. *Journal of Geophysical Research, 114*, C02012. https://doi.org/10.1029/2008JC004949.

Halkides, D. J., Waliser, D. E., Lee, T., Menemenlis, D., & Guan, B. (2015). Quantifying the processing controlling intraseasonal mixed-layer temperature variability in the tropical Indian Ocean. *Journal of Geophysical Research, Oceans, 120*, 692–715. https://doi.org/10.1002/2014JC010139.

Han, W., McCreary, J. P., Anderson, D. L. T., & Mariano, A. J. (1999). Dynamics of the eastern surface jets in the equatorial Indian Ocean. *Journal of Physical Oceanography, 29*, 2191–2209. https://doi.org/10.1175/1520-0485(1999) 029<2191:DOTESJ>2.0.CO;2.

Han, W., Shinoda, T., Fu, L.-L., & McCreary, J. P. (2006). Impact of atmospheric intraseasonal oscillations on the Indian Ocean dipole during the 1990s. *Journal of Physical Oceanography, 36*, 670–690.

Han, W., Yuan, D., Liu, W. T., & Halkides, D. J. (2007). Intraseasonal variability of the Indian Ocean Sea surface temperature during boreal winter: Madden-Julian oscillation verus submonthly forcing and processes. *Journal of Geophysical Research, 112*, C04001. https://doi.org/10.1029/2006JC003791.

Hermes, J., & Reason, C. J. C. (2008). Annual cycle of the South Indian Ocean (Seychelles-Chagos) thermocline ridge in a regional ocean model. *Journal of Geophysical Research, 113*, C04035. https://doi.org/10.1029/2007JC004363.

Hodur, R. M. (1997). The naval research Laboratory's coupled ocean/atmosphere mesoscale prediction system (COAMPS). *Monthly Weather Review, 125*, 1414–1430.

Hood, R. R., Coles, V. J., Huggett, J. A., Landry, M. R., Levy, M., Moffett, J. W., & Rixen, T. (2024a). Chapter 13: Nutrient, phytoplankton, and zooplankton variability in the Indian Ocean. In C. C. Ummenhofer, & R. R. Hood (Eds.), *The Indian Ocean and its role in the global climate system* (pp. 293–327). Amsterdam: Elsevier. https://doi.org/10.1016/B978-0-12-822698-8.00020-2.

Hood, R. R., Kohler, K. E., McCreary, J. P., & Smith, S. L. (2003). A four-dimensional validation of a coupled physical–biological model of the Arabian Sea. *Deep Sea Research Part II: Topical Studies in Oceanography, 50*(22–26), 2917–2945.

Hood, R. R., Rixen, T., Levy, M., Hansell, D. A., Coles, V. J., & Lachkar, Z. (2024b). Chapter 12: Oxygen, carbon, and pH variability in the Indian Ocean. In C. C. Ummenhofer, & R. R. Hood (Eds.), *The Indian Ocean and its role in the global climate system* (pp. 265–291). Amsterdam: Elsevier. https://doi.org/10.1016/B978-0-12-822698-8.00017-2.

Howden, S. D., & Murtugudde, R. (2001). Effects of river inputs into the Bay of Bengal. *JGR Oceans, 106*, 19825–19843.

Iizuka, S., Matsuura, T., & Yamagata, T. (2000). The Indian Ocean SST dipole simulated in a coupled general circulation model. *Geophysical Research Letters, 27*, 3369–3372.

Jaeger, G. S., Lucas, A. J., & Mahadevan, A. (2020). Formation of interleaving layers in the Bay of Bengal. *Deep Sea Research, Part II, 172*, 104717.

Jayakumar, A., Vialard, J., Lengaigne, M., Gnanaseelan, C., McCreary, J. P., & Kumar, B. P. (2011). Processes controlling the surface temperature signature of the Madden-Julian oscillation in the thermocline ridge of the Indian Ocean. *Climate Dynamics, 37*, 2217–2234.

Jensen, T. G. (1991). Modeling the seasonal undercurrents in the Somali Current system. *Journal of Geophysical Research-Oceans, 96*, 22151–22167.

Jensen, T. G. (1993). Equatorial variability and resonance in a wind-driven Indian Ocean model. *Journal of Geophysical Research-Oceans, 98*, 22533–22552.

Jensen, T. G., Shinoda, T., Chen, S., & Flatau, M. (2015). Ocean response to CINDY/DYNAMO MJOs in air-sea coupled COAMPS. *Journal of the Meteorological Society of Japan, 93A*, 157–178.

Kantha, L. H., & Clayson, C. A. (2004). On the effect of surface gravity waves on mixing in an oceanic mixed layer model. *Ocean Modelling, 6*, 101–124.

Karmakar, N., & Misra, V. (2020a). Differences in northward propagation of convection over the Arabian Sea and Bay of Bengal during boreal summer. *Journal of Geophysical Research-Atmospheres, 125*.

Karmakar, N., & Misra, V. (2020b). The fidelity of a regional coupled model in capturing the relationship between intraseasonal variability and the onset/demise of the Indian summer monsoon. *Climate Dynamics, 54*, 4693–4710.

Kido, S., Tozuka, T., & Han, W. (2019a). Anatomy of salinity anomalies associated with the positive Indian Ocean Dipole. *Journal of Geophysical Research: Oceans, 124*, 8116–8139. https://doi.org/10.1029/2019JC015163.

Kido, S., Tozuka, T., & Han, W. (2019b). Experimental assessments on impacts of salinity anomalies on the positive Indian Ocean Dipole. *Journal of Geophysical Research: Oceans, 124*, 9462–9486. https://doi.org/10.1029/2019JC015479.

Kindle, J. C., & Thompson, J. D. (1989). The 26- and 50-day oscillations in the Western Indian Ocean: Model results. *Journal of Geophysical Research, 94*, 4721–4736.

Knauss, J. A., & Taft, B. A. (1964). Equatorial Undercurrent of theIndian Ocean. *Science, 143*, 354–356. https://doi.org/10.1126/science.143.3604.354.

Knox, R. A. (1974). Reconnaissance of the Indian Ocean equatorial undercurrent near Addu Atoll. *Deep Sea Research and Oceanographic Abstracts, 21*, 123–129. https://doi.org/10.1016/0011-7471(74)90069-2.

Koné, V., Aumont, O., Lévy, M., & Resplandy, L. (2009). Physical and biogeochemical controls of the phytoplankton seasonal cycle in the Indian Ocean: A modeling study. *Indian Ocean Biogeochemical Processes and Ecological Variability, 185*, 350.

Krishnamohan, K., Vialard, J., Lengaigne, M., Masson, S., Samson, G., Pous, S., Neetu, S., Durand, F., Shenoi, S., & Madec, G. (2019). Is there an effect of Bay of Bengal salinity on the northern Indian Ocean climatological rainfall? *Deep Sea Research, Part II, 166*, 19–33.

Kumar, S. P., Roshin, R. P., Narvekar, J., Kumar, P. D., & Vivekanandan, E. (2009). Response of the Arabian Sea to global warming and associated regional climate shift. *Marine Environmental Research, 68*(5), 217–222.

Kwiatkowski, L., Torres, O., Bopp, L., Aumont, O., Chamberlain, M., Christian, J. R., & Ziehn, T. (2020). Twenty-first century ocean warming, acidification, deoxygenation, and upper-ocean nutrient and primary production decline from CMIP6 model projections. *Biogeosciences, 17*(13), 3439–3470.

Lachkar, Z., Lévy, M., & Smith, S. (2018). Intensification and deepening of the Arabian Sea oxygen minimum zone in response to increase in Indian monsoon wind intensity. *Biogeosciences, 15*(1), 159–186.

Lachkar, Z., Smith, S., Lévy, M., & Pauluis, O. (2016). Eddies reduce denitrification and compress habitats in the Arabian Sea. *Geophysical Research Letters, 43*(17), 9148–9156.

Lau, N. C., & Nath, M. J. (2004). Coupled GCM simulation of atmosphere-ocean variability associated with zonally asymmetric SST changes in the tropical Indian Ocean. *Journal of Climate, 17*, 245–265.

Lee, T., Fukumori, I., Menemenlis, D., Xing, Z., & Fu, L. (2002). Effects of the Indonesian throughflow on the Pacific and Indian oceans. *Journal of Physical Oceanography, 32*, 1404–1429. https://doi.org/10.1175/1520-0485(2002)032<1404:EOTITO>2.0.CO;2.

Lee, T., & Marotzke, J. (1998). Seasonal cycles of meridional overturning and heat transport of the Indian Ocean. *Journal of Physical Oceanography, 28*, 923–943.

Lévy, M., Resplandy, L., Palter, J. B., Couespel, D., & Lachkar, Z. (2021). The crucial contribution of mixing to present and future ocean oxygen distribution. *Ocean Mixing*, 329–344.

Lévy, M., Shankar, D., André, J. M., Shenoi, S. S. C., Durand, F., & de Boyer Montégut, C. (2007). Basin-wide seasonal evolution of the Indian Ocean's phytoplankton blooms. *Journal of Geophysical Research: Oceans, 112*(C12).

Li, Y., Han, W., Ravichandran, M., Wang, W., Shinoda, T., & Lee, T. (2017a). Bay of Bengal salinity stratification and Indian summer monsoon intraseasonal oscillation: 1. Intraseasonal variability and causes. *Journal of Geophysical Research: Oceans, 122*, 4291–4311. https://doi.org/10.1002/2017JC012691.

Li, Y., Han, W., Shinoda, T., Wang, C., Ravichandran, M., & Wang, J.-W. (2014). Revisiting the wintertime intraseasonal SST variability in the tropical South Indian Ocean: Impact of the ocean interannual variation. *Journal of Physical Oceanography, 44*(7), 1886–1907.

Li, Y., Han, W., Wang, W., Ravichandran, M., Lee, T., & Shinoda, T. (2017b). Bay of Bengal salinity stratification and Indian summer monsoon intraseasonal oscillation: 2. Impact on SST and convection. *Journal of Geophysical Research, 122*, 4312–4328. https://doi.org/10.1002/2017JC012692.

Li, Y., Jourdain, N. C., Taschetto, A. S., Gupta, A. S., Argüeso, D., Masson, S., & Cai, W. (2017c). Resolution dependence of the simulated precipitation and diurnal cycle over the maritime continent. *Climate Dynamics, 48*(11−12), 4009–4028.

Li, H., Kanamitsu, M., & Hong, S.-Y. (2012). California reanalysis downscaling at 10 km using an ocean–atmosphere coupled regional model system. *Journal of Geophysical Research – Atmospheres, 117*, D12118.

Li, Q., Webb, A., Fox-Kemper, B., Craig, A., Danabasoglu, G., Large, W. G., & Vertenstein, M. (2016). Langmuir mixing effects on global climate: WAVEWATCH III in CESM. *Ocean Modelling, 103*, 145–160.

Lighthill, M. J. (1969). Dynamic response of the Indian Ocean to the onset of the southwest monsoon. *Philosophical Transactions of the Royal Meteorological Society A, A265*, 45–92.

Lu, B., Ren, H. L., Scaife, A. A., Wu, J., Dunstone, N., Smith, D., Wan, J., Eade, R., MacLachlan, C., & Gordon, M. (2018). An extreme negative Indian Ocean Dipole event in 2016: Dynamics and predictability. *Climate Dynamics, 51*, 89–100. https://doi.org/10.1007/s00382-017-3908-2.

Luo, J. J., Behera, S., Masumoto, Y., Sakuma, H., & Yamagata, T. (2008). Successful prediction of the consecutive IOD in 2006 and 2007. *Geophysical Research Letters, 35*, L14S02. https://doi.org/10.1029/2007GL032793.

Luo, J. J., Masson, S., Behera, S., & Yamagata, T. (2007). Experimental forecasts of the Indian Ocean dipole using a coupled OAGCM. *Journal of Climate, 20*, 2178–2190. https://doi.org/10.1175/JCLI4132.1.

Luther, M., & O'Brien, J. J. (1985). A model of the seasonal circulation in the Arabian Sea forced by observed winds. *Progress in Oceanography, 14*, 353–385.

Lutjeharms, J. R. E. (2006). *The Agulhas Current*. Springer-Verlag. ISBN 10354042392. 330 pp.

Masson, S., Menkes, C., Delecluse, P., & Boulanger, J.-P. (2003). Impacts of salinity on the eastern Indian Ocean during the termination of the fall Wyrtki Jet. *Journal of Geophysical Research, 108*(C3), 3067. https://doi.org/10.1029/2001JC000833.

Masumoto, Y., & Meyers, G. (1998). Forced Rossby waves in the southern tropical Indian Ocean. *Journal of Geophysical Research, 103*(C12), 27589–27602.

McCreary, J. P., Kundu, P. K., & Molinari, R. L. (1993). A numerical investigation of dynamics, thermodynamics and mixed-layer processes in the Indian Ocean. *Progress in Oceanography, 31*, 181–244.

McCreary, J. P., Kohler, K. E., Hood, R. R., & Olson, D. B. (1996). A four-component ecosystem model of biological activity in the Arabian Sea. *Progress in Oceanography, 37*(3–4), 193–240.

McCreary, J. P., & Kundu, P. K. (1985). Western boundary circulation driven by an alongshore wind: With application to the Somali current system. *Journal of Marine Research, 43*, 493–516.

McCreary, J. P., Yu, Z., Hood, R. R., Vinaychandran, P. N., Furue, R., Ishida, A., & Richards, K. J. (2013). Dynamics of the Indian-Ocean oxygen minimum zones. *Progress in Oceanography, 112*, 15–37.

McPhaden, M. J., Beal, L. M., Bhaskar, T. V. S. U., Lee, T., Nagura, M., Strutton, P. G., & Yu, L. (2024). Chapter 17: The Indian Ocean Observing System (IndOOS). In C. C. Ummenhofer, & R. R. Hood (Eds.), *The Indian Ocean and its role in the global climate system* (pp. 393–419). Amsterdam: Elsevier. https://doi.org/10.1016/B978-0-12-822698-8.00002-0.

Melzer, B. A., Jensen, T. G., & Rydbeck, A. V. (2019). The life cycle and interannual variability of the Great Whirl from altimetry-based eddy tracking. *Geophysical Research Letters, 46*. https://doi.org/10.1029/2018GL081781.

Menezes, V. V., Phillips, H. E., Schiller, A., Bindoff, N. L., Domingues, C. M., & Vianna, M. L. (2014). South Indian Countercurrent and associated fronts. *Journal of Geophysical Research, Oceans, 119*, 6763–6791. https://doi.org/10.1002/2014JC010076.

Menezes, V. V., Phillips, H. E., Vianna, M. L., & Bindoff, N. L. (2016). Interannual variability of the South Indian Countercurrent. *Journal of Geophysical Research: Oceans, 121*(5), 3465–3487.

Metzger, E. J., Hurlburt, H. E., Xu, X., Shriver, J. F., Gordon, A. L., Sprintall, J., Susanto, R. D., & van Aken, H. M. (2010). Simulated and observed circulation in the Indonesian Seas: 1/12° global HYCOM and the INSTANT observations. *Dynamics of Atmospheres and Oceans, 50*, 275–300. https://doi.org/10.1016/j.dynatmoce.2010.04.002.

Miller, A. J., Collins, M., Gualdi, S., Jensen, T. G., Misra, V., Pezzi, L. P., Pierce, D. W., Putrasahan, D., Seo, H., & Tseng, Y.-H. (2017). Coupled ocean-atmosphere-hydrology modeling and predictions. *Journal of Marine Research, 75*, 361–402.

Misra, V., Mishra, A., & Bhardwaj, A. (2017). High-resolution regional-coupled ocean–atmosphere simulation of the Indian summer monsoon. *International Journal of Climatology, 37*, 717–740.

Misra, V., Mishra, A., & Bhardwaj, A. (2018). Simulation of the intraseasonal variations of the Indian summer monsoon in a regional coupled ocean—Atmosphere model. *Journal of Climate, 31*(8), 3167–3185.

Murtugudde, R., & Busalacchi, A. J. (1999). Interannual variability of the dynamics and thermodynamics of the tropical Indian Ocean. *Journal of Climate, 12*, 2300–2326.

Nagura, M., & McPhaden, M. J. (2010). Wyrtki Jet dynamics: Seasonal variability. *Journal of Geophysical Research, 115*, C07009. https://doi.org/10.1029/2009JC005922.

Nagura, M., & McPhaden, M. J. (2018). The shallow overturning circulation in the Indian Ocean. *Journal of Physical Oceanography, 48*, 413–434.

Nyadjro, E., Jensen, T. G., Richman, J., & Shriver, J. (2017). On the interactions between wind, SST and the subsurface ocean in the southwestern Indian Ocean. *IEEE Geoscience and Remote Sensing Letters, 14*(12), 2315–2319.

Ogata, T., & Masumoto, Y. (2010). Interactions between mesoscale eddy variability and Indian Ocean dipole events in the southeastern tropical Indian Ocean—Case studies for 1994 and 1997/1998. *Ocean Dynamics, 60*, 717–730.

Paulmier, A., & Ruiz-Pino, D. (2009). Oxygen minimum zones (OMZs) in the modern ocean. *Progress in Oceanography, 80*(3–4), 113–128.

Pearce, A. F., & Feng, M. (2013). The rise and fall of the "marine heat wave" off Western Australia during the summer of 2010/2011. *Journal of Marine System, 111–112*, 139–156. https://doi.org/10.1016/j.jmarsys.2012.10.009.

Phillips, H. E., Menezes, V. V., Nagura, M., McPhaden, M. J., Vinayachandran, P. N., & Beal, L. M. (2024). Chapter 8: Indian Ocean circulation. In C. C. Ummenhofer, & R. R. Hood (Eds.), *The Indian Ocean and its role in the global climate system* (pp. 169–203). Amsterdam: Elsevier. https://doi.org/10.1016/B978-0-12-822698-8.00012-3.

Pianezze, J., Barthe, C., Bielli, S., Tulet, P., Jullien, S., Cambon, G., et al. (2018). A new coupled ocean-waves-atmosphere model designed for tropical storm studies: Example of tropical cyclone Bejisa (2013–2014) in the South-West Indian Ocean. *Journal of Advances in Modeling Earth Systems, 10*.

Prasad, T. G., & McClean, J. L. (2004). Mechanisms for anomalous warming in the western Indian Ocean during dipole mode events. *Journal of Geophysical Research, 109*, C02019. https://doi.org/10.1029/2003JC001872.

Prasanna Kumar, S., Muraleedharan, P. M., Prasad, T. G., Gauns, M., Ramaiah, N., De Souza, S. N., & Madhupratap, M. (2002). Why is the Bay of Bengal less productive during summer monsoon compared to the Arabian Sea? *Geophysical Research Letters, 29*(24).

Quadfasel, D. (1982). Low frequency variability of the 20°C isotherm topography in the western equatorial Indian Ocean. *Journal of Geophysical Research, 87*(C3), 1990–1996.

Rahman, H., et al. (2020). An assessment of the Indian Ocean mean state and seasonal cycle in a suite of interannual CORE-II simulations. *Ocean Modelling, 145*, 101503.

Ramachandran, S., Tandon, A., Mackinnon, J., Lucas, A. J., Pinkel, R., Waterhouse, A. F., Nash, J., Shroyer, E., Mahadevan, A., Weller, R. A., & Farrar, J. T. (2018). Submesoscale processes at shallow salinity fronts in the Bay of Bengal: Observations during the winter monsoon. *Journal of Physical Oceanography, 48*(3), 479–509.

Rao, S. A., Masson, S., Luo, J. J., Behera, S. K., & Yamagata, T. (2007). Termination of Indian Ocean dipole events in a coupled general circulation model. *Journal of Climate, 20*, 3018–3035. https://doi.org/10.1175/JCLI4164.1.

Renault, L., Masson, S., Oerder, V., Jullien, S., & Colas, F. (2019). Disentangling the mesoscale ocean-atmosphere interactions. *Journal of Geophysical Research: Oceans, 124*, 2164–2178.

Renault, L., McWilliams, J. C., & Penven, P. (2017). Modulation of the Agulhas Current retroflection and leakage by oceanic current interaction with the atmosphere in coupled simulations. *Journal of Physical Oceanography, 47*, 2077–2100. https://doi.org/10.1175/JPO-D-16-0168.1.

Renault, L., Molemaker, M. J., Gula, J., Masson, S., & McWilliams, J. C. (2016). Control and stabilization of the Gulf stream by oceanic current interaction with the atmosphere. *Journal of Physical Oceanography, 46*(11), 3439–3453.

Reppin, J., Schott, F. A., Fischer, J., & Quadfasel, D. (1999). Equatorial currents and transports in the upper central Indian Ocean: Annual cycle and interannual variability. *Journal of Geophysical Research, 104*, 15495–15514. https://doi.org/10.1029/1999JC900093.

Resplandy, L. (2018). Will ocean zones with low oxygen levels expand or shrink? *Nature, 557*.

Resplandy, L., Lévy, M., Bopp, L., Echevin, V., Pous, S., Sarma, V. V. S. S., & Kumar, D. (2012). Controlling factors of the oxygen balance in the Arabian Sea's OMZ. *Biogeosciences, 9*(12), 5095–5109.

Resplandy, L., Lévy, M., Madec, G., Pous, S., Aumont, O., & Kumar, D. (2011). Contribution of mesoscale processes to nutrient budgets in the Arabian Sea. *Journal of Geophysical Research: Oceans, 116*(C11).

Resplandy, L., Vialard, J., Lévy, M., Aumont, O., & Dandonneau, Y. (2009). Seasonal and intraseasonal biogeochemical variability in the thermocline ridge of the southern tropical Indian Ocean. *Journal of Geophysical Research: Oceans, 114*(C7).

Roxy, M. K., Modi, A., Murtugudde, R., Valsala, V., Panickal, S., Prasanna Kumar, S., & Lévy, M. (2016). A reduction in marine primary productivity driven by rapid warming over the tropical Indian Ocean. *Geophysical Research Letters, 43*(2), 826–833.

Roxy, M. K., Ritika, K., Terray, P., & Masson, S. (2014). The curious case of Indian Ocean warming. *Journal of Climate, 27*(22), 8501–8509.

Roxy, M. K., Saranya, J. S., Modi, A., Anusree, A., Cai, W., Resplandy, L., … Frölicher, T. L. (2024). Chapter 20: Future projections for the tropical Indian Ocean. In C. C. Ummenhofer, & R. R. Hood (Eds.), *The Indian Ocean and its role in the global climate system* (pp. 469–482). Amsterdam: Elsevier. https://doi.org/10.1016/B978-0-12-822698-8.00004-4.

Ruppert, J. H., & Johnson, R. H. (2015). Diurnally modulated cumulus moistening in the preonset stage of the Madden–Julian oscillation during DYNAMO. *Journal of the Atmospheric Sciences, 72*(4), 1622–1647. https://doi.org/10.1175/JAS-D-14-0218.1.

Ryabchenko, V. A., Gorchakov, V. A., & Fasham, M. J. R. (1998). Seasonal dynamics and biological productivity in the Arabian Sea Euphotic Zone as simulated by a three-dimensional ecosystem model. *Global Biogeochemical Cycles, 12*(3), 501–530.

Saji, N. H., Goswami, B. N., Vinayachandran, P. N., & Yamagata, T. (1999). A dipole mode in the tropical Indian Ocean. *Nature, 401*, 360–363.

Saji, N. H., Xie, S.-P., & Tam, C.-Y. (2006). Satellite observations of intense intraseasonal cooling events in the tropical south Indian Ocean. *Geophysical Research Letters, 33*, L14704. https://doi.org/10.1029/2006GL026525.

Samson, G., Masson, S., Lengaigne, M., Keerthi, M. G., Vialard, J., Pous, S., Madec, G., Jourdain, N., Julien, S., Menkes, C., & Marchesiello, P. (2014). The NOW regional coupled model: Application to the tropical Indian Ocean climate and tropical cyclone activity. *Journal of Advances in Modeling Earth Systems, 06*. https://doi.org/10.1002/2014MS000324.

Sarma, V. V. S. S., & Udaya Bhaskar, T. V. S. (2018). Ventilation of oxygen to oxygen minimum zone due to anticyclonic eddies in the Bay of Bengal. *Journal of Geophysical Research: Biogeosciences, 123*(7), 2145–2153.

Sasaki, H., Kida, S., Furue, R., Nonaka, M., & Masumoto, Y. (2018). An increase of the Indonesian Throughflow by internal tidal mixing in a high-resolution Quasi-Global Ocean simulation. *Geophysical Research Letters, 45*, 8416–8424.

Schiller, A., Godfrey, J. S., McIntosh, P. C., Meyers, G., & Wijffels, S. E. (1998). Seasonal near-surface dynamics and thermodynamics of the Indian Ocean and Indonesian throughflow in a global ocean general circulation model. *Journal of Physical Oceanography*, 2288–2312. https://doi.org/10.1175/1520-0485(1998)028<2288:SNSDAT>2.0.CO;2.

Schott, F. A., & McCreary, J. P. (2001). The monsoon circulation of the Indian Ocean. *Progress in Oceanography, 51*, 1–123.

Schott, F. A., Swallow, J. C., & Fieux, M. (1990). The Somali Current at the equator: Annual cycle of currents and transports in the upper 1000 m and connection to neighbouring latitudes. *Deep Sea Research Part A, 37*(12), 1825–1848. https://doi.org/10.1016/0198-0149(90)90080-F.

Schrum, C. (2017). Regional climate modeling and air-sea coupling. In *Oxford research encyclopedia of climate science*. From: https://oxfordre.com/climatescience/view/10.1093/acrefore/9780190228620.001.0001/acrefore-9780190228620-e-3.

Seo, H. (2017). Distinct influence of air-sea interactions mediated by mesoscale sea surface temperature and surface current in the Arabian Sea. *Journal of Climate, 30*, 8061–8079.

Seo, H., Miller, A. J., & Norris, J. R. (2016). Eddy-wind interaction in the California current system: Dynamics and impacts. *Journal of Physical Oceanography, 46*, 439–459.

Seo, H., Miller, A. J., & Roads, J. O. (2007). The Scripps Coupled Ocean–Atmosphere Regional (SCOAR) model, with applications in the eastern Pacific sector. *Journal of Climate, 20*, 381–402.

Seo, H., Murtugudde, R., Jochum, M., & Miller, A. J. (2008). Modeling of mesoscale coupled ocean–atmosphere interaction and its feedback to ocean in the Western Arabian Sea. *Ocean Modelling, 25*, 120–131.

Seo, H., Subramanian, A. C., Miller, A. J., & Cavanaugh, N. R. (2014). Coupled impacts of the diurnal cycle of sea surface temperature on the Madden-Julian Oscillation. *Journal of Climate, 27*, 8422–8443.

Seo, H., Subramanian, A. C., Song, H., & Chowdary, J. S. (2019). Coupled effects of ocean current on wind stress in the Bay of Bengal: Eddy energetics and upper ocean stratification. *Deep-Sea Research Part II, 168*(104), 617.

Seo, H., Xie, S.-P., Murtugudde, R., Jochum, M., & Miller, A. J. (2009). Seasonal effects of Indian Ocean freshwater forcing in a regional coupled model. *Journal of Climate, 22*, 6577–6596.

Shenoi, S. S. C., Shankar, D., & Shetye, S. R. (2002). Differences in heat budgets of the near-surface Arabian Sea and Bay of Bengal: Implications for the summer monsoon. *Journal of Geophysical Research, 107*(C6), 3052. https://doi.org/10.1029/2000JC000679.

Shi, Q., & Bourassa, M. A. (2019). Coupling ocean currents and waves with wind stress over the Gulf stream. *Remote Sensing, 11*, 1476.

Shinoda, T. (2005). Impact of the diurnal cycle of solar radiation on intraseasonal SST variability in the western equatorial Pacific. *Journal of Climate, 18*, 2628–2636.

Shinoda, T., Han, W., Jensen, T. G., Zamudio, L., Metzger, E. J., & Lien, R.-C. (2016). Impact of the Madden-Julian Oscillation on the Indonesian Throughflow in the Makassar Strait during the CINDY/DYNAMO field campaign. *Journal of Climate, 29*, 6085–6108. https://doi.org/10.1175/JCLI-D-15-0711.1.

Shinoda, T., Han, W., Metzger, E. J., & Hurlburt, H. (2012). Seasonal variation of the Indonesian Throughflow in Makassar Strait. *Journal of Physical Oceanography, 42*, 1099–1123. https://doi.org/10.1175/JPO-D-11-0120.1.

Shinoda, T., Han, W., Zamudio, L., Lien, R.-C., & Katsumata, M. (2017). Remote Ocean response to the Madden-Julian oscillation during the DYNAMO field campaign: Impact on Somali Current System and the Seychelles-Chagos thermocline ridge. *Atmosphere, 8*(9), 171.

Shinoda, T., Pei, S., Wang, W., Fu, J. X., Lien, R.-C., Seo, H., & Soloviev, A. (2021). Climate process team: Improvement of ocean component of NOAA climate forecast system relevant to Madden-Julian oscillation simulations. *Journal of Advances in Modeling Earth Systems, 13*. https://doi.org/10.1029/2021MS002658.

Small, R. J., et al. (2008). Air–sea interaction over ocean fronts and eddies. *Dynamics of Atmospheres and Oceans, 45*, 274–319.

Smith, R. L., Huyer, A., Godfrey, J. S., & Church, J. A. (1991). The Leeuwin Current off Western Australia, 1986–1987. *Journal of Physical Oceanography, 21*, 323–345. https://doi.org/10.1175/1520-0485(1991)021<0323:TLCOWA>2.0.CO;2.

Soloviev, A. V., & Lukas, R. (1996). Observation of spatial variability of diurnal thermocline and rain-formed halocline in the western Pacific warm pool. *Journal of Physical Oceanography, 26*, 2529–2538.

Song, Q., Vecchi, G. A., & Rosati, A. (2007). Indian Ocean variability in the GFDL CM2 coupled climate model. *Journal of Climate, 20*, 2895–2916.

Song, Q., Vecchi, G. A., & Rosati, A. J. (2008). Predictability of the Indian Ocean Sea surface temperature anomalies in the GFDL coupled model. *Geophysical Research Letters, 35*, L02701. https://doi.org/10.1029/2007GL031966.

Sprintall, J., Biastoch, A., Gruenburg, L. K., & Phillips, H. E. (2024). Chapter 9: Oceanic basin connections. In C. C. Ummenhofer, & R. R. Hood (Eds.), *The Indian Ocean and its role in the global climate system* (pp. 205–227). Amsterdam: Elsevier. https://doi.org/10.1016/B978-0-12-822698-8.00003-2.

Sui, C.-H., Lau, K.-M., Takayabu, Y. N., & Short, D. A. (1997). Diurnal variations in tropical oceanic cumulus convection during TOGA COARE. *Journal of the Atmospheric Sciences, 54*(1990), 639–655.

Swallow, J. (1964). Equatorial undercurrent in the western Indian Ocean. *Nature, 204*, 436–437. https://doi.org/10.1038/204436a0.

Swallow, J. C., & Fieux, M. (1982). Historical evidence for two gyres in the Somali Current. *Journal of Marine Research, 40*, 747–755.

Takaya, Y., Bidlot, J. R., Beljaars, A., & Janssen, P. A. (2010). Refinements to a prognostic scheme of skin sea surface temperature. *Journal of Geophysical Research, 115*, C06009.

Tanizaki, C., Tozuka, T., Doi, T., & Yamagata, T. (2017). Relative importance of the processes contributing to the development of SST anomalies in the eastern pole of the Indian Ocean Dipole and its implication for predictability. *Climate Dynamics, 49*, 1289–1304. https://doi.org/10.1007/s00382-016-3382-2.

Thompson, B., Gnanaseelan, C., Parekh, A., & Salvekar, P. S. (2009). A model study on oceanic processes during the Indian Ocean Dipole termination. *Meteorology and Atmospheric Physics, 105*, 17–27. https://doi.org/10.1007/s00703-009-0033-8.

Thompson, B., Gnanaseelan, C., & Salvekar, P. S. (2006). Variability in the Indian Ocean circulation and salinity and its impact on SST anomalies during dipole events. *Journal of Marine Research, 64*, 853–880.

Thoppil, P. G., Richman, J. G., & Hogan, P. J. (2011). Energetics of a global ocean circulation model compared to observations. *Geophysical Research Letters, 38*, L15607. https://doi.org/10.1029/2011GL048347.

Tozuka, T., Dong, L., Han, W., Lengainge, M., & Zhang, L. (2024). Chapter 10: Decadal variability of the Indian Ocean and its predictability. In C. C. Ummenhofer, & R. R. Hood (Eds.), *The Indian Ocean and its role in the global climate system* (pp. 229–244). Amsterdam: Elsevier. https://doi.org/10.1016/B978-0-12-822698-8.00014-7.

Ummenhofer, C. C., Geen, R., Denniston, R. F., & Rao, M. P. (2024a). Chapter 3: Past, present, and future of the South Asian monsoon. In C. C. Ummenhofer, & R. R. Hood (Eds.), *The Indian Ocean and its role in the global climate system* (pp. 49–78). Amsterdam: Elsevier. https://doi.org/10.1016/B978-0-12-822698-8.00013-5.

Ummenhofer, C. C., Taschetto, A. S., Izumo, T., & Luo, J.-J. (2024b). Chapter 7: Impacts of the Indian Ocean on regional and global climate. In C. C. Ummenhofer, & R. R. Hood (Eds.), *The Indian Ocean and its role in the global climate system* (pp. 145–168). Amsterdam: Elsevier. https://doi.org/10.1016/B978-0-12-822698-8.00018-4.

Vecchi, G. A., Xie, S.-P., & Fischer, A. S. (2004). Ocean–atmosphere covariability in the western Arabian Sea. *Journal of Climate, 17*, 1213–1224.

Vinayachandran, P. N., Iizuka, S., & Yamagata, T. (2002a). Indian Ocean dipole mode events in an ocean general circulation model. *Deep-Sea Research Part II, 49*, 1573–1596.

Vinayachandran, P. N., Kurian, J., & Neema, C. P. (2007). Indian Ocean response to anomalous conditions in 2006. *Geophysical Research Letters, 34*, L15602. https://doi.org/10.1029/2007GL030194.

Vinayachandran, P. N., McCreary, J. P., Jr., Hood, R. R., & Kohler, K. E. (2005). A numerical investigation of the phytoplankton bloom in the Bay of Bengal during Northeast Monsoon. *Journal of Geophysical Research: Oceans, 110*(C12).

Vinayachandran, P. N., Murty, V. S. N., & Ramesh Babu, V. (2002b). Observations of barrier layer formation in the Bay of Bengal during summer monsoon. *Journal of Geophysical Research: Oceans, 107*(C12), SRF–19.

Vinayachandran, P. N., & Nanjundiah, R. S. (2009). Indian Ocean Sea surface salinity variations in a coupled model. *Climate Dynamics, 33*, 245–263. https://doi.org/10.1007/s00382-008-0511-6.

Vinayachandran, P. N., & Saji, N. H. (2008). Mechanisms of South Indian Ocean intraseasonal cooling. *Geophysical Research Letters, 35*, L23607. https://doi.org/10.1029/2008GL035733.

Vinayachandran, P. N., et al. (2018). BoBBLE: Ocean–atmosphere interaction and its impact on the South Asian Monsoon. *Bulletin of the American Meteorological Society, 99*(8), 1569–1587.

Vincent, C. L., & Lane, T. P. (2017). A 10-year austral summer climatology of observed and modeled intraseasonal, mesoscale, and diurnal variations over the Maritime Continent. *Journal of Climate, 30*, 3807–3828.

Wajsowicz, R. C. (2004). Climate variability over the tropical Indian Ocean sector in the NSIPP seasonal forecast system. *Journal of Climate, 17*, 4783–4804.

Wajsowicz, R. C. (2005). Potential predictability of tropical Indian Ocean SST anomalies. *Geophysical Research Letters, 32*, L24702. https://doi.org/10.1029/2005GL024169.

Wajsowicz, R. C. (2007). Seasonal-to-interannual forecasting of tropical Indian Ocean Sea surface temperature anomalies: Potential predictability and barriers. *Journal of Climate, 20*(13), 3320–3343. https://doi.org/10.1175/JCLI4162.1.

Wang, H., McClean, J. L., Talley, L. D., & Yeager, S. G. (2018). Seasonal cycle and annual reversal of the Somali Current in an eddy-resolving global ocean model. *Journal of Geophysical Research: Oceans, 123*, 6562–6580. https://doi.org/10.1029/2018JC013975.

Wang, J., & Yuan, D. (2015). Roles of western and eastern boundary reflections in the interannual sea level variations during negative Indian Ocean dipole events. *Journal of Physical Oceanography, 45*, 1804–1821.

Warner, J. C., Armstrong, B., He, R., & Zambon, J. B. (2010). Development of a coupled ocean–atmosphere–wave–sediment transport (COAWST) modeling system. *Ocean Model, 35*(3), 230–244.

Wei, J., Malanotte-Rizzoli, P., Eltahir, E. A. B., Xue, P., & Xu, D. (2014). Coupling of a regional atmospheric model (RegCM3) and a regional ocean model (FVCOM) over the maritime continent. *Climate Dynamics, 43*(5–6), 1575–1594.

Weller, R. A., & Anderson, S. P. (1996). Surface meteorology and air-sea fluxes in the western Equatorial Pacific warm pool during the TOGA coupled ocean-atmosphere response experiment. *Journal of Climate, 9*(8), 1959–1990. https://doi.org/10.1175/1520-0442(1996)009<1959:SMAASF>2.0.CO;2.

Weller, R. A., Fischer, A. S., Rudnick, D. L., Eriksen, C. C., Dickey, T. D., Marra, J., Fox, C. A., & Leben, R. R. (2002). Moored observations of upper ocean response to the monsoon in the Arabian Sea during 1994–1995. *Deep Sea Research, 49B*, 2195–2230.

Wernberg, T., Smale, D. A., Tuya, F., Thomsen, M. S., Langlois, T. J., De Bettignies, T., … Rousseaux, C. S. (2013). An extreme climatic event alters marine ecosystem structure in a global biodiversity hotspot. *Nature Climate Change, 3*, 78–82. https://doi.org/10.1038/nclimate1627.

Wiggert, J. D., Hood, R. R., Banse, K., & Kindle, J. C. (2005). Monsoon-driven biogeochemical processes in the Arabian Sea. *Progress in Oceanography, 65*(2–4), 176–213.

Wiggert, J. D., & Murtugudde, R. G. (2007). The sensitivity of the southwest monsoon phytoplankton bloom to variations in eolian iron deposition over the Arabian Sea. *Journal of Geophysical Research: Oceans, 112*(C5).

Wiggert, J. D., Murtugudde, R. G., & Christian, J. R. (2006). Annual ecosystem variability in the tropical Indian Ocean: Results of a coupled bio-physical ocean general circulation model. *Deep Sea Research Part II: Topical Studies in Oceanography, 53*(5–7), 644–676.

Wijesekara, H. W., et al. (2016). ASIRI an ocean–atmosphere initiative for Bay of Bengal. *Bulletin of the American Meteorological Society, 97*, 1859–1884.

Wyrtki, K. (1973). An equatorial jet in the Indian Ocean. *Science, 181*, 262–264.

Xie, S. P., Annamalai, H., Schott, F. A., & McCreary, J. P. (2002). Structure and mechanisms of South Indian Ocean climate variability. *Journal of Climate, 15*, 864–878.

Xue, P., Malanotte-Rizzoli, P., Wei, J., & Eltahir, E. A. B. (2020). Coupled ocean–atmosphere modeling over the maritime continent: A review. *Journal of Geophysical Research, Oceans, 125*, e2019JC014978.

Yamagata, T., Behera, S., Doi, T., Luo, J.-J., Morioka, Y., & Tozuka, T. (2024). Chapter 5: Climate phenomena of the Indian Ocean. In C. C. Ummenhofer, & R. R. Hood (Eds.), *The Indian Ocean and its role in the global climate system* (pp. 103–119). Amsterdam: Elsevier. https://doi.org/10.1016/B978-0-12-822698-8.00009-3.

Yang, Y., Xie, S. P., Wu, L., Kosaka, Y., Lau, N. C., & Vecchi, G. A. (2015). Seasonality and predictability of the Indian Ocean dipole mode: ENSO forcing and internal variability. *Journal of Climate, 28*, 8021–8036. https://doi.org/10.1175/JCLI-D-15-0078.1.

Yao, Z., Tang, Y., Chen, D., Zhou, L., Li, X., Lian, T., & Ul Islam, S. (2016). Assessment of the simulation of Indian Ocean Dipole in the CESM—Impacts of atmospheric physics and model resolution. *Journal of Advances in Modeling Earth Systems, 8*, 1932–1952. https://doi.org/10.1002/2016MS000700.

Yokoi, T., Tozuka, T., & Yamagata, T. (2008). Seasonal variation of the Seychelles Dome. *Journal of Climate, 21*, 3740–3754.

Yoneyama, K., & Zhang, C. (2020). Years of the maritime continent. *Geophysical Research Letters, 47*, e2020GL087182.

Yoshida, K. (1959). A theory of the Cromwell current (the equatorial undercurrent) and of the equatorial upwelling—An interpretation in a similarity to a coastal circulation. *Journal of the Oceanographic Society of Japan, 15*, 159–170.

Yu, L. (2019). Global air–sea fluxes of heat, fresh water, and momentum: Energy budget closure and unanswered questions. *Annual Review of Marine Science, 11*(1), 227–248. https://doi.org/10.1146/annurev-marine-010816-060704.

Yu, Z., & McCreary, J. P. (2004). Assessing precipitation products in the Indian Ocean using an ocean model. *Journal of Geophysical Research, Oceans, 109*, C05013. https://doi.org/10.1029/2003JC002106.

Yuan, X., Ummenhofer, C. C., Seo, H., & Su, Z. (2020). Relative contributions of heat flux and wind stress on the spatiotemporal upper-ocean variability in the tropical Indian Ocean. *Environmental Research Letters, 15*, 084047.

Zhao, N., & Nasuno, T. (2020). How does the air-sea coupling frequency affect convection during the MJO passage? *Journal of Advances in Modeling Earth Systems, 12*, e2020MS002058.

Chapter 19

Paleoclimate evidence of Indian Ocean variability across a range of timescales[⍟]

Mahyar Mohtadi[a,b], Nerilie J. Abram[c,d], Steven C. Clemens[e], Miriam Pfeiffer[f], James M. Russell[e], Stephan Steinke[g], and Jens Zinke[h]

[a]*MARUM-Center for Marine Environmental Sciences, University of Bremen, Bremen, Germany,* [b]*Faculty of Geosciences, University of Bremen, Bremen, Germany,* [c]*Research School of Earth Sciences, The Australian National University, Canberra, ACT, Australia,* [d]*ARC Centre of Excellence for Climate Extremes, The Australian National University, Canberra, ACT, Australia,* [e]*Department of Earth, Environmental, and Planetary Sciences, Brown University, Providence, RI, United States,* [f]*Institut für Geowissenschaften, Christian-Albrechts-Universität zu Kiel, Kiel, Germany,* [g]*Department of Geological Oceanography & State Key Laboratory of Marine Environmental Science, College of Ocean and Earth Sciences, Xiamen University, Xiamen, People's Republic of China,* [h]*School of Geography, Geology and the Environment, University of Leicester, Leicester, United Kingdom*

1 Introduction

Our understanding of the Indian Ocean state, dynamics, and variability is limited by the brevity of instrumental records, which are too short to detect forced changes against background natural variability and are particularly scarce in and around the Indian Ocean (e.g., Cai et al., 2018). This knowledge gap of natural variability on the relevant historical timescales also limits the assessment of anthropogenic impact on regional climate and environment. Closing this gap is thus essential to realistically project future changes, and to develop appropriate strategies for mitigating climate change.

There have been significant advances in our ability to describe and model past climate systems, yet our understanding of oceanic and atmospheric processes within the Indian Ocean realm remains rudimentary in many respects. This is largely because on one hand, the Indian Ocean remains under-sampled in both space and time, and on the other hand, the complexity of the highly dynamic climate of the Indian Ocean makes it even more challenging for climate simulations of the past. Most climate models are unable to quantitatively reproduce the Indian Ocean climate in their control runs, particularly over the maritime continent in the eastern Indian Ocean, which is characterized by deep atmospheric convection and weak coupling between wind and temperature fields (see e.g., DiNezio et al., 2016; Flato et al., 2013).

Proxy-based reconstructions of the Indian Ocean beyond the historical records are therefore critical not only to extend the "climate memory" for model simulations but also to better understand the Indian Ocean response to various forcing mechanisms and background conditions before the anthropogenic impact. This chapter synthesizes the available information on the past state and variability of the Indian Ocean, inferred from proxy-based reconstructions and model simulations. Reconstructions of the interannual to decadal climate variability (Section 2) are based on monthly to annually resolved corals and tree rings, with only a few lacustrine and marine sedimentary records, or speleothems. In contrast, sedimentary records and stalagmites are the main climate archives for detecting past climate variability on millennial (Section 3) to orbital (Section 4) timescales.

2 Interannual to decadal variability

2.1 Sea surface temperature (SST) and ocean circulation

One of the most accurate and prominent archives of past Indian Ocean conditions are the tropical corals. Multicore coral SST reconstructions indicate an early onset of anthropogenic warming in the Indian Ocean in the early 19th century, long before instrumental records of SST became available, and show that the most recent decades between 1986 and 1995 CE

[⍟]This book has a companion website hosting complementary materials. Visit this URL to access it: https://www.elsevier.com/books-and-journals/book-companion/9780128226988.

The Indian Ocean and its Role in the Global Climate System. https://doi.org/10.1016/B978-0-12-822698-8.00007-X

were the warmest decades in the past 400 years (Abram et al., 2016; Tierney et al., 2015a). It has been shown that such long-term temperature trends modulate the interannual SST variability (e.g., Pfeiffer & Dullo, 2006; Timm et al., 2005).

Sumatra corals suggest a tight coupling between SSTs in the eastern Indian Ocean to the Indian Ocean Dipole (IOD, Saji et al., 1999) and El Niño-Southern Oscillation (ENSO) variance during the last millennium (Abram et al., 2020b). They further indicate a mid-millennium shift at around 1590 CE from a reduced (~30%) IOD and ENSO variability and a La Niña-like mean state to a more El Niño-like mean state and an increase in IOD and ENSO interannual variance during the "Little Ice Age" (early 14th to mid-19th century, Figs. 1 and 2). This finding is supported by reconstructed thermocline conditions offshore of southern Indonesia based on Mg/Ca and stable oxygen isotopes in foraminifera that indicate a stronger, El Niño-related local upwelling during the Little Ice Age (Steinke et al., 2014b).

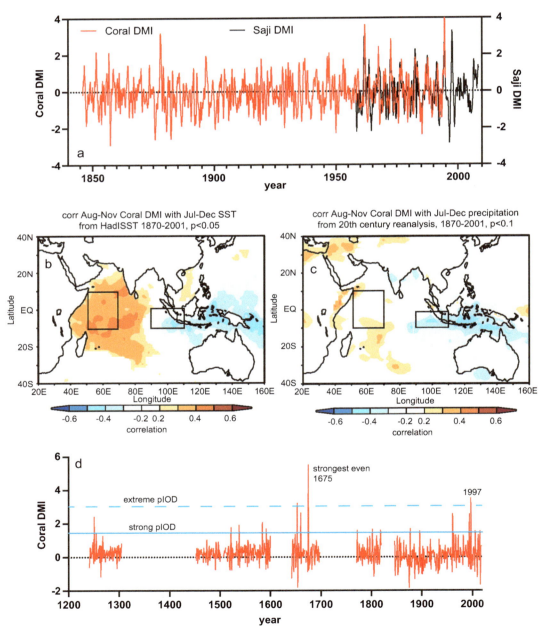

FIG. 1 Reconstruction of the IOD from corals. (a) Coral data compared to instrumental IOD (Saji et al., 1999). (b) Spatial correlation of coral IOD with HadISST for August to November. (c) Same as b but with precipitation from the 20th century reanalysis. Spatial correlations computed in knmi climate explorer, colors indicate correlation at 95% level. (d) Paleoclimatic perspective of the June-December IOD from Sumatra fossil corals (Abram et al., 2020a). Thresholds for strong (*blue* line) and extreme positive IOD events (*blue dashed* line) are indicated, with the strongest event recorded in 1675 CE. *(From Abram et al. (2008).)*

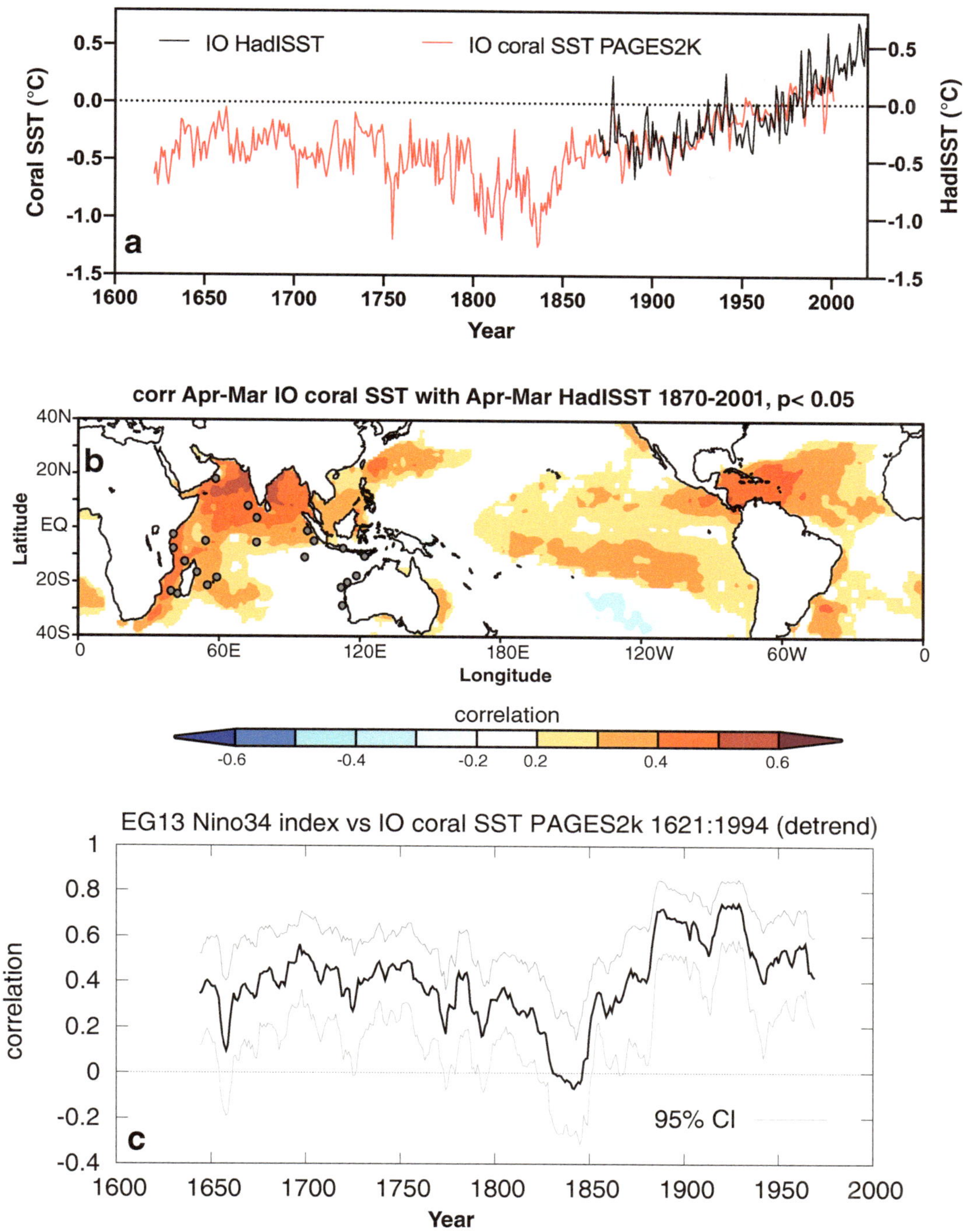

FIG. 2 Coral-based SST reconstructions from the Indian Ocean. (a) Reconstruction of mean annual Indian Ocean SST from corals (*red*) within PAGES2K (Tierney et al., 2015a) compared to instrumental SST from HadISST. The root mean square error is indicated below. (b) Spatial correlation of tropical Indian Ocean SST from corals with HadISST after linear detrending. Spatial correlations computed in KNMI climate explorer, only correlation at 95% level are indicated by colors. Published coral sites are marked by white dots. (c) 31-year running correlation of tropical Indian Ocean SST from corals with an ENSO index (Emile-Geay et al., 2013). Note the stable ENSO relationship with the exception of the mid-1800s.

Coral data from Sumatra also reveal that IOD events during the mid-Holocene at around 6570 to 4100 years ago were up to 50% stronger than the strongest events in the late Holocene and instrumental record (Abram et al., 2008; Fig. 3). They also show that the IOD variability has changed during the 19th and 20th centuries and that the frequency and strength of positive IOD events have increased since the 1960s (Abram et al., 2008; Charles et al., 2003). Yet, the recent increase is not unprecedented in the context of the last millennium (Abram et al., 2020b). However, the ongoing Indian Ocean warming trend projected by 21st-century models accompanied by more frequent positive IOD events imply that the IOD variability will imminently move outside of its natural range of the last millennium (Abram et al., 2020b).

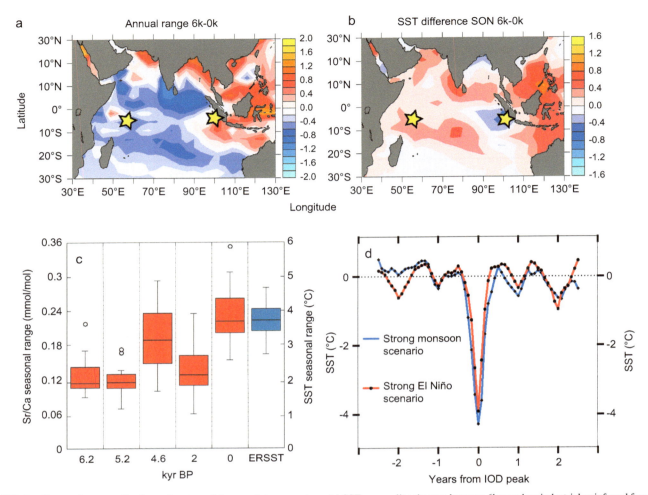

FIG. 3 Changes in seasonality from climate model-proxy data comparison. (a) SST seasonality changes between 6kyr and preindustrial as inferred from the Kiel Climate Model (Park et al., 2009). The location of fossil coral studies indicated by *yellow* dots. (b) September to November IOD season SST difference between 6kyr and preindustrial as inferred from the Kiel Climate Model highlighting a positive IOD-like mean state. (c) SST seasonality boxplots from Seychelles corals (*red*) between 6kyr (lowest) and present (highest), as well as instrumental ERSST3b between 1990 and present (*blue*). The colored box highlights the area with 60% of the data with the median seasonality indicated by a solid line. The whiskers show the maximum range and the open circles are outliers (Zinke et al., 2014). (d) IOD reconstructions from the Middle (*blue* line; strong Monsoon scenario) and Late Holocene (*red* line; strong ENSO scenario) from fossil Sumatra corals (Abram et al., 2007). *(Modified from Zinke et al. (2014).)*

Coral-based SST reconstructions from the western and equatorial Indian Ocean show a stationary ENSO-SST relationship since the peak of the Little Ice Age in the 17th century (Cobb et al., 2001; Cole et al., 2000; Damassa et al., 2006; Hennekam et al., 2018; Leupold et al., 2021; Pfeiffer & Dullo, 2006; Timm et al., 2005; Zinke et al., 2004, 2005, 2016, 2019). In contrast, coral-based SST data from the subtropical southwest Indian Ocean only show an ENSO teleconnection when ENSO variability is strong (Crueger et al., 2009; Zinke et al., 2004, 2005, 2016), i.e., during periods of a cooler mean SST in the late 19th century and the peak of the Little Ice Age (Charles et al., 1997; Leupold et al., 2021; Pfeiffer et al., 2017; Zinke et al., 2004). Interestingly, SST reconstructions from the central Indian Ocean suggest that the ENSO-SST teleconnection is symmetric since 1675 CE, with an equal magnitude of warming during El Niño and cooling during La Niña years (Leupold et al., 2021). This finding is important, as it has been suggested that an asymmetric ENSO teleconnection with El Niño causing greater warming than La Niña, drives the current warming trend in the Indian Ocean (Roxy et al., 2014).

Major Indian Ocean surface currents were also affected by the ENSO, IOD, and the Pacific Decadal Oscillation (PDO) in the past centuries. For instance, the Indonesian Throughflow (ITF) leakage was stronger during negative PDO state and

La Niña years of the past 200 years (Hennekam et al., 2018). Corals from La Réunion reveal that since 1830 CE, variations in salinity advection by the South Equatorial Current were controlled by ENSO and PDO (Pfeiffer et al., 2004b) with a mid-20th-century freshening of surface waters, potentially by the ITF (Pfeiffer et al., 2019). Reconstruction of the southeastern Indian Ocean Leeuwin Current sea-level height variability based on a linear combination of an ENSO index and coral SST records shows a long-term increase of sea-level height in concert with long-term warming of 0.05°C per decade between 1795 and 2010 CE (Zinke et al., 2014). SST and Leeuwin Current strength covaried on interannual and decadal timescales, with the most extreme sea level and SST anomalies of the past 200 years occurring after 1980 CE, and corresponding to ENSO extremes (Zinke et al., 2014).

In summary, paleoclimate data reveal a (non)stationary ENSO-SST relationship in the (subtropical) tropical Indian Ocean (e.g., Leupold et al., 2021; Zinke et al., 2004) with a symmetric warming of the tropical Indian Ocean during both phases of ENSO (Leupold et al., 2021). Further, enhanced ENSO and IOD variability during the Little Ice Age (Abram et al., 2020a, 2020b) indicate that the recent IOD variance is not unprecedented in context of last millennium (Abram et al., 2020b), despite the strong anthropogenic warming trend since the early 19th century (Abram et al., 2016; Pfeiffer et al., 2017; Tierney et al., 2015a).

2.2 Rainfall, monsoon, and the intertropical convergence zone (ITCZ)

Paleoclimate studies suggest that ENSO is the primary driver of interannual rainfall variability in the western and eastern Indian Ocean (Abram et al., 2007; Cole et al., 2000; Wolff et al., 2011). Instrumental data and climate archives from Africa indicate that ENSO drives rainfall variability over East Africa by affecting the "short rains" from October to December, with El Niño (La Niña) resulting in anomalously wet (dry) conditions (Nash et al., 2016; Nicholson, 2017). The close correspondence between SST time series from the western Indian Ocean and terrestrial proxy archives for surface air temperature and rainfall over western India, east and southeast Africa over the last 350 years has also been related to variations in ENSO (Charles et al., 1997; Cole et al., 2000; Therrell et al., 2006; Zinke et al., 2008a) or ENSO-IOD (Abram et al., 2020a; Zinke et al., 2014). A bimonthly seawater stable oxygen isotope ($\delta^{18}O$) reconstruction in corals from Mayotte in the northern Mozambique Channel reveals variations on timescales of 5–6 and 18–25 years (Zinke et al., 2008b), which was found to be related to ENSO that negatively impacts the freshwater balance (Grove et al., 2013). Likewise, coral data from offshore southwest Madagascar indicate a negative impact of ENSO on freshwater balance between 1973 and 1995 and might have done so since 1659 CE (Zinke et al., 2004). However, results from Chagos Island in the central Indian Ocean show a positive ENSO-related freshwater balance during the second half of the last century (Pfeiffer et al., 2006; Timm et al., 2005).

Model simulations support the abovementioned ENSO control of rainfall and show increased precipitation over East Africa during short rains when SST in the western Indian Ocean is high, consistent with the seasonality and processes associated with ENSO-IOD variability (Ummenhofer et al., 2009). Similar to the west Indian Ocean realm, tree-ring (Cook et al., 2010) and lake sediments (Rodysill et al., 2013) from the eastern Indian Ocean indicate that severe droughts during the 19th century occurred during strong El Niño events and suggest a prominent role of ENSO in interannual rainfall variability there.

However, the role of ENSO in driving rainfall patterns in the Indian Ocean is likely complex. For example, coral $\delta^{18}O$ data indicate that the ENSO-rainfall teleconnection is nonstationary and depends on the central Indian Ocean mean SST (Pfeiffer et al., 2004a; Timm et al., 2005). Similarly, coral records from Malindi (Kenya) and Seychelles reveal a better correlation of local SST to Niño 3.4 index than to rainfall over Africa or South Asia on decadal timescales during the past 200 years (Cole et al., 2000) and thus suggest a weak connection between the ENSO-related east African rainfall and local SST variability. Sedimentary records of flooding from the eastern tropical Pacific and Indonesia show increased flood frequency and intensity, interpreted as increased ENSO variability, during the first half of the last millennium (Moy et al., 2002; Rodysill et al., 2019), opposing the coral-based scenario (Fig. 1). Finally, sedimentary records from the eastern Indian Ocean suggest a La Niña-like mean state during the Little Ice Age (Rodysill et al., 2019), which is in contrast to the inferred El Niño-like conditions during the Little Ice Age based on corals (Abram et al., 2020a).

The Australian summer monsoon, ITCZ, and IOD also play an important role in driving rainfall patterns in the eastern Indian Ocean. Proxies of rainfall isotopic composition and sedimentary records generally indicate drying during the early centuries of the last millennium and wetter conditions over the maritime continent during the Little Ice Age (Griffiths et al., 2016; Konecky et al., 2013; Rodysill et al., 2013; Tierney et al., 2010). These changes have been interpreted to reflect a variety of phenomena, including a strengthened Australasian summer monsoon and southward migration of the ITCZ (e.g., Konecky et al., 2013; Tierney et al., 2010), equatorward contraction of the ITCZ during the Little Ice Age (Griffiths et al., 2016; Yan et al., 2015), and changing IOD behavior (Scroxton et al., 2017).

As mentioned earlier, interpretation of rainfall pattern across the Indian Ocean realm during the Little Ice Age remains debated. Speleothem $\delta^{18}O$ and lake sediment records from the Asian mainland provide evidence for the weakening of the Asian summer monsoon during the Little Ice Age (e.g., Zhang et al., 2008), in support of models and theory linking hemispheric cooling to meridional precipitation changes. Wet conditions in southeastern Africa during the Little Ice Age support this theory (Brown & Johnson, 2005). Yet northeastern and equatorial African records indicate much more complex spatiotemporal patterns, with regional drought during the Medieval Warm Period (<1000–1270 CE) giving way to wet conditions extending from the Horn of Africa to easternmost equatorial Africa during the Little Ice Age (Tierney et al., 2015a; Verschuren, 2004; Verschuren et al., 2000) versus drought in central equatorial Africa (Alin & Cohen, 2003; Russell et al., 2007; Russell & Johnson, 2005, 2007). These zonal anomalies during the Little Ice Age have been interpreted to reflect ENSO and/or IOD variations (Russell et al., 2007; Tierney et al., 2013). However, to date, it has proven difficult to demonstrate multidecadal to century-scale precipitation dipole anomalies over the Indian Ocean during the last millennium (Konecky et al., 2014; Scroxton et al., 2017).

Besides ENSO and IOD, other climate phenomena have been proposed to control rainfall variability across the Indian Ocean. Decadal Indian Ocean rainfall cycles of the past 300 years were related to the PDO, with positive (or negative) PDO phases associated with increased (or reduced) SST and rainfall in eastern Madagascar and reduced (increased) precipitation in southern Africa and eastern Australia (Grove et al., 2013). In-phase variations in speleothem $\delta^{18}O$ records from Oman (Burns et al., 2002), Madagascar (Scroxton et al., 2017), and southern Indonesia (Griffiths et al., 2016) suggest that over the past 1600 years, multidecadal rainfall variability was controlled by expansion and contraction of the tropical rain belt, rather than a north-south movement of the ITCZ or zonal shifts associated with the Walker circulation (Yan et al., 2015). However, observations and model simulations suggest that differences between the East African and maritime continent records could be partly due to the nonstationarity of the relationship between precipitation amount and isotopic composition of rainfall ($\delta^{18}O$), particularly over the maritime continent, questioning the applicability of this proxy for reconstructing rainfall amount (Konecky et al., 2014).

Finally, solar activity and volcanic eruptions have been suggested to control rainfall across the Indian Ocean in the past. Tree ring data and model simulations of the last millennium indicate that volcanic eruptions lead to wetter (drier) conditions over southeast (central) Asia (Anchukaitis et al., 2010). Similarly, sunspot peaks during the current warm period correlate with positive rainfall anomalies over the South Asian monsoon (van Loon & Meehl, 2012). Marine sedimentary records from the eastern Arabian Sea and speleothem $\delta^{18}O$ time series from Oman, India, and China suggest that decreased Asian Monsoon intensity corresponds to solar minima during the last two millennia (Agnihotri et al., 2002; Burns et al., 2002; Sinha et al., 2015) and the Holocene (Fleitmann et al., 2003; Gupta et al., 2005; Neff et al., 2001; Tiwari et al., 2015; Zhang et al., 2008). It has been suggested that solar variations influence the South Asian monsoon either directly (Fleitmann et al., 2003; Gupta et al., 2005) or nonlinearly through the changes in ENSO and IOD (Tiwari et al., 2015). Different to the Asian monsoons, rainfall proxy records from the Australian-Indonesian monsoon region (Konecky et al., 2013; Steinke et al., 2014a) and equatorial East Africa (Verschuren et al., 2000) reveal intervals of increased rainfall corresponding to minima in solar irradiance. Model experiments indicate that solar minima cause strong cooling over cloud-free India and northern Australia associated with enhanced subsidence and drier conditions there, while anomalous low-level wind convergence and ascend result in wetter conditions over the maritime continent and equatorial Indian Ocean (Steinke et al., 2014a).

In summary, data and model simulations indicate that variations in the state of ENSO, IOD, PDO, and monsoon/ITCZ control temperature, circulation, and rainfall across the Indian Ocean realm on interannual to decadal timescales. However, deciphering the relative role of these climate phenomena and their forcing mechanisms in shaping the Indian Ocean climate in the past remains debated and appears to depend, in part, on the choice of proxies, sites, and models.

3 Centennial to millennial variability

3.1 Monsoon and the ITCZ

Marine and terrestrial climate archives across the Indian Ocean realm reveal an ocean circulation control on millennial timescales that is mainly related to variations in the North Atlantic Meridional Overturning Circulation (AMOC). Periods of a less-vigorous AMOC during the last glacial period, e.g., during Heinrich Stadials and the Younger Dryas (11,600–12,900 years ago) with cooler conditions in the North Atlantic are generally associated with anomalously dry conditions over the Northern Hemisphere monsoon regions, as documented by different proxies from different archives and sites (Fig. 4) from the South Asian monsoon (Berkelhammer et al., 2012; Deplazes et al., 2013; Dutt et al., 2015), East Asian monsoon (Wang et al., 2001; Zhou et al., 2016), central and western part of the Australian-Indonesian monsoon (Partin et al., 2015; Russell et al., 2014), North African monsoon (Thomas et al., 2012; Tierney & deMenocal, 2013), and the

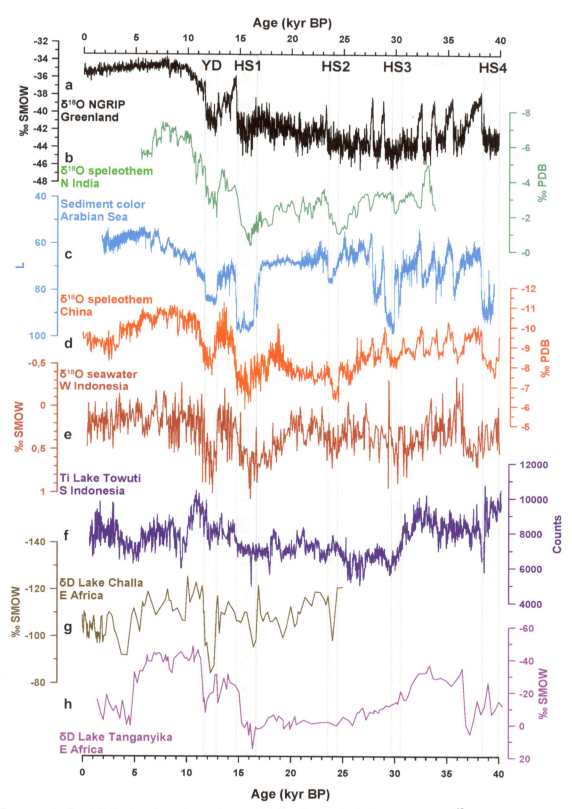

FIG. 4 Proxy records of precipitation from the northern and equatorial Indian Ocean realm (b–h) compared to the $\delta^{18}O$ record of the Greenland ice core NGRIP (a, Svensson et al., 2008). All ages are from original publications and in thousand years before present (kyr BP). Vertical gray bars indicate the North Atlantic cold climate anomalies, the Younger Dryas (YD) and the Heinrich Stadials (HS). Isotopic values are given in per mill (‰) and relative to the Standard Mean Ocean Water (SMOW) or PeeDee Belemnite (PDB). (b) $\delta^{18}O$ of Mawmluh Cave speleothems, north India (Berkelhammer et al., 2012); (c) sediment reflectance/color from site SO130-289KL in the northern Arabian Sea (Deplazes et al., 2013); (d) composite $\delta^{18}O$ of Chinese cave speleothems (Cheng et al., 2016); (e) seawater $\delta^{18}O$ offshore west Sumatra (Mohtadi et al., 2014); (f) Ti counts in Lake Towuti sediments, Sulawesi (Russell et al., 2014); (g, h) deuterium isotopes (δD) in east African lake sediments from Challa (Tierney et al., 2011) and Tanganyika (Tierney et al., 2008), respectively.

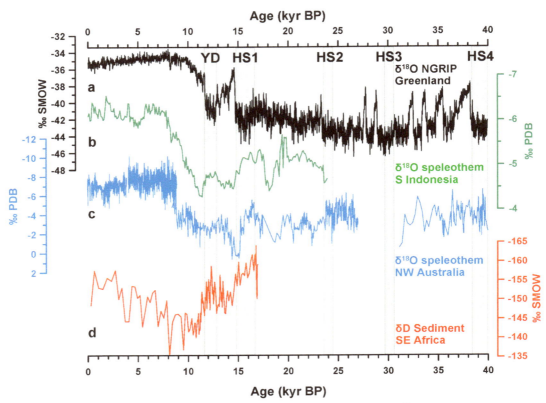

FIG. 5 Proxy records of precipitation from the southern Indian Ocean realm (b–d) compared to the $\delta^{18}O$ record of the Greenland ice core NGRIP (a, Svensson et al., 2008). (b) $\delta^{18}O$ of Liang Luar cave speleothems in Flores Island (Ayliffe et al., 2013); (c) $\delta^{18}O$ of Ball Gown cave speleothems, northwest Australia (Denniston et al., 2013); (d) δD record of GeoB9307-3 offshore Mozambique (Schefuß et al., 2011).

southeast African monsoon (Brown et al., 2007). In contrast, periods of stronger AMOC and a relatively warmer North Atlantic result in generally wetter conditions at or north of the equator (Mohtadi et al., 2014; Wurtzel et al., 2018) and drier conditions over the Southern Hemisphere monsoon regions including the southern part of the Australian-Indonesian monsoon domain (Ayliffe et al., 2013; Denniston et al., 2013; Griffiths et al., 2009) and the South African monsoon system (Fig. 5, Brown et al., 2007; Schefuß et al., 2011; Thomas et al., 2012). Drying during periods of weaker AMOC extends well south of the equator, to ~10°S in East Africa (Tierney et al., 2008), whereas southern Sulawesi experienced weak to no anomalies during AMOC disruptions, potentially due to the proximity of this region to the mean location of the ITCZ (Krause et al., 2019).

Mechanistically, a weaker or disrupted thermohaline circulation results in a reduced northward heat transport by the surface ocean in the Northern Hemisphere (Ganopolski & Rahmstorf, 2001). Consequently, the atmosphere compensates for the reduced oceanic heat transport and the resulting interhemispheric energy imbalance by transferring warm air from the warmer Southern Hemisphere to the cooler Northern Hemisphere, which is accomplished by a southward displacement of the ITCZ (Donohoe et al., 2013; Marshall et al., 2014; Schneider et al., 2014). A more southerly position of the ITCZ and the reorganization of the Hadley circulations result in a general drying of the Northern Hemisphere and the equatorial Indian Ocean, and wetter conditions in the Southern Hemisphere (Fig. 6).

Comparison of the rainfall anomaly records of Heinrich Stadial 1 with a transient simulation of the last deglaciation shows great similarity in terms of the sign of change over most monsoon regions and supports the concept of millennial-scale changes in the AMOC intensity shaping not only the Australasian and African monsoon rainfall but also the rainfall patterns worldwide (Fig. 6, Mohtadi et al., 2016). Most model simulations agree that monsoon rainfall anomalies during Heinrich Stadials stem from a quasi-instantaneous global atmospheric response to the Northern Hemisphere cooling (Gibbons et al., 2014; Marzin et al., 2013; Mohtadi et al., 2014; Otto-Bliesner et al., 2014; Pausata et al., 2011). Over the Indian Ocean, a transient paleoclimate model simulation indicates that cooling over the North Atlantic is rapidly transmitted eastward by the atmosphere, strengthening anticyclonic circulation over the Arabian Peninsula (Otto-Bliesner et al., 2014). The resulting northerly wind anomalies cool the western Indian Ocean, drying much of East Africa and causing

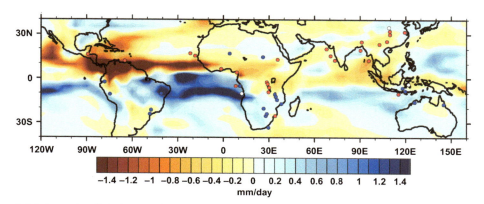

FIG. 6 Changes in rainfall (in mm day^{-1}) during Heinrich Stadial 1 relative to the last glacial maximum (at 21,000 years ago) as indicated by monsoon proxy records (dots) and simulated by the TraCE-21k climate model (annual-mean precipitation anomalies, color shading). *Red (blue)* dots indicate negative (positive) rainfall anomalies, white dots indicate no change or uncertain changes. Positive (negative) anomalies south (north) of the equator indicate a southward displacement of the ITCZ. *(From Mohtadi et al. (2016).)*

weaker precipitation in the Indian monsoon domain (Otto-Bliesner et al., 2014), since cooling in the western Arabian Sea leads to weaker monsoon winds aloft and eventually a weaker South Asian monsoon (Tierney et al., 2015b). Other model results suggest that the overall drying of the Indian monsoon during the North Atlantic abrupt events stems from either a direct atmospheric response through a stationary Rossby wave-train teleconnection that originates in the northern North Atlantic and weakens the vertical shear in the atmosphere (Mohtadi et al., 2014), or indirectly through SST changes in the tropical Atlantic that perturb the subtropical jet over Africa and Eurasia (Marzin et al., 2013).

Taken together, model simulations and reconstructions agree that centennial- to millennial-scale variations in the Indian Ocean climate are controlled remotely by changes in the strength of the AMOC. However, debate exists on how these changes modified ocean and atmosphere circulation patterns and were transferred the Indian Ocean.

3.2 IOD and ENSO

IOD-sensitive archives from the eastern equatorial Indian Ocean reveal that IOD events were characterized by a longer duration of enhanced surface water cooling accompanied by droughts during the middle Holocene (Abram et al., 2007). Climate model simulations suggest that the enhanced cooling and drying caused by strong cross-equatorial winds correspond to a more positive IOD-like mean configuration across the tropical Indian Ocean during the middle Holocene (Fig. 3, Abram et al., 2007; Park et al., 2009). This inference is supported by fossil corals from Seychelles in the western Indian Ocean showing a much lower seasonal SST range in the middle Holocene (6.2–5.2 kyr BP) compared to 1990–2003 CE (Fig. 3, Zinke et al., 2014), and by rainfall reconstructions over Sumatra (Niedermeyer et al., 2014).

Likewise, the increase in Southeast African rainfall during the early/middle Holocene has been attributed to a positive IOD phase along with a stronger South Asian monsoon (Tierney et al., 2008). In contrast, shoaling of the thermocline during the past 3000 years offshore western Sumatra most likely indicates an increased upwelling and a more positive IOD-like mean state during the late Holocene (Kwiatkowski et al., 2015). This finding is supported by model simulations indicating a prolonged configuration of a negative mode of the IOD during the middle Holocene (de Boer et al., 2014), and an insolation-controlled shift from a negative mode of the IOD during the early Holocene to a positive mode of the IOD during the late Holocene (Kwiatkowski et al., 2015). Yet this stands in contrast to lake-level reconstructions suggesting a late Holocene drying in East Africa (e.g., Gasse, 2000).

Varve thickness and organic biomarker reconstructions from sediments of Lake Challa indicate that ENSO variability was weak between 300 BCE and 300 CE, increased from 400 to 750 CE followed by a gradual decline during the Medieval Warm Period and an abrupt breakdown around 1300 CE, and remained low during the Little Ice Age (Wolff et al., 2011). Leaf wax data from the same lake sediments reveal wet conditions and a strong East African monsoon during the early Medieval Warm Period and arid conditions and weak monsoon during the late Medieval Warm Period, followed by increased runoff and an intensified rainfall during the Little Ice Age (Tierney et al., 2011). The connection between wet conditions in East Africa and El Niño during the Little Ice Age has been dismissed since prevailing wet conditions and a strong monsoon over central Indonesia during the same period (Oppo et al., 2009; Tierney et al., 2010) are inconsistent with El Niño patterns (Tierney et al., 2015a). In contrast, thermocline reconstructions from the eastern Indian Ocean imply a substantial contribution of ENSO to the centennial-scale upwelling variations, with a prevalence of an El Niño-like mean state during the Little Ice Age and a La Niña-like mean state during the Medieval Warm Period (Steinke et al., 2014b).

Taken together, paleoclimate studies do not reveal a consistent picture of past IOD and ENSO behavior on centennial to millennial timescales, illustrating the challenge of reconstructing IOD and ENSO for the geological past. This stems from the fact that these timescales are not covered by climate archives capable of resolving single IOD and ENSO events, such as corals.

3.3 Ocean circulation and conditions

On millennial timescales, there is only a subtle response of ocean surface and thermocline temperature ($\sim 1°C$) to abrupt climate changes in the North Atlantic during Heinrich Stadials and the Younger Dryas (Gibbons et al., 2014; Mohtadi et al., 2014; Panmei et al., 2017; Tierney et al., 2015b). In the upwelling regions offshore of southwest Sumatra and south Java, these cold climate events are associated with an increased austral winter upwelling possibly due to stronger cross-equatorial winds (Mohtadi et al., 2011) and supported by alkenone-based SST reconstructions showing a slight decrease in the upwelling SST offshore the Sunda Strait (Lückge et al., 2009; Mohtadi et al., 2010a). However, both data and model results suggest a net mean annual SST warming in the eastern Indian Ocean and the Bay of Bengal during Heinrich Stadials and the Younger Dryas (Gibbons et al., 2014; Mohtadi et al., 2014; Panmei et al., 2017; Setiawan et al., 2015), possibly caused by the global rise in the atmospheric CO_2 during these periods (Moffa-Sanchez et al., 2019). Alternatively, the larger energy demand of the Northern Hemisphere during boreal winter should have strengthened the Hadley circulation and surface westerly/northwesterly winds that combined with an anomalous descend, to create downwelling and higher SST in the eastern Indian Ocean (Mohtadi et al., 2014).

Similar to the eastern Indian Ocean, variations in upwelling and marine productivity in the Arabian Sea reveal great similarity to the abrupt climate changes in the North Atlantic on millennial timescales, which has been attributed to ocean circulation changes, specifically the formation of the Subantarctic Mode Water and Antarctic Intermediate Water (e.g., Böning & Bard, 2009), or weaker South Asian monsoon winds during Heinrich Stadials and the Younger Dryas (Altabet et al., 2002; Deplazes et al., 2014).

Cooling in the western Indian Ocean related to the air-sea interactions outlined before (Otto-Bliesner et al., 2014; Tierney et al., 2015b) may explain the dramatic impacts of North Atlantic stadials in East Africa, including complete desiccation of the region's large lakes (Gasse, 2000) and drought extending well south of the equator to regions dominated by austral summer precipitation (Stager et al., 2011), despite the southward shift of the ITCZ (Otto-Bliesner et al., 2014). Proxy evidence and model simulations indicate a sluggish Agulhas Current transporting warmer and saltier waters during Heinrich Stadials and a reduction of the positive wind stress curl over the subtropical southern Indian Ocean, caused by changes in the southeasterly trades and Southern Hemisphere Westerlies in response to the AMOC slowdown (Simon et al., 2013, 2015).

4 Orbital and glacial-interglacial variability

4.1 Monsoon and the ITCZ

Few summer monsoon reconstructions are suitably resolved and long enough for quantitative time series analysis at orbital timescales (Fig. 7, Day et al., 2015; Hong et al., 2005; Rao et al., 2016; Wang et al., 2003). The relative sensitivity of monsoon proxies to different forcing mechanisms can be derived by assessing coherence (the linear correlation among variables as a function of period when the phase is set to zero) and phase relationships (the timing of proxy maxima or minima at each orbital period) among them at eccentricity, obliquity, and precession; the well-known drivers of insolation (radiative forcing) through time. Spectral analysis is a powerful tool to understand the underlying physics of the climate system, allowing quantification of phase and coherence relationships among climate variables and their potential forcing mechanisms.

A number of proxies have been used to reconstruct summer-monsoon strength in the Bay of Bengal and Arabian Sea where winds are 90% steady from the southwest at $\sim 15\,m\,s^{-1}$ during the summer-monsoon months (Figs. 8 and 9, records #1–4). Although the close association between the onset of summer monsoon rains over India and the abrupt strengthening of the southwesterly Somali Jet over the Arabian Sea is well established in the modern climatology (Boos & Emanuel, 2009), none of the wind-related proxies directly record rainfall. A wide variety of proxies have also been developed to reconstruct variability in the isotopic composition of precipitation, precipitation amount, and runoff from river drainage

Paleoclimate evidence of Indian Ocean variability **Chapter | 19** 455

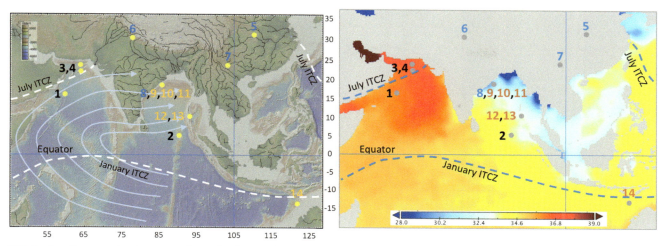

FIG. 7 Location maps showing topography, major rivers, and summer-season salinity (2005–2012; Antonov et al., 2010) as well as schematic summer-season winds and ITCZ location. *Blue* vertical line is at 105°E, approximately separating the East Asian and South Asian monsoon subsystems. Proxy locations are denoted numerically; colors denote proxies considered to represent wind strength (*black*), precipitation isotopic composition (*blue*) and rainfall/runoff (*orange*). Numbers and color designations are carried through Figs. 8 and 9. References as follows: 1 (Clemens & Prell, 2003), 2 (Bolton et al., 2013), 3 (Ziegler et al., 2010), 4 (Caley et al., 2011), 5 (Cheng et al., 2016), 6 (Kathayat et al., 2016), 7 (Cai et al., 2015), 8 (McGrath et al., 2021), 9 (Clemens et al., 2021), 10 (Yamamoto et al., 2022), 11 (Clemens et al., 2021), 12 (Gebregiorgis et al., 2018), 13 (Jöhnck et al., 2020), 14 (Zhang et al., 2020).

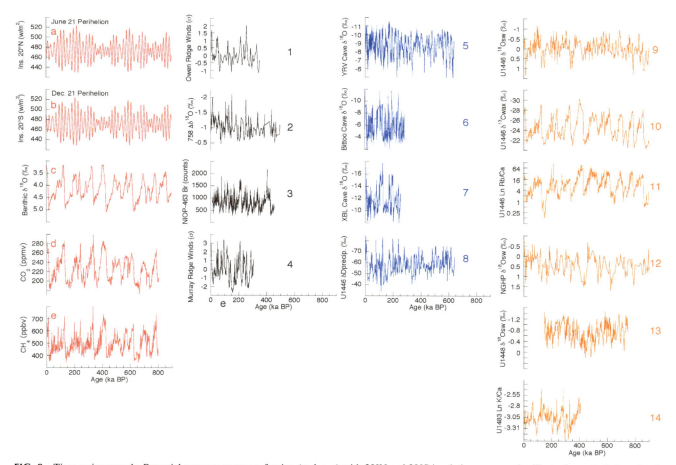

FIG. 8 Time series records. Potential summer-monsoon forcing (*red*; a–e) with 20°N and 20°S insolation as examples illustrating out-of-phase hemispheric relationships and that 23-kyr variance dominates low-latitude insolation. References as follows: a and b (Berger, 1978; Laskar et al., 2004), c (Lisiecki & Raymo, 2005), d (Bereiter et al., 2015; Lüthi et al., 2008), e (Loulergue et al., 2008). Summer-monsoon reconstructions are numbered 1–14. Numbers and colors correspond to those in Figs. 7 and 9. Isotopic records have axes inverted such that all time series represent response maxima in the upward direction. All records span the past 900 ka except for 13 which is Miocene-Pliocene in age. All records represent the South Asian summer monsoon except for 14 (Australian summer monsoon) and 5 (East Asian summer monsoon).

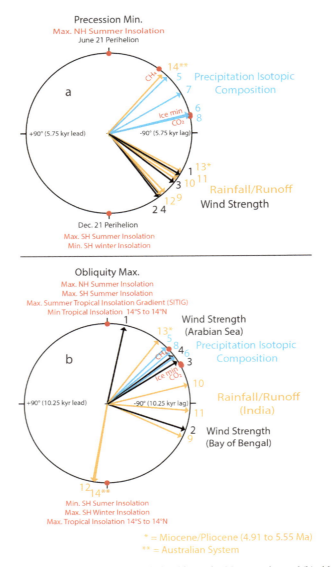

FIG. 9 Phase wheel summaries of cross spectral coherence and phase relationships at the (a) precession and (b) obliquity bands. Zero phase is set at June 21 perihelion for precession and at maximum obliquity. Positive phase (counterclockwise) represents leads and negative (clockwise) represents lags. *Red* text and dots represent the timing of external forcing factors (hemispheric and seasonal insolation maxima and minima; at 0 and 180 degree) and internal forcing factors (global ice volume minima and greenhouse gas maxima). Vectors represent the timing of monsoon proxy responses within the 23-ka precession and 41-ka obliquity cycles (averaged over the number of cycles spanned in a given record). Phase errors (not shown) are typically ±10 to 20 degrees. Vectors are plotted provided coherence (e.g., with the insolation forcing) exceeds a predetermined threshold, typically an 80% confidence threshold; record 7, for example, does not meet the threshold for obliquity variance. The phase of the climate response provides information regarding relative sensitivity to the known external and internal forcing mechanisms. For example, if a climate proxy responds dominantly and directly to Northern Hemisphere summer insolation at the precession band, then its vector would plot at or near 0 degree on the precession phase wheel. Significant leads or lags indicate the influence of other (e.g., internal) drivers such as ice volume and/or greenhouse gases. Proxy response numbers and colors correspond to those in Figs. 7 and 8.

basins (Figs. 8 and 9) including $\delta^{18}O$ of speleothems ($\delta^{18}O_{speleo}$; records #5,6,7), δD of leaf wax (δ^2H_{wax}; record #8), $\delta^{18}O$ of seawater ($\delta^{18}O_{sw}$) from locations sensitive to river runoff (records #9, 12,13), $\delta^{13}C$ of leaf wax ($\delta^{13}C_{wax}$; record #10), and scanning X-ray Fluorescence-derived elemental ratios (XRF), also from locations sensitive to river runoff (records #11,14). All records are dominated by combinations of variance in the eccentricity and/or obliquity bands with the exception of East Asian proxy #5, which is dominated almost exclusively by precession-band variance (Caley et al., 2011; Clemens et al., 2021; Clemens & Prell, 2003; Gebregiorgis et al., 2018; McGrath et al., 2021; Ziegler et al., 2010).

In the following, cross-spectral analysis is used to isolate the variance in each orbital band, determine the coherence (linear correlation) among variables, and the phase (timing) of the maximum response. The results are then summarized on phase wheels, provided they meet established coherence thresholds; proxy groupings are assessed to determine which components of the monsoon system are reflected in the various proxy records.

4.1.1 Eccentricity

All of the monsoon proxy records within the South Asian geographic region have significant components of variance at the 100-ky eccentricity band associated with the coupled dynamics of global ice volume and greenhouse gas, for example, in record #10 (Fig. 8). At the eccentricity period, strengthened monsoon circulation is associated with interglacial intervals characterized by decreased ice volume and increased atmospheric greenhouse gas concentrations, both driving the system toward warmer surface conditions and increased atmospheric moisture content as demonstrated in single-forcing climate simulations (Erb et al., 2015); decreased northern hemisphere ice sheets during interglacials result in northward ITCZ shifts while increased greenhouse gases drive increased equatorial precipitation. Hereafter, these dynamics are referred to as the coupled ice volume-greenhouse gas forcing.

4.1.2 Precession

The Asian proxies yield two distinct clusters (Fig. 9a). Records #5, 6, 7, and 8 cluster at an average phase of −60 degrees on the precession phase wheel. These proxies all reflect the isotopic composition of terrestrial precipitation and their similar response suggests that they share a common, large-scale forcing. Isotopically depleted values are in phase with greenhouse gas maxima and ice volume minima but lag northern-hemisphere summer-insolation maxima by 3800 years or 3.8 ka (60/360 degree × 23 ka). Thus, these proxies are interpreted as responding strongly to coupled ice volume-greenhouse gas dynamics (McGrath et al., 2021). Hence, while external insolation forcing (sensible heating of continental land masses) may be critical to initiating monsoon circulation, internal climatic boundary conditions set the timing of the maximum response (the most depleted terrestrial precipitation isotopic values). In accordance with results of isotope-enabled climate simulations, strong depletion is interpreted as responding to longer transport paths between the oceanic moisture source and the continental moisture sink and/or increased rainout along that transport path, with lesser contribution from increased local rainfall (the amount effect) (Hu et al., 2019; Tabor et al., 2018).

In contrast to the ∼−60 degree timing of terrestrial precipitation isotope minima, South Asian records #1–4 and #9–13 cluster together with an average phase of −130 degrees on the precession phase wheel (Fig. 9a). This indicates that the two proxy groups (#5–8 vs. #1–4 and 9–13) monitor different aspects of the monsoon system with different sets of underlying drivers. The similar phasing of wind strength maxima in the Arabian Sea and Bay of Bengal records (#1–4), including XRF reconstruction of river runoff (#11), $\delta^{13}C_{wax}$ reconstruction of continental vegetation structure (#10), and $\delta^{18}O_{sw}$ reconstructions of rainfall/runoff from continental margins (#9, 12), indicate a coupling of cross-equatorial moisture-transporting winds and South Asian summer-monsoon rainfall over the oceans and surrounding continents. The −130 degree phase indicates that South Asia rainfall maxima are not directly driven by Northern Hemisphere summer insolation forcing. Instead, maximum rainfall/runoff falls between the timing of maximum ice volume-greenhouse gas forcing and precession maxima (Pmax; −180 degrees), interpreted as the timing of maximum latent heat transport from the Southern Hemisphere Indian Ocean into the South Asian system (Bolton et al., 2013; Caley et al., 2011; Clemens & Prell, 2003; Gebregiorgis et al., 2018). Analysis of modern dynamics (Venugopal et al., 2018) as well as current and future climate change simulations (Mei et al., 2015; Wang et al., 2020) underscore the influence of Southern Hemisphere dynamics on Northern Hemisphere summer monsoons via cross-equatorial energy transport. The $\delta^{18}O_{sw}$ reconstruction record (#13) suggests that similar mechanisms were in play as far back as the late Miocene (Jöhnck et al., 2020). Record #14 indicates that Australian summer monsoon maxima occur shortly after Northern Hemisphere summer insolation maxima, lagging Northern Hemisphere summer insolation maxima by 42 degrees. Zhang et al. (2020) interpret this timing as reflecting poleward expansion of the summer ITCZ in both hemispheres.

Idealized (insolation-only) climate simulations indicate a 180-degree difference in the timing of rainfall maxima over the Bay of Bengal relative to the surrounding continents at the precession band (Bosmans et al., 2018). When the climate system is forced only by changes in precession-band insolation, rainfall over the South Asian continents is strongest at precession minima (P_{min}; 0 degree on the precession phase wheel, Fig. 9a) while rainfall over the Bay of Bengal is maximized at Pmax (180 degrees on the precession phase wheel, Fig. 10a). The rainfall/runoff proxy cluster at −130 degrees matches neither result, indicating that insolation-only simulations cannot capture the orbital-scale timing of monsoon rainfall maxima. This highlights a need for long, fully coupled transient climate simulations with realistic ice volume, sea level, and greenhouse gas boundary conditions.

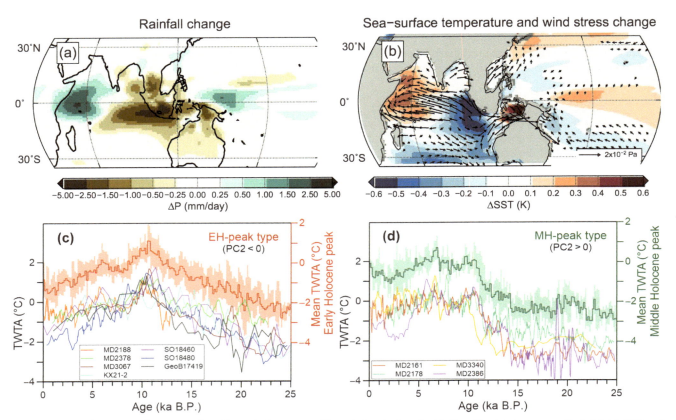

FIG. 10 Simulation *(top)* and reconstructions *(bottom)* of ocean conditions during the last glacial. *Top*: Simulated response of rainfall (a), SST and surface winds (b) to shelf exposure during the Last Glacial Maximum. Shadings and vectors represent anomalies compared to the control run. *Bottom*: Thermocline temperature anomaly (TWTA) records around the maritime continent. (c) The early Holocene (EH) type, with average TWTA (*brown*) and the original TWTA records of the open-ocean sites; (d) the mid Holocene (MH) type for the near-equator sites in the maritime continent with average TWTA (*green*) and the original TWTA record. *(Parts (a) and (b) from DiNezio et al. (2016) and parts (c) and (d) from Dang et al. (2020).)*

4.1.3 Obliquity

The Asian obliquity-band proxy groupings are less distinct (Fig. 9b) indicating more regional responses. The terrestrial precipitation-isotope proxy cluster (Fig. 9b; records #5, 6, and 8) is in phase with coupled ice volume-greenhouse gas as found for precession, indicating similar sensitivity to ice volume-greenhouse gas forcing at both orbital bands. However, the South Asian wind strength proxy records (#1–4) and rainfall/runoff proxy records (#9–13) are not as tightly clustered and have significantly smaller phase lags relative to the precession band results. For example, the Arabian Sea wind proxy records (#1, 3, 4) lead the Bay of Bengal (Indian margin) rainfall/runoff proxy records (#9, 10, 11) indicating an uncoupling of Arabian Sea winds and Bay of Bengal rainfall/runoff at the obliquity band. For obliquity, the Arabian Sea winds are more coupled to ice volume-greenhouse gas forcing and, hence, the terrestrial precipitation isotope records. However, the Bay of Bengal wind strength proxy record (#2) remains consistent with the phase of Bay of Bengal rainfall/runoff proxies from the Indian margin records (#9, 10, and 11) indicating a regional linkage between wind strength and rainfall amount in the Bay of Bengal. The Andaman Sea rainfall/runoff proxy record (#12) has a phase different from all other South Asian proxies, plotting near obliquity minima. This outlier has been interpreted as either (a) a response to increased meridional insolation gradient in the summer hemisphere at obliquity minima, characterized by warming of tropical regions (±14 degrees (Imbrie et al., 1993; Mantsis et al., 2014)) and/or (b) increased isolation of the Andaman Sea as ice volume increases and sea level falls, increasing the retention of fresh water flowing into the basin (Gebregiorgis et al., 2018). Interestingly, the Andaman $\delta^{18}O_{sw}$ record (#12) and Australian XRF record (#14) proxies have the same obliquity-band phase. The Australian phase response (out of phase with the Northern Hemisphere records) has been interpreted in the context of latitudinal migration of the ITCZ, consistent with the traditional explanation of out-of-phase Northern- and Southern-Hemisphere monsoon responses. Alternatively, the similar phasing as Andaman Sea record #12 suggests the possibility of a sea level influence as well, associated with the broad continental shelf at this location.

This summary is largely based on coherence and phase analyses of the 12 records within the South Asian domain, making the case that the $\delta^{18}O_{speleo}$ and δ^2H_{wax} proxies (measured in both cave and marine archives) reflect changes in the isotopic composition of rainfall, predominantly responding to changing source areas and transport path dynamics, with lesser influence of rainfall amount. The tightly clustered phase relationships, spanning both the East Asian and South Asian systems at precession and obliquity timescales, points to large-scale changes in ice volume-greenhouse gas boundary conditions as the primary factors driving changes in the isotopic composition of monsoonal rainfall. A separate multiproxy group composed of $\delta^{18}O_{sw}$, $\delta^{13}C_{wax}$, and elemental XRF ratios located proximal to continental drainages is interpreted to reflect the combined influence of local rainfall/runoff amount. On the basis of phase, summer-monsoon rainfall/runoff maxima are interpreted as being influenced both by large-scale ice volume-greenhouse gas conditions as well as the flux of moisture from the Southern Hemisphere Indian Ocean, at least at the precession band; an inference based on the common phase relationship with wind proxies at precession timescales.

Records from the African monsoon show a regionally consistent picture and suggest a strong control of precession on regional rainfall dynamics until the latter stages of the mid-Pleistocene, when ice volume-greenhouse gas forcing strengthens (deMenocal, 1995, 2004; Rossignol-Strick, 1983; Trauth et al., 2003, 2009). Precessional cycles in monsoonal rainfall are well-documented in East African climate records (e.g., Joordens et al., 2011; Kingston et al., 2007; Lourens et al., 1996; Lupien et al., 2018; Nutz et al., 2017; Rossignol-Strick, 1983, 1985; Trauth et al., 2003) and supported by model simulations (Bosmans et al., 2015), though ice volume-greenhouse gas forcing is also thought to play a critical role in East African climate evolution (Otto-Bliesner et al., 2014). For instance, several terrestrial records (e.g., Gasse, 2000; Tierney et al., 2008) suggest widespread drying through most of northern and equatorial East Africa during the Last Glacial Maximum (19,000–23,000 years ago). It remains unclear when forcings related to ice-volume began to strongly impact the East African climate, due to a paucity of long records from the region spanning the Pleistocene to the present.

Rainfall proxy records from the maritime continent and the Australian-Indonesian monsoon region suggest generally drier conditions during the Last Glacial Maximum (Dam et al., 2001; Krause et al., 2019; Partin et al., 2007; Reeves et al., 2013; Vogel et al., 2015). However, records from the westernmost maritime continent (e.g., Mohtadi et al., 2017) and from southern Indonesia (Ayliffe et al., 2013; Ruan et al., 2019) suggest wetter conditions, at least seasonally. Speleothem $\delta^{18}O$ data from Borneo provided one of the first high-resolution hydroclimate records extending into and beyond the Last Glacial Maximum (Carolin et al., 2013, 2016; Meckler et al., 2012; Partin et al., 2007) that vary primarily in response to boreal fall insolation with minor contribution of ice volume-greenhouse gas. In contrast, a δD record from northwest Sumatra suggests no orbital-scale change in precipitation during the last 24,000 years (Niedermeyer et al., 2014), while other records exhibit little precessional influence but very strong responses to ice volume-greenhouse gas (Krause et al., 2019; Russell et al., 2014; Windler et al., 2019). Some studies hypothesize that seasonal insolation forcing is the dominant control on regional precipitation, whether it is via local (equatorial) seasonal heating and its impacts on the strength of convection within the ITCZ (Carolin et al., 2013), or seasonal insolation changes at higher latitudes which influence the strength of the austral summer monsoon (Mohtadi et al., 2011, 2016).

4.2 Sea-level and ocean conditions

Reconstructions of the thermocline depth and temperature in the Timor Sea suggest that the ITF was stronger at the thermocline depth during interglacials and at the surface during glacials, mainly as a result of sea level changes (Xu et al., 2006, 2008). Likewise, model simulations of the glacial climate suggest a strong control of sea level on the hydroclimate conditions of the Indian Ocean realm (DiNezio et al., 2016; DiNezio & Tierney, 2013), in which atmospheric cooling over the exposed shelf areas, particularly the Sahul Shelf, excites Bjerknes feedback in the eastern Indian Ocean and a more positive IOD-like mean state during sea level lowstands (Fig. 10, DiNezio et al., 2016). However, recent compilation studies suggest that over the past 25,000 years, variations in ENSO have been controlling the ITF variability by changing the precipitation amount and vertical mixing in the upper water column (Zhang et al., 2018), which follow the September insolation changes at the equatorial band (Fig. 10, Dang et al., 2020). These results dismiss the hypothesis of a sea level control on the ITF, since thermocline temperatures remain high in this region until the mid-Holocene, long after the full flooding of the Sunda and Sahul shelves and the complete opening of all the relevant ITF passages at about 10,000 years ago (Hanebuth et al., 2000).

SST reconstructions from the eastern tropical Indian Ocean vary on glacial-interglacial timescale akin to the Antarctic temperatures, both in nonupwelling (Mohtadi et al., 2014) and upwelling (Gibbons et al., 2014; Mohtadi et al., 2010a, 2010b; Setiawan et al., 2015; Wang et al., 2018; Xu et al., 2006, 2008) areas, implying a coherent response of the mean annual SST in the eastern tropical Indian Ocean to changing global ice volume and greenhouse gases on glacial-interglacial timescales (Moffa-Sanchez et al., 2019). Results on ocean conditions during the Last Glacial Maximum are conflicting and suggest either increased interannual variability related to increased IOD activity (Thirumalai et al., 2019), or more stable conditions and a well-stratified water column due to a stronger Walker circulation (Mohtadi et al., 2017), which implies a reduced IOD activity.

5 Conclusions

Proxy data and model simulations of the Indian Ocean SST and monsoon rainfall in the Common Era suggest a strong connection to the state of the IOD and ENSO, solar and volcanic activity, and the position of the ITCZ (see review by PAGES Hydr2k Consortium, 2017, and references therein). However, despite the great amount of data and model simulations with their interpretations being largely consistent with modern observations, uncertainty remains in the relative importance of, and the interaction between, the forcing factors. Disagreement exists on the direction of change and thus the inferred state of ENSO-IOD on centennial scales and seems to depend on the choice of the proxy.

Millennial-scale variability in temperature and productivity of the Indian Ocean and rainfall over the surrounding continents appear to be controlled by changes in the strength of the oceanic global thermohaline circulation. Freshwater perturbations in the North Atlantic are capable of greatly reducing this circulation (Alley & Clark, 1999), as evidenced by iceberg release and freshwater discharge during Heinrich Stadials and the Younger Dryas. The subsequent cooling of the North Atlantic provokes rapid changes in the atmospheric circulation by displacing the ITCZ to the warmer Southern Hemisphere, a weaker Northern Hemisphere monsoon, and a drier and warmer northern Indian Ocean.

Understanding the forcing and response of the Indian Ocean climate on longer timescales is similarly complicated by the overlapping influence of several forcing factors and changing boundary conditions. A growing body of evidence indicates the combined influence of external (orbital) and internal (ice volume and greenhouse gases) factors driving changes in tropical SST, thermocline temperature and depth, and monsoon rainfall. Enhanced spatial coverage and continued proxy development will further elucidate the relative influence of these factors at various temporal and spatial scales. Application of the multiproxy approach helps to resolve divergent interpretations that arise from interpretation of individual proxy records or proxy types. Progress along these lines, coupled with proxy-model comparisons will continue to lead to a deeper understanding of dynamic and thermodynamic components of the Indian Ocean climate.

6 Educational resources

https://www.ncdc.noaa.gov/data-access/paleoclimatology-data
https://www.pangaea.de/

Acknowledgments

MM acknowledges funding from the Deutsche Forschungsgemeinschaft (DFG) grant MO2546/3-1 and the German Ministry of Education and Research grants 03G0806A (CARIMA) and 03G0864F (CAHOL). SCC acknowledges the International Ocean Discovery Program and U.S. National Science Foundation (NSF) grant OCE1634774. MP acknowledges funding by the DFG grant PF676/2-1. JMR acknowledges partial support from the U.S. NFC for his contributions to this manuscript. SS acknowledges financial support from the DFG grant STE1044/4-1 and the Xiamen University President Fund project 20720170071. JZ acknowledges funding by the DFG grant ZI1659-1-1 and the Royal Society Wolfson Foundation grant RSWF-FT-180000.

Author contributions

JZ, MM, MP, and NJA wrote Section 2.1, JMR, JZ, MM, MP, and NJA wrote Section 2.2, and SCC wrote Section 4.1. All the other sections were written by MM Figs. 1–3 were drafted by JZ, MP, and NJA, Figs. 4, 5, and 10 by MM, Fig. 6 by MM and SS, and SCC drafted Figs. 7–9. All authors contributed to the discussion of content and overall structure of the chapter.

References

Abram, N. J., Gagan, M. K., Cole, J. E., Hantoro, W. S., & Mudelsee, M. (2008). Recent intensification of tropical climate variability in the Indian Ocean. *Nature Geoscience*, *1*(12), 849–853.

Abram, N. J., Gagan, M. K., Liu, Z., Hantoro, W. S., McCulloch, M. T., & Suwargadi, B. W. (2007). Seasonal characteristics of the Indian Ocean dipole during the Holocene epoch. *Nature*, *445*(7125), 299–302.

Abram, N. J., Hargreaves, J. A., Wright, N. M., Thirumalai, K., Ummenhofer, C. C., & England, M. H. (2020a). Palaeoclimate perspectives on the Indian Ocean dipole. *Quaternary Science Reviews*, *237*, 106302. https://doi.org/10.1016/j.quascirev.2020.106302.

Abram, N. A., McGregor, H. V., Tierney, J. E., Evans, M. N., McKay, N. P., Kaufman, D. S., & the PAGES 2k Consortium. (2016). Early onset of Industrial-era warming across the oceans and continents. *Nature, 536*, 411–418.

Abram, N. J., Wright, N. M., Ellis, B., Dixon, B. C., Wurtzel, J. B., England, M. H., ... Heslop, D. (2020b). Coupling of Indo-Pacific climate variability over the last millennium. *Nature, 579*(7799), 385–392. https://doi.org/10.1038/s41586-020-2084-4.

Agnihotri, R., Dutta, K., Bhushan, R., & Somayajulu, B. L. K. (2002). Evidence for solar forcing on the Indian monsoon during the last millennium. *Earth and Planetary Science Letters, 198*(3), 521–527. https://doi.org/10.1016/S0012-821X(02)00530-7.

Alin, S. R., & Cohen, A. S. (2003). Lake-level history of Lake Tanganyika, East Africa, for the past 2500 years based on ostracode-inferred water-depth reconstruction. *Palaeogeography, Palaeoclimatology, Palaeoecology, 199*(1), 31–49. https://doi.org/10.1016/S0031-0182(03)00484-X.

Alley, R. B., & Clark, P. U. (1999). The deglaciation of the northern hemisphere: A global perspective. *Annual Review of Earth and Planetary Sciences, 27*, 149–182.

Altabet, M. A., Higginson, M. J., & Murray, D. W. (2002). The effect of millennial-scale changes in Arabian Sea denitrification on atmospheric CO_2. *Nature, 415*(6868), 159–162. https://doi.org/10.1038/415159a.

Anchukaitis, K. J., Buckley, B. M., Cook, E. R., Cook, B. I., D'Arrigo, R. D., & Ammann, C. M. (2010). Influence of volcanic eruptions on the climate of the Asian monsoon region. *Geophysical Research Letters, 37*(22), L22703. https://doi.org/10.1029/2010gl044843.

Antonov, J. I., et al. (2010). World Ocean Atlas 2009, Salinity. *NOAA Atoas NESDIS 69*. U.S. Government Printing Office.

Ayliffe, L. K., Gagan, M. K., Zhao, J.-X., Drysdale, R. N., Hellstrom, J. C., Hantoro, W. S., Griffiths, M. L., Scott-Gagan, H., Pierre, E. S., Cowley, J. A., & Suwargadi, B. W. (2013). Rapid interhemispheric climate links via the Australasian monsoon during the last deglaciation. *Nature Communications, 4*. https://doi.org/10.1038/ncomms3908.

Bereiter, B., Eggleston, S., Schmitt, J., Nehrbass-Ahles, C., Stocker, T. F., Fischer, H., Kipfstuhl, S., & Chappellaz, J. (2015). Revision of the EPICA Dome C CO_2 record from 800 to 600-kyr before present. *Geophysical Research Letters, 42*(2). https://doi.org/10.1002/2014gl061957.

Berger, A. L. (1978). Long-term variations of caloric insolation resulting from the earth's orbital elements. *Quaternary Research, 9*(2), 139–167. https://doi.org/10.1016/0033-5894(78)90064-9.

Berkelhammer, M., Sinha, A., Stott, L., Cheng, H., Pausata, F. S. R., & Yoshimura, K. (2012). An abrupt shift in the Indian monsoon 4000 years ago. In L. Giosan, D. Q. Fuller, K. Nicoll, R. K. Flad, & P. D. Clift (Eds.), *Vol. 198. Climates, landscapes, and civilizations* (pp. 75–88). Washington, DC: Geophysical Monograph Series.

Bolton, C. T., Chang, L., Clemens, S. C., Kodama, K., Ikehara, M., Medina-Elizalde, M., Paterson, G. A., Roberts, A. P., Rohling, E. J., Yamamoto, Y., & Zhao, X. (2013). A 500,000 year record of Indian summer monsoon dynamics recorded by eastern equatorial Indian Ocean upper water-column structure. *Quaternary Science Reviews, 77*, 167–180. https://doi.org/10.1016/j.quascirev.2013.07.031.

Böning, P., & Bard, E. (2009). Millennial/centennial-scale thermocline ventilation changes in the Indian Ocean as reflected by aragonite preservation and geochemical variations in Arabian Sea sediments. *Geochimica et Cosmochimica Acta, 73*(22), 6771–6788.

Boos, W. R., & Emanuel, K. A. (2009). Annual intensification of the Somali jet in a quasi-equilibrium framework: Observational composites. *Quarterly Journal of the Royal Meteorological Society, 135*(639), 319–335. https://doi.org/10.1002/qj.388.

Bosmans, J. H. C., Drijfhout, S. S., Tuenter, E., Hilgen, F. J., & Lourens, L. J. (2015). Response of the North African summer monsoon to precession and obliquity forcings in the EC-Earth GCM. *Climate Dynamics, 44*(1–2), 279–297. https://doi.org/10.1007/s00382-014-2260-z.

Bosmans, J. H. C., Erb, M. P., Dolan, A. M., Drijfhout, S. S., Tuenter, E., Hilgen, F. J., Edge, D., Pope, J. O., & Lourens, L. J. (2018). Response of the Asian summer monsoons to idealized precession and obliquity forcing in a set of GCMs. *Quaternary Science Reviews, 188*, 121–135. https://doi.org/10.1016/j.quascirev.2018.03.025.

Brown, E. T., & Johnson, T. C. (2005). Coherence between tropical East African and South American records of the little ice age. *Geochemistry, Geophysics, Geosystems, 6*(12). https://doi.org/10.1029/2005GC000959.

Brown, E. T., Johnson, T. C., Scholz, C. A., Cohen, A. S., & King, J. W. (2007). Abrupt change in tropical African climate linked to the bipolar seesaw over the past 55,000 years. *Geophysical Research Letters, 34*, L20702. https://doi.org/10.1029/2007gl031240.

Burns, S. J., Fleitmann, D., Mudelsee, M., Neff, U., Matter, A., & Mangini, A. (2002). A 780-year annually resolved record of Indian Ocean monsoon precipitation from a speleothem from South Oman. *Journal of Geophysical Research: Atmospheres, 107*(D20), ACL 9-1–ACL 9-9. https://doi.org/10.1029/2001JD001281.

Cai, Y., Fung, I. Y., Edwards, R. L., An, Z., Cheng, H., Lee, J.-E., Tan, L., Shen, C.-C., Wang, X., Day, J. A., Zhou, W., Kelly, M. J., & Chiang, J. C. H. (2015). Variability of stalagmite-inferred Indian monsoon precipitation over the past 252,000 y. *Proceedings of the National Academy of Sciences, 112*(10), 2954–2959. https://doi.org/10.1073/pnas.1424035112.

Cai, W., Wang, G., Gan, B., Wu, L., Santoso, A., Lin, X., Chen, Z., Jia, F., & Yamagata, T. (2018). Stabilised frequency of extreme positive Indian Ocean dipole under 1.5°C warming. *Nature Communications, 9*(1), 1419. https://doi.org/10.1038/s41467-018-03789-6.

Caley, T., Malaize, B., Zaragosi, S., Rossignol, L., Bourget, J., Eynaud, F., Martinez, P., Giraudeau, J., Charlier, K., & Ellouz-Zimmermann, N. (2011). New Arabian Sea records help decipher orbital timing of Indo-Asian monsoon. *Earth and Planetary Science Letters, 308*(3–4), 433–444. https://doi.org/10.1016/j.epsl.2011.06.019.

Carolin, S. A., Cobb, K. M., Adkins, J. F., Clark, B., Conroy, J. L., Lejau, S., Malang, J., & Tuen, A. A. (2013). Varied response of Western Pacific hydrology to climate forcings over the last glacial period. *Science, 340*(6140), 1564–1566. https://doi.org/10.1126/science.1233797.

Carolin, S. A., Cobb, K. M., Lynch-Stieglitz, J., Moerman, J. W., Partin, J. W., Lejau, S., Malang, J., Clark, B., Tuen, A. A., & Adkins, J. F. (2016). Northern Borneo stalagmite records reveal West Pacific hydroclimate across MIS 5 and 6. *Earth and Planetary Science Letters, 439*, 182–193. https://doi.org/10.1016/j.epsl.2016.01.028.

Charles, C. D., Cobb, K., Moore, M. D., & Fairbanks, R. G. (2003). Monsoon-Tropical Ocean interaction in a network of coral records spanning the 20th century. *Marine Geology, 201*(1–3), 207–222.

Charles, C. D., Hunter, D. E., & Fairbanks, R. G. (1997). Interaction between the ENSO and the Asian Monsoon in a coral record of tropical climate. *Science*, *277*, 925–928.

Cheng, H., Edwards, L. R., Sinha, A., Spötl, C., Yi, L., Chen, S., ... Zhang, H. (2016). The Asian monsoon over the past 640,000 years and ice age terminations. *Nature*, *534*(7609), 640–646. https://doi.org/10.1038/nature18591.

Clemens, S. C., & Prell, W. L. (2003). A 350,000 year summer-monsoon multi-proxy stack from the Owen Ridge, Northern Arabian Sea. *Marine Geology*, *201*, 35–51. https://doi.org/10.1016/S0025-3227(03)00207-X.

Clemens, S. C., Yamamoto, M., Thirumalai, K., Giosan, L., Richey, J. N., Nilsson-Kerr, K., Rosenthal, Y., Anand, P., & McGrath, S. M. (2021). Remote and local drivers of Pleistocene South Asian summer monsoon precipitation: A test for future predictions. *Science Advances*, *7*(23), eabg3848. https://doi.org/10.1126/sciadv.abg3848.

Cobb, K. M., Charles, C. D., & Hunter, D. E. (2001). A central tropical Pacific coral demonstrates Pacific, Indian, and Atlantic decadal climate connections. *Geophysical Research Letters*, *28*(11), 2209–2212. https://doi.org/10.1029/2001GL012919.

Cole, J. E., Dunbar, R. B., McClanahan, T. R., & Muthiga, N. A. (2000). Tropical Pacific forcing of decadal SST variability in the Western Indian Ocean over the past two centuries. *Science*, *287*(5453), 617–619. https://doi.org/10.1126/science.287.5453.617.

Cook, E. R., Anchukaitis, K. J., Buckley, B. M., D'Arrigo, R. D., Jacoby, G. C., & Wright, W. E. (2010). Asian monsoon failure and megadrought during the last millennium. *Science*, *328*(5977), 486–489. https://doi.org/10.1126/science.1185188.

Crueger, T., Zinke, J., & Pfeiffer, M. (2009). Patterns of Pacific decadal variability recorded by Indian Ocean corals. *International Journal of Earth Sciences*, *98*(1), 41–52. https://doi.org/10.1007/s00531-008-0324-1.

Dam, R. A. C., Fluin, J., Suparan, P., & van der Kaars, S. (2001). Palaeoenvironmental developments in the Lake Tondano area (N. Sulawesi, Indonesia) since 33,000 yr B.P. *Palaeogeography, Palaeoclimatology, Palaeoecology*, *171*(3–4), 147–183.

Damassa, T. D., Cole, J. E., Barnett, H. R., Ault, T. R., & McClanahan, T. R. (2006). Enhanced multidecadal climate variability in the seventeenth century from coral isotope records in the western Indian Ocean. *Paleoceanography*, *21*(2). https://doi.org/10.1029/2005PA001217.

Dang, H., Jian, Z., Wang, Y., Mohtadi, M., Rosenthal, Y., Ye, L., Bassinot, F., & Kuhnt, W. (2020). Pacific warm pool subsurface heat sequestration modulated Walker circulation and ENSO activity during the Holocene. *Science Advances*, *6*(42), eabc0402. https://doi.org/10.1126/sciadv.abc0402.

Day, J. A., Fung, I., & Risi, C. (2015). Coupling of south and east Asian monsoon precipitation in July–August. *Journal of Climate*, *28*(11), 4330–4356. https://doi.org/10.1175/jcli-d-14-00393.1.

de Boer, E. J., Tjallingii, R., Vélez, M. I., Rijsdijk, K. F., Vlug, A., Reichart, G.-J., Prendergast, A. L., de Louw, P. G. B., Florens, F. B. V., Baider, C., & Hooghiemstra, H. (2014). Climate variability in the SW Indian Ocean from an 8000-yr long multi-proxy record in the Mauritian lowlands shows a middle to late Holocene shift from negative IOD-state to ENSO-state. *Quaternary Science Reviews*, *86*, 175–189. https://doi.org/10.1016/j.quascirev.2013.12.026.

deMenocal, P. B. (1995). Plio-Pleistocene African climate. *Science*, *270*, 53–59.

deMenocal, P. B. (2004). African climate change and faunal evolution during the Pliocene–Pleistocene. *Earth and Planetary Science Letters*, *220*(1), 3–24. https://doi.org/10.1016/S0012-821X(04)00003-2.

Denniston, R. F., Wyrwoll, K.-H., Asmerom, Y., Polyak, V. J., Humphreys, W. F., Cugley, J., Woods, D., LaPointe, Z., Peota, J., & Greaves, E. (2013). North Atlantic forcing of millennial-scale Indo-Australian monsoon dynamics during the Last Glacial period. *Quaternary Science Reviews*, *72*, 159–168. https://doi.org/10.1016/j.quascirev.2013.04.012.

Deplazes, G., Lückge, A., Peterson, L. C., Timmermann, A., Hamann, Y., Hughen, K. A., Rohl, U., Laj, C., Cane, M. A., Sigman, D. M., & Haug, G. H. (2013). Links between tropical rainfall and North Atlantic climate during the last glacial period. *Nature Geoscience*, *6*(3), 213–217.

Deplazes, G., Lückge, A., Stuut, J.-B. W., Pätzold, J., Kuhlmann, H., Husson, D., Fant, M., & Haug, G. H. (2014). Weakening and strengthening of the Indian monsoon during Heinrich events and Dansgaard-Oeschger oscillations. *Paleoceanography*, *29*(2), 99–114. https://doi.org/10.1002/2013pa002509.

DiNezio, P. N., & Tierney, J. E. (2013). The effect of sea level on glacial Indo-Pacific climate. *Nature Geoscience*, *6*(6), 485–491. https://doi.org/10.1038/ngeo1823.

DiNezio, P. N., Timmermann, A., Tierney, J. E., Jin, F.-F., Otto-Bliesner, B., Rosenbloom, N., Mapes, B., Neale, R., Ivanovic, R. F., & Montenegro, A. (2016). The climate response of the Indo-Pacific warm pool to glacial sea level. *Paleoceanography*, *31*(6), 866–894. https://doi.org/10.1002/2015pa002890.

Donohoe, A., Marshall, J., Ferreira, D., & McGee, D. (2013). The relationship between ITCZ location and cross-equatorial atmospheric heat transport: From the seasonal cycle to the last glacial maximum. *Journal of Climate*, *26*(11), 3597–3618. https://doi.org/10.1175/jcli-d-12-00467.1.

Dutt, S., Gupta, A. K., Clemens, S. C., Cheng, H., Singh, R. K., Kathayat, G., & Edwards, R. L. (2015). Abrupt changes in Indian summer monsoon strength during 33,800 to 5500 years B.P. *Geophysical Research Letters*, *42*(13), 5526–5532. https://doi.org/10.1002/2015gl064015.

Emile-Geay, J., Cobb, K. M., Mann, M. E., & Wittenberg, A. T. (2013). Estimating central equatorial pacific SST variability over the past millennium. Part II: Reconstructions and implications. *Journal of Climate*, *26*(7), 2329–2352. https://doi.org/10.1175/jcli-d-11-00511.1.

Erb, M. P., Jackson, C. S., & Broccoli, A. J. (2015). Using single-forcing GCM simulations to reconstruct and interpret quaternary climate change. *Journal of Climate*, *28*(24), 9746–9767. https://doi.org/10.1175/JCLI-D-15-0329.1.

Flato, G., Marotzke, J., Abiodun, B., Braconnot, P., Chou, S. C., Collins, W., Cox, P., Driouech, F., Emori, S., Eyring, V., Forest, C., Gleckler, P., Guilyardi, E., Jakob, C., Kattsov, V., Reason, C., & Rummukainen, M. (2013). Evaluation of climate models. In T. F. Stocker, D. Qin, G.-K. Plattner, M. Tignor, S. K. Allen, J. Boschung, ... P. M. Midgley (Eds.), *Climate change 2013: The physical science basis. Contribution of working group I to the fifth assessment report of the intergovernmental panel on climate change*. Cambridge, United Kingdom, New York, NY, USA: Cambridge University Press.

Fleitmann, D., Burns, S. J., Mudelsee, M., Neff, U., Kramers, J., Mangini, A., & Matter, A. (2003). Holocene forcing of the Indian monsoon recorded in a stalagmite from southern Oman. *Science*, *300*, 1737–1739.

Ganopolski, A., & Rahmstorf, S. (2001). Rapid changes of glacial climate simulated in a coupled climate model. *Nature*, *409*(6817), 153–158. https://doi.org/10.1038/35051500.

Gasse, F. (2000). Hydrological changes in the African tropics since the last glacial maximum. *Quaternary Science Reviews*, *19*, 189–211.

Gebregiorgis, D., Hathorne, E. C., Giosan, L., Clemens, S., Nürnberg, D., & Frank, M. (2018). Southern hemisphere forcing of South Asian monsoon precipitation over the past ~1 million years. *Nature Communications*, *9*(1), 4702. https://doi.org/10.1038/s41467-018-07076-2.

Gibbons, F. T., Oppo, D. W., Mohtadi, M., Rosenthal, Y., Cheng, J., Liu, Z., & Linsley, B. K. (2014). Deglacial $\delta^{18}O$ and hydrologic variability in the tropical Pacific and Indian oceans. *Earth and Planetary Science Letters*, *387*, 240–251. https://doi.org/10.1016/j.epsl.2013.11.032.

Griffiths, M. L., Drysdale, R. N., Gagan, M. K., Zhao, J. X., Ayliffe, L. K., Hellstrom, J. C., Hantoro, W. S., Frisia, S., Feng, Y. X., Cartwright, I., Pierre, E. S., Fischer, M. J., & Suwargadi, B. W. (2009). Increasing Australian-Indonesian monsoon rainfall linked to early Holocene sea-level rise. *Nature Geoscience*, *2*(9), 636–639.

Griffiths, M. L., Kimbrough, A. K., Gagan, M. K., Drysdale, R. N., Cole, J. E., Johnson, K. R., Zhao, J.-X., Cook, B. I., Hellstrom, J. C., & Hantoro, W. S. (2016). Western Pacific hydroclimate linked to global climate variability over the past two millennia. *Nature Communications*, *7*. https://doi.org/10.1038/ncomms11719.

Grove, C. A., Zinke, J., Peeters, F., Park, W., Scheufen, T., Kasper, S., Randriamanantsoa, B., McCulloch, M. T., & Brummer, G. J. A. (2013). Madagascar corals reveal a multidecadal signature of rainfall and river runoff since 1708. *Climate of the Past*, *9*(2), 641–656. https://doi.org/10.5194/cp-9-641-2013.

Gupta, A. K., Das, M., & Anderson, D. M. (2005). Solar influence on the Indian summer monsoon during the Holocene. *Geophysical Research Letters*, *32*, L17703. https://doi.org/10.1029/2005GL022685.

Hanebuth, T., Stattegger, K., & Grootes, P. M. (2000). Rapid flooding of the Sunda shelf: A late-Glacial Sea-level record. *Science*, *288*(5468), 1033–1035. https://doi.org/10.1126/science.288.5468.1033.

Hennekam, R., Zinke, J., van Sebille, E., ten Have, M., Brummer, G.-J. A., & Reichart, G.-J. (2018). Cocos (keeling) corals reveal 200 years of multidecadal modulation of Southeast Indian Ocean hydrology by Indonesian throughflow. *Paleoceanography and Paleoclimatology*, *33*(1), 48–60. https://doi.org/10.1002/2017PA003181.

Hong, Y. T., Hong, B., Lin, Q. H., Shibata, Y., Hirota, M., Zhu, Y. X., Leng, X. T., Wang, Y., Wang, H., & Yi, L. (2005). Inverse phase oscillations between the east Asian and Indian Ocean summer monsoons during the last 12 000 years and paleo-El Nino. *Earth and Planetary Science Letters*, *231*(3–4), 337. https://doi.org/10.1016/j.epsl.2004.12.025.

Hu, J., Emile-Geay, J., Tabor, C., Nusbaumer, J., & Partin, J. (2019). Deciphering oxygen isotope records from Chinese speleothems with an isotope-enabled climate model. *Paleoceanography and Paleoclimatology*, *34*(12), 2098–2112. https://doi.org/10.1029/2019PA003741.

Imbrie, J., Berger, A., & Shackleton, N. J. (1993). Role of orbital forcing: A two-million-year perspective. In J. A. Eddy, & H. Oeschger (Eds.), *Global changes in the perspective of the past* (pp. 263–277). New York: Wiley.

Jöhnck, J., Kuhnt, W., Holbourn, A., & Andersen, N. (2020). Variability of the Indian monsoon in the Andaman Sea across the Miocene-Pliocene transition. *Paleoceanography and Paleoclimatology*. https://doi.org/10.1029/2020PA003923.

Joordens, J. C. A., Vonhof, H. B., Feibel, C. S., Lourens, L. J., Dupont-Nivet, G., van der Lubbe, J. H. J. L., Sier, M. J., Davies, G. R., & Kroon, D. (2011). An astronomically-tuned climate framework for hominins in the Turkana Basin. *Earth and Planetary Science Letters*, *307*(1), 1–8. https://doi.org/10.1016/j.epsl.2011.05.005.

Kathayat, G., Cheng, H., Sinha, A., Spötl, C., Edwards, L. R., Zhang, H., … Breitenbach, S. F. M. (2016). Indian monsoon variability on millenial-orbital timescales. *Scientific Reports*, *6*. https://doi.org/10.1038/srep24374.

Kingston, J. D., Deino, A. L., Edgar, R. K., & Hill, A. (2007). Astronomically forced climate change in the Kenyan Rift Valley 2.7–2.55 Ma: Implications for the evolution of early hominin ecosystems. *Journal of Human Evolution*, *53*(5), 487–503. https://doi.org/10.1016/j.jhevol.2006.12.007.

Konecky, B. L., Russell, J. M., Rodysill, J. R., Vuille, M., Bijaksana, S., & Huang, Y. (2013). Intensification of southwestern Indonesian rainfall over the past millennium. *Geophysical Research Letters*, *40*(2), 386–391. https://doi.org/10.1029/2012gl054331.

Konecky, B., Russell, J., Vuille, M., & Rehfeld, K. (2014). The Indian Ocean zonal mode over the past millennium in observed and modeled precipitation isotopes. *Quaternary Science Reviews*, *103*, 1–18. https://doi.org/10.1016/j.quascirev.2014.08.019.

Krause, C. E., Gagan, M. K., Dunbar, G. B., Hantoro, W. S., Hellstrom, J. C., Cheng, H., Edwards, R. L., Suwargadi, B. W., Abram, N. J., & Rifai, H. (2019). Spatio-temporal evolution of Australasian monsoon hydroclimate over the last 40,000 years. *Earth and Planetary Science Letters*, *513*, 103–112. https://doi.org/10.1016/j.epsl.2019.01.045.

Kwiatkowski, C., Prange, M., Varma, V., Steinke, S., Hebbeln, D., & Mohtadi, M. (2015). Holocene variations of thermocline conditions in the eastern tropical Indian Ocean. *Quaternary Science Reviews*, *114*, 33–42. https://doi.org/10.1016/j.quascirev.2015.01.028.

Laskar, J., Robutel, P., Joutel, F., Gastineau, M., Correia, A. C. M., & Levrard, B. (2004). A long-term numerical solution for the insolation quantities of the earth. *Astronomy and Astrophysics*, *428*. https://doi.org/10.1051/0004-6361:20041335.

Leupold, M., Pfeiffer, M., Watanabe, T. K., Reuning, L., Garbe-Schönberg, D., Shen, C. C., & Brummer, G. J. A. (2021). El Niño–Southern Oscillation and internal sea surface temperature variability in the tropical Indian Ocean since 1675. *Climate of the Past*, *17*(1), 151–170. https://doi.org/10.5194/cp-17-151-2021.

Lisiecki, L. E., & Raymo, M. E. (2005). A Pliocene-Pleistocene stack of 57 globally distributed benthic $\delta^{18}O$ records. *Paleoceanography*, *20*(1). https://doi.org/10.1029/2004PA001071.

Loulergue, L., Schilt, A., Spahni, R., Masson-Delmotte, V., Blunier, T., Lemieux, B., Barnola, J.-M., Raynaud, D., Stocker, T. F., & Chappellaz, J. (2008). Orbital and millennial-scale features of atmospheric CH_4 over the past 800,000 years. *Nature*, *453*(7193), 383–386.

Lourens, L. J., Antonarakou, A., Hilgen, F. J., van Hoof, A. A. M., Vergnaud-Grazzini, C., & Zachariasse, W. (1996). Evaluation of the Plio-Pleistocene astronomical timescale. *Paleoceanography*, *11*, 391–413.

Lückge, A., Mohtadi, M., Rühlemann, C., Scheeder, G., Vink, A., Reinhardt, L., & Wiedicke, M. (2009). Monsoon versus ocean circulation controls on paleoenvironmental conditions off southern Sumatra during the past 300,000 years. *Paleoceanography, 24*, PA1208. https://doi.org/10.1029/2008PA001627.

Lupien, R. L., Russell, J. M., Feibel, C., Beck, C., Castañeda, I., Deino, A., & Cohen, A. S. (2018). A leaf wax biomarker record of early Pleistocene hydroclimate from West Turkana, Kenya. *Quaternary Science Reviews, 186*, 225–235. https://doi.org/10.1016/j.quascirev.2018.03.012.

Lüthi, D., Le Floch, M., Bereiter, B., Blunier, T., Barnola, J. M., Siegenthaler, U., Raynaud, D., Jouzel, J., Fischer, H., Kawamura, K., & Stocker, T. F. (2008). High-resolution carbon dioxide concentration record 650,000-800,000 years before present. *Nature, 453*(7193), 379–382. https://doi.org/10.1038/nature06949. ISSN 0028-0836.

Mantsis, D. F., Lintner, B. R., Broccoli, A. J., Erb, M. P., Clement, A. C., & Park, H.-S. (2014). The response of large-scale circulation to obliquity-induced changes in meridional heating gradients. *Journal of Climate, 27*(14), 5504–5516. https://doi.org/10.1175/jcli-d-13-00526.1.

Marshall, J., Donohoe, A., Ferreira, D., & McGee, D. (2014). The ocean's role in setting the mean position of the inter-tropical convergence zone. *Climate Dynamics, 42*(7), 1967–1979. https://doi.org/10.1007/s00382-013-1767-z.

Marzin, C., Kallel, N., Kageyama, M., Duplessy, J.-C., & Braconnot, P. (2013). Glacial fluctuations of the Indian monsoon and their relationship with North Atlantic climate: New data and modelling experiments. *Climate of the Past, 9*, 2135–2151. https://doi.org/10.5194/cp-9-2135-2013.

McGrath, S. M., Clemens, S. C., Haung, Y., & Yamamoto, M. (2021). Greenhouse gas and ice volume drive Pleistocene Indian summer monsoon precipitation isotope variability. *Geophysical Research Letters*. https://doi.org/10.1029/2020GL092249.

Meckler, A. N., Clarkson, M. O., Cobb, K. M., Sodemann, H., & Adkins, J. F. (2012). Interglacial hydroclimate in the tropical west pacific through the late Pleistocene. *Science, 336*(6086), 1301–1304. https://doi.org/10.1126/science.1218340.

Mei, R., Ashfaq, M., Rastogi, D., Leung, L. R., & Dominguez, F. (2015). Dominating controls for wetter South Asian summer monsoon in the twenty-first century. *Journal of Climate, 28*(8), 3400–3419. https://doi.org/10.1175/JCLI-D-14-00355.1.

Moffa-Sanchez, P., Rosenthal, Y., Babila, T. L., Mohtadi, M., & Zhang, X. (2019). Temperature evolution of the Indo-Pacific Warm Pool over the Holocene and the last deglaciation. *Paleoceanography and Paleoclimatology, 34*(7), 1107–1123. https://doi.org/10.1029/2018pa003455.

Mohtadi, M., Lückge, A., Steinke, S., Groeneveld, J., Hebbeln, D., & Westphal, N. (2010a). Late Pleistocene surface and thermocline conditions of the eastern tropical Indian Ocean. *Quaternary Science Reviews, 29*(7–8), 887–896. https://doi.org/10.1016/j.quascirev.2009.12.006.

Mohtadi, M., Oppo, D. W., Steinke, S., Stuut, J.-B. W., De Pol-Holz, R., Hebbeln, D., & Lückge, A. (2011). Glacial to Holocene swings of the Australian-Indonesian monsoon. *Nature Geoscience, 4*(8), 540–544. https://doi.org/10.1038/ngeo1209.

Mohtadi, M., Prange, M., Oppo, D. W., De Pol-Holz, R., Merkel, U., Zhang, X., Steinke, S., & Lückge, A. (2014). North Atlantic forcing of tropical Indian Ocean climate. *Nature, 509*(7498), 76–80. https://doi.org/10.1038/nature13196.

Mohtadi, M., Prange, M., Schefuß, E., & Jennerjahn, T. C. (2017). Late Holocene slowdown of the Indian Ocean Walker circulation. *Nature Communications, 8*(1), 1015. https://doi.org/10.1038/s41467-017-00855-3.

Mohtadi, M., Prange, M., & Steinke, S. (2016). Palaeoclimatic insights into forcing and response of monsoon rainfall. *Nature, 533*(7602), 191–199. https://doi.org/10.1038/nature17450.

Mohtadi, M., Steinke, S., Lückge, A., Groeneveld, J., & Hathorne, E. C. (2010b). Glacial to Holocene surface hydrography of the tropical eastern Indian Ocean. *Earth and Planetary Science Letters, 292*, 89–97.

Moy, C. M., Seltzer, G. O., Rodbell, D. T., & Anderson, D. M. (2002). Variability of El Niño/Southern oscillation activity at millennial timescales during the Holocene epoch. *Nature, 420*(6912), 162–165.

Nash, D. J., De Cort, G., Chase, B. M., Verschuren, D., Nicholson, S. E., Shanahan, T. M., Asrat, A., Lézine, A.-M., & Grab, S. W. (2016). African hydroclimatic variability during the last 2000 years. *Quaternary Science Reviews, 154*, 1–22. https://doi.org/10.1016/j.quascirev.2016.10.012.

Neff, U., Burns, S. J., Mangini, A., Mudelsee, M., Fleitmann, D., & Matter, A. (2001). Strong coherence between solar variability and the monsoon in Oman between 9 and 6 kyr ago. *Nature, 411*(6835), 290–293. https://doi.org/10.1038/35077048.

Nicholson, S. E. (2017). Climate and climatic variability of rainfall over eastern Africa. *Reviews of Geophysics, 55*(3), 590–635. https://doi.org/10.1002/2016RG000544.

Niedermeyer, E. M., Sessions, A. L., Feakins, S. J., & Mohtadi, M. (2014). Hydroclimate of the western Indo-Pacific warm Pool during the past 24,000 years. *Proceedings of the National Academy of Sciences, 111*(26), 9402–9406. https://doi.org/10.1073/pnas.1323585111.

Nutz, A., Schuster, M., Boës, X., & Rubino, J.-L. (2017). Orbitally-driven evolution of Lake Turkana (Turkana depression, Kenya, EARS) between 1.95 and 1.72 Ma: A sequence stratigraphy perspective. *Journal of African Earth Sciences, 125*, 230–243. https://doi.org/10.1016/j.jafrearsci.2016.10.016.

Oppo, D. W., Rosenthal, Y., & Linsley, B. K. (2009). 2,000-year-long temperature and hydrology reconstructions from the Indo-Pacific warm pool. *Nature, 460*(7259), 1113–1116.

Otto-Bliesner, B. L., Russell, J. M., Clark, P. U., Liu, Z., Overpeck, J. T., Konecky, B., deMenocal, P., Nicholson, S. E., He, F., & Lu, Z. (2014). Coherent changes of southeastern equatorial and northern African rainfall during the last deglaciation. *Science, 346*(6214), 1223–1227. https://doi.org/10.1126/science.1259531.

PAGES Hydr2k Consortium. (2017). Comparing proxy and model estimates of hydroclimate variability and change over the common era. *Climate of the Past, 13*, 1851–1900. https://doi.org/10.5194/cp-13-1851-2017.

Panmei, C., Naidu, D. P., & Mohtadi, M. (2017). Bay of Bengal exhibits warming trend during the younger dryas: Implications of AMOC. *Geochemistry, Geophysics, Geosystems, 18*(12), 4317–4325. https://doi.org/10.1002/2017GC007075.

Park, W., Keenlyside, N., Latif, M., Stroeh, A., Redler, R., Roeckner, E., & Madec, G. (2009). Tropical Pacific climate and its response to global warming in the Kiel climate model. *Journal of Climate, 22*, 71–92.

Partin, J. W., Cobb, K. M., Adkins, J. F., Clark, B., & Fernandez, D. P. (2007). Millennial-scale trends in West Pacific warm pool hydrology since the last glacial maximum. *Nature, 449*(7161), 452–455.

Partin, J. W., Quinn, T. M., Shen, C. C., Okumura, Y., Cardenas, M. B., Siringan, F. P., Banner, J. L., Lin, K., Hu, H. M., & Taylor, F. W. (2015). Gradual onset and recovery of the Younger Dryas abrupt climate event in the tropics. *Nature Communications*, 6. https://doi.org/10.1038/ncomms9061.

Pausata, F. S. R., Battisti, D. S., Nisancioglu, K. H., & Bitz, C. M. (2011). Chinese stalagmite $\delta^{18}O$ controlled by changes in the Indian monsoon during a simulated Heinrich event. *Nature Geoscience*, 4(7), 474–480.

Pfeiffer, M., & Dullo, W.-C. (2006). Monsoon-induced cooling of the western equatorial Indian Ocean as recorded in coral oxygen isotope records from the Seychelles covering the period of 1840–1994AD. *Quaternary Science Reviews*, 25(9), 993–1009. https://doi.org/10.1016/j.quascirev.2005.11.005.

Pfeiffer, M., Dullo, W.-C., & Eisenhauer, A. (2004a). Variability of the intertropical convergence zone recorded in coral isotopic records from the Central Indian Ocean (Chagos archipelago). *Quaternary Research*, 61(3), 245–255. https://doi.org/10.1016/j.yqres.2004.02.009.

Pfeiffer, M., Reuning, L., Zinke, J., Garbe-Schönberg, D., Leupold, M., & Dullo, W.-C. (2019). Multidecadal oscillations of $d^{18}O$ seawater reconstructed from paired $d^{18}O$ and Sr/Ca measurements of a La Reunion coral. *Paleoceanography and Palaeoclimate*. https://doi.org/10.1029/PA2019003770.

Pfeiffer, M., Timm, O., Dullo, W.-C., & Garbe-Schönberg, D. (2006). Paired coral Sr/Ca and $\delta^{18}O$ records from the Chagos archipelago: Late twentieth century warming affects rainfall variability in the tropical Indian Ocean. *Geology*, 34(12), 1069–1072. https://doi.org/10.1130/g23162a.1.

Pfeiffer, M., Timm, O., Dullo, W.-C., & Podlech, S. (2004b). Oceanic forcing of interannual and multidecadal climate variability in the southwestern Indian Ocean: Evidence from a 160 year coral isotopic record (La Réunion, 55°E, 21°S). *Paleoceanography*, 19(4). https://doi.org/10.1029/2003PA000964.

Pfeiffer, M., Zinke, J., Dullo, W. C., Garbe-Schönberg, D., Latif, M., & Weber, M. E. (2017). Indian Ocean corals reveal crucial role of World War II bias for twentieth century warming estimates. *Scientific Reports*, 7(1), 14434. https://doi.org/10.1038/s41598-017-14352-6.

Rao, Z., Li, Y., Zhang, J., Jia, G., & Chen, F. (2016). Investigating the long-term palaeoclimatic controls on the δD and $\delta^{18}O$ of precipitation during the Holocene in the Indian and East Asian monsoonal regions. *Earth-Science Reviews*, 159, 292–305. https://doi.org/10.1016/j.earscirev.2016.06.007.

Reeves, J. M., Bostock, H. C., Ayliffe, L. K., Barrows, T. T., De Deckker, P., Devriendt, L. S., Dunbar, G. B., Drysdale, R. N., Fitzsimmons, K. E., Gagan, M. K., Griffiths, M. L., Haberle, S. G., Jansen, J. D., Krause, C., Lewis, S., McGregor, H. V., Mooney, S. D., Moss, P., Nanson, G. C., ... van der Kaars, S. (2013). Palaeoenvironmental change in tropical Australasia over the last 30,000 years—A synthesis by the OZ-INTIMATE group. *Quaternary Science Reviews*, 74, 97–114. https://doi.org/10.1016/j.quascirev.2012.11.027.

Rodysill, J. R., Russell, J. M., Crausbay, S. D., Bijaksana, S., Vuille, M., Edwards, R. L., & Cheng, H. (2013). A severe drought during the last millennium in East Java, Indonesia. *Quaternary Science Reviews*, 80, 102–111. https://doi.org/10.1016/j.quascirev.2013.09.005.

Rodysill, J. R., Russell, J. M., Vuille, M., Dee, S., Lunghino, B., & Bijaksana, S. (2019). La Niña-driven flooding in the Indo-Pacific warm pool during the past millennium. *Quaternary Science Reviews*, 225, 106020. https://doi.org/10.1016/j.quascirev.2019.106020.

Rossignol-Strick, M. (1983). African monsoons, an intermediate climate response to orbital insolation. *Nature*, 304(7), 46–49.

Rossignol-Strick, M. (1985). Mediterranean Quaternary sapropels, an immediate response of the African monsoon to variation of insolation. *Palaeogeography, Palaeoclimatology, Palaeoecology*, 49, 237–263.

Roxy, M. K., Ritika, K., Terray, P., & Masson, S. (2014). The curious case of Indian Ocean warming. *Journal of Climate*, 27(22), 8501–8509. https://doi.org/10.1175/jcli-d-14-00471.1.

Ruan, Y., Mohtadi, M., van der Kaars, S., Dupont, L. M., Hebbeln, D., & Schefuß, E. (2019). Differential hydro-climatic evolution of East Javanese ecosystems over the past 22,000 years. *Quaternary Science Reviews*, 218, 49–60. https://doi.org/10.1016/j.quascirev.2019.06.015.

Russell, J. M., & Johnson, T. C. (2005). A high-resolution geochemical record from Lake Edward, Uganda Congo and the timing and causes of tropical African drought during the late Holocene. *Quaternary Science Reviews*, 24(12), 1375–1389. https://doi.org/10.1016/j.quascirev.2004.10.003.

Russell, J. M., & Johnson, T. C. (2007). Little ice age drought in equatorial Africa: Intertropical convergence zone migrations and El Niño–Southern oscillation variability. *Geology*, 35(1), 21–24. https://doi.org/10.1130/g23125a.1.

Russell, J. M., Verschuren, D., & Eggermont, H. (2007). Spatial complexity of 'little ice age' climate in East Africa: Sedimentary records from two crater lake basins in western Uganda. *The Holocene*, 17(2), 183–193. https://doi.org/10.1177/0959683607075832.

Russell, J. M., Vogel, H., Konecky, B. L., Bijaksana, S., Huang, Y., Melles, M., Wattrus, N., Costa, K., & King, J. W. (2014). Glacial forcing of central Indonesian hydroclimate since 60,000 y B.P. *Proceedings of the National Academy of Sciences*, 111(14), 5100–5105. https://doi.org/10.1073/pnas.1402373111.

Saji, N. H., Goswami, B. N., Vinayachandran, P. N., & Yamagata, T. (1999). A dipole mode in the tropical Indian Ocean. *Nature*, 401, 360–363.

Schefuß, E., Kuhlmann, H., Mollenhauer, G., Prange, M., & Pätzold, J. (2011). Forcing of wet phases in Southeast Africa over the past 17,000 years. *Nature*, 480(7378), 509–512.

Schneider, T., Bischoff, T., & Haug, G. H. (2014). Migrations and dynamics of the intertropical convergence zone. *Nature*, 513(7516), 45–53. https://doi.org/10.1038/nature13636.

Scroxton, N., Burns, S. J., McGee, D., Hardt, B., Godfrey, L. R., Ranivoharimanana, L., & Faina, P. (2017). Hemispherically in-phase precipitation variability over the last 1700 years in a Madagascar speleothem record. *Quaternary Science Reviews*, 164, 25–36. https://doi.org/10.1016/j.quascirev.2017.03.017.

Setiawan, R. Y., Mohtadi, M., Southon, J., Groeneveld, J., Steinke, S., & Hebbeln, D. (2015). The consequences of opening the Sunda Strait on the hydrography of the eastern tropical Indian Ocean. *Paleoceanography*, 30(10), 1358–1372. https://doi.org/10.1002/2015pa002802.

Simon, M. H., Arthur, K. L., Hall, I. R., Peeters, F. J. C., Loveday, B. R., Barker, S., Ziegler, M., & Zahn, R. (2013). Millennial-scale Agulhas Current variability and its implications for salt-leakage through the Indian–Atlantic Ocean Gateway. *Earth and Planetary Science Letters*, 383, 101–112. https://doi.org/10.1016/j.epsl.2013.09.035.

Simon, M. H., Gong, X., Hall, I. R., Ziegler, M., Barker, S., Knorr, G., van der Meer, M. T. J., Kasper, S., & Schouten, S. (2015). Salt exchange in the Indian-Atlantic Ocean gateway since the last glacial maximum: A compensating effect between Agulhas Current changes and salinity variations? *Paleoceanography*, 30(10), 1318–1327. https://doi.org/10.1002/2015PA002842.

Sinha, A., Kathayat, G., Cheng, H., Breitenbach, S. F. M., Berkelhammer, M., Mudelsee, M., Biswas, J., & Edwards, R. L. (2015). Trends and oscillations in the Indian summer monsoon rainfall over the last two millennia. *Nature Communications*, 6. https://doi.org/10.1038/ncomms7309.

Stager, J. C., Ryves, D. B., Chase, B. M., & Pausata, F. S. R. (2011). Catastrophic Drought in the Afro-Asian monsoon region during Heinrich event 1. *Science*, *331*(6022), 1299–1302. https://doi.org/10.1126/science.1198322.

Steinke, S., Mohtadi, M., Prange, M., Varma, V., Pittauerova, D., & Fischer, H. W. (2014a). Mid- to Late-Holocene Australian–Indonesian summer monsoon variability. *Quaternary Science Reviews*, *93*, 142–154. https://doi.org/10.1016/j.quascirev.2014.04.006.

Steinke, S., Prange, M., Feist, C., Groeneveld, J., & Mohtadi, M. (2014b). Upwelling variability off southern Indonesia over the past two millennia. *Geophysical Research Letters*, *41*(21). https://doi.org/10.1002/2014gl061450.

Svensson, A., Andersen, K. K., Bigler, M., Clausen, H. B., Dahl-Jensen, D., Davies, S. M., Johnsen, S. J., Muscheler, R., Parrenin, F., Rasmussen, S. O., Röthlisberger, R., Seierstad, I., Steffensen, J. P., & Vinther, B. M. (2008). A 60 000 year Greenland stratigraphic ice core chronology. *Climate of the Past*, *4*, 47–57. https://doi.org/10.5194/cp-4-47-2008.

Tabor, C. R., Otto-Bliesner, B. L., Brady, E. C., Nusbaumer, J., Zhu, J., Erb, M. P., Wong, T. E., Liu, Z., & Noone, D. (2018). Interpreting precession-driven $\delta^{18}O$ variability in the south Asian monsoon region. *Journal of Geophysical Research: Atmospheres*, *123*(11), 5927–5946. https://doi.org/10.1029/2018JD028424.

Therrell, M. D., Stahle, D. W., Ries, L. P., & Shugart, H. H. (2006). Tree-ring reconstructed rainfall variability in Zimbabwe. *Climate Dynamics*, *26*(7), 677. https://doi.org/10.1007/s00382-005-0108-2.

Thirumalai, K., DiNezio, P. N., Tierney, J. E., Puy, M., & Mohtadi, M. (2019). An El Niño mode in the glacial Indian Ocean? *Paleoceanography and Paleoclimatology*, *34*(8). https://doi.org/10.1029/2019pa003669.

Thomas, D. S. G., Burrough, S. L., & Parker, A. G. (2012). Extreme events as drivers of early human behaviour in Africa? The case for variability, not catastrophic drought. *Journal of Quaternary Science*, *27*(1), 7–12. https://doi.org/10.1002/jqs.1557.

Tierney, J. E., Abram, N. J., Anchukaitis, K. J., Evans, M. N., Giry, C., Kilbourne, K. H., ... Zinke, J. (2015a). Tropical Sea surface temperatures for the past four centuries reconstructed from coral archives. *Paleoceanography*, *30*(3), 226–252. https://doi.org/10.1002/2014pa002717.

Tierney, J. E., & deMenocal, P. B. (2013). Abrupt shifts in horn of Africa hydroclimate since the last glacial maximum. *Science*, *342*(6160), 843–846. https://doi.org/10.1126/science.1240411.

Tierney, J. E., Oppo, D. W., Rosenthal, Y., Russell, J. M., & Linsley, B. K. (2010). Coordinated hydrological regimes in the Indo-Pacific region during the past two millennia. *Paleoceanography*, *25*(1), PA1102. https://doi.org/10.1029/2009PA001871.

Tierney, J. E., Pausata, F. S. R., & deMenocal, P. (2015b). Deglacial Indian monsoon failure and North Atlantic stadials linked by Indian Ocean surface cooling. *Nature Geoscience*, *9*, 46–50. https://doi.org/10.1038/ngeo2603.

Tierney, J. E., Russell, J. M., Huang, Y., Damste, J. S. S., Hopmans, E. C., & Cohen, A. S. (2008). Northern hemisphere controls on tropical Southeast African climate during the past 60,000 years. *Science*, *322*(5899), 252–255. https://doi.org/10.1126/science.1160485.

Tierney, J. E., Russell, J. M., Sinninghe Damsté, J. S., Huang, Y., & Verschuren, D. (2011). Late quaternary behavior of the East African monsoon and the importance of the Congo Air Boundary. *Quaternary Science Reviews*, *30*(7–8), 798–807. https://doi.org/10.1016/j.quascirev.2011.01.017.

Tierney, J. E., Smerdon, J. E., Anchukaitis, K. J., & Seager, R. (2013). Multidecadal variability in East African hydroclimate controlled by the Indian Ocean. *Nature*, *493*(7432), 389–392.

Timm, O., Pfeiffer, M., & Dullo, W.-C. (2005). Nonstationary ENSO-precipitation teleconnection over the equatorial Indian Ocean documented in a coral from the Chagos Archipelago. *Geophysical Research Letters*, *32*(2). https://doi.org/10.1029/2004GL021738.

Tiwari, M., Nagoji, S. S., & Ganeshram, R. S. (2015). Multi-centennial scale SST and Indian summer monsoon precipitation variability since the mid-Holocene and its nonlinear response to solar activity. *The Holocene*, *25*(9), 1415–1424. https://doi.org/10.1177/0959683615585840.

Trauth, M. H., Deino, A. L., Bergner, A. G. N., & Strecker, M. R. (2003). East African climate change and orbital forcing during the last 175 kyr BP. *Earth and Planetary Science Letters*, *206*(3), 297–313. https://doi.org/10.1016/S0012-821X(02)01105-6.

Trauth, M. H., Larrasoaña, J. C., & Mudelsee, M. (2009). Trends, rhythms and events in Plio-Pleistocene African climate. *Quaternary Science Reviews*, *28*(5), 399–411. https://doi.org/10.1016/j.quascirev.2008.11.003.

Ummenhofer, C. C., Gupta, A. S., England, M. H., & Reason, C. J. C. (2009). Contributions of Indian Ocean sea surface temperatures to enhanced east African rainfall. *Journal of Climate*, *22*(4), 993–1013.

van Loon, H., & Meehl, G. A. (2012). The Indian summer monsoon during peaks in the 11 year sunspot cycle. *Geophysical Research Letters*, *39*(13), L13701. https://doi.org/10.1029/2012gl051977.

Venugopal, T., Ali, M. M., Bourassa, M. A., Zheng, Y., Goni, G. J., Foltz, G. R., & Rajeevan, M. (2018). Statistical evidence for the role of southwestern Indian Ocean heat content in the Indian summer monsoon rainfall. *Scientific Reports*, *8*(1), 12092. https://doi.org/10.1038/s41598-018-30552-0.

Verschuren, D. (2004). Decadal and century-scale climate variability in tropical Africa during the past 2000 years. In R. W. Battarbee, F. Gasse, & C. E. Stickley (Eds.), *Past climate variability through Europe and Africa* (pp. 139–158). Dordrecht: Springer Netherlands.

Verschuren, D., Laird, K. R., & Cumming, B. F. (2000). Rainfall and drought in equatorial East Africa during the past 1,100 years. *Nature*, *403*(6768), 410–414. https://doi.org/10.1038/35000179.

Vogel, H., Russell, J. M., Cahyarini, S. Y., Bijaksana, S., Wattrus, N., Rethemeyer, J., & Melles, M. (2015). Depositional modes and Lake-level variability at Lake Towuti, Indonesia, during the past ~29kyr BP. *Journal of Paleolimnology*, *54*(4), 359–377. https://doi.org/10.1007/s10933-015-9857-z.

Wang, Y. J., Cheng, H., Edwards, R. L., An, Z. S., Wu, J. Y., Shen, C. C., & Dorale, J. A. (2001). A high-resolution absolute-dated Late Pleistocene monsoon record from Hulu Cave, China. *Science*, *294*(5550), 2345–2348. https://doi.org/10.1126/science.1064618.

Wang, B., Clemens, S. C., & Liu, P. (2003). Contrasting the Indian and east Asian monsoons: Implications on geologic timescales. *Marine Geology*, *201*, 5–21. https://doi.org/10.1016/S0025-3227(03)00196-8.

Wang, X., Jian, Z., Lückge, A., Wang, Y., Dang, H., & Mohtadi, M. (2018). Precession-paced thermocline water temperature changes in response to upwelling conditions off southern Sumatra over the past 300,000 years. *Quaternary Science Reviews, 192*, 123–134. https://doi.org/10.1016/j.quascirev.2018.05.035.

Wang, B., Jin, C., & Liu, J. (2020). Understanding future change of global monsoon projected by CMIP6 models. *Journal of Climate*. https://doi.org/10.1175/JCLI-D-19-0993.1.

Windler, G., Tierney, J. E., DiNezio, P. N., Gibson, K., & Thunell, R. (2019). Shelf exposure influence on Indo-Pacific warm Pool climate for the last 450,000 years. *Earth and Planetary Science Letters, 516*, 66–76. https://doi.org/10.1016/j.epsl.2019.03.038.

Wolff, C., Haug, G. H., Timmermann, A., Damsté, J. S. S., Brauer, A., Sigman, D. M., Cane, M. A., & Verschuren, D. (2011). Reduced interannual rainfall variability in East Africa during the last ice age. *Science, 333*(6043), 743–747. https://doi.org/10.1126/science.1203724.

Wurtzel, J. B., Abram, N. J., Lewis, S. C., Bajo, P., Hellstrom, J. C., Troitzsch, U., & Heslop, D. (2018). Tropical Indo-Pacific hydroclimate response to North Atlantic forcing during the last deglaciation as recorded by a speleothem from Sumatra, Indonesia. *Earth and Planetary Science Letters, 492*, 264–278. https://doi.org/10.1016/j.epsl.2018.04.001.

Xu, J., Holbourn, A., Kuhnt, W., Jian, Z., & Kawamura, H. (2008). Changes in the thermocline structure of the Indonesian outflow during terminations I and II. *Earth and Planetary Science Letters, 273*(1–2), 152–162.

Xu, J., Kuhnt, W., Holbourn, A., Andersen, N., & Bartoli, G. (2006). Changes in the vertical profile of the Indonesian throughflow during termination II: Evidence from the Timor Sea. *Paleoceanography, 21*, PA4202. https://doi.org/10.1029/2006PA001278.

Yamamoto, M., Clemens, S. C., Seki, O., Tsuchiya, Y., Huang, Y., O'ishi, R., & Abe-Ouchi, A. (2022). Increased Interglacial atmospheric CO_2 levels followed the mid-Pleistocene transition. *Nature Geoscience, 15*. https://doi.org/10.1038/s41561-022-00918-1.

Yan, H., Wei, W., Soon, W., An, Z., Zhou, W., Liu, Z., Wang, Y., & Carter, R. M. (2015). Dynamics of the intertropical convergence zone over the western Pacific during the Little Ice Age. *Nature Geoscience, 8*(4), 315–320. https://doi.org/10.1038/ngeo2375.

Zhang, P., Cheng, H., Edwards, R. L., Chen, F., Wang, Y., Yang, X., Liu, J., Tan, M., Wang, X., Liu, J., An, C., Dai, Z., Zhou, J., Zhang, D., Jia, J., Jin, L., & Johnson, K. R. (2008). A test of climate, Sun, and culture relationships from an 1810-year Chinese cave record. *Science, 322*(5903), 940–942. https://doi.org/10.1126/science.1163965.

Zhang, P., Xu, J., Holbourn, A., Kuhnt, W., Beil, S., Li, T., Xiong, Z., Dang, H., Yan, H., Pei, R., Ran, Y., & Wu, H. (2020). Indo-Pacific hydroclimate in response to changes of the intertropical convergence zone: Discrepancy on precession and obliquity bands over the last 410 kyr. *Journal of Geophysical Research: Atmospheres, 125*(14). https://doi.org/10.1029/2019JD032125.

Zhang, P., Xu, J., Schröder, J. F., Holbourn, A., Kuhnt, W., Kochhann, K. G. D., Ke, F., Wang, Z., & Wu, H. (2018). Variability of the Indonesian throughflow thermal profile over the last 25-kyr: A perspective from the southern Makassar Strait. *Global and Planetary Change, 169*, 214–223. https://doi.org/10.1016/j.gloplacha.2018.08.003.

Zhou, X., Sun, L., Chu, Y., Xia, Z., Zhou, X., Li, X., Chu, Z., Liu, X., Shao, D., & Wang, Y. (2016). Catastrophic drought in east Asian monsoon region during Heinrich event 1. *Quaternary Science Reviews, 141*, 1–8. https://doi.org/10.1016/j.quascirev.2016.03.029.

Ziegler, M., Lourens, L. J., Tuenter, E., Hilgen, F., Reichart, G.-J., & Weber, N. (2010). Precession phasing offset between Indian summer monsoon and Arabian Sea productivity linked to changes in Atlantic overturning circulation. *Paleoceanography, 25*. https://doi.org/10.1029/2009PA001884.

Zinke, J., D'Olivo, J. P., Gey, C. J., McCulloch, M. T., Bruggemann, J. H., Lough, J. M., & Guillaume, M. M. M. (2019). Multi-Trace-Element Sea surface temperature coral reconstruction for the southern Mozambique Channel reveals teleconnections with the tropical Atlantic. *Biogeosciences, 16*(3), 695–712. https://doi.org/10.5194/bg-16-695-2019.

Zinke, J., Dullo, W. C., Heiss, G. A., & Eisenhauer, A. (2004). ENSO and Indian Ocean subtropical dipole variability is recorded in a coral record off Southwest Madagascar for the period 1659 to 1995. *Earth and Planetary Science Letters, 228*(1–2), 177–194.

Zinke, J., Pfeiffer, M., Park, W., Schneider, B., Reuning, L., Dullo, W. C., Camoin, G. F., Mangini, A., Schröder-Ritzrau, A., Garbe-Schönberg, D., & Davies, G. R. (2014). Seychelles coral record of changes in sea surface temperature bimodality in the western Indian Ocean from the Mid-Holocene to the present. *Climate Dynamics, 43*(3–4), 689–708. https://doi.org/10.1007/s00382-014-2082-z.

Zinke, J., Pfeiffer, M., Timm, O., Dullo, W. C., & Brummer, G. J. A. (2008a). Western Indian Ocean marine and terrestrial records of climate variability: A review and new concepts on land–ocean interactions since AD 1660. *International Journal of Earth Sciences, 98*(1), 115. https://doi.org/10.1007/s00531-008-0365-5.

Zinke, J., Pfeiffer, M., Timm, O., Dullo, W. C., & Davies, G. R. (2005). Atmosphere-ocean dynamics in the Western Indian Ocean recorded in corals. *Philosophical Transactions of the Royal Society A: Mathematical, Physical and Engineering Sciences, 363*(1826), 121–142. https://doi.org/10.1098/rsta.2004.1482.

Zinke, J., Pfeiffer, M., Timm, O., Dullo, W. C., Kroon, D., & Thomassin, B. A. (2008b). Mayotte coral reveals hydrological changes in the western Indian Ocean between 1881 and 1994. *Geophysical Research Letters, 35*, L23707. https://doi.org/10.1029/2008GL035634.

Zinke, J., Reuning, L., Pfeiffer, M., Wassenburg, J. A., Hardman, E., Jhangeer-Khan, R., Davies, G. R., Ng, C. K. C., & Kroon, D. (2016). A sea surface temperature reconstruction for the southern Indian Ocean trade wind belt from corals in Rodrigues Island (19° S, 63° E). *Biogeosciences, 13*(20), 5827–5847. https://doi.org/10.5194/bg-13-5827-2016.

Chapter 20

Future projections for the tropical Indian Ocean

M.K. Roxy[a], J.S. Saranya[a,b], Aditi Modi[a,c], A. Anusree[a], Wenju Cai[d,e], Laure Resplandy[f], Jérôme Vialard[g], and Thomas L. Frölicher[h,i]

[a]Centre for Climate Change Research, Indian Institute of Tropical Meteorology, Ministry of Earth Sciences, Pune, India, [b]School of Earth and Environmental Sciences, College of Natural Sciences, Seoul National University, Seoul, Republic of Korea, [c]Interdisciplinary Programme in Climate Studies, Indian Institute of Technology Bombay, Mumbai, India, [d]Frontier Science Centre for Deep Ocean Multispheres and Earth System and Physical Oceanography Laboratory, Ocean University of China, Qingdao, China, [e]Center for Southern Hemisphere Oceans Research (CSHOR), CSIRO Oceans and Atmosphere, Hobart, TAS, Australia, [f]Department of Geosciences and High Meadows Environmental Institute, Princeton University, Princeton, NJ, United States, [g]LOCEAN-IPSL, Sorbonne Universités (UPMC, Univ Paris 06)-CNRS-IRD-MNHN, Paris, France, [h]Climate and Environmental Physics, Physics Institute, University of Bern, Bern, Switzerland, [i]Oeschger Centre for Climate Change Research, University of Bern, Bern, Switzerland

1 Introduction

The tropical Indian Ocean (40–120°E, 30°S–30°N) underwent basin-wide warming during the last 150 years (1871–2020). From an average sea surface temperature (SST) of 26.44°C in the 1870s, the basin recorded an average 0.76°C increase, raising the basin-average SSTs to 27.2°C by the 2010s (Fig. 1). Recent research and the Intergovernmental Panel on Climate Change (IPCC) reports point out that the fastest ocean surface warming since the 1950s has occurred in the Indian Ocean and western boundary currents, while ocean circulation has slowed down the warming or even slightly cooled the surface in parts of the Southern Ocean, equatorial Pacific, North Atlantic, and coastal upwelling systems (Collins et al., 2019; Fox-Kemper et al., 2021; IPCC, 2021). The SST warming trend in the Indian Ocean was strongest during the last seven decades (1950–2020), at a rate of 0.12°C per decade. These SST changes are dwarfed by the projected surface warming of 3°C between 2020 and the end of the century (at a rate of 0.38°C per decade), if anthropogenic emissions continue to increase at the current rate (Fig. 1, SSP5-8.5 scenario).

The rapid warming in the Indian Ocean is not limited to the surface. The heat gain in the Indian Ocean represents about one-quarter of the global ocean heat gain since 1990, primarily due to a redistribution of heat from the Pacific to the Indian Ocean (Beal et al., 2020; Cheng et al., 2017). There are uncertainties regarding monitoring the change in the total heat content and exchange in and out of the basin (e.g., the heat exchange via the Indonesian Throughflow and the Agulhas Current (Sprintall et al., 2024; Tozuka et al., 2024). However, it is certain that the Indian Ocean exhibits the fastest surface warming among all the other tropical oceans, in recent decades (Beal et al., 2020; Gnanaseelan et al., 2017; Hermes et al., 2019; Roxy et al., 2020). The IPCC Special Report on Ocean and Cryosphere in a Changing Climate indicates that global ocean warming in the upper 2000 m would be 5–7 times higher than the warming recorded since 1970 under the business-as-usual scenario by 2100, and 2–4 times higher under the low emission scenario (Collins et al., 2019).

SST variations mediate heat exchange across the air-sea interface, with high SSTs over the tropics accompanied by changes in atmospheric convection and circulation. A large part of the Indian Ocean is covered by the tropical warm pool, characterized by permanently warm SSTs greater than 28°C, and is therefore often called the heat engine of the globe (De Deckker, 2016; Rao et al., 2012; Roxy et al., 2019). Warming SSTs in the Indian Ocean imply a ramping up of this heat engine through intensification and expansion of the warm pool, thereby impacting the local and global climate (Beal et al., 2020; Roxy et al., 2020).

Indian Ocean warming contributes to increasing monsoon droughts and floods, and premonsoon heatwaves over South Asia (Li et al., 2022; Rohini et al., 2016; Roxy et al., 2015, 2017; Wang et al., 2021). Warming of the tropical Indian Ocean

---

[*]This book has a companion website hosting complementary materials. Visit this URL to access it: https://www.elsevier.com/books-and-journals/book-companion/9780128226988.

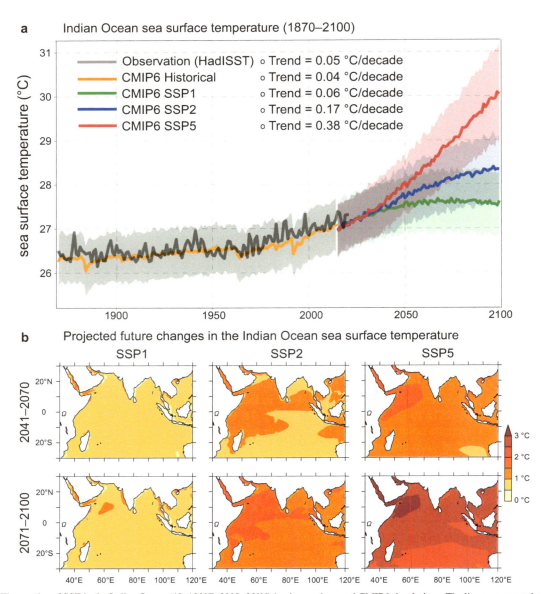

FIG. 1 (a) Time series of SST in the Indian Ocean (40–120°E, 30°S–30°N) in observations and CMIP6 simulations. The lines represent the observations (HadISST, 1870–2020, *black*) and CMIP6 multimodel ensemble mean of historical simulations (1870–2014, orange) and the future projections (2015–2100) under SSP1-2.6 (*green*), SSP2-4.5 (*blue*), and SSP5-8.5 (*red*) emission scenarios. Shading represents the intermodel uncertainty (intermodel standard deviation). (b) The projected multimodel mean changes in SST over the Indian Ocean for near (2041–2070) and far future (2071–2100) periods in different emission scenarios (SSP1-2.6, SSP2-4.5, SSP5-8.5), with respect to the historical simulations for the period 1985–2014.

in the recent decades has led to frequent droughts and occasional locust outbreaks in eastern Africa, threatening food security in this region (Funk et al., 2008; Salih et al., 2020). The effect of long-term warming in the Indian Ocean is reinforced by the occasional Indian Ocean Dipole (IOD) events, which preconditions and exacerbates bushfires over Australia since 1950 (Cai et al., 2009). The increase in ocean heat content has resulted in a rise in sea level via thermal expansion of seawater (Swapna et al., 2020), a potential increase in extremely severe cyclones and their rapid intensification (Bhatia et al., 2018; Deshpande et al., 2021; Murakami et al., 2017; Singh & Roxy, 2022) and consistent rise in the frequency and intensity of marine heatwaves in the Indian Ocean (Oliver et al., 2018; Qi et al., 2022; Saranya et al., 2022). The warming of the Indian Ocean also has far-reaching global impacts on intraseasonal-to-climate timescales. It modulates the Madden Julian Oscillation (MJO) and monsoon intraseasonal oscillation, which alters regional rainfall patterns (Rodrigues et al., 2019; Roxy et al., 2019; Sabeerali et al., 2014), and strengthens the Atlantic meridional overturning circulation thereby influencing the global climate (Hu & Fedorov, 2019, 2020).

The increase in atmospheric CO_2 and the associated ocean warming have very likely contributed to biogeochemical changes in the tropical Indian Ocean. These biogeochemical changes include the observed decreasing trends in pH (Piontkovski & Queste, 2016), dissolved oceanic oxygen (O_2) concentrations (Helm et al., 2011; Ito et al., 2017; Lévy et al., 2021; Stramma et al., 2008), and marine phytoplankton distribution (Piontkovski & Queste, 2016; Prakash et al., 2012; Roxy et al., 2016) in the tropical Indian Ocean. Combined, the more frequent marine heatwaves and biogeochemical changes potentially impact the marine ecosystem and fisheries in the tropical Indian Ocean (e.g., do Rosário Gomes et al., 2014; Frölicher & Laufkötter, 2018; Hood et al., 2024a, 2024b; Marsac et al., 2024; Naqvi et al., 2009; Piontkovski & Queste, 2016).

Given the magnitude and impact of the rapid warming in the tropical Indian Ocean, it is important to quantify and assess the future evolution of this warming under different climate change scenarios. This chapter discusses the future projections of the physical and biogeochemical changes in the Indian Ocean using existing literature based on Coupled Model Intercomparison Project (CMIP) Phase 5 (CMIP5) and available Phase 6 (CMIP6) simulations. The CMIP6 simulations are used to prepare the analysis and figures for historical simulations and future projections in the current chapter. The future scenarios are represented by Shared Socioeconomic Pathways (SSPs) of projected socioeconomic global changes up to 2100, based on greenhouse gas emissions scenarios with different climate policies (O'Neill et al., 2016; Riahi et al., 2017). Here we utilize three pathways: a world of sustainability-focused growth and equality where the radiative forcing is limited to $2.6\,W\,m^{-2}$ by the end of the 21st century (SSP1-2.6, low-forcing scenario); a "middle of the road" world where trends broadly follow their historical patterns and the radiative forcing is limited to $4.5\,W\,m^{-2}$ (SSP2-4.5, medium-forcing scenario); and the high road—a world of rapid and unconstrained fossil fuel-driven growth in economic output and energy use where the radiative forcing is high at $8.5\,W\,m^{-2}$ (SSP5-8.5, high-forcing scenario).

2 Projected changes in sea surface temperature, Indian Ocean Dipole, and heat content

The CMIP6 projections of future SSTs in the Indian Ocean show basin-wide warming but with substantial regional and seasonal variations as documented in previous studies (Cai et al., 2013). Fig. 1 shows the projected changes in SST over the Indian Ocean for the near (2041–2070) and far future (2071–2100) in different CMIP6 scenarios with respect to the historical period 1985–2014. All the CMIP6 future projections, regardless of the specific scenario, show maximum warming in the northwestern Indian Ocean, including the Arabian Sea, and reduced warming off the Sumatra and Java coasts in the southeast Indian Ocean (Fig. 1). These patterns are also simulated by the CMIP5-type of models (Zhao & Zhang, 2016; Zheng et al., 2010). While the CMIP6 SSP1-2.6 projects a basin-wide Indian Ocean warming of 0.06°C per decade, SSP2-4.5 and SSP5-8.5 projects an increased rate at 0.17°C and 0.38°C per decade, respectively (Fig. 1).

The strong warming pattern in the northwestern Indian Ocean and relatively weaker warming over the southeastern Indian Ocean in the future projections is consistent with a corresponding increase and decrease in precipitation over these regions, respectively, and strong easterly winds over the tropical Indian Ocean (Li et al., 2016). The changes in the easterlies along the equator and the faster warming in the west than the east in the Indian Ocean accompany a reduced strength of the Walker circulation in response to global warming (Vecchi et al., 2006). These environmental conditions create a conducive condition for the formation of the IOD pattern through the shoaling of thermocline over the eastern equatorial Indian Ocean (Zheng et al., 2010), and ease at which the atmospheric convergence moves to the west (Cai et al., 2014).

Regardless of the skewness in warming patterns in the Indian Ocean, instrumental records do not show any significant trends in IOD behavior. Also, the projected changes in the frequency and intensity of future IOD events remain uncertain in terms of the amplitude of the traditional dipole mode index using SSTs (Hui & Zheng, 2018; McKenna et al., 2020; Saji et al., 1999). Meanwhile, in terms of IOD-induced rainfall anomalies, climate models project that the frequency of extreme positive IOD events could increase by almost a factor of three, from a one-in-seventeen-year event in the 20th century to a one-in-six-year event in the 21st century (Cai et al., 2014; Collins et al., 2019). However, the frequency of IODs flattens if the global mean temperature is maintained below 1.5–2.0°C warming by 2050 (relative to the preindustrial level), as in the Paris Climate Agreement (Cai et al., 2018).

However, it is important to note that there is a debate about the dynamic processes related to the projected changes in IOD under global warming. The IOD simulations have a significant bias and intermodel spread that could lead to an overestimation of the projected increase in extreme positive IOD events (Li et al., 2016) while other studies suggest the bias could lead to an under-estimate of the projected increase (Wang et al., 2017). In fact, the CMIP historical simulations fail in reproducing the observed changes in the zonal SST gradients over the Indian Ocean in response to greenhouse gas forcing (Roxy et al., 2014).

Recent studies show that dynamics for extreme IOD events as in 1997 and 2019 and for moderate IOD events as in 1982 and 2015 are different, with nonlinear subsurface ocean dynamics playing a more important role in the extreme

IOD events (Cai et al., 2014, 2021; Yang et al., 2020). As such, extreme and moderate IOD feature vastly different SST anomaly patterns, and two indices are required to represent their difference. Examining the response of the two types of IOD events in terms of IOD SST anomalies finds that the frequency of extreme IOD increases by 66% whereas the frequency of moderate IOD events decreases by 52% in the 21st century climate projections, as compared to the 20th century model historical simulations (Cai et al., 2021).

An increase in greenhouse gas emissions will not only raise the SSTs on interannual to decadal timescales but will also change its variability and seasonal cycle. On annual and longer time scales, the seasonal cycle is responsible for around 90% of the total surface temperature variance (Dwyer et al., 2012). CMIP6 models project a large increase in the magnitude of the seasonal cycle, in response to increased emissions under various SSPs (Fig. 2). The magnitude shift is largest during March–May when the mean temperatures are also at the highest (Fig. 2b, SST change of 2.75°C under high emission scenario). It is important to note that while the maximum basin mean temperatures in the Indian Ocean during 1980–2020 (observations and historical simulations) remained below 28°C (26–28°C) throughout the year, the minimum temperatures under SSP5-8.5 by the end of the 21st century is above 28°C (28.5–30.7°C) year around. This could have a potential response on the intensity of convection and the genesis and evolution of cyclones in the Indian Ocean region, where SSTs above 28°C are generally conducive for deep convection and cyclogenesis (Gadgil et al., 1984; Roxy, 2014; Singh et al., 2021). Heavy rainfall events and extremely severe cyclones have already increased since the 1950s (Deshpande et al., 2021; Roxy et al., 2017) and are projected to increase further with increasing SSTs in the tropical Indian Ocean region (Collins et al., 2019; Murakami et al., 2017; Wang et al., 2021).

The rapid warming in the Indian Ocean is not limited to the surface. Similar to the SSTs, the ocean heat content (OHC) also exhibits an increasing trend in the Indian Ocean since the 1950s (Han et al., 2014), though somewhat modulated by multidecadal variations (Ummenhofer et al., 2020, 2021). The paucity of long-term in situ observations makes it difficult to accurately quantify the relative contributions of warming trends and multidecadal variations to the changes in OHC and sea level (Beal et al., 2020; Nidheesh et al., 2017). The heat gain in the Indian Ocean represents about one-quarter of the global ocean heat gain since 1990, primarily due to a redistribution of heat from the Pacific to the Indian Ocean (Beal et al., 2020;

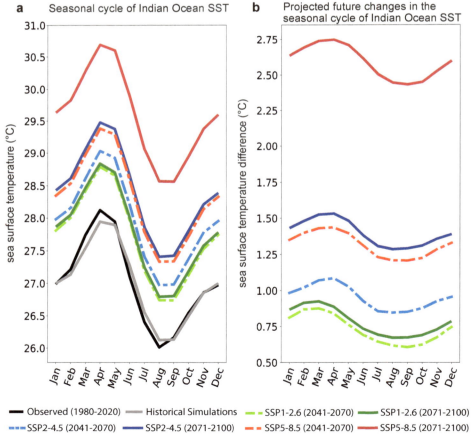

FIG. 2 (a) The seasonal cycle of SSTs over the Indian Ocean in the historical simulations (1985–2014) and as projected in different emission scenarios (SSP1-2.6, SSP2-4.5, SSP5-8.5) in CMPI6, for the periods 2041–2070 and 2071–2100. (b) The change in SSTs in near and far future projections, with respect to the historical simulations (1985–2014).

Cheng et al., 2017; Sprintall et al., 2024; Tozuka et al., 2024). The long-term trends in the upper (0–700 m) ocean heat content (OHC$_{700}$) exhibit decadal variations, associated mainly with the transport from the Pacific to the Indian Ocean via the Indonesian throughflow, which is tightly linked to the decadal variability of the El Niño Southern Oscillation (ENSO) (Han et al., 2014).

The rate of increase in the tropical Indian Ocean heat content was at 3.7 zetta-joules (1 zetta-joule = 10^{21} J) per decade for OHC$_{700}$, and 4.5 zetta-joules per decade for deep (0–2000 m) ocean heat content (OHC$_{2000}$), during 1960–2016 (Fig. 3). While the OHC$_{700}$ is projected to increase at a rate of 7 zetta-joules per decade for the low emission scenario, a trend of 11–17 zetta-joules per decade is projected under a mid-to-high emission scenario, between 2020 and the end of the 21st

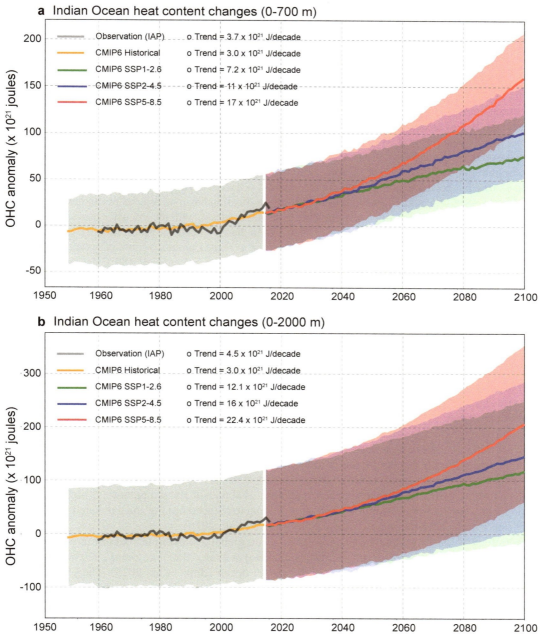

FIG. 3 Time series of ocean heat content anomaly (OHC) in the Indian Ocean (40–120°E, 30°S–30°N) in observation (Institute of Atmospheric Physics (IAP) ocean temperature analysis) and CMIP6 simulations relative to period 1960–2016, for (a) 0–700 m and (b) 0–2000 m. The *black, orange, green, blue,* and *red* lines represent the observations (Cheng et al., 2017) and CMIP6 multimodel ensemble mean of historical simulations (1950–2014) and future projections (2015–2100) under different emission scenarios (SSP1-2.6, SSP2-4.5, SSP5-8.5), respectively. The trend (slope) is estimated for the period represented by the respective lines in the figure. The shaded region represents the intermodel uncertainty (intermodel standard deviations).

century. The OHC$_{2000}$ is projected to increase at a rate of 16–22 zetta-joules per decade under mid-to-high emission scenarios during the same period. The IPCC Special Report on the Ocean and Cryosphere in a Changing Climate and Sixth Assessment Report projects a global OHC increase at about 256 zetta-joules/decade for high emission scenarios (Bindoff et al., 2019; Fox-Kemper et al., 2021), which indicates that the Indian Ocean OHC increase may contribute to about 1/9th of the global OHC rise.

3 Projected changes in marine heatwaves

The basin-wide warming in the Indian Ocean is a significant factor contributing to the increased occurrence of marine heatwaves (Frölicher et al., 2018; Oliver et al., 2018; Saranya et al., 2022). Marine heatwaves (Hobday et al., 2016) are periods of extremely high temperatures (i.e., SSTs exceeding the seasonally-varying 90th percentile threshold based on the 1970–1999 reference period in our case) in the ocean that can significantly impact marine organisms and ecosystems (Collins et al., 2019; Hughes et al., 2017; Smale et al., 2019). Oliver et al. (2018) show that globally, the frequency of marine heatwaves has increased on average by 34% and the duration by 17% during the last century. CMIP5 and CMIP6 projections indicate a significant further increase in the frequency, intensity, and duration of marine heatwaves in the future due to global warming (Frölicher et al., 2018; Plecha & Soares, 2020). These projections indicate that by the end of the century, global oceans will be in a near-permanent marine heatwave state (Frölicher et al., 2018; Oliver et al., 2019). Notably, the number of marine heatwave days in the Indian Ocean is projected to increase from 20 days per year (in historical simulations, 21 days per year in observations) to 250 days per year by 2100 (SSP 8.5). The maximum intensity of marine heatwaves (i.e., SST anomaly exceeding the seasonally varying 90th percentile threshold) in the Indian Ocean is projected to increase from 0.60°C per year (in historical simulations, 0.96°C per year in observations) to 3.4°C per year by 2100 (Fig. 4, SSP 8.5).

Fig. 4a shows that for both the near and far future, SSP1-2.6 has an average number of marine heatwave days <200, while SSP2-4.5 and SSSP5-8.5 have higher values >250 days, indicating basin-wide near-permanent marine heatwave conditions (Fig. 4c). The maximum intensities also indicate a similar contrast, with the intensity projected below 1.2°C for the SSP1-2.6 while it is around 1.2–1.4°C for SSSP5-8.5 (Fig. 4b). In the far future, the entire Indian Ocean basin is projected to have marine heatwaves of above 1.8°C maximum intensity. The average cumulative intensity (sum of temperature anomalies over the duration of a marine heatwave, in °C) in SSP2-4.5 and SSP5-8.5 exceeds 480°C days in the near and far future for most of the tropical Indian Ocean (Fig. 4).

4 Projected changes in the biogeochemistry of the Indian Ocean

The western Indian Ocean, including the Arabian Sea, is one of the most prominent marine productivity hotspots within the tropical Indian Ocean, where upwelling of cold nutrient-rich water by the seasonally reversing monsoon winds promotes large phytoplankton blooms (Hood et al., 2024a; Prasanna Kumar et al., 2001; Wiggert et al., 2005). Ocean color measurements suggest that the surface chlorophyll—a proxy for estimating net primary productivity—has declined by up to 30% in the Arabian Sea during 1998–2013 (Roxy et al., 2016). Historical simulations from the CMIP5 Earth system models also agree with the observations, suggesting a 20% decline in this region, during 1950–2015 (Roxy et al., 2016). These strong declining trends in phytoplankton production are mainly due to the increased stratification of the oceanic water column caused by the warming of the ocean surface (Behrenfeld et al., 2006; Roxy et al., 2016). Global ocean salinity changes have contributed to enhanced ocean stratification in the high-latitude regions (Cheng et al., 2020), amplifying the effects of ocean warming, while such an impact is not evidenced for most of the tropical Indian Ocean (Li et al., 2020). In the low-latitude regions—that are primarily nutrient-limited as sunlight is abundant and temperature is high—an increase in ocean surface warming increases the water column stratification leading to reduced mixing and a reduced supply of nutrients from the subsurface into the surface (Behrenfeld et al., 2006; Boyce et al., 2010).

Earlier generations of Earth system models projected a mean global decline in the net primary productivity during the 21st century (Behrenfeld et al., 2006; Bopp et al., 2001, 2013; Boyce et al., 2010; Fung et al., 2005; Steinacher et al., 2010), associated with a net decrease in carbon export into the deeper ocean (Bopp et al., 2013). The CMIP6 models generally outperform the CMIP5 predecessors in simulating the observed ocean biogeochemistry (Séférian et al., 2020). The CMIP6 ensemble mean projects a further significant reduction in depth-integrated net primary production in the western Indian Ocean and Arabian Sea by the end of the 21st century (Kwiatkowski et al., 2020).

The CMIP6 projections indicate a reduction in surface chlorophyll concentrations in both the near-future (2041–2070) and far-future (2071–2100), with respect to the reference period (1976–2005) in most parts of the tropical Indian Ocean, particularly the western Arabian Sea and western Bay of Bengal (Fig. 5). Under medium-to-high emission scenarios, an 8%–10% decrease in surface chlorophyll is projected for the western Arabian Sea region (50–65°E, 5–25°N) by the end of

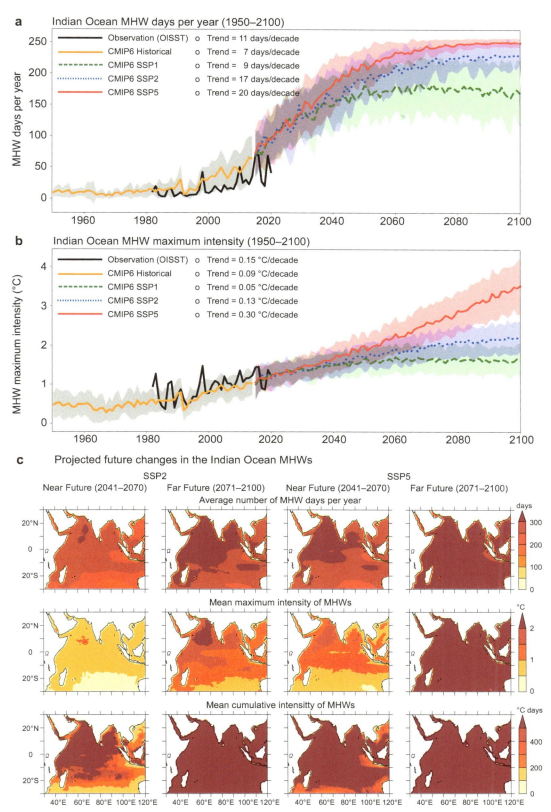

FIG. 4 (a) Marine heatwave (MHW) days and (b) maximum intensity (maximum temperature anomaly exceeding the seasonally-varying 90th percentile threshold during a marine heatwave, °C) in observations (National Oceanic and Atmospheric Administration Optimum Interpolation Sea Surface Temperature V2 data, NOAA OISST; Banzon et al., 2014), and CMIP6 simulations, for the tropical Indian Ocean. The baseline climatology for calculating the anomalies for historical and future simulations is based on the period 1970–2000. The trend (slope) is estimated for the period represented by the respective lines in the figure. (c) Future changes (with respect to 1970–2000) in the marine heatwave metrics—average number of marine heatwave days, mean maximum intensity, and mean cumulative intensity—using CMIP6 simulations (ensemble mean) under medium (SSP2-4.5) and high (SSP5-8.5) emission scenarios.

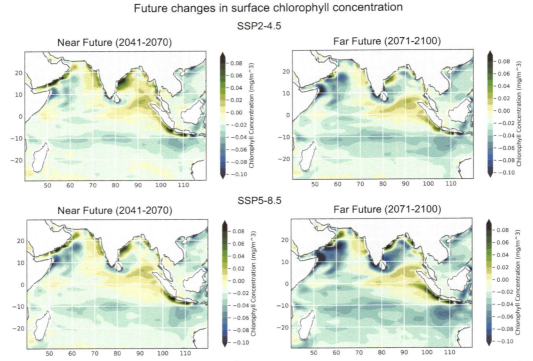

FIG. 5 Projected changes in ocean surface chlorophyll (in $mg\,m^{-3}$) during the summer monsoon (June–September) in (a) SSP2-4.5 for 2041–2070, (b) SSP2-4.5 for 2071–2100, (c) SSP5-8.5 for 2041–2070, and (d) SSP5-8.5 for 2071–2100, with respect to the reference period 1976–2005, in CMIP6 (ensemble mean) simulations.

the 21st century (Fig. 5). In contrast, the models project an increase in surface chlorophyll in the southeast Indian Ocean off the Sumatran and Java coasts and along most coastal regions in the northern Indian Ocean (Oman, western and eastern India, Myanmar).

This decline in chlorophyll is likely tied to the warming-driven increase in stratification (less-mixed), which limits the supply of subsurface nutrient-rich waters to the surface. Therefore, the tropical Indian Ocean may face a net negative impact on marine primary production (Marinov et al., 2010). Diatoms, which form a major component of the marine phytoplankton and account for more than 40% of the global CO_2 biological pump (Treguer & Pondaven, 2000), are projected to undergo a rapid decrease as compared to other phytoplankton types in the Indian Ocean, owing to stronger nitrate limitation (Bopp et al., 2005; Cermeño et al., 2008).

Although changes in stratification contribute to the observed and projected changes in marine primary production here, alterations to winds may also be crucial to understand. While the observed impacts of wind changes on the seasonal phytoplankton blooms are yet to be delineated (Roxy et al., 2016), the potential impact of future changes in monsoon winds on these blooms cannot be ruled out and requires an in-depth investigation (Parvathi et al., 2017; Praveen et al., 2016). As of now, the significant biases in CMIP6 models in representing the Indian Ocean variability and the monsoon system (Halder et al., 2021; Singh et al., 2019), and uncertainties and intermodel spread in the projections of biogeochemical changes limit a definitive understanding of future changes in marine primary production in the Indian Ocean (Tagliabue et al., 2021).

As the atmospheric carbon dioxide increases due to continued carbon emissions, the amount of carbon dioxide absorbed by the ocean also increases (IPCC Special Report on Ocean and Cryosphere in a Changing Climate), resulting in an increased concentration of hydrogen ions in the seawater, reducing its pH (Orr et al., 2005). The mean global ocean pH has already decreased from 8.16 to about 8.07 since the industrial revolution (Dore et al., 2009; Orr et al., 2005). Ocean biogeochemical model simulations show that the western Arabian Sea has undergone rapid acidification from 8.12 to 8.05 since 1961 (Sreeush et al., 2019). The western Arabian Sea has acidified more than the rest of the tropical Indian Ocean basin by drawing up anthropogenic CO_2 embedded into the deeper ocean during the process of upwelling, particularly during the southwest monsoon season.

The ocean acidity is projected (based on CMIP5 models) to increase, with the pH decreasing by an average of about -0.04 units per decade by the end of the 21st century (Dunne et al., 2013; Jiang et al., 2019; Kwiatkowski et al., 2020). Bopp

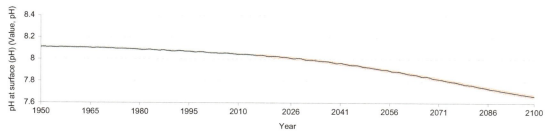

FIG. 6 Observed and projected surface pH in the tropical Indian Ocean under high emission scenario (SSP5-8.5), in CMIP6 (ensemble mean) simulations. The light shading indicates the spread between the 10th–90th percentiles. *(Adapted from the IPCC WGI Interactive Atlas (IPCC, 2021).)*

et al. (2013) and Frölicher et al. (2016) find a very low intermodel spread among the CMIP5 models and internal variability in their projections of a decline in pH over the course of the 21st century, increasing the confidence in these results. The projections of global mean surface pH suggest that it is likely to fall to around 7.67 by the year 2100, under a high-emissions scenario in case of no mitigation (Bernie et al., 2010). This would be likely five times the current amount of acidification. The projected change in surface pH over the Indian Ocean by CMIP6 models is on par with the globally projected change (Fig. 6). The corals that form reefs using aragonite and the phytoplankton, which requires calcites to form their shells, are projected to undergo a decline in the calcification rates in the low-latitude regions under reducing pH conditions (Feely et al., 2009). In the Arabian Sea, the planktonic foram *Globigerinoides ruber* already shows a significant reduction in the calcification rates used in their shell formation (de Moel et al., 2009).

The volume of oxygen minimum zones (OMZs) in the northern Indian Ocean has not dramatically changed over past decades, with observation suggesting that parts of the OMZ in the Arabian Sea have shrunk while others have expanded over this timeframe (Banse et al., 2014; Piontkovski & Al-Oufi, 2015; Queste et al., 2018). Yet there is growing observational evidence that oxygen concentrations are declining in most of the tropical Indian Ocean (e.g., Hood et al., 2024b; Lévy et al., 2021). In the northern Arabian Sea, dissolved oxygen concentrations measured during 1960–2010 show a decreasing trend in the surface mixed layer, potentially triggering large harmful algal (Noctiluca) winter blooms since the 2000s (do Rosário Gomes et al., 2014). Global ocean oxygen concentrations are very likely to decrease further in response to anthropogenic warming (IPCC SROCC), and this deoxygenation is expected to persist for thousands of years (Frölicher et al., 2020; IPCC, 2021; Oschlies, 2021). However, in contrast to the future projections of phytoplankton and pH, the future evolution of subsurface oxygen concentration in the tropical Indian Ocean is inconsistent across models from both the CMIP5 (Bopp et al., 2013; Busecke et al., 2019; Cabré et al., 2015; Lévy et al., 2021) and CMIP6 generations (Kwiatkowski et al., 2020). These projections are impeded by strong biases in oxygen distribution, with most Earth system models overestimating the oxygen concentration in the Arabian Sea and about half of them lacking an OMZ core in this region (Fig. 6, Rixen et al., 2020). Regional ocean model sensitivity experiments suggest, however, that the volume of the OMZ in the Arabian Sea could expand in response to anthropogenically-driven warming and wind intensification in the region (Lachkar et al., 2018, 2019). Recent work using Earth system models suggests that in response to the ocean warming, the outer OMZ expands while the OMZ core contracts, and in between the oxygen is redistributed with little effect on the OMZ volume (Ditkovsky et al., 2023).

5 Summary and discussion

The rapid warming of the Indian Ocean (at a rate of 0.12°C per decade between 1950 and 2020) has altered the weather and climate over the densely populated Indian Ocean rim nations, threatening the food, water, and energy security of the region (Anwar et al., 2023). Climate models project an increased basin-wide surface warming in the Indian Ocean, at the rate of 0.17–0.38°C per decade during 2020–2100, under medium-to-high greenhouse gas emission scenarios. The models project a large shift in the amplitude of the seasonal cycle, with the largest change during March–May, when the mean temperatures are also highest. The seasonal SST cycle ranging between 26°C and 28°C during 1980–2020 is projected to shift to 28.5–30.7°C by the end of the 21st century, under medium-to-high emission scenarios. The ocean heat content (OHC_{2000}) is also projected to consistently increase basin-wide in the future, at the rate of 16–22 zetta-joules per decade under mid-to-high emission scenarios. The frequency and intensity of future IOD events are projected to increase in the 21st century while the significant model bias and substantially large intermodel spread could be a factor potentially overestimating the projected increase in extremely positive IOD events.

The basin-wide warming in the Indian Ocean is a significant factor contributing to an increased frequency and intensity of marine heatwaves. The frequency, intensity, and the area covered by marine heatwaves are projected to increase substantially, driving a basin-wide near-permanent marine heatwave condition by the end of the 21st century, under medium and high emission scenarios. Rapid surface warming results in a stratified ocean, preventing the mixing of the surface with subsurface waters, which reduces the exchange of nutrients, oxygen, carbon, and heat. Under medium-to-high emission scenarios, an 8%–10% decrease in surface phytoplankton abundance is projected for the Arabian Sea region by the end of the 21st century (Fig. 5), potentially driven by enhanced thermal stratification. There is, however, a projected increase in surface phytoplankton in the southeast Indian Ocean off the Sumatran-Java coasts, and a few coastal regions in the north Indian Ocean.

With rising atmospheric CO_2 levels, ocean acidification is also increasing at a rapid pace. The surface pH of the tropical Indian Ocean is projected to decrease below 7.7 by the end of the 21st century, compared to a pH above 8.1 during the early 20th century. Since the pH scale is logarithmic, even a drop of 0.1 pH units represents approximately a 30% increase in the relative acidity of ocean water. The change may be easier to fathom when we realize that a 0.1 fall in human blood pH can result in rather profound health consequences and multiple-organ failure. The projected changes in pH may be detrimental to the marine ecosystem since many marine organisms—particularly corals and organisms that depend on calcification to build and maintain their shells—are sensitive to the change in ocean acidity (Doney et al., 2009).

Earth system models project inconsistent future changes in subsurface ocean oxygen concentrations in the tropical Indian Ocean. The high uncertainties in these projections are likely due to model limitations in simulating the biophysical processes that control oxygen concentrations in the Indian Ocean (Bopp et al., 2017; Lachkar et al., 2016; Resplandy et al., 2012; Rixen et al., 2020; Séférian et al., 2020). At the same time, gaps in high-quality long-term ocean observations limit our ability to separate multidecadal variability and anthropogenic warming in tropical oceans, particularly in terms of the observed changes in the biogeochemistry (Beal et al., 2020; Hermes et al., 2019; Hood et al., 2009; Stammer et al., 2019). Improved simulations of upper-ocean biophysical processes in Earth system models and urgent investment in high-resolution ocean observations are necessary to address the limitations of current projections, which exhibit high uncertainties in future changes in ocean biogeochemistry in the tropical Indian Ocean.

6 Educational resources

IPCC WGI Interactive Atlas: An interactive tool for spatial and temporal analyses of the observed and projected climate change information based on the IPCC Working Group I contribution to the Sixth Assessment Report (IPCC, 2021).

Website: https://interactive-atlas.ipcc.ch
Code with the data: https://github.com/IPCC-WG1/Atlas

Author contributions

MKR conceived and led the chapter. JSS and AA contributed to the analysis and figures in Sections 2 and 3, and AM contributed to Section 4. All authors contributed to the discussion of content and overall chapter structure and provided feedback on the entire chapter.

References

Anwar, F., De, S., Durbarry, A., Fozdar, F., Hermes, J., Khan, H., ... Mohee, R. (2023). *Indian Ocean futures: Prospects for shared regional success*. UWA Public Policy Institute.

Banse, K., Naqvi, S., Narvekar, P., Postel, J., & Jayakumar, D. (2014). Oxygen minimum zone of the open Arabian Sea: Variability of oxygen and nitrite from daily to decadal timescales. *Biogeosciences, 11*(8), 2237–2261.

Banzon, V. F., Reynolds, R. W., Stokes, D., & Xue, Y. (2014). A 1/4-spatial-resolution daily sea surface temperature climatology based on a blended satellite and in situ analysis. *Journal of Climate, 27*(21), 8221–8228.

Beal, L., Vialard, J., Roxy, M., Li, J., Andres, M., Annamalai, H., Feng, M., Han, W., Hood, R., & Lee, T. (2020). A road map to IndOOS-2: Better observations of the rapidly warming Indian Ocean. *Bulletin of the American Meteorological Society, 101*(11), E1891–E1913.

Behrenfeld, M. J., O'Malley, R. T., Siegel, D. A., McClain, C. R., Sarmiento, J. L., Feldman, G. C., Milligan, A. J., Falkowski, P. G., Letelier, R. M., & Boss, E. S. (2006). Climate-driven trends in contemporary ocean productivity. *Nature, 444*(7120), 752–755.

Bernie, D., Lowe, J., Tyrrell, T., & Legge, O. (2010). Influence of mitigation policy on ocean acidification. *Geophysical Research Letters, 37*(15).

Bhatia, K., Vecchi, G., Murakami, H., Underwood, S., & Kossin, J. (2018). Projected response of tropical cyclone intensity and intensification in a global climate model. *Journal of Climate, 31*(20), 8281–8303.

Bindoff, N. L., Cheung, W. W., Kairo, J. G., Arístegui, J., Guinder, V. A., Hallberg, R., ... Williamson, P. (2019). Changing ocean, marine ecosystems, and dependent communities. *IPCC Special Report on the Ocean and Cryosphere in a Changing Climate* (pp. 447–587). Cambridge, UK and New York, NY, USA: Cambridge University Press.

Bopp, L., Aumont, O., Cadule, P., Alvain, S., & Gehlen, M. (2005). Response of diatoms distribution to global warming and potential implications: A global model study. *Geophysical Research Letters, 32*(19).

Bopp, L., Monfray, P., Aumont, O., Dufresne, J. L., Le Treut, H., Madec, G., Terray, L., & Orr, J. C. (2001). Potential impact of climate change on marine export production. *Global Biogeochemical Cycles, 15*(1), 81–99.

Bopp, L., Resplandy, L., Orr, J., Doney, S., Dunne, J., Gehlen, M., Halloran, P., Heinze, C., Ilyina, T., & Séférian, R. (2013). Multiple stressors of ocean ecosystems in the 21st century: Projections with CMIP5 models. *Biogeosciences, 10*(10), 6225–6245.

Bopp, L., Resplandy, L., Untersee, A., Le Mezo, P., & Kageyama, M. (2017). Ocean (de) oxygenation from the Last Glacial Maximum to the twenty-first century: Insights from Earth System models. *Philosophical Transactions of the Royal Society A: Mathematical, Physical and Engineering Sciences, 375*(2102), 20160323.

Boyce, D. G., Lewis, M. R., & Worm, B. (2010). Global phytoplankton decline over the past century. *Nature, 466*(7306), 591–596.

Busecke, J. J., Resplandy, L., & Dunne, J. P. (2019). The equatorial undercurrent and the oxygen minimum zone in the Pacific. *Geophysical Research Letters, 46*(12), 6716–6725.

Cabré, A., Marinov, I., Bernardello, R., & Bianchi, D. (2015). Oxygen minimum zones in the tropical Pacific across CMIP5 models: Mean state differences and climate change trends. *Biogeosciences, 12*(18), 5429–5454.

Cai, W., Cowan, T., & Raupach, M. (2009). Positive Indian Ocean dipole events precondition Southeast Australia bushfires. *Geophysical Research Letters, 36*(19).

Cai, W., Santoso, A., Wang, G., Weller, E., Wu, L., Ashok, K., Masumoto, Y., & Yamagata, T. (2014). Increased frequency of extreme Indian Ocean Dipole events due to greenhouse warming. *Nature, 510*(7504), 254–258.

Cai, W., Wang, G., Gan, B., Wu, L., Santoso, A., Lin, X., Chen, Z., Jia, F., & Yamagata, T. (2018). Stabilised frequency of extreme positive Indian Ocean Dipole under 1.5 C warming. *Nature Communications, 9*(1), 1–8.

Cai, W., Yang, K., Wu, L., Huang, G., Santoso, A., Ng, B., Wang, G., & Yamagata, T. (2021). Opposite response of strong and moderate positive Indian Ocean Dipole to global warming. *Nature Climate Change, 11*(1), 27–32.

Cai, W., Zheng, X.-T., Weller, E., Collins, M., Cowan, T., Lengaigne, M., Yu, W., & Yamagata, T. (2013). Projected response of the Indian Ocean Dipole to greenhouse warming. *Nature Geoscience, 6*(12), 999–1007.

Cermeño, P., Dutkiewicz, S., Harris, R. P., Follows, M., Schofield, O., & Falkowski, P. G. (2008). The role of nutricline depth in regulating the ocean carbon cycle. *Proceedings of the National Academy of Sciences, 105*(51), 20344–20349.

Cheng, L., Trenberth, K. E., Fasullo, J., Boyer, T., Abraham, J., & Zhu, J. (2017). Improved estimates of ocean heat content from 1960 to 2015. *Science Advances, 3*(3), e1601545.

Cheng, L., Trenberth, K. E., Gruber, N., Abraham, J. P., Fasullo, J. T., Li, G., Mann, M. E., Zhao, X., & Zhu, J. (2020). Improved estimates of changes in upper ocean salinity and the hydrological cycle. *Journal of Climate, 33*(23), 10357–10381.

Collins, M., Sutherland, M., Bouwer, L., Cheong, S.-M., Frölicher, T., Jacot Des Combes, H., ... Tibig, L. (2019). Extremes, abrupt changes and managing risk. *IPCC Special Report on the Ocean and Cryosphere in a Changing Climate* (pp. 589–655). Cambridge, UK and New York, NY, USA: Cambridge University Press.

De Deckker, P. (2016). The indo-Pacific warm Pool: Critical to world oceanography and world climate. *Geoscience Letters, 3*(1), 20.

de Moel, H., Ganssen, G., Peeters, F., Jung, S., Kroon, D., Brummer, G., & Zeebe, R. (2009). Planktic foraminiferal shell thinning in the Arabian Sea due to anthropogenic ocean acidification? *Biogeosciences, 6*(9), 1917–1925.

Deshpande, M., Singh, V. K., Ganadhi, M. K., Roxy, M., Emmanuel, R., & Kumar, U. (2021). Changing status of tropical cyclones over the North Indian Ocean. *Climate Dynamics, 57*(11), 3545–3567.

Ditkovsky, S., Resplandy, L., & Busecke, J. (2023). Unique ocean circulation pathways reshape the Indian Ocean oxygen minimum zone with warming. *Biogeosciences, 20*, 4711–4736. https://doi.org/10.5194/bg-20-4711-2023.

do Rosário Gomes, H., Goes, J. I., Matondkar, S., Buskey, E. J., Basu, S., Parab, S., & Thoppil, P. (2014). Massive outbreaks of *Noctiluca scintillans* blooms in the Arabian Sea due to spread of hypoxia. *Nature Communications, 5*.

Doney, S. C., Fabry, V. J., Feely, R. A., & Kleypas, J. A. (2009). Ocean acidification: The other CO_2 problem. *Annual Review of Marine Science, 1*, 169–192.

Dore, J. E., Lukas, R., Sadler, D. W., Church, M. J., & Karl, D. M. (2009). Physical and biogeochemical modulation of ocean acidification in the central North Pacific. *Proceedings of the National Academy of Sciences, 106*(30), 12235–12240.

Dunne, J. P., John, J. G., Shevliakova, E., Stouffer, R. J., Krasting, J. P., Malyshev, S. L., Milly, P., Sentman, L. T., Adcroft, A. J., & Cooke, W. (2013). GFDL's ESM2 global coupled climate–carbon earth system models. Part II: Carbon system formulation and baseline simulation characteristics*. *Journal of Climate, 26*(7), 2247–2267.

Dwyer, J. G., Biasutti, M., & Sobel, A. H. (2012). Projected changes in the seasonal cycle of surface temperature. *Journal of Climate, 25*(18), 6359–6374.

Feely, R. A., Doney, S. C., & Cooley, S. R. (2009). Ocean acidification: Present conditions and future changes in a high-CO_2 world. *Oceanography, 22*(4), 36–47.

Fox-Kemper, B., et al. (2021). *Ocean, Cryosphere and Sea Level Change*. Rep. IPCC.

Frölicher, T. L., Aschwanden, M., Gruber, N., Jaccard, S. L., Dunne, J. P., & Paynter, D. (2020). Contrasting upper and deep ocean oxygen response to protracted global warming. *Global Biogeochemical Cycles, 34*(8), e2020GB006601.

Frölicher, T. L., Fischer, E. M., & Gruber, N. (2018). Marine heatwaves under global warming. *Nature, 560*(7718), 360–364.

Frölicher, T. L., & Laufkötter, C. (2018). Emerging risks from marine heat waves. *Nature Communications, 9*(1), 1–4.

Frölicher, T. L., Rodgers, K. B., Stock, C. A., & Cheung, W. W. (2016). Sources of uncertainties in 21st century projections of potential ocean ecosystem stressors. *Global Biogeochemical Cycles, 30*(8), 1224–1243.

Fung, I. Y., Doney, S. C., Lindsay, K., & John, J. (2005). Evolution of carbon sinks in a changing climate. *Proceedings of the National Academy of Sciences, 102*(32), 11201–11206.

Funk, C., Dettinger, M. D., Michaelsen, J. C., Verdin, J. P., Brown, M. E., Barlow, M., & Hoell, A. (2008). Warming of the Indian Ocean threatens eastern and southern African food security but could be mitigated by agricultural development. *Proceedings of the National Academy of Sciences, 105*(32), 11081–11086.

Gadgil, S., Joshi, N. V., & Joseph, P. V. (1984). Ocean-atmosphere coupling over monsoon regions. *Nature, 312*, 141–143.

Gnanaseelan, C., Roxy, M. K., & Deshpande, A. (2017). Variability and trends of sea surface temperature and circulation in the Indian Ocean. In M. Rajeevan, & S. Nayak (Eds.), *Observed climate variability and change over the Indian region* (p. 382). Springer.

Halder, S., Parekh, A., Chowdary, J. S., Gnanaseelan, C., & Kulkarni, A. (2021). Assessment of CMIP6 models' skill for tropical Indian Ocean Sea surface temperature variability. *International Journal of Climatology, 41*(4), 2568–2588.

Han, W., Vialard, J., McPhaden, M. J., Lee, T., Masumoto, Y., Feng, M., & De Ruijter, W. P. (2014). Indian Ocean decadal variability: A review. *Bulletin of the American Meteorological Society, 95*(11), 1679–1703.

Helm, K. P., Bindoff, N. L., & Church, J. A. (2011). Observed decreases in oxygen content of the global ocean. *Geophysical Research Letters, 38*(23).

Hermes, J. C., Masumoto, Y., Beal, L., Roxy, M., Vialard, J., Andres, M., Annamalai, H., Behera, S., d'Adamo, N., & Feng, M. (2019). A sustained ocean observing system in the Indian Ocean for climate related scientific knowledge and societal needs. *Frontiers in Marine Science, 6*, 355.

Hobday, A. J., Alexander, L. V., Perkins, S. E., Smale, D. A., Straub, S. C., Oliver, E. C., Benthuysen, J. A., Burrows, M. T., Donat, M. G., & Feng, M. (2016). A hierarchical approach to defining marine heatwaves. *Progress in Oceanography, 141*, 227–238.

Hood, R. R., Coles, V. J., Huggett, J. A., Landry, M. R., Levy, M., Moffett, J. W., & Rixen, T. (2024a). Chapter 13: Nutrient, phytoplankton, and zooplankton variability in the Indian Ocean. In C. C. Ummenhofer, & R. R. Hood (Eds.), *The Indian Ocean and its role in the global climate system* (pp. 293–327). Amsterdam: Elsevier. https://doi.org/10.1016/B978-0-12-822698-8.00020-2.

Hood, R. R., Rixen, T., Levy, M., Hansell, D. A., Coles, V. J., & Lachkar, Z. (2024b). Chapter 12: Oxygen, carbon, and pH variability in the Indian Ocean. In C. C. Ummenhofer, & R. R. Hood (Eds.), *The Indian Ocean and its role in the global climate system* (pp. 265–291). Amsterdam: Elsevier. https://doi.org/10.1016/B978-0-12-822698-8.00017-2.

Hood, R. R., Wiggert, J. D., & Naqvi, S. W. A. (2009). Indian Ocean research: Opportunities and challenges. *Indian Ocean Biogeochemical Processes and Ecological Variability, 185*, 409–429.

Hu, S., & Fedorov, A. V. (2019). Indian Ocean warming can strengthen the Atlantic meridional overturning circulation. *Nature Climate Change, 9*(10), 747–751.

Hu, S., & Fedorov, A. V. (2020). Indian Ocean warming as a driver of the North Atlantic warming hole. *Nature Communications, 11*(1), 1–11.

Hughes, T. P., Kerry, J. T., Álvarez-Noriega, M., Álvarez-Romero, J. G., Anderson, K. D., Baird, A. H., Babcock, R. C., Beger, M., Bellwood, D. R., & Berkelmans, R. (2017). Global warming and recurrent mass bleaching of corals. *Nature, 543*(7645), 373–377.

Hui, C., & Zheng, X.-T. (2018). Uncertainty in Indian Ocean dipole response to global warming: The role of internal variability. *Climate Dynamics, 51*(9), 3597–3611.

IPCC. (2021). Climate change 2021: The physical science basis. In *Contribution of Working Group I to the Sixth Assessment Report of the Intergovernmental Panel on Climate Change (Rep.)*.

Ito, T., Minobe, S., Long, M. C., & Deutsch, C. (2017). Upper ocean O_2 trends: 1958–2015. *Geophysical Research Letters, 44*(9), 4214–4223.

Jiang, L.-Q., Carter, B. R., Feely, R. A., Lauvset, S. K., & Olsen, A. (2019). Surface Ocean pH and buffer capacity: Past, present and future. *Scientific Reports, 9*(1), 1–11.

Kwiatkowski, L., Torres, O., Bopp, L., Aumont, O., Chamberlain, M., Christian, J. R., Dunne, J. P., Gehlen, M., Ilyina, T., & John, J. G. (2020). Twenty-first century ocean warming, acidification, deoxygenation, and upper-ocean nutrient and primary production decline from CMIP6 model projections. *Biogeosciences, 17*(13), 3439–3470.

Lachkar, Z., Lévy, M., & Smith, S. (2018). Intensification and deepening of the Arabian Sea oxygen minimum zone in response to increase in Indian monsoon wind intensity. *Biogeosciences, 15*(1), 159–186.

Lachkar, Z., Lévy, M., & Smith, K. S. (2019). Strong intensification of the Arabian Sea oxygen minimum zone in response to Arabian Gulf warming. *Geophysical Research Letters, 46*(10), 5420–5429.

Lachkar, Z., Smith, S., Lévy, M., & Pauluis, O. (2016). Eddies reduce denitrification and compress habitats in the Arabian Sea. *Geophysical Research Letters, 43*(17), 9148–9156.

Lévy, M., Resplandy, L., Palter, J. B., Couespel, D., & Lachkar, Z. (2021). *The crucial contribution of mixing to present and future ocean oxygen distribution*. Elsevier.

Li, G., Cheng, L., Zhu, J., Trenberth, K. E., Mann, M. E., & Abraham, J. P. (2020). Increasing ocean stratification over the past half-century. *Nature Climate Change, 10*(12), 1116–1123.

Li, G., Xie, S.-P., & Du, Y. (2016). A robust but spurious pattern of climate change in model projections over the tropical Indian Ocean. *Journal of Climate, 29*(15), 5589–5608.

Li, B., Zhou, L., Qin, J., & Murtugudde, R. (2022). Increase in intraseasonal rainfall driven by the Arabian Sea warming in recent decades. *Geophysical Research Letters, 49*(20), e2022GL100536.

Marinov, I., Doney, S., & Lima, I. (2010). Response of ocean phytoplankton community structure to climate change over the 21st century: Partitioning the effects of nutrients, temperature and light. *Biogeosciences, 7*(12), 3941–3959.

Marsac, F., Everett, B., Shahid, U., & Strutton, P. G. (2024). Chapter 11: Indian Ocean primary productivity and fisheries variability. In C. C. Ummenhofer, & R. R. Hood (Eds.), *The Indian Ocean and its role in the global climate system* (pp. 245–264). Amsterdam: Elsevier. https://doi.org/10.1016/B978-0-12-822698-8.00019-6.

McKenna, S., Santoso, A., Gupta, A. S., Taschetto, A. S., & Cai, W. (2020). Indian Ocean Dipole in CMIP5 and CMIP6: Characteristics, biases, and links to ENSO. *Scientific Reports, 10*(1), 1–13.

Murakami, H., Vecchi, G. A., & Underwood, S. (2017). Increasing frequency of extremely severe cyclonic storms over the Arabian Sea. *Nature Climate Change, 7*(12), 885–889.

Naqvi, S. W. A., Naik, H., Jayakumar, A., Pratihary, A. K., Narvenkar, G., Kurian, S., Agnihotri, R., Shailaja, M., & Narvekar, P. V. (2009). Seasonal anoxia over the western Indian continental shelf. *Indian Ocean Biogeochemical Processes and Ecological Variability, 185*, 333–345.

Nidheesh, A., Lengaigne, M., Vialard, J., Izumo, T., Unnikrishnan, A., Meyssignac, B., Hamlington, B., & de Boyer Montégut, C. (2017). Robustness of observation-based decadal sea level variability in the Indo-Pacific Ocean. *Geophysical Research Letters, 44*(14), 7391–7400.

Oliver, E. C., Burrows, M. T., Donat, M. G., Sen Gupta, A., Alexander, L. V., Perkins-Kirkpatrick, S. E., Benthuysen, J. A., Hobday, A. J., Holbrook, N. J., & Moore, P. J. (2019). Projected marine heatwaves in the 21st century and the potential for ecological impact. *Frontiers in Marine Science, 6*, 734.

Oliver, E. C., Donat, M. G., Burrows, M. T., Moore, P. J., Smale, D. A., Alexander, L. V., Benthuysen, J. A., Feng, M., Gupta, A. S., & Hobday, A. J. (2018). Longer and more frequent marine heatwaves over the past century. *Nature Communications, 9*(1), 1–12.

O'Neill, B. C., Tebaldi, C., van Vuuren, D. P., Eyring, V., Friedlingstein, P., Hurtt, G., Knutti, R., Kriegler, E., Lamarque, J.-F., & Lowe, J. (2016). The scenario model intercomparison project (ScenarioMIP) for CMIP6. *Geoscientific Model Development, 9*(9), 3461–3482.

Orr, J. C., Fabry, V. J., Aumont, O., Bopp, L., Doney, S. C., Feely, R. A., Gnanadesikan, A., Gruber, N., Ishida, A., & Joos, F. (2005). Anthropogenic Ocean acidification over the twenty-first century and its impact on calcifying organisms. *Nature, 437*(7059), 681–686.

Oschlies, A. (2021). A committed fourfold increase in ocean oxygen loss. *Nature Communications, 12*(1), 1–8.

Parvathi, V., Suresh, I., Lengaigne, M., Izumo, T., & Vialard, J. (2017). Robust projected weakening of winter monsoon winds over the Arabian Sea under climate change. *Geophysical Research Letters, 44*(19), 9833–9843.

Piontkovski, S., & Al-Oufi, H. (2015). The Omani shelf hypoxia and the warming Arabian Sea. *International Journal of Environmental Studies, 72*(2), 256–264.

Piontkovski, S. A., & Queste, B. Y. (2016). Decadal changes of the Western Arabian Sea ecosystem. *International Aquatic Research, 8*(1), 49–64.

Plecha, S. M., & Soares, P. M. (2020). Global marine heatwave events using the new CMIP6 multi-model ensemble: From shortcomings in present climate to future projections. *Environmental Research Letters, 15*(12), 124058.

Prakash, P., Prakash, S., Rahaman, H., Ravichandran, M., & Nayak, S. (2012). Is the trend in chlorophyll-a in the Arabian Sea decreasing? *Geophysical Research Letters, 39*(23).

Prasanna Kumar, S., Madhupratap, M., Dileepkumar, M., Muraleedharan, P., DeSouza, S., Gauns, M., & Sarma, V. (2001). High biological productivity in the central Arabian Sea during the summer monsoon driven by Ekman pumping and lateral advection. *Current Science, 81*(12), 1633–1638.

Praveen, V., Ajayamohan, R., Valsala, V., & Sandeep, S. (2016). Intensification of upwelling along Oman coast in a warming scenario. *Geophysical Research Letters, 43*(14), 7581–7589.

Qi, R., Zhang, Y., Du, Y., & Feng, M. (2022). Characteristics and drivers of marine heatwaves in the western equatorial Indian Ocean. *Journal of Geophysical Research: Oceans, 127*(10), e2022JC018732.

Queste, B. Y., Vic, C., Heywood, K. J., & Piontkovski, S. A. (2018). Physical controls on oxygen distribution and denitrification potential in the north west Arabian Sea. *Geophysical Research Letters, 45*(9), 4143–4152.

Rao, S. A., Dhakate, A. R., Saha, S. K., Mahapatra, S., Chaudhari, H. S., Pokhrel, S., & Sahu, S. K. (2012). Why is Indian Ocean warming consistently? *Climatic Change, 110*(3–4), 709–719.

Resplandy, L., Lévy, M., Bopp, L., Echevin, V., Pous, S., Sarma, V., & Kumar, D. (2012). Controlling factors of the oxygen balance in the Arabian Sea's OMZ. *Biogeosciences, 9*(12), 5095–5109.

Riahi, K., Van Vuuren, D. P., Kriegler, E., Edmonds, J., O'neill, B. C., Fujimori, S., Bauer, N., Calvin, K., Dellink, R., & Fricko, O. (2017). The shared socioeconomic pathways and their energy, land use, and greenhouse gas emissions implications: An overview. *Global Environmental Change, 42*, 153–168.

Rixen, T., Cowie, G., Gaye, B., Goes, J., do Rosário Gomes, H., Hood, R. R., Lachkar, Z., Schmidt, H., Segschneider, J., & Singh, A. (2020). Reviews and syntheses: Present, past, and future of the oxygen minimum zone in the northern Indian Ocean. *Biogeosciences, 17*(23), 6051–6080.

Rodrigues, R. R., Taschetto, A. S., Gupta, A. S., & Foltz, G. R. (2019). Common cause for severe droughts in South America and marine heatwaves in the South Atlantic. *Nature Geoscience, 12*(8), 620–626.

Rohini, P., Rajeevan, M., & Srivastava, A. (2016). On the variability and increasing trends of heat waves over India. *Scientific Reports, 6*.

Roxy, M. (2014). Sensitivity of precipitation to sea surface temperature over the tropical summer monsoon region—And its quantification. *Climate Dynamics, 43*(5–6), 1159–1169.

Roxy, M., Dasgupta, P., McPhaden, M. J., Suematsu, T., Zhang, C., & Kim, D. (2019). Twofold expansion of the Indo-Pacific warm pool warps the MJO life cycle. *Nature, 575*(7784), 647–651.

Roxy, M., Ghosh, S., Pathak, A., Athulya, R., Mujumdar, M., Murtugudde, R., Terray, P., & Rajeevan, M. (2017). A threefold rise in widespread extreme rain events over Central India. *Nature Communications, 8*(1), 708.

Roxy, M., Gnanaseelan, C., Parekh, A., Chowdary, J. S., Singh, S., Modi, A., Kakatkar, R., Mohapatra, S., Dhara, C., & Shenoi, S. (2020). Indian Ocean warming. In *Assessment of climate change over the Indian region* (pp. 191–206). Springer.

Roxy, M. K., Modi, A., Murtugudde, R., Valsala, V., Panickal, S., Prasanna Kumar, S., Ravichandran, M., Vichi, M., & Lévy, M. (2016). A reduction in marine primary productivity driven by rapid warming over the tropical Indian Ocean. *Geophysical Research Letters, 43*(2), 826–833.

Roxy, M. K., Ritika, K., Terray, P., & Masson, S. (2014). The curious case of Indian Ocean warming. *Journal of Climate, 27*(22), 8501–8509.

Roxy, M. K., Ritika, K., Terray, P., Murtugudde, R., Ashok, K., & Goswami, B. N. (2015). Drying of Indian subcontinent by rapid Indian Ocean warming and a weakening land-sea thermal gradient. *Nature Communications, 6*, 7423.

Sabeerali, C., Rao, S. A., George, G., Rao, D. N., Mahapatra, S., Kulkarni, A., & Murtugudde, R. (2014). Modulation of monsoon intraseasonal oscillations in the recent warming period. *Journal of Geophysical Research: Atmospheres, 119*(9), 5185–5203.

Saji, N. H., Goswami, B. N., Vinayachandran, P. N., & Yamagata, T. (1999). A dipole mode in the tropical Indian Ocean. *Nature, 401*(6751), 360–363.

Salih, A. A., Baraibar, M., Mwangi, K. K., & Artan, G. (2020). Climate change and locust outbreak in East Africa. *Nature Climate Change, 10*(7), 584–585.

Saranya, J. S., Roxy, M. K., Dasgupta, P., & Anand, A. (2022). Genesis and trends in marine heatwaves over the tropical Indian Ocean and their interaction with the Indian summer monsoon. *Journal of Geophysical Research: Oceans, 127*(2), 1–16. https://doi.org/10.1029/2021JC017427.

Séférian, R., Berthet, S., Yool, A., Palmieri, J., Bopp, L., Tagliabue, A., Kwiatkowski, L., Aumont, O., Christian, J., & Dunne, J. (2020). Tracking improvement in simulated marine biogeochemistry between CMIP5 and CMIP6. *Current Climate Change Reports, 6*(3), 95–119.

Singh, D., Ghosh, S., Roxy, M. K., & McDermid, S. (2019). Indian summer monsoon: Extreme events, historical changes, and role of anthropogenic forcings. *Wiley Interdisciplinary Reviews: Climate Change, 10*(2), e571. https://doi.org/10.1002/wcc.571.

Singh, V. K., & Roxy, M. K. (2022). A review of the ocean-atmosphere interactions during tropical cyclones in the North Indian Ocean. *Earth System Reviews, 226*(2022), 103967. https://doi.org/10.1016/j.earscirev.2022.103967.

Singh, V. K., Roxy, M., & Deshpande, M. (2021). Role of warm ocean conditions and the MJO in the genesis and intensification of extremely severe cyclone Fani. *Scientific Reports, 11*(1), 1–10.

Smale, D. A., Wernberg, T., Oliver, E. C., Thomsen, M., Harvey, B. P., Straub, S. C., Burrows, M. T., Alexander, L. V., Benthuysen, J. A., & Donat, M. G. (2019). Marine heatwaves threaten global biodiversity and the provision of ecosystem services. *Nature Climate Change, 9*(4), 306–312.

Sprintall, J., Biastoch, A., Gruenburg, L. K., & Phillips, H. E. (2024). Chapter 9: Oceanic basin connections. In C. C. Ummenhofer, & R. R. Hood (Eds.), *The Indian Ocean and its role in the global climate system* (pp. 205–227). Amsterdam: Elsevier. https://doi.org/10.1016/B978-0-12-822698-8.00003-2.

Sreeush, M. G., Rajendran, S., Valsala, V., Pentakota, S., Prasad, K., & Murtugudde, R. (2019). Variability, trend and controlling factors of Ocean acidification over Western Arabian Sea upwelling region. *Marine Chemistry, 209*, 14–24.

Stammer, D., Bracco, A., AchutaRao, K., Beal, L., Bindoff, N. L., Braconnot, P., Cai, W., Chen, D., Collins, M., & Danabasoglu, G. (2019). Ocean climate observing requirements in support of climate research and climate information. *Frontiers in Marine Science, 6*, 444.

Steinacher, M., Joos, F., Frölicher, T., Bopp, L., Cadule, P., Cocco, V., Doney, S., Gehlen, M., Lindsay, K., & Moore, J. (2010). Projected 21st century decrease in marine productivity: A multi-model analysis. *Biogeosciences, 7*(3), 979–1005.

Stramma, L., Johnson, G. C., Sprintall, J., & Mohrholz, V. (2008). Expanding oxygen-minimum zones in the tropical oceans. *Science, 320*(5876), 655–658.

Swapna, P., Ravichandran, M., Nidheesh, G., Jyoti, J., Sandeep, N., Deepa, J., & Unnikrishnan, A. (2020). Sea-level rise. In *Assessment of climate change over the Indian region* (pp. 175–189). Springer.

Tagliabue, A., Kwiatkowski, L., Bopp, L., Butenschön, M., Cheung, W., Lengaigne, M., & Vialard, J. (2021). Persistent uncertainties in ocean net primary production climate change projections at regional scales raise challenges for assessing impacts on ecosystem services. *Frontiers in Climate, 149*.

Tozuka, T., Dong, L., Han, W., Lengainge, M., & Zhang, L. (2024). Chapter 10: Decadal variability of the Indian Ocean and its predictability. In C. C. Ummenhofer, & R. R. Hood (Eds.), *The Indian Ocean and its role in the global climate system* (pp. 229–244). Amsterdam: Elsevier. https://doi.org/10.1016/B978-0-12-822698-8.00014-7.

Treguer, P., & Pondaven, P. (2000). Silica control of carbon dioxide. *Nature, 406*(6794), 358–359.

Ummenhofer, C. C., Murty, S. A., Sprintall, J., Lee, T., & Abram, N. J. (2021). Heat and freshwater changes in the Indian Ocean region. *Nature Reviews Earth & Environment*, 1–17.

Ummenhofer, C. C., Ryan, S., England, M. H., Scheinert, M., Wagner, P., Biastoch, A., & Böning, C. W. (2020). Late 20th century Indian Ocean heat content gain masked by wind forcing. *Geophysical Research Letters, 47*(22), e2020GL088692.

Vecchi, G. A., Soden, B. J., Wittenberg, A. T., Held, I. M., Leetmaa, A., & Harrison, M. J. (2006). Weakening of tropical Pacific atmospheric circulation due to anthropogenic forcing. *Nature, 441*(7089), 73–76.

Wang, B., Biasutti, M., Byrne, M. P., Castro, C., Chang, C.-P., Cook, K., Fu, R., Grimm, A. M., Ha, K.-J., & Hendon, H. (2021). Monsoons climate change assessment. *Bulletin of the American Meteorological Society, 102*(1), E1–E19.

Wang, G., Cai, W., & Santoso, A. (2017). Assessing the impact of model biases on the projected increase in frequency of extreme positive Indian Ocean dipole events. *Journal of Climate, 30*(8), 2757–2767.

Wiggert, J., Hood, R., Banse, K., & Kindle, J. (2005). Monsoon-driven biogeochemical processes in the Arabian Sea. *Progress in Oceanography, 65*(2–4), 176–213.

Yang, K., Cai, W., Huang, G., Wang, G., Ng, B., & Li, S. (2020). Oceanic processes in ocean temperature products key to a realistic presentation of positive Indian Ocean Dipole nonlinearity. *Geophysical Research Letters, 47*(16), e2020GL089396.

Zhao, Y., & Zhang, H. (2016). Impacts of SST warming in tropical Indian Ocean on CMIP5 model-projected summer rainfall changes over Central Asia. *Climate Dynamics, 46*(9), 3223–3238.

Zheng, X.-T., Xie, S.-P., Vecchi, G. A., Liu, Q., & Hafner, J. (2010). Indian Ocean dipole response to global warming: Analysis of ocean–atmospheric feedbacks in a coupled model. *Journal of Climate, 23*(5), 1240–1253.

Index

Note: Page numbers followed by *f* indicate figures and *t* indicate tables.

A

Abbasid Caliphate, 36
Abyssal circulation, 187–190
Acidification, 16, 267–268, 271, 275–276, 284, 330–331, 342, 377, 408, 476–478
Aerosols, 65, 231, 234–236, 240, 329, 334, 337–342, 337*f*, 341–342*t*
Agulhas current system, 181, 209, 213–216, 219–220, 408
Agulhas leakage, 181–182, 184, 190–191, 206*f*, 209, 213–216, 219, 394
Air-sea coupling, 11, 110, 124–125, 127*f*, 128, 135, 155–156, 428, 429*f*, 430–431
Air-sea feedback, 109, 135, 421
Air-sea fluxes, 103–104, 402–404, 408, 428
Air-sea gas exchange, 330–334, 336
Amount effect, 58–59, 61, 457
Andaman Sea, 45, 123, 126–127, 306–307, 332, 365–368, 366*f*, 381, 458
Anoxia, 265, 271–274, 359–360, 407
Antarctic Oscillation, 107–108
Arabian/Persian Gulf, 365, 369–372, 381–382
Arakan Empire, 37
Asian monsoon, 10–11, 49–58, 63–67, 113, 205, 450
Atlantic Meridional Overturning Circulation (AMOC), 11, 159, 209, 216, 394, 397, 450–452, 469–470
Atmospheric bridge, 103–104, 113, 152–153, 155–156, 233, 240, 400

B

Barrier layer, 8–9, 9*f*, 13, 20, 79, 84–87, 394, 402–403, 432–433
Basra, 43
Batavia, 39–40
Biogeochemistry, 13–20, 208, 276, 308, 331, 337, 339–340, 351, 407, 425, 431–434, 432*f*, 434*f*, 474–477, 476–477*f*
Biological pump, 16, 273, 275, 284, 330, 433–434, 476
Bjerknes feedback, 103–104, 109, 113, 124, 180, 459
Boreal summer intraseasonal oscillation, 53, 56, 65, 431

C

Cape Town, 2–3, 40–41
Carbon, 16, 208, 265–285, 303, 305, 310, 340, 351, 353–354, 367, 371, 474

Carbonate pump, 367
Carbon dioxide (CO_2), 16, 208, 267–268, 274–276, 284–285, 329–331, 330*f*, 331*t*, 342, 476
Carbon export flux, 268–271, 281–282
Carbon monoxide, 329, 335–337
Chlorophyll-*a*, 233, 280, 298, 303*f*, 377, 379–381, 474
Chola Kingdom, 36
CLAW hypothesis, 334
Climate mode, 5, 10–11, 66, 107, 110, 124–125, 126*f*, 146–147, 159–160, 215–216, 231, 233–234, 248
Climate variations, 103, 106, 155–156, 230–231, 356–357, 404, 425–427
Coastal upwelling, 1, 9, 113, 281, 283, 298–299, 303–304, 312–313, 315, 330–334, 341, 431–433, 469
Cold pool, 79, 404
Convective quasi-equilibrium, 51
Coral bleaching, 121, 126–127, 276
Corals, 267–268, 275–276, 368, 372, 375, 377, 446, 446*f*, 448–449
Coupled Model Intercomparison Project (CMIP), 218–219, 234–236, 240, 427, 430, 471
Cyanobacteria, 308, 314, 352–355, 356*f*
Cyclogenesis, 128, 130, 366–367, 472

D

Decadal variability, 17–18, 56, 58, 63, 156–158, 183, 216, 219, 229–236, 330, 397, 406–407, 445–450, 478
Decolonization, 45
Delagoa Bay, 44
Denitrification, 16, 265, 271–273, 284, 294, 297, 305, 354, 354*f*, 356–359, 374, 433
Deposition, wet/dry, 339–340
Dhow, 34, 41*f*, 43*f*, 251
Diabatic heating, 81–83, 109, 150, 153, 158–160
Diapycnal, 186–188, 191, 207
Diazotrophs, 341, 342*t*, 355–356, 355*f*, 358
Dibromomethane (CH_2Br_2), 334
Dimethylsulfide (DMS), 329, 334–335, 335*f*, 337–338
Dissolved organic carbon (DOC), 16–17, 268–271, 269*f*, 274, 276–284, 367, 397
Diurnal warm layer, 86
Dust storm, 371, 376*f*, 377

E

Earthquake, 122–123, 131–132, 133*f*, 134–135
East India Company (Dutch; English; French), 39–42, 41*f*
Eccentricity, 454–457
Eddy kinetic energy (EKE), 257, 422
El Niño, 10–11, 49–50, 56–57, 66, 106–110, 113, 124, 129–130, 148–156, 152*f*, 155*f*, 158, 206–208, 215–216, 255, 257, 393–395, 400, 426, 446, 449, 453–454
El Niño-Southern Oscillation (ENSO), 3, 49–50, 103, 121–122, 145, 179, 206–207, 233, 248, 275–276, 331, 367, 393–394, 446, 453–454, 472–473
Emissions, 63–65, 235–236, 329, 332–338, 335*f*, 337*f*, 341, 471–474, 477–478
Endemism, 377, 381
Endosymbiotic dinoflagellates of corals, 377
Equatorial upwelling, 113, 234, 257
Equatorial waves, 79, 89–90, 91*f*, 181, 190, 395–396, 398, 401–402, 424
Estuary, 303, 331–332
Euphotic zone, 208, 276, 279–281, 283, 294, 304–305, 315, 330, 367, 407
Eutrophic, 298–299, 303, 341, 358
Exclusive Economic Zone, 252–253, 408

F

Free troposphere, 79, 82–83

G

General circulation model (GCM), 65, 67
Goa, 38–40, 39*f*
Grazing, 13, 15, 303, 307, 312, 314–315
Greenhouse gasses, 16, 64*f*, 218–219, 232–237, 240, 265, 267, 271, 284, 329–334, 330*f*, 331*t*, 332–333*f*, 354, 471–472
Gulf of Aden, 35*f*, 43, 178, 295–296, 365, 372–375, 372*f*, 374*f*, 377, 382

H

Hadley circulation, 51–52, 51*f*, 56, 65, 107–108, 452, 454
Halocarbons, 329, 334–335
Halocline, 84, 371, 425
Harappan civilization, 34, 67, 277
Holocene Climate Optimum, 61, 67
Hormuz, 34, 38–40, 40*f*, 45, 369, 371, 377–379

H

Hypersaline, 369, 375
Hypoxia, 265, 266t, 272–273, 371

I

Independence movements, 45
Indian monsoon, 55f, 57–58, 105, 147–149, 157–158, 234, 353, 404, 433–434, 452–453
Indian Ocean basin mode, 10–11, 56, 106, 111f, 124, 125f, 145
Indian Ocean Observing System (IndOOS), 3, 179, 241, 393–395, 398–409
Indian Ocean Dipole (IOD), 7f, 8–9, 81, 103–106, 110–112, 121–122, 126f, 129f, 145, 147, 179, 233, 245, 273, 293, 367, 393–395, 405f, 421, 427, 446, 453–454, 469–474
Indian Ocean subtropical dipole, 103, 105–108, 110–112, 112f, 129–130, 145, 233–234, 240
Indian Ocean Tuna Commission, 250f, 254f, 255–256, 256f
Indonesian Throughflow (ITF), 1, 7f, 92, 125–126, 169, 205–209, 206f, 229, 268f, 281–282, 282f, 293, 394, 396–397, 424, 427f, 448–449, 469, 472–473
Indo-Pacific Warm Pool, 7f, 11–12, 12f, 56, 58, 145, 157–158
Interbasin exchanges, 170, 205–206, 219, 394
Interbasin interactions, 114, 145–146, 157, 160, 240, 421, 435
Interdecadal Pacific Oscillation (IPO), 56, 110, 125–126, 206–207, 229, 231–233
Intergovernmental Panel on Climate Change (IPCC), 63–66, 469, 473–474, 478
International Indian Ocean Expedition (IIOE), 2–3, 114, 310, 351
Intertropical Convergence Zone (ITCZ), 11, 56–57, 59f, 61, 89, 128, 146–147, 336–337, 449–459
Intraseasonal oscillation (ISO), 7f, 11, 53, 56, 65, 79, 81–84, 82–83f, 87, 114, 145–147, 189, 426, 469–470
Intraseasonal variability (ISV), 11, 20, 66, 79, 82, 87–90, 92, 145–147, 175, 245, 331, 421, 428, 430f
Iron (Fe), 13, 44–45, 275, 295–298, 303, 315, 341, 356, 371, 432, 432f
Islam, 34, 36
Isoprene, 329, 334–335
Isotopes, 61, 275–276, 446, 449, 451f

K

Kelvin wave, 9, 81
Kerguelen Plateau, 216–218, 331

L

Lacustrine sediments, 61, 445
La Niña, 10, 56–57, 105–110, 113, 124–126, 125f, 129–130, 134, 148, 152–153, 206–207, 257, 394, 400–401, 405, 448–449
Large marine ecosystem, 249, 251–254, 381
Leeuwin current, 1–2, 5, 9–10, 12–15, 20, 109, 169, 181, 183–185, 190, 205, 206f, 209–210, 212, 218–220, 247, 257, 281–282, 282f, 284–285, 293–294, 310–312, 315–316, 332, 421–423, 448–449
Little Ice Age, 60f, 61, 446

M

Madden-Julian Oscillation (MJO), 7f, 80f, 81, 121–122, 129f, 145, 293, 424
Majapahit Kingdom, 36
Mangroves, 249–251, 255, 375, 377
Mare clausum, 38–39
Mare liberum, 40
Marine heatwave (MHW), 108, 121, 123–127, 145, 151, 394, 469–471, 474, 475f, 478
Marine sediments, 272–273
Maritime continent, 12–13, 56–57, 81, 84, 87, 90–92, 109, 153, 219, 293, 407, 428, 431, 449–450, 458f, 459
Mascarene High, 105–110, 113, 234
Melaka Strait, 36
Mercury (Hg), 335–337
Meridional overturning circulation, 6, 8f, 169–170, 185–187, 190–191, 205, 209, 216, 218–219, 397–398
Mesopotamia, 33–34, 61
Mesoscale, 10–11, 19, 63, 79, 178, 189–190, 210, 214–216, 219, 257, 259–260, 278–279, 311, 314–315, 351, 371–372, 379, 428, 431
Mesoscale eddies, 10, 19, 178, 190, 210, 214, 257, 278–279, 281–282, 377, 379, 407, 428, 431–432
Methane (CH_4), 329, 331–332, 332f
Methanogenesis, 331–332
Methyliodide (CH_3I), 334
Methylmercury (MeHg), 337
Microbial loop, 303, 307, 377
Ming Dynasty, 36–37
Mixed-layer, 107–108
Mode of variability, 121–122, 145, 234, 245
Moist static energy, 51–52, 51f, 65, 89, 150, 153
Moisture mode, 82–83, 89, 91–92
Monsoon, 1, 33–34, 49, 169
Monsoon depression, 53–54, 56, 65–66
Monsoon intraseasonal oscillation, 7f, 53, 469–470
Mozambique Channel, 5, 9–10, 14–15, 42, 44, 187–188, 213–214, 219, 248, 257, 282–283, 312–313, 315–316, 449
Mughal Sultanate, 37–38

N

Net primary productivity (NPP), 246–247, 246f, 408, 474
N_2 fixation, 15, 273, 314, 354–356, 354f, 358–359, 359f
Ningaloo Niño/Niña, 103, 108–113, 123–124, 126–127, 145, 233–234, 394, 400–401, 435
Nitrate (NO_3), 1, 267f, 271, 276, 294, 295f, 297–298f, 298, 304f, 308f, 310–312, 354, 357–358, 371, 397
Nitrification, 276, 294, 333, 354, 354f, 356–358
Nitrogen oxides (NO, NO_2, NO_x), 329, 335–338, 342
Nitrous oxide (N_2O), 16, 265, 271, 329, 332–334, 333f, 354, 354f, 356
North Atlantic Oscillation (NAO), 57–58, 67, 150–151, 159
Nutrients, 1, 13–15, 208, 304–305, 307, 310–312, 314, 339–340, 374

O

Obliquity, 456f, 458–459
Observing system, 170, 275, 394–395, 397, 408
Ocean-atmosphere coupling, 79–81, 92–93, 103–104, 147
Ocean heat content (OHC), 79, 90, 209, 219, 236, 402, 405, 469–470, 472–474, 473f, 477
Oceanic bridge, 152–153, 233
Oligotrophy, 246, 301–302, 314–315
Orbital parameters, 454–460
Orography, 52–53, 65, 67, 373–374
Oxygen (O_2), 16–17, 265, 303f, 305
Oxygen isotope, 58–61, 67, 446
Oxygen minimum zone (OMZ), 2–3, 265–267, 351, 433, 477
Ozone (O_3), 329, 335–337

P

Pacific Decadal Oscillation (PDO), 19, 56, 229, 257–258, 393, 448–449
Particulate organic carbon (POC), 16–17, 18f, 268–271, 270f, 276–285, 280f, 352f
Pelagic fish, 249–251, 253, 254f
pH, 16–17, 274–276, 353, 476–478, 477f
Phenological analysis, 374
Phosphate (PO_4), 1, 271, 276, 294, 295–298f, 303f, 304, 310, 312, 371
Phytoplankton, 13–15, 251, 271, 276, 294–315, 337–338, 340–342, 353–354, 379, 431–433, 474, 476–477
Pollution, atmospheric, 329, 337, 342
Precession, 454, 456f, 457–459
Predictability, 66, 103, 110–112, 124–125, 155–156, 240, 427
Primary production, 13–15, 17–19, 208, 277, 294, 298, 300–303, 302f, 305–306, 308, 311, 313, 315–316, 352–354, 358–359, 359f, 377

R

RAMA (Research Moored Array for African-Asian-Australian Monsoon), 3, 85–86, 169, 179–180, 395–396, 398–400, 402, 403f, 404–405, 407–408, 424
Rectifier effect, 81
Red Sea, 34–36, 178–179, 214f, 268f, 331, 365
Regional modeling, 10–11, 16
Rossby wave, 9, 56–58, 79, 81, 85, 89–90, 91f, 104–105, 107–108, 124, 129, 150–151, 154–155, 158–160, 174–176, 181, 207–208, 213, 215, 234, 367, 401–402, 405, 422, 424, 426

S

Salinity, 13, 20, 86, 206–207, 206f, 212, 237–238, 268–269f, 293, 373, 399f, 407, 424–425, 427f
Scleractinian corals, 377
Sea breeze, 10, 49–52, 67, 89
Sea level rise, 2, 130, 219, 237–238, 239f, 398, 405, 407
Sea of Oman, 365, 366f, 371–372, 377–381, 380–381f
Sea surface height (SSH), 5–6, 89–90, 91f, 181, 183, 210, 211f, 218–219, 294, 367, 380f, 395, 397, 400–401, 405–407
Sea surface salinity (SSS), 12–13, 12f, 86, 171–172, 237, 380f, 395, 435
Sea surface temperature (SST), 2f, 7–8, 79–81, 145, 156f, 230–231, 230f, 282f, 380f, 393, 445–449, 471–474
Seychelles-Chagos Thermocline Ridge (SCTR), 1, 7–9, 9f, 13, 81, 85, 90, 146–147, 150–151, 183, 186, 191, 231, 246–248, 281, 293, 307–310, 424–425
Shared Socioeconomic Pathways (SSP), 471
Silicate (SiO_4), 1, 13, 276, 294, 295–298f, 304, 304f, 310, 312, 314, 367, 371, 432
Slavery/slave trade, 44
Solar forcing, 54, 92
Solubility pump, 16, 275, 284, 330
Somali current, 7f, 90, 171–174, 186, 190–191, 216–218, 396–399, 408, 422, 434f
Song Dynasty, 36–37
South Asian monsoon, 10–11, 20, 49–67, 59f, 148–149, 450–453, 455f
South Australia current system, 184–185

Southern Annular Mode, 107–108, 124–125, 126f, 215–216, 257
Southern Ocean, 6, 33, 169–170, 181–183, 188, 190–191, 211–212, 216–219, 217f, 230–231, 241, 276, 331–332, 353, 469
Speleothem, 60f, 63, 445, 450, 451–452f, 454–456, 459
Sri Lanka Dome, 1, 171, 178, 281, 305
Storm surge, 121, 132, 397
Stratification, 1, 8–9, 17–18, 20, 79–81, 87–88, 135, 206–208, 247–248, 272, 280, 285, 293, 304–305, 316, 367, 379–381, 402, 425, 429–430, 433–435, 474, 476
Subduction, 5, 122–123, 131–132, 131f, 186, 191, 275, 283, 310–311, 314, 395–396
Submesoscale, 86, 184, 373
Subsidence, 56–58, 79, 82–83, 132–133, 152–153, 157–158, 169, 450
Subtropical, 103, 106–108, 182–183, 283, 314–315, 398, 400
Suez Canal, 44–45, 375
Sulfur dioxide (SO_2), 335–337

T

Tamil Nadu, 33, 50
Tang Dynasty, 36
Tasman leakage, 10, 20, 205, 206f, 209, 212–213, 219
Teleconnection, 11, 57–58, 66, 79, 81, 121–122, 124, 145, 148, 150–151, 151f, 158–160, 219–220, 231, 427, 448–449, 452–453
Thermocline, 9, 104–105, 179, 183, 206–207, 306–307, 453
Thermohaline circulation, 3, 10–11, 20, 205, 212, 377, 452, 460

Trace gases, 329, 334–335, 341–342
Tropical cyclone (TC), 2, 53–54, 84, 121, 127–130, 149, 393, 403, 407
Troposphere, 11, 57, 79, 82–83, 88–89, 127–128, 150, 153, 158–160, 329, 428
Tsunami, 5, 131–134
Tuna, 19, 255–259, 256f

U

Undercurrent, 172, 175, 179–180, 184–185, 210, 212, 400
Upwelling, 1, 158, 169, 255, 273, 275–276, 293, 373, 426, 432–433, 454

W

Walker circulation, 10–13, 55–57, 81, 82f, 83, 124, 148–149, 151f, 152–153, 155–156, 158, 206–207, 229, 233, 240, 394, 400, 450, 459–460, 471
Whalers/whaling, 42, 44
World Ocean Circulation Experiment (WOCE), 3, 274, 394–395
Wyrtki jet, 5, 169, 172–175, 179–181, 190, 293, 308, 398–402, 424

Y

Yongle Emperor, 36–37

Z

Zanzibar, 38–39, 41, 41f, 43–44, 406–407, 406f
Zheng He, 36–37
Zooplankton, 13–15, 294–316